D0161243

Collin College Library
WYLIE CAMPUS
Wylie, Texas 75098

WITHDRAWN

The Handbook of Medicinal Chemistry
Principles and Practice

Registered charity number: 2078

ROYAL SOCIETY
OF CHEMISTRY

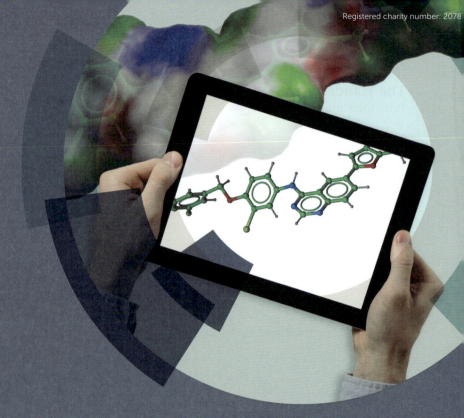

In meetings? At the lab? On the move?

Medicinal Chemistry Toolkit app

- Calculating pharmacologically relevant physicochemical properties based on chemical structure
- Helping you to translate target affinity into expected cellular potency
- Simplifying the process of calculating clinical dose

Take advantage of easy to access functions whenever and wherever you are

Get the 2014 Medicinal Chemistry Toolkit app today

AVAILABLE NOW!

Download on the
App Store

The Handbook of Medicinal Chemistry
Principles and Practice

Edited by

Andrew Davis
AstraZeneca, Mölndal, Sweden
Email: andy.davis@astrazeneca.com

Simon E Ward
University of Sussex, Brighton, UK
Email: Simon.Ward@sussex.ac.uk

THE QUEEN'S AWARDS
FOR ENTERPRISE:
INTERNATIONAL TRADE
2013

Print ISBN: 978-1-84973-625-1
PDF eISBN: 978-1-78262-183-6

A catalogue record for this book is available from the British Library

© The Royal Society of Chemistry 2015

All rights reserved

Apart from fair dealing for the purposes of research for non-commercial purposes or for private study, criticism or review, as permitted under the Copyright, Designs and Patents Act 1988 and the Copyright and Related Rights Regulations 2003, this publication may not be reproduced, stored or transmitted, in any form or by any means, without the prior permission in writing of The Royal Society of Chemistry or the copyright owner, or in the case of reproduction in accordance with the terms of licences issued by the Copyright Licensing Agency in the UK, or in accordance with the terms of the licences issued by the appropriate Reproduction Rights Organization outside the UK. Enquiries concerning reproduction outside the terms stated here should be sent to The Royal Society of Chemistry at the address printed on this page.

The RSC is not responsible for individual opinions expressed in this work.

The authors have sought to locate owners of all reproduced material not in their own possession and trust that no copyrights have been inadvertently infringed.

Published by The Royal Society of Chemistry,
Thomas Graham House, Science Park, Milton Road,
Cambridge CB4 0WF, UK

Registered Charity Number 207890

Visit our website at www.rsc.org/books

Printed and bound by CPI Group (UK) Ltd, Croydon, CR0 4YY

Preface

Medicinal Chemistry sits at the heart of the pharmaceutical industry and the medicinal chemist has one of the most challenging and rewarding jobs imaginable. The medicinal chemist designs the drug which must balance often conflicting demands of a suitable dose, by the chosen delivery route, at a desired dose frequency to provide a therapeutic effect while maintaining margins to adverse effects throughout the dosing period. The drug molecule may be given to millions of patients all of whom may respond to the drug differently, and all of whom must be treated safely and effectively. Whilst drug discovery is undoubtedly an endeavour involving a wide range of scientific disciplines, the medicinal chemists are critical to the design and progression of a drug molecule. It is the medicinal chemist who integrates and balances the diverse inputs into a single chemical structure which has the potential to become a new medicine.

This is an enormously difficult task. Our advances in synthetic organic chemistry mean that we can respond well to the challenges of preparing and purifying new molecules and chemists can be trained in these skills during undergraduate and graduate studies. In contrast, compound design is far harder to control and requires extensive experience and knowledge to take the sometimes subjective decisions to arrive at a potential drug candidate. There are few universal rules in drug design, and barely any universally accepted guidelines, and it sometimes seems success is more a matter of chance. But, as Louis Pasteur said, "chance favours the prepared mind". However, given the current challenges and high attrition during the development phase, and the acceptance that many reasons for failure are directly attributable to the chemical structure of the drug candidate, medicinal chemists have a duty to design the best molecule possible to advance from research into development and beyond.

The aim of this book, through a series of monographs by leading scientists from across the world, from major pharmaceutical companies, biotechnology companies, contract research organisations and academia is to prepare the medicinal chemist to spot the good chances.

The book covers the whole R&D process from target validation through to late stage clinical trials, through descriptions of the background science, the process, learnings, case studies, leading references and even hints and tips.

The foreword has been written by one of our industry's most respected scientists, Simon Campbell CBE FRS, FMedSci. Simon Campbell joined Pfizer as a Medicinal Chemist in 1972, and was a key member of the teams that led to such blockbuster drugs as Cardura, Norvasc and Viagra. He went on to become Pfizer's Senior Vice President for World-wide Drug Discovery and Medicinal R&D in Europe. He was President of the Royal Society of Chemistry from 2004 to 2006

The Handbook of Medicinal Chemistry: Principles and Practice
Edited by Andrew Davis and Simon E Ward
© The Royal Society of Chemistry 2015
Published by the Royal Society of Chemistry, www.rsc.org

and maintains a very active and influential role in our industry. With his considerable experience Simon provides us with his personal learnings, and the undoubted opportunities for medicinal chemistry looking forward.

The early chapters describe the tools of the medicinal chemist's trade such as physical organic chemistry, computational chemistry and QSAR, library design, fragment based lead generation and structure based design.

The middle section of the book covers the supporting scientific disciplines, including assay development, receptor pharmacology and *in vivo* model development, drug metabolism and pharmacokinetics, molecular biology, toxicology and translational science, computational biology and of critical importance, intellectual property.

The later sections of the book describe the overall research and development process from target generation, lead identification and optimisation through to pharmaceutical development, clinical development and chemical development, including the importance of efficient project management.

Due to the high levels of failure faced during drug development, case studies of successful R&D are hard to find, but are invaluable as a touchstone for pathways to success. So the last three chapters provide case studies of drugs that made it into the later stages of clinical development and/or onto the market, Brilinta, Aleglitazar and Lapatinib. Even during the preparation of this book, one of our case studies was unfortunately halted during Phase III trials. As sad as Phase III failure is, few drugs reach this stage of clinical development and there are many lessons to be learnt in this story that justify its esteemed place in this section.

The book began as life as a proposal to update to a 3rd edition the Royal Society of Chemistry's long running publication "Principles and Practice of Medicinal Chemistry". The first edition was published over 20 years ago, and was a spin-out from the biannual Royal Society of Chemistry Medicinal Chemistry Summer Workshop, which itself has been running for over 40 years and has been the training ground for many of our industry's leading medicinal chemists. The 3rd edition proposal retained some distinctive features of its predecessors, being highly practitioner focused, but grew to incorporate a broader context and to reflect the changing reader demographic reflected in the changing industry and drug discovery environments. It also grew to incorporate new opportunities that did not exist 20 years ago.

Paper publishing is as valid today as it has ever been, but mobile computing and e-publishing are changing the way information can be used and presented. E-publications allow interaction with the content which cannot occur with paper. App-stores allow easy access to sophisticated software that can be delivered and updated with ease. Many tools potentially useful to medicinal chemists do not exist in an easily accessible and secure manner. So for the 3rd edition we wanted to develop, as a companion to the print book, a set of useful medicinal chemistry apps to run locally on tablet computers, and also a fully interactive e-book version to complement the paper copy. The apps would bring to life concepts described within the book chapters and allow chemists to quickly and easily find help in their design challenges.

While even 10 years ago protein structure visualisation and small molecule modelling required high-end workstations and costly software, nowadays this can be accomplished on a tablet computer. Indeed, the frontispiece image of this book was designed inside the freeware app iMolview from Molsoft on an Apple iPad3. Similarly static pictures of X-ray crystal structures within the chapters can be brought into high resolution reality, and the reader can interact with the exact data that the original medicinal chemist used in the documented design. Structures can link to ChemSpider or even Wikipedia and other online resources providing deeper context, and hyperlinks to regulatory guidance mean the medicinal chemist has access to primary information sources relevant to each chapter.

So while this 3rd edition was inspired by its predecessors, with the companion apps and the e-book format, it was time to change the book's name. We hoped the book would become an everyday companion for the practicing medicinal chemist, and so the title ''Handbook of Medicinal Chemistry'' seemed appropriate. With both print and electronic format and companion apps we hope that, with this handbook, we can more fully prepare the mind of the medicinal chemist to pick the right chances.

Andrew Davis and Simon E Ward

Contents

The Handbook of Medicinal Chemistry: Principles and Practice
Edited by Andrew Davis and Simon E Ward
© The Royal Society of Chemistry 2015
Published by the Royal Society of Chemistry, www.rsc.org

Chapter 7   Drug Metabolism   **184**
C. W. Vose and R. M. J. Ings

**Chapter 8   Prediction of Human Pharmacokinetics, Exposure and Therapeutic Dose in
             Drug Discovery**   **208**
Dermot F. McGinnity, Ken Grime and Peter J. H. Webborn

Chapter 9   Molecular Biology for Medicinal Chemists   239
Giselle R. Wiggin, Jayesh C. Patel, Fiona H. Marshall and
Ali Jazayeri

Chapter 10   Assays   266
Tim Hammonds and Peter B. Simpson

Chapter 11   *In Vitro* Biology: Measuring Pharmacological Activity   292
Iain G. Dougall

Chapter 12   Animal Models: Practical Use and Considerations   310
Milenko Cicmil and Robbie L. McLeod

Chapter 16   Toxicology and Drug Development   **413**
Mark W. Powley

Chapter 17   Patents for Medicines   **424**
Paul A. Brady and Gordon Wright

Chapter 22 Pharmaceutical Properties—the Importance of Solid Form Selection **566**
Robert Docherty and Nicola Clear

Introduction

I am delighted to have been invited to write an introduction for The Handbook of Medicinal Chemistry: Principles and Practice. The editors and authors have played an outstanding role in covering all of the major components of modern medicinal chemistry in an expert and timely manner, within a comprehensive handbook relevant to newcomers and experienced scientists alike. This volume will be a pleasure to read through, and then to pick out relevant sections for in depth consultation. I am sure Principles and Practice will be in constant use, as different problems arise in drug discovery projects almost on a daily basis, and will become essential reading for medicinal chemists of whatever background and experience.

An overview of the dramatic progress we have made with healthcare quality shows that life expectancy has consistently risen over the past century with an increase from 60 to over 85 years for women in most industrialised nations. Similar trends are evident for men and equally importantly, the developing world is now moving in the same direction. While improved standards of hygiene, nutrition, housing and other factors are obviously important, it is estimated that 40% of the recent increase in life expectancy in the US is due to modern medicines largely discovered by the pharmaceutical industry:[1] powerful antibiotics are available to treat life-threatening bacterial infections; hypertension (the silent killer) can be controlled by any number of once-daily therapies; elevated cholesterol which is a major cardiovascular risk factor, is well managed with statins, while H2 antagonists and even proton pump blockers are available over the counter to treat gastric ulcers. When HIV/AIDS appeared on the scene in the early 1980s, it was considered a death sentence and control was thought to be beyond our reach due to facile transmission and potential for resistance. Today, over thirty drugs from six mechanistic classes are available, and those in the West who contract the virus have enjoyed much improved quality of life and longevity. Importantly, similar benefits are now emerging in the developing world where for example, life expectancy in Kwa Zulu-Natal has risen from 49 in 2003 to 60 years in 2011 as affordable anti-retroviral combinations became available in the public healthcare system. Hopefully, recent headlines from *The Economist* such as: "The end of AIDS? How 5 million lives have been saved and a plague could be defeated" are now within sight, and a fair balance between drug pricing and health benefits will become commonplace.

Despite such outstanding success, there are still tremendous healthcare challenges facing medicinal chemists and the whole drug discovery community. We all know that cardiovascular disease (CVD) is a major risk factor responsible for over four million deaths in Europe each year, but few realise that 80% of global CVD mortality actually occurs in low- to middle-income countries which are disproportionately affected. The prevalence of obesity in US adults will grow

The Handbook of Medicinal Chemistry: Principles and Practice
Edited by Andrew Davis and Simon E Ward
© The Royal Society of Chemistry 2015
Published by the Royal Society of Chemistry, www.rsc.org

to 50% by 2030, and it is also estimated that 92 million Chinese already suffer from Type 2 diabetes. While malaria, TB and HIV are still scourges in many parts of Africa, non-communicable diseases killed 36 million people in 2008 which represents 63% of total deaths, with the majority occurring in emerging nations. In any one year, 40% of Europeans will be affected by some type of brain disorder with total annual costs of care around Euro 800 billion, more than CVD, cancer and diabetes combined. Mental depression is responsible for 38% of all morbidity and 23% of Quality-Adjusted Life Years (QALYs) lost, whereas the corresponding figures for cancer are 3 and 16%, respectively. The WHO has forecast an impending disaster due to unchallenged increases in antimicrobial resistance, but only four new classes of antibiotics have been introduced over the past 40 years. In response to these major threats to health and well being, the demand for new medicines will continue unabated, albeit with different emphasis on quality of life or longevity depending on regional differences in economic and social development. However, new paradigms for research focus, organisation, cooperation and funding will be required, as we adjust to an ever changing scenario of contracting Pharma and withdrawal from major therapeutic areas. This introduction offers a personal perspective on some factors that may influence success and failure in drug discovery, and suggests how the sector might learn from the past and evolve in the future.

Size and organisation are key factors for innovative drug discovery which have been overlooked in endless rounds of mergers and acquisitions over the past decade, and the relentless drive to international research conglomerates. During the period when we were most productive at the Pfizer research laboratories in Sandwich, our total staffing was probably around two to three hundred, but that period witnessed outstanding discoveries such as amlodipine, diflucan, doxazosin and sildenafil. Our research was driven by dedicated scientists working together in multidisciplinary teams towards common objectives within a supportive, but focussed environment. Unusually, drug metabolism experts were also integral members of discovery projects which provided a significant competitive edge, as we did not have to beg, borrow or steal from development which was the norm throughout Pharma at that time. While we fully understood the need to compete internationally, we operated largely on a local and personal scale where a trip to the US was an annual treat, not a weekly routine. We all knew each other, managers and directors walked the job, and we were not distracted by administration. Scientists were constantly in and out of each other's laboratories as we had a hunger to generate, share and exploit new data that would drive our projects forward. Face to face discussions were the norm, and stimulated a level of intellectual challenge far beyond impersonal e-mails and text messages. The current journals section of the library was a focal point for discussions where we swapped ideas as we jostled for the latest articles, but paper copy has largely disappeared and individual online access may not generate the same thought-provoking synergies unless alternative communication networks are established. In addition, we valued our "Tribal Elders" who had "been there, done that" and freely shared their experience, but successful role models have largely disappeared in today's cost cutting climate. However, the added value generated through a mentoring and supportive culture coupled with institutionalised learning cannot be over estimated.

As we grew we had to adapt, and I became drawn to the concept of the Roman Centurion who traditionally leads and cares for 100 soldiers which seemed a sensibly sized unit, particularly in a research environment. When there were 100 chemists in my discovery group, I knew them all and what they were doing, and I was also able to engage at a personal level. However, as the group expanded it became more difficult to maintain that level of interaction, and informal discussions were diluted. Dunbar's number of 150 is an estimate of the social contacts humans can cope with, obviously at differing levels of engagement, which is roughly in line with the Centurion concept. The average size of a village in the Domesday Book of 1086 was also around 150, and any further increase stimulated migration to form new settlements. These numbers intuitively feel right as they reflect the importance of personal contact, and also address the critical mass necessary for survival. Similar considerations should underpin drug discovery organisations where large groups should be broken down into nimble, multidisciplinary units that can be managed and led on a personal scale. Teams should be largely autonomous but accountable, with innovation and a data-driven culture recognised and rewarded, rather than the consensus management and upward decision making that has ossified Pharma in recent years.

Critical mass is probably more important than size *per se* as the ability to respond rapidly to breaking science can make the difference between success and failure. For example, we quickly realised that half a dozen chemists on a lead optimisation project would not be competitive, whereas 12 to 15 could hold their own. However, we could never manage the teams of 20 to 40 that others mobilised as duplication, poor communication and a loss of personal responsibility inherent in such large groups compromised productivity and motivation. Innovative scientists often want to be different, but some can drift into peripheral activities with a lack of focus and commitment to team objectives. Crucially, the concept of critical mass and nimble research units became confused with absolute size in the fruitless drive to build the largest R&D organisations. Even before the merger with Wyeth, Pfizer had an annual R&D budget of nearly $8 billion with thousands of staff spread over eight centres on three continents, which may not be conducive to a personal or nimble approach. The negative impact of mergers and acquisitions on productivity is well documented[2] as research simply cannot be effectively managed in such massive units, nor can innovation survive, particularly with multiple locations, cultures and ever changing leadership. Technology can be expanded in a modular manner and centralised facilities for HTS, gene sequencing and other service operations are efficient and cost effective, but innovation simply does not scale. If readers were to take one key message from this introduction, it would be my strong conviction that drug discovery is a personal and shared experience, not a metrics-focused, mechanical event. So many times, successful projects are driven by a small core of dedicated champions with a burning desire to address particular medical needs, working together in a research-friendly environment not dominated by numbers.

Hype and premature over-investment in new technologies are other examples of how Pharma lost its way with the drive towards "faster, cheaper, better," but quality was lost in the pursuit of numerical goals. Most companies thought that industrialisation of drug discovery was the way of the future and that attrition need not be improved if the number of candidates entering development was significantly increased. Numbers and metrics became key drivers and innovation and personal accountability were lost in the process. Gone were the days of research proposals that laid out a biological rationale and thoughtful chemistry plans that were subject to rigorous challenge, and HTS assumed the default mode for new projects. HTS became a macho competition across Pharma with migration from 96 to 384 to 1536 well plates, and the drive to generate millions of data points over the shortest time frame. However, assays were often not robust, and quality control was poor. Compound collections contained everything chemists had registered, and it took some time to weed out frequent hitters, reactive intermediates and undesirable structural flags that were never intended to included in screening files in the first place. Unfortunately, re-building these collections also became numbers driven as it was easy to impress senior managers with the claim to synthesise millions of peptides overnight, but without adding that these compounds had little utility for drug discovery. Combinatorial libraries constructed from simple, non-peptide building blocks suffered a similar fate as focus on "what we can make" rather than "what we should make" led to large collections of closely related compounds with low value for screening, particularly as mixtures. Some Pharma companies responded by investing up to $100 million in building diverse, multi-million compound collections, but such large files are rarely screened routinely as representative sub-sets usually provide an idea of the relevance of the overall library to a particular target. However, it is encouraging that HTS has matured considerably over the past decade, where greater attention to assay reproducibility and compound quality has been rewarded with viable hit matter identified much of the time. More recently, advances in computational chemistry and structural biology have led to the integration of virtual screening with smart HTS which has further increased success rates. Such improvements are a tremendous advance, but the time-scales and resources required for new technologies to reach maturity are quite sobering.

Of course, the allure of HT-everything drove massive investments in numerous other technologies including every "omics" under the sun, often through fear of losing out to competitors rather than an appreciation of real value or time scales involved. The Gartner Hype Cycle neatly summarises new initiatives passing through a technology trigger, peak of inflated expectation, trough of disillusion, slope of enlightenment and plateau of productivity that we have all experienced. Multiple external collaborations often proved a distraction from drug discovery, and some major investments from the 1990s are still way off delivery. For example, billions have been invested in DNA- and RNA-based therapies as interest shifted from antisense to ribozymes to RNA interference, although it was obvious that delivery would be a common problem that still has not been solved generically. However, the first systemic antisense drug was approved in 2013, some 23 years from "a blank sheet of paper to market" which again brings home the timescales required for new technologies to bed-in and mature. Gene therapy involves the simple concept of introducing a gene into a cell to express a particular protein involved in disease, but the few regulatory approvals to date are limited to niche indications with return on investment still a long way off. Perhaps the highest hopes were raised over the sequencing of the human genome which was announced in draft form in 2001, with ambitious claims that this would revolutionise healthcare diagnosis and treatment. This may turn out to be the case, but more than a decade later, millions of gene sequences are in hand with "the dream" still far from reality. Maybe the fundamental thought processes outlined by James Watson in *The Double Helix* put such numbers-driven approaches into context. However, some genes and SNPs have shown a weak association with disease but there has been little impact on target validation or patient selection, except for particular cancers. In the latter case, identification of genetic markers of drug sensitivity has proved to be extremely powerful in patient stratification for clinical trials and targeted therapies, but there has been little progress with other diseases.

Improved candidate survival is another key issue, as the enormous cost of bringing a new medicine to market is unacceptable since it also includes wasted investment in the numerous failures that occur throughout the drug discovery and development process. Alarmingly, recent surveys suggest that less than 10% of preclinical candidates that enter development reach the market, and it is difficult to imagine that any other business sector would accept such an appalling failure rate. Reducing attrition must be a major priority for Pharma in general and medicinal chemists in particular, since even modest improvements would have a significant impact on the cost-effective output of new medicines

The individual reasons for candidate failure have been well documented, but the dual themes of mechanism- and compound-related attrition are particularly relevant during the discovery phase. Validating a new target in the laboratory is a daunting task even with today's sophisticated technologies and realistically, only a certain level of confidence can be established that a particular pathway or mechanism will be relevant in man. To mitigate risk, mining gene families has received particular attention on the assumption that experience with one clinically-relevant member could be extended to close relatives. While this may apply to druggability, there seems to be a high level of biological redundancy such that seemingly attractive targets may not be involved in physiological or pathophysiological processes. For example, despite convincing rationale for disease relevance and the discovery of potent ligands for numerous members of the adenosine and PDE families, only a handful of drugs have actually resulted. Clearly, animal experiments are still poorly predictive of the clinical situation, particularly for CNS and cancer where above average attrition is par for the course. Mechanism-related failures may also be a consequence of evaluating new drugs in heterogeneous patient groups such that efficacy signals from responsive subsets are lost in the noise.

Reducing mechanism-related failures calls for greater innovation and investment in target validation, but animal models always have limitations and rapid progression of quality candidates to the clinic may be more informative. This will require developing robust biomarkers that confirm drug activity in relevant tissues, identifying patient sub-groups that respond to a particular mechanism of action and establishing definitive clinical end points. Overall, a much better understanding and interpretation of PK/PD relationships will be required, and early enough to influence discovery projects. Some consider that these initiatives will fragment markets, but the cost of clinical trials and attrition will be significantly reduced, and surely targeting patients with a high chance of response must be a key objective? Pre-competitive collaborations for both target validation and patient selection will become more common, and there are encouraging signs that Pharma is moving in this direction.

Given such significant investment in biological and clinical sciences, medicinal chemists have a key role to play in designing high quality candidates capable of completing definitive Phase 2 proof of concept studies where full dose–response relationships can be explored. Their challenge is to optimise the physicochemical and molecular properties they so well understand to eliminate compound-related failures such that decisions on candidate progression can be made on efficacy and safety data alone. 30% of all candidate failures are due to inadequate clinical efficacy, but probing new mechanisms with sub-optimal compounds provides minimal learning at significant cost. Indeed, analysis of 44 Phase 2 programmes at Pfizer[3] confirmed that the majority of failures was due to lack of efficacy, but in 43% of those cases it was not possible to conclude that the mechanism had been properly tested due to limited exposure and target engagement.

The Lipinski Rule of Five is now part of the fabric of drug design since these data-driven guidelines summarise the physicochemical parameters that influence permeability and oral absorption. While there are exceptions, medicinal chemists who push the guidelines to the limit usually bequeath compound-related deficiencies to their colleagues at some stage in the discovery and development process, and which often come home to roost in the clinic. Various analyses have shown that molecular weights of drug candidates decrease along the development pathway which must raise at least an amber flag to those pursuing lead series with molecular weights above 400. Increasing molecular weight and lipophilicity is seductive as this allows introduction of structural diversity and novel substituents that improve potency and allow differentiation from prior art. However, while low oral doses are obviously preferred for clinical candidates, the median target affinity for current small molecule drugs is around 20 nM, so the goal of continually driving down absolute potency may be less important than focussing on ligand efficiency which reflects the average binding energy per heavy atom. Ligand lipophilicity efficiency may be even more relevant for lead optimisation as this provides a constant reminder that SARs should be developed without compromising physicochemical properties.

For compounds with high molecular weight and lipophilicity, solubility is almost invariably compromised and is often not improved during lead optimisation such that bioavailablity may be low and variable. Such compound-related limitations are significant barriers to exploring dose–response relationships in the clinic and may also have a negative impact on eventual commercialisation. Compounds at the fringes of the guidelines tend to be more susceptible to CYP oxidation/induction, which can reduce bioavailability through first pass metabolism, generate biologically active and/or toxic metabolites and lead to significant drug-drug interaction liabilities. Encouragingly, medicinal chemists now have a greater understanding of the scientific principles that control absorption, distribution and metabolism, and failure during development for pharmacokinetic factors has been reduced from 40 to below 10%.[4]

Entropy driven, non-specific interactions are important for binding between small molecules and proteins so compounds with high molecular weight and lipophilicity tend to be promiscuous with significant off-target activities. Given that safety issues in animals and man are responsible for some 30% of candidate losses, medicinal chemists should work within physicochemical parameters associated with success, not failure. Of course, there are exceptions such as natural products and some anti-virals for example, and larger, more complex molecules may be required to block protein-protein interactions, but passive drift outside the guidelines should be avoided.

The challenges to medicinal chemists are clear: physicochemical property inflation should be reduced; compound-related failures eliminated; and attrition significantly improved. We have a unique responsibility for discovering new drugs that will meet future medical needs and to ensure the viability of industry-based research in years to come, but personal accountability can be eroded as the drug discovery processes is broken down into compartments with "experts" assigned to artificial stages from design to candidate selection. Such fragmentation may simplify metrics, but may be personally unrewarding and less productive than a holistic approach where chemists have target laboratory and clinical profiles in mind even as they consider early hit structures.

Phenotypic screens were common in the 1970s when I joined Pfizer, and the rigorous mechanistic approach pioneered by Sir James Black was only just starting to make an impact. I became a member of the antihypertensive project where we were trying to improve on prazosin, a diaminoquinazoline derivative discovered by our colleagues in Groton. It had been suggested that prazosin acted as a PDE inhibitor, but the biological target was unknown so our screening sequence was alarmingly simple: synthesis then oral administration to spontaneously hypertensive rats, which actually was common practice at that time. Of course, a fall in blood pressure confirmed oral availability and perhaps our compounds were hitting a single target, but negative results were difficult to interpret and we abandoned the project. Some time later, Sandwich pharmacologists showed that prazosin was the prototype for a new mechanistic class of post-synaptic α_1-adrenoceptor antagonists and we immediately understood why these compounds lowered blood pressure. Screening switched to functional blockade of noradrenaline-induced vascular contraction through α_1-receptors which enabled us to rapidly identify the basic pharmacophore responsible for affinity and selectivity, while interrogation of the prior art suggested how SARs could be developed in an innovative fashion. Almost immediately, we synthesised UK33,274 (doxazosin), a potent and highly selective α_1-adrenoceptor antagonist that was later marketed as Cardura®, a once-daily antihypertensive agent that attained annual sales of over $1 billion.

When we started our calcium antagonist project to seek a once-daily follow on to nifedipine for the treatment of angina and hypertension, we screened compounds in guinea pig hearts as we thought we should target the cardiac rigour that occurs during an ischeamic attack. We did indeed discover a novel series of anti-rigour agents, but without a trace of calcium antagonist activity. Again, when we moved to specific binding and functional screens, we made rapid progress with the discovery of UK48,340 (amlodipine), a potent and selective calcium antagonist with complete bioavailability and a 30 hour half-life in dogs. This compound is marketed as Norvasc® (Istin® in the UK) for the once-daily treatment of angina and hypertension and became the World's fourth most popular drug with annual peak sales of $5.5 billion. I am convinced that neither amlodipine nor doxazosin would be bringing benefit to patients today if we had not moved from such crude and inappropriate phenotypic screens to defined mechanistic targets, but of course one size does not fit all.

Our attempts at phenotypic screening at the animal or organ level were poorly considered and were not productive, and like most of the industry we became attracted to mechanism-based approaches. This was driven not only by difficulties in rational prosecution of lead matter, but also from our experience that "no mechanism" candidates had higher failure rates in development. In addition, there was always a lingering fear that unexpected side effects might appear in the clinic when biological targets were not defined. Accordingly, one might assume that rational, target-based approaches would dominate today's landscape but surprisingly, in my view at least, 37% of first in class NMEs approved by the FDA over the decade up to 2008 originated from phenotypic screening. A defined mechanism of action may be preferable, but it is not essential for regulatory approval where agencies focus more on efficacy and safety. Some consider that phenotypic assays are more relevant for a complex disease condition than screening against a single molecular target, but follow-up can be challenging as activity reflects multiple parameters such as access, distribution and promiscuity. In addition, Structure-Based Drug Design is not relevant and "ligand efficiency" has limited value, and the richness of prior art is often lost when targets are unknown. Despite these caveats, innovative medicinal chemists have a fine record in overcoming such challenges and translating phenotypic hits into successful clinical drugs. Traditionally, there has been a poor and well-documented return from HTS against single antimicrobial targets, and phenotypic screening has proved more appropriate. For example, the Medicines for Malaria Venture has recently coordinated screening of Pharma libraries in a phenotypic, blood-stage malaria assay where numerous, attractive leads were identified, some of which have been transformed into high-quality clinical candidates. Wider use of carefully defined phenotypic screening should be expected in future as newer technologies such as chemical proteomics have significantly facilitated target identification, and some claim up to a 70% success rate within months or even weeks.

The relative merits of small molecules and biologicals are regularly debated as if it were one class or the other, whereas both will play important roles in meeting future medical needs. It is expected that up to eight of the top ten drugs in 2014 will be biologicals which is taken by some to mark the end of small molecules, but this may be an artefact of timing in that Biotech was initially some way behind Pharma, and these products have taken time to mature. Indeed, several leading biologicals have passed or are near the end of their patent life and "The Cliff" does not respect particular molecules. Generic biosimilars will make an increasing impact, although there are still hurdles particularly in the US, but revenues may not be eroded as rapidly as for small molecules. While biologicals have been outstandingly successful for the treatment of arthritis, cancer and diabetes, for example, these molecules are expensive to make and can cost thousands of dollars each month, without offering the convenience of oral administration for chronic diseases. Dose simplification has been an important driver for the widespread acceptance of statins and for the success of anti-retroviral therapies in the developing world for example, which would be impractical with biologicals. Regenerative medicine and stem cell therapies will also find a place for some diseases, but such approaches are likely to focus on specific patient populations, given potential high cost and specialist administration. Pressures on healthcare budgets will increase as the population ages, but there should be a continuing role for novel, small molecules that provide cost effective therapies that can be conveniently taken by mouth. Indeed, 26 of the 39 NCEs approved by the FDA in 2012 are small molecules with only two monoclonal antibodies which may be a pointer to the future, or simply a reflection of a "one off" mix of research projects initiated some ten or more years ago. Whatever the future holds, medicinal chemists will be key players addressing clinical needs not only through small molecules but also with the design and production of hybrid biological therapies, and full participation in new chemical and synthetic biology initiatives.

Pre-competitive collaborations will become more important in the future in order to reduce cost, risk and duplication. Most pharma portfolios probably share 70–80% similarity with multiple and parallel investments in the same targets, and often molecular scaffolds. For example, several companies took neurokinin and endothelin antagonists to the clinic over similar periods but with little reward, while the cumulative time and effort committed to renin inhibitors was absolutely staggering. Such duplicative failures might be avoided through precompetitive collaborations between industry and academia for target validation, particularly given the alarming claim that far less than 50% of biological publications can be repeated by third parties. Surely, we are past the point where individual Pharma/Biotech companies can continue to make parallel investments to reach the same negative conclusions given the tremendous pressures the industry is facing? Identifying patient populations that respond to new mechanisms of action is also essential, but this will require cooperative investment from industry, academia, health services and regulators. If validated targets and patient sub-sets do enter the public domain earlier than at present, then responsibility for establishing a competitive edge and robust IP will depend largely on innovative medicinal chemistry which will simply become too valuable to contract out. There are signs that the community may be moving towards precompetitive collaborations with the Structural Genomics Consortium championing more open interactions and providing wide access to chemical tools to probe new targets. Medicinal chemists play a central role in such initiatives by designing prototype molecules and developing analytical capability to build our understanding of biological pathways and mechanisms, and for target validation. Strict criteria for compound potency and selectivity should be demanded for proof of concept studies, and a further frame shift in chemical innovation will be required to exploit receptors and enzymes currently considered undruggable, and for those yet to be discovered.

On a broader precompetitive front, the EU Innovative Medicines Initiative has launched a new Euro 224 million programme jointly funded with industry to channel academic and industry partners towards new classes of antibiotics that address antimicrobial resistance. A further EU Public Private Partnership will invest nearly Euro 200 million to bring together multiple partners to create a Lead Factory comprising a European Screening Centre and compound collection. Access to HTS and 0.5 million diverse structures could enhance the rate of lead generation across the community, particularly for academic researchers who have previously had difficulty in identifying tractable chemical matter. In the US, a National Centre for Advancing Translational Sciences has been established with focus on facilitating translation from the laboratory to clinic which could have significant pre-competitive impact, although there are vociferous critics of both mission and budget. Ten pharmaceutical companies have formed a non-profit organisation called TransCelerate BioPharma to accelerate the development of new medicines, while DataShare aims to create a repository of information from cancer trials carried out by Pharma, academia and public institutions that can be shared across the community. More broadly, an international AllTrials initiative is campaigning for industry and regulators to make full Clinical Study Reports available, and GSK has taken the lead amongst large Pharma by agreeing to participate. Such precompetitive collaborations in drug discovery and development not only have the potential to reduce costs and risks, but also to bring significant patient benefit.

Economic conditions will become harsher than in the past with unflinching pressures on budgets at national, regional and local levels. Healthcare costs overall and drug prices in particular will be under the closest scrutiny as we move more towards an ageing society. Continued rises in health investment as a percentage of GDP will simply not be sustainable worldwide. New medicines will have to demonstrate positive outcomes over existing treatments, with hard evidence of reduced mortality and morbidity, improved quality of life and savings in overall healthcare budgets. There will be high expectations, or more likely demands, for innovative and cost- effective medicines that will transform treatment paradigms and justify reimbursement. Although NICE in the UK has led the way in relating treatment benefits and costs to QALYs and

Disability-Adjusted Life Years, such agencies are now commonplace throughout the world and criteria for reimbursement are becoming more stringent. Indeed, 2012 may prove to be a watershed with respect to pricing and reimbursement as five orphan drugs approved by the FDA have annual prices between $100 000–300 000 while several new anti-cancers will cost from $7000–10 000 per month, and there is already significant pushback from oncology experts. Healthcare systems may not be able to offer such expensive new therapies unless significant clinical benefit can be demonstrated, but earlier industry-agency agreement on target efficacy/safety criteria could minimise negative reimbursement decisions currently taken after years of costly investment. Encouragingly, the FDA has introduced a "breakthrough" status for fast tracking innovative new medicines based on Phase 2 data which resulted in the approval of ivacaftor for cystic fibrosis in 2012.

There have been high expectations that the developing world would provide a more welcoming environment as living standards rise, but leading countries such as China and India are driving down drug costs even more aggressively than in the West, and are tending to favour local manufacturers. Bringing cost-effective healthcare to the general population is their first priority, although expanding middle classes may be willing to pay higher prices for some new medicines. However, these markets are currently not robust enough to support investment in R&D at historical levels and few new drugs have emerged from generic companies. Given the mantra that "innovative R&D follows premium priced markets" it is unlikely that high-investment pharmaceutical research will make a major shift eastwards in the near future, particularly given worrying threats to IP that had previously been secured elsewhere. However, China has announced a five-year plan to invest $7 billion in academic projects that might lead to new drugs and eventually spawn an innovative pharmaceutical industry, although the need to build expertise and depth is openly accepted.

So what of the future? Some ten years ago, I suggested to a sceptical audience that the future pharmaceutical industry would be largely located in the US with outposts in Europe and Japan, which may well come to pass. However, even the US is in flux as budget deficits and pressure to reduce healthcare expenditure continue to force down drug costs and R&D investment. Consequently, traditional organisations are consigned to the past as the number of leading pharmaceutical companies in the US has declined from 42 in 1988 to 11 today, and all have undergone significant downsizing with major site closures. In the UK, international players such as AstraZeneca, GSK, Pfizer, Merck, Novartis, Organon (Merck/Schering) and Roche have abandoned modern research facilities, there have been thousands of job losses and the overall situation is probably still meta-stable. Indeed, decentralisation of R&D organisations is in full flow, as Pharma continues to minimise fixed costs by externalising routine research activities to CROs, and by working more closely with the academic community. For example, AstraZeneca has significantly reduced resource on neuroscience research and has moved to a virtual model where a small internal team collaborates with leading academic centres to share reward and risk. Pfizer has established Centres for Therapeutic Innovation in Boston, New York, San Diego and San Francisco to facilitate interactions with academic institutions, and has placed their own staff in collaborator laboratories. While these initiatives should provide early access to new biology, translation to successful drug discovery projects still has to be realised, and there will be the inevitable trade off between publications and IP. In addition, core expertise within Pharma, particularly medicinal chemistry, cannot be eroded too far, as successful collaborations require complementary intellectual contributions from both partners, and coherence on objectives.

Simple arithmetic suggests that given the significant scale of Pharma contraction and reduction in R&D investment, the number of new drugs reaching deserving patients will decrease, and there are also concerns that whole therapeutic areas are being abandoned. Historically, around five First in Class new medicines have been approved each year and any decline would leave major clinical needs unsatisfied. This shortfall will probably not be compensated for by Biotech where investment in early stage companies has been severely scaled back, nor is it clear that continued Pharma investment will be sustainable even at today's levels. Alternative models for R&D funding will be required involving academia, charities, governments, industry and private investors. However, given the time-scales and uncertainties traditionally inherent in drug discovery programmes, there may be pressure from funding bodies to reduce costs and risks through increased emphasis on target validation, attrition, predictive toxicology, and patient selection, and to develop more open collaborations. Funders may also need to be convinced that lessons from the past have been learned, and that cost effective and sustainable models for drug discovery can evolve to provide acceptable returns on investment.

Now would be an opportune time to strengthen drug discovery capabilities in the public sector by co-localising industry-experienced medicinal chemists alongside world class biologists and clinicians with a real commitment to the discovery of new medicines. In many cases, a fundamental change in mind set will be required for medicinal chemists to be accepted as equal partners, rather than as a service function. It will be important to build up chemistry to a critical mass as simply adding a few experienced scientists to established academic groups would not be effective. Of course, there are already research institutes and academic centres focused on drug discovery but not on the scale now required, and integration of Pharma veterans within the wider community will take time as there is little appreciation of the skills base required for medicinal chemistry. However at steady state, barriers between "academic" and "industry" researchers may soften and there would be increased permeability across previously defined disciplines and sectors. Of course, broadening individual skill sets should not be allowed to compromise quality control. Drug discovery centres would be more output-focussed than traditional academia with set objectives and goals, but rigid metrics would not be appropriate; the traditional industry dichotomy of "scientists" and "managers" would disappear, and a culture of innovation and scientific excellence would flourish. Long-term investment in the most challenging disease areas such as antibacterials and neurosciences would be encouraged and supported. There will also be important roles for Public Private Partnerships some of which have attracted significant funding for drug discovery, and have appointed scientists with industry experience who are building real and virtual R&D portfolios with multiple projects ranging from early hits to regulatory approval. These organisations and charities have traditionally focused on diseases of the developing world and cancer, but similar commitments to a wider range of therapeutic areas will be required in the future. Overall, there is a strategic and pressing need to strengthen competitive drug discovery initiatives outside Pharma and Biotech, and concerted efforts from interested parties will be required to ensure research capabilities are commensurate with future medical needs.

Drug discovery organisations will be more heterogeneous in the future, but research units could be roughly scaled in multiples of 50–100, with say a total of 200–300 multidisciplinary scientists providing an optimal balance of critical mass, personal interactions, individual accountability and potential for commercial success. Multidisciplinary teams would have disease and project focus, and would be closely integrated with clinical and academic colleagues. Medical need and scientific excellence would be fundamental drivers for project selection, which would be owned by teams through target validation, hit discovery, lead optimisation, candidate selection, biomarker PK/PD to clinical proof of concept. All team members would be actively involved in science right up to the limit of their abilities, including project leaders and directors. Skilled laboratory scientists would be recognised and rewarded with proper career progression. There would be ready access to the most relevant technologies such as HTS, protein crystallography, computational chemistry, fragment screening *etc.*, which would be expertly exploited as enablers rather than constraints or solutions *per se.* Of course, goals would be defined at group and personal levels and decisions taken with respect to portfolio priorities rather than individual preference, but the driving force would be quality not quantity. This would engender a culture of innovation and realism in which knowledge transfer and training of future generations were also highly valued. "Think global, act local" would recognise a fiercely competitive external environment, but focus on personal interactions and knowledge-based decisions would be much more effective than continual multi-site meetings, transatlantic travel and late night video conferences.

Medicinal chemists have never been in such a strong position to meet the challenges that now face drug discovery given the major scientific advances we have experienced over the past decades. We have unprecedented knowledge to design and synthesise new molecules, understand protein structure and function, and to appreciate the physicochemical factors that control delivery, efficacy and safety. We have the tools we need to exploit the massive worldwide investment in biomedical sciences, and to be more innovative and effective in execution and decision making from idea to proof of concept. Our challenge is to work with biology and clinical colleagues within a research-driven, but sustainable environment to integrate and apply our skills to discover innovative molecules that will meet the medical needs of the twenty-first century.

S. F. Campbell, Kent, UK

Acknowledgements

I am most grateful for the interest, encouragement and insightful comments from Julian Blagg, Duncan Campbell, Manoj Desai, David Fox and Graeme Stevenson.

REFERENCES

1. F. Lichtenberg, *NBER Working Paper*, 2003, 9754.
2. J. L. LaMattina, *Nat. Rev. Drug Discovery*, 2011, **10**, 559.
3. P. Morgan, P. H. Van Der Graaf, J. Arrowsmith, D. E. Feltner, K. S. Drummond, C. D. Wegner and S. D. A. Street, *Drug Discovery Today*, 2012, **17**, 419.
4. I. Kola and J. Landis, *Nat. Rev. Drug Discovery*, 2004, **3**, 711.

CHAPTER 1

Physicochemical Properties and Compound Quality

M. PAUL GLEESON,*[a] PAUL D. LEESON*[b] AND HAN VAN DE WATERBEEMD[c]

[a] Department of Chemistry, Faculty of Science, Kasetsart University, 50 Phaholyothin Rd, Chatuchak, Bangkok 10900, Thailand; [b] GlaxoSmithKline Medicines Research Centre, Gunnels Wood Road, Stevenage, Hertfordshire, SG1 2NY, United Kingdom; [c] 14 Rue de la Rasclose, 66690 Saint Andre, France, E-mail: han.vandewaterbeemd@orange.fr
*E-mail: paul.gleeson@ku.ac.th; paul.d.leeson@gsk.com; paul.leeson@virgin.net

1.1 INTRODUCTION

Currently the pharmaceutical industry is facing a severe productivity crisis. The rate of attrition from candidate drug nomination to the market place is 96% or higher.[1] In the preclinical phase, toxicity is the major cause of termination, and efficacy is the major single cause of Phase II and Phase III clinical failure (Figure 1.1). These top level reasons for failure are informative and useful, but the clinical attrition data fail to tell the full story because there are multiple reasons for efficacy failure. Digging deeper into the underlying causes of Phase II clinical attrition, where the success rate is markedly lower than other phases, revealed that as many as 43% of the molecules entering Phase II in Pfizer during 2005–2009 did not adequately test the clinical hypothesis.[2] These molecules did not have sufficiently convincing evidence that they reached the desired concentrations (exposure) to occupy the biological target in the clinical study. Despite coming from a single organisation, these observations ask some hard questions of the decision-making processes used by the industry. Are robust pharmacokinetic–pharmacodynamic relationships present in the candidate drug and how will they be confirmed in the clinic? And, of most relevance to this review, why take molecules with potential liabilities forward to costly clinical trials, when it would be preferable to further optimise them at the preclinical stage?

In this chapter we will first provide some background definitions to the key physicochemical properties, then look at the evidence for drug-like physicochemical properties as measures of compound quality in drug discovery. It has been clear for some time that molecules patented by medicinal chemists, as well as those in early clinical phases, have physical properties that are

The Handbook of Medicinal Chemistry: Principles and Practice
Edited by Andrew Davis and Simon E Ward
© The Royal Society of Chemistry 2015
Published by the Royal Society of Chemistry, www.rsc.org

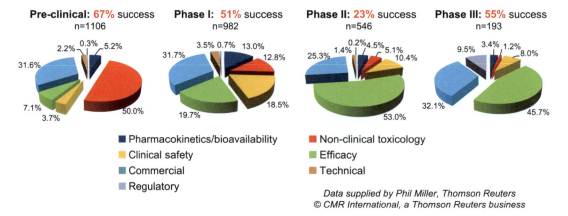

Pre-clinical: **67%** success
n=1106

Phase I: **51%** success
n=982

Phase II: **23%** success
n=546

Phase III: **55%** success
n=193

- ■ Pharmacokinetics/bioavailability
- ■ Clinical safety
- ■ Commercial
- ■ Regulatory
- ■ Non-clinical toxicology
- ■ Efficacy
- ■ Technical

Data supplied by Phil Miller, Thomson Reuters
© CMR International, a Thomson Reuters business

Figure 1.1 Industry success rates and causes of attrition, 2006–2010.
We thank Dr Phil Miller of Thomson Reuters for providing these data.

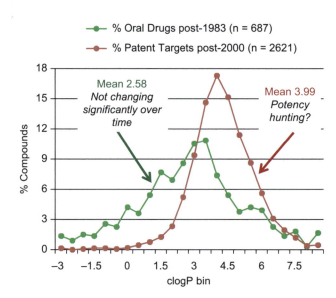

Figure 1.2 The distribution of calculated clogP in oral drugs launched since 1983[84] and in patent targets
filed by 18 major companies in 2000–2010.[95]

distinct from marketed drugs (Figure 1.2). We go on to explore reasons for this discrepancy, and
suggest that the drug discovery processes used may be unknowingly introducing molecular
property bias.

1.2 PHYSICOCHEMICAL PROPERTIES

The fundamental physicochemical properties most often used in defining compound quality
are shown in Table 1.1. Of these, log P, pK_a, log $D_{7.4}$, together with solubility and hydrogen
bonding descriptors are of critical importance.

Table 1.1 Distribution of common molecular properties for a dataset of oral drugs. Based on a dataset of ~2000 oral drugs (data from reference 84 updated with drugs launched since 2011).

Descriptor	Mean	Minimum	Maximum	Std. Dev.	Percentile Range 68%	95%	99%
MWT	341.0	42.0	1526.7	139.6	201.4–480.5	61.8–620.1	0.0–759.7
clogP	2.5	− 9.1	20.4	2.5	− 0.1–5.0	− 2.6–7.5	− 5.2–10.1
log D pH 7.4	1.2	− 15.9	17.2	2.7	− 1.5–3.9	− 4.2–6.6	− 7–9.4
HBA	5.3	0.0	51.0	3.4	1.9–8.7	− 1.6–12.1	− 5–15.6
HBD	1.6	0.0	19.0	1.7	− 0.1–3.3	− 1.8–5.0	− 3.6–6.8
TPSA	72.9	0.0	772.2	53.0	19.9–125.9	− 33.1–178.8	− 86.1–231.8
RotB	4.5	0.0	31.0	3.5	1.0–8.0	− 2.5–11.5	− 6.0–15
Carbo Ar ring	1.1	0.0	7.0	0.9	0.2–2.0	− 0.6–2.8	− 1.5–3.7
Hetero Ar ring	0.4	0.0	6.0	0.7	− 0.2–1.1	− 0.9–1.8	− 1.6–2.5
Ar ring	1.5	0.0	8.0	1.0	0.5–2.6	− 0.5–3.6	− 1.6–4.7
Neg	0.3	0.0	4.0	0.6	− 0.3–0.8	− 0.9–1.4	− 1.4–1.9
Pos	0.5	0.0	6.0	0.6	− 0.1–1.1	− 0.6–1.6	− 1.2–2.2
Chiral atoms	1.7	0.0	21.0	3.0	− 1.3–4.7	− 4.3–7.7	− 7.3–10.7
Fsp3	0.4	0.0	1.0	0.2	0.2–0.7	− 0.1–0.9	− 0.3–1.2
C atom	17.7	1.0	67.0	7.7	10.0–25.4	2.2–33.1	− 5.5–40.9
Csp3	7.8	0.0	48.0	6.3	1.5–14.1	− 4.8–20.4	− 11–26.7
Ar atom	8.7	0.0	46.0	5.8	3.0–14.5	− 2.8–20.3	− 8.6–26.1
HA	23.7	3.0	102.0	9.7	14.1–33.4	4.4–43.1	− 5.3–52.8

1.2.1 Lipophilicity

The terms lipophilicity and hydrophobicity are often used interchangeably, but IUPAC provides distinct definitions:[3]

Hydrophobicity is the association of non-polar groups or molecules in an aqueous environment, which arises from the tendency of water to exclude non-polar molecules.

Lipophilicity represents the affinity of a molecule for a lipophilic environment.

For many years the standard system in which to measure lipophilicity has been the n-octanol/water partition system. The equilibrium of a neutral compound between n-octanol and water is measured, normally at 20 °C and the partition coefficient reported on a log_{10} scale.

$$\log P = \log_{10} \left(\frac{[\text{drug}]_{n\text{-octanol}}}{[\text{drug}]_{\text{water}}} \right)$$

The solvent n-octanol became the standard lipophilic phase for the partition experiment, as it is almost non-water miscible, UV-transparent, and due to its hydroxyl group, many drug molecules can dissolve in it, unlike more hydrophobic alkane phases. It was the solvent system chosen by Corwin Hansch in the 1960s in his seminal paper at the birth of QSAR. Since then many compounds have had their log Ps measured, and large compilations exist. These databases provided the basis for prediction algorithms to calculate log P, in particular one of the first, and most popular, CLOGP. Other solvent systems have been used to define a lipophilicity scale including alkanes, chloroform, phospholipid membrane vesicles, and even retention times on various HLPC column stationary phases. But n-octanol is still the dominant system, because of the large and growing database of measurements and the now highly developed predictive

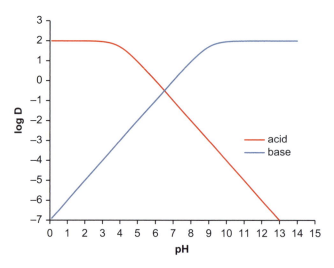

Figure 1.3   Plot of log D *vs* pH for an acid with log P = 2 and pK$_a$ = 4 (red) and a base (blue) with log P = 2 and pK$_a$ = 9.

methods. Note that these databases contain log P values measured for a compound in its neutral state, as well as log D values measured at a selected pH, often pH 7.4 (see further below).

The standard experimental procedure for measuring partition coefficients is known as the "shake-flask" method. Traditionally this would measure the equilibration of one compound, pre-dissolved in either the aqueous phase or lipophilic phase depending on its likely log P shaken in a glass laboratory bottle containing the immiscible partitioning liquids at constant temperature. The equilibration could be left overnight to achieve equilibrium, and was a highly labour-intensive measurement. The method has now been automated to run on modern laboratory automation in 96-well plates removing the throughput bottleneck of the traditional method.[4]

The n-octanol largely supports only the partition of neutral species. When the drug molecule contains an ionisable centre, the distribution of the compound between n-octanol and water becomes dependent upon the aqueous phase pH, and so the equilibrium must be measured at a particular pH.

For an acid:

$$\log D = \log P - (1 + 10^{pH - pK_a})$$

For a base:

$$\log D = \log P - (1 + 10^{pK_a - pH})$$

A theoretical plot of log D *vs* pH for an acid with pK$_a$ = 4 and a base with pK$_a$ = 9, both of log P = 2 is shown in Figure 1.3. When the solution pH equals the compound's pK$_a$, the compound is 50% ionised, and the observed log D at that pH will be approximately 0.3 log units below the log P (log D = log P – log 2).

Above the pK$_a$ of an acid and below the pK$_a$ of a base, for every 1 log unit change in aqueous phase pH, the log D changes by 1 log unit. As a standard point of comparison, the distribution coefficient measured at pH 7.4, representing physiological pH, is chosen. Most often measurements of log D$_{7.4}$ are made rather than measuring the log P. As many drug molecules contain an ionizing centre, measurement of log P would require a compound by compound choice of aqueous phase pH to ensure the compound was in its neutral form in the aqueous phase when the distribution between the phases was measured, which is a complication for an automated screening assay, and risks exposing the compound to extremes of pH during the experiment. The log P can be estimated from a log D and a pK_a measurement.

As we shall see lipophilicity is a primary determinant of compound quality, and log $D_{7.4}$ can be lowered by either decreasing the log P of the molecule, or moving the pK_a of the ionisable centre further away from pH 7.4 (lower for an acid or higher for a base). But lowering log $D_{7.4}$ while maintaining a high log P is not a favourable optimization strategy. As we shall see in this chapter, and apparent throughout the book, high log $D_{7.4}$ *and* high log P are detrimental. So what are good ranges to aim for? For oral drugs log $D_{7.4}$ values between 0–2 would seem a good target range, and for log P 2–3.

1.2.2 Calculating log P and log $D_{7.4}$

Probably the best known log P calculator is the CLOGP algorithm. It was developed at Pomona College[5] and is available through their own software package and also the DAYLIGHT cheminformatics system.[6,7] This empirically derived calculator uses a fragment based approach to estimate log P based on the 2-D graph of the molecule. The program fragments the molecule into polar fragments and isolating carbons (an isolating carbon is one which is not doubly or triply bonded to a polar fragment). The fragmental constants were estimated from simple molecules in the measured log P database. Where the fragment is unknown, it can be estimated, historically this resulting in "missing fragment" error. The additive approach includes correction terms to account for neighbourhood of polar atoms and groups, intramolecular hydrogen bonding and electronic effects.

A simpler but also widely used clogP algorithm uses atom-based functions, such as that proposed by Ghose-Crippen.[8] Many different log P calculators have been proposed and are in common use, many being variants on these two fundamental methods. A recent review compares their performance of over 20 currently available algorithms on two public databases and the Pfizer database of 95809 measurements.[9] While many methods produced reasonable results on the public database, few were successful predicting the in-house dataset. They concluded that a simple equation based on the number of carbon atoms and number of heteroatoms outperformed many methods

$$\log P = 1.46 \ (\pm 0.02) + 0.11 n_{\text{carbons}} - 0.11 \ (\pm 0.001) n_{\text{heteroatoms}}$$

Many companies with their own internal measured log P databases use QSAR approaches to either "tune" the published methods, by having them as input descriptors to a multivariate QSAR model, or calculate log P directly from QSAR models trained on their internal measurement databases using their favourite molecular descriptors.

For chemists working in projects the important question is: "does the algorithm I use predict my chemistry with acceptable accuracy and precision?". The method which is most suitable for your chemistry may differ from project to project.

In order to calculate log $D_{7.4}$ from log P, the ionization constants of the molecule must also be calculated. The physchem suite of ACDLabs has implemented a fully integrated package for calculating log P, pK_a and log D.[10] QSAR methods have been used to estimate log $D_{7.4}$ directly from chemical structure[11] without the need to calculate log P and pK_a separately.

Often within a chemical series, structural changes are being made far enough away from the ionising centre that the ionization constants of the series remain constant. Correlations between measured log $D_{7.4}$ and calculated log Ps can then use used to guide further compound optimization.

1.2.3 Ionisation Constants

Since biological membranes only really support the passive partition of neutral molecules, the ionization state of a molecule is an important property. The ionization constant, K_a is normally recorded as the negative logarithm of the ionization constant, with most drugs with ionisable centres having pK_as in the range 2–12. The pK_a is the pH at which the compound in solution is 50% ionized. Ionisation constants can be measured by a number of methods including potentiometric titration, spectrophotometrically or even by NMR. As already described the ACDLabs software can be used to calculate pK_as. ACDLabs software uses a set of Hammett equations, and an internal database of σ-values together with complex structural perception to identify the electronic environment of the ionizing centre. But pK_as can also be calculated using physics-based approaches of computational chemistry.

Manipulating pK_a is an important strategy in drug design, to optimise potency through direct drug-receptor interactions, manipulation of overall physical properties such as log $D_{7.4}$, improving solubility by introduction of an ionizing centre, controlling other pharmacokinetic properties such as lung retention[12] and modulating off-target activities such as hERG potency.[13]

1.2.4 Hydrogen Bonding

Hydrogen bonds are key drug-receptor interactions driving enthalpic binding, but also a key means of manipulating bulk physicochemical properties. Different functional groups have intrinsic different hydrogen bonding abilities, and various hydrogen bonding scales have been derived. But so far these have found few applications in drug projects. The Δlog P scales, whereby the difference between the log P in two different solvent systems, often n-octanol/water and alkane/water, appear to encode for hydrogen bonding capacity of a solute and its uptake into the brain,[14] and Δlog P measurements have been proposed as a way of describing intramolecular hydrogen bonding.[15]

Maybe one of the reasons why the intrinsic hydrogen bonding ability may be less important is that we are often exchanging hydrogen bonds (between solvent and a protein active site for example), and so increasing the hydrogen binding ability may favour the formation of the new bond but disfavour the breaking of another. The overall benefit gained by the exchange may be difficult to predict. Hydrogen bond counts are, however, widely used, and in particular the number of hydrogen bond donors appears to be a very important compound quality metric, as the number of hydrogen bond donors appears to have a large impact on permeability. The number of hydrogen bond acceptors has a wider tolerated range and is the primary means of manipulating log P.

The topological polar surface area (PSA or TPSA) descriptor is a means of quantifying the overall number of polar hydrogen bonding groups contained in the molecule. It has been suggested that for CNS drugs the PSA should be below 90 $Å^2$ [16] while it can be somewhat higher for peripheral (non-CNS) oral drugs.[17] Polar surface area is a key part of Pfizer's CNS multi-parameter optimization algorithm for identifying drugs with greater probability of success in testing hypotheses in the clinic.[18]

1.2.5 Solubility

In order for a drug to act it must be in solution. Therefore solubility is a key molecular property. For poorly soluble compounds, dissolution rate is also an important factor, although dissolution rate is likely highly correlated to the overall equilibrium solubility, in that poorly soluble compounds are likely slower to dissolve, as has been demonstrated for a series of substituted benzoic acids salts of benzylamine.[19] Modern formulation techniques can improve solubility and dissolution, as will be discussed in the Pharmaceutical Development chapter 22, but adds complexity, time and costs to the development process.

But what is sufficient solubility? Much is dependent upon dose. Estimates for a 1 mg kg^{-1} dose gives a minimum acceptable solubility of 5, 50 and 500 μg ml^{-1} for a low, medium and highly permeable compound, respectively. The concept of maximal absorbable dose (MAD) is a useful ranking tool for potential drug candidates.[20]

$$MAD = (S) \times (K_a) \times (SIWV) \times (SITT)$$

Where S = solubility (mg ml^{-1} at pH 6.5; K_a = trans-intestinal absorption rate constant (min^{-1}); SIWV = small intestinal water volume (approximately 250 ml for man) and SITT = small intestinal transit time (approximately 270 minutes for man).

The Biopharmaceutics Classification System (BCS) of novel chemical entities classifies drugs based on their permeability and solubility,

BCS class 1 being high permeability high solubility
BCS class 2 being high permeability low solubility
BCS class 3 being low permeability high solubility
BCS class 4 being low solubility low permeability

The FDA has issued guidance allowing applications for biowaivers for BCS class 1 immediate release solid dose oral drugs from the need for *in-vivo* bioequivalence testing, as absorption is unlikely to be dependent upon dissolution and gastric emptying time.[21]

1.2.6 Measurement of Solubility

The measurement of solubility, while superficially a simple experiment, has many pitfalls and caveats. The experimentally determined solubility is dependent upon:

Buffer, ionic strength
Temperature
pH
Supersaturation
The starting solid state form, its history and impurities

amongst others. Many experimental protocols try to measure to thermodynamic equilibrium, but for some compounds this can take extended periods of time. The solubility is often measured at a particular pH, often pH7.4 and in a particular buffer, although the solubility of the neutral form can be measured, in which case its known as the intrinsic solubility.

Dissolution rate is an even more involved experiment. In order to control as much of the potential sources of variability as possible, intrinsic dissolution rate requires a spinning disc of the solid to control surface area and fluid flow across the surface.

In order to generate solubility on large number of compounds, solubility from a stock DMSO solution injected in to a buffer, after suitable equilibration can be measured. Solubility can then be determined from the turbidity threshold, or quantified spectrophotometrically or by HPLC. In a validation study at AstraZeneca the measured solubilities of 200 compounds showed a good correlation with a gold standard manual method measured from solid with average-fold deviation of 3-fold.[22] Some compounds showed much larger differences. It was noted that the authors did not expect a perfect correlation within the replicate error of each other's measurements, as they had changed the solid state form.

1.2.7   Calculating Solubility

Solubility is difficult to measure accurately and precisely and more difficult to predict. Most solubility calculators use empirical QSAR equations based on literature data or in-house data-sets. Their predictiveness for your chemical series should always be tested, as described in the QSAR chapter.

The Yalkowsky General Solubility Equation (GSE),[23] which is not in fact a QSAR model, but a physically-derived equation, contains a negative coefficient for log P and melting point, as dissolution would require the breaking and crystal packing interactions and the disruption of water structure by the hydrophobic nature of the drug.

$$\log S = -\log P - 0.01\ mp + 1.2$$

Few QSAR models, nor indeed the Yalkwosky GSE for solubility, do better than an average error of ±0.9 log units, a range so wide as to be of limited utility in optimizing solubility for most projects.

1.2.8   Other Compound Quality Indicators

Alongside molecular weight, and the described physicochemical descriptors, many counted descriptors are used including hydrogen bond acceptor and donor counts (HBA and HBD), rotatable bond counts (RotB), aromatic and aliphatic ring counts and other atom-type counts (Carbo Ar ring, Hetero Ar ring, Ar ring, Chiral atoms, C atom); charge counts (Pos Neg) *etc* as indicated in Table 1.1.

1.3   COMPOUND QUALITY AND DRUG-LIKENESS

The influence of a molecule's physical properties on its behaviour in biological systems has been recognised since the 19th century. The seminal contributions of Corwin Hansch and co-workers since the 1960s established lipophilicity (or hydrophobicity) as a key measure to take account of in drug discovery projects. The lipophilicity measure most frequently employed is the logarithm of the partition coefficient between water (log P) or buffer (log D) and 1-octanol, which can be measured or calculated. The importance of an *optimal* lipophilicity comes from consideration of both the pharmacokinetic and pharmacodynamic properties of drug molecules. In gaining exposure to drug targets *in vivo* (pharmacokinetics), molecules have to be both water soluble and membrane permeable, while binding to drug receptor targets (pharmacodynamics) requires desolvation from bulk water. The ionisation constants and ionic class of molecules are clearly important in these processes.[24] As long ago as 1987, Hansch[25] had proposed the 'minimal hydrophobicity' hypothesis:

"Without convincing evidence to the contrary, drugs should be made as hydrophilic as possible without loss of efficacy."

This hypothesis has withstood an onslaught of later analyses. A new generation of medicinal chemists, now employing pharmacokinetic and toxicity optimisation in the discovery phase, have been busy 'reinventing' the role of lipophilicity, as well as seeking other molecular properties and guidelines that might help to better understand overall drug-likeness. While a convincing definition of drug-likeness *versus* non-drug-likeness has proven elusive, we think that lipophilicity[26] remains the single most important physical property to optimise.

1.3.1 The Rule of Five, and Other Physical Properties

Following the insights provided by Hansch, the 'rule of five' (Ro5) published by Christopher Lipinski in 1997 provided the next step forward.[27] Lipinski noted that marketed drugs and drugs in clinical development had distinctive distributions of physical properties, with 90 percentile values of clogP $\sim$5, molecular weight $\sim$500, OH and NH count (a hydrogen bond donor estimate) $\sim$5, and O plus N atom count (a hydrogen bond acceptor estimate) $\sim$10. Lipinski suggested keeping below these values as a useful property guideline for obtaining good permeability in drug candidate molecules. The Ro5 is indeed a powerful indicator, with about 80% of all marketed oral drugs passing all four rules and, more tellingly, $<$5% failing two or more rules. It is also easy to remember and to calculate. These facts have contributed to its continuing wide acceptance (in February 2013, the Ro5 paper had been cited $>$5400 times according to SciFinder®). There is a danger of improperly using property cut-offs too literally. For example, it makes no sense to reject a molecule only because it has a molecular weight of 510 (the Ro5 was not intended to be used in this way). Conversely, less desirable molecules may only just pass all four rules. One approach to overcome this hard cut-off issue uses a 'quantitative estimate of drug-likeness' (QED), which scores molecules on a continuous scale, using desirability functions that are derived from each particular property distribution.[28] A number of drugs that fail Lipinski criteria have QED scores that overlap with drugs that pass the criteria.

Following the Ro5, there have been numerous studies that support the role of physical properties in ADMET studies, summarised in number of reviews.[26,29–33] Notably, studies using large data sets from pharmaceutical companies reach similar conclusions and insights about the importance of optimising physical properties, especially lipophilicity.[34–38] Additional relevant physical properties include polar surface (PSA) which influences permeability, bioavailability and *in vivo* toxicity; H-bond donors (HBA) which influence permeability; rotatable bond count, thought to influence bioavailability; the fraction of tetrahedral carbon atoms (Fsp3), which is correlated with solubility and is higher in drugs than in compounds in the clinical phase;[39] aromatic ring[40] count, which is related to compound solubility and other 'developability' attributes, independently of lipophilicity;[41] non-aromatic rings;[42] chirality;[43] and ionisation.[24]

It is important to realise that some physical properties are closely inter-correlated. It has been known for some time that lipophilicity is a composite measure, quantitatively dependent on molecular size, hydrogen bonding, and polarity.[44] There is the fascinating observation that three-dimensional plots of related physical properties in organic molecules (for example, log P, molecular weight and PSA) form a plane, dubbed the BC plane,[45] from which the underlying properties, suggested to be 'bulk' (B) and 'cohesiveness' (C) can be calculated using principal component analysis (PCA). A PCA model generated on $\sim$30 000 pre-clinical compounds[34] shows that much of the information contained within commonly used descriptors is redundant (Figure 1.4). Four orthogonal components can describe $\sim$81% of the variation in the 12 descriptors used. The analysis indicated that the most meaningful, non-redundant descriptors, located at the extreme of each axis, can be encompassed by molecular weight, lipophilicity (clogP, or alternatively PSA, HBA or HBD), HBDs and ionisation state indicators. The principal component descriptor approach has clear merit at a fundamental level but has not caught on widely, perhaps because these parameters are more difficult for chemists to calculate, understand and use than log P, molecular weight, PSA, hydrogen bond count, *etc.*, in the design of new molecules.

1.3.2 ADME and Physicochemical Properties

The derivation of rules and guidelines for specific ADME assays have been reported in the literature over many decades and these are commonly used during hit selection and to help guide optimisation. A PSA $<$ 60 Å^2 has been reported to be desirable to ensure complete absorption.[46]

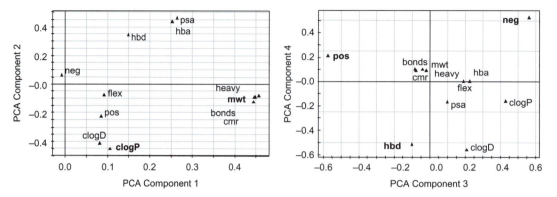

Figure 1.4 PCA loadings plot showing the inter-correlation between common physicochemical parameters. Highlighted are the key descriptors that contribute essentially unique chemical information.
Adapted with permission from (M. P. Gleeson, Generation of a set of simple, interpretable ADMET rules of thumb, *J. Med. Chem.*, 2008, **51**, 817–834.). Copyright 2014 American Chemical Society.

In addition, the importance of ionisation state has been noted in a number of publications on bioavailability.[34,47] Acidic molecules with low to moderate PSA have a high probability of having acceptable exposure. The exposure of neutral and basic compounds was found to be largely controlled by their Ro5 characteristics.[47] More recently, a detailed analysis of exposure data has shed light on the role of log P, MWT and rotatable bonds (RB).[48] It was found that higher MWT significantly impacted the fraction absorbed, while fraction escaping gut wall and hepatic elimination decreased with increasing lipophilicity. It was also found that a parabolic relationship existed between bioavailability and polar descriptors. RB and MWT had an overall negative effect on bioavailability.

The dependence of solubility on lipophilicity has been known for decades.[49] Additional factors such as the presence of functional groups that can provide stabilisation of the crystal lattice are important terms to consider.[50] Large, lipophilic molecules will generally have poor solubility, yet while polar features are preferred, they can lead to reduced solubility by formation of strong intermolecular interactions within the crystal lattice. Ionisation is important, with acid and bases having dramatically higher solubilities.[34,51] Further, Young *et al.* reported on the detrimental effect that aromatic rings can have on solubility, above and beyond that due to their MWT and log P.[41]

Lipophilicity is also recognised as one of the key descriptors controlling permeability.[52] Fichert *et al.* propose log D values between 0 and 3 for good permeability.[53] A complicating factor is that the data from some researchers show a linear relationship between log P_{app} and log P or log D,[52,54] while for others it is parabolic in nature.[53–55] Neutral molecules are expected to be the most permeable on average, followed by bases, zwitterions, and finally acids.[53,55] It has also been noted that as PSA (and MWT) increases, permeability is found to decrease.[34,35,52,53,56]

Like membrane permeability, penetration of the CNS barrier is also inversely affected by log P and PSA. Hansch *et al.* observed that the relationship between CNS penetration and octanol/water log P was parabolic, with a maximum P_{app} at a log P of ~ 2.[25] The most commonly used guideline involves PSA, where values < 60–70 Å^2 are expected to have good CNS penetration.[57] Norinder and Haeberlein[58] have proposed the following simple rules: rule 1, if N + O (the number of nitrogen and oxygen atoms) in a molecule is ≤ 5 it has a high chance of entering the brain; and rule 2, if log P $-$ (N + O) is > 0 then logBB is positive (*i.e.* greater drug concentration in the brain *versus* blood). The degree of permeability of a molecule may also be affected by transporter activity such as at P-gp, causing efflux from the brain. The consensus view is that increasing size and hydrogen bonding capability lead to increased P-gp substrate activity.[34,59,60] These trends are in line with a recent study of the physicochemical properties of CNS space,[61] showing that the physicochemical parameters associated with CNS drugs were considerably more restrictive than drug candidates in current development, particularly with respect to clogP and MWT.

Once a compound has been absorbed into the bloodstream it is also necessary to consider how and where it is distributed. Again, this shows a dependency on simple physicochemical properties. It is found that the volume of distribution of compounds is of the order bases > neutral ~ zwitterions > acids, with lipophilicity playing an additional role.[62–64] The opposite trend exists for plasma protein binding, with acidic compounds being the most bound (and thus least effectively distributed), followed by neutral compounds, bases and zwitterions. Both globally, and within specific series, log P has shown a strong positive correlation with plasma protein binding.[65–68]

Based on a consideration of the ADMET SAR trends described above and elsewhere,[34,51,69–72] it is clear that molecular properties lying outside of those that encompass oral drugs generally lead to poorer overall ADMET properties. A concern for drug development pipelines dominated by molecules within less desirable regions of property space (high MWT and clogP) is that many will fail in more expensive clinical development phases.[32,73] A strategy to reduce late stage risk would be to balance the pipeline with more drug-like compounds with less probability of ADMET issues (see Figure 1.2). Given the relationships between ADMET and physical properties, we hypothesise that restraining basic physicochemical properties within drug-like space should increase the probability of success of a candidate drug portfolio. In Table 1.1 we report the mean and percentile ranges for physicochemical properties of ~2000 compiled oral drugs. In essence, the area of occupation of the majority of oral drugs lies within rather tight boundaries; however drug discovery is increasingly encompassing a much larger physical property space which contains fewer marketed drugs.

1.3.3 Toxicity and Physicochemical Properties

In a study of 245 candidate drugs from Pfizer,[74] a key learning point was the observation that toxic outcomes in rats and dogs were directly proportional to plasma exposure (total or free); a total plasma concentration of 10 μM resulted in equal incidence of toxic and non-toxic outcomes. Using the 10 μM exposure criterion, compounds that had clogP < 3 and total polar surface area (TPSA) > 75 Å^2 were six-fold less likely to show toxicity *versus* compounds with clogP > 3 and total polar surface area (TPSA) < 75 $\text{Å}.^2$ Using clogP alone, there was a 2.4-fold increase in toxic outcomes when clogP > 3 *versus* clogP < 3.[74] A follow-up study from Lilly[75] using 485 compounds in rats provides support, showing that at the 10 μM no adverse effect level, there was a 3.2-fold increase in toxic outcomes with clogP > 3 *versus* clogP < 3, but no significant effect of TPSA. Increasing *in vivo* exposure, as a consequence of increased volume of distribution and reduced clearance, were the strongest surrogate measures reported for the observed lowest no adverse effect levels.[76] These two studies[74,75] are consistent in showing increased toxicity risk as both exposure and lipophilicity increase.

The decision to progress a molecule to development is rarely a consequence of 'no toxicity' in animals, but is dependent on the margins between the highest therapeutic exposure and the no toxic effect exposure (therapeutic index). Calculating these margins is a challenging but necessary step from lead discovery to late phase clinical studies.[77] There are no studies we know of that have looked at safety margins in relation to physical properties. A study from AstraZeneca[76] explored the differences between basic candidates that failed preclinically for animal toxicity *versus* those that progressed to the clinic; it is reasonable to assume that the clinical compounds would have higher safety margins than those that did not proceed. Lipophilicity played no role, perhaps because the mean clogP was lower in AstraZeneca compounds (3.2) *versus* the Lilly compounds (4.0; from the supplementary data in reference 75). A model was derived from principal least square analysis and was supported by a set of recent marketed drugs. The main factors that promoted progression to man are consistent with emerging drug-like themes: small size, increased three-dimensionality and increased polarity.[76]

Toxicity can result from unwanted effects of molecules at other biological targets (promiscuity), as well as from their primary mechanism. A role for lipophilicity in molecule-based *in vivo* toxicity is supported by a number of studies, summarised in a recent review[30] of receptor promiscuity, where several studies show that compound lipophilicity, basicity, and some structural features are important. *In vitro* cellular toxicity is also indicative of *in vivo* toxicity risk, and, consistent with the promiscuity data, cytoxicity increases with increasing pK_a (>5.5) and lipophilicity (>2).[78] Drugs that show idiosyncratic toxicity, and are either withdrawn or have a 'black box' warning, have higher doses and exposures than do the top 200 prescription drugs; keeping the dose <40 mg dramatically lowers the probability of idiosyncratic events.[79] Finally, the risk of drug induced liver injury increases when the dose is >100 mg and clogP is >3 (the 'rule of two').[80]

Since toxicity is the major cause of preclinical attrition (Figure 1.1), and the bulk of compounds pursued by industry have clogP >3 (Figure 1.2), primary recommendations on how to reduce preclinical toxicity attrition are obvious: aim for low efficacious exposure levels, and reduce clogP in drug candidates.

1.3.4 Effect of Time on Oral Drug Properties

Drug properties will obviously depend on the initial selection of the molecules, then on the pressures applied to them throughout the arduous and demanding development process, where differences in physical properties exist between the development phases.[73] One approach to explore the relative influence of different physical properties is to look at trends over time in marketed drugs. The hypothesis is that those physical properties of drugs that do not change over time, or that change the least, are more important 'drug-like' attributes. The first such study looked at FDA approved oral drugs[81] launched pre- and post-1983.[82] In the post-1983 group, there were significant increases in molecular weight, hydrogen bond acceptors, PSA, ring count and rotatable bonds *versus* the pre-1983 group. In contrast, log P, hydrogen bond donors and %PSA showed no differences in the two groups.[82] More detailed studies, using the year of invention or publication as well as launch, were consistent in showing that two of the Lipinski parameters, log P and hydrogen bond donor count, showed much less change over time than molecular weight and hydrogen bond acceptor count.[31,83] Other properties changing little over time in oral drugs are Fsp3 and Ar-sp3, the difference between aromatic atom and sp3 carbon atom counts.[84] The physical properties of oral drugs show a number of differences according to ion class (*i.e.* acids, bases, neutrals or zwitterions); however there has been little change in log P by ion class over 50 years, with the exception of more recent acidic drugs.[84] Our updated analysis of the publication dates of oral drugs versus their molecular properties shows the clear increase in MWT and total AR rings over time, and the increase in clogP since the 1990s (Figure 1.5).

The mean physical properties of the molecules produced by pharmaceutical companies in patents from 2000–2010 show little change over this time period, but on average these have 'inflated' bulk physical properties *versus* post-1983 marketed drugs.[84] Patented molecules also have increased mean aromatic ring count, and are less structurally complex in having notably fewer mean chiral centres and lower mean Fsp3 values than the drug set. Patent data are the most reliable to use in assessing trends in practice, because they are strategic, being published by pharmaceutical companies to protect inventions of potential commercial value. In contrast, literature data, from both industry and academia, are not strategic and are specially selected to illustrate scientific discoveries. Despite this, a year-by-year study[85] of the compounds published over 50 years in the *Journal of Medicinal Chemistry* shows marked increases in most physical properties over time, and is generally consistent with both the changes seen in drug properties and with patent data. In this study, the mean lipophilicity of published compounds and drugs

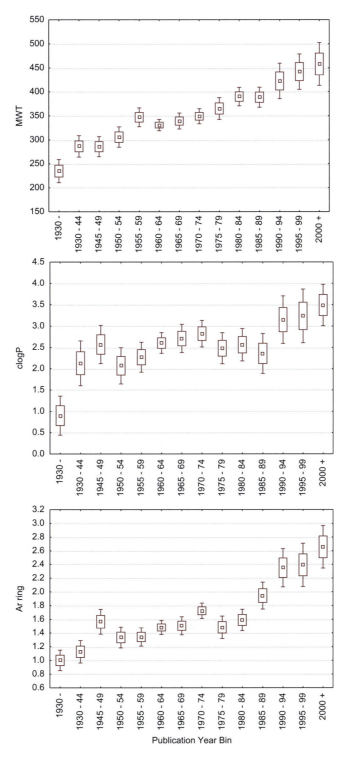

Figure 1.5 Analysis of the variation of three physicochemical properties over time (date of first publication) for a dataset of ~2000 oral drugs (data from reference 84 updated with drugs launched since 2011).

are not notably different until after 1990, when the published compounds start to increase while drugs do not. The synthetic methods commonly used in medicinal chemistry, especially metal mediated cross coupling of aryl moieties, almost certainly contribute to the more aromatic and lipophilic molecules found in patents and in the literature in comparison with drugs.[86–88]

1.3.5 Non-Oral Drug Properties

Drugs given by non-oral routes have differing mean physical properties from oral drugs,[81] although >98% of a group of inhaled, ophthalmic and transdermally administered drugs still comply with the Ro5.[89] Intravenous drugs have lower lipophilicity than oral drugs, which can be explained by the need for the dose to be soluble in relatively small volumes. The mean log molar solubility of a group of 79 intravenous drugs[90] was − 1.57 *versus* − 2.57 for 512 oral drugs (data calculated from supplementary material in reference 90).

Since many drugs are used for both non-oral and oral administration, it is fair to say that the design concepts for *specifically* targeting some non-oral routes lag behind the better understood oral approach. This is perhaps best exemplified by inhaled drugs, for the topical treatment of airways disorders, which in comparison with oral drugs are significantly larger, less lipophilic and have more hydrogen-bonding groups.[91] In this case, designing molecules with low permeability and solubility are strategies proposed to increase lung retention and duration of action. Reducing bioavailability and building in specific soft metabolic sites to limit systemic exposure are further strategies for inhaled molecule design. The consequences of these tactics (which appear to be the opposite of oral drug design) for drug safety have yet to be clarified. There is an extra challenge in inhalation of optimising suitable crystalline physical forms for use in dry powder devices. The current state of inhaled molecule design is nicely illustrated by the efforts to find dual action β_2 agonists and muscarinic antagonists with long duration, for treatment of chronic obstructive pulmonary disease.[92]

1.3.6 Effect of Target Class

Differences between mean physical properties according to their target class have been observed in drugs,[84,93] literature compounds[94] and in patented compounds (Table 1.2).[84,95] In general, these differences reflect the differences in physical properties of the target class endogenous ligands. For example, mean lipophilicity is greatest for drugs acting at nuclear hormone receptors and lipidergic G protein-coupled receptors and least for proteases; there is a similar trend in pharmaceutical patents. In general, the patented molecules also show increased physical properties across most target classes in comparison with oral drugs.[95] An interesting exception is kinase inhibitor drugs, where patents are notably less lipophilic than the first drugs for this target class. With hindsight, this suggests that early less optimised kinase inhibitors were chosen for development, most probably driven by competitive pressures and the urgent therapeutic need in cancer treatment. Increases in aromatic atom or ring count are evident in most target classes *versus* drugs,[84] perhaps a result of favoured synthetic methodology.[86]

Despite the differences in mean physical properties between target classes, there is high variability within each target class and it is not possible to predict target class 'membership' of a molecule based on physical properties alone.[26] The observed differences result from a combination of real differences between binding sites ('druggability'), the selection of chemical starting points, and the subsequent optimisation strategies pursued. If the lead molecules chosen are based on endogenous ligands or related structures from screening, then this will likely lead to corresponding target class differences in drug candidates. *A key point follows*: if the hypothesis is not challenged that a particular receptor will always require a lipophilic or large ligand, then a lipophilic or large drug candidate will be a highly probable result.

Table 1.2 The effect of target class on a range of physicochemical properties. The mean and standard deviation (in parenthesis) values are reported. Care should be taken not to over-interpret this data given the small numbers in certain categories (*i.e.* GPCR and Kinase).

Descriptor	All (N = 2108)	GPCR-amine (N = 278)	Ion Channel (N = 151)	Nuclear Receptor (78)	Protease (N = 50)	Transporter (N = 48)	Kinase (N = 23)	GPCR-peptide (N = 22)	GPCR-lipid (N = 14)
MWT	341 (139.6)	335.5 (85.2)	325.2 (115.1)	379.7 (82.8)	446.2 (138.3)	306.1 (79.8)	467.1 (75)	464.6 (81.3)	394.5 (72.7)
clogP	2.5 (2.5)	2.9 (2)	2.7 (1.7)	4.2 (2)	2.3 (2.3)	3 (1.6)	4.2 (1.3)	4.5 (1.7)	4.2 (2.1)
log D pH 7.4	1.2 (2.7)	1.3 (1.8)	1.8 (1.7)	3.5 (1.5)	0.4 (2.6)	1.4 (1.7)	3.3 (1.1)	2.5 (2.4)	2.7 (2)
O + N	5.3 (3.4)	4.4 (2.1)	5.1 (2.4)	3.8 (1.7)	7.8 (2.6)	3.9 (2.3)	7.1 (1.2)	7.1 (2.5)	5.2 (2.2)
OH + NH	1.6 (1.7)	1.3 (1.2)	1.2 (0.9)	1.2 (1)	2.6 (1.4)	1.3 (1.1)	1.8 (0.8)	1.7 (0.6)	1.9 (0.7)
TPSA	72.9 (53)	52.6 (30.9)	69.4 (34.5)	59.4 (27.3)	110.3 (37.2)	54.4 (39.9)	87.9 (15.6)	93.6 (34.2)	81.7 (25.8)
RotB	4.5 (3.5)	5.1 (2.9)	3.8 (3)	3.2 (2.4)	8.1 (4.4)	4.1 (2.2)	6.1 (2)	6.9 (2.1)	7.9 (3.3)
Carbo Ar ring	1.1 (0.9)	1.4 (0.7)	1.3 (0.8)	0.8 (1.1)	1.2 (0.8)	1.4 (0.6)	1.8 (0.7)	2.1 (0.8)	1.2 (1)
Hetero Ar ring	0.4 (0.7)	0.4 (0.6)	0.3 (0.6)	0.1 (0.3)	0.4 (0.7)	0.2 (0.5)	1.6 (0.7)	1.3 (0.8)	0.5 (0.8)
Ar ring	1.5 (1)	1.8 (0.8)	1.6 (0.9)	0.9 (1.2)	1.6 (1.1)	1.7 (0.7)	3.4 (0.8)	3.5 (1.2)	1.7 (1.6)
Neg	0.3 (0.6)	0 (0.2)	0.1 (0.4)	0.2 (0.4)	0.5 (0.6)	0.1 (0.3)	0.1 (0.3)	1 (0.7)	0.6 (0.5)
Pos	0.5 (0.6)	0.9 (0.3)	0.4 (0.5)	0.1 (0.3)	0.7 (0.6)	0.7 (0.5)	0.5 (0.6)	0.3 (0.5)	0.1 (0.3)
Chiral atoms	1.7 (3)	1.1 (1.4)	0.9 (1.8)	4.1 (3.2)	2.9 (1.8)	1.1 (1.3)	0.2 (0.5)	1 (1.4)	2.1 (2.1)
Fsp3	0.4 (0.2)	0.4 (0.2)	0.4 (0.2)	0.6 (0.2)	0.5 (0.2)	0.4 (0.2)	0.2 (0.1)	0.3 (0.1)	0.5 (0.3)
C atom	17.7 (7.7)	18.8 (5.1)	17.1 (6.8)	22.1 (4.3)	23.2 (7.9)	16.1 (4.9)	23.2 (3.7)	25.4 (4.8)	22.6 (3.3)
Csp3	7.8 (6.3)	8.4 (3.5)	6.2 (4.7)	12.7 (5.9)	11.9 (5.2)	6 (3.4)	5.6 (2.6)	8.1 (3.9)	11 (5)
Ar atom	8.7 (5.8)	10.1 (4.1)	9.2 (5.2)	5.3 (6.8)	9 (6.3)	9.7 (3.9)	18.7 (5)	19 (6)	9.4 (8.4)
HA	23.7 (9.7)	23.8 (6)	22.9 (8.3)	26.6 (4.4)	31.6 (10)	21.2 (5.5)	32.4 (4.7)	33.5 (5.9)	28.3 (5.2)

1.3.7 Effect of the Individual Chemist and the Organisation

Drug discovery is a highly creative activity that has generated an armoury of > 2000 medicines with a massive positive overall effect on human health. There is no prescribed way to invent a drug—the selection of a chemical starting point and the path taken to optimise it are prone to the intuitive, but also subjective, mind-set of the chemists and other scientists involved. Lajiness *et al.*[96] assessed the consistency of medicinal chemistry decisions by asking a group of chemists to select a set of drug-like molecules from lists of 2000 molecules. They found that medicinal chemists were not very consistent in the compounds they rejected as being un-desirable, due to human subjectivity. The authors noted that this has significant implications for project decision making since different teams would presumably take forward different starting points. Kutchukian *et al.*[97] considered the decision making further, going into greater detail to understand what precise factors are guiding medicinal chemists' decision making. The authors assessed what factors (physical properties, scaffold type, functional groups, diversity, *etc.*) were behind the clustering a set of ~ 4000 fragments into lead-like and non-lead-like sets. Interestingly, of the 19 chemists involved, most compound decisions were made taking into account no more than 2–3 parameters (fewer than they realised), with ring topology and functional groups being the key factors. In addition, the complexity of the problem was greatly simplified by each individual to facilitate the selection strategy, and there appears to be greater consensus between chemists in this study on what constitutes an undesirable *versus* a desirable molecule. Further, despite their own assertions, individual chemists are generally unaware of the factors that impact their decision making, depending mostly on their instinctive feeling or prejudice about a molecule's desirability.[97]

An interesting retrospective study on the numbers of compounds made in lead optimisation showed that chemists' activities frequently result in more molecules being synthesised than are needed.[98] In this study, the median numbers of molecules made in candidate-producing chemical series was 147, with the candidate being the median 19th in the series. In five of the 23 series discussed, the candidate was the first or second molecule; in these cases, a single modification provided a 'step-jump' resulting in candidate criteria being met, with the re-mainder of the molecule already optimised by earlier efforts. These observations suggest that work in a series could be stopped if this key progression step is identified. A useful definition of a chemical series was suggested: '*a distinct set of compounds that contain a common structural motif that consistently provides the analogues with the same unique advantage.*'[98]

While individual chemists may have differing perceptions about compound quality, they generally work in teams, with multiple inputs available to select molecules and to solve problems in lead optimisation. In addition, computational tools are now widely available for chemists to use to help predict properties in advance of deciding which molecules to synthesise.[99] These factors might be expected to have had an 'equalising' influence on decision making, but there is strong evidence that a team's thinking processes are influenced by the local environment. Thus, significant differences in physical properties in specific projects are seen in the patents pub-lished by pharmaceutical companies. For example, a comparison of the mean molecular weight and lipophilicity seen in patents from AstraZeneca, GlaxoSmithKline, Pfizer and Merck Inc., showed marked differences between them that were independent of target class.[31] Identical differences were seen at the project level in the prosecution, by all four companies, of the same C–C chemokine receptor 5 antagonist pharmacophore (a phenpropylpiperidine).[31]

A further detailed study[95] of the patents of 18 pharmaceutical companies, over the period 2001–2010, confirmed the presence of an 'organisational' effect influencing drug discovery activities. The specific targets pursued were taken into account by using the physical property differences between all possible pairs of companies working on the same target. In fact, there was a considerable overlap in targets of interest, and target-unbiased property differences in

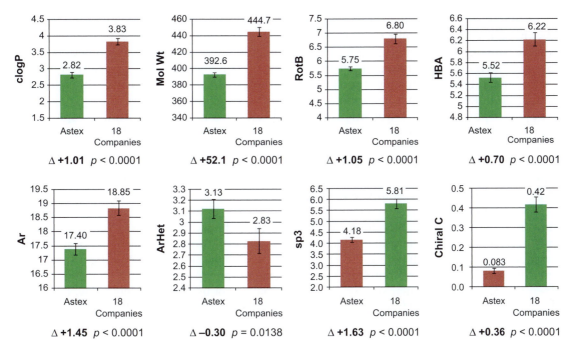

Figure 1.6 Impact of fragment-based drug discovery using data derived from Astex Therapeutics (a company that focuses only on fragment-based drug discovery) and 18 companies.[95]

companies correlated highly with their average property values, especially with lipophilicity. The conclusion is that the differences in physical properties between various organisation's compounds are not influenced very much by the targets or target classes pursued, but rather more by the different strategies and tactics followed.[95] Project teams appear to be more influenced by their local beliefs, knowledge and experience than by external observations. Perhaps a 'not invented here' syndrome also prevails? Compared to Astex, a specialist fragment-based drug discovery organisation, pharmaceutical companies had fewer drug-like patented molecules, except for sp3 and chiral atom counts, on the same shared targets (Figure 1.6).[95] In Astex, the *raison d'etre* is atom-by-atom modification of fragments, driven by X-ray structure determination and ligand efficiency optimisation, with essentially the same approach applied to all targets. It is clearly harder for a large company to drive such a consistent philosophy.

1.3.8 'Exception' Space

Examples of drugs that violate the Ro5 are often cited as examples, accompanied by concerns that over-stringent application of 'rules' will restrict innovation.[100,101] In response to the innovation concerns, it is arguable that pointing drug design towards more restricted drug-like chemical space places a premium on innovative problem solving whereas doing the opposite does not. Physical property 'exception' molecules should enter development if alternatives cannot be found, but equally, if such molecules dominate a pipeline, the risk of failure would be expected to increase. We think that active management of the physical property profile of an early stage drug portfolio is an essential step in the war on attrition.

It has not been easy to extract general insights from 'exception' drugs which lie outside, or at the extremes of, Ro5 space. A number of studies suggest that high lipophilicity will be better tolerated if it is associated with aliphatic moieties (as found in many natural products) rather than aromaticity.[39,41,84] An analysis of $> 30\,000$ GlaxoSmithKline compounds used a combined developability score from high-throughput aqueous solubility, artificial membrane permeability, human serum albumin binding and CYP450 3A4 inhibition screens to specifically investigate sub-optimal chemical space.[102] Compounds were divided into quadrants using molecular weight>or <400 and clogP>or <4.[34] The 'largest and greasiest' quadrant (clogP>400 and molecular weight>400) had the lowest developability score, which was predominantly linked to lipophilicity, not to size. In all clogP/molecular weight regions, 'composite' descriptors [clog D+aromatic ring count],[41] [aromatic atom count minus sp3 carbon count],[84] [clog D+(aromatic atom count minus sp3 carbon count)], and [sp3 carbon count/total atom count][39] were important for developability.

Blocking sites of metabolism can improve pharmacokinetic profiles but this will normally increase bulk physical properties and reduce solubility. Some natural product drugs do exist in a different 'chemical space' compared to synthetic drugs.[103] Steroids, macrocycles, sugars and vitamins are more 'three-dimensional' molecules, with a greater number of tetrahedral (sp3 hybridised) carbon atoms, and fewer aromatic atoms, than other drugs. Cyclosporin is the benchmark for a complex natural product that is an oral drug, and it has been proposed that the intramolecular hydrogen bonding it displays may be of general utility in 'hiding' the effect of high numbers of hydrogen bonding atoms in membrane permeation.[104] Natural product screening has had spectacular success in the past but is not used very widely today. Exploiting the structural motifs found in natural products has much promise in search of novel templates and scaffolds,[105] but progress has been slow because this strategy is burdened by challenging synthetic chemistry.[106]

Non-standard formulations can 'rescue' poorly soluble molecules, a recent example being the HCV protease inhibitor telaprevir (Figure 1.7), which violates three of the Ro5 guidelines. Telaprevir requires a special spray-dried dispersion formulation of an amorphous polymorph,[107] and dosing with a high fat diet improves absorption. Solving such problems can therefore help to advance challenging molecules, but also adds very significantly to early development time

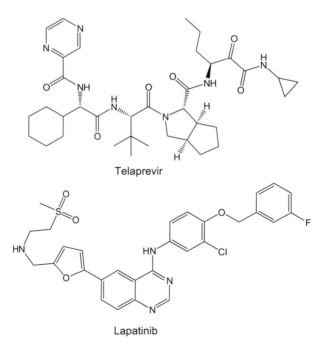

Telaprevir

Lapatinib

Figure 1.7   Examples of oral drugs that violate the rule of five. Top: Telaprevir,[107] molecular weight 680; clogP 5.4; O+N count 13. Bottom: Lapatinib,[108–111,138] molecular weight 581; clogP 5.8.

and cost.[33] In the case of cancer drugs, a higher risk to benefit ratio is acceptable relative to other therapy areas. An example is lapatinib, a significant new drug for treatment of breast cancer, which acts as a specific dual inhibitor of EGFR (endothelial growth factor receptor, ErbB1) and ErbB2 kinases. The molecular weight and lipophilicity of lapatinib break the Ro5 guideline, and it also has five aromatic rings indicating predictable developability risks.[41] Lapatinib binds in a novel mode to EGFR where it unexpectedly displays slow off-rate kinetics, thereby benefiting cellular efficacy.[108] On the other hand it carries a 'black box warning' because of a significant risk of drug induced liver injury, which can be explained by the formation of a number of reactive metabolites.[109] Lapatinib also inhibits the hERG (human Ether-à go-go) ion channel, a well-established cardiovascular risk,[110] and has variable bioavailability because of its high dose (1500 mg) and poor solubility above pH 4.[111] Thus the unexpected kinetic advantage of the structure is offset by a number of disadvantages which would probably be prohibitive in other therapeutic areas. The point here is that if a project is restricted to sub-optimal chemical space, one strategy is to actively identify and then optimise advantageous properties, rather than relying on serendipity.

1.4   THE DRUG DISCOVERY PROCESS: DOES IT UNKNOWINGLY INTRODUCE A BIAS IN MOLECULAR PROPERTIES?

Drug discovery involves the iterative optimisation of a lead series against a diverse panel of assays with different dependencies on the key physicochemical parameters. Thus great care must be taken that the data from particular experimental assays are carefully considered so as to not unknowingly bias the physicochemical properties of a lead series into an undesirable area of property space with respect to other parameters.

For example, it has been demonstrated by the early work of Kuntz *et al.*[112] and Andrews *et al.*[113] that an increase in molecular size will generally lead to an increase in binding affinity. This is primarily due to favourable van der Waals (VDW) and hydrophobic interactions, meaning that large lipophilic compounds will be amongst active compounds in most screens. Without due care, this could lead to the progression of fewer lead-like molecules should potency considerations dominate decision making.[32,114] The recent focus on ligand efficiency provides a means to mitigate against this effect[115,116] and to facilitate the optimisation of potency while keeping the physicochemical parameters to a region of chemical space more likely to realise balanced potency and ADMET properties.

1.4.1   Ligand Efficiency

Using measures of ligand efficiency, which provide quantitative estimates of how much biological potency is obtained for the physical properties of a molecule, have the potential to supplant the use of physical properties alone. Essentially a 'bang for your buck' metric, virtually any property or combination of properties can be used.[116] The most popular and useful ligand efficiency metrics use size and lipophilicity measures, alone or in combination:

- Ligand efficiency (LE) = $pX_{50} \times 1.37$/heavy atom count[116]
- Lipophilic ligand efficiency (LLE or LiPE) = $pX_{50} - \log P$ (or log D)[31]
- LELP = $\log P$/LE[117]
- LLE_{AT} = $0.111 + (1.37 \times LLE)$/heavy atom count[118]

Ligand efficiency metrics are especially useful for placing the potency gained into perspective for the specific target, rather than in an absolute sense. Lipophilic efficiency measures are highly relevant in optimisation,[118] as exemplified by a set of recent drugs showing mean potency improvement with little change in mean lipophilicity, and consequently increased LLE, in moving from lead to drug.[119] What are ideal values of LE and LLE? In a compilation of 261 oral drug potencies,[32] the mean LE is 0.45 and the mean LLE is 4.43 (calculated from the supplementary data provided in ref. 32). However, given the differences in druggability and physicochemical properties of targets, it is appropriate to consider the LE and LLE values of drugs relative to other molecules acting on the same target.[120] A survey of 45 oral drugs acting on 23 targets examined the fraction of compounds reported in the literature, acting on each target, which had higher values of both LE and LLE *versus* the drug molecules.[120] The median was 6.5%; the value for kinases was 22% and other targets 2.7%. Kinase drugs are therefore less fully optimised than other classes; since they are predominantly used for cancer, a higher risk to benefit ratio is acceptable, as discussed earlier. For other targets, the majority of drug molecules are highly optimised, extracting near-maximal potency at their target in relation to their physical properties. A study of seven VEGF2 kinase inhibitors[121] used for treatment of renal carcinoma provides strong support for the application of ligand efficiency measures. In this study, LLE values derived from affinities for VEGF2 complete with the juxtamembrane domain correlated positively with both kinase specificity, and with clinical measures of progression free survival.

Drug efficiency measures, using potencies and estimates of the free concentration of the drug dose that is available for interaction at the site of action, have potential in guiding projects towards drug-like space.[122–124] It is important to note that the method does not involve the optimisation of the plasma protein binding of the drug which is known to be therapeutically less significant.[122] The drug efficiency is estimated from the oral exposure and the *in vitro* binding data, which is then assessed in relation to its *in vitro* target affinity.[123–125] Drug efficiency is closely linked to free drug concentration and correlates well with LLE.

The application of ligand efficiency metrics combined with physical properties is a powerful means of tracking compound quality, as shown by an analysis of Factor Xa inhibitors[126] using a plot of LLE$_{AT}$ *versus* a hydrophobicity metric, the 'property forecast index'[41] (PFI) equal to log D (determined chromatographically) plus aromatic ring count. In this example, the marketed drugs, rivaroxaban, apixaban and edoxaban had markedly higher LLE$_{AT}$ values and lower PFI values than other Factor Xa inhibitors.[126] This type of analysis can easily be applied to any target. Some other examples[127–129] chosen to illustrate LLE optimisation strategies are given in Figure 1.8. Essentially, optimising LLE is increasing the relative contribution of polar *versus* non-polar binding interactions, and has been suggested by one group to be the most important efficiency measure for medicinal chemists to use.[130]

1.4.2 Multi-Objective Parameter Optimisation

The principal issue that can arise during target potency optimisation, which has been shown almost beyond doubt by numerous researchers using different datasets and methodologies, is that as molecular size and lipophilicity increase, ADMET parameters will deteriorate.[27,34,72] It is necessary to consider upfront the 'druggability' and relevance of the target in question,[131] the biological relevance of the assay being used,[132] and what level of affinity is likely to be considered acceptable. This is necessary since the dominance of one particular parameter will make meeting other parameters difficult, given the often confounding effects that each of the physical and biological parameters can have on each other.[32,133,134] For example, a highly potent molecule but with poor physical properties will probably be poorly soluble and permeable, and have a low unbound fraction in blood; such factors frequently render high potency levels redundant.

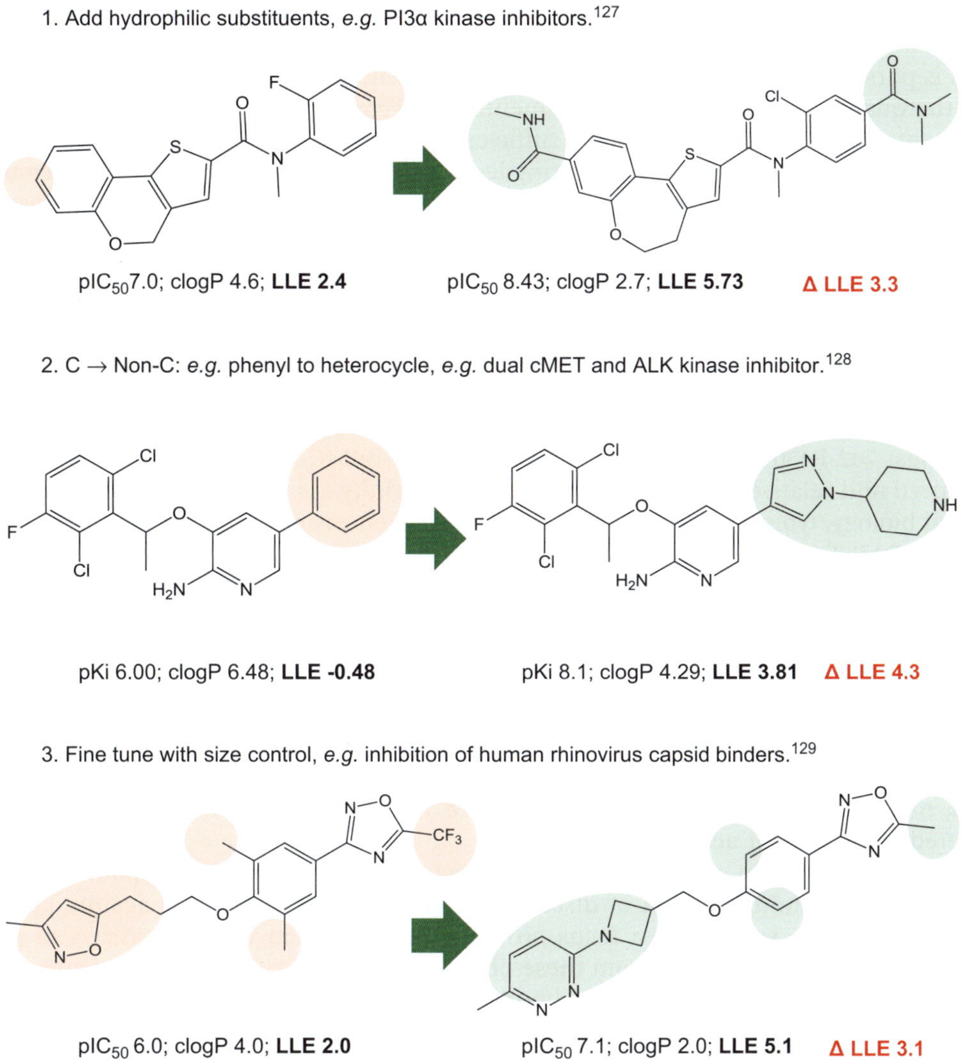

1. Add hydrophilic substituents, *e.g.* PI3α kinase inhibitors.[127]

pIC$_{50}$7.0; clogP 4.6; **LLE 2.4**   pIC$_{50}$ 8.43; clogP 2.7; **LLE 5.73**   Δ LLE 3.3

2. C → Non-C: *e.g.* phenyl to heterocycle, *e.g.* dual cMET and ALK kinase inhibitor.[128]

pKi 6.00; clogP 6.48; **LLE -0.48**   pKi 8.1; clogP 4.29; **LLE 3.81**   Δ LLE 4.3

3. Fine tune with size control, *e.g.* inhibition of human rhinovirus capsid binders.[129]

pIC$_{50}$ 6.0; clogP 4.0; **LLE 2.0**   pIC$_{50}$ 7.1; clogP 2.0; **LLE 5.1**   Δ LLE 3.1

Figure 1.8 Illustrations of three useful design strategies in LLE optimisation, where potency is increased and lipophilicity is successfully reduced at the same time.[127–129]

The ability to effectively manage a large number parameters, some of which can be orthogonal, correlated or inversely correlated, is absolutely critical in lead optimisation.[135] The effective balancing of diverse parameters is central to the successful iteration between successive design-make-test cycles within lead optimisation. The team psychology that is required to achieve this fine balancing act has been discussed in detail by Segal *et al.*[136] The authors have assessed the psychological barriers that can influence decision including putting inappropriate weight on certain parameters, being over confident in the program progress (resulting in delayed termination decisions), or not learning from past experience. The authors note that individuals are often biased towards recent events and sub-consciously ignore longer term trends. This a dilemma often faced in drug discovery, requiring scientists on the one hand to innovate and on the other hand to remember and follow the lessons of past. In fact knowledge is applied in a very inconsistent way by medicinal chemists, as we have discussed in detail in Section 1.3.6.

The inherent preference for optimising *in vitro* potency begins from the outset, at hit identification.[32,95,114] This elevated importance is not only unwarranted, but because the discovery process is pursued serially, it is very likely to be detrimental to the ultimate goal of finding new drugs: the quality of the output is fundamentally affected by the criteria used in the earlier steps. In this case, only hits that reach a sufficient potency threshold are prioritised for further optimisation (typically $< \sim 1\ \mu M$) and during lead optimisation a newly synthesised compound will be assessed in the higher throughput target potency assay first, as other ADMET assays are typically lower throughput or used only with selected compounds. The outcomes of this approach can be seen in any cursory analysis of structure–activity relationship (SAR) studies in medicinal chemistry journals. It is typical for studies to report the target affinity and structures of novel series for a given target but it is less common for *in vitro* ADMET, let alone *in vivo* ADMET parameters, to be reported. This is confirmed from a recent analysis of the ChEMBL database, compiled from the medicinal chemistry literature, which contains $>1\,400\,000$ unique compounds and $>350\,000$ target bioactivities, but just 3644 rat bioavailabilities (quantitative %F) and just 742 P450 3A4 inhibitions measurements (full IC_{50} values).[32]

The speed and relative simplicity of *in vitro* target affinity assays, coupled with insights from structural biology, typically make SAR for the target much easier to pursue than that of *in vitro* or *in vivo* ADMET.[33] ADMET measurements are typically performed on the most potent molecules, take longer to obtain,[137] generally display less clear SAR, and can require changes to the pharmacophore that are incompatible with the target SAR. The fact that efficacy is the single largest failure in the clinic of drug candidates shows that *in vitro* target affinity receives too much weight in design-make-test cycles. What level of potency is really required to test a mechanism? A 1 nM compound in a simple biochemical assay may not necessarily translate to 1 nM in a more relevant cellular, let alone animal or human, model.[138,139] Thus, why insist on obtaining nM potent molecules when there is often no specific evidence that it is required? Instead, nM potency should only be pursued once all other parameters (Table 1.1) are at least considered achievable or acceptable.

Bennani reports that in a typical research project, 20–30% of the time is spent fine tuning a molecule to fit the animal model of disease perfectly.[100] Teague cautions against box-ticking using *in vitro* assays, suggesting a focus on using more predictive *in vivo* assays to progress compounds.[140] The key message from these observations is clear: *don't over optimise molecules based on models known to be imperfect.* The ideal approach would be to test the validity of the disease–target–molecule combination with diverse compounds having (a) different modes of action (not just high potencies), (b) good ADME properties, (c) different structural templates and (d) greater exploration of physical property space.

1.5   HINTS AND TIPS

In Table 1.3 we have compiled suggested solutions for improving medicinal chemical quality and productivity, taken directly from three recent papers by respected scientists in four different pharmaceutical companies. Their suggestions are surprisingly consistent given the different sources of information used, and the specific topics being considered. We add to Table 1.3 by proposing that the human bias displayed in hit selection and lead optimisation can be countered across all drug discovery disciplines by (a) increased education and use of computational tools, (b) involving greater numbers of individuals in the hit selection process, (c) a greater focus on drug-like concepts in terms of both molecular and biological properties (Table 1.4), (d) assessing the validity of the disease target link earlier in development and (e) improving confidence in human dose prediction. A summary of some specific hints and tips to use in optimisation is given in Box 1.1.

Table 1.3 A list of the recommendations taken from three articles whose authors are based at four companies (AstraZeneca, GlaxoSmithKline, Gideon Richter and Vertex Pharmaceuticals). Analysis of different datasets and medicinal chemistry aspects give rise to similar overall conclusions.

	Leeson et al.: The influence of the 'organisational factor' on compound quality in drug discovery[74]	*Hann et al.: Finding the sweet spot: the role of nature and nurture in medicinal chemistry*[12]	*Walters et al.: What Do Medicinal Chemists Actually Make? A 50-Year Retrospective*[64]
Multi-objective optimisation	Full adoption of preclinical ADMET optimisation, problem-solving and the integration of these disciplines into compound design from hit identification onwards, in parallel with traditional optimisation of biological activity.	Do not be afraid to revert to a series of lower potency if it has better physicochemical properties. Extensive optimisation of a scaffold that is not amenable to achieving a desirable balance of potency and ADME (absorption, distribution, metabolism and excretion) properties is likely to be a waste of time and resources	More informative and higher through-put *in vitro* ADME/toxicity assays. Advances in assay technologies have led drug discovery teams to increasingly employ additional *in vitro* assays. Compounds in drug discovery programs are now routinely tested in solubility, CYP and hERG assays. *In vitro* systems such as Caco-2 and parallel artificial membrane permeation are commonly used as surrogates for permeability. A number of cellular systems have been developed to provide an indication of potential adverse outcomes. While many *in vitro* assays provide benefit, the correlation with *in vivo* data can be inconsistent. Increases in the quality and applicability of *in vitro* assays will enable medicinal chemists to gain insights earlier and avoid potential liabilities.
A consideration of molecular properties, particularly when optimising potency	Avoidance of over-reliance on potency optimisation without sufficient consideration of ligand efficiency or lipophilic ligand efficiency which can result in property inflation or 'molecular obesity'. The adoption of strategies to enhance screening collections with fragment-like, lead-like or chemically 'beautiful' compounds, especially removing chemically 'ugly' and highly lipophilic molecules.	Consider the chemical tractability (ligandability) of the target, and if it is poor then investigate different mechanisms of action or different pathways.	More emphasis on ligand efficiency, particularly in the hit-to-lead process. This in turn would support a re-evaluation of screening collections and perhaps the use of fragment-based approaches. Greater ligand efficiency would also tend to lead to lower log P, which has clearly been shown to reduce toxicological liabilities.

Table 1.3 *(Continued)*

	Leeson et al.: The influence of the 'organisational factor' on compound quality in drug discovery[74]	Hann et al.: Finding the sweet spot: the role of nature and nurture in medicinal chemistry[12]	Walters et al.: What Do Medicinal Chemists Actually Make? A 50-Year Retrospective[64]
Multivariate design—improved understanding of progression/ termination strategy	The adoption of lead generation strategies that set challenging criteria for progression, resulting in the timely termination of projects with unpromising hits or leads.	Select appropriate metrics for multi-dimensional optimisation; use ligand efficiency and lipophilic efficiency metrics in hit-to-lead optimisation and change to more complex metrics emphasising dosage to support lead optimisation. Leave suboptimal scaffolds early. Extensive optimisation of a scaffold resisting the balanced optimisation of potency and ADME properties wastes chemistry resources in inappropriate property space.	A return to the mind-set of simultaneous multivariate optimisation. The ultimate goal of a drug discovery program is to generate a compound that is safe and efficacious in humans. In order to develop such a compound, teams must optimise many criteria including affinity, selectivity, activity, properties, and pharmacokinetics. These criteria are often optimised in a serial fashion. Teams will optimise a single criterion like enzyme potency and then sacrifice these gains to optimize a second property like solubility. The process of trading off one property for another can repeat for dozens of cycles. A number of approaches can be applied to support multiobjective optimisation. One is to employ visualisation software that enables teams to appreciate the entirety of the data. A number of groups have recently developed visualisation tools oriented toward drug discovery programs. Another approach is to apply multiobjective optimisation algorithms and attempt to simultaneously optimise multiple criteria. While multiobjective optimisation has been successfully applied in a number of other fields, the primary limitation in drug discovery is the limited accuracy of computational models.

Use of more advanced computational methods and rules to incorporate drug-likeness	– An awareness of and improving understanding of the molecular features associated with toxicity and reactivity risks. – The application of computational models (at a global or project-specific level) and property calculation in advance of chemical synthesis, so that predictably poorer molecules are never synthesised.	Deeper understanding of true drug-likeness. Simple numerical rules of thumb are helpful only to a degree. Rather, we need to develop an improved understanding of what makes some molecules drug-like despite having physical properties outside the usual ranges. These rules can be global (applicable to all molecules) or local (fine-tuned for each chemical class or target tissue) and will be especially valuable for tackling "undruggable" targets.	–
Increasing the diversity of chemical syntheses	A willingness to take on more challenging, complex synthetic chemistry and to introduce new synthetic templates and scaffolds.	Reduced reliance on "easy" chemistry. While Pd-mediated sp2_sp2 couplings and amide bond-forming reactions have their place, we believe that a greater emphasis on the art of synthesis in medicinal chemistry would dramatically improve the physical properties of our molecules.	–
Consider past precedent and avoid unnecessary behavioural biases	Learning from past experiences and from recently marketed drugs; remembering the lessons learnt from the pre-genomic drug discovery period.	– Stay focused on the 'sweet spot' and committed to delivering high-quality compounds, but remain open-minded to the many ways this can be achieved. – Cultural changes are difficult and need supporting throughout an organisation and at all phases. Exchange of views and access to data-driven knowledge are important and could help reduce behavioural 'addictions'. In-depth analysis of drug discovery case studies and success stories contributes substantially to these efforts, and towards the understanding of true compound quality. – Resist timelines that compromise compound quality.	

Table 1.4   Ideal optimised oral physicochemical and ADMET property ranges.

Physical properties	Ligand efficiency	ADMET
• log D 0–2; log P 2–3 • HBD ≤ 2; PSA < 60–70 Å^2 • MWT < 450 • Aromatic rings ≤ 3 • No undesirable functional groups • Lack of overt covalent binding (excluding target-specific design)	• LE ≥ 0.4 • LLE ≥ 5	• Human dose prediction from at least two species < 100 mg (ideally < 40 mg to avoid idiosyncratic toxicity) • PK–PD established in one species • > 100-fold selective in cross screening

---

BOX 1.1   OPTIMISATION HINTS AND TIPS

- Do not increase pX_{50} by increasing lipophilicity or size alone. Optimising LE ($pX_{50} \times 1.37$/heavy atom count; ideally > 0.4) and LLE (pX_{50} – log P/D; ideally > 5) is much more important.
- Chemical starting points should be lead like, with potential for progression, ideally LE > 0.3 and LLE > 1–2.
- In optimisation: use plots of pX_{50} *versus* clogP, log D, or PFI (log D + aromatic ring count) and MWT. Aim to expand and populate the 'north west' boundary (high potency and low log P or log D) by designing compounds with appropriate physical properties.
- In optimisation: use plots of LE *versus* LLE, adding in literature molecules active at the desired target. Aim to expand and populate the 'north east' boundary.
- Measure and optimise solubility and *in vitro* metabolic stability from hit identification onwards.
- Obtain *in vivo* pharmacokinetic data and conduct human dose prediction with early lead compounds.
- Aim to optimise lead compound dose and dose frequency, not just the pX_{50}.

1.6   CONCLUSIONS

It has been shown that oral drug physicochemical properties are rather restrictive. Yet we observe that the industry is pursuing fewer drug-like molecules, which appear to be increasingly failing to progress through clinical development. We acknowledge that drugs can be found outside the space represented by the Ro5 and related guidelines. However, we hypothesise that a portfolio comprised of such molecules would suffer greater attrition than a portfolio with a drug-like property distribution, and its success would require greater serendipity. A rational approach should focus on the area of property space with the highest probability of success, but would not exclude venturing beyond in specific cases, especially if there are strong therapeutic reasons, no solutions can be found in drug-like space, and advantages of being in suboptimal space are identified and optimised.

We have discussed how the serial application of reductionist models of human disease and pharmacokinetics can prove challenging. Each assay system has associated experimental error and variable biological relevance, making its application in progressing individual compounds challenging. It is our belief that a careful balance of properties should be pursued in drug discovery and due weight should be placed on each parameter, based on the best available understanding of the disease. Great care should be taken not to adopt practices that bias the progression of compounds or unfavourably prioritise optimisation of some parameters over others. A primary focus on target affinity is not recommended, since this will risk introducing ADMET-unfriendly physicochemical properties.

KEY REFERENCES

Good general reading on physicochemical properties:
D. Smith, P. Artursson, A. Avdeef, L. Di, G. F. Ecker, B. Faller, J. B. Houston, M. Kansy, E. H. Kerns, S. D. Krämer, H. Lennernäs, H. van de Waterbeemd, K. Sugano and B. Testa, *Mol. Pharmaceutics*, 2014, **11**(6), 1727–1738.
H. van de Waterbeemd, Physicochemical Properties, in *Medicinal Chemistry, Principles and Practice, 2nd Ed*, ed. F. King, Royal Society of Chemistry, Cambridge, 2002, Chapter 10, pp. 195–214.
H. van de Waterbeemd, Physicochemical Approaches to Drug Absorption, in *Drug Bioavailability, 2nd Ed.*, eds. H. van de Waterbeemd and B. Testa, VCH, 2009, Chapter 5, pp. 71–100.
H. van de Waterbeemd, Improving compound quality through in vitro and in silico physicochemical profiling, *Chem. Biodiv.*, 2009, **6**, 1760–1766.

Rule of Five and ADMET:
C. A. Lipinski, F. Lombardo, B. W. Dominy and P. J. Feeney, *Adv. Drug Delivery Rev.*, 1997, **23**, 3–25.

Ligand efficiency concept:
A. L. Hopkins, C. R. Groom and A. Alex, *Drug Discovery Today*, 2004, **9**, 430–431.

The impact of drug-like principles:
P. D. Leeson and B. Springthorpe, *Nat. Rev. Drug Discov.*, 2007, **6**, 881–890.

REFERENCES

1. M. E. Bunnage, *Nat. Chem. Biol.*, 2011, 7, 335–339.
2. P. Morgan, P. H. Van Der Graaf, J. Arrowsmith, D. E. Feltner, K. S. Drummond, C. D. Wegner and S. D. A. Street, *Drug Discovery Today*, 2012, **17**, 419–424.
3. http:/www.iupac.org/reports/1997/6905vanderwaterbeemd/index.html.
4. M. C. Wenlock, T. Potter, P. Barton and R. P. Austin, *J. Biomol. Screening*, 2011, **16**(3), 348.
5. A. J. Leo, *Chem. Rev.*, 1993, **93**, 1281.
6. http://www.biobyte.com/index.html.
7. http://www.daylight.com/.
8. A. K. Ghose and G. M. Crippen, *J. Chem. Inf. Comput. Sci.*, 1987, **B**, 21–35.
9. R. Mannhold, G. I. Poda, C. Ostermann and I. V. Tetko, *J. Harm. Sci.*, 2009, **98**, 862.
10. http://www.acdlabs.com/.
11. P. Bruneau and N. R. McElroy, *J. Chem. Inf. and Model.*, 2006, **46**(3), 1379–1387.
12. A. E. Cooper, D. Ferguson and K. Grime, *Curr. Drug Metab.*, 2012, **13**(4), 457–473.
13. C. Jamieson, E. M. Moir, Z. Rankovic and G. Wishart, *J. Med. Chem.*, 2006, **49**(17), 5029–5046.
14. M. H. Abraham, W. E. Acree, Jr., A. J. Leo, D. Hoekman and J. E. Cavanaugh, *J. Pharm. Sci.*, 2010, **99**, 2492–2501.

15. M. Shalaeva, G. Caron, Y. A. Abramov, T. N. O'Connell, M. S. Plummer, G. Yalamanchi, K. A. Farley, G. H. Goetz, L. Philippe and M. J. Shapiro, *J. Med. Chem.*, 2013, **56**(12), 4870–4879.
16. H. van de Waterbeemd, G. Camenisch, G. G. Folkers, J. R. Chretien, O. A. Raevsky and A. Oleg, *J. Drug Target*, 1999, **6**, 151.
17. H. van de Waterbeemd in *Methods for Assessing Oral Drug Absorption*, ed. J. Dressman, Dekker, New York, 2000, 31.
18. T. T. Wager, X. Hou, P. R. Verhoest and A. Villalobos, *ACS Chem. Neuroscience*, 2010, **1**, 435.
19. H. Parshad, K. Frydenvang, T. Liljefors and C. S. Larsen, *Int. J. Pharaceutics*, 2002, **237**, 193.
20. K. Johnson and A. Swindell, *Pharmaceutical Res.*, 1996, **13**, 1795.
21. http://www.fda.gov/downloads/Drugs/Guidances/ucm070246.pdf.
22. N. Colclough, A. Hunter, P. W. Kenny, R. S. Kittlety, L. Lobedan, K. Y. Tam and M. A. Timms, *Bioorg. Med. Chem.*, 2008, **16**, 6611.
23. N. Jain and S. H. Yalkowsky, *J. Pharm. Sci.*, 2001, **90**(2), 234–252.
24. D. T. Manallack, R. J. Prankerd, E. Yuriev, T. I. Oprea and D. K. Chalmers, *Chem. Soc. Rev.*, 2013, **42**, 485–496.
25. C. Hansch, J. P. Bjoerkroth and A. Leo, *J. Pharm. Sci.*, 1987, **76**, 663–687.
26. M. J. Waring, *Expert Opin. Drug Discov.*, 2010, **5**, 235–248.
27. C. A. Lipinski, F. Lombardo, B. W. Dominy and P. J. Feeney, *Adv. Drug Delivery Rev.*, 1997, **23**, 3–25.
28. G. R. Bickerton, G. V. Paolini, J. Besnard, S. Muresan and A. L. Hopkins, *Nat. Chem.*, 2012, **4**, 90–98.
29. N. A. Meanwell, *Chem. Res. Toxicol.*, 2011, **24**, 1420–1456.
30. Á. Tarcsay and G. M. Keserű, *J. Med. Chem.*, 2013, **56**, 1789–1795.
31. P. D. Leeson and B. Springthorpe, *Nat. Rev. Drug Discov.*, 2007, **6**, 881–890.
32. M. P. Gleeson, A. Hersey, D. Montanari and J. Overington, *Nat. Rev. Drug Discov.*, 2011, **10**, 197–208.
33. M. M. Hann and G. M. Keserü, *Nat. Rev. Drug Discov.*, 2012, **11**, 355–365.
34. M. P. Gleeson, *J. Med. Chem.*, 2008, **51**, 817–834.
35. M. J. Waring, *Bioorg. Med. Chem. Lett.*, 2009, **19**, 2844–2851.
36. T. W. Johnson, K. R. Dress and M. Edwards, *Bioorg. Med. Chem. Lett.*, 2009, **19**, 5560–5564.
37. P. B. Cox, R. J. Gregg and A. Vasudevan, *Bioorg. Med. Chem.*, 2012, **20**, 4564–4573.
38. P. D. Leeson and J. R. Empfield, *Annu. Reports Med. Chem.*, 2010, **45**, 393–407.
39. F. Lovering, J. Bikker and C. Humblet, *J. Med. Chem.*, 2009, **52**, 6752–6756.
40. L. M. Salonen, M. Ellermann and F. Diederich, *Angew. Chem. Int. Ed., Engl.*, 2011, **50**, 4808–4842.
41. R. J. Young, D. V. S. Green, C. N. Luscombe and A. P. Hill, *Drug Discovery Today*, 2011, **16**, 822–830.
42. T. J. Ritchie, S. J. F. Macdonald, S. Peace, S. D. Pickett and C. N. Luscombe, *MedChemComm*, 2012, **3**, 1062–1069.
43. A. G. Leach, E. A. Pilling, A. A. Rabow, S. Tomasi, N. Asaad, N. J. Buurma, A. Ballard and S. Narduolo, *MedChemComm*, 2012, **3**, 528–540.
44. M. J. Kamlet, M. H. Abraham, R. M. Doherty and R. W. Taft, *J. Am. Chem. Soc.*, 1984, 464–466.
45. R. D. I. Cramer, *J. Am. Chem. Soc.*, 1980, **102**, 1837–1849.
46. K. Palm, P. Stenberg, K. Luthman and P. Artursson, *Pharm. Res.*, 1997, **14**, 568–571.
47. Y. C. Martin, *J. Med. Chem.*, 2005, **48**, 3164–3170.
48. M. V. S. Varma, R. S. Obach, C. Rotter, H. R. Miller, G. Chang, S. J. Steyn, A. El-Kattan and M. D. Troutman, *J. Med. Chem.*, 2010, **53**, 1098–1108.

49. C. Hansch, J. E. Quinlan and G. L. Lawrence, *J. Org. Chem.*, 1968, **33**, 347–350.
50. N. Jain and S. H. Yalkowsky, *J. Pharm. Sci.*, 2001, **90**, 234–252.
51. A. G. Leach, H. D. Jones, D. A. Cosgrove, P. W. Kenny, L. Ruston, P. MacFaul, J. M. Wood, N. Colclough and B. Law, *J. Med. Chem.*, 2006, **49**, 6672–6682.
52. T. J. Hou, W. Zhang, K. Xia, X. B. Qiao and X. J. Xu, *J. Chem. Inf. Comput. Sci.*, 2004, **44**, 1585–1600.
53. T. Fichert, M. Yazdanian and J. R. Proudfoot, *Bioorg. Med. Chem. Lett.*, 2003, **13**, 719–722.
54. A. Nordqvist, J. Nilsson, T. Lindmark, A. Eriksson, P. Garberg and M. Kihlén, *QSAR Comb. Sci.*, 2004, **23**, 303–310.
55. R. P. Verma, C. Hansch and C. D. Selassie, *J. Comput. Aided Mol. Des.*, 2007, **21**, 3–22.
56. G. Camenisch, J. Alsenz, H. van de Waterbeemd and G. Folkers, *Eur. J. Pharm. Sci.*, 1998, **6**, 313–319.
57. J. Kelder, P. D. Grootenhuis, D. M. Bayada, L. P. Delbressine and J. P. Ploemen, *Pharm. Res.*, 1999, **16**, 1514–1519.
58. U. Norinder and M. Haeberlein, *Adv. Drug. Deliv. Rev.*, 2002, **54**, 291–313.
59. G. Gerebtzoff and A. Seelig, *J. Chem. Inf. Model.*, 2006, **46**, 2638–2650.
60. P. Crivori, B. Reinach, D. Pezzetta and I. Poggesi, *Mol. Pharm.*, 2006, **3**, 33–44.
61. T. T. Wager, R. Y. Chandrasekaran, X. Hou, M. D. Troutman, P. R. Verhoest, A. Villalobos and Y. Will, *ACS Chem. Neurosci.*, 2010, **1**, 420–434.
62. G. Berellini, C. Springer, N. J. Waters and F. Lombardo, *J. Med. Chem.*, 2009, **52**, 4488–4495.
63. F. Lombardo, R. S. Obach, M. Y. Shalaeva and F. Gao, *J. Med. Chem.*, 2004, **47**, 1242–1250.
64. M. P. Gleeson, N. J. Waters, S. W. Paine and A. M. Davis, *J. Med. Chem.*, 2006, **49**, 1953–1963.
65. G. Colmenarejo, A. Alvarez-Pedraglio and J. L. Lavandera, *J. Med. Chem.*, 2001, **44**, 4370–4378.
66. K. Yamazaki and M. Kanaoka, *J. Pharm. Sci.*, 2004, **93**, 1480–1494.
67. K. Valko, S. Nunhuck, C. Bevan, M. H. Abraham and D. P. Reynolds, *J. Pharm. Sci.*, 2003, **92**, 2236–2248.
68. M. P. Gleeson, *J. Med. Chem.*, 2007, **50**, 101–112.
69. D. F. V. Lewis, S. Modi and M. Dickins, *Drug Metab. Rev.*, 2002, **34**, 69–82.
70. M. L. Lewis and L. Cucurull-Sanchez, *J. Comput.-Aided Mol. Des.*, 2009, **23**, 97–103.
71. U. Norinder and C. A. Bergstrom, *Chem. Med. Chem.*, 2006, **1**, 920–937.
72. P. Gleeson, G. Bravi, S. Modi and D. Lowe, *Bioorg. Med. Chem.*, 2009, **17**, 5906–5919.
73. M. C. Wenlock, R. P. Austin, P. Barton, A. M. Davis and P. D. Leeson, *J. Med. Chem.*, 2003, **46**, 1250–1256.
74. J. D. Hughes, J. Blagg, D. A. Price, S. Bailey, G. A. DeCrescenzo, R. V. Devraj, E. Ellsworth, Y. M. Fobian, M. E. Gibbs, R. W. Gilles, N. Greene, E. Huang, T. Krieger-Burke, J. Loesel, T. Wager, L. Whiteley and Y. Zhang, *Bioorg. Med. Chem. Lett.*, 2008, **18**, 4872–4875.
75. J. J. Sutherland, J. W. Raymond, J. L. Stevens, T. K. Baker and D. E. Watson, *J. Med. Chem.*, 2012, **55**, 6455–6466.
76. T. Luker, L. Alcaraz, K. K. Chohan, N. Blomberg, D. S. Brown, R. J. Butlin, T. Elebring, A. M. Griffin, S. Guile, S. St-Gallay, B.-M. Swahn, S. Swallow, M. J. Waring, M. C. Wenlock and P. D. Leeson, *Bioorg. Med. Chem. Lett.*, 2011, **21**, 5673–5679.
77. P. Y. Muller and M. N. Milton, *Nat. Rev. Drug. Discov.*, 2012, **11**, 751–761.
78. S. Lu, B. Jessen, C. Strock and Y. Will, *Toxicol. In Vitro*, 2012, **26**, 613–620.
79. A. F. Stepan, D. P. Walker, J. Bauman, D. A. Price, T. A. Baillie, A. S. Kalgutkar and M. D. Aleo, *Chem. Res. Toxicol.*, 2011, **24**, 1345–1410.
80. M. Chen, J. Borlak and W. Tong, *Hepatology*, 2012, **3**, 388–396.
81. M. Vieth, M. G. Siegel, R. E. Higgs, I. A. Watson, D. H. Robertson, K. A. Savin, G. L. Durst and P. A. Hipskind, *J. Med. Chem.*, 2003, **47**, 224–232.

82. P. D. Leeson and A. M. Davis, *J. Med. Chem.*, 2004, **47**, 6338–6348.
83. J. R. Proudfoot, *Bioorg. Med. Chem. Lett.*, 2005, **15**, 1087–1090.
84. P. D. Leeson, S. A. St-Gallay and M. C. Wenlock, *MedChemComm*, 2011, **2**, 91–105.
85. W. P. Walters, J. Green, J. R. Weiss and M. A. Murcko, *J. Med. Chem.*, 2011, **54**, 6405–6416.
86. S. D. Roughley and A. M. Jordan, *J. Med. Chem.*, 2011, **54**, 3451–3479.
87. A. Nadin, C. Hattotuwagama and I. Churcher, *Angew. Chem., Int. Ed.*, 2012, **51**, 1114–1122.
88. P. Maclellan and A. Nelson, *Chem. Commun.*, 2013, **49**, 2383–2393.
89. Y. B. Choy and M. R. Mark, *R. Prausnitz, Pharm. Res.*, 2011, **28**, 943–948.
90. L. Benet, F. Broccatelli and T. Oprea, *AAPS J.*, 2011, **13**, 519–547.
91. T. J. Ritchie, C. N. Luscombe and S. J. F. Macdonald, *J. Chem. Inf. Model.*, 2009, **49**, 1025–1032.
92. A. D. Hughes, A. McNamara and T. Steinfeld, *Progress Med. Chem.*, **51**, 71–95.
93. M. Vieth and J. J. Sutherland, *J. Med. Chem.*, 2006, **49**, 3451–3453.
94. R. Morphy, *J. Med. Chem.*, 2006, **49**, 2969–2978.
95. P. D. Leeson and S. A. St-Gallay, *Nat. Rev. Drug Discov.*, 2011, **10**, 749–765.
96. M. S. Lajiness, G. M. Maggiora and V. Shanmugasundaram, *J. Med. Chem.*, 2004, **47**, 4891–4896.
97. P. S. Kutchukian, N. Y. Vasilyeva, J. Xu, M. K. Lindvall, M. P. Dillon, M. Glick, J. D. Coley and N. Brooijmans, *PLoS ONE*, 2012, **7**, e48476.
98. D. R. Cheshire, *Drug Discovery Today*, 2011, **16**, 817–821.
99. S. W. Muchmore, J. J. Edmunds, K. D. Stewart and P. J. Hajduk, *J. Med. Chem.*, 2010, **53**, 4830–4841.
100. Y. L. Bennani, *Drug Discovery Today*, 2011, **16**, 779–792.
101. H. Zhao, *Drug Discovery Today*, 2011, **16**, 158–163.
102. T. J. Ritchie, S. J. F. Macdonald, S. Peace, S. D. Pickett and C. N. Luscombe, *MedChemComm*, 2013, **4**, 673–680.
103. A. Ganesan, *Curr. Opin. Chem. Biol.*, 2008, **12**, 306–317.
104. A. Alex, D. S. Millan, M. Perez, F. Wakenhut and G. A. Whitlock, *MedChemComm*, 2011, **2**, 669–674.
105. D. Camp, R. A. Davis, M. Campitelli, J. Ebdon and R. J. Quinn, *J. Nat. Prod.*, 2011, **75**, 72–81.
106. R. W. Huigens III, K. C. Morrison, R. W. Hicklin, T. A. Flood Jr, M. F. Richter and P. J. Hergenrother, *Nat. Chem.*, 2013, **5**, 195–202.
107. A. D. Kwong, R. S. Kauffman, P. Hurter and P. Mueller, *Nat. Biotech.*, 2011, **29**, 993–1003.
108. E. R. Wood, A. T. Truesdale, O. B. McDonald, D. Yuan, A. Hassell, S. H. Dickerson, B. Ellis, C. Pennisi, E. Horne, K. Lackey, K. J. Alligood, D. W. Rusnak, T. M. Gilmer and L. Shewchuk, *Cancer. Res.*, 2004, **64**, 6652–6659.
109. S. Castellino, M. O'Mara, K. Koch, D. J. Borts, G. D. Bowers and C. MacLauchlin, *Drug Metab. Dispos.*, 2012, **40**, 139–150.
110. H.-A. Lee, E.-J. Kim, S.-A. Hyun, S.-G. Park and K.-S. Kim, *Basic Clin. Pharmacol. Toxicol.*, 2010, **107**, 614–618.
111. K. M. Koch, N. J. Reddy, R. B. Cohen, N. L. Lewis, B. Whitehead, K. Mackay, A. Stead, A. P. Beelen and L. D. Lewis, *J. Clin. Oncol.*, 2009, **27**, 1191–1196.
112. I. D. Kuntz, K. Chen, K. A. Sharp and P. A. Kollman, *Proc. Natl. Acad. Sci. U. S. A.*, 1999, **96**, 9997–10002.
113. P. R. Andrews, D. J. Craik and J. L. Martin, *J. Med. Chem.*, 1984, **27**, 1648–1657.
114. M. M. Hann, *MedChemComm*, 2011, **2**, 349–355.
115. C. Abad-Zapatero and J. T. Metz, *Drug Discovery Today*, 2005, **10**, 464–469.
116. A. L. Hopkins, C. R. Groom and A. Alex, *Drug Discovery Today*, 2004, **9**, 430–431.
117. Á. Tarcsay, K. Nyíri and G. M. Keserű, *J. Med. Chem.*, 2012, **55**, 1252–1260.

118. P. Mortenson and C. Murray, *J. Comput.-Aided Mol. Des.*, 2011, **25**, 663–667.
119. E. Perola, *J. Med. Chem.*, 2010, **53**, 2986–2997.
120. A. L. Hopkins, P. D. Leeson, G. M. Keserű, D. C. Rees and C. H. Reynolds, *Nat. Rev. Drug Discov.*, 2014, **13**, 105–121.
121. M. McTigue, B. W. Murray, J. H. Chen, Y.-L. Deng, J. Solowiej and R. S. Kania, *Proc. Natl Acad. Sci. USA*, 2012, **109**, 18281–18289.
122. L. Z. Benet and B. A. Hoener, *Clin. Pharmacol. Ther.*, 2002, **71**, 115–121.
123. K. Valko, E. Chiarparin, S. Nunhuck and D. Montanari, *J. Pharm. Sci.*, 2012, **101**, 4155–4169.
124. D. Montanari, E. Chiarparin, M. P. Gleeson, S. Braggio, R. Longhi, K. Valko and T. Rossi, *Expert Opin. Drug Discov.*, 2011, **6**, 913–920.
125. S. Braggio, D. Montanari, T. Rossi and E. Ratti, *Expert Opin. Drug Discov.*, 2010, **5**, 609–618.
126. R. J. Young, *Bioorg. Med. Chem. Lett.*, 2011, **21**, 6228–6235.
127. S. T. Staben, M. Siu, R. Goldsmith, A. G. Olivero, S. Do, D. J. Burdick, T. P. Heffron, J. Dotson, D. P. Sutherlin, B.-Y. Zhu, V. Tsui, H. Le, L. Lee, J. Lesnick, C. Lewis, J. M. Murray, J. Nonomiya, J. Pang, W. W. Prior, L. Salphati, L. Rouge, D. Sampath, S. Sideris, C. Wiesmann and P. Wu, *Bioorg. Med. Chem. Lett.*, 2011, **21**, 4054–4058.
128. J. J. Cui, M. Tran-Dube, H. Shen, M. Nambu, P.-P. Kung, M. Pairish, L. Jia, J. Meng, L. Funk, I. Botrous, M. McTigue, N. Grodsky, K. Ryan, E. Padrique, G. Alton, S. Timofeevski, S. Yamazaki, Q. Li, H. Zou, J. Christensen, B. Mroczkowski, S. Bender, R. S. Kania and M. P. Edwards, *J. Med. Chem.*, 2011, **54**, 6342–6363.
129. A. Morley, N. Tomkinson, A. Cook, C. MacDonald, R. Weaver, S. King, L. Jenkinson, J. Unitt, C. McCrae and T. Phillips, *Bioorg. Med. Chem. Lett.*, 2011, **21**, 6031–6035.
130. K. D. Freeman-Cook, R. L. Hoffman and T. W. Johnson, *Future Med. Chem.*, 2013, **5**, 113–115.
131. P. J. Hajduk, J. R. Huth and C. Tse, *Drug Discovery Today*, 2005, **10**, 1675–1682.
132. D. C. Swinney and J. Anthony, *Nat. Rev. Drug Discov.*, 2011, **10**, 507–519.
133. K. H. Grime, P. Barton and D. F. McGinnity, *Mol. Pharmaceutics*, 2012, **10**, 1191–1206.
134. D. A. Smith, L. Di and E. H. Kerns, *Nat. Rev. Drug Discov.*, 2010, **9**, 929–939.
135. M. D. Segall, *Curr. Pharm. Des.*, 2012, **19**, 1292–1310.
136. M. Segall and A. Chadwick, *Future Med. Chem.*, 2011, **3**, 771–774.
137. S. J. F. Macdonald and P. W. Smith, *Drug Discovery Today*, 2001, **6**, 947–953.
138. K. E. Lackey, *Curr. Top. Med. Chem.*, 2006, **6**, 435–460.
139. A. M. Davis, D. J. Keeling, J. Steele, N. P. Tomkinson and A. C. Tinker, *Curr. Top. Med. Chem.*, 2005, **5**, 421–439.
140. S. J. Teague, *Drug Discovery Today*, 2011, **16**, 398–411.

CHAPTER 2

Parallel Synthesis and Library Design

ANDY MERRITT

MRCT Centre for Therapeutics Discovery, 1-3 Burtonhole Lane, Mill Hill,
London NW7 1AD, UK
E-mail: Andy.Merritt@tech.mrc.ac.uk

2.1 INTRODUCTION

Why do companies build and maintain large chemical libraries? Often described as the most important asset of a pharmaceutical company's research arm (the 'crown jewels'[1]) they are typically the product of a large number of man years for internal synthesis plus a significant (multimillion dollar) spend on external compound acquisition from an ever widening range of commercial sources. Significant overhead is spent annually on selecting, acquiring, synthesising, maintaining and analysing compounds and the investment in facilities to curate, protect and distribute collections,[2] with direct equipment costs estimated to fall in the $1–2 million range.[3]

The answer to the initial question in the previous paragraph is clear—to increase the chance of finding something novel. For any company progressing lead discovery and optimisation programmes, if enough is known about a particular target and the type of molecules capable of interacting with it (in a pharmacologically relevant manner) then as long as there is novelty inherent in that knowledge, a curated and diverse collection of compounds is not required. All that is demanded of a compound management process in that situation is the shepherding of new compounds through any required assays to support project progression. However for other targets, often early stage and novel, but also fast follower targets where an organisation is trying to catch and overtake the known state of discovery, a compound collection becomes an invaluable source to potentially find something novel (usually a small molecule start point, but also target validation tools) that can be used to initiate a medicinal chemistry discovery programme. Where little or nothing is known about the requirements of the target active site in terms of preferred interacting molecules then that search may most likely be based on complete sampling of as much variety of chemical space as possible. If there is predetermined knowledge of the target (specific protein structure or knowledge of closely related proteins) then it may be possible to sample the compound collection to produce a set of compounds with a predetermined bias towards that target. However, whether the target structure is known or not, the main

The Handbook of Medicinal Chemistry: Principles and Practice
Edited by Andrew Davis and Simon E Ward
© The Royal Society of Chemistry 2015
Published by the Royal Society of Chemistry, www.rsc.org

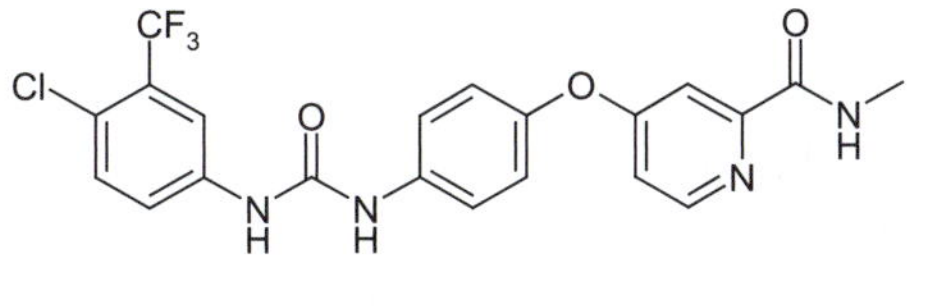

Figure 2.1 Examples of marketed drugs where high throughput screening and optimisation were part of the discovery process.

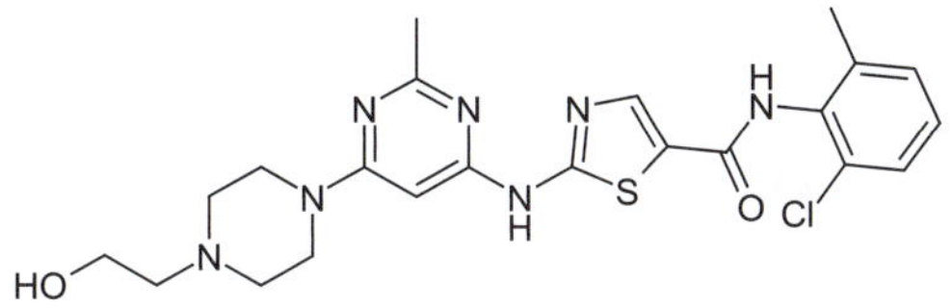

element of the screening exercise is the same—to find something new by accessing and sampling the best possible selection of hit, lead or drug like compounds available. Of course such an approach should not be considered in isolation for any new target, and should be carefully considered alongside other more rational design and established empirical medicinal chemistry approaches that are highlighted in many sections of this book. But it is true to say that there are now many examples where well designed arrays associated with high throughput screening methods have led to compounds that are now into the clinic.[4] For example, Sorafenib (Bayer/Onyx)[5] and Dasatinib (BMS)[6] are both tyrosine kinase inhibitors discovered in part from initial high throughput screening (in 1994 and 1997, respectively) followed by targeted array synthesis (Figure 2.1).

So what could be defined as a well-designed library? Like the old story of economists, put five chemists into a room and ask that question and the likelihood is that you'll get at least six separate answers. Indeed, studies have shown in some cases just asking the same chemist twice on separate occasions can provide differing results. In a 2004 study by Pharmacia[7] when chemists were asked to select/reject compounds from a set of 2000, the average pairwise agreement within the 13 chemists included was only 28%. Moreover nine chemists were subsequently given the same set of compounds to repeat the rejection process, with a result of only 51% consistency. But putting personal subjectivity aside there are some underlying principles that can be applied and are independent of any one favourite algorithm for selecting A over B or grouping X with Y instead of Z. It was during the initial development of combinatorial chemistry approaches to collection design in the 1990s, and the associated explosion in chemical technologies (design, synthesis, purification and analysis) that many key concepts of successful application of library design to drug discovery were learnt (or in many cases relearnt after having been forgotten!). The next sections of this chapter will briefly review the historical development of key approaches to library synthesis and construction to illustrate how we have got to our current stages of compound library and screening collection design. The remainder of the chapter will then focus on design strategies for general compound libraries and larger targeted arrays aimed towards specific protein classes. The development of the concepts and specific technologies of combinatorial chemistry and the application of combinatorial approaches to specific target prosecution (sometimes through equally large libraries) is beyond the scope of this chapter and has been covered in detail elsewhere.[8]

2.2 THE START OF COMBICHEM IN DRUG DISCOVERY

The development of miniaturised screening leading to high throughput approaches was a significant advancement of drug discovery.[9] The standardisation of assay format into microtitre plates, initially 96 well format, alongside the development of automated processing, radically changed the opportunity for screening to deliver new leads into drug discovery programmes. Automation of plate movement, liquid handling and plate reading processes meant that where a few 10s of compounds may have been tested in a day by manual techniques, suddenly 1000s were possible in enzyme, (membrane bound) receptor and even whole cell assay format. Further enhanced by the miniaturisation of wells on the plates, from 96 to 384 (and subsequently 1536), high throughput screening of compound collections of 100 000s or more became clearly feasible, and when run alongside mechanism and knowledge/structural based targeted screening approaches provided much greater opportunity to identify novel lead series and structural classes.

As high throughput screening developed rapidly in the late 1980s and early 1990s attention was turned to the feedstock for such efforts—company compound collections. These had typically built up by a combination of 'file' compounds from previous and ongoing lead optimisation programmes and natural products, sourced either from in house fermentation or through external acquisition of samples, be they soil, microbe or plant derived. A compound collection of one to two hundred thousand such compounds was not atypical, but the potential for further growth through these traditional routes would always be limited. A 'traditional' medicinal chemist was likely to add no more than 40–50 compounds in any year, and perhaps even more significantly any file collection built on past programmes would clearly only represent those chemical areas that had been of interest. Many collections were significantly populated by specific structural classes, for example β-lactams or steroids. Meanwhile natural products were often complex structures, difficult to work with in lead optimisation, and becoming harder to source with exclusivity. International treaties correctly limited the ability to source natural products from countries without due regard to intellectual property ownership[10] and even when novel active natural products were identified, it was possible for more than one company to independently and concurrently identify the same structural series.[11,12]

So if high throughput screening presented the opportunity to screen 100 000s of compounds in a matter of days whilst collection sizes were still limited, alternative mechanisms to grow the collections were targeted. Collection sharing deals were struck between companies[13] and this concept was effectively continued in the mergers of the 1990s[14] where the formation of combinations such as GlaxoWellcome, Smithkline Beecham, AstraZeneca, Novartis, and Aventis, for example, provided immediate increases in corporate collection size. A recent analysis[15] of the combination of compound libraries from Bayer Healthcare AG and Schering AG following the takeover of the latter by Bayer showed a very low direct overlap of chemical structures between the two organisations (0.04% for in house synthesised and 1.5% in total) and reached the conclusion that collaborative screening efforts between companies (either through consortia or the result of more commercial takeovers) would be an effective means of increasing diversity coverage of screening libraries.

In addition, acquisition of compounds from external sources was increased, both from commercial and academic sources. Commercial suppliers provided compounds that could be added to screening collections, though these were available to all companies, thus raising concern over intellectual property control, and at that time were limited to only a few suppliers of fine chemicals. Access to more varied chemistry was available through academic collaborations, and many academic groups found they could fund several aspects of their research with money from compound selling, however a combination of structural integrity, purity, and sustainability of resupply were all potential issues for the pharmaceutical companies using this approach.

The optimum solution for companies appeared to be a combination of the above, but enhanced with an even greater component derived from a significant increase of productivity from their own chemists. Such internally derived compounds would be proprietary, exclusive and could be targeted if necessary to areas of most interest to the company concerned. Knowledge would be retained for further synthesis, follow up and analogue work thus providing confidence downstream of any initial positive results. The rapid development of high throughput screening had demonstrated that technology and rethinking of strategies could in combination provide major increases in productivity, and drug companies began to consider whether this could be also true for chemistry.

Fortunately such ideas and approaches had already been developed, though not in the field of synthetic organic chemistry but in peptide chemistry. The technology and methodology of solid phase chemistry had been developed by Merrifield[16] in the 1960s and subsequent automation of the approach, maximising the advantages of forcing conditions (through excess reagent) and purification (through filtering), was well developed by this time.[17] Indeed, some solid phase work with non-peptide structures had been developed by the 1970s[18] though had not achieved widespread use in mainstream synthetic chemistry.

The ability to carry out peptide chemistry on support in parallel was demonstrated by Geysen[19] with the development of polystyrene coated pins. Using this methodology synthesis could be carried out in spatially addressed arrays so that common steps (deprotection and activation steps for example) could be performed using bulk reagents and reaction vessels. At around the same time Furka[20] was developing the approach of split and mix using resin beads to allow synthesis of large numbers of peptides (albeit as mixtures) in very few reactions (Figure 2.2). Houghten[21] introduced the compartmentalisation of resin beads as "teabags," thus allowing a more efficient and scaled up handling of the process, and introducing the idea that packaged resin could then be traced through the synthetic sequence thus allowing identification of the resulting compound (or compound mixture depending on the approach adopted).

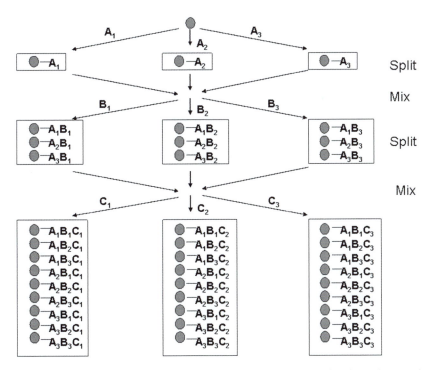

Figure 2.2 Polymer supported strategy of split-mix synthesis in the production of screening compound libraries.

These initial developments focused on manufacturing large numbers of small peptide fragments, used for example to evaluate protein–protein interactions (epitope mapping)[22] or enzyme[23,24] and antibody[25] specificities. The mixtures produced using the split mix approach needed to be deconvoluted to single active compounds, and a number of methods were developed, including iterative deconvolution[26] (fixed positions in mixtures and subsequent sub-library synthesis), positional scanning[27] (replicated synthesis of same library but with a different fixed position in each mixture) and orthogonal pooling strategies[28] (replicated synthesis with orthogonal chemistries allowing different pooling strategies) (Figure 2.3).

For a trimer library A-B-C with 25 monomers at each position (=15625 compounds)

A_1 = specific monomer 1 at position A
A_n = mixture of 25 monomers at position A
A_{Fixed} = 'solved' monomer at position A

Iterative Deconvolution—sequential synthesis and screening, 3 rounds:

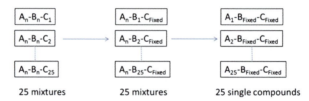

Positional Scanning—concurrent synthesis and screening, 1 round:

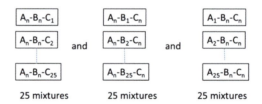

Orthogonal pooling—concurrent synthesis and screening, 1 round:

Two separate pooling strategies for monomers A (similar for B and C):

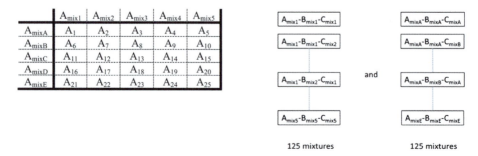

Figure 2.3 Deconvolution strategies to identify single compound hits from pooled samples out of split-mix libraries.

For readers interested in the statistical effectiveness of such strategies and the subsequent development of these approaches for HTS, especially with implementation into post synthesis pooling, the review by Kainkaryam and Woolf is worth reading.[29] The authors explore the concepts of adaptive *versus* non adaptive pooling strategies and provide several further examples of the effectiveness of the approaches.

2.3 FROM PEPTIDES TO SMALL MOLECULES

The ability of combinatorial chemistry to make large numbers of peptides, combined with various new screening approaches did not escape the attention of those involved in early hit identification programmes. Although peptides were not suitable compounds for lead identification, analysis of drug discovery literature confirmed what many practitioners were aware of, that the large majority of drug discovery programmes involved amide bond formation or related reactions (including heterocycle formation through subsequent dehydration). As such, many of the drug discovery compounds should be accessible using similar chemistries to those of peptide synthesis.

The first 'small molecule' combinatorial library was published by Ellman,[30] who demonstrated that a library of 40 benzodiazepines could be produced using solid phase approaches, with three points of diversity, or variation, on the core structure (Scheme 2.1). Ellman expanded this work, using the pin method of Geysen to give 192 compounds,[31] and further expanded this to several thousand compounds in later publications.[32] De Witt described the preparation of array compounds on solid phase using the 'Diversomer' approach,[33] coupled with simple automation that was the first of many automated synthetic approaches to be introduced. That De Witt was based in industry was significant—the approach of combinatorial chemistry was clearly applicable to issues of drug discovery where obtaining data to make the next structural series decisions was the driving component of the research rather than the development of the core discipline.

Over the following few years the two main strategies of split and mix (to generate large libraries using solid phase approaches) and parallel synthesis (focused on smaller libraries) were

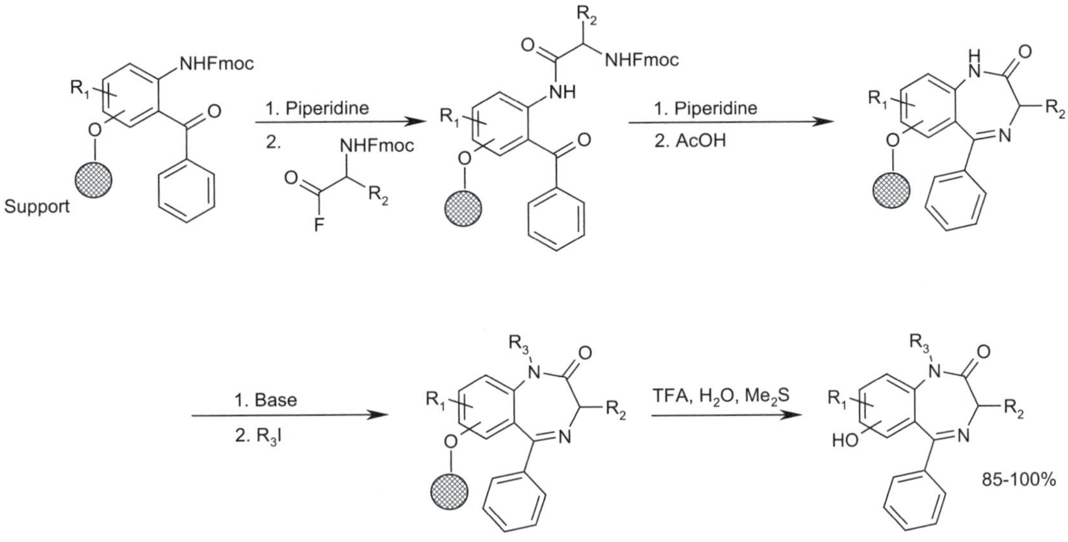

Scheme 2.1 The first published example of a small molecule array synthesised on solid phase—Ellman's benzodiazepine synthesis.[30]

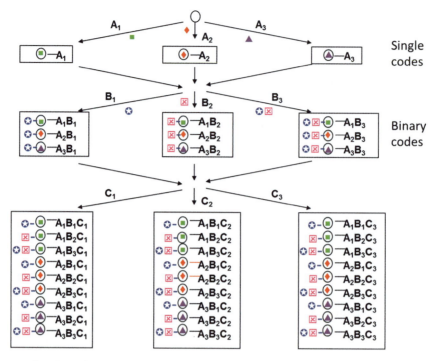

Figure 2.4   Introduction of encoding tags and strategies to split-mix synthesis.

refined and developed. The main focus for lead discovery split and mix approaches was on means of identifying compounds without the need for resynthesis or deconvolution stages, which typically took too long for fast moving lead discovery projects to allow simple mixture libraries to have an impact.[34] Tagging approaches were developed, where the solid phase was orthogonally reacted with molecules that could be 'read', typically using mass spectrometric approaches (Figure 2.4).[35] At the same time the "teabag" concept of Houghten was further developed, both with advancements of the container system, but more importantly with the inclusion of inert radiofrequency tags.[36] These then allowed the synthetic history of any container to be either tracked or directed, thus combining the potential of split and mix with both the potential scale and single product outcome of parallel methods.

At the same time there were also rapid developments in both the range of chemistry applicable to solid phase and in alternative approaches looking to maximise the advantages of solid phase techniques whilst keeping those of solution phase. The range of chemistries on solid phase became almost as broad as traditional solution chemistry,[37-39] though in the context of this review it is worth noting (perhaps discouragingly) that a recent review[40] of the current 'medicinal chemistry toolbox' showed a similar prevalence of amide chemistry in drug programmes. Considering just 'constructive' reactions (excluding protection/deprotection and oxidation/reduction processes) then 24.4% of reactions were simple acylations, whilst an additional 11.3% were N-heterocycle formations primarily through dehydration of intermediate acylated amines. N/O alkylation (including reductive amination methods) accounted for an additional 26% of reactions.

Attempts to get the solid phase 'in solution' included soluble polymers (*e.g.* polyethylene glycol monomethyl ethers,[41] non-cross-linked polystyrenes[42]) that could be precipitated for purification purposes, and the combination of fluorocarbon fluids and perfluorinated substrates[43] to allow separation from both aqueous and organic solution when required. The most applicable development to address the combination of solid and solution phase approaches was in supported reagents, either as scavengers to remove excess reagents or unreacted substrates[44] or as removable reagents to catalyse specific reaction steps.[45] These approaches have achieved widespread use in mainstream synthetic chemistry as well as in the combinatorial research area, and have been extensively reviewed elsewhere.[46,47] Further examples specifically associated with array library design will be discussed later in the chapter in the section on realising a collection.

2.4 MY LIBRARY'S BIGGER THAN YOUR LIBRARY—THE 'UNIVERSAL' LIBRARY

Before considering current best practices and use of high throughput and parallel chemistry in drug discovery and lead optimisation it is important to understand how the initial promise of combinatorial chemistry failed to deliver, and the subsequent backlash against large combinatorial approaches that heralded the start of the 21st century. As has been described above, high throughput screening had rapidly developed as a key component of drug discovery, to be utilised where possible alongside other lead seeking strategies to maximise the chances of new serendipitous results. The need for 'feedstock' for the screening regime was compelling a push to maximise the scale of compound collections. New elements of diversity driven design were exploring a whole range of new ideas on compound structures.[48-53] In this light the power of combinatorial chemistry to generate potentially millions of compounds could not be overlooked. Pharmaceutical companies rapidly followed each other in building in-house combinatorial groups, whilst external new companies were developed to focus on the technology of delivering large numbers of compounds. Many of these were subsequently acquired by pharmaceutical companies, often accompanied with the expressed intent to allow these new technology companies to continue to operate independently of the mainstream world of drug discovery.

Thus by the mid to late 1990s there were many groups using combinatorial chemistry to generate large numbers of compounds, either within pharmaceutical companies or standalone companies operating fee for service provision of libraries. The range of chemistry and structural motifs expanded, and groups were able to make libraries of hundreds of thousands of compounds with a wide variety of structures, extremely rich in functionality.

The pinnacle of such approaches were the 'Universal Libraries', a concept that developed under a range of titles in many groups.[54,55] The hypothesis was a simple and powerful one. By using a set of core templates with several differentially protected functionalities and decorating these in a comprehensive combinatorial fashion with sets of compounds rich in potentially pharmacologically relevant functional groups displayed in directionally controlled manners, it should be possible to devise a single library that would cover all of 'pharmacological space' as relevant to target proteins in drug discovery. Some groups suggested this could be achieved with only a small number of cores series, whilst others argued that greater central variety would be needed. However all had one thing in common—the technology of synthesis, the concepts of spatial design of the molecules and the power of combinatorial numbers had driven the development rather than any real consideration of the nature of the resultant structures, which had to be viable structures for drug discovery optimisation programmes. Indeed at that time the belief was expressed by some that the need for optimisation itself would be mostly eliminated—after all, from such a large and comprehensive library surely the drug itself would be present in the first screening.

2.5 FROM COMBICHEM TO HIGH THROUGHPUT CHEMISTRY—REMEMBERING IT'S ALL ABOUT DRUGS

"The pharmaceutical industry has benefited...from rapid access to a large number of novel compounds and related biological data though combinatorial chemistry and high throughput screening. However this plethora of data has yet to translate into clinical success."

The above extract from Oprea's review[56] of the impact of combinatorial chemistry is just one of many that could be used at this point. Clearly the generation of millions of compounds, not to mention the investment of significant resources into developing technologies, strategies and expertise had not reaped the hyped dividends so readily promised in the early days of combinatorial chemistry. So where did it go wrong?

One of the most fundamental issues was a misconception around the scale of synthetic compound numbers as they related to all of potential chemical (or biological chemistry) space. Traditional medicinal chemistry and drug discovery had been a discipline where, once biological data had pointed the direction, the next compound for test used to take a week to prepare, and a medicinal chemist was seen as prolific if they added 100 test compounds over the lifetime of a particular project. The promise of 100 000 or more compounds from a small team and a few weeks' effort was therefore clearly a step change. Multiply that by concerted planning and the promise of hits every time from a library of maybe 1–2 million compounds appeared to be a reasonable supposition. In short, the naive view was that this step up in compound productivity was bound to yield success in screening campaigns and optimisation work. However, as computational chemists had been pointing out all along, the reality of druggable chemical space was in a completely different dimension. Final numbers vary between advocates of different techniques, but certainly the number of potential compounds to fill that space can be measured in numbers vastly greater than could ever be made (indeed greater than the number of atoms in the universe).[57,58] In a conceptual world of perhaps 10^{70} potential drug molecules then 10^{6} is never going to deliver every time!

Even if the design of a library meant the potential blockbuster drug compound was intended to be in the library, the possibility of it actually being present was limited by the quality of the chemistry of the early libraries, and moreover the means of assessing whether it was in there did not exist. Although analytical (and purification) tools and capabilities have become much more powerful (*vide infra*) in the early days it was only possible to assess quality through extensive validation of the chemistry on sample sets and then build confidence by sampling a subset of final compounds, though even this step was not viable if split mix approaches yielding mixtures of compounds were being pursued. Solid phase methods especially were prone to producing varied yields in parallel steps, and the final cleavage of compounds often could generate unexpected and indefinable products due to the often forcing nature of cleavage conditions.[59]

The combinatorial chemists of the 1990s set themselves up as the new force in drug discovery. Although other areas of chemistry saw and utilised the potential of combinatorial approaches[60] it was in drug discovery that the practitioners viewed their way as revolutionary, leading as it would to a complete change in approaches to lead identification. As such, those who got involved in the field were often excellent scientists who were driven by the development of technology and the strategies of maximising the value of those technologies. Attempts to spread combinatorial approaches into mainstream drug discovery were at best of limited impact.[61] The belief that they were developing a whole new, and more effective, science for drug discovery is well illustrated by the publication challenges and how they were overcome. As the early practitioners of combinatorial chemistry looked to publish work they found the mainstream journals reluctant to accept manuscripts, demanding as they did levels of quality assurance and data than were not only not being gathered but due to the nature of the techniques of the day were not even feasible. Rather than work within the established literature constraints to refine how combinatorial chemistry could be adapted the result was the establishment of new journals dedicated to the science of Combichem.[62]

The separation of combinatorial technology approaches from mainstream drug discovery had a most significant impact on the design of libraries. Driven as it was by the desire to produce large numbers and to make maximum use of the associated technologies, it was almost inevitable that the libraries produced would have large, highly functionalised structures.[63]

In addition, the production of large numbers of compounds around similar core structures created an illusion of diversity but in reality exacerbated the issue identified so much earlier within compound collections being dominated by common core motifs.

The rehabilitation of combinatorial chemistry (as high throughput chemistry) was enabled by a number of analyses of problems associated with the earlier approaches (leading to several strategies such as described below), alongside the more widespread development of understanding of factors critical in limiting attrition in potential drugs across all aspects of drug discovery. Three particular strategies are worth noting here as they have had major impact on the design of combinatorial approaches; the physicochemical properties of drug structures and their ability to cross biological membranes; the size of lead molecules and subsequent optimisation impact; and the incorporation of experience and knowledge into targeted library approaches.

The first of these is the seminal publication of Lipinski,[64] outlining the 'rule of 5' as criteria to determine to likelihood that a particular compound will pass through biological membranes, and therefore have potential to act as a drug substance. Early library structures typically had a profile of properties with mean molecular weight well above the Lipinski limits of 500, and high functionality counts (especially amide bonds) that inevitably leads to too high a level of both H-bond acceptors and donors.[65] Therefore screening such libraries in any lead discovery phase, or using such design templates in lead pursuit and optimisation is fraught with developability issues and, not surprisingly, initial results from such libraries did not become successful development candidates. As all the Lipinski parameters can be calculated from compound structures it was simple to incorporate such factors into any design approach, for example using weighted penalties in a design strategy or just setting hard limits on molecular weight and other properties.

Extending the physicochemical property limitation further, Teague and colleagues from AstraZeneca published an analysis that showed that for lead compounds these parameters needed to be even stricter,[66] as lead optimisation consistently added both molecular weight and lipophilicity to any series as it progressed towards development candidate status. On a similar note, Hann[67] demonstrated that success rate of lead discovery was inversely related to the complexity of the screening structures, and that for more complex designs the likelihood of finding a successful hit against a target were very low.

Finally, the application of knowledge of past success has been brought into the design of libraries, most effectively for large targeted libraries for protein family screening. One example of this is the work of Lewell and Judd[68] where the knowledge of known active compounds against classes of related 7-TM structures was used to design library building block sets incorporating 'privileged' substructures. Computational algorithms looked for common feature motifs across a range of active structures, using chemically intelligent fragmentation approaches to identify real substructures that could be introduced into new designs.

2.6 REALISING A COLLECTION—TECHNOLOGY DEVELOPMENT AND COMMERCIAL OFFERINGS

Alongside the development of strategies of design and selection the development of combinatorial chemistry and subsequent movement to high throughput chemistry approaches has driven a number of technological advances. Many of these have been 'of the moment'; for example a number of high level automation approaches were extremely effective in producing large numbers of compounds but now exist only in archives of scientific equipment. Others however have become commonplace approaches, as have many of the developments in parallel analysis and purification, initially driven by the challenge of large number synthetic approaches.

Synthetic automation is perhaps the most notable example of such short lifetime technologies. As in other sections of this review, fully comprehensive reviews of the wide range of synthetic automation equipment are available elsewhere,[69] and only illustrative examples are used here. For example, three synthetic automated technologies were in use within GlaxoWellcome in the late 1990s, all of which are now 'retired' (and indeed examples of all have been donated to the Science Museum in London). Initial solid phase work was driven by 'Advanced Chemtech' ACT machines.[70] Based around liquid handling robotics, and using proprietary designed reaction blocks there were a number of designs supporting solid phase chemistry. At the same time split mix approaches incorporated through the acquisition of Affymax by GlaxoWellcome were carried out on Encoded Synthetic Library (ESL) synthesisers,[71] automation based around adaption of peptide synthesisers with the ability to mix and redistribute resin to reaction vessels. Finally an arm of solution phase based work was supported by the development of synthetic robotics on a Tecan liquid handling bed with adaption for solvent removal through gas enhanced evaporation.[72] Between these three technologies millions of compounds were synthesised during the late 1990s, however all were to be subsequently overtaken by the development of RF-encoded encapsulated resin in the IRORI system.[73] Using automated directed sorting with capacity for up to 10 000 vessels this became the workhorse of large number synthesis, but was itself superseded by IRORI development of the X-Kan,[74] with 2D bar-coding replacing the RF tag approach. In the period of only 10 years, within just one company therefore we have seen the introduction and subsequent displacement of over four separate automated synthesisers, and in reality several more systems (*e.g.* Myriad,[75] Zinnser Sophas,[76] Argonaut Trident and Quest systems[77]) were also in use during the same period, again most of which are now no longer in use.

The type of automated synthetic equipment outlined above has typically remained as tools of the dedicated diversity chemist, with the development of expertise around synthetic automation technology, and several groups continue to develop extensions to these approaches.[78] Of much greater impact and lasting effect was the development of simpler parallel reaction equipment, much of which was developed in pharmaceutical laboratories and subsequently commercialised through equipment manufacturer partnerships.[72] Many examples are available and in use today, but examples include parallel tube based reaction blocks introduced by companies such as STEM,[79] allowing controlled stirring and heating of arrays of solution based reactions at significant scale, whilst Radleys introduced equipment based on commercialising the common practices of having several reactions on a single stirrer hotplate.[80] The carousel took advantage of the magnetic field created by a stirrer, whilst the greenhouse allowed reactions to be carried out readily under inert conditions. For solid phase chemistry a number of block based clamped filter based systems were introduced, including Bohdan Miniblocks,[81] which took advantage of a layout format identical to microtitre plates, thus facilitating subsequent transfer to assay plates.

As discussed earlier, the development of polymer supported reagents and sequestration agents has made solution phase approaches to parallel chemistry viable, allowing filtration and work up approaches to be used in parallel using filtration reagent blocks. This area has recently been reviewed[82] and includes resin capture and release approaches, tagged reagents and substrates. The following examples illustrate how these approaches have been applied in library syntheses. Strohmeier and Kappe[83] used resin capture and release steps in the preparation of 1,3 thiazine libraries (Scheme 2.2). Parlow[84] reports the use of 2 different tagged reagents to support purification by removal of reagent by-products in Suzuki coupling reactions (Scheme 2.3). Wang[85] describes the use of polymer supported phosphines in the wide ranging syntheses of triazolopyridines (Scheme 2.4). Perhaps the ultimate demonstration of the power and flexibility of polymer supported reagents and reactions is in the synthetic work of the Ley group, who has produced several publications of total syntheses of natural products (Scheme 2.5)[86] as well as a number of approaches to library and array syntheses.[87]

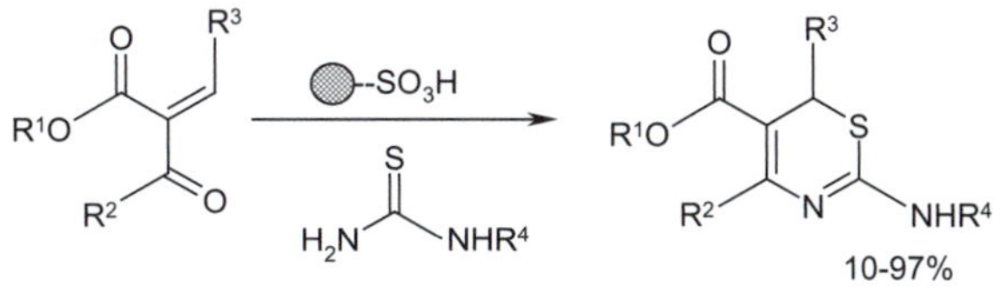

Scheme 2.2 Using polymer bound reagents to capture an intermediate with subsequent release into solution on further reaction.

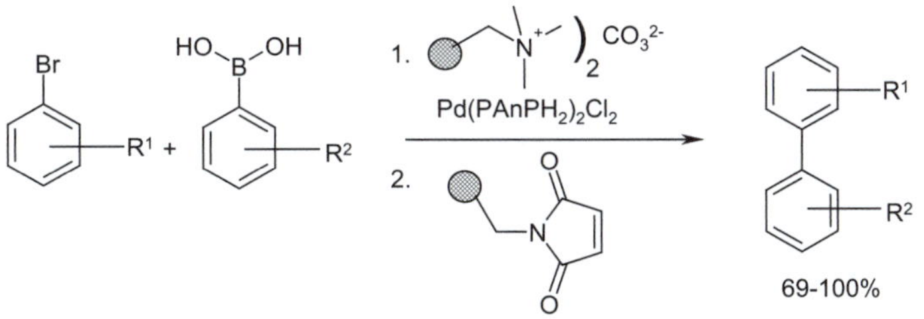

Scheme 2.3 Using polymer bound reagents to scavenge excess reagents and reagent biproducts.

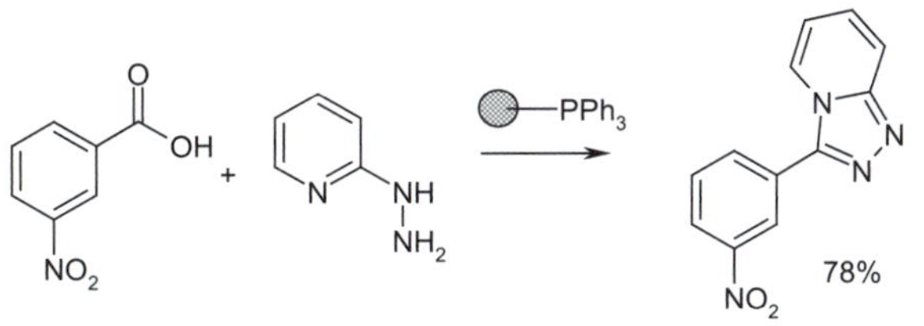

Scheme 2.4 Example of supported reagents in solution phase synthesis.

One now commonplace technique that developed alongside the high throughput chemistry techniques has been the use of microwaves to heat and accelerate reactions.[88] Although it was initially thought that microwaves could have a specific effect on reaction trajectories and rates, it is now generally agreed that the primary impact is the same as thermal acceleration, albeit a much faster and energy efficient one.[89] There are specific exceptions where homogeneous reactions may be affected by localised heating of solid catalysts[90] and recent designs of reaction vessels incorporate microwave absorbing materials to maximise the effectiveness of microwave heating.[91] However generally microwave technology has the main advantage of rapid heating, combined with being linked to automatic processing equipment that allows array chemistry to use this approach as a very specific tool for rapid compound synthesis. For example, a recent synthesis of dihydropyrimidone libraries using stepwise multi component Biginelli chemistry and Pd/Cu mediated cross coupling reactions, both accelerated and in high yield, illustrates some of the range and impact of microwave assisted synthesis (Scheme 2.6).[92]

Alongside parallel synthesis developments, the ability to analyse and purify large number of compounds has also developed extensively. The use of scavenger reagents and supported sequestration approaches, alongside catch and release methodologies certainly improved the purity and quality of combinatorial chemistry reactions. However it has been the development of fast, automated LC-MS analysis systems[93] and the more recent development of fast, parallel, mass directed preparative LC[94] that has allowed the approach of purifying all synthesised compounds to take over from previous triage processes,[95] whereby moderate to good purity compounds were typically progressed into screening without additional purification, and only

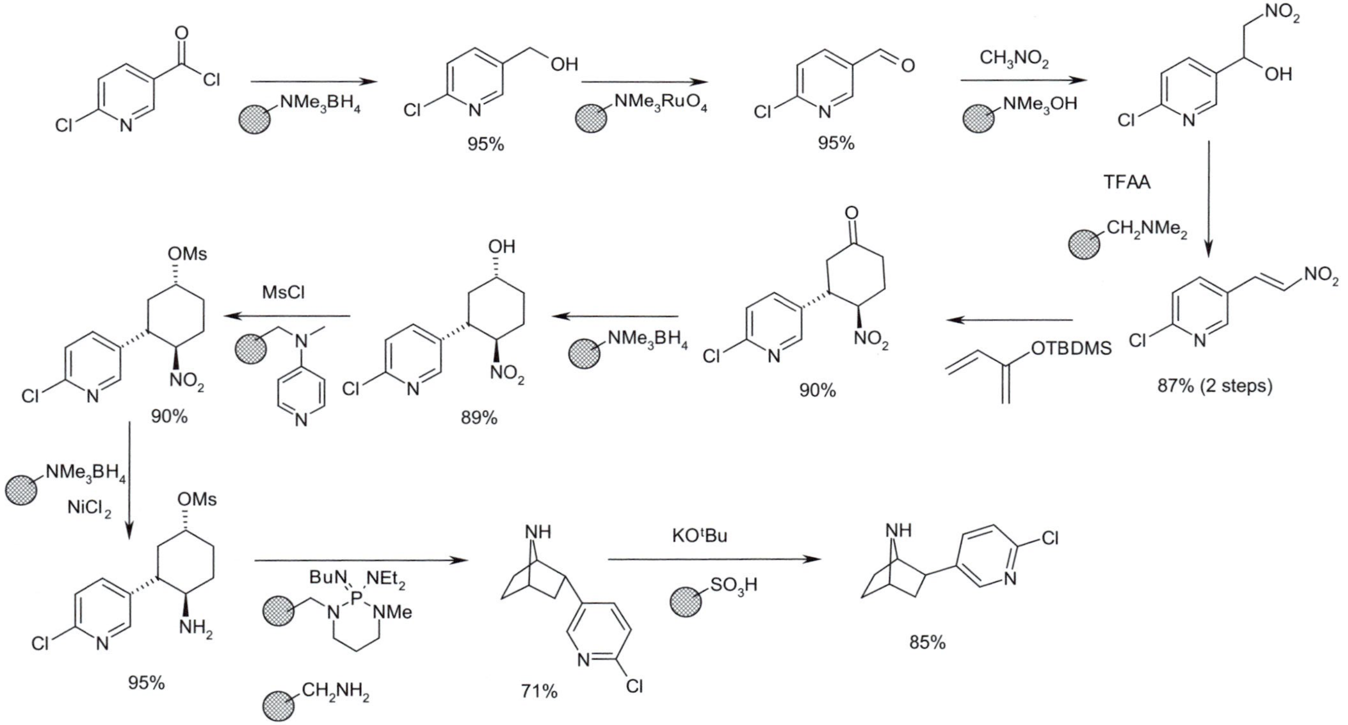

Scheme 2.5 Multi step total synthesis of (±)epibatidine using supported reagents and/or scavengers at every step, and with no additional purification steps.

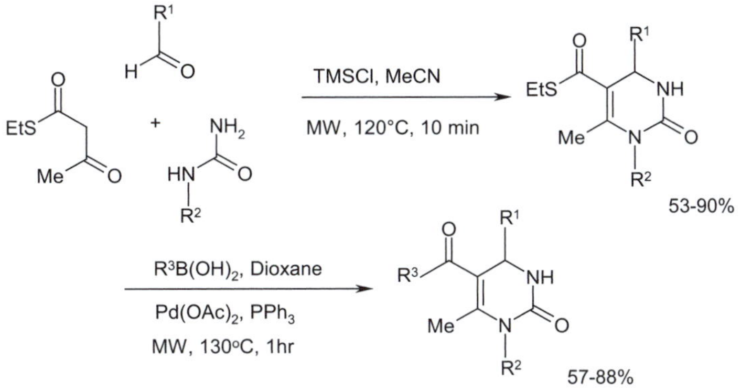

Scheme 2.6 Microwave assisted synthesis of a library of dihydropyrimidones.

the less successful reactions were purified. The ability to estimate concentration using LC methods[96,97] has added a further level of quality into library compound in screening, as assay level concentrations can now also be determined with greater confidence rather than assuming only a single concentration across an entire collection.

The development of equipment and technologies to deliver novel chemical libraries within pharmaceutical companies as illustrated above was driven by the intent to create novelty within a company's collections. However, the lack of a wide variety of chemistry applicable to such approaches, combined with a limited internal resource meant that this could only be partially successful. As indicated above, some companies addressed this further through the acquisition of small synthesis companies, whilst others commissioned the external creation of companies to focus more closely on pure synthesis of collection compounds. However, the opportunity for alternative external support of collection development was not missed by others, and companies offering compounds for sale, often acquired through collaborative approaches with academic institutions, became more prominent. Initial issues of compound quality and of access to material for follow up studies plagued this process, but as more companies became involved and began to offer higher quality assured products the option to build collections though purchase of compounds became more prominent. In particular the opening up of the former Soviet states in Eastern Europe and Russia allowed the rapid development of a number of companies offering compounds, initially brokering academic sourced material but rapidly moving towards commissioned and designed libraries. Today there exists a highly competitive supply market built on a long tradition of good organic chemistry in these regions that offer screening sets targeted to particular proteins, general screening sets with the ability to cherry pick bespoke selections, custom synthesis of novel structures around array formats and full contract research services.[98] This market has developed in response to the needs and quality demands of the customers, typically the large pharma organisations, though many academic grant applications have been built around a component of library purchase for novel target investigation in universities. Many of the companies are now correctly regarded as design leaders in their own right, developing algorithms and approaches to defining novel chemistry and chemical space. As an illustration of the most well-known and valued companies, Table 2.1 shows a breakdown in percentage coverage terms of the top eight suppliers that MRCT used in an exercise in 2011 to build a representative diversity set suitable for screening in academic laboratories with modest HTS capacity.

In addition, the suppliers have continued to take note of the developing understanding of undesirable structural types and properties. Knowledge of many of these has only developed through screening and (failed) follow up across a number of targets, and publications such as

Table 2.1 Top eight suppliers of compounds used in a recent 10K library construction at MRCT.

Supplier	Compounds	% of library
Enamine Ltd.	3117	31
ChemDiv Inc.	1530	15
ChemBridge Corp.	1459	15
Vitas-M Laboratory Ltd.	1026	10
Maybridge (Thermo Fisher Scientific)	960	10
Life Technologies Corp	619	6
Asinex	606	6
InterBioScreen Ltd.	382	4

the PAINS paper from Baell and Holloway[99] are now used to help define desirable chemical space by suppliers. However, the onus still rests with the purchaser (*caveat emptor*) to ensure that any process of compound acquisition has robust mechanisms of analysis and filtering that are maintained to current knowledge to maximise the likelihood of quality hits from any screen, and such approaches will be outlined in the following sections.

At the time of writing the total commercial offering of screening compounds is in the excess of 20 million compounds and the efficiency of compound delivery, cherry pick selection and variable quantity supply at modest cost per unit item means that purchase of compounds is the preferred route for rapid collection development for the majority of parties interested in developing libraries for screening. Sub-selection from the commercial offering for initial library preparation also supports initial follow up of screening results through purchase of similar structures—'analogue by catalogue' (*vide infra*). There is still some question however over the absolute breadth of coverage of commercial offerings, in the same way that combinatorial 'universal libraries' were once believed to represent all of chemical space. In a follow up paper to the PAINS publication Baell has postulated that the commercial offering of millions of compounds is a 'shallow pool' that can be represented with fewer than 350 000 compounds,[100] and undoubtedly there are many more regions of chemical space that could be opened up and explored by alternative chemistries (and indeed technologies, such as the DNA encoding approaches that can increase the compound count to billions—more on that later)—the question beyond this chapter (though picked up in other sections of this book) is how much of that is truly 'drug space'.

2.7 DESIGN STRATEGIES

As highlighted in the introduction to this chapter, a compound collection can be broadly categorised as supporting two types of investigation—either looking for start points for discovery programmes where little or nothing is known about the target or for maximising the chances of finding novel results for targets where we believe we do have some understanding of the underlying requirements for that target (usually but not exclusively structurally derived). The first approach requires a focus on diversity and as wide a selection of compounds exploring chemical (or drug like) space as possible, though there are limitations that need to be considered around structures carrying unwanted liabilities that may potentially limit the developability of a particular series. The second approach needs to rely on structure guided knowledge to allow the library to focus down to advantageous regions of chemical space (often around 'privileged' motifs and structures). A good chemical collection design strategy can and should encompass both components (at least for collections intended for broad usage—an organisation whose whole focus is on one particular protein target class would do best to focus towards maximising coverage around knowledge based design for that target class). However for the sake of clarity the two types of use (and relevant design) will be considered sequentially in this review.

2.8 DIVERSITY COLLECTIONS

Approaches to diversity, with extensive focus on the algorithms behind computational, cheminformatic and mathematical modelling, and the comparative analysis of the effectiveness of such design strategies, have been extensively discussed elsewhere[101–104] and are covered in other chapters of this book; a relatively high level appreciation is all that is necessary to illustrate the issues relevant to this chapter. Indeed, the effectiveness of any particular diversity model and design strategy is difficult to quantify objectively, as by the very nature of the use of such diversity libraries sparse data sets are created as primary outcomes, with the majority of data based on single point biological measurements that even when positive are not exhaustively followed through to confirmation. Potential chemical structures of interest are typically rapidly reduced down to a small number of compounds through cascades of counter screening (again often single point), calculated physical properties, and structural evaluation based on chemist intuition and it is this smaller subset that then may have more detailed data measured and evaluated. The objective of any novel target drug discovery screening campaign is to reach a decision point on 'hit identification' as fast as possible preferably generating a few good structural series for further medicinal chemistry development rather than on statistical validation of the effectiveness of any particular design strategy. A screening campaign is expensive in both time and consumable costs (often under close scrutiny to remain tight to budget and timelines) even when focused as rapidly as possible towards the hit finding goal and it is not surprising that experiments to understand why a design strategy may be effective or otherwise are not considered worth pursuing.

Although objective result based evaluation of a collection design is therefore rare, the importance of applying good design principles, especially around those learnt through the mistakes of the earlier days of chemical libraries, is clearly understandable. A good design must first and foremost try to limit the presence of undesirable compounds (why have them if they will never be followed up) whilst working to the physical constraints of the collection itself and the means of generating such collections (for example, it is easier to achieve a very favourable property profile through individual purchase of compounds but may be more cost effective and efficient to generate compounds through a combinatorial synthesis approach that will generate a wider profile of properties, even though this will produce a small number of undesirable structures). The overall size of a collection should reflect the constraints set by physical capacity (how many compounds can be held and processed by a particular group or organisation), level of automation in both compound handling and screening (what capacity of screening is actually achievable) and the cost per well of particular screening targets (is the focus on high throughput, low cost biochemical assays or higher cost cell based or complex reagent based).

Once the realistic limits of the size of a collection have been decided, then the next stage is to determine the highest quality selection of compounds to use as the potential library selection set. This could be defined from commercially available compounds and/or based on virtual libraries from available and reliable synthetic transformations available to the library designers (again the synthesis of these libraries may be from commercial suppliers as well as in house resources). The use of virtual libraries based on parallel and combinatorial synthesis approaches brings additional constraints of practicality, as sparse synthesis of representatives of a matrix of compounds is often much more labour intensive than blanket synthesis of all, but again this is a balance that needs to be considered for each case in turn and cost/value analysis carried out for each synthesis design and sub-selection strategy. For the sake of this process however this review will focus on the commercial purchase strategy, but all the decision processes can and should be equally applied if using a synthesis driven approach.

All commercial compounds available need to be collated together and then analysed to remove those compounds that would not be wanted in a general diversity screening campaign—that is anything that can be predicted to have toxicity or promiscuity effects, plus all compounds where the molecular properties and physicochemical parameters mean that the compounds would be very unlikely to be successfully progressed to a small molecule drug candidate. Once again the actual values of for example molecular weight cut-off or polar surface area value are subjective decisions based on the strategy and previous experience of the designer and the organisation involved, and many different views will exist, though in a study on using the wisdom of crowds to develop compound libraries Agrafiotis[105] found some strong common held understanding of important properties and parameters. This study involved medicinal chemists from six different Johnson and Johnson Pharmaceuticals R&D sites across the US and Europe, and examples of the findings are shown in Table 2.2. One overall finding was that it appears easier for chemists to agree on molecules that they do not like rather than on those that they may all favour.

Many exclusion filters applied at this stage are well understood and include historic considerations such as those based on known reactive and toxic functionailities;[106,107] these will include structures such as alkylating agents (epoxides, aziridines, activated alkyl halides *etc.*), acylating agents (acyl halides, anhydrides, sulfonyl halides *etc.*) reactive carbon–heteroatom multiple bonds (aldehydes, ketones, imines, 1,2-dicarbonyls *etc.*) and heteroatom–heteroatom single bonds (disulfides, perethers *etc.*). Others are those that have been learnt through several years of high throughput screening[99] (and having been the repeated subject of failed hit to lead prosecutions). The actual mechanisms by which these compounds may interfere with screens are varied, and will include off target effects through promiscuous activity, false screening results due to protein aggregation, non-stoichiometric non-specific binding or interference with assay read out technologies. It should also be remembered that knowledge around problem structural types and functionalities is continually developing and regular reanalysis of a screening collection based on updated filters is to be recommended. A single structural type to illustrate these issues is shown in Figure 2.5.

Rhodanines and related thiohydantoin structures have been identified in a multiple of screens across a wide range of target classes. A comprehensive review[108] of this class of compound concluded that many of the results could be put down to aggregation effects, reactivity to proteins (through a conjugate addition mechanism) and the generation of reactive compounds through photochemistry during assays. However in another study,[109] a large library of rhodanines and other related structures were synthesised and analysed for their propensity to hit multiple targets. Structures containing an exocyclic sulfur double bond and with a benzylidene substitution (thus creating an extended aromatic system) were identified as 'frequent hitters'. However further studies precluded both aggregation and protein reactions as mechanisms

Table 2.2   Preferred compound properties identified by 'crowd sourcing' methodology within Johnson and Johnson.

Property	Preferred	Disliked
Molecular weight	300–400	<250 or >425
Rotatable bonds	4	<3 or >6
H-Bond donors	1	>2
H-Bond acceptor	3	<2 or >4
ALogP	Dependant on target location	>4

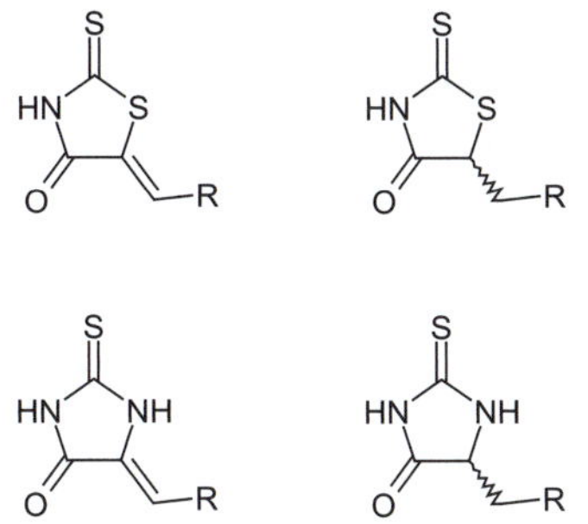

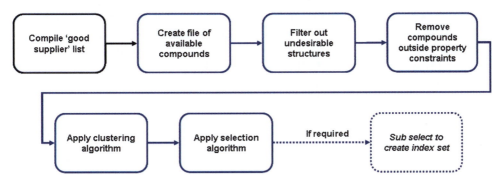

Figure 2.5   Rhodanine and related heterocycles—frequent hitting PAINS or viable hit to lead start points?

Figure 2.6   Typical process flow in the design and selection of a screening library from commercial sources.

involved—the authors concluded that the electronic and hydrogen bonding potential of these structures led to the formation of a wide range of weak/moderate molecular interactions. As such the compounds were acting in a specific manner, but were able to do so with a wide range of proteins.

A typical process to define a diversity based screening collection can therefore be summarised in Figure 2.6.

In brief, an initial collection of commercially available compounds is drawn together from reliable suppliers (another aspect that in reality is developed through experience of ordering compounds—it is important to use suppliers that have the track record of delivering compounds to order, as any design will be compromised if suppliers cannot supply their catalogue compounds). This set is then subjected to filters to remove undesirable and reactive structures and functionalities. Additional property limits will also be applied, such as molecular weight and lipophilicity ranges to ensure the set falls within a desired range of properties (again de-pendant on the type and use of the set—for example, focus on CNS targets may have a different lipophilicity profile), yielding a final set of available and acceptable structures. At this point the application of a selection method is required to reduce the numbers down to a final level that can be accommodated in the physical limitations of the collection housing. This will typically be a two-step process, with firstly some form of structural based algorithm applied which places the available compounds into clusters of 'like' compounds—the definition of like being based on the algorithm applied (most commonly based on Tanimoto similarity indices).[110,111] A second stage sampling method is then applied to the clusters to select representation of the clusters. Statistical methods[112] have been developed to consider the optimum number of compounds that need to be selected to maximise the potential of finding a hit from a cluster (assuming that there is a hit in that cluster), but in simplistic terms the intention is to obtain a sufficient density of coverage of each cluster to maximise the chances of finding a positive from that cluster whilst at the same time allowing reasonable sampling across all the clusters. Within the MRCT collection, for example, the range is variable but an aim during the initial construction and subsequent refinement has always been to achieve between 10 and 20 representative cluster members for each substructure.

The advantage of using the cluster-selection approach is that it allows the design of an optimum representation of a full set of compounds within the constraints applied to the screening set (most typically size and cost driven). Though used typically to generate screening libraries that range from 10's of thousands up to millions of compounds, the approach can be used to support other diversity based strategies. For example, if the ability to carry out a large (100 000 plus compound) HTS campaign is limited for a particular target (possibly by reagent cost or technical capability) then sampling of the clustered diversity screening library can be used to define smaller 'index' sets of compounds, based on a sparser selection of compounds from the original clusters in the initial screening collection build. Clearly such an approach may limit the statistical likelihood of finding hits for any given cluster (as sampling is much more sparse) but it may be a pragmatic necessity that allows for a target to still be explored through screening approaches, and should any hits be found then there is immediate available follow up from the larger screening set which may then allow rapid assessment of the potential of a particular structural series.

Sampling of the wider cluster should also be the preferred first steps for following up any hits resulting from a full diversity HTS. Any hits from a screen should be analysed to identify whether there are multiple hits from a particular structural cluster. Follow up confirmation screening should be applied not only to the compounds that were found in the initial screen but also to other nearest neighbours in the same cluster. Assuming this initial rescreening confirms interest in the structural series then the next stage should be to return to the larger commercially available cluster that the screening set was drawn from and near neighbours should then be purchased for screening and the establishing of any early SAR trends (often referred to as 'analogue by catalogue'). Although there will never be perfect coverage of all the substituted analogues a good SAR design would demand, this approach is a highly efficient and time effective means of rapidly assessing a series' potential for further SAR development and allows for comparison between series rapidly. Testament to the (cost) effectiveness of such an approach (initial screen, confirm close neighbours, explore SAR by accessing wider cluster compounds) is the adoption of similar approaches by many large pharmaceutical companies to sample their in house diversity compound decks. Often several million compounds in size, the economics of screening such large collections has become challenging, and smaller sets representing the large set are now often screened, with the equivalent of 'analogue by catalogue' then being carried out on their in house large set (and also externally with commercial compounds). In one example an exercise based primarily on elimination of molecular redundancy within a screening file has allowed Pfizer to reduce their primary screening deck by almost 1.5 million compounds.[113]

Building a diverse screening set based on commercially available 'lead like' small molecules is the most common approach adopted for building and developing a screening capacity.[114] However alternative approaches towards diverse sets are used and these can be particularly valuable for some classes of targets, especially if the outcome of a screen is focused towards the generation of tool compounds for target validation and biological understanding of a particular target or pathway rather than the specific identification of a potential small molecule drug discovery start point. A number of smallish sets of pharmacologically active compound sets are available, some from commercial sources (*e.g.* Sigma Aldrich Lopac[115] or the Prestwick screening collection[116]) whilst others have been built by charitable groups to support specific areas of disease biology (*e.g.* NINDS set[117]). Such sets can be used to interrogate biological pathways, for example in whole cell or *in vivo* phenotypic screens, as many of the compounds in the set will have well described pharmacology and underlying target information (albeit often developed in alternative indication studies). Such sets can also be used for studies looking at repositioning known compounds for new indications, identifying previous unseen pharmacology in novel systems that have the potential to become therapeutic intervention points.

Sets of natural products can serve a similar purpose to the pharmacologically active sets, and many are available as well defined and characterised sets of isolated compounds (as opposed to earlier natural product screening approaches based on fractionated uncharacterised extracts). It has been argued[118] that natural products should have a greater success rate in screening for biologically active compounds given that they are generated in a 'biological selection environment' and certainly hits found from natural products can often serve as tool compounds for valuable target validation exercises (assuming they have a good level of selectivity) even if they are not obvious start points for drug discovery programmes due to complexity or limited availability. In an extension to just accessing and screening natural products, the biosynthetic pathways themselves can in some cases be used to generate diversity compounds. For example, the polyketide biosynthesis pathways have been studied and modified to generate compound sets for biological evaluation.[119] Taking the idea one step further and into the synthetic chemistry domain, the concept of Diversity Oriented Synthesis (DOS)[120,121] has been developed to allow synthetic chemists to focus on delivering structures similar to natural products in their complexity. This approach allows synthetic chemistry groups to exploit their established chemistry methodology to generate screening compounds, though it is worth noting that to date the majority of these efforts have been limited to screening within the local generating groups.

Before moving on to more focused library design it is worth considering the concept of fragment screening and dynamic combinatorial libraries as parts of the whole spectrum of diversity screening approaches.

Fragment screening represents an alternative approach to lead discovery and subsequent optimisation by accessing the widest possible diversity of chemical space using smaller molecules. Rather than sampling chemical space through a large number of discrete compounds, fragment screening covers wide areas with few compounds in a very nebulous manner, with any interactions likely to be weak but still identifiable, and that can subsequently be optimised through structure guided synthesis and design. Fragment screening has become a significant approach to diversity screening[122] and is covered in much greater detail in the dedicated chapter of this volume.

The related approach of dynamic combinatorial libraries typically uses a fragment like approach, but instead of following lead discovery programmes using iterative structural based design, the concept relies on the protein target itself building directed larger molecules of interest from a cocktail of fragments present in an assay mixture, using reversible bond forming reactions.[123] Astex have described the extension of fragment screening to generate larger lead like molecules bound to CDK2.[124] Mixtures of aryl hydrazines and isatins were soaked into individual crystals of CDK2, and under equilibrating conditions reacted in a condensation reaction to form hydrazones. These were then examined *in situ* using X-ray crystallography, before the most promising compounds were resynthesised and fully profiled in typical assays. The best compound had an IC_{50} of 30 nM (Scheme 2.7).

Another dynamic combinatorial approach using the target protein to template the chemistry was described by Sunesis pharmaceuticals.[125] In this example, the target Aurora Kinase was initially modified by site directed mutagenesis to present a cysteine SH close to the putative binding site. This handle was then exploited in a dynamic combinatorial chemistry strategy using mixtures of disulphide building blocks, which under the equilibrating conditions underwent S–S cleavage and reformation of disulphide bonds. Any building block favoured to fit

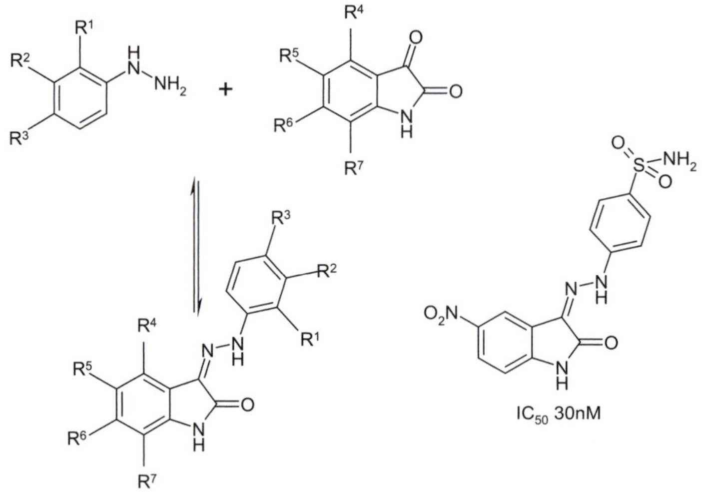

Scheme 2.7   Using reversible dynamic combinatorial chemistry in screening an extended fragment library *in situ* during crystal soaking experiments against the protein kinase CDK2.

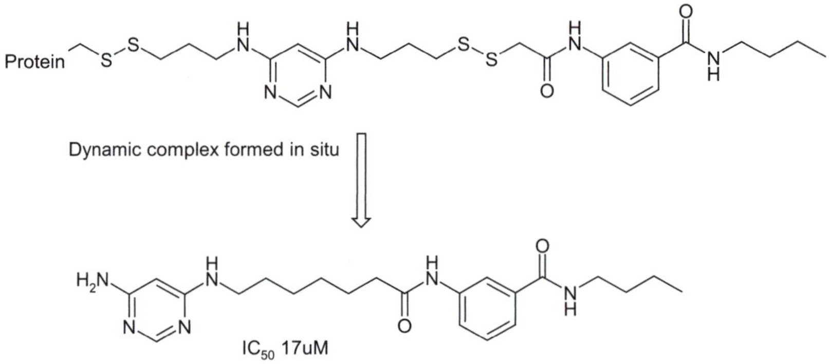

Scheme 2.8   Disulfide equilibration in dynamic combinatorial chemistry and subsequent optimisation of compounds active at Aurora Kinases.

in the binding site of the kinase was therefore held close to the cysteine handle in a favourable position to form a disulphide bridge. The initial series of monomers incorporated a second set of disulphide links, thus allowing a second round of equilibration with another set of monomers, and finally yielding compounds with activity in the single micromolar range (Scheme 2.8).

An alternative approach to performing reactions directly in the presence of a protein is to allow the dynamic mixture to equilibrate before introducing the target protein. Therascope have described such an approach to target novel neuraminidase inhibitors, using reductive amination chemistry on a scaffold related to known inhibitors.[126] In this example the initial imine formation was performed in the absence of the neuraminidase, and the resulting mixture reduced to yield a set of amines that could be profiled by LC-MS. The same reaction sequence was then repeated, this time with the introduction of neuraminidase during the initial imine equilibration and following the reduction step the amine profile was again analysed. A specific number of ketone examples were dramatically amplified by the addition of the enzyme, with all subsequently shown by resynthesis to be potent inhibitors of the enzyme, the most potent having a Ki of 85 nM (Scheme 2.9).

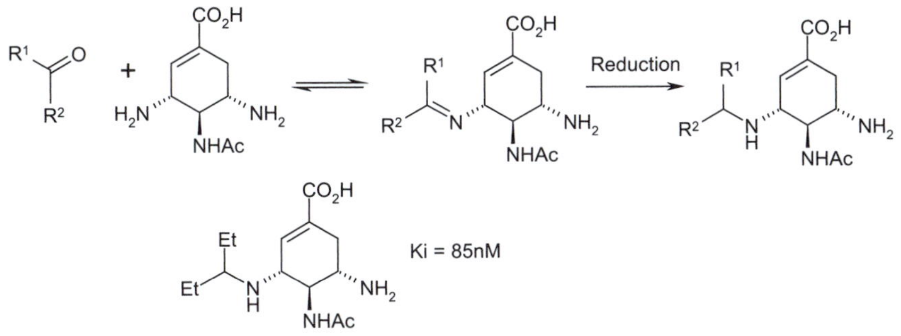

Scheme 2.9 Target protein driven amplification of dynamic combinatorial chemistry to identify favoured constructs for inhibiting neuraminidase following reduction to non-reversible products.

2.9 TARGETED LIBRARIES

The design of targeted sets of compounds against specific protein classes can be approached from two complementary approaches akin to other design methods used in computational approaches to single target drug discovery, either protein structure based or ligand/substrate based. However, with the intent in designing such sets being to target future as yet unknown family members of the same protein family, additional care needs to be taken in selecting the most experimentally validated targets and ligands to build from, and to ensure any alignment of structures (protein or ligand based) is undertaken with an eye to class related common features wherever possible. However, even the best designed arrays can provide surprising results as demonstrated in the design of novel arrays targeting the kinase CDK2[127] and related family members. Based on crystallographic structural data using active oxindole compounds novel structures were designed around an aza indole structure, with the intent to maximise the interaction with the hinge region of the active site by picking up additional binding. However, analysis of the active compounds from the subsequent arrays using crystallography showed that rather than just picking up additional binding, the novel structures had adopted two new binding poses, with one even having reversed the direction of the core scaffold in its binding to the hinge region (Figure 2.7).

Rational design of targeted libraries based on protein structures is likely to be most effective where a significant amount of structural information around the protein family is known. The kinase family[128] and several protease family sub-classes[129] are good candidates for such approaches, sharing as they do significant active site structural homology between members of the same family. Perhaps surprisingly, another class of drug targets that has received similar structural attention is protein–protein interaction targets. In this case it is structural motifs that are the product of the protein secondary structure that are targeted, for example mimetics of the spatial arrangement of alpha helix side chains.[130]

Where limited structural knowledge of the protein class is available then ligand based screening sets have often been designed around trying to define privileged structures and motifs from active ligands and inhibitors of related proteins which can then be incorporated into targeted library designs, approaches that have been outlined by both AstraZeneca[131] and GSK.[68] As a simple illustration of the concept, the selection of a tricyclic biphenyl motif to include as a monomer building block would come from the analysis of known serotonin/adrenergic reuptake inhibitors (Figure 2.8).

There are however some issues with using this approach to define targeted sets and libraries. By designing a library around known structural motifs then there is a built in risk that any results from that library will sit in close structural space to known compounds, which in turn

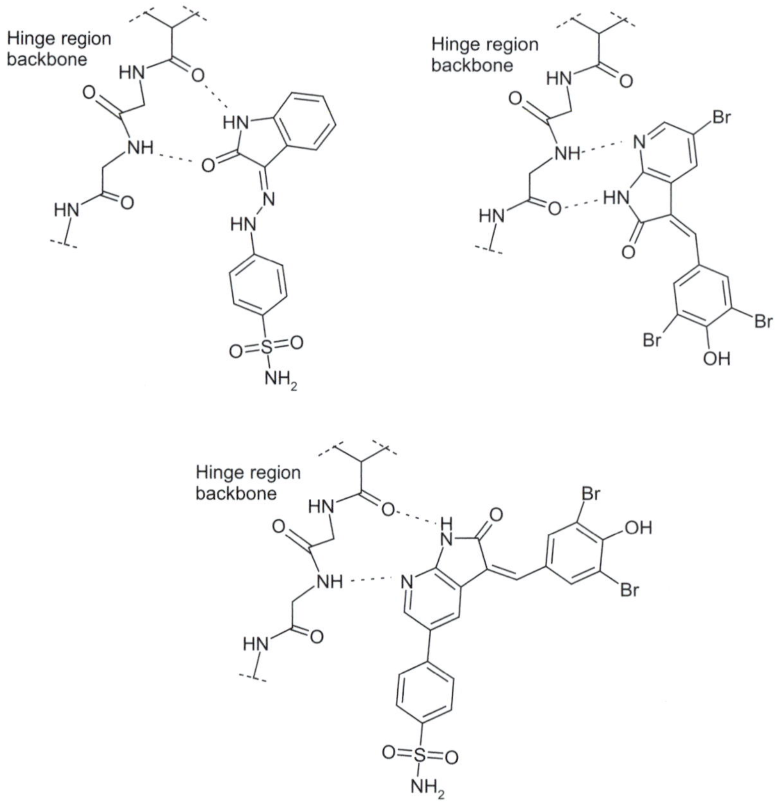

Figure 2.7 Three distinct modes identified for (aza)oxindoles binding to the hinge region of protein kinases.

brings issues around patentability and freedom to operate. Moreover the grouping of some common protein structures is often at a superficial and broad level, leading to very generic designs (for example, libraries purportedly focused for ion channels or GPCRs rather than sub-classes of these targets). A limited number of studies have shown that for many of these 'targeted' library approaches, outside of kinase and some family A GPCR focused libraries, the statistical results of screening campaigns are no different from using general diversity based high throughput screening,[101] which if done at full scale also brings the potential for moving into truly novel chemical space for a target class. There are however published examples where targeted libraries have been used to successfully identify novel start points for drug discovery programmes, including kinases,[129] voltage gated ion channels[132,133] and serine/cysteine proteases.[129]

2.10 COMBINATORIAL POWER IN DESIGN

The original drive behind the development of combinatorial chemistry to support high throughput screening was the potential to access very large numbers of compounds using efficient synthetic paradigms. The limited success of early screening approaches using mixtures, either synthesised as such or through pooling strategies, combined with the increasing capacity of screening campaigns meant the focus shifted significantly to one assay one compound strategies, and this has remained the standard approach to screening in most companies to

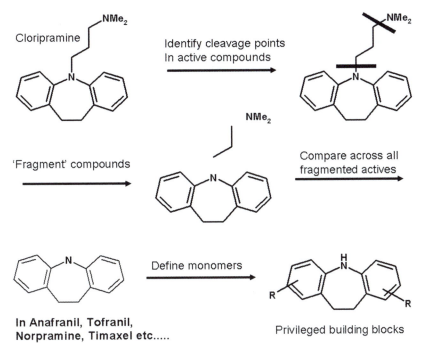

Figure 2.8 Illustration of the process that ReCap uses to identify privileged monomers from known active compounds.

date. However as illustrated in the historical review earlier in this article, a significant proportion of screening has shifted to more targeted approaches, either to specific protein families or to more refined means of interrogating diverse chemical space. This has been the result of a number of factors, with the cost of screening large numbers of compounds being a key component.

The alternative approach to controlling costs whilst still maximising the scale of combinatorial chemistry would be to return to the screening of compounds as mixtures, and approaches to support this have continued to develop in parallel to the mainstream developments in single compound screening. Houghten has continued to use and develop combinatorial libraries screening methodologies to maximise the effectiveness of compound mixture screening,[134] with extensive use of computational[135] and mathematical[136] modelling to support lead identification.

Finally, in an approach that takes the art of combinatorial synthesis back to its initial beginnings and focus on very large numbers, the use of DNA encoding to allow the rapid screening of extremely large mixtures of compounds has been described by several groups,[137,138] illustrated here by the production of 7 million triazine compounds by the Praecis group (Figure 2.9).[139] Making use of the sensitivity of PCR approaches to rapidly amplify a particular code, libraries of multiple millions of compounds have been prepared, screened and deconvoluted, though the range of chemistry associated with the small molecule still carries the same potential restrictions and liabilities associated with previous polymer bound and tagged approaches. Now part of GSK, this group recently published[140] the discovery of highly potent and selective ADAMTS-5 inhibitors using this technology, with the identification of compounds that did not carry the usual type of zinc binding motif (*e.g.* hydroxamic acids) that typically create selectivity issues across other zinc containing metalloproteinases (Figure 2.10).

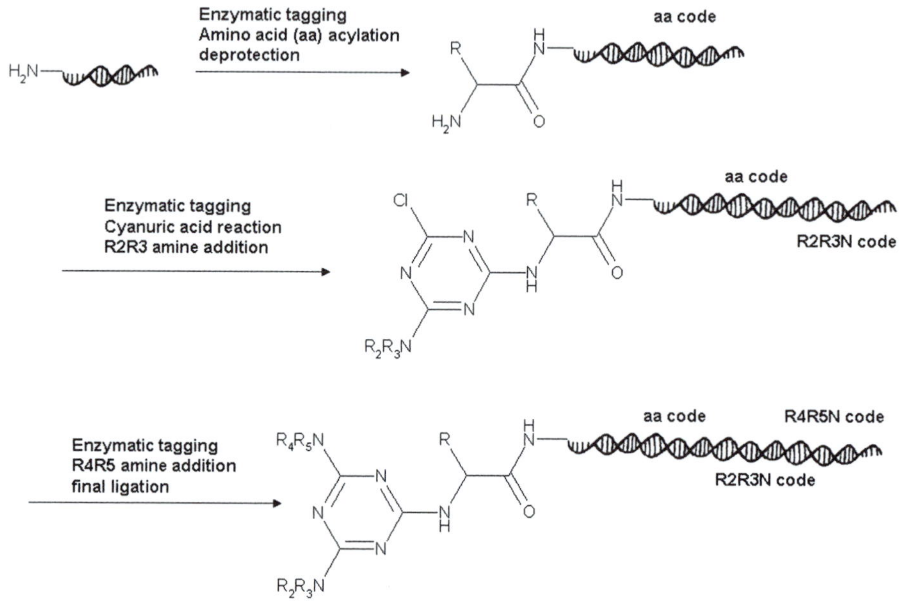

Figure 2.9 Preparation of a three point diversity triazine library with DNA encoding.

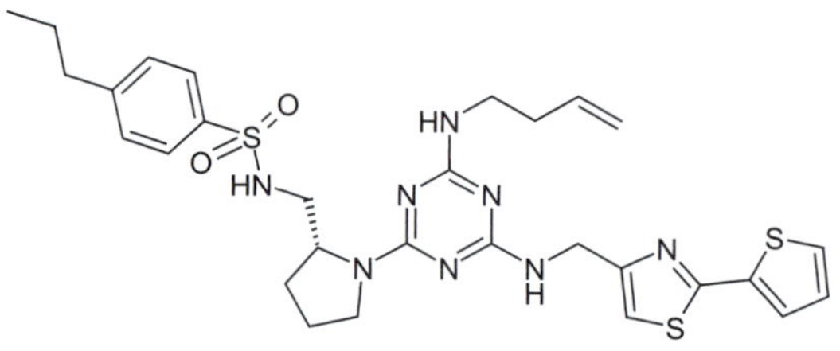

ADAMTS-5 IC$_{50}$ = 0.030uM

ADAMTS-4 IC$_{50}$ = 1.5uM

MMP-13 IC$_{50}$ > 25uM

Figure 2.10 Selective and potent ADAMTS-5 inhibitor identified through the screening of a 4 billion member DNA encoded triazine library.

2.11 CONCLUSION

Combinatorial practices, be they large library purchase, syntheses or focused efforts of parallel chemistry around SAR generation, have become widespread throughout the drug discovery process. The initial promise of Combichem, leading as it did in the 1990s to the development of specialist teams and companies, has gone through a process of expansion, realisation, disappointment and reassessment, to reach a point where it is a valuable tool, part of the overall armoury of drug discovery to be used alongside other approaches. Compound collection numbers, very much the initial driver of the combinatorial explosion, are still significant factors in defining how drug discovery can be prosecuted. However, rather than the in house (or commissioned) combinatorial approach it is as much through purchase of compounds that these numbers are built.[114] Whereas 15 years ago purchasing compounds was very often a lottery of quality, availability and pharmaceutical relevance, it is now possible to build very large high quality diverse screening sets from commercial sources.

So what should be the take home messages from the last 15–20 years of combinatorial approaches to drug discovery? Well established understanding and due reflection on the multi parameter complexity of drug discovery should never be displaced by the technological challenges of a new strategy—after all it is still about finding an active compound that will elicit the correct response in a physiological system, not about the technology. Universal approaches will almost certainly never exist, and application of knowledge about the target proteins can maximise the effectiveness of one design over another. The application of derived drug/lead like properties at the start of any design strategy will save a lot of time at the screening stages of any program.

Combinatorial chemistry began as a tool for understanding biological processes. The application to drug discovery and the generation of small molecule drug compounds became a dream that for many developed into a nightmare of over investment and limited return. But 20:20 hindsight is always right, and we should not be so quick to condemn the work of the earlier combinatorial pioneers. Without those pushing the boundaries of the science we wouldn't now have an approach that when applied correctly can enormously shorten the discovery cycle and maximise the opportunity to optimise in parallel across a wider range of parameters than could ever have been imagined. Pick up any copy of the *Journal of Medicinal Chemistry* or *Biological and Medicinal Chemistry* and randomly open to an article—the odds are now very strong that one of the descriptors 'parallel', 'array', 'high throughput' or even 'combinatorial' will be prominent. The hype came and went but the processes embedded and stayed.

HINTS AND TIPS 1

Typical Equipment for the Parallel Synthesis Chemist

There are many varieties of parallel equipment available on the market, supporting synthesis, analysis, purification and final characterisation, and it would be impossible to provide a comprehensive listing here. The following suggestions are those which the author and his group have found to be useful, reliable and where relevant user friendly. This should not be viewed as a specific endorsement and many equivalent products are available. Moreover there is much to be recommended in a pragmatic approach of adaptation of normally available equipment—the development of some of these pieces of hardware that are now commercial products began life as elastic band and sealing tape prototypes in the author's laboratories back in the early 1990s.

Synthesis equipment: Adaptation of a normal magnetic stirrer to allow parallel reactions with varying levels of heating cooling and inert atmosphere control can be achieved using equipment such as the suite of reaction stations supplied by Radleys (www.radleys.com); including the simple Starfish multiple reaction station, the range of carousel stations and the greenhouse parallel synthesisers. The latter two series also have the advantage of companion workup stations that allow for simple work up procedures in a matching format to the reaction numbers. For more dedicated parallel reaction stations then either the Metz heater shaker system (also available from Radleys) or the STEM RS series reaction stations (www.electrothermal.com) provide dedicated parallel tube systems. More specialist equipment, typically relying on sequential flow reactions rather than parallel reactions, is also available, either for general reaction conditions (see www.syrris.com) or specialist gas reactions (H-Cube; www.thalesnano.com/h-cube). Finally the use of microwave heating has radically changed parallel chemistry capabilities over the past decade, and either CEM (www.cemmicrowave.co.uk) or Biotage (www.biotage.com) equipment is readily usable in the research laboratory.

Work up and purification: Parallel filtration and aqueous extraction processes are available to match the reaction stations as described above, and parallel evaporation systems such as the range of Genevac centrifuge systems (www.genevac.com) or gas evaporation systems such as the FlexiVap work stations (www.glascol.com) allow rapid concentration of multiple reactions. Filtration cartridges can be used either as simple clean up filters or by careful selection of resin content can allow selective removal or sequestration of various functionalities—for the range of solid phase extraction (SPE) and filtration cartridges available see suppliers such as Agilent (www.chem.agilent.com). For more complex purification and separation then parallel column chromatography such as the Isolera system from Biotage (www.biotage.com) can process up to four samples in parallel through automated column chromatography.

HINTS AND TIPS 2

Compounds and Functionalities to Avoid—the PAINS of Hit Discovery

As discussed in the main chapter, there have been a number of publications describing the issues of 'frequent hitters' and trying to understand why some structural motifs are best avoided in any hit to lead follow up program. There are a number of grey areas, and it is not for this author to state categorically that any particular class of molecule, even if pursued carefully, will never yield a viable lead compound for further optimisation. However, there are several classes of compounds that, through repeated evaluation, have clearly been identified as 'problem' structures that should be progressed only with open eyes and an awareness of the history of similar strategies. That list is ever changing and, as more is understood about underlying mechanisms of biological non-specific activity, changes in both increasing and reducing directions. However, the types of core structures highlighted in the PAINS paper[99] are always worth treating with care (Figure 2.11):

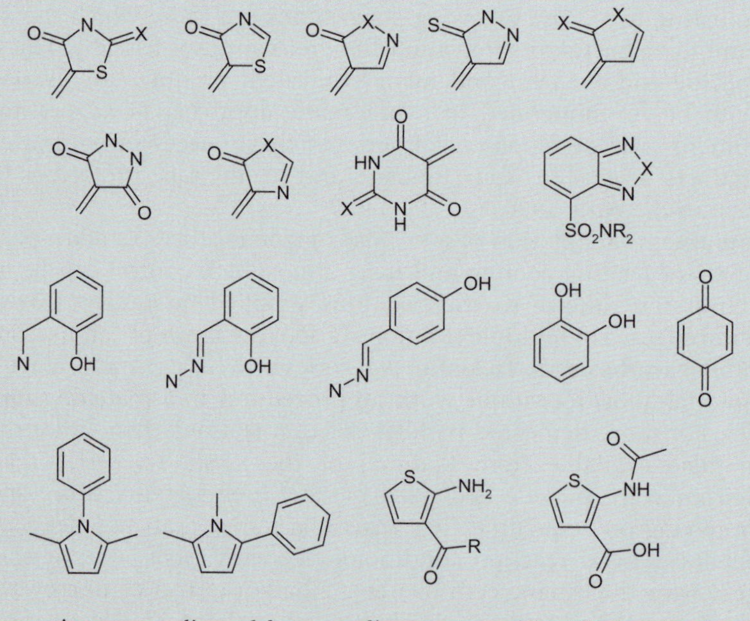

Figure 2.11 Illustrative core cyclic and heterocyclic structures identified as having 'frequent hitter' potential by Baell and Holloway in the PAINS paper.[99]

HINTS AND TIPS 3

Compounds and Functionalities to Avoid—the AZ approach

Perhaps easier to identify and avoid are more specific functional groups and reactive structures. AstraZeneca has identified a wide range of such functionalities in a recent paper,[141] including several of the following motifs:

Class 1: Bland structures
- Compounds containing atoms other than hydrogen, carbon, nitrogen, oxygen, sulfur, fluorine, chlorine, bromine and iodine
- Fewer than four carbon atoms
- Fewer than 12 heavy atoms
- No polar atoms (nitrogen, oxygen, sulfur)
- Straight or unbranched structures
- Positively charged atoms (for example, quaternary nitrogen)
- Compounds with three or more acidic groups
- Alkyl or aryl amine (with no other heteroatom)
- Hydroxyl or thiol (with no other heteroatom)
- Only hetero atom is one acid or derivatives

Class 2: reactive structures
- Michael acceptors: $C=C-C=O$, $C=C-CN$, $C=C-SO_2$, $C=C-NO_2$
- Reactive ester or thioester
- Anhydride
- Alpha halo ketone
- Halo methylene ether
- Acid halide and thio acid halide
- Aliphatic and aromatic aldehyde
- Peroxide
- Epoxide, aziridine, thiirane or oxazirane
- Thiocyanate
- Isocyanate, isothiocyanate
- Isocyanide, isonitrile

Class 3: frequent hitters
- More than two nitro groups
- Dihydroxybenzene
- Nitrophenols

Class 4: dye-like structures
- Two nitro groups on same aromatic ring, including naphthalene
- Diphenyl ethylene cyclohexadiene

Class 5: unlikely drug candidates or unsuitable fragments
- Large ring $\geq C9$
- C9 chain not in any rings
- Crown ethers
- Multi-alkene chain: $C=CC=CC=C$ or $N=CC=CC=C$
- Diyne: $-C\equiv C-C\equiv C-$
- Annelated rings such as phenanthrene, anthracene and phenalene
- Two sulfur atoms (not sulfones) in 5-membered rings or 6-membered rings
- Triphenylmethyl

Class 6: difficult series or natural compounds
- Steroids
- Penicillin or cephalosporin
- Prostaglandins

Class 7: general 'ugly' halogenated structures
- Di- or trivalent halogens
- N–, S–, P– and O–halogens
- Sulfonyl halides
- Triflates: SO_3CX_3

Class 8: general 'ugly' oxygen
- Five or more hydroxyl groups
- p-,p'-dihydroxybiphenyl
- p-,p'-dihydroxystilbene
- Formic acid esters

Class 9: general 'ugly' nitrogen
- Hydrazine (not in ring)
- Three or more guanidines
- Two or more N-oxides
- Azo (N=N) or diazonium (N≡N)
- Carbodiimide
- N-nitroso groups
- Aromatic nitroso groups
- Cyanohydrin or (thio)acylcyanide
- Nitrite
- Nitramine
- Oxime

Class 10: general 'ugly' sulfur
- Five or more sulfur atoms
- Disulfide
- Sulfate
- Sulfonic acid
- Thioketone
- Sulfonic ester (except for aryl or alkyl–SO$_3$–aryl groups)
- Sulfanylamino groups
- 1,2-thiazol-3-one
- Dithiocarbamate
- Thiourea, isothiourea, thiocarbamic acid or thiocarbonate
- Isocyanate or isothiocyanate
- Thiocyanate
- Thiol
- Dithioic or thioic acid

HINTS AND TIPS 4

How to Build a Diverse Collection–The 5 Minute Practical Guide:

The main body of text contains a significant discussion on the steps and processes in designing a collection–but the basic principles and steps can be summarised in a few short bullets (and referring back to Figure 2.6 in the main chapter):

- Identify as many viable commercial compound collections as possible
- Collate into a single data set in a format best suited to downstream analysis
- Apply initial exclusion filters based on strategic considerations to reduce to only compounds that would be acceptable in the collection
- Identify any cost and storage constraints to which the overall collection must comply to determine any size and/or format limitations
- Analyse the compound set using a relevant similarity and clustering protocol to generate a clustered available dataset
- Select representation from the cluster analysis to meet the number constraints of the compound collection whilst maximising coverage of the clustered dataset
- Allow chemist analysis of the selected compounds to ensure best options from cluster selections have been chosen by algorithm
- Spend the money. . . . .

KEY REFERENCES

M. D. Hack, D. N. Rassokhin, C. Buyck, M. Seierstad, A. Skalkin, P. Ten Holte, T. K. Jones, T. Mirzadegan and D. K. Agrafiotis, *J. Chem. Inf. Model.*, 2011, **51**, 3275.
M. A. Snowden and D. V. S. Green, *Curr. Op. Drug Disc. Dev.*, 2008, **11**, 553.
J. B. Baell and G. A. Holloway, *J. Med. Chem.*, 2010, **53**, 2719.

REFERENCES

1. I. Yates, *Drug Discovery World*, 2003, 35.
2. W. S. Fillers, *Drug Discovery World*, 2004, 86.
3. D. Booth, http://www.dddmag.com/articles/2013/04/tackling-challenges-compound-management.
4. R. Macarron, M. N. Banks, D. Bojanic, D. J. Burns, D. A. Cirovic, T. Garyantes, D. V. S. Green, R. P. Hertzberg, W. P. Janzen, J. W. Paslay, U. Schopfer and G. S. Sittampalam, *Nat. Rev. Drug Discovery*, 2011, **10**, 188.
5. S. Willhelm, C. Carter, M. Lynch, T. Lowinger, J. Dumas, R. A. Smith, B. Schwartz, R. Simantov and S. Kelley, *Nat. Rev. Drug Discovery*, 2006, **5**, 835.
6. J. Das, P. Chen, D. Norris, R. Padmanabha, J. Lin, R. V. Moquin, Z. Shen, L. S. Cook, A. M. Doweyko, S. Pitt, S. Pang, D. R. Shen, Q. Fang, H. F. De Fex, K. W. McIntyre, D. J. Shuster, K. M. Gillooly, K. Behnia, G. L. Schieven, J. Wityak and J. C. Barrish, *J. Med. Chem.*, 2006, **49**, 6819.
7. M. S. Lajiness, G. M. Maggiora and V. Shanmugasundaram, *J. Med. Chem.*, 2004, **47**, 4891.
8. A. Merritt, *New Synthetic Technologies in Medicinal Chemistry*, ed. E. Farrant, RSC Publishing, Cambridge, 2012, p. 6.
9. G. R. Nakayama, *Curr. Op. Drug Disc. Dev.*, 1998, **1**, 85.
10. T. D. Mays and K. D. Mazan, *J. Ethnopharm.*, 1996, **51**, 93.
11. J. D. Bergstrom, M. M. Kurtz, D. J. Rew, A. M. Amend, J. D. Karkas, R. G. Bostedor, V. S. Bansal, C. Dufresne, F. L. Van Middlesworth, O. D. Hensens, J. M. Liesch, D. L. Zink, K. E. Wilson, J. Onishi, J. A. Milligan, G. Bills, L. Kaplan, M. N. Omstead, R. G. Jenkins, L.

Huang, M. S. Meinz, L. Quinn, R. W. Burg, Y. L. Kong, S. Mochales, M. Mojena, I. Martin, F. Pelaez, M. T. Diez and A. W. Alberts, *Proc. Natl. Acad. Sci. U. S. A.*, 1993, **90**, 80.

12. A. Baxter, B. J. Fitzgerald, J. L. Hutson, A. D. McCarthy, J. M. Motteram, B. C. Ross, M. Sapra, M. A. Snowden, N. S. Watson, R. J. Williams and C. Wright, *J. Biol. Chem.*, 1992, **267**, 11705.

13. R. Thiericke, *Modern Methods of Drug Discovery*, ed. A. Hillisch and R. Hilgenfeld, Birkhauser, Switzerland, 2003, p. 71.

14. O. Amedee-Manesmee, *Therapie*, 1999, **54**, 419.

15. J. Schamberger, M. Grimm, A. Steinmeyer and A. Hillisch, *Drug Discovery Today*, 2011, **16**, 636.

16. R. B. Merrifield, *J. Am. Chem. Soc.*, 1963, **85**, 2149.

17. B. Merrifield, *Methods Enzymol.*, 1995, **289**, 3.

18. C. C. Leznoff, *Acc. Chem. Res.*, 1978, **11**, 327.

19. H. M. Geysen, R. H. Meloen and S. J. Barteling, *Proc. Natl. Acad. Sci. U. S. A.*, 1984, **81**, 3998.

20. A. Furka, F. Sebestyen, M. Asgedom and G. Dibo, *Int. J. Pept. Prot. Res.*, 1991, **37**, 487.

21. R. A. Houghten, *Proc. Natl. Acad. Sci. U. S. A.*, 1985, **82**, 5131.

22. S. Rodda, G. Tribbick and M. Geysen, *Combinatorial peptide and nonpeptide libraries*, ed. G. Jung, VCH, Weinheim, 1996, p. 303.

23. P. M. St Hilaire, M. Willert, M. A. Juliano, L. Juliano and M. Meldal, *J. Comb. Chem.*, 1999, **1**, 509.

24. E. Apletalina, J. Appel, N. S. Lamango, R. A. Houghten and I. Lindberg, *J. Biol. Chem.*, 1998, **273**, 26589.

25. J. R. Appel, J. Buencamino, R. A. Houghten and C. Pinilla, *Mol. Div.*, 1996, **2**, 29.

26. R. A. Houghten and C. T. Dooley, *Bioorg. Med. Chem. Lett.*, 1993, **3**, 405.

27. C. T. Dooley and R. A. Houghten, *Life Sci.*, 1993, **52**, 1509.

28. B Deprez, X. Williard, L. Bourel, H. Coste, F. Hyafil and A. Tartar, *J. Am. Chem. Soc.*, 1995, **117**, 5405.

29. R. M. Kainkaryam and P. J. Woolf, *Curr. Op. Drug Disc. Dev.*, 2009, **12**, 339.

30. B. A. Bunin and J. A. Ellman, *J. Am. Chem. Soc.*, 1992, **114**, 10997.

31. B. A. Bunin, M. J. Plunkett and J. A. Ellman, *Proc. Natl. Acad. Sci. U. S. A.*, 1994, **91**, 4708.

32. C. G. Boojamra, K. M. Burow, L. A. Thompson and J. A. Ellman, *J. Org. Chem.*, 1997, **62**, 1240.

33. S. H. DeWitt, J. S. Kiely, C. J. Stankovic, M. C. Schroeder, D. M. R. Cody and M. R. Pavia, *Proc. Natl. Acad. Sci. U. S. A.*, 1993, **90**, 6909.

34. C. Barnes and S. Balasubramanian, *Curr. Op. Chem. Biol.*, 2000, **4**, 346.

35. M. H. J. Ohlmeyer, R. N. Swanson, L. W. Dillard, J. C. Reader, G. Asouline, R. Kobayashi, M. Wigler and W. C. Still, *Proc. Natl. Acad. Sci. U. S. A.*, 1993, **90**, 10922.

36. K. C. Nicolaou, X.-Y. Xiao, Z. Parandoosh, A. Senyei and M. P. Nova, *Angew. Chem. Int. Ed. Engl.*, 1995, **34**, 2289.

37. P. H. H. Hermkens, H. C. J. Ottenheijm and D. Rees, *Tetrahedron*, 1996, **52**, 4527.

38. P. H. H. Hermkens, H. C. J. Ottenheijm and D. C. Rees, *Tetrahedron*, 1997, **53**, 5643.

39. S. Booth, P. H. H. Hermkens, H. C. J. Ottenheijm and D. C. Rees, *Tetrahedron*, 1998, **54**, 15385.

40. S. D. Roughley and A. M. Jordan, *J. Med. Chem.*, 2011, **54**, 3451.

41. H. Han, M. M. Wolfe, S. Brenner and K. D. Janda, *Proc. Natl. Acad. Sci. U. S. A.*, 1995, **92**, 6419.

42. S. Chen and K. D. Janda, *J. Am. Chem. Soc.*, 1997, **119**, 8724.

43. A. Studer, S. Hadida, R. Ferritto, S.-Y. Kim, P. Jeger, P. Wipf and D. P. Curran, *Science*, 1997, **275**, 823.

44. S. W. Kaldor, M. G. Siegel, J. E. Fritz, B. A. Dressman and P. J. Hahn, *Tetrahedron Lett.*, 1996, **37**, 7193.

45. S. V. Ley, O. Schucht, A. W. Thomas and P. J. Murray, *J. Chem. Soc., Perkin Trans. 1*, 1999, 1251.
46. A. Solinas and M. Taddei, *Synthesis*, 2007, 2409.
47. S. V. Ley, I. R. Baxendale and R. M. Myers, *Comprehensive Medicinal Chemistry 2*, ed. J. B. Taylor and D. J. Triggle, Elsevier, Amsterdam, 2006, vol. 3, p. 791.
48. E. J. Martin, J. M. Blaney, M. A. Siani, D. C. Spellmeyer, A. K. Wong and W. H. Moos, *J. Med. Chem.*, 1995, **38**, 1431.
49. J. H. Van Drie and M. S. Lajiness, *Drug Discovery Today*, 1998, **3**, 274.
50. J. S. Mason and M. A. Hermsmeier, *Curr. Op. Chem. Biol.*, 1999, **3**, 342.
51. D. H. Drewry and S. S. Young, *Chemom. Intell. Lab. Syst.*, 1999, **48**, 1.
52. D. K. Agrafiotis, J. C. Myslik and F. R. Salemme, *Mol. Div.*, 1999, **4**, 1.
53. A. R. Leach and M. M. Hann, *Drug Discovery Today*, 2000, **5**, 326.
54. M. R. Pavia, S. P. Hollinshead, H. V. Meyers and S. P. Hall, *Chimia*, 1997, **51**, 826.
55. M. J. Sofia, R. Hunter, T. Y. Chan, A. Vaughan, R. Dulina, H. Wang and D. Gange, *J. Org. Chem.*, 1998, **63**, 2802.
56. T. I. Oprea, *Curr. Op. Chem. Biol.*, 2002, **6**, 384.
57. A. W. Czarnik, *ChemtTracts Org. Chem.*, 1995, **8**, 13.
58. Y. C. Martin, *Perspect. Drug Disc. Des.*, 1997, 7, 159.
59. F. Z. Dorwald, *Organic Synthesis on Solid Phase*, Wiley and Sons, Hoboken, 2002.
60. A. Merritt, *Rodd's Chemistry of Carbon Compounds, Topical Volumes, Asymmetric Catalysis*, ed. M Sainsbury, Elsevier, Amsterdam, 2nd edn, 2001, vol. 5, p. 259.
61. A. T. Merritt, *Drug Discovery Today*, 1998, **3**, 505.
62. *J. Comb. Chem.; Comb. Chem. High Throughput Screening; Mol. Diversity.*
63. J. Alper, *Science*, 1994, **264**, 1399.
64. C. A. Lipinski, F. Lombardo, B. W. Dominy and P. J. Feeney, *Adv. Drug Del. Rev.*, 1997, **23**, 3.
65. R. A. Fecik, K. E. Frank, E. J. Gentry, S. R. Menon, L. A. Mitscher and H. Telikepalli, *Med. Res. Rev.*, 1998, **18**, 149.
66. S. J. Teague, A. M. Davis, P. D. Leeson and T. Oprea, *Angew. Chem. Int. Ed. Engl.*, 1999, **38**, 3743.
67. M. M. Hann, A. R. Leach and G. Harper, *J. Chem. Inf. Comput. Sci.*, 2001, **41**, 856.
68. X. Q. Lewell, D. B. Judd, S. P. Watson and M. M. Hann, *J. Chem. Inf. Comput. Sci.*, 1998, **38**, 511.
69. http://www.combinatorial.com/.
70. http://www.peptide.com/.
71. B. Evans, A. Pipe, L. Clark and M. Banks, *Bioorg. Med. Chem. Lett.*, 2001, **11**, 1297.
72. N. Bailey, A. W. J. Cooper, M. J. Deal, A. W. Dean, A. L. Gore, M. C. Hawes, D. B. Judd, A. T. Merritt, R. Storer, S. Travers and S. P. Watson, *Chimia*, 1997, **51**, 832.
73. X.-Y. Xiao, R. Li, H. Zhuang, B. Ewing, K. Karunaratne, J. Lillig, R. Brown and K. C. Nicolaou, *Biotech. Bioeng.*, 2000, **71**, 44.
74. http://www.nexusbio.com/.
75. N. Hird and B. MacLachlan, *Laboratory Automation in the Chemical Industries*, ed. D. G. Cork and T. Sugawara, Marcel Dekker, New York, 2002, p. 1.
76. http://www.zinsser-analytic.com/.
77. http://www.combichemlab.com/website/files/Combichem/Workstations/argonaut.htm.
78. N. Kuroda, N. Hird and D. G. Cork, *J. Comb. Chem.*, 2006, **8**, 505.
79. http://www.stemcorp.co.uk/s_index.cfm.
80. http://www.radleys.com/.
81. http://uk.mt.com/gb/en/home/products/L1_AutochemProducts/L3_Post-Synthesis-Work-up.html.

82. J. J. Parlow, *Curr. Op. Drug Disc. Dev.*, 2005, **8**, 757.

83. G. A. Strohmeier and C. O. Kappe, *Angew. Chem. Int. Ed. Engl.*, 2004, **43**, 621.

84. P. Lan, D. Berta, J. A. Porco, M. S. South and J. J. Parlow, *J. Org. Chem.*, 2003, **68**, 9678.

85. Y. Wang, K. Sarris, D. R. Sauer and S. W. Djuric, *Tetrahedron Lett.*, 2007, **48**, 2237.

86. J. Habermann, S. V. Ley and J. S. Scott, *J. Chem. Soc., Perkin Trans., 1*, 1999, 1253.

87. S. V. Ley, M. Ladlow and E. Vickerstaffe, *Exploiting Chemical Diversity for Drug Discovery*, ed. P. A. Bartlett and M. Entzeroth, RSC Publishing, London, 2006, p. 3.

88. C. O. Kappe and D. Dallinger, *Nat. Rev. Drug Discovery*, 2006, **5**, 51.

89. A. De La Hoz, A. Diaz-Ortiz and A. Moreno, *Chem. Soc. Rev.*, 2005, **34**, 164.

90. B. Desai and C. O. Kappe, *Topics in Curr. Chem.*, 2004, **242**, 177.

91. J. M. Kremsner and C. O. Kappe, *J. Org. Chem.*, 2006, **71**, 4651.

92. L. Pisani, H. Prokopcova, J. M. Kremsner and C. O. Kappe, *J. Comb. Chem.*, 2007, **9**, 415.

93. J. N. Kyranos, H. Lee, W. K. Goetzinger and L. Y. T. Li, *J. Comb. Chem.*, 2004, **6**, 796.

94. W. Leister, K Strauss, D. Wisnoski, Z. Zhao and C. Lindsley, *J. Comb. Chem.*, 2003, **5**, 322.

95. S. J. Lane, D. S. Eggleston, K. A. Brindred, J. C. Hollerton, N. L. Taylor and S. A. Readshaw, *Drug Discovery Today*, 2006, **11**, 267.

96. A. W. Squibb, M. R. Taylor, B. L. Parnas, G. Williams, R. Girdler, P. Waghorn, A. G. Wright and F. S. Pullen, *J. Chromatogr. A*, 2008, **1189**, 101.

97. S. Lane, B. Boughtflower, I. Mutton, C. Paterson, D. Farrant, N. Taylor, Z. Blaxill, C. Carmody and P. Borman, *LC-GC Europe*, 2006, **19**, 161.

98. R. Richardson, *Exploiting Chemical Diversity for Drug Discovery*, ed. P. A. Bartlett and M. Entzeroth, RSC Publishing, Cambridge, 206, p. 112.

99. J. B. Baell and G. A. Holloway, *J. Med. Chem.*, 2010, **53**, 2719.

100. J. B. Baell, *J. Chem. Inf. Model.*, 2013, **53**, 39.

101. D. V. S. Green, *Drug Design Strategies*, ed. D. J. Livingstone and A. M. Davis, RSC Publishing, Cambridge, 2012, p. 367.

102. D. J. Huggins, A. R. Venkitaraman and D. R. Spring, *ACS Chem. Biol.*, 2011, **6**, 208.

103. R. E. Dolle, *Methods Mol. Biol.*, 2011, **685**, 3.

104. D. M. Schnur, B. R. Beno, A. J. Tebben and C. Cavallaro, *Methods Mol. Biol.*, 2011, **672**, 387.

105. M. D. Hack, D. N. Rassokhin, C. Buyck, M. Seierstad, A. Skalkin, P. Ten Holte, T. K. Jones, T. Mirzadegan and D. K. Agrafiotis, *J. Chem. Inf. Model.*, 2011, **51**, 3275.

106. J. Ashby and R. W. Tennant, *Mutation Res.*, 1991, **257**, 229.

107. G. M. Rishton, *Drug Discovery Today*, 1997, **2**, 382.

108. T. Tomasic and L. P. Masic, *Expert Opin. Drug Discov.*, 2012, **7**, 549.

109. T. Mendgen, C. Steuer and C. D. Klein, *J. Med. Chem.*, 2012, **55**, 743.

110. Y. C. Martin, J. L. Kofron and L. M. Traphagen, *J. Med. Chem.*, 2002, **45**, 4350.

111. P. Willett, J. M. Barnard and G. M. Downs, *J. Chem. Inf. Comput. Sci.*, 1998, **38**, 983.

112. G. Harper, S. D. Pickett and D. V. S. Green, *Comb. Chem. High Throughput Screening*, 2004, **7**, 63.

113. G. A. Bakken, A. S. Bell, M. Boehm, J. R. Everett, R. Gonzales, D. Hepworth, J. L. Klug-McLeod, J. Lanfear, J. Loesel, J. Mathias and T. P. Wood, *J. Chem. Inf. Model.*, 2012, **52**, 2937.

114. M. A. Snowden and D. V. S. Green, *Curr. Op. Drug Disc. Dev.*, 2008, **11**, 553.

115. http://www.sigmaaldrich.com/catalog/product/sigma/lo1280?lang = en&region = GB.

116. http://www.prestwickchemical.com/index.php?pa = 26.

117. http://www.ninds.nih.gov/.

118. J. Rosen, J. Gottfries, S. Muresan, A. Backlund and T. I. Oprea, *J. Med. Chem.*, 2009, **52**, 1953.

119. P. Caffrey, J. F. Aparicio, F. Malpartida and S. B. Zotchev, *Curr. Top. Med. Chem.*, 2008, **8**, 639.

120. G. L. Thomas, E. E. Wyatt and D. R. Spring, *Curr. Op. Drug Disc. Dev.*, 2006, **9**, 700.
121. C. J. O'Connor, H. S. G. Beckmann and D. R. Spring, *Chem. Soc. Rev.*, 2012, **41**, 4444.
122. M. Baker, *Nat. Rev. Drug Discovery*, 2013, **12**, 5.
123. P. T. Corbett, J. Leclaire, L. Vial, K. R. West, J.-L. Wietor, J. K. M. Sanders and S. Otto, *Chem. Rev.*, 2006, **106**, 3652.
124. M. S. Congreve, D. J. Davis, L. Devine, C. Granata, M. O'Reilly, P. G. Wyatt and H. Jhoti, *Angew. Chem. Int. Ed.Engl.*, 2003, **42**, 4479.
125. M. T. Cancilla, M. M. He, N. Viswanathan, R. L. Simmons, M. Taylor, A. D. Fung, K. Cao and D. A. Erlanson, *Bioorg. Med. Chem. Lett.*, 2008, **18**, 3978.
126. M. Hochgurtel, R. Biesinger, H. Kroth, D. Piecha, M. W. Hofmann, S. Krause, O. Schaaf, C. Nicolau and A. V. Eliseev, *J. Med. Chem.*, 2003, **46**, 356.
127. E. R. Wood, L. Kuyper, K. G. Petrov, R. N. Hunter III, P. A. Harris and K. Lackey, *Biorg. Med. Chem. Lett.*, 2004, **14**, 953.
128. S. L. Posy, M. A. Hermsmeier, W. Vaccaro, K.-H. Ott, G. Todderud, J. S. Lippy, G. L. Trainor, D. A. Loughney and S. R. Johnson, *J. Med. Chem.*, 2011, **54**, 54.
129. C. J. Harris, R. D. Hill, D. W. Sheppard, M. J. Slater and P. F. W. Stouten, *Comb. Chem. High Throughput Screening*, 2011, **14**, 521.
130. L. R. Whitby and D. L. Boger, *Acc. Chem. Res.*, 2012, **45**, 1698.
131. J. G. Kettle, R. A. Ward and E. Griffin, *Med. Chem. Comm.*, 2010, **1**, 331.
132. S. C. Mayer, J. A. Butera, D. J. Diller, J. Dunlop, J. Ellingboe, K. Y. Fan, E. Kaftan, B. Mekonnen, D. Mobilio, J. Paslay, G. Tawa, D. Vasilyev and M. R. Bowlby, *Assay Drug Dev. Tech.*, 2010, **8**, 504.
133. N. Y. Mok and R. Brenk, *J. Chem. Inf. Model.*, 2011, **51**, 2449.
134. R. A. Houghten, C. Pinilla, M. A. Giulianotti, J. R. Appel, C. T. Dooley, A. Nefzi, J. M. Ostresh, Y. Yu, G. M. Maggiora, J. L. Medina-Franco, D. Brunner and J. Schneider, *J. Comb. Chem.*, 2008, **10**, 3.
135. A. B. Yongye, C. Pinilla, J. L. Medina-Franco, M. A. Giulanotti, C. T. Dooley, J. R. Appel, A. Nefzi, T. Scior, R. A. Houghten and K. Martinez-Mayorga, *J. Mol. Model.*, 2011, **17**, 1473.
136. R. G. Santos, M. A. Giulianotti, C. T. Dooley, C. Pinilla, J. R. Appel and R. A. Houghten, *ACS Comb. Sci.*, 2011, **13**, 337.
137. L. Mannocci, M. Leimbacher, M. Wichert, J. Scheuermann and D. Neri, *Chem. Commun.*, 2011, **47**, 12747.
138. R. E. Kleiner, C. E. Dumelinz and D. R. Liu, *Chem. Soc. Rev.*, 2011, **40**, 5707.
139. M. A. Clark, R. A. Acharya, C. C. Arico-Muendel, S. L. Belyanskaya, D. R. Benjamin, N. R. Carlson, P. A. Centrella, C. H. Chiu, S. P. Creaser, J. W. Cuozzo, C. P. Davie, Y. Ding, G. J. Franklin, K. D. Franzen, M. L. Gefter, S. P. Hale, N. J. V. Hansen, D. I. Israel, J. Jiang, M. J. Kavarana, M. S. Kelley, C. S. Kollmann, F. Li, K. Lind, S. Mataruse, P. F. Medeiros, J. A. Messer, P. Myers, H. O'Keefe, M. C. Oliff, C. E. Rise, A. L. Satz, S. R. Skinner, J. L. Svendsen, L. Tang, K. van Vloten, R. W. Wagner, G. Yao, B. Zhao and B. A. Morgan, *Nat. Chem. Biol.*, 2009, **5**, 647.
140. H. Deng, H. O'Keefe, C. P. Davie, K. E. Lind, R. A. Acharya, G. J. Franklin, J. Larkin, R. Matico, M. Neeb, M. M. Thompson, T. Lohr, J. W. Gross, P. A. Centrella, G. K. O'Donovan, K. L. Bedard, K. Van Vloten, S. Mataruse, S. R. Skinner, S. L. Belyanskaya, T. Y. Carpenter, T. W. Shearer, M. A. Clark, J. W. Cuozzo, C. C. Arico-Muendal and B. A. Morgan, *J. Med. Chem.*, 2012, **55**, 7061.
141. J. G. Cumming, A. M. Davis, S. Muresan, M. Haeberlein and H. Chen, *Nat. Rev. Drug Discovery*, 2013, **12**, 948.

CHAPTER 3

Useful Computational Chemistry Tools for Medicinal Chemistry

DARREN V. S. GREEN

GlaxoSmithKline Medicines Research Centre, Gunnels Wood Road, Stevenage, Hertfordshire SG1 2NY, UK
E-mail: Darren.vs.green@gsk.com

3.1 PHYSICS BASED *VS.* EMPIRICAL MODELS

The underlying physical laws necessary for the mathematical theory of a large part of physics and the whole of chemistry are thus completely known, and the difficulty is only that the exact application of these laws leads to equations much too complicated to be soluble.

P. A. M. Dirac[1]

Key to the application of computational chemistry to medicinal chemistry is the calculation of energy. Accurate calculation of absolute energies, for example the change in free energy upon binding of a small molecule to a protein, remains an unfulfilled aspiration of computational chemists. Figure 3.1. Illustrates the complexity of ligand binding when expressed in free energy terms, involving changes in both enthalpy and entropy, and explains why it is that prediction of binding energy has been found to be so difficult.

This is a scientific Grand Challenge and it is therefore fortunate that components of the free energy of binding that are applicable to drug discovery, for example the relative energies of molecular conformations, or tautomers in heterocyclic rings, are computationally tractable to a degree of accuracy that is useful in the design of new molecules (see Table 3.1 for selected key milestones in the development of Computational Chemistry as a scientific discipline).

The key difference between computational chemistry and QSAR modelling is in the atomistic level of detail that is modelled, with properties of molecules predicted from first principles study of the atoms that comprise them. This chapter will explore the various techniques available, outline the approximations which are made in order to enable the calculation and discuss what these might mean in real life application to medicinal chemistry problems.

The Handbook of Medicinal Chemistry: Principles and Practice
Edited by Andrew Davis and Simon E Ward
© The Royal Society of Chemistry 2015
Published by the Royal Society of Chemistry, www.rsc.org

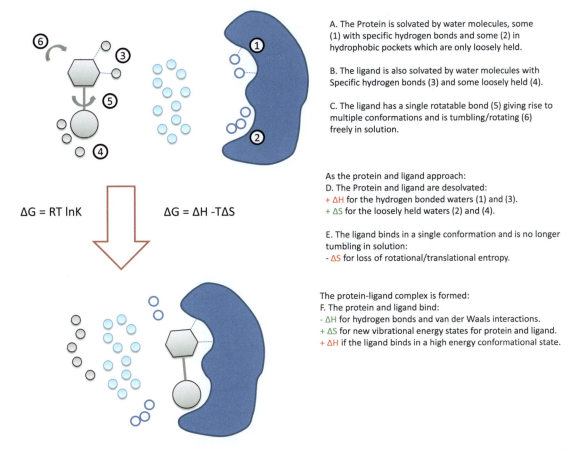

A. The Protein is solvated by water molecules, some (1) with specific hydrogen bonds and some (2) in hydrophobic pockets which are only loosely held.

B. The ligand is also solvated by water molecules with Specific hydrogen bonds (3) and some loosely held (4).

C. The ligand has a single rotatable bond (5) giving rise to multiple conformations and is tumbling/rotating (6) freely in solution.

As the protein and ligand approach:
D. The Protein and ligand are desolvated:
+ ΔH for the hydrogen bonded waters (1) and (3).
+ ΔS for the loosely held waters (2) and (4).

E. The ligand binds in a single conformation and is no longer tumbling in solution:
- ΔS for loss of rotational/translational entropy.

The protein-ligand complex is formed:
F. The protein and ligand bind:
- ΔH for hydrogen bonds and van der Waals interactions.
+ ΔS for new vibrational energy states for protein and ligand.
+ ΔH if the ligand binds in a high energy conformational state.

$$\Delta G = RT \ln K \qquad \Delta G = \Delta H - T\Delta S$$

Figure 3.1 The free energy of binding of a ligand to a protein comprises many components. Although enthalpic binding from direct molecular interactions receives most attention from chemists, the numerous entropic terms can come to dominate the energetics and these need to be considered with as much care.

3.2 MOLECULAR MECHANICS AND MOLECULAR ORBITAL THEORY

Central to all computational chemistry methods is the assumption that the Born–Oppenheimer approximation holds, that is that the mass difference (a proton is 1836 times heavier than an electron) between nuclei and electrons is sufficient, so that their motions can be considered separately.[29,30] This assumption then allows quantum mechanical methods to focus on the electronic state of the molecule given a fixed position for the nuclei, whilst molecular mechanics may focus on the nuclear coordinates of a molecule in its ground electronic state.

3.2.1 Quantum Mechanics

The key concept in molecular quantum mechanics is the wavefunction, most often denoted as Ψ. The wavefunction completely describes the properties of a quantum mechanical system, and has a value at every point in space. The square of the wavefunction yields the electron density at that point, and integration of the electron density over a volume gives the probability of finding an electron in that volume. That the localisation of electron density is not fixed, but can be computed, is an important result from quantum mechanics, introducing concepts for molecular recognition such as partial atomic charges and polarisability.

Table 3.1   Key milestones in the development of computational chemistry as a discipline.

1924	Lennard-Jones proposes the Lennard-Jones[6–12] interatomic potential[2]
1926	Schrödinger develops the "wave equation" that mathematically describes the distribution of an electron through space.[3]
1928	Mulliken develops a "molecular orbital" theory where electrons are assigned to orbitals across an entire molecule[4]
1929	Lennard-Jones introduces the linear combination of atomic orbitals approximation for the calculation of molecular orbitals.[5]
1930	London explains van der Waals forces as due to the interacting fluctuating dipole moments between molecules.[6]
1938	Coulson makes the first accurate calculation of a molecular orbital wavefunction for the hydrogen molecule.[7]
1946	Westheimer publishes calculations on a series of hindered biphenyls which demonstrate that observed molecular properties could be rationalised through the use of computed geometries and energies.[8]
1950	Barton lays down the foundations of Conformational Analysis in a study of steroid chemistry.[9]
1956	The first *ab initio* Hartree–Fock calculations on diatomic molecules.
1961	Hendrickson publishes the first use of a computer for force field calculations in a study of ring conformations.[10]
1965	The semi empirical method CNDO is published by Pople and co-workers.[11]
1965	Wiberg employs a Steepest Descent algorithm for geometry optimisation with a molecular mechanics forcefield.[12]
1965	The molecular structure drawing program ORTEP developed at Oak Ridge National Laboratory.[13]
1970	Gaussian70 made freely available *via* the Quantum Chemistry Program Exchange (QCPE).[14]
1970	The term "Computational Chemistry" first used to describe the new scientific discipline: "It seems, therefore, that 'computational chemistry' can finally be more and more of a reality."[15]
1972	Wiberg and Boyd describe the dihedral driver method for systematic conformational analysis.[16]
1973	Cambridge Crystallographic Data Centre makes its 3D structure files widely available.[17]
1973	N. L. Allinger describes the modelling of hydrocarbons with a new force field, MM1.[18]
1975	The first molecular dynamics simulation of a protein is published by Warshel and Levitt.[19]
1980	The Carbo Index described for the quantitation of molecular similarity based on the 3D electron density distribution in molecules.[20]
1982	Kuntz and co-workers lay down the foundations for the DOCK algorithm, the starting point for a multitude of researchers in the area of virtual screening.[21]
1983	The Connolly Surface algorithm enables real time visualisation of high quality coloured molecular surfaces on desktop computers.[22]
1985	GRID is published by Peter Goodford. A computational procedure for determining energetically favourable binding sites on biologically important macromolecules.[23]
1987	Free Energy Perturbation predictions made for the potency of novel Thermolysin inhibitors.[24]
1987	The Concord automatic 3D structure generation program developed.[25]
1993	Zanamivir, an inhibitor of the influenza neuraminidase enzyme, is discovered with the assistance of GRID in one of the first examples of protein structure based drug discovery.[26]
1997	First geometry optimisation of an entire protein using semi-empirical quantum mechanics.[27]
2010	The first millisecond simulation of a protein is performed using Anton, a computer designed specifically for high speed molecular dynamics.[28]

All properties of the system (molecule, or collection of molecules) may be calculated by the application of various mathematical operators on the wavefunction (Figure 3.2).

The operator of most interest to us in the energy operator, or Hamiltonian $\hat{H}$, which give rise to the familiar Schrödinger equation:

$$\hat{H}\Psi = E\Psi \tag{3.1}$$

Unfortunately it is not possible to analytically solve the Schrödinger equation for systems with more than one electron (the hydrogen atom). It can be argued that this fact spawned the

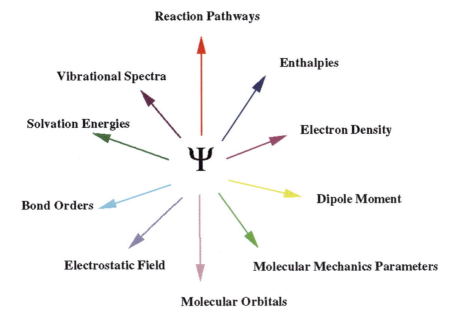

Figure 3.2 The power and versatility of quantum mechanics. All molecular properties of interest to medicinal chemists can be computed from the wavefunction of a molecule.

discipline of what is known as Computational Chemistry and the majority of the key milestones relate to the 50+ year quest for accurate solutions. The most common technique used is the Hartree–Fock method, in which each electron is considered to move in a static field of all the other electrons. This is an iterative calculation which exploits the Variational Principle (the true energy of a system is always lower than that produced when the Hamiltonian is applied to an incorrect wavefunction). The wavefunction is first estimated and then iteratively modified to minimise the energy. In order to make the calculations tractable, the wavefunction itself is comprised of multiple molecular orbitals, ψ, each represented by a Linear Combination of Atomic Orbitals (LCAO) centred on each atom:

$$\psi = \sum_{i=1}^{n} c_i \varphi_i \tag{3.2}$$

Each atomic orbital φ_i is modelled by mathematical functions designed to mimic the shape of s, p, d and f orbitals whilst being mathematically efficient (as solving the Schrodinger equation involves integrating across the overlap of these atomic orbitals). The most accurate atomic orbitals mimic those derived analytically for the hydrogen atom and are known as Slater Type Orbitals (STOs). The STO for a 1s orbital centred on an atom is given by:

$$\varnothing_{1s}^{SF} = \left(\frac{\zeta^3}{\pi}\right)^{1/2} e^{-\zeta r} \tag{3.3}$$

Where ζ is an exponent which varies according to the atom type, the larger the exponent the smaller the 1s orbital (for hydrogen atom $\zeta = 1$, for the helium atom $\zeta = 1.69$). However, Hartree–Fock calculations involve computing the overlap between orbitals, and for STOs this is slow. The best combination of shape and efficiency is achieved by using multiple Gaussian-type orbitals (GTOs) for each atomic orbital, chosen to provide a best fit to an STO. The form of an atom centred GTO is:

$$\emptyset_{1s}^{GF} = \left(\frac{2\alpha}{\pi}\right)^{3/4} e^{-\alpha r^2} \tag{3.4}$$

As GTOs do not describe atomic orbitals well, particularly near the nucleus, normally it requires a combination of at least 3 of these functions to provide a reasonable representation (Figure 3.3).[31] The advantage of using GTOs is that the overlap of two Gaussians is itself a Gaussian, and this overlap can be computed very quickly. How many and what type of Gaussians are applied to each orbital are collectively referred to as a "basis set".

There are many such basis sets to choose from, but the split-valence basis sets of Pople[32] have been the most successful at modelling the type of molecular systems of interest here. As most intermolecular interactions are dominated by the valence electrons, more attention is paid to those atomic orbitals than to the inner shell. Basis sets have a nomenclature all of their own, for example the popular 6-31G** has the inner shell represented by a single basis function constructed from six GTOs, whilst the valence orbitals are represented by two basis functions, one of three GTOS and the other of just one. The "**" indicates that polarisation functions are added to both heavy (so d-type orbitals for C, N and O atoms) and hydrogen atoms (p-type orbitals); these functions allow the electron density on an atom to become anisotropic in response to, for example, a directional hydrogen bond. A " + " would indicate diffuse functions (larger s and p functions which tail off slowly as the distance from the nucleus increases) are added, for example in order to model an ionised carboxylic acid.

These approaches are sufficient to calculate many properties of interest to drug researchers—dipole moments, charge distributions and bond orders—and can be combined with optimisation algorithms to produce energy minimised molecular conformations. A significant deficiency of so-called Restricted Hartree Fock methods (RHF) is the neglect of electron correlation. As the RHF method considers the movement of an electron in an average, static, field on all the other electrons in the molecule it cannot model the observed behaviour that electrons tend to avoid each other where possible. There are many methods developed to overcome this, such as Møller–Plesset Perturbation theory which obtains a more accurate estimation of the true energy by writing the true eigenvalues as a power series of ever more complex wavefunctions, where the basic HF energy is given by the sum of the zeroth- and

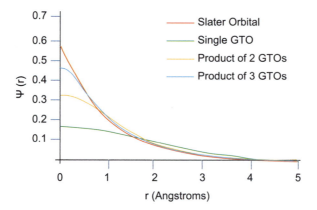

Figure 3.3　In order to fit the mathematical form of a Slater Type Orbital (STO), which is a fine description of an atomic orbital, multiple Gaussian Type Orbitals must be combined. In this case three GTOs are used to produce a close match to a single STO.

first-order energies. To improve on the HF energy it is therefore necessary to add the second order wavefunction and this level of theory is referred to as MP2. MP2 includes the excitations obtained by promoting electrons into higher energy orbitals. It is, however, computationally expensive and is therefore normally used as a single point energy calculation, once the geometry of the molecule has been minimised using HF methods.

The time required for HF calculations scale at the rate of N^4 (where N = number of electrons in the molecule) and therefore quickly consume available computing resources (even today). Therefore, techniques were developed which would allow quantum mechanical calculations whilst reducing the number of integral evaluations. Incorporation of experimental data and parameterisation are key to the success of such methods, and give rise to the term "semi-empirical quantum mechanics" to describe them. For example, the energy of a carbon 1s orbital can be taken from experimental data rather than computed. The most successful methods focus on the reduction of calculations required for differential overlap, that is, the interaction between two orbitals centred on different atoms. The first method, CNDO[11] (Complete Neglect of Differential Overlap) led to a variety of successive improvements (INDO, NDDO), which the most successful being the MNDO (Modified Neglect of Differential Overlap) methods. These neglect interactions between certain orbitals, based on atom and orbital type, but allow inclusion of, for example, the interaction of p orbitals in the C and O of a carbonyl bond. Although conjugation in such systems is therefore included in the model, the lack of a full inclusion of electron interactions across a conjugated bond does lead to lower estimates of, for example, rotational barriers in a molecule such as formamide. There are various forms of MNDO model in use, the most popular being AM1[33] and PM3[34] (or PM6), the difference between these being not in the level of theory used, but in how the empirical parameters are derived. The advantage of these methods is that they scale well—modern codes scale linearly with the number of electrons in the system, whilst providing results which are competitive with moderate basis set HF calculations (due to the inclusion of experimental and parameterisation, semi-empirical calculations also include some consideration of electron correlation).

In recent years, quantum mechanics practice has been heavily influenced by a third class of theoretical approach—Density Functional Theory (DFT). In contrast to HF theory, which is concerned with determination of the set of single electron wavefunctions (from which electron density distributions can be derived), DFT directly calculates the electron density distribution and the total electronic energy. Although the Thomas–Fermi model of 1920s heritage contained many of the key concepts, it was not until the 1960s that Hohenberg and Kohn[35] demonstrated that the ground state energy and other properties of a system were uniquely defined by the electron density distribution. This finding gives rise to the language of DFT—the energy of the molecule is a unique *functional* of the electron density distribution. There are many advantages of DFT, in particular the explicit inclusion of electron correlation and many of the methods used to implement DFT calculations are very similar to those used in HF theory, for example the iterative procedures to achieve self-consistency. Most of the difficulty in implementation is in the derivation of the best mathematical function, and in this regard DFT resembles semi-empirical methods. For example, to compute the electron exchange and correlation contributions to the energy, it is necessary to estimate the electron density, for example the commonly applied local density approximation (LDA) is based on a model of a uniform electron gas.

Chemists are most likely to encounter DFT as a hybrid calculation with HF methods which add a correlation term based on electron density to the electrostatic and configuration terms computed by HF. A popular combination is B3 LYP (Becke's three parameter functional[36] which combines HF and DFT terms for electron configuration; the Lee Yang and Parr correlation functional).[37] These methods are implemented with traditional basis sets, and therefore a calculation will often be described as B3LYP/6-31G*. DFT allows the inclusion of electron correlation at a reasonable computational cost, scaling at $\sim N^3$. However, for the purposes of drug discovery, DFT has some serious deficiencies. In particular, hydrogen bonds can be too short, and van der Waals interactions are poorly handled.[38] In contrast to HF theory, where the solution is to increase the level of theory and increase the basis set, improving the performance of DFT requires the empirical derivation of an improved functional.

3.2.2 Molecular Mechanics

The technique of molecular mechanics (MM) originated with the vibrational spectoscopy community, who determine the forces holding molecules together from knowledge of the molecular structure and the observed vibrational frequencies. In vibrational spectroscopy, electrons are not studied and are assumed to adopt an optimal distribution around the nuclei. Application of the Born–Oppenheimer approximation then allows the energy of the molecule to be defined in terms of the nuclear positions.

Due to its origins, in MM the energy of the molecule has contributions from bond-stretches, angle stretches, out of plane bending and torsional energies with additional terms to describe non-bonded terms for van der Waals and electrostatic interactions (Figure 3.4).

A common mathematical form for a simple MM method is:

$$E(r^N) = \sum_{bonds} \frac{ki}{2}(l_i - l_{i,0}) + \sum_{angles} \frac{ki}{2}(\theta_i - \theta_{i,0})^2 + \sum_{torsions} \frac{Vn}{2}(1 + \cos(n\omega - \gamma))$$
$$+ \sum_{i=1}^{N} \sum_{j=i+1}^{N} \left(4\varepsilon_{ij} \left[\left(\frac{\sigma_{ij}}{r_{ij}} \right)^{12} - \left(\frac{\sigma_{ij}}{r_{ij}} \right)^{6} \right] + \frac{q_i q_j}{4\pi\varepsilon_0 r_{ij}} \right) \tag{3.5}$$

Where r denotes the positions of a set of atoms N. The first term is the bond stretch of two bonded atoms, which increases as the bond length l_i deviates from the reference value l_0, and in this implementation is modelled by a harmonic (symmetrical) potential as described by

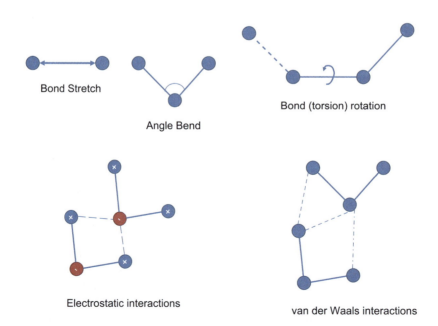

Figure 3.4   The fundamental molecular motions and interactions encoded in a molecular mechanics force field.

Hooke's Law. Similarly, the second term describes the energies associated by deformation of a bond angle from its reference and again is modelled by a harmonic potential. The third, torsional, term describes how the energy changes as a bond rotates and this, necessarily a more complex term, comprises the cosine of the torsional angle. The final term is a composite function to describe the non-bonded interactions between atoms and in this instance uses a 6–12 Lennard-Jones potential for van der Waals energies and a simple Coulomb charge potential, modified by the dielectric constant, ε of the environment (1 for a vacuum, 78.4 for water, approximately 2 for the interior of a protein).

The reference bond lengths, angles and torsional terms, van der Waals radii and atomic charges all need to be provided in order for a calculation to be performed. The combination of the mathematical energy function and the set of required parameters together define a force field. The first accurate and common purpose force field for hydrocarbons was Allinger's MM1.[18] Since then, ever more complex schemes have been invented,[39,40] adding better bond stretch and angle terms and so-called cross terms between them (for example, as a bond angle decreases, adjacent bonds stretch to minimise the 1,3 atomic interactions).

All force fields make use of atom typing. Typically each atom is defined by its hybridisation state (*e.g.* sp$_2$, sp$_3$), number or type of substituents (*e.g.* OH), formal charge (N^{4+}) and sometimes bond type (*N*-amide). Parameters are then assigned to bonds between atom types to reproduce bond lengths, angles and torsions. A force field can therefore contain hundreds of parameters.

For applications in drug discovery, such as conformational analysis and protein–ligand interactions, the most important terms are the non-bonded interactions, which are also the most difficult to parameterise. There is a multitude of force fields available and is not always easy to select the most appropriate for a molecule.[40] There is certainly no "universal" force field that reliably models all molecules. Unlike quantum mechanics, where a result can usually be improved by an increase in theory, poor behaviour with a MM force field is difficult to improve, as it is not easy to identify a single parameter as the "fault" and even harder to rebalance a force field after having made a change. It is most common to swap force fields completely in an attempt to get a better result.

3.2.3 Electronic Distribution and Electrostatic Isopotentials

One of the most challenging set of parameters to derive are the coulombic terms. The energy function itself is straightforward, but requires that there are charges for each atom. Atomic charges are one of the major deficiencies in the MM method. Quantum mechanics tells us that the charge distribution in an atom is anisotropic (uneven); for example, the water molecule has two lone pairs (high electron density) and two hydrogen atoms (low electron density) arranged in a tetrahedral configuration. Many force fields, however, use a single charge centred on each atom, a "monopole". There are a variety of methods for determining partial charges. Quantum mechanics calculations are now the method of choice, with the earliest method due to Mulliken,[41] who proposed that each atom receive half of the electron density where that was due to contributions from two atoms. However, this method neglects differences in electronegativity between atoms, and is also highly dependent on the basis set used. Other, more empirical, methods were adopted, such as the widely adopted Gasteiger–Marsili method.[42] This is an iterative approach, based on electronegativities of atoms. Each atom is allocated a starting partial charge (based on atom types) which is then modified by the difference in electronegativities of the bonded atoms. As this is a pairwise comparison, these changes are iterated until the charges no longer change (the process has reached convergence). Once *ab initio* quantum mechanics calculations were sufficiently accessible, a new method came to dominate: fitting atomic charges to reflect the electrostatic potential of the molecule.

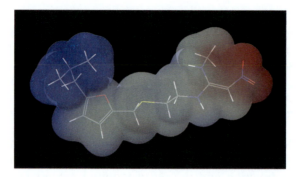

Figure 3.5 The molecular electrostatic potential of Zantac mapped onto a surface of constant electron density.

The valence bond model of molecules has been hugely successful. It allows an orderly depiction of molecules where electrons are assigned to bond orders of whole numbers (single, double, triple), where formal charges are assigned to atoms (*e.g.* N+) and where chemical reactions occur by movements of whole electrons between bonds, rationalised by the drawing of curly arrows. None of this is real, of course, with only the complexity of the—quantum mechanical—truth preventing wholesale rejection of this over-simplified model of chemistry. However, if we wish to model molecular recognition, we are forced to abandon representations that are useful to human minds, and to adopt representations that are closer to what a molecule "sees", and feels, about other molecules. The molecular electrostatic potential[43] of a molecule can be calculated at any point in space, for example, at the location of a protein atom even if the molecule is far away. As electrostatic forces are relatively long range, and are large at short range, the MEP of a molecule is an excellent description of how the molecule is "seen" by other molecules and the magnitude of the MEP at any point describes how forcefully its potential energy might be "felt" by another molecule. The MEP can be calculated from the wavefunction by placing of a unit positive charge at a point in space and summing the coulombic interaction energies between it and the electrons and nuclei in the molecule (therefore, one can also calculate an MEP from a molecular mechanics force field). An MEP at a single point in space is not in itself particularly useful, therefore it is usual to calculate the MEP surface of a molecule on a surface, typically chosen close to the van der Waals radii of the atoms, or at constant electron density. This surface is an excellent visualisation of the molecule (Figure 3.5), but can also be used to derive accurate partial charges for a molecule for use with a force field, which by definition cannot pre-define partial charges for all possible molecules. In the standard CHELPG algorithm,[44] a regular grid of points are placed around the molecule, extending to 3 Å further than the van der Waals radii of atoms in the molecule. A set of atom-centred partial charges are derived using a procedure that minimises the error function (difference between the electrostatic potential values from the quantum mechanics calculation and that produced by the partial charges) subject to constraint that the partial charges sum to the total molecular charge.

3.2.4 Dimensional Molecular Similarity

If the molecular electrostatic potential is a good representation of how a protein sees a ligand, then two molecules with similar MEPs should share similar biological properties. The similarity between two MEPs can be defined using the Carbo index:[20]

$$R_{AB} = \frac{\int P_A P_B dv}{\left(\int P_A^2 dv\right)^{1/2} \left(\int P_B^2 dv\right)^{1/2}} \tag{3.6}$$

Where *P* denotes the MEP for molecule A or B integrated over the 3D space surrounding the two molecules. The potential, P_r at a point r for a molecule of n atoms may be calculated directly from the electron density or from point charges:

$$P_r = \sum_{i=1}^{n} \frac{q_i}{(r - R_i)} \tag{3.7}$$

Where $(r - R_i)$ is distance from atom *i* to the point *r* in 3D space and q_i is the partial charge on the atom. By assigning partial lipophilicity constants to atoms, a molecular lipophilic similarity can also be calculated in the same way.[45]

As computational chemistry makes common use of Gaussian functions to model electron distribution, it was realised that the Carbo index could be used to provide a measure of shape similarity, and this could be made fast enough for general use by use of atom-centred Gaussian functions to fit the curve of electron density against distance from the nucleus.[46] These methods are at the heart of most modern methods of computing 3D shape similarities.[47]

3.2.5 Energy Minimisation

Using a force field allows the calculation of the molecular energy for a particular set of nuclear positions. In the very first studies, these atomic positions were taken from physical molecular models![48] In order to realise the full potential of the methods, it is necessary to be able to compute the atomic positions which give rise to the lowest potential energy. This is termed energy minimisation, and there are a variety of techniques employed. One of the simplest and earliest methods is that of Steepest Descents.[49] Think of the potential energy surface of a molecule as a hill with a ball on it. If we start halfway down the hill, the force on the ball (gravity) will point down the slope. The Steepest Descents algorithm will move the ball parallel to the force on it, *i.e.* straight down the hill. The algorithm does not know where the hill ends and therefore needs to know how far to move the ball down hill before recalculating the potential energy. The most common way of doing this is to use a fixed "step size". If we extend our analogy from a hill slope to a valley, there is now a more complex minimisation. Steepest Descents will "tack" down the valley (Figure 3.6) using a series of orthogonal steps. The fixed step size may not be the most efficient, but avoids overshooting the energy minimum at the bottom of the valley. A more efficient method is Conjugate Gradients,[50] which takes into account the previous step and the current forces on the ball, to produce a much better idea of the true nature of the potential energy surface and will minimise using fewer steps, and therefore fewer energy calculations. Even more complex minimisation methods are the so-called second derivative methods, which take into account not just the gradients (as in Conjugate Gradients), but the curvature of the energy surface. One popular method is the Newton–Raphson,[51] which will find the energy minimum for a quadratic function in one step.

The great benefits of MM over QM methods are the ability to directly incorporate experimental data on molecular structure (bond lengths, angles, torsions) and the sheer speed of the calculations which, with modern computers, allow studies on drug like molecules in seconds. This speed allows the study of large molecules such as proteins, and the motion of molecules in time, an area of computational chemistry generally referred to as molecular simulation.

3.3 MOLECULAR SIMULATION AND DYNAMICS

Energy minimisation allows us to generate individual energy minima on the potential energy surface of a molecule. However, as the laws of thermodynamics tell us, properties of interest (binding energies, solvation energies, movement of a molecule through a membrane, *etc.*) are the product of all of the energy states in a system, including the vibrational and conformational

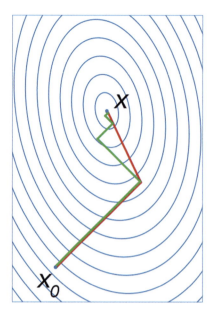

Figure 3.6 A comparison of the convergence of a steepest descent algorithm (green) and the corresponding performance of the conjugate gradients method (in red) for minimizing a quadratic
function associated with a given linear system. Conjugate gradient, assuming exact arithmetic,
converges in at most n steps where n is the size of the matrix of the system (here $n = 2$).

energy minima. For a protein–ligand system, a full quantification of all the possible energy
states of the system is infeasible. Techniques which sample the possible energy states in order
to approximate the true ensemble of states are commonly referred to as molecular simulations.
Simulations are also able to model time-dependent behaviour, providing a detailed representation of, for example, how a molecule changes conformation upon binding to a protein.

The first published simulation method, the Monte Carlo technique,[52] uses a random sampling of states of the system (for example, conformations) to build up a representative ensemble
from which the properties (*e.g.* binding energy) can be calculated. A new conformation is
generated, for example by rotating a random torsional angle by a random number of degrees.
The energy of the conformer is calculated, and then accepted in to the ensemble of representative conformations according to a probability (p) which is modified by the energy (E) of the
conformation:

$$p = e^{-E/kT} \tag{3.8}$$

Therefore, low energy conformers have a much higher probability of being accepted into the
ensemble of states than would a high energy conformer, but these will have at least some
representation in the model.

It was recognised that the equations of motion could be harnessed to describe the behaviour
of a real system, and this begat the technique of molecular dynamics.[53] In contrast to the
sampling strategy of Monte Carlo methods, molecular dynamics is a deterministic method,
which means that any future state of the system can be predicted from the current state. Given a
starting configuration of a molecule, or collection of molecules, the forces on the atoms are
determined, typically using a MM force field.[54] Newtonian mechanics are then applied to the
system in order to move the atoms for a particular time. After each time step, the forces on the
atoms are recalculated and combined with the current positions and velocities of the atoms to
produce new atomic coordinates, directions and velocities. For typical molecular systems,

bearing in mind how quickly non-bonded forces can change as atoms become close to each-other (*e.g.* using the Lennard-Jones 6–12 potential) the time step between calculation must be very short, of the order of 1–10 femtoseconds (10^{-15} to 10^{-14} s). In this way, molecular dynamics produces a trajectory which describes the change on molecular motion over time.

There are subtle theoretical differences between Monte Carlo and molecular dynamics methods. Monte Carlo is typically performed under conditions of constant number of atoms, volume and temperature ("NVT") whilst dynamics uses constant energy but not temperature ("NVE"). Therefore, the temperature in a MD simulation can be varied, and this is often ex-ploited to put more energy in the system to allow, for example, high energy conformational changes, to be seen in a shorter time than would usually be observed, so that the simulation would be possible using available computational resources (CPU time is proportional to the number of time steps required for evaluation in the simulation).

Both MC and MD methods require significant compute resources, as the number of con-figurations of a system, particularly for larger molecules, is extremely large and the number of non-bonded terms to be evaluated in the MM force field scales by the square of the number of atoms in the system. There are a number of common approximations applied in order to make the simulations feasible. A non-bonded cut-off is generally applied: the Lennard-Jones potential declines rapidly with distance (at 2.5 times the vdW radius σ the potential has just 1% of the value at σ). Electrostatic terms, in contrast, have a much longer range and generally require calculation up to at least 10 Å distance. It is also useful to keep a list of atom pairs which are included in the calculation—the non-bonded neighbour lists. This prevents the recalculation of interatomic distances on every atom pair in the system at every time step in the simulation. The neighbour lists are typically updated every 10–20 time steps.

Molecular simulations allow the study of conformational ensembles. As such they allow the inclusion of entropic contributions to calculations, where previously we have spoken only about enthalpic energies. However, particularly for MD, it is unlikely that the sampling of high energy states is always comprehensive.

Simulations hold out the tantalising possibility of being able to predict the free energy of binding for a novel molecule to a protein. There are many reasons why this is not yet, and may never be, possible, but for a time it seemed that the computational community was on the brink of a revolution in how drug discovery was performed. Calculation of the absolute free energy of binding of a ligand to a protein is fraught with difficulty, and to compare calculated free en-ergies for two different molecules would normally produce small differences from very large numbers, where any differences are swamped by the errors in the large numbers. The Free Energy Perturbation (FEP) method[24] uses a non-physical application of the thermodynamic energy cycle to compare the binding energies of two different ligands to the same protein (Figure 3.7). ΔG_1 and ΔG_2 are respectively the free energies of binding of Ligand 1 and Ligand 2 to the protein. As stated above, accurate calculation of this free energy may not be compu-tationally feasible. The key to the method is to calculate the *relative* free energies between the systems. Because the value around a thermodynamic cycle must be zero, the difference in free energies of binding, $\Delta G_2 - \Delta G_1$, must equal $\Delta G_4 - \Delta G_3$. ΔG_3 is the difference in free energy between the ligands in solution, and ΔG_4 the free energy difference of the two ligand–receptor complexes, neither of which can be observed in the laboratory. However, the FEP method, being theoretical, can exploit the non-observable changes ΔG_3 and ΔG_4. The calculation starts with one of the ligands x, and gradually changes it into the other ligand y *via* a gradual perturbation, λ, of the molecular mechanics parameters. For example, the van der Waals energy term in the calculation can be written as the sum of the contributions from the partial ligands x and y:

$$\nu_{ij}^{IJ}(\lambda) = 4(1-\lambda)\varepsilon_x\left(\alpha_x^{LJ6-12}\right) + 4\lambda\varepsilon_y\left(\alpha_y^{LJ6-12}\right) \tag{3.9}$$

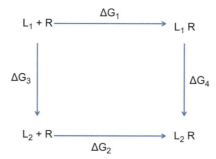

Figure 3.7   The thermodynamic cycle for the binding energies of two protein–ligand complexes which is exploited by the Free Energy Perturbation method. By using the purely theoretical energy changes ΔG_3 and ΔG_4, the difference in free energies of binding between the ligands can be computed.

Where α_x^{LJ6-12} is a modified Lennard–Jones 6–12 potential for ligand X, to ensure that the system behaves well as X disappears to nothing, *i.e.* the mutation from ligand X to ligand Y is complete.

Initial FEP calculations were encouraging—relative free energies of binding could be calculated within experimental error for drug-like molecules. Unfortunately, further experience with the method served to highlight the many difficulties of simulating complex molecular interactions with simple force fields, approximate methods and insufficient computing resources.[55]

A semi-empirical alternative is the Linear Response method,[56] which attempts to compute absolute free energies of binding from just two simulations, one for the ligand in solution and one for the protein–ligand complex. The electrostatic and van der Waals interaction energies from these simulations are then combined to produce an estimate of the free energy of binding:

$$\Delta G = \beta(\langle \gamma_{\text{ligand–protein}}^{\text{el}} \rangle - \langle \gamma_{\text{ligand–solvent}}^{\text{el}} \rangle) + \alpha(\langle \gamma_{\text{ligand–protein}}^{\text{vdw}} \rangle - \langle \gamma_{\text{ligand–solvent}}^{\text{vdw}} \rangle) \tag{3.10}$$

Where β and α are parameters. Initially it had been hoped that some generic values of β and α could be derived from existing data, but it soon became clear that this is not the case.

Although many improvements have been published and new approaches invented, numerous experts doubt that the reliable, high throughput prediction of free energy of binding is a tractable problem for computational chemistry.[57]

3.4   MODELLING SOLVATION

The previous sections concentrate on the energy surface of a single molecule and many computational studies are conducted without consideration of the environment of the molecule—in the active site of an enzyme, in a membrane or in solution. Such studies are termed *in vacuo* or "gas phase" calculations. The environment of a molecule may have a profound effect on the energetics. For example, the preferred tautomer state of pyridone is reversed from gas phase to aqueous phase (Figure 3.8).[58] In the gas phase, the internal hydrogen bond from the hydroxyl proton to the nitrogen lone pair stabilises that tautomer. However, in water, this coulombic attraction is screened by the dielectric, whilst in the keto-form, the resonance structure allows electrons to be drawn from the ring towards the electronegative oxygen, yielding a strong electrostatic interaction with the polar solvent, at the same time increasing the dipole of the molecule and hence adding further to the large solvation energy.

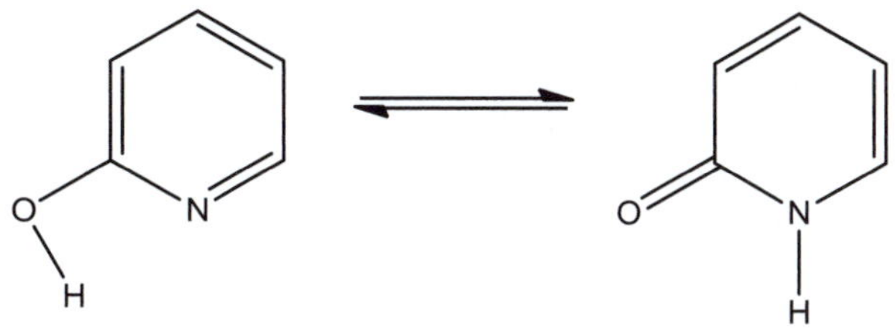

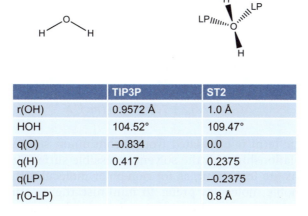

Gas phase relative energy   0 kcal/mol   0.5 kcal/mol

Aqueous phase relative energy   +4.1 kcal/mol   0 kcal/mol

Figure 3.8   Solvent effects can have profound impacts on molecular structure. In the gas phase, and in non-polar solvents, the dominant form of 2-pyridone is the hydroxyl tautomer. However, in water, the large solvation energy of the keto-form makes it the dominant form.

	TIP3P	ST2
r(OH)	0.9572 Å	1.0 Å
HOH	104.52°	109.47°
q(O)	−0.834	0.0
q(H)	0.417	0.2375
q(LP)		−0.2375
r(O-LP)		0.8 Å

Figure 3.9   A comparison of two water models used in molecular simulations. The TIP3P was an early standard, even though it did not reflect the tetrahedral hydrogen binding preference of water. More recent models such as ST2 provide a more accurate representation, albeit at a higher computational cost.

The intuitive method of modelling solvation is to include explicit solvent molecules in the calculation and immerse the ligand in a "water bath". This is the method of choice for modern simulation methods. However, inclusion of explicit solvent molecules is impractical for quantum mechanics calculations (remember the N^4 or N^3 scaling) and even for molecular mechanics, the need to evaluate all the molecular interaction terms for each pair of water molecules adds significantly to the computation required. Hence, simplified models of water were sought, one of the most successful being Jorgensen's TIP3P (Figure 3.9) model.[59] This model has charges centred on the oxygen and both hydrogens. The water molecule is kept rigid during the simulation, and water–water van der Waals terms are computed *via* a single Lennard-Jones term centred on the oxygen of each molecule. In order to reproduce the properties of bulk water (effective dipole moment of 2.6 Debyes), the atomic charges are large. The TIP3P model does not adequately describe the directionality of electron density on a water molecule, which is towards the oxygen lone pairs. There are more realistic models available, such as the ST2 system[60] which uses a more realistic tetrahedral geometry with "lone pair" point charges on the oxygen atom. However, the increased computational effort required to use these has meant that the simpler models are more likely to be used. These models have been proven very effective and can even be polarised by their environment by use of a fluctuating charge model.[61]

A simpler solution is to utilise classical reaction field theory.[62] Here, the solvent is considered as a bulk continuum and characterised by the dielectric constant. The solute model is considered as being embedded in a cavity within the solvent, in the simplest case a spherical cavity is used. The solute charge distribution and dipole cause a reaction field in the solvent (as, for example, water molecules orientate to hydrogen bond with a carboxylic acid solvate) and the interaction of these fields produces the electrostatic energy of solvation. Such models are known as continuum solvation models, are simple to compute and surprisingly accurate for simple solutes. More sophisticated solutions involve the construction of more realistic cavity shapes by interlocking spheres,[63] or using an isoelectronic electron density surface to describe the shape of the cavity.[64] In these models, the solute molecular electrostatic potential is calculated at points on the surface, and these charges then represent the solvent reaction field which is used to compute electrostatic interactions with the solute. An alternate, but related, approach is the use of Poisson–Boltzman methods[65] which allow a better description of the ionic strength of the solvent and the change in dielectric from molecule to bulk solvent.

Other terms, such as the energy to create the cavity in the solvent, and van der Waals interactions, are generally computed in a much more approximate fashion, for example in the GBSA (Generalised Born Surface Area) method:[66]

$$\Delta G_{sol} = \Delta G_{el} + 7.2 \times SA \; cal/mol/\mathring{A}^2 \qquad (3.11)$$

Where the electrostatic term is calculated from a continuum solvent model and the other terms abstracted to a relationship with the solvent accessible surface (SA) of the solute.

The surface area terms are important, as for non-polar molecules they model the observed behaviour that hydrophobic molecules prefer to minimise the amount of surface area that interacts with water, preferring instead to self-associate or adopt appropriate conformations. Although still a contentious topic, this phenomenon known as the Hydrophobic Effect[67] is thought to be entropy driven (Figure 3.10).

This phenomenon is known to affect binding of ligands to proteins, increasing the affinity *via* non-specific binding. The majority of drug molecules are moderately lipophilic, exploiting the effect to help adhere to the target of choice, without incurring penalties associated with too much lipophilicity (poor solubility, low levels of absorption, promiscuous binding to other proteins, toxicity).[68] Recent studies with isothermal calorimetry confirm a relationship between lipophilicity and entropic binding energies,[69] whilst at the same time highlighting that our ability to decompose free energy contributions and use them in prospective design remains imperfect.[70]

Being entropic in nature, this effect is also very hard to model with the methods described here, the most successful methods being those based on measurements and QSAR techniques, for example the many log P models that are available.[71]

3.5 CONFORMATIONS, CONFORMATIONAL ENERGY AND DRUG DESIGN

The biological and physicochemical properties of a molecule are determined to a large extent by the three dimensional structures, or conformations, which are accessible. The scientific

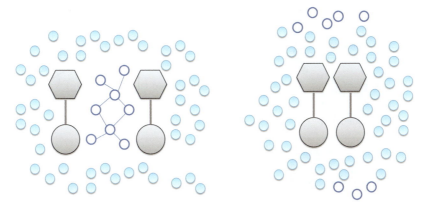

Figure 3.10 A representation of the Hydrophobic Effect. Two organic molecules are dissolved in water. Between them is a water network that cannot hydrogen bond with the lipophilic molecules, and therefore adopt an ordered network of hydrogen bonds. As the two organic molecules come together, reducing the hydrophobic surface area exposed to the solvent, the ordered water molecules are released to bulk solvent, therefore increasing entropy, and lowering the free energy of the system.

field of conformational analysis can be said to have originated in a study by Barton,[9] who related the reactivity of substituted cyclohexanes to the equatorial or axial orientation of the ring substituents. Molecular mechanics calculations allowed theoreticians to take the principles of conformational analysis and reduce it to routine practice.[16] Conformational analysis necessitates a conformational search for accessible conformations of a molecule. The preferred conformations can then be determined through calculation of relative energies of the accessible conformations. For simple molecules, a systematic search is facile: each rotatable bond is systematically rotated *via* a fixed amount (say, 30 degrees) whilst the bond lengths and angles are kept constant. Energies may be computed at each point, or energy minimisation conducted starting at low energy conformers. The objective is to discover all the accessible conformations, which will lie at energy minima. Ideally, the global energy minimum will be determined, which will describe the dominant conformation. This systematic conformational data can be used to produce a map of the potential energy surface for a molecule. Possibly the most famous example of this type of analysis is the Ramachandran plot[72] for the alanine peptide, a model for the accessible conformations of amino-acids in proteins (Figure 3.11).

However, systematic searches become problematic when dealing with drug size molecules, in that the number of conformations to be computed scales by O^N where N is the number of rotatable bonds. For example, a small molecule with three rotational bonds, sampled at $30°$ intervals, has 1728 conformations. A molecule roughly twice the size (six rotatable bonds) would have 2.9 million conformations, and adding a single atom to that molecule could lead to 36 million conformations! At 1 second per conformer, the analysis would take over a year. Fortunately, search algorithms have been developed which reduce the size of the problem by avoiding unproductive areas of conformational space.[73] Search trees allow the early recognition of, say, a high energy torsion, and then ignore all conformers in that branch of the search. As conformational analysis is such a fundamental technique, many other methods have been invented: fragment based building,[74] random and Monte Carlo searches,[75] molecular dynamics/simulated annealing, genetic algorithms[76] and distance geometry.[77] This last method is useful in the generation of conformers which satisfy experimentally determined distance information, for example from NMR experiments. Distance geometry uses matrices of distances between

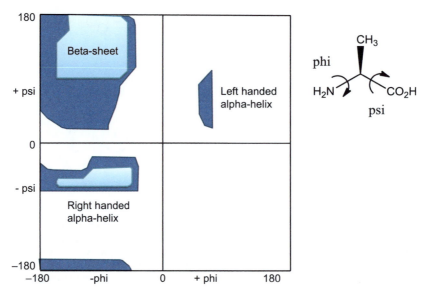

Figure 3.11   The famous Ramachandran plot for peptides. The torsion angles phi and psi (shown here for the amino acid alanine) are systematically rotated and close contacts between atoms noted. Any combination of phi and psi which result in steric clashes are disallowed, and are responsible for all the white space in the graph. These regions are therefore not accessible to any amino acid except for glycine, which is unique in that it does not have a side chain. The darkblue areas are those which can be accessed if the van der Waals radii are slightly reduced, and this allows the emergence of the left handed helix region, which is sometimes observed in protein structures. The light blue areas are the low energy conformers, and correspond to the commonly observed beta-sheet and alpha-helix conformations.

atoms alongside rules from trigonometry to parse randomly generated conformations and weed out those that are geometrically impossible. Those that cannot be ruled out are refined until the conformation better satisfies the distance constraints and finally a force field minimisation may be employed on the smaller numbers of remaining conformers.

Relative conformational energies are crucially important in drug discovery. A difference in energy of only 1.4 kcal/mol results in a ratio of 1 : 10 in the populations of two conformers.

Returning to the relationships of Gibbs free energy introduced in Section 3.1, we can write:

$$\Delta H - T\Delta S = -2.3RT\log K \qquad (3.12)$$

From this we can use conformational properties to increase the enthalpy of binding (ΔH) or reduce the entropic penalty of binding (ΔS). Both of these require us to use the technique of Conformational Constraints. If we are able to constrain a molecule to have a better geometric match of the functionality of the ligand with the complementary groups on the receptor, this will increase ΔH. This requires much precision in design and is challenging even with protein structure data. A more forgiving approach is to increase the concentration of the binding conformation of the molecule in solution, which reduces the entropic penalty upon binding. Take a simple molecule with three accessible conformations arising from one rotatable C–C bond.

We exploit the Boltzman relationship between the number of states in a system and its entropy:

$$S = k \ln W \qquad (3.13)$$

Comparing the change from a molecule with a freely rotating C–C bond (W1, 3 conformational states) with that where the C–C bond has been constrained (W2, 1 state), the change in entropy is:

$$\Delta S = R\ln(W1/W2) \text{ kcal/mol} \tag{3.14}$$

$$T\Delta S = RT\ln(1/3) \text{ kcal/mol} = -0.6 \text{ kcal/mol} \tag{3.15}$$

This is a gain of ~four fold in affinity, and is comparable to that gained from a strong hydrogen bond.[78]

Conformational constraints are a familiar tactic in small molecule drug discovery, offering predictable gains in affinity from small changes to molecules without requiring detailed understanding of the receptor site.

Figure 3.12 shows two successful examples of conformational constraints, one by use of a ring system to constraint the rotation about an important bond vector,[79] whilst the other

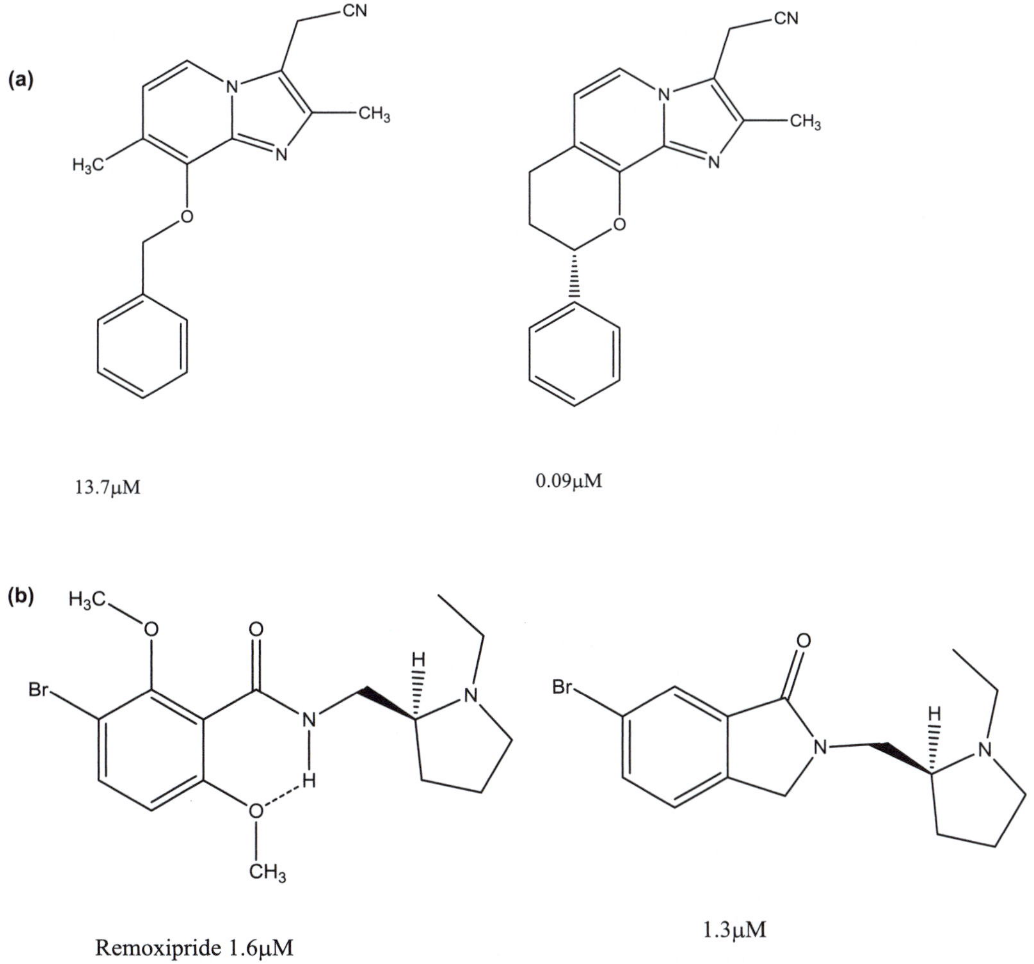

Figure 3.12 (a) The successful introduction of a conformational constraint in Gastric H+/K+ ATPase inhibitors. (b) An example of the use of intramolecular hydrogen bonds to constrain a conformation in D2 receptor antagonists.

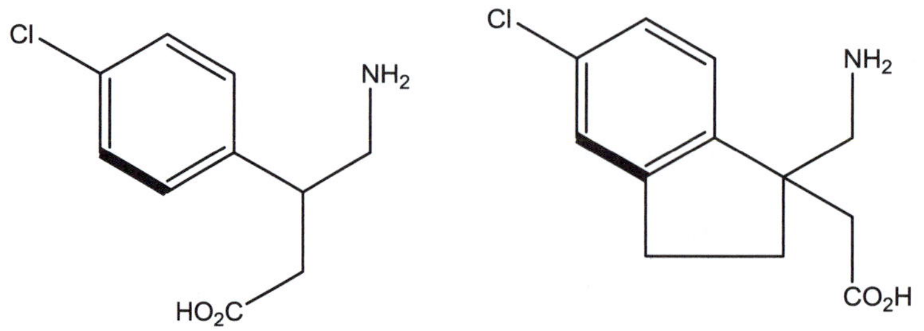

Baclofen (420nM)

Constrained analogue designed from
small molecule crystal structure (>100μM)

Figure 3.13 An unsuccessful application of conformational constraints in GABAb receptor ligands.

example uses intramolecular hydrogen bonding constrain the accessible conformations of the molecule.[80] Modern computational chemistry methods are quite capable of predicting the geometries and energetics of this type of system.

Figure 3.13 illustrates that it is quite possible for conformational constraints to fail if one has insufficient knowledge of the receptor environment or binding conformations of the molecule. The small molecule crystal structure of baclofen was determined and used to design analogues which would lock the molecule into the observed low energy conformation.[81] None of these were active, meaning that either the binding conformation of the molecule is different to that seen in the solid state, or that the atoms providing the conformational constraint may not be tolerated in the active site of the receptor.

3.6 QUANTIFYING MOLECULAR INTERACTIONS FROM EXPERIMENTAL DATA

There is a significant body of experimental data that can be used instead of, in conjunction with, or to train computational approaches. One of the most useful of these is the Cambridge Structural Database (CSD, www.ccdc.cam.ac.uk/) of small molecule crystal structures.[82] Starting from a straightforward repository of data, the Cambridge Crystallographic Data Centre (CCDC) have developed a number of analysis tools which reveal a treasure trove of information about small molecule conformation and functional group interaction geometries.

IsoStar[83] (www.ccdc.cam.ac.uk/Solutions/CSDSystem/Pages/IsoStar.aspx) is a tool that gives access to a library of the intermolecular interactions found within the CSD (and some from the PDB). Searches of the CSD have been carried out for over 12 000 particular intermolecular interactions. The data can be analysed by functional group, and a picture of how that group interacts with others can be displayed in various forms. For example, Figure 3.14 shows the interaction patterns of an ester group with any hydroxyl group (phenolic, aliphatic or water).

These plots can be reduced to a contour plot to reveal underlying distributions. In Figure 3.15 two contours have been computed to represent the density of observed observations per unit volume of space. This method clearly shows the preference of an ester to use the carbonyl to hydrogen bond, and the geometrical preference to have the hydrogen bond directed towards the oxygen lone pairs.

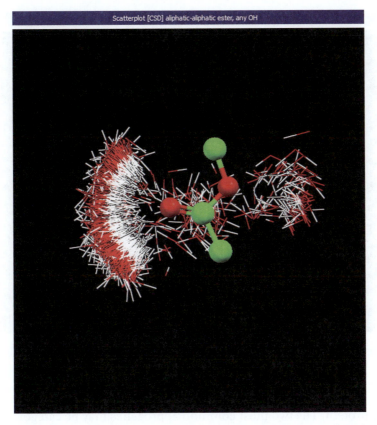

Figure 3.14 Example output from IsoStar for an ester fragment interacting with hydroxyl groups.

The CCDC data provides geometric preference and likelihood statistics, but the observational data does not provide insight into the strength of hydrogen bonds nor of the binding energy that might be afforded by making such interactions with the protein. For many years researchers relied on the elegant work of Abraham *et al.*[84] which yielded a set of standard hydrogen bond strengths for common organic functional groups and rings. It is also possible to use quantum mechanics calculations to produce such values for novel fragments, and to do so using quite straightforward methods such as computing the MEP.[85]

3.7 DOCKING AND SCORING FUNCTIONS

Perhaps the most familiar application of computational chemistry is that of structure based design, where the crystal structure of the protein has been determined. Computational methods offer the possibility of virtual screening to discover leads in an existing collection of molecules, or to select from possible molecules which could be synthesised as part of a lead optimisation program. It is useful to think of this problem in two parts, that of docking (generation of plausible poses of the ligand and protein) and scoring (selecting the correct binding mode and conformation for a single molecule and correctly ranking molecules against each other).

The generation of ligand poses within a protein is a similar problem to that of conformational analysis, with the added complexity that we must also take account of the orientation of

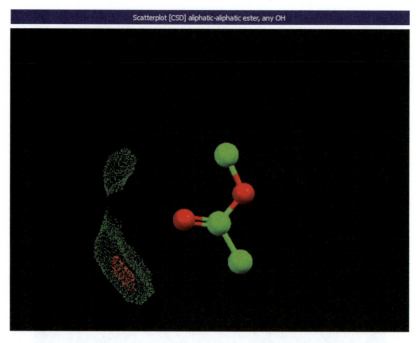

Figure 3.15   The IsoStar output contoured according to the density of observations, illustrating how the ester fragment prefers to make hydrogen bonds with its p orbitals.

the ligand to the protein. The two most common approaches may be categorised as systematic and random. A systematic search seeks to sample all the possible conformations and orientations of the molecule, but faces the same problems as for conformational analysis, multiplied by the fact that calculations are needed on every conformer to test the interaction with the protein. A common strategy, as employed by the popular methods Dock[21] and FlexX,[86] is to fragment the molecule and dock the largest or most rigid part of the molecule into the protein. Other parts of the molecule are grown onto the successfully docked fragments, until the whole set of plausible docking poses have been generated for the molecule. Alternative approaches often employ stochastic (random) methods to generate poses. The popular GOLD algorithm[87] uses a genetic algorithm, with the bond torsions encoded in the chromosome which is allowed to mutate and breed, with better scoring ("fitter") solutions being more likely to pass on their DNA to new poses. Monte Carlo approaches are also effective at exploring possible binding modes.

   Plausible docking poses need to be compared to each other, and to those of other ligands, in order to perform a useful virtual screen. Even if the correct binding poses are produced, this cannot be used unless a scoring function can distinguish it from the other possibilities. Scoring functions are best categorised into three classes.[88] The first, and earliest, are the physics based methods, using the first principles methods described in this chapter, typically using molecular mechanics force fields (although there are some studies using QM). The seminal GRID program[23] employs a molecular mechanics force field, which is used to calculate the potential energy of the protein at points on a grid thrown over the protein active site. GRID uses atom type probes to generate interaction energy contours for visualisation, and the approach is readily adapted and used for whole molecules. AutoDock[89] employs the AMBER force field.[90] A problem with the use of force fields to score docking poses is the "hardness" of the Lennard-Jones 6–12 potential. A small error in positioning of a molecule in close proximity to the protein can result in very large repulsion energies, and therefore some methods such as G-Score[91] utilise a softer potential, for example an 8-4 term which is more forgiving.

Pure force field based methods were not found to be satisfactory, and require significant compute time, and therefore empirical methods emerged. These methods derive scoring functions which reproduce experimental data such as binding energies and observed binding geometries as extracted from the CCDC database. The LUDI scoring function[92] involves the number of hydrogen bonds made (which need to be of an orientation and geometry consistent with those observed in crystal structures), lipophilic contact surface and the number of rotatable bonds:

$$\text{Log } 1/K_i = 1.4 \times \text{ionic hydrogen bonds} + 0.83 \times \text{neutral hydrogen bonds}$$
$$+ 0.030 \times \text{lipophilic contact surface area}$$
$$- 0.25 \times \text{number of rotatable bonds} - 0.91 \tag{3.16}$$

Subsequent empirical schemes utilised different methods to encode the non-bonded interactions, for example ChemScore[93] replaces the hydrophobic surface area with contacts between hydrophobic atom pairs, and the F-Score[86] adds a specific term for aromatic interactions. One advantage of empirical methods is the ability to add fitted terms for phenomena that are difficult to compute from first principles, for example entropic components (often based on the number of rotatable bonds) and desolvation energies.

The third category is knowledge based scoring functions. These attempt to encode observed information from protein–ligand crystal structures in a way that can be employed to good effect on novel systems. Typically, this entails extraction of all pairs of atoms seen to interact in a crystal structure (for example, a ligand carbon atom in an aromatic ring contacting an aromatic carbon atom in the protein). The geometries and configurations of all examples of each atom pair are combined to define the optimal position, and allowable deviation for that type of interaction. A score for a docking pose is then the sum of all the atom-pair interactions. Examples of these methods are the Potential of Mean Force,[94] DrugScore[95] and SMoG.[96] A significant advantage of such methods is the speed by which they can be computed, however like QSAR models, the predictions can be suspect when applied to molecular systems which are unlike those in the training set.

There are dozens of scoring functions in the literature, none of which can be relied on to work for every problem.[97] It is also difficult to predict when a particular method might work well. This problem gave rise to the development of consensus scoring schemes, whereby several scoring functions are employed at the same time and the results combined. Typically, a set of complementary scoring functions are chosen, for example the X-CSCORE[98] model uses force field based (GOLD, DOCK, FlexX), empirical (ChemScore) and knowledge based (PMF) functions. The hope is that by combining the models, consistent high performing docking poses will be raised to the top, and good molecules will be penalised less by a poor prediction in one method.

Although much intellectual firepower has been deployed in the quest for reliable docking and scoring methods, the problem remains stubbornly resistant to the best efforts of the computational community. The same issues that prevent reliable predictions of free energy changes with the most sophisticated computer models reappear, and are amplified, as we try to compare the properties of very diverse ligands. As a general rule of thumb, generation of plausible docking poses is less of a problem than having a scoring function that recognises the right answer when it has been found.

3.8 EXAMPLES OF IMPACTFUL COMPUTATIONAL CHEMISTRY ON DRUG DESIGN

It is rare to see examples of computational design which are primarily responsible for a drug molecule. This is because drug discovery is a very difficult occupation, and although

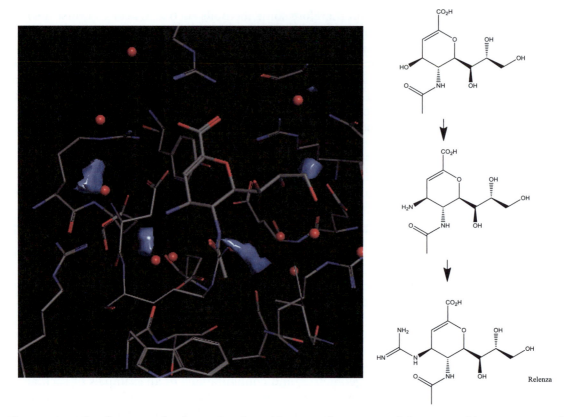

Figure 3.16 The discovery of Relenza. Starting with a crystal structure of the neuramidase enzyme and a the weak inhibitor, 2-deoxy-2,3-didehydro-*N*-acetylneuraminic acid (DANA), the GRID program was used to guide medicinal chemistry. First the 4-amino substituent was introduced which increased affinity for the enzyme. The image shows the crystal structure of the 4-amino analogue, with the blue area directly below the amino group clearly indicating the predicted increase in potency with the introduction of a larger positively charged moiety. Following this prediction, the 4-guanidino group was one of several functional groups introduced and was found to be a potent inhibitor of the enzyme, yielding the anti-influenza drug Relenza.

computational techniques might help with binding affinity or solubility, application to areas such as predictive toxicology is in its infancy.[99] In recent years, the proportion of drugs derived from protein structure based approaches has increased and therefore these have had computational input. One of the cleanest examples of computational impact is the Neuraminidase inhibitor, Relenza (Figure 3.16). The introduction of the key 4-amino and then the 4-guanidine substituent was prompted by the use of the GRID program, which enabled the chemistry team to gain multiple orders of binding affinity whilst making only tens of molecules.[26]

A less successful example, but one which provides many excellent touch-points, is the work of Lam *et al.* on inhibitors of the HIV-1 Protease (Figure 3.17).[100] Starting from a linear peptide which was seen to bind to the protein *via* a water molecule, a 3D database search yielded some ideas that a six membered ring could deliver a water mimetic in the right geometric orientation whilst at the same time positioning a key hydroxyl group to bind with the active site aspartic acid residues. Using knowledge from the peptide SAR, where two hydroxyls provide better binding affinity, a seven membered ring was postulated. As a ketone has poor hydrogen bond

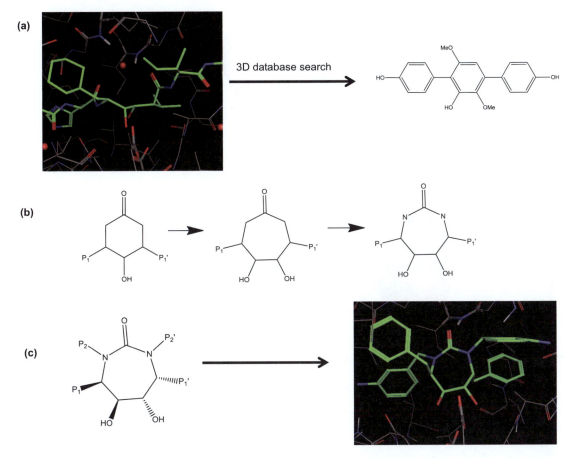

Figure 3.17 The design of non-peptidic inhibitors of HIV Protease. Starting from crystal structures of peptide inhibitors, a 3D database search was used to identify molecular architectures which could replace a key water molecule and position a hydroxyl group which was required to bind to the catalytic aspartic acids in the protein active site (shown in the first crystal structure image). This search yielded the tri-phenyl moiety shown which inspired the team to design a six membered ring which might better position the functionality whilst incorporating some synthetic routes to add substituents which could interact with the protein's other recognition pockets. The initial idea was refined to a seven membered ring (so as to incorporate two hydroxyl groups, known to be better inhibitors in the peptide analogues) and the ketone was replaced by a urea to incorporate a stronger hydrogen bonding group, and also to allow further points of diversification. Only then was ''pure'' computation invoked, to determine the precise stereochemistry required to fit the active site and position the key functional groups with a low energy conformation of the ring template. The design can be seen to be correct in the second protein structure image.

strength, this was substituted for a urea. With the idea taking shape, the challenge was to predict the correct stereoisomers and substituents that would produce a low energy conformation of the ring and deliver all the substituents to the correct binding pockets. This was decided using computational approaches, and the resulting molecule synthesised. The team was successful in designing a non-peptidic ligand, and the subsequent protein–ligand crystal structure confirmed the predictivity of the design. However, the molecular structure and associated substituents is very lipophilic, which gave rise to insurmountable DMPK issues and the series did not deliver a drug.

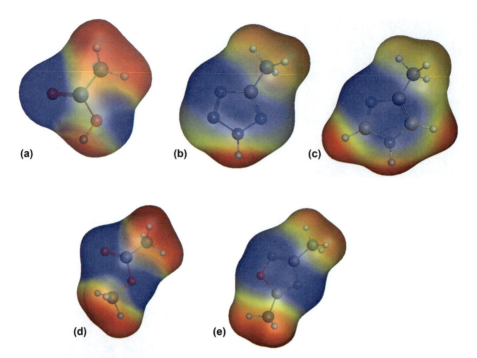

Figure 3.18 The use of molecular electrostatic potential maps to design inhibitors of squalene synthase.[101] Analogues with a carboxylic acid (a) or tetrazole (b) substituent are more potent than the inactive imidazole (c). The MEP maps of (a) and (b) show a similar pattern of charge density which is not shared by (c), but is shared by the active methyl ester (d) and subsequently designed 3,5-dimethyl-1,2,4-oxadiazole (e).

An example of how molecular electrostatic potentials can be applied in lead optimisation is given by Bamford *et al.*[101] in the design of novel inhibitors of squalene synthase (Figure 3.18). The calculations were first used to aid interpretation of the structure activity relationships in a series of carboxylic acid and methyl ester analogues, and then applied to aid design of heterocyclic isosters of the methyl ester.

Finally, a fine example of the successful application of ligand based modelling is the design of selective 5HT-1D agonists for the treatment of migraine by Glen *et al.*[102] The conformationally restricted agonist methysergide was used to map the 3D positions of critical binding motifs, such as the basic amine, using the Active Analog method.[103] By this method, a 3D pharmacophore was developed and used to screen potential ligands. The design process involved conformational analysis of potential ligands using molecular mechanics, with charges derived from AM1 semi-empirical QM calculations. Low energy conformations found from the molecular mechanics calculations were then superimposed onto the 3D pharmacophore using constrained molecular mechanics minimisation with a simplex algorithm. There were concerns over potential intra-molecular hydrogen bonds which might prevent adoption of the preferred conformation, and these were examined with molecular dynamics simulations. One of the key design criteria was selectivity for the 5HT-1D receptor over the related 5HT-2A, and therefore a "selectivity site" was computed by the differential volume occupied by ligands selective for 5HT-1D over 5HT-2A *versus* those with no selectivity, and the volume added to the pharmacophore model. This computationally based design strategy yielded the drug zolmitriptan (Figure 3.19).

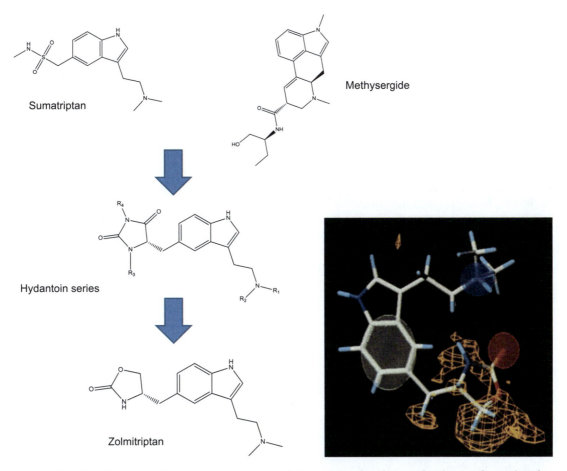

Figure 3.19 The discovery of zolmitriptan employed rigorous molecular mechanics conformational analysis combined with a 3D pharmacophore. Known 5HT-1D agonists such as methysergide and sumatriptan were used to define a 3D pharmacophore and selectivity volume. A series of hydantoins were designed using systematic conformational analysis and superimposition onto the 3D model (shown for zolmitriptan).
The molecular graphics image is adapted with permission from Glen *et al.*[102] Copyright 1995 American Chemical Society.

HINTS AND TIPS

The application of computational chemistry to a drug design problem is not a simple matter of installing a software package and pressing some buttons. Nor is it a case of always running the most complex simulation available to you. What this chapter has sought to impart is that computational chemistry involves approximations, and it is vitally important to have an appreciation of how those approximations impact the ability of a calculation to answer the question to hand. If a molecule has the potential to form an intramolecular hydrogen bond, then it is more useful to include solvation effects in the calculation than to use expensive electron correlation methods. On the other hand, if there is a potential for a ligand to be polarised by adjacent residues in an active site, it may be more useful to run quantum mechanics calculations than to use molecular dynamics simulations of the entire complex.

In particular, with modern molecular graphics packages it is important to remember that what you see on the screen is not real, it is a model. Proteins move, often in surprising ways. Hydrogens are only observed in very high resolution crystal structures, most of the time they are assigned to a position by the crystallographer, as are many water molecules. Indeed, the use of crystallography data is not always as straightforward as it might seem.[104]

Always consider how solvation is treated in your system, especially when thinking of hydrogen bonding. Are you taking entropy into account, particularly if using flexible chains to bridge two apparently important binding motifs?

Hydrophobicity, being entropic in nature, is very hard to model with the physics-based methods, the most successful methods being those based on measurements and QSAR techniques, for example the clogP and alogP algorithms. However, molecular simulations are beginning to show value in understanding of the role of water in structure based design and this is an area full of promise for the future. The physics based methods and empirical based methods therefore both have a complimentary role in trying to understand drug–receptor interactions.

Finally, remember that internal ligand energies have the same contribution to the free energy of binding as does protein–ligand interaction energies. Always model the ligand in solution as well as in the protein.

KEY REFERENCES

Two text books are recommended for students who wish to further understand the theoretical background to computational chemistry applications. The best general text is:
A. R. Leach, *Molecular Modelling: Principles and Applications*, Pearson Education Ltd., Harlow, UK, 2nd Edition, 2001.

Quantum Mechanics demands its own text:
A. Szabo and N. S. Ostlund, *Modern Quantum Chemistry: Introduction to Advanced Electronic Structure Theory*, McGraw-Hill, New York, 1st (Revised) Edition, 1989.

For an example of the systematic use of computational chemistry in the design of a marketed drug the reader is referred to:
R. C. Glen, G. R. Martin, A. P. Hill, R. M. Hyde, P. M. Woollard, J. A. Salmon, J. Buckingham and A. D. Robertson, *J. Med. Chem., J. Med. Chem.*, 1995, **38**, 3566.

REFERENCES

1. P. A. M. Dirac, *Proc. R. Soc. Lond.*, 1929, **123**, 714.
2. J. E. Lennard-Jones, *Proc. R. Soc. Lond.*, 1924, **106**, 463.
3. E. Schrödinger, *Ann. Phys.*, 1926, **79**, 361.
4. R. S. Mulliken, *Phys. Rev.*, 1928, **32**, 186.
5. J. E. Lennard-Jones, *Trans. Farad. Soc.*, 1929, **25**, 668.
6. R. Eisenschitz and F. London, *Z. Physik.*, 1930, **60**, 491.
7. C. Coulson, *Trans. Faraday Soc.*, 1938, **33**, 1479.
8. F. H. Westheimer and J. E. Mayer, *J. Chem. Phys.*, 1946, **14**, 733.

9. D. H. R. Barton, *Experimentia*, 1950, **6**, 316.
10. J. B. Henrickson, *J. Am. Chem. Soc.*, 1961, **83**, 4537.
11. J. A. Pople, D. P. Santry and G. A. Segal, *J. Chem. Phys.*, 1965, **43**, S129.
12. K. B. Wiberg, *J. Am. Chem. Soc.*, 1965, **87**, 1070.
13. C. K. Johnson, *Oak Ridge National Laboratory Report no. ORNL-3794*, 1965. http://www.ornl. gov/ortep/ortep.html.
14. W. J. Hehre, W. A. Lathan, R. Ditchfield, M. D. Newton and J. A. Pople, *Gaussian70*, Quantum Chemistry Program Exchange, Program No. 237, 1970.
15. S. Fernbach and A. H. Taub, in *Computers and Their Role in the Physical Sciences*, ed. S. Fernbach and A. H. Taub, Gordon and Breach, New York, 1970, p. 539.
16. K. A. Wiberg and R. H. Boyd, *J. Am. Chem. Soc.*, 1972, **94**, 8426.
17. F. H. Allen, O. Kannard, W. D. S. Motherwell, W. G. Town and D. G. Watson, *J. Chem. Doc.*, 1973, **13**, 211.
18. N. L. Allinger, *J. Am. Chem. Soc.*, 1973, **95**, 3893.
19. M. Levitt and A. Warshel, *Nature*, 1975, **94**, 253.
20. R. Carbó, L. Leyda and M. Arnau, *Int. J. Quantum Chem.*, 1980, **17**, 1185.
21. T. D. Kuntz, J. M. Blaney, S. J. Oatley, R. Langridge and T. E. Ferrin, *J. Mol. Biol.*, 1982, **161**, 269.
22. M. L. Connolly, *J. Appl. Cryst.*, 1983, **16**, 548.
23. P. J. Goodford, *J. Med. Chem.*, 1985, **28**, 849.
24. P. A. Bash, U. C. Singh, F. K. Brown, R. Langridge and P. A. Kollman, *Science*, 1987, **235**, 574.
25. R. S. Pearlman, *Chem. Des. Autom. New.*, 1987, **2**, 1.
26. M. von Itzstein, W.-Y. Wu, G. B. Kok, M. S. Pegg, J. C. Dyason, B. Jin, T. Van Phan, M. L. Smythe, H. F. White, S. W. Oliver, P. M. Colman, J. N. Varghese, D. M. Ryan, J. M. Woods, R. C. Bethell, V. J. Hotham, J. M. Cameron and C. R. Penn, *Nature*, 1993, **363**, 418.
27. J. J. P. Stewart, *J. Mol. Struct.-THEOCHEM*, 1997, **401**, 195.
28. D. E. Shaw, P. Maragakis, K. Lindorff-Larsen, S. Piana, R. O. Dror, M. P. Eastwood, J. A. Bank, J. M. Jumper, J. K. Salmon, Y. Shan and W. Wriggers, *Science*, 2010, **330**, 341.
29. A. R. Leach, *Molecular Modelling: Principles and Applications*, Pearson Education Ltd., Harlow, UK, 2nd Edition, 2001.
30. A. Szabo and N. S. Ostlund, *Modern Quantum Chemistry: Introduction to Advanced Electronic Structure Theory*, McGraw-Hill, New York, 1st (Revised) Edition, 1989.
31. Gaussian Approximations to 1s Slater Type Orbitals, from the Wolfram Demonstrations Project: http://demonstrations.wolfram.com/GaussianApproximationsTo1sSlaterTypeOrbitals.
32. J. S. Binkley, J. A. Pople and W. J. Hehre, *J. Am. Chem. Soc.*, 1980, **102**, 939.
33. M. J. S. Dewar, E. G. Zoebisch, E. F. Healy and J. J. P. Stewart, *J. Am. Chem. Soc.*, 1985, **107**, 3902.
34. J. J. P. Stewart, *J. Comp. Chem.*, 1989, **10**, 221.
35. P. Hohenberg and W. Kohn, *Phys. Rev.*, 1964, **136**, B864.
36. A. D. Becke, *Phys. Rev. A*, 1988, **38**, 3098.
37. C. Lee, W. Yang and R. G. Parr, *Phys. Rev. B*, 1988, **37**, 785.
38. C. J. Cramer, *Essentials of Computational Chemistry: Theories and Models*, John Wiley & Sons, Hoboken, NJ, 2013, p. 271.
39. N. L. Allinger, *J. Am. Chem. Soc.*, 1977, **99**, 8127.
40. T. Liljefors, K. Gundertofte, P. Norrby and I. Petterson, in *Computational Medicinal Chemistry for Drug Discovery*, ed. P. Bultinck, H. De Winter, W. Langenaeker and J. P. Tollenare, CRC Press, New York, 2003, p. 1.
41. R. S. Mulliken, *J. Chem. Phys.*, 1955, **23**, 1833.
42. J. Gasteiger and M. Marsili, *Tetrahedron*, 1980, **22**, 3219.

43. R. Bonaccorsi, A. Pullman, E. Scrocco and J. Tomasi, *Theoret. Chim. Acta.*, 1972, **24**, 51.
44. C. M. Brenerman and K. B. Wiberg, *J. Comput. Chem.*, 1990, **11**, 361.
45. P. Gaillard, P.-A. Carrupt, B. Testa and A. Boudon, *J. Comp. Aid. Mol. Des.*, 1994, **8**, 83.
46. A. C. Good and W. G. Richards, *J. Chem. Inf. Comput. Sci.*, 1993, **33**, 112.
47. A. Nicholls, G. B. McGaughey, R. P. Sheridan, A. C. Good, G. Warren, M. Mathieu, S. W. Muchmore, S. P. Brown, J. A. Grant, J. A. Haigh, N. Nevins, A. N. Jain and B. Kelley, *J. Med. Chem.*, 2010, **53**, 3862.
48. N. L. Allinger, *J. Am. Chem. Soc*, 1959, **81**, 5727.
49. E. W. Weisstein, Method of Steepest Descent, from MathWorld, a Wolfram Web Resource: http://mathworld.wolfram.com/MethodofSteepestDescent.html.
50. R. Fletcher and C. M. Reeves, *Comput. J.*, 1964, **7**, 149.
51. E. W. Weisstein, Newton's Method, from MathWorld, a Wolfram Web Resource: http://mathworld.wolfram.com/NewtonsMethod.html.
52. N. Metropolis and S. Ulam, *J. Amer. Stat. Assoc.*, 1949, **44**, 335.
53. B. J. Alder and T. E. Wainwright, *J. Chem. Phys.*, 1959, **31**, 459.
54. J. A. McCammon, B. R. Gelin and M. Karplus, *Nature*, 1977, **267**, 585.
55. P. Kollman, *Chem. Rev.*, 1993, **93**, 2395.
56. J. Åqvist, C. Medina and J. E. Samuelsson, *Protein Eng.*, 1994, **7**, 385.
57. J. Tirado-Rives and W. Jorgensen, *J. Med. Chem.*, 2006, **49**, 5880.
58. A. R. Katritzky, *Handbook of Heterocyclic Chemistry*, Pergamon, New York, 1985, pp. 47–50.
59. W. L. Jorgensen, J. Chandrasekhar, J. D. Madura, R. W. Impey and M. L. Klein, *J. Chem. Phys.*, 1983, **79**, 926.
60. F. H. Stillinger and A. Rahman, *J. Chem. Phys.*, 1974, **60**, 1545.
61. S. W. Rick, S. J. Stuart and B. J. Berne, *J. Chem. Phys.*, 1994, **101**, 6141.
62. L. Onsager, *J. Am. Chem. Soc.*, 1936, **58**, 1486.
63. S. Miertus, E. Scrocco and J. Tomasi, *Chem. Phys.*, 1981, **55**, 117.
64. M. Cossi, V. Barone, R. Cammi and J. Tomasi, *Chem. Phys. Lett.*, 1996, **255**, 327.
65. M. K. Gilson and B. Honig, *Proteins: Struct., Funct., Genet.*, 1988, **4**, 7.
66. W. C. Still, A. Tempczyk, R. C. Hawley and T. Hendrickson, *J. Am. Chem. Soc.*, 1990, **112**, 6127.
67. G. Nemethy and H. A. Scheraga, *J. Chem. Phys.*, 1962, **36**, 3382.
68. M. Waring, *Exp. Opin. Drug. Dis.*, 2010, **5**, 235.
69. A. Biela, F. Sielaff, F. Terwesten, A. Heine, T. Steinmetzer and G. Klebe, *J. Med. Chem.*, 2012, **55**, 6094.
70. A. Schön, N. Madani, A. B. Smith, J. M. Lalonde and E. Freire, *Chem. Bio. Drug. Des.*, 2011, **77**, 161.
71. R. Mannhold, G. I. Poda, C. Ostermann and I. V. Tetko, *J. Pharm. Sci.*, 2009, **98**, 861.
72. G. N. Ramachandran, C. Ramakrishnan and V. Sasisekharan, *J. Mol. Biol.*, 1963, **7**, 95.
73. A. R. Leach, K. Prout and D. P. Dolata, *J. Chem. Inf. Comp. Sci.*, 1990, **30**, 316.
74. K. D. Gibson and H. A. Scheraga, *J. Comp. Chem.*, 1987, **8**, 826.
75. Z. Q. Li and H. A. Scheraga, *Proc. Natl. Acad. Sci. U. S. A.*, 1987, **84**, 6611.
76. R. S. Judson, W. P. Jaeger, A. M. Treasurywala and M. L. Peterson, *J. Comp. Chem.*, 1993, **14**, 1407.
77. G. M. Crippen and T. F. Havel, in *Distance Geometry and Molecular Conformation*, Chemometrics Research Studies Series 15, John Wiley & Sons, New York, 1988.
78. P. R. Andrews, D. J. Craik and J. L. Martin, *J. Med. Chem.*, 1984, **27**, 1648.
79. J. J. Kaminski, B. Wallmark, C. Briving and B. M. Andersson, *J. Med. Chem.*, 1991, **34**, 533.
80. C. G. Wermuth, in *The practice of Medicinal Chemistry*, ed. C. G. Wermuth, Academic Press, London, 2nd edition, 2003, p. 219.

81. A. Mann, T. Boulanger, B. Brandau, F. Durant, G. Evrard, M. Heaulme, E. Desaulles and C. G. Wermuth, *J. Med. Chem.*, 1991, **34**, 1307.
82. F. H. Allen, *Acta Cryst.*, 2002, **B58**, 380.
83. I. J. Bruno, J. C. Cole, J. P. M. Lommerse, R. S. Rowland, R. Taylor and M. L. Verdonk, *J. Comput.-Aided Mol. Des.*, 1997, **11**, 525.
84. M. H. Abraham, P. P. Duce, D. V. Prior, D. G. Barratt, J. J. Morris and P. J. Taylor, *J. Chem. Soc., Perkin Trans.*, 1989, **2**, 1355.
85. J. A. Platts, *Phys. Chem. Chem. Phys.*, 2000, **2**, 973.
86. B. Kramer, M. Rarey and T. Lengauer, *Proteins: Struct., Funct., Bioinf.*, 1999, **37**, 228.
87. G. Jones, P. Willett, R. C. Glen, A. R. Leach and R. Taylor, *J. Mol. Biol.*, 1997, **267**, 727.
88. D. B. Kitchen, H. Decornez, J. R. Furr and J. Bajorath, *Nat. Rev. Drug. Disc.*, 2004, **3**, 935.
89. D. S. Goodsell, G. M. Morris and A. J. Olson, *J. Mol. Recog.*, 1996, **9**, 1.
90. P. A. Kollman, *J. Am. Chem. Soc.*, 1984, **106**, 795.
91. R. D. Clark, A. Strizhev, J. M. Leonard, J. F. Blake and J. B. Matthew, *J. Mol. Graph. Mod.*, 2002, **20**, 281.
92. H.-J. Böhm, *J. Comp.-Aid. Mol. Des.*, 1994, **8**, 243.
93. M. D. Eldridge, C. W. Murray, T. R. Auton, G. V. Paolini and R. P. Mee, *J. Comput. Aid. Mol. Des.*, 1997, **11**, 425.
94. I. Muegge and Y. C. Martin, *J. Med. Chem.*, 1999, **42**, 791.
95. H. Gohlke, M. Hendlich and G. Klebe, *J. Mol. Biol.*, 2000, **295**, 337.
96. R. S. DeWitte and E. I. Shakhnovich, *J. Am. Chem. Soc.*, 1996, **118**, 11733.
97. G. L. Warren, C. W. Andrews, A.-M. Capelli, B. Clarke, J. LaLonde, M. H. Lambert, M. Lindvall, N. Nevins, S. F. Semus, S. Senger, G. Tedesco, I. D. Wall, J. M. Woolven, C. E. Peishoff and M. S. Head, *J. Med. Chem.*, 2006, **49**, 5912.
98. R. Wang, L. Lai and S. Wang, *J. Comp.-Aid. Mol. Des.*, 2002, **16**, 11.
99. C. Merlot, *Drug Discovery Today*, 2010, **15**, 16.
100. P. Y. Lam, P. K. Jahdav, C. J. Eyermann, C. N. Hodge, Y. Ru, L. T. Bacheler, J. L. Meek, M. J. Otto, M. M. Rayner, Y. N. Wong, C.-H. Chang, P. C. Weber, D. A. Jackson, T. R. Sharpe and S. Erickson-Viitanen, *Science*, 1994, **263**, 380.
101. M. J. Bamford, C. Chan, A. P. Craven, B. W. Dymock, D. Green, R. A. Henson, B. E. Kirk, M. G. Lester, P. A. Procopiou, M. A. Snowden, S. J. Spooner, A. R. P. Srikantha, N. S. Watson and J. A. Widdowson, *J. Med. Chem.*, 1995, **38**, 3502.
102. R. C. Glen, G. R. Martin, A. P. Hill, R. M. Hyde, P. M. Woollard, J. A. Salmon, J. Buckingham and A. D. Robertson, *J. Med. Chem., J. Med. Chem.*, 1995, **38**, 3566.
103. G. R. Marshall, C. D. Barry, H. E. Bosshard, R. A. Dammkoehler and D. A. Dunn, The Conformational Parameter in Drug-Design: The Active Analogue Approach, Computer Assisted Drug Design: ACS Symposium series 112: American Chemical Society, Washington DC, 1979.
104. A. M. Davis, S. J. Teague and G. J. Kleywegt, *Angew. Chem., Int. Ed.*, 2003, **42**, 2718.

CHAPTER 4

Structure-Based Design for Medicinal Chemists

JEFF BLANEY*[a] AND ANDREW M. DAVIS[b]

[a] Small Molecule Drug Discovery, Genentech, 1 DNA Way, South San Francisco, CA 94080, USA; [b] AstraZeneca Respiratory and Inflammation Innovative Medicines, Pepparedsleden 1 Mölndal, 43183, Sweden
*E-mail: blaney.jeff@gene.com

4.1 INTRODUCTION

Medicinal chemistry is the most expensive part of pre-clinical research.[1] Hit-to-lead optimization to a clinical development candidate is an iterative, empirical process that frequently requires the design, synthesis, and testing of several thousand compounds by large research teams over several years. Identifying the best compounds as early as possible through efficient selection and optimization is critical. Understanding structure–activity relationships (SAR) is the key to designing new analogues with better properties. Structure-based design uses high-resolution X-ray co-crystal structures and models of analogue series members bound to the target protein to interpret SAR and help create new, testable design hypotheses.

Structure-based design is an essential part of medicinal chemistry. The availability of X-ray crystal structures for many drug discovery targets and improvements in molecular modeling software makes it practical for medicinal chemists to do their own modeling and design. X-ray crystal structures for most non-membrane proteins can be routinely accessible with sufficiently high resolution (≤ 2.5 Å). A well-supported project can deliver co-crystal structures within two weeks from request, which is fast enough to help drive iterative design. Structure-based design is used during hit identification, hit-to-lead optimization, and late-stage optimization. Its most obvious applications are for improving potency and selectivity, but it can also help guide later-stage optimization of pharmacokinetic, toxicity, solubility, or other challenges by modeling small molecule scaffold or substituent changes which are likely to retain activity.

We will focus on how we've seen structure-based design performed during many medicinal chemistry projects in industry. Structure-based design relies on a solid understanding of physical organic chemistry, iterative determination of X-ray co-crystal structures, accurate assays to

The Handbook of Medicinal Chemistry: Principles and Practice
Edited by Andrew Davis and Simon E Ward
© The Royal Society of Chemistry 2015
Published by the Royal Society of Chemistry, www.rsc.org

measure binding affinity, and experienced use of molecular modeling software, structure and physical property databases, and associated data-mining and analysis software. The most successful approaches evolved from trying to drive design primarily through computational approaches to focusing instead on iterative, hypothesis-based designs by expert scientists supplemented by computer modeling.[2] The recent review by Bissantz, Kuhn, and Stahl[3] and book chapters by Klebe[4,5] are excellent additional sources to learn about structure-based design.

Predicting the three-dimensional structures of proteins from their primary sequence (the protein folding problem) is neither reliable nor accurate enough for structure-based design. Predicting the structure of a specific target protein from the crystal structure of a highly similar protein (homology modeling), where the sequence identity is $\geq$ 30%, can provide structures that are helpful, but co-crystal structures of relevant ligands bound to the target protein are much more valuable.

Structure-based design can be started based on publicly available structures if you are lucky enough to have relevant structures with similar ligands already in the Protein Data Bank (PDB, www.rcsb.org)[6] but it will be difficult, if not impossible, to make progress without consistent, iterative feedback of new co-crystal structures determined with inhibitors you have designed and synthesized complexed with your protein. This requires a robust crystallographic system and is usually done by soaking pre-formed protein crystals in a solution containing the small molecule inhibitor. Co-crystallization is sometimes required and can be practical if the crystals grow quickly enough. Teague and Davis reviewed issues and recommendations in interpreting protein crystal structures for use in structure-based design.[7,8]

Computational methods are still inadequate to predict potency (related to affinity and binding free energy) with sufficient accuracy to drive medicinal chemistry in most cases, but they provide useful qualitative guidance. We usually rely on computational methods to eliminate designs that are unlikely to succeed, rather than to pick the top compounds. Free-energy perturbation (FEP) approaches have improved to the point where they can predict relative affinities of closely related analogues in some cases,[9] but these are very challenging calculations to perform and are not routinely applied yet in most industrial medicinal chemistry projects. We propose which molecules to synthesize next empirically, guided by observations of preferred non-covalent interactions and supplemented with qualitative docking and other calculations.

Why is predicting protein–ligand affinity so hard? The relationship between the Gibbs free energy and the binding constant shows that each ten-fold change in potency is only a 1.4 Kcal/mol (5.9 kJ/mol) change in binding free energy:

$$\Delta G = \Delta H - T\Delta S \tag{4.1}$$

$$\Delta G = -2.3 \ RT \log K = -1.4 \log K \tag{4.2}$$

Small changes in binding energy cause exponential changes in binding affinity. A medicinal chemist may be thrilled to improve potency by two-fold (0.4 Kcal/mol, 1.7 kJ/mol), but the current practical computational predictive accuracy in the best cases, for closely related analogues, is 10–100-fold (1.4–2.8 kcal/mol, 5.9–10.8 kJ/mol). Affinity does not correlate with the number of hydrogen bonds formed, the buried surface area of hydrophobic interactions, or many other simple first and second-order terms. X-ray crystallography provides high-resolution, time-averaged pictures of how molecules interact, but can't label interactions by their relative importance. For example, a shorter protein–ligand hydrogen bond doesn't necessarily provide higher affinity.

Structure-based design is done by proposing hypotheses based on observed SAR, co-crystal structures of relevant series with the target and related proteins, database searches and analysis, and modeling. A typical hypothesis would be to improve potency by modifying a known ligand to fill a pocket or make a specific interaction. We use X-ray crystallography, biochemical, and biophysical assays to test the hypothesis: did the compound bind as expected and achieve its structural design goal? If so, did it actually improve potency, selectivity, or other desired outcomes? Do unexpectedly active or inactive compounds bind like their close analogues?

4.2 HISTORY

Beddell *et al.* designed small molecules from scratch (*de novo* design) to bind to an allosteric site on haemoglobin in the first structure-based design publication in 1976.[10,11] They performed this impressive feat of *de novo* design with wire Kendrew models of haemoglobin's crystal structure! This is particularly amazing given how difficult it still is to design now with much higher resolution X-ray structures, vastly improved modeling tools, and much greater knowledge of protein–ligand interactions. Enthusiasm for *de novo* design peaked during the 1980s and early 1990s. However, *de novo* design was not reliable: most designs were inactive. We learned that it is relatively easy to design a molecule that fits into a binding site; the challenge is recognizing which apparently complementary molecules actually bind to their target and to rank-order their affinities. Predicting free energies of protein–ligand binding has proven to be far more difficult than most of us realized. We still cannot predict relative free energies of binding reliably and accurately enough to drive design and synthesis during hit-to-lead optimization.

The first protein–ligand docking paper was published in 1982.[12] Docking predicts the pose (the ligand's orientation in the binding site and its conformation) and relative affinity of compounds selected from a database of available or virtual small molecules. Many docking programs have been developed since then, but they all struggle to predict affinity (scoring).[13–17] Docking has had modest success in virtual screening and typically provides hits in the low μM (IC_{50}, K_i, K_d, *etc.*) affinity range, especially if guided with prior SAR knowledge of active molecules.[18,19] Docking is more successful at predicting poses and is commonly used to qualitatively test whether a proposed molecule binds as designed. Shape and feature-based methods are competitive with docking for virtual screening.[16,20] This is surprising since these methods usually ignore the protein and only include the ligand: this suggests there is more *exploitable* knowledge in the bound ligand structure than in the protein site. Swann *et al.* developed a shape/feature protocol focused solely on pose prediction that can provide a more reliable result when co-crystal structures of similar analogues are available during a hit-to-lead optimization project.[21]

The Protein Data Bank[6] has over 75 000 protein structures, of which about 50 000 are protein–ligand complexes. High-resolution (≤ 2.5 Å) X-ray crystal structures are now routinely available for most drug discovery targets, except for membrane-bound proteins (for example, GPCRs and ion channels). They are becoming tractable in many cases, but they are still extremely challenging.[22,23] The Cambridge Structural Database (CSD)[24] contains about 700 000 small molecule crystal structures and increases by about 40 000 molecules per year, with about 140 000 high-quality structures that are relevant for medicinal chemistry.[25] The PDB and CSD crystal structure databases have been studied extensively to understand the geometric preferences of non-covalent interactions[3,26–31] and small molecule conformational preferences.[25,32,33] Protein crystal structures typically contain about 40–60% solvent by volume (mostly water), while small molecule crystal structures are dry (some are solvates, with one to a few molecules of solvent per organic molecule). Small molecule crystal "packing forces" were previously believed to induce conformational strain on small molecules, but recent analysis of the much larger PDB and CSD databases shows that such cases are rare:[34] the CSD does in fact provide a reference set of preferred small molecule conformations (in particular, preferred torsion angles) that are relevant for structure-based design.[25] However, the small molecule X-ray crystal structure conformation is typically only one of many possible preferred conformations and is unlikely to represent the conformation bound to a drug target.

4.3 INTERPRETING X-RAY CRYSTAL STRUCTURES

Teague, Davis, Kleywegt and co-workers, through a number of reviews[7,8,35] reminded us that "An X-ray crystal structure is one crystallographer's subjective interpretation of an observed electron-density map expressed in terms of an atomic model." While the crystallographer will have done his best to fit the atomic structure into the observed electron density map, the atomic model will have regions of high or low confidence, depending upon the quality of the electron density in that region. But a chemist might interpret the atomic coordinates as if they are at perfect resolution, irrespective of the quality of the original electron density. The ambiguities within the model maybe at the level of atoms, residues, or even whole regions of the protein, and without forewarning these ambiguities can misguide design hypotheses. So how can a chemist protect himself from basing design hypotheses on ambiguous regions of the structure? Together with the coordinates, atoms in the model also include a "temperature factor" (also known as the B-factor or atomic displacement parameter), which models the effects of static and dynamic disorder in the crystal. B-factors provide very useful information about the relative reliability of different parts of the model. The isotropic temperature factor of an atom is related to its mean-square displacement. If they are high, *e.g.* for a lysine side chain, this usually means that little or no electron density was observed for the atoms in that side chain, and that the coordinates are therefore less reliable. Except at high resolution (typically, better than ~ 1.5 Å), where there are sufficient observations to warrant refinement of anisotropic temperature factors (requiring six parameters per atom), temperature factors are usually constrained to be isotropic (requiring only one parameter per atom). There are other ambiguities within the model that may be more difficult to identify, even considering the reliability of the model in that region of space. Examination of the atomic model together with its electron density minimizes the ambiguities, but does not remove them. Common ambiguities include the positions of N and O atoms in the side chains of glutamine and asparagine, since they are isoelectronic. Even the position of nitrogens and carbons in histidines, imidazoles and pyridines can be ambiguous. Since the presence of hydrogen atoms is inferred rather than observed, the tautomeric state of bound ligands cannot be determined directly. The protonation state of ionizable residues on the protein or ligand cannot be assigned directly from the structure, as the pK_as of ionizable residues can be drastically shifted from their aqueous values in the microenvironment of the protein. The crystallographer makes informed guesses in these cases when placing molecular features within the density, considering potential hydrogen bonds, but this decision is often made before the water molecules in the model have been added. Waters are particularly difficult. Unless at high resolution (typically ≤ 2.5 Å), the presence of a water molecule in a structure cannot be determined with certainty, and it becomes quite a subjective matter whether a feature in the electron density map is a water molecule or noise. Any small molecule also present within the protein structure poses different problems for the crystallographer. Whereas high quality dictionaries of bond lengths, angles and torsions are available for amino acids and nucleic acids, the same is not true for small molecules. This is critical because the electron density usually lacks sufficient information to completely determine the atomic positions, so a molecular mechanics force field is used to help refine both protein and small molecule atoms. If appropriate terms are lacking or wrong for the small molecule, the fit of these atoms into observed electron density may be wrong. Finally, protein flexibility is often underestimated, and the assumption of a rigid receptor may limit the pharmacophoric space that can be explored. Collaborate closely with your crystallographer, so the uncertainties in the structure are considered within your design hypotheses.

4.4 VISUALIZING SHAPE COMPLEMENTARITY

Focus design on shape complementarity between the binding site and ligand first. Most molecular graphics programs provide a variety of display options. The clearest way to view and interpret protein–ligand interactions is to display the stick model (bonds) of the ligand and the protein binding site, along with the transparent solvent-accessible surface[36] (also called the "interaction surface") of the binding site, for example, by selecting all protein atoms within ≤ 10 Å of any ligand atom. The solvent-accessible surface is calculated 1.4 Å beyond the van der Waals surface of the protein (1.4 Å is the radius of a water molecule if it is treated as a sphere, slightly smaller than the radius of carbon, nitrogen, and oxygen). The resulting smooth surface touches ligand atoms if they are in van der Waals contact. Ligand atoms intersect the surface if they make a hydrogen bond or ionic interaction; otherwise, intersection indicates a steric clash. If you observe this in an X-ray co-crystal structure, inspect the electron density map and consult with your crystallographer to determine whether the structure needs additional refinement. If you observe this in a model, the ligand is unlikely to bind as modeled unless either it or the site can "relax" and open up. The shape and size of unfilled pockets and larger regions in the site is much easier to see with this solvent-accessible surface than with the more familiar Connolly molecular surface.[36] Non-planar ligand features provide opportunities to improve solubility[37] and new vectors for design: solvent-accessible surfaces make it easier to spot where non-planar groups can fit. Figures 4.1 and 4.2 compare the solvent-accessible and Connolly surfaces of a chymotrypsin-inhibitor complex (PDB 3VGC). The solvent-accessible surface makes chymotrypsin's specificity for aromatic rings much more obvious. The Connolly surface is displayed at the actual surface of the protein, so that viewing the complementarity of a ligand in the site also requires displaying the surface of the ligand (or an unusually good three-dimensional imagination). Displaying both protein and ligand surfaces is straightforward in molecular modeling software, but visually interpreting the fit of the two complicated, three-dimensional surfaces is much more difficult than comparing the protein solvent-accessible surface with the stick model of the ligand.

3D stereoscopic display makes it much easier to visualize the shape of a binding site and interpret intermolecular interactions. All molecular modeling programs support a variety of 3D

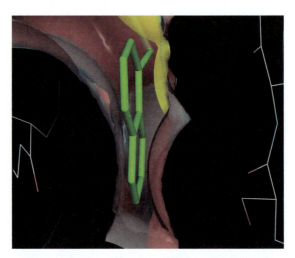

Figure 4.1   Solvent-accessible or interaction surface of a chymotrypsin-inhibitor complex (PDB 3VGC). The protein solvent-accessible surface is color-coded by simple properties: carbon (hydrophobic) = gray, hydrogen bond acceptor = pink, hydrogen bond donor = cyan, Asp/Glu negatively charged side chain oxygens = red, Arg/Lys/His positively charged side chain nitrogens = blue, Cys/Met sulfur = yellow. Divalent sulfur attached to an aromatic ring is "invisible" from a logP perspective (logP $PhSCH_3$ = 2.74, logP $PhCH_3$ = 2.73), but it is hydrophilic in an alkyl chain (log P $CH_3CH_2SCH_3$ = 1.54, log P $CH_3CH_2CH_3$ = 2.36[89]). All molecular graphics were done with PyMOL.[90]

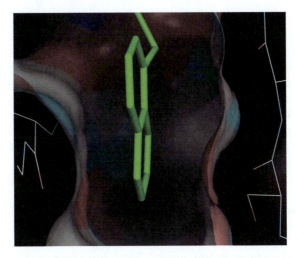

Figure 4.2 Connolly molecular surface of the same chymotrypsin-inhibitor complex.

display methods, including very effective options for laptops or computers with standard graphics cards. Anaglyph red/blue or magenta/cyan 3D work well with inexpensive plastic glasses. Many programs also support FPR (film pattern retarder) 3D, which requires an inexpensive FPR 3D monitor (available from LG and other companies) and uses cheap, passive polarized glasses (FPR 3D is also referred to as 'Zalman 3D', although Zalman no longer makes these monitors).

4.5 WHAT DRIVES BINDING?

Gas-phase non-covalent association is driven by polar and electrostatic interactions; van der Waals interactions are weak. The situation reverses in water, where hydrophobic interactions drive protein folding and intermolecular interactions. Fersht *et al.*[38] used site-directed mutagenesis to show that, in general, binding is driven by shape complementarity and hydrophobicity, while polar interactions drive specificity. Water is 55.5 M and dominates all aspects of aqueous non-covalent binding. Water is an excellent hydrogen bond donor and acceptor: a free small molecule and unbound protein site are solvated prior to binding. Water must be removed from the ligand and protein during binding. The desolvation penalty incurred by a highly polar substituent, such as a hydroxyl group, may cost more than the energy of its interaction in the protein–ligand complex. Polar interactions only improve potency if they improve upon the interactions between the free ligand and water, and the protein binding site and water: it's difficult to form better interactions than water.

High-resolution X-ray crystal structures identify many bound waters. The electron density map only "sees" the oxygen atom of the water; hydrogen atoms are not visible. Hydrogen bonds are inferred based on the hydrogen bond donors and acceptors between water and the site, water and the ligand, and between different water molecules. Some parts of a binding site are not optimal for binding water and waters in these regions will be easier to displace. Some waters are also tightly bound and very difficult to displace. A common problem is to decide whether to design a compound to interact with a bound water molecule or to displace the water. Crystallographic B-factors indicate the relative level of order and disorder for atoms and are sometimes mistakenly interpreted as a surrogate for water affinity, such that waters with low B-factors are deemed to bind more tightly and therefore harder to displace. Well-ordered waters may be trapped in an energetically suboptimal orientation and displaceable. For example, most kinase inhibitors bind to the classic kinase hinge with one to three hydrogen bonds, and most fragment hits bind to the hinge. Why do small molecules bind so frequently to the hinge? Computational analysis suggests that the waters which hydrogen bond to the hinge are thermodynamically unstable: they are trapped by the hydrophobic floor and roof of the planar binding site and cannot make their full set of possible hydrogen bonds,[39] so they are easily displaced by a ligand with the appropriate shape and hydrogen bond donor and acceptor geometry. Recent computational methods provide qualitative guidance about the relative thermodynamic stability of waters in a binding site[26,39–44] and can be used to help decide which waters to interact with or to displace.

Proteins are flexible and their binding sites may change conformation in response to different small molecule ligands, other interacting proteins, temperature, pH, and ionic strength. Ligand binding modes can change dramatically due to changes in crystal conditions. For example, co-crystal structures of trypsin with an inhibitor containing a weakly basic pyridine ($pK_a = 7.5$) were determined in two different crystal forms, at pH 7 and 8. The inhibitor binds with its pyridyl group in the S1 specificity pocket in one crystal form, but completely flips over end-for-end in the other crystal form and places its neutral chloronaphthyl group in the S1 pocket.[45]

A single X-ray crystal structure of a protein with an empty binding site (apo form) is not necessarily a good model for a bound conformation; the active or binding site may change conformation when a ligand binds. Flipped binding modes can occur between similar small molecules, whether they are weakly bound fragments ($K_i > 100$ μM) or tightly bound larger inhibitors ($K_i < 100$ nM). Rather than interpret such observations as evidence of unstable binding leading to confusing SAR, consider viewing these different binding modes as separate starting points for the design of a new series: in structure-based design, series membership should be viewed as a combination of the chemical series (*e.g.* scaffold) along with its binding mode. For example, adding a group to a ligand to fill a pocket in the site can lock in a binding mode, prevent subsequent flipping, and differentiate one binding mode from the other. See Figure 4.9, Section 4.12, for an example, which compares compounds **2** ($IC_{50} = 15$ nM) and **5** ($IC_{50} = 350$ nM) in the active site of iNOS.

4.6   ENTHALPY–ENTROPY COMPENSATION

Biochemical, biophysical, and cell biology assays measure a ligand's K_i or IC_{50} to estimate the Gibbs free energy of binding. Computational chemistry approaches usually estimate enthalpies of binding. X-ray crystal structures tell us about structure, but provide no information about binding enthalpies or free energies. The observation of an interaction tells us nothing about its strength.

Modern biophysical techniques can factor the free energy changes upon binding into their enthalpic and entropic contributions. In principle this provides a greater level of detail for structure-based design, although prospective case studies are lacking. Freire *et al.* analyzed a number of well-known structure-based lead-drug optimizations and suggested enthalpy-driven binders are more likely to make drugs.[46–48] This has prompted the experimental study of enthalpy and entropy within drug discovery programs, and suggested a paradigm of selecting enthalpic-driven binders as far back as the lead selection phase. But it is still unclear how to use enthalpy/entropy data to help drive iterative, structure-based drug design due to the complexity of enthalpy–entropy compensation.

The simple concept that hydrophobic binding is driven by entropy, and polar interactions such as hydrogen bonds are due to enthalpy, is not supported by an increasing number of studies. There are examples of entropically-driven hydrogen bonds, and enthalpically-driven hydrophobic interactions, and while these two parameters are mathematically independent, and potentially independently optimizable, the cooperative nature of protein–ligand binding resulting in enthalpy–entropy compensation complicates the original simple concept.

Klebe used isothermal calorimetry together with X-ray crystallography to analyze the SAR of a series of thrombin inhibitors.[50] Confusingly, cyclopentyl and cyclohexyl analogues, while very similar in ΔG, showed completely opposite behavior in ΔH and ΔS. The cyclopentyl analogue's binding was more entropically driven, while the cyclohexyl analogue's binding was enthalpically driven. The differences in ΔH and ΔS were large, but resulted in nearly identical overall free energy of binding for two analogues, which only differed by one methylene group. Close examination of the X-ray co-crystal structures showed well-resolved density for the cyclohexyl ring, but the cyclopentyl ring was disordered. While a retrospective rationalization can always be made, this level of subtlety is impossible to predict with our current level of understanding of drug–receptor interactions. Klebe also studied the binding of a set of congeneric inhibitors of thermolysin.[51] Through the series H, methyl, i-Propyl, n-Propyl, sec-butyl, i-butyl, and benzyl, the binding became increasingly entropic, although with substantial enthalpic compensation, so overall across the series the change in Gibbs free energy was small, compared to substantial changes in entropy or enthalpy. These changes in thermodynamic signatures were rationalized with changes observed in high-resolution X-ray structures. As they highlighted, the analyses worked at the limit of accuracy in current protein crystallography, and required considerations of details not usually considered in computational chemistry, simply because we are not aware of their importance in protein–ligand structure activity relationships. Klebe, Ladbury, and Freire have suggested that enthalpy measurements could even be used to select leads.[48] Using case studies of statins and HIV protease inhibitors as illustration, they indicate that best-in-class drugs may have enthalpy-driven rather than entropy-driven binding. While the properties of follower drugs depend on much more than affinity, enthalpy-driven potency may be just another indication of the importance of selecting compounds with high ligand lipophilic efficiencies, and a further indicator of compound quality.[49]

4.7 SMALL MOLECULES BIND IN THEIR LOWEST ENERGY, PREFERRED CONFORMATIONS

Small molecules rarely bind to proteins outside of their preferred minimum energy conformations. However, the literature is confused on this subject. Earlier work argued that a significant number of bound ligand conformations in the PDB were strained and had high energy, in some cases, even higher than the total free energy of binding (the total free energy of binding of a protein–ligand complex is $1.4 \times pK_i$, for example, a complex with a $K_i = 1$ nM has a free energy of binding of $1.4 \times 9 = 13.6$ kcal/mol or 56.9 kJ/mol). These studies suffered from ligand refinement errors in PDB X-ray co-crystal structures,[52] molecular mechanics force field issues, and other problems. More recent work with higher resolution, higher quality co-crystal structures and more careful computational analysis concludes that most ligands bind in one of their preferred, low-energy conformations, if not their minimum energy conformation.[53] A protein dissipates strain energy through small changes across its many degrees of freedom much more easily than for the ligand to adopt an unfavorable conformation by straining its few rotatable bonds.

"Soft" or low-energy torsion barriers are important: for example, anisoles ($ArOCH_2R$) and anilines (ArNHR) prefer coplanar conformations, while alkylaryls ($ArCH_2R$), arylsulfonamides and arylsulfones prefer a perpendicular conformation (Figure 4.3). These preferences must be considered and exploited when designing linkers between aryl rings; even small deviations are seldom observed and cost binding free energy. Linker atoms control both distance and direction. Brameld *et al.*[32] provide an excellent survey of conformational preferences for small molecule linkers and groups encountered in medicinal chemistry. Schärfer *et al.*[25] analyzed the

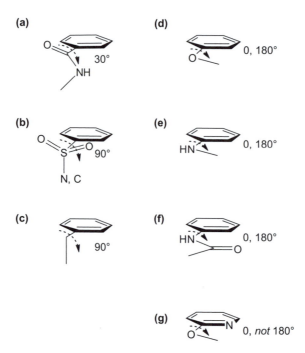

Figure 4.3 The most frequently observed torsions[32] in the CSD and PDB for (a) benzamides, (b) aryl-sulfonamides and arylsulfones, (c) arylalkyls, (d) anisoles, (e) anilines, (f) anilides, and (g) 2-alkoxypyridines, which have a very strong preference for a coplanar torsion with the alkyl group *syn* to the pyridine N, so that the pyridine N and alkoxy O lone pairs are *anti*.

torsion angle distributions from the CSD and PDB to develop an expert system complementary to the CSD's Mogul software.[24]

There are surprisingly large geometric differences between the substituent exit vectors on five-membered heterocycles. Replacing carbon, nitrogen, or oxygen with sulfur obviously alters the ring geometry and orientation of substituent exit vectors, but the various C, N, O heterocycles are typically presumed to be isosteric, with similar substituent vector orientations. Salonen *et al.*[54] probed the SAR of Factor Xa inhibitors in the S1 pocket by comparing three heterocycle linkers in inhibitor **1** (Figure 4.4): 3,5-disubstituted isoxazole **1a**, 2,4-disubstituted oxazole **1b**, and 2,5-disubstituted oxazole **1c**. Oxazoles **1b** (146 nM K_i) and **1c** (1620 nM K_i) were much less potent inhibitors than isoxazole **1a** (9 nM K_i). X-ray co-crystal structures determined at 1.25–1.33 Å resolution revealed that the oxazoles **1b** and **1c** flipped over 180° relative to the isoxazole **1a**. Searching the CSD showed surprising differences between the substituent vectors in these three heterocycles (Figure 4.4). The 15- to 180-fold lower activity of **1b** and **1c** is likely due to a combination of the different steric requirements (due to the different exit vector angles) of the oxazoles *vs.* the parent isoxazole, different preferred conformations of the heterocycle linker relative to the main inhibitor scaffold, and also their different dipole vectors. **1a**, **1b**, and **1c** stack face-to-face against peptide bonds lining the Factor Xa S1 pocket. Harder *et al.*[27] studied the face-to-face packing of aromatic heterocycles to peptide bonds and concluded the optimal interacting dipole vectors should be antiparallel, consistent with the improved activity of **1b** *vs.* **1c**.

4.8 PREFERRED PROTEIN–LIGAND INTERACTIONS

Conquest, Relibase, Isostar, and Superstar programs from the CSD[24] are the best available tools for studying intermolecular interactions in X-ray crystal structures. Conquest performs 2D

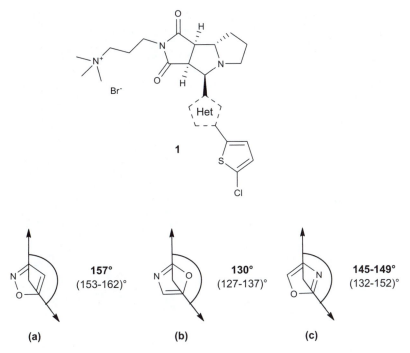

Figure 4.4 CSD search results from Salonen *et al.*[54] for the angle between the two substituent vectors of isoxazole **1a**, and oxazoles **1b** and **1c** in Factor Xa inhibitor **1**. The most frequently observed angles are shown in **bold** and the range of angles is shown in parentheses. These angular preferences are reproduced by MOE[91] with the MMFF94 force field.[92,93]

substructure searches of the CSD with 3D geometric constraints. Relibase performs similar searches for the PDB. IsoStar and SuperStar summarize preferred interactions and their distributions from the CSD and PDB for about 300 functional groups with 48 different contact groups. We use ConQuest and Relibase to search for interactions similar to those observed in co-crystal structures of protein–ligand complexes during structure-based design projects; these searches can provide ideas for how to optimize the ligand.

The comprehensive structure-based design review from Bissantz, Kuhn, and Stahl[3] provides detailed examples with preferred geometries for the common classes of protein–ligand interactions: hydrogen bonds OH···O=C, OH···N(=)-, NH···O=C, NH···N(=)-, halogen bonds, multipolar interactions C=O– amide, C–F–O=C (peptide backbone), cation-Pi, hydrophobic contacts, aromatic face-to-face (also analyzed more recently by Huber *et al.*[55]), aromatic edge-to-face, aromatic ring face-to-face with amide, and sulfur–aryl. Kuhn *et al.*[26] took this qualitative study another step to develop ViewContacts,[56] a software tool to annotate protein–ligand complexes with each of these interaction types and also highlight favorable and unfavorable interactions (Figures 4.5 and 4.6).

Neutral hydrogen bond distances range from 2.7–3.2 Å, while charged hydrogen bonds with carboxylates range from 2.6–3.0 Å. The hydrogen bond angle donor–hydrogen–acceptor is most frequently observed >150°, with the hydrogen bond oriented to approach the lone pair of the acceptor. Sulfonamide and sulfone oxygens are weak hydrogen bond acceptors and prefer to orient their hydrogen bonds linearly along the S=O bond. Hydrogen bonds to acceptors in aromatic rings or carbonyl groups lie within 30° of the plane of the aromatic ring or carbonyl group. Aromatic hydrogens (ArH) in aromatic rings with electron-withdrawing groups or in aromatic heterocycles are weak hydrogen bond donors; ArH hydrogen bonds are frequently observed between kinase inhibitors and one of the kinase hinge backbone carbonyl oxygens.

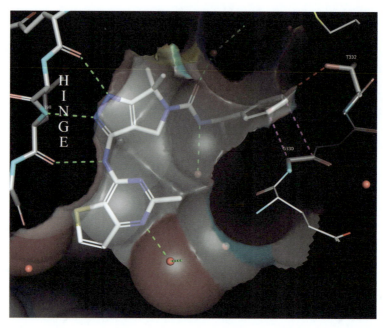

Figure 4.5 PAK4 kinase-inhibitor complex (PDB 2X4Z) interactions highlighted by ViewContacts:[56] three hydrogen bonds (green) between the kinase hinge and an inhibitor, an aromatic face-to-face interaction (magenta) with the peptide backbone, and a probable repulsive interaction (red) between a phenyl ring and the hydroxyl of Thr-332.

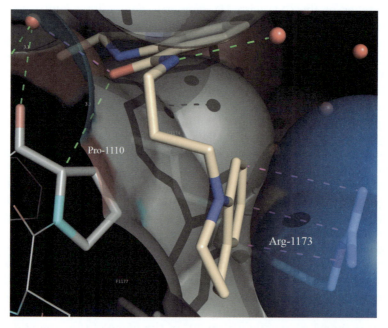

Figure 4.6 Bromodomain CREBBP-inhibitor complex (PDB 4NYW) interactions highlighted by View-Contacts[56] show a hydrogen bond (green) between the inhibitor's amide carbonyl oxygen and the alpha carbon proton of Pro-1110, and an aromatic face-to-face interaction (magenta) with the guanidinium group of Arg-1173.

ArH hydrogen bond lengths vary from about 3.1–3.6 Å. The hydrogen on the alpha carbon of the peptide backbone can also be a weak hydrogen bond donor (Figure 4.6).

Halogen bonds occur between aromatic chlorine, bromine, and iodine substituents in aromatic rings with electron-withdrawing groups, and a carbonyl oxygen from the protein backbone, Asn, Gln, Asp, or Glu side chains. Halogen bonds are counterintuitive: instead of the halogen behaving as though it has a partial negative charge, it acts as a surrogate hydrogen-bond donor. This is due to the "sigma-hole" at the end of the bond between the aromatic carbon and the halogen, which creates a partial positive charge on the halogen for on-axis interactions.[57] Increasing the electron-withdrawing power of substituents on the halogen's aromatic ring (*e.g.* increasing their Hammett sigma constant) enhances the halogen bond. The halogen-oxygen distance is much shorter than the normal van der Waals contact distance and becomes progressively shorter as the halogen gets larger. The preferred halogen bond C–X–O angle is linear. Halogen bond interaction strengths also increase with the size of the halogen. Halogen bonds are frequently observed, but very challenging to design. Hardegger *et al.* designed halogen bonds for inhibitors of MEK1 kinase and cathepsin L, where they compared ArX interacting with a backbone carbonyl oxygen for X = H, F, Cl, Br, and I.[58] Halogen bonds for Cl, Br, and I improved IC_{50} relative to H by 2–13-fold, 7–24-fold, and 26–45-fold for MEK1 and cathepsin L, respectively. Fluorine does not form halogen bonds, but can interact orthogonally with the carbonyl carbons of amides, where the C–F bond is oriented perpendicular to the plane of the amide bond and oriented such that the F contacts the carbonyl carbon with a range of 3.0–3.7 Å. This orthogonal dipolar C–F$\cdots$C=O interaction can contribute several-fold to binding affinity. Aromatic rings and peptide bonds pack against each other with T-shaped, edge-to-face or parallel-displaced, face-to-face interactions (Figure 4.5).[27,31] Aromatic rings also make face-to-face interactions with the guanidinium group of arginine (Figure 4.6).

4.9 HYDROGEN BONDS

Affinity does not correlate with the number of hydrogen bonds. The goal is to identify which observed or potential hydrogen bonds are most important and try to optimize them. Individual protein–ligand hydrogen bonds contribute zero to about 3 kcal/mol (13 kJ/mol) to binding free energy.[59] Quantitative prediction of hydrogen bonding strength is still extremely difficult; empirical rules based on geometry and network coordination through software such as View-Contacts[56] provide qualitative help.[26] The CSD software tools IsoStar and SuperStar provide useful tools for retrieving and analyzing specific geometric preferences and distributions for different hydrogen bond donor–acceptor interactions.[24]

What other clues can we use to design and optimize hydrogen bonding? Bissantz *et al.* suggested that hydrogen bonding is energetically asymmetric: donors are more important to target than acceptors.[3] This is based on affinity data from several protein–ligand systems showing that not satisfying a donor (with an acceptor from an inhibitor or water) costs more than ignoring an acceptor, and analysis of PDB kinase-inhibitor complexes which also shows a strong preference for satisfying the kinase hinge's central backbone NH with an acceptor, but that many inhibitors ignore one of the hinge's backbone carbonyl oxygens, leaving it desolvated. This is also qualitatively consistent with the asymmetry of Lipinski's "Rule of 5", which allows up to 10 acceptors, but only five donors.

Laurence *et al.* developed the experimental pK(BHX) scale to measure the relative strengths of over 1100 different hydrogen bond acceptors (HBA).[60] This is the most self-consistent and comprehensive database available of relative hydrogen bond acceptor strengths. It is an excellent reference for choosing HBA groups in structure-based design. Bissantz *et al.* found a good agreement between pK(BHX) acceptor strength and the frequency of hydrogen bonds to OH groups observed in the CSD.[3] Hydrogen bond-accepting ability does not correlate with proton basicity. HBA strengths tend to decrease in the order amides/sulfoxides/pyridines/ N-methylimidazoles > ketones > esters/alcohols > sulfones/sulfonamides/ethers, but there are large variations depending on cyclization and substituents. Some of the pK(BHX) results seem counter-intuitive. For example, sulfonamide and sulfones are very hydrophilic based on their π values and might also seem similar to amides and ketones, but they are weak hydrogen bond acceptors on the pK(BHX) scale. This is consistent with the observation that only about 30% of the sulfonamides in the CSD or PDB form hydrogen bonds, but 75–80% of them are in close contact with an aliphatic carbon.[3] Green and Popelier recently published a quantum chemical approach to calculate pK(BHX), which should be practical to predict values for those groups missing from the database.[61]

4.10   ELECTROSTATICS

The detailed electrostatics inside a binding site is poorly understood. The local dielectric constant inside a protein binding site is believed to vary from two to four, and increases to bulk water's dielectric (80) outside of the protein.[62] Mutagenesis experiments by Fersht *et al.*[63] showed that the pK_a of side chains in binding sites could vary over 1 pK_a unit due to per-turbations over 10 Å away, which suggests that charged groups at similar distances from a bound ligand could also influence its binding and selectivity. Long-range electrostatics are clearly significant, but are seldom deliberately exploited in structure-based design, where we typically focus on simpler direct and water-mediated protein–ligand contacts. Tidor described an elegant approach to designing inhibitors that includes long-range electrostatics.[64,65] Why does the classic "fluorine walk" work sometimes? Fluorine causes the maximum electronic perturbation with minimal steric change and can empirically probe the local dielectric en-vironment in a binding site.

4.11   HYPOTHESIS-BASED DESIGN

Most structure-based design is iterative, based on co-crystal structures and SAR from previous molecules. We recommend the following guidelines for approaching a structure-based design program. Review all available high quality co-crystal structures and models for your target and similar proteins by aligning them based on structurally conserved regions of their active or binding sites, and displaying them with consistent visualization features (surface, colour-coding, annotation, *etc.*), such that you can easily view and compare each structure super-imposed on one or more other structures. Any observed binding site conformation is a viable target for inhibitor design. Look for conserved features in structures, including bound or computationally predicted waters that may differentiate the target from off-targets, and co-operative motions or other conformational changes, which might be exploited during design. Search the CSD and PDB for molecules that make similar interactions to those observed in your co-crystal structures or models; matches in these databases can provide valuable clues to how to improve existing interactions and also identify similar groups that might also fit your site. Focus on shape first: optimal ligands match the binding site's solvent-accessible (interaction surface) very closely, with few unfilled areas or pockets. Shape-based methods[20,66] find molecules with similar shape to bound ligands can also provide good ideas for "scaffold-hopping": swapping cores or other features of lead molecules to identify new compounds with improved properties.

   As your design-synthesize-test cycle begins, consider each new design as a probe to interro-gate the protein: a well-designed molecule will provide useful feedback even if it is less potent than its reference analogue. Each design should test a specific hypothesis, for example, to fill a pocket, stabilize a conformation, or make a hydrogen bond. Designs should also test protein flexibility by pushing into pockets to learn whether they can expand. Long timescale (milli-second) molecular dynamics simulations are beginning to access significant conformational change,[67–69] but it is still not practical to predict conformational change of a protein in response to ligand binding. An X-ray co-crystal structure can determine whether the structural design objective was met. As the SAR within a chemical series develop, and confidence in the under-standing gained from the structural information grows, compounds whose biological data appear as outliers within the SAR should ideally be followed up by crystallography. These outliers can sometimes identify new binding poses that could lead to improved designs. Structure-based design considerations are part of the medicinal chemistry multiparameter optimization process and need to be balanced against the many other terms covered in Chapter 21 on Lead Optimization.

4.11.1 Polar Interactions

Hydrogen bonds do not necessarily improve binding free energy, but they are critical for spe-cificity. Hydrogen bond donor/acceptor mismatches are usually not tolerated and will cost substantial binding free energy; you might be able to exploit a mismatch with a related off-target protein to increase selectivity. Burying a polar group in a hydrophobic region or *vice-versa* will also hurt binding. However, remember the goal is not to make as many hydrogen bonds or polar interactions as possible, but rather to attempt to identify critical opportunities for polar interactions and optimize them. The observation of Bissantz *et al.* that hydrogen bonding is energetically asymmetric suggests that design should focus first on satisfying hydrogen bond donors.[3] The pK(BHX) scale is valuable for selecting and comparing potential ligand hydrogen bond-accepting groups.[60]

Compounds that are ionized at physiological pH are likely to reduce membrane permeability, so it's usually best to minimize the overall charge of the small molecule to zero or one, avoid strong acids and bases (pK$_a$s < 3 or > 10, respectively), and to focus on using neutral sub-stituents (or at least those with pK$_a$ close to 7.4) to interact with charged side chains on the protein. Where part of your ligand points to solvent, this may be a good region to incorporate a solubilizing group, while not compromising potency.

4.11.2 Interactions at the Entrance to a Binding Site

Surprising changes in SAR are occasionally observed at the entrance to a binding site (the solvent front), where ligand atoms contact protein atoms in a highly exposed region. These interactions usually have little effect on affinity, due to the groups in this region being highly solvated and flexible. However, protein–ligand interactions at the solvent front can be im-portant. Two different groups independently discovered a strong enzyme-inhibitor interaction at the surface of Hepatitis C Viral Polymerase (HCVP). Antonysamy *et al.* optimized a weak 14 µM K$_d$ fragment screening hit using parallel synthesis with small libraries to a 0.46 µM K$_d$ lead compound.[70] The most potent members of the series had a tertiary carboxamide that bound through a face-to-face interaction between the plane of the amide and the imidazole of His-475 on the protein's surface (Figure 4.7). His-475 accepts a hydrogen bond from the ad-jacent Ser-476 OH and donates a hydrogen bond to the backbone carbonyl oxygen of Asp-375. His-475 and Ser-476 are rigid across the many co-crystal structures determined during this project. This face-to-face interaction appears to have been responsible for improving K$_d$ by 30–70-fold. Li *et al.* also discovered a similar face-to-face interaction with an aromatic, fused ring heterocycle from a different chemical series binding at the same site.[71]

There is surprisingly little literature on protein–ligand interactions at the solvent front, but this example shows that the same intermolecular interactions that contribute to affinity in the

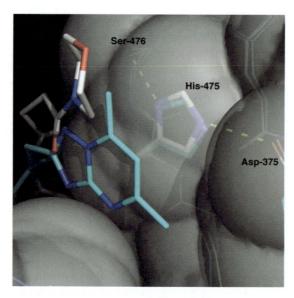

Figure 4.7   The white HCVp inhibitor's tertiary amide[70] makes a face-to-face interaction with His-475 on the surface of HCVp (PDB 3CJ5). The cyan inhibitor's aromatic heterocycle[71] makes a similar face-to-face interaction (PDB 3FRZ). Dashed yellow lines indicate hydrogen bonds between His-475, the adjacent Ser-476, and the backbone carbonyl oxygen of Asp-375. The solvent-accessible surface of the protein is light gray.

interior of a binding site can also make similar contributions on the protein surface. Face-to-face interactions between planar, polarizable groups such as His, Trp, and amide and aromatic rings of inhibitors may be good design targets for interactions at the solvent front: these groups are apolar perpendicular to the plane of their aromatic rings or amide and presumably are poorly solvated. The rigidity of His-475 and Ser-476 may also be a critical factor, as protein side chains on the surface are usually very flexible and frequently disordered. Understanding how to design interactions at the solvent front will improve our ability to design inhibitors for protein–protein interactions.

4.11.3   Self-Fulfilling Prophecy: The Local Minimum Problem

Molecular modeling programs for structure-based design provide features to build, modify, and energy-minimize small molecules within the protein binding site, along with interactive graphics visualization. They include options to constrain the entire protein or the protein outside of the binding site. We recommend the latter in most cases, since allowing limited flexibility in the site can alleviate steric clashes, optimize interactions, and simulate the small conformational changes in the protein backbone, side chains, and ligand that are frequently observed between co-crystal structures of closely related inhibitors. A skilled modeler can usually fit any designed ligand into a binding site using the powerful features in the top modeling programs, which can lead to a self-fulfilling prophecy: designs need to be tested to ensure their designed binding mode is in one of the ligand's preferred conformations. Energy-minimizing the protein–inhibitor complex finds the nearest local minimum conformation of the inhibitor in the site, which may have higher energy than the free inhibitor's preferred conformations. The modeled, bound inhibitor conformation should always be compared to the free inhibitor's preferred conformations, identified by thorough conformational search of the inhibitor without the protein. Start by taking by the modeled, bound inhibitor conformation and minimize its energy without the protein: if its conformation changes substantially, the bound conformation is high-energy and should be modified. Even if this simple test succeeds, you should perform the complete conformational search to ensure that your designed, bound conformation is close to a preferred, low-energy conformation. How close? Bound conformations seldom deviate significantly from preferred, low-energy conformations of the free small molecule, but there is no hard acceptance threshold we can provide for either the geometric (*e.g.* RMSD) or energetic difference, due to differences in force fields, uncertainty in crystal structures, and other factors mentioned in Section 4.3. We recommend as a general guideline that the bound conformation should be within 5 kcal/mol (21 kJ/mol) of the lowest energy conformations found in the conformational search, and that its torsion angles should vary by no more than 10° from the closest conformer. The inhibitor's modeled torsion angles should also be compared with the observed distribution of similar torsion angles in the CSD and PDB using software such as the CSD program Mogul[24,72] and/or the torsion analyzer from Schärfer *et al.*[25]

4.12   CASE STUDY: NITRIC OXIDE SYNTHASE

Structure-based design of inducible nitric oxide (iNOS) synthase inhibitors highlights several key points in this chapter. Nitric oxide synthases (NOS) catalyze the conversion of arginine to nitric oxide (NO) and citrulline. The released nitric oxide is an important signaling and cytotoxic agent. Neuronal NOS (nNOS) and endothelial NOS (eNOS) are constitutively expressed and play important roles in gastrointestinal (GI) motility, memory, and vascular tone, respectively. iNOS is induced by a variety of cells at sites of inflammation and can produce large amounts of NO. The uncontrolled production of NO is thought to contribute to the pathology of a number of inflammatory conditions, which makes iNOS a very interesting anti-inflammatory target. The three NOS isoforms have highly conserved binding sites and very similar structures, so gaining selectivity for iNOS was likely to prove difficult.

Most of the known iNOS inhibitors contained a basic amidine that mimics the substrate arginine's guanidinium group by stacking over the active-site heme and forming a charged, bidentate interaction with the carboxylate side chain of an invariant glutamate, Glu-377. AstraZeneca scientists optimized several amidine series: quinazolines, aminopyridines, or bicyclic thienoxazepines (Figure 4.8).[73,74]

The program was supported by iterative crystallography, with over 70 co-crystal structures determined. Potent inhibitors were synthesized, but all suffered from poor pharmacokinetics. Chemists optimized potency by growing into the iNOS pocket, building in hydrophobicity and attempting to make new polar interactions, for example by interacting with residues important for binding the arginine substrate. Potency was improved and 10 nM IC$_{50}$ iNOS inhibitors became common. Through this process, computational chemistry was able to rule out medicinal chemistry designs that didn't fit the active site, but was rarely able to confidently predict potency within the series. However, detailed structural investigations driven by a close dialog between medicinal chemists, crystallographers, and computational chemists were critically important in unraveling SAR and guiding design hypotheses. As the iNOS active-site pocket was filled, SAR began to identify iNOS inhibitors with selectivity over eNOS and nNOS (Figure 4.8). X-ray structures showed that iNOS-selective inhibitors displaced an invariant glutamine residue, Gln-263, and opened a new pocket (Figure 4.9).

Gln-263, although common to all three NOS isoforms, provided a valuable source of selectivity.[73,74] Its movement induced further movements up to 20 Å away and opened a new iNOS selectivity pocket; sequence differences outside the active site were thought to be the likely cause of selectivity. Gln-263 adopted different conformations, depending upon the ligand. Some conformations introduced n/e NOS selectivity, but at the expense of iNOS activity, while other glutamine conformations increased iNOS potency and increased selectivity over n/e NOS.

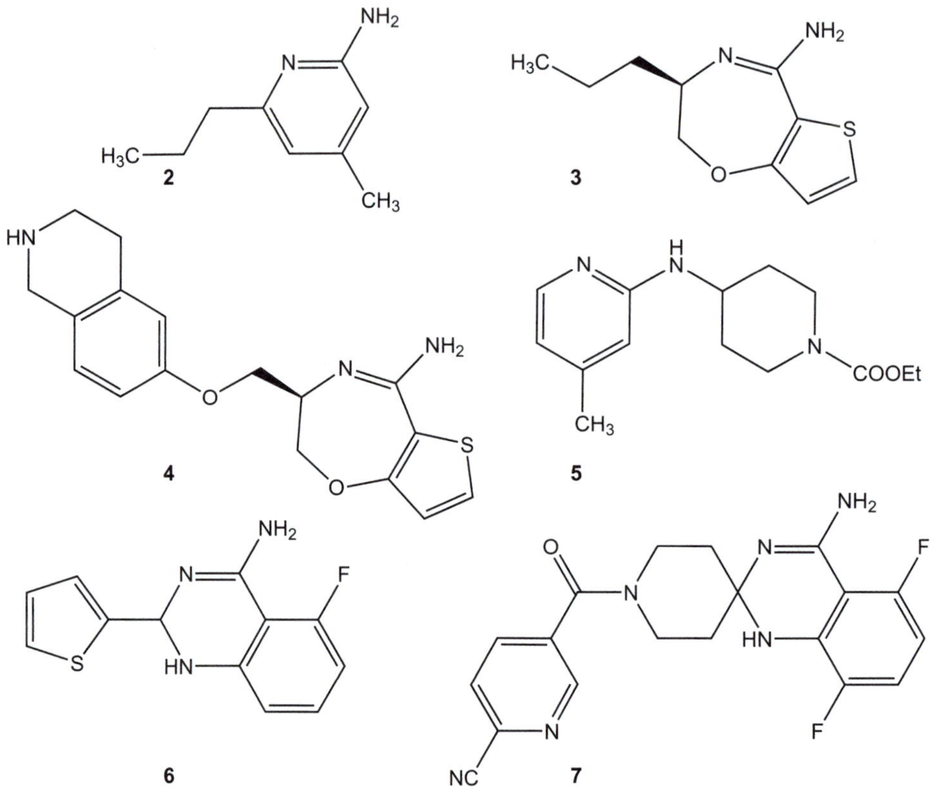

Figure 4.8 Part of the iNOS inhibitor series studied by AstraZeneca, their iNOS IC$_{50}$s, selectivity *vs.* eNOS and nNOS, and PDB codes for their X-ray co-crystal structures.[73]

	iNOS IC$_{50}(\mu M)$	Selectivity		PDB
		eNOS/ iNOS	nNOS/ iNOS	
2	0.015	2.3	1.5	3E6N
3	0.04	5	1.2	3EAH
4	0.4	125	2.5	3EBF
5	0.35	166	57	3E7G
6	0.01	250	100	3E7I
7	0.035	>2860	23	3E7T

Selectivity through perturbation of the invariant Gln-263 residue proved to be a very valuable design lesson, and the "Gln pocket" became design dogma. However, even with potent and selective inhibitors, pharmacokinetic limitations prevented further optimization of the amidine or aminopyridine series. A different starting point was required.

The project team had been aware of a number of published, non-basic iNOS inhibitors[75] similar to compound **8** (Figures 4.10 and 4.11), but they were weak, lacked the charged, bidentate interaction with Glu-377, and were therefore considered non-specific inhibitors since they did not fit the project SAR dogma. They were overlooked for many months, but out of desperation for a new starting point the team decided to use iterative X-ray crystallography to

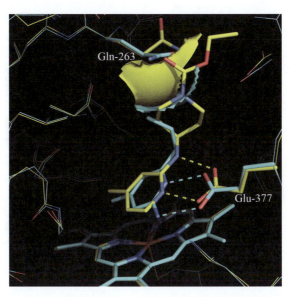

Figure 4.9 Non-selective aminopyridine **2** (cyan, PDB 3E6N, mouse iNOS) makes a charged, bidentate interaction (dashed cyan lines) between its amidine and Glu-377. The 2-aminopyridine of a selective analogue, **5** (yellow, PDB 3E7G, human iNOS), flips over 180°, but still makes a similar bidentate interaction (dashed yellow lines) with Glu-377. The N-carboxyethylpiperidine group of **5** displaces Gln-263, which rotates its side chain to a different conformation and opens a new pocket, shown by comparing the solvent-accessible surfaces of the side chains of the two Gln-263 conformers.

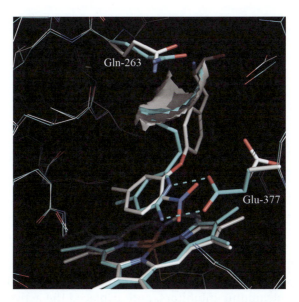

Figure 4.10 The X-ray co-crystal structure of compound **8** (white PDB 4UX6) in mouse iNOS reveals that it stacks its neutral aromatic ring over the heme,[74] similar to the basic amidine **2** (cyan, PDB 3E6N, mouse iNOS),[73] and displaces Glu-377 from its position observed with amidines. The solvent-accessible surfaces of the sidechains of Gln-263 are similar. Both compounds have little selectivity.

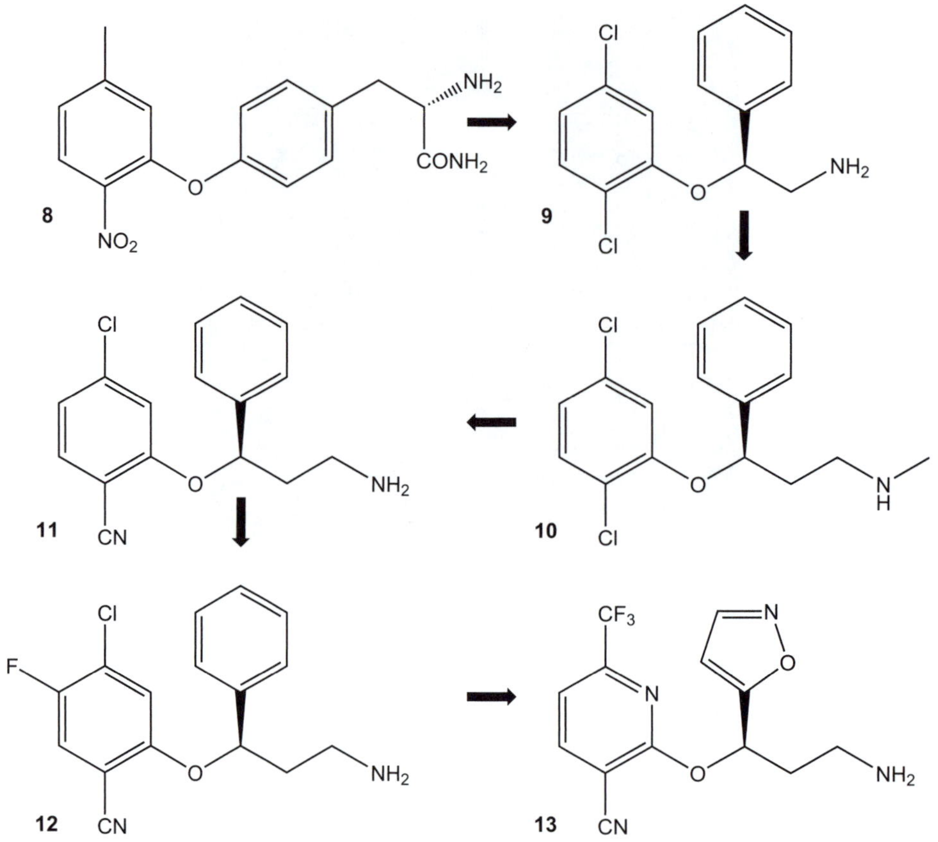

		Selectivity		
	iNOS IC$_{50}(\mu M)$	eNOS/ iNOS	nNOS/ iNOS	PDB
8	1.2			
9	1.9	>533	52	
10	0.022	>4500	210	
11	0.004	>10^4	>270	2Y37
12	0.006	>10^4	35	
13	0.005	>10^4	1.3	

Figure 4.11 Optimization of iNOS inhibitors proceeded rapidly to exploit the "Gln pocket" for potency and selectivity.[74]

explore the surprising SAR of these weak inhibitors. The new X-ray crystal structures were an eye-opener.[74] They showed that an aromatic ring in the non-basic inhibitors stacked over the heme, similar to the basic amidine inhibitors, and displaced the active site Glu-377 instead of interacting with it (Figure 4.10).

This was a turning point for the project. These inhibitors were no longer dismissed as weak, non-specific starting points; they were now a priority opportunity for a new series. Design progressed through a sequence of hybrid structures using previously learned SAR, starting with

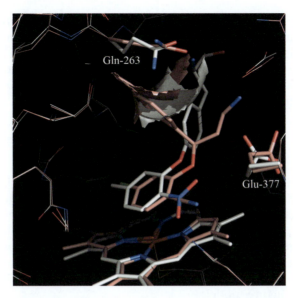

Figure 4.12   X-ray co-crystal structures of poorly selective compound **8** (white) and highly selective compound **11** (pink, PDB 2Y37) in mouse iNOS, showing the different Gln-263 residue positions influenced by the Gln-pocket substituent.[74] Solvent-accessible surfaces for the side chain of Gln-263 highlight the extra volume opened up by compound **11**.

the neutral heme binder, displacing the glutamate, reaching for the arginine's amino acid binding site initially with a zwitterion, building into the glutamine pocket for selectivity, and then replacing the zwitterionic amino acid with a basic group to improve cell permeability (Figures 4.11 and 4.12). Chemistry focused on neutral heme binders that improved potency while decreasing log P to improve ADMET properties. But metabolism was still a problem for this new chemical series. Metabolite identification showed that oxidation *para* to the ether oxygen on the aromatic heme binder was a particular metabolic soft spot. Analysis of the X-ray crystal structure suggested there was enough space to block metabolism by fluorination. Unsurprisingly, N-demethylation of compound **10** was also observed, but the resulting primary amine was still acceptable for potency and selectivity. Through this sequence, the original μM inhibitors were optimized to nM potency with high n/e NOS selectivity and good *in vivo* stability.[74] Drug discovery is seldom this straightforward, and other problems remained to be solved. The series showed some binding to noradrenaline and serotonin transporters, and also inhibited CYP2D6. Additional structure-based design learning led to the removal of these liabilities. The design strategy was to further reduce lipophilicity, while maintaining the important enzyme-inhibitor interactions already identified. The interactions with the heme and Gln pocket were considered to be lipophilic, and so it was felt unlikely that polarity could be introduced successfully in these regions. But while heterocycles are considerably more polar than phenyl rings, they are only polar around their edges, apolar perpendicular to the ring plane, and can form face-to-face stacking interactions. Hence, replacing the phenyl ring stacking over the heme (compounds **8–12**) with a pyridine (compound **13**), and replacing the "Gln pocket" phenyl with isoxazole, resulted in reduced overall lipophilicity and the removal all three off-target liabilities, while maintaining potent iNOS inhibition. The project continued to use these design lessons to further develop the series.[74]

This case study reinforces key points in this chapter. Computational chemistry could not predict potency, but was used extensively to filter design lists for compounds that were unlikely to bind. Amino acid changes distal from the active site had profound effects on potency and selectivity. Determining X-ray co-crystal structures of compounds with unusual SAR, particularly those that challenged the pharmacophore dogma within the project, provided key insights and new pharmacophores. Protein flexibility and induced fit were critical to potency and selectivity, and led to the discovery of a completely new pharmacophore. Finally, combining the structural information with consideration of physicochemical properties and ADMET properties underpinned the whole design strategy.

HINTS AND TIPS

The clearest way to view and interpret protein–ligand interactions is to display the stick model (bonds) of the ligand and the protein-binding site, along with the transparent solvent-accessible surface. Ligand atoms intersect the surface if they make a hydrogen bond or ionic interaction; otherwise, intersection indicates a steric clash.

While the crystallographer will have done his/her best to fit the atomic structure into the observed electron density map, the atomic model will have regions of high or low confidence, depending upon the quality of the electron density in that region. View the electron density map along with the structure model.

Collaborate closely with your crystallographer, so the uncertainties in the structure are considered within your design hypotheses.

Flipped binding modes can occur between similar small molecules, whether they are weakly or tightly bound. Treat these different binding modes as separate starting points for the design of a new series: series membership should be viewed as a combination of the chemical series along with its binding mode.

Each design should test a specific hypothesis. Most successful structure-based design is iterative, based on previous X-ray co-crystal structures and SAR, plus modeling to test the feasibility of a hypothesis.

Designs should occasionally attempt to push into protein pockets to learn whether they can expand.

Compounds that are unexpectedly active or inactive should be prime candidates for structure determination: outliers sometimes contain surprises that lead to new designs.

Focus design on shape complementarity between the binding site and ligand first and polar interactions second.

Polar interactions only improve potency if they improve upon the interactions between the free ligand and water, and the protein binding site and water: it's difficult to form better interactions than water. However, well-placed polar interactions can make very large differences.

Ensure that protein hydrogen-bond donors make good hydrogen bonds with ligand acceptors or water: satisfying protein donors appears to be more important than satisfying acceptors.

The pK(BHX) scale is an excellent reference for comparing and rank-ordering relative hydrogen-bond acceptor strengths for common functional groups: acceptor strengths tend to decrease in the order amides/sulfoxides/pyridines/N-methylimidazoles > ketones > esters/alcohols > sulfones/sulfonamides/ethers.

Computational methods provide useful qualitative guidance to eliminate designs that are unlikely to succeed, but are too unreliable to predict relative activities among proposed, apparently well-designed compounds.

The CSD and PDB databases provide excellent sources of data for preferred conformations and interactions. Ensure your bound conformation is close to a preferred low energy conformation using computational methods to search for the lowest energy conformations and by comparison with molecules with similar torsion types in the CSD.

Small molecules rarely bind to proteins outside of their low energy conformations. A protein dissipates strain energy through small changes across its many degrees of freedom much more easily than for the ligand to adopt an unfavorable conformation by straining its few rotatable bonds.

Minimize the overall charge of the small molecule to zero or one, avoid strong acids and bases (pK_as < 3 or > 10, respectively), and use neutral substituents (or at least those with pK_a close to 7.4) to interact with charged side chains on the protein.

Binding selectivity between protein isoforms can come not only from sequence differences within the active site, but also from differences further away.

4.13 SUMMARY

Many marketed drugs can trace a key design decision in their discovery to a structure-based approach, including the neuraminidase inhibitors zanamivir[76] and oseltamivir,[77] the first marketed renin inhibitor aliskiren,[78] the carbonic anhydrase glaucoma drug dorzolamide,[79] many HIV protease inhibitors,[80] the HCV protease inhibitors telaprevir,[81] boceprevir,[82] and simeprevir,[83] and various kinase inhibitors including imatinib,[84] nilotinib,[85] gefitinib,[86] and vemurafenib,[87] to name but a few examples. Even membrane proteins are now amenable to crystallography, particularly GPCRs, and the first clinical candidates targeting GPCRs designed using X-ray crystal structure information are now emerging.[88]

Almost 40 years since it began, structure-based drug design is still qualitative, with many surprises. We are still learning how to model and understand non-covalent aqueous interactions. Our knowledge and practice have advanced dramatically during the last 10–15 years due to the advent of high-throughput crystallography, which makes it practical to determine many co-crystal structures during a drug discovery project. The PDB and CSD databases and companion software have grown tremendously. They provide comparative structures to test designs for preferred conformations and interactions. A protein–ligand X-ray structure is a highly provocative and valuable tool to stimulate new design hypotheses in the minds of drug discovery scientists. Together with physical organic chemistry concepts, considerations of compound quality, and excellent synthetic chemistry to carry out ambitious designs, structure-based design is now a fully established and validated medicinal chemistry tool.

KEY REFERENCES

The paper by Bissantz, Kuhn, and Stahl from *J. Med. Chem.* is highly recommended, as are the chapters by Klebe.

The reviews on the use of X-ray crystal structures by Davis, Teague and Kleywegt are also recommended reading for any chemist embarking on a structure-based design campaign.

C. Bissantz, B. Kuhn and M. Stahl, *J. Med. Chem.*, 2010, **53**, 5061.
G. Klebe, *Drug Design: Methodology, Concepts and Mode-of-Action*, Springer-Verlag, Berlin Heidelberg, 2013, pp. 61–88 and pp. 429–447.
Davis, S. Teague and G. Kleywegt, *Angew. Chem., Int. Ed.*, 2003, **24**, 2718.
M. Davis, S. A. St-Gallay and G. J. Kleywegt, *Drug Discovery Today*, 2008, **13**, 83.

REFERENCES

1. S. M. Paul, D. S. Mytelka, C. T. Dunwiddie, C. C. Persinger, B. H. Munos, S. R. Lindborg and A. L. Schacht, *Nat. Rev. Drug Discov.*, 2010, **9**, 203.
2. J. Blaney, *J. Comput.-Aided Mol. Des.*, 2012, **26**, 13.
3. C. Bissantz, B. Kuhn and M. Stahl, *J. Med. Chem.*, 2010, **53**, 5061.
4. G. Klebe, *Drug Design: Methodology, Concepts and Mode-of-Action*, Springer-Verlag, Berlin Heidelberg, 2013, pp. 61–88.
5. G. Klebe, *Drug Design: Methodology, Concepts, and Mode-of-Action*, Springer-Verlag, Berlin Heidelberg, 2013, pp. 429–447.
6. H. M. Berman, J. Westbrook, Z. Feng, G. Gilliland, T. N. Bhat, H. Weissig, I. N. Shindyalov and P. E. Bourne, *Nucleic Acids Res.*, 2000, **28**, 235.
7. A. Davis, S. Teague and G. Kleywegt, *Angew. Chem., Int. Ed.*, 2003, **24**, 2718.
8. A. M. Davis, S. A. St-Gallay and G. J. Kleywegt, *Drug Discovery Today*, 2008, **13**, 831.
9. W. L. Jorgensen, *Acc. Chem. Res.*, 2009, **42**, 724.
10. C. R. Beddell, P. J. Goodford, F. E. Norrington, S. Wilkinson and R. Wootton, *Br. J. Pharmacol.*, 1976, **57**, 201.
11. C. R. Beddell, P. J. Goodford, G. Kneen, R. D. White, S. Wilkinson and R. Wootton, *Br. J. Pharmacol.*, 1984, **82**, 397.
12. I. D. Kuntz, J. M. Blaney, S. J. Oatley, R. Langridge and T. E. Ferrin, *J. Mol. Biol.*, 1982, **161**, 269.
13. J. B. Cross, D. C. Thompson, B. K. Rai, J. C. Baber, K. Y. Fan, Y. Hu and C. Humblet, *J. Chem. Inf. Model.*, 2009, **49**, 1455.
14. N. Moitessier, P. Englebienne, D. Lee, J. Lawandi and C. R. Corbeil, *Br. J. Pharmacol.*, 2008, **153**, S7.
15. I. J. Enyedy and W. J. Egan, *J. Comput.-Aided Mol. Des.*, 2008, **22**, 161.
16. G. B. McGaughey, R. P. Sheridan, C. I. Bayly, J. C. Culberson, C. Kreatsoulas, S. Lindsley, V. Maiorov, J.-F. Truchon and W. D. Cornell, *J. Chem. Inf. Model.*, 2007, **47**, 1504.
17. G. L. Warren, C. W. Andrews, A.-M. Capelli, B. Clarke, J. LaLonde, M. H. Lambert, M. Lindvall, N. Nevins, S. F. Semus, S. Senger, G. Tedesco, I. D. Wall, J. M. Woolven, C. E. Peishoff and M. S. Head, *J. Med. Chem.*, 2006, **49**, 5912.
18. P. Kolb, R. S. Ferreira, J. J. Irwin and B. K. Shoichet, *Curr. Opin. Biotechnol.*, 2009, **20**, 429.
19. A. R. Leach, B. K. Shoichet and C. E. Peishoff, *J. Med. Chem.*, 2006, **49**, 5851.
20. A. Nicholls, G. B. McGaughey, R. P. Sheridan, A. C. Good, G. Warren, M. Mathieu, S. W. Muchmore, S. P. Brown, J. A. Grant, J. A. Haigh, N. Nevins, A. N. Jain and B. Kelley, *J. Med. Chem.*, 2010, **53**, 3862.
21. S. L. Swann, S. P. Brown, S. W. Muchmore, H. Patel, P. Merta, J. Locklear and P. J. Hajduk, *J. Med. Chem.*, 2011, **54**, 1223.
22. D. Kozma, I. Simon and G. E. Tusnády, *Nucleic Acids Res.*, 2013, **41**, D524.
23. P. Raman, V. Cherezov and M. Caffrey, *Cell. Mol. Life Sci.*, 2006, **63**, 36.
24. C. R. Groom and F. H. Allen, *Angew. Chem., Int. Ed.*, 2014, **53**, 662.
25. C. Schärfer, T. Schulz-Gasch, H.-C. Ehrlich, W. Guba, M. Rarey and M. Stahl, *J. Med. Chem.*, 2013, **56**, 2016.
26. B. Kuhn, J. E. Fuchs, M. Reutlinger, M. Stahl and N. R. Taylor, *J. Chem. Inf. Model.*, 2011, **51**, 3180.
27. M. Harder, B. Kuhn and F. Diederich, *ChemMedChem*, 2013, **8**, 397.
28. E. A. Meyer, R. K. Castellano and F. Diederich, *Angew. Chem., Int. Ed.*, 2003, **42**, 1210.
29. K. Müller, C. Faeh and F. Diederich, *Science*, 2007, **317**, 1881.
30. R. Paulini, K. Müller and F. Diederich, *Angew. Chem., Int. Ed.*, 2005, **44**, 1788.
31. L. M. Salonen, M. Ellermann and F. Diederich, *Angew. Chem., Int. Ed.*, 2011, **50**, 4808.

32. K. A. Brameld, B. Kuhn, D. C. Reuter and M. Stahl, *J. Chem. Inf. Model.*, 2008, **48**, 1.
33. S. J. Cottrell, T. S. G. Olsson, R. Taylor, J. C. Cole and J. W. Liebeschuetz, *J. Chem. Inf. Model.*, 2012, **52**, 956.
34. A. J. Cruz-Cabeza, J. W. Liebeschuetz and F. H. Allen, *CrystEngComm*, 2012, **14**, 6797.
35. S. J. Teague, *Nat. Rev. Drug Discov.*, 2003, **2**, 527.
36. M. L. Connolly, *Science*, 1983, **221**, 709.
37. F. Lovering, J. Bikker and C. Humblet, *J. Med. Chem.*, 2009, **52**, 6752.
38. A. R. Fersht, J. P. Shi, J. Knill-Jones, D. M. Lowe, A. J. Wilkinson, D. M. Blow, P. Brick, P. Carter, M. M. Waye and G. Winter, *Nature*, 1985, **314**, 235.
39. D. D. Robinson, W. Sherman and R. Farid, *ChemMedChem*, 2010, **5**, 618.
40. T. Young, R. Abel, B. Kim, B. J. Berne and R. A. Friesner, *Proc. Natl. Acad. Sci. U. S. A.*, 2007, **104**, 808.
41. R. Abel, T. Young, R. Farid, B. J. Berne and R. A. Friesner, *J. Am. Chem. Soc.*, 2008, **130**, 2817.
42. Chemical Computing Group Inc., *3D-RISM: Molecular Operating Environment (MOE)*, Montreal, QC, Canada, 2013.
43. OpenEye Scientific Software, *SZMAP*, Santa Fe, NM, 2013.
44. G. Cui, J. M. Swails and E. S. Manas, *J. Chem. Theory Comput.*, 2013, **9**, 5539.
45. M. T. Stubbs, S. Reyda, F. Dullweber, M. Möller, G. Klebe, D. Dorsch, W. W. K. R. Mederski and H. Wurziger, *ChemBioChem*, 2002, **3**, 246.
46. E. Freire, *Drug Discovery Today*, 2008, **13**, 869.
47. E. Freire, *Chem. Biol. Drug Des.*, 2009, **74**, 468.
48. J. E. Ladbury, G. Klebe and E. Freire, *Nat. Rev. Drug Discov.*, 2010, **9**, 23.
49. P. D. Leeson and B. Springthorpe, *Nat. Rev. Drug Discovery*, 2007, **6**, 881–890.
50. C. Gerlach, M. Smolinski, H. Steuber, C. A. Sotriffer, A. Heine, D. G. Hangauer and G. Klebe, *Angew. Chem., Int. Ed.*, 2007, **46**, 8511.
51. A. Biela, N. N. Nasief, M. Betz, A. Heine, D. Hangauer and G. Klebe, *Angew. Chem., Int. Ed.*, 2013, **52**, 1822.
52. G. L. Warren, T. D. Do, B. P. Kelley, A. Nicholls and S. D. Warren, *Drug Discovery Today*, 2012, **17**, 1270.
53. J. Liebeschuetz, J. Hennemann, T. Olsson and C. R. Groom, *J. Comput.-Aided Mol. Des.*, 2012, **26**, 169.
54. L. M. Salonen, M. C. Holland, P. S. J. Kaib, W. Haap, J. Benz, J.-L. Mary, O. Kuster, W. B. Schweizer, D. W. Banner and F. Diederich, *Chem. Eur. J.*, 2012, **18**, 213.
55. R. G. Huber, M. A. Margreiter, J. E. Fuchs, S. von Grafenstein, C. S. Tautermann, K. R. Liedl and T. Fox, *J. Chem. Inf. Model.*, 2014.
56. Desert Scientific Software, *Proasis 3*, Norwest NSW, Australia, 2013.
57. R. Wilcken, M. O. Zimmermann, A. Lange, A. C. Joerger and F. M. Boeckler, *J. Med. Chem.*, 2013, **56**, 1363.
58. L. A. Hardegger, B. Kuhn, B. Spinnler, L. Anselm, R. Ecabert, M. Stihle, B. Gsell, R. Thoma, J. Diez, J. Benz, J.-M. Plancher, G. Hartmann, Y. Isshiki, K. Morikami, N. Shimma, W. Haap, D. W. Banner and F. Diederich, *ChemMedChem*, 2011, **6**, 2048.
59. M. A. Williams and J. E. Ladbury, *Methods Princ. Med. Chem.*, 2003, **19**, 137.
60. C. Laurence, K. A. Brameld, J. Graton, J.-Y. Le Questel and E. Renault, *J. Med. Chem.*, 2009, **52**, 4073.
61. A. J. Green and P. L. A. Popelier, *J. Chem. Inf. Model.*, 2014, **54**, 553.
62. P. Kukic, D. Farrell, L. P. McIntosh, B. García-Moreno E, K. S. Jensen, Z. Toleikis, K. Teilum and J. E. Nielsen, *J. Am. Chem. Soc.*, 2013, **135**, 16968.
63. A. J. Russell and A. R. Fersht, *Nature*, 1987, **328**, 496.
64. K. A. Armstrong, B. Tidor and A. C. Cheng, *J. Med. Chem.*, 2006, **49**, 2470.
65. Y. Shen, M. K. Gilson and B. Tidor, *J. Chem. Theory Comput.*, 2012, **8**, 4580.

66. J. A. Grant, M. A. Gallardo and B. T. Pickup, *J. Comput. Chem.*, 1996, **17**, 1653.

67. A. C. Pan, D. W. Borhani, R. O. Dror and D. E. Shaw, *Drug Discovery Today*, 2013, **18**, 667.

68. S. Piana, K. Lindorff-Larsen and D. E. Shaw, *Proc. Natl. Acad. Sci. U. S. A.*, 2012, **109**, 17845.

69. Y. Shan, E. T. Kim, M. P. Eastwood, R. O. Dror, M. A. Seeliger and D. E. Shaw, *J. Am. Chem. Soc.*, 2011, **133**, 9181.

70. S. S. Antonysamy, B. Aubol, J. Blaney, M. F. Browner, A. M. Giannetti, S. F. Harris, N. Hebert, J. Hendle, S. Hopkins, E. Jefferson, C. Kissinger, V. Leveque, D. Marciano, E. McGee, I. Nájera, B. Nolan, M. Tomimoto, E. Torres and T. Wright, *Bioorg. Med. Chem. Lett.*, 2008, **18**, 2990.

71. H. Li, J. Tatlock, A. Linton, J. Gonzalez, T. Jewell, L. Patel, S. Ludlum, M. Drowns, S. V. Rahavendran, H. Skor, R. Hunter, S. T. Shi, K. J. Herlihy, H. Parge, M. Hickey, X. Yu, F. Chau, J. Nonomiya and C. Lewis, *J. Med. Chem.*, 2009, **52**, 1255.

72. I. J. Bruno, J. C. Cole, M. Kessler, J. Luo, W. D. S. Motherwell, L. H. Purkis, B. R. Smith, R. Taylor, R. I. Cooper, S. E. Harris and A. G. Orpen, *J. Chem. Inf. Comput. Sci.*, 2004, **44**, 2133.

73. E. D. Garcin, A. S. Arvai, R. J. Rosenfeld, M. D. Kroeger, B. R. Crane, G. Andersson, G. Andrews, P. J. Hamley, P. R. Mallinder, D. J. Nicholls, S. A. St-Gallay, A. C. Tinker, N. P. Gensmantel, A. Mete, D. R. Cheshire, S. Connolly, D. J. Stuehr, A. Aberg, A. V. Wallace, J. A. Tainer and E. D. Getzoff, *Nat. Chem. Biol.*, 2008, **4**, 700.

74. D. R. Cheshire, A. Aberg, G. M. K. Andersson, G. Andrews, H. G. Beaton, T. N. Birkinshaw, N. Boughton-Smith, S. Connolly, T. R. Cook, A. Cooper, S. L. Cooper, D. Cox, J. Dixon, N. Gensmantel, P. J. Hamley, R. Harrison, P. Hartopp, H. Käck, P. D. Leeson, T. Luker, A. Mete, I. Millichip, D. J. Nicholls, A. D. Pimm, S. A. St-Gallay and A. V. Wallace, *Bioorg. Med. Chem. Lett.*, 2011, **21**, 2468.

75. M. Cowart, E. A. Kowaluk, J. F. Daanen, K. L. Kohlhaas, K. M. Alexander, F. L. Wagenaar and J. F. Kerwin, *J. Med. Chem.*, 1998, **41**, 2636.

76. M. von Itzstein, W. Y. Wu, G. B. Kok, M. S. Pegg, J. C. Dyason, B. Jin, T. Van Phan, M. L. Smythe, H. F. White and S. W. Oliver, *Nature*, 1993, **363**, 418.

77. C. U. Kim, W. Lew, M. A. Williams, H. Liu, L. Zhang, S. Swaminathan, N. Bischofberger, M. S. Chen, D. B. Mendel, C. Y. Tai, W. G. Laver and R. C. Stevens, *J. Am. Chem. Soc.*, 1997, **119**, 681.

78. N. C. Cohen, *Chem. Biol. Drug Des.*, 2007, **70**, 557.

79. J. J. Baldwin, G. S. Ponticello, P. S. Anderson, M. E. Christy, M. A. Murcko, W. C. Randall, H. Schwam, M. F. Sugrue, J. P. Springer and P. Gautheron, *J. Med. Chem.*, 1989, **32**, 2510.

80. Y. Mehellou and E. De Clercq, *J. Med. Chem.*, 2010, **53**, 521.

81. A. D. Kwong, R. S. Kauffman, P. Hurter and P. Mueller, *Nat. Biotechnol.*, 2011, **29**, 993.

82. K. X. Chen and F. G. Njoroge, *Prog. Med. Chem.*, 2010, **49**, 1.

83. Å. Rosenquist, B. Samuelsson, P.-O. Johansson, M. D. Cummings, O. Lenz, P. Raboisson, K. Simmen, S. Vendeville, H. de Kock, M. Nilsson, A. Horvath, R. Kalmeijer, G. de la Rosa and M. Beumont-Mauviel, *J. Med. Chem.*, 2014, **57**, 1673.

84. R. Capdeville, E. Buchdunger, J. Zimmermann and A. Matter, *Nat. Rev. Drug Discov.*, 2002, **1**, 493.

85. E. Weisberg, P. W. Manley, S. W. Cowan-Jacob, A. Hochhaus and J. D. Griffin, *Nat. Rev. Cancer*, 2007, **7**, 345.

86. A. J. Barker, K. H. Gibson, W. Grundy, A. A. Godfrey, J. J. Barlow, M. P. Healy, J. R. Woodburn, S. E. Ashton, B. J. Curry, L. Scarlett, L. Henthorn and L. Richards, *Bioorg. Med. Chem. Lett.*, 2001, **11**, 1911.

87. G. Bollag, J. Tsai, J. Zhang, C. Zhang, P. Ibrahim, K. Nolop and P. Hirth, *Nat. Rev. Drug Discov.*, 2012, **11**, 873.

88. S. P. Andrews, J. S. Mason, E. Hurrell and M. Congreve, *MedChemComm*, 2014, **5**, 571.

89. C. Hansch, A. Leo and D. Hoekman, *Exploring QSAR: Hydrophobic, Electronic, and Steric Constants*, American Chemical Society, Washington, D.C., 1995.

90. Schrödinger, LLC, *The PyMOL Molecular Graphics System, Version 1.5.0.4*, 2010.

91. Chemical Computing Group Inc., *Molecular Operating Environment (MOE)*, Montreal, QC, Canada, 2013.

92. T. A. Halgren, *J. Comput. Chem.*, 1999, **20**, 730.

93. T. A. Halgren, *J. Comput. Chem.*, 1996, **17**, 490.

CHAPTER 5

Fragment Based Lead Discovery

RODERICK E. HUBBARD[a,b]

[a] YSBL, University of York, Heslington, York, YO10 5DD, UK; [b] Vernalis (R&D) Ltd,
Granta Park, Abington, Cambridge, CB21 6GB, UK
E-mail: roderick.hubbard@york.ac.uk

5.1 INTRODUCTION

The starting point for small molecule drug discovery programs is to identify initial "hit" compounds that bind to the target and which have the potential for optimisation to clinical candidates with the desired therapeutic effect. For many organisations, the preferred way of identifying such compounds has conventionally been by screening large numbers (often millions) of compounds in some cellular, biochemical, or target binding assay (High Throughput Screening, HTS). Usually, these compounds are of molecular weight 300 to 450 Da and are selected to have appropriate drug-like[1] or lead-like[2] properties. There are two major issues with such approaches. The first is that these initial hits usually need substantial and often challenging modification to introduce the required properties. The second is that HTS often generates no tractable hits, particularly for new target classes. Even with many millions of compounds, HTS only samples a very small fraction of available chemical space and the collection is usually dominated by compounds from previous drug discovery campaigns.[3]

The essential feature of fragment based methods is that drug discovery begins with screening of a relatively small library (typically 1000s) of compounds of low molecular weight (average 190 Da) that are more likely to bind.[4] The fragment hits are then evolved, usually guided by the structure of the compounds bound to the therapeutic target, to larger lead compounds which can then be optimised by conventional medicinal chemistry methods. There have been great advances in the methods of fragment based lead discovery over the past 15 years, and there are now many compounds in the clinic (reviewed[5–8]) and the first fragment derived drug has now been approved and is on the market.[9]

This chapter provides an overview of the current practice of fragment based lead discovery (see also previous reviews[6,10–14]). The emphasis is on practical aspects of the methods. Although there are some examples of fragments binding to other targets (such as nucleic acids[15]), the methods have been mostly applied to protein targets, which will be the focus of this discussion.

The Handbook of Medicinal Chemistry: Principles and Practice
Edited by Andrew Davis and Simon E Ward
© The Royal Society of Chemistry 2015
Published by the Royal Society of Chemistry, www.rsc.org

In addition, I include a section describing some of the history of the evolution of fragment based discovery.

5.2 THE GENERAL FEATURES OF FBLD

Figure 5.1 is a summary of the general approach to fragment based discovery that has been developed at Vernalis over the past 10 years.[12,16] Most practitioners have a similar approach: a fragment library is screened, usually by a biophysical method. Information about fragments is combined with other hits from virtual or experimental screening of conventional libraries (such as by HTS) and from the literature. The binding of the various hits is usually validated by one or more orthogonal biophysical methods while attempts are made to determine a structure, preferably by X-ray crystallography, but if necessary, using NMR and other methods to generate models of ligand binding.

Finding fragments that bind to most binding sites on most targets is relatively straightforward: the challenge is knowing what to do with the fragments—both in choosing which fragments to progress and how to evolve them to compounds which can be advanced like any other compound in lead optimisation. An important first step is to optimise the fragment core itself, exploring SAR by purchase (often called SAR by catalog) or limited synthesis to test binding hypotheses and to identify the key vectors for optimisation. There are two key characteristics that fragments bring to the fragment to hit to lead optimisation process. The first is that for most targets, fragment screening provides choice in the number and diversity of chemical starting points. The second is that the medicinal chemist has a small start point; with

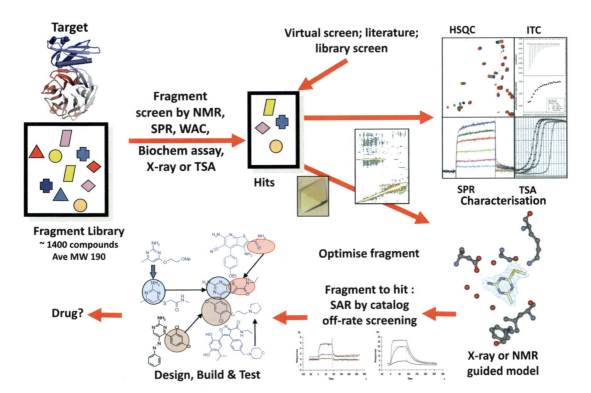

Figure 5.1 A schematic representation of the overall fragment based drug discovery process at Vernalis. See text for details.

careful design, this gives opportunities for developing highly efficient, lower molecular weight lead compounds while optimising the required drug-like properties.

The following four sections summarise the key features of these different steps.

5.3   FRAGMENT LIBRARY

The main difference between a fragment and any other hit compound in drug discovery is that the fragment is small (and thus needs to evolve) and binds with a low affinity to the target. The low affinity means that sensitive methods are needed (discussed later) to detect such weak binding. Both the screening technology and the method for growing the fragments need to be taken into account in designing the library.

As with any screening approach, the quality of the hits obtained depend on the quality of the compounds in the fragment library. There has recently been a considerable amount of literature published on library design.[17-22] Most practitioners have adopted a similar approach which is summarised in Figure 5.2. The available compounds (either commercially available or from the corporate collection) are filtered to identify compounds with the desired molecular weight, excluding those that contain functional groups that are predicted to be toxic or reactive and excluding compounds that do not have identified sites for synthetic elaboration. This step also includes a filter on predicted solubility. Some libraries have an additional step, selecting compounds predicted to be suitable for a particular target class, either through requiring a particular pharmacophore or through virtual screening. The remaining set of compounds is then usually assessed on some measure of chemical diversity—in the case of the Vernalis fragment library, we have used the three point 2D pharmacophore method. The final selection of compounds usually involves a manual inspection step, where medicinal chemists triage the list of compounds to identify the chemotypes that they feel should be progressed. The compounds are then purchased and checked for purity and solubility.

The maintenance and curation of a fragment library takes considerable effort. The fragments are screened at relatively high concentrations (100s of μM for most biophysical screens; many mM for crystallographic soaks), so the solubility and purity of the compounds needs to be regularly checked, particularly when the libraries are maintained in high concentration stock solutions. This is not only to ensure that the compounds are still in solution, but also to check for aggregation or stability issues. A small percentage of some contaminants can have a marked effect in a high concentration assay. The physicochemical requirements of a library are a major factor that constrains the contents of a fragment library, in particular the solubility which is usually linked to the lipophilicity.

At Vernalis, the library is assessed and updated about every 9 months. Compounds are removed if they show lack of solubility or stability, or when stocks are depleted and the compound is no longer available. In addition, fragments that have proven particularly difficult to progress or demonstrated liabilities (metabolic *etc.*) are removed to make way for new fragments. All compounds synthesised within the company that meet the general criteria for fragments are considered for inclusion; in addition, novel scaffolds are often synthesised based on ideas from the literature or understanding of a particular target class. The overall size of the library is maintained between 1100 to 1500 compounds—primarily because of the overhead of library QC and maintenance.

There are a number of published variations on these general principles. The first of these are libraries that have been constructed to include particular elements to aid the screening method. Dalvit *et al.*[23-25] are enthusiastic proponents of using ^{19}F as a probe; the main advantage is that the ^{19}F nucleus is particularly sensitive in the NMR experiment and also the isotope is 100% abundant. This markedly increases the sensitivity compared to ^{1}H NMR spectra, such that as

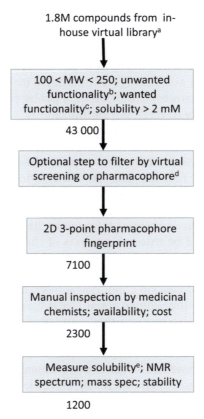

Figure 5.2 A typical workflow to design a fragment library. The numbers are the number of compounds which remain to be considered and are taken from a fragment design protocol followed at Vernalis in 2003.[20]

[a]The database of available compounds rCAT.[103]

[b]SMARTs strings were used to remove compounds that had:

- Four aliphatic carbons except if also contains X–C–CC–X, X–C–C–X, X–C–X with X=O or N
- Any atom different from H, C, N, O, F, Cl, S
- –SH, S–S, O–O, S–Cl, N–halogen
- Sugars
- Conjugated system: R=C=C=O, with R different from O, N, or S or aromatic cycle
- (C=O)–halogen, O–(C=O)–halogen, SO_2–halogen, N=C=O, N=C=S, N–C(=S)–N
- Acyclic C(=O)–S, Acyclic C(=S)–O, Acyclic N=C=N
- Anhydride, aziridine, epoxide, *ortho* ester, nitroso
- Quaternary amines, methylene, isonitrile
- Acetals, thioacetal, N–C–O acetals
- Nitro group
- >1 chlorine atom

[c]SMARTs strings were used to accept compounds that had:

- R–COOMe, R–COOH; R–NHMe, R–N(Me)$_2$, R–NH$_2$
- R–CONHMe, R–CON(Me)$_2$, R–CONH$_2$; R–SO$_2$NHMe, S–SO$_2$N(Me)$_2$, R–SO$_2$NH$_2$
- R–OMe, R–OH; R–Sme
- ≥ 1 ring system

[d]This filter was not included for the example numbers used here.

[e]For NMR screening (see later), the compounds are stored at 200 mM in d$_6$-DMSO and need to be soluble aqueous at 500 µM ; the requirements differ for other methods of screening.

low as 5 μM fragment is required, dramatically reducing the solubility constraint. One of the early proponents of structure-based design, the company SGX, designed a library containing Br atoms as electron dense atoms to aid identification in crystallographic screening;[26] one group went as far as designing a paired set of libraries where a fragment with a Br could be used for crystallographic screening with a non-Br fragment available for other assays. An issue with both of these approaches is that the atom introduced as a probe can itself be important for interaction, complicating the pathway to fragment optimisation.

A second variation is to enrich the fragment library with fragments derived from known drugs (so-called privileged scaffolds) or to use virtual screening or a pharmacophore (as described above) for a particular target class. Usually, these approaches will increase the hit rate, but by definition, are not providing any new insights into the chemotypes which will interact with the target. For these reasons, most practitioners aim for the library to be generic and suitable for most targets.[17,20–22]

A final variation to note is to construct a fragment library that is representative of the available compounds. This could be through analysis of the available compounds[18] or to synthesise libraries of compounds based on the fragments; this type of approach ensures there are near neighbours of fragments available for rapid generation of SAR.

There has recently been some discussion about how important it is for drug molecules to include sp^3 carbon atoms giving a more three dimensional shape.[27] Such considerations have led to the proposal that 3D fragments may also be important, particularly for some of the new target classes (protein–protein interactions, carbohydrate processing proteins). To date, there have been two main challenges for such ideas. The first is that such complex shapes usually require more atoms; in addition, the chemical suppliers have not focussed on this area. For these reasons, there are only a small number of such 3D compounds commercially available which meet the MW criteria. The second is that the subsequent evolution of molecules with (usually) quite complex stereochemistry is synthetically challenging. It remains to be seen whether the current interest in the idea of 3D fragments really leads to any new chemical entities and thus useful lead compounds.

A concern expressed by some is that many drug discovery campaigns will begin from screening the same fragment library and so the same fragment hits will be considered. Anecdotally, there is remarkably little overlap in the exact chemical structures selected to be in different fragment libraries—it appears that there is enough diversity in the commercially available compounds less than 250 Da that the chance of choosing exactly the same fragment for a particular chemotype is quite low. However, it is the case that rather similar fragments are found by different project teams; what is important is the optimisation of the fragments. As we will see later in the analysis of various HSP90 projects in Section 5.9, this is the difficult part of fragment-based discovery and where real discovery is made.

Finally, a comment on the molecular weight of the compounds in the fragment library. A key feature of fragment screening is that a small number of compounds samples a potentially huge chemical space—*i.e.* a small fragment is representative of many more compounds that can incorporate or be inspired by that fragment. The scaling of the chemical space with number of compounds is nicely illustrated by the work of Reymond.[28] His group has developed algorithms to enumerate the compounds that can be generated using standard chemical protocols and which fit within what can be termed drug-like space. Although there are many approximations in the analysis, the main message is that chemical space increases approximately 8.3-fold for the addition of each heavy atom. This means that a library of 1000 fragments of molecular weight 190 is equivalent to 10^7 compounds of molecular weight 280 and 10^{18} compounds of molecular weight 450. This emphasises the need to keep the fragments within a library as small as possible, while still having sufficient mass and functionality to register in an assay.[4] For these reasons, it is usual to think of fragments as less than 250 Da, and calling compounds between 250 to 350 Da scaffolds.

5.4 FRAGMENT SCREENING APPROACHES

For a simple binding event of a ligand (L) to a protein (P), the equilibrium can be represented as shown in Equation 5.1.

$$\text{(a)} \quad PL_{(aq)} \xrightleftharpoons[k_{on}]{k_{off}} P_{(aq)} + L_{(aq)}$$

$$\text{(b)} \quad K_D = \frac{k_{off}}{k_{on}} = \frac{[P_{(aq)}][L_{(aq)}]}{[PL_{(aq)}]}$$

$$\text{(c)} \quad \Delta G^{\circ} = \Delta H^{\circ} - T\Delta S^{\circ} = -RT \ln K_D$$

Equation 5.1   Equation 5.1 For the simple equilibrium in Equation 5.1(a), the dissociation constant (K_D) is the ratio of the off rate (k_{off}) divided by the on rate (k_{on}), as shown in Equation 5.1(b). K_D is related to the standard Gibbs free energy (ΔG°) and the enthalpy (ΔH°) and entropy (ΔS°) of binding as shown in Equation 5.1(c) at temperature (T); R is the molar gas constant.

There are now many examples where small fragments that bind to a target with a dissociation constant in the mM range can be progressed to lead compounds. The need to detect binding at such low affinity has inspired the development of many techniques for screening fragments. These have relied not only on the development of more sensitive equipment but also an increased understanding of how to set up and validate the assays and the limitations and advantages of the different methods.

Figure 5.3 summarises the range of affinities which are usually accessible by these techniques, based on experience at Vernalis.[12] The following is a brief summary of the main characteristics and considerations for each of the screening approaches.

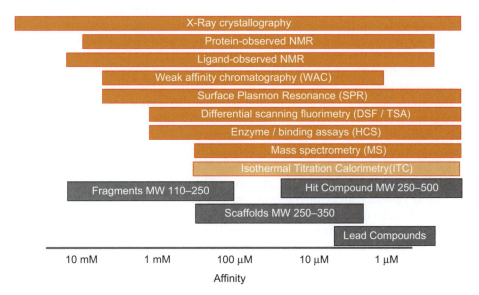

Figure 5.3   A schematic representation of the sensitivity range of the different techniques presented in the text for detecting binding of compounds to protein targets. The affinity scale is for the dissociation constant for the equilibrium PL = P + L. For all techniques, there will be exceptions to these sensitivity limits; the diagram summarises the general experience in most projects, presuming that all components are soluble and do not interfere with the detection method.

5.4.1 Protein-Observed NMR

This is the screening method used by the NMR group at Abbott, used in the first published fragment based discovery project[29] and used in most of the subsequent fragment campaigns at that company.[7,30] The main NMR pulse sequence used is known as HSQC (Heteronuclear Single Quantum Coherence) which is a 2D NMR experiment, most often where transfer of signal between ^1H and ^{15}N in the isotopically labelled protein results in a spectrum where each amide yields a peak; the position of each peak depends on the local chemical environment and is sensitive to ligand binding (see typical spectrum in the top right panel of Figure 5.1). Although more sophisticated NMR experiments are possible for larger proteins, the usual limitation is for proteins less than 35 kDa (otherwise the peaks are too broad due to the slower tumbling time). Usually, the experiments are performed with protein at a concentration of 20–50 μM to give sufficient signal. This solubility requirement increases the amount of protein needed for screening; as it is the protein that is being observed, this also makes screening with mixtures more complicated (see ligand-observed NMR below). The dynamic range of HSQC measurements is quite broad, mainly limited by compound solubility. Hits (that is some change) can be seen for compounds up to mM in affinity.

There has been limited use of the method for screening of fragments at other organisations; to some extent, this is because of the patent that Abbott gained on some of the methods, but mainly because of the limitations outlined above. However, protein-observed NMR is used more widely for characterising fragment binding. An advantage of the method is that it can identify fragments that bind to alternate regions of the protein—this would be seen in a different pattern of amide shifts in the HSQC spectrum.[31]

5.4.2 Ligand-Observed NMR

This is regarded by many as the most robust of the biophysical screening methods, because of the dynamic range of the assay (hits can be identified with K_D between 100 nM and 10 mM) and because there is quality control within each experiment—the NMR signals report that the ligand is intact and in solution and that the protein is still folded and in solution.

Figure 5.4 summarises the experimental design. Typically, these experiments are performed with 10 μM protein and with mixtures of up to 10 fragments at 500 μM each; as long as there is not one fragment which has a particularly strong residence time on the target, the experiment should be able to detect when more than one fragment in a mixture binds to the target. The selection of fragments for each mixture is not too onerous—the only checks are that there is one peak for each fragment in the mixture that is distinctive and that the fragments do not associate with each other in the mixture. A number of different NMR experiments are used with the most widely applied being Saturation Transfer Difference.[32] In this experiment, the protein sample is irradiated at the chemical shift of the protons of the hydrophobic core; this radiation is transferred to any ligand which touches the protein during the timescale of the experiment, which is measured as the difference seen in the 1D spectrum recorded for the ligand. Additional experiments that are used by some are the water-logsy[33] (which involves transfer of radiation from solvent molecules) and CPMG[34] (measuring differential relaxation of the ligand in solution and bound). Each of these experiments has a different physical basis, so a hit in all three of the experiments is more reliable. The general experience at Vernalis is that a hit in all three NMR experiments has a 70% success rate in giving a crystal structure, falling to 40% if only a hit in two of the experiments and rarely if a hit in only one experiment. The main limitation for detection is the solubility of the ligand; at Vernalis (see Figure 20.5 in reference 12), fragments have been reliably identified between 100 nM and 5 mM binding. The high concentration of fragments needed to observe such low affinity binding can sometimes result in non-specific

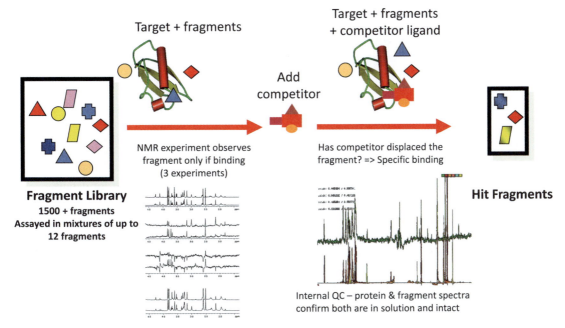

Target + fragments

Add competitor

Target + fragments + competitor ligand

NMR experiment observes fragment only if binding (3 experiments)

Has competitor displaced the fragment? => Specific binding

Fragment Library
1500 + fragments
Assayed in mixtures of up to 12 fragments

Hit Fragments

Internal QC – protein & fragment spectra confirm both are in solution and intact

Figure 5.4   A schematic representation of fragment screening by competitive ligand-observed NMR spectroscopy. See text for details.

binding. To control for this, a competitive step is used where a known ligand (which can be a peptide or a tool compound) is added and the spectra reacquired to see if the signal disappears.

Such a competitive experiment identifies fragments whose binding is effected by such a competitor ligand and so, in principle, would identify a true allosteric site as well. However, many alternate, cryptic sites (sometimes functional, often not) elsewhere on the protein active site can only be probed if the experiment is constructed to look for them. For example, the primary active site can be blocked by incubation with a high concentration of a known tight binding (*i.e.* slow kinetics) inhibitor during screening.

A variation of the ligand-observed NMR screening is the proprietary TINS method. Here, the protein is immobilised on a column and ligands flowed past.[35–37]

5.4.3   Surface Plasmon Resonance (SPR)

SPR is a technique which measures changes in molecular weight. The experiment is based on immobilising one component (either protein or ligand) on a surface; the change in properties of an optical beam shining on this surface reflects the mass of what is attached. It is then possible to observe the change in mass as the binding partner is flowed over the surface. There are two main configurations: direct mode and indirect mode.

Figure 5.5 is a sketch of the direct binding mode which provides the most information on the kinetics of binding (see Equation 5.1). The protein is immobilised to the surface and the ligand flowed over. As the ligand binds, there is a change in the response of the detector; the on rate or k_{on} for binding can be calculated from this initial slope. When the ligand is no longer injected over the surface, the ligand dissociates from the surface and the off rate or k_{off} can be measured from the resulting change in the response. For most targets, the fragments bind with a low affinity, with rapid on and off rates, so it is not possible to make a great deal from the kinetics (see the SPR trace in the panel at the top right of Figure 5.1). However, the off rate is usually

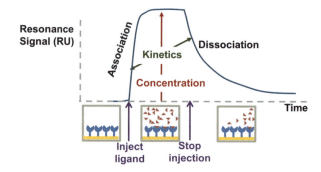

Figure 5.5   A schematic representation of surface plasmon resonance (SPR). See text for details.

what changes as a ligand is optimised. SPR can thus be an effective way of detecting progress in compound optimisation. In addition, slow off-rate compounds can be of great benefit in some therapeutic indications (such as oncology), as a long half-life on the target (or residence time) can be associated with increased efficacy.[38]

In indirect mode, a good ligand is immobilised to the surface, usually through an extended linker which needs careful design so as not to impede binding to the protein. The protein is then injected over the surface. As the protein binds, there is a large change in signal—inclusion of a high concentration of a fragment that binds to the protein will prevent binding. In this case, it is not possible to directly measure kinetics of binding but binding affinities can be extracted.

This technique has been available for many decades following the introduction of the first instrument (the initial manufacturer/trade name is Biacore). Although more sensitive equipment now allows weak fragment binding to be observed, the major advances have come with experience in the attachment methods, in the software that analyses the response traces, and importantly, a wider understanding of how to validate the binding and how to recognise when there are anomalies.[39-42] It is now generally found that fragment hits of 120 Da can be seen; although some practitioners have pushed the limit on affinity into the mM range, most find that it is challenging to reliably measure fragments binding with an affinity worse than 500 μM, mainly for reasons of solubility and interference with the chip surface and equivalent failures.

There are a variety of attachment strategies—many use kits that biotinylate free lysine residues on the surface of the protein[43] or use direct amine coupling reactions. There are protocols for establishing the best conditions, but this can result in a chip where the protein is randomly oriented; in addition, it is not possible to regenerate the chip once it has been used with a tight binding compound (and they are quite expensive consumables). Another alternative is to couple *via* the histidine tag usually attached to most over-expressed proteins for purification purposes—and immobilise on a Ni^{2+} surface. The standard hexa-his tag can be too weak to give a stable chip surface (the protein leaches off) but a double his tag[44] can provide a stable surface which has the advantage that it can be cleared to regenerate the rather expensive chips for different proteins.

5.4.4   Thermal Shift Analysis (TSA) or Differential Scanning Fluorimetry

In this method, a solution containing the target protein and the possible ligand or fragment is slowly heated up in the presence of a fluorescent dye. As the protein unfolds, there is a change in fluorescence as the dye binds to the hydrophobic surfaces; a change in the so-called melting temperature (or T_m) indicates that the ligand has bound and stabilised the protein. The top right panel of Figure 5.1 shows some representative traces.

The method was initially developed for screening of compound libraries[45] and very powerfully applied to select ligands which could increase the chance of crystallisation[46] or improving the stability of proteins.[47] For such ligands (usually better than a K_D of 1 μM) the method is reasonably robust as the ligands will tend to provide a clear stabilisation of the protein. A number of groups (particularly academics) have exploited the technique for fragment screening. As documented by many (*e.g.* reference 48), there are challenges in recognising the very small T_m shift seen for fragments. Because such small molecules give little stabilisation of the protein (if any) there can be many false negatives if TSA is used as the primary screening method.[12,48]

The main advantage of the technique is the relatively small amount of protein material required for screening and that standard laboratory equipment (such as a PCR machine) can be used.

5.4.5 Biochemical Assay

This category includes any assay that depends on readout of enzyme activity or any assay where a spectrophotometric readout (such as fluorescence or UV/visible light) monitors binding or the displacement of a labelled ligand. The advent and investment in High Throughput Screening (HTS) has stimulated the development of a range of robust screening approaches and platforms for measuring such activity or binding. The main issues for such assays for screening fragments are their detection limits and their sensitivity to the high concentrations needed to observe such low affinity binding; this has led to a wide literature on the origins of frequent hitter compounds and false positives.[49,50] For some classes of target (such as kinases), such assays are reasonably robust, particularly where the binding site being inhibited has a good ligandable shape and chemistry. In such cases, it is possible to screen a fragment library with reasonable confidence and minimal false positives and negatives. However, for some targets, the assays either lack sufficient sensitivity or are susceptible to interference (*e.g.* fluorescence). A recent article[12] summarises experience with fragment screening for a number of different targets. Validated hits identified by NMR screening and for which crystal structures are available were assessed for binding in the relevant wet assay. Whereas the false negatives in the wet assay were minimal for kinases and ATPases, there were some targets for which very few of the fragment hits would have been identified if only a wet assay had been used (see Figure 20.4 in reference 12).

5.4.6 Crystallography

An early emphasis in fragment-based methods was to use X-ray crystallography to identify fragment hits (see the discussion later). To some extent, this was because many of the early adopters and adapters of the methods had a strong background in structural biology and crystallography (see section 5.7) but the main advantages are that an immediate picture is obtained of how the fragment is binding and it is possible to obtain discrete structures for very weak ligand binding. However, the main challenge is whether the protein crystal is suitable, that is whether the crystal will withstand soaking in high concentrations of ligands and still diffract and whether crystal packing makes it difficult for low affinity ligands to readily soak into the binding site. The latter constraint emphasises that observing a ligand binding from a crystal soaking experiment is as much about kinetics as thermodynamics. It can take a number of attempts (varying soaking time, temperature *etc.*) before a structure of a complex is obtained.

Generating a suitable crystal system for ligand soaking can be a real challenge. Sometimes, modifications to the protein construct and/or crystallisation conditions will generate a suitable crystal form; however, requiring this can limit the types of projects for which fragment screening can be performed. In addition, as many protein crystals are formed with protein produced from truncated or mutated constructs, the cryptic sites are not always natural or functional. However, there are a number of examples where soaking into a crystal identifies unexpected cryptic (occasionally allosteric) binding sites on the protein (see examples reference 51–53).

Notwithstanding these caveats, a crystal structure of a fragment bound to the target protein validates fragment binding and provides important information to aid the design of improved compounds.

5.4.7   Mass Spectrometry

There have been great advances in mass spectrometry instrumentation for detection and characterisation of non-covalent complexes and this has delivered real insight into the composition of multi-protein assemblies (*e.g.* reference 54) and have been successfully used in detecting protein–ligand binding when the ligand binds with affinities better than 100 µM (*e.g.* reference 15). However, this is not an appropriate range for fragment screening, although the technique has found application when there is covalent fragment binding,[55] such as in the tethering approaches of Sunesis.[56]

5.4.8   Isothermal Titration Calorimetry (ITC)

This technique is based on measuring the heat (enthalpy in Equation 5.1(b)) that is taken up or released on binding of a ligand to a protein. There are two main limitations to the dynamic range of affinities that can be monitored—the amount of heat that is released and the solubility of one or other of the protein or ligand. The usual configuration is to inject small amounts of a concentrated solution of the ligand into a sample of protein; if the ligand is poorly soluble and the protein very soluble, then the protein can be injected into the ligand sample. The instrument compares the heat difference between the cell containing the protein (or ligand) and a blank sample; this is recorded as spikes of heat; the total heat is obtained by integration. As increasing amounts of ligand and protein bind, so the amount of heat reduces, and a sigmoidal curve such as seen in the top right panel of Figure 5.1 is observed. The concentration at which equilibrium is achieved allows the equilibrium constant to be calculated; it is then possible to calculate the entropy change on binding.

The detection limit is generally for a K_D in the 500 µM to 5 nM range; however the main limitation is the amount of protein required to generate sufficient heat. This means the method has limited application to fragment screening, but is a reliable method for validating fragment hits and subsequent optimisation of compounds.

5.4.9   Other Ideas and Approaches

The success and investment in fragment-based discovery has coincided with recognition of the importance of compound properties in lead optimisation. Taken together, these have encouraged the development of a number of other technologies for detecting and characterising protein–ligand interactions. Some of these methods are proprietary such as the fluorescence based interferometry method from the company Evotec[57] and the capillary electrophoresis methods from the company Selcia.[58] A recent method from an academic group in Sweden may be more accessible—weak affinity chromatography or WAC.[59] In this technique, the protein is attached to a silica gel and loaded onto a chromatography column; a mixture of fragments is flowed over the column. The retention time reflects the binding to the protein and the ligand can be identified if the column is connected to a mass spectrometer. As with any immobilisation technique, the main challenge is to fix the protein to the support, while retaining its folding and binding integrity.

5.4.10 Validating Fragment Hits—Comparing Methods

It is important to validate any fragment hits found in a screening campaign. Although the false positive rate for some targets (such as kinases) is relatively low for well conducted, competitive binding screens, it is useful to remove any false hits before investing too much time and resource. Validation is crucial for more challenging targets—such as protein–protein interaction targets—where the screening techniques can report incorrectly for all sorts of reasons.

There has been some debate (and reports) of the differences between the biophysical techniques both in hit rates and the nature of the hits. In principle—all the methods should give the same results, provided that the conditions are equivalent. However, it is the case that the sensitivity between the methods varies and in particular the differences in buffer conditions required for the different assays may have a differential effect on compound solubility or aggregation, or on the binding affinity itself.

5.5 FRAGMENT HIT RATES

5.5.1 Hits *vs.* Non-Hits

35% of the Vernalis fragment library is a hit against at least one target whereas 65% of the library has not yet been a hit against any target. To date, it has not been possible to find any combination of physicochemical or other properties that differentiates a hit from a non-hit sufficiently to be used as a criteria for removal of that class of compounds from the library. An analysis of hit rates and properties[17] showed that the hits are slightly more hydrophobic (higher lipophilicity, calculated as SlogP[60]) than non-hits, with slightly fewer rotatable bonds and more rings. However, these differences are subtle. The description of the Pfizer library design process[21] included a preliminary report on hit rates for the first library of 2592 fragments (so-called GFI-I) screened against 13 different targets by STD-NMR. 766 fragments hit at least on target, with only one fragment hitting all 13 targets—about 30%. The hit rates reported by AstraZeneca are more difficult to compare, as a number of different screening strategies were used.[22]

5.5.2 Hits for Different Types of Target

The Vernalis analysis[17] also found slight variation in the nature of the hits found for conventional targets such as kinases compared to the hits found for protein–protein interaction (PPI) targets which tend to have larger, flatter and more hydrophobic binding cavities. The hits for PPIs were of higher molecular weight, with more rotatable bonds and rings compared to those for other targets.

5.6 DETERMINING STRUCTURES OF FRAGMENTS BOUND

X-ray crystallography is the most information-rich method for providing a detailed picture of how a fragment is binding to a target. In some organisations, a fragment does not qualify as a true "hit" to be considered for progression until a crystal structure of the target is obtained with the fragment bound. Although other structural information (such as from NMR or from molecular docking) can provide enough information, the details of solvent and protein conformation provided by an X-ray structure gives a more robust framework for designing modifications to evolve the fragment. There are some caveats on the detailed interpretation of the structures—nearly all structures are determined from crystals immersed in cryoprotectant and frozen at liquid nitrogen temperature to protect against damage from the x-ray beam. In addition, some crystal packing can affect the detail of the conformation.

There are two ways of obtaining a protein–ligand crystal structure—either soaking or co-crystallisation. Co-crystallisation can be resource consuming—the crystallisation conditions may vary from ligand to ligand and it can take time and protein to identify new conditions for each ligand. In general, soaking provides the same structures (though there are exceptions, where the ligand can induce a conformational change). In most projects, the approach is to do soaking for speed, with occasional co-crystallisation attempts to cross-check the structures.

There are variants in the soaking protocols. The ligand can be added to the crystallisation solution or added to the cryoprotectant. For some crystals, it is not possible to obtain an apo structure (that is unliganded); one approach is to generate crystals with a moderately weak ligand binding (an example would be ADP-PNP binding to a kinase) and then soak out the ligand with the fragment solution which then binds. This approach is how the PDPK1 example in Figure 5.19 was achieved.

With the exception of the SAR by NMR approaches of Abbott (see Figures 5.7 to 5.9) all of the successful fragment optimisation campaigns have relied on X-ray crystal structures. In general, determining the structure of individual complexes by NMR is very time consuming—both in the time taken to collect the various NMR spectra, but also in assigning the spectra.

5.7 THE EVOLUTION OF THE IDEAS AND METHODS—A HISTORICAL PERSPECTIVE

As with most areas of science, it is possible to track the emergence over time of the main results that underpin current practice. The following is a selective discussion of the origin of some of the underlying principles and early applications of fragment-based methods, see also reference 14.

5.7.1 Some Early Ideas

Most discussions of fragment-based discovery begin with the papers from Jencks.[61] He reminded the community that for the simple equilibrium describing the dissociation of a protein–ligand complex (Equation 5.1), the Gibbs free energy relationship means that a doubling of the energy of interaction between protein and ligand will square the binding constant, K_D.

This means that if two functional groups each bind with a 1 mM K_D (equivalent to -17 kJ mol^{-1}) then a molecule containing both groups making the same interaction will have a binding energy of -34 kJ mol^{-1}, equivalent to a K_D of 1 µM. The Jencks paper also discusses how the binding of one group will overcome the rotational and translational entropy for the interaction, so any subsequent interactions are contributing to binding affinity alone.

The notion of different functional groups making a distinct contribution to binding was explored by Andrews in 1984.[62] In this paper, he performed a simple analysis of the average energy of interaction that is achieved by a particular functional group. This paper was to a large extent ignored on first publication; it was not until the mid-2000s that the ideas of functional group efficiency were formalised and more widely accepted.

One of the central features of many fragment screening campaigns is the determination of large numbers of crystal structures of the fragments and subsequent evolved hits binding to the protein. Perhaps the first recorded example of multiple crystal structures with potential ligands bound was the work by Perutz and Abraham[63] who looked at the structure of haemoglobin with various compounds such as bezafibrate bound, to understand the structural basis of their potential to affect the conformation of the protein and be used as anti-sickling agents.

5.7.2 The Emergence of De Novo Structure-Based Design

During the 1980s and early 1990s, the computational chemists developed series of ideas on how to exploit the structure of a protein to identify potential interaction sites for ligands. The first widely used example was the GRID program.[64] Here, a virtual grid of points is placed over the binding site of interest and a simple energy of interaction (van der Waals and electrostatic) computed for a particular probe—say a methyl group or a water molecule. The resulting grid of points can then be contoured to generate a very graphical representation of the regions of the binding site which would favour such an interaction. An early demonstration of the simplicity and power of this approach was the optimisation of the sialic acid analogue, Neu5Ac2en for binding to the influenza enzyme neuraminidase, to introduce a guanidinium group, guided by GRID calculations.[65] The resulting compound became zanamivir, marketed by GSK as Relenza. Further examples of such computational mapping approaches were the programs MCSS (Multiple Copy Simultaneous Search[66]) —where molecular dynamics calculations identify binding poses (that is a conformation, position and orientation) for larger functional groups (such as acetamide or benzene) and LUDI[67]—which used the pattern of functional group interactions seen in small molecule structures to predict hot spots for binding on the surface of a protein. Together, these computational approaches reinforced the idea that there are discrete binding sites for small functional groups on the surface of proteins.

The development of the MSCS approach (Multiple Solvent Crystal Structures) by Ringe[68] as extended by others[69,70] was the first attempt to map the binding site of a protein experimentally. These structures emphasised that the binding sites of proteins have evolved to be chemically attractive and that discrete binding poses can be observed even for very weak binding molecules.

As well as identifying functional group binding sites, the computational community also developed a number of different programs that attempted to link the observed or computed functional groups together—examples include the Caveat approach,[71] Hook[122] and the linking components in LUDI.[67] Although intellectually attractive and computationally elegant, these approaches did not gain widespread acceptance. The approaches can generate many millions of possible molecules and the main issue (which in many respects remain today) is the inability to accurately predict the energy of interaction of an evolved molecule. In addition, the programs would often generate compounds that were synthetically intractable. Ironically, the maturing of fragment based discovery has generated many thousands of crystal structures of fragments bound to a protein—a rich dataset on which these computational approaches could be revisited and refined.

5.7.3 The Emergence of Fragment-Based Lead Discovery

The first published example of the successful use of fragment-based methods to generate potent lead compounds is the pioneering work of the Abbott group in the mid-1990s. The first paper[29] coined the phrase "SAR by NMR" and described the discovery of nanomolar inhibitors of the FK506 binding protein. The approach relies on NMR spectroscopy for detecting fragment binding (using HSQC measurements, see above) and for determining the structure of protein–fragment complexes. Five separate stages can be identified as sketched out in Figure 5.6(a):

- Screen 1: a library of fragments (in Abbott's case, up to 10 000 of average MW 210) is screened for binding to the target by HSQC NMR;
- Optimise 1: selected hits are optimised for binding by limited chemistry;
- Screen 2: a second screen is performed of the protein with the optimised fragment bound, usually with smaller molecular weight fragments;

- Optimise 2: this second fragment is then optimised for binding to the target. The structure(s) of the target with the fragment(s) bound is determined by NMR and this information is used to guide linking;
- Linking: the linking of the fragments together to give higher affinity hits.

This initial paper on FK506 inhibitor discovery was followed by further work from this group, including stromelysin,[72] E2 from papilloma virus,[73] urokinase[74] and metalloproteases.[75] However, there were few reports of successful application of this same SAR by NMR approach from other organisations; this perhaps reflects the challenge of organising the appropriate medicinal chemistry resource around the biophysical methods, ensuring that such low affinity starting points are exploited by what can be quite challenging chemistry. In addition, there were not many targets where this strategy of linking multi-site fragments was appropriate.

As well as NMR, Abbott were the first (and sometimes forgotten) proponents of the use of X-ray crystallography for screening of fragments; the work of Nienaber *et al.*[76] established the approach which many others then went on to exploit.

The late 1990s saw variations of the methods developed in various large pharma such as the work at Roche[77] on crystallographic screening—the so-called Needles approach exemplified on DNA gyrase. However, the most significant growth in development and use of the methods, as well as increasingly vigorous campaigning for the approach, came from within the small, structure-based discovery companies such as Astex, Vernalis, SGX and Plexxikon that were established in the late 1990s and early 2000s. The core rationale for these companies was to build medicinal chemistry around a structure-based platform. Fragments (or in the case of Plexxikon, scaffolds focussed on a particular chemogenomic space[78,79]) provided a way for these smaller companies to obtain the initial hits without the investment in multi-million member compound libraries and the automation required for HTS. Some of the important method developments and examples from the early days of these companies were:

- Development of high throughput crystallography for screening and structure determination. Astex focus on developing a crystal system suitable for soaking of fragment mixtures and have invested heavily in developing the crystallographic and data analysis tools to analyse the resulting structures.[80,81] SGX (and also a company called Syrrx)[82–84] developed and exploited high throughput crystallisation coupled with dedicated synchrotron beamlines.
- Development of fragment libraries—Vernalis published one of the first detailed descriptions of how to select fragments for a library.[20] Astex developed some ideas on the characteristics of compounds that were hits in fragment screens and coined the "rule of 3".[85,86] Although most libraries are now closer to 180 Da average weight, the phrase was a useful marketing tool in establishing the field, though not without its critics,[87] as acknowledged in a recent perspective.[85] In addition, various analyses of the characteristics of fragments emerged which helped to guide further library design.[17,88,89]

5.7.4 Some Important Underpinning Concepts

An important early concept for fragments comes from the discussion of molecular complexity by Hann and co-workers.[4] They argued that a compound needs to be of a certain size (and complexity—that is number of features capable of interaction) to bind to a target, but if it gets too large then it is more likely to have features that prevent it from binding. A second important concept which has had a major impact on medicinal chemistry in general, and fragment evolution in particular, is that of ligand efficiency,[90] which built on some earlier ideas about the maximal affinity attainable by ligands.[91] These ideas have merged with others from retrospective analyses of the properties of successful drug-like compounds[92–94] to give a whole raft of different efficiency metrics.[95] There is a danger that these metrics can be applied too rigorously—for example, an early fragment may have much potential once optimised but low ligand efficiency to begin with. However, they do provide a useful monitor of compound properties to advise the chemist/modeller and a useful metric with which to persuade the fragment sceptic that these low affinity starting points are efficient.

5.8 FRAGMENT EVOLUTION

Quite a lot of time (and space) has been used in this article describing the design of fragment libraries and the methods for detecting fragment binding. However, it is relatively straight-forward to find fragment hits for most sites on most protein targets. The real challenge is deciding which fragments to choose for optimisation and evolving those fragments to effective hits and leads. There have been a number of reviews of published fragment based evolution.[11,96,97]

Figure 5.6 is a schematic summary of the three main approaches to fragment evolution. The initial ideas were to identify fragments binding to different sites on a protein and link them together using SAR by NMR as discussed above. Although there are some striking examples from the Abbott group, there are few other examples of fragment linking from other groups (some examples are reference 98, 99). Most examples in the literature are of fragment growing—where the structure of the initial fragment guides the addition of functional groups to achieve the required affinity and selectivity. Fragment merging is where information is used from a number of different sources (fragments, virtual screening, literature compounds) to design new scaffolds.

Figures 5.7–5.19 provide a summary of a set of fragment evolution stories, chosen to reflect these different approaches. The captions to the individual examples include notes on some of

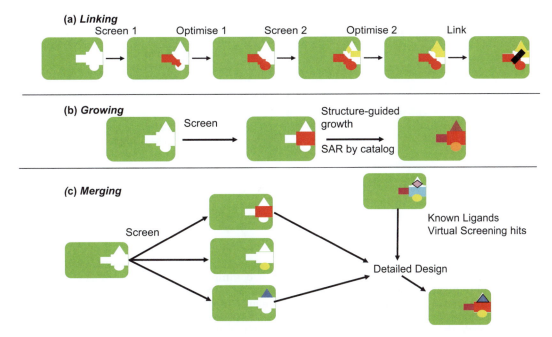

Figure 5.6 A schematic representation of the different strategies for fragment evolution.

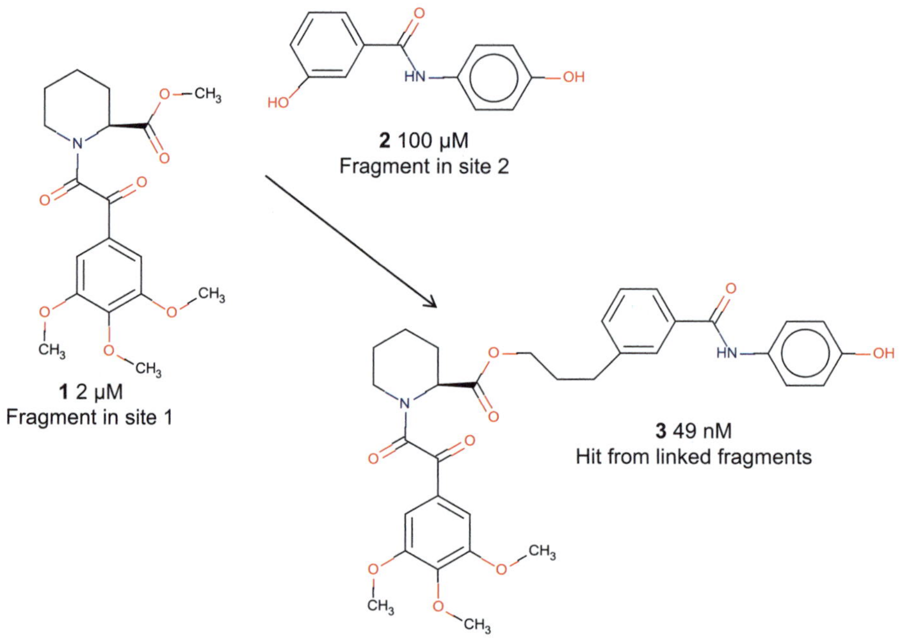

Figure 5.7   SAR by NMR: FKBP.[29] Compound **1** was identified by HSQC NMR from a screen of a fragment library as binding to a ^{15}N labelled sample of the protein FKBP. A subsequent screen by HSQC NMR of FKBP with a library of fragments in the presence of compound **1** identified compound **2** binding to an alternate site. Structures determined by NMR identified the appropriate vectors for linking the two fragments and compound **3** was identified as one of the more potent hits. The activity numbers are the reported K_D from the NMR HSQC titrations.

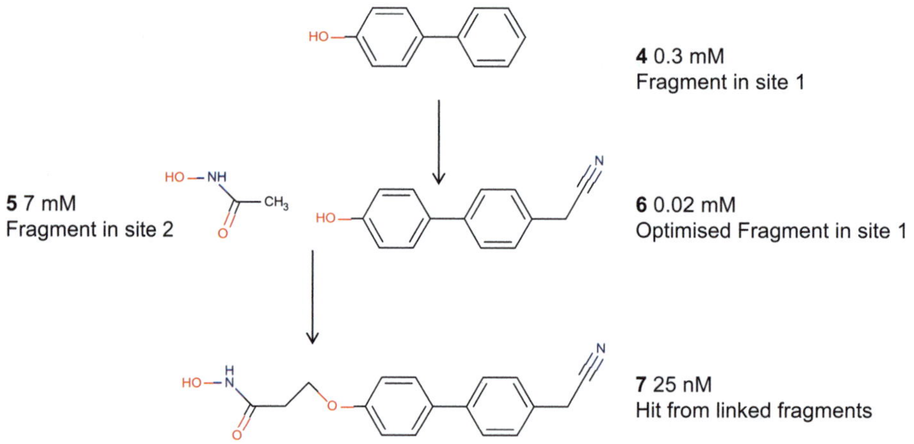

Figure 5.8   SAR by NMR: Stromelysin.[72] Compound **4** was identified by HSQC NMR from a screen of a fragment library as binding to a ^{15}N labelled sample of the metalloprotease, stromelysin. Subsequent optimisation led to compound **5**; a subsequent screen by HSQC NMR of stromelysin with a library of smaller fragments in the presence of compound **5** identified compound **6** binding to an alternate site. Structures determined by NMR identified the appropriate vectors for linking the two fragments and compound 7 was identified as one of the more potent hits.

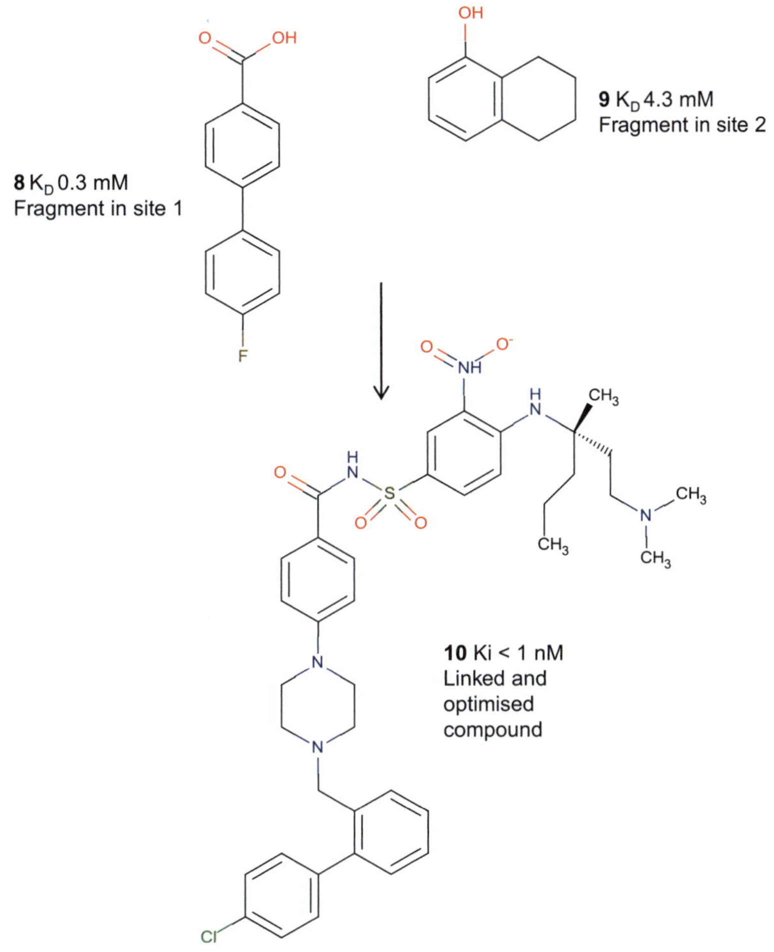

Figure 5.9 SAR by NMR: Bcl-2/Bcl-x$_L$.[104,105] Compound **8** was identified by HSQC NMR from a screen of a fragment library as binding to a ^{15}N labelled sample of the pro-apoptotic protein, Bcl-2. A subsequent screen by HSQC NMR of Bcl-2 with a library of smaller fragments in the presence of compound **8** identified compound **9** binding to an alternate site. A combination of NMR structure determination and substantial chemistry effort led to the clinical candidate, ABT737, compound **10**. The K$_D$ values were reported from an HSQC titration; the Ki value estimated from inhibition of BH3 domain binding in a wet assay. The projects summarised in Figures 5.7 and 5.8 are striking examples of linking two fragments where the initial fragments are very clearly present in the linked compounds. In this Bcl-2 example, there has been substantial evolution. However, the fragment methods provided starting points for chemistry which were not obtained from high throughput screening. Although considerable chemistry effort (and time) is needed to generate leads against such targets, the fragments at least provide a starting point.

the lessons that can be learnt. A more general comment is on ligand efficiency. As discussed above, this idea plays an important role in fragment and lead optimisation. The calculation keeps the design focussed on achieving the maximal potency possible with the addition of heavy atoms. This principle is nicely illustrated in the CHK1 example (Figure 5.11) where the fall in ligand efficiency highlighted that a particular modification was not optimal—and could be corrected. The CDK2 example (Figure 5.12) emphasises another aspect of fragment evolution: if the ligand efficiency is kept as high as possible during the early optimisation, then

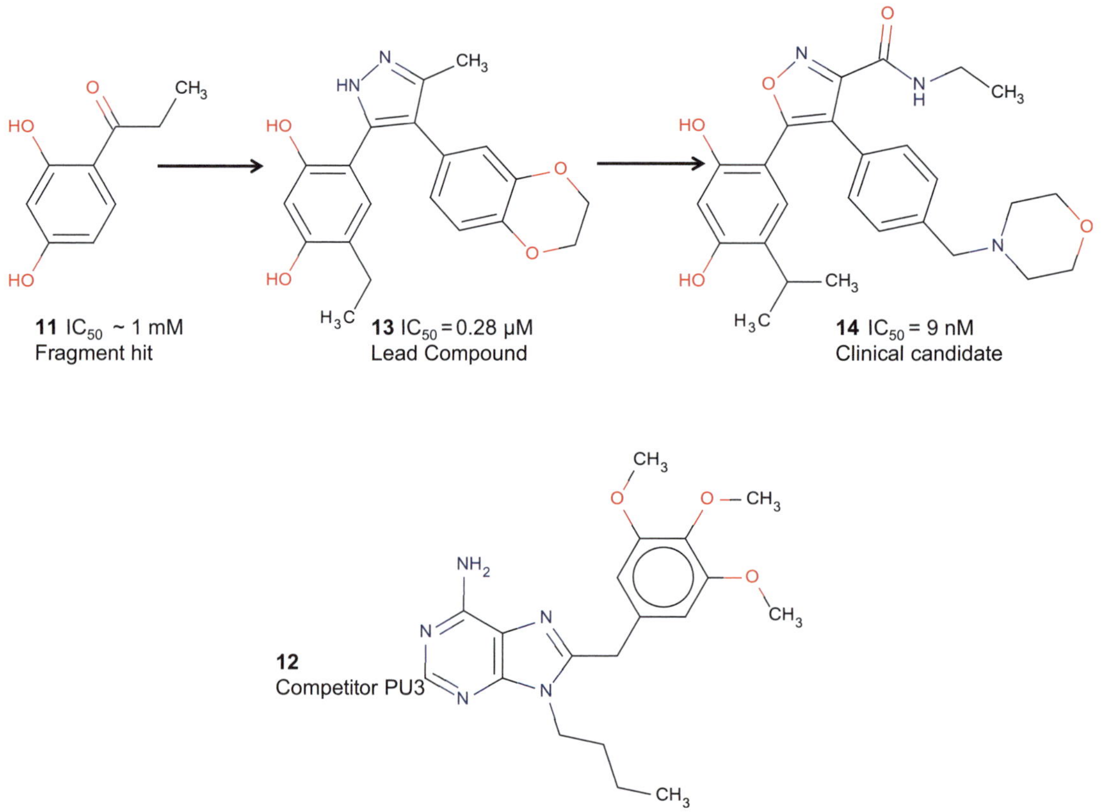

Figure 5.10   SAR by Catalog: HSP90.[102,106–109] Compound **11** was identified from a fragment library by ligand-observed NMR as competing with PU3[110] (compound **12**) for binding to N terminal domain of the molecular chaperone, HSP90. The resorcinol sub-structure was used to search a database of commercially available compounds;[103] the resulting compounds were assessed by computational docking for fit into the binding site. Compound **13** was identified as a sub-micromolar hit; this compound was also found by a medium throughput screen and by virtual screening. Subsequent structure-based optimisation led to the compound **14** (AUY922) which is currently in Phase II trials for various cancers. The activity numbers are the reported IC_{50} values from fluorescence polarisation measurements made for displacing a fluorescein labelled resorcinol-pyrazole.[111]

some of that efficiency can be "spent" in making modifications to improve drug-like potency and efficacy.

5.9   FRAGMENTS AND CHEMICAL SPACE

An early concern in fragment based discovery was whether selecting hits from such a small library would lead to problems in issues with IP for the optimised compounds. This fear has not been realised. There is some evidence that certain targets are attractive for certain scaffolds,[100] and it is clear that some companies have very similar fragments in their libraries. This becomes apparent when comparing fragment and other drug discovery campaigns from different companies on the same target. A details analysis has been carried out for HSP90 projects[101]—this analysis also emphasises the point that a fragment screen can quite effectively assess the accessible chemical space for a target. The first fragment screen on this target in 2002 identified 17 hits from a ligand-observed screen of a library of 729 fragments.[102] An analysis in 2011

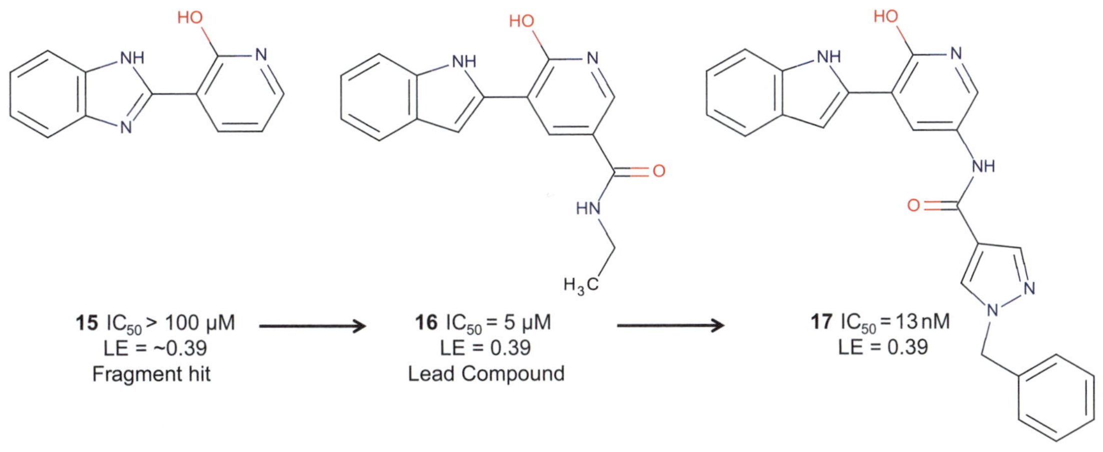

Figure 5.11 Fragment growth: CHK1. Compound **15** was identified from a fragment library as inhibiting the enzyme activity of the kinase, CHK1. The crystal structure of the fragment bound to the ATP pocket of CHK1 identified suitable vectors for growth—addition of the amide to give compound **16** gave an improvement in affinity while maintaining ligand efficiency. Subsequent optimisation led to compound **17** which retains the ligand efficiency. Earlier analogues of compound **17** had a reversed amide; the affinity was 200 μM with a ligand efficiency of 0.33. Analysis of the crystal structure identified sub-optimal hydrogen bonding to the solvent which was corrected in compound **17**. The activity numbers are the IC_{50} values from an activity assay for CHK1 which monitors the ATP driven phosphorylation of a CHK1 substrate.

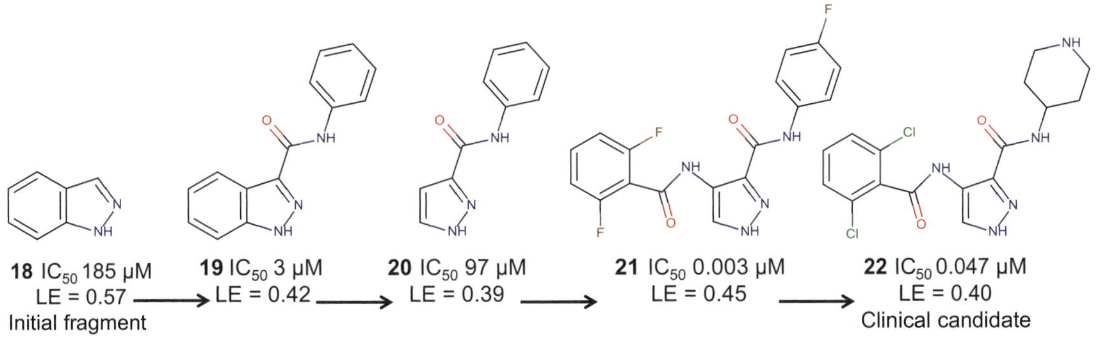

Figure 5.12 Fragment growth: CDK inhibitor.[112] Compound **18** was identified from an X-ray crystallographic screen of a fragment library as binding to CDK2. Iterative, detailed structure-based design led to compounds **19** to **21** in which the crystal structures of the growing compounds was used to identify regions of the compound for modification and addition. The final compound **22** is AT7519 which is being assessed in clinical trials as a therapy for solid tumours. The activity numbers are from a kinase activity assay. As discussed in the text, the high ligand efficiency achieved for compound **21** provided opportunities to tune the properties in compound **22** by "spending" some of that ligand efficiency.

HSP90 showed that just five of the fragments recapitulate the key features of the published lead and candidate compounds. In some cases, very similar fragments had been chosen for evolution by different companies; although there are sometimes some echoes of one company's compound in another, all of the compounds have achieved sufficient novelty.

23 K$_D$ 550 µM
LE = 0.37
Initial fragment

24 KD 15 µM
LE = 0.39

25 K$_D$ 200 µM
K$_i$ >500 µM
LE = 0.39

26 Ki 3.7 µM
LE = 0.45

27 Ki 2 nM
LE = 0.48

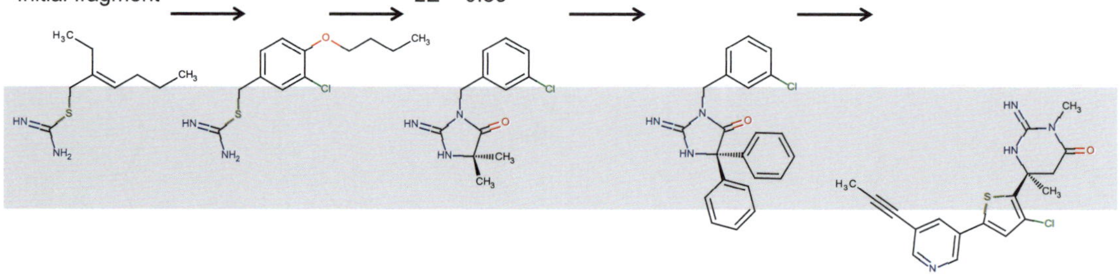

Figure 5.13 Fragment growth: BACE1 inhibitor.[113] Compound **23** was identified as binding to BACE1 from an HSQC based screen of a fragment library. 204 close analogues were synthesised from which compound **24** was identified which was co-crystallised with BACE1. The structure led to the design of the iminohydantoin scaffold (compound **25**) which presented vectors into substrate recognition pockets; subsequent structure based design led to compound **26** which has the right balance of physicochemical properties for *in vivo* efficacy. The grey square in the background emphasises the development of the core of this series. K$_D$ values were from NMR HSQC titrations; Ki values from an activity assay. This project demonstrated the importance of taking time to identify and optimise the initial scaffold and the need for careful and persistent structure-guided chemistry. Another example of designing a fragment using multiple crystal structures of different chemotypes is the work at Vertex on ltk kinase inhibitors.[114]

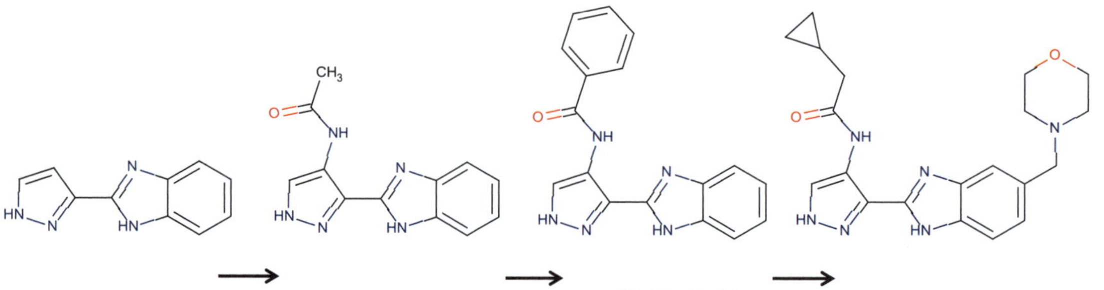

28 IC$_{50}$ 0.9 µM
LE = 0.59
CDK2 = 28 µM
Initial fragment

29 IC$_{50}$ 7 nM
LE = 0.54
CDK2 = 1 µM

30 IC$_{50}$ 6 nM
LE = 0.49
CDK2 = 52 nM

31 IC$_{50}$ 3 nM
LE = 0.41
CDK2 = 510 nM

Figure 5.14 Fragment growth: Aurora inhibitor.[115] Compound **28** was initially identified from an X-ray crystallographic screen of a fragment library as binding to CDK2 with an affinity of 28 µM; it was subsequently found to be a highly efficient inhibitor of Aurora A. This practise is variously called target hopping or chemogenomics, where a library of compounds which has some affinity for a class of proteins can be used as starting point for generation of selective inhibitors for specific members of the family. This is the same approach as used for B-raf (see Figure 5.15[116]). Iterative, detailed structure-based design led to compounds **29** to **31** in which the crystal structures of the growing compounds was used to identify regions of the compound for modification and addition. Of particular concern with this series was selectivity; as with the CDK2 example in Figure 5.12, the high ligand efficiency of compound **29** has been "spent" to improve the selectivity (and improve other physicochemical properties). The final compound **31** is AT9283 which is being assessed in clinical trials as a therapy for solid tumours and haematological malignancies. The activity numbers are from a kinase activity assay.

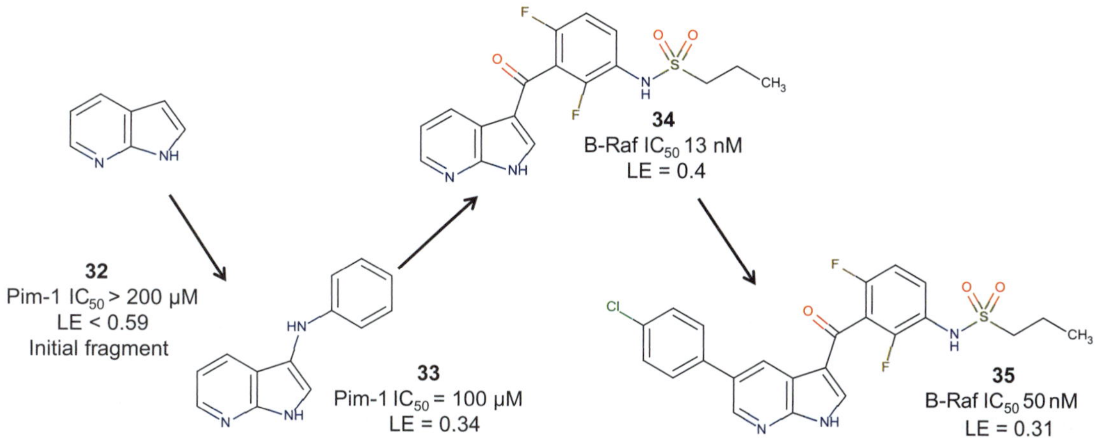

Figure 5.15 Fragment growth: B-raf kinase.[9,116] Compound **32** was initially identified by biochemical screening and confirmed by X-ray crystallography as binding to the kinase Pim-1 and early structure-guided optimisation gave compound **33** and the compound **34** (PLX4720). This design was based on binding to first Pim-1 and then Fgfr1 kinase, both of which were more readily crystallised. Such a surrogate approach (including introducing mutations into the binding pocket of a homologous protein which crystallises) can provide the necessary structural guidance for fragment and hit optimisation. A crystal structure is only a model and this approach can be successful as long as the rationale from the structures in the surrogate correlate with the inhibitory data generated in wet assays. Eventually, structures of B-Raf were generated leading to compound **35** now on the market as Vemurafenib for the treatment of mutant-B-Raf driven melanoma. Although the activity on the enzyme (and the targeted mutant B-Raf with V600E) is modest, it shows remarkable selectivity in cellular and *in vivo* models. The activity numbers are the reported activity from a kinase activity assay.

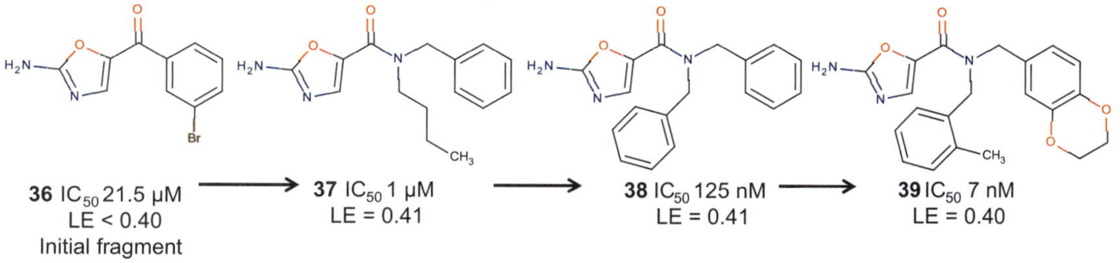

Figure 5.16 Fragment growth: Biotin carboxylase.[117] The initial fragment **36** was identified by biochemical screening of a library of 5200 fragments with validation by STD NMR. The crystal structure bound to biotin carboxylase identified potential regions for optimisation and led to the redesigned amide containing template **37**. Subsequent structure-guided growth led to **38** and then **39**. Additional hit series were also generated for this target by merging features of fragments and also merging fragments with functional groups identified from hit compounds discovered by virtual screening. The activity numbers are the reported IC_{50} from an activity assay for acetyl-CoA carboxylase (biotin carboxylase is a subunit of this enzyme).

5.10 CONCLUDING REMARKS

This article has covered most of the practical aspects of fragment based discovery. It will hopefully be useful in providing an introduction to the ideas and methods, with some practical

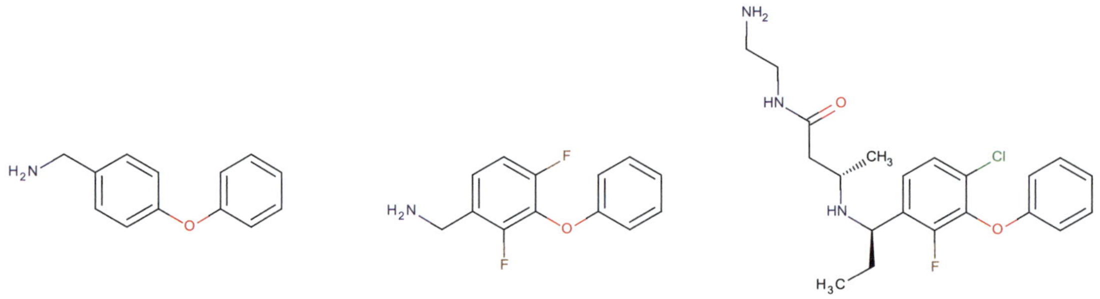

Figure 5.17 Fragment growth: HCV protease/helicase.[51] Crystals of the HCV protease/helicase were screened by x-ray crystallography; a novel, pre-existing cavity was identified at the interface between the two domains, into which fragments such as **40** bound. Such compounds stabilise the inactivated conformation of the protein; the functional significance of this allosteric effector was realised early on in optimisation of the compounds. The initial fragment was optimised to compound **41** before structure-guided growth led to compound **42**; this inhibits the virus *in vivo*. The activity numbers are the reported IC_{50} from a protease FRET based assay; K_D was measured by ITC.

examples of methods of fragment library design, screening and fragment evolution. There are many opportunities for future development of the methods and their effective integration with medicinal chemistry. In this concluding section, I consider some of them.

It is clear from the examples presented here, that most of the published examples of successful fragment based discovery campaigns are for "conventional" targets, such as kinases, where there are well defined binding sites and where structure-based design can rapidly progress fragments to leads. Although other screening methods such as HTS will identify many hit compounds, the advantage that fragments bring is in identifying novel scaffolds and binding motifs that can be exploited to give selectivity, novel IP, and most importantly, choice in the lead optimisation process. The main challenge in effective use of fragment based methods on such targets within established organisations is to persuade the medicinal chemist that starting from such a small starting point can be successful, rather than trying to fix the issues with a large hit from HTS. The opportunity in large organisations is using the fragments alongside HTS—the fragment hits can give insights into the important motifs and interactions that can be made and a deconstruction of the HTS hits into the constituent parts can lead to new ideas.

There is real potential for fragments to provide starting points against "non-conventional" targets, such as protein–protein interactions, large multi-subunit complexes or intrinsically disordered proteins; the challenge here is that although it is relatively straightforward to find fragment hits against such targets, it can take time (a number of years) and commitment to progress these fragments to suitable leads or even respectable hit compounds. Little has been published, as yet, on using fragments for such targets. There is anecdotal evidence of success, but that extreme care needs to be taken in evolving the fragments to hits—not only because of the difficulty of establishing robust models of how the fragments bind, but also because there are often issues with the assays for such targets that take some time to resolve.

In conclusion—the main advantage of fragment based discovery is that the methods give the medicinal chemist choice. They provide a large choice of starting points, the opportunity to start small and grow carefully to do "good" medicinal chemistry, and they provide starting points for targets for which more conventional screening approaches fail. An important message is that

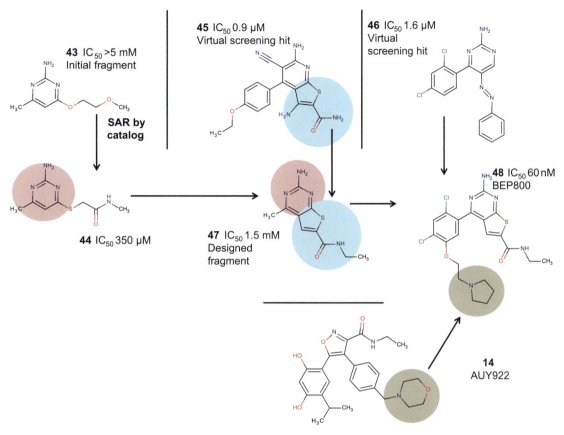

Figure 5.18 Merging fragments and other hits: HSP90.[102,118] Compound **43** was one of the 17 fragments identified as binding competitively with PU3[119] (compound **12**) by ligand-observed NMR screening of the N terminus of HSP90 with the initial Vernalis fragment library of 729 compounds[20] (SeeDs1) in 2002; subsequent SAR by catalog (guided by the crystal structure) led to compound **44**. Compounds **45** and **46** were identified by virtual screening of the ATPase site of the N terminus of HSP90. When these hits were initially considered, there was a concern that the thioether of **44** has potential metabolic liabilities; in addition, the cyano moiety of **45** was making unfavourable interactions with the binding site. The crystal structures of these various compounds bound to HSP90 suggested the design of the fragment **47**; this was synthesised and, although relatively weak binding, the crystal structure showed it bound as predicted. Subsequent structure-guided evolution of **48** was rapid, with incorporation of the dichloro-benzene suggested by **46** and the incorporation of a solubiliser at the position suggested by **14**, AUY922, the clinical candidate previously. The resulting compound **48** (BEP800) is orally bioavailable and shows efficacy in tumour regression models. The activity numbers are the reported IC_{50} from a fluorescence polarisation assay.[111] This example illustrates the power of combination of compound properties understood from multiple crystal structures; it also emphasises that the weaker binding fragments should be considered—the decision of which fragment to evolve is a balance of synthetic opportunity, the vectors that are available and the physicochemical properties of the resulting compounds.

fragment based discovery is not necessarily any faster—most projects spend some time exploring fragment SAR before embarking on optimisation chemistry. But the evidence suggests that the fragment based methods can deliver better quality of compounds and a higher chance of success in reaching clinical candidates.

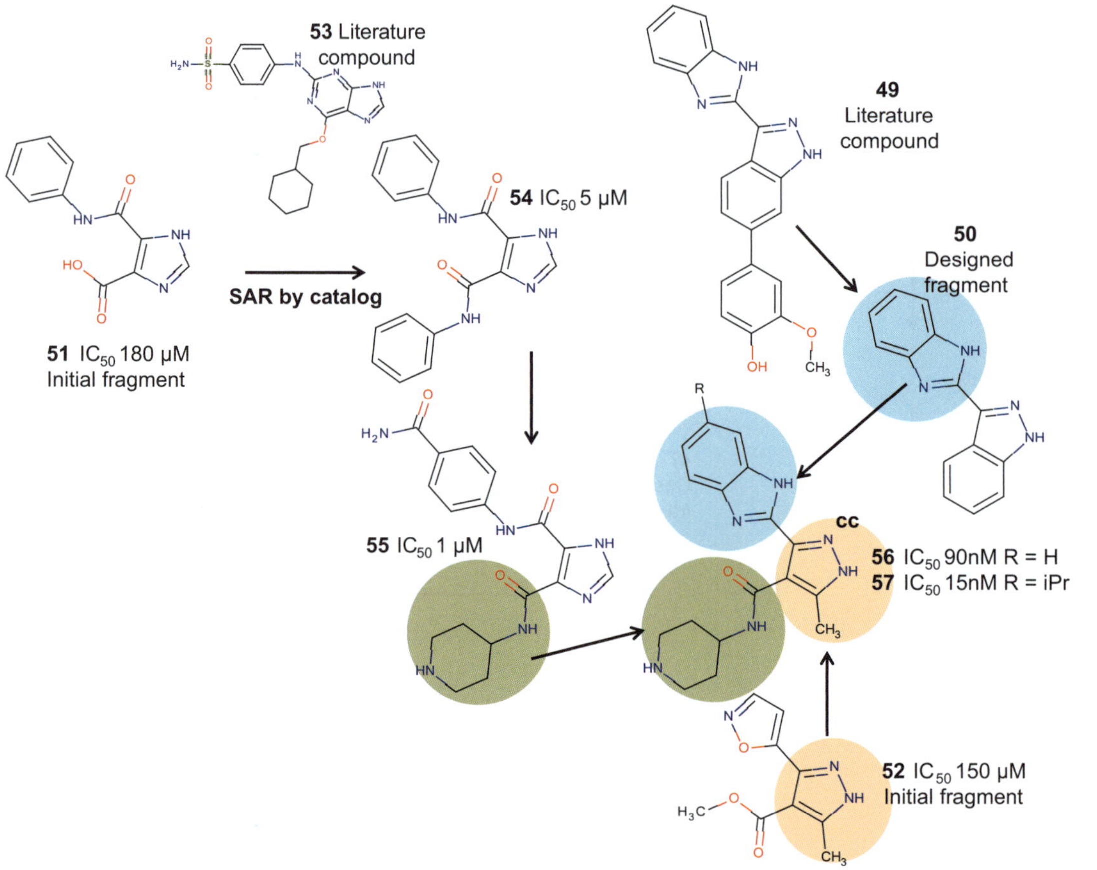

Figure 5.19 Merging fragments and other hits: PDPK1.[120] Compound **49** was a known promiscuous kinase inhibitor; the crystal structure bound to CHK1 kinase identified a core scaffold which gave the designed fragment **50**. A ligand observed NMR screen identified over 80 fragments that bound competitively with staurosporine to PDPK1; compounds **51** and **52** were selected. The publication of the CDK2 inhibitor compound **53** listed PDPK1 activity; a crystal structure bound to PDPK1 identified a key hydrophobic interaction made by the cyclohexyl ring. The available compounds were searched (SAR by catalog) for compounds that included a hydrophobic group in a similar position relative to the imidazole; compound **54** was identified which with minor structure-guided optimisation gave compound **55**. An overlay of compounds **50, 52, 54** and **55** suggested the merger of features into the compound **56** (R=H) which was optimised to **57** (R = iPr). This compound was relatively selective over other kinases, cell active and gave the expected PD marker responses when administered *in vivo*. This tool compound had appropriate properties to investigate the therapeutic potential of PDPK1 inhibitors in various models. The activity numbers are the reported IC_{50} from a kinase activity assay. Another example of fragment based discovery against this target is work at GSK.[121]

HINTS AND TIPS

- Fragments are small (MW < 250 Da, optimally around 190 Da) compounds that have suitable physicochemical properties for high concentration screening. The crucial property of a fragment is low molecular weight—remember that the size of chemical space increases by 8x per heavy atom added, so the smaller the compounds the greater the chemical space sampled when screening.

- Most methods for detecting fragment binding require high concentration screening; this means that most fragment libraries consist of highly soluble compounds
- It is vital that the fragment library is regularly curated to check for aggregation, solubility, or decomposition; this will save a lot of time and effort chasing false positives in screening.
- Most practitioners find that a fragment library of around 1000 fragments is sufficient to provide hits for most targets—it is resource intensive to curate much larger libraries. Key is to maintain a chemically diverse library.
- The most widely applicable and robust method for fragment screening is ligand-observed NMR; however, if sufficient protein is not available, then SPR is a good general purpose screening method. Key is gaining experience and expertise in the method, so that anomalies can be identified and the screening library optimised for that technique.
- Most successful fragment to lead campaigns involve structure-guided growth of the fragment.
- As fragment methods are increasingly deployed, there are opportunities for using the fragment hits to dissect key features of HTS hits.

Acknowledgements

I thank all at Vernalis and York who have made such important contributions to the development of my ideas about fragment-based drug discovery. In particular for this article, I have valued the enthusiasm and input to discussions from Ben Davis and James Murray and the analysis from Ijen Chen on fragment hit rates.

KEY REFERENCES

S. B. Shuker, P. J. Hajduk, R. P. Meadows and S. W. Fesik, *Science*, 1996, **274**, 1531–1534.

The very first publication on fragment based discovery from Abbott.

M. N. Schulz and R. E. Hubbard, *Curr. Opin. Pharmacol.*, 2009, **9**, 615–621.

A concise overview of the methods and applications.

Volume 593 of *Methods in Enzymology*: there are many papers from key practitioners in the field, with good commentaries on the methods and reviews. For example: R. E. Hubbard and J. B. Murray, *Methods Enzymol.*, 2011, **493**, 509–531.

REFERENCES

1. A. Ajay, W. P. Walters and M. A. Murcko, *J. Med. Chem.*, 1998, **41**, 3314–3324.
2. M. M. Hann and T. I. Oprea, *Curr. Opin. Chem. Biol.*, 2004, **8**, 255–263.
3. R. Macarron, *Drug Discovery Today*, 2006, **11**, 277–279.
4. M. M. Hann, A. R. Leach and G. Harper, *J. Chem. Inf. Comput. Sci.*, 2001, **41**, 856–864.
5. M. Orita, K. Ohno, M. Warizaya, Y. Amano and T. Niimi, *Methods Enzymol.*, 2011, **493**, 383–419.
6. M. Congreve, G. Chessari, D. Tisi and A. J. Woodhead, *J. Med. Chem.*, 2008, **51**, 3661–3680.

7. P. J. Hajduk and J. Greer, *Nat. Rev. Drug. Discov.*, 2007, **6**, 211–219.
8. G. E. de Kloe, D. Bailey, R. Leurs and I. J. de Esch, *Drug Discovery Today*, 2009, **14**, 630–646.
9. G. Bollag, J. Tsai, J. Zhang, C. Zhang, P. Ibrahim, K. Nolop and P. Hirth, *Nat. Rev. Drug. Discov.*, 2012, **11**, 873–886.
10. M. Fischer and R. E. Hubbard, *Mol. Interv.*, 2009, **9**, 22–30.
11. M. N. Schulz and R. E. Hubbard, *Curr. Opin. Pharmacol.*, 2009, **9**, 615–621.
12. R. E. Hubbard and J. B. Murray, *Methods Enzymol.*, 2011, **493**, 509–531.
13. D. C. Rees, M. Congreve, C. W. Murray and R. Carr, *Nat. Rev. Drug. Discov.*, 2004, **3**, 660–672.
14. G. Chessari and A. J. Woodhead, *Drug Discovery Today*, 2009, **14**, 668–675.
15. P. P. Seth, A. Miyaji, E. A. Jefferson, K. A. Sannes-Lowery, S. A. Osgood, S. S. Propp, R. Ranken, C. Massire, R. Sampath, D. J. Ecker, E. E. Swayze and R. H. Griffey, *J. Med. Chem.*, 2005, **48**, 7099–7102.
16. R. E. Hubbard, B. Davis, I. Chen and M. J. Drysdale, *Curr. Top. Med. Chem.*, 2007, **7**, 1568–1581.
17. I. J. Chen and R. E. Hubbard, *J. Comput. Aided Mol. Des.*, 2009, **23**, 603–620.
18. M. N. Schulz, J. Landstrom, K. Bright and R. E. Hubbard, *J. Comput. Aided Mol. Des.*, 2011, **25**, 611–620.
19. R. E. Hubbard, I. Chen and B. Davis, *Curr. Opin. Drug. Discov. Devel.*, 2007, **10**, 289–297.
20. N. Baurin, F. Aboul-Ela, X. Barril, B. Davis, M. Drysdale, B. Dymock, H. Finch, C. Fromont, C. Richardson, H. Simmonite and R. E. Hubbard, *J. Chem. Inf. Comput. Sci.*, 2004, **44**, 2157–2166.
21. W. F. Lau, J. M. Withka, D. Hepworth, T. V. Magee, Y. J. Du, G. A. Bakken, M. D. Miller, Z. S. Hendsch, V. Thanabal, S. A. Kolodziej, L. Xing, Q. Hu, L. S. Narasimhan, R. Love, M. E. Charlton, S. Hughes, W. P. van Hoorn and J. E. Mills, *J. Comput. Aided Mol. Des.*, 2011, **25**, 621–636.
22. J. S. Albert, N. Blomberg, A. L. Breeze, A. J. Brown, J. N. Burrows, P. D. Edwards, R. H. Folmer, S. Geschwindner, E. J. Griffen, P. W. Kenny, T. Nowak, L. L. Olsson, H. Sanganee and A. B. Shapiro, *Curr. Top. Med. Chem.*, 2007, **7**, 1600–1629.
23. C. Dalvit and A. Vulpetti, *Magn. Reson. Chem.*, 2012, **50**, 592–597.
24. A. Vulpetti and C. Dalvit, *Drug Discovery Today*, 2012, **17**, 890–897.
25. C. Dalvit, N. Mongelli, G. Papeo, P. Giordano, M. Veronesi, D. Moskau and R. Kummerle, *J. Am. Chem. Soc.*, 2005, **127**, 13380–13385.
26. J. Blaney, V. Nienaber and S. K. Burley, in *Fragment-based approaches in drug discovery*, ed. W. Jahnke and D. A. Erlanson, Wiley-VCH.Verlag GmbH & Co, KGaA, 2006, pp. 215–248.
27. F. Lovering, J. Bikker and C. Humblet, *J. Med. Chem.*, 2009, **52**, 6752–6756.
28. T. Fink and J. L. Reymond, *J. Chem. Inf. Model.*, 2007, **47**, 342–353.
29. S. B. Shuker, P. J. Hajduk, R. P. Meadows and S. W. Fesik, *Science*, 1996, **274**, 1531–1534.
30. P. J. Hajduk, *Mol. Interv.*, 2006, **6**, 266–272.
31. W. Jahnke, R. M. Grotzfeld, X. Pelle, A. Strauss, G. Fendrich, S. W. Cowan-Jacob, S. Cotesta, D. Fabbro, P. Furet, J. Mestan and A. L. Marzinzik, *J. Am. Chem. Soc.*, 2010, **132**, 7043–7048.
32. M. Mayer and B. Meyer, *Angew. Chem., Int. Ed.*, 1999, **38**, 1784–1788.
33. C. Dalvit, P. Pevarello, M. Tatò, M. Veronesi, A. Vulpetti and M. Sundström, *J. Biomol. NMR*, 2000, **18**, 65–68.
34. S. G. Meiboom and D. Gill, *Rev. Sci. Instrum.*, 1958, **29**, 688–691.
35. M. Kobayashi, K. Retra, F. Figaroa, J. G. Hollander, E. Ab, R. J. Heetebrij, H. Irth and G. Siegal, *J. Biomol. Screening*, 2010, **15**, 978–989.
36. G. Siegal and J. G. Hollander, *Curr. Top. Med. Chem.*, 2009, **9**, 1736–1745.

37. S. Vanwetswinkel, R. J. Heetebrij, J. van Duynhoven, J. G. Hollander, D. V. Filippov, P. J. Hajduk and G. Siegal, *Chem. Biol.*, 2005, **12**, 207–216.

38. R. A. Copeland, D. L. Pompliano and T. D. Meek, *Nat. Rev. Drug. Discov.*, 2006, **5**, 730–739.

39. A. M. Giannetti, in *Fragment-Based Drug Design: Tools, Practical Approaches, and Examples*, ed. L. C. Kuo, 2011, vol. 493 , pp. 169–218.

40. R. L. Rich, G. A. Papalia, P. J. Flynn, J. Furneisen, J. Quinn, J. S. Klein, P. S. Katsamba, M. B. Waddell, M. Scott, J. Thompson, J. Berlier, S. Corry, M. Baltzinger, G. Zeder-Lutzi, A. Schoenemann, A. Clabbers, S. Wieckowski, M. M. Murphy, P. Page, T. E. Ryan, J. Duffner, T. Ganguly, J. Corbin, S. Gautam, G. Anderluh, A. Bavdek, D. Reichmann, S. P. Yadav, E. Hommema, E. Pol, A. Drake, S. Klakamp, T. Chapman, D. Kernaghan, K. Miller, J. Schuman, K. Lindquist, K. Herlihy, M. B. Murphy, R. Bohnsack, B. Andrien, P. Brandani, D. Terwey, R. Millican, R. J. Darling, L. Wang, Q. Carter, J. Dotzlaf, J. Lopez-Sagaseta, I. Campbell, P. Torreri, S. Hoos, P. England, Y. Liu, Y. Abdiche, D. Malashock, A. Pinkerton, M. Wong, E. Lafer, C. Hinck, K. Thompson, C. Di Primo, A. Joyce, J. Brooks, F. Torta, A. B. B. Hagel, J. Krarup, J. Pass, M. Ferreira, S. Shikov, M. Mikolajczyk, Y. Abe, G. Barbato, A. M. Giannetti, G. Krishnamoorthy, B. Beusink, D. Satpaev, T. Tsang, E. Fang, J. Partridge, S. Brohawn, J. Horn, O. Pritsch, G. Obal, S. Nilapwar, B. Busby, G. Gutierrez-Sanchez, R. Das Gupta, S. Canepa, K. Witte, Z. Nikolovska-Coleska, Y. H. Cho, R. D'Agata, K. Schlick, R. Calvert, E. M. Munoz, M. J. Hernaiz, T. Bravman, M. Dines, M. H. Yang, A. Puskas, E. Boni, J. J. Li, M. Wear, A. Grinberg, J. Baardsnes, O. Dolezal, M. Gainey, H. Anderson, J. L. Peng, M. Lewis, P. Spies, Q. Trinh, S. Bibikov, J. Raymond, M. Yousef, V. Chandrasekaran, Y. G. Feng, A. Emerick, S. Mundodo, R. Guimaraes, K. McGirr, Y. J. Li, H. Hughes, H. Mantz, R. Skrabana, M. Witmer, J. Ballard, L. Martin, P. Skladal, G. Korza, I. Laird-Offringa, C. S. Lee, A. Khadir, F. Podlaski, P. Neuner, J. Rothacker, A. Rafique, N. Dankbar, P. Kainz, E. Gedig, M. Vuyisich, C. Boozer, N. Ly, M. Toews, A. Uren, O. Kalyuzhniy, K. Lewis, E. Chomey, B. J. Pak and D. G. Myszka, *Anal. Biochem.*, 2009, **386**, 194–216.

41. M. Elinder, M. Geitmann, T. Gossas, P. Kallblad, J. Winquist, H. Nordstrom, M. Hamalainen and U. H. Danielson, *J. Biomol. Screening*, 2011, **16**, 15–25.

42. U. H. Danielson, *Curr. Top. Med. Chem.*, 2009, **9**, 1725–1735.

43. G. Papalia and D. Myszka, *Anal. Biochem.*, 2010, **403**, 30–35.

44. M. Fischer, A. P. Leech and R. E. Hubbard, *Anal Chem*, 2011.

45. J. K. Kranz and C. Schalk-Hihi, *Methods Enzymol.*, 2011, **493**, 277–298.

46. F. H. Niesen, H. Berglund and M. Vedadi, *Nat. Protoc.*, 2007, **2**, 2212–2221.

47. M. Vedadi, F. H. Niesen, A. Allali-Hassani, O. Y. Fedorov, P. J. Finerty, Jr., G. A. Wasney, R. Yeung, C. Arrowsmith, L. J. Ball, H. Berglund, R. Hui, B. D. Marsden, P. Nordlund, M. Sundstrom, J. Weigelt and A. M. Edwards, *Proc. Natl. Acad. Sci. U. S. A.*, 2006, **103**, 15835–15840.

48. M. N. Schulz, J. Landstrom and R. E. Hubbard, *Anal. Biochem.*, 2013, **433**, 43–47.

49. K. Babaoglu, A. Simeonov, J. J. Irwin, M. E. Nelson, B. Feng, C. J. Thomas, L. Cancian, M. P. Costi, D. A. Maltby, A. Jadhav, J. Inglese, C. P. Austin and B. K. Shoichet, *J. Med. Chem.*, 2008, **51**, 2502–2511.

50. K. E. Coan and B. K. Shoichet, *J. Am. Chem. Soc.*, 2008, **130**, 9606–9612.

51. S. M. Saalau-Bethell, A. J. Woodhead, G. Chessari, M. G. Carr, J. Coyle, B. Graham, S. D. Hiscock, C. W. Murray, P. Pathuri, S. J. Rich, C. J. Richardson, P. A. Williams and H. Jhoti, *Nat. Chem. Biol.*, 2012, **8**, 920–925.

52. J. D. Bauman, D. Patel, C. Dharia, M. W. Fromer, S. Ahmed, Y. Frenkel, R. S. Vijayan, J. T. Eck, W. C. Ho, K. Das, A. J. Shatkin and E. Arnold, *J. Med. Chem.*, 2013, **56**, 2738–2746.

53. A. L. Perryman, Q. Zhang, H. H. Soutter, R. Rosenfeld, D. E. McRee, A. J. Olson, J. E. Elder and C. D. Stout, *Chem. Biol. Drug Des*, 2010, **75**, 257–268.

54. S. E. Rollauer, M. J. Tarry, J. E. Graham, M. Jaaskelainen, F. Jaeger, S. Johnson, M. Krehenbrink, S.-M. Liu, M. J. Lukey, J. Marcoux, M. A. McDowell, F. Rodriguez, P. Roversi, P. J. Stansfeld, C. V. Robinson, M. S. P. Sansom, T. Palmer, M. Hogbom, B. C. Berks and S. M. Lea, *Nature*, 2012, **492**, 210–214.

55. M. K. Ameriks, S. D. Bembenek, M. T. Burdett, I. C. Choong, J. P. Edwards, D. Gebauer, Y. Gu, L. Karlsson, H. E. Purkey, B. L. Staker, S. Q. Sun, R. L. Thurmond and J. A. Zhu, *Bioorg. Med. Chem. Lett.*, 2010, **20**, 4060–4064.

56. D. A. Erlanson, J. A. Wells and A. C. Braisted, *Annu. Rev. Biophys. Biomol. Struct.*, 2004, **33**, 199–223.

57. T. Hesterkamp, J. Barker, A. Davenport and M. Whittaker, *Curr. Top. Med. Chem.*, 2007, 7, 1582–1591.

58. C. Austin, S. N. Pettit, S. K. Magnolo, J. Sanvoisin, W. Chen, S. P. Wood, L. D. Freeman, R. J. Pengelly and D. E. Hughes, *J. Biomol. Screening*, 2012, **17**, 868–876.

59. E. Meiby, H. Simmonite, L. le Strat, B. Davis, N. Matassova, J. D. Moore, M. Mrosek, J. Murray, R. E. Hubbard and S. Ohlson, *Anal. Chem.*, 2013, **85**, 6756–6766.

60. S. A. Wildman and G. M. Crippen, *J. Chem. Inf. Comput. Sci.*, 1999, **39**, 868–873.

61. W. P. Jencks, *Proc. Natl. Acad. Sci. U. S. A.*, 1981, **78**, 4046–4050.

62. P. R. Andrews, D. J. Craik and J. L. Martin, *J. Med. Chem.*, 1984, **27**, 1648–1657.

63. D. J. Abraham, M. F. Perutz and S. E. Phillips, *Proc. Natl. Acad. Sci. U. S. A.*, 1983, **80**, 324–328.

64. P. J. Goodford, *J. Med. Chem.*, 1985, **28**, 849–857.

65. M. Vonitzstein, W. Y. Wu, G. B. Kok, M. S. Pegg, J. C. Dyason, B. Jin, T. V. Phan, M. L. Smythe, H. F. White, S. W. Oliver, P. M. Colman, J. N. Varghese, D. M. Ryan, J. M. Woods, R. C. Bethell, V. J. Hotham, J. M. Cameron and C. R. Penn, *Nature*, 1993, **363**, 418–423.

66. A. Miranker and M. Karplus, *Proteins*, 1991, **11**, 29–34.

67. H. J. Bohm, *J. Comput. Aided Mol. Des.*, 1992, **6**, 61–78.

68. K. N. Allen, C. R. Bellamacina, X. C. Ding, C. J. Jeffery, C. Mattos, G. A. Petsko and D. Ringe, *J. Phys. Chem.*, 1996, **100**, 2605–2611.

69. A. C. English, C. R. Groom and R. E. Hubbard, *Protein Eng.*, 2001, **14**, 47–59.

70. A. C. English, S. H. Done, L. S. Caves, C. R. Groom and R. E. Hubbard, *Proteins*, 1999, **37**, 628–640.

71. P. A. Bartlett, G. T. Shea, S. Waterman and S. J. Telfer, *Abstr. Pap. Am. Chem. S.*, 1991, **202**, 44–COMP.

72. E. T. Olejniczak, P. J. Hajduk, P. A. Marcotte, D. G. Nettesheim, R. P. Meadows, R. Edalji, T. F. Holzman and S. W. Fesik, *J. Am. Chem. Soc.*, 1997, **119**, 5828–5832.

73. P. J. Hajduk, J. Dinges, G. F. Miknis, M. Merlock, T. Middleton, D. J. Kempf, D. A. Egan, K. A. Walter, T. S. Robins, S. B. Shuker, T. F. Holzman and S. W. Fesik, *J. Med. Chem.*, 1997, **40**, 3144–3150.

74. P. J. Hajduk, S. Boyd, D. Nettesheim, V. Nienaber, J. Severin, R. Smith, D. Davidson, T. Rockway and S. W. Fesik, *J. Med. Chem.*, 2000, **43**, 3862–3866.

75. S. W. Fesik, P. J. Hajduk, G. Sheppard, E. T. Olejniczak, D. G. Nettesheim, R. P. Meadows, P. A. Marcotte, S. B. Shuker, D. H. Steinman, G. M. Carrera, J. Severin, K. Walter, H. Smith, E. Gubbins, R. Simmer, T. F. Holzman, D. W. Morgan, S. K. Davidsen and J. B. Summers, *Abstr. Pap. Am. Chem. S.*, 1997, **214**, 109–MEDI.

76. V. L. Nienaber, P. L. Richardson, V. Klighofer, J. J. Bouska, V. L. Giranda and J. Greer, *Nat. Biotechnol.*, 2000, **18**, 1105–1108.

77. H. J. Boehm, M. Boehringer, D. Bur, H. Gmuender, W. Huber, W. Klaus, D. Kostrewa, H. Kuehne, T. Luebbers, N. Meunier-Keller and F. Mueller, *J. Med. Chem.*, 2000, **43**, 2664–2674.

78. D. R. Artis, J. J. Lin, C. Zhang, W. Wang, U. Mehra, M. Perreault, D. Erbe, H. I. Krupka, B. P. England, J. Arnold, A. N. Plotnikov, A. Marimuthu, H. Nguyen, S. Will, M. Signaevsky, J. Kral, J. Cantwell, C. Settachatgull, D. S. Yan, D. Fong, A. Oh, S. Shi, P. Womack, B. Powell, G. Habets, B. L. West, K. Y. Zhang, M. V. Milburn, G. P. Vlasuk, K. P. Hirth, K. Nolop, G. Bollag, P. N. Ibrahim and J. F. Tobin, *Proc. Natl. Acad. Sci. U. S. A.*, 2009, **106**, 262–267.

79. G. L. Card, L. Blasdel, B. P. England, C. Zhang, Y. Suzuki, S. Gillette, D. Fong, P. N. Ibrahim, D. R. Artis, G. Bollag, M. V. Milburn, S. H. Kim, J. Schlessinger and K. Y. Zhang, *Nat. Biotechnol.*, 2005, **23**, 201–207.

80. W. T. Mooij, M. J. Hartshorn, I. J. Tickle, A. J. Sharff, M. L. Verdonk and H. Jhoti, *ChemMedChem*, 2006, **1**, 827–838.

81. T. L. Blundell, H. Jhoti and C. Abell, *Nat. Rev. Drug. Discov.*, 2002, **1**, 45–54.

82. P. Kuhn, K. Wilson, M. G. Patch and R. C. Stevens, *Curr. Opin. Chem. Biol.*, 2002, **6**, 704–710.

83. Z. Zhang, N. K. Sauter, H. van den Bedem, G. Snell and A. M. Deacon, *J. Appl. Crystallogr.*, 2006, **39**, 112–119.

84. D. Hosfield, J. Palan, M. Hilgers, D. Scheibe, D. E. McRee and R. C. Stevens, *J. Struct. Biol.*, 2003, **142**, 207–217.

85. H. Jhoti, G. Williams, D. C. Rees and C. W. Murray, *Nat. Rev. Drug. Discov.*, 2013, **12**, 644.

86. M. Congreve, R. Carr, C. Murray and H. Jhoti, *Drug Discovery Today*, 2003, **8**, 876–877.

87. H. Koster, T. Craan, S. Brass, C. Herhaus, M. Zentgraf, L. Neumann, A. Heine and G. Klebe, *J. Med. Chem.*, 2011, **54**, 7784–7796.

88. P. J. Hajduk, M. Bures, J. Praestgaard and S. W. Fesik, *J. Med. Chem.*, 2000, **43**, 3443–3447.

89. P. J. Hajduk, *J. Med. Chem.*, 2006, **49**, 6972–6976.

90. A. L. Hopkins, C. R. Groom and A. Alex, *Drug Discovery Today*, 2004, **9**, 430–431.

91. I. D. Kuntz, K. Chen, K. A. Sharp and P. A. Kollman, *Proc. Natl. Acad. Sci. U. S. A.*, 1999, **96**, 9997–10002.

92. J. R. Empfield and P. D. Leeson, *IDrugs*, 2010, **13**, 869–873.

93. P. D. Leeson and B. Springthorpe, *Nat. Rev. Drug. Discov.*, 2007, **6**, 881–890.

94. A. P. Hill and R. J. Young, *Drug Discovery Today*, 2010, **15**, 648–655.

95. M. D. Shultz, *Bioorg. Med. Chem. Lett.*, 2013, **23**, 5980–5991.

96. G. G. Ferenczy and G. M. Keseru, *J. Med. Chem.*, 2013, **56**, 2478–2486.

97. D. A. Erlanson, *Curr. Opin. Biotechnol.*, 2006, **17**, 643–652.

98. J. J. Barker, O. Barker, S. M. Courtney, M. Gardiner, T. Hesterkamp, O. Ichihara, O. Mather, C. A. Montalbetti, A. Muller, M. Varasi, M. Whittaker and C. J. Yarnold, *ChemMedChem*, 2010, **5**, 1697–1700.

99. N. Howard, C. Abell, W. Blakemore, G. Chessari, M. Congreve, S. Howard, H. Jhoti, C. W. Murray, L. C. Seavers and R. L. van Montfort, *J. Med. Chem.*, 2006, **49**, 1346–1355.

100. P. J. Hajduk, M. Bures, J. Praestgaard and S. W. Fesik, *J. Med. Chem.*, 2000, **43**, 3443–3447.

101. S. D. Roughley and R. E. Hubbard, *J. Med. Chem.*, 2011.

102. S. Roughley, L. Wright, P. Brough, A. Massey and R. E. Hubbard, *Top. Curr. Chem.*, 2011.

103. N. Baurin, R. Baker, C. Richardson, I. Chen, N. Foloppe, A. Potter, A. Jordan, S. Roughley, M. Parratt, P. Greaney, D. Morley and R. E. Hubbard, *J. Chem. Inf. Comput. Sci.*, 2004, **44**, 643–651.

104. T. Oltersdorf, S. W. Elmore, A. R. Shoemaker, R. C. Armstrong, D. J. Augeri, B. A. Belli, M. Bruncko, T. L. Deckwerth, J. Dinges, P. J. Hajduk, M. K. Joseph, S. Kitada, S. J. Korsmeyer, A. R. Kunzer, A. Letai, C. Li, M. J. Mitten, D. G. Nettesheim, S. Ng, P. M. Nimmer, J. M. O'Connor, A. Oleksijew, A. M. Petros, J. C. Reed, W. Shen, S. K. Tahir, C. B. Thompson, K. J. Tomaselli, B. Wang, M. D. Wendt, H. Zhang, S. W. Fesik and S. H. Rosenberg, *Nature*, 2005, **435**, 677–681.

105. A. M. Petros, J. Dinges, D. J. Augeri, S. A. Baumeister, D. A. Betebenner, M. G. Bures, S. W. Elmore, P. J. Hajduk, M. K. Joseph, S. K. Landis, D. G. Nettesheim, S. H. Rosenberg, W. Shen, S. Thomas, X. Wang, I. Zanze, H. Zhang and S. W. Fesik, *J. Med. Chem.*, 2005, **49**, 656–663.

106. X. Barril, P. Brough, M. Drysdale, R. E. Hubbard, A. Massey, A. Surgenor and L. Wright, *Bioorg. Med. Chem. Lett.*, 2005, **15**, 5187–5191.

107. P. A. Brough, W. Aherne, X. Barril, J. Borgognoni, K. Boxall, J. E. Cansfield, K. M. Cheung, I. Collins, N. G. Davies, M. J. Drysdale, B. Dymock, S. A. Eccles, H. Finch, A. Fink, A. Hayes, R. Howes, R. E. Hubbard, K. James, A. M. Jordan, A. Lockie, V. Martins, A. Massey, T. P. Matthews, E. McDonald, C. J. Northfield, L. H. Pearl, C. Prodromou, S. Ray, F. I. Raynaud, S. D. Roughley, S. Y. Sharp, A. Surgenor, D. L. Walmsley, P. Webb, M. Wood, P. Workman and L. Wright, *J. Med. Chem.*, 2008, **51**, 196–218.

108. B. W. Dymock, X. Barril, P. A. Brough, J. E. Cansfield, A. Massey, E. McDonald, R. E. Hubbard, A. Surgenor, S. D. Roughley, P. Webb, P. Workman, L. Wright and M. J. Drysdale, *J. Med. Chem.*, 2005, **48**, 4212–4215.

109. S. A. Eccles, A. Massey, F. I. Raynaud, S. Y. Sharp, G. Box, M. Valenti, L. Patterson, A. de Haven Brandon, S. Gowan, F. Boxall, W. Aherne, M. Rowlands, A. Hayes, V. Martins, F. Urban, K. Boxall, C. Prodromou, L. Pearl, K. James, T. P. Matthews, K. M. Cheung, A. Kalusa, K. Jones, E. McDonald, X. Barril, P. A. Brough, J. E. Cansfield, B. Dymock, M. J. Drysdale, H. Finch, R. Howes, R. E. Hubbard, A. Surgenor, P. Webb, M. Wood, L. Wright and P. Workman, *Cancer Res.*, 2008, **68**, 2850–2860.

110. L. Wright, X. Barril, B. Dymock, L. Sheridan, A. Surgenor, M. Beswick, M. Drysdale, A. Collier, A. Massey, N. Davies, A. Fink, C. Fromont, W. Aherne, K. Boxall, S. Sharp, P. Workman and R. E. Hubbard, *Chem. Biol.*, 2004, **11**, 775–785.

111. R. Howes, X. Barril, B. W. Dymock, K. Grant, C. J. Northfield, A. G. Robertson, A. Surgenor, J. Wayne, L. Wright, K. James, T. Matthews, K. M. Cheung, E. McDonald, P. Workman and M. J. Drysdale, *Anal. Biochem.*, 2006, **350**, 202–213.

112. P. G. Wyatt, A. J. Woodhead, V. Berdini, J. A. Boulstridge, M. G. Carr, D. M. Cross, D. J. Davis, L. A. Devine, T. R. Early, R. E. Feltell, E. J. Lewis, R. L. McMenamin, E. F. Navarro, M. A. O'Brien, M. O'Reilly, M. Reule, G. Saxty, L. C. Seavers, D. M. Smith, M. S. Squires, G. Trewartha, M. T. Walker and A. J. Woolford, *J. Med. Chem.*, 2008, **51**, 4986–4999.

113. Z. Zhu, Z. Y. Sun, Y. Ye, J. Voigt, C. Strickland, E. M. Smith, J. Cumming, L. Wang, J. Wong, Y. S. Wang, D. F. Wyss, X. Chen, R. Kuvelkar, M. E. Kennedy, L. Favreau, E. Parker, B. A. McKittrick, A. Stamford, M. Czarniecki, W. Greenlee and J. C. Hunter, *J. Med. Chem.*, 2010, **53**, 951–965.

114. J. D. Charrier, A. Miller, D. P. Kay, G. Brenchley, H. C. Twin, P. N. Collier, S. Ramaya, S. B. Keily, S. J. Durrant, R. M. Knegtel, A. J. Tanner, K. Brown, A. P. Curnock and J. M. Jimenez, *J. Med. Chem.*, 2011, **54**, 2341–2350.

115. S. Howard, V. Berdini, J. A. Boulstridge, M. G. Carr, D. M. Cross, J. Curry, L. A. Devine, T. R. Early, L. Fazal, A. L. Gill, M. Heathcote, S. Maman, J. E. Matthews, R. L. McMenamin, E. F. Navarro, M. A. O'Brien, M. O'Reilly, D. C. Rees, M. Reule, D. Tisi, G. Williams, M. Vinkovic and P. G. Wyatt, *J. Med. Chem.*, 2009, **52**, 379–388.

116. J. Tsai, J. T. Lee, W. Wang, J. Zhang, H. Cho, S. Mamo, R. Bremer, S. Gillette, J. Kong, N. K. Haass, K. Sproesser, L. Li, K. S. Smalley, D. Fong, Y. L. Zhu, A. Marimuthu, H. Nguyen, B. Lam, J. Liu, I. Cheung, J. Rice, Y. Suzuki, C. Luu, C. Settachatgul, R. Shellooe, J. Cantwell, S. H. Kim, J. Schlessinger, K. Y. Zhang, B. L. West, B. Powell, G. Habets, C. Zhang, P. N. Ibrahim, P. Hirth, D. R. Artis, M. Herlyn and G. Bollag, *Proc. Natl. Acad. Sci. U. S. A.*, 2008, **105**, 3041–3046.

117. I. Mochalkin, J. R. Miller, L. Narasimhan, V. Thanabal, P. Erdman, P. B. Cox, J. V. Prasad, S. Lightle, M. D. Huband and C. K. Stover, *ACS Chem. Biol.*, 2009, **4**, 473–483.

118. P. A. Brough, X. Barril, J. Borgognoni, P. Chene, N. G. Davies, B. Davis, M. J. Drysdale, B. Dymock, S. A. Eccles, C. Garcia-Echeverria, C. Fromont, A. Hayes, R. E. Hubbard, A. M. Jordan, M. R. Jensen, A. Massey, A. Merrett, A. Padfield, R. Parsons, T. Radimerski, F. I. Raynaud, A. Robertson, S. D. Roughley, J. Schoepfer, H. Simmonite, S. Y. Sharp, A. Surgenor, M. Valenti, S. Walls, P. Webb, M. Wood, P. Workman and L. Wright, *J. Med. Chem.*, 2009, **52**, 4794–4809.

119. B. Dymock, X. Barril, M. Beswick, A. Collier, N. Davies, M. Drysdale, A. Fink, C. Fromont, R. E. Hubbard, A. Massey, A. Surgenor and L. Wright, *Bioorg. Med. Chem. Lett.*, 2004, **14**, 325–328.

120. R. E. Hubbard, *J Synchrotron Radiat*, 2008, **15**, 227–230.

121. J. R. Medina, C. J. Becker, C. W. Blackledge, C. Duquenne, Y. Feng, S. W. Grant, D. Heerding, W. H. Li, W. H. Miller, S. P. Romeril, D. Scherzer, A. Shu, M. A. Bobko, A. R. Chadderton, M. Dumble, C. M. Gardiner, S. Gilbert, Q. Liu, S. K. Rabindran, V. Sudakin, H. Xiang, P. G. Brady, N. Campobasso, P. Ward and J. M. Axten, *J. Med. Chem.*, 2011, **54**, 1871–1895.

122. M. B. Eisen, D. C. Wiley, M. Karplus and R. E. Hubbard, *Proteins*, 1994, **19**, 199–221.

CHAPTER 6

Quantitative Structure–Activity Relationships

ANDREW M. DAVIS

Respiratory Inflammation and Immunity Innovative Medicines, AstraZeneca R&D Mölndal, Pepparedsleden 1, Mölndal 43183, Sweden
E-mail: andy.davis@astrazeneca.com

6.1 QUANTITATIVE STRUCTURE–ACTIVITY RELATIONSHIPS IN DRUG DESIGN SAR

Quantitative structure activity relationships (QSARs) are common tools for medicinal chemists, used to help decide what molecule to make next. QSARs are statistical empirical models that relate a quantitative description of chemical structure features of a series of molecules to the responses those molecules show in an experimental system. QSARs are empirical models, that is, they are based on observed trends and correlations between the chemical descriptors and response variables, rather than on an *a priori* physics-based model. Although empirical in nature, a robust QSAR model can be used to propose a physical hypothesis that underlies an observed relationship

The medicinal chemistry approach to structure activity relationships is based on serial pairwise comparisons of structural changes with activity changes. QSAR takes a complementary approach and tries to identify how structural changes across a series of molecules are related to their activity.

6.2 BRIEF HISTORY OF QSAR

The first QSAR model is often said to have come from the work of Alexander Crum-Brown and Thomas Fraser, which was read before the Royal Society of Edinburgh on January 6 1868.[1] They were studying the muscle paralysing activities of quaternised alkaloid "poisons". They focused on differences in activity and chemical structure. As the constitution of the substances was not known, they thought "there is more hope of arriving at some definite conclusion by studying the changes produced in the action of physiologically active substances, by performing on them certain well-defined chemical operations which introduce known changes to their composition." They treated natural alkaloids strychnia, brucia, thebaia, codeia, morphia and nicotia

The Handbook of Medicinal Chemistry: Principles and Practice
Edited by Andrew Davis and Simon E Ward
© The Royal Society of Chemistry 2015
Published by the Royal Society of Chemistry, www.rsc.org

with methyl iodide and studied their effects in rabbits and dogs. In their paper, they expressed the relationship between structural changes (Δ) and changes activity as a mathematical equation (Equation (6.1)).

$$\Delta(\text{physiological activity}) = f(\Delta\text{constitution}) \tag{6.1}$$

Modern QSAR still follows this formalism, where chemical descriptors are used to numerically quantify chemical constitution, and the mathematical tool, sometimes called a machine learning algorithm, is used to define the function "f", that relates structure to activity.

In 1869 Richardson[2] showed that the narcotic effect of primary alcohols was inversely related to their molecular weight, and in 1893 Richet[3] showed that the toxicity of simple ethers, alcohols, and ketones was inversely related to their solubility in water. A major step was taken towards our modern approach to QSAR when in 1899, Hans Horst Meyer[4] first proposed and demonstrated that anaesthetic potency was related to lipid solubility, and two years later a similar proposal was made by Charles Overton.[5] They showed the narcotic effect on tadpoles was proportional to the partition coefficient of the compounds from water into olive oil, as a model for the partition of the narcotics into the biological membrane. Further, their correlation prompted them to hypothesise that it was the compounds partitioning into nerve cells that led to the narcosis. Today we still use partition coefficients as a measure of a compound's hydrophobicity, although n-octanol has superseded olive oil as a hydrophobic receptor phase.

In the 1930s, quantitative structure reactivity relationships defined the field of physical organic chemistry, through the work of Hammett.[6] Hammett was inspired by the work of Brönsted and Pederson, who had found that the logarithm of rate constants for acid and base catalysed reactions could be related to the pKas of the general acids and bases participating in the reaction.[7] Hammett proposed that the effect of substituents attached to the benzene ring, *meta* or *para* to a reaction centre, upon the reaction rate or position of the a chemical equilibrium in which the reaction centre participates, is a property of the electronic effect of the substituent alone. Different reactions may have different sensitivities to the effect of this substituent. Hammett defined a descriptor, σ, to quantify the electronic effect of a substituent. The σ scale used the ionization of benzoic acids as the calibrating reaction, and σ_m or σ_p represented the difference in pKa between the different m- or p-substituted benzoic acids and benzoic acid itself. The sensitivity of a particular reaction to the electronic effect of a substituent was described by ρ, the slope of the correlation between the logarithm of the equilibrium (K) or rate constant (k) for that reaction and the sigma values (Equation (6.2)).

$$\log k = \rho\sigma + \text{constant} \tag{6.2}$$

In Hammett's initial publication he tabulated σ values for 31 substituents, and linear energy correlations for 39 reactions. While Hammett's observations were empirical, the number of reactions he was able to find that conformed to the equation led him to the "...unavoidable conclusion that a substituent affects rates and equilibria essentially by a single internally transmitted mechanism. This may be either because the substituent directly attracts or repels electrons more than does the hydrogen atom it replaces, or because the substituent permits the construction of alternative electronic distributions of a highly polar nature which resonate with the non-polar one, or from some combination of these effects."

Hammett's work was extended by many others, notably in the 1950's by Robert Taft, who defined an electronic substituent constant scale for aliphatic systems.[8] Taft used the acid and base catalysed hydrolysis of esters as the calibrating reaction. He proposed that as the acid and base catalysed hydrolysis of esters proceed through a very similar tetrahedral transition state

structure, and under identical conditions, both reactions would experience similar steric influences. As the base catalysed reaction went from a neutral ground state to a negatively charged
transition state, it would be much more sensitive to electronic effects of appended substituents
than the acid catalysed hydrolysis reaction which goes from a positively charged ground
state (rapid pre-protonation of the ester carbonyl) to a similarly positively charged transition
state. Hence acid-catalysed hydrolysis of esters would be largely insensitive to the electronic
effect of substituents, but still sensitive to their size, while base catalysed hydrolysis would
be sensitive to both size and electronic effects. Based on this differential sensitivity, Taft
was able to define a new a steric substituent constant scale, termed Es. To differentiate the
new σ-scale from Hammett's σ-scale derived from ionization of benzoic acids, Taft gave the
symbol σ* to the electronic effect of the aliphatic substituent. The sensitivity of a particular
reaction to the electronic and steric effects of a substituent was described by ρ* and Δ,
respectively, the regression coefficients from the correlation between the logarithm of the
equilibrium (K) or rate constant (k) for that reaction and the σ* and Es values of the substituents (Equation (6.3)).

$$\log (k - R/k - Me) = \rho^*\sigma^* + (\Delta \times Es) + constant \tag{6.3}$$

Hammett had already alluded to both induction and resonance participating in transmitting
the electronic effect of the substituent to the reaction centre. In the 1960's Swain and Lupton
went on to factor the two effects into their Field and Resonance (F & R) scales for aromatic
substituents.[9]

Taft, Swain and Lupton's work was able to extend Hammett's single descriptor approach to
more complex systems, and analysis by using regression to fit multiple parameter equations.

The leap of QSARs from physical-organic chemistry to medicinal chemistry occurred in the
early 1960's through the work of Corwin Hansch and Toshio Fujita (For a more detailed description of the emergence of this very first QSAR equation see reference 11).[10,11] They were
studying the effect of auxins as growth regulators (Figure 6.1). Their hypothesis was that the

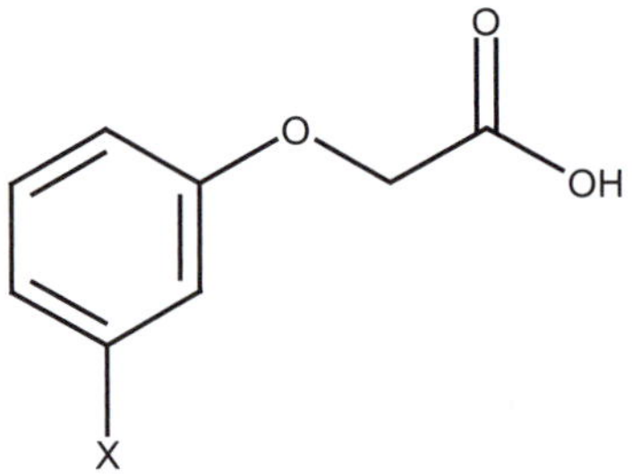

Figure 6.1 Auxin plant growth regulators—the first Hansch–Fujita QSAR dataset.

$$\log 1/C = -1.97\pi^2 + 3.24\pi + 1.86\sigma + 4.16 \tag{6.4}$$

$n = 21$
$s = 0.484$
$r^2 = 0.776$
$C =$ molar concentration producing 10% increase above control in the length of 3 mm oat
   seedlings
$n = 21$ (3-CF_3, 4-Cl, 3- I, 4-F, 3-Br, 3-SF_5, 3-Cl, 3-NO_2, 3-SMe, 3-Et, 3-SCF_3, 3,4-$(CH)_4$, 3-OMe,
   3-Me, 3-CN, 3-Pr, 4-OMe, 3-Ac, 3-F, H)
$s =$ residual standard deviation of the regression
$r^2 =$ coefficient of variation, the fraction of the variation in log1/C explained by the regression
   equation

auxin mode of action was more chemical in nature and that a covalent bond may be involved in their activity. They were therefore seeking to apply the Hammett equation to describe biological activity, but their lack of success led them to re-evaluate their hypothesis. Their new hypothesis was that transport through biological membranes to the site of activity may contribute to the rate determining step. By defining a new substituent constant scale π, representing the substituent's contribution to the overall lipophilicity. The π parameter was defined as the difference in logarithm of the partition coefficient between n-octanol and water of the substituted compound to the logarithm of the partition coefficient of the unsubstituted compound. Hansch and Fujita were able to merge the work of Overton and Meyer, and Hammett to extend the Hammett equation to describing biological activity (Equation (6.4)).

They modified the Hammett equation to incorporate both a linear and non-linear dependence upon this new substituent constant, π, incorporating both a π and π^2 term. The work was extended in a follow-up paper to show the general application in a number of other systems.[12]

Hansch and colleagues took a multiple variable (multivariate) approach to describe biological activity and used multiple linear regression to fit the relationship between a set of descriptors and biological activity. The n-octanol/water system used to define π has become the standard system on which to base the measure of lipophilicity, and is the defining system for the CLOGP algorithm, used to predict log P, which was also a product of the Pomona College group. The descriptors of Hammett, Taft, Swain and Lupton have all been used, together with Hansch and Fujita's π in the derivation of classical QSAR equations to describe electronic, steric and lipophilic contributions to biological activity.

While QSAR by definition uses a numerical description of chemical constitution, we should also mention at this point the Free–Wilson methods.[13] Free–Wilson analysis directly correlates biological activity with the presence or absence of a substituent at a particular position. The presence of absence of the substituent in the Free–Wilson method is represented as a numerical "indicator" descriptor, where 1 = presence of a substituent and, 0 = absence of a particular substituent. It has remained a popular approach since its introduction in the early 1960's and has recently inspired the current interest in matched molecular pairs analysis.

QSAR has maintained its popularity and impact since its introduction in the 1960s. It brings together a number of key disciplines, biology, computational chemistry and statistics. Its power to predict what molecule to make next is undoubted if applied in an appropriate fashion.

6.3 QSAR MODEL QUALITY

The literature is full of QSAR models, but not all QSAR models are born equal. There are a number of pitfalls that await the unprepared QSAR scientist. QSAR models with apparently appealing statistics can in reality encode no more than chance relationships, identified through brute-force searches for correlations between chemical descriptors and biological responses in large data matrices. Models that make accurate predictions for the 'training set' compounds, used to define the model, may provide poor predictions for the structurally distinct query compounds that are applied to the model in real life usage.[14] Superficially appealing model statistics can easily mislead the unwary and suggest the use of models that merely describe the structure of the training data. Such models, if used prospectively, can waste time and money on the synthesis of compounds that are subsequently found not to possess improved properties.[15] This generates some uncertainty in medicinal chemists' minds about the application of QSAR models to their own problems. A good QSAR model, however, can very quickly guide a chemist towards optimal compounds. For example, the hERG models[16] used by AstraZeneca chemists have contributed to the reduction in compounds synthesized showing unwanted "red-flag" hERG potency <1 µM from 25.8% of all compounds tested in 2003 to only 6% in 2010.

The problem of differentiating a robust QSAR model from one that is not robust is so great, and the need for predictive models so high, that the Organisation for Economic Co-operation and Development (OECD) has defined good practice in the validation of QSAR models, with the aim of increasing regulatory body acceptance of QSAR models for toxicity and environmental assessment.[17] They have also published a "QSAR toolbox" for use by regulatory authorities and companies to allow look-up, near neighbour analysis and robust prediction of toxicity and environmental properties of new compounds based on its database of structures and responses. Indeed, the European Community in its Registration Evaluation Authorisation and Restriction of Chemicals (REACH) now accepts validated QSARs as part of the required complete dossier on the physicochemical, biological and technological properties for every chemical circulating in Europe.[18] The OECD guidelines, although focused on environmental toxicology, are recommended reading for any scientist involved in the derivation of QSAR models. The guidelines state that, to facilitate consideration of a QSAR model for regulatory purposes, the model should be associated with:

1. An unambiguous endpoint;
2. An unambiguous algorithm;
3. A defined domain of applicability;
4. Appropriate measures of goodness-of-fit, robustness and predictivity;
5. A mechanistic interpretation, if possible.

These points will be highlighted in the next sections. The guidelines include a recommendation to seek a mechanistic interpretation, if possible, and this brings the QSAR equation full circle. In Hammett's original work, he was able to propose an underlying physical model from the observation of many similar structure–property correlations for chemical equilibria and reactions conforming to the Hammett equation. Predictive QSAR models must identify underlying controlling physics of the system, and can bring new physical understandings to the systems being studied.

The publication policies of journals have also changed and many now explicitly require all data and molecular structures used to carry out a QSAR/QSPR study to be reported in the publication, or they should be readily available without infringements or restrictions. This allows the reader to reproduce and validate the analysis in the paper and helps to improve the quality of published work. See for instance the ACS guidelines on publication in Journal of Medicinal Chemistry (reference 19). With care in their generation and use, QSAR models can be a very powerful tool in the medicinal chemist's armoury.

6.4   THE LANGUAGE OF QSAR: DESCRIPTORS, MACHINE LEARNING METHODS AND STATISTICS

The QSAR discipline has its own language like any other. A QSAR model building process requires:

- A set of response data (y-data) for the series of molecules being studied.
- A set molecular descriptors, each descriptor a column/vector of numbers, a row for each molecule, quantifying a molecular property for each molecule (x-data).
- A mathematical tool or algorithm to correlate the response data to one or more of the molecular descriptors, sometimes called a machine learning method.

The outputs from QSAR modelling include the model, where the mathematical relationship between one or more of the descriptors to the response data is defined, and a number of statistical diagnostics which aim to help a user assess the quality of a model, its utility in prediction, and property-space covered by the model.

6.4.1 An Unambiguous Endpoint—the Biological Response

The OECD guideline 1^{17} requires an unambiguous endpoint. Data from a single well-defined and controlled assay is the best starting point for QSAR model development. Data can be combined from different assays measuring the same endpoint, if the relationship between those assays is understood, by the use of cross-screening of standard compounds. Where uncertainties exist about data from different assays, it may be better to build separate models in the first instance, or at least to look for systemic errors in prediction of the compounds in the test set which came from the differing assays. Replicate measurements in the assay can give you an estimate of the assay experimental error. A QSAR model should not be able to predict with a resolution better than the experimental error of the assay.

Once the dataset has been selected, next begins the pre-processing stage of the analysis. The response data is often "transformed" into a scale where the experimental errors follow "a normal distribution". This is an advantage for some of the statistical tools used to form the correlations between descriptors and response. This is often, but not always, a logarithm scale (*e.g.* log IC_{50}, $-$ log IC_{50}, pA_2, log (solubility)). The negative logarithm is often used, as it makes higher potency compounds take larger positive values, but the response data being modelled could be categorized data, where the response data may only exist as active/inactive which are represented in the QSAR model by a set of 1 and 0's, indicator variables.

Where response data is missing for some compounds, or if some of the compounds only contain qualitative estimates of activity (*e.g.* less than or greater than categorizations), these can be used as test set compounds. Sometimes these $> <$ compounds contain the most interesting data, for example when modelling *in vitro* metabolic stability assay data. Often the most stable compounds show no detectable metabolism and the metabolic stability is reported as less than a limiting value. Excluding these compounds from the model development would exclude the exact compounds we wish to emulate. In this case they can be left in the training set and included as continuous data, given a minimum or maximum value. Classification methods like random forests are insensitive to using continuous data alongside categorical data.

6.4.2 The Numerical Descriptors of Chemical Constitution

How you describe changes in molecular structure across a series of molecules within a vector of numbers is one of the critical steps in setting up a QSAR model. Many different types of molecular descriptors are available, representing molecular size, volume, lipophilicity, hydrogen bonding counts, shape, and electronic distribution. Representing 3-dimensional shape within a single number vector is challenging, although many shape descriptors have been suggested. Herein lays one of the problems of QSAR. While some descriptors like H-bond donors counts are intuitively understood by medicinal chemists, some descriptors, in an attempt to encode complex properties such as shape, molecular connectivity or charge distribution, although mathematically precise and derived from molecular structure, can be difficult for the medicinal chemist to deconvolute. There are so many descriptors that can be calculated that there are whole books just devoted to them. The DRAGON descriptor suite alone contains 4885 molecular descriptors for use in QSAR.[20] The sheer number available causes QSAR another problem: which descriptors to include in the analysis? Many of the mathematical tools used to define QSAR equations are very sensitive to the number of descriptors included in the analysis. Including more descriptors increases the probability of finding correlations just by chance, particularly when there are more descriptors than the number of molecules with biological activity being modelled. If this is the case regression tools, such as partial least squares analysis, which extract summary variables from a larger descriptor matrix and use "cross validation" (see later) to assess the significant contribution of variables added to the model, are less sensitive to finding spurious correlations.

Comparative molecular field analysis (CoMFA) and related 3-dimensional QSAR methods attempt to describe both the 3-dimensional shape and charge distribution of molecules by recording the interaction between the target molecules and a small probe molecule over a 3-dimensional grid surrounding the molecules in the test set. The molecules have to be overlayed in 3-dimensional space so common molecular features occupy common volumes in space. The overlay can be thought of as a putative pharmacophore. The result of the QSAR analysis is the importance of the interactions the probe felt at each point in space to describing the response variable. These coefficients can be mapped back onto their respective points in 3-dimensional space to form a 3-dimensional regression map. These can be interpreted by the medicinal chemist. For CoMFA the probe is usually a positively charged methyl group which perceives electrostatics and volume. The maps tell the medicinal chemists where in 3D space to place or remove charge and molecular volume to increase or decrease potency.[21]

A number of descriptor calculators are available on the internet.[22] ACD physical property suite calculates log P, log $D_{7.4}$, and ionization constants. ACDLabs also makes available iLab, a pay-for-service web-based system for calculated physical properties.[23] The CDK toolkit is an open-source cheminformatics package, developed by more than 50 researchers worldwide, and widely used to provide chemical descriptors for QSAR modelling.[24] A number of other cheminformatics toolkits also exist including RDkit and OpenBabel.[25] CINFONY is an open-source application program interface to a number of cheminformatics toolkits including CDK, RDkit,[26] and OpenBabel. A number of molecular descriptor calculators are also available within the Bioclipse software system.[27]

In most QSAR approaches it is not necessary to know the exact nature of the interactions between the drugs and its molecular target. Where such structural information is known, whether from experimental X-ray structures, NMR or computational models, descriptors can be calculated directly from the observed drug receptor interactions (CoMFA,[28] GRID-GOLPE,[29,30] VALIDATE[31]). In fact, some of the scoring functions inside docking programs use methods based on QSAR equations to rank the affinity of molecules from the observed drug–receptor interactions (*e.g.* LUDI[32]).

6.4.3   Preparation of the Dataset

Whatever machine learning method you choose, a similar overall approach is taken. Some machine learning methods are sensitive to the numerical range of the data in each descriptor. Descriptors which happen to have a larger range because of the units chosen can unduly influence the fitting the descriptors to the y response data using your chosen statistical learning method. Therefore the dataset is often "normalized" or "unit scaled" by subtracting the mean value of each descriptor vector from each descriptor value, and dividing by the standard deviation of the descriptor column. Hence, each descriptor will have a mean of zero and standard deviation of 1.0. In those methods sensitive to scale, this process gives each descriptor an equal weight to contribute to the model. Their contributions will then be adjusted in model building.

Next the dataset is split into a training set and a test set. The training set is used to define the mathematical relationship, or correlation, between the y-response and the x-descriptors using the chosen machine learning method. The test set is used to assess the quality of the QSAR model by its ability to predict a set of molecules that were not part of the training phase. Because many of the machine learning methods used to define QSAR models are very good at fitting the x-descriptors to the y-response data, the quality of the model is usually best assessed on how well it predicts an independent test set, rather than the fit to the training set compounds.

Some machine learning methods require the dataset to be split into three: a training set, a test set/tuning set and a validation set. In this case the test set is used to stop the model training phase, or to select between different models, while the validation set is an independent test of model performance. The disadvantage of this approach is that it limits the number of compounds available to train the machine learning algorithm.

The test set/validation set can be selected from the original dataset in a number of ways, including random selection, by clustering and selecting representative examples from the original dataset, or by selecting in time order, so the latest compounds make the test set. When the model is to be used in a design-make-test environment, a time-ordered test set most closely reflects how the model will be used to predict future compounds. The size of the test set selected often depends on the amount of compounds on which you have activity data. The compounds selected for the test set limit the amount of information you make available to the machine learning method to build a model from. A common approach is to select 20–30% of your dataset as a test set.

6.4.4 Exploring the Dataset

A key step in QSAR analysis is getting to know your dataset. It is common to examine plots of response *vs.* variables, as it may inform your choice of statistical tool, and it may allow you to refine your dataset. Examining how the compounds distribute over the property space defined by the descriptors may allow outlying points or clusters to be observed, which may affect how the training set for the model is selected.

In Figure 6.2, we have plotted the activity of series of molecules as –log (act) *vs.* log P. We can see examples of a) strong linear relationships with key variables, and b) strong non-linear

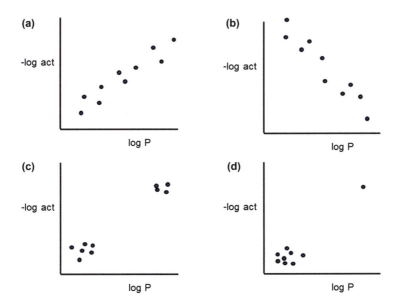

Figure 6.2 Example of examining relationships in your dataset. An activity variable, − log (activity), is plotted *vs.* log P, showing a) strong linear relationships, b) non-linear relationships, c) highly clustered data which would drive a high correlation, and finally d) a dataset containing a large outlier, which also would drive a high correlation. How does the distribution of data points in each graph influence your analysis?

relationships which may necessitate using a non-linear QSAR approach. Apparent non-linearity within the dataset from such simple examination may bias you away from choosing inherently linear machine learning methods such as multiple linear regression or Partial Least Squares analysis in favour of a more non-linear method such as neural networks or random forests, or it may encourage you to manage the non-linearity by introducing derived descriptors raised to a power term. For example, in Hansch's very first QSAR equation, a parabolic π^2 term was incorporated, as there appeared to be an optimum value of π for biological activity. Figure 6.2 also contains a dataset where c) distinct clustering occurs, which may mean the space between the clusters is not well described, or the response being largely categorised into two groups. One may wish to explore the groups in separate QSAR models or accept a categorized model. The final dataset d) shows a very strong outlier which may unduly bias the QSAR modelling method, and produce a trivial model which just describes the different behaviour of that one compound. Compounds with extreme values may be removed from the data matrix, as they may unduly bias the analysis or maybe they can be left in, to examine their influence on the overall model. Either way, a key part of building a QSAR model is to understand its sensitivity to individual pieces of data.

Often such visual explorations of the dataset are enough to understand important underlying QSAR relationships. A lot of information can be extracted by such a simple exploration, and this approach is very complementary to the traditional medicinal chemist pairwise approach to exploring SAR. In the following case study, which describes the design of long-acting dual-acting β_2 adrenoceptor and dopamine D_2 receptor agonists, a simple visual analysis of such plots provided a powerful platform for building a very predictive model.

6.4.5   Case Study: D_2/β_2 Agonists

For the treatment of COPD and asthma, topically administered β_2-adrenoceptor agonists are the most commonly prescribed anti-bronchoconstrictor agents, and topically applied steroids are used to reduce inflammation of airways, but there has been relatively little attention paid to the investigation of mechanisms that could specifically reduce airway hyper-reactivity.

This project's aim was to control bronchoconstriction through β_2 agonism, and to control underlying hyper-reactivity of the lung through modulation of sensory nerve traffic controlled through dopamine receptor activation.[33] Our working hypothesis was that the stimulation of D_2-receptors on afferent nerves should lead to the inhibition of nerve activity in the lung. Therefore, a dual acting D_2 receptor and β_2 adrenoceptor (D_2/β_2) agonist would combine both activities and have a desirable therapeutic profile. We discovered a series of benzothiazolone dual acting D_2/β_2 agonists that provided a starting point for optimisation towards long duration compounds, while controlling the primary pharmacologies and off-target selectivities. The discovery of this series and the optimisation of the primary pharmacologies has been described elsewhere.[34]

In the project, an *ex vivo* electrically stimulated guinea pig tracheal strip was used to measure the functional duration of action of the agonists on bronchial relaxation. Such an experiment was used in the discovery of the archetypal long acting β_2 agonist, salmeterol.

In brief, a piece of guinea pig tracheal strip was held suspended in air in a force balance so the force of contraction of the tissue could be measured when electrically stimulated to contract. The tissue was kept viable by being bathed in oxygenated Krebs solution. The tissue was made to contract using an electrical stimulation and the force of contraction measured against time. Into the bathing Krebs solution, β_2 agonists (and β_2 antagonists to check recovery of the tissue) could be added to elicit relaxation of the tissue. The infusion of compound could be stopped and the relaxation of the tissue measured against time. The experiment could be run for 12 hours or more and experience taught that compounds that maintained relaxation for 3 hours with no sign of recovery would likely still be active at 12 hours, so the experiment was run for 3 hours routinely, and compounds showing no recovery in tissue tone after 3 hours were recorded with durations >180 minutes. Compounds were optimised for their D_2 receptor and β_2 adrenoceptor agonism, and then progressed to this *ex vivo* duration of action screen to check for long-acting agonism.

Salmeterol was supposed to utilize binding to an "exosite" on the β_2 receptor close to the agonist binding site, to gain duration of action. This suggested duration to be very structurally dependent.[35] But our own emerging structure–activity relationships for duration of action in these D_2/β_2 agonists suggested that duration was controlled by lipophilicity, and not by specific drug–receptor interactions. There were also literature challenges to the salmeterol "exosite" theory[36] which gave us confidence in pursuing lipophilicity control. We determined the lipophilicity experimentally by measuring the n-octanol–water distribution coefficients at a fixed pH of 7.4, as the compounds contained multiple ionising centres which made the measured partition coefficients pH dependent. We were also able to calculate logD$_{7.4}$ for new compounds using CLOGP and pKa calculations, and the growing database of our own measurements. For early compounds, their β_2 duration of action followed a simple log $D_{7.4}$ relationship, suggesting that log $D_{7.4} > 2.0$ gave long-acting compounds (Figure 6.3). The apparent break-point in the graph was an artefact of the screen only running for 3 hours routinely, and many compounds having recorded durations >180 minutes.

However, as the chemistry evolved the simple log $D_{7.4}$ relationship disappeared, and we lost confidence in our model and hypothesis (Figure 6.4).

Close examination of the data plots, facilitated by the graphing package SPOTFIRE,[37] revealed that short acting lipophilic compounds were members of a subseries where the pKa of the secondary amine was reduced to below 8.0. In fact, all these compounds contained a sulphone group two carbons from the secondary amine (Figure 6.5).

Our new model therefore depended upon lipophilicity and the pKa of the secondary amine, but the dependence on pKa was a classified response as all compounds with an amine pKa > 8.0 behaved in a similar manner. Because we measured duration in a screen that was capped at 3 hours, the overall structure–activity relationship was inherently non-linear. Because of the classified dependence on pKa (pKa $\neq 8$) and the inherently nonlinear relationship of duration on logD$_{7.4}$, most QSAR approaches failed to identify the underlying model. The complete dataset is shown in Table 6.2.

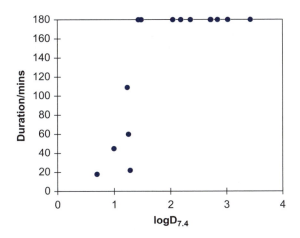

Figure 6.3 We hypothesized that duration of action was simply a function of how much compound could partition into the tissue and was described by log $D_{7.4}$.

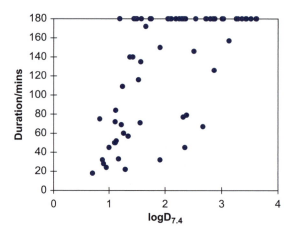

Figure 6.4   The early log D$_{7.4}$ relationship disappeared as the chemistry evolved.

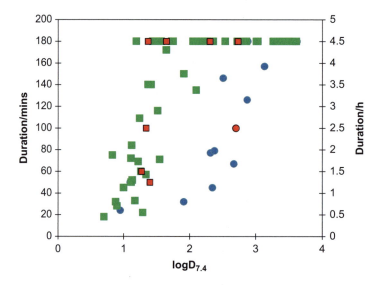

Figure 6.5   Plot of β$_2$ duration *vs.* log D$_{7.4}$, coloured by the amine basicity. Green squares: secondary amine pKa > 8. Blue circles: secondary amine pKa < 8. Red points were predictions for sub-sequently synthesized compounds in each series.

   Table 6.3 showed compounds subsequently made based on the deduced QSAR model. The duration predictions were in fact so robust, that the β$_2$ duration screen was subsequently only run infrequently, allowing chemistry to progress much more rapidly based on duration pre-dictions using estimates of log D$_{7.4}$ and pKa.

   So while QSAR as a discipline appears synonymous with statistical tools and complicated mathematics to identify underlying relationships between the response variables and chemical descriptors, visual data exploration with a keen medicinal chemist's eye is probably the most powerful QSAR tool, and indeed allows a more focused QSAR exploration to be undertaken, if only to evaluate the robustness of the visually identified pattern.

   The robustness of the β$_2$ duration log D$_{7.4}$/pKa model caused us to look for a mechanistic explanation. Our hypothesis was that basic amines could partition in an interfacial way into

Table 6.1 Various machine learning tools used in QSAR analyses—advantages and disadvantages.

Statistical tool	Supervised/unsupervised	Linear	Advantages	Disadvantages
Principal component analysis (PCA)	Unsupervised	Linear	Data projection Reducing the number of variables	Often hard to interpret summary variables
SIMCA	Unsupervised/supervised	Linear	Classification using PCA analysis	Often hard to interpret summary variables
Cluster analysis (CA)	Unsupervised	Linear	Looking for groupings within datasets	
Non-linear mapping (NLM)	Unsupervised	Nonlinear	To reduce high dimensionality data to 2–3 dimensions	Can be hard to interpret map as point proximity does not necessarily imply "nearness"
Multiple linear regression (MLR)	Supervised	Linear	When linear underlying relationships are expected When number of compounds $\gg$ number variables Produces simple models selecting most important variables	Susceptible to chance correlations when allowed algorithm is allowed to choose between multiple variable
Partial least squares (PLS)	Supervised	Linear	When linear underlying relationships are expected When descriptors are highly correlated When number of variables $\gg$ number compounds Useful when multiple response data exists—common controlling features and unique features in describing each response are obtained from a single analysis	Inherently linear
Support vector machines (SVM)	Supervised	Nonlinear	Particularly good for classified responses	Hard to interpret the underlying meaning like PCA
Regression trees	Supervised	Nonlinear	Robust method which can be used in many circumstances Particularly good with large datasets (100s of compounds)	Susceptible to fitting signal as well as noise Prediction is a consensus from many trees, therefore harder to interpret the model
Neural networks (NN)	Unsupervised/supervised	Nonlinear	When an underlying nonlinear relationship is expected	Susceptible to overtraining (model fitting the noise)
Gaussian processes	Supervised	Nonlinear		Highly complex modelling method

Table 6.2 Physicochemical properties and β_2 potency and *in vitro* duration data for compounds studied.

No	Sidechain	log $D_{7.4}$	B = pK_a > 8 N = pK_a < 8	β_2 p[A]$_{50}$	Int. Act.	β_2 duration
1	NH(CH$_2$)$_6$NH(CH$_2$)$_2$benthiazolone	1[a]	B	7.86	0.4	45
2	NH(CH$_2$)$_6$O(CH$_2$)$_2$Ph	2.73[a]	B	8.23	0.5	>180
3	NH(CH$_2$)$_5$O(CH$_2$)$_2$Ph	2.19	B	7.80	0.32	>180
4	NHCH(CH$_3$)(CH$_2$)$_5$O(CH$_2$)Ph	3.03	B	8.08	0.45	>180
5	NH(CH$_2$)$_5$O(CH$_2$)$_3$Ph	2.72	B	7.00	0.30	>180
6	NH(CH$_2$)$_3$SO$_2$(CH$_2$)$_2$NH(CH$_2$)$_2$Ph	0.7	B	7.88	0.37	18
7	NH(CH$_2$)$_6$O(CH$_2$)$_2$-Ph(-4-OH)	2.05	B	8.81	0.30	>180
8	NH(CH$_2$)$_5$CONH(CH$_2$)$_2$Ph	1.29[a]	B	7.18	0.35	22
9	NHC(CH$_3$)$_2$(CH$_2$)$_5$O(CH$_2$)$_2$Ph	3.43	B	8.42	0.46	>180
10	NH(CH$_2$)$_6$NHCONHPh	1.26[a]	B	7.01	0.20	60
11	NH(CH$_2$)$_6$SO$_2$(CH$_2$)$_2$Ph	1.44[a]	B	7.00	0.21	>180
12	NH(CH$_2$)$_6$O(CH$_2$)$_2$-2-pyridyl	1.24	B	8.12	0.45	109
13	NH(CH$_2$)$_6$O(CH$_2$)$_2$-2-thiazole	2.36	B	7.78	0.33	>180
14	NHCH$_2$C(CH$_3$)$_2$(CH$_2$)$_4$O(CH$_2$)-Ph(-4-OH)	2.85	B	9.03	0.44	>180
15	NH(CH$_2$)$_6$O(CH$_2$)$_2$-Ph(4-NH$_2$)	1.49	B	9.40	0.40	>180
16	NH(CH$_2$)$_6$O(CH$_2$)$_2$-Ph(4-NHCONH$_2$)	1.42	B	9.20	0.49	140
17	NH(CH$_2$)$_3$S(CH$_2$)$_2$O(CH$_2$)$_2$Ph	2.31[a]	B	8.02	0.39	>180
18	NHCH$_2$C(CH$_3$)$_2$(CH$_2$)$_4$O(CH$_2$)-Ph(-4-NO$_2$)	3.26	B	6.35	0.23	>180
19	NH(CH$_2$)$_6$NHCH$_2$CF$_2$Ph	1.57[a]	B	6.84	0.17	>180
20	NHCH$_2$C(CH$_3$)$_2$(CH$_2$)$_4$O(CH$_2$)$_2$-Ph(-4-NH$_2$)	2.29	B	8.86	0.35	>180
21	NH(CH$_2$)$_3$SO$_2$(CH$_2$)$_2$O(CH$_2$)$_2$Ph	1.37[a]	B	7.93	0.27	140
22	NH(CH$_2$)$_3$O(CH$_2$)$_2$O(CH$_2$)$_2$Ph	1.19[a]	B	8.72	0.53	>180
23	NH(CH$_2$)$_2$O(CH$_2$)$_3$O(CH$_2$)$_2$Ph	2.06[a]	B	7.16	0.31	>180
24	NHCH$_2$C(CH$_3$)$_2$(CH$_2$)$_4$O(CH$_2$)$_2$-Ph(-2-NH$_2$)	2.25	B	7.58	0.32	>180
25	NHCH$_2$C(CH$_3$)$_2$(CH$_2$)$_4$O(CH$_2$)$_2$-Ph(-3-NH$_2$)	2.29	B	8.39	0.51	>180
26	NH(CH$_2$)$_2$NHCO(CH$_2$)$_2$O(CH$_2$)$_2$Ph	1.34[a]	B	9.27	0.58	57
27	NH(CH$_2$)$_2$SO$_2$(CH$_2$)$_3$O(Ch$_2$)$_2$Ph	2.51[a]	N	8.09	0.51	146
28	NH(CH$_2$)$_2$NH(CH$_2$)$_3$O(CH$_2$)$_2$Ph	0.88[a]	B	7.01	0.54	32
29	NH(CH$_2$)$_2$CF$_2$(CH$_2$)$_3$O(CH$_2$)$_2$Ph	3.26[a]	B	7.77	0.49	>180
30	NH(CH$_2$)$_2$S(Ch$_2$)$_3$O(CH$_2$)$_2$Ph	2.86[a]	B	8.36	0.54	>180
31	NH(CH$_2$)$_2$NHSO$_2$(CH$_2$)$_2$O(CH$_2$)$_2$Ph	1.56[a]	B	7.39	0.43	135
32	NHCH$_2$C(CH$_3$)$_2$CH$_2$S(CH$_2$)$_2$O(CH$_2$)$_2$-Ph(-2-NH$_2$)	3.44[a]	B	7.13	0.39	>180
33	NH(CH$_2$)$_3$SO$_2$(CH$_2$)$_2$O(CH$_2$)$_2$-Ph(-2-NH$_2$)	1.17[a]	B	7.46	0.26	33
34	NH(CH$_2$)$_2$NHCO(CH$_2$)$_3$O(CH$_2$)$_2$Ph	1.55[a]	B	7.56	0.27	71
35	NH(CH$_2$)$_3$SO$_2$(CH$_2$)$_2$O(Ch$_2$)$_2$-Ph(-2-NO$_2$)	1.37[a]	B	7.45	0.28	50
36	NH(CH$_2$)$_3$SO$_2$(CH$_2$)$_2$NH(CH$_2$)$_2$OPh	0.9[a]	B	6.73	0.45	28
37	NH(CH$_2$)$_3$S(CH$_2$)$_2$O(CH$_2$)2-pyridyl	0.83[a]	B	7.32	0.40	75
38	NH(CH$_2$)$_2$SO$_2$(CH$_2$)$_3$O(CH$_2$)$_2$-2-thiazolyl	2.35[a]	N	7.72	0.51	45
39	NH(CH$_2$)$_2$SO$_2$(CH$_2$)$_3$O(CH$_2$)$_2$-Ph(-4-F)	2.67[a]	N	7.92	0.49	67
40	NH(CH$_2$)$_2$NHCO(CH$_2$)$_2$O(CH$_2$)$_3$Ph	1.91[a]	B	8.63	0.45	150
41	NH(CH$_2$)$_3$SO$_2$(CH$_2$)$_3$O(CH$_2$)$_2$Ph	1.65[a]	B	8.24	0.41	172
42	NH(CH$_2$)$_2$S(CH$_2$)$_3$O(CH$_2$)$_2$-Ph(-4NO$_2$)	3.29[a]	B	6.78	0.13	>180
43	NH(CH$_2$)$_3$S(CH$_2$)$_2$O(CH$_2$)$_2$-1-naphthyl	3.36	B	5.80	0.54	>180
44	NH(CH$_2$)$_2$SO$_2$(CH$_2$)$_3$O(Ch$_2$)$_2$-Ph(-4-NO$_2$)	2.32[a]	N	6.73	0.26	77
45	NH(CH$_2$)$_2$S(CH$_2$)$_3$O(CH$_2$)$_2$-2-pyridyl	1.75[a]	B	7.51	0.55	>180
46	NH(CH$_2$)$_3$SO$_2$(CH$_2$)$_2$O(CH$_2$)$_2$-1-naphthyl	2.54[a]	B	7.73	0.55	>180
47	NH(CH$_2$)$_2$N(CH$_3$)CO(CH$_2$)$_2$O(CH$_2$)$_2$Ph	1.52[a]	B	8.29	0.75	116
48	NH(CH$_2$)$_3$S(CH$_2$)$_3$O(CH$_2$)$_2$Ph	3.01	B	7.32	0.35	>180

Table 6.2   (*Continued*)

No	Sidechain	log $D_{7.4}$	B = $pK_a > 8$ N = $pK_a < 8$	β_2 p[A]$_{50}$	Int. Act.	β_2 duration
49	NH(CH$_2$)$_3$S(CH$_2$)$_2$O(CH$_2$)$_2$OPh	2.11[a]	B	7.06	0.15	>180
50	NH(CH$_2$)$_3$SO$_2$(CH$_2$)$_2$O(Ch$_2$)$_2$OPh	1.13[a]	B	7.44	0.48	52
51	NH(CH$_2$)$_2$SO$_2$(CH$_2$)$_3$O(CH$_2$)$_2$-2-pyridyl	0.95[a]	N	7.49	0.79	24
52	NH(CH$_2$)$_2$SO$_2$(CH$_2$)$_3$O(CH$_2$)$_2$-Ph(-4-Br)	3.36[a]	N	7.40	0.52	>180
53	NH(CH$_2$)$_2$S(CH$_2$)$_3$O(CH$_2$)$_2$Ph(4-NHSO$_2$Ph)	3.52[a]	B	7.57	0.53	>180
54	NH(CH$_2$)$_2$SO$_2$NH(CH$_2$)$_2$O(CH$_2$)$_2$Ph	2.38[a]	B	6.48	0.33	79
55	NH(CH$_2$)$_3$S(CH$_2$)$_2$O(CH$_2$)$_3$Ph	2.8[a]	B	7.86	0.30	>180
56	NH(CH$_2$)$_2$S(CH$_2$)$_3$O(CH$_2$)$_3$Ph	3.61[a]	B	7.27	0.47	>180
57	NH(CH$_2$)$_3$SO(CH$_2$)$_2$O(Ch$_2$)$_2$Ph	1.12[a]	B	8.47	0.57	84
58	NH(CH$_2$)$_3$SO$_2$(CH$_2$)$_2$O(CH$_2$)$_3$Ph	2.1[a]	B	8.69	0.50	>180
59	NH(CH$_2$)$_3$O(CH$_2$)$_2$S(CH$_2$)$_2$Ph	1.73[a]	B	7.03	0.53	>180
60	NH(CH$_2$)$_2$SO$_2$(CH$_2$)$_3$O(CH$_2$)$_2$Ph(4-NHSO$_2$Ph)	2.88[a]	N	7.70	0.70	>180
61	NH(CH$_2$)$_2$SO$_2$(CH$_2$)$_3$O(CH$_2$)4Ph	3.62[a]	N	7.76	0.17	>180
62	NH(CH$_2$)$_2$SO$_2$(CH$_2$)$_3$O(CH$_2$)$_3$Ph	3.13[a]	N	7.35	0.36	157
63	NH(CH$_2$)$_2$SO$_2$(CH$_2$)$_3$O(CH$_2$)$_2$-Ph(-4-OH)	1.91[a]	N	8.65	0.44	32
64	NH(CH$_2$)$_2$SO$_2$(CH$_2$)$_2$S(CH$_2$)$_2$Ph	2.34[a]	B	7.57	0.41	>180
65	NH(CH$_2$)$_2$SO$_2$(CH$_2$)$_3$O(Ch$_2$)$_2$Ph(4-CONH$_2$)	1.1[a]	N	7.54	0.40	50
66	NH(CH$_2$)$_2$SO$_2$(CH$_2$)$_3$O(CH$_2$)$_2$-2- (-5-methylthiazole)	2.87[a]	N	7.61	0.69	126
67	NH(CH$_2$)$_3$NHSO$_2$(CH$_2$)$_2$O(CH$_2$)$_2$Ph	1.22[a]	N	7.51	0.36	69
68	NH(CH$_2$)$_3$SO$_2$NH(CH$_2$)$_2$OCH$_2$Ph	1.11[a]	B	7.40	0.59	72
69	NH(CH$_2$)$_3$SO$_2$NH(CH$_2$)$_2$O(CH$_2$)$_2$Ph	1.46[a]	B	7.54	0.59	>180

[a]Measured log $D_{7.4}$.

membrane phospholipids of the tracheal strip, causing them to be more "membrane-philic" than their n-octanol/water log $D_{7.4}$ measurements would predict. In fact a better relationship existed with log P—we were able to study this in detail by following the dependence of membrane partition upon pH, and even probe the thermodynamics of membrane partition for basic and neutral amine series of the dual D_2/β_2 agonists. The more basic compounds showed a more enthalpically-driven partition. Due to the commercial sensitivities of the ongoing pharmaceutical development program around chosen D_2/β_2 agonists, we were able to publish the underlying physicochemical understanding[38] 5 years before we could publish the dataset that led to that understanding.[39]

We now know that the efficient membrane partitioning we were identifying in this guinea pig tracheal strip duration model was indicating compounds that would have high volumes of distribution *in vivo*, as the basic compounds had much longer half-lives *in vivo* than the weakly basic compounds. Volume of distribution of bases is also governed by the same affinity for ordered phospholipids, and it seems this even governs the sites of disposition of neutral or basic drugs. Neutrals and acids appear to favour depoting into adipose tissue, while bases favour lean tissue.[40]

6.4.6 Building the QSAR Model

The OECD QSAR guideline 2 requires an unambiguous algorithm. If the algorithm is unambiguously defined and all its parameterisations and settings accurately recorded, results can be replicated. It builds confidence in the method if the quality of the model is not too sensitive to parameterisation. Investigations to automate QSAR model building have explored the effects of model and algorithm parameterization and found methods such as PLS and random forests are robust to changes in model parameterization.[41]

Table 6.3 Physicochemical properties, predicted β_2 duration and observed β_2 duration data for compounds subsequently synthesised.

No	Sidechain	log $D_{7.4}$	B = pK_a > 8 N = pK_a < 8	Predicted β_2 duration/ mins	Measured β_2 duration mins
71	$NH(CH_2)_2NHSO_2(CH_2)_3O(CH_2)_2Ph$	1.83[a]	B	180	151
71	$NH(CH_2)_3SO_2(CH_2)_3O(CH_2)_2Ph$	1.34[a]	B	80	45
73	$NH(CH_2)_3SO_2(CH_2)_2OCH_2Ph$	1.15[a]	B	60	74
74	$NH(CH_2)_2SO_2(CH_2)_3O(CH_2)_2(1\text{-isoquinolyl})$	2.35[a]	N	50	60
75	$NH(CH_2)_2NHSO_2(CH_2)_2O(CH_2)_2Ph$	2.22[a]	B	>180	>180
76	$NH(CH_2)_3SO_2(CH_2)_2O(CH_2)_2(2\text{-(5-methylthiophenyl)}$	1.63[a]	B	150	>180
77	$N(CH_2)_3N(CH_3)SO_2(CH_2)_2O(CH_2)_2Ph$	1.24[a]	B	60	48
78	$NH(CH_2)_2SO_2(CH_2)_3O(CH_2)2(5\text{-(4-methyl-1-3-thiazolyl)}$	1.44[a]	N	40	31
79	$NH(CH_2)_2SO_2(CH_2)_3O(CH_2)_2NHPh$	1.98[a]	N	40	32
80	$NH(CH_2)_2SO_2(CH_2)_3O(CH_2)_2(\text{-2-(5-methylpyridyl)}$	1.56[a]	N	40	37
81	$NH(CH_2)_2SO_2(CH_2)_3O(CH_2)_2(4\text{-methoxyphenyl})$	2.42	N	60	56
82	$NH(CH_2)_2SO_2(CH_2)_3O(CH_2)_2\text{-(4-methanesulphamidophenyl)}$	1.34[a]	N	40	30
83	$NH(CH_2)_2SO_2(CH_2)_3O(CH_2)_2\text{-1-benzofuran}$	2.86[a]	N	160	>180
84	$NH(CH_2)_2SO_2(CH_2)_3O(CH_2)_2(4\text{-cyanophenyl})$	2.00[a]	N	40	40
85	$NH(CH_2)_2S(CH_2)_2O(CH_2)_2Ph$	3.53[a]	B	>180	>180
86	$NH(CH_2)_3SO_2NH(CH_2)_2S(CH_2)_2Ph$	1.77[a]	B	180	>180
87	$NH(CH_2)_3NHSO_2(CH_2)_2OCH_2(1\text{-naphthalene})$	1.69[a]	B	180	>180
88	$NH(CH_2)_3SO_2NH(CH_2)_2O(CH_2)_2(4\text{-fluorophenyl})$	1.57[a]	B	150	164
89	$NH(CH_2)_2SO_2(CH_2)_3)O(CH_2)_2(2\text{methylphenyl})$	2.84[a]	N	160	144
90	$NH(CH_2)_2SO_2(CH_2)_3O(CH_2)_2SPh$	2.68[a]	N	120	155
91		2.32[a]	N	60	135
92	$N(CH_3)(CH_2)_3O(CH_2)_2O(CH_2)_2Ph$	2.44[a]	B	>180	>180
93	$NH(CH_2)_3SO_2(CH_2)_2O(CH_2)_3(5\text{-methylthiophene}$	2.41[a]	B	>180	>180
94	$N((CH_2)_2CH_3)(CH_2)_3O(CH_2)_2O(CH_2)2Ph$	2.74[a]	B	>180	>180
95	$NH(CH_2)_3SO_2NH(CH_2)_2O(CH_2)_2(3\text{-trifluoromethylphenyl})$	2.62[a]	B	>180	>180
96	$NH(CH_2)_2NHSO_2(CH_2)_2O(CH_2)_2(3,5\text{-dimethylphenyl})$	2.59[a]	B	>180	>180
97	$N(CH_3)(CH_2)_2N(CH_3)CO(CH_2)_2O(CH_2)_2Ph$	2.59[a]	B	>180	>180
98	$N(CH_2CH_3)(CH_2)_3O(CH_2)_2O(CH_2)_2Ph$	2.45	B	>180	>180
99	$NH(CH_2)_3SO_2NH(CH_2)O(CH_2)_2(3,5\text{-dimethylphenyl})$	2.92	B	>180	>180
100		1.81[a]	B	180	>180
101	$NH(CH_2)_3SO_2NH(Ch_2)_2O(Ch_2)_2(3\text{-chlorophenyl})$	1.92[a]	B	180	>180
102	$NH(CH_2)_3SO_2NH(CH_2)_2O(CH_2)_2SPh$	1.84[a]	B	180	>180
103	$NH(CH_2)_3SO_2NH(CH_2)_2O(CH_2)_2\text{-(4-methylphenyl})$	2.07[a]	B	>180	>180

[a]Measured log $D_{7.4}$.

There are a number of statistical methods widely used in QSAR analysis, and many that have been explored, Table 6.1. They can be divided into two categories, unsupervised and supervised methods. Unsupervised methods are used to explore the distribution of the compounds

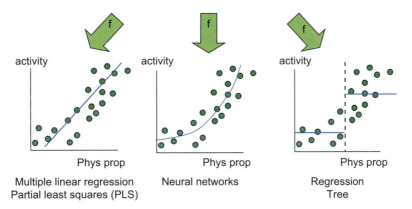

Figure 6.6 How different QSAR tools model data. MLR and PLS apply a linear model. Neural networks can fit non-linear relationships. Regression tree approaches classify data to form nodes enriched in one class over another. Multiple classes can be used to model continuous data.

through property space, and do not use the response variable to choose or weight variables. Supervised methods attempt to find correlations between the responses and descriptors.

There are many supervised methods that have been used with some success in QSAR. They have been extensively described previously and here we will focus on three very popular methods that exemplify different approaches to building a correlation between x descriptors and the y response (Figure 6.6).

6.4.6.1 Multiple Linear Regression (MLR)

Hansch used multiple linear regression, and this is still a common method today. It is limited to datasets where there are many more compounds than descriptors, as it is particularly sensitive to chance correlations where many descriptors are available. The principle on which multiple linear regression was built assumes all variables are relevant in the model. But MLR algorithms allow "forward and backward stepping", where descriptor variables are sequentially added or excluded from the models, facilitating QSAR scientists to pick and choose from a larger array of descriptors those that are relevant to the model to find a good model. The statistics which govern the selection were not designed to allow such choice, and due to the combinatorial nature of selection of a few relevant descriptors from a matrix of many more, MLR is particularly susceptible to chance correlations.[42] But MLR does produce simpler to interpret models, that is if the descriptors are meaningful to a medicinal chemist.

Multiple linear regression assumes all variables are uncorrelated of each other, but this is rarely the case. Where variables are correlated, then the coefficients of the regression equation become dependent upon each other, and so their size and even sign cannot be used independently of all the other descriptors in the model to deduce that variable's unique contribution.

6.4.6.2 Partial Least Squares Analysis or Projection to Latent Structures (PLS)

Partial least squares analysis (PLS), introduced to Physical Sciences by H Wold and co-workers in 1979, is a linear correlation method that is less sensitive to chance correlations in datasets where there are more descriptors than compounds.[43] It uses the inter-correlation structure of the descriptor set to form new underlying summary descriptors (components) as weighted linear combinations of the original variables. PLS forms these summary variables to maximize the correlation with the response variable. The fact the method uses the intercorrelation structure of the descriptor set makes it less sensitive to a chance correlation in any one variable. The method principal components analysis (PCA) is similar in that it forms summary variables as weighted linear combinations of the original variables, but in PCA no response variable is defined to supervise the extraction of components. In PCA analysis each component is chosen in decreasing order of its ability to summarise the information on the x-descriptor set. PLS is a development from principal component analysis.

The influence of the original variables on the derivation of these new summary descriptors is described by the "loadings" or weightings of the variables on the new summary descriptor, and the summary descriptor's numerical values are the scores.

Because all the original variables contribute more or less to forming each summary variable, PLS and other multivariate models are harder to interpret their underlying physical meaning.

6.4.6.3   Neural Networks

Various flavours of neural networks exist. Their structure has a set of layered nodes inter-connected to each other, which prompts the analogy to the structure of the human brain. They often consist of an input layer, one node per descriptor, one or more hidden layers, on which the input descriptor nodes are projected, and an output layer. There are non-linear "transfer functions" that operate between the nodes of the input layer and the hidden layers and mathematically transform the original data as it is combined onto the "hidden layer" which allow the neutral network to model non-linear data. They usually run in an iterative fashion where either a random, or a distribution (Bayesian approach) of weighting parameters are start points for linking the input nodes to the hidden and output nodes. These weights are iteratively optimized, to drive a correlation between descriptors and response. The fitting routines together with the non-linear transfer functions in neural networks are very powerful, and if unchecked, will fit both signal and noise in the training set. So neural networks often maintain a calibration set/tuning set, the prediction of which is used to stop the training phase. Neutral networks can generate a single QSAR equation, or in the case of Bayesian approaches, an ensemble of models.

6.4.6.4   Regression Trees and Random Forests

Regression Trees and Random Forests are a form of classification method. In regression trees, the method attempts to classify the response into "bins" that represent ranges of activity. The method attempts to find critical values of x-variables that result in an improvement in the purity of y-bins with correctly classified compounds. Random Forests is an extension where instead of one tree, an ensemble of trees is formed, each built on a subset of the original dataset by sampling rows and columns. Each tree in the "forest" is a decision tree. When a prediction is made, all models participate in the prediction and the predicted value is an average from across the forest of trees. The dispersion of predictions from across the ensemble of models can give an indication of the confidence in the prediction. Like PLS, Random Forests appear less sensitive to the size of the descriptor matrix, as spurious models influence becomes averaged out across the ensemble of models. Because the prediction is made from across the forest of trees, it is more difficult to interpret which are the truly important underlying descriptors, although the method provides diagnostics to indicate important underlying variables.

6.4.7   Appropriate Measures of Goodness of Fit—Model Diagnostics

OECD guideline 4 states one should have appropriate goodness of fit statistics and model diagnostics.[17] Familiar statistical diagnostics from knowledge of simple linear regression are also used with the more intricate machine learning methods. The coefficient of variation in fit to training set r^2, the fraction of the variation in response described by the model, and residual root mean square error—the variation in the response variable unexplained by the model—are often reported. But the fit to the training set is not a good indication of a model's ability to generalize. Spurious models can have very appealing fit statistics, but will no ability to predict. Rather, most analyses focus on the performance of the model in prediction of a chosen test set. Similar quality of fit statistics to r^2 can be defined in this prediction mode, but to indicate the assessment is based on prediction rather than fit, the r^2 diagnostic is replaced with Q^2 and root mean square error with root mean square error in prediction (RMSE/RMSEP). Unlike r^2, which is bounded between 0–1, Q^2 can take values less than zero, meaning the model predicts the test set worse than if one had used the mean of the model as the prediction for every compound.

r^2 for the fitted model can be written as Equation (6.5).

$$r^2 = 1 - \frac{\sum (y_{obs} - y_{fitted})^2}{\sum (y_{obsd} - y_{mean})^2} \qquad (6.5)$$

Q^2, the r^2 in prediction, can be defined in a similar way (Equation (6.6)).

$$Q^2 = 1 - \frac{\sum (y_{obs} - y_{predicted})^2}{\sum (y_{obsd} - y_{mean\ train\ set})^2} \qquad (6.6)$$

The root mean square error in prediction (RMSE) value is particularly useful as it is in the same units as the experiment, and can be directly compared to the experimental error of the response variable. The RMSE of the test set gives the medicinal chemist an indication of the likely error associated with a prediction (Equation (6.7)).

$$RMSE = \frac{\sum (y_{predicted} - y_{obs})^2}{n - 1} \qquad (6.7)$$

The average error in prediction tells you if there is any systematic bias in the predictions (Equation (6.8)).

$$mean\ absolute\ error/bias = \frac{\sum_{i=1}^{n_{ext}} (y_{obs} - y_{predicted})}{n} \qquad (6.8)$$

An external set of compounds may be predicted with a systematic bias, which would inflate RMSE and decrease Q^2. A non-zero mean absolute error indicates the predictions may not be accurate, but may still be useful in ranking compounds. It also may indicate a structural or substructural feature present in the test set but not encoded in the training set or its description.

Assessment of performance based on a test set only using Q^2 or even RMSE can still be problematic. Figure 6.7 shows the performance of four project test sets compared to the test set used to validate a QSAR model. For project a (Figure 6.7(a), the test set predictions are good, and will show a high Q^2 and a low RMSE lower than the standard deviation of the test set, and no mean error (bias) in predictions. For project b (Figure 6.7(b)), this project's compounds are predicted with a high mean error, which will inflate RMSE and reduce Q^2, but it is clear that the project's predictions rank well. Project c (Figure 6.7(c)) is an interesting case. The project compounds are predicted with a low RMSE but Q^2 will also be low as the compounds themselves do not cover a range of response.

An ambiguity occurs with the Q^2 statistic when the test set data is not evenly distributed over the range of the training set. As the variance of the external test set approaches the RMSE of the fitted model, the Q^2 measure would approach zero—even though it would appear the predictions are in accord with the model. Consonni[43] defined a new Q^2 statistic that expresses the mean predicted error sum of squared deviations between the observed and predicted values for the test set over the mean training set sum of squared deviations from the mean value (Equation (6.9)).

$$Q^2 = 1 - \frac{\left[\sum_{i=1}^{n_{ext}} (\hat{y}_i - y_i)^2\right]/n_{ext}}{\left[\sum_{i=1}^{n_{TR}} (y_i - \bar{y}_{TR})^2\right]/n_{TR}} = 1 - \frac{PRESS/n_{ext}}{TSS/n_{TR}} \tag{6.9}$$

n_{ext} = number of compounds in external set
$\hat{y}_i$ = the prediction estimate of y
y_i = the observed value
n_{TR} = the number of compound in the training set,
$\bar{y}_{TR}$ = the mean of the traiing set observed values,
PRESS is predicted error sum of squares deviations between the observed and measured y values
TSS is the training set sum of squared deviations from the mean

Consonni demonstrated that this formulation of Q^2 is stable with test sets of different variances.[44]

For project c (Figure 6.7c) we would say the predictions are in line with the model, although within the range studied we cannot rank them. For project d (Figure 6.7(d)), the project is predicted with a high RMSE, not driven by high mean error, and one would deduce that the project is poorly predicted by this model.

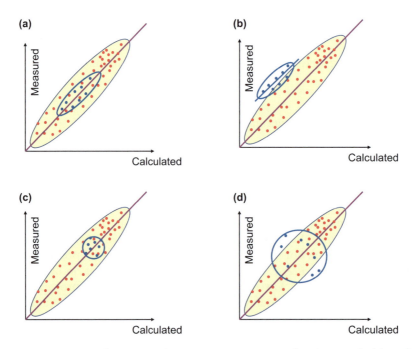

Figure 6.7   How a QSAR model (in red dots) may predict a test set (in blue dots). (a) Well predicted; (b) predicted with a large bias; (c) predicted in-line with the model, although the model cannot rank the predictions; and (d) test set is poorly predicted.

Even if your project behaves like Figure 6.7(a), this model may still not be useful to your project. The assessment of the quality of the model can only be made in the context of what you are trying to achieve. The RMSE of the predictions needs to be compared to the resolution in prediction you are looking for. If you wish to move your prediction by 0.3 log units, and the RMSE of the test set is still 0.5 log units, the model does not have enough resolution to confidently predict such a small change.

The use of a test set in the model evaluation process protects you from using a spurious QSAR model as it is less likely that a model that randomly fits the training set would also predict well compounds that the model has not seen. The use of a test set in the model building process excludes valuable chemical information from the model building process, and so some methods attempt to utilize all the data and perform an internal validation procedure known as cross validation. In cross validation a number of cases from the training set are left out during the model fitting process and then predicted from the resulting model. Multiple models are built with a fraction of the compounds excluded at each build. Enough models are built so that each compound is predicted from a model from which it was excluded. A prediction statistic can be defined, as in Equation 6.6. To differentiate this cross-validated measure of predictiveness from a measure of predictiveness calculated from an external test set, a lower-case q^2 symbol is often used to delineate a cross-validated statistic. A $q^2 > 0.3$ is often taken as a rule of thumb for a potentially useful model.

Leave-one-out cross validation is the simplest and least challenging validation method, and has been widely criticised[45] as being too easy a test of the model. Leaving out compounds in multiple groups is a tougher test for the model. The archetypal PLS program, SIMCA,[46] uses seven subsets as default, so at each rebuild $1/7^{th}$ of the training set is left out at each model build.

Cross-validation can only give an unbiased estimate of future predictive ability if future compounds come from the same population as those in the cross-validation sample. This is not a safe assumption as the nature of drug design is that compounds evolve with time.

The assessment of the significance of the model can also be explored by performing a y-randomisation experiment. The y-randomisation procedure compares the performance of models built for a randomly shuffled y-response. When repeated multiple times, the distribution of r^2 values obtained is an assessment of the significance of the original model, without assuming the data follow a particular statistical distribution (non-parametric). Y-randomisation has been referred to as probably the most powerful validation procedure,[47] although only if the full model generation procedure is repeated including variable selection, rather than just permuting the y-variables of the final model.[48]

6.4.8 A Defined Domain of Applicability

A critical question for medicinal chemists using a model for the first time is, will this model make good predictions for my compound series. One approach to providing that confidence is to assess the "chemical space" covered by the QSAR model and how "far" the compounds being predicted are from that space. The chemical space encompassed by the model is termed the domain of applicability. The hope is that molecules being predicted that are "near" the applicability domain of the training set compounds will be predicted with lower error and therefore with more confidence than compounds further away.

Confidence in the model can be gained by examining the error in prediction of compounds in the test set most similar to your compounds.

Definition of domain of applicability of the QSAR model and "distance" of a compound being predicted to the defined domain is an active area of research in QSAR. The OECD QSAR Guideline 3 states that one should have a defined Domain of Applicability.[17] The principles are

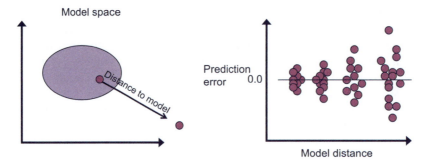

Figure 6.8 Schematic showing the domain of applicability of a QSAR model and the distance of prediction compounds from it. One might expect as compounds get "further" from the model space, the confidence in the predictions may decline. This can be assessed by plotting the prediction error *vs.* distance to model of an independent test set of molecules.

Training set	Test set	RMSE
AZ train	AZ test	0.78
Literature train	Literature test	0.84
AZ train	Literature test	1.88
Literature train	AZ test	1.00
AZ-literature train	AZ test	0.79
AZ-literature train	Literature test	0.82

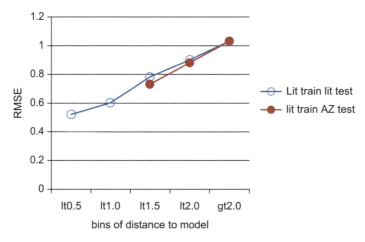

Figure 6.9 Evaluation of solubility models built on AstraZeneca in-house data and literature data, predicting their respective test sets. The literature model does worse on the AZ test set because the average distance of the AZ compounds from the literature based model was higher. It actually performed equally well on the AZ compounds when the distance to model was considered. When a mixed training set was used this model was equally able to predict both in-house and literature compounds.
Data from ref. 49.

easy to grasp, although the mathematics associated with some of the methods can be quite complex (Figure 6.8).

A suitable distance to model measure should show a dependence of the distance upon the error in prediction of the test set compounds. By knowing the distance to the model space of the compound being predicted, the confidence in the prediction can be inferred based on the error in prediction of the test set compounds similarly distance from the model space. Unfortunately, many of the distance measures only a produce a weak relationship to error in prediction which limits the confidence that can be extracted from the statistic.

The effect of distance to model on prediction precision was clearly shown by Bruneau in an assessment of an internal AstraZeneca (AZ) solubility dataset compared to a model built on public domain solubility data (Figure 6.9).[49] Bruneau found that when using the internal AZ model to predict internal AZ compounds or the model built on the public domain data to predict the public domain test set, the models gave comparable prediction statistics with similar root mean square errors in prediction. But when used to cross-predict each other's test set, the predictions were worse. Interestingly, while the solubility model built on external data did a worse job at predicting the AZ compound test set than its own test set, the predictions were not as bad as the AZ model predicting the public domain test set. When Bruneau examined how the prediction error depended on distance to model, the results became understandable. The AZ compounds were more distant from the public domain compounds than their own training set, which corresponded to an increase in the prediction error, but their error in prediction was similar to public domain compounds at a similar distance to the public domain training set compounds (Figure 6.9).

A number of different numerical approaches are taken to defining the domain of applicability and the distance of compounds being predicted from it (distance to model). Euclidian distance (geometric distance) assumes the model space is spherically distributed, and all compounds in multivariate space are "equally distant". It is like taking a ruler and measuring the separation of points in the multivariate space. Alternatively, one can use a Mahalanobis distance (probability distance) which takes note of the intercorrelation structure of the dataset and hence the shape of the model space to be non-spherical and more "ellipsoid". Distances can also be measured in structure space using molecular fingerprints, amongst many others.[50]

So far there appears to be no one measure that performs better than another across many datasets. Suffice to mention here that medicinal chemists should request information on the distance to the model space of the compound being predicted, and this distance can be calibrated based on the relationship of the error in prediction to the distance to the model of the test set compounds. With this information, one has some guidance as to what degree of confidence to place in the resulting prediction. With often such limited confidence in these potentially very powerful diagnostics, what advice can we give medicinal chemists? Firstly work closely with your QSAR scientist when assessing the model's utility to your compounds. Medicinal chemists who do commit to synthesis based on the predictions from a model will get real-time feedback on the quality of the model, once their compounds are tested. Maybe this is a more useful confidence build for the practicing medicinal chemist than any QSAR diagnostic.

6.4.9 Trying To Interpret Your Model—What Are The Controlling Descriptors?

The OECD QSAR Guideline 5 states that one should attempt to gain a mechanistic interpretation, if possible.[17] A predictive QSAR model must identify some of the underlying physical controlling features of the response variable being modelled, although in the case of multivariate models it can be difficult to disentangle the primary controlling properties when many variables contribute to the overall model. The examples and case studies described so far led to mechanistic interpretations which guided further insights, from Overton and Meyer's theory of narcosis, Hammett's hypothesis for electronic induction and resonance, Hansch's "random walk" of drugs to their receptors, or understanding of the affinity of charged drugs for biological membranes, from the duration of action of dual-acting D_2/β_2 agonists. The QSAR expert can work with the medicinal chemist in interpreting the model, so the model can be used to rationally design the next compound, rather than just using the QSAR model as a black-box filter. Particularly with complex multivariate models, trying to interpret the controlling properties so new compounds can be rationally designed, so called inverse QSAR, can be challenging. There has therefore been a renewed interest in Free–Wilson type methods, which does not use derived descriptors, and this inspired the recently introduced method of matched molecular pairs.

6.5 MATCHED MOLECULAR PAIRS ANALYSIS

Matched molecular pairs analysis is a complementary approach to QSAR modelling. It is often termed an inverse QSAR approach, as it attempts to predict the structural change required to bring about a particular change in the response variable. Matched molecular pairs are pairs of molecules that differ only by a particular well defined structural modification, the rest of the molecule being common. Where that same variation is repeated over a number of pairs of molecules, the average change in the response variable and associated standard deviation can provide a statistically meaningful description of the effect of that structural variation on the response. As a worked example, in a set of Pfizer CDK4 inhibitors, six matched pairs were found where –Br was substituted for acetyl.[51] In the six matched pairs, where bromo was changed to acetyl, the mean change in CDK4 log(potency) was 1.2 logunits with a standard error of 0.32 (Figure 6.10).

Matched molecular pairs analysis bears similarities to Free–Wilson methods, although matched molecular pairs analysis focuses on the transformation, and the associated change in biological activity rather than attempting to quantify the contribution of a particular substituent *per se*. Modern cheminformatics tools allow the automated perception of matched molecular pairs, even in very large datasets. So far commercial or publically available code to undertake automated matched molecular pairs analysis is only just emerging. KNIME, the Konstanz Information Miner, is an open-source data analytics platform, widely used in the chemoinformatics community. A version of matched molecular pairs exists for the KNIME platform. The open-source chemoinformatics toolkit RDKIT also contains an implementation of matched

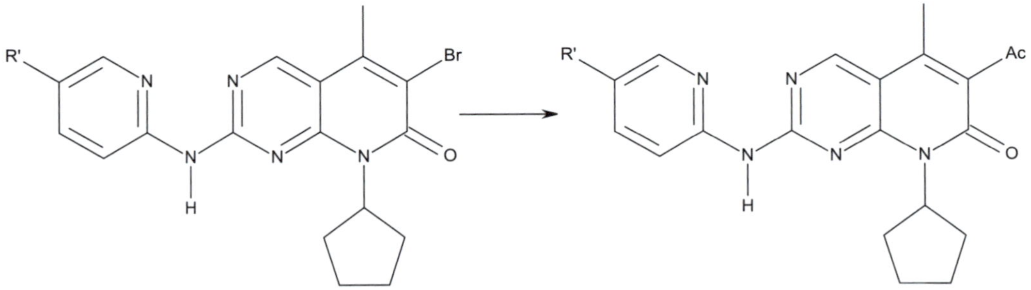

| | pIC$_{50}$ | | pIC$_{50}$(Br)– |
R'	Br	Ac	pIC$_{50}$(Ac)
Piperazine	6.8	8.0	+1.2
(CH$_3$OCH$_2$CH$_2$)$_2$N	6.0	7.3	+1.3
3,5-dimethylpiperazine	7.2	7.4	+0.2
N-methylpiperazine	6.9	8.3	+1.4
4-hydroxypiperazine	7.1	7.7	+0.6
Morpholine	5.7	8.4	+2.7

Figure 6.10 Example of matched molecular pairs analysis from a set of Pfizer CDK4 inhibitors. Data taken from Ref. 51.

molecular pairs analysis. RDKIT is the chemoinformatics engine in MyChEMBL, allowing RDKIT matched molecular pairs analysis to be undertaken within the ChEMBL database. These are two implementations, although computational procedures to undertake matched molecular pairs analysis have been described in detail by a number of groups.

Dosseter and colleagues found 1826 matched pairs in the AstraZeneca database of 135,000 human microsomal metabolic stability measurements for 24 common transforms of methylene groups, common sites for oxidative metabolism.[52] Within this database they found 4 transforms that increased significantly metabolic stability. These were replacement of both Ar-CH$_2$-Ar and Ar-CH$_2$-Alk with ether, introduction of a basic side-chain into an aliphatic chain, or replacement of -CH$_2$- with sulphone. They also identified transforms that resulted in significant decreases in stability. Unsurprising to experienced medicinal chemists, a significant correlation with log P was found across all identified transforms affecting metabolic stability. While this may seem a trivial result, the power of QSAR and matched molecular pairs is apparent when the structural and bulk property effects of molecular changes can be deconvoluted.

Gleeson has published a comprehensive matched pairs analysis from the GlaxoSmithKline database of 500,000 data points across 8 absorption distribution metabolism and elimination (ADMET) endpoints.[53] They studied matched molecular pairs for hydrogen replacements with a list of predefined substituents where the total number of pairs $>= 20$ that came from >5 structural families identified by Daylight fingerprint clusters, to access (hopefully) generalizable rules across chemical series. Their conclusion was there was no perfect substituent which has significant beneficial effects across all ADMET endpoints; the choice is always a compromise. For example, a substituent chosen to improve solubility may have a detrimental effect on permeability. Nevertheless the tables do provide useful guidance, as for your particular optimisation problem, one may be more concerned with one endpoint rather than another and may be able to tolerate changes which are detrimental to one property if it is not so critical as another. Figure 6.11 shows a matched molecular pair analysis for changes in P450 3A4 *vs.* logD

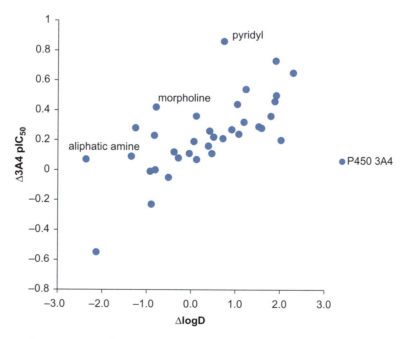

Figure 6.11 Plot of matched molecular pairs analysis from GSK database showing Δ3A4 pIC$_{50}$ *vs.* ΔlogD7.4 for hydrogen being substituted by various functional groups.

from the GSK ADMET database.[54] The change in property associated with each substituent is plotted *vs.* clogD$_{7.4}$. This indicates whether the contribution of the substituent to changes in P450 3A4 is more or less than a bulk hydrophobic effect as defined by the rest of the substituents studied. When a pyridine, morpholine and even an aliphatic amine are substituent for hydrogen, the increase in P450 3A4 is greater than that expected considering the effect on lipophilicity or ionisation. This is in agreement with patterns identified previously by QSAR analysis.[55] The QSAR suggested an additional mechanism of 3A4 inhibition operating, which was chelation of the nitrogen lone-pair to haem iron of cytochrome P450 3A4,[56] subsequently observed by X-ray crystallography.[57,58] The QSAR analysis also identified a method to overcome this, by blocking the nitrogen acceptor through *ortho* substitution.

6.6 EXAMPLES OF INFLUENTIAL QSAR MODELS

Many marketed drugs have depended, at least in part, on classical QSAR approaches in their discovery, including drugs as diverse as norfloxacin, lomerazine, flobufen[59] and cimetidine, amongst many others.

Because QSAR models attempt to find patterns across a series of molecules in a dataset, they are particularly suited to ADMET endpoints, which are more bulk property-controlled than structurally dependent. Drug-receptor interactions inherent to primary potency endpoints are often fraught with "activity cliffs" – where activity falls away very rapidly with small structural variations – which are harder for QSAR models to identify without knowledge of the nature of the drug–receptor interactions, as can be gained from X-ray crystallography and docking studies. Nevertheless, QSAR models are often sought for primary potency endpoints, as they are easy to look for, and can at least help the medicinal chemist explore the SAR within a dataset, even if they are not refined enough to be used for accurate prediction. The literature is replete with examples of QSAR models being applied in drug design.

The fundamental understanding of drug phospholipid interactions driving long duration of action of D$_2$/β$_2$ agonists in the earlier case study was later used to aid the design of β$_2$ agonists in projects at Pfizer[60] and was cited in the design of indacaterol, the recently introduced long acting β$_2$ agonist from Novartis.[61] The AstraZeneca global hERG QSAR model has contributed to the reduction in the synthesis of "red-flag" compounds (compound that are measured to have a hERG potency <1μM) from 25.8% of all compounds tested in 2003 to only 6% in 2010.

The false negative prediction rate of potentially genotoxic impurities by *in silico* models was recently surveyed across 8 companies. The methods for prediction were given and the approaches across the companies were very similar. The true negative prediction rate was found to be 94%, and when expert evaluation of the results was included in the decision this increased to 99%.[65] The results of this analysis are currently being written into an ICH guideline M7 on genotoxic impurities.

The control of lipophilicity is of fundamental importance to drug design. While the archetypal CLOGP algorithm from Pomona College was manually constructed as a fragmental based system from a combination of first principles and empirical observations, a number of the most widely used ClogP and ClogD algorithms are QSAR based including ALOGP,[62] XLOGP[63] and MLOGP.[64] Many pharmaceutical companies have derived their own log P and log D calculators using QSAR approaches, which outperform the standard CLOGP method.[65]

The scoring functions used inside many structure-based design docking programs are QSAR derived. The archetypal scoring function in LUDI, was derived from regression analysis of a set of pre-determined physics-based descriptors calculated from the 82 protein-ligand X-ray crystal structures together with the ligands reported affinity.[32] The initial dataset defining scoring function SCORE1 of 45 complexes[66] (Equation (6.10)) was extended to 82 complexes, and was named SCORE2 (Equation (6.11)). The SCORE 2 function has a lower residual standard deviation of 8.8 kJ/mol, (relative to 9.5 kJ/mol for SCORE1) suggesting SCORE2 fits the experimental data for $\Delta G_{binding}$ better than SCORE1. The prediction of an independent test set of 12 complexes showed a root mean square error in prediction RMSE = 8.8kJ/mol, in good agreement with the unexplained error in fitting the training set data. It is interesting that the magnitude of some of the coefficients change significantly, which is probably due to the changing intercorrelation structure of the descriptors within the larger dataset. As mentioned earlier in the multiple linear regression section, the regression equation can be used to predict affinity, but the magnitude of the coefficients may not be quantitatively interpreted, as the intercorrelation of the descriptor set makes them dependent upon each other

$$SCORE1 \; \Delta G_{binding} = 5.4G_0 - 4.7G_{HB} - 8.3G_{ionic} - 0.17G_{apolar} + 1.4G_{rot \, kJ/mol}$$
$$N = 45, \; R = 0.835 \; s = 9.5$$

$$(6.10)$$

$$SCORE2 \; \Delta G_{binding} = -1.4G_0 - 3.1G_{HB} - 6.6G_{ionic} - 0.15G_{apolar} + 1.0G_{rot \, kJ/mol}$$
$$N = 85 \; R = 0.841 \; s = 8.8$$

$$(6.11)$$

LUDI scoring function has inspired many other docking scoring functions.

6.7   ACCESSING QSAR TOOLS AND MODELS

For visual dataset exploration, SPOTFIRE[67] has already been mentioned, and both SPOTFIRE and VORTEX from Dotmatics[68] enable data exploration through graphs, linked to chemical structure visualisation. But even Microsoft EXCEL, or indeed any graph-drawing package can reveal exciting details when structure–activity data is plotted. The statistical package JMP from SAS Institute[69] has a very useful multiple linear regression package as well as the ability to make simple graphs for data exploration. It is often this author's first start point for a QSAR analysis.

Many commercial computational chemistry packages contain QSAR model building implementations. They bring together molecular descriptor calculators, machine learning methods, statistical diagnostics and visualization into seamless packages.

A number of online systems are available for the automatic building of QSAR models, for example OCHEM,[70] and OECD QSAR toolbox.[71] Submitting datasets to online QSAR engines, outside corporate firewalls may be problematic for some medicinal chemists, as this potentially compromises corporate confidentiality. Therefore local implementations are often preferable. The R statistics suite[72] is commonly used at the core of computational chemistry QSAR implementations. Common computational chemistry software contain useful QSAR implementations with descriptor calculation and machine learning tools. But freely available integrated QSAR toolkits are less common.

The Bioclipse platform[27] is an open source QSAR and bioinformatics platform, developed as a collaboration between the Dept. of Pharmaceutical Biosciences, Uppsala University, Sweden, and the European Bioinformatics Institute. Bioclipse can be freely downloaded to private computers, and will run under Windows, Linux and Macintosh operating systems. Bioclipse contains a fully integrated QSAR toolkit that can be used for datasets of up to 2000 compounds. It contains a choice of descriptor sets including Signature descriptors, which are a type of fingerprint descriptor, which allows the results of the QSAR analysis to be mapped back onto the training set structure, highlighting areas of the molecule most significantly contributing to the response. Bioclipse may be a useful starting point for anyone wishing to build their own QSAR models from scratch.

For matched molecular pairs analysis, QSAR, as well as a plethora of other cheminformatics tasks, the open-source KNIME cheminformatics platform is available. A commercial version of KNIME also exists offering technical support and features for enterprise level use.[73–74]

Many medicinal chemists will want to use QSAR models, rather than build their own. Many QSAR models for ADMET endpoints are available in the public domain, either as freeware or in commercial packages. These are based on literature datasets, and in the absence of any data from their own chemistry, or expertise to build their own QSAR models, these are all many medicinal chemists may have access to. How confident one can be in the prediction from such models requires careful thought. If the compounds you are predicting are unlike the compounds within the training set of the model, one may be concerned with the quality of the prediction. One should remember the OECD guidelines when assessing the utility of these models for your intended use. Before you use the model to prioritise your synthetic targets and commit valuable resources to synthesizing your chosen targets, consider the hints and tips given below as guidance.

HINTS AND TIPS FOR USING QSAR MODELS

For a medicinal chemist using a QSAR model, the following steps are recommended:

- Look at the distribution of compounds in the training set and test set over the range of responses. Are the data evenly distributed or is there a region of activity or property space that contains better coverage—indicating where better predictions are likely to be made?
- Examine the Q^2 and RMSE of the test set used to validate the model. Does the model have the resolution you require to solve your project problem? A high r^2 or q^2/Q^2 looks impressive, but if the range of the training set or test set is very large, the unexplained residual error in model can still be large. So examine the RMSE of the model. For example, if the RMSE of predictions of the test set is 0.7 log units and you only wish to move your property by 0.3 log units, it's likely you are within the noise of the model.
- Examine the error in prediction of compounds in the test set that are "closest" to the compound you are predicting (based on distance to model measures). You may find this gives you more or less confidence in the model. You may find that within the test set, compounds that are more similar to the training set are predicted better than the average across the test set. If these compounds are similar to your target compound this will give you more confidence in the model's ability to predict your compound.
- The QSAR model can be used as a filter for a virtual library, or interpreted to influence compound design. Examine the controlling features of the QSAR model; this may give you an indication of how you can rationally change your molecules to improve the property you are trying to optimise.
- Examine the QSAR model visually. Plot the important controlling variables against the response variable; this may allow you see patterns in the dataset that can guide further design.
- Are the important variables similar as new compounds are added to the model, or if compounds are removed from the model, or if sub-models are built during cross validation? Instability in selected variables can point to a poor or spurious QSAR model.
- Are new compounds well predicted by your model? This is the ultimate test of any model.

- As new compounds are synthesized and measured in your assay, ask your computational chemist to update the QSAR model to minimize the distance the compounds you will design will be from the domain of applicability of the QSAR model. Keep monitoring your predictions as compounds synthesised are subsequently measured in the assay your QSAR model is supposed to predict. This may give you increased confidence in the model, or highlight new features in your molecules not identified in the model (*e.g.* it is surprisingly badly predicted where previous compounds had been well predicted). QSAR models may be automatically updated, ensuring your model always contains your latest compounds (AutoQSAR).[75]

KEY REFERENCES

www.QSARtoolbox.org Homepage of the OECD QSAR toolbox and associated documentation, including the OECD.

www.orchestra-qsar.eu Homepage of Orchestra, an EU funded project to disseminate recent research on *in silico* methods for evaluating toxicity of chemicals. Includes an online course and e-book *Theory, Guidance and Applications on QSAR and REACH*.

Drug Design Strategies – Quantitative Methods, ed. D. Livingstone and A. M. Davis, Royal Society of Chemistry, Cambridge, 2011. Useful compendium of monographs on quantitative methods including QSAR models written by practitioners in the QSAR field.

H. Van de Waterbeemd and S. Rose, in *The Practice of Medicinal Chemistry*, ed. C. G. Wermuth, Elsevier, London, 3rd edn, 2008, pp. 491–513.

REFERENCES

1. A. Crum-Brown and T. R. Fraser, *J. Anat. Physiol.*, 1868, **2**(2), 224.
2. B. J. Richardson, *Medical Times and Gazette*, 1869, **2**, 703.
3. M. C. Richet, *Compt. Rend. Soc. Biol.(Paris)*, 1893, **45**, 775.
4. H. Meyer, *Arch. Exp. Path. Pharm.*, 1899, **42**, 109.
5. C. E. Overton, *Studien ber die Narkose, zugleich ein Beitragzur allgemeinen Pharmakologie*, Fischer, Jena, 1901 (English translation by R. L. Lipnick, *Studies on Narcosis*, Chapman and Hall, London, 1991).
6. L. P. Hammett, *J. Am. Chem. Soc.*, 1937, **59**, 96.
7. P. Bronsted, *Physik. Chem.*, 1924, **108**, 185.
8. (a) R. W. Taft, *J. Am. Chem. Soc.*, 1952, **74**, 2729; (b) R. W. Taft, *J. Am. Chem. Soc.*, 1952, **74**, 3120; (c) R. W. Taft, *J. Am. Chem. Soc.*, 1953, **75**, 4538.
9. C. G. Swain and E. C. Lupton, Jr., *J. Am. Chem. Soc.*, 1968, **90**, 4328.
10. C. Hansch, P. P. Maloney, T. Fujita and R. M. Muir, *Nature*, 1962, **194**, 178.
11. T. Fujiita, *J. Comput.-Aided Mol. Des.*, 2011, 25, 509.
12. T. Fujita, J. Iwasa and C. Hansch, *J. Med. Chem.*, 1964, **86**(23), 5175.
13. S. M. Free and J. W. Wilson, *J. Med. Chem.*, 1964, 395.
14. T. R. Stouch, J. R. Kenyon, S. R. Johnson, X.-Q. Chen, A. Doweyko and Y. Li, *J. Comput.-Aided Mol. Des.*, 2003, **17**(2-4), 83.
15. A. M. Davis, in *Drug Design Strategies: Quantitative Approaches*, ed. A. M. Davis and D. J. Livingstone, RSC Publishing, London, 2012, pp. 242–266.
16. C. L. Gavaghan, C. H. Arnby, N. Blomberg, G. Strandlund and S. Boyer, *J. Comput.-Aided Mol. Des.*, 2007, **21**(4), 189.
17. http://www.oecd.org/env/hazard/qsar.

18. Anon, Regulatory Use of (Q)SARs under REA*CH*, 2009 http://echa.europa.eu/doc/press/webinars/regulatory_use_of_qsars_under_reach_doris_hirmann_echa.pdf.

19. http://pubs.acs.org/paragonplus/submission/jmcmar/jmcmar_authguide.pdf.

20. I. V. Tetko, J. Gasteiger, R. Todeschini, A. Mauri, D. Livingstone, P. Ertl, V. A. Palyulin, E. V. Radchenko, N. S. Zefirov, A. S. Makarenko, V. Y. Tanchuk and V. V. Prokopenko, *J. Comput.-Aided Mol. Des.*, 2005, **19**, 453.

21. R. D. Cramer III, D. E. Patterson and J. D. Bunce, *J. Am. Chem. Soc.*, 1988, **110**(18), 5959.

22. http://crdd.osdd.net/descriptors.php.

23. Advanced Chemistry Development, Inc., Toronto, ON, Canada, www.acdlabs.com, 2013.

24. http://sourceforge.net/apps/mediawiki/cdk/index.php?title = Main_Page.

25. http://openbabel.org/wiki/Main_Page.

26. http://www.rdkit.org/.

27. O. Spjuth, L. Carlsson, J. Alvarsson, V. Georgiev, E. Willighagen and M. Eklund, *Curr. Top. Med. Chem. (Sharjah, United Arab Emirates)*, 2012, **12**(18), 1980.

28. R. D. Cramer III, D. E. Patterson and J. D. Bunce, *J. Am. Chem. Soc.*, 1988, **110**(18), 5959.

29. P. J. Goodford, *J. Med. Chem.*, 1985, **28**, 849.

30. J. Nilsson, H. Wikstrom, A. Smilde, S. Glase, T. Pugsley, G. Cruciani, M. Pastor and S. Clementi, *J. Med. Chem.*, 1997, **40**(6), 833.

31. R. D. Head, M. L. Smythe, T. I. Oprea, C. L. Waller, S. M. Green and G. R. Marshall, *J. Am. Chem. Soc.*, 1996, **118**(16), 3959.

32. H.-J. Böhm, *J. Comput.-Aided Mol. Des.*, 1998, **12**, 309.

33. R. V. Bonnert, R. C. Brown, D. Chapman, D. R. Cheshire, J. Dixon, F. Ince, E. C. Kinchin, A. J. Lyons, A. M. Davis, C. Hallam, S. T. Harper, J. F. Unitt, I. G. Dougall, D. M. Jackson, K. McKechnie, A. Young and W. T. Simpson, *J. Med. Chem.*, 1998, **41**, 4915.

34. R. V. Bonnert, R. C. Brown, D. Chapman, D. R. Cheshire, J. Dixon, F. Ince, E. C. Kinchin, A. J. Lyons, A. M. Davis and C. Hallam, *J. Med. Chem.*, 1998, **41**(25), 4915.

35. J. Bradshaw, R. T. Brittain, R. A. Coleman, D. Jack and I. Kennedy, *Br. J. Pharmacol.*, 1993, **108**, 507.

36. G. P. Anderson, A. Linden and K. F. Rabe, *Eur. Respir. J.*, 1994, **7**, 569.

37. Spotfire, TIBCO Software Inc. 212 Elm StreetSomerville, MA, 02144 United States.

38. R. P. Austin, P. Barton, A. M. Davis, C. N. Manners and M. C. Stansfield, *J. Pharm. Sci.*, 1998, **87**(5), 599.

39. R. P. Austin, P. Barton, R. V. Bonnert, R. C. Brown, P. A. Cage, D. R. Cheshire, A. M. Davis, I. G. Dougall, F. Ince, G. Pairaudeau and A. Young, *J. Med. Chem.*, 2003, **46**(15), 3210.

40. P. Barton, A. M. Davis, D. J. McCarthy and P. Webborn, *J. Pharm. Sci.*, 1997, **86**(9), 1034.

41. S. L. Rodgers, A. M. Davis, N. P. Tomkinson and H. Van de Waterbeemd, *Mol. Inf.*, 2011, **30**, 256.

42. D. Livingston, D. J. Salt, R. Critchton and S. Ajmani, *J. Chem. Inf. Mod.*, 2007, **47**, 143.

43. R. W. Gerlach, B. R. Kowalski and H. O. A. Wold, *Herman Analytica Chimica Acta*, 1979, **112**(4), 417.

44. V. Consonni, D. Ballabio and R. Todeschini, *J. Chemom.*, 2010, **24**(3–4), 194.

45. A. Golbraikh and A. Tropsha, *J. Mol. Graphics Modell.*, 2002, **20**(4), 269.

46. SIMCA, Umetrics, AB Box 7960, SE-907 19, Umeå, Sweden.

47. H. Kubinyi, in *Handbook of Chemoinformatics*, ed. J. Gasteiger, Wiley-VCH, Weinheim, 2003, vol. 4, pp. 1532–1554.

48. C. Ruecker, G. Ruecker and M. Meringer, *J. Chem. Inf. Model.*, 2007, **47**(6), 2345.

49. P. Bruneau, *J. Chem. Inf. Model.*, 2001, **41**(6), 1605.

50. S. Weaver and M. P. Gleeson, *J. Mol. Graphics Mod.*, 2008, **26**, 1312.

51. A. G. Leach, H. D. Jones, D. A. Cosgrove, P. W. Kenny, L. Ruston, P. MacFaul, J. M. Wood, N. Colclough and B. Law, *J. Med. Chem.*, 2006, **49**(23), 6672.

52. A. Dosseter, *MedChemComm*, 2012, **3**, 1518.
53. M. P. Gleeson, A. Hersey and S. Hannongbua, *Curr. Top. Med. Chem.* (Sharjah, United Arab Emirates), 2011, **11**, 358.
54. M. P. Gleeson, G. Bravi, S. Modi and D. Lowe, *Biorg. Med. Chem.*, 2009, **17**, 5906.
55. R. J. Riley, A. J. Parker, S. Trigg and C. N. Manners, *Pharm. Res.*, 2001, **18**, 652.
56. D. A. Smith, S. M. Abel, R. Hyland and B. Jones, *Xenobiotica*, 1998, **28**, 10957.
57. P. A. Williams, J. Cosme, D. M. Vinkovic, A. Ward, H. C. Angove, P. J. Day, C. Vonrhein, I. J. Tickle and H. Jhoti, *Science*, 2004, **305**, 683.
58. M. Ekroos and T. Sjogren, *Proc. Natl. Acad. Sci. U. S. A.*, 2006, **103**, 13682.
59. T. Fujita, *Quant. Struct.-Act. Relat.*, 1997, **16**, 10.
60. A. D. Brown, M. E. Bunnage, P. A. Glossop, M. Holbrook, R. D. Jones, C. A. L. Lane, R. A. Lewthwaite, S. Mantell, C. Perros-Huguet and D. A. Price, *Bioorg. Medchem. Lett.*, 2007, **17**, 6188.
61. F. Baur, D. Beattie, D. Beer, D. Bentley, M. Bradley, I. Bruce, S. J. Charlton, B. Cuenoud, R. Ernst, R. A. Fairhurst, B. Faller, D. Farr, T. Keller, J. R. Fozard, J. Fullerton, S. Garman, J. Hatto, C. Hayden, H. He, C. Howes, D. Janus, Z. Jiang, C. Lewis, F. Loeuillet-Ritzler, H. Moser, J. Reilly, A. Steward, D. Sykes, L. Tedaldi, A. Trifilieff, M. Tweed, S. Watson, E. Wissler and D. Wyss, *J. Med. Chem.*, 2010, **53**, 3675.
62. A. K. Ghose and G. M. Crippen, *J. Comput. Chem.*, 1986, **7**(4), 565.
63. A. K. Ghose, V. N. Viswanadhan and J. J. Wendoloski, *J. Phys. Chem. A*, 1998, **102**(21), 3762.
64. I. Moriguchi, S. Hirono, Q. Liu, I. Nakagome and Y. Matsushita, *Chem. Pharm. Bull.*, 1992, **40**(1), 127.
65. P. Bruneau and N. R. McElroy, *J. Chem. Inf. Model.*, 2006, **46**(3), 1379.
66. H. J. Bohm, *J. Comput.-Aided Mol. Des.*, 1998, **8**, 243.
67. http://spotfire.tibco.com/.
68. www.dotmatics.com/products/vortex/.
69. JMP®, Version 10, SAS Institute Inc., Cary, NC, 1989–2007.
70. www.ochem.edu.
71. www.QSARtoolbox.org.
72. R Development Core Team, *R: A language and environment for statistical computing*, R Foundation for Statistical Computing, Vienna, Austria, http://www.R-project.org, 2006.
73. www.knime.org.
74. M. R. Berthold, N. Cebron, F. Dill, T. R. Gabriel, T. Kotter, T. Meinl, P. Ohl, C. Sieb, K. Thiel and B. Wiswedel, in *Proceedings of the 31st Annual Conference of the Gesellschaft für Klassifikation e.V.. (Studies in Classification, Data Analysis, and Knowledge Organization)*. Berlin, Germany: Springer, 319, 2007.
75. D. J. Wood, D. Buttar, J. G. Cumming, A. M. Davis, U. Norinder and S. L. Rodgers, *Mol. Inf.*, 2011, **30**(11–12), 960.

CHAPTER 7

Drug Metabolism

C. W. VOSE*[a] AND R. M. J. INGS[b]

[a] CVFV Consulting. Address at time of writing: Centre for Integrated Drug Development, Quintiles Ltd, 500 Brook Drive, Green Park, Reading, Berkshire RG2 6UU, United Kingdom; [b] RMI-Pharmacokinetics, 1317, Bulrush Ct, Carlsbad, CA 92011, USA
*E-mail: cvfv7@tiscali.co.uk

7.1 INTRODUCTION

This chapter is intended as a general overview of drug metabolism for the medicinal chemist. Key references and a bibliography have been included at the end to provide sources of further more detailed information.

Drug metabolism is an important elimination pathway. It may be defined as 'The chemical alteration of a drug by a biological system with the principal purpose of eliminating it from the system'. Mammals use exogenous compounds for the synthesis of their essential components and the maintenance of life. When a foreign compound cannot be assimilated into these pathways it will be eliminated. Drug elimination may occur directly by excretion in urine or bile for intrinsically water soluble drugs, indirectly by metabolism followed by the excretion of the metabolites in urine or bile, or by a combination of these processes. Metabolism generally produces products that are more water soluble and more easily excreted. The metabolic fate of a drug can influence its pharmacodynamics and toxicology.

7.2 DRUG METABOLISM PATHWAYS

The metabolism of drugs may be classified into two types: Phase I and Phase II pathways.

Phase I pathways are those which produce or introduce a new chemical group into a molecule (Table 7.1). There is a wide range of Phase I reactions which generally yield a product (metabolite) more water soluble and thus more easily excreted than the drug and may also produce metabolites which are substrates for the Phase II pathways. However, some Phase I pathways may produce reactive, potentially toxic metabolites, *e.g.* epoxides, quinone-imines, nitrosamines.

The Handbook of Medicinal Chemistry: Principles and Practice
Edited by Andrew Davis and Simon E Ward
© The Royal Society of Chemistry 2015
Published by the Royal Society of Chemistry, www.rsc.org

Table 7.1 Phase I metabolic pathways.

Reaction type	Pathway
Oxidation	Aliphatic or aromatic hydroxylation
	N- or *S*-oxidation
	N-, *O*-, or *S*-dealkylation
Reduction	Nitro reduction to hydroxylamine, amine
	Carbonyl reduction to alcohol
Hydrolysis	Ester to acid and alcohol
	Amide to acid and amine
	Hydrazides to acid and substituted hydrazine

Oxidations, which are the most common Phase I reactions, include hydroxylation, *e.g.* propranolol, debrisoquine, oxidation at nitrogen or sulfur atoms, *e.g.* nicotine, sulindac, and N- or O-dealkylation, *e.g.* imipramine, misonidazole (Figure 7.1). The dealkylation pathway is oxidative, as the initial step is hydroxylation in the alkyl group adjacent to the heteroatom with subsequent cleavage of the C-heteroatom bond.

Reduction of nitro or carbonyl groups leads to amines and alcohols, respectively (Figure 7.1), with a consequent increase in water solubility. Similarly, hydrolysis of esters, amides or hydrazides yields the more water soluble acids, alcohols, amines and hydrazines respectively.

Phase II or conjugation pathways differ from Phase I pathways in that they link the drug and/or a Phase I metabolite of a drug with an endogenous molecule, *e.g.* glucuronic acid, sulfate, an amino acid (Table 7.2). The prerequisite for conjugation reactions is that the molecule (drug or Phase I metabolite) has a suitable chemical group, *e.g.* OH, NH_2, COOH to which the endogenous substrate can be attached. The conjugation of a drug or Phase I metabolite with glucuronic acid, sulfate, an amino acid or glutathione (Figure 7.2) generally results in a more water soluble product for excretion. Additional conjugation pathways are N-acetylation of amines and O-, N-, and S-methylation. These, unlike other Phase II reactions, generally result in a more lipophilic product.

Some products of Phase II conjugation reactions may contribute to drug toxicity, *e.g.* glucuronide or sulfate conjugates of certain substituted N-hydroxyamides, have been implicated in the induction of bladder cancer and acyl glucuronides of some carboxylic acids have been associated with hepatotoxicity. The endogenous reagent and xenobiotic substrates are shown in Table 7.2 for the conjugation pathways.

7.3 SITES OF DRUG METABOLISM

Drug metabolism can occur in most tissues and organs of the body, *e.g.* liver, kidneys, gut, blood, plasma. The liver is probably the most efficient metabolizing organ, having a high capacity for most metabolic reactions. The kidneys and gut wall are important sites for Phase II or conjugation reactions and the mucosal cells of the small intestine express significant levels of cytochrome P450 3A4 and 2C9, and peptidases. The enzymes in the gut wall and liver contribute to the extensive metabolism that occurs with some drugs after absorption from the GI tract leading to low oral bioavailability. This is the source of the well-known "first-pass metabolism" of orally administered drugs. Hydrolysis of esters and amides may occur in most tissues as well as blood and plasma.

Within the cell, the complex metabolic pathways may occur in the endoplasmic reticulum (the microsomes used in *in vitro* metabolism studies are derived from disruption of this membrane), mitochondria, and the cell cytosol. Many oxidative reactions are carried out by the membrane-bound mixed function oxidases or CYP450 enzymes in the microsomes. These enzymes are

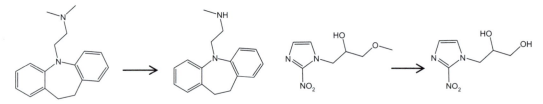

AROMATIC HYDROXYLATION

Propranolol: β-blocker

ALIPHATIC HYDROXYLATION

Debrisoquine: anti-hypertensive

N-DEALKYLATION

imipramine: anti-depressant

O-DEALKYLATION

misonidazole: radiosensitizer

NITRO-REDUCTION

nitrazepam: hypnotic

CARBONYL REDUCTION

warfarin: anti-coagulant

ESTER HYDROLYSIS

aspirin: analgesic

AMIDE HYDROLYSIS

procainamide: anaesthetic

Figure 7.1 Common Phase I metabolic pathways.

dominant in the Phase I metabolic pathways and within each animal species including man a number of different isozymes are expressed, each showing some substrate specificity. Thus inter-species differences in metabolism reflect in part the properties of the P450 isozymes expressed. Some 60% of marketed drugs are metabolised by CYP450.[1] Many marketed drugs are metabolised by more than one CYP450 enzyme and these CYP450 enzymes or close analogues of them are also expressed in other mammalian species, *e.g.* rat, dog, monkey.[2,3] Two human

Table 7.2 Phase II conjugation pathways.

Conjugation reaction	Endogenous reagent or substrate	Xenobiotic substrate
Glucuronidation	Uridine Diphosphate glucuronic acid (UDPGA)	Carboxylic acid, alcohol, phenol, amine
Sulphation	3′-Phosphoadenosine-5′-phosphosulphate (PAPS)	Alcohol, phenol, amine
Acetylation	Acetyl-CoA	Amine
Amino acid	Glycine, glutamine	Carboxylic acid
Glutathione conjugation	Glutathione	Epoxides, arene oxides, Chloro compounds, Quinone-imines
Methylation	*S*-adenosyl methionine	Phenols, amines, thiols

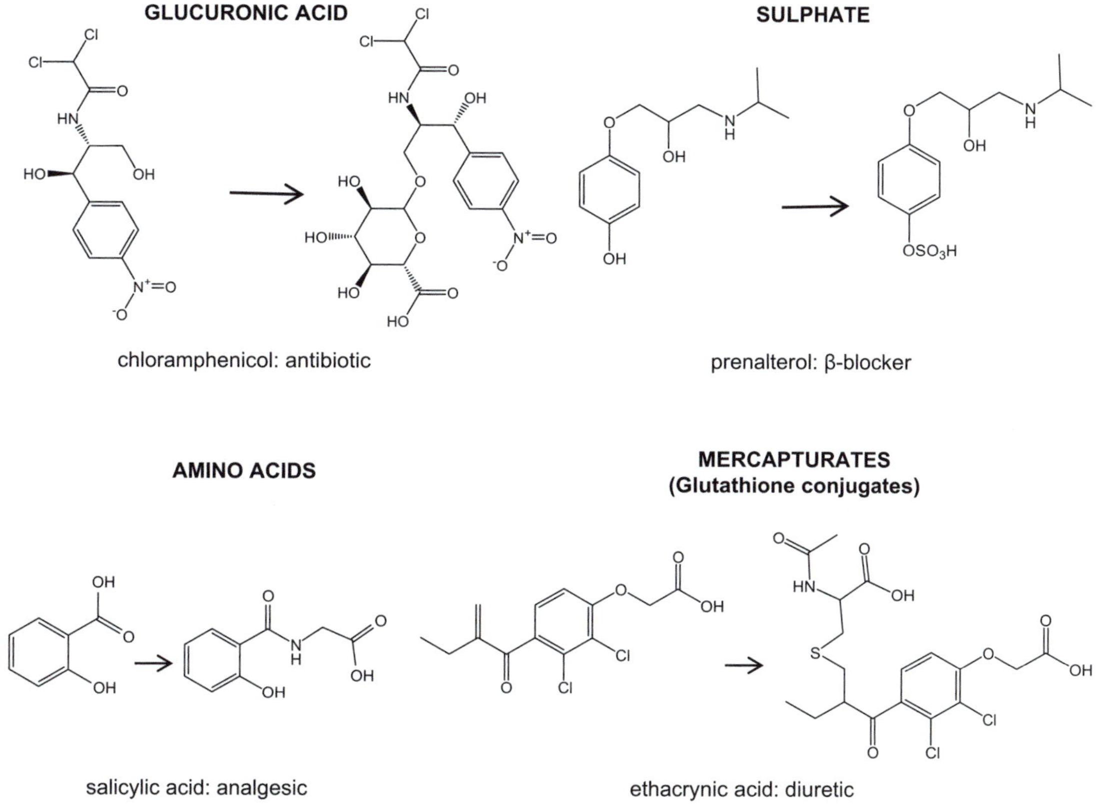

GLUCURONIC ACID

chloramphenicol: antibiotic

SULPHATE

prenalterol: β-blocker

AMINO ACIDS

salicylic acid: analgesic

MERCAPTURATES
(Glutathione conjugates)

ethacrynic acid: diuretic

Figure 7.2 Common Phase II conjugation pathways.

CYP450 isozyme families comprising 3A4/5 and 2C9/19 and the isozyme 2D6 appear to be most important in drug metabolism, being responsible for some 40-50%, 14% and 30%, respectively, of the hepatic CYP450 mediated metabolism of marketed drugs.[4] In contrast, CYP450 isozymes 1A2, 2E1 and 2A6 metabolise only about 6%, 5% and 4% of marketed drugs. Table 7.3 shows the characteristics of several human CYP450 isozymes involved in the metabolism of xenobiotics including drugs, and shows the overlaps in substrate specificity even between these major isozymes. Many of these enzymes are known to be inhibited and/or induced by drugs and other xenobiotics *e.g.* dietary components, and such effects can modulate the metabolism and thus the pharmacokinetics and pharmacodynamics of a drug that has metabolism by these enzymes as a

Table 7.3 Human Cytochrome P-450 isozyme characteristics.

Isozyme	Hepatic CYP450 (%)	Substrate Examples	Reaction	Inhibition	Induction	Polymorphic
1A2	18	PAH Caffeine Paracetamol Tacrine	Epoxidation Hydroxylation	√	√	√
2A6	6	Basic, Neutral 7-Ethoxycoumarin Nicotine Cyclophosphamide	De-ethylation Hydroxylation	√	√	N/A
2C9	20	Acidic Phenytoin Tolbutamide Retinol Ethylmorphine Lidocaine	Hydroxylation Hydroxylation Hydroxylation Dealkylation Dealkylation	√√√	√	√√ Poor metabolisers 5–10% Cauc 23% Oriental
2C19	2–5	Basic Phenytoin Tolbutamide Retinol Ethylmorphine Lidocaine	Hydroxylation Hydroxylation Hydroxylation Dealkylation Dealkylation	√√√	√	√√ Poor metabolisers 5–10% Cauc 23% Oriental
2D6	2–5	β-Blockers SSRIs	Hydroxylation Dealkylation	√√√	–	√√√ Poor metabolisers 5–10% Caucasians
2E1	9	Paracetamol	Quinone-imine formation	√	√	√
3A4	30	Basic, Acidic, Neutral Dihydropyridines Cyclosporine Ethylmorphine Benzodiazepines Steroids Macrolides	Aromatisation Dealkylation Dealklation Hydroxylation Hydroxylation	√√√	√√	N/A High inter- individual variability

major elimination route. A review in 2004 by Ingelman-Sundberg[5] provides comparative information on the CYP450 enzymes in animals and man.

Some metabolic pathways, such as N-acetylation or β-oxidation, occur in the mitochondria, and conjugation (Phase II) reactions such as glucuronidation may occur in the cytosol or be membrane-bound. The enzymes involved in most other drug metabolism pathways, *e.g.* glucuronyl transferases, also exist in multiple forms (isoenzymes, isozymes), which also show differing substrate specificities.

7.4 RELATIONSHIP BETWEEN STRUCTURE AND EXTENT OF METABOLISM

The extent of drug metabolism is essentially dependent on the chemical structure of the drug. This determines the drug's physicochemical characteristics, *e.g.* pKa (acid, base or neutral), its lipophilicity (log P and log D) and molecular weight. These will influence the potential for the drug to be excreted unchanged in urine or bile, eliminated by metabolism alone or by a

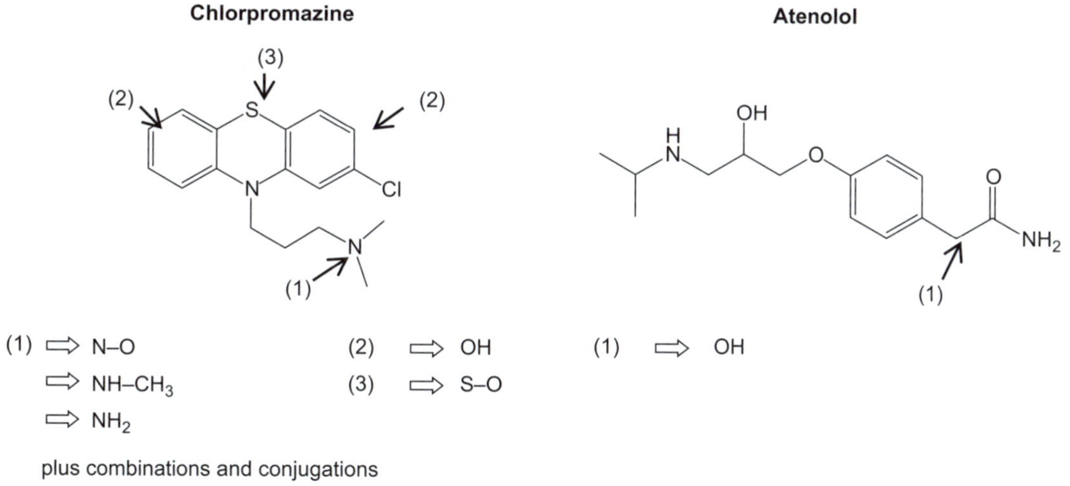

Chlorpromazine

Atenolol

(1) ⇨ N–O
⇨ NH–CH$_3$
⇨ NH$_2$

(2) ⇨ OH
(3) ⇨ S–O

(1) ⇨ OH

plus combinations and conjugations

Figure 7.3 Comparison of metabolic pathways for chlorpromazine and atenolol.

combination of excretion and metabolism. Thus, the antidepressant drug chlorpromazine (Figure 7.3) is very lipophilic (log P = 5.14), poorly water soluble, and has many potential sites of metabolism. It undergoes N-dealkylation, aromatic hydroxylation, sulfoxidation, N-oxidation and a combination of these Phase I processes. Some of the resulting Phase I metabolites are also substrates for Phase II conjugation reactions, *e.g.* glucuronidation, sulfation. This produces many metabolites that are excreted in urine and bile with limited excretion of unchanged drug. In contrast, the much less lipophilic atenolol (log P = 0.16) is excreted predominantly unchanged in urine. Only a small amount (≤5% dose) is metabolised to its monohydroxy metabolite.

The extent of metabolism of many drugs lies between these two extremes, with a mixture of parent drug and several metabolites being excreted. Within a series of compounds, *e.g.* β-blockers, the importance of metabolism as an elimination pathway generally increases with increasing lipophilicity (log P) of the drug.[6]

7.5 HOW IS DRUG METABOLISM STUDIED?

The metabolic fate of a drug is generally studied using a combination of techniques including solid phase or liquid–liquid extraction of biological samples, chromatography (most commonly HPLC or UHPLC) and mass spectrometry and/or proton NMR. Most commonly, the eluent from the chromatograph is delivered directly to the spectrometer, *e.g.* LC-MS, UHPLC-MS, LC-MS/MS, LC-NMR, enabling metabolite identification without the need for separate extraction, concentration and purification techniques. LC-MS or LC-MS/MS are also the most commonly used specific quantitative methods for a drug and its metabolite(s) in biological samples, *e.g.* plasma, blood, urine.

The use of drugs radiolabelled with ^{14}C, ^{3}H, or in some cases, ^{35}S provides a technique to detect and quantify all drug-related material in the complex mixture of endogenous compounds in biological samples in animals and man. Measurement of total radioactivity in blood, plasma, urine and faeces by liquid scintillation counting (LSC), following oxidation for blood and faecal samples, gives information on the absorption and the routes, rates and extent (excretion balance) of excretion of drug related material and the tissue distribution of drug related material. However, measurement of total radioactivity alone in plasma, urine, faeces, tissues or other samples does not give an accurate assessment of drug concentration or pharmacokinetics of drug and/or metabolites. Total radioactivity is the sum of the concentrations of the drug (if present) and any radiolabelled metabolites, as shown for example in plasma (Figure 7.4).

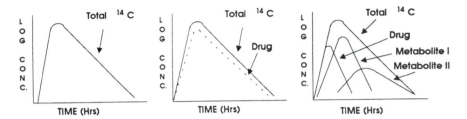

Figure 7.4   Comparison of concentration–time profiles for total radioactivity, parent drug and metabolites.

Over the past 15 years, accelerator mass spectrometry (AMS), which has been extensively used in radio-carbon dating of archaeological artefacts, has been applied to drug metabolism studies in man. AMS is approximately 10^6 times more sensitive than liquid scintillation counting thereby allowing the determination of pharmacokinetics, excretion balance, PK of parent drug and metabolite profile proportions of total radioactivity using microdoses (nCi to pCi) of radioactivity whereas µCi doses are normally used for such studies in man.[7,8] Thus such investigations if appropriate can be incorporated into clinical studies in healthy volunteers and/or patients as part of single and multiple ascending doses in First in Man studies and/or in Phase II or III registration studies.

Thus the appropriate combination of extraction procedures, LC-MS, LC-MS/MS and LC-NMR with radiolabelled drug enables the characterisation of the absorption, distribution, metabolism and excretion of a drug. In particular, it provides information on the extent of metabolism in the biological samples, metabolite identification and, where necessary, isolation and purification of additional quantities of metabolites for further investigation, *e.g.* identification, activity, reactivity.

It may also be possible to carry out direct NMR analysis of biological samples (*e.g.* urine) with minimal sample preparation. This can provide very rapid metabolic information when relatively high doses have been administered and urinary excretion of drug and metabolites is rapid and extensive. Thus it may be possible to identify metabolites and estimate their concentrations in small urine volumes by direct analysis of the samples or following freeze-drying and dissolution in D_2O using proton NMR spectroscopy. Alternatively, solid-reverse phase extraction and step gradient elution with water or aqueous buffer containing increasing proportions, *e.g.* 0%, 20%, 40%, 80%, 100% of methanol or acetonitrile, may provide sufficient purification and concentration of metabolite fractions from biological samples to enable identification by proton and/or ^{13}C-NMR analysis.

The detection and identification of metabolites of drugs containing fluorine may, in some cases, be obtained by direct ^{19}F-NMR analysis of urine samples or deproteinised plasma samples, because in contrast to proton NMR, there is essentially no background signal from endogenous components in biological samples, as shown by Wade *et al.* with trifluoromethylaniline[9] as a model compound. This approach was also used with imirestat to identify its metabolites in dog urine, including a metabolite resulting from an NIH fluorine shift.[10]

Drug metabolism may be studied *in vivo* or *in vitro*. Most commonly preparations from liver, *e.g.* S9 supernatants, microsomes (fragmented endoplasmic reticulum), hepatocytes or liver slices, are used to assess the *in vitro* metabolism of a drug and identify metabolites. S9 supernatants and microsomes can also be prepared from other tissues *e.g.* small intestine, lung. Centrifugation of the appropriate organ or tissue homogenate from the relevant species, *e.g.* rat, dog, monkey or man, at 9000 to 12 000×g removes the nuclei and mitochondria to give the tissue S9 supernatant. Centrifuging the S9 fraction at about 136 000×g produces microsomes from the relevant tissue.

S9 supernatants in phosphate buffer (pH 7.4) incubated with the drug of interest with added NADPH (reduced nicotinamide-adenosine diphosphate), UDPGA (uridine diphosphate glucuronic acid) or PAPS (phosphoadenosine phosphosulfate) enables CYP450 Phase I metabolism, Phase II conjugation with glucuronic acid or with sulfate, respectively. Addition of all three factors allows all three pathways and their interactions to be investigated.

Microsomes in phosphate buffer (pH7.4) are most commonly used to investigate CYP450 metabolism when incubated with the drug of interest and added NADPH. However, in the presence of added UDPGA and/or PAPS and a pore forming agent, *e.g.* alamethicin, allows investigation of Phase II conjugation with glucuronic acid and/or with sulfate, respectively.

Addition of glutathione in such systems also allows investigation of its conjugation with any reactive metabolites resulting from CYP450 oxidation.

In drug discovery these systems allow comparison of rates, extents and pathways of metabolism within a series of compounds to select the compound(s) with the preferred metabolic characteristics, *e.g.* low first-pass metabolism to allow selection of the most appropriate candidate drug. *In vitro* studies also allow early investigation of species differences/similarities in drug metabolism including man to support the interpretation of the relevance of non-clinical pharmacology and toxicology for man and to predict possible metabolic fate in man. It may also provide an early indication of any marked differences in metabolite profile in animals and man, *e.g.* absence of a major human metabolite in animals. This information will influence the development programme, as in a case where a significant human metabolite is not found in the species used in toxicology studies specific studies will be required as part of the development programme to investigate the potential toxicity of such a metabolite.

Once a metabolite(s) has been identified a specific quantitative analytical method can be developed where appropriate, to allow measurement of its concentrations in biological fluids and thus its pharmacokinetics.

7.6 WHY DO WE STUDY DRUG METABOLISM?

7.6.1 The Industry Perspective

Drug metabolism information provides a link between the animals used in pharmacology and toxicology studies and man. A comparison of the drug's fate in all species studied allows the interpretation of the relevance to man. A drug may be pharmacologically active, inert or toxic, as may a drug metabolite. A metabolite's pharmacological activity may be the same as that of its precursor drug or be completely different.

If the drug itself is inert and one or more of its metabolites are the source of its activity the drug is acting as a prodrug. In some cases the prodrug is the result of a conscious drug design decision, *e.g.* the antibiotic pivampicillin, an ester prodrug which yields improved systemic exposure to ampicillin as the prodrug is more efficiently absorbed and then rapidly hydrolysed to ampicillin. For another drug, leflunomide, an immunomodulator used to treat rheumatoid arthritis (RA), its prodrug character was realised during early development when it was shown to be converted to teriflunomide. This major plasma metabolite is an inhibitor of dihydro-orotate dehydrogenase and thus of *de novo* pyrimidine biosynthesis. This suppresses growth of rapidly dividing cells, including activated T cells, and may explain the activity of leflunomide in RA and of teriflunomide which was approved by the FDA in 2012 as an oral treatment for multiple sclerosis.

Non-clinical evaluation of drug safety in animal toxicology studies usually includes assessment of parent drug plasma concentrations. These are used to aid selection of starting and maximum doses in First in Man studies and to assess the relevance of potential risks suggested by non-clinical findings and are used to guide monitoring in clinical trials. This is normally sufficient when the metabolic profile in humans is similar to that in at least one of the animal species used in non-clinical studies. However, metabolic profiles can vary across species both quantitatively and qualitatively, and in some cases clinically relevant metabolites may not have been identified or not been adequately evaluated during non-clinical safety studies. This situation can occur if the metabolite is formed only in humans and is absent in the animal test species or if the metabolite is present at disproportionately higher levels in humans than in the animal species used in the standard toxicity testing with the parent drug. For zoniporide (Figure 7.5) rat and dog encompass the parent drug metabolites seen in man with no evidence

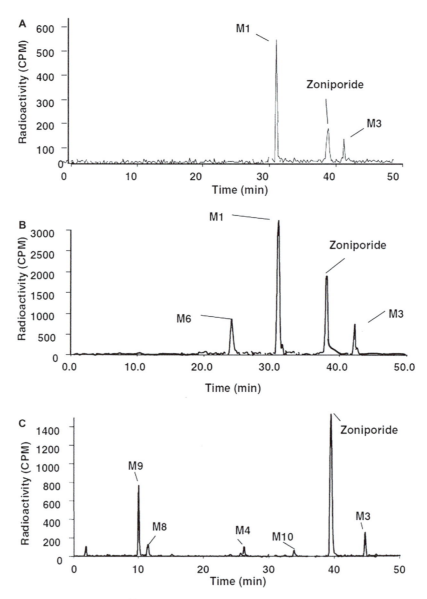

Figure 7.5   Species comparison of [^{14}C]-zoniporide metabolism in plasma of (A) healthy male volunteer, (B) Sprague-Dawley rat and (C) Beagle dog (reprinted by permission of American Society of Pharmacology and Experimental Therapeutics from D. Dalvie, C. Zhang, W. Chen, T. Smolarek, R. Scott Obach and C. M. Loi, Cross-species comparison of the metabolism and excretion of zoniporide: Contribution of aldehyde oxidase to interspecies differences, *Drug Metab. Dispos.*, **38**, 641–654, 2010[35]).

of disproportionate exposure to metabolites in man compared to rat and dog. A number of other metabolites were also detected in rat and dog. If a significant drug metabolite in man is not observed in the animal species used in pharmacology or toxicology studies, the potential pharmacologic or toxicologic effects of the metabolite in man may not be reflected in the non-clinical safety evaluation.

In *in vitro* metabolism studies, apart from qualitative and quantitative species comparisons, measurement of drug concentrations at various times during the incubation with human liver microsomes or hepatocytes will allow the estimation of the rate of metabolism as reflected in the *in vitro* half-life and also allows determination of the *in vitro* intrinsic clearance of the drug. The *in vitro* intrinsic clearance can be used to rank the metabolic stability of a series of drugs and to estimate the *in vivo* clearance of a selected development candidate.[11] This approach allows the ranking of compounds in a series of active ones as being low, moderate or high clearance drugs, and thus having the potential to have high, moderate or low oral bioavailability, respectively.

7.6.2 Guidance on Safety Testing of Metabolites

In 2008 the FDA issued guidance on the safety testing of metabolites[12] and an updated guidance on this topic was issued in 2010 as part of the ICH M3 (R2) document on non-clinical safety studies[13] focused on the importance of metabolite profiling and metabolite quantification, in supporting the safety evaluation in R and D, and ultimately marketing approval of a drug. These so-called MIST (Metabolites in Safety Testing) guidances focused on metabolites in human plasma that were either unique, or were present at disproportionately higher concentrations to those of the parent drug in man than in animals. In the original FDA guidance non-clinical characterisation would be required for a metabolite if at steady-state the metabolite plasma AUC was > 10% of parent drug in man and a similar or greater relative exposure was not observed in at least one toxicology species to provide coverage of its potential toxicity.

The ICH guidance (M3(R2)) broadened the criterion for non-clinical safety characterisation of a human metabolite was only warranted when it was observed at exposures > 10% of *total* drug related exposure and was at significantly greater levels in humans than the maximum exposure seen in the toxicity studies. The FDA recently modified their guidance to reflect the change to ≥ 10% of drug-related material. Measurement of total drug related exposure may be based on comparison with total radioactivity or based on non-radioisotope methods to estimate total drug-related material and metabolite exposures.[14,15] This may also be expedited by using plasma sample pooling methods to reduce the number of samples that have to be analysed to provide the required exposure data.[16,17]

Clearly these guidances have implications for drug R and D activities. If there are human metabolites not covered by the toxicology studies then specific toxicology studies on the metabolite(s) may be required, *e.g.* 3 month toxicity, genotoxicity, reprotox studies. An early understanding of the level of coverage for potential human metabolites during and following candidate selection will be important in assessing the need for a more extensive toxicology programme on metabolites to enable the clinical investigation and ultimately marketing of a drug candidate.

Drug metabolism information can assist drug discovery programmes. Thus, compounds which showed high *in vitro* potency may yield low *in vivo* potency because of poor oral bioavailability due to low or incomplete absorption and/or extensive 'first-pass' metabolism. Many potent antibiotics, *e.g.* cephalosporins, are polar compounds with low log D at physiological pH and thus are poorly absorbed from the gastrointestinal tract. This problem may be avoided by formation of appropriate ester prodrugs which are much better absorbed, and then hydrolysed during first-pass metabolism in gut wall and/or liver to release the active species.

Similarly if rapid metabolism is limiting the pharmacodynamics of a compound, modification of the structure may be used to reduce the effect. This can produce a more effective drug candidate for development. This approach was used to design the β_1-adrenoceptor antagonist betaxolol, which had improved bioavailability (80%) and half-life in man (14–22 h) resulting from modifying the *p*-substituent in metoprolol (bioavailability about 50%; $t_{1/2}$ 3.5 h). This work used *in vitro* metabolism studies in liver preparations (*e.g.* microsomes, homogenates) to select the most metabolically stable compounds.[18]

A series of imidazolyl and aryl substituted propan-1-one compounds were being investigated as potential antibiotics, targeted at anaerobic bacteria, *e.g. B. fragilis*. However, the *in vitro* activity did not correlate with that found *in vivo*. One compound that was very active *in vitro* was essentially inactive after i.v. administration to rat or mouse. A metabolism study in the mouse using the i.v. [^{14}C]-imidazolyl labelled drug (40 mg/kg) showed rapid and extensive conversion to [^{14}C]-imidazole. Although the drug was stable in phosphate buffered saline at room temperature, it was rapidly converted to [^{14}C]-imidazole in mouse plasma *in vitro* and under the *in vitro* conditions, agar broth at 37 °C, used to test its activity against *B. fragilis*.[19] Further investigations showed that compounds less active *in vitro* did show activity *in vivo* but no really active compounds could be found. The conclusions were that *in vitro* activity was due to formation of reactive aryl substituted propen-1-one products formed by a retro-Michael loss of the imidazole group. However, presumably *in vivo* the reactive bactericidal intermediates were being removed by reaction with tissue nucleophiles *e.g.* albumin, and thus were unable to achieve adequate concentrations to inhibit bacterial growth.

7.7   WHAT FACTORS CAN MODIFY DRUG METABOLISM?

7.7.1   Dose Level

As the dose of drug increases, the capacity of the metabolic enzyme systems may be saturated. This can lead to alternative pathways coming into operation and/or to a disproportionately high concentration of drug or of a toxic or active metabolite being present. This can result in exaggerated pharmacology and/or toxicity. Thus paracetamol (Figure 7.6) is eliminated by conjugation with glucuronic acid and sulfate at normal therapeutic doses. At higher doses, sulfate and glucuronide conjugation become saturated and formation of a mercapturic acid by conjugation of reactive Phase I metabolites with glutathione is also observed. Intentional overdose saturates glucuronidation and sulfation and depletes the glutathione. The reactive intermediates accumulate and cause damage to cell macromolecules with resultant liver and kidney toxicity. Thus saturation of metabolic elimination pathways at high doses can modify the exposure to parent drug and/or metabolites with consequent changes in pharmacology and toxicology. These effects should be considered when assessing the relevance of results from non-clinical toxicology studies to man. This is why toxicokinetics of drug and where appropriate metabolites are studied in toxicology studies.

7.7.2   Route of Administration

Drugs given orally have to pass the gut microflora and digestive enzymes in the gut lumen and drug metabolising enzymes of the intestinal wall and liver before reaching the systemic circulation.

Metabolism may occur at any or all of these sites and can reduce the drug concentration (amount) in the systemic circulation. If the drug itself is the active compound, this will affect the intensity of its pharmacodynamic effects. When this pre-systemic metabolism occurs in gut wall or liver, it is called "first-pass" metabolism or the "first-pass" effect.

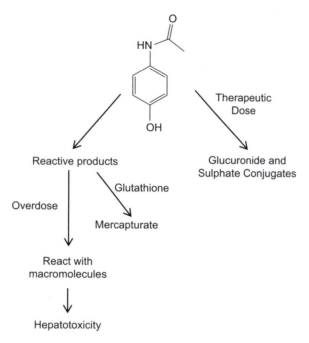

Figure 7.6   Metabolic pathways for paracetamol.

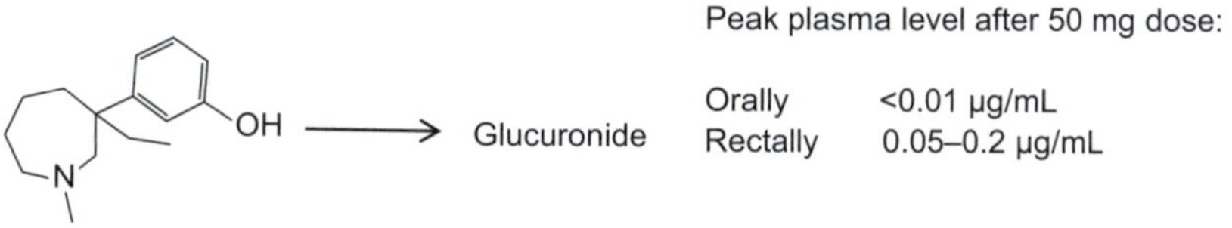

Peak plasma level after 50 mg dose:

| | Orally | <0.01 µg/mL |
| | Rectally | 0.05–0.2 µg/mL |

Glucuronide

Figure 7.7   Effects of route of administration on Meptazinol metabolism and bioavailability.

Meptazinol (Figure 7.7) is an analgesic drug subject to a very high first-pass effect *via* conjugation with glucuronic acid or sulfate. This is markedly reduced if the drug is administered rectally since blood drainage from the rectum predominantly avoids the hepatic portal vein. Thus, much less drug is eliminated before reaching the systemic circulation, higher concentrations are obtained for a given dose and therefore it was a more effective analgesic when given rectally because of the decreased first-pass metabolism in gut wall and liver.[20] However, this was a route and formulation modification to avoid a problem inherent in the chemical structure of the drug. Currently a discovery DMPK optimisation project would attempt to avoid such a problem using early *in vitro* screening for intrinsic clearance and estimated *in vivo* clearance and potential bioavailability to guide selection of development candidates. Drug administration by transdermal, sub-lingual, buccal or inhaled pulmonary routes has also been used to avoid "first-pass" effects.

The "first-pass" effect for extensively metabolised drugs may contribute to between- and within-subject variability.

7.7.3   Species Differences in Metabolism

There are frequently quantitative differences in the Phase I pathways between species. These are at present in some cases difficult to predict and interpret. The general conclusion is that in man the rates of these pathways is slower than in mouse, rat and dog, in part reflecting the changes in metabolic rate with changes in body weight and surface area although this may vary for particular compounds owing to species differences in the enzymes involved.

A number of qualitative differences are seen for Phase II pathways. Dogs are unable to acetylate aromatic amines and are more sensitive to the pharmacologic and toxic effects of such compounds. Cats lack the ability to form glucuronide conjugates and only primates form amino-acid conjugates with glutamine. These differences can have implications for the screening of compounds, in animal species other than man. Differences in biliary excretion also occur. Biliary excretion is a complex process involving active transport of the drug or metabolite into bile against the concentration gradient. The molecular weight and a lipophilic moiety linked to an anionic or cationic structure influence this elimination pathway. As a broad generalisation, compounds (drug or metabolites) with appropriate structures and MW > 325 will undergo extensive biliary excretion in rat and dog, whereas this is more common in man at MW > 500.

7.7.4 Gender-Related Differences

There are significant differences in the capacity of certain metabolic pathways in male and female rodents, particularly rats. Thus, male rats required higher doses of hexobarbital to induce sleep than female rats because of their greater ability to inactivate the drug by aliphatic hydroxylation. Such differences tend to be less important in other species. They are related to the effect of sex differences in the expression of cytochrome P-450 isozymes in the rat. However, such effects have been reported for several compounds in man, *e.g.* fluvoxamine, and should be investigated as part of the development of a new drug in man.

7.7.5 Age

This is an important determinant of drug metabolism and is best studied in man. At either end of the age range, the liver (and other tissues) is (are) generally less capable of metabolic reactions, than in the subjects aged 18–45 used in early clinical studies. For example, neonates are essentially unable to conjugate chloramphenicol with glucuronic acid (Figure 7.2). This resulted in an accumulation of drug leading to toxic cardiovascular effects (grey baby syndrome). Similar problems occur in the aged liver (Table 7.4) with the reduced clearance of imipramine in the liver requiring changes in the administered dose, to avoid excessive drug accumulation and hence increased side-effects.

The potential for age-related changes in metabolism and thus pharmacokinetics and pharmacodynamics needs to be investigated early in drug development for compounds ultimately intended to treat paediatric or elderly patients. These effects can be investigated *in vitro* with

Table 7.4 Age-dependent effects on imipramine pharmacokinetic parameters.

	Young	Old (>70)
Clearance (mL/min)	950	570
Half-life (h)	17	30
C_{max} (ng/mL)	10–20	40–45

Volume of distribution and protein binding were unchanged: the decreased clearance is due to the reduced ability of the old liver to *N*-dealkylate.

cadaver liver tissue from elderly subjects and in healthy elderly subjects as part of the First in Man single and multiple dose safety, tolerability and pharmacokinetics studies.

7.7.6   Disease Effects on Metabolism

Drugs which are mainly eliminated by metabolism can show changes in their kinetics in patients with liver disease. Propranolol is more bioavailable in cirrhotic patients because of a much reduced first-pass effect in the damaged liver. Similarly, conversion of a pro-drug to active compound may be decreased in liver disease. However, it is difficult to predict the effect of a particular liver disease on the fate of a specific drug but based on the ADME characteristics of a drug such investigations in patients may be needed to support the safety evaluation of the drug.

7.7.7   Drug Interactions

As many drugs are primarily eliminated by metabolism and the enzymes mediating these reactions can be subject to inhibition or induction by other compounds, the potential always exists for metabolism-mediated drug–drug interactions with potential clinical safety and efficacy repercussions. This was highlighted by the discovery of a case of *torsade de pointes*, resulting from QT interval prolongation, in a patient receiving the antihistamine, terfenadine, with the antifungal agent, ketoconazole.[21] In normal circumstances, relatively lipophilic terfenadine is almost completely metabolised to the much more hydrophilic active species fexofenadine on its first pass through the liver *via* CYP3A4 so there are very low or no circulating plasma levels of the intact drug. However, in the presence of a potent CYP3A4 inhibitor, such as ketoconazole, this reaction is blocked resulting in detectable plasma levels of terfenadine which interacts with the cardiac hERG receptor causing a prolongation of the QT interval with the potential for *torsades de pointes* resulting in reported fatalities, ultimately leading to the withdrawal of terfenadine as a marketed product. Grapefruit juice has also been shown to inhibit CYP3A4, thereby decreasing the clearance and thus increasing steady-state plasma levels of drugs cleared extensively by 3A4 metabolism, *e.g.* terfenadine and statins, thus modifying their wanted and unwanted pharmacodynamic effects. Hence the detailed guidances or guidelines[22,23] now issued by most major Regulatory Authorities on testing for drug–drug interactions and QT interval prolongation.

The understanding of the number and type of metabolic drug–drug interactions has exploded over recent years with the introduction of *in vitro* technologies such as recombinant human cytochromes, as well as the use of microsomes and hepatocytes with specific and selective substrates and inhibitors. For many, a metabolic drug–drug interaction appears to be simply the inhibition of a cytochrome by a candidate compound, often identified in very early screening assays for cytochrome inhibition in the candidate drug selection process. However, it is much more complex than that and caution is advised on rejecting any compound based on the results of such tests alone, since potentially good compounds could be unnecessarily discarded and compounds with other types of metabolic drug–drug interaction liabilities missed, as is discussed later when your drug is a target of a drug–drug interaction rather than being a causative agent. To make a rational decision there needs to be data on the expected clinically relevant drug concentrations and even the extent of plasma protein binding. Also, if an interaction occurs, it needs to be established whether it can be successfully managed, such as with the anti-viral drugs where inhibitory interactions are even used to boost efficacy.[24]

The more common human cytochromes involved with the metabolism of xenobiotics are listed in Table 7.5, together with examples of substrates, inhibitors and inducers. A diverse range of chemical structures is seen in inhibitors and in inducers. Inhibitors tend to have a

Table 7.5 Examples of substrates, inhibitors and inducers of different human CYP450 isozymes.

CYP450	Substrate	Inhibitor	Inducer
CYP1A2	Alosetron/duloxetine/ tizanidine	Ciprofloxacin/cimetidine/ fluvoxamine	Omeprazole/montelukast/ phenytoin
CYP2A6	Coumarin/nicotine	Ritonavir/tranylcypromine	Phenobarbital/rifampicin
CYP2B6	Bupropion/efavirenz	Clopidogrel/ticlopidine	Phenobarbital/efavirenz/rifampin
CYP2C8	Paclitaxel/repaglinide	Anastrozole/gemfibrozil/ trimethoprim	Phenobarbital/primidone/ rifampin
CYP2C9	Celecoxib/phenytoin/ warfarin	Amiodarone/metronidazole/ troglitazone	Rifampin/secobarbital/ carbamazepine
CYP2C18	S-mephenytoin	cimetidine	
CYP2C19	Clobazam/omeprazole/ S-mephenytoin	Fluvoxamine/esomeprazole/ ticlopidine	Carbamazepine/prednisone/ rifampin
CYP2D6	Desipramine/ dextromethorphan/ metoprolol	Buproprion/quinidine/ terbinafine	
CYP2E1		Disulfiram/ritonavir	Ethanol/isoniazid
CYP3A4/5	Cyclosporine/lovastatin/ sildenafil	Clarithromycin/ketoconazole/ saquinavir	Carbamazepine/rifampin/ St. John's wort

basic nitrogen able to strongly bind to the haem iron in the CYP450 active site, *e.g.* imidazole group as in ketoconazole, quinoline group as in quinidine, thiazole group as in ritonavir. Inducers vary from ethanol, phenobarbital, anti-epileptics drugs, *e.g.* carbamazepine, to a complex cyclic antibiotic like rifampicin. A lot of emphasis has been placed on inhibitory metabolic drug–drug interactions since these tend to be the cause of safety concerns, but just as important can be metabolic drug–drug interactions resulting from enzyme induction that often lead to a reduced, or a lack of efficacy. Also, there are two distinct directions in which the identification of metabolic drug–drug interaction has to be approached. The first is as described above, where the drug or candidate compound is the causative agent or perpetrator of a metabolic drug–drug interaction. The second is when a drug or candidate compound is the target or a victim of a metabolic drug–drug interaction. Each of these aspects is equally important but requires completely different strategies for their investigation.

A perpetrator of a metabolic drug–drug interaction can be by two routes, inhibition and induction. If we first consider inhibition interactions, the most common studies, particularly in the early discovery screening phase, are for direct inhibition of the major cytochrome P450s (CYP1A2, CYP2B6, CYP2C8, CYP2C9, CYP2C19, CYP2D6 and CYP3A4/5), usually determining the percent inhibition at a single concentration, although sometimes an attempt is also made to estimate an IC_{50}. These can be performed using either recombinant human cytochromes or human microsomes with specific substrates for each cytochrome. If a compound proceeds into development, these data are usually insufficient, especially once the compound enters clinical development, and more detailed studies are required to accurately determine the IC_{50} and K_i for those cytochromes where inhibition is observed. For such studies it is advisable not only to examine for direct inhibition, but also for time-dependent and mechanism-dependent inhibition. Direct inhibition involves the binding of the drug candidate to the cytochrome, and depending on the binding site, can be classified as competitive, noncompetitive, uncompetitive or mixed inhibition. For this type of inhibition, usually human liver microsomes are used with specific model substrates, such as those recommended in the FDA Drug Interaction guidance (2012)[22] to determine the IC_{50} and K_i (inhibition constant) of inhibitors, with different concentrations of the candidate compound used, and if the type of inhibition is being investigated, different substrate concentrations as well. If there is good evidence that the inhibition is competitive, the use of different substrate concentrations may be omitted since the K_i is calculated by half the IC_{50}, but if

Table 7.6 Criteria to assess the clinical impact of a direct cytochrome inhibition.

[I]/Ki ratio	Prediction
[I]/Ki > 1	Likely
1 > [I]/Ki > 0.1	Possible
0.1 > [I]/Ki	Remote

not, the K_i has to be calculated using the equations cited in the FDA and European guidances.[22,23] These guidances also offer a method using K_i to assess whether the observed inhibition should be followed up with clinical interaction studies (Table 7.6).

Time-dependent inhibition occurs when the potential inhibitor slowly complexes with the cytochrome and is investigated by preincubating (*e.g.* 30 min) the microsomes with the compound prior to adding NADPH to initiate metabolism. Metabolism- or mechanism-based inhibition involves the candidate compound acting as a suicide substrate by being metabolised to a product that either binds tightly to the ferrous ion in the haem moiety of the cytochrome or forms a reactive intermediate that binds covalently to the enzyme. This type of inhibition can make management of any resultant drug–drug interaction more difficult, since the enzyme is effectively 'killed', so the inhibition is prolonged, requiring the synthesis of new enzyme to replace that which has been lost. Thus, the duration of inhibition will be dependent not on the pharmacokinetics of the candidate drug but on the speed of formation of new enzyme. A pre-incubation stage (*e.g.* 30 min) is incorporated into a study to investigate metabolism-dependent inhibition but unlike the time-dependent inhibition, this is performed in the presence of NADPH to enable the metabolite to be formed. With metabolism-dependent inhibition it is necessary not only to determine K_i but also K_{inact}, the inactivation constant.

Enzyme induction is another possible means where a candidate drug could be a perpetrator of a metabolic drug–drug interaction. Some xenobiotic compounds are able to interact with nuclear receptors such as aromatic hydrocarbon receptor (AhR), constitutive androstane receptor (CAR) and pregnane X receptor (PXR), each inducing sets of drug metabolising enzymes and transporters including the CYP1As, CYP2Bs, CYP2Cs, CYP3As, glucuronyl transferases and P-gp, depending on the receptor. Activation of these receptors, however, is species specific so when investigating induction drug–drug interactions, it is recommended to always use human systems and not to be tempted to extrapolate animal induction data to the human situation. An initial screening assay using human CAR and PXR binding assays or cell-based reporter gene assays can be invaluable in alerting to potential induction issues and the risk of drug–drug interactions. The more definitive study, however, uses a functional assay where different concentrations of the test compound are incubated with human hepatocytes for 2 to 3 days and the microsomal activities of the enzymes of interest compared before and after incubation using specific, model substrates. The E_{max} and EC_{50} for induction are determined and compared to those of positive controls run at the same time to assess the possible clinical impact of the induction. Induction normally results in a reduction of efficacy, due to increased metabolism of drugs, and of particular concern is the induction of CYP3A4, since this can result in a reduction of plasma levels of estrogens and progestins found in oral contraceptives, possibly resulting in unplanned pregnancy, and reduction in cyclosporine plasma levels resulting in transplant rejection due to increased cyclosporine clearance.

A victim of a metabolic drug–drug interaction is when our candidate compound is the target, and requires a completely different strategy to that described above for perpetrators of a drug–drug interaction. When establishing if a test compound could be a victim of a drug–drug interaction, *i.e.* another co-administered compound alters its clearance, it is necessary to establish and quantify every route of elimination of the compound, including non-metabolic routes of elimination such as renal. The rationale is that the fewer the routes of elimination, the greater the potential for a compound to be a victim of a metabolic drug–drug interaction. The rule of thumb is that if any one route of elimination exceeds 25% of the total clearance of a compound, there is a risk of a clinically meaningful drug–drug interaction. To evaluate this is no simple process, since, for instance, if there is a substantial renal component, the extent of clearance by glomerular filtration and each of the transporters involved with tubular secretion (see Section 7.9) will need to be quantified.

To identify the metabolic component, often termed as reaction phenotyping, the extent of non-cytochromal (*e.g.* hydrolases, FMO, MAO, glucuronyl transferases, carbonyl reductases) and cytochromal metabolism will need to be established, usually by preincubating human hepatocytes or human S9 fractions (these preparations contain the non-cytochromal enzymes whereas microsomes may not) with 1-aminobenzotriazole (1-ABT) to inhibit all the cytochromal activity then determining how much of the test compound is still metabolised. If metabolism occurs in the presence of 1-ABT, the enzymes responsible should be identified, although the risk of a compound largely metabolised non-cytochromally showing a clinically relevant drug–drug interaction tends to be lower than those primarily metabolised by cytochromes. The FDA and EMA recommend that reaction phenotyping for the cytochromal enzymes is performed by at least two assays of the three offered. The first is incubation of the test compound with a panel of recombinant major human cytochromes to ascertain which metabolises it. The rates of metabolism by each cytochrome should be normalised based on the average specific content of the cytochrome in human liver microsomes since the expression of the recombinants can show wide variation. This is probably the simplest of the assays and can be introduced early in a development program to alert of potential risks for the future. The second method compares the rate of metabolism of a candidate compound, either by its disappearance or by the production of a specific metabolite, with and without either a selective chemical inhibitor or inhibitory antibody for the each of the cytochrome P450s of interest. The final method utilises the natural individual variation in human metabolism with a correlation analysis using a panel of human microsomes from individual subjects where the activity of each of enzymes of interest is fully characterised. The test compound is considered a substrate of an enzyme when there is a good correlation of the metabolism rate with the activities of the respective enzyme in the microsomes of the individual subjects making up the panel.

As can be seen from the above discussion any understanding of the risk of a potential metabolic drug–drug is much more than performing a quick CYP inhibition screen. It must be put into context of whether it is viewed as a perpetrator or a victim of an interaction as well as the final use of the drug. The therapeutic index of the drugs involved, together with the plasma concentrations of the candidate drug obtained at anticipated clinical doses, plasma protein binding, type of inhibition/induction, *etc.*, must all be brought into these considerations. The FDA and European guidelines on drug–drug interactions provide a series of very useful decision trees to help in the interpretation of the *in vitro* inhibition and induction data. Many of the interactions, once identified and quantified, can be very manageable and sometimes may even be of benefit, such as ritonavir boosted protease inhibitors when treating HIV and probenecid with some β-lactam antibiotics.

Before embarking on expensive clinical studies, it is always possible to simulate the possible clinical consequences of any of these metabolic drug–drug interactions using physiological based pharmacokinetic models with programs such as Simcyp (www.simcyp.com). This program has the advantage that it is population based incorporating the variability of the distribution of different human cytochromes so it is possible to simulate not just average effect but also the extremes where often the real clinical problems lie. Also, if there are multiple major elimination pathways ($\geq 25\%$ of total clearance), it is possible to simulate the effect of different drug–drug interaction scenarios.

7.7.8 Genetics

Individual differences in expression of different isoenzymes involved in drug metabolism may lead to wide population differences in the metabolic fate of a compound. Some 10% of the Caucasian population are unable to hydroxylate debrisoquine (Figure 7.1) and other basic drugs *e.g.* β-blockers, SNRIs metabolised by the same P450 isoenzyme (P450 2D6) with potential effects on their activity and side effects. Similarly, there are genetic differences in the ability to N-acetylate certain classes of drug, *e.g.* dapsone, isoniazid. Some 50% of Caucasians have a much lower capacity to carry out this reaction which in turn influences the activity and side-effects of such compounds. If a polymorphic metabolising enzyme is the major elimination pathway there will be increased exposure to the drug resulting from the reduced clearance in "poor metabolisers", *i.e.* those having low expression levels of the drug, or reduced exposure to the drug in "fast metabolisers", *i.e.* those having very high expression levels of the enzyme. This can result in significantly higher adverse events and/or toxicity particularly for drugs with a narrow therapeutic window in "poor metabolisers". Thus 2D6 polymorphism has influenced the adverse event profiles of CNS drugs such as venlafaxine[25] and haloperidol[26] with "poor metabolisers" showing increased incidences of adverse events compared to normal metabolisers. Based on the identification of the relative importance of the metabolising enzymes for a given drug in early non-clinical studies such effects in man are investigated early in drug development in Phase I studies and in ethnicity studies.

7.8 REACTIVE METABOLITES

Metabolites as previously mentioned may be pharmacologically/toxicologically inert, show similar pharmacology and/or toxicology to the parent drug or may have different pharmacological and/or toxicological effects to the parent drug.

Some metabolites are able to chemically react with endogenous components which in certain circumstances produce toxicity. Thus CYP450 mediated aromatic hydroxylation involves potentially reactive arene oxide intermediates. These can non-enzymatically convert to phenols, be metabolised to dihydrodiol metabolites by epoxide hydrolase and/or undergo glutathione-S-transferase catalysed conversion to mercapturic acid. Saturation of the enzyme catalysed pathways or glutathione depletion may in some circumstances result in covalent binding to cellular macromolecules. However, many drugs are metabolised *via* the arene oxide intermediate without significant associated toxicity.

Similarly carboxylic acid drugs can undergo conjugation *via* the acid moiety with glucuronic acid to form 1β-D-acyl glucuronides or form acyl-CoA thioesters. Such metabolites can in some circumstances react with cellular proteins. This can occur with acyl glucuronides by transacylation transfer of the acyl group from the glucuronic acid to amino or hydroxyl groups in proteins. It is also possible for the acyl group of the acyl 1β-D-glucuronide to migrate to form the 2β, 3β or 4β isomers. These isomers have the ability to react with proteins *via* the 1β hemi-acetal group. Again there are few examples where toxicity has been clearly linked to acyl glucuronide metabolite formation.

Various pharma companies have used *in vitro* microsomal metabolism screens, *e.g.* glutathione trapping, or the extent of covalent binding of ^{14}C-labelled drugs to protein, *e.g.* ≥ 50 pM drug equivalents/mg liver protein, as the basis for rejecting candidates in early development due to concerns for their possible toxicity risks. These tests have overall been poorly predictive of the potential for toxicity $\geq 40\%$ false positives or poor correlation with observed liver toxicity in non-clinical safety studies. Most companies appear to have abandoned such approaches because of these issues. Reviews by Skonberg *et al.*[27] (2008) and Park *et al.*[28] (2011) provide significant insight into the challenges of interpreting the safety implications of such metabolites. This indicated that the risk of such toxicity appears to be greater, the higher the dose administered. More recently R. A. Thompson *et al.*, at AstraZeneca, published a detailed comparison between *in vitro* covalent binding burden of ^{14}C-labeled drugs and/or metabolites in human hepatocytes determined as the fraction of metabolism leading to covalent binding, corrected for the maximum prescribed clinical dose with the aggregated *in vitro* toxic effects of five cellular test systems. The results for 36 drugs known to exhibit different idiosyncratic adverse events in man indicated this more detailed approach may offer the potential to select drug candidates with a lower risk of inducing such toxicity.[29]

7.9 TRANSPORTERS

Although not strictly drug metabolism, drug transporters can greatly impact the metabolism of a drug *via* their role in drug absorption, distribution and elimination, especially with respect to potential drug–drug interactions since they can be inhibited and induced, sometimes by the same compounds that inhibit or induce specific cytochrome P450s. Probably the best known and studied transporter is the efflux transporter P-gp (MDR1) which has been blamed for many a drug failing in drug development, often quite unjustly, but we shall come back to that later.

The understanding of transporters and their role in the overall ADME of drugs has gathered momentum over recent years with the identification of many different transporter proteins, both for xenobiotics and endogenous compounds, together with their genes. This, combined with the development of experimental systems to identify xenobiotic substrates, inhibitors and inducers of the specific transporters has provided a much better awareness of their impact on the ADME of different drugs. The subject has now grown to an extent that it would be impossible to provide a detailed and comprehensive overview in this chapter and the interested reader should refer to the many excellent reviews that are available including those by Russel (2010),[30] Giacomini and Sugiyama (2002)[31] and The International Transporter Consortium (2010).[32]

The transporters of particular interest to the drug developer can be divided into four broad super-families. There are two categories of solute carrier transporters (SLC), one of which acts mainly on transporting anionic compounds such as benzylpenicillin, frusemide, ibuprofen and the statins (OATs and OATPs) and the other which transports mainly cations such as cimetidine, ranitidine and metformin (OCTs, OCNTs and MATEs). Then there are the ATP binding cassette (ABC) efflux transporters that transport a wide range of anionic, cationic and neutral compounds from digoxin to doxorubicin and, as the name suggests, rely on ATP hydrolysis to actively move substrates across membranes. The SLC transporters can undertake both drug uptake and efflux whereas the ABC type of transporter only undertakes efflux. Finally, there are the peptide transporters which can be especially important for the absorption and elimination of certain peptide-like compounds such as the β-lactam antibiotics, lisinopril and oseltamivir (Tamiflu). For convenience, the major human transporters are listed in Table 7.7, together with some known substrates, but it will be obvious from this list that many of the transporters are quite promiscuous, with a large amount of crossover between substrates and with substrates outside the class that they are attributed to.

The main tissues where these transporters are found include the intestine, liver, kidney and brain as summarised in Table 7.7 and Figure 7.8. The brain has been highlighted, in addition to the excretory organs, because it can be especially important when developing CNS acting drugs, with the efflux transporters forming an integral component of the notorious blood brain barrier.

Much of a small molecule's absorption, distribution and elimination rely to some extent on the process of passive diffusion with a molecule passing through a biological membrane from an area of high concentration to an area of low concentration. This requires no energy but depends on the molecule possessing some degree of lipophilicity so that it can dissolve in the bimolecular lipid layer of the cell membrane to enable it to diffuse through. For molecules with

Table 7.7 Summary of the major human drug transporters and their substrates (data collated from Russel, 2010;[30] International Transporter Consortium, 2010;[32] FDA guidance on drug–drug Interactions, 2012[22]).

Transporter	Typical tissue distribution	Membrane distribution	Substrate examples
PEPT1	Intestine/Kidney	BBM	Ampicillin/Amoxicillin/Cefaclor/Cefadroxil/Enalapril
PEPT2	Kidney	BBM	Amoxacillin/Bestatin/Cefaclor/Valganciclovir
OCT1	Intestine/Liver	BLM/SM	Acyclovir/Ganciclovir/Metformin/Cimetidine/Quinine
OCT2	Kidney/Brain	BLM	Cimetidine/Ranitidine/Metformin/Propanolol/Zidovudine
OCT3	Liver	BLM	Cimetidine
OCTN1	Intestine/Kidney	BBM	Mepyramine/Quinidine/Verapamil/Gabapentin
OCTN2	Intestine/Kidney	BBM	Mepyramine/Quinidine/Valproate/Cephaloridine/Ematine
OAT1	Kidney	BLM	Adefovir/Methotrexate/Furosemide/Ibuprofen/Cimetidine
OAT2	Liver	SM	Erythromycin/Cimetidine/Taxol/Bumetanide/Salicylate
OAT3	Kidney/Liver/Muscle/Heart	BLM	Benzylpenicillin/Tetracycline/Ranitidine/Ketoprofen/Pravastatin
OAT4	Kidney	BBM	Tetracycline/Zidovudine/Bumetanide/Ketoprofen/Salicylate
MATE1	Liver/Kidney	CM/BBM	Cimetidine/Procainamide/Metformin/Cephalexin/Fexofenadine
MATE2	Kidney	BBM	Cimetidine/Procainamide/Metformin/Fexofenadine/Oxaliplatin
OATP1A2	Intestine/Kidney	BBM	Fexofenadine/Enalapril/Rosuvastatin/Imatinib/Saquinavir
OATP1B1	Liver	BBM	Benzylpenicillin/Rifampicin/Pravastatin/Valsartan/Troglitazone
OATP1B3	Liver	BLM	Digoxin/Rifampicin/Enalapril/Fluvastatin/Valsartan/Paclitaxel
OATP2B1	Liver/Intestine	SM/BBM	Benzylpenicillin/Bosentan/Pravastatin/Glibenclamide
OATP4C1	Kidney	BLM	Digoxin/Ouabain/Methotrexate
MDR1 (P-gp)	Intestine/Liver/Kidney/Brain	BBM/CM/BBM	Vincristine/Doxorubicin/Paclitaxel/Digoxin/Ritonavir/Cyclosporin
MDR3	Liver	CM	Digoxin/Paclitaxel/Vinblastine
MRP2	Intestine/Liver/Kidney/Brain	BBM/CM/BBM	Vinblastine/Doxorubicin/Ritonavir/Glutathione conjugates
MRP3	Intestine/Liver/Kidney	BLM/SM/BLM	Glucuronide conjugates/Methotrexate
MRP4	Intestine/Liver/Kidney	BBM/SM/BBM	Topotecan/6-Thioguanine/Tenovir/Cefazolin/Furosemide
BCRP	Intestine/Liver/Kidney	BLM/CM/BBM	Topotecan/SN38/Imatinib/Nelfinavir/Rosuvastatin/Ofloxacin

BLM: Basal lateral membrane; BBM: Brush border membrane; SM: Sinusoidal membrane; CM: Canalicular membrane.

very limited lipophilicity, their diffusion across cell membranes becomes severely restricted so their absorption, distribution and elimination is constrained to paracellular transport through loose junctions between cells, hydrophilic (aqueous) pores in the cell membrane or drug transporters. It is for this type of compound that transporters have the most impact since they don't have passive diffusion to fall back on. Also, it must be remembered that a freely permeable lipophilic compound, although less impacted by transporters, will probably be eliminated by

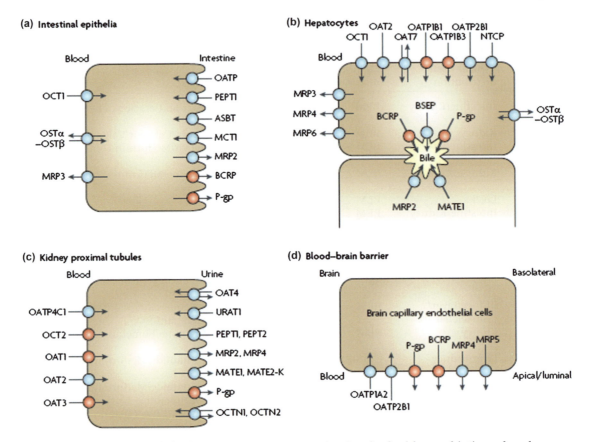

Figure 7.8 A summary of the human transporter proteins involved with xenobiotic and endogenous compound transport across cell walls (reprinted by permission of Macmillan Publishers Ltd: International Transporter Consortium, *Nat. Rev. Drug Discov.*, 9, 215–236, 2010[32]).

metabolism, but the resultant metabolites generally will be more polar. Thus, as the lipophilicity of the metabolites decreases with some, such as glucuronides and sulfates, completely losing their lipophilicity, they become increasingly reliant on transporters to remove them from the body.

The movement of compounds across membranes by different transporters is saturable and follows Michaelis–Menten kinetics in the same way as enzyme mediated reactions (Equation 7.1).

$$v = \frac{V_{max}*C}{K_m + C} \tag{7.1}$$

Where v is the rate of transport; V_{max} is the maximum transport rate; K_m is the Michaelis constant; C is the drug concentration.

Thus if the concentration of the xenobiotic compound being transported by a specific transporter is low compared to its K_m as defined above, the rate of transport will be approximately proportional to the concentration of the xenobiotic, but as the concentration increases, the rate of increase will slow and asymptote to the V_{max}. This leads into a frustrating topic, since because the technology is well established, compounds are often routinely tested as a substrate for MDR1 (P-gp) very early in the discovery process, using an *in vitro* bidirectional Caco-2 monolayer screen to determine the efflux ratio. The rationale for performing this test so early in the screening process is that P-gp is thought to restrict oral absorption and/or prevent compounds from entering the brain. Thus, if the efflux ratio for a compound is ≥2, it is considered to be a substrate for P-gp and quite often, discarded because of that. However, P-gp is easily saturated, mitigating its effect on compounds with a high flux into cells such as highly permeable, lipophilic compounds. P-gp will only impact absorption, remembering there will be large amounts of the drug in the intestinal tract, when the diffusion through the cells becomes severely restricted such as with very restricted permeability (hydrophilic), high molecular weight and/or severe dissolution limitation. Moreover, P-gp does not usually restrict high permeability, lipophilic compounds from entering the brain unless there is high plasma protein binding effectively restricting the unbound drug from crossing the cell membrane and keeping the effective concentration to P-gp low. Thus, it is highly premature to discard any compound early in the discovery process based solely on Caco-2 efflux data, without considering the other attributes of the compound.

Since the kinetics of transporter flux are Michaelis–Menten, it is not surprising that they can be inhibited or induced, much like the cytochromal enzymes. Inhibition can be competitive, noncompetitive or uncompetitive which can lead to potential drug–drug interactions either directly (*e.g.* transporters involved in renal or biliary excretion) or indirectly (*e.g.* transporters presenting a compound to a site where they can be eliminated by biliary excretion or metabolism). Induction of drug transporters is also a possible cause of drug–drug interactions but its understanding is less advanced than that of inhibition. Interestingly, some of the same nuclear receptors (CAR and PXR) that induce cytochromal enzymes also induce P-gp (MDR1), the MRPs and OATP2 (Xu *et al.*, 2004;[33] Mottino and Catania, 2008[34]) further illustrating the inter-relationship between drug transporters and drug metabolising enzymes. The impact of drug transporters on potential drug–drug interactions has been reviewed in detail by the International Transporter Consortium (2010)[32] and was subsequently followed up by regulatory guidances from the FDA (2012)[22] and EMEA (2012).[23]

7.10 CONCLUSIONS

Drug metabolism is an important elimination route for many compounds. An understanding of drug metabolism pathways and the factors which influence them provides information about mechanisms underlying changes in the kinetics, toxicity, and pharmacodynamics of drugs in animals and man. Information on the log $D_{7.4}$ or log P pK_a and structural complexity of a drug allows some prediction of metabolic fate. Extrapolation across species is not precise for metabolic pathways. However, in general, the small (shorter life span) species are more adept at Phase I oxidative pathways, producing a wider range of metabolites. Thus in man metabolite profiles are often less complex than in animals. These differences tend to reflect the decrease in basal metabolic rate with increasing body weight. An understanding of the relationship between drug metabolism, pharmacology, and toxicology can be applied to aid the design of drug candidates.

Preliminary information on metabolism can be of value in selecting drug development candidates and in designing compounds with improved kinetic and thus pharmacodynamics profiles.

DRUG METABOLISM HINTS AND TIPS

- Within a series of compounds, *e.g.* β-blockers, the importance of metabolism as an elimination pathway generally increases with increasing lipophilicity of the drug.[4]
- If a significant drug metabolite in man is not observed in the animal species used in pharmacology or toxicology studies or is only seen at much lower levels than those in man, the potential pharmacologic or toxicologic effects of the metabolite in man may not be reflected in the non-clinical safety evaluation. *In this case a separate evaluation of that metabolite may be necessary. This is described in the FDA MIST guidance document.*

- Drug administration by transdermal, sub-lingual, subcutaneous, intramuscular, buccal, inhaled pulmonary, or rectal routes have been used for some drugs to avoid "first pass" effects.
- A number of qualitative species differences are seen for Phase II pathways:
  - ○ Dogs are unable to acetylate aromatic amines and are more sensitive to the pharmacologic and toxic effects of such compounds.
  - ○ Cats lack the ability to form glucuronide conjugates and only primates form amino-acid conjugates with glutamine.
- As a broad generalisation, compounds (drug or metabolites) with appropriate structures and MW > 325 will undergo extensive biliary excretion in rat and dog, whereas this is more common in man at MW > 500.
- The FDA and European guidelines on drug–drug interactions provide a series of very useful decision trees to help in the interpretation of the *in vitro* inhibition and induction data.
- P-gp, a transmembrane drug transporter, is easily saturated, mitigating its effect on compounds with a high flux into cells such as highly permeable, lipophilic compounds. P-gp will only impact absorption, remembering there will be large amounts of the drug in the intestinal tract, when the diffusion through the cells becomes severely restricted such as with very restricted permeability (hydrophilic), high molecular weight and/or severe dissolution limitation.
- P-gp efflux can have a much greater impact on the net permeability of compounds attempting to cross blood brain barrier.

KEY REFERENCES

Pharmacokinetics and Drug Metabolism in Drug Design, 3rd Edition, eds. D. A. Smith, C. Allerton, A. Kalgutkar, H. van de Waterbeemd and D. K. Walker, Wiley-VCH, 2012.
Comprehensive Medicinal Chemistry, 2nd Edition, Vol. 5, ADME-Tox Approaches, eds. B. Testa and H. van de Waterbeemd, Elsevier, 2007.
J. H. Lin and A. Y. H. Lu, *Pharmacol. Rev.*, 1997, **49**(4), 403–449.
FDA guidance to Industry M3(R2) nonclinical safety studies for the conduct of human clinical trials & marketing authorization for pharmaceuticals, 2010.
S. C. Khojasteh, H. Wong and C. E. C. A. Hop, *Drug Metabolism and Pharmacokinetics*, 1st Edition, Springer, 2011.

REFERENCES

1. R. J. Bertz and G. R. Granneman, *Clin. Pharmacokin.*, 1997, **32**(3), 210–258.
2. D. A. Smith, *Drug Metab. Rev.*, 1991, **23**, 355–373.
3. M. Matignoni, G. M. M. Groothuis and R. de Kanter, *Exp. Opin. Drug Metab. Toxicol.*, 2006, **6**, 875–894.
4. T. Shimada, H. Yamazaki, M. Minura, Y. Inui and F. P. Guengerich, *J. Pharmacol. Exptl. Ther.*, 1994, **270**, 414–423.
5. M. Ingelman-Sundberg, *Naunyn-Schmiedeberg's Arch. Pharmacol.*, 2004, **369**, 89–104.
6. G. R. Bourne, in *Progress in Drug Metabolism*, ed. J. W. Bridges and L. F. Chasseaud, Wiley, London, 1981, Vol. 6, pp. 77–110.
7. G. Lappin and L. Stevens, *Expert Opin. Drug Metab. Toxicol.*, 2008, **4**, 1021–1033.

8. G. Lappin, M. Rowland and R. C. Garner, *Expert Opin. Drug Metab. Toxicol.*, 2006, **2**, 419–427.

9. K. R. Wade, J. Troke, C. M. Macdonald, I. D. Wilson and J. K. Nicholson, in *Methodological Surveys in Biochemistry and Analysis, Bioanalysis of Drugs and Metabolites especially Anti-inflammatory and cardiovascular*, ed. E. Reid, J. D. Robinson and I. D. Wilson, Plenum Press, New York and London, 1987, **18**, F-5, pp. 383–388.

10. P. J. Gilbert, T. E. Hartley, J. A. Troke, R. G. Turcan, C. W. Vose and K. Watson, *Xenobiotica*, 1992, **22**, 775–787.

11. J. B. Houston, *Biochem. Pharmacol.*, 1994, **47**, 1469–1479.

12. *FDA Guidance for Industry: safety testing of drug metabolites*, 2008.

13. *ICH M3 (R2) Non-clinical safety studies for conduct of human clinical trials and marketing authorisation for pharmaceuticals*, 2010.

14. C. Yu, C. L. Chen, F. L. Gorycki and T. G. Neiss, *Rapid Commun. Mass Spectrom.*, 2007, **21**, 497–502.

15. L. Leclercq, F. Cuyckens, G. S. Mannens, R. de Vries, P. Timmerman and D. C. Evans, *Chem. Res. Toxicol.*, 2009, **22**, 280–293.

16. R. A. Hamilton, W. R. Garrett and B. J. Kline, *Clin. Pharmacol. Ther.*, 1981, **29**, 408–413.

17. C. E. C. A. Hop, Z. Wang, Q. Chen and G. Kwei, *J. Pharm. Sci.*, 1998, **87**, 901–903.

18. P. M. Manoury, J. L. Binet, J. Rousseau, F. M. Lefevre-Boug and I. G. Cavero, *J. Med. Chem.*, 1987, **30**, 1003–1011.

19. G. Dean and C. W. Vose, *Methodological Surveys in Biochemistry and Analysis*, ed. E. Reid and I. D. Wilson, Royal Society of Chemistry, Cambridge, 1990, **20**, pp. 207–210.

20. R. A. Franklin, *Xenobiotica*, 1988, **18**(1), 105–112.

21. B. P. Monahan, C. L. Ferguson, E. S. Killeavy, B. K. Lloyd, J. Troy and L. R. Cantilena, *J. Am. Med. Assoc.*, 1990, **254**, 2788–2790.

22. *FDA Guidance for Industry Drug Interaction Studies: Study Design, Data Analysis, Implications for Dosing and Labeling Recommendations*, 2012.

23. *EMA Guideline on the Investigation of Drug Interactions*, CPMP/EWP/560/95/Rev1 2012.

24. J. G. Gerber, *Clin. Infect. Dis*, 2000, **30**(Suppl 2), S123–129.

25. M. E. Shams, B. Arnath, C. Hiemke, A. Dragicevic, M. J. Muller, R. Kaiser, K. Lachner and S. Hartter, *J. Clin. Pharmacol. Therap.*, 2006, **31**(5), 493–502.

26. J. G. Brockmoller, J. Kirchheiner, W. Silke, C. Sachse and B. Muller-Oerlinghausen, *Clin. Pharmacol. Therap.*, 2002, **72**, 438–452.

27. C. Skonberg, J. G. Olsen, K. G. Madsen, S. H. Hansen and M. P. Grillo, *Expert Opin. Drug Metab. Toxicol.*, 2008, **4**, 425–438.

28. B. K. Park, A. Boobis, S. Clarke, C. E. Goldring, D. Jones, J. G. Kenna, C. Lambert, H. G. Laverty, D. J. Nesbitt, S. Nelson, D. A. Nicholl-Griffith, R. S. Obach, P. Routledge, D. A. Smith, D. J. Tweedie, N. Vermeulen, I. D. Wilson and T. A. Baillie, *Nature Rev. Drug Discov.*, 2011, **10**, 292–306.

29. R. A. Thompson, E. M. Isin, Y. Li, L. Weidolf, K. Page, I. D. Wilson, S. Swallow, B. Middleton, S. Stahl, A. J. Foster, H. Dolgos, R. Weaver and J. G. Kenna, *Chem. Res. Toxicol.*, 2012, **25**, 1616–1632.

30. F. G. M. Russel, in *Enzyme- and Transporter-based drug–drug interactions. Progress and future challenges*, ed. K. S. Pang, A. D. Rodrigues and R. M. Peter, Springer, New York, 2010, 27–49.

31. K. M. Giacomini and Y. Sugiyama, in *Membrane transporters and drug response in Goodman and Gilman's the Pharmacological Basis of Therapeutics*, 11th Edition, eds. L. L. Brunton, J. S. Lasko and K. L. Parker, McGraw-Hill, 2002, pp. 41–70.

32. The International Transporter Consortium, *Nat. Rev. Drug Discovery*, 2010, **9**, 215–236.

33. C. Xu, C. Y. T. Li and A.-N. T. Kong, *Arch. Pharm. Res.*, 2005, **28**, 249–268.

34. A. D. Mottino and V. A. Catania, *World J. Gastroenterol*, 2008, **14**, 7068–7074.

35. C. Zhang, W. Chen, T. Smolarek, R. Scott Obach and C. M. Loi, *Drug Metab. Dispos.*, 2010, **38**, 641–654.

CHAPTER 8

Prediction of Human Pharmacokinetics, Exposure and Therapeutic Dose in Drug Discovery

DERMOT F. McGINNITY,*[a] KEN GRIME[b] AND PETER J. H. WEBBORN[c]

[a] DMPK, Drug Safety and Metabolism, AstraZeneca, Mölndal, Sweden; [b] DMPK, Respiratory, Inflammation and Autoimmune IMed, AstraZeneca, Mölndal, Sweden; [c] DMPK, Drug Safety and Metabolism, AstraZeneca, Alderley Park, United Kingdom
*E-mail: Dermot.F.McGinnity@astrazeneca.com

8.1 INTRODUCTION

The application of pharmacokinetic principles in drug design is now widespread due primarily to the recognition of the role of plasma concentrations as a surrogate for measuring drug effects. A basic tenet of pharmacology is that the magnitude of a pharmacological response is a function of the concentration of drug at the site of action. Thus the objective of therapy can be achieved by maintaining an adequate concentration of drug at the site of action for the necessary duration. In clinical pharmacology the objective is to maintain a therapeutic concentration high enough to give the desired response, but not so high so as to elicit an undesired response; this is the 'therapeutic window' and immediately brings to the fore the concept of the management, or design, of drug concentration–time profiles. As the concentration of drug cannot usually be determined at the site of action, the concentration is usually measured in blood or plasma. The relevance of this surrogate site, and the drivers for the observed changes over time, will depend upon a number of factors and assumptions. Our understanding of such factors are best understood through the science of pharmacokinetics (PK), which is defined as the study of change of drug concentration over time and describes a systematic approach to relating dose to amount of drug in the body, typically as viewed from plasma. Pharmacodynamics (PD) is the study of how drug concentration relates to effect. Quantitative pharmacology or pharmacokinetics–pharmacodynamics (PKPD) through modelling the mechanism of drug action (*e.g.* agonism or antagonism) describes the relationship between dose, concentration and the intensity and duration of response.

The Handbook of Medicinal Chemistry: Principles and Practice
Edited by Andrew Davis and Simon E Ward
© The Royal Society of Chemistry 2015
Published by the Royal Society of Chemistry, www.rsc.org

This chapter is designed to be used by the practicing medicinal chemist and attempts to provide both the context and some specific advice to apply best practice for PK and efficacious dose prediction. Hints and tips are suggested throughout the chapter and methods to predict human PK parameters outlined. Worked examples of both proprietary candidate and approved drugs are used to highlight particular themes. This chapter can be used in conjunction with the associated Dose to Man mobile app which allows medicinal chemists to simulate therapeutic doses and exposures in Man for their drug discovery projects *via* the approaches and equations presented herein. General definitions of the key PK parameters and their inter-dependencies, are included here, but for a comprehensive analysis of pharmacokinetics and full derivation of equations the reader is directed towards the authoritative reference *Clinical Pharmacokinetics* by Rowland and Tozer.[1]

A key question, the answer to which, to a degree, defines a project strategy is "Why conduct PK studies in animals, as part of a drug discovery program?" There was a time when "good" PK properties in animals *per se* were seen as a sign of quality or value in molecules. As animal data may poorly predict human kinetics, the real value of animal data is to validate, or otherwise the *in vivo*, *in vitro* and *in silico* approaches that are used to predict PK in Man. The authors seek to explain how such assessments are made and how application of such an understanding can lead to robust predictions of human PK.

Three key PK parameters are derived following intravenous (IV) administration. A measured dose is administered and drug levels in plasma are determined over time. Early concentrations are back-extrapolated to estimate the concentration at $t = 0$, and simple mass balance considerations allow estimation of the volume that the dose would have to be dissolved in to yield that concentration. This yields the initial volume of distribution (with units of L or L/Kg). If there is a mono exponential decline in plasma concentrations, this is the volume of distribution (V_d). The same assessment (amount of drug in the body relative to the plasma concentration) is used to derive other volume terms, whose derivations are beyond the scope of this article. V_d is most usefully thought of as a measure of the relative affinity of the compound for tissues and plasma. The volume term that most accurately describes this measure, for multi exponential profiles is the volume of distribution at steady-state (V_{ss}).

The area under the plasma concentration–time profile (AUC) for a given dose is a measure of how efficiently a compound is removed from the body (*i.e.* compounds yielding low AUCs are eliminated by efficient processes). This ratio (Dose/AUC) is used to derive the clearance (CL) of a compound and has units of flow (commonly mL/min/Kg). CL is the second key PK parameter derived which describes drug elimination (see Hints and Tips: Clearance (CL)).

The third key parameter obtained from a simple fit of IV data is the half-life $(T_{1/2})$. Multi-exponential declines are more complex, but generally the terminal $T_{1/2}$ is the most important, as it describes the decrease in plasma concentration during the elimination phase. $T_{1/2}$ does not describe the efficiency of eliminating processes (CL does this) and is actually a secondary PK parameter, dependent upon fundamental parameters. In the case of a one compartment system $T_{1/2} = \ln 2 \times V_d/CL$. The significance of this, from a drug design perspective, is that $T_{1/2}$ can be modulated by changing the V_d or CL.

Following oral administration, a key parameter is oral bioavailability (F). This describes the fraction of the dose that reaches the systemic circulation, and is a measure of a compound's ability to avoid the processes that protect the body from exposure to foreign compounds, *i.e.* metabolism in the liver, metabolism in the gut, biliary elimination and the physical barrier of the gut wall (poor dissolution is also a factor that reduces F). Thus F is a product of the fraction of dose that is absorbed unchanged through the gut and the fraction of drug that escapes 'first-pass' extraction in the liver. High extraction compounds (*e.g.* metoprolol, diltiazem) that are rapidly metabolised and have a high hepatic CL will have low systemic exposure following an oral dose and low F. Low extraction compounds (*e.g.* atenolol, cetirizine) that are relatively stable to metabolism and have a low hepatic CL, assuming they are absorbed across the gut wall will have high systemic exposure following an oral dose and high F. Bioavailability is estimated from $F = AUC_{oral}/AUC_{IV}$, for an equivalent dose.

PK is typically separated into distinct physiological processes of Absorption, Distribution, Metabolism and Excretion (ADME). PK can be studied using different mathematical models broadly classified as non-compartmental models, compartmental models and physiologically based (PBPK) models. All three approaches have use within modern drug discovery, determined by the questions being asked. Compartmental PK modelling represents the body as a minimum number of empirical compartments in equilibrium, with the number of compartments usually defined by the number of exponential phases in the plasma concentration–time profile. Indeed for many drugs, a mono-exponential fit, and hence a one compartment model may reasonably describe the bulk of the data. However, for some drugs, whose kinetics are characterised by a slow redistribution of a reasonable fraction of the dose from tissues, a multi-compartment model may be required. In PBPK modelling, compartments are chosen to represent physiological compartments such as tissues and organs of the body (although they may be grouped), connected by anatomically correct blood flows. Elimination rates and partition coefficients are estimated for relevant compartments. PBPK models can be powerfully applied to explain the disposition of drugs in tissues and to generate testable hypotheses, relating efficacy and extrapolation of drug behaviours across species. The advantage of compartmental and PBPK models is that once the experimental data has been fitted to provide estimates of the model parameters, drug concentration–time profiles can be simulated, to explore a variety of dose regimens. Simulations are a powerful method for aiding the mechanistic investigation of drug disposition, for dose setting and for aiding clinical study design.

Both compartmental and PBPK models require some assumptions about the kinetics of the compound. A non-compartmental analysis, although not enabling simulation, is able to be used to determine fundamental parameters that describe the kinetics of compounds: CL, V_{ss}, $T_{1/2}$ and F. Such a non-compartmental approach forms the foundation of human PK and dose prediction method ("Dose to Man") that is the focus of this Chapter.

Effective therapies require potency against the relevant target, sufficient selectively against unwanted pharmacological responses and an adequate PK profile to sustain the drug at the site of action, from an acceptable dose size and frequency. Thus, the medicinal chemist attempts to optimise many parameters simultaneously whilst attempting to avoid impairing any desired properties. What can assist in this difficult, and at times seemingly insurmountable challenge, is that the PK properties of drugs are often strongly associated with their inherent physicochemical properties. Understanding these relationships is a key element of drug design. The most important physicochemical properties, in this context, are molecular size, hydrophobicity, aqueous solubility and ionisation state at physiological pH. It is beyond the scope of this chapter to provide a comprehensive review of this topic but the authors can strongly recommend *Metabolism, Pharmacokinetics and Toxicity of Functional* groups edited by Dennis Smith.[3] These properties, to a greater or lesser extent, affect dissolution in the gastrointestinal tract and membrane permeability (and therefore oral absorption), susceptibility to drug metabolising enzymes and other elimination processes.

Likely human PK and clinical dose can be estimated from *in vitro* experiments utilising human derived material and supported by *in vivo* animal studies. Defining clinically efficacious dose is a very useful parameter in the optimisation of compounds as a holistic measure of 'quality' that relies on both prediction of human PK and of the anticipated therapeutic concentration. Starting from target identification and continuing throughout drug discovery and into the clinic, the prediction and refinement of therapeutic concentration in patients is key and should be based on a quantitative understanding of the target and relevant biomarkers. This requires an integrated quantitative assessment, in which the pharmacology, target, tissue and disease are all considered. However such knowledge is not always available *a priori* and it is common that uncertainty around likely effective therapeutic concentration is a major risk factor.

The underlying assumption to the paradigm represented in Figure 8.1 assumes that efficacy at the target tissue is related directly to the target free plasma concentration at steady state ($C_{ss,min}$) and that the target concentration must be maintained for the whole of the dosing interval. For many targets, this will be a cautious assumption since efficacy may only require partial occupancy for a fraction of this period. Clearly, where a PD or disease model considered representative of the human disease exists, data from such models should be used to define the target plasma or tissue concentration and level of target occupancy required for efficacy. It should be stressed that the "Dose to Man" approach outlined here is very much a default starting position in the absence of other information and should be refined during the lifetime of the project.

A test of the predictive capability of the "Dose to Man" approach was to predict the human PK and therapeutic dosage for a number of marketed oral drugs. A database was collated which contained human PK data and *in vitro* potency data for marketed drugs against a range of targets from different classes including G-protein coupled receptors, enzymes including kinases and ion channels.[4] This work demonstrated that setting the minimum efficacious concentration at $3 \times$ the relevant free *in vitro* potency is a reasonable conservative starting position, in the absence of other information. This default assumption has been shown to hold for a number of G-protein coupled receptors antagonists including antimuscarinics, antihistaminics and β-adrenoceptor blockers.[5]

Similarly, establishing the PK drivers and profile that is likely to deliver clinical efficacy is also a feature of successful drug hunting projects. This PKPD relationship is underpinned by the nature of the drug–target interaction and the role of the biological target in the pathophysiology of the disease. PKPD relationships can be derived from clinical precedents, preclinical models and mechanistic studies of the drug target interaction, and are refined throughout the life of a project. Early in a project, when critical improvements in compound properties are required to establish that a chemical series is optimisable and to generate initial data supporting the biological hypothesis, it is useful to have a simple model that integrates optimisable properties to show overall progress. Thus, for oral drug delivery, a simplistic one compartment PK model can be used, in which the target plasma (unbound drug) concentration at steady state is directly related to efficacy and that the PK elimination $T_{1/2}$ is the effective $T_{1/2}$ (the phase of the drug concentration–time profile responsible for maintaining drug concentrations above the required level to achieve efficacy). With such assumptions, Equation 8.1

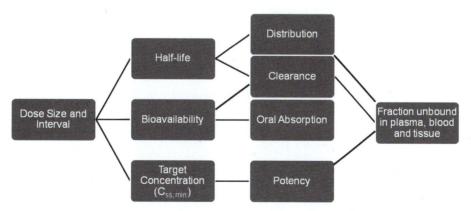

Figure 8.1 Paradigm for assessing human PK, minimum effective concentration and therapeutic dose.

which assumes rapid (instantaneous) absorption can be used to estimate the likely daily dose of any given compound.

$$\text{Dose (mg/kg/day)} = \frac{(24/\tau) \times \text{MEC} \times V_{ss}\left(\exp(k_{el} \times \tau) - 1\right)}{F} \quad (8.1)$$

Where MEC is minimum effective concentration, V_{ss} is volume of distribution at steady state, F is oral bioavailability, k_{el} is the elimination rate constant and τ is the dosing interval. The same model but with a first order absorption rate constant (k_a) included is given by the Equation 8.2.

$$\text{Dose (mg/kg/day)} = \frac{(24/\tau) \times \text{MEC} \times (k_a - k_{el})}{F \times k_a \times \left[\dfrac{1}{1 - e^{-k_{el} \times \tau}} - \dfrac{1}{1 - e^{-k_a \times \tau}}\right]} \quad (8.2)$$

The parameter estimates required are thus F, k_a and k_{el}. The accuracy of this approach depends on the validity of the underlying assumptions and the ability to estimate these parameters. F is a function of the fraction of dose absorbed (F_{abs}), the fraction of drug escaping intestinal metabolism (F_g) and the fraction of the drug escaping hepatic CL (F_h) such that $F = F_{abs} \times F_g \times F_h$. The absorption rate constant is commonly rapid enough not to significantly impact on dose (authors unpublished observation based on an analysis of marketed oral drugs). The most influential parameter is the elimination rate constant (k_{el}). This is estimated from the V_{ss} and CL as $k_{el} = 0.7 \times V_{ss}/CL$.

As a worked example, assume a drug of MW 450 has an unbound potency (pX) against the target of 8 and the fraction unbound in plasma, $fu_p = 0.05$, blood : plasma ratio, B:P = 1 and for efficacy requires a total MEC equivalent to $3 \times pX = 270$ ng/mL. PK parameters are CL = 1 mL/min/kg, $F_{abs} = 0.5$ (and rapid absorption), F = 0.48, $V_{ss} = 1$ L/kg, $T_{1/2} = 12$ h so $K_{el} = 0.06$ h^{-1}. Using Equation 8.1 the single daily dose equates to 1.8 mg/kg.

It is clear from Equation 8.1 that there are essentially only four parameters to optimise: potency against the target, CL, V_{ss} and F_{abs}. Alteration of one of these parameters whilst fixing the others at a single set value demonstrates that potency and F_{abs} have a linear impact on dose whilst CL and V_{ss} can have a much greater impact (Figure 8.2). It is worth noting that these parameters are rarely, if ever, independent, and that the challenge of drug discovery is overcoming the confounding SARs of these parameters. However, for any given target, in the absence of a more thorough understanding of the human PKPD relationship, this equation can set the foundation for a rational drug discovery optimisation strategy.

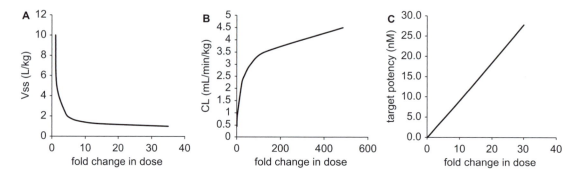

Figure 8.2 Interrelationships between dose, CL, V_{ss} and potency.

8.2 PK IN DRUG DISCOVERY: A HISTORICAL OVERVIEW

The fundamentals of what we would now recognise as PK theory and analysis were established by 1960,[6] but it took another 20 years before PK optimisation became integral to the drug discovery process. By 1960, the key concepts such as volume of distribution, elimination rates, compartmental models, the essentials of drug absorption and bioavailability, as well as free-drug and steady-state considerations, were used to describe and understand the *in vivo* behaviour of a limited number of drugs.

During the 1960's enhanced bioanalytical methodologies enabled a pharmacokinetically driven understanding of the behaviour of many drugs in clinical use, describing such phenomena as drug accumulation, effects of renal impairment and non-linear kinetics. However, at this time PK considerations had little influence on compound design. In an era of phenotypic screening, where tissue baths and animal experimentation were the drivers of medicinal chemistry, any PK considerations were only notional components in the rationalisation of efficacious dose and duration of effect, relative to potency.

Three advances, one bioanalytical, one based on the use of human tissue, and the other a development of a novel PK parameter, made PK optimisation an achievable Medicinal Chemistry ambition and thrust drug metabolism and pharmacokinetics (DMPK) to the centre of drug discovery. The bioanalytical advance was the adaption of what was a historically a qualitative tool for elucidation of molecular structure, into a high capacity quantitative tool. Advances that enabled the coupling of HPLC to triple quadrupole mass spectrometers around 1980[7] with appropriate software and sample handling systems became the industry standard.[8] The ability to rapidly develop sensitive, selective methods and deliver high volumes of data in parallel with assays of activity, meant *in vitro* based predictions and validation through subsequent *in vivo* PK studies became a driving force in discovery projects.

The earliest use of PK assessments within the pharmaceutical industry was in the development phase, with a focus on understanding and correctly quantifying safety risks associated preclinical observations, inter-subject variability,[9,10] inter-species differences[11] and drug–drug interactions.[12] It was natural that the frustration of selecting compounds subsequently shown to have PK deficiencies, such as low F or short $T_{1/2}$, that consequently limited clinical utility, would translate into a desire to select, and later design, compounds based on PK properties.

The influential PK innovation was the development of CL concepts by Rowland *et al.*[13] and others. CL concepts impacted many aspects of PK and enabled explanation of several hitherto unexplained observations, based in the different behaviours of high and low CL compounds to the effects of enzyme inducers and inhibitors. However from a drug design point of view, the realisation that it was this new parameter—CL, and not $T_{1/2}$—that best described the efficiency of an eliminating process, was fundamental. Pharmacokineticists were able to develop a much clearer understanding of drug elimination processes and to produce the kind of single parameter SAR and optimisation sought by medicinal chemists.

As CL has units of flow, it can be compared to the blood flow to an eliminating organ to establish the extraction ratio, thereby enabling a readily understandable calibration of measured CL. A key feature of CL is that it can be derived without having to assume any underlying PK model, for example, a one or two compartment model. It is remarkable to think that the true value of one of the simplest, most powerful equations in PK, CL = Dose/AUC, did not start to be fully exploited until 1973.

A major component of the early publications was the development of models of eliminating organs, notably the liver, but also the kidney. For hepatic CL, such models yielded a mechanism for extrapolating *in vitro* data to *in vivo*[14–16] enabling development of the third key advance—robust prediction of human metabolic CL based on enzyme activity. This key step forward meant that human predictions could account for the well-recognised interspecies differences in metabolic capabilities. Until this point, variants on simple allometric scaling, where predictions are based on some physiological scaling factor (*e.g.* body weight), were the primary methods available.[17]

The science of designing drugs based on PK principles was a new challenge for the pharmaceutical industry that would not only lead to generation of structure–activity relationships for many of the processes that drive drug disposition, but also to a transformation in the understanding of these processes. The ability to describe and predict rates of metabolism based on physicochemical properties was described by 1970,[18] and the succeeding years saw extension of this work leading to models that described drug distribution,[19] renal CL,[19] plasma protein binding (ppb)[20] and intestinal absorption,[21] the key elements governing PK properties. Accurate predictions of CL purely from physicochemical properties, beyond the narrow bounds of a single chemical series, remain elusive, partly because such predictions require an estimate of rate, rather than just affinity, but also because of the relatively narrow dynamic range in CL estimates (generally 2 log units). From this time, it was common in the pharmaceutical industry to build physiologically based PK models of human kinetics, based on physicochemical properties and an *in vitro* assessment of rate of elimination, with refinement as preclinical data became available. These approaches were ultimately formalised and commercialised in such products as SimCYP™ and Gastroplus™.

The early application, learning and impact of DMPK principles in the industry have been captured in a series of publications by the group at Pfizer in Sandwich, which was arguably the leading group in developing the thinking that brought an insightful and rigorous use of data.[22–24]

The learning by the pharmaceutical industry and its ability to translate this into more successful candidates was a major factor in the development of the large number of blockbuster drugs in the 1990's. Furthermore not only were first in class compounds commonly once daily (QD) due to optimisation of CL, but discovery programs could be initiated based on a PK goal with some confidence. Similarly by the 1990's best-in-class compounds started to be approved, where the key advantages were essentially PK. Since 2000, there have been a number of reviews of the state-of the art integration of PK into the drug discovery process, including extensive analysis of the prediction of human kinetics from preclinical data from our own laboratories.[25]

8.3   OPTIMISING PHARMACOKINETICS IN DRUG DISCOVERY

The ability to predict human PK and likely efficacious clinical dose are essential elements of any 'drug hunting' strategy. They allow the focus of optimisation to be on the influential parameters, potency against the target, F_{abs}, CL, and V_{ss} and enable a rational assessment of when a credible candidate compound has been synthesised. The next sections outline the authors' experience and view of current best practice for predicting these parameters for Man.

8.3.1   Absorption

For orally administered drugs, it is important not just to determine F in preclinical species but also to understand the relevance of the contribution of first pass hepatic and intestinal metabolism, and of intestinal permeability. What is acceptable in terms of preclinical and clinical absorption is a question that in our experience can confound drug discovery teams and lead to time and effort being unnecessarily wasted. Fundamentally, drug absorption impacts in two areas: limiting exposure in preclinical safety testing and driving excessive variability in systemic drug levels in patients. As inter-patient variability in drug exposure is inversely related to bioavailability,[26] absolute bioavailability in man of greater than 30% is an appropriate target, and it is therefore judicious to target a minimum predicted human absorption of at least 50%, as F is a product of the fraction of absorbed drug (F_{abs}), the fraction escaping intestinal metabolism (F_g) and liver extraction (F_h) as it passes from the portal vein to the systemic circulation

($F = F_{abs} \times F_g \times F_h$). The 30% target for F should not be viewed as a hard cut-off, since several marketed oral drugs sit in this category, but simply as an area where inter-subject variability and the impact on dose size may hamper drug development.

An assessment of the role of intestinal drug extraction is a component of the human F prediction. The human and preclinical animal intestinal drug metabolizing enzymes are well characterised with CYP3A, CYP2C9 and UGT dominating.[27–32] Mathematical models enabling *in vitro* data to be used in the prediction of F_g have also been described.[33] Despite intestinal CYP content being extremely low compared to that of the liver[34,35] and intestinal intrinsic CL values being similar to hepatic once corrected for expression levels,[36] extraction by the gut can in some cases be similar to or exceed hepatic extraction.[33,37] The reasons for this include the efficient location of the drug metabolizing enzymes and the p-glycoprotein (P-gp) drug efflux transporter (which often shares substrate specificity with CYP3A4) in the villus tip of the enterocytes, facilitating cycling of drugs and prolonged exposure and thus, intestinal metabolism.[33,34,38] Significant intestinal extraction is commonly associated with highly metabolically unstable drugs[33,39] so for a drug to be efficiently extracted by intestinal drug metabolising enzymes, it would need not only to have sufficient exposure to those enzymes in the intestine but also to be relatively rapidly metabolised. However, oral drug discovery programs typically optimise towards compounds with moderate to high permeability and solubility, high intrinsic metabolic stability and low involvement of intestinal efflux transporters. Such new chemical entities (NCEs) are unlikely to carry a significant risk of intestinal metabolic extraction, making gut metabolism a minor consideration in human dose prediction.

Intestinal efflux by intestinal drug efflux transporters such as P-gp is unlikely to limit the absorption of compounds with moderate to high permeability, and good solubility.[40,41] However, if a drug falls outside this category and the K_m describing the affinity of the substrate–transporter interaction is relatively high (tens of micromolar or above) and/or the dose is considerably less than 1 mg/kg, *in vivo* P-gp dependent efflux is more likely.[42] In the absence of more sophisticated simulations incorporating solubility and dissolution rate measures, the intestinal drug concentration range sufficient for interacting with P-gp may be estimated from the maximum dose taken /250 mL (intestinal fluid volume) or alternatively, the maximum concentration in the enterocyte can be estimated from ($F_{abs} \times k_a \times Dose/Q_{ent}$) where Q_{ent} is enterocyte blood flow.[33,39] It should be noted that this Q_{Gut} equation generally gives 100-fold lower estimations of concentration than using the intestinal fluid volume approach, such that a 1 mg/kg oral dose may be differentially described as having a relevant intestinal concentration of approximately 6 or 600 µM.[28] Caution should also be used when using K_m values from *in vitro* assays such as the Caco-2 assay since they may over-estimate the *in vivo* value and are sensitive to the expression level of P-gp.[42] Thus, whilst for compounds with moderate to low permeability an assessment of the role of P-gp can be made using the Q_{Gut} equation and P-gp K_m estimated from basolateral to apical Caco-2 drug concentration data, it is advisable when in a 'risk zone' to use simulation software, such as Gastroplus™ that has been validated with known clinical data for converting *in vitro* experimental values (from the particular laboratory where the novel data is being generated) to those that can be used for *in silico* prediction.

Although variables such as intestinal transit time and rate of dissolution from a tablet can control the extent of oral absorption in some situations, targeting compounds with appropriate physicochemical properties that ensure sufficient solubility, of crystalline material, and trans-membrane permeability should maximise the oral absorptive potential, and avoid pharmaceutical complexities. The apparent permeability measure (P_{app}) is typically obtained from assessment of drug flux across a monolayer of cells intended to mimic the intestinal barrier. The most commonly used are the Caco-2 and MDCK cell lines.[33,43] Relationships between human absorption and *in vitro* P_{app} are required to put the *in vitro* data into context but P_{app} data must first be transformed into effective permeability (P_{eff}) which describes intestinal permeability per unit surface area.[33,44] It is important to assess the impact of the key variables (measured *in vitro* permeability and solid crystalline solubility data) through PK modelling/simulation using tools such as Gastroplus™ or SimCYP™.[42,45,46] To facilitate rapid and effective decision making early in the life of a NCE it is possible to use such modelling tools to generate a solubility–permeability heat map.[48] From the analysis shown it is evident that for a 1 mg/kg dose, a crystalline solubility of 100 μM and a Caco-2 P_{app} of 5×10^{-6} cm/second, this will likely result in a fraction absorbed in man of greater than 50%, with lower values on solubility or P_{app} putting the NCE in a risk category for lower human F_{abs}.[48] This illustrates how to effectively use *in vitro* data in a powerful way to make robust decisions in drug discovery.

Rat oral absorption is, in the main, not a good predictor of human absorption.[4] This is most likely explained by the fact that once normalised for body surface area, the rat small intestine has a four times lower surface area compared to human.[48] Nonetheless, the rat absorption data is important for determining if safety assessment studies can be adequately performed. It is therefore necessary to identify whether it is possible to achieve high enough exposures in the rat to allow suitable margins over the predicted human exposure to be attained. Human exposure can be predicted using the available data or the generic target profile for the candidate drug. Predicting human absorption from preclinical *in vivo* data is more achievable using dog as a model, since the dog to human absorption correlation appears strong for high molecular weight (>325) compounds, when absorption is not by the paracellular route (Figure 8.3). In our experience, for drugs absorbed by the transcellular route, not only does dog PK data indicate a systematically greater absorption than in rat, but also that dog absorption, when using crystalline material and studied with an appropriate formulation, better represents the clinical situation.

To summarise, *in vitro* crystalline solubility and P_{app} data allow an early estimation of human absorption which can then be supported by *in vivo* dog absorption data. Rat oral absorption data should be treated with caution as our experience suggests this may be an underestimate of absorption in higher species including human and poor oral absorption in the rat should not

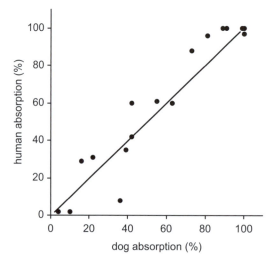

Figure 8.3 Relationship showing correlation between human absorption and dog absorption for compounds with MW>325. Shown is the line of unity.
Figure adapted with permission from McGinnity *et al.*, 2007.[4] Copyright 2007 Bentham Science Publishers.

per se preclude progression of compounds into development assuming the requisite oral exposure margins in safety studies are achievable.

HINTS AND TIPS: ABSORPTION

- $F = AUC_{po}/AUC_{iv}$ (if equivalent doses given).
- Target $F > 30\%$ to avoid excessive inter-individual variation in man.
- Intestinal PGP efflux is unlikely for compounds with high permeability and solubility.
- Rat oral absorption is not a good predictor of human but vital for determining if safety studies can be adequately performed. Dog is a better model for human absorption.
- A crystalline solubility measure of 100 μM and a Caco-2 P_{app} of 5×10^{-6} cm/sec will likely result in F_{abs} in man $>50\%$; note absolute values will depend on assays in any given laboratory.

8.3.2 Volume of Distribution

V_d can be viewed as a measure of the relative affinity of a drug for tissues and for plasma. This is expressed in the equation: $V_d = [fu_p / fu_T] \times V_T + V_P$, with units of L/Kg, where fu_p is the unbound drug fraction in the plasma; fu_T is the unbound drug fraction in the tissue; V_T is the volume of the tissue (~ 1 L/Kg) and V_P is the volume of the plasma (0.03 L/Kg). As V_T is considerably greater than V_P changes in fu_p will directly affect V.[1] Due to the physiological similarity, inter-species differences in tissue binding are assumed to be minimal and the key driver of interspecies differences in V are differences in ppb as outlined by Øie and Tozer.[49] Another way of expressing this is $V \times F_u = fu_T$ or $V_u = fu_T$. Therefore if unbound volumes are consistent across species, human volume can be predicted from animal data, by estimating V_u (Figure 8.4).

For drug-like chemotypes one should expect characteristically low (acids), moderate to low (neutrals) or moderate to high (bases) steady state V_d, where the low, moderate and high labels can be assigned as less than 1 L/kg, 1–3 L/kg and greater than 3 L/kg (Figure 8.5). In a Drug Discovery setting, knowledge of the expected boundaries for a given chemical class can be translated into a strategy for obtaining the necessary human elimination $T_{1/2}$ through an understanding of the extent to which CL will need to be reduced and what parameters are available in order to make such a change. The overriding influence on the distribution of acidic drugs is that of extensive binding to plasma albumin. Apparent distribution volumes thus approach that of albumin, approximately 0.1 L/kg,[1] and in our experience do not exceed 0.3 L/kg unless active hepatic uptake or entero-hepatic re-circulation is a determining factor, increasing the amount of drug in tissues. As such, an acidic NCE of interest with a V_{ss} greater than this value merits closer attention, starting with scrutiny of the PK profile itself. The V_d for neutral drugs is governed by hydrophobic interactions with plasma proteins and tissue membranes. Increasing lipophilicity raises tissue affinity but has the opposite effect of restricting tissue distribution due to an increase in ppb. Consequently, V_{ss} tends to be confined to the range 1 to 3 L/kg for a high proportion of neutral compounds (Figure 8.5). Basic drugs tend to have similar ppb to neutral drugs, in contrast to acids which show higher binding even for a given $logD_{7.4}$.[50] Being positively charged at physiological pH, bases have favourable interactions with acidic phospholipid head groups leading to higher tissue affinity and therefore for the same ppb as a neutral compound, a base will tend to have a higher tissue distribution.[51] Ion trapping in such sub-cellular acidic organelles as lysosomes also has a large effect, particularly in lysosome rich organs such as the liver and lung.[52] Even a weakly basic drug may have an apparent V_d three times the physical blood volume and dibasic drugs can have V_d values greatly in excess of its monobasic analogues.[50,53]

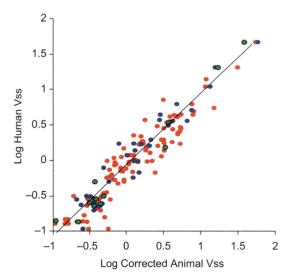

Figure 8.4 Using preclinical species to predict human V_{ss}. Relationship between human V_{ss} values and preclinical animal V_{ss} values corrected for human/animal differences in ppb as detailed in McGinnity *et al.*[4] Red circles—rat, green—mouse and blue—dog. The dataset includes both data generated in this laboratory and literature data from rat, dog and mouse and compared to clinical V_{ss}. Line of regression shown with the equation $\log y = 1.\log x + 0.06$, $r^2 = 0.93$, $p = 4.6 \times 10^{-101}$, afe = 1.36. Thus human V_{ss} can be predicted from preclinical animals using Equation 8.3. Figure adapted with permission from McGinnity *et al.*, 2007.[4] Copyright 2007 Bentham Science Publishers.

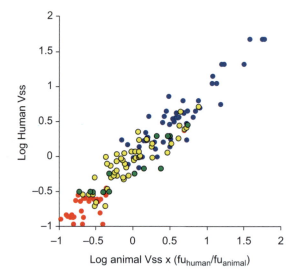

Figure 8.5 Using preclinical species to predict human V_{ss} for compounds with a range of physicochemical properties. Relationship between human V_{ss} values and preclinical animal V_{ss} values corrected for human/animal differences in ppb. Red circles—acids, green—zwitterions, yellow—neutrals and blue—bases. The same data is included in Figure 8.4 and McGinnity *et al.*, 2007.[4]

In summary, V_{ss} is reasonably predictable from *in silico*,[54] *in vitro* data,[54] and physicochemical properties.[6,55] Once preclinical *in vivo* PK data is available, it can be considered the most robust and predictable of PK parameters with the Øie and Tozer method superior for making human predictions, as demonstrated by several laboratories.[56-58] Thus, human V_{ss} can be predicted using preclinical animal data and Equation 8.3.

$$\log\left(Vss_{\text{human}}\right) = \log\left(Vss_{\text{animal}}\frac{fu_{\text{human}}}{fu_{\text{animal}}} + 0.1\left(1 - \frac{fu_{\text{human}}}{fu_{\text{animal}}}\right)\right) - 0.06 \qquad (8.3)$$

HINTS AND TIPS: VOLUME OF DISTRIBUTION

- V_d correlates with the relative affinity of compound for tissues. V_{ss} values for acids are low (0.1–0.3 L/kg), moderate to low for neutrals (1–3 L/kg) and moderate to high for bases (> 3 L/kg).
- Bases have higher V_d due to favourable interactions with acidic phospholipid heads and trapping in acidic sub-cellular organelles.
- Using preclinical *in vivo* PK data and correcting for species ppb differences, V_{ss} is the most predictable of PK parameters: Unbound V_{ss} should be conserved within two-fold across species—if not reasons need to be understood.
- Recirculation of parent drug after elimination in bile (of parent or glucuronide that can be hydrolysed to parent in intestines) can lead to extended PK profile and therefore larger calculated V_{ss}. This can yield incorrect predictions if extrapolating V_{ss} to a species where re-circulation does not occur to same extent.

8.3.3 Clearance

Optimisation of CL is typically one of the more significant challenges for a drug discovery project. Identification of the elimination route and rate in preclinical species and optimisation in human are major goals in most projects. The major drug elimination routes in humans and preclinical species are metabolic, renal and biliary. As there is currently no reliable way to predict human elimination pathways from purely *in silico* or *in vitro* methods, a combination of establishing CL routes in preclinical species and use of *in vitro* human tools is required to predict human CL.

Until relatively recently it has been common practice to predict human CL by cross-species allometric scaling irrespective of the elimination route.[59-61] However, for hepatic metabolic CL, it is now widely accepted that the use of *in vitro* data (*in vitro–in vivo* extrapolation, IVIVE) can be relied on to make more accurate predictions.[62-66] This science came to the fore almost 20 years ago following the seminal publication of an IVIVE strategy based on *in vitro* estimates of rates of turnover and a mathematical model of the liver. The models describing hepatic metabolic CL (CL_H) incorporated terms for liver blood flow (Q_h), hepatic intrinsic clearance (CL_{int}) and blood binding (fu_b).[14] Over the subsequent years, a significant increase in studies and awareness of this area led to gains in understanding and refinement.[14,63-66] Isolated hepatocytes are regarded as the most useful *in vitro* system for predictive studies since they contain the full complement of enzymes a compound is likely to encounter during first pass metabolism and transporter proteins, which can be key determinants of hepatic CL[62,67,68] and should therefore form the basis of IVIVE for CL. In simple terms the system uses an estimate of compound turnover, scaling factors, non-specific binding terms (*in vitro* and *in vivo*) and a model in which delivery to the liver can restrict the amount of turnover *in vivo*. Perhaps the most controversial elements have been the absence or inclusion of *in vitro* and *in vivo* non-specific binding terms and the fact that the approach yields a systematic under-prediction of *in vivo* CL_{int}, for reasons not yet understood. Many articles have discussed this subject and it is now commonly agreed that incorporation of drug binding terms is necessary.[47,62–64,69–71] The prediction method for *in vivo* hepatic metabolic CL involves initially building *in vivo–in vitro* CL_{int} models for each species using test sets of drugs (acid, basic and neutral) with known *in vivo* hepatic metabolic CL values. Prediction of unbound *in vivo* CL_{int} rather than CL affords a fuller understanding of predictive accuracy[62,65,66] as there is no limit on values by blood flow. Hepatocyte CL_{int} is measured and used with the incubational binding term, fu_{inc} (fraction unbound in the incubation) to calculate unbound *in vitro* CL_{int} which is corrected, based on the number of cells used in the test, up to the whole liver unbound CL_{int} using scaling factors.[66] To establish prediction accuracy, the unbound *in vivo* CL_{int} of the reference compounds is calculated from *in vivo* hepatic metabolic CL (using total CL and subtracting non-hepatic metabolic CL values) adjusted for *in vivo* blood binding (using ppb and blood to plasma (B:P) ratios). The derived unbound *in vitro* and *in vivo* CL_{int} values form a line of correlation from which future predictions of unbound *in vivo* CL_{int} values for NCEs can be made once the *in vitro* CL_{int} is determined.[62,63,65,69] If the unbound *in vivo* CL_{int} has been accurately predicted to within two-fold in two preclinical species (typically rat and dog), the approach can be used in human with a degree of confidence that a similar relationship may apply. Figure 8.6 and the regression equation in the legend demonstrate how this is done, using all the *in vitro* measured terms on the X-axis and then reading off predicted in total (bound) *in vivo* CL_{int} from the Y-axis. This *in vivo* CL_{int} is then put back through the Well Stirred Model equation ($CL_H = CL_{int} \times Q/CL_{int} + Q$) to predict the human *in vivo* hepatic metabolic clearance.

Besides elimination *via* metabolism, drugs can be directly excreted in urine or bile. Renal excretion of drugs is typically complex and may involve active secretion into the proximal tubules and/or passive filtration in the glomerulus. The high pressure and relatively large pores in the glomerulus result in almost all the free drug in plasma entering the proximal tubule and beginning the first stage of renal elimination. Mathematically, the filtration rate, as a CL, is the product of the fraction unbound in plasma and glomerular filtration rate (GFR),[1] and is readily

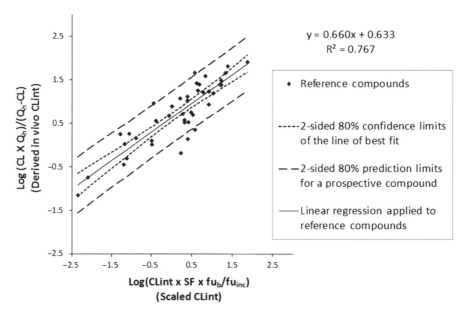

Figure 8.6   Human hepatocyte regression line comparing *in vitro* CL_{int} with *in vivo* CL_{int} with all *in vitro* variables grouped together on x-axis.
Figure adapted from Sohlenius-Sternbeck *et al.*, 2012.[63] Reproduced with permission of Informa Healthcare.

estimated if fu_p is known. However, as human GFR is only approximately 1.7 mL/min/kg and the majority of drugs are highly bound to plasma proteins, CL by this mechanism is relatively low, particularly relative to hepatic CL. Moreover, the physiology of the kidney tubules favours passive re-absorption into the blood of molecules of sufficient lipophilicity to readily permeate cells. As a consequence, passive renal CL is highly correlated with lipophilicity, such that only compounds with negative $logD_{7.4}$ values are passively renally cleared to any significant extent.[72] Uptake transporter proteins, designed to salvage important endogenous compounds, may also recover drugs, if sufficient structural similarities exist.

Active renal secretion can be considered a two-step process consisting of uptake across the basolateral membrane of the proximal tubule followed by exit across the apical membrane. Different sets of transporters polarised to either the apical or basolateral membrane are involved: In man the organic anion transporters OAT1 and OAT3 and the organic cation transporter OCT2 are the predominant transporters in the basolateral membrane whilst the apical step can involve MDR1 (P-glycoprotein), multidrug resistance protein 2 (MRP2), MRP4 or breast cancer resistance protein (BCRP) along with organic cation transporters including OCTN1, OCTN2 and MATE-1 and organic anion transporters such as OAT4 or URAT1.[73] Intuitively, estimating active human renal CL *via* transporters should be less predictable than when just passive processes are involved. However, an accurate prediction method using dog renal CL corrected for ppb and kidney blood flow species differences has emerged for a diverse set of 36 actively secreted drugs.[74] Male rat renal CL correlates less well with human possibly due to poor species cross-over of OAT substrates or male/female differences for rat organic anion transporting polypeptide (OATP) substrates.[74–76]

Biliary excretion of drugs is also commonly a two-step process involving the hepatic uptake transporters OATP, OAT or OCT and bile canalicular efflux transporters BCRP, MRP-1, P-glycoprotein or MRP2.[75,76] Biliary excretion can be an important route of elimination, but there has not been a wealth of literature on the subject of predicting human biliary CL,[77,78] perhaps because of the scarcity of relevant clinical data.[64] A variety of inter-species allometric scaling approaches have been assessed,[78–80] but given the low number of drugs used in the analyses, the fact that some of the examples used "all drug related material" rather than just parent drug[81] and that allometry under-predicts human biliary CL for some drugs but not others,[82,83] a more extensive analysis has been required. Morris and co-workers recently demonstrated that from a database of eighteen drugs with known rat and human biliary clearances, that when unbound CL is considered, simple allometry using an exponent of 0.66 gave the best predictions. However, for only about two-thirds of the compounds did the human predictions fall within three-fold of those observed, and in agreement with previous studies, some drugs were shown to have human biliary CL over-estimated by one to two orders of magnitude. Multiple species allometry using biliary CL data corrected for ppb yielded much improved predictions.[84]

High biliary CL is most often associated with acidic compounds, probably due to the effective synergistic actions of OATP and MRP transporters. Inability to de-risk human predictions of biliary CL can effectively exclude a major area of chemistry from exploitation, and therefore this area remains an important challenge. In the authors' laboratory, a comprehensive analysis of 22 drugs of all charge types and several different therapeutic classes has been compiled in order to compare rat and human biliary CL data.[85] For 86% of the drugs, rat unbound biliary CL values, when normalised for body weight, exceeded those for man by factors ranging from nine- to over 2500-fold. Hepatic uptake and efflux transporter involvement was defined for many of the drugs and the findings suggested that, regardless of the biliary efflux transporters implicated, when drugs do not require active hepatic uptake to access the liver, the differences in rat, dog and human biliary CL may be insignificant. Conversely, when the organic anion-transporting polypeptide drug transporters are involved, one may expect at least a ten-fold underestimate of human biliary CL from the rat. Perhaps such observations are not surprising given that biliary CL is defined, in these studies, as the amount of drug in the bile relative to the plasma AUC.

Consequently, removal from plasma is a significant factor in biliary excretion[86] and functionally the rat hepatic uptake transporters appear more efficient than their human and dog counterparts.[67,87] A recent study of 123 NCEs showing significant overlap in physicochemical space between rat biliary excretion data and human OATP/rat Oatp substrate definition appears to support this hypothesis[88] but more studies are warranted in this space.

Extra-hepatic metabolic CL can also be important. A starting point for consideration of such should include a combination of *in cerebro* identification of metabolically labile functional groups and subsequent experimental determination of the major metabolites and the enzymes responsible in order to elucidate potential mechanisms *via*, for example, amidases/esterases, amine oxidases, and transferases. Subsequent experiments using different subcellular fractions such as plasma, cytosol and non-hepatic microsomes can be illuminatory. Plasma hydrolysis of drugs can be scaled to whole body CL from *in vitro* data by multiplying the measured *in vitro* elimination rate constant $(\ln(2)/T_{1/2})$ by the volume of plasma. Unless the rate is rapid, drug hydrolysis in the plasma alone is unlikely to be a major CL pathway. For example, an *in vitro* measured plasma hydrolysis $T_{1/2}$ of 5 min results in a rat and human CL of approximately 7 mL/min/kg since blood volumes are approximately 50 mL/kg. More important perhaps is the indication that hydrolysis can occur at a number of sites throughout the body and the cumulative CL can thus be both challenging to predict and high. Moreover, if a drug is unstable in plasma, instability post sampling from the *in vivo* PK experiment or in the prepared analytical standards increases the risk of inaccurate measurements.

Because of the large numbers of measured input parameters for human CL predictions (hepatocyte CL_{int}, extra hepatic CL_{int}, ppb, incubational binding, B:P ratio, renal and biliary CL in rat and dog), CL predictions are open to more uncertainty than absorption or V_{ss}. Nevertheless, the strategy presented provides an effective set of experiments to facilitate drug optimisation and reduces the risk of an incorrect prediction of human CL as far as currently possible. In summary, total human CL can be predicted as follows: hepatic metabolic CL from IVIVE involving human hepatocyte CL_{int} determination (Figure 8.6), renal CL predicted from dog renal CL and biliary CL predicted from rat biliary CL.

HINTS AND TIPS: CLEARANCE (CL)

CL is the parameter that best describes the efficiency of an eliminating process

- CL = Dose/AUC.
- CL relates the rate of elimination (ng/min) to the substrate concentration (ng/mL) *i.e.* CL = Rate of elimination/[Substrate].
- CL has units of flow, and can be related to physiological parameters such as organ blood flows and glomerular filtration rate (GFR). A CL:organ blood flow ratio of 1:2 means 50% of drug is extracted in a passage through that organ.
- Clearances are additive, if the processes are parallel.
- As human GFR is only ~ 1.7 mL/min/kg and as the majority of drugs are highly bound to plasma proteins, CL *via* renal elimination is typically low.
- Biliary excretion can be an important route of elimination and remains challenging to predict to man.
- CL is influenced by plasma protein binding. Relating the rate of removal to the concentration of unbound drug in plasma, gives 'intrinsic clearance' (see Hints and Tips: Intrinsic Clearance (CL_{int}))
- CL concepts are used to rationalise bioavailability (F) and observed $T_{1/2}$, and to enable predictions from *in vitro* data.

HINTS AND TIPS: INTRINSIC CLEARANCE (CL$_{INT}$)

- Assessment of predictive accuracy should use *in vivo*/*in vitro* CL$_{int}$ (linear relationship), not predicted/observed CL because of the hyperbolic function linking CL to CL$_{int}$ and liver blood flow.
- Confidence in predicting human hepatic metabolic *in vivo* CL$_{int}$ is increased if *in vivo*/*in vitro* CL$_{int}$ is <2 for rat and dog.
- Hepatocytes are the optimal *in vitro* system for predicting *in vivo* CL, as they contain the full complement of drug metabolising enzymes. The requirement of drugs to cross membranes through passive or active transport means that sub-cellular fractions such as human liver microsomes (HLM) are removed from the *in vivo* situation.
- HLM CL$_{int}$ defines the oxidative metabolic liability and has a larger dynamic range than human hepatocyte CL$_{int}$. Comparing HLM and human hepatocyte CL$_{int}$ values is valuable in increasing mechanistic understanding.
- For poor preclinical predictions of metabolic CL from *in vitro* data, investigate the following variables:
  - *In vitro* CL$_{int}$, incubational binding, ppb and B:P ratio.
  - Hepatocyte CL$_{int}$ incubations are artificial so consider presence and concentration of organic solvent; compound concentration *in vitro* and *in vivo*—if there is a difference, is it important? Does the assay allow for active transport? Incubation conditions including pH and oxygenation and impact on enzyme viability/rate of metabolism should be considered.
- Table 8.1 lists the standard physiological parameters and *in vitro* scaling factors for mouse, rat, dog and human.[2]

Table 8.1   Table of standard physiological and *in vitro* scaling factors.[2]

Parameter	Mouse	Rat	Dog	Human
Q$_h$ (mL/min/kg)	152	72	55	20
Liver weight (g/kg)	60	40	32	24
Standard body weight (kg)	0.025	0.25	12	70
Microsomal protein yield (mg/g liver)	45	61	55	40
Hepatocellularity (10^6 cells/g liver)	125	163	169	120

Q$_h$: hepatic blood flow.

8.4   STRATEGIC USE OF PK PARAMETERS

As outlined above, it is useful to focus optimisation on the three key PK parameters that underpin a PK profile CL, V$_{ss}$ and F$_{abs}$, only adding complexity as necessary. Inconsistencies between experimental data and predictions based on physicochemical properties, or between *in vitro* and *in vivo* systems, are the primary drivers for embarking on more detailed studies at an early stage. CL and V$_{ss}$ have the biggest impact on dose and C$_{max}$:C$_{min}$ ratios and because of the logarithmic relationship between trough concentration and T$_{1/2}$. A basic tenet of clinical PK is that the magnitude of both the desired response and toxicity are functions of the drug concentration. Accordingly, for a relevant drug target, therapeutic failure results when either the concentration is too low, giving ineffective therapy, or when the concentration is too high, producing unacceptable off-target effects. Between these limits lies a concentration range associated with therapeutic success—the 'therapeutic window'. This fluctuation in drug concentration depends both on frequency of dosing and the $T_{1/2}$.[50] Unless presented with evidence to the contrary, a default assumption maybe that target plasma (unbound) concentration at steady state is directly related to efficacy and PK elimination $T_{1/2}$ is the effective $T_{1/2}$. Time to attain required effect can influence this assumption as can the effective $T_{1/2}$ not being the elimination $T_{1/2}$, *e.g.* extended PK profile driven by biliary re-circulation.

For QD dosing, it is prudent to target a predicted human $T_{1/2}$ of between 16–20 h, if drug effect is related to trough concentration, since being two-fold out in either direction (*e.g.* 16 h turns out to be 8 h or 32 h) elevates the dose and C_{max} by only a factor of 2 or results in an accumulation from single dose to steady state of only 4-fold. On the other hand to be wrong by a factor of 2 in the prediction of a $T_{1/2}$ of only 8 h (*e.g.* 8 h to 4 h) for a QD drug would lead to a dose and C_{max} elevation of eight-fold. Although a somewhat empirical assessment, it is a constructive way to view the risks. Of course one may take the view that appropriate compounds are potentially being discarded and that drug projects are asked to chase excessive goals. Our experience is that predictions are commonly too optimistic, due to biased assessments of uncertainty and risk.[89] Projects can attempt to de-risk this, by allowing for a margin of error, such that projects are viable even when what were genuinely perceived as the most likely outcomes, turn out to be overly optimistic. Even then it should be noted that the predictive methodologies outlined in this review do not offer a panacea for successful PK prediction (subjectively characterised by correctly estimating a parameter within two-fold), since successful predictions of individual PK parameters may not equate to a successful prediction of clinical dose and exposure due to the cumulative impact of two-fold errors in CL and V_{ss} predictions on $T_{1/2}$, exposure and dose.[4]

HINTS AND TIPS: PREDICTION OF PK AND DOSE

- Human hepatic metabolic CL can be predicted from human hepatocyte CL_{int} determination, human renal CL predicted from dog renal CL and human biliary CL from rat biliary CL when drugs do not require active hepatic uptake to access the liver. Prediction of biliary CL remains an important challenge.
- Investigation of poor IVIVE of CL should begin with the *in vivo* PK profile itself.
- Once satisfied with the quality and relevance of existing data, investigation of the *in vitro* measured terms should follow: B:P, ppb, CL_{int} and fu_{inc}.
- Publications from Pfizer, J&J and GSK[90–92] confirm that allometric scaling approaches offer fairly poor predictivity of human clearance. However, they may offer a 'sense-check' on human CL and $T_{1/2}$ predictions made from more a mechanistic approach, *e.g.* Human elimination $T_{1/2}$ often approximates to $4\times$ rat $T_{1/2}$ and $2\times$ dog $T_{1/2}$.
- For once daily (QD) dosing, a human $T_{1/2}$ prediction of between 16–20 h is appropriate, since being two-fold out in either direction (*e.g.* 16 h to 8 h or 32 h) elevates the dose and C_{max} only by two-fold.

8.4.1 Acidic Compounds

Understanding how PK parameters influence dose, allows differentiation between chemical series, that due to an undesirable combination of PK properties, are likely to be challenging to optimise, and those that have a favourable property profile. For example, an awareness of the likely limited range of V_{ss} values for a given chemotype (Figure 8.5) allows an approach for obtaining acceptable $T_{1/2}$ in man through an *a priori* assessment of the risks associated with achieving an acceptable CL in that chemotype. For example, if optimizing acidic compounds, the upper-limit for V_{ss} is likely to be 0.3 L/kg, without uptake transporter involvement (Figure 8.5). Using the equation elimination $T_{1/2} = \ln 2 \times V_{ss}/CL$, it is clear that a CL value of 0.15 mL/min/kg must be obtained if a 16 h $T_{1/2}$ is to be achieved (assuming a rapid enough absorption that the elimination $T_{1/2}$ is dependent on CL and V_{ss}). Using AstraZeneca prediction methods[63] and assuming that the acidic drug in question is 99% bound to plasma proteins and has a B:P ratio of 0.6, a human hepatocyte CL_{int} of 1 µL/min/million cells would not only be required, but would need to be accurately determined. Indeed, if the ppb was only 90%, a human hepatocyte CL_{int} of 0.1 µL/min/million cells would be required which would be challenging experimentally to determine with confidence. A postulated strategy for optimizing acidic drugs is to manipulate the CL of such compounds through increasing ppb, provided that free blood levels can be maintained to provide efficacy at the target receptor.[62] Typically ppb is not a suitable parameter for optimisation in drug discovery[51] but this example provides a good illustration that a quantitative understanding of the interplay between the different PK and PD parameters can facilitate the right strategy to identify an acceptable clinical candidate within a defined area of chemical space. Historical approaches may have involved a little more serendipity, albeit guided by *in vivo* pharmacology results, but the end-point is the same.

Only 3% of 60 marketed oral acidic drugs have a ppb of less than 99% with $T_{1/2}$ of more than 8 h.[93,94] Reservations over using modulation of ppb as a design strategy within drug discovery have been justifiably raised since free drug concentration is not influenced by ppb (See Hints and Tips: Plasma Protein Binding).[51] If the aim of a drug project is to occupy a set fraction of receptors/enzymes/ion channels *etc.* to achieve a required efficacy, targeting a minimum free drug concentration (C_{min}) at the end of the dosing interval is key. The period prior to the drug concentration reaching that value can give a greater occupancy for a much less than linear increase in efficacy. In this very common scenario, for acidic drugs with distribution volumes *fixed* at a lower limiting value, attenuation of the $T_{1/2}$ through increasing ppb (impacting on the CL but not V_{ss}) lowers the dose for the required C_{min}. An example of how such a strategy was used in an AstraZeneca project is detailed in Case Examples (Section 8.5.3).

8.4.2 Neutral and Basic Compounds

It is worth stressing, that in most cases, ppb is not a suitable parameter for optimisation in drug discovery since changing ppb, in isolation of any other change, will equally alter the V_d and CL. Thus $T_{1/2}$ will remain unchanged except in the specific case of lower V_d limited drugs described above. A common misconception is that decreasing ppb should result in lower dose since lower total drug concentrations can achieve the same free concentration required for efficacy. However, this overlooks the fact that increasing the free fraction also results in higher CL. Thus for most drugs, decreasing ppb has no effect on the free drug concentration, because free drug concentrations are governed by CL_{int}. This is elegantly and comprehensively outlined by Smith *et al.*[51] This article stresses the focus should be on changing fundamental physiochemical parameters such as logP, pKa and PSA to effect a change in CL and absorption to modulate free drug concentration.

HINTS AND TIPS: PLASMA PROTEIN BINDING

- Read Smith *et al.*'s 2010 paper[51] which outlines common misconceptions about the relationships between ppb and *in vivo* efficacy and provides guidance on the use of ppb in drug discovery.
- *In vivo* efficacy is determined by *unbound drug concentration* at the target, which is modulated primarily by CL_{int}, *not ppb*.

- Decreasing ppb has no effect on the *in vivo* unbound drug concentration, AUC_u or $T_{1/2}$ for most compounds.
- Thus ppb is typically not a suitable parameter for optimisation in drug discovery.
- Acidic drugs are typically highly bound to albumin with low V_{ss} (~ 0.1 L/kg) which are in common with the distribution of albumin. *For these compounds*, plasma $T_{1/2}$ may be lengthened by increasing ppb *via* attenuating CL but not V_{ss} (see case example Section 8.5.3).
- Focus on physiochemical parameters such as logP, pKa and PSA to effect a change in CL and absorption, in order to optimise free drug concentration.

Neutral drugs have moderate V_{ss} values and as with acidic drugs above, a simple assessment indicates that if V_{ss} is 1 L/kg and a $T_{1/2}$ of 20 h is required, CL must be 0.55 mL/min/kg. If ppb is 90% (typical for a neutral drug of moderate lipophilicity), the required human hepatocyte CL_{int} is 1 µL/min/million cells. An appropriate strategy to optimise free drug concentration and $T_{1/2}$ of neutrals may be to lower lipophilicity and *via* metabolic blocking control unbound CL_{int}. A similar strategy can be considered for basic drugs, although the fact that bases have high V_{ss} can be used as an advantage if other issues such as safety considerations allow, raising the pKa to facilitate increased $T_{1/2}$ through higher V_{ss}. An AstraZeneca project highlighting some of the issues associated with optimizing basic drugs is detailed in Case Examples (Section 8.5.4).

Table 8.2 describes a typical screening cascade for oral projects. Not described are *in silico* approaches to guide both design and screening strategies. All synthesised project compounds are screened in assays synchronous with the generation of primary pharmacology, selectivity and safety data. Based on predefined criteria for each subsequent assay set, compounds are progressed into subsequent DMPK, efficacy and safety assays/models towards a goal of shortlisting candidate drugs for detailed profiling. From primary screens (including target potency data) an early prediction of human PK and dose can be made from just five DMPK assays and by the end of the second round of screening more confidence is gained on the validity of the predictions.

8.5 CASE EXAMPLES

In this section we document several select case examples of using the Dose to Man algorithms for both candidate and marketed drugs studied in our laboratory or using data from literature sources. Here we show the usefulness, but also the limitations of this approach.

8.5.1 H₁ Receptor Antagonists

Second-generation H_1 receptor antagonists were developed to provide effective oral treatment of conditions such as allergic rhinitis but without the CNS related sedative effects of the first generation, and with a longer duration of action. The most commonly used second generation antagonists include desloratadine, fexofenadine, cetirizine and levocetirizine. It is illuminatory to compare and contrast the pharmacological and PK properties of desloratadine and fexofenadine and to retrospectively 'predict' the therapeutic concentrations and dose given their *in vitro* potencies and PK properties (Figure 8.7).

Desloratadine, which is the major active metabolite of loratadine, is an antagonist of the H_1 receptor with a measured pK_i of 8.9. Fexofenadine, the active metabolite of terfenadine, has an approximately ten-fold lower potency, pK_i 7.8.[4] Indeed, of the second generation antihistamines, desloratadine has the greatest binding affinity for the H_1 receptor.

The standard clinical daily dose of desloratadine is 5 mg once-daily whilst fexofenadine is 120 mg, 2 × 60 mg tablets taken 12 h apart. Significant symptom reduction is observed 1–3 h

Table 8.2 A suggested generic DMPK screening cascade for oral projects.

Screening set	Assays/Models	Decisions/comments
1	LogD$_{7.4}$	Is the compound in the correct property space with respect to CL$_{int}$, ppb, permeability, CYP inhibition?
	Solubility	With measured *in vitro* Papp (or predicted once correlation of observed data with relevant physicochemical properties established) human absorption can be estimated
	Rat hepatocyte CL$_{int}$	Rat hepatic metabolic clearance can be predicted with assumed B:P and ppb[a]
	Human liver microsomal (HLM) CL$_{int}$	Initial estimate of human hepatic metabolic clearance with ppb[b]
	Human plasma protein binding (ppb)	Allows prediction of human clearance and calculation of relevant potency to base human dose prediction on.
2	Human hepatocyte CL$_{int}$	Estimate of human hepatic metabolic clearance with assumed or measured B:P and ppb
	Ppb (rat)	See comments in Table notes[a]
	Caco-2 AB Papp (pH 6.5/7.4)	For predicting human absorption[c]
	Rat PK	Can *in vivo* CL$_{int}$ be predicted accurately to within two-fold—good *in vitro–in vivo* extrapolation (IVIVE)?[d]
3	Dog PK IV/PO with urine collection	Confirm two species IVIVE, human absorption and renal CL prediction
	Reversible CYP inhibition	Drug–Drug Interaction (DDI) prediction[e]
	Time dependent CYP inhibition assay	DDI prediction
4	Biliary clearance assessment	Considered as candidate drug profiling assays
	DDI assays (enzyme identification, drug transporter inhibition assays),	
	Caco-2 AB/BA efflux and drill down	
	Reactive metabolite assays (*e.g.* cyanide and reduced glutathione trapping following metabolic activation)	

[a]For a streamlined screening approach, assaying only for human ppb in primary screens is recommended. Rat ppb is often predictable to a degree of accuracy (particularly when one knows the value of human ppb) that allows acceptable decisions on whether to go to rat PK experiment or not in the second set of screens—rat CL predictions can be made using estimations of rat ppb from human ppb or logD based predictions and when rat ppb is measured an accurate understanding of the predictivity of rat CL is gained.

[b]HLM CL$_{int}$ has a larger dynamic range than human hepatocyte CL$_{int}$ due to the scaling factors (*i.e.* 120 million hepatocytes/g liver compared to 45 mg microsomal protein/g liver) and offers an understanding of the oxidative metabolism liability for the compound in the absence of complications such as membrane permeability and active transport considerations associated with hepatocytes. In the authors' experience, obtaining HLM and human hepatocyte CL$_{int}$ values has on many occasions proved valuable in identifying issues with the latter that require investigation. Obviously if phase II routes of metabolism (directly on the parent molecule) are dominant, the use of HLM is obviated.

[c]Early in a project it may be necessary to obtain this data in primary screening, but since apical to basolateral Caco-2 permeability data is highly predictable from such parameters as logD, it can quickly be relegated down the screening cascade. It is not necessary to generate Caco-2 P$_{app}$ efflux ratio data early in the screening cascade unless there is a concern over lack of penetration to the central nervous system. However, when apical to basolateral permeability values give concern over predicted extent of human absorption, it may be useful to obtain apical to basolateral permeability data at higher compound concentrations than the typical standard screening concentration of 10 μM.

[d]Early on in projects it is necessary to obtain IV rat CL data to understand if there is a good IVIVE. Oral rat absorption data is also useful early to understand if there will be genuine concerns over rat F limiting safety or PD studies. However, once IVIVE understanding is established, obtaining rat IV data can be relegated down the screening cascade and oral data need only be generated in planning for PD studies.

[e]For a given chemical series an understanding of how CYP inhibition relates to logD$_{7.4}$ can be gained early and therefore testing is appropriate once there is an understanding of whether the compound is likely to have acceptable human PK properties.

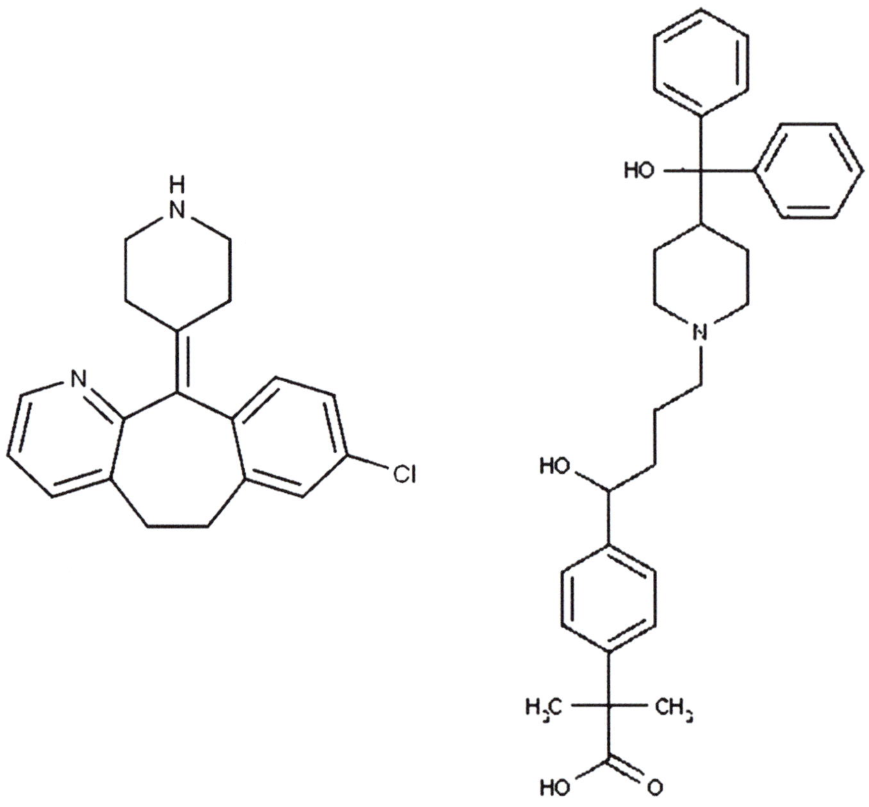

Desloratadine   **Fexofenadine**

Figure 8.7 Chemical Structure of H1 antagonists desloratadine and fexofenadine.

following dosing of both desloratadine and fexofenadine and as do they not readily cross the blood–brain barrier[95] the incidence of drowsiness is similar to placebo. Using the pKi of 8.9 and fraction unbound in plasma of 0.4, the predicted minimum effective concentration (MEC) based on 50–75% receptor occupancy (1–3 × pKi) is between 1–3 ng/mL, which is comparable to the reported clinical MEC of ~2 ng /mL. Using the Dose to Man approach *via* Equation 8.1, as outlined above, using an MEC range of 1–3 ng/mL and estimates of human PK parameters (V_d/F of 49 L/kg, $T_{1/2}$ of 27 h, F ~ 30%) the predicted clinical daily dose would be between 3–9 mg and this compares well to the actual clinical dose of 5 mg. For fexofenadine, using the pKi of 7.8 and fraction unbound of 0.4 the predicted MEC (again based on 1–3 × pK_i) is 22–65 ng/mL which again well approximates the clinic value of ~30 ng /mL. A MEC of 22–65 ng/mL and an estimate of human PK (V_d of 5.6 L/kg, $T_{1/2}$ of 14 h, F of 33%, assume B:P = 1) results in an estimated clinical daily dose of between 43–125 mg *vs.* the actual dose of 120 mg.

Thus, for these two GPCR inhibitors, measurements of therapeutic concentrations translate very well to clinical efficacy *via* good estimations of receptor occupancy. Also, despite the human PK parameters for both compounds being estimates due to the lack of derived parameters from an IV dose, 'successful' dose prediction outcomes such as these exemplify the semi-quantitative use of this approach. Moreover, albeit from a relatively simplistic view, a key differentiator between desloratadine and fexofenadine, which otherwise show similar PK properties (high V_d, moderate-high CL, long $T_{1/2}$, moderate F) is the ten-fold greater potency of desloratadine which results in a concomitant (24-fold) lower therapeutic dose. This quantitative comparison of antihistamines supports the common understanding that optimising potency against a relevant target protein, all other parameters being equal lowers the dose proportionally. In the last 10–15 years there has been a significant rise in the early use of DMPK and Safety data to reduce attrition in the Development phase. Whilst these have, by and large, been driven by increased knowledge of key risk factors, clinical observations and regulatory requirements, they create additional hurdles for medicinal chemistry efforts and potentially lead to a plethora of parameters to optimise. All optimisable parameters in drug discovery are not equal. A primary focus on optimising potency against the relevant target whilst maintaining drug like physico-chemical properties, which maximise the chances of an *acceptable* PK and safety profile is, in the authors' view, a strategy which maximises the likelihood of success of a candidate drug becoming a successful therapy. In our view, the Dose to Man paradigm outlined here facilitates, in a quantitative manner, the optimisation strategy of design teams.

8.5.2 Brilinta/Brilique (Ticagrelor)

Brilinta/Brilique (Ticagrelor) is an oral antiplatelet treatment for acute coronary syndromes in a new chemical class called cyclopentyltriazolopyrimidines. Ticagrelor is the first direct acting, reversibly-binding oral adenosine diphosphate receptor antagonist of the purigenic receptor $P2Y_{12}$ thus preventing blood clots and reducing risk of atherothrombotic events and cardio-vascular related morbidity. It was discovered at AstraZeneca Charnwood in the authors' laboratories in the late 1990's and represents one of the first examples in our company of modern drug discovery with Chemistry, Pharmacology, Safety and DMPK combining effectively. As such it is a good case example to examine the utility of the Dose to Man approach (Figure 8.8).

Ticagrelor is an antagonist of human $P2Y_{12}$ receptor with a measured pKi of 8.7[96] and demonstrates a rapid onset of pharmacological effect as demonstrated by inhibition of platelet aggregation of >70% (for 90% of patients) by 2 h post loading dose of 180 mg.[97] The clinical oral maintenance dose is 90 mg twice daily. Preclinical data from the dog arterial thrombosis model suggests a high level of platelet inhibition is required to induce an antithrombotic effect[98] and therefore for this target a receptor occupancy of $10 \times K_i$ (equivalent to pIC_{90}) may be an appropriate minimum value for clinical benefit. A pIC_{90} value of 6.4 against platelet aggregation was determined for ticagrelor in human whole blood (AstraZeneca data on file) and assuming a minimum of 90% receptor occupancy is required for clinical benefit, this equates to a predicted MEC of 208 ng/mL. Use of human whole blood assays can be useful in estimating

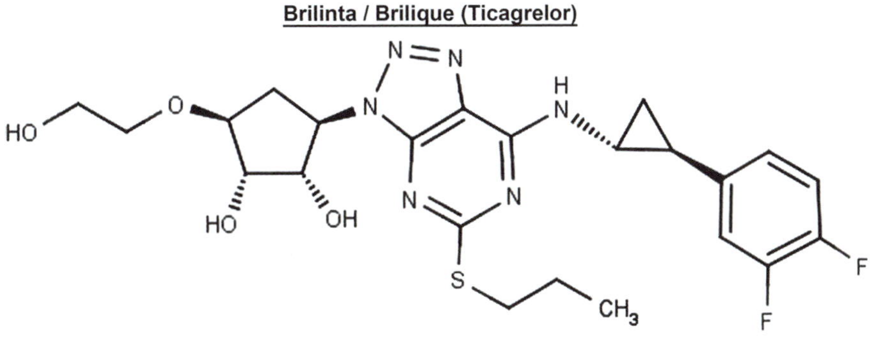

Brilinta / Brilique (Ticagrelor)

Figure 8.8 Chemical Structure of Brilinta/Brilique (Ticagrelor).

MECs especially for compounds such as ticagrelor which has a very high measured ppb (>99%). Following multiple dosing at 90 mg twice daily the clinically determined MEC was ~150 ng/mL.[99] Using the Dose to Man approach *via* Equation 8.1, as outlined above, and using the *a priori* predicted MEC of 208 ng/mL and estimates of human PK parameters (V_d 1.3 L/kg, $T_{1/2}$ 7 h, F 36%) the predicted clinical daily dose would be 230 mg and this compares to the actual clinical dose of 180 mg. However it should be noted that ticagrelor has an active metabolite AR-C124910XX with similar potency *vs.* the P2Y$_{12}$ receptor. The systemic exposure to the active metabolite is approximately 30–40% of that observed for ticagrelor and it is thought to contribute to the antiplatelet effect.[97,100] This may explain in part why the clinical dose is lower than that predicted based on 90% receptor occupancy of parent alone. Indeed this case exemplifies the utility but also the limitation of the Dose to Man approach. Establishing PK–PD relationships in preclinical species and importantly anticipating their translation to the clinic (including the presence and quantification of active metabolites) should be a key goal of the preclinical phase, to provide a key element of assessment of likely clinical dose. The approaches outlined in this chapter do not offer a panacea for successful prediction of dose of candidate drugs. However quantitative approaches such as these, applied consistently across different candidates and a portfolio facilitates a comparative evaluation of compound quality and facilitates design and selection of those candidates most likely to become successful therapies.

8.5.3 Acidic Compounds

An AstraZeneca drug discovery project with a chemical series containing an acidic moiety observed that rat CL after IV dosing correlated very well with rat ppb, with rat hepatocyte metabolic CL$_{int}$ (RH CL$_{int}$) having a much lesser effect (Figure 8.9A). It was also observed that rat V_{ss} had a reasonable relationship with ppb (Figure 8.9B). Thus it was evident that decreasing ppb from 99.8 to 98% bound (ten-fold increase in free fraction) could gain a maximum of four-fold in V_{ss} whereas the gain in rat clearance could be approximately ten-fold, when comparing compounds with similar RH CL$_{int}$.

The project optimisation approach was to increase acidity with the intention of increasing potency and ppb at the same time as decreasing logD$_{7.4}$ to reduce human hepatocyte metabolic CL$_{int}$ (HH CL$_{int}$). Additionally minor structural changes to the molecules were investigated as a possible way to increase ppb without a gain in lipophilicity. This strategy aimed to maintain whole blood potency despite an increase in ppb, maintain or lower hepatic metabolic intrinsic clearance and lower clearance through higher ppb.

A comparison of three leading compounds describes the approach from a PK perspective. All three had similar RH CL$_{int}$, a three-fold range in primary potency (antagonist binding against the target, pIC$_{50}$) and critically have a range of approximately 25-fold in rat ppb.

8.5.3.1 Rat In Vitro and In Vivo Profile

Compound 1: RH 15 µL/min/million cells, rPPB = 99.91% bound (0.09% free), CL = 0.23 mL/min/kg, V_{ss} = 0.1 L/kg, $T_{1/2}$ = 5.7 h

Compound 2: RH = 11 µL/min/million cells, rPPB = 99.6% bound (0.4% free), CL = 2.4 mL/min/kg; V_{ss} = 0.2 L/kg, $T_{1/2}$ = 0.9 h

Compound 3: RH = 12 µL/min/million cells, rPPB = 97.6% bound (2.4% free), CL = 4.2 mL/min/kg; V_{ss} = 0.3 L/kg, $T_{1/2}$ = 0.8 h

Compound 1 demonstrated that even in rat, a long elimination $T_{1/2}$ was attainable even with lower limiting V_{ss}. However, the compound did not have sufficient primary potency to carry the strategy,

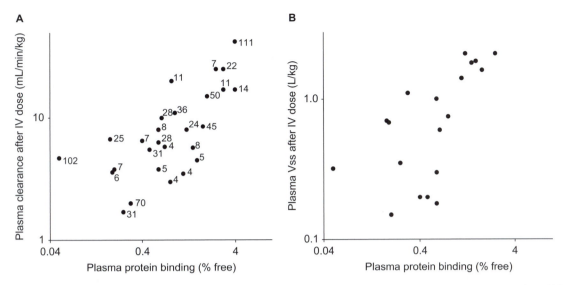

Figure 8.9   The relationship of (A) rat CL and (B) V_{ss} with ppb for an acidic chemical series. Each solid circle represents the data for a single compound in the chemically related chemical series. Numbers next to solid circles represent the determined RH CL_{int} in $\mu L/min/million$ cells.

since the potency assayed in human whole blood (hWBP) was 10 μM, almost ten-fold lower than compound 2 and the HH Cl_{int} was over three-fold higher than compounds 2 and 3.

Compound 2 and compound 3 had very similar elimination $T_{1/2}$ in the rat and in isolation these were not long enough to be confident in attaining an acceptable human $T_{1/2}$. However, compound 1 indicated that the strategy could be successful. Moreover, RH Cl_{int} values were not low and yet it was still possible to achieve a rat $T_{1/2}$ approaching 6 h. In dog PK studies, both compounds had similar ppb to rat, but the two compounds had quite different dog hepatocyte Cl_{int} (DH Cl_{int}) values. The results clearly indicated that when hepatic metabolic clearance was low and ppb high, it was possible to achieve reasonable $T_{1/2}$.

8.5.3.2   Dog In Vitro and In Vivo Profile

Compound 2: $V_{ss} = 0.15$ L/kg, CL $= 0.5$ mL/min/kg, $T_{1/2} = 4$ h, DH $= 4$ $\mu L/min/million$ cells, dPPB $= 99.5\%$ bound (0.45% free)
Compound 3: $V_{ss} = 0.21$ L/kg, CL $= 6.5$ mL/min/kg, $T_{1/2} = 0.5$ h, DH $= 11$ $\mu L/min/million$ cells, dPPB $= 98\%$ bound (2% free)

The human PK and dose predictions indicated that a combination of low HH Cl_{int} and high hPPB coupled with good potency selected compound 2 as the lead compound for the project. Although compound 3 was ten-fold more potent in whole blood (hWBP) than compound 1 and the dose predictions were similar, the longer predicted human $T_{1/2}$ for compound 1 afforded less risk resulting from variability in human PK.

8.5.3.3   Human In vitro and Predicted In vivo Profile

Compound 2: HH $= 2.6$ $\mu L/min/million$ cells, hPPB $= 99.9\%$ bound (0.1% free), predicted CL $= 0.04$ mL/min/kg, $V_{ss} = 0.1$ L/kg, $T_{1/2} = 30$ h, hWBP $= 5.8$, predicted dose $= 0.67$ mg/kg/day
Compound 3: HH $= 3.5$ $\mu L/min/million$ cells, PPB $= 99.41\%$ bound (0.59% free), predicted CL $= 0.28$ mL/min/kg, Vss $= 0.2$ L/kg, $T_{1/2} = 8$ h, hWBP $= 6.8$, predicted dose $= 0.75$ mg/kg/day

8.5.4 Basic Compounds

The following project example with basic compounds as the chemical series highlights some of the easier gains, compared to acidic drug optimisation, and some of the challenges. V_{ss} values were typically in the 2 to 5 L/kg range with compounds having one piperidine basic centre with pKa in the range of 7.5 to 8.5. However, with molecular weights in excess of 550, Caco-2 apparent permeability (P_{app}) values generally only exceeded 2 at $logD_{7.4}$ values >2. Aqueous solubility values were typically several hundred micromolar when measured at pH 7.4 but acceptable absorption in rat, dog and predicted human (based on permeability and solubility measurements) was evident only in relatively lipophilic compounds. Human liver microsomal (HLM) CL_{int} was also shown to have a positive relationship to lipophilicity. Three major sites of metabolism were identified and metabolic blocking was possible on two of these to reduce CL_{int}. However, metabolism around the pharmacophore (central piperidine ring) was not possible to control and the overall trend with lipophilicity dominated. Thus CL still was a central issue in controlling elimination $T_{1/2}$. Attempts to raise the basicity to increase V_{ss} and thus decrease reliance on CL for $T_{1/2}$ were tempered by concerns over cardiac safety issues related to hERG potency. Balancing several issues like this is of course common in drug discovery and compromises are inevitable to achieve an acceptable rather than necessarily an optimal profile. From the chemical series described, a leading compound was identified with a $logD_{7.4}$ of 2.2, pKa of 7.5, aqueous solubility at pH 7.4 of 160 µM and primary antagonist binding IC_{50} of 0.1 nM. The compound properties and human PK and dose predictions are as shown below.

HH = 7 µL/min/million cells, HLM = 18 µL/min/mg, hPPB = 69 %, Predicted human CL = 8 ml/min/Kg, Vss = 6 L/kg, $T_{1/2}$ = 9 h, F = 30%, primary binding = 0.1 nM, predicted dose = 0.1 mg/kg/day (once daily)

8.5.5 Inhaled PK

Inhaled drug therapy has been established practice over many years. Optimising drug delivery, desired efficacy, PK and associated duration of action does bring different challenges and perspectives compared to oral drug therapy. Below the fundamental principles and issues are addressed.

In contrast to oral drug delivery, an inhaled lung dose is often limited by the capacity of the delivery device and is generally low, typically <1 mg per day in man. Material is also required to have low particle size (typically 1–5 µm) to increase the chance of deposition to the deeper airways and involves generation of a dry powder or aerosol cloud to deliver this.[101,102] With human inhaled delivery devices, the fraction reaching the lung can vary but good lung targeting has been reported to be in the 30–50% range.[103] The lung is known to be naturally permeable to small molecule drugs and other therapeutic agents such as peptides and proteins, more so than the intestines.[104,105] A large proportion of the dose can be absorbed within minutes and absorption half-lives are typically minutes rather than hours, particularly when the compound is delivered in solution. Even low permeability compounds (Caco-2 P_{app} of approximately 0.1×10^{-6} cm/sec) may give absorption half-lives of only up to 1 h, indicating that reducing permeability alone may be insufficient to achieve duration consistent with once a day dosing.

For antagonist drugs, minimum duration of pharmacological activity should be at least proportionate to the PK $T_{1/2}$. Thus, following inhaled delivery, understanding the drivers for drug duration in the lung over the dosing interval is important. Lung PK profiles are typically biphasic over a 24 h period post dosing, with an initial phase primarily due to absorption through the lung and a secondary phase describing the likely effective duration of the drug. An understanding the shape of the drug concentration profile in the lung should be used to facilitate effective *in vivo* pharmacology study designs, for example by performing dose response experiments at the start of the PK phase (typically 1–4 h post dosing) likely to be associated with the duration of action. Attenuating dissolution rate of a dry powder inhaled drug can slow lung absorption enough and directly give rise to appropriately long effective $T_{1/2}$ in the lung. Indeed, poor aqueous solubility (lower than 1 µM) can lead to duration of many hours as inhaled particles dissolve slowly within the lung tissue as a prelude to absorption into the systemic circulation.

Once inhaled particles are delivered into the trachea and deeper airways, subsequent PK profiles can be influenced by a number of factors including the form of the material (salt form and particle size) and the inherent physicochemical properties of the molecule. Ion class can have a profound influence on lung retention. Mono-bases, intermediate and dibasic compounds have much longer duration than acidic or neutral drugs.[106] This ranking is perhaps not surprising given that it is well established that bases have higher general affinity for tissues which is often reflected by large V_{ss} values. Increasing basicity may therefore be considered as one strategy for optimizing lung retention and duration for inhaled drugs. Lysosomal trapping of basic drugs, as a result of the pH difference between cytosol (pH 7.2) and lysosomes (pH 5), offers a reasonable mechanistic explanation for the observations, since the lung is known to be a lysosomal rich organ.[107,108] Modulating pulmonary drug disposition through the highly predictable mechanism of distribution volume (lung half-lives tend to be well conserved across species,[107] has some possible advantages over that of relying on slow dissolution rate to drive prolonged lung retention, since this may lead to variable and unpredictable human PK particularly in the patient populations due to increased mucociliary clearance and clearance of drug due to cough and phagocytic elimination.[109] Additionally, the particulate material may lead to an increased risk of an irritant response which in turn may hamper safety testing and interpretation of pharmacological data. Another consideration worthy of note is that this approach is highly dependent on the physical characteristics of the crystal form of the compound and in early discovery programs material properties are generally poorly characterised and not optimised. Nevertheless, dissolution rate driven lung PK duration is a viable strategy when the pharmacological target cannot tolerate basic molecules. It is vital for projects to develop good PKPD relationships as this builds confidence in predicting to man as well as identifying drivers for optimisation. As with oral PKPD there are a number of challenges including understanding the impact of the target mechanism (agonist or antagonist and possible drug–receptor binding kinetics and/or downstream cascade events), the level of efficacy required and the relevance to human disease. Traditionally PKPD relationships have come from understanding the circulating blood (or plasma) concentrations since the unbound concentrations are believed to be in equilibrium with the effect site and therefore reflect the exposures that are needed to drive a pharmacological effect ('free' drug hypothesis).[51] Whilst this is likely to be true in the lung, the unbound concentration may be different than that in the systemic circulation because, as described above, lung retention mechanisms may rely on slow 'delivery' of drug into the 'effective bio-phase' of the lung tissue either by tissue distribution or slowed absorption. Not only is this local unbound drug concentration dependent on the balance between dissolution rate, mucociliary clearance, absorption into the lung tissue across the airway epithelium, absorption into the systemic circulation, affinity for lung tissues and trapping of drug in sub-cellular organelles within the lung, but it is also technically very difficult to assess, but whether *in vitro* or *in vivo* estimations are made, the key is to understanding drug concentrations at steady state. *In vitro* techniques that have been used to estimate the unbound drug concentration in organs other than the lung can also be applied to provide an assessment for inhaled drugs.[110] Of course it is possible to establish and understand PKPD relationships using total drug concentrations and preclinical animal models and these can be helpful in making human PK and dose predictions.[106]

For human dose predictions it is useful to focus on the required lung dose, rather than total inhaled dose as this avoids the complication of needing to know the efficiency of delivery from the final device used in man. However, it is necessary to correct this dose for any significant differences in animal and human potency, ideally derived from relevant cellular or *ex vivo* tissue models. Since preclinical *in vivo* pharmacological models are likely to play a significant role in human dose prediction, any differences in preclinical to clinical effective dose for drugs of the same or similar pharmacology with known human dose should also be accounted for. In addition it is of course appropriate to correct the predicted human dose for any differences in pharmacokinetics across species to ensure that the human dose will provide sufficient lung exposure over the desired dosing interval.

8.6 SUMMARY

In summary, the purpose of this chapter has been to demonstrate that based on a robust understanding of how physicochemical properties drive PK, relatively little experimental data is required establish an drug discovery optimisation strategy and gain confidence that credible clinical candidates are selected, with a high likelihood of achieving the PK goals in the clinic. A system has been described for integrating information to rank and select compounds, which as a minimum, is a sophisticated scoring function, and at best is an effective tool for assessing clinical utility. Additionally, in an age when centralised screening functions or external contract research organisations provide much routine DMPK data for Pharma, internal project facing scientists can focus on maximizing the chances of success through building predictions, knowledge and mechanistically examining potential inconsistencies, and instigating the right experiment at the right time for compounds.

KEY REFERENCES

M. V. Varma, G. Chang, Y. Lai, B. Feng, A. F. El-Kattan, J. Litchfield and T. C. Goosen, *Drug Metab. Dispos.*, 2012, **40**, 1527.

Provides key insights into the interplay between physicochemical space and optimal key DMPK properties, thus facilitating Medicinal Chemists to use the current book chapter with compound properties to maximise the chances of success.

D. A. Smith, L. Di and E. H. Kerns, *Nat. Rev. Drug. Discov.*, 2010, **9**, 929.

A very good assessment of how not understanding impact of ppb can lead you astray.

P. Ballard, P. Brassil, K. H. Bui, H. Dolgos, C. Petersson, A. Tunek and P. J. Webborn, *Drug Metab. Rev.*, 2012, **44**, 224.

Outlines tactics for applying principles in projects and the use of decision trees.

K. H. Grime, P. Barton and D. F. McGinnity, *Mol. Pharm.*, 2013, **10**, 1191.

Provides the learning from many years of Fisons, Astra and AstraZeneca Charnwood DMPK and out of that, tactics on how to effectively approach Drug Discovery DMPK.

D. A. Smith, *Curr. Top. Med. Chem.*, 2011, **11**, 467.

Gives a detailed overview of all aspects of Drug Discovery DMPK and the relationship to drug properties, covering a historical perspective, what we as an industry have learned and relevant tactics and strategies for 21st century Drug Discovery.

REFERENCES

1. M. Rowland and T. N. Tozer, *Clinical Pharmacokinetics: Concepts and Applications*, Williams and Wilkins, Philadelphia, 3rd edn, 1989.
2. A.-K. Sohlenius-Sternbeck, C. Jones, D. Ferguson, B. J. Middleton, D. Projean, E. Floby, J. Bylund and L. Afzelius, *Xenobiotica*, 2012, **42**, 841.
3. *Metabolism, Pharmacokinetics and Toxicity of Functional groups*, ed. D. A. Smith, RSC Drug Discovery Series No. 1, Cambridge, U.K., 2010.
4. D. F. McGinnity, J. Collington, R. Austin and R. Riley, *Curr. Drug. Metab.*, 2007, **8**, 463.
5. D. A. Smith, B. C. Jones and D. K. Walker, *Med. Res. Rev.*, 1996, **16**, 243.
6. J. G. Wagner, *Pharmacol. Ther.*, 1981, **12**, 537.
7. J. D. Henion and G. A. Maylin, *Biomed. Mass Spectrom.*, 1980, **7**, 115.
8. C. Lindberg, J. Paulson and A. Blomqvist, *J. Chromatogr. A.*, 1991, **554**, 215.
9. C. C. Peck, *J. Pharmacokinet. Pharmacodyn.*, 2010, **37**, 617.
10. D. D. Breimer, *Clin. Pharmacokinet.*, 1983, **8**, 371.
11. J. H. Lin, *Drug Metab. Dispos.*, 1995, **23**, 1008.
12. B. Clark and D. A. Smith, *Crit. Rev. Toxicol.*, 1984, **12**, 343.
13. M. Rowland, L. Z. Benet and G. G. Graham, *J. Pharmacokinet. Biopharm.*, 1973, **1**, 123.
14. J. B. Houston, *Biochem. Pharmacol.*, 1994, **47**, 1469.
15. R. J. Riley, D. F. McGinnity and R. P. Austin, *Drug Metab. Dispos.*, 2005, **33**, 1304.
16. T. Lavé, K. Chapman, P. Goldsmith and M. Rowland, *Exp. Opin. Drug Metab. Toxicol.*, 2009, **9**, 1039.
17. H. J. Boxenbaum, *Pharmacokinet. Biopharm.*, 1982, **10**, 201.
18. C. Hansch, *J. Med. Chem.*, 1970, **13**, 964.
19. S. Toon and M. Rowland, *J. Pharmacol. Exp. Ther.*, 1983, **225**, 752.
20. M. Láznícek, J. Kvĕtina, J. Mazák and V. Krch, *J. Pharm. Pharmacol.*, 1987, **39**, 79.
21. C. A. Lipinski, F. Lombardo, B. W. Dominy and P. J. Feeney, *Adv. Drug Delivery Rev.*, 1997, **23**, 3.
22. M. J. Humphrey and D. A. Smith, *Xenobiotica*, 1992, **22**, 743.
23. D. A. Smith, K. Beaumont, N. J. Cussans, M. J. Humphrey, S. G. Jezequel, D. J. Rance, D. A. Stopher and D. K. Walker, *Xenobiotica*, 1992, **22**, 1195.
24. D. A. Smith, B. C. Jones and D. K. Walker, *Med. Res. Rev.*, 1996, **16**, 243.
25. P. Ballard, P. Brassil, K. H. Bui, H. Dolgos, C. Petersson, A. Tunek and P. J. Webborn, *Drug Metab. Rev.*, 2012, **44**, 224.
26. E. T. Hellriegel, T. D. Bjornsson and W. W. Hauck, *Clin. Pharmacol. Ther.*, 1996, **60**, 601.
27. H. Komura and M. Iwaki, *Drug Metab. Rev.*, 2011, **43**, 476.
28. A. Galetin, M. Gertz and J. B. Houston, *Drug Metab. Pharmacokinet.*, 2010, **25**, 28.
29. K. Thelen and J. B. Dressman, *J. Pharm. Pharmacol.*, 2009, **61**, 541.
30. M. F. Paine, H. L. Hart, S. S. Ludington, R. L. Haining, A. E. Rettie and D. C. Zeldin, *Drug Metab. Dispos.*, 2006, **34**, 880.
31. Q. Y. Zhang, D. Dunbar, A. Ostrowska, S. Zeisloft, J. Yang and L. S. Kaminsky, *Drug Metab. Dispos.*, 1999, **27**, 804.
32. T. Prueksaritanont, L. M. Gorham, J. H. Hochman, L. O. Tran and K. P. Vyas, *Drug Metab. Dispos.*, 1996, **24**, 634.
33. J. Yang, M. Jamei, K. R. Yeo, G. T. Tucker and A. Rostami-Hodjegan, *Curr. Drug. Metab.*, 2007, **8**, 676.
34. M. M. Doherty and W. N. Charman, *Clin. Pharmacokinet.*, 2002, **41**, 235.
35. M. F. Paine, M. Khalighi, J. M. Fisher, D. D. Shen, K. L. Kunze, C. L. Marsh, J. D. Perkins and K. E. Thummel, *J. Pharmacol. Exp. Ther.*, 1997, **283**, 1552.
36. A. Galetin and J. B. Houston, *J. Pharmacol. Exp. Ther.*, 2006, **318**, 1220.

37. A. Galetin, *Curr. Drug. Metab.*, 2007, **8**, 643.

38. Y. Zhang and L. Z. Benet, *Clin. Pharmacokinet.*, 2001, **40**, 159.

39. *Guidance for Industry Drug Interaction Studies—Study Design, Data Analysis, Implications for Dosing and Labeling Recommendations*, U.S. Department of Health and Human Services Food and Drug Administration Center for Drug Evaluation and Research (CDER), 2012.

40. X. Lin, S. Skolnik, X. Chen and J. Wang, *Drug Metab. Dispos.*, 2011, **39**, 265.

41. T. Murakami and M. Takano, *Expert Opin. Drug Metab. Toxicol.*, 2008, **4**, 923.

42. M. B. Bolger, V. Lukacova and W. S. Woltosz, *AAPS J.*, 2009, **11**, 353.

43. I. J. Hidalgo, T. J. Raub and R. T. Borchardt, *Gastroenterology*, 1989, **96**, 736.

44. D. Sun, H. Lennernas, L. S. Welage, J. L. Barnett, C. P. Landowski, D. Foster, D. Fleisher, K. D. Lee and G. L. Amidon, *Pharmacol. Res.*, 2002, **19**, 1400.

45. M. Jamei, S. Marciniak, K. Feng, A. Barnett, G. Tucker and A. Rostami-Hodjegan, *Expert Opin. Drug Metab. Toxicol.*, 2009, **5**, 211.

46. V. K. Sinha, J. Snoeys, N. V. Osselaer, A. V. Peer, C. Mackie and D. Heald, *Biopharm. Drug Dispos.*, 2012, **33**, 111.

47. K. H. Grime, P. Barton and D. F. McGinnity, *Mol. Pharm.*, 2013, **10**, 1191.

48. J. M. DeSesso and C. F. Jacobson, *Food Chem. Toxicol.*, 2001, **39**, 209.

49. S. Øie and T. N. Tozer, *J. Pharm. Sci.*, 1979, **68**, 1203.

50. R. J. Riley, I. J. Martin and A. E. Cooper, *Curr. Drug. Metab.*, 2002, **3**, 527.

51. D. A. Smith, L. Di and E. H. Kerns, *Nat Rev Drug Discov*, 2010, **9**, 929.

52. A. C. MacIntyre and D. J. Cutler, *Biopharm. Drug Dispos.*, 1988, **9**, 513.

53. A. E. Cooper, D. Ferguson and K. Grime, *Curr. Drug. Metab.*, 2012, **13**, 457.

54. G. Berellini, C. Springer, N. J. Waters and F. Lombardo, *Med. Chem.*, 2009, **52**, 4488.

55. M. P. Gleeson, N. J. Waters, S. W. Paine and A. M. Davis, *J. Med. Chem.*, 2006, **49**, 1953.

56. P. Zou, N. Zheng, Y. Yang, L. X. Yu and D. Sun, *Exp. Opin. Drug Metab. Toxicol.*, 2012, **8**, 855.

57. R. D. Jones, H. M. Jones, M. Rowland, C. R. Gibson, J. W. Yates, J. Y. Chien, B. J. Ring, K. K. Adkison, M. S. Ku, H. He, R. Vuppugalla, P. Marathe, V. Fischer, S. Dutta, V. K. Sinha, T. Björnsson, T. Lavé and P. Poulin, *J. Pharm. Sci.*, 2011, **100**, 4074.

58. F. Lombardo, N. J. Waters, U. A. Argikar, M. K. Dennehy, J. Zhan, M. Gunduz, S. P. Harriman, G. Berellini, I. Liric Rajlic and R. S. Obach, *J. Clin. Pharmacol.*, 2013, **53**, 178.

59. I. Mahmood, *J. Pharm. Sci.*, 2010, **99**, 2927.

60. K. W. Ward and B. R. Smith, *Drug Metab. Dispos.*, 2004, **32**, 603.

61. D. B. Campbell, *Ann. NY Acad. Sci.*, 1996, **801**, 116.

62. K. Grime and R. J. Riley, *Curr. Drug Metab.m*, 2006, **7**, 251.

63. A.-K. Sohlenius-Sternbeck, C. Jones, D. Ferguson, B. J. Middleton, D. Projean, E. Floby, J. Bylund and L. Afzelius, *Xenobiotica*, 2012, **42**, 841.

64. T. Lavé, K. Chapman, P. Goldsmith and M. Rowland, *Exp. Opin. Drug Metab. Toxicol.*, 2009, **9**, 1039.

65. K. Ito and J. B. Houston, *Pharm. Res.*, 2005, **22**, 103.

66. R. S. Obach, *Curr. Top. Med. Chem.*, 2011, **11**, 334.

67. M. G. Soars, D. F. McGinnity, K. Grime and R. J. Riley, *Chem. Biol. Interac.*, 2007, **168**, 2.

68. Y. Shitara, A. P. Li, Y. Kato, C. Lu, K. Ito, T. Itoh and Y. Sugiyama, *Drug Metab. Pharmacokinet.*, 2003, **18**, 33.

69. R. J. Riley, D. F. McGinnity and R. P. Austin, *Drug Metab. Dispos.*, 2005, **33**, 1304.

70. H. C. Rawden, D. J. Carlile, A. Tindall, D. Hallifax, A. Galetin, K. Ito and J. B. Houston, *Xenobiotica*, 2005, **35**, 603.

71. Y. Naritomi, S. Terashita, A. Kagayama and Y. Sugiyama, *Drug Metab. Dispos.*, 2003, **31**, 580.

72. D. A. Smith, B. C. Jones and D. K. Walker, *Med. Res. Rev.*, 1996, **16**, 243.

73. C. D. A. Brown, R. Sayer, A. S. Windass, I. S. Haslam, M. E. De Broe, P. C. D'Haese and A. Verhulst, *Toxicol. App. Pharmacol.*, 2008, **233**, 428.

74. S. W. Paine, K. Menochet, R. Denton, D. F. McGinnity and R. J. Riley, *Drug Metab. Dispos.*, 2011, **39**, 1008.

75. H. Tahara, M. Shono, H. Kusuhara, H. Kinoshita, E. Fuse, A. Takadate, M. Otagiri and Y. Sugiyama, *Pharm. Res.*, 2005, **22**, 647.

76. Y. Kato, K. Kuge, H. Kusuhara, P. J. Meier and Y. Sugiyama, *J. Pharmacol. Exp. Ther.*, 2002, **302**, 483.

77. D. A. Smith, *Curr. Top. Med. Chem.*, 2011, **11**, 467.

78. Y. Shitara, T. Horie and Y. Sugiyama, *Eur. J. Pharm. Sci.*, 2006, **27**, 425.

79. I. Mahmood, *J. Pharm. Sci.*, 2005, **94**, 883.

80. I. Mahmood and C. Sahajwalla, *J. Pharm. Sci.*, 2002, **91**, 1908.

81. Y. Kato, K. Kuge, H. Kusuhara, P. J. Meier and Y. Sugiyama, *J. Pharmacol. Exp. Ther.*, 2002, **302**, 483.

82. J. A. Scatina, D. R. Hicks, M. Kraml and M. N. Cayen, *Xenobiotica*, 1989, **19**, 991.

83. I. Påhlman, M. Edholm, S. Kankaanranta and M. L. Odell, *Pharm. Pharmacol. Com.*, 1998, **4**, 493.

84. M. Morris, Y. Yang, Y. A. Gandhi, S. G. Bhansalia and L. J. Benincosa, *Biopharm. Drug Dispos.*, 2012, **33**, 1.

85. K. H. Grime and S. W. Paine, *Drug Metab. Dispos.*, 2013, **41**, 372.

86. M. Yamazaki, S. Akiyama, R. Nishigaki and Y. Sugiyama, *Pharm. Res.*, 1996, **13**, 1559.

87. P. Gardiner and S. W. Paine, *Drug Metab. Dispos.*, 2011, **39**, 1930.

88. M. V. Varma, G. Chang, Y. Lai, B. Feng, A. F. El-Kattan, J. Litchfield and T. C. Goosen, *Drug Metab. Dispos.*, 2012, **40**, 1527.

89. A. Tversky and D. Kahneman, *Science*, 1974, **185**, 1124.

90. K. W. Ward and B. R. I. Smith, *Drug Metab. Dispos.*, 2004, **32**, 603.

91. G. W. Caldwell, J. A. Masucci, Z. Yan and W. Hageman, *Eur. J. Drug Metab. Pharmacokinet.*, 2004, **29**, 133.

92. F. Lombardo, N. J. Waters, U. A. Argikar, M. K. Dennehy, J. Zhan, M. Gunduz, S. P. Harriman, G. Berellini, I. Liric Rajlic and R. S. Obach, *J. Clin. Pharmacol.*, 2013, **53**, 178.

93. W. Taeschner and S. Vozeh, in *Pharmacokinetic Drug Data, Clinical Pharmacokinetics. Drug Data Handbook*, ed. N. Holford, Adis International Limited, New Zealand, 3rd edn, 1998, pp. 1–48.

94. L. Z. Benet and R. L. Williams, in *The Pharmacological Basis of Therapeutics*, ed. A. G. Gilman, L. S. Goodman, T. W. Rall, A. S. Nies and P. Taylor, Macmillan Publishing Co., New York, 8th edn, 1990, pp. 1650–1735.

95. P. Devillier, N. Roche and C. Faisy, *Clin. Pharmacokinet.*, 2008, **47**, 217.

96. B. Springthorpe, A. Bailey, P. Barton, T. N. Birkinshaw, R. V. Bonnert, R. C. Brown, D. Chapman, J. Dixon, S. D. Guile, R. G. Humphries, S. F. Hunt, F. Ince, A. H. Ingall, I. P. Kirk, P. D. Leeson, P. Leff, R. J. Lewis, B. P. Martin, D. F. McGinnity, M. P. Mortimore, S. W. Paine, G. Pairaudeau, A. Patel, A. J. Rigby, R. J. Riley, B. J. Teobald, W. Tomlinson, P. J. Webborn and P. A. Willis, *Bioorg. Med. Chem. Lett.*, 2007, **17**, 6013.

97. Brilique product information sheet. www.ema.europa.eu/docs/en_GB/document_library/EPAR_-_Product_Information/human/001241/WC500100494.pdf.

98. J. J. J. van Giezen and R. G. Humphries, *Semin. Thromb. Hemost.*, 2005, **31**, 195.

99. R. Teng, P. Mitchell and K. Butler, *Eur. J. Clin. Pharmacol.*, 2012, **68**, 1175.

100. R. Teng and K. Butler, *Eur. J. Clin. Pharmacol.*, 2010, **66**, 487.

101. N. Labris and M. Dolovich, *Br. J. Clin. Pharmacol.*, 2003, **56**, 588.

102. N. Labris and M. Dolovich, *Br. J. Clin. Pharmacol.*, 2003, **56**, 600.

103. S. Rohataki, G. Rhodes and P. Chaikin, *J. Clin. Pharmacol.*, 1999, **39**, 661.

104. J. Patton, S. Fishburn and J. Weers, *Proc Am. Thorac. Soc.*, 2004, **1**, 338.

105. A. Tronde, B. Norden, H. Marchner, A. Wendel, H. Lennernäs and U. Bengtsson, *J. Pharm. Sci.*, 2003, **92**, 1216.
106. A. E. Cooper, D. Ferguson and K. Grime, *Curr. Drug. Metab.*, 2012, **13**, 457.
107. F. Boer, *Br. J. Anaesth.*, 2003, **91**, 50.
108. K. Yokogawa, J. Ishizahi, S. Ohkuma and K. Myamoto, *Methods Find. Exp. Clin., Pharmacol.*, 2002, **24**, 81.
109. B. Olsson, E. Bondesson, L. Borgström, S. Edsbäcker, S. Eirefelt, K. Ekelund, L. Gustavsson and T. Hegelund-Myrbäck, in *Controlled Pulmonary Drug Delivery*, ed. H. D. C Smyth and A. J. Hickey, New York, Springer, 2011, ch. 2, pp. 21–50.
110. M. Fridén, F. Bergström, H. Wan, M. Rehngren, G. Ahlin, M. Hammarlund-Udenaes and U. Bredberg, *Drug Metab. Dispos.*, 2011, **39**, 353.

CHAPTER 9

Molecular Biology for Medicinal Chemists

GISELLE R. WIGGIN, JAYESH C. PATEL, FIONA H. MARSHALL AND
ALI JAZAYERI*

Heptares Therapeutics Limited, BioPark, Broadwater Road, Welwyn Garden City, AL7 3AX, UK
*E-mail: ali.jazayeri@heptares.com

9.1 BRIEF HISTORY OF MOLECULAR BIOLOGY

As the name suggests, molecular biology pertains to a branch of biology that is concerned with understanding life on a molecular level. It is difficult to accurately ascertain when molecular biology was born as an independent scientific approach, but the general consensus is the early part of the twentieth century, when attempts to isolate and characterise molecules present in organisms became tractable with the advent of novel technologies. Technologies such as X-ray crystallography and electron microscopy acted as tremendous catalysts in the progression of molecular biology and allowed scientists for the first time to visualise and gain a mechanistic understanding of the organisation of the molecular world that had remained in obscurity until then. Electron microscopy was invented in 1931 by German electrical engineer Max Knoll and the physicist Ernst Ruska, while the contributions of the father and son British physicists, Henry and Lawrence Bragg, allowed solution of X-ray diffraction of crystals to determine molecular structures. The fact that the main discoverers of these early molecular biology techniques were primarily physicists is not coincidental, and is a reminder that at its origin, other disciplines such as physics and chemistry played a major contributory role. This early and organic cross-contribution was formalised in 1931 by Warren Weaver at the influential Rockefeller Foundation, who supported research to pursue the following aim:

"...to encourage the application of the whole range of scientific tools and techniques, and especially those which had been so superbly developed in the physical sciences, to the problems of living matter."

It is claimed that Weaver coined the term molecular biology, which at least symbolically ushered in an extremely fertile period of scientific discovery with huge consequences across a multitude of areas, ranging from human and animal health to agriculture whilst raising serious

The Handbook of Medicinal Chemistry: Principles and Practice
Edited by Andrew Davis and Simon E Ward
© The Royal Society of Chemistry 2015
Published by the Royal Society of Chemistry, www.rsc.org

ethical and moral questions about the role of science in society as well as the relationship between humanity and nature. In the process, molecular biology has placed what was once deemed to be the realm of science fiction firmly in the realm of science reality; Dolly the first cloned sheep being a prominent example. Apart from such sensational examples, molecular biology has had a profound effect on various aspects of drug discovery and this is the main topic of this chapter. In this section we will highlight the basics of molecular biology described through a historical perspective and go on to outline how the molecular techniques are shaping the drug discovery pipeline.

The first major landmark in the development of molecular biology was the discovery that DNA constituted the genetic material. In a set of classical experiments, Fred Griffith in 1928 provided the experimental framework for this discovery. In these experiments mice were infected with two different strains of the bacteria *Pneumococcus*: the so-called S and R strains. The S strain possesses smooth capsular glycoproteins and is fatally virulent, whereas the R strain exhibits a rough cell surface and is harmless. Griffith demonstrated that although mice survived infection with the heat-inactivated form of the virulent S strain, co-infection of this form of the S strain with the normally harmless R strain resulted in infection and death. It was thus concluded that an unknown factor present in the heat-inactivated S strain was capable of "transforming" the inert R strain.[1] In the following decades Oswald Avery, Colin MacLeod, and Maclyn McCarty followed up on these experiments and demonstrated that only isolated DNA from the S strain preserved this transformative property, in contrast to purified protein, lipid or RNA.[2] A few years later, these findings were confirmed by Alfred Hershey and Martha Chase using bacteriophage T2, a virus capable of infecting bacteria and transforming them into virus factories by attaching to and injecting their DNA inside the bacterial cells. In their experiments, Hershey and Chase labelled the bacteriophage protein and DNA with ^{35}S and ^{32}P, respectively. Following infection of bacteria, the fate of each radioisotope was carefully followed and it was shown that whilst ^{32}P was taken up by the host cells and was detected inside the infected cells, ^{35}S remained in culture medium and was separated from the infected cells. These experiments together fully established the central position of DNA as the hereditary material.[3]

Soon after the publication of the Hershey–Chase experiments one of the most ground breaking discoveries in the history of molecular biology was published in *Nature*. In 1953, James Watson and Francis Crick presented the three-dimensional structure of DNA that despite its apparent simplicity of two intertwined helical chains showed novel features that hinted at "considerable biological importance".[4] The key feature was the way that the purine and pyrimidine bases paired and held the two chains together. Based on the assumption that these nitrogen bases are found in their most common keto rather than enol configuration, then only adenine (purine) can pair with thymine (pyrimidine) and guanine (purine) can pair with cytosine (pyrimidine). This direct observation from the model of the X-ray diffraction was experimentally supported by the observation that in natural DNA of multiple organisms the ratios of guanine to cytosine and adenine to thymine were always close to unity. The implication of this specific base pairing rule was not lost on Watson and Crick; they correctly suggested that it might be a possible copying mechanism of the genetic material.[4] A few years later Matthew Meselson and Franklin Stahl showed very elegantly that DNA was indeed replicated in a semiconservative fashion; whereby each helix acts as the template for the synthesis of the new strand.[5]

Armed with these discoveries, the field of molecular biology moved towards understanding the flow of information from DNA to proteins in the next half of the twentieth century. Various experiments lead to the hypothesis that a linear sequence of nucleotides in DNA specifies the linear sequence of amino acids in proteins. Early in the second half of the twentieth century, Francois Jacob, Mathew Meselson and Sydney Brenner showed that the information encoded in DNA is initially transcribed into messenger RNA (mRNA) before being translated into amino acid sequences.[6] Furthermore, intriguing experiments by Francis Crick, Sidney Brenner, Leslie Barnett and R. J. Watts-Tobin showed that the language of DNA is coded in triplets.[7] This discovery propelled a number of scientists to decode the genetic code and reveal how the cellular machinery reads DNA and translates it into protein sequence. Chief amongst them was the chemist Har Gorbind Khorana who set out to apply chemistry to biological problems and in effect started the field of chemical biology. Khorana took a purely chemical approach to deciphering the language of DNA and in effect created the technology required to crack the DNA code. Taking advantage of recently developed cell-free protein expression systems and his method of generating synthetic polynucleotides of defined sequences, Khorana managed to decode the entire genetic code. He started off by synthesising a long DNA polymer consisting of two repeating guanine and thymine nucleotides (GT)n, which would constitute two triplets of GTG and TGT. Such a polymer was predicted to direct synthesis of two amino acids in strictly alternating sequence. The subsequent cell-free protein synthesis reaction showed that the resulting sequence led to the generation of a polypeptide with repeating valine and cysteine sequence. Expanding this strategy to generate triplet and quadruplet repeats allowed Khorana to work out the entire combination of the genetic code in a systematic way. It is worth mentioning that the initial deciphering steps were made possible by the results of Nirenberg and Leder who deciphered a number of the codes through development of a technique to detect direct binding of different aminoacyl–tRNAs (adapter transfer RNA molecules carrying different amino acids) to pre-defined triplets. By 1966 these joint efforts resulted in the complete genetic code being declared.[8,9] The final version revealed that the genetic code exhibits degeneracy, which means that most of the amino acids are coded for by more than one codon. Codon degeneracy arises as there are more codon combinations than the number of amino acids; the total possible of triplet combinations out of the four bases of DNA is 64 (4^3) that need to code for 20 common amino acids. It turns out that all 20 common amino acids are coded for by 61 codons out of the possible 64, with the remaining three encoding for stop codons that signal the termination point of protein synthesis. It is thought that the genetic redundancy is a mechanism to increase the potential of the genomes to absorb errors by increasing the chances of silent mutations (for a detailed review of how the genetic code was cracked see www.nobelprize.org/educational/medicine/gene-code/history.html).

By 1970, the big pieces of the molecular biology puzzle were in place and there was good mechanistic understanding of how cells use the genetic information as a recipe for synthesising proteins. These insights allowed the field of molecular biology to enter a new phase of technology development that resulted in the creation of a set of tools that have collectively ushered in the era of recombinant technology. One of the earliest examples of these is the discovery and characterisation of restriction enzymes. These are groups of enzymes that recognise a specific sequence of DNA and create a double-stranded cut in the DNA. The majority of the restriction enzyme recognition sites are between four to eight nucleotides long and most are palindromic. These enzymes have evolved in bacteria as a defence mechanism against viral infection; about 3800 restriction enzymes have been identified, with about 600 available commercially as purified protein preparations. These reagents were originally used as a way of generating crude molecular maps; nowadays, they are an indispensable part of the cloning kit that allow easy and rapid cloning of DNA fragments. However, the full potential of these enzymes as a molecular biology tool was not fully realised until the invention of polymerase chain reaction (PCR), arguably the most powerful and most utilised technique in molecular biology. Developed in the early 1980s by Kary Mullis,[10] this technique allows exponential amplification of a specific region of DNA (for a review see www.dnalc.org/resources/animations/pcr.html). The method relies on thermal cycling that oscillates between heating to melt the template double stranded DNA and cooling to allow DNA polymerase to copy each strand at specific positions determined by short primers (Figure 9.1). The key feature of the polymerases employed in PCRs is that they are

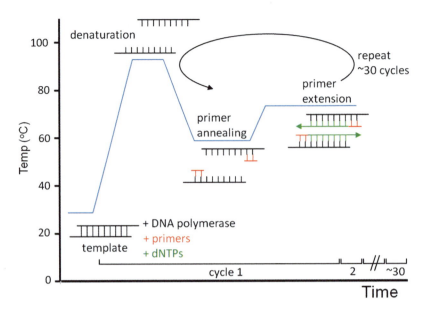

Figure 9.1   Graphical representation of the Polymerase Chain Reaction (PCR). The reaction mix contains the double-stranded template DNA, primers to direct amplification of both DNA strands, deoxynucleotides (dNTPs) and a thermostable DNA polymerase, in addition to buffer containing magnesium, which is an essential cofactor of the enzyme. There are three main steps the reaction, which is carried out in an automated thermal cycler. The first stage involves heating to approximately 95 °C to separate the DNA strands, by disrupting the hydrogen bonds between complementary bases. The temperature is then reduced to about 5 °C lower than the T_m of the individual primers, to allow the primers to anneal to the template. The reaction conditions are then increased to around 70 °C, dependant on the optimal operating temperature of the polymerase, to allow primer extension to occur, synthesising DNA strands complementary to the template. The length of time of this extension step is determined by the length of DNA to be amplified and the processivity of the polymerase, with modern enzymes being able to amplify 2–4 kilobases (thousand base pairs) of DNA in under 1 min. Under optimal conditions the amount of DNA target is doubled at each extension step, leading to exponential amplification of the specific DNA fragment. For a standard amplification reaction, this cycle is repeated between 25 and 30 times.

heat-stable, such as Taq polymerase, an enzyme originally isolated from the thermophilic bacterium *Thermus aquaticus*. Kary Mullins was awarded the Nobel Prize in Chemistry for his invention; he shared the prize with Michael Smith who invented an equally powerful technology called site-directed mutagenesis.[11] This technology allows introduction of a specific mutation in a specific location in DNA, thereby enabling investigators to study the effect of single mutation in a single gene. The final piece of technological advancement that was developed in the late 1970s was the DNA sequencing technology developed by Fredrick Sanger and colleagues that allowed determination of the nucleotide order in DNA using *in vitro* DNA synthesis reaction terminated by incorporation of dideoxynucleotides (Figure 9.2).[12] Together this slew of technological developments placed the field of molecular biology in a position to effectively and accurately manipulate the genetic material and provide detailed mechanistic answers to specific questions. This eventually led to a number of large initiatives designed to provide large scale datasets, most famously the Human Genome Project (HGP; see section 9.2.2 for further detail). Inspired by the success of the HGP and its technical legacy, the first post-genome decade has seen further large data-driven molecular biology efforts such as International HapMap Project,[13] which charted the points at which human genomes commonly differ, and the Encyclopedia of

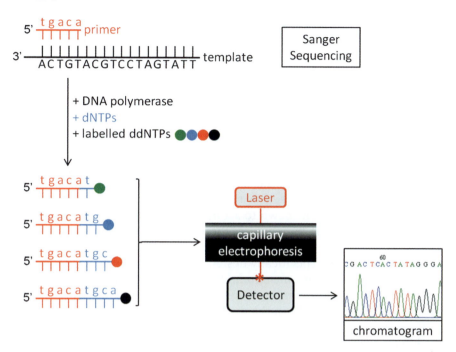

Figure 9.2 Sanger sequencing utilises a modified version of the basic PCR. Only one complementary primer is added to amplify a single strand of DNA. In addition to the usual dNTPs, fluorescently-labelled dideoxynucleotides (ddNTPs) are included at a concentration that allows one ddNTP to be incorporated per base pair of amplified DNA. The incorporation of a ddNTP causes chain-termination, as these modified nucleotides lack the 3'-hydroxyl group required for the formation of a phosphodiester bond between two nucleotides. This results in the amplification of a series of DNA fragments, each differing by one base pair in length. These fragments can be separated by capillary electrophoresis on the basis of length. The fluorescent tag on the incorporated ddNTP of each DNA fragment is detected, following activation by a laser. The sequence of DNA bases is then determined from the resulting chromatogram.

DNA Elements (ENCODE),[14] which aims to identify every functional element in the human genome. The effect of the human genome project on human health is more difficult to assess and appears to have had limited effect to date. The main reason for this is the complexity of the post-genome biology that has been brought into sharp focus by the recent publications from the ENCODE project, indicating what was previously designated as "non-coding junk DNA" appears to have functional properties in gene regulation. Moreover, single nucleotide polymorphisms associated with disease by genome-wide association studies are enriched within these non-coding, but functional regions. These results have served as a reminder that despite the revolutionary advances made in a relatively short period of time, the field of molecular biology can, and will continue, to influence human health and the process of drug discovery. In subsequent sections we will explore the wider impact of molecular biology on specific areas of drug discovery.

9.2 IMPACT OF MOLECULAR BIOLOGY ON TARGET IDENTIFICATION AND VALIDATION

9.2.1 From Disease to Gene—A Genetic Approach

The advent of molecular biology has revolutionised the path to drug discovery, changing the process from a largely trial-and-error based assessment of whether a compound can change a phenotype or outcome, to specifically designing a drug to fit a hypothesis-driven, rationally-selected target. The availability of entire genome sequences allows a target of interest to be identified, along with mutations that may cause disease, and potential homologues that may affect target specificity. The cloning and recombinant expression of a target allows for routine *in vitro* and *in vivo* screening of compounds, structural determination and downstream phenotypic analysis in cell lines or whole organisms, without having to rely on native expression and complex purifications. Techniques such as mutagenesis of targets can allow for identification of ligand binding sites or the determination of the effect of disease-causing mutations.

A comparison of an ancient drug, salicylic acid, and a modern treatment, the B-Raf kinase inhibitor vemurafenib, illustrates how the use of molecular biology techniques has changed the drug discovery process. Salicylic acid has been used to reduce pain and inflammation for thousands of years, but it has only been in the last few decades that the mode of action has begun to be elucidated. It has been shown to act by reducing the synthesis of pro-inflammatory prostaglandins,[15] at least in part by reducing the expression of the gene encoding inducible cyclooxygenase 2.[16]

By contrast, vemurafenib was specifically designed to fulfil a particular mode of action. The sequencing of DNA from malignant melanoma biopsies identified mutations within the B-Raf kinase in 80% of cases. The most common alteration is a single point mutation of an adenine instead of a thymine at position 1799 in the nucleotide sequence. This leads to the substitution of valine by glutamate at amino acid 600 (known as the V600E mutation). This causes the activation of the kinase, which stimulates the growth-promoting mitogen activated protein kinase (MAPK) pathway. Vemurafenib was identified using a structure-guided discovery approach, initiated with the screening of a fragment-based library; the first approved drug to originate from this method. Hits were co-crystallised with the wild-type and mutant kinase, and the compound optimised further to preferentially target the V600E form. Vemurafenib inhibits the activation of the MAPK pathway by mutant Raf, but not the wild-type kinase. The drug leads to tumour shrinkage and improves survival in patients with V600E-positive melanoma. It was approved by the FDA in August 2011.[17,18]

The following sections will consider molecular biology techniques used in the initial stages of the drug discovery process, from target identification to validation.

9.2.2   Human Genome Project and Beyond

Being able to map every gene in the genome has been a fundamental breakthrough in being able to identify and validate novel drug targets. The Human Genome Project was an international scientific research project with the primary aims of determining the complete sequence of the human genome, to physically and functionally map all the estimated 20 000–25 000 human genes, and to make them accessible for further study. It followed on from and utilised many technological developments made in the sequencing of the genomes of several species of bacteria, in addition to model organisms including the yeast *Saccharomyces cerevisiae*, the nematode worm *Caenorhabditis elegans*, the fruit fly *Drosophila melanogaster*, and the mouse *Mus musculus*.

The human program was initiated in October 1990 and completed in 2003. It was co-ordinated by the US Department of Energy and National Institutes of Health. The UK Wellcome Trust became a major partner, with further contributions coming from Japan, France, Germany and China, amongst others (www.ornl.gov/sci/techresources/Human_Genome/home.shtml). Alongside these governmental programs, human genome sequencing was also undertaken by private companies, such as Celera Genomics (www.celera.com/).

Working drafts of the sequence were published in 2001,[19] and by April 2003 the project was considered complete, with 99% coverage of the gene-containing regions to an accuracy of

99.99%. In order to sequence the 3 billion bases of the genome, the DNA of each chromosome was first subcloned into smaller fragments and amplified in bacteria, in the form of bacterial artificial chromosomes (BACS) or cosmids, and sequencing proceeded on a clone-by-clone basis. By contrast, Celera Genomics used a method known as "shotgun sequencing", where the genomic DNA was broken into much smaller random fragments, which were assembled post-sequencing *in silico*.[20] This shotgun method is now routinely used for genome sequencing. Sanger sequencing was used for both the public and private sequencing efforts (Figure 9.2, www.pbs.org/wgbh/nova/body/sequence-DNA-for-yourself.html; www.wiley.com/college/pratt/0471393878/student/animations/dna_sequencing/).

The massive amount of data generated from the sequencing of the human genome is a source of potentially thousands of novel drug targets.[21,22] The risk of this avalanche of information is that a slew of targets that are not properly validated will clog up development pipelines and actually lead to a decline in progress and productivity. In order to identify and validate the most promising of these requires multiple technologies and successful integration of this knowledge. Initial data mining of the genome sequence data may involve looking for genes that have similarity to known gene families implicated in disease or known drug targets. This helps limit targets to those which are "druggable" and can assist in determining target specificity. Many bioinformatic tools for analysing sequence homology are available on public databases, such as those curated by the National Centre for Biotechnology Information, including BLAST (Basic Local Alignment Search Tool), HMMER (Hidden Markov Model profiler) and SAM (Sequence Alignment Modeller). Such studies have been used to identify all members of the G-protein coupled receptor superfamily. TRAIL (tumour necrosis factor (TNF)-related apoptosis–inducing ligand) isoforms, members of the TNF family, were identified from sequence databases by searching for proteins with homology to the most conserved region of TNF.[23] Forms of TRAIL have since been developed as cancer therapeutics.[24]

Comparing the human genome with that of other model organisms can give clues to gene function in humans. Gene function can be inferred by removing a particular gene from an organism or cell and observing the behaviour or phenotype. Large scale gene-silencing screens carried out in *C. elegans*, *Drosophila* or mice can suggest novel targets. The human genome can then be searched to find whether homologues of these genes exist. This is explored further in the following section on model organisms (Section 9.2.3).

Expression profiling monitors the expression of the genes encoded in the genome. Databases of the partial sequence of expressed genes (expressed sequence tags or 'ESTs') begun to be assembled prior to the sequencing of the whole genome. These could be searched to identify genes that had a specific expression pattern in diseased tissue. Microarray technology has developed from the availability of whole genome sequence information, and, at least in theory, offers the ability to compare the expression of all genes within a single experiment. Many different comparisons can be made to assist in the identification of genes whose expression profile suggests that they may be potential drug targets. For example, the samples compared may be from normal or diseased tissues, from cells treated with different drugs or where a gene has been silenced or knocked out, or from patients with different clinical manifestations of a disease. DNA microarrays make use of the inherent property of single-stranded DNA to specifically bind their complementary sequences in order to detect and quantify mRNA expression. There are two main types, oligonucleotide DNA microarrays and cDNA microarrays. In the case of oligonucleotide DNA microarrays, short sequences of 25–70 bases in length are either chemically synthesised or printed onto solid surfaces. For cDNA microarrays, 200–2000 nucleotide long cDNA fragments are PCR-generated and spotted onto slides. Total RNA is isolated from the relevant cell lines or tissue, amplified by PCR, fluorescently labelled and hybridised to the microarray. The fluorescent signals are then analysed and software tools utilised to analyse the different gene expression profiles (www.bio.davidson.edu/courses/genomics/chip/chip.html).

Specific examples are often confirmed using real-time quantitative PCR. Such methods have been used to identify genes upregulated upon metastasis in melanoma cell lines, including the small GTPase RhoC.[25] Many large scale microarray projects have made their results publically available so the data can be interrogated by other investigators. This includes the Connectivity Map (C-Map), which is a collection of genome-wide transcriptional expression data from cultured human cancer cells that have been treated with bioactive molecules (www.broadinstitute.org/cmap).[26] Further improvements in sequencing technology may make it possible to identify and quantify mRNA sequences directly, without having to rely on hybridisation.

One drawback of analysing gene expression to identify novel targets is that mRNA levels do not always correlate with amounts of expressed protein, which can be regulated by numerous translational and post-translational mechanisms. Proteomics methods can be used to address these issues. Such methods include mass spectrometry and protein microarrays, which are used to identify, quantify and compare protein levels in different samples. Protein microarrays can be used to quantify protein levels by measuring binding to an array of immobilised antibodies, or vice versa, by probing an array of immobilised cell lysate mixtures with antibodies. Protein microarrays can also map and quantify the network of protein–protein interactions by immobilising purified proteins onto a chip and probing with mixtures of other proteins to pull out binders. By assessing protein expression profiles and post-translational modifications (PTMs) in healthy and diseased, or drug-treated samples, proteomics has the potential to discover, identify and quantify novel targets for therapeutic intervention of disease.[27]

Single nucleotide polymorphisms (SNPs) can be used as markers to map genes that modify disease susceptibility or those related to drug responsiveness. Databases of single nucleotide polymorphisms (SNPs) can be analysed to determine if a SNP in a particular protein is associated with a disease phenotype (*e.g.* www.ncbi.nlm.nih.gov/SNP/; http://snp.ims.u-tokyo.ac.jp/). The International HapMap consortium has created a catalogue of common genetic variants that occur in humans. The map describes what these variants are, where they occur in our DNA, and how they are distributed among people within populations, and among populations in different parts of the world. The project is designed to provide information that other researchers can use to link genetic variants to the risk for specific illnesses, which will lead to new methods of preventing, diagnosing, and treating diseases, particularly those resulting from multiple genetic and environmental factors, such as cancer, stroke, heart disease, diabetes or depression (http://hapmap.ncbi.nlm.nih.gov).

Access to the complete human genome sequence and these resulting bioinformatic tools has revolutionised genetic studies by providing novel markers and more rapid means of gene mapping. Genome-wide association studies (GWAS) for a disease involves collecting and genotyping the DNA from patients and control subjects. The samples are genotyped on the basis of the presence of selected marker SNPs, and statistical analysis is used to determine whether a particular genomic region is linked to the disease. A GWAS was carried out by deCODE Genetics on Icelandic patients with atherosclerosis. This implicated the gene encoding the EP3 prostaglandin receptor in peripheral arterial obstructive disease. DG-041, a potent EP3 receptor antagonist was developed and is in the early stages of clinical development. In order to ensure that targets identified were druggable, the company Oxagen restricted patient genotyping to GPCRs. This methodology identified the gene CRTH2, the receptor for prostaglandin PGD2, as being linked to asthma. OC-000459 was developed as a CRTH2 antagonist, which is currently in Phase II trials as a novel oral treatment for respiratory and gastrointestinal inflammatory disorders. However, there has been considerable criticism of the benefit of GWAS.[28,29] Despite the fact that more than 1500 publications have reported more than 2000 associations of particular SNPs with certain traits, the majority of these show a very small elevation in the risk of disease, with little predictive value. Early studies were affected by flaws such as a lack of well-defined case and controls groups, and insufficient sample sizes. Even when such errors can be overcome, some believe that the GWAS concept is fundamentally flawed, due to the fact that it typically identifies commonly occurring genetic variants, thus failing to pick up lower frequency, but potentially more significant polymorphisms. GWAS also are less able to detect associations cause by large scale genomic variation, such as insertions and deletions. In addition, many of the SNPs identified as being associated with diseases are located in non-coding regions, making dissection of their role more difficult. Developments in sequencing technology will no doubt lead to improvements in these association studies, and eventually high-throughput whole genome sequencing will enable direct identification of disease-causing mutations on a population level. Indeed, the ability to obtain the genome sequence of patients and to identify mutations associated with disease phenotypes is already turning the concept of personalised medicine into a reality. There are many examples where the ability of a particular drug to treat a condition is dependent on the genotype of the patient. The case of vemurafenib in treating melanomas with the V600E B-Raf mutation was illustrated earlier. Other examples include a poor response to long-acting beta-2 agonists (LABAs) in asthma patients with arginine rather than glycine at amino acid 16 of the beta-2 receptor.[30] In breast cancer therapy, the monoclonal antibody therapy Herceptin is given to patients that have been shown to overexpress the HER2 receptor. Kalydeco is a cystic fibrosis treatment that targets one (G551D) of the potential 1500 mutations found in the ion channel CFTR, found in 4% of patients.[31]

Having the sequence of all human genes enables researchers to generate or purchase cDNAs encoding any gene of interest for further study. This can be used to analyse the effect of gene overexpression on a genome wide basis. 27 000 cDNAs were overexpressed in a human cancer cell line to test which had the ability to overcome the anti-proliferative effect of the compound aprotoxin A. Overexpression of fibroblast growth factor receptors was shown to partially rescue the effect of the drug, suggesting that these proteins may have a role in mediating resistance to this chemotherapeutic agent.[32]

Downstream of obtaining the human genome sequence, further developments include ENCODE, the Encyclopedia of DNA Elements. This National Human Genome Research Institute funded initiative aims to identify all regions of transcription, transcription factor association, chromatin structure and histone modification in the human genome sequence. This will provide new insights into the organisation and regulation of the genome, including potential new drug targets.

9.2.3 Model Organisms

9.2.3.1 Bacteria

Bacteria have been long studied as a cause of human disease. They have also been key to the fields of genetics and molecular biology, with many major discoveries being made in bacterial systems. In addition, the use of bacteria to express large quantities of a target protein, rather than having to rely on extensive purification from native sources, has revolutionised the screening of new drugs.

The first non-viral microorganism to undergo complete genome sequencing was the bacterium *Haemophilus influenza*, with the sequence published in 1995 (reviewed in ref. 33). This set a benchmark for all subsequent sequencing projects, in terms of the quality of sequence to be obtained and how the data should be made accessible to the research community. Many of the technological lessons learnt from this program were key to the progress of the human genome project. Over 400 prokaryote genomes have now been sequenced, leading to a glut of sequencing data and many bioinformatic tools have been developed in order to annotate this data and determine gene function. Identification of genes present in pathological prokaryotes that do not have close homologues in humans may provide potential targets for much needed novel antibiotics or vaccines. Mapping the mutations that allow bacteria to become resistant to current therapies may allow such resistance to be overcome or avoided in future. This strategy has been employed in viruses, where identification of the mutations in influenza that confer resistance to certain antivirals has led to the conclusion that resistance is less likely to develop against drugs that most closely resemble the natural ligand.[34] Microarray techniques have been utilised in an attempt to identify novel genes which may regulate different aspects of pathogenesis, for example the host-pathogen interaction.[35]

9.2.3.2 Yeast

Sacchromyces cerevisiae and *Sacchromyces pombe* have both proven to be invaluable eukaryotic model organisms. These unicellular fungi can be grown rapidly and economically in culture. They can be easily transformed with exogenous DNA, and homologous recombination occurs with high efficiency, to allow specific gene deletions or replacements to be made. *S. cerevisiae* was the first eukaryote to be sequenced, with the sequence being published in 1996 by an international consortium.[36] It was found to consist of 12 Mb of DNA, with approximately 6000 protein-coding genes. The *S. pombe* sequence was completed in 2002 and has same quantity of DNA, and about 5000 genes.[37] The research community has compiled and annotated this genomic information into publically accessible databases, which has allowed yeast research to follow an integrated, systems biology approach.

Although the divergence of yeast from the mammalian lineage occurred hundreds of millions of years ago, there is a high degree of sequence conservation between these fungi and humans. The ease of genetics in yeast has allowed extensive characterisation of pathways regulating metabolism and signal transduction, which includes many targets for human disease, such as the cell cycle regulating cyclin-dependent kinases,[38,39] components of the cholesterol synthesis pathway,[40-42] GPCRs[43,44] and MAPKs.[45-47] There are numerous examples of human and yeast proteins being able to complement for one another, which allow yeast to be used a model system for the human protein function.

Whole genome microarrays, which provide a complete profile of gene expression, were first developed for yeast, and yeast has proven to be a model for the use of this technology. Using a PCR-based approach, a deletion strain has been generated for almost every protein-coding gene in *S. cerevisiae* (www-sequence.stanford.edu/group/yeast_deletion_project/deletions3.html). This has enabled genome-wide phenotypic screens to be carried out. The effect of point mutations can also be analysed in the knockout background. Combining gene deletions has been used to carry out synthetic lethal screens to further characterise pathways and overcome redundancy. The ability to target a protein that selectively kills cancer cells is the fundamental aim of cancer chemotherapy, and synthetic lethal screens can be used to identify genes that can cause cytotoxicity only in cancer cells that lack a particular protein function, for example repair of DNA damage.[48,49] The combination of these classic genetics with chemical screening has been termed "chemical genetics", and yeast is a particularly powerful organism for such studies.[50]

Yeast knockouts have also been used to carry out haploinsufficiency analysis.[51,52] Lowering the dosage of a single gene from two copies to one copy in diploid yeast results in a heterozygote that is sensitised to any drug that acts on the product of this gene. This haploinsufficient phenotype thereby identifies the gene product of the heterozygous locus as the likely drug target. This was shown to be successful for six targets in a small scale screen, including the confirmation of cytochrome P450 lanosterol 14a-demethylase (ERG11) as the target of the antifungal fluconazole.[51] A larger screen confirmed known targets and furthermore identified novel interactions. The corollary of this assay can also be used, in that overexpression of the target protein can grant resistance to a compound of interest. Overexpression analysis can also identify novel components of pathways and be used in synthetic lethal screens on sensitised backgrounds.

Yeast has been an invaluable tool for two-hybrid analysis, which identifies protein–protein interactions. It can often be the case that a protein directly associated with a disease proves to be an "undruggable" target, whereas regulatory binding partners for these proteins may prove to be much more suitable; for example Raf kinase binding to the oncogenic G protein Ras.[53] Variations on the traditional two-hybrid methodology can be used to directly assay the binding of small molecules or biotherapeutics. Mass spectrometry is proving to be a complementary technique to two-hybrid analysis, allowing interactions to be mapped on a higher-throughput scale.[54,55]

As an additional benefit, the study of non-pathogenic *S. cerevisiae* and *S. pombe* has also proven useful in the development of treatments for pathogenic fungi.

9.2.3.3 Caenorhabditis elegans

The nematode *C. elegans* is a highly attractive model for the identification of novel targets for a number of reasons, in particular, the high conservation of biochemical pathways between worms and humans, the ease and speed of genetic methods designed to identify and characterise new mutants, and the relatively low cost associated with *C. elegans* research. They are optically clear so can be easily screened by microscopy. The worm only has 959 somatic cells, with each cell division being invariant and therefore able to be mapped. Despite this, *C. elegans* forms complex structures such as a nervous system and a digestive tract. A key reason for the suitability of *C. elegans* as model organism in target identification and validation is due to the ease of carrying out gene silencing by RNA interference—this is described in more detail in the section on RNAi (Section 9.2.4).

C. elegans was the first multicellular organism to be fully sequenced, the sequence being published in 1998. The genome consists of 100 million base pairs and approximately 20 000 protein coding genes. An annotated database of the genome together with bioinformatic analysis tools, can be found on web servers such as WormBase (www.wormbase.org). These repositories and tools have themselves served as models for analysing other genomes. Comparison with the human genome suggests that around 75% of human genes have homologues in *C. elegans*, illustrating the high degree of sequence, and therefore most likely, functional conservation between these organisms. Indeed, components of pathways that are deregulated in human disease and are targeted by many drugs, such as the oncogenic Ras-MAPK pathway, the insulin signalling pathway that regulates growth and metabolism, the presenilin pathway mutated in familial forms of Alzheimer's disease, and apoptotic pathways implicated in cancer and degenerative diseases, are all highly conserved between *C. elegans* and humans. Studies in *C. elegans* have proven invaluable in rapidly gaining an understanding of the normal biological functions of these proteins that are mutated in human disease. The relative simplicity of the nervous system in *C. elegans* makes it attractive as a means of assessing homologous neuronal targets. The anti-depressant fluoxetine was already known to function as a selective serotonin uptake inhibitor (SSRI), but other modes of action remained unclear. The application of the drug to *C. elegans* mutants lacking serotonin signalling allowed other targets to be identified, including a novel family of transmembrane proteins.[56]

Loss-of-function deletions of *C. elegans* genes are available, but the tractability of RNAi in this organism, means that this is the preferred means of gene silencing (see RNAi section). Microarray techniques to analyse gene expression, and mass spectrometry to analyse protein–protein interactions are being utilised, in combination with gene silencing, to elucidate many of the regulatory pathways in *C. elegans*.

9.2.3.4   Drosophila melanogaster

The fruit fly has been used as the standard model organism for genetic research for over 100 years. *Drosophila* was selected due to its short life cycle, together with straightforward and economical maintenance. Classical genetic mapping was enabled in *Drosophila* by the presence of polytene chromosomes in the salivary gland. These show a reproducible banding pattern, which allows a physical map to be generated that can be linked to a genetic map. The modern era of *Drosophila* genomics has been facilitated by the completion of the sequencing of the genome in 2000, which was found to consist of approximately 13 000 genes.[57] This information has allowed the generation of genome-wide libraries of full-length clones and genetic mutations, although many of the traditional genetic tools (such as mutation sets generated by radiation or chemicals), developed over the past century, are still routinely used. The genetic information is collated into databases such as Flybase (http://flybase.harvard.edu).

There is a high degree of overall homology to the human genome, with an analysis of human disease-linked genes identifying 77% that have clear counterparts in *Drosophila*. The Homophila (http://homophila.sdsc.edu) database compares information between the human and *Drosophila* genomes in order to identify such links. Often a single *Drosophila* gene has multiple counterparts in mammals, for example a single p53 isoform exists in the fly, in comparison to the three mammalian p53, p63, and p73 isoforms. This simplicity can be of benefit in target identification as it removes a layer of redundancy that may mask observation of a phenotype. Mutations in human proteins have been shown to have orthologous effects in *Drosophila*; for example the oncogenic Ras V12 mutant leads to an overgrowth phenotype when expressed in the developing eye of the fly.[58] Complementation of *Drosophila* mutants with human proteins again demonstrates the high level of functional conservation. This enables *Drosophila* to be used to assess even complex phenotypes with relevance to human disease, such as behavioural and psychiatric disorders.

The development of technologies to generate targeted gene knockouts in *Drosophila* has enabled genome-wide deletions to be screened. Modifier screens examining the combined effect of mutation, deletion, or overexpression can be used to assign genes to the same pathway or process. Such screens were used to identify genes involved in Ras signalling, by enhancement or suppression of the eye overgrowth phenotype. Chemical genetic screens can also be carried out to identify mutants that are resistant or hypersensitive to compounds of interest.

In order to speed up the analysis of *Drosophila* mutants, automated animal sorters have been developed that can collect mutant embryos, based for example on the expression of a readily detectable marker such as green fluorescent protein. Whole genome microarrays are available for *Drosophila* and have been used to assess global transcriptional status. Mapping of the interaction network in the *Drosophila* proteome has been carried out using mass spectrometry and yeast-two-hybrid analysis. The use of RNAi-mediated gene silencing in *Drosophila* is considered in the RNAi section.

9.2.3.5   Danio rerio *(Zebrafish)*

Danio rerio is a vertebrate model organism, traditionally used for the study of development and embryogenesis. The optical clarity of embryos, ability to assay in multiwell formats, speed of development and economical maintenance costs all serve to make zebrafish highly useful for genetic analysis and drug discovery.[59,60]

The genome was sequenced by the Sanger Centre, and despite the fact that zebrafish diverged from humans more than 450 million years ago, the sequence similarity between zebrafish and human genes is very high, with 70% of protein coding human genes related to genes found in the zebrafish.[61] One disadvantage of the use of zebrafish in genetic screens is that they possess two copies of many mammalian genes due to an evolutionary gene duplication event. This can lead to redundancy, which may prevent a phenotype being revealed in a genetic or chemical screen. An online database, the Zebrafish Information Network (ZFIN), compiles the genetic, genomic, and developmental information (zfin.org).

Forward genetic screens on mutants induced by chemicals or random insertions, have identified many mutations that cause disease phenotypes similar to those seen in humans, such as polycystic kidney disease, heart disease, leukaemia and nervous system disorders. These mutants could be used to carry out genetic suppressor screens, to identify interacting partners that may prove to be drug targets, or chemical screens for disease modulators.

Reverse genetic screens can be carried out in zebrafish using morpholinos. These are antisense oligonucleotides that have been chemically modified to increase stability, by increasing resistance to nuclease-mediated degradation. Antisense oligonucleotides hybridise to the translation-initiation or splicing sites of specific mRNAs to knockdown expression. Antisense technology can be utilised in other model systems and has been developed for therapeutic purposes, but in more recent times has been superseded by RNAi. However, its high level of effectiveness in zebrafish makes it a robust method of generating gene knockdowns in this system. Morpholinos can be used in suppressor screens to identify genes that prevent or slow the development of a disease phenotype, induced by genetic mutations, pharmacological agents or infection.

The massive number of potential targets made available as a consequence of whole genome sequencing has, in some ways, made the process of target identification even more difficult. This has led some groups to revert to phenotype-based screening of small compounds as the first stage of drug discovery. Zebrafish are an ideal model for such screens, as they offer a high-throughput platform, equivalent to that used *in vitro* or in cells, with the huge benefits of an *in vivo* system, which allows issues such as non-specific or toxic effects to be identified at a preliminary stage. Zebrafish that have the gridlock mutation, which causes a vascular defect, or crb (crash and burn), a cell-cycle mutation, were used to screen small-molecule libraries. In both screens, compounds that can reverse the phenotypic effects of the mutations were identified.[62–65] For the gridlock mutation, two structurally distinct classes of compound were found to suppress the phenotype. These function to inhibit pathways downstream of the vascular endothelial growth factor (VEGF) receptor, thus revealing the importance of these signalling networks in endothelial development.[62,66] The suppression of the crb mutation by persynthamide, aphidicolin or hydroxyurea enabled the identification of pathways that contribute to the specific cell cycle phenotypes observed.[63]

Zebrafish are of particular use in the later stages of drug development, such as ADME, pharmacokinetics, safety and toxicology, providing whole organism data much more economically than using mammalian model organisms (see section 9.3.3).

9.2.3.6 Mouse

The mouse serves as the most well characterised mammalian model system, having been used for many years, even prior to the molecular biology revolution. There are several disease models that have arisen in mice through natural breeding, including NOD (non-obese diabetic) mice, which develop Type I diabetes, and SCID (severe combined immune deficiency) mice, with an almost complete absence of an immune system. The murine genome was first published in 2002.[67] Comparison with the human genome illustrates a striking similarity, with 95% of the genes being highly conserved. The sequencing of whole genomes has identified hundreds of genes with unknown function, many of which have the potential to be future drug targets. Mouse knockouts are amongst the most powerful means of identifying gene function in a mammalian system.

A mouse knockout is traditionally generated by designing a targeting vector that contains two arms of homology to the gene of interest, separated by a cassette (encoding, for example, neomycin resistance) that disrupts the gene and also functions as a selectable marker. This construct is linearised and introduced into mouse embryonic stem (ES) cells by electroporation. The regions of homology in the targeting vector allow the construct to insert into the target gene by homologous recombination. The ES cells are grown *in vitro* in the presence of G418 (neomycin) to select for those that have taken up the targeting vector. The colonies of selected cells are then picked, expanded, and screened by PCR and/or Southern blotting to ensure correct insertion has occurred (illustrated in www.ncbi.nlm.nih.gov/books/NBK22002/figure/A2286/). Positive clones are then combined with fertilised mouse oocytes, by microinjection or aggregation. These eggs are then introduced into pseudo-pregnant foster mothers, and the resultant offspring are chimeric mice, originating from cells derived from both the original oocyte and the targeted embryonic stem cells. Strong chimeras are bred further, and, if cells derived from the targeted ES cells have contributed to the germline, the gene deletion can be passed to future generations. Progeny are PCR genotyped from DNA isolated from tail biopsies to confirm that the gene knockout is present. The process of generating knockouts is being continuously advanced to improve efficiency and reduce the time required; for example, the novel CRISPR/Cas system allows multiple genes to be targeted simultaneously.[68,69]

Gene trapping is an alternative strategy that doesn't target a specific gene, instead a cassette is inserted at random, and the position of insertion determined retrospectively using markers present on the inserted cassette. ENU mutagenesis is an alternative forward genetics approach, whereby the alkylating agent N-ethyl-N-nitrosourea (ENU) is used to generate random mutations in mice. The mutagenised mice are then screened for disease phenotypes and the disease-causing gene is then identified. Several large screens have carried been carried out using this method. A collaboration between the Medical Research Council and GlaxoSmithKline identified 500 novel phenotypes using this method.

The power of mouse knockout technology in drug discovery was illustrated in a retrospective analysis by Zambrowicz and Sands in 2003.[70] The 100 best-selling drugs at that time were identified to target 43 host proteins (14 of the drugs were anti-infectives which do not have human targets). 34 of the 43 host targets had been knocked out in mice, and 29 of these knockouts yielded useful information on gene function and pharmaceutical relevance, in most cases leading to a direct correlation between the phenotype of the knockout and the therapeutic utility of the drug. A selection of these is listed in Table 9.1.

Table 9.1 Correlation between the therapeutic utility of various drug targets and the mouse knockout phenotype of the target gene [from Zambrowicz and Sands, 2003.[70]]

Drug target	Drug utility	Mouse knockout phenotype
$H+/K+$ ATPase	Gastroesophageal reflux disease	No or decreased stomach acid
COX2	Arthritis	Reduced inflammation, reduction in collagen-induced arthritis
Serotonin transporter	Depression	Altered open-field behaviour
PPAR-γ	Diabetes	Increased insulin sensitivity in heterozygotes, embryonic lethality in homozygotes
Angiotensin receptor AT_1	Hypertension	Low blood pressure
$P2Y_{12}$	Atherosclerosis	Decreased platelet aggregation

This confirms that mouse knockouts serve as good models for human disease despite concerns as to differences in murine and human physiology, whether gene compensation prevents analysis of true gene function, whether knocking out a gene throughout development is relevant to inhibiting a target in the adult animal, and whether embryonic lethality prevents identification of the best targets.

These latter concerns have been overcome in recent years by more advanced gene targeting technology that allows genes to be knocked out conditionally in specific tissues or at certain stages of development, using recombination-based techniques. A targeting vector is generated where the gene of interest (or a region of it) is flanked by recognition sites for the Cre- or Flp-recombinase enzymes. These recombinases function to excise the DNA between the recognition sites, creating the gene knockout. Mice that express the recombinase under the control of a tissue- or developmental stage-specific promoter are crossed with mice carrying the targeting vector. This results in the gene only being deleted in the selected tissue or at a specific stage in development. It is also possible to generate "knock-in" animals which, instead of failing to express a gene, express a mutant form of the encoded protein, for example, a protein with a known disease-causing mutation, such as B-Raf V600E.

Mouse knockouts also have the potential to reveal risks or side-effects of targeting a particular protein. Many COX-2 inhibitors have had to be withdrawn from use due to cardiovascular risks. Subsequent analysis of COX-2 knockout mice identified a potential mechanism for this issue, through the reduction in levels of a cardio-protective prostacyclin.[71]

Gene knockouts and gene trapping have been carried out on an industrial scale by both pharmaceutical companies and public consortia. Lexicon Pharmaceuticals initiated a program in 2001, called Genome5000, designed to discover the function of nearly 5000 mammalian genes through comprehensive phenotyping of gene knockout mice (www.lexicon-genetics.com/research). Efforts were focused on genes from the human genome that encode "druggable" targets. 4650 complete gene knockouts were generated in mice using gene knockout or gene trapping technologies. More complex and subtle mutations including conditional alleles (as described above), point mutations, and humanised alleles were also generated. The program was completed in 2007, with more than 100 drug discovery targets identified across a wide range of therapeutic areas.

The Sanger Institute Gene Trap Resource was a major project that isolated and characterised gene trap mouse embryonic stem (ES) cell lines to generate reporter-tagged, loss of function mutations. The project produced more than 10 000 characterised gene trap insertions in ES cells that are stored frozen and are freely available to the research community.

As opposed to knockouts or knock-ins which target endogenous murine genes, transgenic mice overexpress exogenous DNA. This technique can be used examine the effect of over-expressing a wild-type or mutant protein to mimic human disease. For example, there are multiple transgenic mouse lines that exhibit symptoms of Alzheimer's disease. These animals overexpress proteins implicated in disease progression, such as amyloid precursor protein, presenilin 1 or apolipoprotein E.[72] Transgenic models of this type have a key role of validating the suitability of targets for the treatment of disease and in the testing of novel therapeutics.

9.2.4 RNA Interference (RNAi)

RNAi-based gene silencing is becoming an increasingly important tool in target identification and validation. This system has revolutionised functional genomics due to its relative ease and ability to provide a powerful reverse genetic approach, particularly in organisms where classical genetic approaches are difficult, or have not been developed.[73]

In 1998, Fire and Mello were the first to describe the gene silencing effect of double-stranded RNA in *C. elegans.*[74] Similar processes were subsequently shown to operate in plants and other

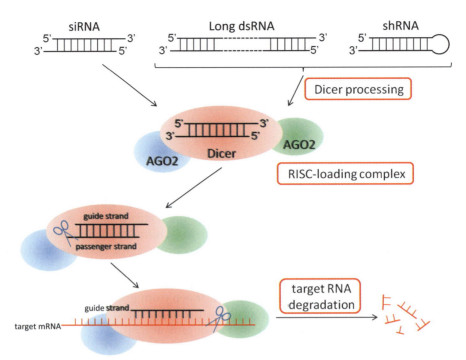

Figure 9.3 The process of RNA interference (RNAi). The process commences with the addition of small interfering RNAs (siRNAs). These are either generated synthetically, or by endogenous Dicer enzymes from long double stranded RNA (dsRNA) or short hairpin RNA (shRNA). The siRNA duplex is then loaded onto the Argonaute (AGO2) component of the RNA-induced silencing complex (RISC), by Dicer and the RNA-binding protein TRBP. AGO2 then selects one strand as a "guide" strand, with the other "passenger" strand being cleaved and ejected from the complex. The guide strand, while still associated with AGO2, then pairs with complementary target mRNAs. AGO2 then cleaves the mRNA target, and the cleaved product is released. The RISC is then recycled, using the same guide strand for several other rounds of target mRNA cleavage. For further details please see the text.

animals. RNAi is initiated with the processing of long, double-stranded RNA (dsRNA) molecules into small (~21–25 nt) interfering RNA (siRNA) complexes, by the endonuclease Dicer. The siRNA complexes have distinctive termini, with the 3′ end carrying a dinucleotide overhang, while the 5′ end has a monophosphate group. Both the length and termini are critical for recognition of the siRNA by the RNA-induced silencing complex (RISC). Once generated, the siRNA duplex is loaded onto the Argonaute (AGO2) component of RISC, by Dicer and the RNA-binding protein TRBP (Figure 9.3, www.nature.com/nrg/multimedia/rnai/animation/index.html). The termini of the siRNA are critical for anchoring the complex onto AGO2. AGO2 then selects the strand of the siRNA duplex with the less thermodynamically stable 5′ end as a "guide" strand, with the other "passenger" strand being cleaved and ejected from the complex. The guide strand, while still associated with AGO2, then pairs with complementary target mRNAs. AGO2 then cleaves the mRNA target, and the cleaved product is released. Base-pairing between the residues 2–11 of the siRNA and the target mRNA has been shown to be critical for recognition and determination of the cleavage site. The RISC is then recycled, using the same guide strand for several other rounds of target mRNA cleavage.

Screening for the effects of the knockdown of gene expression by RNAi is now a widely used tool in the pharmaceutical industry for the target identification and validation. It can be undertaken at a high-throughput level, borrowing many of the techniques previously developed for undertaking large-scale chemical screens. The most powerful RNAi screens cover the whole genome of model organisms or cell lines. The use of this system has therefore been completely dependent on the genome sequencing projects of *C. elegans, Drosophila*, mouse, and human. Today, genome-wide RNAi screening is possible *in vivo* in *C. elegans*, in both tissue culture cells and *in vivo* in *Drosophila*, and in cell lines from mice, rats, and humans.[75]

For screening purposes, various RNAi reagents are used to exploit the endogenous processing mechanisms, including synthetic siRNAs, small hairpin RNAs (shRNAs), small hairpin and long dsRNAs. The use of synthetic siRNAs was first described by Elbashir *et al*. in 2001.[76] These are composed of RNA duplexes of 19 complementary base pairs (bp) and 2-nucleotide 3′ overhangs. They are transfected into cells or injected into animals. On entering cells, one strand of the siRNA duplex is incorporated into the RISC, leading to degradation of the target mRNA, as described above. The effects of the siRNAs are transient, especially in actively dividing cells. This is in contrast to shRNA-synthesising vectors, which allow for controlled or continuous expression of small 50–70 bp single-stranded RNA transcripts that contain both the sense and antisense strand complementary to the selected mRNA target. They are either transfected into cells as plasmid DNAs or delivered using viral particles to cells that are difficult to transfect, and are maintained as extra-chromosomal copies or stably integrated in the genome as transgenes. The RNA transcript folds back to form a stem-loop structure. shRNAs are then processed in the cytoplasm by the ribonuclease Dicer to generate siRNAs (Figure 9.3). These vectors can be placed under the control of inducible promoters, restricting expression temporally or spatially.[77]

Long dsRNAs are usually 200–500 nucleotides (nts) in length. They can be injected into animals, delivered *via* bacteria, expressed as transgenes, or delivered into cultured cells by transfection or bathing. Long dsRNAs are not used in mammalian systems as they trigger the interferon response. This is an endogenous antiviral mechanism triggered by dsRNA, which is a by-product in the replication cycle of virtually all viruses. Viral dsRNA is recognised and binds to endogenous proteins, such as Toll Like Receptor 3, which stimulate production of interferon and other cytokines. These in turn activate numerous signalling pathways that aim to combat viral infection, for example by decreasing protein synthesis and degrading RNA, eventually leading to cell death. This can obviously mask the gene-specific effect of any applied dsRNA. Recently it has been shown that even siRNA of 23 nt can activate the interferon response, as can the presence of certain GU-rich sequences.[78–83] A fuller understanding of the response of cells to such RNA molecules is critical to the use of this method in screening.

9.2.4.1 Cell-Based RNAi Assays

RNAi screens are carried out routinely in *Drosophila* and mammalian cultured cells. *Drosophila* S2 cells are particularly amenable to RNAi, being able to take up dsRNA even in the absence of transfection reagents, and obtaining knockdown efficiencies of 95%. Rapidly evolving technological developments in mammalian systems are allowing similar levels of effectiveness to be achieved. Such screening builds upon instrumentation, assays and other methods previously developed for chemical screening in cells. These high-throughput screens may measure the level of a metabolite or protein, or the rate of transcription, utilising fluorescent dyes or luciferase. Each RNAi, or pool of RNAis against the same gene, is contained in a separate well of a 96- or 384-well plate, or for further miniaturisation, can be printed onto a microarray slide. Screens have been carried out to investigate signal transduction pathways, differentiation and pathogen interactions.[77]

RNAi can be combined with chemical screening to provide additional information, or a more focused screen. For example, cells can be sensitised with a drug prior a whole-genome RNAi screen. This was undertaken with neratinib, a small molecule tyrosine kinase inhibitor of the ErbB receptor family currently in Phase III clinical trials for breast cancer. A genome-wide functional RNAi screen was combined with a lethal dose of neratinib in the SKBR-3 breast cancer cell line, to discover chemoresistant interactions with the drug. This screen identified novel mediators of cellular resistance to the kinase inhibitor that could lead in the identification of drug targets to be used in conjunction with neratinib to help prevent drug resistance.[84]

Alternatively, a single gene can be knocked down using RNAi in all samples, prior to a high-throughput chemical screen being undertaken. This has been carried out in *Drosophila* cells to identify novel modulators of the Wnt signalling pathway, which has a clear role in tumour development, particularly in colorectal cancers. RNAi was used to target the protein axin and thus activate the Wnt pathway, before cells were screened with a ~15 000 compound library. This identified three compounds that inhibited Wnt-mediated transcription.[85] Synthetic lethal screens, which have proven to be so powerful in yeast to uncover genetic redundancy and network interactions, can also be undertaken using combinations of RNAi.

9.2.4.2   In Vivo *RNAi Screening*

In vivo genome-wide RNAi screening has been carried out extensively in *C. elegans*, which are particularly amenable to this mechanism. It is very straightforward to apply the silencing RNA to these organisms; dsRNAs can be introduced by soaking the animal in a solution of dsRNA, by feeding the worms bacteria that express long dsRNAs, by injection of dsRNA into animals or eggs, or by generating transgenic shRNA-expressing animals. Moreover, in *C. elegans*, RNAi has been shown to be both systemic, whereby RNAi initiated in one tissue can spread to another, and transitive, in that the original RNAi signal is amplified and results in the production of siRNA to regions upstream of the sequence to which the original dsRNA was complementary.

Many whole genome screens have been carried out on all the ~20 000 genes in *C. elegans*, focusing on systems that are highly conserved between the worm and mammals, including metabolism, aging and synaptic transmission. Ashrafi *et al.* carried out an RNAi screen to identify novel components involved in lipid metabolism.[86] The use of Nile Red, a fluorescent dye that can be fed to the worms and is taken up into fat, allowed for a relatively simple screen of the visualisation and quantification of lipid droplets. Knockdown of 305 of the genes tested increased the accumulation of fat, whilst 112 genes decreased fat levels when inactivated. Many of these genes have human homologues, some of which are known to be involved in lipid metabolism. Others represented genes not previously implicated in fat regulation, which could prove to be targets for the treatment of obesity.

Genome-wide RNAi screens have also been carried out *in vivo* in *Drosophila*, on systems such as pain perception, obesity, heart function, bacterial infections of the gut, neural stem cell self renewal, and neurological disease. RNAi knockdown is induced *via* injection or expression of dsRNAs, and acts cell-autonomously in *Drosophila*, facilitating tissue- and life cycle stage-specific studies. Expression of long or short dsRNA hairpins *via* a transgene is also an option. RNAi approaches have also begun to be used *in vivo* in mice. RNAi can be introduced into ES cells from which transgenic mice can be generated. Lentiviral-based RNAi constructs can also be applied to tissues such as the skin. An *ex vivo* approach has proved to be useful in the study of cancer development, whereby RNAi can be added to tumour cells in culture, these cells can then be introduced into mice to assess the effect of the gene knockdown on tumourogenesis.[77]

A comparison of RNAi with other gene knockdown techniques is highlighted in Table 9.2.

9.3   IMPACT OF MOLECULAR BIOLOGY ON HIT IDENTIFICATION TO LEAD OPTIMISATION

The role of molecular biology on the development of lead series from the initial hits has gained further significance with recent changes in the source of the hit series.[87,88] The greatest impact has been in the development of hit series derived from fragment screening approaches; given

Table 9.2 Advantages and disadvantages of various gene knockdown techniques.

Gene knockdown technique	Advantages	Disadvantages
Gene knockout	• Total elimination of expression of a specific gene • Can be carried out in cells or whole organisms	• Generation of knockout can be time-consuming, especially in mice • Can result in lethality, preventing analysis of knockout • Compensation (*e.g.* upregulation of another gene isoform) can occur, obscuring the effect of knockout
Conditional gene knockout	• Can generate a knockout in a tissue-specific or developmental-stage specific fashion • Can be used to overcome lethality or compensation associated with total knockout	• Generation can be more time-consuming than total gene knockout • Can result in incomplete loss of gene expression • May not overcome issues of lethality or compensation
Knock-in	• Can generate a specific mutation in gene rather than a loss of gene expression, which may more closely resemble a disease phenotype than gene knockout • Can be used to overcome lethality or compensation associated with total knockout	• Generation can be more time-consuming than total gene knockout
Morpholino	• Very rapid generation of knockdown • Stable, not susceptible to nucleases • Can be used as probes for visualisation	• Off-target effects • Delivery can be challenging and laborious (*e.g.* microinjection) • Incomplete loss of gene expression can occur
RNAi	• Widely applicable to many cell types and model organisms • Very rapid generation of knockdown • Delivery of RNAi can be very straight-forward (*e.g.* in *C. elegans*) • Can be carried out on a high-throughput scale	• Sequence-dependent and independent ("interferon response") off-target effects • Incomplete loss of gene expression can occur • Delivery of RNAi can be challenging (*e.g.* neurons) • Can be susceptible to nucleases • Knockdown can be transient

their small size, deciphering the structure activity relationship (SAR) can be particularly conducive to molecular approaches. One such approach is site-directed mutagenesis (SDM), whereby specific mutations that modify the steric/chiral centre, charge, hydrophobicity and hydrogen bonding potential of key residue side chains can yield significant insights into binding interactions. In some cases, scanning mutagenesis can be used to systematically change all the residues at the putative drug-binding interface. Alanine is the favoured residue for scanning SDM given that it eliminates side chain interactions without altering the main chain conformation (in contrast to glycine or proline), nor does it impose extreme electrostatic or steric effects. These SDM methods are particular powerful when coupled to lower throughput secondary screening techniques such as Surface Plasmon Resonance (SPR), NMR and X-ray crystallography.

9.3.1 Surface Plasmon Resonance

Surface Plasmon Resonance (SPR) is an optical method used to monitor label-free molecular interactions in real time.[89] It measures changes in the refractive index of light as compounds, solvated in buffer, flow over and bind to target molecules immobilised on a sensing surface. In addition to use as a screening tool, SPR can reveal much about the binding interaction, such as kinetics, affinity, concentrations and stoichiometry. More recently, it has been coupled with targeted SDM to probe the topology of the ligand-binding pocket of the Adenosine A_{2A} receptor ($A_{2A}R$, a GPCR drug target for Parkinson's disease and neurodegeneration). This approach, termed Biophysical Mapping (BPM), generates a matrix of binding data for multiple ligands, of diverse chemotypes, against a panel of putative active site mutations. It provides a novel way to overcome the common constraints of needing labelled compounds for binding measurements. Using BPM, Zhukov *et al.* analysed eight mutants of $A_{2A}R$, each containing a single amino acid change predicted to alter ligand binding according to literature and in-house SDM data. Each mutant was screened against an array of 21 ligands by SPR to identify mutation-dependent differences in binding affinity and compile a map of binding site interactions within $A_{2A}R$. Together the data demonstrated clear SAR trends and binding poses for a series of xanthine ligands and proprietary compounds to $A_{2A}R$.[90] These predictions were very close to the experimental binding modes observed through subsequent high-resolution X-ray crystallography data and facilitated lead optimisation to identify a preclinical candidate for the treatment of Parkinson's Disease.

9.3.2   X-Ray Crystallography

X-ray crystallography has traditionally been viewed as a resource-intensive, low throughput method with the major bottleneck being generation of diffraction grade crystals. However, advances in molecular biology, specifically protein engineering, have significantly benefited structure determination.[91]

Today, heterologous expression systems are routinely used to generate recombinant proteins in quantities far superior to that usually found in native sources. Bacteria, typically *E. coli*, are often the expression host of choice given their speed, genetic tractability and low costs. For example, bacterial mutant strains lacking proteases or co-expressing rare codon tRNAs and molecular chaperones may be exploited to generate difficult or low-expressing target proteins. Despite these benefits, many eukaryotic proteins can prove intractable, given their need for complex folding or post-translational modifications, and hence require alternative expression systems based on yeast, insect or mammalian cells.

Molecular biology techniques may also be employed to reduce the inherent heterogeneity within the target protein and maximise its propensity to crystallise. With the aid of bioinformatics, target proteins can be engineered at the cDNA level to lack unstructured domains or flexible regions such as N- and C-termini or loop residues. In addition, SDM is often used to eliminate sites for post-translational modifications such as glycosylation and lipidation. The resulting constructs can be benchmarked for optimal truncation boundaries and monodispersity by chromatography based screening techniques, *i.e.* fluorescence size exclusion chromatography (fSEC), before progressing.

One very important class of drug targets that have proved notoriously difficult to crystallise are membrane proteins, particularly cell surface receptors and channels. This is primarily attributed to their instability outside cellular membranes and their inherent flexibility resulting from their need to transition between different signalling conformations. G Protein Coupled Receptors (GPCRs) are a classic example, wherein Rhodopsin remained the only GPCR structure solved up until 2007. Since then protein engineering techniques have been instrumental in uncovering many more GPCR structures, with two approaches in particular proving successful. The first, involves using SDM to engineer a minimal number of conformationally selective point mutations into GPCRs that enhance their thermostability. These stabilised receptors (StaRs) are more amenable to detergent extraction from cellular membranes and facilitate purification of corrected folded receptors for structural studies and/or screening applications such as BPM and TINS.[92] This approach has been used successfully to generate the high-resolution structure of β1-adrenergic receptor (β1AR), a key regulator of cardiac function. The second approach is to utilise fusion proteins or monoclonal antibodies. Fusion proteins have long been used in crystallisation to enhance protein folding and solubility and assist in purification. However, Cherezov *et al.* demonstrated that replacing the poorly structured third intracellular loop (ICL3) of β2-Adrenergic receptor (β2AR) with T4 lysozyme (T4L) generated a fusion protein more amenable to crystallography.[93] The T4L served to impose protein rigidity by restricting the movement of the flanking trans-membrane helices and increase polar surfaces for better crystal contacts. A similar plan to constrain the flexible ICL3 region was employed by Rasmussen *et al.* using monoclonal antibodies.[94] Together, these approaches have helped deliver eight different GPCR structures and provided evidence for a common architecture and conformational trigger within this protein superfamily. More pertinent for drug design, the data also alluded to the presence of a druggable cavity for small molecule binding in potentially all GPCRs. Another major benefit of structure-based drug discovery (SBDD) is the potential to improve drug selectively between homologous targets and thereby minimise side effects.[95] In the case of GPCRs, these difficulties are highlighted by the many antipsychotic drugs available today, such as clozapine and olanzepine, which despite acting at their intended target also hit additional GPCRs leading to side effects such as weight gain and cognitive defects. The increasing number of GPCR structures promises to significantly improve our understanding of this protein super-family and should help expose their huge therapeutic potential.[96,97]

The contributions of molecular biology techniques have helped transform X-ray crystallography into a high throughput strategy for drug discovery. With the aid of computational methods, the high resolution structural data can be used to model target proteins for docking and refinement of potential lead compounds. In addition, these 3D models also offer a viable platform for *in-silico* based fragment screening of libraries (*i.e.* virtual screening) for hit identification.

9.3.3 Safety and Clinical Efficacy

Most drugs that reach pre-clinical and clinical trials fail due to safety and toxicity concerns. In recent times, drug discovery has adopted a systems biology approach to evaluate the adverse effects of drug exposure to cells and minimise the rates of attrition. This approach typically involves a combination of genomics (*i.e.* genome-wide gene expression profiling using microarray methods), proteomics (*i.e.* cell and tissue-wide protein profiling using gel electrophoresis and mass spectroscopy techniques) and metabolomics (*i.e.* profiling drug induced metabolites typically using NMR and mass spectroscopy technology) to monitor the effects of drugs on a global scale.[98,99] Biomarkers that provide signatures of disease or successful treatment are also increasingly being employed as molecular reporters of drug safety.[100] A classic example is the Her-2 receptor which was found to be over-expressed in 20–30% of tumours from breast cancer patients. Despite indicating poor prognosis, this biomarker provided researchers with a new target for novel therapies. It resulted in the development of Herceptin, the first FDA approved antibody-based therapeutic to a cancer-related molecular marker (Her-2) effective at reducing proliferation of cancer cells in patients.[101]

Model organisms provide another means for testing toxicity and efficacy in late stage drug development. They offer a more relevant way to address issues such as metabolic inactivation, failure to reach target tissues and off-target effects. In support of this approach, genome-sequencing efforts have made it easier to generate transgenic and knock-out animals that better mimic human pathology. The benefits can be model systems that respond to toxicity, *i.e.* carcinogens, more rapidly (6–9 months instead of 2 years). In addition, whole model organisms, such as the nematode *Caenorhabditis elegans*, the fruit fly *Drosophilia melanogaster* and the zebra fish *Danio rerio* can be used for large-scale therapeutic screening, even in cases when the target is unknown.[102] For example, transgenic *C. elegans* models for Alzheimer's, Parkinson's and Huntington's diseases are available and can be used to screen for compounds that reverse the disease phenotype. These organisms can be used to evaluate drug safety whilst combining the genetic tractability, low cost and culture conditions compatible with HTS campaigns. Another advantage is that they can also flag the presence of pro-drugs, compounds metabolised from inactive to active drugs, which fail to register in *in vitro*/cell-based screens.

Progress in molecular and cellular biology has also helped address liabilities associated with drug absorption, distribution, metabolism and excretion (ADME). In particular, cloning and characterisation of the superfamily of cytochrome P450 enzymes (CYPs) has enabled *in vitro* drug screening for possible inhibitors of this important class of drug metabolising enzymes.[103] CYPs mediate the oxidative metabolism and inactivation of most (>75%) drugs and their inhibition can have significant implications for clearance, both of the parent drug and any co-administered drugs. Other important biological assays for drug toxicity include monitoring cardiac liability (hERG test) and genotoxic/carcinogenic potential (Ames test). Collectively, these approaches should help triage the best drug candidates for tolerance and efficacy in human clinical trials.

HINTS AND TIPS

Gene expression profiling is useful to identify novel targets, but mRNA levels do not always correlate with amounts of expressed protein, which can be regulated by numerous translational and post-translational mechanisms.

The ability to obtain the genome sequence of patients and to identify mutations associated with disease phenotypes is already turning the concept of personalised medicine into a reality.

Identification of the mutations in influenza that confer resistance to certain antivirals has led to the conclusion that resistance is less likely to develop against drugs that most closely resemble the natural ligand.[34]

These site directed mutagenesis coupled to techniques such as Surface Plasma Resonance (SPR), NMR and X-ray crystallography, are powerful tools to uncover structure–activity relationships.

KEY REFERENCES

J. Watson and F. Crick, *Nature*, 1953, **171**, 737–738.
R. Saiki, D. Gelfand, S. Stoffel, S. Scharf, R. Higuchi, G. Horn, K. Mullis and H. Erlich, *Science*, 1988, **239**, 487–491.
E. S. Lander *et al.*, *Nature*, 2001, **409**, 860–921.

REFERENCES

1. F. Griffith, *J. Hyg. (Lond).*, 1928, **XXVII**, 2–159.
2. O. Avery, C. McLeod and M. McCarty, *J. Exp. Med.*, 1944, **79**, 137–58.
3. A. Hershey and M. Chase, *J. Gen. Physiol.*, 1952, **36**, 39–56.
4. J. Watson and F. Crick, *Nature*, 1953, **171**, 737–738.

5. M. Meselson and F. W. Stahl, *Proc. Natl. Acad. Sci. U. S. A.*, 1958, **44**, 671–682.

6. S. Brenner, F. Jacob and M. Meselson, *Nature*, 1961, **190**, 576–581.

7. F. Crick, L. Barnett, S. Brenner and R. Watts-Tobin, *Nature*, 1961, **192**, 1227–1232.

8. D. Nirenberg and P. Leder, *Science*, 1964, **145**, 1399–1407.

9. H. Khorana, H. Büchi, H. Ghosh, N. Gupta, T. Jacob, H. Kössel, R. Morgan, S. Narang, E. Ohtsuka and R. Wells, *Cold Spring Harb. Symp. Quant. Biol.*, 1966, **31**, 39–49.

10. R. Saiki, D. Gelfand, S. Stoffel, S. Scharf, R. Higuchi, G. Horn, K. Mullis and H. Erlich, *Science*, 1988, **239**, 487–491.

11. C. A. Hutchison 3rd, S. Phillips, M. H. Edgell, S. Gillam, P. Jahnke and M. Smith, *J. Biol. Chem.*, 1978, **253**, 6551–6560.

12. F. Sanger, S. Nicklen and A. R. Coulson, *Proc. Natl. Acad. Sci. U. S. A.*, 1977, **74**, 5463–5467.

13. T. I. H. Consortium, *Nature*, 2003, **426**, 789–796.

14. T. E. P. Consortium, *Nature*, 2012, **489**, 57–74.

15. J. R. Vane, *J. Allergy Clin. Immunol.*, 1976, **58**, 691–712.

16. X. M. Xu, L. Sansores-Garcia, X. M. Chen, N. Matijevic-Aleksic, M. Du and K. K. Wu, *Proc. Natl. Acad. Sci. U. S. A.*, 1999, **96**, 5292–5297.

17. G. Bollag, J. Tsai, J. Zhang, C. Zhang, P. Ibrahim, K. Nolop and P. Hirth, *Nat. Rev. Drug Discov.*, 2012, **11**, 873–886.

18. J. Tsai, J. T. Lee, W. Wang, J. Zhang, H. Cho, S. Mamo, R. Bremer, S. Gillette, J. Kong, N. K. Haass, K. Sproesser, L. Li, K. S. M. Smalley, D. Fong, Y.-L. Zhu, A. Marimuthu, H. Nguyen, B. Lam, J. Liu, I. Cheung, J. Rice, Y. Suzuki, C. Luu, C. Settachatgul, R. Shellooe, J. Cantwell, S.-H. Kim, J. Schlessinger, K. Y. J. Zhang, B. L. West, B. Powell, G. Habets, C. Zhang, P. N. Ibrahim, P. Hirth, D. R. Artis, M. Herlyn and G. Bollag, *Proc. Natl. Acad. Sci. U. S. A.*, 2008, **105**, 3041–3046.

19. E. S. Lander *et al.*, *Nature*, 2001, **409**, 860–921.

20. J. C. Venter *et al.*, *Science*, 2001, **291**, 1304–1351.

21. R. Kramer and D. Cohen, *Nat. Rev. Drug Discov.*, 2004, **3**, 965–972.

22. J. Hall, P. Dennler, S. Haller, A. Pratsinis, K. Säuberli, H. Towbin, K. Walther, K. Walthe and J. Woytschak, *Nat. Rev. Drug Discov.*, 2010, **9**, 988.

23. S. R. Wiley, K. Schooley, P. J. Smolak, W. S. Din, C. P. Huang, J. K. Nicholl, G. R. Sutherland, T. D. Smith, C. Rauch and C. A. Smith, *Immunity*, 1995, **3**, 673–682.

24. H. Walczak, R. E. Miller, K. Ariail, B. Gliniak, T. S. Griffith, M. Kubin, W. Chin, J. Jones, A. Woodward, T. Le, C. Smith, P. Smolak, R. G. Goodwin, C. T. Rauch, J. C. Schuh and D. H. Lynch, *Nat. Med.*, 1999, **5**, 157–163.

25. G. Gremel, M. Rafferty, T. Y. K. Lau and W. M. Gallagher, *Crit. Rev. Oncol. Hematol.*, 2009, **72**, 194–214.

26. G. Roti and K. Stegmaier, *Br. J. Cancer*, 2012, **106**, 254–261.

27. R. Savino, S. Paduano, M. Preianò and R. Terracciano, *Int. J. Mol. Sci.*, 2012, **13**, 13926–13948.

28. T. A. Manolio, *Nat. Rev. Genet.*, 2013, **14**, 549–558.

29. P. M. Visscher, M. A. Brown, M. I. McCarthy and J. Yang, *Am. J. Hum. Genet.*, 2012, **90**, 7–24.

30. I. Sayers, *Clin. Sci. (Lond).*, 2013, **124**, 517–519.

31. J. W. Hanrahan, H. M. Sampson and D. Y. Thomas, *Trends Pharmacol. Sci.*, 2013, **34**, 119–125.

32. H. Luesch, *Mol. Biosyst.*, 2006, **2**, 609–620.

33. P. M. Carroll, B. Dougherty, P. Ross-Macdonald, K. Browman and K. FitzGerald, *Pharmacol. Ther.*, 2003, **99**, 183–220.

34. J. N. Varghese, P. W. Smith, S. L. Sollis, T. J. Blick, A. Sahasrabudhe, J. L. McKimm-Breschkin and P. M. Colman, *Structure*, 1998, **6**, 735–746.

35. M. Kato-Maeda, Q. Gao and P. M. Small, *Cell. Microbiol.*, 2001, **3**, 713–719.

36. A. Goffeau, B. G. Barrell, H. Bussey, R. W. Davis, B. Dujon, H. Feldmann, F. Galibert, J. D. Hoheisel, C. Jacq, M. Johnston, E. J. Louis, H. W. Mewes, Y. Murakami, P. Philippsen, H. Tettelin and S. G. Oliver, *Science*, 1996, **274**, 546, 563–567.

37. V. Wood, R. Gwilliam, M.-A. Rajandream, M. Lyne, R. Lyne, A. Stewart, J. Sgouros, N. Peat, J. Hayles, S. Baker, D. Basham, S. Bowman, K. Brooks, D. Brown, S. Brown, T. Chillingworth, C. Churcher, M. Collins, R. Connor, A. Cronin, P. Davis, T. Feltwell, A. Fraser, S. Gentles, A. Goble, N. Hamlin, D. Harris, J. Hidalgo, G. Hodgson, S. Holroyd, T. Hornsby, S. Howarth, E. J. Huckle, S. Hunt, K. Jagels, K. James, L. Jones, M. Jones, S. Leather, S. McDonald, J. McLean, P. Mooney, S. Moule, K. Mungall, L. Murphy, D. Niblett, C. Odell, K. Oliver, S. O'Neil, D. Pearson, M. A. Quail, E. Rabbinowitsch, K. Rutherford, S. Rutter, D. Saunders, K. Seeger, S. Sharp, J. Skelton, M. Simmonds, R. Squares, S. Squares, K. Stevens, K. Taylor, R. G. Taylor, A. Tivey, S. Walsh, T. Warren, S. Whitehead, J. Woodward, G. Volckaert, R. Aert, J. Robben, B. Grymonprez, I. Weltjens, E. Vanstreels, M. Rieger, M. Schäfer, S. Müller-Auer, C. Gabel, M. Fuchs, A. Düsterhöft, C. Fritzc, E. Holzer, D. Moestl, H. Hilbert, K. Borzym, I. Langer, A. Beck, H. Lehrach, R. Reinhardt, T. M. Pohl, P. Eger, W. Zimmermann, H. Wedler, R. Wambutt, B. Purnelle, A. Goffeau, E. Cadieu, S. Dréano, S. Gloux, V. Lelaure, S. Mottier, F. Galibert, S. J. Aves, Z. Xiang, C. Hunt, K. Moore, S. M. Hurst, M. Lucas, M. Rochet, C. Gaillardin, V. A. Tallada, A. Garzon, G. Thode, R. R. Daga, L. Cruzado, J. Jimenez, M. Sánchez, F. del Rey, J. Benito, A. Domínguez, J. L. Revuelta, S. Moreno, J. Armstrong, S. L. Forsburg, L. Cerutti, T. Lowe, W. R. McCombie, I. Paulsen, J. Potashkin, G. V. Shpakovski, D. Ussery, B. G. Barrell, P. Nurse and L. Cerrutti, *Nature*, 2002, **415**, 871–880.

38. P. Nurse, *Nat. Cell Biol.*, 2012, **14**, 776.

39. M. Moorthamer, M. Panchal, W. Greenhalf and B. Chaudhuri, *Biochem. Biophys. Res. Commun.*, 1998, **250**, 791–797.

40. G. W. Robinson, Y. H. Tsay, B. K. Kienzle, C. A. Smith-Monroy and R. W. Bishop, *Mol. Cell. Biol.*, 1993, **13**, 2706–2717.

41. C. Summers, F. Karst and A. D. Charles, *Gene*, 1993, **136**, 185–192.

42. P. Y. Lum, C. D. Armour, S. B. Stepaniants, G. Cavet, M. K. Wolf, J. S. Butler, J. C. Hinshaw, P. Garnier, G. D. Prestwich, A. Leonardson, P. Garrett-Engele, C. M. Rush, M. Bard, G. Schimmack, J. W. Phillips, C. J. Roberts and D. D. Shoemaker, *Cell*, 2004, **116**, 121–137.

43. S. J. Dowell and A. J. Brown, *Methods Mol. Biol.*, 2009, **552**, 213–229.

44. J. S. Gutkind, *Sci. STKE*, 2000, **2000**, re1.

45. R. E. Chen and J. Thorner, *Biochim. Biophys. Acta*, 2007, **1773**, 1311–1340.

46. M. V Metodiev, D. Matheos, M. D. Rose and D. E. Stone, *Science*, 2002, **296**, 1483–1486.

47. J. S. Gutkind, *Sci. STKE*, 2000, **2000**, re1.

48. J. A. Simon, P. Szankasi, D. K. Nguyen, C. Ludlow, H. M. Dunstan, C. J. Roberts, E. L. Jensen, L. H. Hartwell and S. H. Friend, *Cancer Res.*, 2000, **60**, 328–333.

49. D. Canaani, *Br. J. Cancer*, 2009, **100**, 1213–1218.

50. J. M. Enserink, *Molecules*, 2012, **17**, 9258–9273.

51. G. Giaever, D. D. Shoemaker, T. W. Jones, H. Liang, E. A. Winzeler, A. Astromoff and R. W. Davis, *Nat. Genet.*, 1999, **21**, 278–283.

52. G. Giaever, P. Flaherty, J. Kumm, M. Proctor, C. Nislow, D. F. Jaramillo, A. M. Chu, M. I. Jordan, A. P. Arkin and R. W. Davis, *Proc. Natl. Acad. Sci. U. S. A.*, 2004, **101**, 793–798.

53. A. Hamdi and P. Colas, *Trends Pharmacol. Sci.*, 2012, **33**, 109–118.

54. U. Rix and G. Superti-Furga, *Nat. Chem. Biol.*, 2009, **5**, 616–24.

55. M. Schenone, V. Dančík, B. K. Wagner and P. A. Clemons, *Nat. Chem. Biol.*, 2013, **9**, 232–240.

56. R. K. Choy and J. H. Thomas, *Mol. Cell*, 1999, **4**, 143–152.

57. M. D. Adams, S. E. Celniker, R. A. Holt, C. A. Evans, J. D. Gocayne, P. G. Amanatides, S. E. Scherer, P. W. Li, R. A. Hoskins, R. F. Galle, R. A. George, S. E. Lewis, S. Richards,

M. Ashburner, S. N. Henderson, G. G. Sutton, J. R. Wortman, M. D. Yandell, Q. Zhang, L. X. Chen, R. C. Brandon, Y. H. Rogers, R. G. Blazej, M. Champe, B. D. Pfeiffer, K. H. Wan, C. Doyle, E. G. Baxter, G. Helt, C. R. Nelson, G. L. Gabor, J. F. Abril, A. Agbayani, H. J. An, C. Andrews-Pfannkoch, D. Baldwin, R. M. Ballew, A. Basu, J. Baxendale, L. Bayraktaroglu, E. M. Beasley, K. Y. Beeson, P. V Benos, B. P. Berman, D. Bhandari, S. Bolshakov, D. Borkova, M. R. Botchan, J. Bouck, P. Brokstein, P. Brottier, K. C. Burtis, D. A. Busam, H. Butler, E. Cadieu, A. Center, I. Chandra, J. M. Cherry, S. Cawley, C. Dahlke, L. B. Davenport, P. Davies, B. de Pablos, A. Delcher, Z. Deng, A. D. Mays, I. Dew, S. M. Dietz, K. Dodson, L. E. Doup, M. Downes, S. Dugan-Rocha, B. C. Dunkov, P. Dunn, K. J. Durbin, C. C. Evangelista, C. Ferraz, S. Ferriera, W. Fleischmann, C. Fosler, A. E. Gabrielian, N. S. Garg, W. M. Gelbart, K. Glasser, A. Glodek, F. Gong, J. H. Gorrell, Z. Gu, P. Guan, M. Harris, N. L. Harris, D. Harvey, T. J. Heiman, J. R. Hernandez, J. Houck, D. Hostin, K. A. Houston, T. J. Howland, M. H. Wei, C. Ibegwam, M. Jalali, F. Kalush, G. H. Karpen, Z. Ke, J. A. Kennison, K. A. Ketchum, B. E. Kimmel, C. D. Kodira, C. Kraft, S. Kravitz, D. Kulp, Z. Lai, P. Lasko, Y. Lei, A. A. Levitsky, J. Li, Z. Li, Y. Liang, X. Lin, X. Liu, B. Mattei, T. C. McIntosh, M. P. McLeod, D. McPherson, G. Merkulov, N. V Milshina, C. Mobarry, J. Morris, A. Moshrefi, S. M. Mount, M. Moy, B. Murphy, L. Murphy, D. M. Muzny, D. L. Nelson, D. R. Nelson, K. A. Nelson, K. Nixon, D. R. Nusskern, J. M. Pacleb, M. Palazzolo, G. S. Pittman, S. Pan, J. Pollard, V. Puri, M. G. Reese, K. Reinert, K. Remington, R. D. Saunders, F. Scheeler, H. Shen, B. C. Shue, I. Sidén-Kiamos, M. Simpson, M. P. Skupski, T. Smith, E. Spier, A. C. Spradling, M. Stapleton, R. Strong, E. Sun, R. Svirskas, C. Tector, R. Turner, E. Venter, A. H. Wang, X. Wang, Z. Y. Wang, D. A. Wassarman, G. M. Weinstock, J. Weissenbach, S. M. Williams, T. Woodage, K. C. Worley, D. Wu, S. Yang, Q. A. Yao, J. Ye, R. F. Yeh, J. S. Zaveri, M. Zhan, G. Zhang, Q. Zhao, L. Zheng, X. H. Zheng, F. N. Zhong, W. Zhong, X. Zhou, S. Zhu, X. Zhu, H. O. Smith, R. A. Gibbs, E. W. Myers, G. M. Rubin and J. C. Venter, *Science*, 2000, **287**, 2185–2195.

58. M. E. Fortini, M. A. Simon and G. M. Rubin, *Nature*, 1992, **355**, 559–561.

59. C. Delvecchio, J. Tiefenbach and H. M. Krause, *Assay Drug Dev. Technol.*, 2011, **9**, 354–361.

60. L. I. Zon and R. T. Peterson, *Nat. Rev. Drug Discov.*, 2005, **4**, 35–44.

61. K. Howe, M. D. Clark, C. F. Torroja, J. Torrance, C. Berthelot, M. Muffato, J. E. Collins, S. Humphray, K. McLaren, L. Matthews, S. McLaren, I. Sealy, M. Caccamo, C. Churcher, C. Scott, J. C. Barrett, R. Koch, G.-J. Rauch, S. White, W. Chow, B. Kilian, L. T. Quintais, J. A. Guerra-Assunção, Y. Zhou, Y. Gu, J. Yen, J.-H. Vogel, T. Eyre, S. Redmond, R. Banerjee, J. Chi, B. Fu, E. Langley, S. F. Maguire, G. K. Laird, D. Lloyd, E. Kenyon, S. Donaldson, H. Sehra, J. Almeida-King, J. Loveland, S. Trevanion, M. Jones, M. Quail, D. Willey, A. Hunt, J. Burton, S. Sims, K. McLay, B. Plumb, J. Davis, C. Clee, K. Oliver, R. Clark, C. Riddle, D. Eliott, G. Threadgold, G. Harden, D. Ware, B. Mortimer, G. Kerry, P. Heath, B. Phillimore, A. Tracey, N. Corby, M. Dunn, C. Johnson, J. Wood, S. Clark, S. Pelan, G. Griffiths, M. Smith, R. Glithero, P. Howden, N. Barker, C. Stevens, J. Harley, K. Holt, G. Panagiotidis, J. Lovell, H. Beasley, C. Henderson, D. Gordon, K. Auger, D. Wright, J. Collins, C. Raisen, L. Dyer, K. Leung, L. Robertson, K. Ambridge, D. Leongamornlert, S. McGuire, R. Gilderthorp, C. Griffiths, D. Manthravadi, S. Nichol, G. Barker, S. Whitehead, M. Kay, J. Brown, C. Murnane, E. Gray, M. Humphries, N. Sycamore, D. Barker, D. Saunders, J. Wallis, A. Babbage, S. Hammond, M. Mashreghi-Mohammadi, L. Barr, S. Martin, P. Wray, A. Ellington, N. Matthews, M. Ellwood, R. Woodmansey, G. Clark, J. Cooper, A. Tromans, D. Grafham, C. Skuce, R. Pandian, R. Andrews, E. Harrison, A. Kimberley, J. Garnett, N. Fosker, R. Hall, P. Garner, D. Kelly, C. Bird, S. Palmer, I. Gehring, A. Berger, C. M. Dooley, Z. Ersan-Ürün, C. Eser, H. Geiger, M. Geisler, L. Karotki, A. Kirn, J. Konantz, M. Konantz, M. Oberländer, S. Rudolph-Geiger, M. Teucke,

K. Osoegawa, B. Zhu, A. Rapp, S. Widaa, C. Langford, F. Yang, N. P. Carter, J. Harrow, Z. Ning, J. Herrero, S. M. J. Searle, A. Enright, R. Geisler, R. H. A. Plasterk, C. Lee, M. Westerfield, P. J. de Jong, L. I. Zon, J. H. Postlethwait, C. Nüsslein-Volhard, T. J. P. Hubbard, H. Roest Crollius, J. Rogers and D. L. Stemple, *Nature*, 2013, **496**, 498–503.

62. R. T. Peterson, S. Y. Shaw, T. A. Peterson, D. J. Milan, T. P. Zhong, S. L. Schreiber, C. A. MacRae and M. C. Fishman, *Nat. Biotechnol.*, 2004, **22**, 595–599.

63. H. M. Stern, R. D. Murphey, J. L. Shepard, J. F. Amatruda, C. T. Straub, K. L. Pfaff, G. Weber, J. A. Tallarico, R. W. King and L. I. Zon, *Nat. Chem. Biol.*, 2005, **1**, 366–370.

64. C. C. Hong, Q. P. Peterson, J.-Y. Hong and R. T. Peterson, *Curr. Biol.*, 2006, **16**, 1366–1372.

65. G. J. Lieschke and P. D. Currie, *Nat. Rev. Genet.*, 2007, **8**, 353–367.

66. C. C. Hong, Q. P. Peterson, J.-Y. Hong and R. T. Peterson, *Curr. Biol.*, 2006, **16**, 1366–1372.

67. R. H. Waterston, K. Lindblad-Toh, E. Birney, J. Rogers, J. F. Abril, P. Agarwal, R. Agarwala, R. Ainscough, M. Alexandersson, P. An, S. E. Antonarakis, J. Attwood, R. Baertsch, J. Bailey, K. Barlow, S. Beck, E. Berry, B. Birren, T. Bloom, P. Bork, M. Botcherby, N. Bray, M. R. Brent, D. G. Brown, S. D. Brown, C. Bult, J. Burton, J. Butler, R. D. Campbell, P. Carninci, S. Cawley, F. Chiaromonte, A. T. Chinwalla, D. M. Church, M. Clamp, C. Clee, F. S. Collins, L. L. Cook, R. R. Copley, A. Coulson, O. Couronne, V. Curwen, T. Cutts, M. Daly, R. David, J. Davies, K. D. Delehaunty, J. Deri, E. T. Dermitzakis, C. Dewey, N. J. Dickens, M. Diekhans, S. Dodge, I. Dubchak, D. M. Dunn, S. R. Eddy, L. Elnitski, R. D. Emes, P. Eswara, E. Eyras, A. Felsenfeld, G. A. Fewell, P. Flicek, K. Foley, W. N. Frankel, L. A. Fulton, R. S. Fulton, T. S. Furey, D. Gage, R. A. Gibbs, G. Glusman, S. Gnerre, N. Goldman, L. Goodstadt, D. Grafham, T. A. Graves, E. D. Green, S. Gregory, R. Guigó, M. Guyer, R. C. Hardison, D. Haussler, Y. Hayashizaki, L. W. Hillier, A. Hinrichs, W. Hlavina, T. Holzer, F. Hsu, A. Hua, T. Hubbard, A. Hunt, I. Jackson, D. B. Jaffe, L. S. Johnson, M. Jones, T. A. Jones, A. Joy, M. Kamal, E. K. Karlsson, D. Karolchik, A. Kasprzyk, J. Kawai, E. Keibler, C. Kells, W. J. Kent, A. Kirby, D. L. Kolbe, I. Korf, R. S. Kucherlapati, E. J. Kulbokas, D. Kulp, T. Landers, J. P. Leger, S. Leonard, I. Letunic, R. Levine, J. Li, M. Li, C. Lloyd, S. Lucas, B. Ma, D. R. Maglott, E. R. Mardis, L. Matthews, E. Mauceli, J. H. Mayer, M. McCarthy, W. R. McCombie, S. McLaren, K. McLay, J. D. McPherson, J. Meldrim, B. Meredith, J. P. Mesirov, W. Miller, T. L. Miner, E. Mongin, K. T. Montgomery, M. Morgan, R. Mott, J. C. Mullikin, D. M. Muzny, W. E. Nash, J. O. Nelson, M. N. Nhan, R. Nicol, Z. Ning, C. Nusbaum, M. J. O'Connor, Y. Okazaki, K. Oliver, E. Overton-Larty, L. Pachter, G. Parra, K. H. Pepin, J. Peterson, P. Pevzner, R. Plumb, C. S. Pohl, A. Poliakov, T. C. Ponce, C. P. Ponting, S. Potter, M. Quail, A. Reymond, B. A. Roe, K. M. Roskin, E. M. Rubin, A. G. Rust, R. Santos, V. Sapojnikov, B. Schultz, J. Schultz, M. S. Schwartz, S. Schwartz, C. Scott, S. Seaman, S. Searle, T. Sharpe, A. Sheridan, R. Shownkeen, S. Sims, J. B. Singer, G. Slater, A. Smit, D. R. Smith, B. Spencer, A. Stabenau, N. Stange-Thomann, C. Sugnet, M. Suyama, G. Tesler, J. Thompson, D. Torrents, E. Trevaskis, J. Tromp, C. Ucla, A. Ureta-Vidal, J. P. Vinson, A. C. Von Niederhausern, C. M. Wade, M. Wall, R. J. Weber, R. B. Weiss, M. C. Wendl, A. P. West, K. Wetterstrand, R. Wheeler, S. Whelan, J. Wierzbowski, D. Willey, S. Williams, R. K. Wilson, E. Winter, K. C. Worley, D. Wyman, S. Yang, S.-P. Yang, E. M. Zdobnov, M. C. Zody and E. S. Lander, *Nature*, 2002, **420**, 520–562.

68. R. Kühn and W. Wurst, *Methods Mol. Biol.*, 2009, **530**, 1–12.

69. D. Li, Z. Qiu, Y. Shao, Y. Chen, Y. Guan, M. Liu, Y. Li, N. Gao, L. Wang, X. Lu, Y. Zhao and M. Liu, *Nat. Biotechnol.*, 2013, **31**, 681–683.

70. B. P. Zambrowicz and A. T. Sands, *Nat. Rev. Drug Discov.*, 2003, **2**, 38–51.

71. Y. Yu, E. Ricciotti, R. Scalia, S. Y. Tang, G. Grant, Z. Yu, G. Landesberg, I. Crichton, W. Wu, E. Puré, C. D. Funk and G. A. FitzGerald, *Sci. Transl. Med.*, 2012, **4**, 132ra54.

72. K. Duff, *Biochem. Soc. Symp.*, 2001, 195–202.

73. K. Gavrilov, W. M. Saltzman, 2012, **85**, 187–200.

Assays

TIM HAMMONDS*[a] AND PETER B. SIMPSON[b]

Cancer Research Technology Ltd, Wolfson Institute for Biomedical Research, Cruciform Building, Gower Street, London WC1E 6BT, UK; [b] Discovery Sciences, AstraZeneca, Alderley Park, Cheshire SK10 4TG, UK
E-mail: thammonds@cancertechnology.com

10.1 USE OF ASSAYS IN DRUG DISCOVERY

From a drug discovery perspective, an assay is an estimate of the ability of a project compound to inhibit, activate or modulate a biological process, and the development of appropriate robust (i.e. accurate and reproducible) assays is essential to project progression. The word 'estimate' is used very deliberately here—it is important to understand from the outset that all assays performed in drug discovery are estimates, be they single point estimates of percentage activity or multi-point estimates of compound potency. The error in assays is brought about by their multiparametric nature, and the more variables an assay has, then in principle the less exact it may become. Given the desired outputs and known variables, the principles of assay development are relatively simple. They are to build an assay that reliably estimates the activity of compounds whilst retaining a link to the biological target of the discovery project, and to have this assay run as often as is required or is possible to enable quick decision making. In drug discovery projects, compounds are tested thorough a series of assays (assay cascades) to determine their suitability for progression through the discovery process. Project assay cascades typically begin with simple biochemical tests and as compounds show promising properties the assays become more relevant to the disease target. This chapter gives an overview of commonly used assay formats in drug discovery cascades, highlighting their strengths and weaknesses and where they might be most effectively used and interpreted.

10.2 ASSAY TECHNOLOGIES

There are a large number of assay techniques available to the drug discovery bioscientist; each is designed to generate information from measurements such as optical spectroscopy, mass spectroscopy, radiometric techniques, nuclear magnetic resonance and surface plasmon

The Handbook of Medicinal Chemistry: Principles and Practice
Edited by Andrew Davis and Simon E Ward
© The Royal Society of Chemistry 2015
Published by the Royal Society of Chemistry, www.rsc.org

74. A. Fire, S. Xu, M. K. Montgomery, S. A. Kostas, S. E. Driver and C. C. Mello, **391**, 806–811.
75. N. Perrimon, J.-Q. Ni and L. Perkins, *Cold Spring Harb. Perspect. Biol.*, 2010
76. S. M. Elbashir, J. Harborth, W. Lendeckel, A. Yalcin, K. Weber and T. Tuschl **411**, 494–498.
77. S. E. Mohr and N. Perrimon, *Wiley Interdiscip. Rev. RNA*, **3**, 145–158.
78. A. J. Bridge, S. Pebernard, A. Ducraux, A.-L. Nicoulaz and R. Iggo, *Nat. Ge* 263–264.
79. C. A. Sledz, M. Holko, M. J. de Veer, R. H. Silverman and B. R. G. Williams, 2003, **5**, 834–839.
80. D.-H. Kim, M. Longo, Y. Han, P. Lundberg, E. Cantin and J. J. Rossi, *Nat. Bio* **22**, 321–325.
81. L. Wang, J.-Y. Zhou, J.-H. Yao, D.-R. Lu, X.-J. Qiao and W. Jia, *Biochem. Commun.*, 2010, **391**, 1363–1368.
82. P. Svoboda, *Curr. Opin. Mol. Ther.*, 2007, **9**, 248–257.
83. A. Reynolds, E. M. Anderson, A. Vermeulen, Y. Fedorov, K. Robinson, D. Leak W. S. Marshall and A. Khvorova, *RNA*, 2006, **12**, 988–993.
84. A. A. Seyhan, U. Varadarajan, S. Choe, W. Liu and T. E. Ryan, *Mol. Biosyst.*, 2012,
85. F. C. Gonsalves, K. Klein, B. B. Carson, S. Katz, L. A. Ekas, S. Evans, F T. Cardozo, A. M. C. Brown and R. DasGupta, *Proc. Natl. Acad. Sci. U. S. .* 5954–5963.
86. K. Ashrafi, F. Y. Chang, J. L. Watts, A. G. Fraser, R. S. Kamath, J. Ahringer an *Nature*, 2003, **421**, 268–272.
87. G. M. Keseru and G. M. Makara, *Drug Discov. Today*, 2006, **11**, 741–748.
88. R. A. Goodnow, *Drug Discov. Today Technol.*, 2006, **3**, 367–375.
89. C. Boozer, G. Kim, S. Cong, H. Guan and T. Londergan, *Curr. Opin. Biotechr* 400–5.
90. A. Zhukov, S. P. Andrews, J. C. Errey, N. Robertson, B. Tehan, J. S. Mason, F. M. Weir and M. Congreve, *J. Med. Chem.*, 2011, **54**, 4312–4323.
91. Z. S. Derewenda, *Acta Crystallogr. D. Biol. Crystallogr.*, 2010, **66**, 604–615.
92. N. Robertson, A. Jazayeri, J. Errey, A. Baig, E. Hurrell, A. Zhukov, C. J. Langm and F. H. Marshall, *Neuropharmacology*, 2011, **60**, 36–44.
93. V. Cherezov, D. M. Rosenbaum, M. a Hanson, S. G. F. Rasmussen, F. S. Kobilka, H.-J. Choi, P. Kuhn, W. I. Weis, B. K. Kobilka and R. C. Stevens, **318**, 1258–1265.
94. S. G. F. Rasmussen, H.-J. Choi, J. J. Fung, E. Pardon and P. Casarosa *et al.*, **469**(7329), 175–180.
95. M. Congreve, C. J. Langmead, J. S. Mason and F. H. Marshall, *J. Med. Che* 4283–4311.
96. J. S. Mason, A. Bortolato, M. Congreve and F. H. Marshall, *Trends Pharmacol. .* 249–260.
97. M. Congreve and F. Marshall, *Br. J. Pharmacol.*, 2010, **159**, 986–996.
98. Y. Cui and R. S. Paules, 2011, **11**, 573–585.
99. J. K. Nicholson, J. Connelly, J. C. Lindon and E. Holmes, *Nat. Rev. Drug Disc* 153–161.
100. D. A. Lewin and M. P. Weiner, *Drug Discovery Today*, 2004, **9**, 976–983.
101. D. Slamon, G. Clark, S. Wong, W. Levin, A. Ullrich and W. McGuire, *Science*, 1987,
102. J. Giacomotto and L. Ségalat, *Br. J. Pharmacol.*, 2010, **160**, 204–216.
103. D. C. Lamb, M. R. Waterman, S. L. Kelly and F. P. Guengerich, *Curr. Opin. Biote* **18**, 504–512.

resonance. Whilst there is utility in all of these methods, the design of assays for routine prosecution in drug discovery ultimately utilises three optical signal measurements: absorbance, fluorescence and luminescence. Within these three categories there are subcategories defined by the nature of photon generation, and each technology has advantages and disadvantages.

Absorbance is the oldest spectrophotometric method used to measure the concentration of a substance in solution. Absorbance-based assays are extremely useful for measuring the activity of enzymes and countless methods have been described utilising the absorbance of natural metabolites (*e.g.* NADH absorbs at 340 nm) or surrogate enzyme substrates such as *para*-nitrophenol phosphate (pNPP) for phosphatase enzymes.[1] For drug discovery purposes, absorbance assays have the attraction of being relatively simple (*i.e.* few components) and cheap, but they have fallen out of favour in recent years and have in general been superseded for HTS by fluorescence or luminescence based technologies. The reason for this is that typical wavelengths used to measure the absorbance of a product or substrate are in the 330–450 nm range. At this wavelength many small molecules in compound collections will also absorb and hence the potential for assay interference is very high. Absorbance assays that avoid this wavelength range still have their place in HTS. For example, the malachite green assay for inorganic phosphate reads absorbance at a wavelength of 590–650 nm.[2] Absorbance assays in the lower spectral region may still have a use in assays outside of HTS, such as secondary validation or mechanism of inhibition studies where fewer, more characterised compounds are used or in multi-step assays where compounds are washed out before the signal is generated (*e.g.* ELISA assays, described in Section 10.3.4).

Fluorescence assays have revolutionised the HTS process since they began to be introduced in the late 1980s. Fluorescent molecules with excitation and emission wavelengths above 450 nm have been available for many years, and many have been used as labelling moieties (tags) for other, non-fluorescent molecules (for a good insight into what is available, search vendor catalogues such as the Molecular Probes range from Invitrogen Life Sciences). With time, reagent suppliers have marketed fluorescent tags with higher wavelengths and increased emission ratios to avoid compound interference and increase assay sensitivity. The parallel development of more sensitive plate readers with the ability to pick up weaker fluorescence signals led to the development of high throughput assays not only for fluorescence intensity (FI), but also for florescence resonance energy transfer (FRET), time resolved fluorescence (TRF and TR-FRET) and fluorescence polarisation (FP). The principles involved in these assays are illustrated in Figure 10.1.

These techniques allow the assay scientist not only to measure the intrinsic fluorescence intensity of a molecule, but also its proximity to another molecule *via* FRET or its apparent size in solution by FP. This ability to measure proximity by fluorescence ultimately has all but replaced radioactive proximity assays such as scintillation proximity assays (SPA). Time resolved fluorescence assays bring an extra dimension in that the emission fluorescence of one assay component is relatively long lived, *i.e.* in the order of μs compared to ns. By adjusting the time at which the fluorescence is read, potentially interfering auto-fluorescent molecules are ignored by the assay system, decreasing the rate of false data. A combination of both TR fluorescence and FRET is known as TR-FRET.

Luminescence assays rely on a chemiluminescent reaction to generate an excited intermediate molecule, which decays to a ground state and emits a photon of defined wavelength. There are many enzymes that create chemiluminescence, but for drug discovery assays the most commonly used are firefly luciferase, Renilla luciferase and aequorin.[3] The reaction for firefly and Renilla luciferase is described in Figure 10.2.

Like Renilla luciferase, aequorin also binds coelenterezine, but only activates it to emit luminescence upon calcium binding. All reagents for these luminescence reactions are

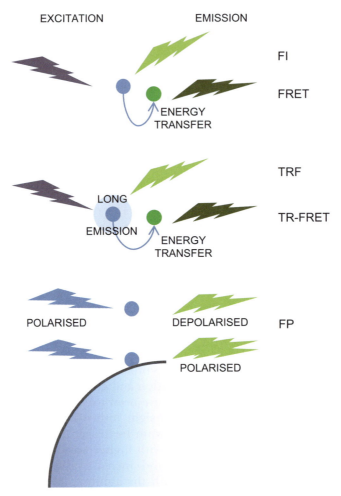

Figure 10.1 Fluorescence techniques for drug discovery assays. Five mechanisms of measurement are shown, each emission (green flash) is the direct or indirect product of and excitation (blue flash). Fluorescent molecules (fluors) are shown as spheres and resonant energy transfer by arrows. FI = Fluorescence intensity is directly measured from a single fluor. FRET = Fluorescence energy resonance transfer, where energy is transferred between fluors before emission. TR = Time resolved, the lifetime of decay and fluorescence of the primary (fluor) is relatively long, allowing for measures of FI or FRET to be made after background or interfering signals have decayed. FP = Fluorescent polarisation, a polarised mixture of excitation is used and the ratio of polarised emitted light is proportional to the apparent size of a fluor in solution, binding to a large protein or affinity bead can be detected by a change in the FP measure.

commercially available, therefore if an assay can be developed that generates any component of a luminescence reaction then that assay can be supplemented with the remaining reagents and the resulting luminescence is a measure of the initial reaction. Luminescent assays have the major advantage that they do not rely upon any other source of light. As such, very low levels of luminescence can be detected and these assays have an excellent limit of sensitivity. Luciferase assays are very robust and amenable to HTS and so they have a key role in biochemical and cell based assays. Luciferases and aequorin can also be genetically engineered into cellular assays to allow the measurement of intracellular processes such as gene expression (Molecular Biology

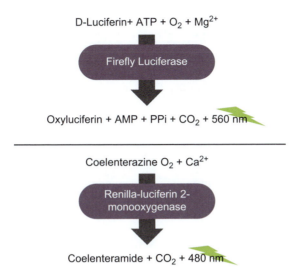

D-Luciferin+ ATP + O_2 + Mg^{2+}

Firefly Luciferase

Oxyluciferin + AMP + PPi + CO_2 + 560 nm

Coelenterazine O_2 + Ca^{2+}

Renilla-luciferin 2-monooxygenase

Coelenteramide + CO_2 + 480 nm

Figure 10.2 Common luminescence reactions for drug discovery assays. The two most common reactions used to detect luminescence, each enzyme utilises a number of cofactors to give an emitted photon (green flash). Measurement of any of the cofactors or the concentration of the enzyme itself can be achieved using these systems. Emitted light is in the blue/green region of the spectrum.

Chapter 9 and Section 10.3.6) or as a probe for calcium flux associated with GPCR signalling (Section 10.3.2).

As they perform a reaction in their own right, luminescence assay systems will attract a number of false positives *via* direct inhibition of the luciferase enzyme.[4] These can easily be detected by running test compounds against the reconstituted assay, but care should be taken to ensure identical conditions are created.

10.2.1 Assay Designs

Assays come in various 'shapes and sizes' and assay developers often describe assays by parameters concerning their build and output. In order to better understand the design, function and potential issues of an assay it is useful to know these terms and in doing so, one can better understand the workings of assays and thus better interpret project data; these are summarised in Table 10.1.

Although the list of terms in Table 10.1 looks complicated, complex assays soon become known by acronyms and trade names. ELISAs (Section 10.3.4) are in fact stopped reactions measured by a multiple wash antibody-based binding/proximity system coupled to a direct product formation absorbance assay, but we know them as ELISAs. Similarly Kinase-Glo™ from Promega[5] which measures ATP remaining after a kinase reaction is a stopped, coupled, cofactor depletion luminescence assay.

As alluded to in Table 10.1, the simplest and theoretically easiest assays to work with are homogenous, direct product accumulation assays, *i.e.* the substrate is directly transformed to product by the target in question and the reaction can be read once all of the reagents are added to the sample well. For HTS, this assay would also be stopped to ensure an even signal across plates tested at various times of the automation process. In reality, assays have to be developed to various designs utilising either entire commercial kits, or by combining in house and

Table 10.1   Common assay nomenclature. Combinations of these parameters and descriptors will be used to describe the format and readout of assays.

Operational Descriptor	Definition	Pros	Cons
Homogenous	All components can be added sequentially	Easy for HTS	Compounds may interfere
Washed	Components must be removed and replaced during the assay operation	Low compound interference Allows technically difficult assays	Cumbersome for HTS
Stopped	The assay is designed to have a fixed signal endpoint.	Easy for HTS Ideal for potency	Must be well timed Not ideal for kinetics
Kinetic	The reaction/signal continues unabated during the testing process	Gold standard for kinetics	Difficult for HTS
Readout value			
Accumulation (Bottom Up)	Measures product generated from zero	Best for kinetic analysis	
Depletion (Top Down)	Measures the loss of one reagent	Poor for kinetics if enzyme activity is weak	Often more noisy May require high % conversion
Ratiometric	Measures the balance between two species	Dispensing errors are adjusted for in ratio Often very robust	Poor linearity. May be limited for kinetics
Binding/ Displacement	Detects two molecules interacting	Direct measures of drug affinity possible	Binding may not be a direct measure of inhibition
Readout type			
Number	Cell number, number of mitochondria, cell protrusion *etc.*	Robust measure allows for easy statistics	Cell assays are biology and as such more error prone.
Direct	The reaction of interest generates the signal	Ideal reaction to measure	
Coupled	A series of other reactions generate assay output from the initial reaction	May increase the rate of product inhibited enzymes Allows measurement of any common cell metabolite	Prone to interference from compounds and contaminating enzymes. May be complex and expensive reagents involved
Proximity	The signal relies upon two objects being close to each other in space	Robust Allow for many non-enzymatic measurements	Poor for kinetics
Measure			
Product	The reaction product that one is interested in the cell	Ideal measure	
Cofactor	Any other reaction component that goes to make product	Many assays exist for cellular cofactors	More prone to contaminant protein interferences
Probe	Labelled versions of any assay component	Ideal for binding reactions	Not a direct measure of enzyme activity

purchased reagents. Signal generation is the one area where recent advances have expanded the number of assays that can be developed. Aside from the ongoing development of novel direct substrates and coupling systems, proximity assays are of particular importance as they have allowed for a far more flexible approach to assay development. In theory the development of an antibody or affinity media that specifically binds a substrate or product of a reaction allows for that component to be measured *via* associated off the shelf FP, FRET or TR-FRET reagents from a variety of manufacturers.

10.2.1.1 Assay Development

Assay development is first a discussion of what is required to be measured and how that will inform the project. One should always start with the end in mind and remember that an assay will be only be considered developed when it is routinely delivering data to the project team that can be used to make decisions on project progress. Achieving an assay signal is only the first stage of assay development, and much effort is then spent developing the basic assay. When an assay is fully developed and producing robust and useful data it is described as 'fit for purpose'.

Many discovery projects begin in high throughput screening and whilst assays for HTS must retaining a functional link to the cellular process to be targeted, the requirements for these assays are largely dictated by the ability of the assay to withstand the automated HTS process. The HTS assay, or a variant thereof, is typically used for routine potency assessment (IC_{50}) as these also require a high throughput. Alongside the primary assay, a series of bespoke biochemical assays will also be needed to elucidate hit compound specificity, selectivity and mechanism of inhibition. Later stage cascade assays are typically cell based, and here assay speed may be sacrificed for relevance to target biology. With all of these requirements in mind, assay development for drug discovery is the process of defining each assay in the cascade and making it fit for purpose. The primary goal is to analyse the reaction at hand and decide upon a format that will detect one of the assay components. Using a mixture of purchased and bespoke reagents, the assays scientist will then attempt to establish an assay signal. Once a stable signal has been established, then assay development proper can begin. Table 10.2 shows a list of considerations for an established assay to be developed at Cancer Research Technology (CRT) along with specific considerations if that assay is to be used for automated HTS.

Whilst this may seem an exhaustive list, it should be noted that each parameter could significantly affect the assay performance and thus project progression. The assay scientist may spend many experiments adjusting the buffer conditions in order to obtain optimal robustness of signal performance, spend time miniaturising an existing assay to allow for a more cost effective HTS, or adjust conditions in an assay to give reproducible IC_{50} data on a regular basis. In reference to the latter, in the case of protein kinase, the IC_{50} value can (and does) vary in response to a vast array of parameters, including a change in the concentration of; ATP, ADP, substrate, kinase protein, divalent cation, or a change in ionic strength, pH, temperature, substrate type, assay type, % product conversion, automation, mode of serial compound dilution, compound batch, compound reweighing from the same batch or the source of protein supplied to the assay. As a consequence, each and every one of these must be accounted for and/or measured in order to maximise the long term accuracy of IC_{50} values throughout the lifetime of a project. This is by no means a scenario unique to kinases; similar variables exist for every assay developed, however in practice many of these variables can be effectively fixed and many can be optimised in parallel. However, faced with a novel reaction, experience tells us that assays will take anything from 4 weeks to 6 months to fully develop as fit for purpose depending on their complexity, past precedence in measuring one of the assay components and the availability of reagents either in house or from commercial suppliers.

Table 10.2 Assay development parameters at CRT and their application to an HTS assay development project.

Assay Parameter Considered	Specific Requirement for HTS
Technology	Suitable for plate reader with stackers
Observed activity derives from the protein in question	Uncertainty is a NO GO for HTS
Availability of reference compound(s)	Potential NO GO if observed activity is uncertain.
DMSO tolerance	Ideally 0.1–5% DMSO.
Assay duration	Incubations >10 mins, read time per plate <10 mins
Kinetic or stopped mode *Kinetic for MOI experiments*	Stopped only, signal stable over a minimum of 2 hours after stopping—at least 4 hours preferable
Format	Must be 384 well
Process steps	Maximum 5 steps; addition volumes ≥ 2 µL
Final assay volume	First choice is 5–10 µL.
Reaction temperature	Must be performed at room temperature
Reaction/total assay time to allow for additions to large number of plates	Minimum 60 mins reaction time, total time <7 hours.
Plate layout	Automation friendly layout with minimal edge effects
Is it for a 1 mM fragment library HTS?	Calculate potential interference rate
Reagents	Available or to be synthesised in house?
	Stable for at least the estimated daily screen duration.
	Stocks stable for the duration of the HTS.
	Single homogenous batch for the entire HTS
Inclusion of detergent *e.g.* Triton X-100	Concentration must be <CMC
Buffer composition	Avoid viscous reagents *e.g.* BSA
Substrate and co-factor concentration	Maximise sensitivity to find desired inhibitor type. (Usually K_M at site of desired action).
	Possible effect of order of reagent additions.
Robustness	Z>0.5 and <3× observed variation of reference compound IC_{50} per run
Interference with assay readout	Artefact assay must be available
Cost	Aim for <15c /well, nominal <35c upper limit for full HTS

HINTS AND TIPS: WHEN ASSAYS FAIL

So, this will happen to you: An assay whose data you are depending on for the next round of chemistry decisions, doesn't deliver. Sometimes assays 'fall over' for lengthy periods and the assay scientist will have to re-optimise conditions to restore the assay to a robust state. Sometimes, the assay scientist will know why it failed – the robot dropped the plate, their student didn't put the detection reagent into the wells, the cells are infected. Often, though, an immediate answer and resolution is less satisfactory than those examples. As the chemist partner to the bioassay scientist, there can be helpful (and, unhelpful) questions to ask. Here are our thoughts, as bioassay scientists, on what those might be:

Helpful:
1. What is the process you follow to identify the root cause of assay failures? *'Learner' statements can make it feel like a joint project issue rather than the individual feeling 'blamed'. And you will be able to see how on top of the issue the bioassay scientist is, by their response.*
2. Is this assay failure due to fundamental issues with the assay design? If so what do you think needs done to rebuild the assay. How long will that take? *I have been in*

a project where the lead chemist readily agreed with me that we should build a more robust, more informative assay, and stopped synthesis for 2 months—which created the breathing space for us to do this well. By the way, the project reached its preclinical candidate on time and the new assay played a key role to get the candidate compound accepted...

3. If this was human error, well ok, we all make mistakes. But are there QC steps you can instigate to make a repeat of this unlikely? *The highly repetitive nature of many assay tasks inevitably leads to the odd dropped ball, as a simple guideline at CRT we rarely intervene unless someone make multiple mistakes or exactly the same mistake twice. Intervention should be training or process orientated.*

4. If this was due to technology issues—are they readily resolvable? If not, do you need my support to strengthen the business case for capital investment? *If these issues are not resolvable in the short term, again think about the usefulness of an ideal, but low robustness assay to the project. Should it be swapped for a less ideal more robust assay*

5. Do you have all the people and help you need to fix the assay in the timeframe you describe? What might being the assay back online quicker? *Your assay scientist should think of this, but junior staff often do not. As project leader you should explore the idea of asking for more resource in any area to speed the fix of a critical path assay.*

Unhelpful:
1. What did you do wrong this time?
2. Why are you always letting me down? Why can't I rely on you?
3. How hard is putting compounds in a well and testing them, for goodness sake?
4. When will this be fixed, I need the data by...
5. Can I have someone else to do this please?

And so on. But I am sure you, informed reader, would never behave like that anyway.

10.3 EXAMPLES OF COMMON DRUG DISCOVERY ASSAYS

This section describes assays that one would typically encounter in drug discovery projects. These assays are in the approximate order one would find them in a typical target based drug discovery project cascade, beginning with methods to measure the activity or binding of specific targets and ending with cellular readouts of both biological pathways and gross cellular effects. However, it should be appreciated that all of the assay types listed below could be used for high throughput screening of libraries if necessary.

10.3.1 Enzyme Assays

A simple enzyme reaction scheme is shown below:

$$E + S + C \leftrightarrow ESC \leftrightarrow E + Sp + Cp \tag{10.1}$$

Where E = Enzyme, S = Substrate and C = Cofactor. The suffix p represents product associated with turnover of cofactor or substrate. A co-factor has been added to this example to highlight the availability of the cofactor in the design of assays. The production or loss of each item on the scheme can theoretically be assayed as a measure of enzyme activity as can direct binding to the enzyme of interest.

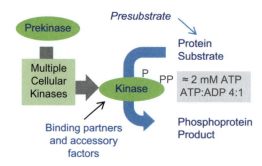

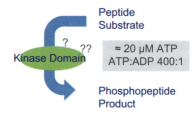

37 °C, cell pH and salt, molecularly crowded cytoplasm

25 °C, 'optimal' pH and salt, artificial buffer

Figure 10.3   Schematic of cellular and HTS kinase reactions. Whilst both produce a phospho product and ADP, cellular kinases and their substrates are often the products of pre-reactions and take place in a high viscosity cytoplasmic environment surrounded by many other proteins and cofactors. For the purposes of HTS this reaction is often simplified to a single domain of the protein and a peptide from the substrate in buffer conditions optimised for HTS and not for comparison to the cellular environment.

It is not possible to extensively cover all of the enzymes one might encounter in this small section, so we will focus on the protein kinases as an example and illustrate the variety of assays available for this enzyme class by way of example of what can be achieved to measure any enzyme reaction. Protein kinases play major roles in multiple disease relevant cell signalling pathways[6] and as a consequence reagent manufacturers have produced multiple assay formats to enable HTS and downstream cascade assays. A generic protein kinase reaction in a cellular context is summarised in Figure 10.3 alongside the typical kinase assay developed for HTS.

Herein is a good example of how HTS arrives at a much simpler assay based on pragmatic considerations, with the assumption that an inhibitor of the phosphorylation of a short substrate peptide by the isolated kinase domain of a protein will also inhibit the full length kinase protein in a cellular context. Whilst this may seem unrealistic, the development of many kinases inhibitors from outputs of HTS prosecuted *via* these assays has to some extent borne out this model of drug discovery.[7]

Figure 10.4 shows a schematic of a kinase reaction along with the points at which it is commonly assayed. There are over 20 kinase assay formats currently in the market.[7,8] Table 10.2 shows examples of each intervention type listed with brief details on assay design, also served by hyperlinks in the Table 10.3.

The range of assays available for kinases is a good example of the breadth of assays that can be developed for an enzyme class using tools that target specific components of the enzyme reaction. This basic principle of assay design applies equally to any other enzyme system as they do the kinase system. If there is a way to specifically measure the depletion or accumulation of a cofactor, substrate or product then assay development can be attempted.

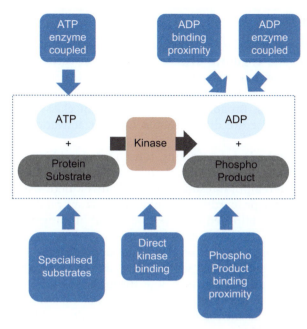

Figure 10.4 Common points of intervention for protein kinase assay design and development.

Table 10.3 Example kinase assays and their mechanistic basis.

Assay Measure	Trade name	Manufacturer	Mechanism	Readout
ATP detection	Kinase Glo	Promega	Direct couple	L
ADP coupled	ADP Glo	Promega	Enzyme coupled	L
	ADP hunter	DiscoveRx	Enzyme coupled	FI
ADP binding proximity	TranScreener	Bellbrook	Ab displacement	FP
	HTRF TranScreener	CisBio	Ab displacement	HTRF
Direct Kinase Binding	Lanthascreen	Invitrogen	Probe displacement	FP
	HitHunter	DiscoveRx	Enzyme complementation	L
Phospho product binding	IMAP	Molecular devices	Metal affinity binding	FP
	Lance	Perkin Elmer	Ab Proximity	TR-FRET
	Delphia	Perkin Elmer	Ab Proximity	TR-FRET
	Kinease	CisBio	Ab proximity	HTRF
Specialised substrates	Z-lyte	Life Technologies	Specific product cleavage	FRET
	Omnia	Life Technologies	Substrate rearrangement	FRET

As drug discovery targets from different enzyme areas emerge in the literature, reagent companies are usually swift to see opportunities and begin to mass produce tools for assay development. Assays developed in these and emerging fields will most likely utilise all of the platform technologies and designs that have been previously validated, such as the generation of reagents that bind specifically to a product, *e.g.* trimethylated histone specific antibodies for histone deacetylases, that can be coupled to a proximity readout,[9] or the creation of bespoke non-native substrates for the enzyme that change their fluorescent properties when processed by the enzyme of interest, *e.g.* fluorescent ubiquitin analogues to measure deubiquitinating enzymes.[10] The plethora of enzyme assays and specific antibodies now available means that in general most enzyme assays can now at least be attempted in assay development for HTS and beyond. Along with extensive use in enzyme assays, the fluorescence proximity/polarisation binding assays are also of value in measuring direct binding interactions such as protein–protein binding and receptor–ligand interactions.

10.3.2 Ion Channel Assays

Ion channels are a drug target class of importance to a range of therapeutic areas, notably within pain, psychiatry, and cognitive diseases, and within cardiovascular conditions.[11] It is often desirable to avoid activity against certain ion channels (*e.g.* hERG block, which was linked to QTc syndrome), while being highly desirable to inhibit, potentiate, activate or modulate activity at other ion channels. Broadly there are two main classes of ion channels—voltage gated, and ligand gated. These are differentiated by their activation mechanism and require different assay approaches.

Originally, the main methods for assessing ion channels were either radioligand binding assays, or patch clamp electrophysiology. Patch clamp electrophysiology remains a gold standard assay, giving high temporal resolution on channel kinetics, and informs the scientists on a wide range of channel properties. Automated electrophysiology also allows control of holding and test voltages, not afforded by FLIPR or ligand-binding assays. This control can be manipulated to better study the biophysical properties and kinetics of an ion channel, *e.g.* current–voltage relationships, inactivation, or state dependence. These are important ion channel properties that can be exploited within drug discovery programs.

The limitations of these techniques (*e.g.* low throughput and inconvenience) led to the emergence of kinetic fluorescence based methods.[11-14] For ligand gated channels, it became relatively straightforward to pre-incubate cells with molecules, add the ligand, then measure changes in intracellular ions (Ca^{2+}, or pH changes linked to Cl^-) or membrane voltage, using these kinetic plate readers. Some attempts have been made to convert these readers to activate and measure voltage gated channels,[15] with mixed success. Second generation voltage dyes have become the normal fluorescence screening approach, on the latest generation of Fluorometric Imaging Plate Reader FLIPR™ or other similar readers.

More recently, technology advances have enabled projects to revert to having the patch clamp at the front of the assay cascade. Sophion, Nanion, Molecular Devices and others have released increasingly sophisticated automated platforms for patch clamping[16-18] enabling a direct rather than surrogate screening endpoint. The throughput enables large scale profiling or hit finding on these platforms, as well as project structure–activity-relationship screening. The Molecular Devices IonWorks™ Barracuda™ reader has the claimed ability to assay voltage and ligand gated channels on a single platform.[19] These platforms can be used by ion channel project scientists, and also by safety pharmacologists screening compound liability against panels of cardiac or brain ion channels linked to cardiovascular adverse events, or seizure risks. Continual improvement in the scope throughput and accuracy of these machines should see previously laborious assays performed in high throughput as a matter of routine.

10.3.3 GPCR Assays

G protein-coupled receptors (GPCRs) are a large superfamily of seven transmembrane domain proteins which sense changes in concentration of ligands outside of cells and transduce this into a variety of intracellular signals. The binding of a ligand to a GPCR results in conformational changes, which in turn leads to the activation of an associated G protein heterotrimer. The activated G protein modulates the activity of intracellular enzymes, which control the production of key intracellular second messengers. These second messengers then act on many downstream targets that regulate gene transcription and cell functions. There are four subclasses of G protein to which different GPCRs may couple; most GPCRs can couple to more than one type although may show a preference. Coupling to the different types of G protein ($G_{\alpha s}$, $G_{\alpha q/11}$, $G_{\alpha i}$, $G_{\alpha 12/13}$) can lead to multiple, often parallel and interacting, signal pathways being activated,[20] as shown in Figure 10.5. While this can make the biological consequences of GPCR inhibition

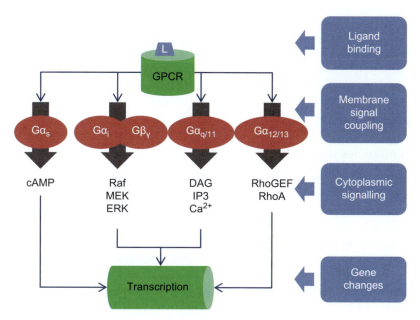

Figure 10.5 Common points of intervention for GPCR assay design and development. As assays move further from the GPCR itself and into downstream signalling events, so they become more prone to off target false positives.

challenging to predict, it opens up a wide variety of types of assays which screeners can build for this drug target family.

Some assay classes discussed elsewhere, such as AlphaScreen™ (Perkin Elmer), and FLIPR™ (Molecular Devices), are suitable for detecting major signalling pathways activated by GPCRs. For example, GPCRs can couple *via* $G_{\alpha q/11}$ proteins to convert lipid phosphates to inositol trisphosphate, which releases Ca^{2+} from the endoplasmic reticulum. Ca^{2+}-sensitive dyes lead to an increasing fluorescence signal, detected by high throughput kinetic plate readers such as FLIPR,[21] or lower cost readers such as FlexStation. This is a relatively transient signal, so for screeners with a lower budget or a preference for less time-dependent readouts, alternative detection can rely on the breakdown of IP_3 back to inositol monophosphate (IP_1). Lithium can prevent recycling of IP_1, allowing accumulation of IP_1 and endpoint detection by immunoassay. One popular kit for IP_1 detection is IPOne™ (CisBio), a Time-Resolved Fluorescence-based immunoassay detection kit that involves competition between the cell lysate and labelled IP_1 for an anti-IP_1 monoclonal antibody.[22]

Coupling to $G_{\alpha s}$ leads to activation of adenylate cyclase, which catalyses conversion of adenosine trisphosphate to the signalling molecule cyclic adenosine monophosphate (cAMP). Conversely, coupling to $G_{\alpha i}$ proteins leads to inhibition of adenylate cyclase. Immunoassay detection kits are commonly used for detection of cAMP levels. Alternatively, multiple G proteins lead to downstream signalling consequences on a widespread protein called ERK. A change in the phosphorylation level of ERK is a relatively generic indicator of GPCR activity that can be measured in various assay kits, from AlphaScreen™ to HTRF and beyond.

For so-called 'orphan' GPCRs where the ligand and intracellular coupling mechanisms are unknown, efforts to 'de-orphanise' these receptors (*i.e.* identify native or synthetic ligands) require the ability to detect activity regardless of signalling pathway. Some assays are now available which do not specifically rely upon the type of G protein or signal transduction pathway activated. For example, Corning's Epic platform detects shape changes, or mass redistribution within cells, using resonant waveguide grating (RWG) technology. Changes in cell confluence, adhesion and activation status in response to addition of a compound will be detected in a generic change in readout. This approach, or conceptually similar impedance-based detection, can be used for screening for ligands of the GPCR, or for detection of modulators or inhibitors.[23]

There are also high content based approaches for GPCR assays, using for example the coupling properties of G proteins to beta-catenin. This provides information on the effects of compounds on multiple aspects of cell biology responses to compound addition, though with often slower and more complex assay protocols.

10.3.4   Immunoassays and ELISA-type Assays

An enzyme-linked immunosorbent assay (ELISA) uses an enzyme coupled to an antibody, or antigen, to detect a specific protein. This can be either a biochemical *in vitro* assay, or a cellular assay typically measuring protein abundance in a cell lysate. There are multiple versions of the basic ELISA principle shown in Figure 10.6.[24] The assay sample, containing an unknown amount of the protein of interest, is immobilized on a surface: either non-specifically *e.g. via* adsorption, or specifically *i.e.* captured by the surface being coated in an antibody. Then, a second, target-specific antibody is applied over the surface so it can bind to the protein. Importantly, this antibody is coupled directly or indirectly to an enzyme. In the final step, a known amount of a substrate of the enzyme is added, producing a detectable signal *e.g.* a change in colour, which is used as the assay readout.

ELISA type assays were commonplace in HTS before the advent of fluorescent proximity binding assays; they are now rarely used for primary HTS of isolated proteins, but have found a significant niche in the analysis of cellular proteins. In many fields of drug discovery, notably in oncology and in inflammation, detection of changes in the level or phospho-form of a protein within signalling pathways can be a key project objective.[6] Cellular target engagement or 'biomarker' assays become mandatory for most discovery cascades and as a consequence, a variety of assay formats may be evaluated in order to accurately, specifically, and sensitively, quantify that change. A key reason for this is to ensure that compounds identified from a primary screen are acting in the expected and desired way, rather than interfering with the assay detection process or acting on an unanticipated alternative protein, before the investment of extensive time and effort in optimising that compound into a lead series.

In cell based systems, the major potential disadvantage of the absorption-based, or indirect, ELISA format is assay specificity. Many or all proteins in the sample may stick to the microtitre

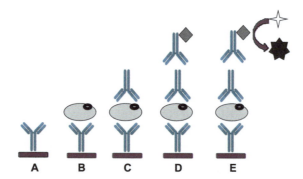

Figure 10.6   Outline of the sandwich ELISA protocol. (A) The microplate well is coated with capture antibody. (B) The assay sample is added and antigen binds to the capture antibody. Non-binding components are washed away. (C) The detection antibody, which binds captured antigen, is added. Unbound antibody is washed off. (D) An enzyme conjugated secondary antibody is added. This binds to the secondary antibody. Unbound antibody is washed off. (E) Enzyme substrate is added, converted to a detectable product and the signal is produced.

well surface—so small concentrations of the protein of interest are in competition with other proteins when binding to the well surface. The sandwich, or direct, ELISA format addresses this issue, by using a capture antibody which specific for the antigen of interest to facilitate its adherence to the surface.[23,25]

Performing an ELISA assay therefore involves having at least one antibody with high specificity for the particular antigen in the target protein, as a very high level of specificity is important for building a reliable screen. For example, in kinase signalling there are thousands of proteins, including hundreds of kinases and substrates; and consequently there are many thousands of different phosphorylation sites.[24,26] If the assay format does not have antibody tools that detect only the target of interest, the project team can be readily misled as to the activity of their lead compounds in cells. To deliver a cellular ELISA whose data can be correctly interpreted by the project team therefore requires highly specific assay tools, and well validated cell assays, to prevent such misinterpretation of screening data.[27]

A key challenge is, then, in selecting generating, and then validating, the appropriate antibodies for development of ELISAs. (Indeed, many non assay development scientists have an active or semi-active interest in the sensitivity and specificity of antibodies, particularly specificity, as they are at the core of home pregnancy testing kits and as such must be fit for purpose!) Most screeners have learnt the painful lesson that not all commercially available antibodies are anything like as target-specific as the vendor claims.[28] Western blots, combined with cell imaging assays, can build confidence that the antibody is binding only (or predominantly) to a protein of the expected molecular weight, located in the expected cellular component.

ELISAs are very specific assays when performed correctly. However, as a preferred technology they are relatively laborious in nature. There are multiple addition steps, and washes, required over the course of an extended period of time.[30] For screeners, the more steps there are in an assay, the more things there are to go wrong or to increase data variability between testing occasions such as automation failures, reagents going off, batch-to-batch differences, or screener error. The major drawback with ELISA assays is that they require repeat wash steps— which even using current high quality automated equipment can increase assay noise, length of time required for plate processing, and assay failure rates.

Homogeneous, no-wash protein detection and quantitation assays are available. AlphaScreen™ is a method for detecting cell signals[29] which contains two bead types; donor beads and acceptor beads. Donor beads contain phthalocyanine, which converts ambient oxygen to an excited form, singlet oxygen, when illuminated at 680 nm. Within its 4 μs half-life, singlet oxygen can diffuse approximately 200 nm in solution. If an Acceptor bead is within that proximity, energy is transferred from the singlet oxygen to thioxene derivatives within the Acceptor bead, producing light emission. The proximity of Acceptor to Donor beads depends on the interaction between the biomolecules that are bound to them. For example, if a receptor is attached to Donor beads and the relevant ligand to Acceptor Beads, then when the ligand binds its receptor, energy is transferred from Donor to Acceptor beads and a signal is produced. So, addition of compounds which displace the ligand will reduce the light signal and allow measurement of the affinity of that compound. AlphaScreen can be used to replace ELISAs by adding capture and detection antibodies to each bead and thus recapitulating the sandwich ELISA format in Figure 10.6 in solution. Compared to ELISAs, AlphaScreen assay times are typically shorter, and the assays are homogenous, however, AlphaScreen has disadvantages too,[29] and its use in HTS has declined in recent years. Limitations in assay sensitivity restrict its use to fairly abundant proteins, assays can be affected by long exposure to ambient light, compounds in screening libraries can scavenge radical oxygen thereby interfering with the detection, and donor beads photo bleaching limits the assay format to a single read.

Some new types of immunoassay have been developed attempting to address this by delivering simpler, quicker, and potentially more robust protocols. A variant of the basic ELISA assay involves the detection antibody being conjugated to a fluorophore. This assay can be read in a plate-based fluorescent reader, or for higher information content, a microscopy high content imaging detection system can be utilised. This version enables multiple assays within each well, as different detection antibodies can be coupled to fluorophores which emit at different wavelengths.[31] Indeed there are now a range of detection approaches available to the assay development scientist, that provide different levels of sensitivity, specificity, and multiplexing options. For example the AlphaLISA™ platform is an evolution of the AlphaScreen™ technology in which the beads contain Europium as the Acceptor fluor,[29] which is reportedly less prone to interferences from assay components. AlphaLISA™ assays require small sample volumes yet they have over a 100-fold greater analytical range than ELISAs. Uptake of this technology has been mixed so far, perhaps reflecting the multitude of options now available to the screener (Fluorescence Polarisation, Time Resolved Fluorescence, Meso Scale Detection) that all can be satisfactory for many drug target assays—though they may well detect overlapping but certainly non-identical sets of active compounds.[32] Ongoing advances in detection technologies are expected to continue to drive improvements in specificity and sensitivity.[33]

10.3.5 Mass Spectrometry

For many years, liquid chromatography-mass spectrometry (LC-MS) has been an important, but late, component of drug testing cascades.[34] It has the advantages of unambiguous identification of minor changes in cellular or assay components and in proteins that may not be amenable to conventional assays.[35] In many cases, labelling of proteins is not required, and sensitivity is sufficiently high for assays to be performed in native or primary cells. However, implementation of LC-MS in routine screening was limited by assay complexity and technological speed. However, recent technological advances in sensitivity and speed of commercial LC-MS systems, *e.g.* RapidFire 360 High-throughput MS System (Agilent),[35,36] and the simplification of the pre-processing of samples has enabled processing speeds of under 10 seconds per sample.

Perhaps the most common use for LC-MS in new drug discovery is the human colon adenocarcinoma cell line (Caco-2), which is used for the measurement of the permeability potential of a compound.[36,37] Another widely used higher throughput *in vitro* mass spectrometry assay is to assess the potential of a compound to inhibit of one or more of the human cytochrome P450 isoforms in parallel.[36,37] Mass spectrometry can be highly desirable as it is highly specific detection and suitable for target molecules that are not amenable to most other assay types. For example minor changes in a metabolite can be detected *via* mass spectrometry where it would be impossible to generate a specific, selective antibody-based assay.

An exciting ongoing challenge is to broaden the applicability of higher throughput mass spectrometry into cellular assays. Separation and quantitation of multiple parallel readouts from interfering cellular components in a timely way is becoming highly enabling for complex metabolism drug targets. Novel, high performance UPLC systems such as the Waters Acquity I-Class are continuing to improve the ability of researchers to derive complex separations at increasing throughput.[38] Cell LC-MS screening assays are an attractive option, *e.g.* for metabolism assays in which multiple pathway components may be affected by a compound, and inhibition at one point can be counteracted by compensation elsewhere in the pathway or network. It is practical to measure multiple metabolites in LC-MS in parallel at reasonable levels of throughput, which should ensure more sophisticated understanding of true compound mechanism of action and downstream pathway consequences.

10.3.6 Cell Reporter Gene

Cell reporter genes (CRG) assays measure the production of an intercellular enzyme known as a reporter enzyme. The basic principle of reporter gene assays is outlined in Figure 10.7. Cellular

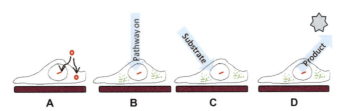

Figure 10.7 Cell reporter gene assay principle. (A) The gene encoding the reporter enzyme is genetically engineered into the host cell, either as a transient plasmid, or *via* incorporation into the genome. (B) Upon pathway stimulation reporter enzyme (green dots) are produced, the strength of pathway signal is proportional to the amount of enzyme produced. (C) The enzyme is exposed to a substrate and (D) converts this substrate to a product that can be quantitated by standard technologies (*e.g.* luminescence). Stages (C) and (D) may be performed after a cell lysis step depending on the type of reporter enzyme produced.

production of the reporter enzyme is coupled to the intracellular activity of a target protein, or target cellular pathway. After pathway stimulation, by for example a receptor ligand, the resulting reporter enzyme is quantitated by adding back the components of the reporter enzyme reaction either to whole cells or after a cell lysis step. Positive data from CRG assays will often result in a compound being described as exhibiting 'on pathway' activity, *i.e.* capable of blocking or activating the target signalling pathway of interest in a cellular setting.

Typical intracellular reporter enzymes used for compound analysis are firefly luciferase, Renilla luciferase and beta galactosidase. The luciferases give rise to luminescence signals as previously described, whilst beta-galactosidase can be used to give an absorbance or fluorescence based signal *via* modified substrates.[39] Reporter genes can also be secreted into the media, *e.g.* placental alkaline phosphatase (Phospha-Light™), or be fluorescent proteins in their own right such as green fluorescent protein (GFP). The signal from enzyme-based CRG assays is very high compared to the number of molecules produced as the product of the assay is an enzyme, which in turn allows for a degree of signal amplification during the detection reaction.

With major suppliers offering ready to use kits and contract services for the creation of bespoke cell reporter assays these assays have become common in discovery cascades. There are two main types of CRG assays; transient transfection, where a plasmid is directly into the cell during the assay process and stable transfection, where the reporter gene is permanently incorporated into the host genome of a cell. Transient transfection assays may be difficult to make robust as the transfection process is prone to variation, whereas stable transfection assays can be extremely robust and as such can be used for compound profiling and for 'target agnostic' or 'pathway' HTS.[40]

The main drawbacks of CRG assays are the potential for off target activity of test compounds. Inhibition of any component of the transcriptional or translational complexes in the cell, cell toxicity or inhibition of the reporter enzyme will all give false positives. All of these false positives are easily removed *via* a parallel, non-target based CRG assay of similar readout and a counterscreen assay of this type is therefore a pre-requisite of any CRG cascade assay.

10.3.7 High Content Cell Assays

High content screening (HCS) methods for drug discovery employ fluorescent labelling of cellular components, and microscopy-based cell image acquisition, combined with quantitation of the resultant fluorescent image by specific analytical algorithms. Stains for cell nuclei and membranes are used as marker for individual cells, and each cell is automatically counted and scored individually for a response in the assay. This enables visualization and measurement of specific protein changes, such as phosphorylation, translocation between compartments, or altered abundance, within individual cells. HCS can be used to determine the mode of action,

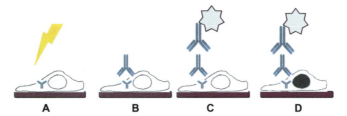

Figure 10.8   Outline of a typical fixed cell HCS protocol. (A) Adherent cells are fixed and permeabilised to ensure antigens are in position. Cells are washed. (B) The antigen detection antibody is added and binds to antigen. Cells are washed to remove unbound antibody. (C) The secondary antibody, labelled with a fluorophore is added and binds to the antigen detection antibody. Cells are washed to remove unbound antibody. (D) DNA is visualised using a fluorescent DNA intercalating dye. The cells are visualised at both the antibody and DNA flour wavelengths. The images are combined to show the amount of antigen present and its localisation relative to the cell nucleus.

potency, and selectivity of potential drug molecules within a biologically relevant cellular context, and also provides information on off-target liabilities and cytotoxicity.

There are currently two predominant types of high content assay: fixed cell antibody and stain based detection systems; and live imaging systems based on fluorescent protein expression and non-toxic cell stains. In the fixed cell format, shown in Figure 10.8, the assay is similar to the ELISAs detailed above. After treatment of cells with compounds, the antigens within cells are fixed *in situ* by an agent such as paraformaldehyde. An agent to permeabilise cells may also be used. An antibody against the antigen of interest is added to the well, incubated, then excess unbound antibody is washed off. A second antibody that is coupled to a fluorophore is then added that binds to the first antibody.[41] By varying the fluorophore used, it is possible to differentially fluorescently label multiple cellular components to build up a more sophisticated view of the cell and therefore of the impact of compound treatment on the cell.

In live cell HCS, cell lines are used in which a protein target or targets are co-expressed with fluorescent labels, *e.g.* GFP. Live imaging of the fluorescent proteins allow for assays that monitor the appearance, longevity, localization of the proteins and also the translocation of these protein targets between cellular compartments upon external cell stimuli. These live cell assays can be very powerful biomarker or pathway assays, but in some cases, tagging with a large label such as GFP may interfere with protein behaviour. In addition, these cell lines can be more time-consuming and costly to obtain.

Because individual cells are labelled imaged and analysed in each assay well, HCS assays have the ability to analyse a subpopulation of cells selectively within a complex network or co-culture of cells. This is increasingly important in areas such as stem cell-based assays in which multiple types of cells are present. One fluorescent antibody is used as a specific marker of each cell type of interest (*e.g.* stem cells) and all analysis can be related back to only those cells expressing that marker, thereby avoiding interference from other cells within the well (*e.g.* feeder cells or differentiated cells).

HCS also enables a primary endpoint to be multiplexed in individual cells in a single well with parallel or downstream events.[42] For example, for kinase targets HCS kinase assays can track, in parallel, the primary target-drug interaction event within cells, *e.g.* autophosphorylation of a kinase, and in the same well, track off-target effects *e.g.* phosphorylation of substrates mediated by other related kinases, or downstream events within the pathway of interest. This combination of both proof of principle and proof of mechanism in a single experiment can provide evidence that a compound enters the cell and inhibits the target, and also that the appropriate downstream consequences are delivered. Multi-endpoint HCS-based safety screens are now frequently performed prior to evaluating compound efficacy *in vivo*.

The use of HCS within drug discovery cascades has increased rapidly over the last 10 years. The technology of HCS instruments has matured in terms of robustness and ease of use offering a range of image resolution, speed, built-in informatics capability, across a wide range of prices. Complex HCS assays are therefore attractive to projects, but unfortunately they are relatively burdensome on screening and compound profiling teams. Expert staff can be required to provide expert guidance to maximize impact from HCS. Novel, multiplexed HCS assays can take longer to build, particularly if novel analysis algorithms are required. Automating complex cell shape and translocation assays and maintaining throughput without compromising quality are also challenging and this can be a significant drain of resources and can delay overall project timelines. Simpler, two-fluorophore cell imaging assays provide lower information content, but can prove to be more robust in routine screening settings and if a HCS assay exists, it may be worth using the simpler assay as a triage. However, it should be mentioned that in hit finding, the investment in HCS may have its merits. HTS groups often report a relatively high success rate in providing progressable leads from cell imaging HTS campaigns,[43] and of course the hits which are found are by definition already active in the cell and may come with extra data which allows for removal of off target and toxic compound structures from the outset.

10.3.8 Cell Phenotypic Assays

A cellular phenotype is based on an observation of something that a cell does as a whole entity, rather than an estimate of protein or pathway modulation. In these assays we are only concerned with counting whole cell phenomena that are part of or recapitulate the disease state to be targeted. Phenotypic assays typically represent the final *in vitro* assay before an animal model of disease is utilised and compounds that show activity are said to have cellular phenotypic efficacy. Phenotypic assays give rise to a measure of cell number, cell morphology, cellular chemistry or cell movement. These generic categories can be further subdivided to more specific phenotypes such as cell cycle status, apoptosis induction, cell differentiation, neurite outgrowth, pseudopodia formation, cell invasion, cell migration, cellular metabolic status *etc*. Each of these subtypes are suitable for more detailed analysis that will also show an associated number, shape chemistry (including biochemistry) or movement phenotype.

The simplest cell phenotype is a measure of cell number, this can give rise to a measure of direct toxicity *i.e.* how many of the original cells died? Or cell growth *i.e.* how many more cells do I have than when I started? Early assays relied on manual microscopy cell counting, *e.g.* by trypan blue exclusion, but these were not amenable to HTS. HTS assays began with assays such as the sulfarhodamine blue (SRB) assay which stain adhered monolayer and use the amount of adsorbed stain as a measure of biomass (cells) remaining after compound treatment.[44] As these involve a washing step to remove unbound stain, advances were made when specific reagents were developed that reacted with intracellular dehydrogenases to give an absorbance or fluorescence signal (*e.g.* MTT, MTS) allowing for homogenous assays for viability.[45] Assays based on cellular enzyme levels give a good estimate of cell number with adherent or non-adherent cells but suffer from the drawback that dead cells may still release active dehydrogenases into the media and thus give a slightly false reading. The CellTitre-Glo assay (Promega) is similar in that it measures the amount of ATP in each well as a surrogate for cell number *via* the luciferin–luciferase system. This assay is also a homogenous mix and read system and has been reported to accurately measure as few as 100 cells per well and is one of the most accurate and robust high throughput measures of cell number available. Along with simple estimates of cell number, biochemical reagents are now available for cellular assays that can estimate the number of live and dead cells and the induction of specific cellular pathways. Assays are often multiplexed *via* kits, *e.g.* the ApoTox-Glo™ caspase-live-dead from Promega which utilises three different fluorescent substrates to gives a simultaneous measure of all three listed outputs.

The advent of high content plate readers that take images of multiple cells in each well has brought the cell survival assay full circle, automating the staining and counting of cells stained as live or dead. This method can be the most accurate estimate of live cell number, also allowing for co-cultures and specific labelling of cell types to be targeted or spared. Another recent innovation in cell growth assays is the ability to measure cells growing in real time. Bright-field imaging systems, such as IncuCyte (Essen Instruments), now facilitate a range of experimental studies on long-term cellular behaviour, as they can be left in the cell incubator for extended time periods, and are now available with fluorescent options enabling combined use of cell shape changes with fluorescent detection of biological events within the cell. Devices such as the IncuCyte can give growth curves over many days. This is particularly useful when looking at compounds which have more subtle growth inhibitory effects or which may have a delay in cell toxicity dependent upon their mechanism of action.

It is worth noting here that in all assays to measure cell number and compound toxicity, the number of cells that one starts with and the duration of incubation can have marked effects on the apparent potency of compounds. Cells in culture will rarely survive happily in microplates for more than 4 or 5 days without showing some signs of nutrient deprivation or edge effects due to evaporation in outer plate wells. As a consequence the number of cells seeded and the duration of incubation should always be tightly controlled.

The most sensitive assays for cell survival are clonogenic assays, where cells are treated as individual cell suspensions and plated for colony formation. These slow, long-term assays can often reveal subtle long term toxicities missed in monolayer experiments, but they are tedious to prepare in a reasonable throughput. The 'blunt' nature and potential for off target false positives in simple direct toxicity measurements often mean that they are replaced by project specific surrogates such as measures of cell replication (*e.g.* Bromovinyldeoxyuridine (BrdU) incorporation) or apoptosis (caspase induction)) with the assumption that this is the desired mode of cell death or arrest and that a positive in a shorter duration, possibly more robust assay will give rise to the desired gross phenotype in a longer term clinical setting.

All of the assays for cell number described above are designed to be used in monolayers, or 2D cultures. Aside from colony formation assays, they are relatively straightforward to design and build as fit for purpose and can be run routinely in project cascades. Growth of cells in 3D cultures is thought to more accurately recapitulate the *in vivo* environment; cells are suspended in a semi-solid media comprising either agarose or cellular polymers such as collagen and grow as distinct clumps and clusters. The 3D nature of the assay means that cells are more difficult to accurately visualise or solubilise for high content or biochemical analysis, although some assays claim to be able to do this.[46] Counting of cells is often done by eye, or possible by dedicated hardware, but the difficulty of setting up robust routine 3D growth assays means that they are often reserved for projects where their need is absolute. There are specially coated plates available that allow for non-adherent 2D growth, said to more closely resemble 3D conditions[47] and these may be a valuable option if the output can be validated against a 3D model for the cell type in question.

Assays for cell morphology, such as cell shape or neurite outgrowth, have become significantly easier to run in high density by using HCS analysis (Section 10.3.7). By differentially labelling appropriate cellular components such as the cell membrane, microtubules and nucleus, the images from HCS can be subjected to complex image analysis and effects on individual cell volume and shape readily quantitated.[48]

Assays for cellular metabolites can also be classed as phenotypic and here the advent of mass spectrometry has increased the scope and range of metabolites that can be measured in reasonable throughput (Section 10.3.5). Whilst bespoke biochemical assays can measure a change in the ratios of individual, or pairs of key cellular co-factors, *e.g.* ATP:ADP, Glutamine: Glutamate, or NAD:NADH, to indicate a phenotypic metabolic switch or stress response, metabolomics assays by mass spec can rapidly measure many tens or hundreds of individual metabolites and generate phenotypic response associated with disease states and responses to chemical agents.[49]

Assays for cell movement are exemplified by chemotaxis assays, cell migration assays and cell invasion assays. Chemotaxis and cell migration monitor the movement of cells in 2D and have been adapted from low throughput experiments to higher throughput modalities. The advent of high content imaging has led to these assays becoming relatively easy to measure in reasonable throughput. The Platypus technologies Oris™ system is designed to leave spaces for migrating cells using a preformed template and the Bellbrook Iuvo™ assay plate system utilises a capillary for chemotaxis experiments. Quantitation of these assays ranges from simply measuring the size of area 'closed' or the distance migrated, to in-depth analysis of the movement of individual cells.[50] However, there is a large degree of noise associated with these assays due to the essentially random movement of the cells and the overlapping effects of toxicity, cell growth and true (anti-)migratory effects on the measured signal. Whilst experimentally they are becoming easier to run, this inherent variability often leads to issues in gaining robustness for day to day use. Invasion assays present all the issues of migration assays multiplied by the issues of working in and assaying cells in 3D cultures. There are companies that sell kits and specialist plates for invasion assays but even with these, design and robust prosecution of invasion assays on a regular basis is not for the faint-hearted and should be given due time and consideration. It is often better to run more robust high content or pathway assays as a triage and reserve the best compounds for bespoke, high information 3D invasion assays before moving to animal models of *e.g.* metastatic disease.

HINTS AND TIPS: ASSAY TROUBLESHOOTING

When an assay fails there will follow a period of time whereby the root cause is established and fixed, this may be lengthy because as discussed in Section 10.2.1.1 there may be many parameters to re-optimise. Typical troubleshooting ideas that assay scientists go through to fix broken assays will include:

1) Check whether the SOP (standard operating protocol) for the assay was followed correctly?
2) Check whether the dispensers and reader were functional and correctly calibrated. *Or if they have just been serviced, if the old data were generated on pre-calibrated equipment data can (rarely) shift slightly after an annual service, or (more commonly) upon using a new piece of hardware for the same task.*
3) What is the recent historical pattern of the assay in terms of signal window, assay robustness? *In AstraZeneca we track this each week so we can see downward trends in assay performance and address them before the assay reaches 'failure' point.*
4) How old were the reagents being used? *Cell lines for example do change behaviour/assay performance after a certain number of passages in culture; also proteins 'go off' over time and become no longer usable in a biochemical assay.*
5) Has any component of the assay involved a recent new batch (protein, detection reagent, *etc.*)? Has this batch been properly demonstrated to perform as well as the previous batch? *Failing an easy answer to this and the previous question, it is not uncommon for assay scientists to start afresh with new batches of every reagent. If the assay behaves again, this is quicker than a laborious examination of each component and data can again be generated. In the long term it is good to know which reagent is most likely to be the cause and this can be followed up separately.*

10.4 ASSAY OUTPUTS

Once assays are in place and are part of the project it is essential to correctly analyse data that they produce and to continually monitor assay robustness. Data analysis is the estimate of a value assignable to a compound activity; this is done objectively *via* statistical analysis and curve fitting. Robustness is estimated from the analysed data *via* a series of quality control measurements designed to constantly monitor the 'fit for purpose' nature of the assay. This is therefore an objective collection of statistics followed by a subjective analysis of their importance for data accuracy. Where these two come together is in the interpretation of individual datasets against large collections of assay data collated over the lifetime of a project, robustness indicators will often explain data anomalies and allow useable data from otherwise unusual values.

10.4.1 Data Analysis

Most assays will arrive at a figure indicating the percentage activity or inhibition of an assay compared to the maximum assay signal, or the concentration of compound at which that assay gives half maximum output. Depending on the assay type and the preferences of discovery teams, there are differences in the way these data are calculated. Common measures of compound effects are summarised in Table 10.4.

Because assays data can be presented in various formats, misreading, mislabelling and subsequent misunderstanding of the data is possible unless a standard data format is agreed upon for each assay. Though it seems obvious, if a team is used to seeing percentage inhibition data, when confronted with a table of percentage of control data (*i.e.* 100% – % inhibition) confusion often ensues. There are also many other measurements that are presented as indicators of compound potency, particularly in GPCR inhibitor discovery[12] and it is worth asking your assay scientist exactly what is being measured and presented.

To generate any of these types of data, the assay must have a series of internal controls and reference compounds that the assay scientist uses to set estimates of full reaction signal and no reaction signal. For maximum accuracy of compound testing, the design of controls and blanks should be such that the positive control (100% signal) recapitulates the activity of the assay in the absence of only the test compound and the negative control (0% signal) recapitulates a fully compound inhibited system. Commonly used positive and negative controls are solvent only and commonly used negative controls are reactions lacking enzyme, or reactions with a saturating amount of a known inhibitor. The number of control well replicates is kept high per assay so as to accurately estimate the 100% and 0% values. With accurate estimates of each of these all of the associated, often high throughput, test compounds can be more robustly

Table 10.4 Typical measure reported in the make test cycle of drug discovery projects.

Measure	Definition
% of control	The percentage of IR remaining
% inhibition	Percentage of IR that is inhibited
% or Fold stimulation	Increase in signal above the positive control as a function of IR
IC_{50}	Concentration required to give 50% of the IR
EC_{50}	Concentration at which an agonist gives 50% of the maximum possible induced response for that agonist
GI_{50}	Concentration at which organisms grow to 50% of the control value
LD_{50}	Concentration at which 50% of organisms are dead

IR = inhibitable response, *i.e.* the difference between the maximum and minimum assay signals.

analysed for significant effects on the assay. A standard measure of robustness for an assay is the Z′ factor (Equation 10.2).[51]

$$1 - \frac{3(STDEVcontrol) + 3(STDEVblank)}{(VALUEcontrol - VALUEblank)} \tag{10.2}$$

Where STDEV = standard deviation of the control or blank and VALUE = the measured mean value of each parameter. The Z′ factor takes into account the variation in the controls and blanks and it is set against a basic hypothesis that if you cannot accurately and repeatedly distinguish between control and blank wells, then you cannot designate intermediate test reading as significantly different from either. The Z′ factor has added to or supplanted measures such as signal to background and signal to noise as measures of assay robustness. Indeed assays with traditionally 'poor' signal to background values of <1.5 are now commonplace for HTS as the very low errors associated with readouts give very robust Z′ scores and negate the need for a large signal window. Z′ factors of 0.5 or above are typically classed as acceptable but it should be remembered that the Z′ factor is a guide and is ultimately used subjectively; assays with Z′ scores of 0.4 are commonly passed if inspection of the data suggests that potency estimates are most likely robust. Once an assay Z′ is calculated and the assay is declared as having run robustly, then values can be assigned to compounds. Outside of HTS where % inhibition is the commonest value calculated, the most common measure reported is compound potency as the IC_{50}. Dose response curves are plotted in a log-linear fashion and the IC_{50} is typically estimated from a fit of the curve to 7–12 datapoints corresponding to different compound concentrations, usually at three-fold dilutions or thereabouts. The most common equation used to estimate the IC_{50} is the four parameter logistic model, or Hill slope model.[12] This generates an IC_{50} and a fitting error associated with this value (based not on multiple IC_{50} values, but on the variance of individual values from the fitted curve). It also generates a Hill slope; in an ideal system (*i.e.* 'ideal' in the sense of the gas equation, a kinetically perfect assay with a perfectly competitive inhibitor) the Hill slope should have a value of 1. However, not all reactions are 'ideal' and whilst shallower or steeper Hill slope gradients can be an indicator of inappropriate compound mechanism[52] this value has an acceptable range of 0.5–2.0. Assay scientists will look at each individual IC_{50} curve and employ standard curve fitting rules and regulations before passing each curve as adequately fitted. For example, there will be a minimum number of points on the curve that may be designated as outlier errors and ignored, the Hill slope and maximum and minimum values will be investigated and IC_{50} values should not be extrapolated and at least one (preferably more than two) datapoints should lie on either side of the calculated IC_{50} value. It is important to realise that IC_{50} values generated are not absolute measures of how potent a compound is, as previously discussed, they are estimated values highly dependent upon the reaction conditions under which they are gathered. As such no two IC_{50} experiments will ever give truly identical data and the role of the IC_{50} is to provide a relative ranking of compound potency against other compounds tested in a particular assay.

10.4.2 Robustness Analysis and Data Comparison

Potency data are generated on a regular basis and as such generate large datasets that require constant monitoring for drift and error. To ensure a reliable IC_{50} ranking system, standard compounds are included in each assay run and compared with historic data to ensure data robustness with time. The hypothesis here is that if the standard compounds are grossly inaccurate, so are the novel compounds tested in that run. To guard against false alarms, more than one standard compound per plate is advised and both have to shift significantly to warrant wholesale data rejection. Once a batch of assays is passed and a compound has a validated IC_{50}, it can be compared to the historic dataset and ranked accordingly. In comparing compound IC_{50} data, it is essential to look at the ratio of IC_{50} values and not the absolute difference.[12] Equations based on the variance of fit error and controls can give degrees of confidence that data are different or similar, but as rule of thumb, with average assay robustness, if two values of IC_{50} are within threefold of each other they cannot be definitively classed as different. So a compound with an IC_{50} at 60 nM is not more potent than to one at 100 nM, whereas an IC_{50} of 1 nM is more potent than 6 nM despite only being "5 nM different" it is in fact six times lower. For an in-depth mathematical analysis of IC_{50} measurement and comparison the Assay Guidance Manual[12] is recommended.

As well as tracking compound potency, all assay performance parameters are tracked with time. Analysis of all historical assay data gives an indication of 'assay drift' towards a lack of robustness and rather than have whole runs failed on the basis of gross variance of the assay the assay developers can intercept and fix the assay to ensure continuous performance.

When assay data are robust and reliable, there are few issues surrounding the relationship between the project chemist and the assay scientist, mutual trust prevails. But there will be times when data are not making sense and at this time it is critical for the project chemist and assay scientist to use their acquired knowledge of the each other's roles to help to determine whether a problem exists and where it does, to assist each other in resolving it such that the project can once again function effectively.

HINTS AND TIPS: WHAT CAN YOU DO WHEN YOUR DATA LOOKS ODD?

As a senior project chemist who has read this chapter you should by now expect assays to give somewhat variable data, but there will be times when things start to make less and less sense *vs.* the hypotheses being tested. Forearmed is forewarned in these circumstances, get to know and the inherent variability of each of the assays in the project cascade and keep an eye on the robustness as the project progresses. Simply asking how the assay went each time is a reasonably good way to keep on top of any changes in assay with time. Flagging poorly soluble, potentially reactive or unstable series in advance and presenting the assay history at each project meeting is also good practise. This is by no way a chemistry only problem, two way communications are critical; at one memorable project meeting at CRT the assay scientist feedback on a (quickly deprioritised) problem compound was that "it was yellow/green in solution this morning and colourless two hours later". It is also worth noting here that whilst it is rare that an assay goes wrong for a specific compound amongst other compounds, it can happen (misfiring liquid dispensers, empty compound wells, *etc*). Once you have established that there is an issue with data then retesting is always the first port of call, explain your concerns and ask for retests and see if the data was indeed just a blip.

If every compound in the assay 'shifts' in potency then the assay or automation conditions will most likely need to be redeveloped somewhat to bring the numbers back on track, similar to the situation in the Assay Troubleshooting section. If an isolated compound or series repeatedly gives non-robust or unexpected data then it is most likely a compound issue and this can include compound handling as well physical and chemical properties of the compound at hand. Solubility is by far the most common reason that compounds do not give reproducibly accurate data in assays, couple this to a variable assay and the situation can quickly become unreadable. It is therefore also helpful to learn to read and understand IC_{50} data and hill slopes and to look for signs of shift (all data points move) or poor solubility (top concentration data points are poor quality) as you go along.

If an assay apparently works but the data are confusing, here are some thoughts on good questions to ask and things to think of when data do not match hypotheses.

1) Did the control compound values match previous values? *Sometimes assays will misbehave for certain compounds or compound series. A major variable here is compound solubility, but it may be that compounds stray into reactive or protein aggregating space. If the assay and standards are accurate but the test compounds are not, look to the sample management and MOA and then to the assay.*

2) Is the synthesis batch of the test compounds the same as that evaluated in the assay previously? *The commonest reason for changes in potency is batch to batch variation or variance due to reweighing from the same batch.*

3) What is the inherent occasion-to-occasion variability on this assay? *It may be that the differences you see in potency between test occasions for the compounds are within the expectations of a 'noisy' assay—and the bioassay scientist should know how variable their assay is.*

4) How steep was the Hill slope and how good was the curve fit? *It is possible the raw data is fine, but the uploaded IC_{50} value is inaccurate due to a poor curve fit. Hopefully the bioscientist has noticed this, but in a large screening run it may have been missed. In AstraZeneca we upload all data but annotate in the database where and why data we do not think should be aggregated due to factors such as poor fit. We rely on chemists to use and interpret these data with their annotations.*

5) Is the hypothesis correct? *Theory is by definition malleable, but data are what they are. If you understand what the assays was designed to do and you know that it performed appropriately then use the data it tells you to change the theory, rather than try to change or repeat the data to fit the theory.*

KEY REFERENCES

Assay Guidance Manual, ed. S. Sittampalam, N. Gal-Edd, M. Arkin, *et al.* 2012, Online. http://www.ncbi.nlm.nih.gov/books/NBK53196/

This assay guidance manual is a web-based resource that is produced by Eli Lilly and is a comprehensive overview of assay development and data analysis. It is a more detailed one stop shop for further information on all of the areas covered in this chapter. We would particularly recommend Chapter 13: "Basics of Enzymatic Assays for HTS".

J. H. Zhang, T. D. Chung and K. R. Oldenburg, *J. Biomol. Screen.*, 1999, **4**, 67.

The original paper to describe the Z-factor, essential reading to understand what assay scientists mean when they talk about assay robustness and how they measure it.

R. Zhang and X. Xie, *Acta Pharmacol. Sin.*, 2012, **33**, 372.

An excellent overview on the workings of and assays available for GPCRs, a major class of cell surface proteins recognised as one of the most successful therapeutic targets for a broad spectrum of diseases.

REFERENCES

1. C. MacKintosh, in *Protein Phosphorylation: A Practical Approach*, ed. D. G. Hardie, IRL Press, New York, 1993, p. 221.
2. J. Feng, Y. Chen, J. Pu, X. Yang, C. Zhang, S. Zhu, Y. Zhao, Y. Yuan, H. Yuan and F. Liao, *Anal. Biochem.*, 2011, **409**(1), 114.
3. L. Rowe, E. Dikici and S. Daunert, *Anal. Chem.*, 2009, **81**(21), 8662.
4. D. S. Auld, N. T. Southall, A. Jadhav, R. L. Johnson, D. J. Diller, A. Simeonov, A. Austin and J. M. Inglese, *J. Med. Chem.*, 2008, **51**(8), 2372.
5. M. Koresawa and T. Okabe, *Assay Drug Dev. Technol.*, 2004, **2**, 153.
6. P. Cohen, *Nat. Rev. Drug Discov.*, 2002, **1**, 309.
7. H. Ma, S. Deacon and K. Horiuch, *Expert Opin. Drug Discov*, 2008, **3**, 607.
8. Y. Jia, Y. C. Quinn, C. S. Kwak and R. Talanian, *Curr. Drug Discovery Technol.*, 2008, **5**, 59.
9. N. Gauthier, M. Caron, L. Pedro, M. Arcand, J. Blouin, A. Labonté, C. Normand, V. Paquet, A. Rodenbrock, M. Ro, N. Rouleau, L. Beaudet, J. Padrós and R. Rodriguez-Suarez, *J. Biomol. Screen.*, 2012, **17**(1), 49.
10. P. Geurink, E. F. Oualid, A. Jonker, D. S. Hameed and H. Ovaa, *ChemBioChem*, 2012, **13**(2), 293.
11. G. J. Kaczorowski, O. B. McManus, B. T. Priest and M. L. Garcia, *J. Gen. Physiol.*, 2008, **131**, 399.
12. *Assay Guidance Manual*, ed. G. S. Sittampalam, N. Gal-Edd, M. Arkin, D. Auld, C. Austin, B. Bejcek, M. Glicksman, J. Inglese, V. Lemmon, Z. Li, J. McGee, O. McManus, L. Minor, A. Napper, T. Riss, O. J. Trask, Jr., and Jeff Weidner, 2012, http://www.ncbi.nlm.nih.gov/books/NBK53196/.
13. A. J. Smith and P. B. Simpson, *Anal. Bioanal. Chem.*, 2003, **377**, 843.
14. J. E González and M. P. Maher, *Recept. Channels*, 2002, **8**, 283.
15. C. J. Huang, A. Harootunian, M. P. Maher, C. Quan, C. D. Raj, K. McCormack, R. Numann, P. A. Negulescu and J. E González, *Nat. Biotechnol.*, 2006, **24**, 439.
16. S. Friis, M. H. Holmqvist, L. D. Løjkner, M. K. Jensen and M. Sunesen, *Biophys. J.*, 2011, **100**, 270.
17. M. H. Bridgland-Taylor, A. C. Hargreaves, A. Easter, A. Orme, D. C. Henthorn, M. Ding, A. M. David, B. G. Small, C. G. Heapy, N. Abi-Gerges, F. Persson, I. Jacobson, M. Sullivan, N. Albertson, T. G. Hammond, E. Sullivan, J. P. Valentin and C. E. Pollard, *J. Pharmacol. Toxicol. Methods*, 2006, **54**, 189.
18. C. Farre, A. Haythornthwaite, C. Haarmann, S. Stoelzle, M. Kreir, M. George, A. Brüggemann and N Fertig, *Comb. Chem. High Throughput Screening*, 2009, **12**, 24.
19. D. J. Gillie, S. J. Novick, B. Donavan, L. A. Payne and C. Townsend, *J. Pharmacol. Toxicol. Methods*, 2013, **67**, 33.
20. R. Zhang and X. Xie, *Acta Pharmacol. Sin.*, 2012, **33**, 372.
21. P. B. Simpson, A. Woollacott, R. Hill and G. Seabrook, *Eur. J. Pharmacol.*, 2000, **392**, 1.
22. E. Trinquet, M. Fink, H. Bazin, F. Grillet, F. Maurin, E. Bourrier, H. Ansanay, C. Leroy, A. Michau, T. Durroux, D. Maurel, F. Malhaire, C. Goudet, J. P. Pin, M. Navale, O. Hernoute, F. Chrétiene, Y. Chapleure and G. Mathis, *Anal. Biochem.*, 2006, **358**, 126.
23. K. Dodgson, L. Gedge, D. C. Murray and M. Coldwell, *J. Recept. Signal Transduction*, 2009, **29**, 163.
24. R. M. Lequin, *Clin. Chem.*, 2005, **51**, 2415.
25. C. K. Dixit, S. K. Vashist, F. O'Neil, B. O'Reilly, B. MacCraith and R. O'Kennedy, *Anal. Chem.*, 2010, **82**, 7049.
26. S. Pearlman, Z. Serber and J. Ferrell, *Cell*, 2011, **117**, 934.
27. G. Høyer-Hansen, M. Hamers, A. N. Pedersen, H. J. Nielsen, N. Brünner, K. Danø and R. W. Stephens, *J. Immunol. Methods*, 2001, **235**, 91.

28. J. Benicky, R. Hafko, E. Sanchez-Lemus, G. Aguilera and J. M. Saavedra, *Cell Mol. Neurobiol.*, 2012, **32**, 1353.
29. R. M. Eglen, T. Reisine, P. Roby, N. Rouleau, C. Illy, R. Bossé and M. Bielefeld, *Curr. Chem. Genomics*, 2008, **1**, 2.
30. M. A. Macmillan, J. P. Orme and K. Roberts, *J. Biomol. Screening*, 2011, **16**, 967.
31. G. R. Richards, J. E. Kerby, G. K. Y. Chan and P. B. Simpson, *Methods Mol. Biol.*, 2006, **356**, 109.
32. M. A. Sills, D. Weiss, Q. Pham, R. Schweitzer, X. Wu and J. J. Wu, *J. Biomol. Screening*, 2002, **7**, 191.
33. R. De La Rica and M. M. Stevens, *Nat. Nanotechnol.*, 2012, **7**, 821.
34. W. A. Korfmacher, *Drug Discovery Today*, 2000, **10**, 1357.
35. E. F. Langsdorf, A. Malikzay, W. A. Lamarr, D. Daubaras, C. Kravec, R. Zhang, R. Hart, F. Monsma, T. Black, C. C. Ozbal, L. Miesel and C. A. Lunn, *J. Biomol. Screening*, 2010, **15**, 52.
36. X. Wu, J. Wang, L. Tan, J. Bui, E. Gjerstad, K. McMillan and W. Zhang, *J. Biomol. Screening*, 2012, **17**, 761.
37. M. S. Lee, *LC/MS Applications in Drug Development*, Wiley, New York, 2002.
38. H. Tsutsui, T. Mochizuki, T. Maeda, I. Noge, Y. Kitagawa, J. Z. Min and T. Toyo'oka, *Anal. Bioanal. Chem.*, 2012, **404**, 1925.
39. K. R. Gee1, W. C. Sun, M. K. Bhalgat, R. H. Upson, D. H. Klaubert, K. A. Latham and R. P. Haugland, *Anal. Biochem.*, 1999, **273**, 41.
40. E. Siebring-van Olst, C. Vermeulen, R. X. de Menezes, M. Howell, E. F. Smit and V. W. van Beusechem, *J. Biomol. Screening*, 2013, **18**(4), 453–461.
41. G. R. Richards, J. E. Kerby, G. K. Y. Chan and P. B. Simpson, *Methods Mol. Biol.*, 2006, **356**, 109.
42. E. H. Mouchet and P. B. Simpson, *IDrugs*, 2008, **11**, 422.
43. P. Alcock, C. Bath, C. Blackett and P. B. Simpson, *Eur. Pharm. Rev.*, 2010, **15**, 13.
44. E. Vega-Avila and M. K. Pugsley, *Proc. West. Pharmacol. Soc.*, 2011, **54**, 10.
45. M. V. Berridge, P. M. Herst and A. S. Tan, *Biotechnol. Annu. Rev.*, 2005, **11**, 127.
46. J. Comley, *Drug Discovery World*, Summer; p. 25, 2010.
47. H. Fukazawa, S. Nakano, S. Mizuno and Y. Uehara, *Int. J. Cancer*, 1996, **67**, 876.
48. A. Shariff, J. Kangas, L. P. Coelho, S. Quinn and R. F. Murphy, *J. Biomol. Screening*, 2010, **15**, 726.
49. G. J. Patti, O. Yanes and G. Siuzdak, *Nat. Rev. Mol. Biol.*, 2012, **13**, 263.
50. J. Huth, M. Buchholz, J. M. Kraus, M. Schmucker, G. von Wichert, D. Krndija, T. Seufferlein, T. M. Gress and H. A. Kestler, *BMC Cell Biology*, 2010, **11**, 24.
51. J. H. Zhang, T. D. Chung and K. R. Oldenburg, *J. Biomol. Screening*, 1999, **4**, 67.
52. B. K. Shoichet, *J. Med. Chem.*, 2006, **49**, 7274.

CHAPTER 11

In Vitro Biology: Measuring Pharmacological Activity

IAIN G. DOUGALL

IGD Consultancy Ltd, Loughborough, Leicester, LE11 3JR, UK
E-mail: igdconsultancy@btinternet.com

11.1 INTRODUCTION

The purpose of this chapter is to describe how molecules synthesised by medicinal chemists are assessed for pharmacological activity *in vitro*. The biological systems and assay readouts typically employed are discussed in detail in Chapter 10 but the primary aim of these experiments is to deliver robust estimates of parameters such as affinity, potency, intrinsic activity (IA) or efficacy as well as providing insights into the mechanism of action of the compounds. Such analysis represents the first step in the evaluation of the biological activity of novel compounds and the information generated is crucial in generating structure activity relationships and ultimately in guiding the rational design of new medicines.

11.2 AGONISTS AND ANTAGONISTS

The majority of drugs exert their biological effects by interacting with proteins, which have the capacity to convert chemical information into biological information. These proteins include plasma membrane bound receptors such as G protein coupled receptors (GPCRs) and tyrosine kinase receptors, ion channels (both ligand gated and voltage operated), enzymes, transporters and transcription factors such as the nuclear hormone receptors (NHRs), which bind to specific consensus sequences of DNA and modulate gene transcription.

Broadly speaking there are two classes of pharmacological agents, agonists and antagonists (or inhibitors). Agonists are capable of inducing a pharmacological response, that is, their chemical information is transduced into a biological response. Antagonists by contrast do not elicit a biological response but are able to block the response of agonists. The following sections will describe the different types of agonists and antagonists and the parameters that define their activity. The discussion and analyses presented focus largely on membrane bound receptors

The Handbook of Medicinal Chemistry: Principles and Practice
Edited by Andrew Davis and Simon E Ward
© The Royal Society of Chemistry 2015
Published by the Royal Society of Chemistry, www.rsc.org

and the classical occupancy model of receptor activation[1-6] but can also be applied to ion channels, NHRs and enzymes (with some modifications).

11.2.1 Agonist Concentration–Effect (E/[A]) Curves

The agonist concentration–effect (E/[A] or dose–response) curve has become one of the hall-marks of modern pharmacology. The generation of such data requires an assay system which can deliver robust and reproducible functional responses elicited solely *via* activation of the target under investigation. Under these conditions, E/[A] curves are typically sigmoidal when plotted in semi-logarithmic form (E/$\log_{10}$[A]) and are described by four parameters: 1) a lower asymptote (β) which represents the basal state of the system; 2) an upper asymptote (α) which represents the maximum effect that the agonist produces in the system; 3) a location or potency ([A_{50}] or EC_{50}), which represents the concentration of agonist that produces an effect equal to 50% of α-β; and 4) a slope parameter (n) which is a measure of the gradient of the curve at the [A_{50}] level. Estimates of these parameters are typically made by fitting experimental E/[A] curve data to the following form of the Hill equation (a saturable function that adequately describes curves of varying gradients):

$$E = \beta + \frac{(\alpha - \beta)[A]^n}{[A]^n + [A_{50}]^n} \tag{11.1}$$

In practice, in the majority of cases $\beta = 0$, that is, the basal effect level is ascribed a value of zero, and therefore most E/[A] curve data can be adequately described by a three-parameter Hill equation as illustrated in Figure 11.1. It is the analysis of how these three curve parameters (α, [A_{50}] and n) are affected by the efforts of the medicinal chemist that drives the optimisation of the pharmacological properties of new compounds. As antagonists do not elicit functional responses, their activity is derived from analysis of how they affect agonist responses (see 11.2.4).

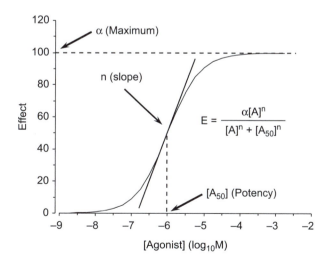

Figure 11.1 Agonist concentration–effect curves. A simulated curve highlighting the parameters that define it, α (the maximum effect), [A_{50}] (potency) the concentration of agonist that produces 50% of α and n, a measure of the slope. [A_{50}] values are often quoted as pA_{50} ($-\log_{10}$[A_{50}]). In this example the [A_{50}] is 10^{-6} M and the pA_{50} is 6.0. Real examples of agonist E/[A] curves are shown in many of the following figures.

11.2.2 Full Agonists, Partial Agonists and Inverse Agonists

The first step in agonist action is the formation of a reversible agonist–receptor (AR) complex, a process that is generally assumed to be governed by the Law of Mass Action. Accordingly, the equilibrium concentration of agonist occupied receptors is a rectangular hyperbolic (a special case of the Hill function where $n = 1$) function of the agonist concentration. This curve is defined by a maximal value of $[R_{tot}]$, the total receptor concentration, and a midpoint value of K_A, the agonist dissociation constant. In theory, K_A is a purely drug-dependent parameter and it determines how well the agonist binds; that is, it is a measure of the affinity (the reciprocal of the dissociation constant) of the agonist for its receptors. Agonist occupancy is subsequently converted into functional effect by the biochemical/biophysical machinery of the cell/tissue and this is what is measured experimentally in the form of an E/[A] curve. The efficiency of this transduction process can vary between agonists and across systems, that is, it is both drug and tissue dependent. Agonist efficacy is a measure of the efficiency of the transduction process. Full agonists have high efficacies and therefore can elicit the maximum effect (E_{max}) that the test system is capable of generating. Partial agonists by contrast have low efficacy and cannot elicit a maximum response (Figure 11.2A).

Measuring the efficacy (and affinity) of full agonists is not straight-forward. This is because it requires an experimental manipulation that decreases the efficacy of the agonist to a level where it behaves as a partial agonist. Irreversible antagonists have been used for this purpose as they covalently modify receptors thereby decreasing $[R_{tot}]$.[4] Their utility in estimating agonist efficacy and affinity is described in section 11.2.4.1. An important consequence of efficacy being both a drug and system dependent parameter is that an agonist can demonstrate different behaviours in different systems. Thus a drug that exhibits partial agonism in one system may be a full agonist in another (with higher $[R_{tot}]$ or more efficient transduction machinery) or effectively an antagonist in yet another (with lower $[R_{tot}]$ or less efficient transduction machinery) (Figure 11.3).

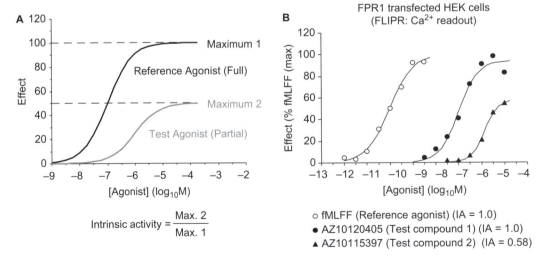

Figure 11.2 Full agonists, partial agonists and intrinsic activity. Panel A shows simulated curves for a reference full agonist and a test partial agonist and how intrinsic activity is calculated. In this example IA of the partial agonist is 0.5. Panel B shows experimental data (increases in intracellular Ca^{2+} levels) for a reference full agonist, fMLFF and two test agonists in HEK cells stably expressing the human FPR1 receptor. AZ10120405 is a full agonist (IA = 1.0) but AZ10115397 exhibits partial agonism (IA = 0.58). Unpublished data.

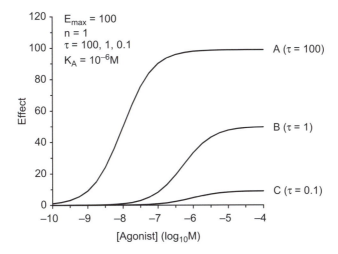

Figure 11.3 System dependence of drug effects. Simulated curves showing how the curve parameters of an agonist change in systems with varying receptor expression. The receptor expression range is 1000-fold from curve A to curve C. In system A, the drug exhibits full agonism (IA = 1.0 and high efficacy (τ = 100)); in system B, it shows partial agonism (IA = 0.5 and low efficacy (τ = 1)) and in system C, it shows very weak partial agonism (IA = 0.09 and very low efficacy (τ = 0.1)). In system C the drug effectively behaves as an antagonist as this level of IA is difficult to detect in most assay systems. The Operational Model of Agonism[6] was used to simulate the data. In this model τ is a measure of the efficacy of the agonist and incorporates both drug (intrinsic efficacy) and system (receptor number ([R_{tot}]) and coupling efficiency) parameters.

As alluded to above, it can be difficult to measure the affinity and efficacy of agonists and typically the information reported to the chemist is the potency ([A_{50}] or more often, pA_{50} ($-\log_{10}[A_{50}]$)) and the intrinsic activity (IA) of the compound. The latter is a measure of the maximal activity of the test compound relative to a reference full agonist.[2] If the test agonist produces a maximum response less than the reference agonist then the IA will be <1.0, for example, if it produces a maximum effect that is 50% of the reference, it will have an IA of 0.5 and exhibit partial agonism (Figure 11.2A). Such compounds are useful to the medicinal chemist as they help direct efforts to optimise the efficacy of compounds for therapeutic benefit, for example, identification of partial agonists were important staging posts in the development of the antagonists, propranolol[7] and cimetidine[8] (see also section 11.3). Finally, it is important to emphasise that the IA scale does not discriminate between full agonists, that is, all full agonists will have an IA of 1.0 (Figure 11.2B) but they may have different efficacies.

Until relatively recently agonist efficacy was considered only as a positive vector associated with increased receptor activity. This dogma was challenged by the discovery of the phenomenon of constitutive receptor activation and compounds that showed inverse agonism,[9,10] that is, they decreased the level of constitutive activation, demonstrating negative efficacy (Figure 11.4). The most likely mechanism for inverse agonism is that such compounds have a selectively higher affinity for the inactive state of the receptor and thereby uncouple spontaneously coupled (active) receptor species.[11] The identification of receptor conformations that display clear preference for different ligands supports this hypothesis and may allow the functional characterisation of ligands by measuring their affinity for a pair (or more) of G protein-coupled receptor conformations.[12] To date inverse agonism has largely been a property detected in genetically engineered systems where receptors (or modified receptors) can be expressed at very high levels. Many of the compounds that exhibit inverse agonism in such systems behave as competitive (neutral) antagonists with zero efficacy in more physiologically

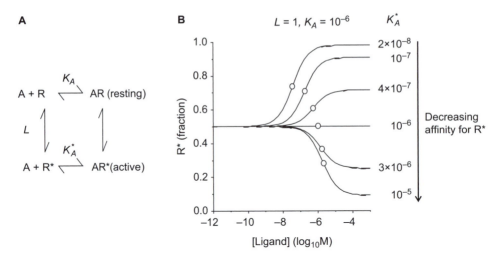

Figure 11.4 Inverse agonism. Panel A shows a simple two-state model in which the receptor exists in R (resting) or R* (active) states. In the absence of agonist (A) the distribution of the two states is governed by the equilibrium constant *L*. The agonist has affinities for the two states governed by the dissociation equilibrium constants, K_A and K_A*. Panel B shows a simulation that represents the effects of six ligands with different K_A:K_A* values. Effect is determined by the fraction of receptors in the R* form. *L* is set to 1 in this example so under basal conditions the receptors are equally distributed between the two states. In such a system inverse agonism is easily detectable. For agonists with higher affinity for R*, positive agonism is observed; for agonists with higher affinity for R, inverse agonism is observed. Ligands with equal affinity for the two states are neutral and behave as antagonists of positive and inverse agonists. Reproduced with permission from Leff, 1995.[11]

relevant assays. As such, the therapeutic relevance of inverse agonism remains largely unknown but this now well documented phenomenon has changed the way pharmacologists view drug–receptor interactions as well as resulting in the reclassification of drugs that were formerly thought to be competitive antagonists (*e.g.* Ranitidine and Propranolol). Moreover, as more diseases are identified that are linked to constitutive receptor activation it seems likely that inverse agonists will become increasingly important tools in addressing unmet clinical needs.

11.2.3 Optimising Agonists

As discussed above, agonists both bind and activate receptors. The optimisation of agonist properties relies on designing compounds with both good affinity and appropriate efficacy. These needs dictate the assays used in agonist focused drug discovery programmes. Affinity can be measured in ligand binding assays, similar to those illustrated in Figure 11.7. Such assays have the benefit of confirming that the compounds do indeed bind to the target of interest but in most instances give no information on efficacy. Functional assays are required to provide estimates of IA or efficacy. These often take the form of simple second messenger readouts such as changes in intracellular calcium or cyclic adenosine monophosphate (cAMP) levels and are nowadays routinely conducted in cell lines expressing the human version of the target receptor. Such systems have the advantage of high throughput and are generally very robust. A disadvantage is that the agonist responses observed may poorly reflect those observed in the biologically relevant cells/tissues due to differences in receptor number and/or coupling and the readout. Testing of compounds in more relevant cellular and tissue systems is therefore essential before progressing to *in vivo* testing. An example of the assays employed and the data

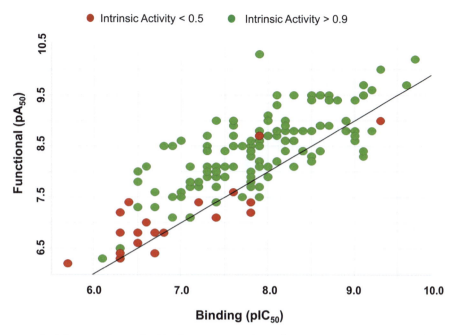

Figure 11.5 Optimising agonists. The binding activities (pIC_{50}) and functional potencies (pA_{50}) of a range of β_2-agonists are shown. The binding activities were measured by displacement of I^{125} labelled cyanopindolol in HEK293 cells stably transfected with the human β_2-adrenoceptor and the functional potencies were assessed by measuring increases in cAMP levels in H292 cells, a human lung epithelial cell line that endogenously expresses the β_2-adrenoceptor. In the latter assay, the intrinsic activity of compounds was assessed relative to the full agonist, formoterol. Compounds are grouped into those with an intrinsic activity <0.5 (●) and those with an intrinsic activity >0.9 (●). As expected compounds largely lie above the "equi-activity" lines, that is, the functional activities are greater than the binding activities and this discrepancy is generally larger for compounds with higher intrinsic activity.

generated in a typical agonist programme is shown in Figure 11.5. Here, both binding and functional assays were used to assess activity of β_2-receptor agonists and as expected compounds with higher intrinsic activity (and therefore efficacy) showed a greater discrepancy between their binding activity and their functional potency (pA_{50}) (as highlighted in section 11.2.2 the binding affinity of full agonists is a poor measure of their functional potency, as pA_{50} is dependent on both agonist affinity and efficacy).

In most cases, the aim of agonist based projects is to identify high potency, high efficacy agonists so that the drug dose ultimately administered will be small and the effect large. In some instances however partial agonists can have therapeutic advantages. Thus if the desirable therapeutic effect is observed in a tissue with high receptor number/coupling but an undesirable side-effect is mediated in a tissue with low receptor number/coupling, a partial agonist of appropriate efficacy could produce agonism in the former but be "silent" in the latter (compare curves A and C in Figure 11.3).

11.2.4 Antagonists

Several different classes of antagonists with distinct mechanisms of action including irreversible competitive, reversible competitive, non-competitive and allosteric have been identified. Their blockade of agonist induced effects can be surmountable (rightward displacement of the

E/[A] curve with no depression of the maximum (α)) or insurmountable (depression of the maximal agonist response (α)). It is important to realise that the profile of antagonism observed can show system dependence, that is, an antagonist can exhibit surmountable activity in one assay system and insurmountable activity in another, despite having the same mechanism of action (see Figure 11.10A and B). A common example of this phenomenon is the behaviour of high affinity competitive antagonists in FLIPR® assays in which the changes in intracellular calcium levels measured are typically transient in nature. In these circumstances, true equilibrium is not reached and the agonist cannot access antagonist bound receptors resulting in non-surmountable antagonism.[13] This contrasts with the behaviour of such antagonists in systems where the agonist's responses are sustained (for example isolated tissues), true equilibrium is reached and the antagonism is surmountable.

The interaction of an antagonist with its receptors is described by a single parameter, affinity, which equates to potency (unlike agonists where potency is dependent on both affinity and efficacy). By definition antagonists have an IA = 0 in functional assays, in which their affinity is measured by studying their interaction with an agonist. The affinity of antagonists can also be measured in binding assays (as can agonists; see Chapter 10) although if such systems are used it is essential to confirm lack of efficacy by subsequent testing in functional assays. The following sections discuss the properties and analysis of the various classes of antagonists.

11.2.4.1 *Irreversible Competitive Antagonists*

Irreversible competitive antagonists are relatively rare but they are worthy of discussion as they form the basis of the receptor inactivation method developed by Furchgott[4] that allows the estimation of the efficacy and affinity of full agonists. These agents (for example phenoxybenzamine) bind covalently to receptors and thereby decrease [R_{tot}] by making a portion of the receptor pool unavailable for agonist binding. As outlined in section 11.2.2, agonist efficacy is partly dependent on [R_{tot}] so this chemical inactivation of receptors decreases agonist efficacy and converts full agonists into partial agonists. Under these conditions the [A_{50}] (or pA_{50}) of the partial agonist curve yields a good estimate of the affinity (K_A or pK_A) of the agonist; the lower the maximum effect (α) the better the [A_{50}] of the E/[A] curve serves as an estimate of agonist affinity. The [A_{50}] estimate obtained under the "low efficacy" condition divided by the [A_{50}] of the control curve (no irreversible antagonist) serves as an approximate measure of the agonist efficacy. In practice, such data can be fitted to mathematical models[4,6,14] that give accurate estimates of the parameters of affinity and efficacy (Figure 11.6).

Unfortunately, the lack of appropriate irreversible antagonists for many receptors precludes this sort of analysis in the majority of cases. The "potency" of the antagonists themselves in such experiments is somewhat meaningless as occupancy does not reach a steady state but rather increases with exposure time and unbound antagonist is removed before assessment of the agonist effects.

11.2.4.2 *Reversible Competitive Antagonists*

Reversible competitive antagonists are probably the most important class of antagonists and a large number of clinically used drugs fall into this class. As outlined above for agonists, the first step in the action of these drugs is the formation of a reversible, relatively short-lasting, drug–receptor complex governed by the Law of Mass Action. This can be measured directly in binding studies used labelled (for example, radioactively or fluorescently) compounds with the binding describing a saturable rectangular hyperbolic function. The concentration of antagonist that binds 50% [R_{tot}] defines the antagonist dissociation constant (K_D or K_B in functional studies)

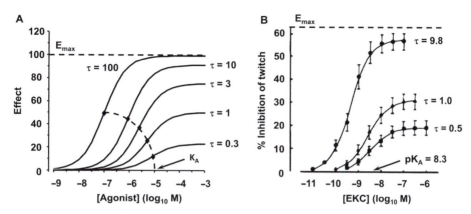

Figure 11.6 Irreversible antagonism. Panel A shows a simulation of the effects of an irreversible antagonist on the response of a full agonist. Increasing concentrations of the irreversible antagonist cause rightward displacement of the control agonist E/[A] curve ($\tau = 100$) with depression of the maximum response (α). As the curves become increasingly depressed the $[A]_{50}$ approaches the K_A affording an estimate of the agonist affinity. The data was generated using the Operational Model of Agonism[6] by varying $[R_{tot}]$ which has the effect of reducing agonist efficacy (τ). Panel B shows an example of irreversible antagonism of the effects of the κ-opioid agonist ethylketocyclazocine by the alkylating agent β-chlornaltrexamine (β-CNA) in isolated, coaxially stimulated guinea-pig ileum. E/[A] curves are shown in the absence of (●) or following 30 minutes exposure to 10 nM (▲) or 30 nM (■) β-CNA. The lines drawn through the data are the results of fitting the data to the Operational Model of Agonism.[6] The estimates of pK_A (affinity) and τ (efficacy) for EKC were 8.3 and 9.8 respectively. Panel B data was reproduced with permission from Leff & Dougall, 1989.[38]

(Figure 11.7A). Binding of a competitive antagonist precludes the binding of other drugs to the same sites on the receptor. This mutually exclusive binding allows test compounds to be assessed for their ability to displace a labelled compound (L), a practice that is routinely used in drug screening to derive an estimate of potency. Such indirect analysis is important as it is impractical to label all test compounds. These displacement experiments typically yield a sigmoidal curve (which can be fitted to a Hill equation similar to Equation (11.1)) from which the IC_{50} (concentration of the test compound that produces 50% displacement of the specific binding of the labelled compound) can be measured (Figure 11.7B).

Assuming that the interaction between the labelled compound and the test compound is competitive then the dissociation constant (K_I) of the test compound can be calculated using a modified Cheng–Prusoff equation:[15]

$$K_I = \frac{IC_{50}}{1 + [L] / K_D} \tag{11.2}$$

Proof of the assumption that the interaction is competitive requires further experimentation, such as studying the displacement by the test compound, of different concentrations of the labelled ligand. As is evident from the Cheng–Prusoff equation, at concentrations of L in excess of K_D the IC_{50} estimate will increase proportionately, that is, higher concentrations of L will require higher concentrations of test compound to displace it. It is therefore imperative that the activities of different test compounds are compared under identical conditions, that is, $[L]/K_D$ should be constant. Since IC_{50} values do not infer a particular mechanism of action, they are routinely used to compare the activities of compounds in binding assays rather than calculated K_I values.

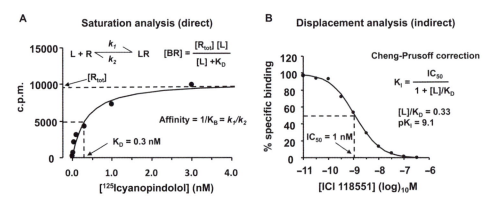

Figure 11.7 Competitive antagonism: Binding assays. Panel A shows a direct binding assay using ^{125}I labelled cyanopindolol as a β_2-adrenoceptor ligand. The curve describes a rectangular hyperbola which saturates at high ligand concentration. The ligand dissociation constant (K_D) was estimated as 0.3 nM and is a measure of the ligand affinity ($1/K_D$). Panel B shows a typical displacement analysis in the same system using 0.1 nM labelled cyanopindolol. The displacing ligand, the selective β_2-adrenoceptor antagonist ICI 118551 produces complete displacement of the specific binding yielding an IC_{50} of 1 nM. In this instance $[L]/K_D$ is < 1.0 so the IC_{50} is a good estimate of the K_I as calculated by the Cheng–Prusoff equation. Unpublished data.

The affinity of competitive antagonists can also be measured in functional assays. In this case, the presence of the antagonist (B) decreases the probability that an agonist–receptor interaction will occur. To achieve the same degree of agonist occupancy and therefore the same effect, in the presence of the antagonist as in its absence, the agonist concentration must be increased. The factor (r) by which it must be increased depends on both the concentration of antagonist ($[B]$) used and on how well it binds (K_B). This relationship which was first described by Schild[16] is shown in Equation (11.3) below:

$$r - 1 = [B]^n/K_B \qquad (11.3)$$

where $r = [A_{50}]/[A_{50}]^c$ (location parameter of the E/[A] curve in the presence of the antagonist/ location parameter of the E/[A] curve in the absence of the antagonist), K_B is the antagonist equilibrium dissociation constant and n represents the order of reaction between the antagonist and the receptors (for a first order reaction $n = 1$, that is, one molecule of antagonist binds to one receptor molecule).

Experimentally, a K_B is estimated by studying the interaction of an agonist and antagonist over a wide range of antagonist concentrations (the wider, the better). This is necessary because drugs which are not reversible competitive antagonists may appear to be so within a narrow range of concentrations (see Figures 11.10A and 11.11A). If the antagonist is truly competitive then it should produce parallel rightward displacement (that is, no change in midpoint slope (n) occurs) of the E/[A] curves with no change in the maximal response (α) (Figure 11.8A), that is, surmountable antagonism is observed. This is intuitively obvious, since the antagonist is merely decreasing the probability that an agonist–receptor interaction will occur, an effect that can always be overcome by increasing the agonist concentration. The analysis involves fitting experimentally derived values of r at different concentrations of antagonist to the following form of Equation (11.3); (Figure 11.8B).[17]

$$\log_{10}(r - 1) = n\log_{10}[B] - \log_{10}K_B \qquad (11.4)$$

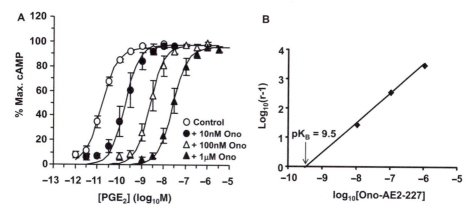

Figure 11.8 Competitive antagonism: functional assays. Antagonism of the PGE_2 mediated induction of cAMP by the competitive antagonist Ono-AE2-227, in CHO cells stably transfected with the human EP_4 receptor (Panel A). Note the concentration-dependent parallel rightward displacement of the control curves. Panel B illustrates the displacements (r values) in Schild plot form derived for one of the five experiments that make up the data shown in panel A. The plot has a slope of unity and the intercept on the x-axis yields an estimate of 9.5 for the pK_B $(-\log_{10}K_B)$.
Modified with permission from Ratcliffe *et al.*, 2007.[39]

Adherence of the data to this equation is judged by the finding of a linear plot with unit slope. Under these conditions the intercept on the x-axis $(\log_{10}[B])$ gives an estimate of K_B. When n is significantly different from 1 then the intercept gives an estimate of pA_2 $(-\log_{10}K_B/n)$. The pA_2 is an empirical estimate of antagonist affinity and equates to the negative logarithm of the concentration of antagonist that produces a two-fold rightward shift $(r=2)$ of the control E/[A] curve. Non-linearity and slopes other than unity can result from many causes. For example, a slope of greater than 1 may indicate incomplete equilibration with the antagonist or removal of the antagonist from the biophase (receptor compartment). A slope that is significantly less than 1 may indicate removal of the agonist by a saturable uptake process, or it may result from the interaction of the agonist with more than one receptor. In the latter case the Schild plot may be non-linear with a clear inflexion.[18] All of these potential complicating factors have been described in detail previously by Kenakin.[19]

Although, Schild type analysis is the most robust method of assessing antagonist behaviour in functional assays, the needs of modern high throughput drug discovery programmes dictate that it is used sparingly to assess the mechanism of action for priority compounds. Routine screening of antagonist properties will more likely be assessed by doing a functional Cheng–Prusoff type experiment (Figure 11.9) in which the effects of several concentrations of the test compound on the response to a single concentration of agonist are studied. The experimental data can then be fitted to the following equation:[20]

$$K_B = \frac{IC_{50}}{(2 + ([A]/[A_{50}])^n)^{1/n} - 1}$$ (11.5)

As was outlined above for binding studies, the estimated IC_{50} is dependent on the concentration of ligand employed. In this case the concentration of agonist (A) relative to its $[A_{50}]$ dictates the IC_{50} (and hence the estimated K_B). Practically, the experimenter usually employs a concentration of agonist that is as close to the $[A_{50}]$ as possible so that the IC_{50} is a good estimate of the K_B. The shape of the agonist E/[A] curve is also important as evidenced by the

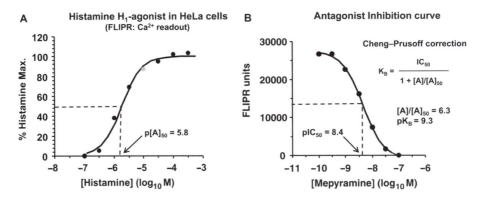

Figure 11.9   Cheng–Prusoff analysis of antagonism in functional assays. Panel A shows a histamine concentration effect curve generated in HeLa cells which endogenously express the human H_1-receptor. 10 µM histamine (grey symbol) was chosen as the concentration of agonist to be used to assess the inhibitory effects of the H_1-receptor antagonist mepyramine (Panel B). Mepyramine completely inhibited the Ca^{2+} induced histamine response and yielded a pIC_{50} value of 8.4. As $[A]/[A_{50}]$ was significantly greater than 1, the affinity (pK_B) estimate (9.3) is considerably greater than the pIC_{50}. In this case the slope of the histamine E/[A] curve was 1 so equation 11.5 simplifies to the equation show in panel B. Unpublished data.

inclusion of the slope parameter (n) in this form of the Cheng–Prusoff equation. When $n = 1$ the equation simplifies to a form equivalent to Equation (11.2). Such analysis, although higher throughput, does not discriminate different modes of action of test compounds, for example, it will not differentiate competitive from non-competitive compounds. Without additional proof that the interaction of agonist and antagonist is competitive it is more appropriate to use the measured IC_{50} as a measurement of antagonist potency rather than calculating a K_B.

11.2.4.3   Non-Competitive Antagonists

Non-competitive antagonists bind to receptors and make them functionally inoperative either by preclusion of agonist binding or through some other biochemical mechanism that obviates agonist effect. Under these circumstances, increasing the agonist concentration cannot over-come the effect of the antagonist and therefore a distinctive feature of non-competitive an-tagonists is the depressive effect they have on the maximal agonist response, that is, they exhibit insurmountable antagonism (Figure 11.10). This mechanism has the potential therapeutic advantage that high concentrations of endogenous agonists, (for example those associated with tumours such as prolactinomas) are less likely to overcome the antagonist effects. The mag-nitude of the depression will however depend on the agonist under study and the system used. This relates to the concept of receptor reserve whereby maximum agonist effects can be achieved at low levels of receptor occupancy (binding). For example, if an agonist can elicit a maximum response by activation of 10% of the receptor population, then there will be a 90% receptor reserve. Receptor reserve depends on both the receptor number ($[R_{tot}]$) and the effi-ciency of stimulus–response coupling as well as the intrinsic efficacy of the agonist, hence non-competitive antagonists will have differing capabilities to depress the maximal response to the same agonist in different systems (compare Figures 11.10A and 11.10B). The same will be true for different agonists in the same system. In terms of measuring the potency of insurmountable antagonists the data can be fitted to a number of models to yield estimates of K_B.[21]

Slowly dissociating reversible competitive antagonists exhibit non-competitive behaviour when their rate of offset prevents correct re-equilibration of agonist, antagonist and receptors during

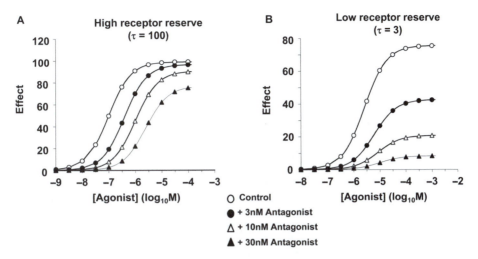

Figure 11.10 Non-competitive antagonism. Simulations showing the effect of a non-competitive antagonist on responses to the same agonist in a system with high receptor reserve (A) or low receptor reserve (B). Increasing concentrations of the antagonist (3, 10, 30 nM) cause more marked depression of the agonist maximum effect in the low reserve system. Data was simulated using a form of the Operational Model of agonism that assumes that antagonist binding precludes binding of the agonist.[37] The model parameters used were $E_m = 100$, $n = 1$, $\tau = 100$ (high reserve) or $\tau = 3$ (low reserve), $pK_A = 5.0$, $pK_B = 9.0$.

the period of agonist response measurement. Under these conditions, a pseudo-irreversible blockade of the receptors occurs whereby the agonist cannot access antagonist-bound receptors. Antagonists with very slow off-rates offer a means of extending the duration of action (pharmacodynamic effect) of therapeutic agents distinct from (or complementary to) optimising their pharmacokinetic profiles.[22,23] In practice this requires the identification of very high potency (nanomolar–picomolar range) compounds although factors such as the rate of free diffusion away from the receptor compartment are also important in determining duration of action.

11.2.4.4 Allosteric Antagonists

The modes of antagonism described above are orthosteric, that is, the antagonist blocks access of the agonist to its binding site through steric hindrance. Allosteric antagonists in contrast bind to their own site on the receptor to induce a change in conformation of the receptor which in turns alters the affinity or efficacy of the receptor for the agonist. It is now clear that allosteric ligands can both increase (for example benzodiazepines at $GABA_A$ receptors) and decrease the affinity and efficacy of other ligands so allosteric modulators is a more appropriate term. One of the key properties of allosteric modulators is their saturability of effect which can be evidenced in functional experiments such as Schild analysis where a curvilinear plot is observed (Figure 11.11; contrast this with the competitive antagonist in Figure 11.8).[24] This results from the fact that while the allosterically modified receptor may have diminished affinity (and/or efficacy) for the agonist, the agonist can still produce receptor activation in the presence of the modulator. A detailed discussion of the complex behaviour of allosteric modulators is beyond the scope of this chapter but the interested reader is referred to some recent reviews.[25–30]

Allosteric modulation offers a number of potential advantages over orthosteric antagonists. Firstly, they can modify (that is, reduce or increase by a small amount) endogenous agonist signals without completely blocking them thus allowing fine-tuning of responses. Secondly, there is the potential to increase the duration of allosteric effect by loading the receptor

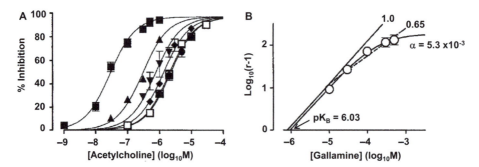

Figure 11.11   Allosteric antagonism. Panel A shows the effects of acetylcholine (Ach) on the electrically evoked contractions of the guinea pig left atrium in the absence (■) or presence of the allosteric modulator gallamine at the following concentrations: 10 μM (▲), 30 μM (▼), 100 μM (◆), 300 μM (□), and 500 μM (●). Panel B shows the Schild plot of the data shown in A. The solid line (slope = 1) denotes the behaviour expected for a competitive antagonist, whereas the dashed line shows the best fit linear regression (and associated slope factor) through the points. The curve through the points and associated parameter estimates represent the fit to an allosteric model. The estimated pK_B was 6.03 and the α value of 5.3×10^{-3} equates to a gallamine-induced decrease in the affinity of ACh of 189-fold. Reproduced with permission from Christopoulos and Kenakin, 2002.[24]

compartment with high concentrations of modulator. Such large concentrations will have no further effect than to prolong the saturated allosteric effect, that is, the saturability of the allosteric ligand can be used to limit effect but increase duration. Another potential advantage of allosterism is increased selectivity. Orthosteric antagonists often have limited selectivity across receptor subtypes, for example, most muscarinic receptor antagonists exhibit poor selectivity between the five known subtypes (M_1–M_5) presumably because they are competing with acetylcholine for very similar recognition sites. However, the surrounding protein structure of the receptors are sufficiently different to offer the potential for selective stabilization of receptor conformations by allosteric modulators. The identification of a number of allosteric antagonists with selectivity for the M_2-receptor supports this hypothesis.[28] Allosteric ligands also exert effects that are probe dependent, meaning that their effects are not the same towards all agonists (in contrast to orthosteric antagonists which typically show the same degree of blockade of all agonists). This may offer further advantages if different agonists induce different responses (or degrees of response) following receptor activation, that is, allosteric antagonists in theory could block the adverse effect of one agonist without affecting the beneficial effect of another. An example of where this approach may bear fruit is the HIV co-receptor CXCR4, as there is evidence of dissociation of virus binding and chemokine peptide activity.[31] This raises the possibility that allosteric molecules that block HIV entry but do not interfere with SDF-1α-mediated chemokine function could be found, thus avoiding the undesirable side-effects of blocking the important physiological roles of CXCR4. A corollary of the above is that it is important to use the physiological/pathophysiological agonist(s) in experiments where possible rather than synthetic agonists. All of the above mentioned potential therapeutic advantages of allosteric modulators remain largely theoretical as very few such agents have reached the market. Nevertheless, the approval of the CCR5 antagonist Maraviroc for the treatment of HIV demonstrates the feasibility of the approach. This compound inhibits the entry of the virus into cells by binding to a receptor site distinct from where the viral gp120 envelope protein binds.[32,33]

Finally, although the discussion above focuses on receptors, allosteric modulation of enzyme function is perhaps a better known phenomenon. The availability of binding sites distinct from those for the substrate again offers the potential for increased selectivity, for example, compounds designed to bind to an allosteric site in a particular protein kinase are likely to have improved selectivity over compounds targeting the well conserved ATP binding site.

11.3 APPLICATION TO DRUG DISCOVERY

The above sections have hopefully given the reader an understanding of the type of experiments conducted and the analyses carried out during the early evaluation of the biological activity of compounds. This section aims to illustrate the utility of this process in drug discovery by describing a series of experiments that led ultimately to the development of AZD1981, a selective DP_2 (CRTh2) receptor antagonist. In 2001, this receptor was discovered, somewhat unexpectedly to respond to prostaglandin D_2 (PGD_2).[34] This finding together with its high expression on Th2 cells, eosinophils and basophils rekindled interest in the role of PGD_2 in allergic diseases and prompted many companies to initiate projects aimed at the identification of selective DP_2 receptor antagonists. At AstraZeneca, during development of the primary assay, a common finding was a biphasic rather than monophasic PGD_2 E/[A] curve (Figure 11.12A). Reasoning that the biphasic curve may be the result of PGD_2-induced release of PGE_2 and subsequent activation of an endogenous EP receptor in the HEK cells in which we had overexpressed the human DP_2 receptor, we attempted to "clean up" the assay by blocking the endogenous production of PGE_2. For that purpose we used the cyclo-oxygenase inhibitor, indomethacin, which

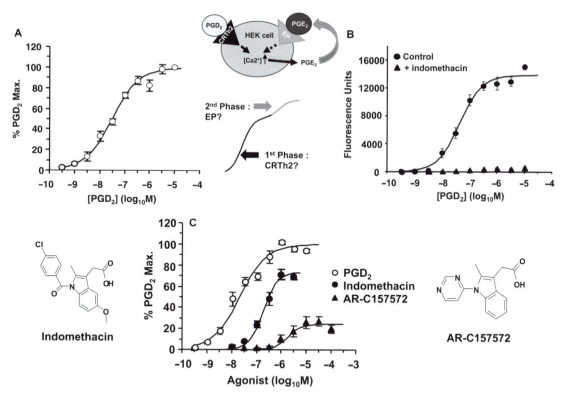

Figure 11.12 Application to drug discovery I: Identification of DP_2-receptor Partial Agonists. Panel A shows typical PGD_2 E/[A] curves in HEK cells stably transfected with the human DP_2-receptor as assessed by measurement of increases in intracellular $[Ca^{2+}]$. The responses appear to plateau at 1 μM then increase again at higher concentrations. The inset shows the logic for using indomethacin to try and make the PGD_2 curve monophasic. Panel B shows the effects of preincubating 3 μM indomethacin on responses to PGD_2—complete ablation was observed. Panel C shows that indomethacin exhibited DP_2 receptor partial agonism and also highlights the discovery of another compound AR-C157572 which had lower potency and lower IA. Panel C data, redrawn with permission from Birkinshaw *et al.*, 2006.[35] Other data unpublished.

we pre-incubated with the cells before measuring the response to PGD_2. To our surprise, this did not make the PGD_2 E/[A] curve monophasic but rather ablated it completely (Figure 11.12B). This suggested that indomethacin was either an antagonist of DP_2 or an agonist that had desensitised the response to PGD_2. Further experimentation confirmed the latter to be true and identified indomethacin as a partial agonist with relatively high IA (0.72) (Figure 11.12C).[35] This finding was subsequently confirmed by other investigators.[36]

As highlighted in section 11.2.2, partial agonists are good staging points on the way to identifying antagonists so we next moved on to test a range of indomethacin analogs and quickly identified a number of moderately potent antagonists (exemplified by AR-C157573, Figure 11.13A and B) as well as some additional partial agonists. These antagonists had a mechanism of action consistent with competitive antagonism in both the primary assay and in more physiologically relevant cells such as human eosinophils (Figure 11.13B). Further optimisation of potency, as well as all the other properties essential for a safe, oral treatment, ultimately led to AZD1981which progressed to clinical development. Interestingly, this compound had a mechanism of action inconsistent with competitive antagonism (Figure 11.13C) as evidenced by the finding of similar pIC_{50} values in a radioligand binding assay, using concentrations of $[^3H]PGD_2$ that were 100-fold different.[37]

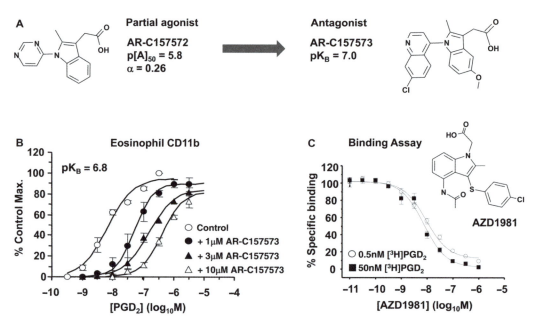

Figure 11.13   Application to drug discovery II: Turning DP_2-receptor Partial Agonists into Antagonists. Panel A illustrates structural modifications that converted the partial agonist AR-C157572 into a moderately potent antagonist, AR-C157573. Panel B shows the effects of AR-C157573 on PGD_2 induced up-regulation of the adhesion molecule CD11b on human eosinophils, as assessed by flow cytometry. The antagonist produced apparently competitive antagonism yielding a pK_B value of 6.8 which was consistent with the value (7.0) obtained in the primary (Ca^{2+}) assay. Panel C shows the effects of the clinical candidate (AZD1981) from this research programme in a radioligand binding assay, employing DP_2 expressing membranes. The measured IC_{50} value for AZD1981was similar against two different concentrations of $[^3H]PGD_2$ indicating that this compound is not a DP_2 receptor competitive antagonist. Panel C data reproduced with permission from Schmidt *et al.*, 2013,[37] other data unpublished.

11.4 CONCLUDING REMARKS

The application of the principles and the various analyses described in this chapter are the cornerstone of structure–activity relationships. They have also significantly enhanced our understanding of the mechanism of action of agonists and antagonists, and ultimately facilitated the discovery of new therapeutic agents. Whereas the measurement of the biological activities of compounds synthesised by medicinal chemists is only one component of a successful drug discovery screening cascade, the information generated influences all future project stages, including clinical testing. Accordingly, this initial interface between the disciplines of chemistry and biology remains one of the most exciting and rewarding areas of medicinal chemistry.

HINTS AND TIPS

Agonist ligand binding assays have the benefit of confirming that the compounds bind to the target of interest, but in most instances give no information on efficacy. Functional assays are required to provide estimates of IA or efficacy.

As drug efficacy is both a drug and system dependent parameter, an agonist can demonstrate different behaviours in different systems.

Testing agonists in recombinant cell systems over-expressing the target of interest offers the advantage of high throughput and robustness, but a disadvantage is that the agonist responses observed may poorly reflect those in the biologically relevant cells/tissues due to differences in receptor number and/or coupling and the readout. Testing of compounds in more relevant cellular and tissue systems is therefore desirable before progressing to *in vivo* testing.

Partial agonists are good staging posts on the way to the design of antagonists.

The Cheng–Prusoff ("IC$_{50}$ type") analyses typically employed in screening cascades provide very limited information on the mechanism of action of the test compounds.

To allow meaningful comparison of IC$_{50}$ values derived from Cheng–Prusoff type analyses it is imperative that the ratio of $[L]/K_D$ or $[A]/[A_{50}]$ remains constant in the assay.

Allosteric modulators offer the potential of increased selectivity over orthosteric modulators.

As allosteric modulators can block the effects of one agonist and not another, it is important to use the physiological/pathophysiological agonist(s) in experiments where possible rather than synthetic agonists.

Acknowledgements

The author would like to thank the AstraZeneca colleagues past and present who generated much of the data presented in this manuscript

---

KEY REFERENCES

R. P. Stephenson, *Br. J. Pharmacol.*, 1956, **11**, 379.
A landmark in modern pharmacology: the introduction of the concept of pharmacological efficacy.
P. Leff and I. G. Dougall, *Trends Pharmacol. Sci.*, 1993, **14**, 110.

Cheng–Prusoff type experimental design is "the norm" in modern day screening campaigns: this article highlights its use and limitation in functional assays.
T. P. Kenakin, in *A Pharmacology Primer: Theory, Applications and Methods*, ed. T. Kenakin, Elsevier Academic Press, Burlington, MA, USA, 3rd Edition, 2009, Chapter 6, pp. 116–119. Essential reading for everyone engaged in drug discovery.

REFERENCES

1. A. J. Clark, in *Handbuch der Experimentellen Pharmakologie Ergansungswerk*, ed. A. Heffter, Springer-Verlag, Berlin, 1937, 4, pp. 1–223.
2. E. J. Ariëns, *Arch. Int. Pharmacodyn. Ther.*, 1954, **99**, 32.
3. R. P. Stephenson, *Br. J. Pharmacol.*, 1956, **11**, 379.
4. R. F. Furchgott, in *Advances in Drug Research*, ed. N. J. Harper and A. B. Simmonds, Academic Press, New York, 1966, 3, pp. 21–55.
5. R. R. Ruffolo, *J. Autonomic Pharmacol*, 1982, **2**, 277.
6. J. W. Black and P. Leff, *Proc. R. Soc. Lond. [Biol.]*, 1983, **220**, 141.
7. M. P. Stapleton, *Texas Heart Institute Journal*, 1997, **24**, 336.
8. ME Parsons, in *Receptor Based Drug Design*, ed. P. Leff, Marcel Dekker, Inc, New York, 1998, 89, pp 195–206.
9. T. Costa and A. Herz, *Proc. Natl. Acad. Sci. U. S. A.*, 1989, **86**, 7321.
10. P. Chidiac, T. E. Hebert, M. Valiquette, M. Dennis and M. Bouvier, *Mol. Pharmacol.*, 1994, **45**, 490.
11. P. Leff, *Trends Pharmacol. Sci.*, 1995, **16**, 89.
12. K. A. Bennett, B. Tehan, G. Lebon, C. G. Tate, M. Weir, F. H. Marshall and C. J. Langmead, *Mol. Pharmacol.*, 2013, **83**, 949.
13. T. R. Miller, D. G. Witte, L. M. Ireland, C. H. Kang, J. M. Roch, J. N. Masters, T. A. Esbenshade and A. A. Hancock, *J. Biomol. Screen.*, 1999, **4**, 249.
14. M. K. James, P. H. Morgan and H. F. Leighton, *J. Pharmacol. Exp. Ther.*, 1989, **249**, 61.
15. Y. C. Cheng and W. H. Prusoff, *Biochem. Pharmacol.*, 1973, **22**, 3099.
16. H. O. Schild, *Br. J. Pharmacol.*, 1949, **4**, 277.
17. O. Arunlakshana and H. O. Schild, *Br. J. Pharmacol.*, 1959, **14**, 48.
18. S. J. Lydford, K. C. W. McKechnie and P. Leff, *Prostaglandins*, 1996, **52**, 125.
19. T. P. Kenakin, *Can. J. Physiol. Pharmacol.*, 1982, **60**, 249.
20. P. Leff and I. G. Dougall, *Trends Pharmacol. Sci.*, 1993, **14**, 110.
21. T. P. Kenakin, in *A Pharmacology Primer: Theory, Applications and Methods*, ed. T. Kenakin, Elsevier Academic Press, Burlington, MA, USA, 3rd Edition, 2009, Chapter 6, pp. 116–119.
22. D. C. Swinney, *Curr. Opin. Drug Discovery Dev*, 2009, **12**, 31.
23. G. Vauquelin and S. J. Charlton, *Br. J. Pharmacol.*, 2010, **161**, 488.
24. A. Christopoulos and T. Kenakin, *Pharmacol. Rev.*, 2002, **54**, 323.
25. M. De Amici, C. Dallanoce, U. Holzgrabe, C. Tränkle and K. Mohr, *Med. Res. Rev.*, 2010, **30**, 463.
26. J. P. Changeux, *Annu. Rev. Biophy*, 2012, **41**, 103.
27. P. Keov, P. M. Sexton and A. Christopoulos, *Neuropharmacol*, 2011, **60**, 24.
28. C. E. Müller, A. C. Schiedel and Y. Baqi, *Pharmacol. Ther.*, 2012, **135**, 292.
29. S. Urwyler, *Pharmacol. Rev.*, 2011, **63**, 59.
30. D. J. Scholten, M. Canals, D. Maussang, L. Roumen, M. J. Smit, M. Wijtmans, C. de Graaf, H. F. Vischer and R. Leurs, *Br. J. Pharmacol.*, 2012, **165**, 1617.

31. N. Heveker, M. Montes, L. Germeroth, A. Amara, A. Trautmann, M. Alizon and J. Schneider-Mergener, *Curr. Biol.*, 1998, **8**, 369.
32. C. Watson, S. Jenkinson, W. Kazmierski and T. Kenakin, *Mol. Pharmacol.*, 2005, **67**, 1268.
33. V. M. Muniz-Medina, S. Jones, J. M. Maglich, C. Galardi, R. E. Hollingsworth, W. M. Kazmierski, R. G. Ferris, M. P. Edelstein, K. E. Chiswell and T. P. Kenakin, *Mol. Pharmacol.*, 2009, **75**, 490.
34. H. Hirai, K. Tanaka, O. Yoshie, K. Ogawa, K. Kenmotsu, Y. Takamori, M. Ichimasa, K. Sugamura, M. Nakamura, S. Takano and K. Nagata, *J. Exp. Med,.*, 2001, **193**, 255.
35. T. N. Birkinshaw, S. J. Teague, C. Beech, R. V. Bonnert, S. Hill, A. Patel, S. Reakes, H. Sanganee, I. G. Dougall, T. T. Phillips, S. Salter, J. Schmidt, E. C. Arrowsmith, J. J. Carrillo, F. M. Bell, S. W. Paine and R. Weaver, *Bioorg. Med. Chem. Lett.*, 2006, **16**, 4287.
36. H. Hirai, K. Tanaka, S. Takano, M. Ichimasa, M. Nakamura and K. Nagata, *J. Immunol.*, 2002, **168**, 981.
37. J. A. Schmidt, F. M. Bell, E. Akam, C. Marshall, I. A. Dainty, A. Heinemann, I. G. Dougall, R. V. Bonnert and C. A. Sargent, *Br. J. Pharmacol.*, 2013, **168**, 1626.
38. P. Leff and I. G. Dougall, *Br. J. Pharmacol.*, 1989, **96**, 702.
39. M. J. Ratcliffe, A. Walding, P. A. Shelton, A. Flaherty and I. G. Dougall, *Eur. Respir. J.*, 2007, **29**, 986.

Animal Models: Practical Use and Considerations

MILENKO CICMIL* AND ROBBIE L. McLEOD

Merck Research Laboratories, 33 Avenue Louis Pasteur, Boston, MA, USA
*E-mail: milenko.cicmil@merck.com

12.1 INTRODUCTION

It is commonly accepted that in general animals, particularly mammalian species, share many common aspects that engender and sustain life. For example, mammals have a complex and dynamic central and autonomic nervous system that regulates multifaceted physiological processes such as digestion, body temperature regulation, respiration, circulation and so on. Thus, the advancement of biomedical sciences is and has been substantially predicated on the thesis that animal physiology can be exploited to learn human biology. While this may be generally true, the application of pharmacological treatments in order to prevent pathophysiological mechanisms across species adds greater opportunity for discord to humans. Historically, the use of animal models in drug discovery has been integral to the development of novel therapies. Unfortunately, the translatability of animal model data into the clinical arena continues to be subject to seemingly everlasting debate (Figure 12.1). Nonetheless, due to pipeline attrition of new drugs, pharmaceutical companies are utilizing several strategies, including pharmacokinetic-pharmacodynamic (PKPD) approaches, to enhance the probability of successfully advancing novel compounds from the bench to human patients. Central to the PKPD modeling is a deep understanding of the utility and limitations of animal models.

In this chapter we make a distinction between pathway biology PD models and disease mechanism models. PKPD modeling can be applied to both types. We define a *pathway biology PD model* as being a model with a relatively quick turn around and a relatively simple measurable output, or endpoint, that informs on drug action and engagement along a specific biological process or pathway. While this type of model can be used in any phase of drug discovery (*i.e.* target validation, lead identification or lead optimization), it is more often used in lead optimization where there are efforts by a scientific team to advance the "best" compound from an inventory of related drugs that may share similar physiochemical and pharmacological

The Handbook of Medicinal Chemistry: Principles and Practice
Edited by Andrew Davis and Simon E Ward
© The Royal Society of Chemistry 2015
Published by the Royal Society of Chemistry, www.rsc.org

- Inconsistent clinical pharmacological translatability
- Acute uniform models of acute responses in adults animals
- Major anatomical differences
- Comparable human endpoint measurements are often difficult in animals
- Animal models do not capture the chronicity or the heterogeneity of human diseases

- Animal models are the most relevant available for modeling in vivo processes
- Historically used to inform on physiology/pathology
- Useful for PK/PD modeling
- Integrates "efficacy" with safety and disease biology

- **No *in vivo* model can encapsulate all the aspects of the symptoms or pathologies of a disease**
- **We can only model what we understand (pathway biology)**

Figure 12.1 Balancing the rational use of animal models in drug discovery. Listed are common arguments against and in support of the use of animal models to support development of novel pharmacological drug entities. The current thinking around the utility of animal models is based on the belief that "the predictive value of an animal model is only as useful as the context in which it is interpreted."

properties. Afterwards, there may be the desire to profile the lead molecule(s) in a disease related model. No animal model encapsulates the complexities (*i.e.* chronicity, heterogeneity, *etc.*) of human illness. Indeed, these assays typically only share some limited feature(s) of human diseases. In short, there are no animal models of asthma, rheumatoid arthritis (RA), Alzheimer's, and so forth. Consequently, we refer to this second type of model as being a *disease mechanism* model. The aim of this chapter is to provide information on the basic principles on the effective use of both pathway biology and disease mechanism models. We provide high level guidance on the use of animal models in the context of *in vitro* pharmacology and PK data, focusing on preventing the over-interpretation of integrative data. We discuss practical aspects of experimental design for *in vivo* studies and offer useful considerations for the medicinal chemist and other biomedical scientists. Ethics, relevant species choice, group size, statistics, pharmacodynamic (PD) end point and PKPD relationship are addressed. Additionally, some practical examples of pathway biology PD (LPS-mediated cytokine liberation in rat serum) and disease mechanism animal (rheumatoid arthritis; asthma) models are provided.

It is worth remembering that in order for scientists to successfully model relevant aspects of a disease, they must have sufficient understanding of the pathological changes associated with the disease itself. Often there is incomplete understanding of the clinical phenotype and this is reflected in the restraints of the current repertoire of animal models. As a consequence, *in vivo* researchers can only model pathophysiological mechanisms that are currently available to them—the implication being that *one can only model what one knows*.

12.2 BASIC PRINCIPLES AND MAJOR CONSIDERATIONS OF ANIMAL MODELING

As we indicated above, there are legitimate arguments against and in support of the use of animal models in pharmaceutical sciences. A recent review by McGonigle and Ruggeri[1] concluded that, "the predictive value of an animal model is only as useful as the context in which it is interpreted." However, "rather than dismissing animal models as not very useful in the drug discovery process, there must be focus on improving existing and developing new animal models." This begs the question, what does the ideal animal model look like? While there is expected to be some understandable uniqueness (among animal models) depending on the pharmacological inquiry being studied, we believe that there are some mutual features that may

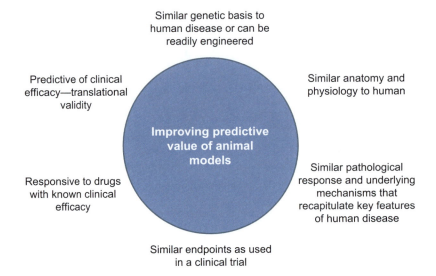

Figure 12.2   The ideal animal model. The figure lists characteristics that if aligned may improve the translatability of animal models. Additionally, "translatability between preclinical and clinical models is improved by having some historical precedence and/or an established blazed path to progress compounds through a pharmaceutical pipeline."

improve translatability to clinical settings (Figure 12.2). For example, these include characteristics such as having similar anatomy, physiology and pathology linking the animal model to the clinical condition, measuring the same or equivalent endpoints or biomarkers in both animal and human experiments and showing comparable pharmacological responses between animal assays and proof of concept studies in man. While obvious, it is important to point out that translatability between preclinical and clinical models are improved by having some historical precedence and/or an established blazed path to progress compounds through a pharmaceutical pipeline. For example, if we were interested in developing a new antihistamine (anti-allergy drug for allergic rhinitis), given the successful development of drugs like azelastine, cetirizine, desloratadine, fexofenadine, levocetirizine, loratadine and olopatadine over the past two decades it may be reasonable to use analogous tactics, including animal experiments, to advance your new drug entity. Indeed, we would be likely to use several of these drugs as positive comparators in our pathway biology PD and disease mechanism models. Nevertheless, before we conduct any *in vivo* experiments it is important to address some the following basic regulatory requirements and milestones for successful animal experimentation.

12.2.1   Ethics, Legal Requirements and 3Rs

Both the European Union and the USA legally require researchers planning any experiment using animals (vertebrates) to consider whether the objective of that experiment can be achieved using an alternative approach. Animal research in the UK is regulated by law. These laws cover the people carrying out the research, the research project itself and the location of the research. Primates, cats, dogs and horses are given special status under the Animals (Scientific Procedures) Act 1986. They can only be used if no alternative research model would suffice and if their use can be justified by the scientist. Breeding facilities are also regulated by law and are formally licensed by the Home Office. Russell and Burch[74] (1959) developed an effective tool to assist the researcher in planning the use of animal models called the '3Rs'—*Replacement*, *Reduction* and *Refinement*.

1. *Replacement*—consider achieving goals using modeling, tissue culture, or lower organisms such as *C. elegans or Drosophila*.
2. *Refinement*—if is not feasible to replace animals, researchers must try to minimize suffering and pain to individual animals.
3. *Reduction*—minimize the number of animals used in each experiment. In order to achieve this goal, researchers need to have a good understanding of their model; they must set clear objectives for the study and understand biological variability, in order to apply appropriate statistical tests.

Although the use of open access journals has revolutionized biomedical research and contributed enormously to data sharing across the research scientific community, there is still substantial evidence showing that the reporting of *in vivo* data for biomedical research often fails to accurately describe research methods and, in some cases, to report results appropriately. This in turn can have a range of implications for the entire research process and in some cases the reputation of the investigators involved. Kilkenny *et al.* highlighted major unreported characteristics of animal experiments ranging from details about species, *i.e.* strain, sex, age, weight, to lack of clarity as to which statistical method has been used. It is vitally important that investigators create and then follow well designed experimental protocols, to ensure the use of a minimum number of animals in providing valid results. It is becoming increasingly recognized that the number of animals used does not necessarily lead to an improvement of the output, but in many cases merely illustrates the bad design of the experiment.[2]

12.2.2 Define Objective of the Study and Readouts

The aims and objectives of the experiment must be clearly defined and the species must be carefully selected (Table 12.1). An experiment is usually designed to test a hypothesis, or differences between two variables, such as dose of a chemical used and an observed response. Alternatively, the objective of the experiment may be to explore a particular hypothesis, to formulate a new hypothesis, or simply to gather data for information purposes. Therefore, although formal statistical testing may be performed, it may not be the main objective of the study. No researcher should design an experiment without having a clear understanding of how the resulting data will be analyzed. In order to perform *in vivo* experiments, researchers need to assign an *'experimental unit'*, which is a unit of replication that can be randomized and receive different treatment. Individual animals, or a group of animals, are often considered to be an experimental unit. However, if for example the objective of the experiment is to compare different diets and the animals in the same cage have the same diet, the cage as a whole will be deemed a unit and not any individual animal. Conversely, if a particular cell type is isolated from animal blood or tissue and placed on culture dishes for different treatment, than each culture dish will be the experimental unit and not the animal the cell(s) originated from. Failure to correctly identify the experimental unit is the most common mistake in the design of *in vivo* experiments. To enable appropriate statistical analysis, experimental unit readouts must be clearly defined, robust and reproducible.

12.2.3 Controlling for Variability

Controlling for variability within *in vivo* experiments is an essential part of the experiment. To design efficient experiments that produce meaningful quantification, researchers must have sufficient understanding of the biological variability between effect (signal) and variability (noise). Variability can be separated into two groups: 'random' variability, which is mainly due to individual variability between animals; and 'fixed' variability, such as animal sex, strain, diet,

Table 12.1 Typical species used in research settings.

Species	Adult weight (g)	Weight at birth (g)	Average longevity (years)
Mice	25–30	1–2	1–3
Rats	250–600	5–6	2–3
Hamster	30–40	1–2	2.5–3
Ferret	750–800	10	5
Guinea pig	500–800	85–90	6–8
Rabbit	1000–7000	100	5–6
Cat	35000–4500	110	13–17
Dog	10000–30000	200–500	13–17
Rhesus Monkey	5000–12000	400–450	15–20
Marmoset	250–600	25–35	8–12

age and housing. The latter can be controlled by researchers. Increasing the signal to noise experimental window will result in an improvement in statistical significance and lead to a reduction in the number of animals used per experiment. The statistical significance test is mainly driven by the signal to noise ratio, hence if 'noise' within the experiment is significant, the biological effect (of) the 'signal' can be masked, making interpretation challenging.

12.2.4   Animal Housing

Animals housed under optimal environmental conditions usually exhibit greater uniformity than animals housed under poor conditions. Bedding, diet (particularly for long term studies), and physical environment can influence the outcome of experiments. Always procure bedding from reputable vendors that have been regularly vetted. Housing mice individually appears to increase variability in comparison to group housing.[3]

12.2.5   Animal Weight

Animals should be stratified by weight using randomized experimental design. Failing to do so can result in variability and create bias within the experiment.

12.2.6   Treatment

Applying treatment uniformly across all animals is an essential part of any *in vivo* experimental design. Although this can be achieved in the majority of cases in experiments using transgenic animals, it may be challenging in some cases due to the limited availability of the animals. Although not ideal, this can be overcome by using a single animal for each experimental group over a period of time.

12.3   ADDITIONAL CONSIDERATIONS WHEN WORKING WITH ANIMALS

12.3.1   Controls

Wherever possible, researchers should always include one or more control groups which are either untreated (satellite), or treated with placebo, or sham-treated animals. It is good practice to consider controls as treatment groups. If experiments are repeated frequently over time, controls should be monitored in order to understand fluctuation and trends. Analyzing the data with time as a control chart (Manhattan plot or Shewhart chart) enables unusual results to be identified from natural variability and is generally considered a useful monitoring tool.

12.3.2 Choice of Animals

Given that diseased animals are usually more variable than healthy animals, it is desirable to use, "Specific Pathogen Free" (SPF) animals. This approach reduces the number of animals required and allows greater statistical precision. SPF animals are widely available from commercial breeders. Care needs to be taken to avoid infection within the animal unit as infections can potentially have a detrimental effect on animal models or readouts.

12.3.3 Species Choice

In general, species choice is mainly governed by: physiology; relevant translatability to humans; availability of the model across different species; model robustness; and model reproducibility. For example, if research is performed for cancer therapy and the objective is to investigate the role of a particular gene or protein within disease-like settings, the most likely strategy would be to use a mouse model. In addition, their smaller size and the abundance of genetic variances make mice often the preferred choice to rats. Generating complete knockout, or conditional knockout models is economical and a relatively straightforward process these days. Rats are the next obvious choice as laboratory species, mainly because of their size and the existence of a large bank of current experimental evidence regarding their physiology, pharmacology and toxicology. Other rodents such as guinea pigs, gerbils, ferrets and hamsters are also frequently used, depending on their anatomical and physiological similarities with humans. A good example of this is the use of guinea pigs, particularly within the respiratory therapy, where similarities with humans in the anatomical and physiological composition of their lungs make this species an attractive choice for animal modelling. Larger species such as rabbits, dogs and sheep are also frequently used in toxicology studies, as well as to investigate particular physiological questions that cannot be addressed using mice and rats. It is also common to use non-human primates in biomedical research (Table 12.2). In general, the larger the animal, the fewer the numbers used in the experiment. Although in drug discovery settings more rats are used than mice, there is no particular statistical justification for this. A more likely explanation is that rats, being larger, allow for more parameters to be measured with greater precision. Rats are also a popular choice for safety and toxicology studies within the drug discovery environment, hence the current attraction of PKPD relationship assessment. The cross-over activity of compound series across the different species is also an important determinant in selecting species for animal model experiments. Ideally test compounds should exhibit similar pharmacology to humans in the chosen species.

12.3.4 Genetic Definition of Strain

There is a large range of genetic types currently used for biomedical research, which can be divided into three main types:

- Genetically undefined 'outbred' animals—each animal in this stock should be considered generally unique (*i.e.* there will be limited knowledge about its genotype), unless determined. Still used in some disciplines, but the use of these animals is considered less desirable by researchers these days due to their inherent variability and potential for genetic drift.
- Genetically defined 'isogenic' animals—the most important factor regarding these strains is that their genotype is replicated across all individuals. Inbred strains are generally produced by more than 20 generations of mating between so called "brother and sisters"

Table 12.2 Attributes and limitations of common mammalian species used in biomedical research.

Species	Advantage	Drawback	Current use
Mouse	Small in size, reproduce quickly, amiable for genetic manipulations.	Share some degree of similarities to human physiology and anatomy but not all.	Majority of biomedical research is currently conducted in mice. Leading mammal for experimental genetics.
Rats	Larger than mice and frequently used as a standard physiological and toxicological species. Popular in cardio-vascular, hypertension and immunological modeling. Used in behavior studies, rats' cognition and memory is superior to mice.	Lack of complete genome data.	Second most cited animal model species. Cardio-vascular, hypertension, brain and nervous system, cancer research, toxicology, respiratory, infection and immunity, psychology and behavior.
Ferrets	Share many anatomical and metabolic features with humans: Estrous cycle similar to human menstrual cycle, develop similar flus symptoms to humans. Alternative to dogs and primates in some toxicology studies	Require specific handling and housing conditions. Lack of genetic information and reagents.	Reproductive, heart, brain and digestive system research. Vaccine and anti-emetics development.
Guinea Pig	Similar lung anatomy and mediators to human air-way disease pathology, similar receptor pharma-cology to human. Allergic reactions—anaphylaxis. Anatomical similarity to human airways. Dietary requirement for vitamin C.	Influence of the axon reflex over airways which is limited in humans. No published reports on transgenic guinea pigs. Compared to mice, substantially less of the guinea pig genome is known. Low arterial blood pressure. High baseline airway hyper-responsiveness.	Research for: asthma and allergies; nutrition, safety and hearing.
Rabbit	Easy to handle and widely available compared to larger animals. Some similarities to bone mineral density and remodeling with humans. Spontaneously develop some form of cystic fibrosis, cholera and hypercholesterolemia.	Diverse genetic back-ground in comparison to inbreed or outbreed mice which can lead to increase in variability if single gene is studied.	Frequently used for antibody production. Research for: lipid metabolism, atherosclerosis hypercholesterolemia, orthopedics and medical device testing prior to larger species, dental implantation.
Dog	Share many anatomical and pathological features with humans.	Larger species requiring larger compound amounts and special li-cense from Home Office. Fewer animals per group limits statistical robustness.	Research use: neuroscience, cardiovascular, diabetes, endocrinology, respiratory, bone and joint studies, multiple sclerosis, safety and toxicology.

Table 12.2 *(Continued)*

Species	Advantage	Drawback	Current use
Non-Human Primates	Closest to human in terms of anatomy, physiology, metabolisms and pathology.	Larger species, fewer animals/group limits statistical robustness and ethical considerations.	Research use: HIV, neurology, behavior, cognition, reproduction, Parkinson's disease, stroke, malaria, respiratory viruses, infectious disease, genetics, xenotransplantation, drug abuse, vaccine and safety drug testing.

and in many cases they represent clones of genetically identical material. These isogenic animals tend to exhibit phenotype uniformity when compared to outbred animals. There are over 600 inbred strains of mice and rats available worldwide. Nevertheless, about 80% of research is conducted using only the most popular 10 strains of isogenic animals (Table 12.3). It is important to emphasize that each strain has its own unique characteristics. Certain strains develop cancer naturally, such as (AKR/J) mouse strains, others are more prone to immunological phenotype (Table 12.3). A common approach is to choose a strain based on a particular research project. However, the most popular strains used in general research are BALB/c or C57BL/6 for mice or Lewis and F344 for rats.

• Partially defined strains of animals (mutant and transgenic with an unidentified genetic background)—only the genotype of the particular locus in these strains is usually defined. This is typical for transgenic animals that are bred with an undefined background in order to improve their survival rate. Potentially, use of these strains can create substantial challenges for data interpretation, since these animals can have genetic drift in their background strain genotypes and this can lead to modification of expression or functional redundancy of the gene of interest.

It is important to note that within each type there are individual strains and stocks of animals with specific characteristics. This increased complexity makes the investigator's task in selecting the correct strain quite a challenge. However, in some specific cases a particular strain, transgenic strain or even mutant is chosen due to its potentially desirable features or characteristics. For example, it is quite common for hypertension studies to use spontaneously hypertensive (SHR) rats since they have a propensity to develop hypertension and therefore provide a good phenotypic research tool.[4,5] Furthermore, certain mouse strains seem to develop a particular type of tumor, or grow tumors at a higher rate than other strains.[6,7] A very popular mouse strain in the asthma research area is BALB/C, because of its propensity to develop robust and reproducible Th2 responses, which are considered one of the pathologies of the human disease. As good practice, however, it is generally desirable to use animals from defined isogenic stock.

12.3.5 Statistical Analysis

In the majority of cases, the purpose of an animal experiment is to allow the researcher to assess whether a particular treatment causes changes in an outcome/biomarker (readout) of interest.

Table 12.3 Most popular mouse and rat strains.

Mouse strains	Comments	Rat strains	Comments
BALB/C	Used for immunology and oncology models. Produce robust Th2 responses frequently used for asthma models. Resistant for diet induced atherosclerosis. Frequently utilized for generating antibodies.	F344	Used for general purpose, experimental autoimmune encephalomyelitis (EAE), oncology and autoimmunity.
C3H	Used in cancer and infectious disease area.	Lewis	Used for transplantation, induced arthritis/inflammation EAE and streptozotocin (STZ)-induced diabetes.
C57BL/6	Most popular strain. Sensitive to pain and cold. Whole genome published. Convenient to create transgenic.	SHR	Hypertension research.
CBA	Used to study autoimmunity. Commonly used for leukemia research also develop mammary tumors.	WKY	Control for SHR rats.
129	Many sub-strains available—used for targeted mutations. High incidence of teratomas.	Brown Norway rats	Used for allergic respiratory and immunological research.
DBA/1	Widely used for arthritis models. Susceptible to develop atherosclerosis once on artherogenic diet.	Dark Agouti	Used for EAE, arthritis and cardiovascular research.
C57BL/10	Used in immunological research. Susceptible to ovalbumine but not to DNP-KLH. Susceptible to TSNB colitis and EAE.	WAG	Used in neuroscience, cancer and immunology research.
AKR/J	Used for cancer research and immunology.	PVG	Immunology research.
SJL	Used for multiple sclerosis. Susceptible to EAE and muscular dystrophy. Immunocompetent but have elevated T cells number.	Sprague Dawley	Most widely used. Used for toxicology, safety reproduction and developmental testing.

This will be followed by statistical analysis, which helps calculate the probability of error within the experiment. In other words, to assess the probability that the observed outcome occurred due to chance, (known as a false positive), instead of the real effect. The proof of good experimental design is to be able to determine beyond reasonable doubt that the causative effect is produced by the treatment itself rather than experimental variability. There is an important relationship between experimental design and the statistical methods used for data analysis. In the majority of situations where the efficacy of either one, or both of these, are called into question, one or two possible scenarios have occurred:

a. *In vivo* scientist possesses good knowledge about *in vivo* models, but often lacks basic principles of statistical design. "We have always performed experiments this way..."

b. Statisticians have extensive theoretical knowledge of statistical design, but often very little understanding of the *in vivo* model.

Table 12.4   Examples of certain statistical analysis tests.

Test	Comments
Tukey's	Methods used for comparing several treatments. Tukey's test is essentially a t-test, except that it corrects for multiple comparisons, therefore more suitable than t-tests.
Dunnett's test	Multiple comparison procedure comparing each of the number of treatments with a single control.
Bonferronni's test	Most conservative multiple comparison method. "Bonferroni adjustment" suggests that the "p" value for each test must be equal to or less than alpha, the critical level of p used to decide on statistical significance, divided by the number of tests.

The analysis of variance (ANOVA) is a contemporary versatile statistical method, commonly employed by researchers to quantitatively analyze experimental data. The validity of ANOVA relies on the following assumptions that:

1. The deviations of each observation from their mean group have normal distributions;
2. The variances within each group are approximately equal; and
3. The observations are independent of one another.

It is common to use ANOVA to compare several treatments within an experiment. Using this approach, ANOVA can indicate whether there is sufficient evidence that the treatment across the groups is different. In addition to analyzing the whole experiment (using all collected data), researchers often like to be able to differentiate controls, or compare individual groups to each other. This can be achieved using post-hoc multiple comparison treatment (Table 12.4). However, the more tests that are performed on the same data set, the greater the possibility, or risk, that one of the tests will produce statistically significant results purely by chance, leading to a false positive. This is mainly due to the reduction in sensitivity caused by multiple comparisons. Tukey's, Dunnett's and Bonferroni's, are the most frequently used multiple comparison tests as briefly described in Table 12.4. These tests should only be used if there is a significant overall treatment difference in ANOVA. It is good practice for experiments to test simple hypotheses, as well as avoiding multiple *post hoc* comparison.

12.3.6   Unexplained Data Exclusion

It can be tempting to exclude an observation without any explanation, because it spoils an otherwise good result. In fact, excluded data points should be noted as well as the reason for the omission. Experimental outliers should be investigated in order to fully understand and improve the model.

12.4   BUILDING A PLATFORM OF EVIDENCE TO ADVANCE THE PHARMACOLOGICAL PIPELINE USING ANIMAL MODELS

A major use of animal models is to explore the relationship between the drug's pharmacokinetic profile, the timing and magnitude of a biomarker and/or a clinically meaningful effect in the animal model. Ideally, the data from the animal model should support an integrative platform of evidence, building with a line of sight to clinical proof of concept (Figure 12.3). Mathematical modeling and simulation experiments are critical parts in formulating the PKPD hypothesis,

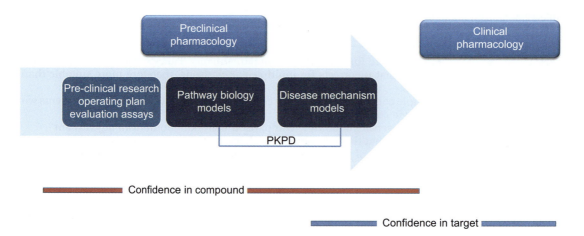

Figure 12.3   Animal models: building platform of evidence to bridge preclinical pharmacology to clinical success. Animal studies are not conducted in isolation. This work should be considered, in the context of *in vitro* and PK data, looking towards clinical proof of concept.

which can then inform the design of later pre-clinical and clinical studies. Control of the pharmacokinetic profile builds confidence that modulation of the chosen biological mechanism is indeed driving the pharmacology, rather than some as yet undefined off-target effect.

Important questions that can be answered with PKPD models include:

a. What degree of "occupancy" of the biological target mechanism is required to generate a meaningful pharmacodynamic response?
b. Does the drug have to be present at high enough concentrations to engage the biological mechanism throughout the dosing period, or can the drug use a hit-and-run mechanism (*i.e.* C_{max} driven PKPD relationship?
c. How long does the target need to be engaged before a meaningful pharmacodynamic effect can be measured?
d. How does the pharmacokinetic and pharmacodynamic profile relate to any adverse events that can also be measured (maybe at higher doses)?

It is good practice to consider the following guiding principles before embarking on designing PKPD *in vivo* models. Although some practical aspects of animal modeling may vary across different research areas, there is unanimous agreement on how to build the animal model or a tool, in order to study and understand the PKPD relationship and achieve the end product.

12.4.1   Specificity

Models should be induced with specific stimulus that is relevant to the target/pathway of interest. In order to understand PKPD, researchers must understand their model and select a biomarker that links to the clinical end point, (*i.e.* which cells or tissue produce biomarkers/ readouts of interest). It is worth remembering some basic principles: A) identify the tissue of interest; B) understand the target expression of the tissue; and C) understand compound exposure at the site of action. The combination of all these parameters can enable assessment of both target engagement and the PKPD relationship.

12.4.2 Robustness

The induction of the model needs to be sufficient in magnitude (assay window) in order to allow differentiation between the compound doses (*i.e.* treatment groups). Although this is often assay dependent, any assay windows less than three-fold are generally considered insufficient for detailed pharmacological evaluation and PKPD assessment, particularly if there is a high degree of experimental variance around the treatment means.

12.4.3 Reproducibility

It is essential to have a reproducible response in the model, enabling direct comparison across different experiments. Monitoring baseline and induction responses longitudinally is essential.

12.4.4 Simplicity

Whenever possible, models should be simplified. Complex model design with multiple or cocktail drug induction challenges tends to result in a large number of variables, which in turn often leads to a reduction in reproducibility.

12.4.5 Tools and Reagents

It is important to have adequate tools in order to generate a robust PKPD model. For example, the use of supra-maximal doses of human specific reagents (*i.e.* cytokines), in order to overcome non specificity of the reagents. The implications of using non-specific tools can often lead to an off target effect or response within the model.

In addition to the aforementioned, it is also very important that *in vivo* pharmacology and PK data are captured within a designated data base, in order to be available for further analysis, monitoring reproducibility and legal/regulatory purposes. Currently, pharmaceutical and biotech companies have different approaches in capturing *in vivo* data, ranging from fully integrated data capturing system available to all scientists, all the way to data being stored on share drives or individual folders being available only to scientists involved in the experiments. This often leads to misinterpretation of the data and difficulties comparing different experimental conditions.

12.5 EXAMPLES OF PATHWAY BIOLOGY PD AND DISEASE MECHANISM MODEL

12.5.1 Pathway Biology PD Model

12.5.1.1 Lipopolysaccharide (LPS) Challenge Model

LPS is a component of bacterial cell membranes and is an endotoxin that through binding to CD14/TLR4/mdm2 receptor complex promotes innate immune responses. The use of LPS is frequently employed in experimental settings to elicit robust, reproducible cytokine release and subsequent inflammatory cell recruitment, (and pathological consequences), in various targeted organ systems. Thus, a LPS model represents a good showcase example to highlight some very basic aspects of PKPD modeling. LPS has been administered by systemic or the inhaled routes to an assortment of animals, including guinea pigs,[8,9] rabbits,[10,11] dogs[12] and pigs.[13] However, rat and mouse LPS models are the most common.[14,15] Given the fact that LPS is also used to study human biology (*i.e.* from *ex vivo* samples or clinical studies), an opportunity is presented to link preclinical PKPD relationships with successful drug action in man. Nonetheless, taking a step back to animal models, there are a few aspects that researchers need to be aware of in order to use LPS vivo models appropriately for PKPD. The first is having a clear understanding of what are the desired essential supporting data sets (*in vitro* and *ex vivo*), that provide mechanistic insights into the basic pharmacology of the test drug. As simple as this may appear to be, many well-meaning scientists have arbitrarily selected doses of LPS, or compounds, for *in vivo* experiments, perhaps sometimes from published reports, without a clear understanding of how their pharmacological agent interacts with the target (receptor, enzyme, *etc.*). Using *in vitro* studies, (receptor binding assay or a cell based functional assay) to establish an IC_{50} (the molar concentration of an antagonist or agonist that elicits 50% of its maximum effect) and the maximum achievable efficacy is a good starting point. As a sidebar, the concept of maximum achievable efficacy is exceptionally important when dealing with agonists, as they can vary in the responses they produce. For example, some agonists can produce their maximum efficacy while only stimulating a low number of receptors (*i.e.* "super agonist"). On the other hand, low potency agonists may interact with the receptor only to produced partial efficacy (*i.e.* partial agonist). Secondly, it is important to have an understanding of "off target" interactions, as this may allow better explanations of the *in vivo* data, particularly around understanding potential side effect liabilities and if the drug is engaging a second target that may also potentially contribute to whole animal efficacy. In the example below, data from a typical LPS PD assay is provided. In this experiment, different doses of LPS were administered by the i.p. route to mice (Figure 12.4). At 1.5 and 3.0 hours post LPS treatment, blood was collected and IL-6 and TNF concentrations in the serum were determined. Although this type of LPS model is robust and reproducible, there are salient points that require mentioning: (1) Cytokines measured in the blood can originate from different sources (liver cells, Kuppffer cells, PBMC or epithelial cells) which can potentially complicate interpretation of the results if target expression is restricted to a particular tissue or cell population. (2) The kinetics of cytokine release is very important. Note the differences in the relative concentrations of TNF and IL-6 as a factor of time. TNF is a short lived cytokine that in the current experiment was only marginally detected after 3 hours. After understanding the temporal profile of LPS mediated cytokine release, the impact of the test compounds can be evaluated in the assay to generate data similar to the simulated graph in Figure 12.5. Shown is a hypothetical effect of an orally administered anti-inflammatory agent on LPS-evoked serum TNF release. Given our understanding of the temporal characteristics of TNF, the plot is simulated to describe the plasma concentrations of Compound A relative to the percent inhibition of serum TNF at 1.5 hours post LPS challenge. Appropriate curve fitting could be allied to the graph and the IC_{50} of approximately 9 μM can be estimated (Figure 12.5).

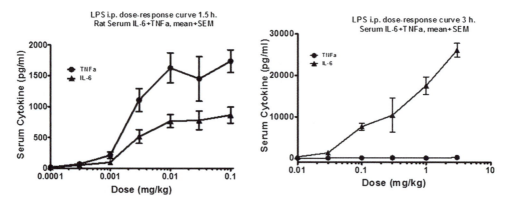

Figure 12.4 Impact of LPS on serum TNF and IL-6 Concentrations. Figure displays the effect of increasing doses of LPS (0.0001–0.3 mg/kg, i.p.) on TNF and IL-6 release at 1.5 and 3.0 hours post administration. Each point represents the MEAN±SEM (n = 8 per treatment group). Byford A., (unpublished).

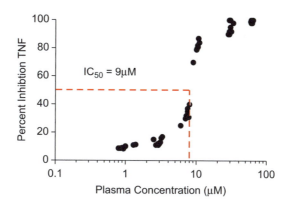

Figure 12.5 Simulated concentration *vs.* effect curve for the effect of a hypothetical drug "Compound A" on TNF concentrations after challenge with LPS. Figure displays the effects of increasing concentrations of Compound A on TNF at 1.5 hours post LPS treatment. Shown are MEAN±SEM with an estimated IC_{50} value.

Neutrophils are quickly recruited to a site of inflammation, or in tissue, following infection or injury.[16] Consequently, the aforementioned pathway biology PD model would be useful in research focused on understanding the impact of drugs on neutrophil trafficking, or upstream cytokines that drive influx of these cells into damaged tissues. One example, is that some in the scientific community have used LPS-mediated release of cytokines and/or pulmonary inflammation to support preclinical—clinical translatability of targets for Chronic Obstructive Pulmonary Diseases (COPD), where neutrophils are believed to play a prominent role in the disease. Drug mechanisms like PDE4 antagonists and CXCR2 inhibitors are good examples of targets that have undergone testing in LPS-challenged animal models and human studies.[17–20] While CXCR2 inhibitors to date have a lackluster success in clinical settings, roflumilast (a PDE4 inhibitor) was approved by the FDA in 2011 as a general anti-inflammatory drug for COPD.

12.5.2 Disease Mechanism Models

While animal models have improved our understanding of the mechanisms driving various diseases such as asthma, rheumatoid arthritis and inflammatory bowel disease, they are not without substantial limitations. While it may be somewhat safe to state that many of the currently available clinical drugs do work in many of the preclinical animal models, it is also accurate to affirm there have been many examples of novel pharmacological agents displaying efficacy in animals and producing little or no effects in man. As an example, the asthma field provides a unique perspective of the challenges in bridging the vast chasm between animal pharmacological assays and efficacy in man. In the early 1990s, a very exciting emerging area of asthma research was around understanding the pathology around tachykinins.[21,22] In brief, tachykinins are a family of neuropeptides (*i.e.* substance P, neurokinin A (NKA) and neurokinin B (NKB) that mediate through their respective receptors (NK1, NK2 and NK3) inflammation, airway constriction and cough in the lungs. Many pharmaceutical companies invested significant resources into developing tachykinergic antagonists. Unfortunately, the animal models, most notably the guinea pig, over predicted the actions of tachykinin inhibitors in the clinic. Sadly there would be numerous clinical failures driven by validation conducted in Th2 driven experimental models, particularly the ovalbumin sensitized and challenged mouse.[23] A bleak history such as this has underscored the translatability of the preclinical and clinical efficacy data. However, it is also important to note that the approaches inducing some of the disease-like changes in small animals only reflect our current advances in understanding the pathogenesis and mechanisms driving these diseases, which is still very limited. It is widely recognized that the term asthma is considered an umbrella for a heterogeneous group of disorders representing different phenotypes which share common symptoms of partially reversible difficulties in breathing.[24] The ultimate goal is to identify the phenotypes, in order to optimize treatment for the particular patient group. To date, the range of severe asthma phenotypes have been identified based on a patient's cellular or inflammatory components. These include high or low airway eosinophila, or neutrophila phenotype; or Th2 high and Th2 low cellular component.[25-27] This high Th2 severe asthma phenotype is particularly interesting since it has been shown to respond to treatment with inhaled corticosteroids, or IL-13 antibody, significantly better than low Th2 phenotype, leading to reduction in exacerbations.[28-30] Similar findings were confirmed using IL-5 Ab in a Hyper-Eosinophilic Syndrome (HES) patient population, compared to normal asthmatics.[31-33] Taking into account the complexity and heterogeneity of asthma, it would be naïve to think that it can be accurately reproduced in any single animal model. Therefore, depending on the mechanisms or the question being asked, a particular animal model can be useful tool to help investigators elucidate the pathway or establish a PKPD relationship for the compound of interest. The recent clinical success of emerging therapeutic antibodies to Th2 inflammatory cytokines like mepolizumab (IL-5), lebrikizumab (IL-13) and the utilization of clinical biomarkers, such as periostin in severe eosinophilic asthmatics, may offer an opportunity to better align animal models with humans. Finally, it is important to note that both IL-5 and IL-13 blocking antibodies have been shown to be efficacious in both the OVA and the House Dust Mite model.[34-39]

A description of the breadth of disease animal models across all therapy areas is beyond the scope of this chapter, for illustration purposes, we will focus on asthma and rheumatoid arthritis animal disease models

12.5.2.1 *Arthritis Like Model*

Rheumatoid arthritis (RA) is a chronic inflammatory disease of unknown etiology, associated with disability and pain, and which significantly affects quality of life.[40] If left untreated, RA can ultimately lead to joint destruction, systemic bone loss, increased risk of fractures and other comorbidities.[41] The pathogenesis of RA comprises a complex inflammatory response involving macrophages, synoviocytes, T/B lymphocytes, proinflammatory cytokines and autoantibodies, causing joint damage as a result of the erosion of bone and cartilage.[42,43]

In RA, the release of numerous proinflammatory mediators, such as cytokines IL-6, IL-1β and TNF, results in increased sensitivity to pain.[44,45] It has also been shown that mechanical thresholds for pain and pressure are decreased in the affected joints of patients with RA.[46] Most anti-rheumatic therapeutics are effective in controlling inflammation, however, further investigation is required in order to identify novel anti-rheumatic agents that can simultaneously inhibit inflammation and pain. Consequently, the investigation of pain is critical for any effective treatment paradigm for RA.[47]

Animal models for RA have played a major role in our understanding of the mechanisms of disease pathophysiology and have supported drug discovery leading to novel therapies.[48] Preclinical models of arthritis share many immunological, clinical and histological characteristics with human RA, however, none of them capture all the facets of the human disease.[49] Today, the most popular arthritis models used are in mice and rats. This is mainly because of cost, genetic homogeneity and the ability to use genetically modified strains. In general, arthritis animal models are generated by treatment or induction with a specific treatment. Even "spontaneous", models can be considered to be induced, since gene expression or deletion is under specific control. Examples of genetic mouse models include, K/BxN and IL-1 receptor knockout mice. Several preclinical models of arthritis, such as Adjuvant Induced Arthritis (AIA) and Collagen Induced Arthritis (CIA) are widely used in drug discovery (Table 12.5).[48] These models are

Table 12.5 Examples of selected animal models utilized in Immunological Research.

Model	Disease-like mechanism	Advantage	Disadvantage	Mechanism/target evaluation
Ovalbumin Challenge Model (OVA)	Asthma	Robust eosinophilia. IL-5/13 induction, IgE, mucus production.	Sensitization required. Antigen not related to human asthma. Lack of "allergic" component of asthma.	Corticosteroids, IL-5/13/33, TSLP, Abs, CRTh2 antagonists.
House Dust Mite model	Asthma	Human relevant allergen. Induce allergic events. No sensitization required. Robust eosinophilia. IL-5/13 induction, IgE, mucus production.	HDM extract contains many proteins in addition to Der p1/f1) that can elicit additional allergic responses. Batch to batch variability.	Corticosteroids, IL-5/13/33, TSLP, Abs, CRTh2 antagonists, TLR4.
Adjuvant Induced Arthritis (AIA)	RA	Captures RA pathology. Robust and reproducible disease suitable for screening compounds. Multiple SOC are efficacious in the model.	Disease is chronic progressive unlike the flares and remission observed in RA. Severe bone loss can lead to cartilage damage due to loss of supportive tissue unlike RA. Does not have the humoral component. Antigen not related to RA. Antigen can cause granuloma in the liver.	T cell, Cox2 and macrophage mediated inflammation Cox2, IL-1, IL-6, IL-12, IL-23, Syk, pan-Jak inhibitor, CCr2.
Collagen Induced Arthritis model (CIA)	RA	Captures RA pathology. Robust and reproducible disease suitable for screening compounds. Does not have the severe bone loss observed in AIA. Cartilage erosion leads to bone loss. Multiple SOC are efficacious in the model.	Disease is chronic progressive unlike the flares and remission observed in RA.	T, B and macrophage mediated inflammation. P38, IL-1, IL-12, IL-23, Syk and pan-Jak inhibitor.
Sreptococcal Cell Wall Model	RA	Captures RA pathology. Disease progression occurs as flares and remission similar to RA. Robust and reproducible disease suitable for screening compounds.	Antigen not related to RA. Does not have the humoral component.	T cell and macrophage mediated inflammation. TNF, Abatacept, anti-IL-1, Corticosteroids, ICAM-1, P_selectin, MIP-2, IL-4/10, INFg, p38.

Table 12.5 (*Continued*)

Model	Disease-like mechanism	Advantage	Disadvantage	Mechanism/target evaluation
		Disease mild compared to AIA or CIA so the animals maintain a healthy status over the course of the study. Multiple SOC are efficacious in the model.		
T Cell Adoptive Transfer Colitis	IBD	Closely resembles human pathology. Multiple SOC are validated in the model.	*In vitro* cell purification adds difficulty and variability. 4–10 weeks for disease development.	T cell mediated inflammation. TNF, IL-12, IL-23, MyD88 in Tcell, NKG2D, integrin, CTLA-4 Ig have been evaluated in the model.
DSS-Colitis	IBD	Convenient induction of intestinal inflammation (DSS in water). Mortality rate is easy to manage. High reproducibility.	Disease phenotype is difficult to suppress.	Epithelial damage. Innate inflammation. Cyclosporin A, Sulfasalazine (5-ASA), TNF blocker, Anti-IL-12 have been evaluated in the model.
TNBS-Colitis	IBD	Involves T cell component. Potential use the hapten as the antigen in *ex vivo* assays.	Highly dependent on the facility microbial environment. Mortality rate highly variable. Short disease duration.	Acute injury + T cell mediated inflammation. TNF, IL-12/23, IL-18, Stat 3, NFKB, antibiotics.
MLRL/lpr Mouse Model	SLE	Spontaneous disease model. Commonly used for compound evaluation, and many pathways have been studied and evaluated in this model. Relatively fast disease development (3–6 month). In addition to nephritis, it has skin, lung, joint, CNS disease phenotype.	Disease development can be heterogeneous and variable. Mortality starts relatively early.	Fas mutation. Lymphoproliferative. Accelerate lupus-like disease development. Four susceptibility locus (Lmb1-Lmb4) are linked to ANA and GN. B cell, T cell, complement, inflammatory cytokine, TLR7 dependent. FcRg independent. Syk, PI3Kg, BTK, IL-6 R, IFNb, IFNg, IL-17, IL-18, IL-21R.Fc, iNO. MyD88, B cell pathway (TACI-Fc, CD20 B depleting mAb, Anti-CD79a or b), CD 4 T cells, Complement C3 inhibitor, Proteasome, CCL2/MCP-1 have been evaluated.

Table 12.5 (*Continued*)

Model	Disease-like mechanism	Advantage	Disadvantage	Mechanism/target evaluation
NZB/NZW F1	SLE	Spontaneous disease model. Commonly used for compound evaluation, and many pathways have been studied and evaluated in this model.	Slow disease development (6–12 months).	Multiple susceptible loci. Autoreactive B and T cell activation. ANA. GN. Complement and FcRg dependent. Syk, BTK, IL-6 R/IL-6, TNF, IFNg. B cell pathway (TACI-Ig, CD20 B depleting mAb, Anti-CD22 B depleting), T cell pathway (CTLA-4 Ig, anti-CD3), proteasome, TLR7/TLR9 have been evaluated.

poly-arthritic, involving multiple joints and the disease phenotype is chronic and progressive, unlike the flares and remissions observed in RA.[50] This review will focus on rat induced arthritis models, with a specific emphasis on the streptococcal cell wall model.

12.5.2.2 Streptococcal Cell Wall (SCW)

SCW is an animal model of arthritis that effectively captures the repeated remission and flaring phenotype, similar to RA.[51] Previous studies have shown that a single intra-peritoneal injection of SCW extract, PeptidoGlycan-PolySaccharide (PG-PS 10s), induces inflammation in the peripheral joints of Lewis rats, with repeated phases of self-reactivating flares resembling RA.[52] However, the recurrence of reactivation is unpredictable and often difficult to control, hence this model was modified by Schwab *et al.*[53] in order to synchronize the flares. The modified SCW model is induced by a local intra-articular (i.a.) injection of SCW extract PG-PS 100p in the hind ankle (Flare 1), followed by a systemic intra-venous (i.v.) challenge (Flare 2) (Figure 12.6). The model is characterized by a mono-arthritic multi-flare phenotype of two distinct remissions and flares. One of the advantages of this model, is that inflammation is only limited to the sensitized joint and there is no detectable involvement of other joints, unlike other preclinical arthritis models.[54] Therefore the contralateral joint can be used as a control.

The first flare, induced by an intra-articular injection of SCW, results in mild paw swelling, peaks 24 hours post sensitization (Flare 1) and resolves over 72 hours. The second flare, induced by an intra-venous challenge with SCW (typically three weeks later), results in a pronounced onset of paw swelling, reaches its peak on day 3 after the i.v. challenge (Flare 2) and resolves over a 7 day period (see Figure 12.7). Previous studies by Schimmer *et al.*[55] show that the early phase of the model can be triggered by neutrophils and Th2 cells. They have also demonstrated that, in addition to T cells, neutrophils were involved in the reactivation of flares. In addition, the model is dependent on multiple proinflammatory cytokines including TNF and IL-1, as assessed by specific anti-cytokine therapy and gene expression analysis.[56,57] Mechanisms that block neutrophil activation and recruitment, Th1/Th2 cell activation and recruitment, or block the production of TNF or IL1 all have shown a degree of efficacy in this model.[56] Furthermore, the glucocorticoid

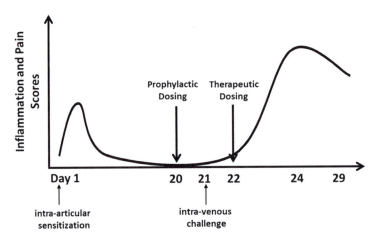

Figure 12.6 Induction of the arthritic phenotype: Inflammation of the ankle joint following intra-articular sensitization (day 1; Flare 1), the first intra-venous challenge (day 21; Flare 2).

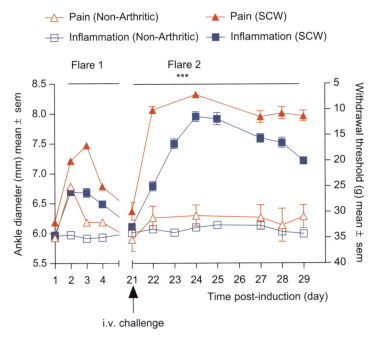

Figure 12.7 Simultaneous evaluation of inflammation and pain in mono-arthritic SCW model. A composite of four independent studies showing inflammation (ankle diameter; Y axis) and pain
(withdrawal threshold; Z axis) over time (days; X axis) following systemic i.v. challenge.[58]

receptor agonists, which down regulates the inflammatory response in many cell types and TNF
antagonists, also demonstrated efficacy in this model.[58] Conversely, therapies targeting ICAM-1,
P-selectin, IL-10 or INFg have shown no effect in this model. On the other hand, mechanisms like
CTLA4, IL-4 and p38 have shown partial effect in this model, but had no advantage in the clinical
settings demonstrated.[55,58,59] In addition to measuring ankle diameter, demonstrating efficacy of
pain in animal models of arthritis is an important step in identifying novel anti-rheumatic agents
that can effectively target inflammation and pain in the clinic.[44] As mentioned above, in addition

to inflammation, many researchers have used this mono-arthritic model to evaluate paw withdrawal threshold readouts (von Frey) as a surrogate for pain.[46,60] The mechanisms leading to pathogenesis in the model can be further investigated by histopathological evaluation, cytokine profiling and immune cell phenotyping in the arthritic joint.

12.5.2.3 Asthma-Like Models

Mice and rats don't develop asthma spontaneously. Only a very low percentage of cats and horses seem to spontaneously develop some symptoms of the asthma phenotype, including allergic airway responses. Asthma, like many diseases, may be better described as a non-genetic syndrome, a group of patients linked by associated symptoms, rather than a common phenotype. Due to the complexities that characterize the majority of asthmatics, such as multiple disease manifestation features, heterogeneity, different cellular pathologies and susceptibility to genetic and environmental factors, it is extremely challenging to create animal models that can capture the full range of disease components and mirror the human situation.

In general, immune responses can be divided into two groups. One group is immunosuppression, arising from an insufficient ability of the immune system to defend itself against invading pathogens, or cancer cells. The second group of immune responses is considered to be an enhancement or hyper-responsiveness of immune system activity. This latter response is thought to be triggered by an agent, or allergen, that can mobilize the humoral or the cellular arm of the immune system. These processes can lead to autoimmunity and hypersensitivity. The most common causality of asthma is exposure to an allergen. This is generally described as a condition characterized by persistent airway inflammation, airway hyperactivity, airway remodeling and reversible airflow limitation. Asthmatic responses can be divided into an induction phase (sensitization) and an effector phase (challenge). The induction phase tends to mark the exposure to an allergen leading to specific immune responses. The second phase is triggered by re-exposure of the same allergen leading to the effector phase, or the display of asthma symptoms. In some patients, these asthma symptoms can be further separated into early asthmatic responses (EAR), mainly consisting of bronchospasm and mucus secretion due to the involvement of mast cell and mediators such as histamine, leukotrienes, proinflammatory cytokines and late asthmatic responses (LAR), which occur 24 hours after the allergen challenge. LARs are thought to be manifested by inflammation of the airway (*i.e.* Th2 cytokine production, increased eosinophilia and CD4+ T cells infiltration). Researchers historically utilized this knowledge in order to devise *in vivo* models that could resemble some of the features of the disease and therefore enable development of drugs for respiratory diseases including asthma.[61,62]

12.5.2.4 The Ovalbumin Model (OVA) of Allergic Lung Inflammation

For a considerable period of time, this model has been considered to be the standard approach for modelling some of the features of the asthma disease phenotype. The approach utilized in this model relies on the ability of antigen in egg white ovalbumin (OVA) to induce immunological responses in the lungs of sensitized rodents. Currently several variations of the protocol exist, however, the principle for all such protocols is relatively similar. Animals are sensitized with OVA in the presence of an adjuvant, usually aluminum hydroxide, which primes immune responses towards the Th2 phenotype. The sensitization, usually 2 weeks, is followed with an aerosol challenge. Traditionally, certain species have been preferred for modeling some aspects of the asthmatic responses, due to the anatomical and physiological features of their lungs (Table 12.2).[63] Although OVA does not induce asthma symptoms in humans, this model can be used to successfully replicate some features of asthma. These include Th2 mediated inflammation, which is very rich in eosinophils, IgE production, edema and mucus secretion, as well as EAR/LAR bronchoconstriction responses. Some of the readouts were shown to be sensitive to standard care treatments such as glucocorticoids and the β_2-adrenergic agonist.[64–66] There is a large body of evidence that links the nature of the allergic Th2 inflammation in the OVA model with clinical phenotypes. In addition to inflammation readouts (eosinophila and Th2 cytokines), airway hyper-responsiveness (AHR) to methacholine was often measured as an end point., although researchers do need to be careful not to use these readouts in isolation when making decisions on the efficacy of their compounds. A good example is measuring AHR in a mouse OVA model. Mice airways have substantially less smooth muscle than human and do not respond readily to pharmacologic agents affecting the bronchial tone.[67] Therefore, high doses of bronchoconstriction agents are needed to elicit responses, which in turn can complicate interpretation of the data. Recently, whole body unrestrained plethysmography and the measurement of "enhanced pause", or PenH, as an artificial parameter of airway function, has gained popularity. PenH is a non-invasive method of measuring AHR in mice, although not without significant concerns regarding the readout and its interpretation.[68,69] A more comprehensive approach assessing methacholine challenge in 29 different strains of mice concluded that changes in PenH appeared to be regulated by underlying genetic factors.[70] Furthermore, using computer simulation to define AHR in OVA-mice models suggests that AHR is measuring something other than the abnormalities in smooth muscle, which is fundamentally different to the response in human asthma and thus requires further investigation.[68]

Currently, asthma drugs are assessed by their ability to improve lung function, as measured by forced expired volume in 1 second (FEV1), or their ability to improve incidences of severe exacerbations in the disease. The majority of animal models that are currently used for asthma research do not measure those parameters and therefore there is a significant gap between measures of efficacy in animal models and humans.

Conversely, researchers have successfully used guinea pigs to assess AHR as guinea pig anatomy and physiology resembles many features seen in humans, making this species very useful for studying respiratory diseases. Several important processes, mediators, regulators of airways disease pathogenesis and breakthroughs in measuring lung mechanics were discovered or demonstrated first in guinea pigs, including the immediate type hypersensitivity reaction, the actions of histamine, the cysteinyl-leukotrienes and their two receptors. Additionally, contractile and relaxant agonists of airway smooth muscle have been shown to be very similar in potency and efficacy between humans and guinea pigs, building confidence in the translatability of this model, at least for bronchodilator mechanisms such as leukotriene D4 receptor antagonists, β_2 agonists and muscarinic M3 antagonists.[71,72]

While mice models of allergic airway diseases have proven to be invaluable for illustrating various aspects and pathways relating to respiratory biology and pathology, they are not necessarily the preferred species from a drug development perspective. Additionally, a recent report by Seok *et al.*[73] suggested that some genomic responses in mice poorly mimic human inflammatory diseases, thereby casting an additional shadow on the translatability of mice models to human inflammatory diseases. Because of these reasons, many researchers use rats as their species of choice in studying lung airways inflammation. The Ovalbumin challenge in Brown Norway (BN) rats is a frequently used rat model for airway inflammation. One of the reasons for the popularity of this rat strain is their ability to produce a substantial Th2 cytokine response, followed by a substantial eosinophilic recruitment in the bronchoalveolar lavage fluid (BALf). It is also interesting to note that these rats develop spontaneous granulomatous pneumonia under normal husbandry conditions. In addition to Th2 cytokines and eosinophilic infiltration, other asthmatic features include: damage to the airway epithelium, thickening of the basement membrane, acute and late-phase airflow obstruction and the presence of airway hyper-responsiveness after the antigen challenge.

12.6 ADDITIONAL ANIMAL MODELS FOR CONSIDERATION

12.6.1 Non-Human Primate Models

In addition to anatomical and structural similarities with humans, in the majority of cases non-human primate models also offer similarities in their response to various challenges. For example, the allergen challenge in human asthmatics and non-human primates is very similar, with an early and a late-phase bronchoconstrictor response, an increase in airway eosinophilia and AHR, unlike when compared to mice models. This approach enables researchers to study human protein and monoclonal antibody therapeutics—a feature often difficult to perform in mice. The major drawbacks to using primate models are the cost and the extended timelines to develop model/phenotype and produce data. For example, house dust mite-induced sensitization develops over 6–12 months and requires frequent exposure to the allergen over the whole period. However, a big advantage is that a colony of non-human primates can be re-studied several times if patient-like endpoints are utilized.

This chapter primarily focused on models recapitulating some aspects of autoimmune disease such as RA and asthma. It is worth mentioning that genetically modified animal models have provided significant insights into the function of genes associated with cancer and neurodegenerative disorders, such as Alzheimer's and Parkinson's disease. Although these animal models have been critical in understanding the underlying biology of the disease processes, none of the existing models faithfully recapitulate all aspects of the pathology. Moreover, there has been poor translatability of efficacy in preclinical animal models to human clinical trials. As these animal models are based on genetic associations that account for a minority of the cases, continued development of new models that incorporate aspects of cancer and neurodegenerative processes involved in sporadic disease can be helpful in treating this rapidly growing unmet medical need.

12.7 SUMMARY

We began this chapter by suggesting that the advancement of biomedical sciences is and has been substantially predicated on the thesis that animal physiology can be exploited to learn human biology. On the other hand, the use of animal models to advance novel pharmacological drug entities continues to be a challenge branded with a multitude of past failures. Thus the question should be asked, do animal models have a role in the drug discovery process at all? Some would default to a negative response, supporting a "glass half empty" philosophy with the mantra, "these animal models are not predictive".

We propose that pharmacology results from animal models needs to be placed into context of both the specific scientific questions being asked and the conditions under which the data was generated. Moreover, we strongly advocate that these models, both pathway biology PD or disease mechanism, need to be integrated into a drug discovery platform geared at building confidence in the novel chemical entities and the pathophysiological target. As described in this chapter, gaining a greater assurance around a new drug can be attained by an understanding of PKPD relationships and its target engagement. Fundamental in establishing a PKPD concept is an appreciation for the challenges of working with animals, which we have discussed in this chapter. Additional confidence in the target will be gained when disease mechanism models are anchored to a human translation platform, which incorporates the use of *in vitro* and/or *ex vivo* human diseased tissues, proteomics and genomics to confirm its impact in patient population. Caution should be used so as not to over interpret results from animal models, given that they only serve as surrogates for complex and typically chronic human conditions. Despite years of research, some diseases such as RA, IBD, COPD and severe asthma still remain without curable treatment or known etiology, perhaps suggesting that no single model will ever provide all the answers. It is challenging to speculate how animal models will be used in the drug discovery process in the future. Regrettably, an ever diminishing number of *in vivo* scientists are currently being trained by universities, which may slow the current movement towards developing an effective, integrative, *in vivo* approach to drug discovery, as proposed in this report.

HINTS AND TIPS FOR ANIMAL MODEL USES

General Housekeeping Hints and Tips

- Always consider ethics and legal requirements before conducting *in vivo* experiments.
- Clearly define objective of the study and readouts.
- Control for variability (animal husbandry and treatment).
- Relevance, Robustness and Reproducibility of the read out are essential components of a good animal model.
- Use appropriate controls and satellite animals.
- Consider use of relevant species and strains based on anatomy, physiology and biological mechanism studied.
- Utilize appropriate statistical input during experimental design and data analysis.

Selecting the Relevant Model, Data Interpretation and Translatability

- Animal models are not accurate representations of human disease and at best can be considered as disease mechanism models. They can mimic some aspects of biological mechanisms present in human disease.
- PKPD modeling provides information linking the desired pharmacological response and the pharmacokinetic profile of the compound.
- Select appropriate induction stimulus and its magnitude (dose) in your model (ED_{80}).
- Investigate temporal profile of the PD response measured and establish relevant time point for desired PD readout.
- Establish target engagement readout and duration of action in the animal model and its relationship to the desired efficacy.
- Understanding the relationship between drug exposure in animals and target engagement *vs. in vitro* potency is essential.
- Consider the translatability of your animal model to the human disease situation. Explore tool compounds or standard of care compounds shown to modulate the same or similar mechanism tested in man, these provide invaluable comparators which can be used to validate your chosen model.

KEY REFERENCES

P. McGonigle and B. Ruggeri, *Biochem. Pharmacol.*, 2014, **87**, 162–171.
B. J. Canning, *Curr. Opin. Pharmacol.*, 2003, **3**, 244–250.
P. McGonigle and B. Ruggeri, *Biochem. Pharmacol.*, 2014, **87**, 162–171.

REFERENCES

1. P. McGonigle and B. Ruggeri, *Biochem. Pharmacol.*, 2014, **87**, 162–171.
2. C. Kilkenny, N. Parsons, E. Kadyszewski, M. F. Festing, I. C. Cuthill, D. Fry, J. Hutton and D. G. Altman, *PLoS One*, 2009, **4**, e7824.
3. M. Chvedoff, M. R. Clarke, J. M. Faccini, E. Irisarri and A. M. Monro, *Arch. Toxicol. Suppl.*, 1980, **4**, 435–438.
4. S. A. Doggrell and L. Brown, *Cardiovasc. Res.*, 1998, **39**, 89–105.
5. C. H. Conrad, W. W. Brooks, J. A. Hayes, S. Sen, K. G. Robinson and O. H. Bing, *Circulation*, 1995, **91**, 161–170.
6. H. Hiai, *Pathol. Int.*, 1996, **46**, 707–718.
7. N. Haran-Ghera, *Adv. Cancer Res.*, 1994, **63**, 245–293.
8. M. Kuno, K. Nemoto, N. Ninomiya, E. Inagaki, M. Kubota, T. Matsumoto and H. Yokota, *Clin. Exp. Pharmacol. Physiol.*, 2009, **36**, 589–593.
9. H. A. Baarsma, S. Bos, H. Meurs, K. H. Visser, M. Smit, A. M. Schols, R. C. Langen, H. A. Kerstjens and R. Gosens, *Respir. Res.*, 2013, **14**, 113–122.
10. J. L. Liang, G. M. Yang, T. Li and L. M. Liu, *J. Trauma Acute Care Surg.*, 2014, **76**, 762–770.
11. Z. Wu, J. N. Li, Z. Q. Bai and X. Lin, *Clin. Exp. Pharmacol. Physiol.*, 2014, **41**(7), 502–508.
12. F. De Vries, J. Leuschner, B. Jilma and U. Derhaschnig, *Int. J. Immunopathol. Pharmacol.*, 2013, **26**, 861–869.
13. E. Wirthgen, M. Tuchscherer, W. Otten, G. Domanska, K. Wollenhaupt, A. Tuchscherer and E. Kanitz, *Innate Immun.*, 2014, **20**, 30–39.
14. X. L. Wu, P. Wang, Y. H. Liu and Y. X. Xue, *J. Mol. Neurosci.*, 2014, **53**, 1–9.
15. G. B. Lieber, X. Fernandez, G. G. Mingo, Y. Jia, M. Caniga, M. A. Gil, S. Keshwani, J. D. Woodhouse, M. Cicmil, L. Y. Moy, N. Kelly, J. Jimenez, Y. Crawley, J. C. Anthes, J. Klappenbach, Y. Ma and R. L. McLeod, *Eur. J. Pharmacol.*, 2013, **718**, 290–298.
16. S. D. Kobayashi and F. R. Deleo, *Wiley Interdiscip. Rev. Syst. Biol. Med.*, 2009, **1**, 309–333.
17. R. W. Chapman, M. Minnicozzi, C. S. Celly, J. E. Phillips, T. T. Kung, R. W. Hipkin, X. Fan, D. Rindgen, G. Deno, R. Bond, W. Gonsiorek, M. M. Billah, J. S. Fine and J. A. Hey, *J. Pharmacol. Exp. Ther.*, 2007, **322**, 486–493.
18. R. Aul, S. Patel, S. Summerhill, I. Kilty, J. Plumb and D. Singh, *Int. Immunopharmacol.*, 2012, **13**, 225–231.
19. R. Virtala, A. K. Ekman, L. Jansson, U. Westin and L. O. Cardell, *Clin. Exp. Allergy*, 2012, **42**, 590–596.
20. B. R. Leaker, P. J. Barnes and B. O'Connor, *Respir. Res.*, 2013, **14**, 137.
21. R. W. Chapman, S. J. Sehring, C. G. Garlisi, A. Falcone, T. T. Kung, D. Stelts, M. Minnicozzi, H. Jones, S. Umland, R. W. Egan and W. Kreutner, *Arzneimittelforschung*, 1998, **48**, 384–391.
22. R. Ramalho, R. Soares, N. Couto and A. Moreira, *BMC Pulm. Med.*, 2011, **11**, 41.
23. K. Mullane, *Biochem. Pharmacol.*, 2011, **82**, 567–585.
24. E. H. Bel, *N. Engl. J. Med.*, 2013, **369**, 2362.
25. M. L. Fajt and S. E. Wenzel, *J. Asthma*, 2014, 1–8.
26. K. F. Chung, *Drugs*, 2014, **74**, 719–728.
27. S. E. Wenzel, *Pulm. Pharmacol. Ther.*, 2013, **26**, 710–715.
28. J. Corren, R. F. Lemanske, N. A. Hanania, P. E. Korenblat, M. V. Parsey, J. R. Arron, J. M. Harris, H. Scheerens, L. C. Wu, Z. Su, S. Mosesova, M. D. Eisner, S. P. Bohen and J. G. Matthews, *N. Engl. J. Med.*, 2011, **365**, 1088–1098.
29. J. Corren, *Curr. Allergy Asthma Rep.*, 2013, **13**, 415–420.
30. E. H. De Boever, C. Ashman, A. P. Cahn, N. W. Locantore, P. Overend, I. J. Pouliquen, A. P. Serone, T. J. Wright, M. M. Jenkins, I. S. Panesar, S. S. Thiagarajah and S. E. Wenzel, *J. Allergy Clin. Immunol.*, 2014, **133**, 989–996.

31. P. Haldar, C. E. Brightling, B. Hargadon, S. Gupta, W. Monteiro, A. Sousa, R. P. Marshall, P. Bradding, R. H. Green, A. J. Wardlaw and I. D. Pavord, *N. Engl. J. Med*, 2009, **360**, 973–984.

32. I. D. Pavord, P. Haldar, P. Bradding and A. J. Wardlaw, *Thorax*, 2010, **65**, 370.

33. M. E. Wechsler, P. C. Fulkerson, B. S. Bochner, G. M. Gauvreau, G. J. Gleich, T. Henkel, R. Kolbeck, S. K. Mathur, H. Ortega, J. Patel, C. Prussin, P. Renzi, M. E. Rothenberg, F. Roufosse, D. Simon, H. U. Simon, A. Wardlaw, P. F. Weller and A. D. Klion, *J. Allergy Clin. Immunol.*, 2012, **130**, 563–571.

34. Y. Wang and C. T. McCusker, *Clin. Exp. Allergy*, 2005, **35**, 1104–1111.

35. G. Yang, L. Li, A. Volk, E. Emmell, T. Petley, J. Giles-Komar, P. Rafferty, M. Lakshminarayanan, D. E. Griswold and P. J. Bugelski, *J. Pharmacol. Exp. Ther.*, 2005, **313**, 8–15.

36. K. L. Tomlinson, G. C. Davies, D. J. Sutton and R. T. Palframan, *PLoS One*, 2010, **5**.

37. E. Hamelmann, G. Cieslewicz, J. Schwarze, T. Ishizuka, A. Joetham, C. Heusser and E. W. Gelfand, *Am. J. Respir. Crit. Care Med.*, 1999, **160**, 934–941.

38. E. Hamelmann, A. Oshiba, J. Loader, G. L. Larsen, G. Gleich, J. Lee and E. W. Gelfand, *Am. J. Respir. Crit. Care Med.*, 1997, **155**, 819–825.

39. I. Akutsu, T. Kojima, A. Kariyone, T. Fukuda, S. Makino and K. Takatsu, *Immunol. Lett.*, 1995, **45**, 109–116.

40. D. L. Scott, F. Wolfe and T. W. Huizinga, *Lancet*, 2010, **376**, 1094–1108.

41. I. B. McInnes and G. Schett, *N. Engl. J. Med.*, 2011, **365**, 2205–2219.

42. I. B. McInnes and G. Schett, *Nat. Rev. Immunol.*, 2007, **7**, 429–442.

43. S. Karmakar, J. Kay and E. M. Gravallese, *Rheum. Dis. Clin. North Am.*, 2010, **36**, 385–404.

44. B. Bingham, S. K. Ajit, D. R. Blake and T. A. Samad, *Nat. Clin. Pract. Rheumatol.*, 2009, **5**, 28–37.

45. J. M. Zhang and J. An, *Int. Anesthesiol. Clin.*, 2007, **45**, 27–37.

46. J. A. Hendiani, K. N. Westlund, N. Lawand, N. Goel, J. Lisse and T. McNearney, *J. Pain*, 2003, **4**, 203–211.

47. Y. C. Lee, J. Cui, B. Lu, M. L. Frits, C. K. Iannaccone, N. A. Shadick, M. E. Weinblatt and D. H. Solomon, *Arthritis Res. Ther.*, 2011, **13**, R83.

48. M. Hegen, J. C. Keith, Jr., M. Collins and C. L. Nickerson-Nutter, *Ann. Rheum. Dis.*, 2008, **67**, 1505–1515.

49. G. Kollias, P. Papadaki, F. Apparailly, M. J. Vervoordeldonk, R. Holmdahl, V. Baumans, C. Desaintes, J. Di Santo, J. Distler, P. Garside, M. Hegen, T. W. Huizinga, A. Jungel, L. Klareskog, I. McInnes, I. Ragoussis, G. Schett, B. Hart, P. P. Tak, R. Toes, W. van den Berg, W. Wurst and S. Gay, *Ann. Rheum. Dis.*, 2011, **70**, 1357–1362.

50. B. Bolon, M. Stolina, C. King, S. Middleton, J. Gasser, D. Zack and U. Feige, *J. Biomed. Biotechnol.*, 2011, **2011**, 569068.

51. K. Kannan, R. A. Ortmann and D. Kimpel, *Pathophysiology*, 2005, **12**, 167–181.

52. W. J. Cromartie, J. G. Craddock, J. H. Schwab, S. K. Anderle and C. H. Yang, *J. Exp. Med.*, 1977, **146**, 1585–1602.

53. J. H. Schwab, S. K. Anderle, R. R. Brown, F. G. Dalldorf and R. C. Thompson, *Infect. Immun.*, 1991, **59**, 4436–4442.

54. L. Bevaart, M. J. Vervoordeldonk and P. P. Tak, *Arthritis Rheum.*, 2010, **62**, 2192–2205.

55. R. C. Schimmer, D. J. Schrier, C. M. Flory, K. D. Laemont, D. Tung, A. L. Metz, H. P. Friedl, M. C. Conroy, J. S. Warren, B. Beck and P. A. Ward, *J. Immunol.*, 1998, **160**, 1466–1471.

56. R. C. Schimmer, D. J. Schrier, C. M. Flory, J. Dykens, D. K. Tung, P. B. Jacobson, H. P. Friedl, M. C. Conroy, B. B. Schimmer and P. A. Ward, *J. Immunol.*, 1997, **159**, 4103–4108.

57. I. Rioja, C. L. Clayton, S. J. Graham, P. F. Life and M. C. Dickson, *Arthritis Res. Ther.*, 2005, **7**, R101–R117.

58. K. Chakravarthy, R. Faltus, G. Robinson, R. Sevilla, J. Shin, M. Zielstorff, A. Byford, E. Leccese, M. J. Caniga, S. Hseih, S. Zhang, C. S. Chiu, J. Z. Hoover, L. Y. Moy, R. L. McLeod, D. Stoffregen, W. Zhang, A. Murtaza and M. Cicmil, 2014, in press.
59. I. Rioja, K. A. Bush, J. B. Buckton, M. C. Dickson and P. F. Life, *Clin. Exp. Immunol.*, 2004, **137**, 65–73.
60. S. R. Chaplan, F. W. Bach, J. W. Pogrel, J. M. Chung and T. L. Yaksh, *J. Neurosci. Methods*, 1994, **53**, 55–63.
61. C. S. Stevenson and M. A. Birrell, *Pharmacol. Ther.*, 2011, **130**, 93–105.
62. B. J. Canning, *Curr. Opin. Pharmacol.*, 2003, **3**, 244–250.
63. G. R. Zosky and P. D. Sly, *Clin. Exp. Allergy*, 2007, **37**, 973–988.
64. M. A. Birrell, C. H. Battram, P. Woodman, K. McCluskie and M. G. Belvisi, *Respir. Res.*, 2003, **4**, 3.
65. D. Wyss, O. Bonneau and A. Trifilieff, *Br. J. Pharmacol.*, 2007, **152**, 83–90.
66. S. Y. Eum, K. Maghni, Q. Hamid, D. H. Eidelman, H. Campbell, S. Isogai and J. G. Martin, *J. Allergy Clin. Immunol.*, 2003, **111**, 1049–1061.
67. J. A. Boyce and K. F. Austen, *J. Exp. Med.*, 2005, **201**, 1869–1873.
68. J. H. Bates, M. Rincon and C. G. Irvin, *Am. J Physiol. Lung Cell. Mol. Physiol.*, 2009, **297**, L401–L410.
69. J. K. Walker, M. Kraft and J. T. Fisher, *Front. Physiol.*, 2012, **3**, 4911–4914.
70. A. Berndt, A. S. Leme, L. K. Williams, R. Von Smith, H. S. Savage, T. M. Stearns, S. W. Tsaih, S. D. Shapiro, L. L. Peters, B. Paigen and K. L. Svenson, *Physiol. Genomics*, 2011, **43**, 1–11.
71. K. Mullane, *Biochem. Pharmacol.*, 2011, **82**, 586–599.
72. B. J. Canning and Y. Chou, *Pulm. Pharmacol. Ther.*, 2008, **21**, 702–720.
73. J. Seok, H. S. Warren, A. G. Cuenca, M. N. Mindrinos, H. V. Baker, W. Xu, D. R. Richards, G. P. McDonald-Smith, H. Gao, L. Hennessy, C. C. Finnerty, C. M. Lopez, S. Honari, E. E. Moore, J. P. Minei, J. Cuschieri, P. E. Bankey, J. L. Johnson, J. Sperry, A. B. Nathens, T. R. Billiar, M. A. West, M. G. Jeschke, M. B. Klein, R. L. Gamelli, N. S. Gibran, B. H. Brownstein, C. Miller-Graziano, S. E. Calvano, P. H. Mason, J. P. Cobb, L. G. Rahme, S. F. Lowry, R. V. Maier, L. L. Moldawer, D. N. Herndon, R. W. Davis and W. Xiao, *Proc. Natl. Acad. Sci. U. S. A.*, 2013, **110**, 3507–3512.
74. W. M. S. Russell and R. L. Burch, Methuen, London, 1959.

CHAPTER 13

Bioinformatics for Medicinal Chemistry

NIKLAS BLOMBERG,[a] BRYN WILLIAMS-JONES[b] AND
JOHN P. OVERINGTON*[c]

[a] ELIXIR Hub, Wellcome Trust Genome Campus, Hinxton, Cambridge CB10 1SD, UK;
[b] ConnectedDiscovery Ltd., 27 Old Gloucester Street, London WC1N 3AX, UK; [c] European
Bioinformatics Institute (EMBL-EBI), European Molecular Biology Laboratory, Wellcome
Trust Genome Campus, Hinxton, Cambridge CB10 1SD, UK
*E-mail: jpo@ebi.ac.uk

13.1   INTRODUCTION

When the first draft of the human genome was finished in 2000 in a close race between the team at the Sanger Center[1] and Craig Venter's Celera Genomics[2] and announced in a joint press-conference by Bill Clinton and Tony Blair, there was a genuine expectation that having the blueprint for humans laid out for inspection would transform medicine and, in particular, drug discovery.[3] Around the same time "array" technologies, chip-based techniques to read the binding of cellular DNA and RNA, became accessible for routine application in laboratories. Based on the assumption that the finite number of human genes would necessitate intellectual property protection of drug targets, most pharmaceutical companies made large investments in genetically driven target discovery and validation programs. With the benefit of hindsight we now know that there was no massive unfolding of new targets amenable to medicinal chemistry.[4,5] While many of the large genetic disease association studies generated strong signals[6] a large fraction of the associations did not map to protein coding genes or, when there were candidate gene associations, their role in disease biology often remained unclear. We have learned that microarray studies of gene expression, while invaluable as a tool to understand the transcriptional response to cellular stimuli, have significant limitations in quantitating RNA response and that the statistical analyses of studies are subject to limitations.[7–9] Understandably this generated an "omics-fatigue" in many industrial, and academic, drug discovery units and attention shifted to "target-class" approaches—concerted efforts to transfer the medicinal chemistry knowledge between homologous targets where chances of successful drug development were felt to be better understood. There is an argument that the development of "omics" and high-throughput screening has actually increased the expense of drug discovery without

The Handbook of Medicinal Chemistry: Principles and Practice
Edited by Andrew Davis and Simon E Ward
© The Royal Society of Chemistry 2015
Published by the Royal Society of Chemistry, www.rsc.org

tangibly improving success. Whether new technologies in drug discovery are open to the same challenges of technical innovation in other fields is debatable, but hype cycles in drug discovery seem to have steeper peaks, deeper troughs, and can take a very long time to plateau.

When bioinformatics surfaced in a drug discovery programme, it was usually to answer questions such as the transferability of assay results between model organisms and humans by analysis of active site mutations, or perhaps to ascertain that the *in vitro* assays were based on a clone from one of the major variants in the human population. These are important issues which will be covered in this chapter, but as nucleotide sequencing technologies have matured and costs have plummeted, many drug discovery programmes will involve large scale patient sequencing either as part of translational medicine efforts or as part of patient stratification for the clinical development programmes. Similarly, metabolomics and proteomics approaches have matured. Many programmes, in particular in oncology, will be using a broad suite of complex multivariate biomarkers. Over the last five years improved experimental and analysis methods have clearly demonstrated that gene expression is an important regulatory factor, and genetic variation of expression levels is associated with many disease states, as evidenced by the *expression QTLs* (quantitative trait loci)—hereditary changes in gene expression caused by polymorphisms in gene promoters and other regulatory elements.[10,11] Indeed, as demonstrated by the ENCODE project, up to 80% of the genome can be assigned a biochemical function[12,13] where cell-type specific regulation is an important aspect of signalling response.[14] Genomic medicine now plays such an important role in biomarker discovery and translational medicine that, while few medicinal chemists or molecular modellers will aspire to expertise in the increasingly complex field of bioinformatics, it is indispensable to have a basic understanding of the technologies, tools and opportunities to fully understand the approaches to target discovery, validation and translation.

As the number of protein encoding genes in the human genome indeed is finite, and continuously revised down,[15] it is increasingly clear that the complexity of human biology does not come from the number of genes, microRNAs and other transcripts but rather in their regulation.[12] The ability to generate, integrate and analyse large scale genomic, proteomic, expression, and increasingly imaging data-sets, has led to an appreciation of the cellular regulation networks and how the same actors can signal in many contexts depending on for example the spatial location of signalling or accessibility of chromatin states.[16,17] Critically, the networks and interactions of these factors are the key to understanding human biology and how multiple subtle changes ultimately drive disease.[18] Network and systems analysis is now an integral part of computational drug discovery.

This chapter aims to give a broad introductory overview of a computational biology toolbox that covers the fields outlined above, with particular emphasis on the many open resources available. Although there is a bewildering array of bioinformatics data resources available, the latest update of the Nucleic Acids Research (NAR) Molecular Biology Database Collection lists 1552 resources,[19] many of these are highly specialized and focussed on the needs of a single community or biological function. Unfortunately many resources are also ephemeral with short funding horizons.[20] Nevertheless, the major primary archives for biological data have been sustained for decades; the data bank for protein structures, PDB, started in 1971 and the EMBL Data Library, the predecessor to today's sequence repositories, was founded in 1981. Many bioinformatic resources are secondary resources that provide value-added services, curation or analysis on top of the primary archives—UniProt, mentioned below, is a prime example, with extensive manual curation of protein sequences and function. Given the plethora of resources and archives, the "integration hubs", services that integrate and summarize data, are increasingly important. Thus there is extensive crossover in content between bioinformatics databases and it is always important to note the provenance—input source—of the underlying data.

Bioinformatics is more than finding a gene sequence and then using that sequence to find similar genes—this chapter aims to give a broad overview of the applicability in drug discovery by highlighting key databases and tools. For a drug discovery programme there are multiple facets which must be investigated to determine the suitability of a drug target for a particular disease, and the validation of that target's relevance in a particular disease. There are large pre-existing sources of evidence, and knowledge can be carefully repurposed from the efforts of others to support project decision-making, but it is vital to take into account the advantages and pitfalls of bioinformatics in experimental design. Computational analysis of large datasets is not a magic wand that makes sense of poorly designed experiments through sheer numbers. This is particularly true for patient stratification and data integration of the increasingly common multi-omics driven biomarker studies. Maintaining the provenance and context for large amounts of data is critical to successful data integration and downstream analysis. Close attention to the final use cases is critical: the understanding of how to select patients based on relevant, measurable and actionable criteria. Bioinformatics also has an important role in the daily project cycle, supporting the trouble-shooting of science issues, supporting and improving experimental design, and finding evidence to progress ideas either into the lab, or into the waste bin.

Bioinformatics and computational biology has a very active training community with a strong ethos of professional training. There are many excellent web-based resources available for most of the key resources. Thus, the emphasis here is to give basic concepts and directions for further studies, investing a few additional hours on the many excellent webinars and on-line training resources is time well-spent. Thoughout this chapter links to relevant on-line training resources are provided, and we can only recommend reaching out to the friendly next-door bioinformatician!

13.2   THE TARGET DOSSIER

The traditional starting point for a medicinal chemist foraging into bioinformatics has been the target dossier or bioinformatics report, a somewhat lengthy document that compiles information from a range of public resources. Today Wikipedia is the first, and in many cases excellent, starting point, and manually curated target reports are, or can be, superseded by public databases and data integration services.

UniProt (www.uniprot.org), or more accurately UniProt KnowledgeBase, is an excellent starting point to explore the known biology around a new protein target. UniProt was formed in 2002 through a merger of the three major protein databases, Swiss-Prot, PIR, and TrEMBL, and is the central protein hub for bioinformatic analysis. UniProt entries are, as far as possible, manually curated, but an automated functional annotation process based on homology complements this to cover the whole protein universe. While the quality of manual annotation is generally very high and the curation process synthesizes multiple lines of evidence, it ultimately depends on the primary literature and will be sensitive to non-reproducibility and bias of experiments.[21,22] It is useful to have a brief understanding of the manual annotation process for effective use of the database; the UniProt/Swiss-Prot curators compile reports based on the data from the scientific literature and will, by necessity, make judgements on the strength of evidence for protein function. It is important to note that the functional assignment will make inferences across cross-species. For instance, annotation of human protein function will take experiments from animal models into account, but *e.g.* posttranslational modifications would only be recorded in the species for which there are experimental data. Thus, it is important to review entries from *orthologous* proteins, functional homologs from related species. The UniProt team recently published an illustrative case-story on the annotation of SIRT-5 which gives a very good overview of the process and assumptions involved.[23]

UniProt functions as an integration hub and provides a large set of links to other resources. For medicinal chemistry the most important aspect of UniProt is perhaps the useful summaries of known drugs from DrugBank, links to Chembl entries (which also have summaries of marketed drugs) as well as an excellent overview of protein structural data from the PDB databank. Again it is important to remember that for crystal structures UniProt will not link across species (*e.g.* the entry human will link to crystal structures of the human protein in the PDB database). As discussed further below under homology modelling many proteins are modular with several distinct, independently folding domains. UniProt provides a summary of the domain annotations, across databases, and has links out to the more detailed assignments in *e.g.* InterPro[24] and SMART.[25] Post-translational modifications, as well as single nucleotide polymorphisms, can be of major importance for drug discovery projects and GeneCards[26] (www.genecards.org, free for academic research, commercial licence required) provides an excellent summary of protein as well as the associated gene data from a very large number of underlying sources, including links to suppliers of reagents and tool compounds.

Understanding of tissue distribution of a target, both at the gene and protein level, is of course a critical aspect of both target validation and risk assessment but can be a major challenge, in particular for targets with low or variable expression levels. GeneCards, mentioned above, provide a high level summary of gene expression but the ExpressionAtlas[27] (www.ebi.ac.uk) is an excellent resource for browsing public functional genomics data. Analysis of gene expression, in particular across experiments and platforms, is non-trivial with many pitfalls. Often there is discordance between experiments but integration of data across resources and the application of simple heuristics markedly increases the quality of results[28]—the recent review by Rung and Brazma[29] provides an excellent overview of public functional genomics resources and is a good guide to the challenges in collating and analysing public expression data-sets.

13.3 PROTEIN STRUCTURE RESOURCES AND HOMOLOGY MODELLING

Structure based design is a mainstay of medicinal chemistry. Detailed understanding of protein ligand interactions requires high-resolution crystal structures but for a new target with unknown 3-dimensional structure much can also be learned from studies on related proteins. Since the seminal observation by Chotia and Lesk[30] that the structural resemblance of two proteins is broadly related to their sequence similarity, there has been intense efforts to model the structures of new proteins by homology—*homology modelling*.

The main use of model structures in drug discovery is perhaps in the understanding of potential selectivity issues, *e.g.* for selection of the appropriate counter screens in a screening cascade, but in the initial phases modelled structures are also useful for selection of peptides as antibody epitopes and understanding the location of disease causing single nucleotide polymorphisms.[31,32]

Nevertheless one of the first success stories of structure based drug design, the development of captopril (www.drugbank.ca/drugs/DB01197) as an inhibitor of angiotensin converting enzyme (www.uniprot.org/uniprot/P12821), ACE in 1977 used homology modelling.[33] Biochemical studies on ACE had established that this enzyme was a zinc-dependent metalloprotease, and 3D structural information from the closely related pancreatic carboxypeptidase A[34] could be used to understand the binding geometry of the succinyl-L-proline (www.chemspider.com/Chemical-Structure.168469.html?rid=ebbd3b1a-1f40-4b2e-bcf8-ec07deb02ba3) lead. Observing that ACE is a dipeptidase and, in contrast to carboxypeptidase A that releases single amino acids, cleaves off the two last amino acids of a peptide chain led to the conclusion that the distance from the C-terminus of the peptide chain to the catalytic zinc ion should be approximately one amino acid longer in ACE. The hypothesis that the carboxy-alkanoyl group of the succinyl-L-proline bound to the active site zinc ion led to a search for better zinc binding substituents. While amine, guanine or amide functionalities did not improve potency, a series mercaptoalkanoyls gave a breakthrough in potency and ultimately led to the synthesis of D-3-mercapto-2-methylpropanoyl-L-proline, captopril.

The use of homology models has remained popular in drug discovery and while protein structure models clearly have significant limitations, blind predictions shows that the models are more similar to the template structure than structures of the target protein,[35,36] they remain an important tool to understand the details of new targets. Modelling of protein structures is a well-established field with robust and mature tools available on-line; the Swiss-Model (www.swissmodel.expasy.org/) server and workbench is a good starting point and the recent review by Schwede[37] gives a good overview of the field and points to many excellent resources. A number of these resources also provides extensive on-line training on protein structures (www.ebi.ac.uk/training/online/) as well as training in the use of specific tools. As usual, Wikipedia (http://en.wikipedia.org/wiki/Homology_modelling) offers a starting point outlining the technical details of a modelling workflow.

The tools and resources provided online give easy access to modelling tools for non-specialists but there are some aspects worth noting. The most critical issue in homology modelling is the quality of the alignment; this establishes the equivalent amino acids and hence geometries of the template structure to the target sequence. For homologous proteins without large insertions or deletions in the sequence the online resources give an adequate result; however care must be taken to ascertain that this is indeed the case. A closely related issue is handling of the modular architecture of proteins; many, if not most, cell-signalling proteins consist of a distinct set of structurally independent domains that are often connected through intrinsically disordered or flexible linkers.[38] The *src* family of kinases[39] is a good example how this modular architecture is used to regulate functional signalling both through spatial location[16] and through the domain rearrangements that drive kinase activation: in the inactive state access to the kinase active site is sterically blocked through binding of the N-terminal SH2 domain to a C-terminal motif. Binding of a cognate cellular partner to the SH2 domain opens up the structure and the kinase domain can access the downstream targets. Analysis of protein structural families and use of multiple different templates when modelling a drug discovery target is critical. In particular it is important to decide on what aspects of the target that will be modelled—sequence similarity in distinct domains can be quite different from the overall homology.

Through large structural genomics consortia aiming to systematically determine protein structures for human drug targets (www.thesgc.org) or *e.g.* tuberculosis targets (www.webtb.org) we now have at least a basic understanding of the structure of over 50% of human proteins.[37] This, together with the recent developments in super-resolution light microscopy,[40] X-ray scattering[41] and electron tomography,[42] is leading to a renewed interest in protein modelling. Combination of structural and volume data from different techniques and resolutions with geometric and structural features from detailed modelling is now starting to give detailed understanding of large macromolecular assemblies, such as the structural organization of genomes[43] or detailed reconstructions of the 26S proteasome.[44] As these techniques mature they are likely to find wide applicability in drug discovery and open up the possibility of structure driven approaches for large signalling complexes or assemblies.

13.4 THE GENOMICS EXPLOSION

The rapid decreasing cost of sequencing, with the $1000 genome rapidly approaching,[45] is driving profound changes in genomic medicine and translational research. On a weekly basis large international collaborations are now unravelling the genomic basis of cancer as well as complex diseases from trans-national patient cohorts.

While the bioinformatics of genome sequencing is a rapidly developing specialist research field beyond the scope of this chapter, there are aspects, particularly around pharmacogenomics, the study of the influence of genetic variation on patient drug response, that are of high relevance to medicinal chemistry and that likely will impact not only patient stratification and clinical trials but also drug design considerations.

Genetic data is an important part of target selection; strong genetic links between molecular targets and human disease phenotypes provides unambiguous support and validation of targets. Examples such as the discovery that loss-of-function alleles of the NaV1.7 α-subunit made homozygous individuals unable to experience pain[46] provided direct evidence of the relevance of this target to pain research and triggered intense drug discovery efforts.[47] Similarly the finding that gain-of-function mutations in the PCSK9 gene led to high LDL-cholesterol level and increased risk of heart disease led to the development of PCSK9 antibodies, currently in late-stage clinical trials.[48] A recent review by Altshuler and colleagues provides a very good overview of target validation *via* genomics[49] although it should be noted that to detect the rare variants involved in complex diseases such as type II diabetes and asthma, very large cohorts, in the excess of 100 000 patients, will be required.[50] For inherited diseases the *Online Mendelian Inheritance in Man*, (OMIM; http://omim.org/) resource provides manual annotations of diseases and genes and the Orphanet (www.orpha.net) database links rare diseases, drugs and genes through manual annotations. While both OMIM and Orphanet focus on the clinical community both resources are invaluable as research tools and can provide strong mechanistic links for disease biology.

The other major impact of the genomics revolution is the growth of personalized medicine, or rather targeted treatments developed together with a companion (genomic) biomarker diagnostic. It is rapidly being established for cancer drug discovery with crizotinib for ALK-positive lung cancers,[51] and vemurafenib[52] for melanoma patients carrying the B-Raf V600E mutation is a prime example. Currently there are well over a hundred compounds with a pharmacogenomics biomarker in the label (www.fda.gov/drugs/scienceresearch/researchareas/pharmacogenetics/ucm083378.htm).

While a growing number of these compounds represent targeted treatments, the understanding of ADME variation on the basis of genetics clearly will have a major impact on medicinal chemistry. Warfarin (www.ebi.ac.uk/chembl/compound/inspect/CHEMBL1464) is an illustrative example; the compound is widely prescribed as an anti-thrombotic, despite having a narrow therapeutic index that requires frequent monitoring and is among the leading causes of emergency room visits due to drug adverse-events. Warfarin is a racemic compound where the most potent isomer, *S*-warfarin, has a shorter half-life than the *R*-isomer (21–43 h compared to 37–89 h) and CYP2C9 driven metabolism. Patients that have CYP2C9 variants CYP2C9*2 (R144C) and CYP2C9*3 (I359L) metabolize the *S*-isoform more slowly and often require a lower dose; the association between these polymorphisms and warfarin dose, as well as risk, has now been established in several studies. Although the FDA label now indicates the pharmacogenomics association, the path to establishing this link demonstrate the complexities taking this into clinical practice.[53]

Similarly the $P2Y_{12}$ antagonist clopidogrel, a prodrug dependent upon CYP2C19 for activation, now carries a black box warning for patients with a common CYP2C19 variant that leads to a truncated, inactive enzyme. Although this link has been established in several studies CYP2C19 genotyping did not affect outcomes in clinical trials and the value of genetic testing in clinical practice remains unclear.[53]

Although the UniProt and GeneCards resources mentioned above annotate known polymorphisms in a target, the Pharmacogenomics KnowledgeBase[54] (www.pharmgkb.org) is the key resource for understanding the effects of genomic variation in drug discovery. In addition to annotated pages of drugs and targets, PharmGKB also provides illustrated pathways for many drugs that summarize known pharmacogenetic effects and provides lists of the clinical evidence.

13.5 SMALL MOLECULE RESOURCES AND DATA INTEGRATION FOR DRUG DISCOVERY

Data integration and systems modelling are rapidly becoming standard tools the analysis of biological systems and while the need for extensive integration of biological data to serve drug discovery, in particular target discovery and target validation support, has been long noted—the seminal paper by Searls[55] remains an excellent overview with many key issues well covered—the main driver today is the integration of genomics, functional genomics and phenotype data for translational research and large stratified medicine studies. Tools, technologies and approaches for integration of chemical, biological and clinical data have emerged as an intensive research field and there are now rapidly maturing public platforms accessible to medicinal chemists. Integrative informatics at the interface between chemistry and biology is a good starting point for medicinal chemists venturing into bioinformatics.

A good recent example of how systematic integration of biological and chemical data resources can drive medicinal chemistry efforts is the platform developed by Bornot *et al.*[56] for systematic investigation of dual-acting receptor modulators within a disease area. They defined a set of medicinal chemistry and bioinformatics workflows that systematically explored data resources and literature for evidence (or indications) of previous dual-acting compounds. Recognizing that many disease areas are mature with a significant history of drug discovery research, their starting point is not a *de novo* search of compound–disease associations but a manually curated list of potential opportunities. Of particular interest in this study is the alignment of medicinal chemistry risk assessment and the supporting target validation evidence (and risks) from detailed mapping of known bioprocesses and molecular mechanisms of disease.

The study from Bornot *et al.* relied on extensive integration of chemistry and *in vitro* pharmacology data; this is an extensively explored area over the last ten years that built on the availability of data-resources from systematic cross screening of compounds against a large panel of *in vitro* assays. Computational scientists in the pharmaceutical industry soon realized that the utility of these data-sources reached far beyond the original purpose of assisting safety assessments.[57–59] As the initial data resources to a large extent were commercial, the intense research efforts into tools to analyse and predict *polypharmacology* for compounds to potentially interact with a range of targets were mostly confined to industrial researchers. This has changed dramatically since the launch of the NIH *Molecular Libraries Initiative*—a large scale molecular screening programme that led to the formation of PubChem. Other resources such as Chembank[60] from the Broad institute screening programme followed and there are also plans underway for a European repository, linked to the EU-Openscreen infrastructure.[61] Of particular interest is the Chembl database which provides compound–target associations abstracted from the literature,[62] and with the addition of the chemistry patent source in SureChem[63] together with integration of bioinformatics resources from EMBL-EBI, will be an important future platform.

While an extensive review of chemogenomics data resources is beyond the scope of this chapter, and extensively covered elsewhere,[61,64–66] the familiar challenge of normalizing chemical structures from different sources, such that charges, stereochemistry, tautomeric and ionization states are consistently treated, is a good illustration to the fundamental challenge of data integration—*equivalence:* What aspects of two different data-sets should be considered "the same" and hence joined together? In many cases this is of course straightforward—starting with the UniProt identifier for a single-chain target one can quickly assemble an overview of small molecule ligands from Chembl, tissue expression from ExpressionAtlas or BioGPS and protein expression data from immunostaining in Human Protein Atlas and ProteomicsXchange.

For a single-chain protein the link to genetic and gene expression data is direct and the assumption of gene–drug and gene–disease pairing will hold. But consider, for instance, 5-HT$_3$ antagonists that bind to large oligomeric ion channels with variable sub-unit composition.[67] In this case there is a one-to-many mapping of the compound to genes, and this is further complicated by a complex tissue variability of subunit composition.[68] In general, for any large scale data integration effort there will be many complicating factors that can undermine analysis efforts. Linking genomic variations (such as single nucleotide polymorphisms) to compound binding should be viewed as a hypothesis rather than a fact derived by deductive reasoning. Simplifications such as: "ion channel isoforms are considered equivalent" allows effective summary and analysis of data-sets, but it is critical to be aware of the assumptions underlying the combination of different data sets. For service-focussed integration efforts, catering to a broad range of users and use-cases, the ability to tune and focus results depending on the expectations of the user is vital. This is the concept of a "scientific lens" where queries can be tuned to return differentially complex results depending on the use-case the user has in mind. The complexity is hidden from the user by the use of comprehensive and flexible data models.[69]

The need for complex integration over multiple data-sources is often driven by seemingly simple user questions: "are there any published data-sets where kinase inhibitors influence expression changes in an immune cell?" Humans are very apt at summarizing data in convenient categories—in the example above "kinase inhibitor" and "immune cell"—that require extensive classification systems, *ontologies*, for any computational mapping and search.[70,71] This is an active area of research and the development of ontologies for human disease and phenotypes as well as similar efforts for model organisms have enabled large scale data integration to support lead generation,[72] compound repurposing[73] as well as translational efforts between mouse models and human disease.[74] The development of high-quality vocabularies and classifications for many of the key concepts in drug discovery research: compounds, targets, diseases, mechanisms and companies, can also be applied to analysis of commercial data and competitive intelligence. The SciBite system (www.scibite.com) is an interesting, open system based on public resources that tags and classifies data in real time. An interesting aspect of the system is the semantic capabilities that allow integration of information across news-sources, molecular databases as well as grants into easily digestible categories.

Recently a systematic analysis of scientific questions or use-cases for preclinical drug discovery across eight pharmaceutical companies and a range of academic laboratories showed a remarkable consistency in priorities and core concepts.[75] This analysis was used to drive the development of a large-scale data integration platform (www.openphacts.org) with an accessible interface and a range of associated public services focused on the early drug discovery process. Platform usage and training is supported by a YouTube channel (www.youtube.com/user/ OpenPHACTS)—again demonstrating the importance and availability of online training resources for bioinformatics services.

Moving further in the drug discovery value chain compound repurposing has been a favourite use-case for large data integration efforts with a number of interesting case-studies published[18,73,76,77] but clearly the most successful examples of repurposing have not come through extensive computational analysis but rather through clinical reasoning—for example, losartan, an anti-hypertensive AT-1 blocker, has been used to treat patients with Marfan's syndrome[78] on the basis of its ability to block TGF β signalling. Much of the current focus is on rare diseases, and as our understanding of these conditions improves through application of genomic medicine, this is likely to be a fruitful area for computational drug discovery.[79]

A second area where genomic medicine and large-scale data integration, including chemistry, will have a transformative impact is oncology. Large scale cancer genomics projects[80] allow an increasingly detailed resolution of the molecular mechanisms of tumour and tumour subtypes; similarly there are now also large-scale data-sets of small molecule effects on cancer cell-types[81,82] linked to the underlying genetic aberrations. Clearly linking and integrating data over "multi-omic" studies for many thousand patients with a broad range of different *in vitro* pharmacologies will be a challenge—highlighted by the lack of correlation between the different assays and conditions[83]—but individualised treatments on the basis of integrating genomic profiles, metabolite biomarkers and chemical screening of tumour samples are moving into clinical practice.[84] For oncology research, multi-omics biomarkers and extensive profiling is rapidly becoming the norm and the medicinal chemistry and drug discovery community will need to develop approaches to rationally design against complex end-points.

Medical and translational research, in both public and commercial organisations, increasingly depend on this integrated analysis of complex data-sets and in particular the rapid advances and plummeting costs in sequencing based assays will have a profound impact on drug discovery research—at the time when the current projects hit the market genomic profiling will be standard practice in most clinical settings. Human genetic variation will likely be an important component of clinical decision-making both in terms of compound efficacy as well as in the assessment of compound safety and pharmacokinetics. The rapid development of genomic medicine is also mirrored in bioinformatics; the current data glut in biology is changing the field to a critical infrastructure service provider. The challenge for medicinal chemistry is to assess which of these aspects that can be built into a programme today. The opportunity is to use the understanding of systems responses to open up new targets and approaches for small molecule drug discovery.

HINTS AND TIPS

Genomic medicine now plays such an important role in biomarker discovery and translational medicine that medicinal chemists need to have to have at least a basic understanding of the technologies, tools and opportunities to appreciate the approaches to target discovery and validation.

Many the key bioinformatics resources are freely available on the internet and these are hyperlinked throughout this chapter.

UniProt functions as an integration hub and provides a large set of links to other resources. For example summaries of known drugs from DrugBank, links to Chembl entries as well as an excellent overview of protein structural data from the PDB databank.

The Chembl database provides compound-target associations abstracted from the literature,[62] and with the addition of the chemistry patent source in SureChem[63] together with integration of bioinformatics resources from EMBL-EBI, will be an important future platform.

The pharmacogenomics database is the key resource for understanding the effects of genomic variation in drug discovery.

Currently there are well over a hundred compounds with a pharmacogenomics biomarker in the label.

The understanding of ADME variation on the basis of genetics is growing, and will have a major impact on medicinal chemistry.

With sequencing plummeting costs, by the time current research compounds hit the market genomic profiling will be standard practice in most clinical settings.

KEY REFERENCES

For a refreshing take on the limitations of computational analysis in Biology based on unparalleled understanding of molecular interactions read T. J. Gibson, *Trends Biochem. Sci.*, 2009, **34**, 471–482.

For a comprehensive review of current protein modelling including the emerging opportunities from imaging and electron microscopy try T. Schwede, *Structure*, 2013, **21**, 1531–1540.

For a very good review of clinical pharmacogenomics as a basis for personalising drug therapy we recommend L. Wang, H. L. McLeod and R. M. Weinshilboum, *N. Engl. J. Med.*, 2011, **364**, 1144.

REFERENCES

1. E. S. Lander, L. M. Linton, B. Birren, C. Nusbaum, M. C. Zody, J. Baldwin, K. Devon, K. Dewar, M. Doyle, W. FitzHugh, R. Funke, D. Gage, K. Harris, A. Heaford, J. Howland, L. Kann, J. Lehoczky, R. LeVine, P. McEwan, K. McKernan, J. Meldrim, J. P. Mesirov, C. Miranda, W. Morris, J. Naylor, C. Raymond, M. Rosetti, R. Santos, A. Sheridan, C. Sougnez, N. Stange-Thomann, N. Stojanovic, A. Subramanian, D. Wyman, J. Rogers, J. Sulston, R. Ainscough, S. Beck, D. Bentley, J. Burton, C. Clee, N. Carter, A. Coulson, R. Deadman, P. Deloukas, A. Dunham, I. Dunham, R. Durbin, L. French, D. Grafham, S. Gregory, T. Hubbard, S. Humphray, A. Hunt, M. Jones, C. Lloyd, A. McMurray, L. Matthews, S. Mercer, S. Milne, J. C. Mullikin, A. Mungall, R. Plumb, M. Ross, R. Shownkeen, S. Sims, R. H. Waterston, R. K. Wilson, L. W. Hillier, J. D. McPherson, M. A. Marra, E. R. Mardis, L. A. Fulton, A. T. Chinwalla, K. H. Pepin, W. R. Gish, S. L. Chissoe, M. C. Wendl, K. D. Delehaunty, T. L. Miner, A. Delehaunty, J. B. Kramer, L. L. Cook, R. S. Fulton, D. L. Johnson, P. J. Minx, S. W. Clifton, T. Hawkins, E. Branscomb, P. Predki, P. Richardson, S. Wenning, T. Slezak, N. Doggett, J. F. Cheng, A. Olsen, S. Lucas, C. Elkin, E. Uberbacher, M. Frazier, R. A. Gibbs, D. M. Muzny, S. E. Scherer, J. B. Bouck, E. J. Sodergren, K. C. Worley, C. M. Rives, J. H. Gorrell, M. L. Metzker, S. L. Naylor, R. S. Kucherlapati, D. L. Nelson, G. M. Weinstock, Y. Sakaki, A. Fujiyama, M. Hattori, T. Yada, A. Toyoda, T. Itoh, C. Kawagoe, H. Watanabe, Y. Totoki, T. Taylor, J. Weissenbach, R. Heilig, W. Saurin, F. Artiguenave, P. Brottier, T. Bruls, E. Pelletier, C. Robert, P. Wincker, D. R. Smith, L. Doucette-Stamm, M. Rubenfield, K. Weinstock, H. M. Lee, J. Dubois, A. Rosenthal, M. Platzer, G. Nyakatura, S. Taudien, A. Rump, H. Yang, J. Yu, J. Wang, G. Huang, J. Gu, L. Hood, L. Rowen, A. Madan, S. Qin, R. W. Davis, N. A. Federspiel, A. P. Abola, M. J. Proctor, R. M. Myers, J. Schmutz, M. Dickson, J. Grimwood, D. R. Cox, M. V. Olson, R. Kaul, C. Raymond, N. Shimizu, K. Kawasaki, S. Minoshima, G. A. Evans, M. Athanasiou, R. Schultz, B. A. Roe, F. Chen, H. Pan, J. Ramser, H. Lehrach, R. Reinhardt, W. R. McCombie, M. de la Bastide, N. Dedhia, H. Blocker, K. Hornischer, G. Nordsiek, R. Agarwala, L. Aravind, J. A. Bailey, A. Bateman, S. Batzoglou, E. Birney, P. Bork, D. G. Brown, C. B. Burge, L. Cerutti, H. C. Chen, D. Church, M. Clamp, R. R. Copley, T. Doerks, S. R. Eddy, E. E. Eichler, T. S. Furey, J. Galagan, J. G. Gilbert, C. Harmon, Y. Hayashizaki, D. Haussler, H. Hermjakob, K. Hokamp, W. Jang, L. S. Johnson, T. A. Jones, S. Kasif, A. Kaspryzk, S. Kennedy, W. J. Kent, P. Kitts, E. V. Koonin, I. Korf, D. Kulp, D. Lancet, T. M. Lowe, A. McLysaght, T. Mikkelsen, J. V. Moran, N. Mulder, V. J. Pollara, C. P. Ponting, G. Schuler, J. Schultz, G. Slater, A. F. Smit, E. Stupka, J. Szustakowski, D. Thierry-Mieg, J. Thierry-Mieg, L. Wagner, J. Wallis, R. Wheeler, A. Williams, Y. I. Wolf, K. H. Wolfe, S. P. Yang, R. F. Yeh, F. Collins, M. S. Guyer, J. Peterson, A. Felsenfeld, K. A. Wetterstrand, A.

Patrinos, M. J. Morgan, P. de Jong, J. J. Catanese, K. Osoegawa, H. Shizuya, S. Choi, Y. J. Chen and International Human Genome Sequencing Consortium, *Nature*, 2001, **409**, 860–921.

2. J. C. Venter, *Science*, 2001, **291**, 1304–1351.
3. E. S. Lander, *Nature*, 2011, **470**, 187–197.
4. A. L. Hopkins and C. R. Groom, *Nat. Rev. Drug Discov.*, 2002, **1**, 727–730.
5. M. Wehling, *Nat. Rev. Drug Discov.*, 2009, **8**, 541–546.
6. S. F. Kingsmore, I. E. Lindquist, J. Mudge, D. D. Gessler and W. D. Beavis, *Nat. Rev. Drug Discov.*, 2008, **7**, 221–230.
7. R. Breitling, *Biochim. Biophys. Acta, Gene Struct. Expression*, 2006, **1759**, 319–327.
8. A. A. Alsheikh-Ali, W. Qureshi, M. H. Al-Mallah and J. P. A. Ioannidis, *PLoS ONE*, 2011, **6**, e24357.
9. Z. Wang, M. Gerstein and M. Snyder, *Nat. Rev. Genet.*, 2009, **10**, 57–63.
10. V. Emilsson, G. Thorleifsson, B. Zhang, A. S. Leonardson, F. Zink, J. Zhu, S. Carlson, A. Helgason, G. B. Walters, S. Gunnarsdottir, M. Mouy, V. Steinthorsdottir, G. H. Eiriksdottir, G. Bjornsdottir, I. Reynisdottir, D. Gudbjartsson, A. Helgadottir, A. Jonasdottir, A. Jonasdottir, U. Styrkarsdottir, S. Gretarsdottir, K. P. Magnusson, H. Stefansson, R. Fossdal, K. Kristjansson, H. G. Gislason, T. Stefansson, B. G. Leifsson, U. Thorsteinsdottir, J. R. Lamb, J. R. Gulcher, M. L. Reitman, A. Kong, E. E. Schadt and K. Stefansson, *Nature*, 2008, **452**, 423–428.
11. W. Cookson, L. Liang, G. Abecasis, M. Moffatt and M. Lathrop, *Nat. Rev. Genet.*, 2009, **10**, 184–194.
12. The ENCODE Project Consortium, *Nature*, 2012, **489**, 57–74.
13. K. R. Rosenbloom, C. A. Sloan, V. S. Malladi, T. R. Dreszer, K. Learned, V. M. Kirkup, M. C. Wong, M. Maddren, R. Fang, S. G. Heitner, B. T. Lee, G. P. Barber, R. A. Harte, M. Diekhans, J. C. Long, S. P. Wilder, A. S. Zweig, D. Karolchik, R. M. Kuhn, D. Haussler and W. J. Kent, *Nucleic Acids Res.*, 2013, **41**, D56–63.
14. S. John, P. J. Sabo, R. E. Thurman, M. H. Sung, S. C. Biddie, T. A. Johnson, G. L. Hager and J. A. Stamatoyannopoulos, *Nat. Genet.*, 2011, **43**, 264–268.
15. M. Pertea and S. L. Salzberg, *Genome Biol.*, 2010, **11**, 206.
16. T. J. Gibson, *Trends Biochem. Sci.*, 2009, **34**, 471–482.
17. S. C. Parker, M. L. Stitzel, D. L. Taylor, J. M. Orozco, M. R. Erdos, J. A. Akiyama, K. L. van Bueren, P. S. Chines, N. Narisu, NISC Comparative Sequencing Program, B. L. Black, A. Visel, L. A. Pennacchio and F. S. Collins, National Institutes of Health Intramural Sequencing Center Comparative Sequencing Program Authors and NISC Comparative Sequencing Program Authors, *Proc. Natl. Acad. Sci. U. S. A.*, 2013, **110**, 17921–17926.
18. E. E. Schadt, S. H. Friend and D. A. Shaywitz, *Nat. Rev. Drug Discov.*, **20098**, 286–295.
19. X. M. Fernandez-Suarez, D. J. Rigden and M. Y. Galperin, *Nucleic Acids Res.*, 2013, **42**, D1–D6.
20. Z. Merali and J. Giles, *Nature*, 2005, **435**, 1010–1011.
21. F. Prinz, T. Schlange and K. Asadullah, *Nat. Rev. Drug Discov.*, 2011, **10**, 712–712.
22. C. G. Begley and L. M. Ellis, *Nature*, 2012, **483**, 531–533.
23. S. Poux, M. Magrane, C. N. Arighi, A. Bridge, C. O'Donovan, K. Laiho and UniProt Consortium, *Database (Oxford)*, 2014, **2014**, bau016.
24. S. Hunter, P. Jones, A. Mitchell, R. Apweiler, T. K. Attwood, A. Bateman, T. Bernard, D. Binns, P. Bork, S. Burge, E. de Castro, P. Coggill, M. Corbett, U. Das, L. Daugherty, L. Duquenne, R. D. Finn, M. Fraser, J. Gough, D. Haft, N. Hulo, D. Kahn, E. Kelly, I. Letunic, D. Lonsdale, R. Lopez, M. Madera, J. Maslen, C. McAnulla, J. McDowall, C. McMenamin, H. Mi, P. Mutowo-Muellenet, N. Mulder, D. Natale, C. Orengo, S. Pesseat, M. Punta, A. F. Quinn, C. Rivoire, A. Sangrador-Vegas, J. D. Selengut, C. J. Sigrist, M. Scheremetjew, J. Tate, M.

Thimmajanarthanan, P. D. Thomas, C. H. Wu, C. Yeats and S. Y. Yong, *Nucleic Acids Res.*, 2012, **40**, D306–12.

25. I. Letunic, T. Doerks and P. Bork, *Nucleic Acids Res.*, 2011, **40**, D302–D305.
26. G. Stelzer, I. Dalah, T. I. Stein, Y. Satanower, N. Rosen, N. Nativ, D. Oz-Levi, T. Olender, F. Belinky, I. Bahir, H. Krug, P. Perco, B. Mayer, E. Kolker, M. Safran and D. Lancet, *Hum. Genomics*, 2011, **5**, 709–717.
27. R. Petryszak, T. Burdett, B. Fiorelli, N. A. Fonseca, M. Gonzalez-Porta, E. Hastings, W. Huber, S. Jupp, M. Keays, N. Kryvych, J. McMurry, J. C. Marioni, J. Malone, K. Megy, G. Rustici, A. Y. Tang, J. Taubert, E. Williams, O. Mannion, H. E. Parkinson and A. Brazma, *Nucleic Acids Res.*, 2014, **42**, D926–32.
28. J. Guo, M. Hammar, L. Oberg, S. S. Padmanabhuni, M. Bjareland and D. Dalevi, *PLoS One*, 2013, **8**, e70568.
29. J. Rung and A. Brazma, *Nat. Rev. Genet.*, 2012, **14**, 89–99.
30. C. Chothia and A. M. Lesk, *EMBO J.*, 1986, **5**, 823–826.
31. T. A. de Beer, R. A. Laskowski, S. L. Parks, B. Sipos, N. Goldman and J. M. Thornton, *PLoS Comput. Biol.*, 2013, **9**, e1003382.
32. J. L. Lahti, G. W. Tang, E. Capriotti, T. Liu and R. B. Altman, *J. R. Soc., Interface*, 9, 2012, 1409–1437.
33. M. Ondetti, B. Rubin and D. Cushman, *Science*, 1977, **196**, 441–444.
34. G. N. Reeke, J. A. Hartsuck, M. L. Ludwig, F. A. Quiocho, T. A. Steitz and W. N. Lipscomb, *Proc. Natl. Acad. Sci. U. S. A.*, 1967, **58**, 2220–2226.
35. J. L. MacCallum, A. Perez, M. J. Schnieders, L. Hua, M. P. Jacobson and K. A. Dill, *Proteins*, 2011, **79**(Suppl 10), 74–9.
36. J. Moult, K. Fidelis, A. Kryshtafovych and A. Tramontano, *Proteins*, 2011, **79**, 1–5.
37. T. Schwede, *Structure*, 2013, **21**, 1531–1540.
38. R. P. Bhattacharyya, A. Reményi, B. J. Yeh and W. A. Lim, *Annu. Rev. Biochem.*, 2006, **75**, 655–680.
39. R. Roskoski Jr., *Biochem. Biophys. Res. Commun.*, 2004, **324**, 1155–1164.
40. L. Schermelleh, R. Heintzmann and H. Leonhardt, *J. Cell Biol.*, 2010, **190**, 165–175.
41. H. D. T. Mertens and D. I. Svergun, *J. Struct. Biol.*, 2010, **172**, 128–141.
42. V. Lučić, A. Rigort and W. Baumeister, *J. Cell Biol.*, 2013, **202**, 407–419.
43. J. Dekker, M. A. Marti-Renom and L. A. Mirny, *Nat. Rev. Genet.*, 2013, **14**, 390–403.
44. K. Lasker *et al.*, *Proc. Natl. Acad. Sci. U. S. A.*, 2012, **109**, 84.
45. E. R. Mardis, *Genome Med.*, 2010, **2**, 84.
46. J. Cox, F. Reimann, A. K. Nicholas, G. Thornton, E. Roberts, K. Springell, G. Karbani, H. Jafri, J. Mannan, Y. Raashid, L. Al-Gazali, H. Hamamy, E. M. Valente, S. Gorman, R. Williams, D. P. McHale, J. N. Wood, F. M. Gribble and C. G. Woods, *Nature*, 2006, **444**, 894–898.
47. A. G. Hayes, L. Arendt-Nielsen and S. Tate, *Curr. Opin. Pharmacol.*, 2014, **14**, 11–17.
48. N. G. Seidah, *Curr. Pharm. Des.*, 2013, **19**, 3161–3172.
49. R. M. Plenge, E. M. Scolnick and D. Altshuler, *Nat. Rev. Drug Discov.*, 2013, **12**, 581–594.
50. V. Agarwala, J. Flannick, S. Sunyaev, GoT2D Consortium and D. Altshuler, *Nat. Genet.*, 2013, **45**, 1418–1427.
51. A. T. Shaw, U. Yasothan and P. Kirkpatrick, *Nat. Rev. Drug Discov.*, 2011, **10**, 897–898.
52. G. Bollag, J. Tsai, J. Zhang, C. Zhang, P. Ibrahim, K. Nolop and P. Hirth, *Nat. Rev. Drug Discov.*, 2012, **11**, 873–886.
53. L. Wang, H. L. McLeod and R. M. Weinshilboum, *N. Engl. J. Med.*, 2011, **364**, 1144–1153.
54. M. Whirl-Carrillo, E. M. McDonagh, J. M. Hebert, L. Gong, K. Sangkuhl, C. F. Thorn, R. B. Altman and T. E. Klein, *Clin. Pharmacol. Ther.*, 2012, **92**, 414–417.

55. D. B. Searls, *Nat. Rev. Drug Discov.*, 2005, **4**, 45–58.

56. A. Bornot, U. Bauer, A. Brown, M. Firth, C. Hellawell and O. Engkvist, *J. Med. Chem.*, 2013, **56**, 1197–1210.

57. J. S. Mason, J. Migeon, P. Dupuis, and A. Otto-Bruc (2008) Use of Broad Biological Profiling as a Relevant Descriptor to Describe and Differentiate Compounds: Structure–In Vitro (Pharmacology–ADME)–In Vivo (Safety) Relationships, in *Antitargets: Prediction and Prevention of Drug Side Effects* (eds R. J. Vaz and T. Klabunde), Wiley-VCH Verlag GmbH & Co. KGaA, Weinheim, Germany.

58. A. L. Hopkins, J. S. Mason and J. P. Overington, *Curr. Opin. Struct. Biol.*, 2006, **16**, 127–136.

59. A. Bender, D. W. Young, J. L. Jenkins, M. Serrano, D. Mikhailov, P. A. Clemons and J. W. Davies, *Comb. Chem. High Throughput Screen*, 2007, **10**, 719–731.

60. K. P. Seiler, G. A. George, M. P. Happ, N. E. Bodycombe, H. A. Carrinski, S. Norton, S. Brudz, J. P. Sullivan, J. Muhlich, M. Serrano, P. Ferraiolo, N. J. Tolliday, S. L. Schreiber and P. A. Clemons, *Nucleic Acids Res.*, 2008, **36**, D351–9.

61. A. Gaulton and J. P. Overington, *Future Med. Chem.*, 2010, **2**, 903–907.

62. A. Gaulton, L. J. Bellis, A. P. Bento, J. Chambers, M. Davies, A. Hersey, Y. Light, S. McGlinchey, D. Michalovich, B. Al-Lazikani and J. P. Overington, *Nucleic Acids Res.*, 2012, **40**, D1100–7.

63. Digital Science transfers SureChem patent chemistry data to EMBL-EBI. https://www.ebi.ac.uk/about/news/press-releases/SureChEMBL.

64. J. Chambers, M. Davies, A. Gaulton, A. Hersey, S. Velankar, R. Petryszak, J. Hastings, L. Bellis, S. McGlinchey and J. P. Overington, *J. Cheminform.*, 2013, **5**, 3-2946-5-3.

65. M. Eriksson, I. Nilsson, T. Kogej, C. Southan, M. Johansson, C. Tyrchan, S. Muresan, N. Blomberg and M. Bjareland, *Mol. Inform.*, 2012, **31**, 555–568.

66. E. Jacoby, *Wiley Interdisciplinary Reviews: Computational Molecular Science*, 2011, **1**, 57–67.

67. A. J. Thompson and S. C. Lummis, *Expert Opin. Ther. Targets*, 2007, **11**, 527–540.

68. A. Jensen, P. Davies, H. Braunerosborne and K. Krzywkowski, *Trends Pharmacol. Sci.*, 2008, **29**, 437–444.

69. A. J. Gray, P. Groth, A. Loizou, S. Askjaer, C. Brenninkmeijer, K. Burger, C. Chichester, C. T. Evelo, C. Goble, L. Harland, S. Pettifer, M. Thompson, A. Waagmeester and A. J. Williams, *Semantic Web*, 2014, **5**(2), 101–113.

70. R. Hoehndorf, M. Dumontier and G. V. Gkoutos, *Briefings Bioinf.*, 2013, **14**, 696–712.

71. R. Hoehndorf, M. Dumontier, J. H. Gennari, S. Wimalaratne, B. de Bono, D. L. Cook and G. V. Gkoutos, *BMC Syst. Biol.*, 2011, **5**, 124-0509-5-124.

72. Q. Zhu, Y. Sun, S. Challa, Y. Ding, M. S. Lajiness and D. J. Wild, *BMC Bioinformatics*, 2011, **12**, 256-2105-12-256.

73. T. I. Oprea, S. K. Nielsen, O. Ursu, J. J. Yang, O. Taboureau, S. L. Mathias, L. Kouskoumvekaki, L. A. Sklar and C. G. Bologa, *Mol. Inform.*, 2011, **30**, 100–111.

74. C. K. Chen, C. J. Mungall, G. V. Gkoutos, S. C. Doelken, S. Kohler, B. J. Ruef, C. Smith, M. Westerfield, P. N. Robinson, S. E. Lewis, P. N. Schofield and D. Smedley, *Hum. Mutat.*, 2012, **33**, 858–866.

75. K. Azzaoui, E. Jacoby, S. Senger, E. C. Rodriguez, M. Loza, B. Zdrazil, M. Pinto, A. J. Williams, V. de la Torre, J. Mestres, M. Pastor, O. Taboureau, M. Rarey, C. Chichester, S. Pettifer, N. Blomberg, L. Harland, B. Williams-Jones and G. F. Ecker, *Drug Discov. Today*, 2013, **18**, 843–852.

76. M. J. Keiser, J. J. Irwin and B. K. Shoichet, *Biochemistry*, 2010, **49**, 10267–10276.

77. Y. Pan, T. Cheng, Y. Wang and S. H. Bryant, *J. Chem. Inf. Model.*, 2014, **54**, 407–418.

78. R. V. Lacro, H. C. Dietz, L. M. Wruck, T. J. Bradley, S. D. Colan, R. B. Devereux, G. L. Klein, J. S. Li, L. L. Minich, S. M. Paridon, G. D. Pearson, B. F. Printz, R. E. Pyeritz, E. Radojewski, M.

J. Roman, J. P. Saul, M. P. Stylianou, L. Mahony and Pediatric Heart Network Investigators, *Am. Heart J.*, 2007, **154**, 624–631.

79. R. Muthyala, *Drug Discovery Today: Therapeutic Strategies*, 2012, **8**, 71–76.
80. Cancer Genome Atlas Research Network, J. N. Weinstein, E. A. Collisson, G. B. Mills, K. R. Shaw, B. A. Ozenberger, K. Ellrott, I. Shmulevich, C. Sander and J. M. Stuart, *Nat. Genet.*, 2013, **45**, 1113–1120.
81. M. J. Garnett, E. J. Edelman, S. J. Heidorn, C. D. Greenman, A. Dastur, K. W. Lau, P. Greninger, I. R. Thompson, X. Luo, J. Soares, Q. Liu, F. Iorio, D. Surdez, L. Chen, R. J. Milano, G. R. Bignell, A. T. Tam, H. Davies, J. A. Stevenson, S. Barthorpe, S. R. Lutz, F. Kogera, K. Lawrence, A. McLaren-Douglas, X. Mitropoulos, T. Mironenko, H. Thi, L. Richardson, W. Zhou, F. Jewitt, T. Zhang, P. O'Brien, J. L. Boisvert, S. Price, W. Hur, W. Yang, X. Deng, A. Butler, H. G. Choi, J. W. Chang, J. Baselga, I. Stamenkovic, J. A. Engelman, S. V. Sharma, O. Delattre, J. Saez-Rodriguez, N. S. Gray, J. Settleman, P. A. Futreal, D. A. Haber, M. R. Stratton, S. Ramaswamy, U. McDermott and C. H. Benes, *Nature*, 2012, **483**, 570–575.
82. J. Barretina, G. Caponigro, N. Stransky, K. Venkatesan, A. A. Margolin, S. Kim, C. J. Wilson, J. Lehar, G. V. Kryukov, D. Sonkin, A. Reddy, M. Liu, L. Murray, M. F. Berger, J. E. Monahan, P. Morais, J. Meltzer, A. Korejwa, J. Jane-Valbuena, F. A. Mapa, J. Thibault, E. Bric-Furlong, P. Raman, A. Shipway, I. H. Engels, J. Cheng, G. K. Yu, J. Yu, P. Aspesi Jr, M. de Silva, K. Jagtap, M. D. Jones, L. Wang, C. Hatton, E. Palescandolo, S. Gupta, S. Mahan, C. Sougnez, R. C. Onofrio, T. Liefeld, L. MacConaill, W. Winckler, M. Reich, N. Li, J. P. Mesirov, S. B. Gabriel, G. Getz, K. Ardlie, V. Chan, V. E. Myer, B. L. Weber, J. Porter, M. Warmuth, P. Finan, J. L. Harris, M. Meyerson, T. R. Golub, M. P. Morrissey, W. R. Sellers, R. Schlegel and L. A. Garraway, *Nature*, 2012, **483**, 603–607.
83. B. Haibe-Kains, N. El-Hachem, N. J. Birkbak, A. C. Jin, A. H. Beck, H. J. Aerts and J. Quackenbush, *Nature*, 2013, **504**, 389–393.
84. T. Pemovska, M. Kontro, B. Yadav, H. Edgren, S. Eldfors, A. Szwajda, H. Almusa, M. M. Bespalov, P. Ellonen, E. Elonen, B. T. Gjertsen, R. Karjalainen, E. Kulesskiy, S. Lagstrom, A. Lehto, M. Lepisto, T. Lundan, M. M. Majumder, J. M. Marti, P. Mattila, A. Murumagi, S. Mustjoki, A. Palva, A. Parsons, T. Pirttinen, M. E. Ramet, M. Suvela, L. Turunen, I. Vastrik, M. Wolf, J. Knowles, T. Aittokallio, C. A. Heckman, K. Porkka, O. Kallioniemi and K. Wennerberg, *Cancer. Discov.*, 2013, **3**, 1416–1429.

CHAPTER 14

Translational Science

ALASDAIR J. GAW

Department of Health + Care, Technology Strategy Board, North Star House, North Star, Swindon, SN2 1UE, UK
E-mail: Alasdair.Gaw@tsb.gov.uk

14.1 INTRODUCTION

The term "translational science" has been defined in different ways in different places. For the purpose of this article we will follow the definition as described by the Medical Research Council of the UK, "Translational Science is the process of the bidirectional transfer of knowledge between basic scientific work (in the laboratory and elsewhere) with that of the person, in health or disease."

This field of science has been brought to the fore in recent years, as the successful development of new medicines has increasingly failed at the point of experimental efficacy testing in the patient, to increase the awareness of the relevance of academic research to society. Translational science encompasses all of the key biomedical disciplines such as genetics, biochemistry, drug metabolism, cellular sciences, physiology, pharmacology, clinical pharmacology and clinical development that are part of the established drug discovery process. Translational science demands the integration of this knowledge from target protein to clinical outcome, rather than the isolationist approach between disciplines that has developed in recent years, as the field of biology and medicine has expanded and become increasingly specialised. This isolation between disciplines has contributed to the high failure rate of clinical efficacy. In this chapter we shall try to describe approaches that can aid in the improvement of the translation of basic science from target to delivering expected clinical outcomes, and provide a perspective for medicinal chemists and where their skills can contribute.

The ultimate aim of any medicinal chemist is to create a compound that will treat disease in patients. Diagnosis is the foundation of determining treatment and is principally based on the use of biomarkers, which are core to the field of translational science and can link the drug discovery process from target identification to clinical testing. A biomarker has been defined by the FDA as "A characteristic that is objectively measured and evaluated as an indicator of normal biologic processes, pathogenic processes, or pharmacologic responses to a therapeutic

The Handbook of Medicinal Chemistry: Principles and Practice
Edited by Andrew Davis and Simon E Ward
© The Royal Society of Chemistry 2015
Published by the Royal Society of Chemistry, www.rsc.org

intervention.''[1] They are essentially quantitative measures that allow us to diagnose and assess the disease process and monitor response to treatment. Strictly speaking, all clinical outcomes are biomarkers, however, those that are exclusively applicable to, or useful in, human beings are usually excluded from the definition as they do not lend themselves to preclinical translational purposes. In such cases alternative biomarkers will need to be sought to establish confidence in a new mechanism, compound or drug prior to the moment in clinical trials that such a clinical outcome can be introduced. For example, in oncology targets tumour size or blood biomarkers may be used in preclinical studies but clinical outcomes will still be around mortality rates. In respiratory disease anti-inflammatory biomarkers may be measured through animal models and early phase clinical trials but reduction in exacerbation rates and improvement in patient reported outcomes will still be required for drug registration by regulators.

The use of biomarkers has become synonymous with translational science as they allow preclinical and clinical effects to be measured in an experimental setting and allow comparison of preclinical observation to observation in the clinic. As a consequence, the use of biomarkers has become essential in modern drug discovery and guidelines have been issued by the FDA[2] and the EMEA[3] for the process of qualifying a biomarker as demonstration of efficacy.

Essentially the qualification process of a biomarker for use in disease requires:

- Context of the disease/condition in which the biomarker will be applied:
  - Symptoms, pathophysiology, risk factors and epidemiology related to the biomarker.
- Context of the use of the biomarker:
  - To demonstrate target engagement or to claim clinical efficacy.
- Influence of current treatments against the biomarker.
- Technical aspects of proposed biomarker measurement:
  - Limits of detection, complexity, intra- and inter-test variability, within subject variability, sensitivity/specificity, positive and negative predictive value, ROC (Receiver Operating Characteristic) curve, change between pre- and post-test.
- Suitability for use in patient care and clinical trial settings.

There are several pre-competitive activities involving industry, academia, patient groups and charities currently underway that interact with regulatory authorities to drive faster acceptance and qualification of biomarkers for uses such as safety biomarkers for medicines (SAFE-T consortium)[4] and new biomarkers of COPD.[5] There are also several initiatives such as those by the MRC and ABPI that bring together academia and industry to extensively phenotype patients in several therapeutic areas such as rheumatoid arthritis, chronic obstructive pulmonary disease and diabetes.[6]

The term translational science frequently refers to the stage at the end of preclinical work progressing into clinical testing (Figure 14.1). Different organisations refer to clinical studies and biomarkers with varying definitions of Proof of Concept, Proof of Principle and Proof of Mechanism (or Target Engagement) studies and biomarkers.

In this article we will define the use of biomarkers in three categories:

- Proof of Concept (PoC) biomarkers: These are biomarkers of presumed direct clinical relevance that can be used to register a drug as effective with regulatory authorities, *e.g.* blood pressure for anti-hypertensives, FEV1 for bronchodilators, cholesterol for statins. These are used if possible from Phase 2A onwards for the first clinical studies in patients; however, if longer duration and greater patient number are required to show an effect they will more commonly be used from Phase 2B studies. As they are required for registration, they are essential for Phase 3 studies and, as confirmation is the main objective of Phase 3 studies, it is highly advisable to use the same biomarkers in both Phase 2b and Phase 3.

Translational science in drug discovery

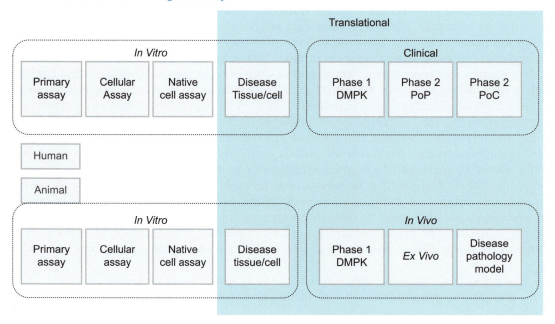

Figure 14.1   The role of translational science in drug discovery.

Occasionally, PoC biomarkers can also be used in Phase1 studies where patients are used, *e.g.* in oncology.

- Proof of Principle (PoP) biomarkers: These are biomarkers that are closely linked to the mechanism of action and closely linked to the clinical symptoms, but are not sufficient to define clinical efficacy for registration due to lack of validation, *e.g.* inflammatory cytokines levels for anti-inflammatory treatments, reduction in amyloid for Alzheimer's treatment, reduction of TNF production by p38 inhibitors, inhibition of cell infiltration to joint with CCR1 inhibitors. These are commonly used in Phase 2 clinical studies that are of short duration and provide confidence of obtaining a Proof of Concept Signal in longer studies. The potential of studying such markers in healthy volunteers, in human challenge models (*e.g.* an LPS challenge in a Phase 1b study), in volunteers with a clear expression of the disease process of interest (*e.g.* testing treatments for allergic asthma in otherwise healthy patients with allergies) should be considered as this could provide an early decision point for a compound or mechanism. Many PoP markers are also suitable for animal model studies and allow Pharmacokinetic/Pharmacodynamic (PK/PD) comparison from animal models to be used to model expected human PK/PD and effect which can then be verified in the Phase 2 studies.
- Proof of Mechanism or Target Engagement (PoM/TE) biomarkers: These are directly linked to activation of the test mechanism normally in an *ex vivo* environment clinically or as a tertiary assay pre-clinically, *e.g.*:
  ○ Reduction in LPS-induced TNF production from human blood by p38 inhibitors, reduction in STAT phosphorylation by JAK inhibitors.
  ○ Binding of a radiolabelled PET ligand or displacement of the label with test compound demonstrating occupancy of the target binding site.

These are commonly used in Phase1 clinical studies that are of short duration and provide confidence of PK/PD relationships. These allow exposure levels to be compared to effect levels and aid dose selection to ensure blockade or activation of the target in the clinical study. These are also suitable for animal model studies and allow PK/PD comparison from animal models to be used to model expected human PK/PD and effect. These PoM markers are fundamental as they can be used in both preclinical and clinical testing and provide the key data that will translate compound effect from bench science models to man.

Within drug discovery and translational science perhaps the greatest role of a biomarker is to provide evidence of target engagement in the various stages of testing of a drug candidate (Figure 14.2).

Assuming that initial DMPK predictions of exposure to engage and block the target were achieved then you have a variety of potential outcomes:

- No exposure for target engagement: No effect on PoM markers; do not progress.
    - Re-evaluate drug metabolism absorption models or stop programme.
- Target engagement: No effect on POM markers; do not progress.
    o Re-evaluate exposure levels required or stop programme.
- Target engaged, effect on POM markers.
    o No effect on PoP markers; do not progress.
        - Re-evaluate link of mechanism to PoP marker or stop programme.
    o Unexpected effect on PoP markers:
        - Re-evaluate hypothesis on link, re-establish platform of evidence.
            • Progress with caution or stop programme.
- Target engaged, effect on PoM, PoP markers.
    o No effect on PoC markers.
        - Re-evaluate hypothesis and trial design or stop programme.

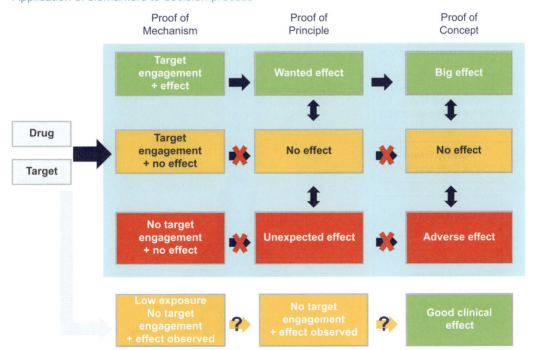

Application of biomarkers to decision process

Figure 14.2 Decision tree for translational testing.

  ○ Unexpected effect on PoC markers.
    ■ Positive effect.
      ● Re-evaluate hypothesis and trial design; progress.
    ■ Adverse effect.
      ● Re-evaluate risk–benefit, progress or stop programme.
- Target exposure too low for expected engagement.
  ○ Effect on PoM markers and/or PoP markers.
    ■ Re-evaluate pharmacodynamic characteristics of compound.
    ■ Consider active metabolite that acts on same target.
    ■ Consider off target mechanism of activity.
      ● If mechanism identifiable then progress.
    ■ If adverse event seen then stop programme.

Only very few compounds progress from screening to man (Figure 14.3), and the demonstration of efficacy in patients is also the point of greatest, and costliest, failure in the drug discovery process. It should always be remembered that this is called Research for a reason: the answers are not yet known therefore failure when a new mechanism is not sufficiently important in disease is to some extent unavoidable, despite the growing focus within the industry on the linkage of target to disease and translation of preclinical hypotheses into the clinic. However, failure because of insufficient target engagement when good markers of target engagement are available should be rare. Certain conditions may well prevent accurate assessment

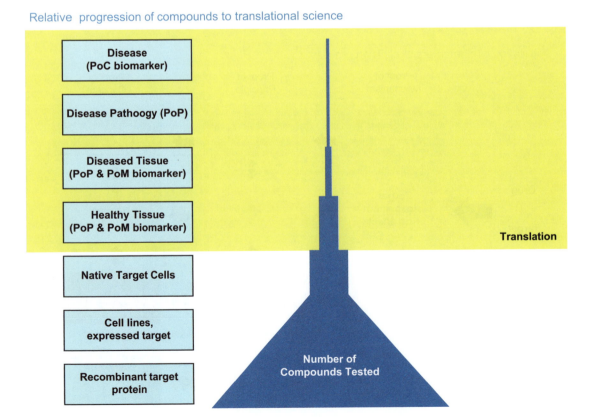

Relative progression of compounds to translational science

Disease (PoC biomarker)

Disease Pathoogy (PoP)

Diseased Tissue (PoP & PoM biomarker)

Healthy Tissue (PoP & PoM biomarker)

Native Target Cells

Cell lines, expressed target

Recombinant target protein

Translation

Number of Compounds Tested

Figure 14.3   Relative progression of compounds into translational screening.

of target engagement, *e.g.* when testing inhaled or topical drugs, but in these cases one should aim to use PoP markers as proof of target engagement at the earliest opportunity. Translational research allows creation of a platform of evidence to improve decision making and increase the expectation of demonstrating an effect in patients.

14.2 LINKING HYPOTHESIS TO DISEASE

Although translational science concentrates on measuring efficacy in man it extends through the target validation process and is important in designing your hypothesis and establishing a screening cascade. Diseases may be defined clearly on a clinical outcome basis, however they are likely to consist of several pathologies that may occur at the organ, tissue and cellular level. At the core of choosing any biological target for Drug Discovery Research you must first know the disease that you wish to treat, the target pathology you wish to change, the clinical impact of changing that pathology and then generate a hypothesis of how the chosen target will achieve that. An example of three potential pathologies within the disease of Chronic Obstructive Pulmonary Disease is shown in Figure 14.4. Maintaining this line of sight from the disease to the target and back is a key element of any translational strategy.

There are five key questions necessary to determine if a target is involved in a disease:

1. Is the target (pathway) present in disease pathology?
2. Is the target (pathway) activated in disease pathology?
3. Will activation of the target (pathway) induce the disease pathology?
4. Will inactivation of the target (pathway) prevent the disease pathology?
5. Will agents which prevent inactivation of the target induce the disease pathology?

This allows you to identify a platform of evidence that will support a hypothesis of expected effect in disease and enables creation of a focused screening strategy that tests the hypothesis (Figure 14.5).

14.3 CREATING A SCREENING STRATEGY FROM MOLECULE TO MAN

The ultimate end point of drug discovery is to show efficacy in man therefore it makes sense to start with the patient and identify the effect you wish to achieve and then establish a translational strategy which will allow you to screen and progress compounds while creating a platform of evidence that will support the expectation of achieving that end point in man. A suggested screening cascade that will provide a platform of evidence suitable for compound progression through to man is shown in Figure 14.6. It is important to remember that the vast majority of compounds synthesised in a drug discovery project will not progress beyond the initial primary screening phases, however the latter assays must be as accurate, rigorous and indicative of compound activity in a physiological environment as possible (Figure 14.3). At the end of the programme molecules will be tested in living human beings at physiological pH, body temperature and in physiological solutions. Therefore a screening strategy must include physiological assay systems that mimic the human situation before progressing into man. When developing novel compounds there is always more that is unknown about the compounds' activity than is believed to be known. It is therefore critical to ensure that the effect being observed can be attributed to the mechanisms that these compounds are targeting. For this reason the measurement of affinity, and efficacy in the case of agonists, is critical at all stages to ensure that the effect occurs at the expected exposure to the molecule.

Chronic Obstructive Pulmonary Disease

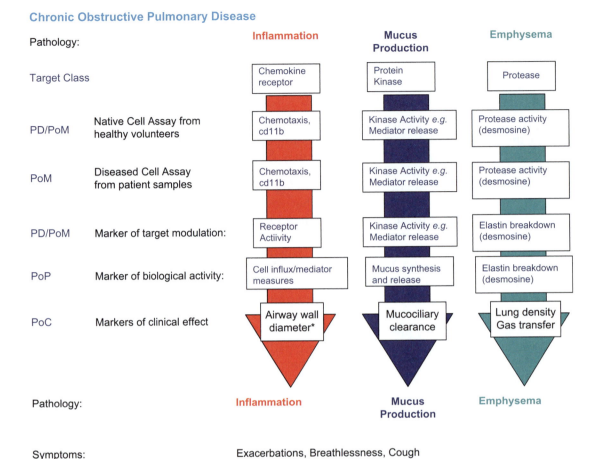

Figure 14.4   Linkage of mechanism to pathology in Chronic Obstructive Pulmonary Disease. Translational depiction of how three separate pathologies that contributes to the symptoms of COPD can be broken down to separate target mechanisms with biomarkers relevant to the individual pathologies.

Assays become more complex the closer they get to mimicking aspects of human disease therefore only a few key compounds should really progress into complex tertiary assays. For primary assays, where the majority of compounds are tested, then a high to medium throughput system for selection of the most potent and stable compounds and determining structure–activity relationships is critical.

With the advent of molecular biology it is now possible and standard practice that targets are generated as human recombinant proteins which are expressed and used by generating cells that express membrane proteins such as GPCRs or enzymes. The target can then be used in broken cell assays for enzyme activity measurement or membrane binding assays. These are the standard approaches for establishing primary assays and are suitable for ranking compounds to develop a structure–activity relationship.

In primary assay systems, where human recombinant target proteins are tested in biochemical based molecular assays, the assay matrix is generally a very simple salt solution buffered to a physiological pH that can be more easily controlled and optimised for a larger signal to noise ratio for screening higher throughput of compounds. However, as these assay

Platform of evidence from target to clinic

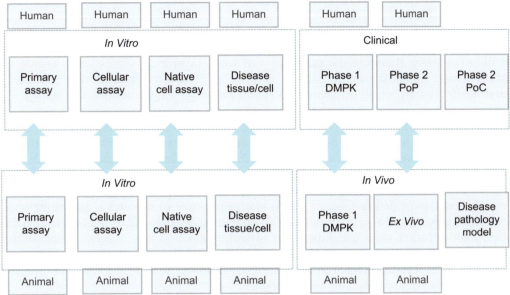

Figure 14.5 Generic screening strategy to generate platform of evidence for translation.

CXCR2 receptor blockade to reduce exacerbation
blocking neutrophil chemotaxis as anti-inflammatory

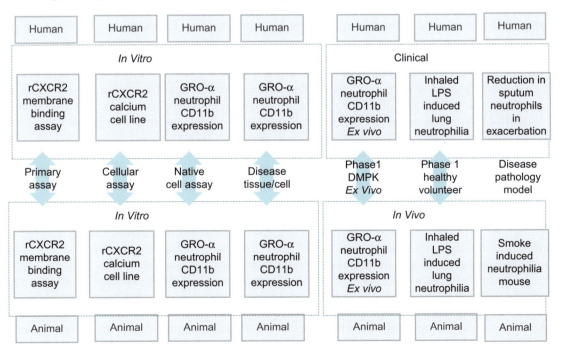

Figure 14.6 Example of a screening strategy to test a chemokine CXCR2 antagonist for Chronic Obstructive Pulmonary Disease.

conditions are non-physiological then issues may occur when progressing the compounds to more physiological conditions required in secondary cellular assays, for example:

- Some assays may perform better at a pH of 6, however the molecule may demonstrate very different biological effect at pH7.4 as pH can affect the amount of charged/neutral molecule in an assay therefore the proportion of active form may vary between assays.
- Many assays have little or no protein present which allows good affinity estimates but underestimates the concentration required in physiological situations (see below).

The human recombinant proteins can also be tested in intact cells to determine the ability to block functional responses. The strength of these secondary assays is that they can demonstrate both affinity and efficacy of compounds as agonist effects of compounds can be seen. In the case of GPCRs whole cell assays can use intracellular calcium signalling as a functional endpoint. By adding test compounds it is possible to observe if compounds have an agonist response in their own right (increasing calcium release) and it is also possible to determine functional calculations of affinity by use of Schild analysis or Cheng–Prussoff analysis. It is always advisable to calculate affinity rather than use functional measures such as IC_{50} values (concentration of inhibitor necessary to reverse a challenge by 50%) as these will vary depending on the challenge level use. This is especially relevant in estimating exposures required for man where the level of challenge agent is variable and can be unknown.

When moving to intact cell assays the assay matrices have to become more physiological to ensure the cells remain viable through the experiment. A common element in secondary cell assays that can cause problems in establishing affinity is the presence of plasma proteins from added serum. Because many assay systems have different requirements for plasma protein content, it is important that affinity estimates in each assay are estimated by the free concentration of molecule in any assay system. This is more apparent in secondary cell systems and tertiary testing systems that use human cells and tissues as there is a greater requirement for more physiological conditions that mimic the human response, for example:

- If an assay has 4% human serum albumin present (similar to human plasma) then a molecule with 99% plasma protein binding will appear to be 100 times less potent than a comparable affinity compound with 1% plasma protein binding. When compounds of different structural and physicochemical properties are being tested (*e.g.* converting from a neutral compound to a weak acid) then it is difficult to determine a strong SAR unless you account for the loss of free molecule due to protein binding.
- Many primary assays control the substrate to standardise signal to noise and screening data, however in intact cells for example the substrate concentration may be considerably greater; this is true for many kinase inhibitors that act at the ATP site. The excess amount of ATP competes with the compound and can reduce the apparent cellular potency by 100 fold.

Therefore, primary and secondary assays can be used to estimate the binding affinity of compounds for the target but do not necessarily demonstrate the concentration of the compound required to block the functional response in the physiological situation. They are useful for affinity measurements but not necessarily for ensuring efficacious exposure to engage and block the target. It is feasible that a compound with an affinity of 1 nM may require an exposure of 10 to 100 nM (or even greater) in blood to achieve a similar degree of target blockade. Thus the importance of using clinically relevant markers of target engagement and compound efficacy come to the fore. If you introduce secondary and tertiary assays and markers that can be utilised in the clinic then the exposure required in man to block the target and observe efficacy can be better determined.

Increasingly, primary human cells, sometimes from disease, are used in the secondary or tertiary screening setting, increasing the relevance of the assay but also likely impacting cost, throughput and complexity. When used in primary screening as "Black Box" or "Phenotypic" screens, the translational relevance of the screening may be very high, but the challenges for the medicinal chemist will increase too. Developing SAR may be difficult and any structure-based design is impossible until the molecular nature of the target is elucidated. Despite great progress in different methodologies of target deconvolution, it may not happen within a relevant timeframe to impact chemistry (indeed it may never happen but whilst it will slow down development it is not necessarily a show-stopper as it is not required for registration of a new drug). The authors recommend that assays based on cells or tissues from patients are used to determine phenotypic effects of key compounds rather than to develop SAR.

It is necessary in establishing any translational screening cascade that cross-species assays are available for the main toxicology species such as dog and rat but also for the species that any animal models are planned in, *e.g.* guinea pig, mouse, rabbit, mini pig, or primate. These assays are necessary to ensure that appropriate species are used for toxicology studies (to demonstrate any side effects of blocking the mechanism) and to be able to identify appropriate dose selection and dosing regimes in animal models to establish PK/PD relationships for effect. It is important that the rank order of potency across several compounds holds up between assays to ensure that the observed effect in cell systems or animal models are a consequence of the expected mechanism and not due to off target mechanisms or active metabolites.

14.4 TRANSLATIONAL SCIENCE AND STRATIFIED MEDICINE

The PoM/TE and PoP biomarkers assays may be carried out on healthy volunteers or blood or tissue samples taken from healthy volunteers, however the critical PoC test has to be carried out on patients. It is true that the PoC marker can sometimes be measured in healthy volunteers in particular with agonistic mechanisms (FEV1 will change in response to beta2 agonists in healthy individuals) but whilst a negative effect may well lead to termination of the project and positive data may be very encouraging, they may not be definitive as it is unlikely to indicate whether the level of response has therapeutic relevance. In such studies the PoC marker functions as a PoM/TE or PoP marker.

As demonstrated (Figure 14.4), within a single disease indication such as COPD there are different PoP/PoC biomarkers for each of the relevant pathologies. For this reason it is important that patients are stratified within the overall disease to ensure that the most relevant patients are tested and to provide the best opportunity to demonstrate efficacy.[7,8]

By selecting the patients into sub-populations for Phase 2 it also allows planning for the future and the development of the candidate drug as a targeted drug that may require a companion diagnostic. This has the potential to allow shorter and more efficacious patient studies enabling faster, more successful progression to market and the earlier generation of a companion diagnostic can enhance rate of adoption by healthcare providers.

Classification of patients into these sub-populations using diagnostic tests that identify the status of a particular biomarker (for example the presence of a genetic mutation or protein) and tailoring the clinical approach accordingly is referred to as stratified medicine. Stratification is desirable for patients, healthcare systems, and pharmaceutical and diagnostic companies as it has both health economic and clinical benefits.

Biomarkers that can be used as PoP can also be used to stratify patients for specific target mechanisms an example being interleukin-13 which is a mediator of allergic asthma.[9] Agents which block or reduce IL-13 could be considered as anti-asthmatic, however to identify asthmatic individuals with high IL-13 levels that may benefit from this treatment, an alternative diagnostic, periostin, a protein whose abundance in blood are influenced by the level of IL-13, has been developed.[10]

Stratification represents a more targeted approach, with the potential for greater efficacy of treatments and minimisation of their side effects as non-responders are not prescribed ineffective medication. A recent example from the oncology field is crizotinib, a drug indicated for the treatment of locally advanced or metastatic non-small cell lung cancer (NSCLC) that is anaplastic lymphoma kinase (ALK)-positive (this occurs in approximately 5% of patients). This patient population has a 10% response to standard chemotherapy, yet 55% respond to the targeted therapeutic.[11]

The necessity of stratification in the development of new medicines is increasingly recognised. Stratification as a factor in drug response is now sufficiently appreciated by regulators and payers that within the last few years the US Food and Drug Administration (FDA) and UK National Institute for Health and Clinical Excellence (NICE) have both rejected drugs due to the lack of a companion diagnostic to identify a responder subpopulation.[12]

A full report on the current importance of stratified medicine for future healthcare has been released in the UK by the Academy of Medical Sciences.[13] The growing need for stratified medicine clarifies the need to include translational science early in the drug discovery process as it not only contributes to the creation of a screening strategy but allows future planning for faster clinical trials and a better estimation of the market opportunities available for companion diagnostics and for improved disease diagnosis. A roadmap describing the challenges associated with the development of stratified medicine and introduction of diagnostics within the UK has been generated by the Technology Strategy Board[14] who have established a Stratified Medicine Innovation Platform to partner across government, industry and the research community to develop this area.

Translational Science and stratified medicine are creating a large opportunity for new and novel diagnostics in the areas of unmet clinical need. In many diseases the probability of achieving a positive clinical outcome from treatment is far greater the earlier a disease is detected and the earlier appropriate treatment can be delivered. There is a considerable unmet clinical requirement for early and detailed characterisation of disease in inaccessible areas of the body such as the brain (dementia, Alzheimer's, oncology), abdominal and thoracic cavities and the deep viscera/soft tissues such as the pancreas and liver. There is a growth in demand for detector and biosensor technologies that can be used on patients for diagnosis or during medical procedures to address these challenges which provides chemical opportunities for development of novel fluorescent or radiolabelled tracers, imaging contrast agents, PET ligands *etc*. Thus, translational science and stratified medicine are not only relevant to aid the design and testing of novel chemical entities but they require the invention of novel tool compounds. This can range from tools for animal testing and model validation to novel diagnostic markers such as PET ligands to enable measurement of target interaction in animal and human.[15,16] The drug discovery process is an integrated multidisciplinary approach and requires the inventiveness of chemists across the entire process from compounds to diagnostics.

Stratified medicine raises challenges in translational medicine and it can impact a drug development programme in a number of ways. Early evidence will be required that the chosen target population is indeed the most likely to respond to therapy and this may make relevant animal and other preclinical models even more difficult to identify. In planning clinical studies when less certain of the target population, an alternative is to include different groups of patients and perform (previously agreed) *post hoc* analysis to determine the best responders. Also whilst you may need fewer patients to see a signal, you may need to recruit a lot more than a traditional study to find such well-defined patients, which will increase the cost and potentially the duration of the study. Within a therapeutic indication it means only a subpopulation will be treatable so the commercial potential may not be as obvious, although regulatory bodies are adopting new reimbursement models as a consequence of this. On balance though, when there is both a strong rationale for stratification and good methods to do so, it is the preferred option. A greater clinical response, not having to treat people who do not respond well or would suffer side effects, are all things that will help drive adoption and reimbursement of a new medicine.

HINTS AND TIPS—POTENTIAL PITFALLS

- Compounds vary in metabolism across species; ensure exposure relative to affinity is relevant when concluding a PK–PD relationship. Try to emulate the human C_{max}–C_{min} range as much as possible.
- Calculate exposure by free active concentrations when estimating human doses relative to animal PK/PD as inter-species protein binding can vary greatly.
- Test selectivity profile of compounds against key targets in animal model species and man. Try to ensure the effect is due to the tested mechanism.
- Ensure that there is evidence of comparable physiology and target expression/distribution between animal models and humans. Use *in vitro* assays where possible to determine the strength of the mechanistic link between species before using models to predict PK/PD in man.
- PoP and PoC markers may have been established for an existing drug for a given indication, but this doesn't mean it will have relevance to a different class of drugs for the same indication… Every drug that is effective in Rheumatoid Arthritis (RA) lowers serum CRP levels so that we now believe that lowering CRP is a good predictor for efficacy in RA, but whilst the very effective inhaled corticosteroids that are commonly used in mild to moderate asthma reduce sputum eosinophils in asthmatics, other drugs that do so equally dramatically such as anti-interleukin 5 antibody were not effective at all.
- Always ask for preclinical evidence linking the marker and pathological improvement for the new mechanism; do not rely on what has been found for different mechanisms.
- Try to determine the level of target engagement (antagonism, agonism, inhibition…) that is required for a relevant therapeutic effect in pre-clinical models and aim to achieve at least such levels when going into man. Effective levels may vary greatly from target to target and despite various analyses, there are no hard and fast rules! It is better to work on free plasma concentration to determine exposure as varying protein concentrations and binding affinities across species can be very misleading. Always try to get relevant species and strain plasma protein binding data for the screening cascade being used.

HINTS AND TIPS—CHEMISTS CONTRIBUTE TO TRANSLATIONAL SCIENCE IN MANY WAYS

- Designing and making compounds of different chemical classes, minimising risk of unknown, off-target activity.
- Designing and making compounds which can be used for both animal and man are most valuable.
- Designing and making selective compounds suitable for testing in animals if required.
  - The majority of screening is now human target based and therefore animal selectivity can be lost, so compounds with suitable DMPK and affinity may have to be made for hypothesis testing in an animal model of a specific species for specific disease pathology. As mouse, rat and dog are the principal species for toxicology and disease pathology models these are the key species to consider.

> - Ensuring species and strain variation in protein binding and physical properties related to tissue penetration are considered.
> - Making selective compounds specifically designed for testing target engagement.
> - Designing and making novel detection agents such as PET ligands that can be essential for areas such as neuroscience where it is difficult to sample the test organ to test if there is target occupancy from the test compound.
> - Developing detection techniques to identify metabolites and in some cases new biomarkers such as with the use of LC-MS/MS.
> - Ensure all members of the project team and various decision-makers are aware of the principles discussed above.
> - Influencing all project team members to ensure that only compounds that are good enough to test the hypothesis are progressed to human studies.

KEY REFERENCES

MRC, Translational Research, www.mrc.ac.uk/Ourresearch/ResearchInitiatives/Translationalresearch/index.htm.
European Medicines Agency, Biomarkers Qualification: Guidance To Applicants, 2008, www.rsihata.com/updateguidance/emea2/2008/7289408en.pdf.
http://issuu.com/acmedsci/docs/realising_the_potential_of_stratifi

REFERENCES

1. Biomarkers Definitions Working Group, *Clin. Pharmacol. Ther.*, 2001, **69**, 89.
2. F. Goodsaid and F. Frueh, *AAPS J.*, 2007, **9**(1), E105.
3. European Medicines Agency, *Biomarkers Qualification: Guidance To Applicants*, 2008, www.rsihata.com/updateguidance/emea2/2008/7289408en.pdf.
4. IMI SAFE-T Consortium, http://www.imi.europa.eu/content/safe-t.
5. R. Casaburi, B. Celli, J. Crapo, G. Criner, T. Croxton, A. Gaw, P. Jones, N. Kline-Leidy, D. A. Lomas, D. Merrill, M. Polkey, S. Rennard, F. Sciurba, R. Tal-Singer, R. Stockley, G. Turino, J. Vestbo and J. Walsh, *COPD*, 2013, **10**(3), 367.
6. MRC, *Translational Research*, www.mrc.ac.uk/Ourresearch/ResearchInitiatives/Translationalresearch/index.htm.
7. A. L. Lazaar, L. E. Sweeney, A. J. MacDonald, N. E. Alexis, C. Chen and R. Tal-Singer, *Br. J. Clin. Pharmacol.*, 2011, **72**(2), 282.
8. P. J. Barnes and R. A. Stockley, *Eur. Respir. J.*, 2005, **25**, 1084.
9. M. Wills-Karp, J. Luyimbazi, X. Xu, B. Schofield, T. Y. Neben, C. L. Karp and D. D. Donaldson, *Science*, 1998, **282**(5397), 2258.
10. J. Corren, R. F. Lemanske, N. A. Hanania, P. R. Korenblat, M. V. Parsey, J. R. Arron, J. M. Harris, H. Scheerens, L. C. Wu, Z. Su, S. Mosesova, M. D. Eisner, S. P. Bohen and J. G. Matthews, *New Engl. J. Med.*, 2011, **365**, 1088.
11. D. E. Gerber and J. D. Minna, *Cancer Cell*, 2010, **18**(6), 548.
12. PricewaterhouseCoopers, *From vision to decision*, 2012, www.pwc.com/gx/en/pharma-life-sciences/pharma2020/vision-to-decision.jhtml.
13. Academy of Medical Sciences, 2013, http://issuu.com/acmedsci/docs/realising_the_potential_of_stratifi.

14. Technology Strategy Board, *Stratified Medicine Innovations Platform*, 2011, https://ktn. innovateuk.org/web/stratified-medicines-innovation-platform.

15. L. Farde, B. Andrée, N. Ginovart, G. Halldin and S. Thorberg, *Neuropsychopharmacol.*, 2000, **22**, 422.

16. K. Varnä, A. Jučaitė, D. J. McCarthy, P. Stenkrona, M. Nord, C. Halldin, L. Farde and S. Kanes, *Cephalagia*, 2013, **33**(10), 853.

CHAPTER 15

Discovery Toxicology In Lead Optimisation

SIMONE BRAGGIO*, MAURO CORSI, ALDO FERIANI, STEFANO FONTANA,
LUCIANA MAROCCHIO AND CATERINA VIRGINIO

Aptuit Srl, Via Fleming 4, 37135 Verona, Italy
*E-mail: simone.braggio@aptuit.com

15.1   INTRODUCTION

The cost impact of late-stage failures of drug candidates has motivated the pharmaceutical industry to develop, validate, and implement a more proactive testing paradigm, including an emphasis on conducting predictive *in vitro* and *in vivo* studies earlier. The goal of drug discovery toxicology is not to reduce or eliminate attrition, as is often misstated as such, but rather to reprioritise efforts to shift attrition of future failing molecules upstream in discovery. This shift in attrition requires additional studies and investment earlier in the candidate evaluation process in order to avoid spending resources on molecules with soon-to-be-discovered development-limiting liabilities.

15.2   *IN SILICO* TOXICOLOGY

In silico toxicology can be defined as the application of computer technologies to analyse and model available data in order to predict toxicological activity of a substance from the chemical structure. It can be strictly a computer-based analysis of the relationship of chemical structure and toxicological activity (SAR), or a simple structural alert identification and association with toxicity.[1–3] *In silico* toxicology methods such as computational toxicology, predictive quantitative structure-activity relationship (QSAR) modelling of toxicity and predictive ADME-Tox are currently used in the pharmaceutical industry at the design stage, to help at identifying lead compounds with reduced toxicological potential risk.[4] In fact, being aware of potential toxicological issues early in the discovery process can be very useful to prioritise chemical series at the hit discovery stage or hopefully to fix these liabilities during lead optimisation (just as potency and pharmacokinetics properties are being optimised), before significant investment of time and financial resources are spent in clinical trials. The advantages of these methods are low costs, standardisation, minimal equipment needs, and short time of execution (*i.e.* high

The Handbook of Medicinal Chemistry: Principles and Practice
Edited by Andrew Davis and Simon E Ward
© The Royal Society of Chemistry 2015
Published by the Royal Society of Chemistry, www.rsc.org

throughput) and can hence be used as a pre-screen and to prioritise compounds for *in vitro* or *in vivo* assay testing.[5–7]

15.2.1 *In Silico* Toxicology Tools

A number of chemical–biological informatics-based tools and related software programs for toxicity prediction along with details of the algorithms have been described in various reviews, published in the literature.[8–15] Some of the most used systems are listed in Table 15.1, along with a general brief description, availability and the biological endpoint they predict. Computational toxicology approaches are typically aimed at building toxicity databases, QSAR models, rule/descriptor-based methods, as well as classical ligand and target-based techniques. Visualisation tools combining QSAR models and systems biology and pharmacology pathway analysis[16] are also available. The choice of the modelling method to be used can be influenced by many factors including: the size and the type of the dataset (few congeneric or many, chemically diverse compounds) the endpoint to be modelled and the type of answers required. The common objective of all methods, however, is to attribute a toxicological indication/effect to a chemical structure.

15.2.2 Databases

Data collection is the most critical and time-consuming task. Toxicology heavily relies on the use of *in vivo* data obtained from animal models. If, however, the mechanism leading to toxicity is known, for example, in the case of the hERG potassium ion channel inhibition and subsequent QT interval prolongation, high-throughput screening (HTS) and high-content screening (HCS) methodologies can be used to replace *in vivo* studies.[17–19] All the computational models employed in the field of toxicity prediction are based on the structure of existing, annotated ligands. Typically, scientists gather data from different sources: literature, patents and, when available, from in-house programs. Following several initiatives aiming at collecting information from legacy, unstructured systems (*e.g.* National Toxicology Program (NTP) reports and literature) and integrate them into organised and searchable open-access databases,[20–26] more and more data are becoming available. Table 15.2 provides an overview of some of the publicly available and commercial databases and electronic data sources which include human health effects of substances useful in risk assessment, safety evaluation, and hazard characterisation. Accordingly, open sources of toxicity data are the first to be exploited in computational modelling and *in silico* toxicology software programs. Furthermore, proprietary toxicity databases tend to offer promise for enhanced "predictive accuracy" or expanded chemical space of the representative molecular structures they contain compared to public databases. In addition, the aforementioned high-throughput capabilities developed and used by the pharmaceutical industry enable the attractive toxicological profiling approach to be pursued, using whole-cell and phenotypical assays (*i.e.* genotox and metabolism modulation, hERG, cytochrome P450 3A4, Ames assay and others).[27–31] Finally, the application of genomics techniques in toxicology, due to the ease of sequencing technology which allows whole-genome expression analyses for organisms used in toxicological tests, has led to the creation of a new discipline called toxicogenomics[32–34] that has proved to be very useful in prediction and interpretation of off-target effects.[35–37]

15.2.3 QSARs and Statistical Modelling

A toxicological QSAR model is a mathematical equation used to predict the toxicity of a new substance from a training set of chemicals with known toxicity and spanning a defined

Table 15.1 Most used computation systems for predicting toxicology alerts.

System/developer	Availability	Description	Toxicity endpoint prediction	Ref.
MDL QSAR (MDL)	Commercial	Quantitative Structure-Activity and Structure-Property modelling system	Acute (oral) tox, hepatoxicity, nephrotoxicity, urinary tract toxicity, mutagenicity, carcinogenicity, skin sensitisation and irritancy	90
HazardExpert (CompuDrug)	Commercial	Fragment, rule-based system	Mutagenicity, carcinogenicity, skin sensitisation and irritancy, immunotoxicity and neurotoxicity	91
Derek (Lhasa Ltd)	Commercial	Knowledge-based expert system	Mutagenicity, carcinogenicity, skin sensitisation and irritancy	78
ADMET Predictor (Simulations Plus Inc.)	Commercial	Modelling based on Neural network	Hepatoxicity, mutagenicity, carcinogenicity, cardiac toxicity, allergenic skin and respiratory sensitisation, phospholipidosis	92
Lazar (Freiburg University)	Freely available	Uses data mining algorithms to derive predictions for untested compounds from experimental training data	Chronic (oral) tox, hepatoxicity, mutagenicity, carcinogenicity	93
Leadscope (Leadscope)	Commercial	Integrate disparate sources of toxicity data, mine historical collections of public and proprietary toxicity data, build predictive models, identify analogs with toxicity data	Hepatoxicity, nephrotoxicity, urinary tract toxicity, neurotoxicity, genetic toxicity, carcinogenicity, reproductive toxicity, cardiac toxicity	94
MCASE/MC4PC (MultiCASE)	Commercial	Identifies molecular fragment with high probability of being associated with a toxic effect	Acute (oral) tox, hepatoxicity, nephrotoxicity, urinary tract toxicity, cytotoxicity, mutagenicity, carcinogenicity, teratogenicity, irritancy, maximum tolerated dose.	56
TOPKAT (Accelrys)	Commercial	Uses cross-validated Quantitative Structural Toxicology Relationships (QSTR) models to assess various measures of toxicity	Mutagenicity, carcinogenicity teratogenicity, lethal dose, skin sensitisation, chronic (oral) toxicity.	53
OncoLogic (US EPA)	Freely available	Knowledge based expert system	Carcinogenicity	77

chemical space. It is usually specific to a single defined endpoint[38] (*i.e.* a specific biological target like hERG) or a defined toxic effect.[39,40] Accordingly, each model has a specific applicability domain which is defined as a calculated region of chemical space of the training data set used to make the model[41] where molecular features of compounds to be predicted are adequately represented. Consequently, toxicity of compounds falling outside the applicability domain cannot be reliably predicted. If the training set is large enough and diverse in molecular

Table 15.2 Publicly available and commercial databases and electronic data sources.

Database/Developer	Availability/access	Description
Acutoxbase ("A-Cute-Tox" EU FP6 project)	Access restricted to project partners https://acubase.amwaw.edu.pl	For ~100 chemicals (50% drugs) available ~100 *in vitro* assays including general acute cytotoxicity, metabolism-mediated toxicity, biokinetics, and organ-specific toxicity; *in vivo*: ~2200 LD_{50} values in rodents (rat and mouse) and other animals (*e.g.* guinea pig, dog) with various administration routes (oral, intravenous, *etc.*) compiled from published literature.
ChemIDplus (US NLM)	Freely available http://chem.sis.nlm.nih.gov/chemidplus/	Structure searchable toxicity data for ~130 000 chemicals retrieved from TOXNET (Toxicology Data Network; http://toxnet.nlm.nih.gov
CEBS (US NIEHS)	Freely available http://cebs.niehs.nih.gov/	*In vivo* data and acute dose on rat for a small number of known hepatoxicants.
RTECS (Symyx Technologies)	Commercial: searchable through the Symyx Toxicity Database http://www.symyx.com/products/databases/bioactivity/rtecs/index.jsp and Leadscope Toxicity Database http://www.leadscope.com/databases/	Structure searchable rat acute oral toxicity (LD_{50}) and acute inhalation toxicity (LC_{50}) data from published literature for ~7000 compounds (~4000 organic).
TerraBase databases	Commercial http://www.terrabase-inc.com/	Several databases include rat and mouse LD_{50} values for different product types (natural compounds, drugs, pesticides). Includes rat or mouse (LD_{50}) and cytotoxicity (IC_{50}) data for 347 compounds from the literature.
ZEBET (BfR ZEBET)	Freely available http://www.dimdi.de	
MRTD Database (US FDA)	Freely available http://www.fda.gov/AboutFDA/CentersOffices/CDER/ucm092199.htm	MRTD (Maximum Recommended Therapeutic Dose) values for 1215 drugs from clinical trials, mostly by oral administration and daily treatments, (3–12 months), 5% of the drugs administered intravenously and/or intramuscularly). Includes structures. Available from FDA and EPA DSSTOX.
DSSTox	http://www.epa.gov/ncct/dsstox/index.html	Distributed Structure-Searchable Toxicity Database Network of downloadable, structure-searchable, standardised chemical structure files associated with toxicity.
RepDose (Fraunhofer Toxicology and Experimental Medicine)	Freely available http://www.fraunhofer-repdose.de/	Sub-acute to chronic, oral and inhalation data for ~700 publicly available chemicals in rat, mouse and dog studies; also including structures, physicochemical properties and study designs.
AERS (US FDA/CDER)	Freely available http://www.fda.gov/Drugs/GuidanceComplianceRegulatoryInformation/Surveillance/AdverseDrugEffects/	US FDA/CDER Adverse Effects Reporting system of post-market safety surveillance for all approved drug and therapeutic biologic products.
CEBS US NIH/NIEHS	Freely available http://cebs.niehs.nih.gov/	Chemical Effects in Biological Systems Knowledgebase; integrates genomic and biological data (including dose–response studies) in toxicology and pathology.
CERES (FDA)	Freely available http:www.fda.gov/	Chemical Evaluation and Risk Estimation System Contain databases on toxicity of food ingredients, drugs, agro, and industrial chemicals, compound profiling, structural alerts, and QSAR-based toxicity predictions.

chemotypes, covering a huge chemical space, the model is termed global, If it is restricted to a set of congeneric compounds, it is termed local.[38-42] Global models are used especially by regulatory authorities that have to assess diverse compounds that may present new structural classes,[43,44] or in drug discovery to broadly evaluate different chemical lead series in order to select the potentially cleanest one. Local models are built to address and hopefully solve toxicity issues in lead optimisation as they are centred on the chemotype(s) of interest and hence are likely to be more accurate and specific. For example, in global hERG models lipophilic and positively charged compounds are in general predicted to be hERG inhibitors with respect to more hydrophilic neutral or negatively charged substances. However, most CNS drugs are lipophilic, carrying an ionisable amine. A useful model is hence the one able to identify, within this unfavourable chemical space, more specific features that hinder hERG interaction in order to reduce or hopefully remove this liability.

The most commonly modelled QSAR endpoints in toxicology are carcinogenicity and genetic toxicity as these findings for a new drug product under development (and its metabolites) can impede regulatory approval. In addition, genotox and carcinogenic potential of a pharmaceutical is a critical regulatory milestone that cannot usually be tested in humans unless using very costly and time consuming assays. In general the aim of statistical commercial modelling software, such as Topkat 53 and Multicase[45,46] (see also Table 15.1), is to analyse available data and automatically build models that, however, have to be carefully reviewed by the scientist. Modelling techniques and structural descriptors also have to be carefully selected. A large number of independent investigations have been widely conducted by experts to assess the commercial and non-commercial *in silico* QSAR/statistical modelling software products in order to determine their performance for predicting carcinogenicity and mutagenicity endpoints. Those based on external validation methods are considered the most scientifically rigorous.[47-56] In general the quality of predictions varies depending on many factors, among the most critical are applicability domain[41,57] and the complexity of the toxicological endpoint (*i.e.* the number of underlying mechanisms of action). Clearly, the more complex an endpoint, the more difficult is the prediction.[58,59] For example, the Ames test (salmonella mutagenicity assay that measures genetic toxicity; one of the mechanisms by which compounds can induce cancer) is a relatively simple assay because it involves a low number of underlying mechanisms, and many models with respectable accuracy exist.[60] By contrast, carcinogenicity, liver toxicity and developmental toxicity are complex endpoints that can be caused by a wide variety of events, and the prediction of these endpoints is usually far from acceptable.[58,61] Hence, although they are useful in a few cases, the broad application of such predictions has been prevented by their lack of accuracy. Clearly from a regulatory review perspective of protecting public health, there is a need for models tuned for sensitivity (*i.e.* accuracy in predicting toxic compounds with low number of false negatives) not specificity (accuracy in predicting non-toxic compounds with low number of false positive). The focus, therefore, is on modelling more simple endpoints, such as off-target activity, to increase accuracy in prediction and hence application usefulness.

15.2.4 Human Knowledge-Based Methods

Among the several approaches to predicting the effects of small molecules from the structure, the human knowledge-based systems, as distinct from statistical-based/machine learning approaches, rely on a set of predefined rules derived from specific examples, widely accepted within the scientific community, to predict toxicity risk. These rules may suggest that the presence of a chemical feature in a molecule or a certain chemical class (*e.g.*, heterocyclic amines, highly reactive groups such as aldehydes) provides a concern about the toxicity risk of a queried chemical.[62-66] Two of the most recognised systems using this modelling approach are the EPA's OncoLogic™ freeware[67] and Lhasa Ltd.'s (non-profit) Derek for Windows

(DFW) program. While the former only evaluates carcinogenicity, DFW has a knowledge base for predicting many toxicological endpoints, including carcinogenicity, mutagenicity, genotoxicity, skin sensitisation and irritancy, teratogenicity, hepatotoxicity, neurotoxicity, and ocular toxicity, and it is fully integrated with a qualitative xenobiotic metabolism prediction program Meteor.[68] One of the problems associated with methods providing such structural alerts is the difficulty to use the identified fragments in medicinal chemistry for drug design purposes. Furthermore, the fragments are sometimes too small to unambiguously map onto a molecule. In addition these structural rules are dependent on the set they are derived from and may, therefore, not be transferable from one chemical series to another. Consequently, although the information contained in these systems is considered reliable enough, these methods suffer from a lack of sensitivity (*i.e.* low accuracy in predicting toxicity issues).[26,56] The direct consequence is that many side effects are likely to be missed.

15.2.5 ADME-Tox Modelling

ADME-Tox modelling may provide a useful indication in order to address metabolic, pharmacokinetic, and toxicological issues related to drug disposition and fate, through *in silico* evaluation of individual or classes of compounds. These methods, such as expert systems and QSAR analysis,[69,70] enable millions of virtual molecules to be examined across a broad range of desired properties in lead identification and optimisation. Some of these methods are based on simple and well-known rules for drug/lead/metabolite likeness, derived from physicochemical properties.[69,71-73] As part of these computational applications, either "off-the-shelf" products or custom built models, there is potential to identify properties of molecules that can be predicted to have "off-target" effects that can potentially have an impact on the human health (*i.e.* hERG). Today the role of computational ADMET in helping to reduce the number of safety issues in the whole discovery phases is well recognised, however, the successful application of these simple approaches are often limited by the paucity of high quality experimental data,[74] and by the use of descriptors neglecting direct structural information about the ADMET proteins and hence of limited impact on drug design, especially in lead optimisation. Accordingly, *in silico* methods based on the 3D structures of relevant proteins might add extra benefits to rule-based and/or statistical 2D methodologies,[75-77] as they provide understanding at a molecular level gained from analysing the ligand and/or protein interactions. So far, structure-based techniques have been rare in computational toxicology due to the lack of structural information for many drug targets, despite the increasing number of x-ray coordinate available in the public databases especially in recent years. In fact, docking experiments relying on target homology models and experimental structure–activity relationships (SARs) to propose binding poses of ligands, can be inaccurate. Furthermore, application of docking-scoring methods to ADMET proteins is a challenging process because they usually have a large and flexible binding cavity; however, promising results relating to metabolising enzymes have been recently reported. An excellent review on structure-based ADMET, including a list of ADMET proteins along with the available 3D structures from the PDB, has recently been published.[75] Finally it is worth mentioning other promising *in silico* ADME-Tox approaches based on quantum mechanical/molecular mechanical (QM/MM) methods, that are used to extract descriptors (*i.e.* HOMO, LUMO) to model the reactivity of compounds towards CYP isoforms and for predicting the rates of reactions in drug metabolism.[78,79]

15.2.6 Application of *In Silico* Tools in Lead Optimisation

Knowledge/rule based high throughput *in silico* "global" tox and structural/physicochemical models are widely applied as filters in the lead identification phase, usually resulting in the selection of developable and potentially clean chemical series. However, it is not guaranteed that these compounds are devoid of toxic effect as these methods have low sensitivity (*i.e.* only know fragments/chemical groups responsible for toxicity well represented in the training set see above).[56] On the other hand, a too restrictive application of these filters may result (especially for poorly tractable targets) in no or too few chemotype compounds left, with the risk of killing valuable leads;[74] moreover, toxicity can build up or hopefully be abolished during lead optimisation,[80] even if in this phase chemical modifications are usually not major. However it is advisable to start a lead optimisation program with multiple, different chemotypes/chemical series, even if some may have some toxicology flags that clearly have to be carefully considered in the optimisation process. Normally little is known about toxicity issues of a new chemical class. Ideally, one would test all new compounds in every available *in vitro* profiling assay to identify potential off-target activities. However, because of cost and practicality aspects, only a few representative compounds (depending on the throughput of the assay) can be tested on a subset of assays, selected according to the knowledge of the on-target relations with the rest of the proteome (in particular with targets associated to adverse events) determined by sequence homology similarity and chemical connectivity (*i.e.* ligands in common). The latter information is provided by chemogenomics databases which contain cross-screening data of collections of compounds tested on a diverse panel of assays spanning the druggable proteome. Interestingly, the two approaches (system biology and chemoproteomics) are not always completely overlapping, rather they provide complementary information. As for the *in vitro* safety pharmacological profiling, only the major potential issues are usually studied (*e.g.* genotoxicity and hERG blockade). ADMET targets are finally added to the panel, among the most common are CYPs interaction, PPB, and P-gp. Once a single liability is identified, computational models are built using the most appropriate approach depending on the available structural information on the target and data on ligands. As soon as a model has been built, any number of structures can be tested *in silico* in a very short time frame. For some of these targets, global models are also commercially available but caution has to be used for their correct application (*i.e.* appropriateness of the applicability domain, see above). Local models, built using in house data, around chemotypes of interest usually provide the best results. Clearly structure-based (docking) and pharmacophore approaches are preferred by medicinal chemists because they provide a better understanding of the recognition processes at a molecular level. "Local" QSAR models (only for congeneric series of compounds) can also be very useful (provided that the chemical descriptors used are simple and meaningful from the physicochemical point of view, rather than a combination of sometimes "exotic" parameters). MLRA (Multi Linear Regression Analysis) and PLS[81] (Partial Least Squares regression), the latter most suitable with a high number of correlated dependent variables, or PLS-DA (Partial Least Squares regression-Discriminant Analysis) when the independent variable is categorical, are among the most used statistical methodologies. Model predictions have to be clearly verified by testing key compounds and, in case of lack of predicitivity, subsequently modified accordingly.

15.3 TARGET SELECTIVITY

The regulatory request to the pharmaceutical industry to improve preclinical safety testing has increased due to recent drug withdrawals.[82] Defining the risk of exposing humans to new drug candidates still depends on preclinical testing which, in many, but not all, cases predicts outcomes in humans accurately. Accordingly, understanding mechanisms of drug toxicity is an essential step toward improving drug safety testing by providing the basis for mechanism-based risk assessments. In a recent paper by Bowes and co-authors[83] it was reported that about of 75% of all adverse reactions produced by novel drugs are dose dependent and they can be predicted on the basis of their pharmacological profile. Drug interaction with off-targets might often be the reason for the adverse reactions produced by a candidate compound when tested in animal or/and clinical studies. Therefore, the precise identification of the off-targets and their deep characterisation are essential steps in modern drug discovery. *In vitro* pharmacological profiling involves the testing of compounds against a panel of targets dissimilar from the intended therapeutic one with the aim of identifying specific interactions responsible for the adverse reactions. A perfect example is the case of fenfluramine. This molecule was originally marketed as an appetite suppressant for weight control and its efficacy was attributed to its agonism to the 5-HT receptor. However, in 1997 fenfluramine was withdrawn from the market since it also induced heart valve disease and pulmonary hypertension. Indeed, it was discovered that whereas the clinically desired effect on appetite was related to the 5-HT2C receptor (primary target), the heart and pulmonary diseases were produced by agonism of the molecule at the 5-HT2B receptor (secondary target or off-target).

Currently, a regulatory guidance describing which targets should constitute the *in vitro* pharmacological selectivity panel does not exist with the exception of the human voltage gated potassium channel subfamily H member 2, also known as hERG, which is now required by the FDA together with other cardiac ion channels.[84]

15.3.1 The Targets Panel for *In Vitro* Selectivity Evaluation

Specialised companies to screen for selectivity of novel chemical candidates exist and all have different types of solutions and different costs by offering small (fewer than 30 targets) and very large (more than 100 targets) target panels. It seems, therefore, that the decision on the panel to be used for off-target characterisation of a novel discovery compound would depend only on economic factors. However, in recent years a new way of thinking has been introduced by pharmaceutical companies with an approach that takes into account if the characterisation of the novel compound takes place in the lead generation/optimisation or in the clinical phase of the drug discovery process. Indeed, in the lead generation and optimisation phases the identification of off-targets has a different consequence with respect to the same request made for a novel candidate ready to be tested in clinical studies. Consequently, the composition of the panel should be different in the two conditions. In addition, the therapeutic area of interest and the primary target family always play an important role in the choice of the target panels to be considered.

Hence, in the lead generation and optimisation phases pharmaceutical companies have now accepted that the screening for potential off-target activity can be carried out by a customised targets panel made of approximately 30–50 targets.[83–85] Its composition may include exemplars of GPCRs, ion channels, enzymes, nuclear receptors and transporters. As pointed out by other authors, the included targets become sentinels for potential targets of liability and the observed positivity would require further evaluation within the target family. The advantage of such an approach in an early discovery phase is that the findings can immediately be transformed by the project using a dedicated SAR strategy to eliminate the liability of the off-target activity. An example of a customised target panel used for projects dealing with CNS indications is listed below:

- Central benzodiazepine receptors (GABA-A),
- Opioid receptors (mu, k, delta),
- Serotonin receptors (5-HT2A, 5-HT2B, 5-HT3),
- NET, DAT, SERT,
- Cannabinoids receptors (CB1, CB2),
- MAO-B,
- Histamine H1 receptor,

- Dopamine D1 and D2 receptors,
- Phosphodiesterase PDE4,
- Neurokinin NK1 receptor,
- Alpha adrenergic receptors (alpha-1 and alpha-2),
- Beta adrenergic receptors (beta 1),
- Adenosine receptors (A2),
- Cycloxygenase COX2,
- Muscarinic receptors (M1 and M2),
- L-type Ca channel,
- Nav1.5 channel,
- hERG channel.

The selected targets can potentially be indicators of drug abuse potential, convulsion, cognitive impairment, emetic and cardiovascular liabilities for compounds interacting with them (Table 15.3).

Two relevant factors still require consideration. First, whichever customised target panel is selected, the target assays should be robust enough for detecting both agonism and antagonism. Second, pharmacological species difference should be kept in mind when developing the assay and translating the results in animal or human studies (*e.g.* a clear pharmacological species difference exists for instance between human and mouse or rat NK1 receptors[86]).

Finally, for a novel candidate ready to be tested in clinical studies there is the general consensus that a large and sometime very large panel of hundreds of potential off-targets is required to widely explore the selectivity of the compound. Projects targeting kinases generally require a large off-targets panel, since more than five hundred different kinases have been identified in humans.

15.3.2 Testing Strategies

Different technical approaches have been adopted for testing compounds on these targets. As discussed in depth by Bowes,[83] binding assays utilise purified preparations of membranes or proteins from recombinant or tissue sources expressing the target and a labelled high affinity ligand that incorporates either a radioisotope or fluorescent probe. Test compounds are assayed for their ability (*e.g.* affinity) to displace the labelled ligand. Binding assays in their general configuration do not indicate if the test compound behaves as an agonist or antagonist.

Functional assays measure activation, inhibition, or modulation of the activity of the target either expressed in a host cell or in a purified preparation. Assays for GPCRs rely on the measurement of secondary messengers like calcium, cAMP of 35-labelled-GTPγS-binding. For ion channels, technologies allowing automated electrophysiological measurement of channel activity when expressed in a host cell are used. Enzymes, including kinases, are assayed by measuring the product formed after their incubation with a specific substrate. Nuclear receptors are usually expressed in cells and tested in combination with gene reporters or by biochemical assays using time resolved fluorescent resonance energy transfer (TR-FRET) technologies.

IC_{50} and K_i values are determined in binding studies (concentration of the test compound that displaces by 50% the labelled ligand; affinity constant of the test compound, respectively). K_i is derived from the IC_{50} value by applying the Cheng–Prussoff equation.[87] In functional studies, EC_{50} values (agonist concentration that produces 50% of the maximal response) or IC_{50} and K_B values (concentration of the antagonist/inhibitor that produces 50% inhibition; antagonist dissociation constant, respectively) are calculated. The K_B parameter can be derived by the use of competitive antagonist affinity from functional inhibition curves using the Gaddum, Schild and Cheng–Prusoff equations.[88]

Table 15.3 Potential hazards produced by compounds acting on the following off-targets panel.

Target	Family	System	Agonism	Antagonism
Adenosine receptors A2	GPCR	CNS / CV	Sedation / Vasodilatation	
Adrenergic α1, α2 receptors	GPCR	CV	Hypertension	Hypotension
Adrenergic β1 receptor	GPCR	CV	Tachycardia	Hypotension
D1 and D2 receptors	GPCR	CNS	Dyskinesia / Hypotension / Hallucination / Psychosis / Increase orgasmic intensity	Extra-pyramidal side effects
Muscarinic M1 and M2 receptors	GPCR	CNS / Peripheral / CV	CNS excitation / Gastrointestinal motility	Agitation / Hallucination / Memory impairment / Tachycardia
Opioid mu, k, delta receptors	GPCR	CNS	Dysphoria / Drug abuse	
H1 receptor	GPCR	CNS		Sedation
5-HT2A, 5-HT2B receptors	GPCR	CV / Peripheral	Vasoconstriction / Heart valve disease / Pulmonary hypertension	
NK1 receptor	GPCR	CNS	Nausea	
Central benzodiazepine receptors (GABA-A)	Ion channel	CNS / Peripheral	Sedation	Muscular spasm seizures, Convulsions
L-type Ca channel	Ion channel	CV		Hypotension
hERG channel	Ion channel	CV		Arrhythmia *Torsade de Pointes* risk
Nav1.5 channel	Ion channel	CV		Arrhythmia Brugada syndrome
5-HT3	Ion channel	CV	Anxiety / Hypotension	
NET, DAT, SERT	Transporters	CNS / CNS		Tachycardia / Drug abuse / Seizures / Serotonin syndrome
MAO-B	Enzymes	CNS		Dyskinesia / Hallucination / Hypotension / Nausea
PDE4	Enzyme	CNS / Peripheral		
COX2	Enzyme	CV	Hypertension / Cardiovascular risks	

A point which generally creates discussion within the discovery project is the choice of the concentration of the compound to be tested against the off-targets panel. In the early profiling performed in the lead generation and optimisation phases it is always recommended that the testing be performed at multiple concentrations to allow a direct and precise estimation of the compound activity (*e.g.* compound affinity, potency). This provides a faster and more reliable result than testing a single concentration. Of course this approach is more expensive if the off-targets list is extensive but the benefit of the approach is the early introduction in the project of a SAR strategy to reduce the potential liability risk of the lead series.

15.3.3   Data Interpretation

Potential leads and candidates should be prioritised based on *in vitro* pharmacological profile, namely the selectivity for the primary target with respect to off-targets (*e.g.* secondary targets). It is generally accepted that the greater the ratio between compound activity for the primary target with respect to the off-target, the better it is for compound safety. However, it is very difficult to assess safety only from *in vitro* data. Complicating factors include species differences in receptor pharmacology, receptor distribution and density, access of the test compound to the active site, compensatory and synergic mechanisms, redundancy and plasticity of response systems. All of these can affect the outcome. As rule of thumb it is suggested to calculate the therapeutic concentrations of the test compound (non-protein bound C_{max}) and if the ratio with the off-target activity measured *in vitro* is greater than 100-fold, it generally represents no safety concern. Leads and candidates displaying lower values should be considered with prudence.

15.4   CELL VIABILITY ASSESSMENT

The determination of cell viability has a fundamental role in all forms of cell culture. In toxicity assays cell viability is the main purpose of the experiment; however, it can also be used to correlate cell behaviour to cell number. One of the most used cell viability assay using single endpoints such is the 3-(4,5-dimethyl-2-thiazolyl)-2,5-diphenyl-2*H*-tetrazolium bromide (MTT) test.

The MTT cytotoxicity assay is a colorimetric method for determining the number of viable cells based on mitochondrial dehydrogenase activity measurement.[89] The MTT tetrazolium compound (a yellow tetrazole) is bioreduced to formazan (coloured) by the intracellular dehydrogenase in living cells. The quantity of formazan that is directly proportional to the number of living cells in culture is measured spectrophotometrically. The assay is performed by adding test compounds at different concentrations to PC-3 cells (or other cell lines) in a 96 well format. The MTT reagent is added directly to the cultured cells and after that an organic solvent is added to dissolve the insoluble purple formazan product into a coloured solution. The absorbance at 490 nm of this solution is quantified by a spectrophotometer. An internal standard is included in each assay for quality control. In Table 15.4 the effect of a series of standard compounds is reported.

As an alternative test to MTT, the Adenosine Triphosphate (ATP) monitoring system based on firefly (*Photinus pyralis*) luciferase can be used.[90] This luminescence assay is the alternative to colorimetric, fluorometric and radioisotopic assays for the quantitative evaluation of proliferation and cytotoxicity of cultured mammalian cells. ATP monitoring can be used to assess the cytocidal, cytostatic and proliferative effects of a wide range of drugs, biological response modifiers and biological compounds. The major advantages are high sensitivity, linearity, simplicity, fast results and the lack of cell harvesting or separation steps. The assay is performed by adding test compounds at different concentrations to PC-3 cells (or other cell lines).

Table 15.4 Compounds when tested in the MTT assay in PC-3 cells. Compound potency, expressed as negative logarithm of the compound concentration producing 50% of reduction of cell viability (pIC_{50}) when tested in the MTT assay in PC-3 cells. The incubation time was 72 h at 37 °C.

Test compound	pIC_{50}
Taxol	8.30
Doxorubicin	7.35
5-Fluoracil	5.15
Cisplatin	4.60

15.5 CARDIAC LIABILITY

Cardiovascular toxicity remains a major cause of concern during preclinical and clinical development of drugs as well as contributing to post-approval withdrawal of marketed drugs.[91] From 1998 to 2008 40 drugs were withdrawn from the US, European, or Asian markets for safety reasons. Most of these drugs are either cardiotoxic, hepatotoxic or neurotoxic.[92] Amongst cardiac related post-approval adverse events reported over the last 40 years, the majority are cardiac arrhythmias. Such data is likely reflecting the increased scrutiny around drug-induced QT interval prolongation and associated arrhythmias over the last decade.[93] Indeed, drug-related *Torsade de Pointe* (TdP), a potentially fatal arrhythmia characterised in the electrocardiogram (ECG) by a prolongation of the QT interval followed by a "twisting" of the waveform around the isoelectric line, accounted for one-third of all drug withdrawals between 1990 and 2006.[94]

There is a need to accurately predict the risk of drug-induced cardiotoxicity in preclinical and early clinical stages in order to avoid progressing drug candidates with a high risk for cardiovascular toxicity into late clinical development and marketing approval phases.

15.5.1 Cardiac Function and Ion Channels

The heart electrical activity determines its ability to beat and pump oxygenated blood successfully, and the overall electrical activity is determined by action potentials (APs) at the level of the cardiomyocyte. The sequential propagation of APs from the sinoatrial node to the ventricular muscle of the heart can be detected at the skin *via* the ECG (Figure 15.1).

The cardiac AP is the electrical change across the membrane of the cardiac myocyte, and results from the sum of the activity of various ion channels. An ion channel is an integral transmembrane protein responsible for the ion's conduction across the cell membrane. One ion channel underlines one ionic current that, together with other ionic currents, contributes to determining the AP. In summary, the AP results from the balance between inward currents that depolarise (Na^+ and Ca^{2+} currents) and outward currents that repolarise (K^+ currents) the cardiac myocytes. In the cardiac AP five phases can be recognised (Figure 15.2):

- Upstroke (phase 0); large sodium current (INa) due to a rapid influx of Na^+ *via* opening of Nav1.5 channels.
- Early repolarisation (phase 1); initiated by closure of Nav1.5 channels and activation of transient outward (Ito) and ultra-rapid (IKur) potassium currents *via* Kv4.2/4.3 and Kv1.5 channels, respectively.
- Plateau (phase 2); sustained by inward Ca^{2+} current (ICa, L) due to L-type Ca^{2+} channels (Cav1.2) and outward delayed rectifier K^+ currents *via* rapid hERG channel (IKr) and in atrial cells *via* ultra-rapid Kv1.5 (IKur).

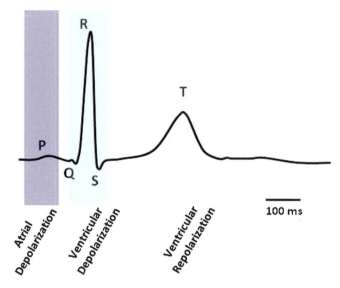

Figure 15.1 Typical human ECG tracing of the cardiac cycle. The P wave, the QRS complex and the T wave are indicated by letters. The baseline of the electrocardiogram corresponds to the isoelectric line. Coloured shaded areas underline the phases of the electrical impulse propagation through the heart.

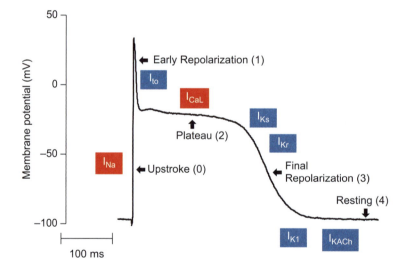

Figure 15.2 Cardiac action potential recorded from a rabbit Purkinje fibre. The five phases are indicated by arrows. Action potentials are generated by the movement of ions through the ion channels present in the cardiac cells and indicated by red (inward currents) and blue (outward currents) boxes.

- Late or final repolarisation (phase 3); closure of Cav1.2 channels and activation of rapid (IKr) and slow outward rectifier (IKs) potassium currents *via* hERG and KCNQ1/mink channels, respectively, which brings the myocyte to the next phase.
- Resting state (phase 4); the resting potential is determined by inward rectifier K^+ current (IK1) *via* Kir2.1 channels and is modulated by acetylcholine-regulated K^+ current (IK, ACh).

15.5.2 Channels With Safety Liabilities

Functional alterations of one or more of the channels contributing to AP lead to an abnormal ECG. Several mutation-associated cardiac diseases, such as atrial fibrillation, Brugada syndrome, cardiac conduction defects, long QT and short QT syndromes, affecting Na^+, K^+ or Ca^{2+} ion currents, have been described. Such cardiac channelopathies may be clinically identified only by the presence of some characteristic ECG abnormalities.[95] Therefore it is evident that drugs that modify the ECG parameters represent a major safety concern for pharmaceutical companies and regulatory agencies. Many compounds interact with cardiac ion channels potentially leading to abnormal propagation of electrical action potentials through the heart tissue. This action is 'designed in' to anti-arrhythmic drugs, but is often an unwanted side effect of non-cardiac drugs.

Historically, for drug development programs that must establish cardiac safety profiles to achieve regulatory approval, the QT interval remains one of the most important ECG parameters. The prolongation of QT interval is the consequence of an increase duration of AP due to a delay in repolarisation that theoretically could arise from an increase in inward (depolarising) current, or alternatively, from a decrease in outward (repolarising) current so that the interaction of the drug with even one ion channel contributing to the AP can result in a cardiac adverse event.

The cellular electrophysiological testing of drugs for cardiac safety tends to involve potassium channels, sodium channels and the L-type calcium channel, because the association of these channels with long QT syndromes is well established. The hERG channel is the most commonly tested ion channel for cardiac liability of new drugs. The reason many drugs are linked to long QT syndrome mediated by hERG is presumed to be due to the structure of the channel which allows the making of promiscuous interactions with many different small molecules.[96] Examples of drugs withdrawn from the market between 1997 and 2002 due to an unacceptable TdP risk are, for example, astemizole and terfenadine (antihistamine drugs), cisapride (gastrokinetic drug), and thioridazine (antipsychotic drug). They were found to be hERG channel blockers.

For many years the only test of drug-induced channel blockade that was viewed as critical for cardiac safety by drug regulators was hERG (Figure 15.3), despite the fact that an ensemble of currents contribute to the AP (and the ECG) and extrapolating the net drug-induced effects on a range of channels based on measuring only one type of current can give a false picture. Some drugs that are not proarrhythmic, such as verapamil (hypertension drug), phenobarbital (anticonvulsant drug) and ranolazine (antianginal drug), show positive results on hERG assay. A clear example of this is verapamil, a potent hERG blocker that does not cause QT prolongation, likely due to its concomitant blockade of the calcium current. Amiodarone is an example of a drug that affects multiple calcium and sodium cardiac currents (with higher sensitivity for late component over the peak component of INa) providing protection from proarrhythmia despite hERG block.[97,98]

In 2005, the International Conference on Harmonisation (ICH) released the guidelines S7B[99] and E14[100] that currently govern cardiac safety landscape. S7B provides a non-clinical testing strategy to evaluate the potential for human pharmaceuticals to affect cardiac electrophysiology, with specific attention focused on ventricular repolarisation. E14 addresses clinical evaluation of QTc (the QT interval "corrected" for the effects of heart rate) prolongation. These guidelines are focused on the hERG (as a surrogate of proarrhythmia) and QTc interval, namely, the measurement of hERG current, the *in vivo* QT assay in an animal model and a trial examining the drug's effects on healthy volunteers' QT interval (a thorough QT study). With evolving nonclinical and clinical data, it is appreciated that the hERG current block is one part of the whole picture related to the proarrhythmic potential of non-antiarrhythmic drugs. S7B and E14 guidelines have been successful since there have not been any withdrawals of marketed drugs for torsadogenic concerns since their adoption. However, they have had the consequence of

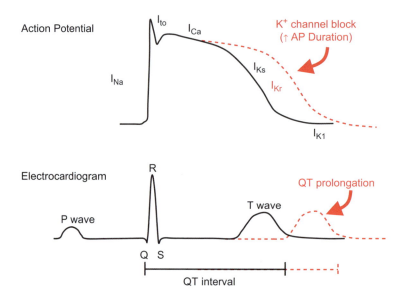

Figure 15.3   Schematic representation of the effect of a hERG channel blocker on the cardiac action potential and on the ECG trace. The potassium channel block has the consequence of increasing the action potential duration and consequently prolonging the QT interval in the ECG recording.

promoting the perception that detection of even a small effect on hERG or mild QTc prolongation will result in adverse regulatory and commercial implications impacting the pharmaceutical drug discovery pipeline by stopping the development of potentially valuable therapeutics.[97]

In 2013 the US Food and Drug Administration (FDA), the Health and Environmental Sciences Institute (HESI) and the Cardiac Safety Research Consortium (CSRC) presented a new paradigm for cardiotoxicity testing,[101] the Comprehensive *in vitro* Proarrhythmia Assay (CiPA). The proposal is based on an established mechanistic understanding of TdP. To assess overall proarrhythmic risk, CiPA relies upon the testing of drugs on multiple human cardiac currents measured in heterologous expression systems, whose electrophysiological effects will then be integrated *in silico* by computer models reconstructing human cellular ventricular electrophysiology and confirmation of effects in a myocyte assay such as human stem-cell derived cardiomyocytes. Evaluation on non-clinical *in vivo* models still remains part of the new paradigm.

A comprehensive *in vitro* set of ion current assays could explore IKr, IKs, IK1, INa (fast and late) and ICaL for drug effects. Potency of current block (IC$_{50}$ values, the concentration of compound that inhibits 50% of the current) would be of key importance and additional blocking mechanisms characterisation (for example voltage-, state-, use- or time-dependency) may be critical.[102]

Human stem-cell derived cardiomyocytes are suggested to be used to provide a cell-based integrated electrophysiological drug response where the set of endogenously expressed channels produce a cardiac AP.[103,104]

15.5.3   Binding *vs.* Functional Assays

The gold standard for studying ion channel function and modulation is the manual patch clamp,[105,106] however it is too slow and too expensive to be adopted in early phases of drug discovery. Alternative approaches with higher throughput and lower costs are binding, fluorescence, or flux assays which represent an indirect readout of ion channel activity and as such are prone to false positives/negatives. The binding assay is not as informative as functional measurements because unless the sites at which test compounds interact with a channel are identical or strongly allosterically coupled to the region where the radiolabel probe binds, potential channel interactions can be missed. In a binding assay there is no possibility to identify compounds that interact with different functional states of the ion channel so binding assay results do not necessarily correlate with functional activity.[107]

Although hERG binding assays (for example [³H]-dofetilide or [³H]-astemizole binding assays) have been used for many years, this approach is conceptually flawed when compared with profiling actives directly on hERG function. Many instances of hERG binding by drug candidates have yielded both false positive and false negative data during testing of novel structural series. False positives can be sorted out by subsequent evaluation in a functional assay. However, false negatives can have disastrous program consequences if the hERG liability is not detected until a potential preclinical candidate is profiled *in vivo*. The time wasted and lack of SAR information on the newly discovered hERG activity usually results in termination of development efforts. The use of hERG binding assays as the only support for medicinal chemistry is too risky since alternatives exist for employing medium- and high-capacity functional hERG assay formats.[107]

Fluorescence assays by using ion sensitive or voltage sensitive fluorescent dyes can detect changes in intracellular concentrations of ions such as calcium or thallium (a permeable ion through potassium channels) or detect changes in membrane potential. Such assays show good correlation with patch clamp results with the risk of false positives in case of fluorescent compounds.

It is clear that assays that directly measure ion channel function are prone to fewer artefacts than those using an indirect measure so that in the last decade automated patch clamp instruments were developed in order to merge the quality of the manual patch clamp technique and the throughput of the other mentioned techniques. As a matter of fact it is now widely applied to reduce the liability of compound series towards the hERG channel as part of safety pharmacology assessment early during medicinal chemistry compound development[108] as well as for the other aforementioned cardiac ion channels.

15.5.4   Integrated Cardiovascular Risk Assessment Strategy

A strategy to progress compounds and discharge cardiovascular risk can be put in place during the lead optimisation phase and in later phases (Figure 15.4).

The identification of early warning signals is suggested by exploiting the knowledge on the target and on the desired mechanism of action of compounds. Early during lead identification the submission of compounds to *in vitro* pharmacological profiling against known cardiovascular targets including cardiac ion channels helps the discharge of cardiac risk and to decrease drug discovery attrition. The testing of compounds in human stem cells-derived cardiomyocytes, where the cardiac set of channels are expressed, allows the evaluation of compound action on cardiac action potentials and to bridge human to preclinical species such as guinea pig and rodents.

The functional assays for ionic currents in recombinant systems, for action potential in human stem cells derived cardiomyocytes and *in vivo* models, provide an integrated screening strategy to discharge cardiovascular risk for novel compounds in the drug discovery and development process.

15.6   DRUG–DRUG INTERACTION

Frequently, patients are given more than one medication and adverse effects have been observed when one drug interacts with a co-administered drug. For the pharmaceutical industry, a significant challenge in early drug discovery is the successful prediction of potential drug–drug

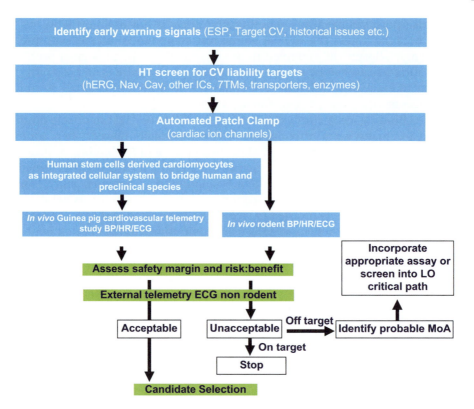

Figure 15.4 Example of an integrated risk assessment strategy. Please note that the high throughput HT screen for liability targets can be expanded outside the cardiovascular field as described in section 15.5. ESP = Early Safety Prediction, CV = Cardio Vascular, HT = High Throughput, BP = Blood Pressure, HR = Heart Rate.

interactions in man (DDI) using *in vitro* CYP450 inhibition data. There have been a number of drugs notably withdrawn from the market because of DDI and this can cause significant ethical and economic problems for the pharmaceutical industry.[109–111] Additionally, it has been reported that the most common cause of deaths in hospitalised patients in the United States is from adverse drug reactions (ADRs) (76 000–137 000 cases a year), making these reactions between the fourth and sixth leading cause of death.[112] It has been found that the incidence of DDI-related ADRs was significant, and most of the events presented important clinical consequences. Because clinicians still have difficulty managing this problem, minimising the factors that increase the risk of DDI-related ADRs is essential.

15.6.1   Drug–Drug Interaction Mechanisms

Of all the human enzymes involved in drug metabolism, CYP450s have been the most extensively characterised and are regarded as the most important in DDI.[113] Cytochrome P450s are mainly concentrated in the liver,[114,115] the major organ for metabolism of oral drugs. Cytochrome P450 (CYP) enzymes are haem-thyolate proteins that are responsible for the oxidative metabolism of a wide variety of xenobiotics (Table 15.5). They comprise a superfamily of related enzymes that are grouped into families and subfamilies based on similarities in amino acid sequences. Currently, there are 18 superfamilies, 32 subfamilies, and 57 isoforms found in human P450s.[116,117] The five major human CYP enzymes responsible for the metabolism of xenobiotics are CYP1A2, CYP2C9, CYP2C19, CYP2D6, and CYP3A4 and account for 70% of all

Table 15.5 Known cytochrome P450 substrates.

Enzyme	Substrates
CYP1A2	Amitriptyline, betaxolol, caffeine, clomipramine, clozapine, chlorpromazine, fluvoxamine, haloperidol, imipramine, olanzapine, ondansetron, propranolol, tacrine, theophylline, verapamil, (R)-warfarin
CYP2A6	Coumarin, betadiene, nicotine
CYP2C9	Amitriptyline, diclofenac, demadex, fluoxetine, ibuprofen, losartan, naproxen, phenytoin, piroxicam, tolbutamide, (S)-warfarin
CYP2C19	Amitriptyline, citalopram, clomipramine, diazepam, imipramine, omeprazole
CYP2D6	Amitriptyline, betaxolol, clomipramine, codeine, clozapine, desipramine, fluoxetine, haloperidol, imipramine, methadone, metoclopramide, metoprolol, nortriptyline, olanzapine, ondansetron, paroxetine, propranolol, risperidone, sertraline, timolol, venlafaxine
CYP2E1	Acetaminophen, caffeine, chlorzoxazone, dextromethorphan, ethanol, theophylline, venlafaxine
CYP3A4/5	Alprazolam, amiodaron, amitriptyline, astemizole, bupropion, buspirone, caffeine, carbamazepine, cerivastatin, cisapride, clarithromycin, clomipramine, codeine, cyclosporine, dexamethasone, dextromethorphan, DHEA, diazepam, diltiazem, donepezil, doxycycline, erythromycin, estradiol, felodipine, fluoxetine, imipramine, lansoprazole, lidocaine, loratadine, lovastatin, midazolam, nicardipine, nifedipine, omeprazole, orphenadrine, paroxetine, progesterone, quinidine, rifampin, sertraline, sibutramine, sildenafil, simvastatin, tacrolimus, tamoxifen, terfenadine, testosterone, theophylline, verapami, vinblastine, (R)-warfarin

Table 15.6 Known inhibitors and inducers of CYP isozymes.

Enzyme	Inducers	Inhibitors
CYP1A2	Cigarette smoke, phenobarbital, ritonavir, carbamazepine, charbroiled foods, vegetables, omeprazole	Enoxacin, ciprofloxacin, grepafloxacin, fluvoxamine, fluoxetine, nefazodone
CYP2A6	Barbiturates	
CYP2C9	Rifampin, carbamazepine, ethanol, phenytoin	Amiodarone, fluvastatin, fluvoxamine, fluoxetine, fluconazole, miconazole, metronidazole, ritonavir, sulfamethoxazole
CYP2C19	Rifampin	Fluvoxamine, fluoxetine, ticlopidine, ritonavir
CYP2D6	Pregnancy	Quinidine, fluoxetine, paroxetine, sertraline, thioridazine, cimetidine, diphenhydramine, haloperidol, ticlopidine (ticlid), ritonavir
CYP2E1	Ethanol, isoniazid, ritonavir	Cimetidine, watercress
CYP3A4/5	Carbamazepine, dexamethasone, rifapentine, prednisone, growth hormone, rifampin, phenobarbital, phenytoin, troglitazone	Ketoconazole, itraconazole, erythromycin, grapefruit juice, fluvoxamine, fluoxetine, diltiazem, verapamil, clarithromycin, omeprazole, ritonavir, indinavir

drug clearance.[118,119] It is estimated that these five CYP enzymes are responsible for approximately 99% of CYP-mediated drug metabolism.[120,121]

Major mechanisms of DDI are:

1. A compound may affect the effective plasma concentrations of other drugs taken concomitantly with the compound of interest, by inhibiting the metabolism of a co-administered drug (Table 15.6). The affected drug thereby might have plasma concentrations higher than intended, leading to toxicity. A well-known case of inhibitory drug–drug interactions is the inhibition of the metabolism of terfenadine by the antifungal

drug ketoconazole, a well-known potent inhibitor of CYP3A4, the P450 isoform responsible for metabolism of terfenadine. A number of patients developed fatal cardiac arrhythmia after they were administered both terfenadine and ketoconazole owing to the elevated level of terfenadine.

2. If the parent drug is a CYP inducer (Table 15.6), it may increase the clearance rate of concomitantly administered drugs, which are metabolised by these CYPs. This may result in a decrease in the effective plasma concentrations of these drugs, thus decreasing their pharmacologic effects. Rifampin–birth control pill interaction is an example of this type of drug–drug interactions. Rifampin, the bacteriocidal antibiotic, is now known to induce CYP3A4 and oestrogen sulfotransferase, the two major pathways of the metabolism of birth control pill ingredients. Women of child-bearing age on birth control pills experienced pregnancy owing to the induction of the drug-metabolising enzymes for the active ingredients of the birth control pills.

3. Metabolites formed *via* CYP metabolism may be responsible for undesirable side-effects, such as organ toxicity. Formation of reactive metabolic intermediates is one of the causes of drug toxicity. Oxidation to electrophilic intermediates or reduction to nucleophilic radicals that can attack DNA or RNA and induce carcinogenicity are two major reactions by which toxicity is exerted. Although many leads are abandoned early on in drug discovery stage due to toxic metabolite formation, presence of a toxic metabolite does not always imply toxicity in a given drug candidate since there are other factors that can make the metabolite toxic or non-toxic. Presence of a toxic metabolite however raises a red flag, which must be extensively examined in animal toxicity studies.

Several of the drug metabolising enzymes, for example the CYP2 family, are polymorphic (having more than one variant of the gene). Although the CYP isozymes generally have similar functional properties, each one is different and has a distinct role (Table 15.7). This polymorphism forms a basis for inter-individual differences in the efficacy of drug treatment, side effects of drugs and the toxic and carcinogenic action of xenobiotics. There is a wide variation in the expression, activity and concentrations of different isozymes among individuals, species and ethnic groups. The expression or the activity of these enzymes is influenced by factors such as species specificity, genetic polymorphism, gender-hormonal control, age, disease and environmental inducers (caffeine, cigarette smoke).

The variability associated with the CYP450 enzymes in each individual results in marked differences in response when the same drug and dose is administered to different individuals. Genetic polymorphism of CYP450 enzymes characterises the general population into three groups:

A. Extensive metabolizers (EM): normal population.
B. Poor metabolisers (PM): Individuals who inherit two inactive alleles (alternative forms of the gene) showing complete absence of enzyme activity.
C. Ultra extensive metabolisers (UEM): Individuals with one common allele and one amplified allele showing enhanced enzyme expression.

Ultra extensive metabolism can cause therapeutic failure due to reduced bioavailability or lack of activation of the drug whereas poor metabolism can lead to drug toxicity and sometimes death. For optimal drug therapy, the prescribing physician should have the knowledge of the genetic makeup of the CYP enzymes in the patient.

The determination of the CYP enzymes responsible for the metabolism of new chemical entities (NCEs) and the identification of interactions with a specific CYP isozyme (*e.g.* inhibition or induction of that isozyme) can aid in predicting clinical drug interactions and *in vitro* methods are commonly used to determine the CYP interaction potential of NCEs.

Table 15.7 Examples of human polymorphic CYPs.

Enzyme	Major Variant Alleles	Mutation	Consequence	% occurrence Caucasians	% occurrence Asian
CYP2A6	CYP2A6*2	L160H	Inactive enzyme	1–3	0
	CYP2A6*3	2A6/2A7 conversions	Not known	0	0
	CYP2A6*4	Gene deletion	No enzyme	1	15
	CYP2A6*5	G479L	Defect enzyme	0	1
CYP2C9	CYP2C9*2	R144C	Reduce affinity for P450 reductase	8–13	0
	CYP2C9*3	I359L	Altered substrate specificity	7–9	2–3
CYP2C19	CYP2C19*2	Aberrant splice site	Inactive enzyme	13	23–32
	CYP2C19*3	Premature stop codon	Inactive enzyme	0	6–10
CYP2D6	CYP2D6*2xn	Gene duplication/ multiduplication	Increased enzyme activity	1–5	0–2
	CYP2D6*4	Defective splicing	Inactive enzyme	12–21	1
	CYP2D6*5	Gene deletion	No enzyme	4–6	6
	CYP2D6*10	P34S, S486T	Unstable enzyme	1–2	50
	CYP2D6*17	T107I, R296C, S486T	Reduce affinity for substrates	0	n.d.
CYP2E1	CYP2E1*2	R76H	Less enzyme expressed	0	1
	CYP2E1*3	V389I	No effects	<1	0
	CYP2E1*4	V179I	No effects	<1	n.d.
CYP34A4	CYP3A4*2	S222P	Higher K_m for substrates	3	0
	CYP3A4*3	M445T	unknown	0	<1

n.d.: not determined (has a very high frequency among Black Africans and African Americans).

15.6.2 CYP Driven DDI Test Systems

In vitro studies are best suited to determining the types of clinical trials needed to assess potential DDIs. FDA guidelines suggest first using *in vitro* studies to assess the effect of drugs on metabolic pathways and if the results indicate possible DDIs, to follow up with *in vivo* assays. Well established protocols are commonly used to determine the CYP driven DDI potential of NCEs and there is a universally accepted strategy to use *in vitro* human liver-based experimental systems.[122,123]

The most widely used systems are discussed below in detail.

15.6.2.1 Expressed Enzymes

Advances in molecular biology have enabled the identification and characterisation of a large number of individual CYP genes. Specific cDNA sequences for particular CYP isozymes have been cloned and expressed heterologously. These enzymes, which can be produced in large amounts to meet the increasing demand of screening assays for drug metabolism research, show catalytic properties comparable to those of HLMs. Resulting recombinant human CYPs are isolated in microsomal forms and are commercially available for CYP phenotyping, metabolic stability screening and inhibitory potential evaluation. Moreover, heterologously engineered cells expressing individual CYPs can be used for metabolism or toxicity studies. Major drawbacks inherent to recombinant models are that concentrations of CYPs are far in excess of their relative amount in the human liver, and that the secondary metabolism cannot be identified.

15.6.2.2 Liver Microsomes

Microsomes can be prepared easily from frozen liver tissues. They contain most of the oxidative drug metabolising enzymes. Their easy preparation and good long-term stability at −80 °C make microsomes the most frequently used *in vitro* system in drug metabolism studies. Microsomes are isolated from liver cells by disrupting the cellular contents and centrifugation at 100 000×g. The ability to phenotype microsomes greatly increases the utility of this system in the identification of specific isozymes responsible. Metabolic information such as metabolic profiles, stability, metabolite identification and kinetics can be obtained from microsomal systems. Microsomal incubations are most often used to obtain information on Phase I reactions. One disadvantage is that the information is not complete as from the cellular systems.

15.6.2.3 Isolated Hepatocytes

Cell cultures or cell suspensions can be used to study multiple aspects of drug metabolism, drug transport across cell membranes, cytotoxicity and enzyme induction in an environment where enzymes and co-factors are present in normal physiological concentrations and cellular integrity is maintained. Hepatocytes are used to study both Phase I and Phase II reactions. Cells can be either primary or permanent cell cultures. Primary cell lines are most often used for drug metabolism studies because permanent cell lines possess very little or no enzyme activity. Primary cells are isolated from fresh liver tissue and can be used immediately after isolation or culture for long-term studies. With the increased availability of fresh human tissues from various commercial and non-profit institutions, human hepatocytes have become the most widely used and preferred *in vitro* system.

15.6.2.4 Tissue Slices

Tissue slices have certain advantages over other systems. With intact cell–cell junctions, normal hepatic cellular architecture is retained in the tissue. Since they contain the complete complement of drug metabolising enzymes with all the cofactors present in relevant concentrations, complete information on the metabolism reactions can be obtained. Liver slices can be easily and rapidly produced. In addition, liver slices are not exposed to proteolytic enzymes that can destroy important membrane receptors of the cell. Although liver slices are increasingly used now in drug metabolism studies, they have certain disadvantages. One drawback is the inadequate penetration of the medium. Liver slices cannot be cryopreserved and they have a limited useful experimental period.

15.6.3 Drug-Metabolising Enzyme Inhibition

Enzyme inhibitors function in different ways. The competitive inhibitors compete with the substrate for the active site, *e.g.* fluvoxamine and caffeine for CYP1A2. The non-competitive inhibitors bind to the enzyme–substrate complex or to the haem group, *e.g.*, ketoconazole. The third type, irreversible inhibitors, inactivate the enzyme either by haem binding or protein binding.

  Primary *in vitro* metabolic systems used in drug metabolism, to evaluate whether the drug in question would inhibit drug-metabolising enzymes, involve hepatic enzymes or tissue preparations. The drug to be studied is incubated with recombinant enzymes, liver microsomes or hepatocytes in the presence of substrates of specific drug-metabolising enzymes. The metabolic rate of these substrates in the presence and absence of the drug would allow one to estimate its inhibitory potential. Results are in general expressed as IC_{50}, a concentration leading to a 50% decrease in activity, or K_i, the inhibitory constant. Low IC_{50} or K_i values as compared with the intended plasma concentration would suggest a high potential of the drug to cause drug interactions with co-administered drugs, which are substrates of the affected enzyme.

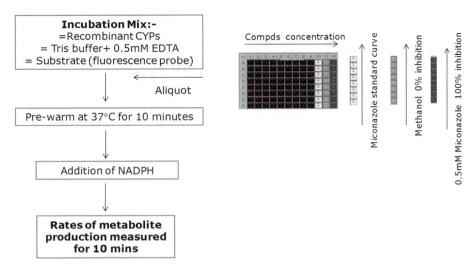

Figure 15.5 Incubation scheme for *in vitro* CYP inhibition fluorescence assay.

In early drug discovery, *in vitro* CYP inhibition fluorescence assays[124] that are more amenable to high throughput are used widely in the pharmaceutical industry for initial screening purposes (Figure 15.5).

Such assays provide critical information for evaluating, ranking and/or eliminating compounds at the earliest stage of drug development. Validation of *Escherichia coli* expressed CYP450 enzymes as surrogates for their counterparts in pooled human liver microsomes[125] and also the comparison of the kinetic properties of the *E. coli* expressed enzymes with liver microsomes with appropriate substrates have been reported.[126] These studies have compared fluorescence-based CYP450 inhibition data generated using CYP450 enzymes (1A2, 2C9, 2C19, 2D6 and 3A4) expressed in *E. coli* (Cypex bactosomes) or human lymphoblastoid cells (Gentest cDNA microsomes), and have analysed the prediction of these data sets toward DDIs using published clinical data on 68 marketed drugs.

Human liver microsomes are the most widely used *in vitro* metabolic models to assess the inhibition potential of drug candidates and, with the increased in number of NCEs due to combinatorial chemistry and high-throughput screening techniques, novel high-throughput liquid chromatography-mass spectrometry (LC-MS)[127,128] assays have been developed. A sensitive and rugged LC-MS/MS method was developed for a comprehensive *in vitro* metabolic interaction cocktail screening.[129] A cocktail, consisting of ten cytochrome P450 (CYP)-selective probe substrates with known kinetic, metabolic and interaction properties *in vivo*, was incubated in a pool of human liver microsomes and then 13 metabolites and internal standard phenacetin were analysed in multiple reaction mode. Recently,[130] by integrating a RapidFire system with an API4000 mass spectrometer (RF-MS), a cytochrome P450 inhibition assay was developed and fully optimised on the system. Compared with the classic liquid chromatography-mass spectrometry method, the RF-MS system generates consistent data with an approximately 20-fold increase in throughput. The lack of chromatographic separation of compounds, substrates, and metabolites can complicate data interpretation, but this occurs in a small number of cases that are readily identifiable. Overall, this system has enabled a real-time and quantitative measurement of a large number of ADME samples, providing a rapid evaluation of clinically important drug–drug interaction potential.

When a drug converted to a reactive metabolite(s) by P450 enzymes interacts with them irreversibly and inactivates their function, this is called metabolism dependent inhibition (MDI).

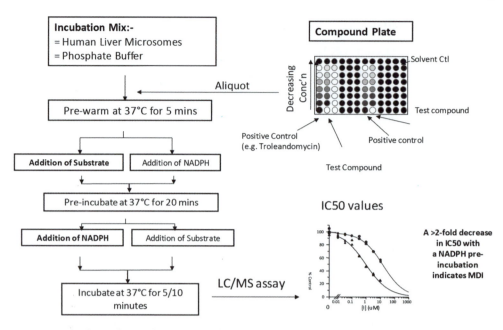

Figure 15.6 Incubation scheme for "IC$_{50}$ shift" experiment for MDI assessment.

Metabolic drug–drug interactions resulting from MDI can display a delayed onset due to the time-dependence in inhibition and can persist even after the inhibitor has been eliminated as enzymatic activity is only restored by *de novo* protein synthesis. Of those drugs currently on the market from various clinical categories that are associated with metabolism-dependent inhibition, most suffer from DDI.[131–133] Indeed, drugs such as astemizole, cerivastatin, cisapride, mibefradil, nefazodone and terfenadine have been withdrawn from the market because of P450 related DDIs.[134] All of these drugs inhibit CYP3A4, with the exception of cerivastatin, which inhibits CYP2C8. Given the importance of CYP inhibition to the developability and safety of new medicines, many pharmaceutical companies have implemented high-throughput assays to evaluate metabolism dependent inhibition (MDI) early in the drug discovery process, in order to enhance the quality of development candidates and reduce attrition.[135–138] In general in drug discovery studies there are three different analytical approaches that are used for the study of P450 inhibition *in vitro*: liquid scintillation counting of radioactivity liberated during site-specific metabolism,[139] selective analysis of fluorescent metabolism,[140–142] and mass spectrometry.[143–145] Traditionally, the extent of inactivation of the test enzyme is determined from a decrease in the IC$_{50}$ after pre-incubation with the test compound. In an "IC$_{50}$ shift" experiment the IC$_{50}$ is determined for a CYP marker activity before and after the test compound has been incubated with enzyme and NADPH for a set pre-incubation time. Generally, IC$_{50}$ shift experiments can be conducted by running the pre-incubation at a higher concentration and then diluting into the activity assessment incubation, or conducting the incubation by adding the probe substrate with no dilution step (Figure 15.6).

15.6.4 Pathway Identification

Understanding which drug metabolism pathways are involved in the metabolism of a drug enables prediction of potential clinical drug–drug interactions (DDIs), as well as prediction of potential variability in patient pharmacokinetics or pharmacodynamics due to genetic polymorphism or different expression levels. Some well-documented examples include the CYP3A4-catalysed oxidation of terfenadine to carboxyterfenadine,[146] the CYP2D6-catalysed *O*-demethylation of codeine to morphine,[147] CYP2C19-catalysed oxidation of proguanil to cycloguanil,[148,149] and the CYP2C9-catalysed oxidation of losartan to carboxylosartan.[150] In recent years, *in vitro* methods for determining which forms of CYP are involved in the metabolism of a given drug, so-called "CYP reaction phenotyping", have become firmly established. Reaction phenotyping can be generally described as a set of *in vitro* experiments with the overall goal of identifying the pathways responsible for the clearance of a new chemical entity (NCE) and, more specifically, to provide information that aids in characterising an NCE as a victim of drug–drug interactions. Generally, pharmaceutical companies are looking for drugs with multiple routes of elimination, in fact, if any one route is impaired, the others can compensate.

In addition to CYP mediated metabolism, several other enzymes, the most significant of which are the flavin-containing monooxygenases (FMO), can mediate the oxidative metabolism of various drugs and xenobiotics.[151] Less extensively studied but potentially just as significant are polymorphisms and DDI involving other drug metabolism routes, such as UDP-glucuronosyltransferase (UGT) mediated glucuronidation pathways.[152–154]

Currently, drug discovery capabilities in predicting clinical DDI for a drug candidate are very adequate where the candidate drug is the perpetrator of DDIs (*e.g.* inhibitor of a drug metabolising enzyme like CYP3A4); while quantitative predictions of risk of DDI as victims are usually conducted relatively late in the drug development process. This is because definitive phenotyping of a given reaction requires a thorough understanding of metabolic pathways and radiolabelled drug or synthetic metabolite standards are usually not available. However, it is often desirable to "screen" for desirable properties (*e.g.* multiple clearance pathways to avoid higher risks of DDI) in drug discovery and lead optimisation. Because radiolabelled material and authentic metabolite standards are not readily available in discovery, the approach of relative quantitation by assessing "substrate depletion" is typical at this stage. In a substrate depletion experiment, the amount of metabolism is assessed after various time points (*e.g.* 0, 5, 10, 20, 30, 45 and 60 minutes) by measuring how much substrate is remaining relative to that found at the initiation of the incubation. Since only the substrate is quantified, the amount of metabolism is equivalent to the sum of all metabolites formed. The rate of disappearance of parent or the rate of appearance of the metabolite(s), if known, is then monitored by appropriate analytical tools, frequently LC-MS/MS. By comparing the rates of metabolism across cell lines, one can then identify which enzymes are capable of metabolising the lead compound.

Compared to measuring the formation of metabolites, this approach is prone to potential underestimation of the determined parameters, especially for compounds with limited clearance. However, it has been shown that data generated from determination of parent disappearance *in vitro*, once processed with an extrapolation tool like SimCyp, can generate an acceptable prediction of likely *in vivo* pharmacokinetic (about 85% success in the prediction).[155]

In drug discovery, perhaps the most common approach to reaction phenotyping is the use of cDNA expressed enzyme systems, each expressing a different enzyme, to evaluate metabolism of lead compounds. For oxidative studies, the test compound is incubated with a panel of individually expressed recombinant human CYP enzymes (CYP1A2, CYP2A6, CYP2B6, CYP2C8, CYP2C9, CYP2C19, CYP2D6, CYP2E1, CYP3A4, and CYP3A5 in a typical panel) expressed in baculovirus-infected insect cell membranes. The incubation mixture typically contains the test compound at a final concentration of 1 μM or 10 μM. Ideally, pilot studies should be conducted to establish the concentration range for the test compound to ensure that substrate concentrations are in the linear range (well below the K_m of the targeted enzyme). In practice, however, the evaluations are more commonly conducted in a screening format with two or three initial starting concentrations of test compound.

Various methods for scaling the results to predict *in vivo* drug clearance have been described, but for screening purposes, the usual approach is to use the following equation to calculate intrinsic clearances (CL_{int}) for each of the individually expressed enzymes based on the estimated turnover rate constants (Equation (15.1)).

$$CLint\ CYPx = \left(\frac{k}{CYPx}\right) \times \text{pmol of CYPx} \times \text{mg microsomal protein} \qquad (15.1)$$

Where CYPx is the concentration of recombinant CYP enzyme in the incubation.

The contribution of an individual CYP enzyme to the overall oxidative metabolism of a drug candidate can be estimated as follows (Equation 15.2):

$$\% \ Contribution\ by\ enzyme_x = \frac{CLint\ x\ (\mu L/min/pmolrCYPx)\ x\ Abundance\ x\ \left(\dfrac{pmol}{mg\ protein}\right)}{\sum CLint\ i\ (\mu L/min/pmolrCYPi)\ x\ Abundance\ i\ \left(\dfrac{pmol}{mg\ protein}\right)} \qquad (15.2)$$

A second approach to reaction phenotyping involves conducting incubations in hepatocytes, microsomes, or some other *in vitro* preparation using normal tissues as the enzyme source, but including selective chemical or immuno-inhibitors of specific enzymatic pathways. By performing a battery of incubations with various inhibitors (Table 15.8), and comparing the relative rates of metabolism, one can identify which inhibitor reduces the overall metabolism to the greatest extent and thereby uncover the metabolic pathway that contributes the most to the clearance of a compound.

The extent of inhibition is determined by comparing the extent of metabolism observed in the presence of inhibitors with the extent of metabolism observed in comparable incubations without inhibitors. This approach is generally favoured for experiments intended to deconvolute the contributions from various enzymes when multiple pathways are involved in metabolism of a compound. For compounds exhibiting low turnover, the use of specific inhibitors presents an additional challenge since it may be very difficult to detect a decrease in an already very low metabolic rate. In such cases, monitoring for appearance of metabolite(s) is an option if the metabolic products are known.

15.6.5   Drug-Metabolising Enzyme Induction

Several drug-metabolising enzymes, for example CYP1A2, CYP2A6, CYP2B6, CYP2C9 and CYP3A4, can be induced. Enzyme induction is a relatively slow adaptive process that increases metabolic capacity, unlike direct enzyme inhibition which is an immediate effect that decreases metabolic capacity. A drug-induced change in the expression of CYP genes is a key cause of issues in pre-clinical development and clinical drug–drug interactions (DDIs). In particular, CYP

Table 15.8 Inhibitors used for microsomal reaction phenotyping incubations.

CYP	Inhibitor
CYP1A1/2	Furafylline
CYP2C9	Sulfaphenazole
CYP2C19	3-Benzylnirvanol
CYP2D6	Quinidine
CYP3A4	Ketoconazole, Troleandomycin

induction can impact pre-clinical safety studies by reducing the exposure of the drug candidate, thus limiting the achievement of the exposure multiples expected by the regulatory agencies for the approval of a drug. P450 induction may increase the pharmacological effect of a drug. For example, upon co-administration of phenobarbital with alprenolol, the oral bioavailability of alprenolol is reduced by 45%. Despite the marked change in bioavailability, there is only a modest reduction in the pharmacological response. This observed effect can be attributed to β-adrenergic receptor activity of an alprenolol metabolite, 4-hydroxylprenolol, which is as potent as the parent. P450 inducers can indirectly cause cellular toxicity by increasing the rate of formation of toxicologically active products, such as stable metabolites or reactive intermediates. For example, CYP2E1 is responsible for the formation of *N*-acetyl-*p*-benzoquinoneimine (NAPQI), a toxic metabolite of paracetamol. Long-term ethanol consumption increases paracetamol hepatotoxicity, whereas acute or simultaneous ethanol and paracetamol consumption has limited or negligible toxicological effects. The time interval between ethanol consumption and paracetamol ingestion is key in terms of the observed toxicological consequences as ethanol is both an inducer and inhibitor of CYP2E1.

The major mechanism for CYP enzyme induction is *via* increased rates of transcription, which is mediated predominantly by the intracellular 'nuclear hormone receptors' Ah, CAR, PXR and PPAR.

The mechanism of induction of CYP1A is based on increased protein synthesis, primarily initiated by the binding of the inducer to the Ah (aryl or aromatic hydrocarbon) cytosolic receptor (AhR). The activation of the constitutive androstane receptor (CAR) by translocation from the cytosol into the nucleus mediates induction of CYP2B genes by phenobarbital (PB)-type inducers. The regulation of the CYP3A family of enzymes is complex, with marked species differences in the capacity of compounds to activate PXR and induce CYP3A in the liver (Figure 15.7). In addition, PXR agonists may induce intestinal CYP3A and the transcription of various genes for transporters.

Quantification of CYP induction in pre-clinical species is tested systematically as part of the toxicological safety evaluation and can be measured by ligand binding, reporter gene assays, mRNA levels (TaqMan®), protein or catalytic end points (Figure 15.8).[156] These assays involve the culturing of human hepatocytes in the presence of the drug in question at a range of concentrations, in parallel with the known positive control compounds, such as rifampicin for CYP3A4, CYP2C9 and CYP2C19, Phenobarbital specific for CYP2B6 and omeprazole for CYP1A2. Induction potential can be assessed by measuring the enzyme activity of model substrates

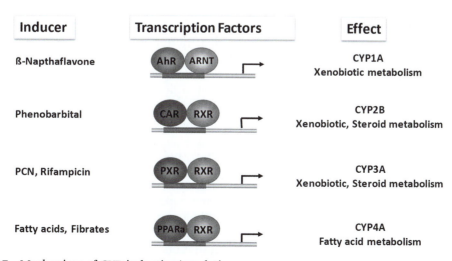

Figure 15.7 Mechanism of CYP induction/regulation.

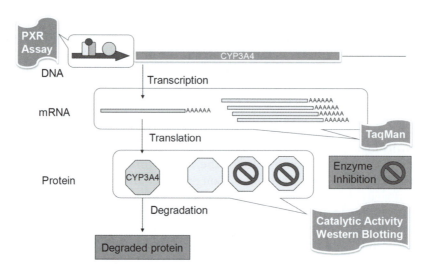

Figure 15.8   Steps of regulation and type of measurements.

determined by HPLC or LC-MS/MS, or by determining the protein or messenger RNA levels of drug-metabolising enzymes, the latter using real-time quantitative reverse transcriptase polymerase chain reaction (qRT-PCR), commonly referred to as TaqMan®.

As these assays require gene and protein expression, human hepatocyte primary cultures represent the most appropriate experimental system. Cultured primary human hepatocytes represent the only proven *in vitro* experimental system for predicting the induction potential of metabolic enzymes and are considered by the FDA as the more realistic system for *in vitro* studies on P450 induction.

Several reporter cell lines[157,158] have also been developed as rapid screening assays allowing P450 induction to be evaluated early in drug development. It is generally believed that the results with these screening assays require to be confirmed by that with primary human hepatocytes.

15.7   TRANSPORTER-MEDIATED DRUG INTERACTIONS

In addition to the effects of drug metabolising enzymes on the pharmacokinetics of drugs, increasing attention is being given to transporters where emerging evidence indicates their important role in modulating drug absorption, distribution, metabolism, and elimination as well as the historical importance of transporters in the development of drug resistant tumours. Transporters, acting alone or in concert with drug metabolising enzymes, can affect the pharmacokinetics and/or pharmacodynamics of a drug and drug–drug interactions involving drug transporters and genetic polymorphisms of drug transporters have been described.[159,160]

Accordingly, methods allowing the rational prediction and extrapolation of *in vivo* drug disposition from *in vitro* data are also essential.

Many different drug transporters are expressed in various tissues, such as the epithelial cells of the intestine and kidney, hepatocytes, and brain capillary endothelial cells.[161–163] In recent years, a number of important transporters have been cloned, and considerable progress has been made in understanding the molecular characteristics of individual transporters. It has now become clear that some of these are responsible for drug transport in various tissues and they may be key determinants of the pharmacokinetic characteristics of a drug as far as its intestinal absorption, tissue distribution, and elimination are concerned.

15.7.1 Most Relevant Transporters for DDIs

Various transporters expressed in the small intestine are involved in the absorption of nutrients or endogenous compounds. Influx transporters expressed in gut, such as oligopeptide transporters (PEPT1), apical sodium-dependent bile acid transporters (ASBT) or organic anion transporting polypeptide-B (OATP), improve drug absorption;[164,165] while primary active efflux transporters, such as P-glycoprotein (P-gp) encoded by multidrug resistance gene (MDR1), multidrug resistance associated protein 2 (MRP2) or the breast cancer resistance protein (BCRP), are expressed on the brush border membrane of enterocytes and excrete their substrates into the lumen, resulting in a potential limitation of net absorption. P-gp contributes to the absorption of many drugs because of its broad substrate specificity.[166] Multispecific transporters are expressed in the liver and kidney and play an important role in the elimination of many xenobiotics, acting as a detoxification system. Many drugs are excreted into the urine *via* organic anion and cation transport systems, expressed on brush border and basolateral membranes of renal tubular cells (PEPT2, OATs, OCT2, OCTNs).[167–169] As far as the liver is concerned, a wide variety of transporter families are known to be present at the sinusoidal and canalicular membranes and play a significant role in hepatobiliary excretion (Figure 15.9).[170–172]

Active transporters expressed on the sinusoidal membrane are responsible for the uptake of drugs from the blood into hepatocytes.[173] Primary active transporters (MDR1, MRP2, BSEP, BCRP) expressed on the canalicular membrane are involved in the biliary excretion of both parent drugs and their metabolites.[174,175]

Due to the broad substrate specificity of drug transporters, drug–drug interactions involving these transporters are very likely. Recently, both inhibition and induction of transporters have been implicated as one mechanism responsible for certain drug–drug interactions. P-gp inhibitors, such as quinidine and verapamil, are known to increase plasma concentrations of digoxin because they block biliary and/or urinary excretion of digoxin *via* P-gp.[176,177] The hepatobiliary

☐ OAT/Oat = organic anion transporter belonging to the SLC22 family

☐ OCT/Oct = organic cation transporter belonging to the SLC22 family

☐ NTCP/Ntcp = Na+/taurocholate cotransporting polypeptide belonging to the SLC10 family

☐ OATP/Oatp = organic anion transporting polypeptide belonging to the SLCO family (previously called SLC21)

☐ MDR1/Mdr1 = multidrug resistance protein (ABCB1)

☐ BCRP/Bcrp = breast cancer resistance protein (ABCG2)

☐ MRP2/Mrp2 = multidrug resistance-associated protein (ABCC2)

☐ BSEP/Bsep = bile salt export pump (ABCB11) all belonging to ABC-carrier proteins;

☐ *OA– = organic anion*
☐ *BS– = bile salt*
☐ *OC+ = organic cation*
☐ *DC– = dicarboxylate*
☐ *GSH = glutathione*

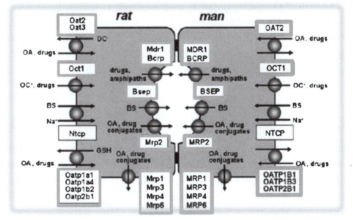

Figure 15.9 Individual carriers in the hepatocyte which are involved in drug uptake and secretion.

Table 15.9 Examples of transporter-mediated drug interactions.

Interacting Drug	Affected Drug	Fold Changes in Substrate Plasma AUC	Consequence
Quinidine	Digoxin	Digoxin exposure 1.7-fold increase	P-glycoprotein (P-gp, MDR1) inhibition
Rifampin	Digoxin	Digoxin exposure 30% decrease	P-gp induction
Dronedarone	Digoxin	Digoxin exposure 2.6-fold increase	P-gp inhibition
Probenecid	Cephradine	Cephradine exposure 3.6-fold increase	Organic Anion Transporter (OAT) inhibition
Cimetidine	Metformin	Metformin exposure 1.4-fold increase	Organic Cation Transporter (OCT) inhibition
Cyclosporine	Rosuvastatin	Rosuvastatin exposure 7-fold increase	Organic Anion Transporting Polypeptide (OATP) Inhibition and Breast Cancer Resistance Protein (BCRP) inhibition
Lopinavir/ Ritonavir	Rosuvastatin	Rosuvastatin exposure 2-fold increase	OATP inhibition

transporter of pravastatin has been shown to be carrier-mediated (OATP-C and MRP2).[178] This transporter is also expected to be involved in distribution to the liver, the main site of cerivastatin distribution, resulting a drug–drug interaction between cerivastatin and cyclosporine during hepatic uptake. Recently, cerivastatin was withdrawn from the market because of severe rhabdomyolysis associated with cerivastatin–gemfibrozil combination therapy (Table 15.9).[179]

P-gp may be involved in human drug–drug interactions associated with absorption. For example, the plasma concentration of talinolol is increased by co-administration of erythromycin.[180] In fact erythromycin inhibits talinolol secretion from enterocytes into the lumen *via* P-gp, resulting in increased net absorption because of the lack of any significant metabolism of talinolol. Grapefruit juice and orange juice reduce oral bioavailability of fexofenadine, an OATP substrate, in healthy volunteers,[181] *via* inhibition of OATP-mediated drug uptake at the intestinal wall. Rifampin induces intestinal P-gp, resulting in reduce oral bioavailability of digoxin.[182] Since rifampin also induces intestinal MRP2, co-administration of rifampin is expected to increase secretion into the lumen of MRP2 substrates, such as glucuronide conjugates.[183]

15.7.2 *In Vitro* Models to Study Transporter Related Drug–Drug Interactions

Many methods and models to study drug–transporter interactions *in vitro* have been and still are being developed. The optimal method to use depends on the specific research question, as well as on the physicochemical nature of the drug that is under investigation. *In vitro* transporter assays are usually carried out with either intact cells or isolated cell membranes overexpressing the transporter of interest. The advantage of *in vitro* assays is that they can be performed with relatively high throughput and low costs as compared to *in vivo* studies. In bi-directional transport assays polarised epithelial cells are cultured on so-called Transwell filter inserts that separate an apical and basolateral compartment (Figure 15.10). Using this approach, the effect of a specific transporter on the actual flux of a drug from the apical to the basolateral side of the cell monolayer and *vice versa* can be studied in detail. This experimental set up is the most widely used *in vitro* assay to investigate whether a drug is a substrate for a specific efflux transporter such as the ABC transporters P-gp, breast cancer resistance protein (BCRP), and the multidrug resistance proteins (MRPs).

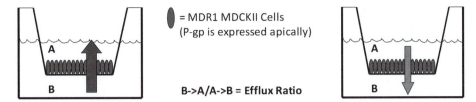

Figure 15.10 Bi-directional transport P-gp assay.

Alternatively, the influence of the drug on the transport of a model substrate can be assessed to evaluate potential drug–drug interactions. These assays are routinely performed using polarising MDCKII (Madine-Darby Canine Kidney) cells that over-express the human version of the transporter of interest, and comparing the translocation of the drug in these cells to that in mock-transfected MDCKII cells, which usually clearly indicates whether a drug is a substrate for the transporter of interest.

In addition to the cell lines transfected with a single human transporter, the human carcinoma cell line Caco-2 is used to study the transport of drugs in combination with inhibitors of the transporter proteins of interest. The advantage of this set-up is that Caco-2 cells are considered a good model for the translocation of drugs over the human intestinal epithelium, expressing considerable levels of a wider range of transporters (thus mimicking the *in vivo* combined effects of various transporters and cellular processes on drug uptake). The fact that Caco-2 cells express multiple transporters can in certain cases also be a disadvantage since inhibitors available for use in these studies show limited specificity.[184] However, for high-throughput screening of drugs for their transporter interaction potential this is a highly valuable and widely used model. A disadvantage of polarising epithelial cells which express human transporters is that to study drug efflux, the drug should first enter the cells. Especially with very hydrophilic compounds this is often a problem, as passive diffusion over the cell membrane will hardly occur and active uptake transporters are necessary. Therefore, to study transporter interactions of drugs that cannot enter the cells described above, it is usually more useful to perform a study with inside-out vesicles expressing the efflux protein of interest. These vesicles can be prepared from almost any cell type (usually the insect cell line Sf-9 or, more preferably, human HEK293 cells) as well as from tissues such as liver or kidney that over-express the efflux transporter of interest. With this assay the active transport of a drug into the inside-out vesicles is studied (as the membranes are "inside-out", the studied efflux protein now transports its substrates towards the inside of the vesicles, where they are captured and can be detected). The tendency for transported compounds to diffuse back across the vesicular membrane limits their usefulness in substrate determination assays to low permeability compounds; however, because the cytoplasmatic side of the transporter is now directly accessible, the selection of an appropriate probe substrate allows the inhibition of the transporter to be measured for compounds of any permeability class. A big advantage of this transporter assay is that it is easily performed in a high-throughput setting. Furthermore, it is relatively fast (and less costly), as no cell culturing is needed for this assay. One limitation to the vesicular transport assay is the batch-to-batch variation. This can result from differences in transporter expression in the host cell line as well as the percentage of inside-out vesicles formed during the membrane isolation procedure. Furthermore, the ability to cryopreserve a single large batch of membrane vesicles over a long time period allows for minimisation of this variability across multiple transport experiments.

To study interactions of a drug with uptake transporters various cell based methods can be used. These models are ideally based on cell lines from human origin (*e.g.* HEK 293 cells), as plasma membrane lipid composition can affect transporter function.[185] However, if not available, cells from other origins (*e.g.* CHO cells, MDCKII cells or Xenopus oocytes) can also be used. The cells over-expressing the uptake transporter of interest (and their mock transfected controls) are plated on 24- or 96-well plates and incubated with the drug under investigation. The time and concentration dependent uptake of the drug and/or the effects of the drug on the uptake of a model substrate are subsequently analysed. This is a very convenient way to study drug interactions with uptake transporters. One should, however, take into account that uptake transporters may need co-transport, anti-transport or exchange of ions for their activity.

Although many *in vitro* assays are available nowadays to study drug transporter interactions, it is obvious that the effect of a transporter on the pharmacokinetics of a drug will not only depend on the direct interaction between a transporter and a drug. In the body many factors (*e.g.* metabolism, tissue distribution of the transporter, presence of endogenous compounds *etc.*) can influence drug–transporter interactions. To gain insight in the *in vivo* effect of the drug transporter interactions, various models can be used. Combined with a wide range of analytical techniques such as imaging, microdialysis and gall bladder/urinary bladder cannulations, these models provide highly useful tools to study the interactions between drugs and transporters *in vivo*. To study the *in vivo* influence of transporters on drug pharmacokinetics, pharmaco-dynamics, and toxicity, many mouse models have been generated recently. The most widely used model is the Mdr1a/1b (P-gp) knock-out mouse which has already been shown to be of great value for studies on the *in vivo* effect of the ABC transporter Pgp, for example to study the effect of P-gp on the brain penetration of drugs.[186] Especially in combination with cell lines expressing human or murine versions of this transporter, it appears quite possible to predict the effect of a transporter in the human situation. As there is a large overlap in substrate specificity between transporters, recently many transporter combination knockout mice (mice lacking more than one transporter) have been generated, and can be used to study the relative *in vivo* effects of the transporters on the drug.[187] Still, although the tissue distribution and substrate specificities of ABC transporters in mice are in general quite comparable to those in humans, there are clearly species differences. To tackle this problem "humanised" transporter mice (mice in which the murine transporter is replaced by the human homologue) are currently being generated.[188] In combination with transporter knockout mice, these are very useful models to study the interaction of drugs with human transporters *in vivo*. If the drug of interest has been shown to interact with a transporter *in vitro*, the effect of known modulators of this transporter on the pharmacokinetics of the drug under investigation (drug–drug interactions) can be determined in various animal models. A disadvantage of this type of interaction study is that the inhibitors are usually not very specific and may also influence other processes in the body. However, these studies are very useful to get an idea of possible transporter related drug–drug interactions *in vivo*. A big advantage is that these studies can be performed in practically every animal model and therefore are easy to incorporate in the ADME studies routinely performed during drug development. With such studies one will not only add highly valuable data to standard ADME study results, but also often provide better interpretations of the pharmacoki-netic profiles found in animals. Furthermore, in combination with the *in vitro* studies described above, this will lead to better predictions of the situation in patients. However, despite all the available knowledge and techniques, it remains difficult to predict drug–transporter effects in the human body with absolute certainty. Research is currently ongoing to describe all these trans-porter-related processes in detail in an *in silico* model[189] and to investigate how data from the *in vitro* or *in vivo* transporter assays can be used as input data for this purpose. This will likely be highly useful for the prediction of internal drug levels and related effects in humans.

15.8 PHOSPHOLIPIDOSIS

Phospholipidosis is a lysosomal storage disorder characterised by the excess accumulation of phospholipids in tissues. Many cationic amphiphilic drugs (CADs), including anti-depressants,

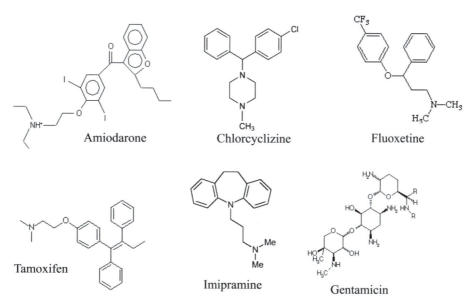

Figure 15.11 Chemical structure of compounds inducing PLD. PLD is induced by drugs with cationic amphiphilic (cationic lipophilic) structure (hydrophobic ring, hydrophilic side-chain and charged amine group).

antihistamines, antianginals, antibiotics, antimalarials, cholesterol-lowering agents and other prescription drugs, are reported to cause drug-induced phospholipidosis (DIPL) in humans, animals and cell culture models (Figure 15.11 and Table 15.10).[190] The mechanisms of DIPL involve trapping or selective uptake of DIPL drugs within the lysosomes and acidic vesicles of affected cells. Drug trapping is followed by a gradual accumulation of drug–phospholipid complexes within the internal lysosomal membranes. The increase in undigested materials results in the abnormal accumulation of multi-lamellar bodies (myeloid bodies) in tissues. DIPL does not necessarily constitute toxicity, and can resolve by itself, but it is predictive of drug or metabolite accumulation in affected tissues that have led to liver, kidney or respiratory failure. For these reasons, DIPL is of concern to the FDA, which in 2004 formed a DIPL Working Group to study the problem.[191] Currently, there are two prevailing hypotheses for the putative mechanisms involved in DIPL:

1) Inhibition of phospholipases due to binding of drug to substrate phospholipases.
2) Drugs bind to the phospholipids resulting in the formation of a complex that cannot be broken down by phospholipases.

The traditional method to evaluate DIPL is visual confirmation of myeloid bodies in tissues by electron microscopy, a time-consuming and labour intensive technique, and/or quantitative PCR. More recently, fluorescent dyes have been employed to assess phospholipidosis in a high throughput manner in cell lines (*e.g.* HepG2 cells).[192] The assay determines DIPL by measuring the accumulation of a specific fluorescent phospholipid (NBD-PE) in HepG2 cells treated with increasing drug concentrations. NBD-PE is broken down and metabolised by untreated cells, and so fluorescence does not accumulate. When phospholipidosis is induced, the phospholipid does not get broken down and fluorescence accumulates into the cells. A cytotoxicity marker is run in parallel to normalise the amount of accumulation to cell viability.

A proposed screening strategy to progress compounds with phospholipidosis liabilities is shown in Figure 15.12.

Table 15.10 Examples of compounds positively or negatively associated with DIPL inducing potential.

No DIPL	Diazepam, 3-OH gepirone, gepirone, buspirone, valproic acid, 5-phenoxybenzamine, ketasarin, almitrine, haloperidol, bufetolol, tetracycline
Positive in cell culture with low DIPL potency in animals	Mainserin, propranolol, clociguanil, noxiptiline, amitriptiline, disobutamide, promazine, mesoridazine, nortriptyline, chlorpromazine, maprotiline, thioridazine
Positive in cell culture with PLD demonstrated in animals	Chlorcyclazine, citalopram, chlorphenteramine, phentermine, fenfluramine, imipramine, tilorone, fluoxetine, iprindole, clomipramine, triparanol, mepacrine, gentamycin, erythromycin, netilimicin, azithromycin
Positive in cell culture with PLD demonstrated in humans and animals	Chloroquine, amiodarone, perhexiline, desethylamiodarone, tamoxifen, gentamycin

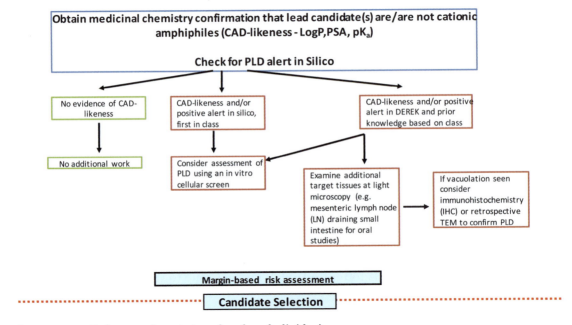

Figure 15.12 Early screening strategy for phospholipidosis.

Phospholipidosis is not considered to be a toxicity, but more likely an adaptive response, however its presence should not be ignored as it is related to secondary damage to tissue structure or impaired function. The decision to progress a compound bearing phospholipid accumulation potential should be taken on a case by case basis. For candidate selection purposes the traditional standard margins-based risk assessment is used to reduce the attrition of safe medicines. Because the incidence and severity of DIPL increases with time, if present in short term toxicology studies, a decision to anticipate longer term toxicology studies should be taken, in order to assess the number of species affected, define the NOAEL, reversibility and safety margins for clinical dosing. It should be finally noted that the occurrence of phospholipid accumulation in preclinical species does not definitely predict for its occurrence in human and, conversely, phospholipid accumulation can occur in human without any preclinical signal.

15.9 PHOTOTOXICITY

Phototoxicity (photoirritation) is a light-induced skin response to a photoreactive substance. Photoallergy is an immunologically mediated reaction to a chemical, initiated by the formation of photoproducts (*e.g.* protein adducts) following a photochemical reaction.

Photosafety testing may include an assessment of acute phototoxicity (photoirritation), photoallergy, 'photogenotoxicity' and photocarcinogenicity.[193]

A photoreactive substance is defined as a chemical (or mixture of chemicals) which absorbs light within the range of natural sunlight (290–700 nm) and generates reactive species following absorption of UV-visible light and distributes sufficiently to light-exposed tissues (*e.g.*, skin, eye). Clinically, phototoxicity results in exaggerated sunburn (erythema, increased skin temperature, pruritus and oedema).

Drugs associated with photosensitivity include tetracyclines, fluoroquinolones, sulfonamides, macrolides, betalactams, nitroimidazoles, and nitrofuranes.

The initial consideration for assessment of photoreactive potential is whether a compound absorbs photons at any wavelength between 290 and 700 nm. Light absorption is characterised by the Molar Extinction Coefficient (MEC, also called molar absorptivity), that reflects the efficiency with which a molecule can absorb a photon at a particular wavelength (typically expressed as $L\ mol^{-1}\ cm^{-1}$). This coefficient is influenced by several factors, including solvents. A compound that does not have a Molar Extinction Coefficient (MEC) greater than $1000\ L\ mol^{-1}\ cm^{-1}$ at any wavelength between 290 and 700 nm is not considered to be sufficiently photoreactive to result in direct phototoxicity. Excitation of molecules by light can lead to generation of Reactive Oxygen Species (ROS), including superoxide anion and singlet oxygen *via* energy transfer mechanisms. Although photoreactivity can result in other molecular outcomes (*e.g.* formation of photoadducts or cytotoxic photoproducts), even in these cases, it appears that ROS are typically generated as well. Thus, ROS generation following irradiation with UV-visible light can be an indicator of phototoxicity potential.[194]

A proposed phototoxicity screening paradigm to identify potentially phototoxic substances is described in Figure 15.13.

The most widely used *in vitro* assay for phototoxicity is the 3T3 Neutral Red Uptake Phototoxicity Test (3T3 NRU-PT) for which an Organisation for Economic Co-operation and Development (OECD) guideline is available.[195] This is currently considered the most appropriate *in vitro* screen for soluble compounds.

In terms of photogenotoxicity testing, the main objective is to make an assessment of the potential of a compound to turn into a photochemical carcinogen upon activation with UV or visible (*sic* solar simulated) radiation. Several *in vitro* photogenotoxicity assays, such as the photo-Ames, photo-chromosome aberration (CA) and photo-comet assays have been described in the literature and are based on standard 'dark' versions of regulatory assays used for genotoxicity assessment.[196]

However, it is recognised that *in vitro* photosafety assays (phototoxicity and photogenotoxicity) are substantially over predicting human hazard. As far as *in vivo* testing is concerned, no standardised study designs have been established.

It should also be mentioned that there are different regional regulatory views with regard to the strategy flow for photoreactive substances; an international effort for addressing different approaches, inconsistencies or gaps in the multiple guidances or papers published from different international bodies is ongoing by the International Harmonisation Conference (ICH), whose S10 guideline is currently the fourth draft.

In conclusion, the evaluation of a candidate with photoreactive potential should be based on experimental results as well as a thorough risk assessment.

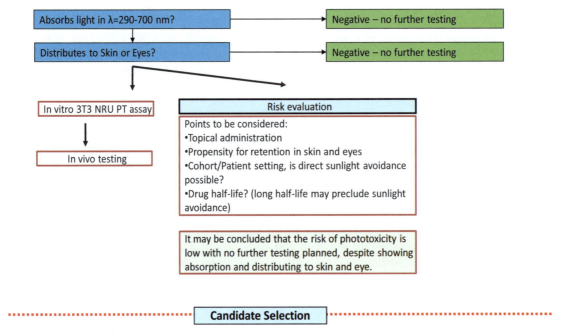

Figure 15.13   Screening paradigm for phototoxicity.

15.10   GENOTOXICITY

Genotoxicity is a term that refers to the ability of an agent to interact with DNA and/or the cellular apparatus that regulates the fidelity of the genome (*i.e.* DNA, spindle apparatus and enzymes involved in the maintenance of genome). Mutagenicity refers to the induction of permanent transmissible changes in the structure of the genetic material of cells or organisms. These changes (mutations) may involve a single gene or a block of genes. Mutations can arise spontaneously or may be induced by a variety of physical and chemical agents. The permanent, hereditary changes can affect either somatic cells of the organism or germ cells and be passed onto progeny. Chemicals that exert their adverse effects through interactions with the genetic material of cells, *i.e.* DNA, and by altering its structure and function are referred to as genotoxins. Note that all mutagens are genotoxic, however not all genotoxins are mutagens as they may not cause retained alterations in DNA sequence.

Cells possess efficient mechanism to prevent expression of the genotoxic mutation such as mechanisms of DNA repair or programmed cell death (apoptosis); mutations occur when the damage may not be fixed.

Type of mutations can be classified as:

- Gene mutations: a detectable permanent change within a single gene (point mutations, insertions, deletions);
- Chromosomal mutations (structural aberrations): morphological alterations in the structure of chromosomes (deletions, inversions, translocations);
- Genomic mutations (numerical aberrations): changes in the number of chromosomes (aneuploidy, polyploidy).

Mutations may induce abortions, congenital malformations or tumours. In addition, diseases such as hemophilia, deafness or Down's syndrome are due to mutations.

Bacterial and mammalian tests have been developed to detect the range of the possible DNA damage.

The most common tests for detection of mutations are the *in vitro* Ames test (that reveals point mutations) and the *in vitro* Chromosome Aberration test in human lymphocytes (that reveals structural chromosomal mutations). *In vivo*, the Micronucleus test in rodents reveals structural mutation as well as numerical aberration.

The most common tests for detection of genotoxicity (*i.e.* DNA damage without evidence of mutation) are for example the Unscheduled DNA Synthesis, the Sister Chromatid Exchange and the Comet assay.

During the early phase of research, high throughput assays such us Fluctuation Test, SOS umu Test or Greenscreen, are used to help screening high number of molecules. At the stage of lead optimisation, more sophisticated assays with lower throughput, although still preliminary, are introduced including Ames Assay and mammalian *in vitro* and *in vivo* tests.

Importantly, at these early stages, the test compound should be at the highest level of purity, to discharge any doubt that positive results were due to the presence of genotoxic impurities or contaminant.

15.10.1 Bacterial Tests

A bacterial test (Ames test) detects gene mutation. The Ames test uses amino-acid dependent strains of *Salmonella typhimurium* and *Escherichia coli*, each carrying different mutations in various genes in the histidine (*Salmonella*) or tryptophan (*Escherichia*) operon.

In the standard plate incorporation assay at least five strains of bacteria are used. These should include four strains of *S. typhimurium* (TA1535, TA1537 or TA97a or TA97, TA98 and TA100) and the strain *E. coli* WP2 uvrA, or *E. coli* WP2 or *E. coli* WP2 (pkM101). The experiment is conducted in the absence and in the presence of metabolic activation (+/-S9).

To accommodate the need for carrying out this critical test as soon as possible, industry has elaborated several versions of the Ames test (see Table 15.11), in order to reduce the requirement of test compound.[197,198]

However, the best approach is to use the five strains, in order to reduce to a minimum the risk of facing mutagenicity issues during the GLP studies. Pros and cons should be evaluated for selecting the reduced tests; *e.g.* knowledge of the chemical class *vs.* timelines.

15.10.2 *In Vitro* Mammalian Tests

There are several mammalian cell systems and each of them can address different end points. In Table 15.12 the tests that are most used for screening purposes are listed; mammalian systems that are used for investigative purposes are not mentioned here. Experiments are performed both in the presence and in the absence of metabolic activation.[199–203]

There are no preferred criteria to select the mammalian test to be included in the lead optimisation phase, however the human peripheral blood lymphocytes (HPLA) or the *in vitro* micronucleus tests, coupled with the bacterial test, offer the most comprehensive preliminary assessment.

15.10.3 Evaluation of Results

Positives results in an Ames test are always highly alerting and should suggest deselecting the candidate. However, some considerations can be made dissecting the result. Positivity can

Table 15.11 Different Ames test versions.

Bacterial test	Strain	Notes	Compound
Ames 5 strains	*S. typhimurium* TA1535, TA1537, TA98, TA100 and *E. coli* WP2 *uvrA* (pKM101)	First choice, same strains that will be used in the GLP studies.	350 mg
Ames 2 strains	*S. typhimurium* TA98 and TA100	Usually two strains of *S. typhimurium* are used, TA98 and TA100 able to detect the most frequent mutations, *i.e.* frameshift and base pair substitution, respectively. This test is also used in the genotoxic evaluation of impurities and of synthesis intermediates.	100 mg
Ames II	TA7001-TA7006 and TA98	It is offered as a standardised kit with quality controlled bacterial strains, however use different strains of the GLP studies.	250 mg

Table 15.12 *In vitro* mammalian genotoxicity tests.

Assays	System	Endpoint
Mouse lymphoma	Mouse lymphoma cell line	Gene mutation and chromosome aberration
In vitro mammalian chromosome aberration	Cell line, *e.g.*, CHO (Chinese Hamster Ovary cell); CHL (Chinese Hamster line)	Chromosome aberration
	Primary human peripheral blood lymphocytes (HPLA)	Structural chromosomal damage and indications of numerical variations (*e.g.* polyploidy, aneuploidy)
Micronucleus test	Primary human peripheral lymphocytes or cell line (*e.g.* CHO)	Structural chromosomal damage
Comet assay	Primary human peripheral lymphocytes or cell line (*e.g.* CHO, MLA)	Single and double-strand breaks and alkali labile sites

be both in the presence and in the absence of metabolic activation, or in one of the two conditions only. A favourable outcome could be derived in case that the positive result is obtained in the presence of metabolic activation only, where the responsible agent could be a metabolite. In this case an analysis of the metabolite profile in the genotoxicity test incubations, for comparison with known metabolite profiles in preclinical species or in human preparations can help determine the relevance of test results. To explore this hypothesis, supplementary *in vivo* genotoxicity tests in liver can be considered (*e.g. in vivo* Comet assay in liver or unscheduled DNA synthesis in the rat). Moreover, a new chemical entity that gives positive results *in vitro* in the presence of metabolic activation might not induce genotoxicity *in vivo* because the metabolite is not formed, is formed in very small quantities, is metabolically detoxified or rapidly excreted, indicating a lack of risk *in vivo*.

As far as *in vitro* mammalian cell assay is concerned, the scientific literature reports a number of conditions that can lead to a positive result of questionable relevance. Any mammalian *in vitro* positive result should be evaluated based on an assessment of the weight of evidence. Factors to consider are, for example, the conditions of the experiments that do not occur *in vivo* (*e.g.* pH, osmolality, precipitates) or genotoxicity that occur only at the most toxic concentrations. If one of these is the case, the weight of evidence indicates a lack of genotoxic potential and the subsequent battery of regulatory tests is not affected, and will include, as standard, a single *in vivo* test only. If there is not sufficient weight of evidence or mechanistic information to rule out the genotoxic potential, in the subsequent battery of regulatory tests, two *in vivo* studies are generally required, with appropriate endpoints, in appropriate tissues and with an emphasis in obtaining sufficient systemic exposure to the test compound.

15.11   EARLY *IN VIVO* TOXICOLOGY

There is a progression of complexity of the *in vivo* studies that are performed in lead optimisation. They are aimed, step by step, to narrow down the candidate's properties. Each study is designed to answer precise questions we pose and are essentially aimed to identify potential hazard and limit toxicity early on. Moreover, the knowledge of the chemical class and the insight from the alerts resulting from *in silico* and *in vitro* work should also direct to the inclusion of studies to explore potential pharmacodynamic issues.

The doses selected at this stage should be multiples of the pharmacologically predicted efficacies, in order answer the question: are there severe toxicities or issues that preclude progression?

The selection of doses is driven by the pharmacological and pharmacokinetic properties of the candidates.

An example of a safety assessment cascade is below exemplified:

1. Preliminary PK in rat and in a second species;
2. *In vivo* tox study, generally performed in one species (rodent) and in one sex (male);
3. Safety pharmacology studies targeted to investigate liabilities previously flagged or inherent to the chemical or pharmacological class (for example cardiovascular issues or convulsions).

Pharmaceuticals must be tested, except in exceptional circumstances, in rodents and non-rodents before and throughout the clinical phases of drug development programmes. For the safety assessment purposes, species are chosen based on similarity to humans with regard to pharmacokinetic profile and, wherever possible, selected species should respond to the primary pharmacodynamic effect of the substance.

The first species used for PK or toxicology studies is rodent, namely the rat. Rodents fall in the low neurophysiological evolutionary scale (*i.e.* have the least capacity to experience pain, distress or lasting harm), are small in size, easy to handle and require a limited amount of compound to be administered. The mouse is sometimes considered; however this species is seldom used for regulatory toxicology studies, due to its small size and the high number of individuals necessary to allow adequate blood collection. During the lead optimisation stage only the rodent species is generally explored. Non-rodents are rarely used, unless driven by previous experience, *e.g.* if the pharmacological receptor is only present in a particular species or previous experience has shown the rodent as non-predictive for a specific toxicity within a particular drug class.[204]

Should a second species be introduced, the default species is the dog, particularly the beagle, because of its size, tractability, and the historical experience which makes the interpretation of toxicological findings easier.[205] Alternatively to dog, the minipig, or rarely the ferret, is sometimes selected. Non-human primates (NHP) should be selected only as a last instance, being the nearest species to man in the evolutionary scale, which implies ethical considerations. For this reason any use of non-human primates must be specifically justified. Justification is based primarily on pharmacological basis when the target is known to be different in non-human primates *versus* other species. Other considerations for selecting NHP are on previous toxicological experience with compounds of that class, ability to achieve the required exposure, similarity to human of important metabolic/biochemical processes (*e.g.* CYP450 structures or activities), specific pharmacological targets (*e.g.* COX-2 distribution in the kidney is similar in non-human primate and man, but different in dog), or poor tolerability in other species (*e.g.* if emesis precludes achieving adequate systemic exposure).

15.11.1 Preliminary Pharmacokinetics

Pharmacokinetic (PK) studies are essential to translate the dose administered in terms of mg/kg to the systemic exposure (AUC) following the process of absorption, distribution, metabolism and excretion. Early *in vitro* and *in vivo* DMPK studies, typically available at the stage when a toxicity study is planned, are always used to help select the species, the doses and the preliminary tox study design.

15.11.2 *In Vivo* Tox Study

Early tox study designs are usually simple, although able to ensure that potential hazard can be detected. A typical study design is reported in Table 15.13.

The active substance is typically administered for 7 days *via* the intended route of administration in human that is, in the majority of the cases, oral, intravenous or inhaled.

A repeated administration of 7 days predicts most of the dose limiting target organ toxicities observed in subsequent pivotal 4-week rodent studies. Target organs that may be missed are musculoskeletal toxicities observed with broad-spectrum matrix metalloproteinase (MMP) inhibitors, which may take 10–14 days to develop. Roberts *et al.*[206] report that at least of 50% of the historical failures were attributable to target organ toxicities or to unexplained death that emerged within 14 days of repeat dosing.

Mortality or severe target organ toxicity at pharmacological exposure precludes progression. The inclusion of three different doses is aimed to explore a range of margins, however the preclinical safety margin (*i.e.* the ratio of the NOAEL—no observable adverse effect level—divided by the predicted human efficacious exposure level or exposure at the maximum anticipated dose in human) will be defined with pivotal (GLP) toxicity studies that will be conducted at a subsequent stage.

In analysing the findings of the studies, discrimination should be made with regard to effects due to target organ toxicity, exaggerated pharmacology (primary pharmacology) or effects mediated by other (secondary) pharmacological targets.[207]

For each of these effects a risk assessment considering the nature of the finding and the safety margin should be made. As examples, alosetron (highly selective serotonin 5-HT3

Table 15.13   Early toxicity study design.

Species	Rat
Sex	Male
Number of animals	4/group
Groups	Vehicle, low, intermediate and high dose
Duration	7 to 14 days
Observations	Clinical signs, body weight, clinical chemistry
Necropsy	Macroscopic examination
Microscopic examination of major organs	Adrenals; liver; lung; heart; kidneys; thymus; adrenals; testes; stomach if oral administration; injection site if IV administration; larynx if inhalation admin

antagonist), cerivastatin (statin), flosequinan (vasodilator) and encainide (member of the class 1C of anti-arrhythmic agents) and rofecoxib (COX-2 selective inhibitor) can be mentioned as agents posing issues related to primary pharmacology. Fenfluramine and dexfenfluramine (for obesity treatment), rapacuronium (neuromuscular blocking agent of nicotinic receptors), astemizole and terfenadine (non-sedating H1 antihistamines), cisapride (gastrointestinal prokinetic motility agent) and mibefradil (long acting calcium channel antagonist) can be quoted as agents posing issues related to secondary pharmacology.[208]

The need for toxicity studies in dog or in other non-rodent species (see species selection) is less frequent and may be of help in case of a new class of agent with unknown toxicity. It is advisable to get toxicokinetic (TK) data at the maximum dose tolerated (MTD) to confirm that the selected species can guarantee the exposure needed. These studies can be non-terminal and aimed only to identify the MTD and the associated TK or may reflect the study design of the rat as described before.

15.11.3 Early Safety Pharmacology Evaluation

The scope of the safety pharmacology studies is to investigate the potential undesirable pharmacodynamic effects of a substance on physiological functions in relation to exposure in the therapeutic range and above. The regulatory core battery includes studies on central nervous system, cardiovascular system and respiratory system. Due to the rate of attrition, at early stages focus is given to the cardiovascular system and to the central nervous system, while the reparatory system is investigated within the regulatory package of studies.

15.11.3.1 *Cardiovascular (CV) Functionality*

One of the major causes of attrition during development is cardiovascular issues, particularly QT prolongation, and a big effort has been devoted to design early studies to set reliable stopping criteria. QT prolongation is an accessible, although imperfect, surrogate biomarker to gauge potential fatal arrhythmias, notably *Torsades de Pointes* (TdP), an extremely rare but potentially lethal adverse drug effect.[209]

A CV attrition reduction strategy to progress compounds and discharge cardiovascular risk which can be put in place during lead optimisation phase and in later phases has been discussed in paragraph 15.5.4 and shown in Figure 15.4.

The telemetry study in rodent is performed after single dose administration in rats, typically males, at three dose levels.

The endpoints that can be collected are reported in Table 15.14.

15.11.3.2 *Nervous System Functionality*

Early assessment of the Nervous System functionality is of value with highly potent compound crossing the blood–brain barrier (BBB) or compounds that are flagged with potential

Table 15.14 Telemetry rat study endpoints.

Hemodynamics	• Systolic, diastolic and mean arterial pressures and heart rate
	• Pulse pressure (SBP-DBP) (assessment of cardiac and vascular function)
	• Rate Pressure Product (HR×SBP) (an index of cardiac work)
	• QA interval (an index of cardiac contractility derived from ECG and BP signals)
ECG intervals	• PR (assessment of Ca channel block/conduction)
	• QRS (assessment of Na channel block/conduction)

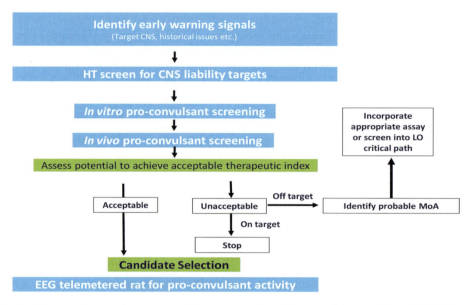

Figure 15.14 Proposed screening strategy for compounds with potential convulsant liabilities.

convulsant properties (*e.g.* cannabinoid-1 antagonists, activators of the excitatory amino acids, GABAA receptor antagonists, 5-HT1B, 1D receptor antagonists). Together with the assessment of the chemical and therapeutic class, early target selectivity assays are able to identify potential hazards produced by compounds acting on off-targets receptors (see paragraph 15.3).

A proposed screening strategy to progress compounds with convulsant liabilities is shown in Figure 15.14.

Hippocampal brain slice assay for assessment of seizure liability[210] is an example of *in vitro* test; *in vivo* studies such us pentylenetetrazol-induced seizure in rats (PTZ) or Maximal electroshock seizure threshold (MEST) test are performed. PTZ is based on the continuous infusion of Pentylenetetrazol and a behavioural assessment in rats. MEST is based on a single corneally delivered electroshock and observation of tonic hind limb extension convulsions.

After candidate selection, telemetric electroencephalographic recording (EEG) in order to correlate PK/PD and interpret *in vivo* signs complementary to the time-right collection of TK metrics will complete the picture.

Undesired effects of drugs on the central and the peripheral nervous system can also be detected with the neurobehavioral observation battery adapted from that first described by Irwin in mice[211] and subsequently modified as the neurologically based Functional Observational Battery (FOB). Behaviour represents the integration and integrity of the nervous system and it is generally considered a sensitive indicator, and perhaps the ultimate assay, of neuronal function.[212] A neurobehavioral observation battery consists of a series of observation/measurements, including home cage and open-arena observation, neuromuscular and sensory motor tests and physiological/neurological measurements. As an example of read out, amphetamine induces a stimulation of the nervous system activities and marked hyperthermia. Chlorpromazine, diazepam and clonidine induce depressive, anxiolytic or sedative effects associated with hypothermia.

In early stages this type of study is performed after single dose administration in male rats or mice.

HINTS AND TIPS

A number of freeware QSAR and rule-based programs are available for toxicity prediction. These global models have limitations, however once a liability in a lead from a series has been identified, local models built using in house data around the chemotypes of interest usually provide the best results.

It is advisable to start lead optimisation with multiple different chemotypes, even if some carry toxicity flags, provided these are considered during the lead optimisation phase.

Drug interaction with off-targets might often be the reason for adverse effects, therefore the precise identification of the off-targets and their characterisation are essential steps in drug discovery.

As a rule of thumb it is suggested that if the ratio of the free C_{max} : off-target activity is >100-fold, it generally represents no safety concern.

Drugs that modify the cardiac action potential represent a major safety concern, and establishing a cardiac safety profile is required to achieve regulatory approval. A comprehensive *in vitro* set of ion current assays could include IKr, IKs, IK1, Ina (fast and late) and ICaL for drug effects.

It is too risky to rely on hERG binding assays as the only support for medicinal chemistry, and a number of medium to high throughput functional hERG assay formats exist.

Clinicians have difficulty in managing drug–drug interactions, hence minimising the factors that increase DDI risks is essential. Well established *in vitro* protocols are used to determine the CYP driven DDI potential, and guide clinical trial design to assess potential DDI's.

Understanding drug metabolism pathways enables prediction of potential DDI's as well as likely variability in patients due to genetic polymorphism or different expression levels.

Victim drugs that have multiple clearance pathways reduce risks of DDIs due to a perpetrator drug.

Drug induced phopholipidosis does not necessarily constitute toxicity, and can resolve by itself, but it is predictive of drug or metabolite accumulation in affected tissue that have led to liver, kidney or respiratory failure, and is therefore a concern for the FDA.

An initial consideration of photoreactive potential is whether the compound absorbs light at any wavelength between 290–700 nm, and a molar extinction coefficient less than $1000 \text{ L mol}^{-1}\text{cm}^{-1}$ within this range is not considered sufficiently photoreactive to result in direct phototoxicity.

The most common tests for detection of mutagenicity are the *in vitro* AMES assay and chromosome aberration test in human lymphocytes and *in vivo* the rat micronucleus test. The compound tested should be of the highest purity to discharge any doubt that positive results are due to impurities or contaminants.

KEY REFERENCES

C. A. Lipinski *et al.*, *Adv. Drug Deliv. Rev.*, 2001, **46**, 3.
L. C. Wienkers and T. G. Heath, *Nat. Rev. Drug Discovery*, 2005, **4**, 825.
J. A. Kramer *et al.*, *Nat. Rev. Drug Discovery*, 2007, **6**(8), 636.

REFERENCES

1. J. Ashby, *Environ. Mutagen*, 1985, **7**, 919.
2. J. Ashby and R. W. Tennant, *Mutat. Res.*, 1988, **204**, 17.
3. J. Ashby and R. W. Tennant, *Mutat. Res.*, 1991, **257**, 229.
4. C. Merlot, *Drug Discovery Today*, 2010, **15**, 16.
5. J. R. Rabinowitz, M. Goldsmith, S. Little and M. Pasquinelli, *Environ. Health Perspect.*, 2008, **116**, 573.
6. D. Jacobson-Kram and J. F. Contrera, *Toxicol. Sci.*, 2007, **96**, 16.
7. P. McGee, *Drug Discov. Dev.*, 2005, **8**, 24.
8. G. M. Pearl, S. Livingston-Carr and S. K. Durham, *Curr. Top. Med. Chem.*, 2001, **1**, 247.
9. N. Greene, *Adv. Drug Deliv. Rev.*, 2002, **54**, 417.
10. S. Ekins, *Biochem. Soc. Trans.*, 2003, **31**, 611.
11. R. Benigni and C. Bossa, *Curr. Comput. Aid. Drug Discov.*, 2006, **2**, 169.
12. C. Gopi Mohan, T. Gandhi, D. Garg and R. Shinde, *Mini Rev. Med. Chem.*, 2007, 7, 499.
13. W. Muster, A. Breidenbach, H. Fischer, S. Kirchner, L. Müller and A. Pähler, *Drug Discovery Today*, 2008, **13**, 303.
14. F. Nigsch, N. J. Maximilan Macaluso, J. BO Mitchell and D. Zmuidinavicius, *Expert Opin. Drug Metab. Toxicol.*, 2009, **5**, 1.
15. P. Szymanski, M. Markowicz and E. Mikiciuk-Olasik, *Int. J. Mol. Sci.*, 2012, **13**, 427.
16. S. Ekins, S. Andreyev, A. Ryabov, E. Kirillov, E. A. Rakhmatulin, S. Sorokina, A. Bugrim and T. Nikolskaya, *Drug Metab. Dispos.*, 2006, **34**, 495.
17. M. Jalaie and D. Holsworth, *Mini Rev. Med. Chem.*, 2005, **5**, 1083.
18. E. Kerns and L. Di, *Drug Discovery Today*, 2003, **8**, 316.
19. J. Bajorath, *Nat. Rev. Drug Discovery*, 2002, **1**, 882.
20. A. M. Richard, *Chem. Res. Toxicol.*, 2006, **19**, 1257.
21. C. Yang, R. D. Benz and M. A. Cheeseman, *Curr. Opin. Drug Discov. Dev.*, 2006, **9**, 124.
22. A. M. Richard, L. Swirsky Gold and M. C. Nicklaus, *Curr. Opin. Drug Discov. Dev.*, 2006, **9**, 314.
23. R. Judson, *Toxicol. Appl. Pharmacol.*, 2008, **233**, 7.
24. R. Judson, A. Richard, D. J. Dix, K. Houck, M. Martin, R. Kavlock, V. Dellarco, T. Henry, T. Holderman, P. Sayre, S. Tan, T. Carpenter and E. Smit, *Environ. Health Perspect.*, 2009, **117**, 685.
25. R. D. Benz, *Expert Opin. Drug Metab. Toxicol.*, 2007, **3**, 109.
26. C. Yang, C. H. Hasselgren, S. Boyer, K. Arvidson, S. Aveston, P. Dierkes, R. Benigni, R. D. Benz, J. Contrera, N. L. Kruhlak, E. J. Matthews, X. Han, J. Jaworska, R. A. Kemper, J. F. Rathman and A. M. Richard, *Toxicol Mech. Method*, 2006, **18**, 277.
27. W. Suter, *Curr. Opin. Chem. Biol.*, 2006, **10**, 362.
28. K. A. Giuliano, P. A. Johnston, A. Gough and D. L. Taylor, *Methods Enzymol.*, 2006, **414**, 601.
29. A. P. Li, *Chem. Biol. Interact.*, 2007, **168**, 16.
30. A. P. Li, *Alternativen zu Tierexperimenten*, 2008, **25**, 43.
31. C. Parng, N. M. Roy, C. Ton, Y. Lin and P. McGrath, *J. Pharmacol. Toxicol. Methods*, 2007, **55**, 103.
32. A. Oberemm, L. Onyon and U. Gundert-Remy, *Toxicol. Appl. Pharmacol.*, 2005, **207**, 592.
33. L. Suter, L. E. Babiss and E. B. Wheeldon, *Chem. Biol.*, 2004, **11**, 161.
34. W. B. Mattes, S. D. Pettit, S. Sansone, P. R. Bushel and M. D. Waters, *Environ. Health Perspect.*, 2004, **112**, 495.
35. A. Bender, J. Scheiber, M. Glick, J. W. Davies, K. Azzaoui, J. Hamon, L. Urban, S. Whitebread and J. L. Jenkins, *ChemMedChem*, 2007, **2**, 861.

36. C. D. Scripture and W. D. Figg, *Nat. Rev. Cancer*, 2006, **6**, 546.
37. A. L. Hopkins, J. S. Mason and J. P. Overington, *Curr. Opin. Struct. Biol.*, 2006, **16**, 127.
38. R. Benigni, T. I. Netzeva, E. Benfenati, C. Bossa, R. Franke, C. Helma, E. Helzebos, C. Marchant, A. Richard, Y. T. Woo and C. Yang, *J. Environ. Sci. Health C. Environ. Carcinog. Ecotoxicol. Rev.*, 2007, **25**, 53.
39. W. Muster, A. Breidenbach, H. Fischer, S. Kirchner, L. Müller and A. Pähler, *Drug Discovery Today*, 2008, **13**, 303.
40. N. L. Kruhlak, J. F. Contrera, R. D. Benz and E. J. Matthews, *Adv. Drug Deliv. Rev.*, 2007, **59**, 43.
41. S. Weaver and M. P. Gleeson, *J. Mol. Graph. Model*, 2008, **26**, 1315.
42. M. G. Knize, F. T. Hatch, M. J. Tanga, E. V. Lau and M. E. Colvin, *Environ. Mol. Mutagen*, 2006, **47**, 132.
43. E. J. Matthews, N. L. Kruhlak, R. D. Benz, J. F. Contrera, C. A. Marchant and C. Yang, *Toxicol. Mech. Methods*, 2008, **18**, 189.
44. C. Yang, C. H. Hasselgren, S. Boyer, K. Arvidson, S. Aveston, P. Dierkes, R. Benigni, R. D. Benz, J. Contrera, N. L. Kruhlak, E. J. Matthews, X. Han, J. Jaworska, R. A. Kemper, J. F. Rathman and A. M. Richard, *Toxicol. Mech. Methods*, 2008, **18**, 277.
45. G. Klopman, *Quant. Struct.*, 1992, **11**, 176.
46. E. J. Matthews and J. F. Contrera, *Regul. Toxicol. Pharmacol.*, 1998, **28**, 242.
47. S. K. Durham and G. M. Pearl, *Curr. Opin. Drug Discov. Dev.*, 2001, **4**, 110.
48. G. M. Pearl, S. Livingston-Carr and S. K. Durham, *Curr. Top. Med. Chem.*, 2001, **1**, 247.
49. M. J. Prival, *Environ. Mol. Mutagen*, 2001, **37**, 55.
50. P. D. Mosier and P. C. Jurs *et al.*, *Chem. Res. Toxicol.*, 2003, **16**, 721.
51. R. D. Snyder, *Environ. Mol. Mutagen.*, 2009, **50**, 435.
52. R. D. Snyder, G. S. Pearl, G. Mandakas, W. N. Choy, F. Goodsaid and I. Y. Rosenblum, *Environ. Mol. Mutagen.*, 2004, **43**, 143.
53. R. Benigni, *Chem. Rev.*, 2005, **105**, 1767.
54. K. B. Arvidson, L. G. Valerio, M. Diaz and R. F. Chanderbhan, *Toxicol. Mech. Methods*, 2008, **18**, 229.
55. J. Mayer, M. A. Cheeseman and M. L. Twaroski, *Regul. Toxicol. Pharmacol.*, 2008, **50**, 50.
56. L. G. Valerio and K. P. Cross, *Toxicol. Appl. Pharmacol.*, 2007, **222**, 1.
57. I. V. Tetko, P. Bruneau, H. W. Mewes, D. C. Rohrer and G. I. Poda, *Drug Discovery Today*, 2006, **11**, 700.
58. B. Simon-Hettich, A. Rothfuss and T. Steger-Hartmann, *Toxicology*, 2006, **224**, 156.
59. E. J. Matthews, N. L. Kruhlak, R. Daniel Benz, J. Ivanov, G. Klopman and J. F. Contrera, *Regul. Toxicol. Pharmacol.*, 2007, **47**, 136.
60. E. J. Matthews, N. L. Kruhlak, M. C. Cimino, R. D. Benz and J. F. Contrera, *Regul. Toxicol. Pharmacol.*, 2006, **44**, 83.
61. F. P. Guengerich and J. S. Macdonald, *Chem. Res. Toxicol.*, 2007, **20**, 344.
62. M. Von Korff and T. Sander, *J. Chem. Inf. Model*, 2006, **46**, 536.
63. P. Foggia, A. Limongiello, F. Tufano and M. Vento, *Int. J. Pattern Recog. Artif. Intell.*, 2006, **20**, 883.
64. R. Kho, J. A. Hodges, M. R. Hansen and H. O. Villar, *J. Med. Chem.*, 2005, **48**, 6671.
65. O. Llorens, J. J. Perez and H. O. Villar, *Int. J. Quant. Chem.*, 2002, **88**, 107.
66. B. Simon-Hettich, A. Rothfuss and T. Steger-Hartmann, *Toxicology*, 2006, **224**, 156.
67. Y.-T. Woo and D. Y. Laiand, in *Predictive Toxicology*, ed. C. Hemla, CRC Press, Boca Raton, FL, USA, 2005, pp. 385–413.
68. C. A. Marchant, K. A. Briggs and A. Long, *Toxicol. Mech. Methods*, 2008, **18**, 177.
69. L. Michielan and S. Moro, *J. Chem. Inf. Model.*, 2010, **50**, 961.
70. M. P. Gleeson, A. Hersey and S. Hannongbua, *Curr. Top. Med. Chem.*, 2011, **11**, 358.

71. H. van de Waterbeemd and E. Gifford, *Nat. Rev. Drug Discovery*, 2003, **2**, 192.

72. P. D. Dobson, Y. Patel and D. B. Kell, *Drug Discovery Today*, 2009, **14**, 31.

73. C. A. Lipinski, F. Lombardo, B. W. Dominy and P. J. Feeney, *Adv. Drug Deliv. Rev.*, 2001, **46**, 3.

74. N. Bhogal, C. Grindon, R. Combes and M. Balls, *Trends Biotechnol.*, 2005, **23**, 299.

75. G. Moroy, *Drug Discovery Today*, 2012, **17**, 44.

76. A. Vedani and M. Smiesko, *Altern. Lab. Anim.*, 2009, **37**, 477.

77. L. Chen, *Drug Discovery Today*, 2012, **17**, 343.

78. C. M. Bathelt, A. J. Mulholland and J. N. Harvey, *J. Phys. Chem.*, 2008, **112**, 13149.

79. R. Lonsdale, J. Oláh, A. J. Mulholland and J. N. Harvey, *J. Am. Chem. Soc.*, 2011, **133**, 15464.

80. H. Kubinyi, *J. Braz. Chem. Soc.*, 2002, **13**, 717.

81. S. Wold, M. Sjostroma and L. Eriksson, *Chemom. Intell. Lab. Syst.*, 2001, **58**, 109.

82. *CPMP/ICH/539/00. Note for guidance on safety pharmacology studies for human pharmaceuticals. S 7 A: Safety pharmacology studies—Step 5.*

83. J. Bowes, A. J. Brown, J. Hamon, W. Jarolimek, A. Sridhar, G. Waldron and S. Whitebread, *Nat. Rev. Drug Discovery*, 2012, **11**, 909.

84. *CPMP/ICH/423/02 EMA. Note for guidance on the nonclinical evaluation of the potential for delayed ventricular repolarization (QT interval prolongation) by human pharmaceuticals.*

85. S. Whitebread, J. Hamon, D. Bojanic and L. Urban, *Drug Discovery Today*, 2005, **10**, 1421.

86. S. Engberg, I. Ahlstedt, A. Leffler, E. Lindström, E. Kristensson, A. Svensson, I. Pahlman, A. Johansson, T. Drmota and B. von Mentzer, *Biochem. Pharmacol.*, 2007, **73**, 259.

87. Y. Cheng and W. H. Prusoff, *Biochem Pharmacol.*, 1973, **22**, 3099.

88. S. Lazareno and N. J. Birdsall, *Br. J. Pharmacol.*, 1993, **109**(4), 1110.

89. W. T. Bellamy, *Drugs*, 1992, **44**(5), 690.

90. J. J. Cali, A. Niles, M. P. Valley, M. A. O'Brien, T. L. Riss and J. Shultz, *Expert. Opin. Drug. Metab. Toxicol.*, 2008, **4**(1), 103.

91. N. Ferri, P. Siegl, A. Corsini, J. Herrmann, A. Lerman and R. Benghozi, *Pharmacol. Ther.*, 2013, **138**, 470.

92. J. S. MacDonald and R. T. Robertson, *Toxicol. Sci.*, 2009, **110**(1), 40.

93. H. Laverty, C. Benson, E. Cartwright, M. Cross, C. Garland, T. Hammond, C. Holloway, N. McMahon, J. Milligan, B. Park, M. Pirmohamed, C. Pollard, J. Radford, N. Roome, P. Sager, S. Singh, T. Suter, W. Suter, A. Trafford, P. Volders, R. Wallis, R. Weaver, M. York and J. Valentin, *Br. J. Pharmacol.*, 2011, **163**, 675.

94. R. R. Shah, *Pharmacogenomics*, 2006, 7, 889.

95. O. Campuzano, P. Beltrán-Alvarez, A. Iglesias, F. Scornik, G. Pérez and R. Brugada, *Genet. Med.*, 2010, **12**, 260.

96. J. S Mitcheson, *Chem. Res. Toxicol.*, 2008, **21**, 1005.

97. P. T. Sager, G. Gintant, J. R. Turner, S. Pettit and N. Stockbridge, *Am. Heart J.*, 2013, **3**, 292.

98. C. A. Remme and A. A. M. Wilde, *Curr. Opin. Pharmacol.*, 2014, **15**, 53.

99. ICH. *Harmonised Tripartite Guideline S7B. Non-clinical evaluation of the potential for delayed ventricular repolarization (QT interval prolongation) by human pharmaceuticals.* Step 4 Version, May 2005.

100. ICH. *Harmonised Tripartite Guideline E14. Clinical evaluation of QT/QTc interval prolongation and proarrhythmic potential for nonantiarrhythmic drugs.* Step 4 Version, May 2005.

101. K. Rae Chi, *Nat. Rev. Drug Discovery*, 2013, **12**, 565.

102. J. Heijman, N. Voigt, L. G. Carlsson and D. Dobrev, *Curr. Opin. Pharmacol.*, 2014, **15**, 16.

103. S. Peng, A. E. Lacerda, G. E. Kirsch, A. M. Brown and A. Bruening-Wright, *J. Pharmacol. Toxicol. Methods*, 2010, **61**, 277.

104. J. Ma, L. Guo, S. J. Fiene, B. D. Anson, J. A. Thomson, T. J. Kamp, K. L. Kolaja, B. J. Swanson and C. T. January, *Am. J. Physiol. Heart Circ. Physiol.*, 2011, **301**, 2006.
105. E. Neher and B. Sakmann, *Nature*, 1976, **260**, 799.
106. O. P. Hamill, A. Marty, E. Neher, B. Sakmann and F. J. Sigworth, *Pflugers Arch.*, 1981, **391**, 85.
107. G. J. Kaczorowski, M. L. Garcia, J. Bode, S. D. Hess and U. A. Patel, *Front. Pharmacol.*, 2011, **78**, 1.
108. C. Möller and H. Witchel, *Front. Pharmacol.*, 2011, **2**, 1.
109. C. D. Furberg and B. Pitt, *Curr. Controlled Trials Cardiovasc. Med.*, 2001, **2**, 205.
110. F. Estelle and R. Simons, *Ann. Allergy Asthma Immunol.*, 1999, **83**, 481.
111. R. Diasio, *Br. J. Clin. Pharmacol.*, 1998, **46**, 1.
112. J. Lazarou, B. H. Pomeranz and P. N. Corey, *J. Am. Med. Assoc.*, 1998, **279**, 1200.
113. S. A. Wrighton, M. Vandenbranden, J. C. Stevens, L. A. Shipley, B. J. Ring, A. E. Rettie and J. R. Cashman, *Drug Metab. Rev.*, 1993, **25**, 453.
114. R. Stringer, P. L. Nicklin and J. B. Houston, *Xenobiotica*, 2008, **38**(10), 1313.
115. R. J. Riley, D. F. McGinnity and R. P. Austin, *Drug Metab. Dispos.*, 2005, **33**(9), 1304.
116. B. S. Kalra, *Indian J. Med. Sci.*, 2007, **61**(20), 102.
117. D. Mitschke, A. Reichel, G. Fricker and U. Moenning, *Drug Metab. Dispos.*, 2008, **36**(6), 1039.
118. J. A. Willian, *Drug Metab. Dispos.*, 2004, **32**(11), 1201.
119. S. Fowler, *AAPS J.*, 2008, **10**(2), 410.
120. R. J. Bertz and G. R. Granneman, *Clin. Pharmacokinet.*, 1997, **32**, 210.
121. S. A. Testino Jr. and G. Patonay, *J. Pharm. Biomed. Anal.*, 2003, **30**, 1459.
122. A. P. Li, *Drug Discovery Today*, 2001, **6**, 357.
123. A. P. Li, *Curr. Top. Med. Chem.*, 2004, **4**, 701.
124. C. L. Crespi, V. P. Miller and B. W. Penman, *Anal. Biochem.*, 1997, **248**, 188.
125. D. F. McGinnity, S. J. Griffin, G. C. Moody, M. Voice, S. Hanlon, T. Friedberg and R. J. Riley, *Drug Metab. Dispos.*, 1999, **27**, 1017.
126. D. F. McGinnity, A. J. Parker, M. Soars and R. J. Riley, *Drug Metab. Dispos.*, 2000, **28**, 1327.
127. M. Yao, M. Zhu, M. W. Sinz, H. Zhang, W. G. Humphreys, A. D. Rodrigues and R. Dai, *J. Pharm. Biomed. Anal.*, 2007, **44**(1), 211.
128. J. Ayrton, R. Plumb, W. J. Leavens, D. Mallett, M. Dickins and G. J. Dear, *Rapid Commun. Mass Spectrom.*, 1998, **12**(5), 217.
129. A. Tolonen, A. Petsalo, M. Turpeinen, J. Uusitalo and O. Pelkonen, *J. Mass Spectrom.*, 2007, **42**(7), 960.
130. X. Wu, J. Wang, L. Tan, J. Bui, E. Gjerstad, K. McMillan and W. Zhang, *J. Biomol. Screen.*, 2012, **17**(6), 761.
131. S. Zhou, E. Chan, L. Y. Lim, U. A. Boelsterli, S. C. Li, J. Wang, Q. Zhang, M. Huang and A. Xu, *Curr. Drug Metab.*, 2004, **5**, 415.
132. S. Zhou, S. Yung Chan, B. Cher Goh, E. Chan, W. Duan, M. Huang and H. L. McLeod, *Clin. Pharmacokinet.*, 2005, **44**, 279.
133. K. M. Bertelsen, K. Venkatakrishnan, L. L. Von Moltke, R. S. Obach and D. J. Greenblatt, *Drug Metab. Dispos.*, 2003, **31**, 289.
134. L. C. Wienkers and T. G. Heath, *Nat. Rev. Drug Discovery*, 2005, **4**, 825.
135. A. Atkinson, J. R. Kenny and K. Grime, *Drug Metab. Disp.*, 2005, **33**, 1637.
136. E. S. Perloff, A. K. Mason, S. S. Dehal, A. P. Blanchard, L. Morgan, T. Ho, A. Dandeneau, R. M. Crocker, C. M. Chandler, N. Boily, C. L. Crespi and D. M. Stresser, *Xenobiotica*, 2009, **39**(2), 99.
137. M. Kajbaf, E. Palmieri, R. Longhi and S. Fontana, *Drug Metab. Letters*, 2010, **4**(2), 104.

138. N. Sekiguchi, A. Higashida, M. Kato, Y. Nabuchi, T. Mitsui, K. Takanashi, Y. Aso and M. Ishigai, *Drug Metab. Pharmacokinet.*, 2009, **24**(6), 500.

139. G. C. Moody, *Xenobiotica*, 1999, **29**, 53.

140. Z. Yan, B. Rafferty, G. W. Caldwell and J. A. Masucci, *Europ. J. Drug Metab. Pharmacokin.*, 2002, **4**, 281.

141. C. L. Crespi and D. M. Stresser, *J. Pharmacol. Toxicol. Methods*, 2000, **44**, 325.

142. T. Yamamoto, A. Suzuki and Y. Kohno, *Xenobiotica*, 2004, **34**, 87.

143. H. Yin, J. Racha, S. Y. Li, N. Olejnik, H. Satoh and D. Moore, *Xenobiotica*, 2000, **30**, 141.

144. R. Weaver, K. S. Graham, I. G. Beattie and R. J. Riley, *Drug Metab. Dispos.*, 2003, **31**, 955.

145. H. K. Lim, N. Duczak Jr, L. Brougham, M. Elliot, K. Patel and K. Chan, *Drug Metab. Dispos.*, 2005, **33**, 1211.

146. A. D. Rodrigues, D. J. Mulford, R. D. Lee, B. W. Surber, M. J. Kukulka, J. L. Ferrero, S. B. Thomas, M. S. Shet and R. W. Estabrook, *Drug Metab Dispos.*, 1995, **23**, 765.

147. K. Hedenmalm, M. Sundgren, K. Granberg, O. Spigset and R. Dahlqvist, *Ther. Drug Monit.*, 1997, **19**, 643.

148. C. Funck-Brentano, L. Becquemont, A. Lenevu, A. Roux, P. Jaillon and P. Beaune, *J. Pharmacol. Exp. Ther.*, 1997, **280**, 730.

149. D. J. Kazierad, D. E. Martin, R. A. Blum, D. M. Tenero, B. Ilson, S. C. Boike, R. Etheredge and D. K. Jorkasky, *Clin. Pharmacol. Ther.*, 1997, **62**, 417.

150. K. A. Youdim, A. Zayed, M. Dickins, A. Phipps, M. Griffiths, A. Darekar, R. Hyland, O. Fahmi, S. Hurst, D. R. Plowchalk, J. Cook, F. Guo and R. S. Obach, *Br. J. Clin. Pharnmacol.*, 2008, **65**(5), 680.

151. S. B. Koukouritaki, *Pediatr. Res.*, 2002, **51**, 236.

152. C. Guillemette, *Pharmacogenomics J.*, 2003, **3**, 136.

153. E. Lévesque, R. Delage, M. O. Benoit-Biancamano, P. Caron, O. Bernard, F. Couture and C. Guillemette, *Clin. Pharm. Ther.*, 2007, **81**, 392.

154. T. K. Kiang, M. H. Ensom and T. K. Chang, *Pharmacol. Ther.*, 2005, **106**, 97.

155. K. A. Youdim, A. Zayed, M. Dickins, A. Phipps, M. Griffiths, A. Darekar, R. Hyland, O. Fahmi, S. Hurst, D. R. Plowchalk, J. Cook, F. Guo and R. S. Obach, *Br. J. Clin. Pharnmacol.*, 2008, **65**(5), 680.

156. T. D. Bjornsson, J. T. Callaghan, H. J. Einolf, V. Fischer, L. Gan, S. Grimm, J. Kao, S. P. King, G. Miwa, L. Ni, G. Kumar, J. McLeod, R. R. Obach, S. Roberts, A. Roe, A. Shah, F. Snikeris, J. T. Sullivan, D. Tweedie, J. M. Vega, J. Walsh and S. A. Wrighton, *J. Clin. Pharm.*, 2003, **43**, 443.

157. J. Raucy, L. Warfe, M. F. Yueh and S. W. Allen, *J. Pharmacol. Exp. Ther.*, 2002, **303**, 412.

158. G. Lemaire, G. de Sousa and R. Rahmani, 2004, **68**, 2347.

159. H. Kusuhara and Y. Sugiyama, in *Drug–drug interactions*, ed. A. D. Rodrigues, Marcell Dekker, Inc., New York, 2001, p. 123.

160. A. Ayrton and P. Morgan, *Xenobiotica*, 2001, **31**, 469.

161. M. Muller and P. L. Jansen, *Am. J. Physiol. Gastrointest. Liver Physiol.*, 1997, **272**, G2185.

162. H. Koepsell, *Annu. Rev. Physiol.*, 1998, **60**, 243.

163. D. K. Meijer, G. J. Hooiveld, A. H. Schinkel, J. E. van Montfoort, M. Haas, D. de Zeeuw, F. Moolenaar, J. W. Smit and P. J. Meier, *Nephrol. Dial. Transplant*, 1999, **14**, 1.

164. V. H. Lee, J. L. Sporty and T. E. Fandy, *Adv. Drug Deliv. Rev.*, 2001, **50**, S33.

165. G. A. Kullak-Ublick, M. G. Ismair, B. Stieger, L. Landmann, R. Huber, F. Pizzagalli, K. Fattinger, P. J. Meier and B. Hagenbuch, *Gastroenterology*, 2001, **120**, 525.

166. H. Kusuhara, H. Suzuki and Y. Sugiyama, *J. Pharm. Sci.*, 1998, **87**, 1025.

167. G. Burckhardt and N. A. Wolff, *Am. J. Physiol. Renal. Physiol.*, 2008, **278**, F853.

168. K. I. Inui, S. Masuda and H. Saito, *Kidney Int.*, 2000, **58**, 944.

169. F. G. Russel, R. Masereeuw and R. A. van Aubel, *Annu. Rev. Physiol.*, 2002, **64**, 563.

170. R. P. Oude Elferink, D. K. Meijer, F. Kuipers, P. L. Jansen, A. K. Groen and G. M. Groothuis, *Biochim. Biophys. Acta*, 1995, **1241**, 215.
171. M. Yamazaki, H. Suzuki and Y. Sugiyama, *Pharm. Res. (NY)*, 1996, **13**, 497.
172. K. N. Faber, M. Müller and P. L. Jansen, *Adv. Drug Deliv. Rev.*, 2003, **55**, 107.
173. P. J. Meier, U. Eckhardt, A. Schroeder, B. Hagenbuch and B. Stieger, *Hepatology*, 1997, **26**, 1667.
174. H. Kusuhara, H. Suzuki and Y. Sugiyama, *J. Pharm. Sci.*, 1998, **87**, 1025.
175. G. J. Hooiveld, J. E. van Montfoort, D. K. Meijer and M. Müller, *Eur. J. Pharm. Sci.*, 2001, **12**, 525.
176. J. M. Kovarik, L. Rigaudy, M. Guerret, C. Gerbeau and K. L. Rost, *Clin. Pharmacol. Ther.*, 1999, **66**, 391.
177. A. Hedman, B. Angelin, A. Arvidsson, R. Dahlqvist and B. Nilsson, *Clin. Pharmacol. Ther.*, 1990, **47**, 20.
178. B. Hsiang, Y. Zhu, Z. Wang, Y. Wu, V. Sasseville, W. P. Yang and T. G. Kirchgessner, *J. Biol. Chem.*, 1999, **274**, 37161.
179. G. Alexandridis, G. A. Pappas and M. S. Elisaf, *Am. J. Med.*, 2000, **109**, 261.
180. U. I. Schwarz, T. Gramatté, J. Krappweis, R. Oertel and W. Kirch, *Int. J. Clin. Pharmacol. Ther.*, 2000, **38**, 161.
181. M. Cvetkovic, B. Leake, M. F. Fromm, G. R. Wilkinson and R. B. Kim, *Drug Metab. Dispos.*, 1999, **27**, 866.
182. A. Johne, J. Brockmöller, S. Bauer, A. Maurer, M. Langheinrich and I. Roots, *Clin. Pharmacol. Ther.*, 1999, **66**, 338.
183. M. F. Fromm, H. M. Kauffmann, P. Fritz, O. Burk, H. K. Kroemer, R. W. Warzok, M. Eichelbaum, W. Siegmund and D. Schrenk, *Am. J. Pathol.*, 2000, **157**, 1575.
184. N. H. Cnubben, H. M. Wortelboer, J. J. van Zanden, I. M. Rietjens and P. J. van Bladeren, *Expert Opin. Drug Metab. Toxicol.*, 2005, **1**, 219.
185. C. Hegedus, G. Szakács, L. Homolya, T. I. Orbán, A. Telbisz, M. Jani and B. Sarkadi, *Adv. Drug Deliv. Rev.*, 2009, **61**, 47.
186. A. Aszalos, *Drug Discovery Today*, 2007, **12**, 838.
187. J. S. Lagas, *Mol. Interv.*, 2009, **9**, 136.
188. L. A. Stanley, B. C. Horsburgh, J. Ross, N. Scheer and C. R. Wolf, *Drug Metab. Rev.*, 2009, **41**, 27.
189. F. Fenneteau, J. Turgeon, L. Couture, V. Michaud, J. Li and F. Nekka, *Theor. Biol. Med. Model.*, 2009, **6**, 2.
190. J. A. Shayman and A. Abe, *Biochim. Biophys. Acta*, 2013, **1831**(3), 602.
191. B. R. Berridge, L. A. Chatman, M. Odin, A. E. Schultze, P. E. Losco, J. T. Meehan, T. Peters and S. L. Vonderfecht, *Toxicol. Pathol.*, 2007, **35**, 325.
192. P. Nioi, B. K. Perry, E. J. Wang, Y. Z. Gu and R. D. Snyder, *Toxicol Sci.*, 2007, **99**(1), 162.
193. T. Maurer, *Food Chem. Toxicol.*, 1987, **25**, 407.
194. ICH S10, *Photosafety Evaluation of Pharmaceutical*, Step 4 version, 23 November 2013.
195. *OECD Test No. 432: In Vitro 3T3 NRU Phototoxicity Test*.
196. A. M. Lynch, S. A. Robinson, P. Wilcox, M. D. Smith, M. Kleinman, K. Jiang and R. W. Rees, *Mutagenesis*, 2008, **23**, 111.
197. N. Flamand, J. Meunier, P. Meunier and C. Agapakis-Caussé, *Toxicology in Vitro*, 2001, **15**, 105.
198. M. Kamber, S. Flückiger-Isler, G. Engelhardt, R. Jaeckh and E. Zeiger, *Mutagenesis*, 2009, **24**(4), 359.
199. *Basic mutagenicity tests: UKEMS recommended procedures*, ed. David Kirkland, 1990.
200. O. Östling and K. J. Johanson, *Biochem. Biophys. Res. Commun.*, 1984, **123**, 291.
201. N. P. Singh, M. T. McCoy, R. R. Tice and E. L. Schneider, *Exp. Cell Res.*, 1988, **175**, 1840.

202. E. Lorge, V. Thybaud, M. J. Aardema, J. Oliver, A. Wakata, G. Lorenzon and D. Marzin, *Mutat. Res.*, 2006, **607**, 13.

203. M. Kirsch-Volders, *Mutagenesis*, 2011, **26**(1), 177.

204. *Non-Rodent Selection in Pharmaceutical Toxicology A "Points to Consider" document*, developed by the ABPI in conjunction with the UK Home Office. August 2002.

205. M. Pellegatti, *Expert Opin. Drug Metab. Toxicol.*, 2013, **9**(10), 161.

206. R. A. Roberts, S. L. Kavanagh, H. R. Mellor, C. E. Pollard, S. Robinson and S. J. Platz, *Drug Discovery Today*, 2014, **19**(3), 341.

207. J. A. Kramer, J. E. Sagartz and D. L. Morris, *Nat. Rev. Drug Discovery*, 2007, **6**(8), 636.

208. D. S. Smith and E. F. Schmid, *Curr. Opin. Drug Discovery Dev.*, 2006, **9**(1), 38.

209. W. S. Redfern, L. Carlsson, A. S. Davis, W. G. Lynch, I. MacKenzie, S. Palethorpe, P. K. Siegl, I. Strang, A. T. Sullivan, R. Wallis, A. J. Camm and T. G. Hammond, *Cardiovasc. Res.*, 2003, **58**, 32.

210. A. Easter, M. E. Bell, J. R. Damewood Jr, W. S. Redfern, J. P. Valentin, M. J. Winter, C. Fonck and R. A. Bialecki, *Drug Discovery Today*, 2009, **14**(17–18), 876.

211. S. Irwin, *Psychopharmacologia (Berl)*, 1968, **13**, 222.

212. E. Moscardo, A. Maurin, R. Dorigatti, P. Champeroux and S. Richard, *J. Pharmacol. Toxicol. Methods*, 2007, **56**, 239.

CHAPTER 16

Toxicology and Drug Development

MARK W. POWLEY

U.S. Food and Drug Administration, Division of Antiviral Products, WO22 RM6373, 10903 New Hampshire Ave., Silver Spring, MD 20993, USA
E-mail: mark.powley@fda.hhs.gov

16.1 INTRODUCTION AND BACKGROUND

Drug development is a lengthy, multi-step process designed to characterize the safety and efficacy of drug candidates before being approved for marketing. Phase 1 clinical trials typically provide safety, tolerability, and pharmacokinetic data for a drug following short-term administration in a small number of healthy subjects or patients. During a Phase 2 clinical trial, activity (*e.g.*, proof of concept) and safety are evaluated in a small number of patients. Finally, Phase 3 clinical trials involving many patients are conducted to provide more definitive efficacy and safety information. Progression through the range of clinical trials and ultimately to product approval depends on a series of risk *vs.* benefit analyses. A critical component of these analyses is the establishment of a toxicity profile. In addition to the human safety database generated during clinical trials, it is necessary to also consider non-clinical data to support decision making. Providing this supporting information is the primary focus of non-clinical safety assessment. This safety assessment covers a broad range of scientific disciplines including safety pharmacology, general toxicology, genetic toxicology, carcinogenicity, and developmental and reproductive toxicology. An understanding of drug metabolism and kinetics is also fundamental to non-clinical safety assessment.

A key assumption in drug development is that data obtained from non-clinical studies will be useful to inform human risk. When assessing the various safety related endpoints, *in vitro* and *in vivo* data is submitted to regulatory authorities to facilitate the decision making process. In some cases, non-clinical studies may supply the sole source of safety data available to support the initiation of a Phase 1 clinical trial. As the clinical safety database is generated during development, non-clinical data helps support decision making in regards to conduct of additional clinical trials as well as the approval for marketing. Non-clinical studies will generally serve as the only source of information for endpoints that are not feasible and/or ethical to study in humans (*i.e.*, genotoxicity, carcinogenicity, developmental and reproductive toxicity).

The Handbook of Medicinal Chemistry: Principles and Practice
Edited by Andrew Davis and Simon E Ward
© The Royal Society of Chemistry 2015
Published by the Royal Society of Chemistry, www.rsc.org

To assist both regulators and sponsors of drug development, there are numerous guidances/ guidelines that provide recommendations related to non-clinical drug development. Globally recognized guidelines are provided by the International Conference for Harmonisation (ICH). Additional recommendations are provided by regional authorities such as the European Medicines Agency (EMA), Japanese Pharmaceuticals and Medical Devices Agency (PMDA), and the United States Food and Drug Administration (FDA). While guidances/guidelines provide recommendations on drug development and highlight current regulatory thinking, decisions are made on a case by case basis and require consideration of the best available science.

16.2   TOXICOLOGY TESTING

Standardized protocols exist for many studies conducted during drug development (*e.g.*, OECD guidelines); however, sound scientific judgment must be exercised to maximize the utility of toxicology studies. When toxicology studies are conducted according to regulatory expectations (*e.g.*, good laboratory practices or GLP), there is added confidence that data are generated under appropriate conditions. However, exploratory or non-GLP studies are sometimes considered "fit for purpose" and, therefore, acceptable for regulatory purposes.

There are several important considerations for designing non-clinical studies. Of particular interest for *in vivo* evaluations is selection of appropriate animal models to maximize clinical relevance. Knowledge of a drug's *in vitro* metabolite profile or additional ADME information as well as pharmacological target characteristics (*e.g.*, tissue expression profile) can provide a basis for comparing humans with the various non-clinical species. Historical experience for a particular drug class may also be informative in aiding species selection.

The route of administration used in non-clinical studies should, in general, mimic the expected clinical route. There may be instances where data from a second route of administration is needed. For instance, intravenous administration of a drug with low oral bioavailability can provide critical systemic exposure resulting in detection of a potentially important clinical toxicity. Data from such a study will potentially provide safety data that could not be obtained otherwise.

Another critical factor in the design of non-clinical safety studies is the selection of doses. *In vitro* studies may include exposures to very high concentrations of questionable clinical relevance. Such a strategy is appropriate for assays intended to identify potential hazards *vs.* defining quantitative risk. *In vivo* studies may include doses yielding systemic exposure comparable to those expected in the clinic. However, these studies should also include higher doses in an attempt to increase the likelihood of detecting toxicity with potential clinical relevance.

In all non-clinical studies designed to characterize drug safety, it is important to determine whether effects are related to the drug or fall within the range of expected observations for a particular experimental model. The characterization of an effect as drug-related should take into account dose–response as well as the statistical and/or biological relevance. An evaluation of relevance will be based on comparison with treated groups and appropriate control groups. Concurrent controls, cells or animals treated with a vehicle or inactive control article, are routinely included in the study design. In addition, historical control values are also an important source of data. When a value is deemed to be statistically significant but falls within the range of historical control values, the effect may be described as drug-related but possessing limited biological relevance.

The following sections are intended to provide brief general descriptions of non-clinical toxicology studies routinely conducted during drug development. Focus is placed on the overall goal, endpoints evaluated, timelines for reporting, key sources of regulatory recommendations, and ultimate utility of the various non-clinical studies. A summary of the recommended non-clinical toxicology studies and timing for submission is provided in Table 16.1. Note that non-clinical development is conducted on a case by case basis so the actual studies conducted

Table 16.1 Recommended non-clinical studies.[a,b]

Study	< Phase 1	< Phase 2	< Phase 3	Marketing Application
Safety Pharmacology				
– cardiovascular				
in vitro	√			
in vivo	√			
– neurological	√			
– respiratory	√			
Genetic Toxicology				
– Option 1				
bacterial reverse mutation assay	√			
in vitro mammalian cell assay	√			
in vivo assessment		√		
– Option 2				
bacterial reverse mutation assay	√			
in vivo assessment #1	√			
in vivo assessment #2	√			
General Toxicology	√	√	√	
Developmental/Reproductive Toxicology				
– fertility			√	
– embryofetal development			√	
– pre/post-natal development				√
Carcinogenicity				√
Miscellaneous Studies				
– immunotoxicity			√	
– photosafety			√	
– abuse liability			√	
– local tolerance			√	

[a]Per ICH M3(R2).[3]
[b]Timing and scope of non-clinical studies may change during drug development.

and timing may vary. For instance, sponsors may be asked to conduct embryofetal evaluations early for drugs in a class known to be associated with teratogenicity.

16.2.1 Safety Pharmacology

The goal of safety pharmacology studies is to characterize drug effects on vital physiological functions. Per ICH S7A,[1] a safety pharmacology testing battery should include extensive evaluations of central nervous system (CNS), cardiovascular, and respiratory parameters. Other organ systems (*e.g.*, kidney, gastrointestinal, *etc.*) may also be evaluated when appropriate. Measurement of multiple endpoints (*e.g.*, CNS assessments include numerous behavioral and functional endpoints) should be included to maximize the ability to detect adverse effects. *In vivo* studies are designed to identify effects over a dose range yielding exposures that exceed the expected clinical exposure. Each study is typically conducted in a single species. Cardio-vascular assessments utilize telemeterized animals allowing continuous measurement over a prolonged period (*e.g.*, 24 hr). Telemetry is a powerful tool for identifying effects that may escape detection by less intensive sampling. In addition to the core battery of *in vivo* tests, an *in vitro* assay should be conducted to assess potential risks for inducing QT prolongation, a potentially serious effect associated with lethal cardiotoxicity. Regulatory recommendations concerning *in vitro* safety pharmacology studies are described in ICH S7B.[2]

Data from safety pharmacology studies can be used to calculate safety margins based on comparison with expected clinical exposure. The results can also be used to help select doses for additional non-clinical evaluations and identify parameters of concern that should be monitored during clinical trials. According to ICH M3(R2), [3] safety pharmacology studies should be submitted prior to Phase 1.

16.2.2 Genetic Toxicology

Genetic toxicology studies are designed to establish a drug's potential to induce changes in the DNA sequence (*i.e.*, mutations) and consequently contribute to carcinogenicity in humans. Because of the correlation between genotoxicity and carcinogenicity, results from genetic toxicology assays are useful for understanding potential carcinogenic risks during the development phases. Similar to the safety pharmacology testing strategy, genetic toxicology testing utilizes a battery approach as described in ICH S2(R1).[4] As the recommended assays are intended to identify potential genotoxic hazards, both *in vitro* and *in vivo* studies are performed using doses yielding exposures that may greatly exceed clinical exposure.

A bacterial reverse mutation assay, also known as the Ames assay, is performed to evaluate the potential to induce gene mutations *in vitro*. Additional *in vitro* studies in mammalian cells are used to detect structural chromosomal damage (*e.g.*, clastogenicity) and/or numerical chromosomal damage (*e.g.*, aneugenicity). Assays capable of investigating these critical endpoints include the mouse lymphoma assay and cytogenetic evaluations designed to identify chromosomal aberrations or micronuclei. An *in vivo* evaluation of chromosomal damage, most often micronuclei in peripheral blood or bone marrow from rodents, is also part of the testing battery performed. The combination of the Ames assay, an *in vitro* mammalian cell assay, and *in vivo* evaluation is referred to as Option 1. A second testing strategy (*i.e.*, Option 2) allows a sponsor to replace the *in vitro* mammalian cell assay with a 2nd *in vivo* endpoint. An example of a 2nd endpoint is the single cell gel electrophoresis assay (*i.e.*, Comet assay) in the liver. When positive results are encountered in any of the recommended assays, additional follow-up testing may be needed to further characterize genotoxic potential.

Genotoxicity data is used to determine whether or not clinical trials are safe to proceed. From a practical standpoint, the concept of safety margins does not apply to drugs that cause gene mutations or are considered true clastogens. In contrast, aneugenic compounds may be associated with a threshold (*i.e.*, exposure below which effects are not observed) and comparison with clinical exposure may allow for margin of safety to be identified. Results from the Ames assay and an assessment of chromosomal damage in a mammalian system is typically submitted to support a Phase 1 study. Results from all components of the testing battery must be provided in advance of Phase 2 clinical trial. When a drug is approved for marketing, the genetic toxicology data is included in the label.

16.2.3 General Toxicology

The goal of general toxicology testing is to identify drug-related toxicity occurring in the whole animal. As such, these studies use high doses sometimes well above expected clinical exposures. Examples of criteria to limit the high-dose include observations of a dose-limiting toxicity, maximum feasible concentration based on limits of solubility, saturation of systemic exposure, large differences in non-clinical exposure *vs.* the expected clinical exposure, or a limit dose (*e.g.*, 1000 or 2000 mg/kg/day).

The studies are wide in scope with regards to duration of dosing as well as the number and diversity of endpoints evaluated. Studies range from single doses to chronic studies of up to 9 months. To facilitate the use of a drug in clinical trials, general toxicology studies are evaluated throughout the development process. The studies are completed in a step-wise manner with initial clinical studies supported by short-term repeat dose non-clinical studies. Extended duration non-clinical general toxicology studies are required to support the initiation of longer term clinical trials and submission of a marketing application (Tables 16.2 and 16.3). Specific recommendations regarding duration and timing of submission for general toxicology studies are provided in ICH M3(R2).[3] Following recommendations in ICH M3(R2)[3] assures that the expected duration of non-clinical data is sufficient to cover proposed clinical dosing.

To maximize the potential to predict clinical risk, multiple endpoints are routinely evaluated in a rodent and non-rodent species. Observations and measurements routinely include changes in behavior/appearance, body weight parameters, food consumption, ophthalmoscopy, and electrocardiograms. Clinical pathology data is collected to help identify target organs of toxicity by monitoring changes in hematology, coagulation, clinical chemistry, and urinalysis parameters. Evaluations of organ weights and macroscopic tissue changes are useful; however, microscopic changes detected by histopathologic evaluation are often the definitive indicators of toxicity. Results of the various assessments are not viewed independently but are instead integrated. Overall, the most convincing evidence of target organ toxicity is consistency in effects detected by multiple endpoints (*e.g.*, hepatotoxicity indicated by both changes in clinical chemistry and histopathology).

Important outcomes from these studies include identification of target organs as well as establishment of the no-observable adverse effect level (NOAEL), no-observable effect level (NOEL), lowest observable effect level (LOEL), and/or maximum tolerated dose (MTD). Ultimately, the specific effect, severity, incidence, and reversibility (*i.e.*, persistence of an effect following a treatment-free period) will be considered in order to determine whether an effect is adverse or non-adverse. Keller *et al.*[5] recently defined an adverse effect as "A change in morphology, physiology, growth, development, reproduction, or life span of a cell or organism, system, or

Table 16.2 Non-clinical general toxicology studies needed to support clinical development[a]

Clinical Trial	Non-Clinical Studies
≤2 weeks	2 weeks[b]
>2 weeks to ≤6 months	Equivalent to duration of clinical trial[b]
>6 months	6 month study in rodent and 9 month study in non-rodent

[a]Modified table from ICH M3(R2).[3]
[b]Studies should be conducted in both rodent and non-rodent.

Table 16.3 Non-clinical general toxicology studies needed to support marketing approval[a]

Clinical Use	Non-Clinical Studies
≤2 weeks	1 month[b]
>2 weeks to ≤1 month	3 months[b]
>1 month to ≤3 months	6 months[b]
>3 months	6 month study in rodent and 9 month study in non-rodent

[a]Modified table from ICH M3(R2).[3]
[b]Studies should be conducted in both rodent and non-rodent.

(sub)population that results in an impairment of functional capacity, an impairment of the capacity to compensate for additional stress, or an increase in susceptibility to other influences."

Results from general toxicology studies can be used to guide clinicians when selecting end-points to monitor in clinical trials. Another primary use of the data is generation of safety margins based on comparison of non-clinical and clinical doses or exposures. To support an initial clinical trial, safety margins are routinely based on body surface area conversion of the non-clinical NOAEL dose to the proposed clinical trial starting dose. Clinical trials are generally considered reasonably safe if the converted NOAEL dose is ≥10-fold the proposed clinical starting dose.[6] Note that larger or smaller safety margins may be appropriate depending on non-clinical toxicity profile observed. Larger safety margins may be warranted when a drug induces severe toxicity, irreversible toxicity, or a non-clinical effect that is difficult to monitor clinically (*e.g.*, certain types of cardiotoxicity). Lower margins of safety are sometimes acceptable for reversible toxicities and those that are more easily monitored in the clinical trial (*e.g.*, liver toxicity). A more refined method of calculating safety margins is to compare systemic exposures once both non-clinical and clinical data are available. Regardless of whether safety margins are calculated using dose, body surface area conversion, or systemic exposure, the information is important the for risk assessment used in making decisions about the acceptability of clinical trials or marketing. General toxicity data is rarely included in the approved drug label.

16.2.4 Developmental and Reproductive Toxicology

Developmental and reproductive toxicology studies evaluate the ability of a drug to impact reproductive function in adult animals as well as development of offspring. While general toxicology studies can provide some limited information (*e.g.*, histopathological changes in reproductive organs), more focused studies are often needed. Regulatory recommendations for these specific evaluations are described in ICH S5(R2).[7]

As part of the evaluation of developmental/reproductive toxicity testing, a study of fertility and early embryonic development (*i.e.*, Segment 1 study) is typically performed in the rat. The Segment 1 study is designed to assess changes in mating behavior and fertility of adult animals as well as effects on implantation and early stages of embryo development. An embryofetal development study (*i.e.*, Segment 2 study) is conducted to assess effects on maternal health as well as embryofetal development. Due to the severity of effects detected during embryofetal development (*e.g.*, malformations associated with teratogenicity), these studies should be evaluated in both a rodent and non-rodent species, most likely the rabbit. Segment 3 evaluations, also called pre-/post-natal development studies, are designed to address potential effects on the pregnant female as well as embryofetal development and pups. Most often these studies are performed in the rat. In all studies, the timing and duration of dosing must correlate with the appropriate reproductive or developmental stage being evaluated. The doses evaluated in developmental and reproductive toxicology studies are determined by similar criteria to those used in general toxicology studies (*e.g.*, dose-limiting toxicity, maximum feasible concentration, saturation of exposure, *etc.*).

Developmental and reproductive toxicology data is used to guide design of clinical trials. While males and females can be included in early clinical trials without a specific assessment of fertility, clinical administration of drugs in women of child bearing potential typically requires appropriate measures (*e.g.*, contraception) until embryofetal development data is available. The data from developmental and reproductive toxicology studies is also used to support regulatory decisions at the marketing stage and, if approved, is listed in the drug label. In general, evaluations of fertility and embryofetal development are needed prior to initiating Phase 3 clinical trials while pre-/post-natal development studies can be submitted with the marketing application.

16.2.5 Carcinogenicity

Carcinogenicity studies are conducted to investigate the ability of a drug to induce tumors arising through both genotoxic and non-genotoxic mechanisms. Because carcinogenicity is a multi-step process requiring relatively long periods of time to develop, the studies are typically conducted following lifetime exposure to a drug. An alternative method is to use transgenic mouse models (*e.g.*, Tg.rasH2, p53$^{+/-}$, *etc.*) genetically modified to increase their response to carcinogens. Lifetime exposure studies include dosing periods of 2 years while the alternative transgenic mouse study requires only 6 months of dosing. Regulatory recommendations for the need and conduct of carcinogenicity studies are included in ICH guidelines S1A,[8] S1B,[9] and S1C(R2).[10]

Endpoints of primary concern include animal health and tumor formation. Tumors are characterized through observation and palpation as well as macroscopic and microscopic pathology. To increase the likelihood of detecting a carcinogenic response, studies are conducted at doses that include the MTD, maximum feasible dose based on limits of solubility, saturation of systemic exposure, or at systemic exposures greatly exceeding the expected clinical exposure.

Because genetic toxicology data serves as a surrogate to rodent carcinogenicity data during development, carcinogenicity studies, when required, are most often submitted prior to marketing approval. Therefore, the data is not often used for regulatory decision making during development. While carcinogenicity data can impact drug approval, the data is primarily used to inform prescribing physicians through the drug label.

16.2.6 Miscellaneous Studies

Drug development sometimes requires additional studies beyond the routine toxicity testing described above. The need to conduct special toxicity studies is dependent on both the drug development strategy as well as empirical non-clinical data collected in early development. When warranted, ICH M3(R2)[3] recommends the following studies should be conducted prior to Phase 3:

- Immunotoxicity studies to further evaluate the effects identified during standard toxicity testing.[11]
- Photosafety studies for drugs that absorb sunlight, distribute to tissues of concern (*i.e.*, skin and/or eye).[12]
- Evaluations of abuse liability for drugs affecting the central nervous system.
- Local tolerance studies for drugs administered through parenteral routes.

Routine toxicology studies may also be needed to demonstrate the safety of novel excipients,[13] impurities,[14–15] unique or important clinical metabolites,[3] and in some cases drugs administered in combination.[3] For drugs intended for use in pediatric populations, non-clinical juvenile toxicity studies in an appropriate species are sometimes informative.[3]

Bridging studies are useful to support a change in the route of administration or a new formulation. For instance, general toxicity should be evaluated for exposures occurring by the new route of administration. Bridging toxicity studies may also be recommended if reformulation results in significant changes to systemic exposure. A change in drug synthesis will likely have qualitative and/or quantitative effects on the impurity profile. As a result, bridging studies may be needed to qualify new impurities or those present at higher levels than previously encountered.

The recommended timing for submitting data from these evaluations varies; however, the studies may be relevant at all stages of development.

16.2.7   Toxicokinetics

Toxicokinetic (TK) data is collected in order to characterize systemic exposures to drugs and relevant metabolites. Measurements of blood/plasma drug concentrations at various time points are routinely included in *in vivo* studies and provide a basis for understanding relationship of exposure with administered dose. Following quantitative analysis, an area under the curve (AUC) can be obtained by applying non-compartmental modeling of the concentration *vs.* time data. Other examples of toxicologically useful parameters obtained during TK analysis are the maximum plasma concentration (C_{max}), time of maximum plasma concentration (T_{max}), and half-life ($t_{1/2}$).

TK data can add needed perspective to dose–response by helping differentiate between linear and non-linear dose–response or identifying saturation of systemic exposure. Systemic exposure data is also useful for detecting potential differences between males and females as well as decreases (*e.g.*, from enzyme induction) or increases (*e.g.*, from accumulation) in exposure over time. TK data from developmental and reproductive toxicology studies can help understand the extent to which drug crosses the placenta or is transmitted during lactation. In addition to TK evaluations of parent drug, characterizing systemic exposure to metabolites may also be justified. In most cases, metabolite measurements are limited to those accounting for a significant fraction of systemic exposure (*e.g.*, >10% of exposure to total drug-related material).[3] ICH S3A[16] provides regulatory recommendations on the conduct of TK evaluations.

16.3   SMALL MOLECULE DRUGS *VS.* BIOPHARMACEUTICALS

Small molecule drugs are primarily organic chemicals that can cause toxicity through both target effects (*e.g.*, exaggerated pharmacology) and off-target effects. Biopharmaceuticals include proteins, peptides, and monoclonal antibodies derived from biotechnological processes. Because biopharmaceuticals have greater specificity than small molecules, toxicity concerns are generally restricted to target effects. Therefore, toxicity testing of biopharmaceuticals should be limited to animal models where the drug is active and have relevant tissue expression of the target. The use of two species is recommended but one may be adequate if no other relevant species exist.

Another important difference between small molecules and biopharmaceuticals that impacts toxicity testing is the potential for immunogenicity. Immunogenicity occurs when a protein elicits an immunologic response potentially resulting in unintended clearance of the drug and/or toxicity. Rapid clearance of the drug can limit the utility of long-term dosing. In addition to evaluating toxicokinetics of the drug, monitoring antibody response is critical to understand exposure–response.

All standard non-clinical safety studies described above are generally considered appropriate to support clinical development and marketing for small molecule drugs. This is not the case for biopharmaceuticals. As a result of the special concerns previously mentioned, some studies are not likely to provide relevant information. Unless an organic linker is included in the molecule, standard genetic toxicology testing is not appropriate. Likewise, carcinogenicity studies are not recommended unless a mechanistic concern exists. Of primary interest are *in vivo* studies that include safety pharmacology, general toxicity, and immunotoxicity endpoints. Developmental/reproductive studies may be appropriate in some cases. ICH S6(R1)[17] provides regulatory recommendations for biopharmaceutical drug development.

16.4   REGULATORY DECISION MAKING

Extensive time and resources are spent characterizing hazards and risk associated with non-clinical toxicology studies. However, empirical safety data is not the only component of regulatory decision making as the potential benefit of the drug must also be considered.

Risk:benefit comparisons are fundamental to decision making and occur throughout development. As drugs are developed for indications ranging from those intended to maintain quality of life to those that will treat life-threating conditions, decisions are made on a case by case basis. Certain non-clinical toxicities may be acceptable for serious and life-threatening indications but would not be acceptable for drugs indicated for less serious conditions. The patient population who will be administered a drug(s) during a clinical trial is another important consideration. A safety liability may be deemed inappropriate for clinical trials involving healthy subjects but may not be viewed as unfavorably for trials involving patients. Similarly, adult patients may be viewed differently than a pediatric population. The duration of dosing will also be considered when making regulatory decisions.

A proposal to initiate a first in human trial must be supported by the appropriate non-clinical data. *In vitro* and/or *in vivo* data describing the activity of a drug are usually submitted to provide an initial indication of potential efficacy. However, the most important consideration at this stage of development is non-clinical safety studies. Using non-clinical data, the regulatory agency will determine whether a proposed clinical trial is appropriate based on the toxicities observed and the duration of non-clinical testing. In the worst case scenario, a drug will be deemed inappropriate for administration in humans under any conditions resulting in a full clinical hold. In other cases, dosing may be allowed in humans but under more restrictive conditions than those proposed by the sponsor's protocol. This partial clinical hold may include a reduction in the starting dose, overall dose range, duration of dosing, number of subjects, *etc.* If deficiencies resulting in a clinical hold are related to non-clinical data, additional studies may be required to remove the hold and allow clinical development to proceed. Throughout the drug development process, non-clinical data helps supplement the clinical experience gained in clinical trials and may also contribute to a drug being placed on full or partial clinical hold.

Non-clinical toxicology data is also considered at the marketing application stage (*e.g.*, NDA, new drug application, or BLA, biologic license application) and plays a role in the regulatory decision to approve a drug and under what conditions. During the review phase of a marketing application, the sponsor and regulatory Agency collaborate on the drug label. The drug label provides details on use of the drug (*e.g.*, indication, recommended dose and duration, potential or known interactions with other drugs) as well as clinical and non-clinical safety data. Critical considerations for the non-clinical data include effects on pregnancy and lactation, mutagenicity, carcinogenicity, fertility, and other significant findings from animal studies. The non-clinical information serves to inform prescribing physicians of the hazards and risks associated with the drug.

16.5 DISCLAIMER

The views expressed are those of the author. No official support or endorsements by the United States Food and Drug Administration are provided.

HINTS AND TIPS

1. Safety Pharmacology: Characterize effects on vital physiological functions including extensive evaluations of central nervous system, cardiovascular, and respiratory parameters. Additional organ systems (*e.g.*, kidney, gastrointestinal, *etc.*) may also be evaluated when appropriate.
2. Genetic Toxicology: Characterize potential to induce changes in DNA sequence (*i.e.*, mutations). Recommended testing includes an *in vitro* assay to evaluated gene mutations (*i.e.*, Ames assay) and assays to evaluate chromosomal damage *in vitro* (*e.g.*, chromosomal aberrations) as well as *in vivo* (*e.g.*, micronuclei).

3. General Toxicology: Characterize drug-related toxicity occurring in the whole animal. Studies are conducted in both a rodent and non-rodent species to maximize the predictive potential. Toxicity is identified through observations and measurements such as changes in behavior/appearance, body weight parameters, food consumption, ophthalmoscopy, and electrocardiograms. Clinical pathology data is collected to help identify target organs of toxicity by monitoring changes in hematology, coagulation, clinical chemistry, and urinalysis parameters.

4. Developmental and Reproductive Toxicology: Characterize effects on reproductive function in adult animals as well as development of offspring. Important evaluations include fertility and early embryonic development, embryofetal development, and pre-/post-natal development.

5. Carcinogenicity: Characterize induction of tumors arising through both genotoxic and non-genotoxic mechanisms. Studies in wild-type rodents include dosing periods of 2 years while the alternative transgenic mouse study requires only 6 months of dosing.

6. Miscellaneous Studies: In some cases, it may be necessary to further characterize immunotoxicity, photosafety, abuse liability, local tolerance, as well as juvenile toxicity. Testing of important clinical metabolites and combination drugs may also considered.

7. Toxicokinetics: Characterize systemic exposure to drugs and relevant metabolites. Measurements of blood/plasma drug concentrations at various time points are routinely included in *in vivo* studies and provide a basis for understanding relationship of exposure with administered dose.

8. Adverse Effects: Important conclusions from animal studies include the no-observable adverse effect level (NOAEL), no-observable effect level (NOEL), lowest observable effect level (LOEL), and/or maximum tolerated dose (MTD). According to Keller *et al.*,[5] adverse is defined as "A change in morphology, physiology, growth, development, reproduction, or life span of a cell or organism, system, or (sub)population that results in an impairment of functional capacity, an impairment of the capacity to compensate for additional stress, or an increase in susceptibility to other influences."

9. Safety Margins: Initial clinical trials are generally considered reasonably safe if the converted NOAEL dose is $\geq$10-fold the proposed clinical starting dose. A more refined method of calculating safety margins is to compare systemic exposure values once both non-clinical and clinical data are available. Larger or smaller safety margins may be appropriate depending on non-clinical toxicity profile observed.

KEY REFERENCES

ICH, *M3(R2) Nonclinical safety studies for the conduct of human clinical trials and marketing authorization for pharmaceuticals*, 2009.

Provides comprehensive summary of regulatory recommendations addressing non-clinical development of small molecules.

ICH, *S6(R1) Preclinical safety evaluation of biotechnology-derived pharmaceuticals*, 2011.

Provides comprehensive summary of regulatory recommendations addressing non-clinical development of large molecules.

FDA, *Guidance for industry: estimating the maximum safe starting dose in initial clinical trials for therapeutics in adult healthy volunteers*, 2005.

Provides description of methods for calculating safety margins and rationale supporting more conservative approaches to clinical trials.

REFERENCES

1. ICH, *S7A Safety pharmacology studies for human pharmaceuticals*, 2001.
2. ICH, *S7B Nonclinical evaluation of the potential for delayed ventricular repolarization (QT interval prolongation) by human pharmaceuticals*, 2005.
3. ICH, *M3(R2) Nonclinical safety studies for the conduct of human clinical trials and marketing authorization for pharmaceuticals*, 2009.
4. ICH, *S2(R1) Genotoxicity testing and data interpretation for pharmaceuticals intended for human use*, 2012.
5. D. A. Keller, D. R. Juberg, N. Cartlin, W. H. Farland, F. G. Hess, D. C. Wolf and N. G. Doerrer, *Tox. Sci.*, 2012, **126**, 291.
6. FDA, *Guidance for industry: estimating the maximum safe starting dose in initial clinical trials for therapeutics in adult healthy volunteers*, 2005.
7. ICH, *S5(R2) Detection of toxicity to reproduction for medicinal products and toxicity to male fertility*, 2005.
8. ICH, *S1A Guideline on the need for carcinogenicity studies of pharmaceuticals*, 1995.
9. ICH, *S1B Testing for carcinogenicity of pharmaceuticals*, 1997.
10. ICH, *S1C(R2) Dose selection for carcinogenicity studies of pharmaceuticals*, 2008.
11. ICH, *S8 Immunotoxicity studies for human pharmaceuticals*, 2005.
12. ICH, *S10 (step 2 draft) Photosafety evaluation of pharmaceuticals*, 2012.
13. FDA, *Guidance for industry: nonclinical studies for the safety evaluation of pharmaceutical excipients*, 2005.
14. ICH, *Q3A(R2) Impurities in new drug substances*, 2006.
15. ICH, *M7 (step 2 draft) Assessment and control of DNA reactive (mutagenic) impurities in pharmaceuticals to limit potential carcinogenic risk*, 2013.
16. ICH, *S3A Toxicokinetics: the assessment of systemic exposure in toxicity studies*, 1995.
17. ICH, *S6(R1) Preclinical safety evaluation of biotechnology-derived pharmaceuticals*, 2011.

CHAPTER 17

Patents for Medicines

PAUL A. BRADY*[a] AND GORDON WRIGHT[b]

[a] Abel & Imray Patent Attorneys, 20 Red Lion Street, London WC1R 4PQ, United Kingdom;
[b] Elkington and Fife LLP, Prospect House, 8 Pembroke Road, Sevenoaks, Kent TN13 1XR, United Kingdom
*E-mail: paul.brady@patentable.co.uk

17.1 INTRODUCTION

"Without patents, the pharmaceutical industry would not exist." So said Jean-Pierre Garnier when chief executive of GSK. This comment is reinforced by headlines in the financial pages of newspapers along the lines of "Loss of drug patents delivers more pain for AstraZeneca"[1] and "Pfizer shares closed down 1.4 percent on the New York Stock Exchange" as "U.S. court invalidates Celebrex patent; generics loom."[2]

This last comment is key to the importance of patents. The costs of researching and developing a new medicine before it can be authorised for marketing are huge. Estimates for the cost of getting a medicine from the bench to the market place range from \$1.5 billion to in excess of \$1.8 billion.[3,4] In contrast, the cost of manufacturing that medicine is small, particularly when the active ingredient is a low molecular weight chemical entity—maybe the order of few dollars or even a few cents per unit dose. Profit margins on the sales price needs to be high, in order to enable the massive cost of drug development to be recouped.

These profit margins can only be maintained whilst the originator (or its licensee) is the exclusive source of the medicine. Once there is competition from a "copycat" or generic version of the medicine, the price falls dramatically: a generic competitor who has not had to make the investment in the drug development can operate profitably at a much lower sale price. The profit margins on the medicine fall, and the originator's market share of the sales of the medicine decrease dramatically.

A typical illustration of the originator's sales figures for a medicine is shown in Figure 17.1. Annual sales grow steadily in the time after the product is placed on the market. However, once the patent expires, the originator's sales drop dramatically. It is not uncommon for sales to fall by over 80% within 3 months of the patent expiry.

The Handbook of Medicinal Chemistry: Principles and Practice
Edited by Andrew Davis and Simon E Ward
© The Royal Society of Chemistry 2015
Published by the Royal Society of Chemistry, www.rsc.org

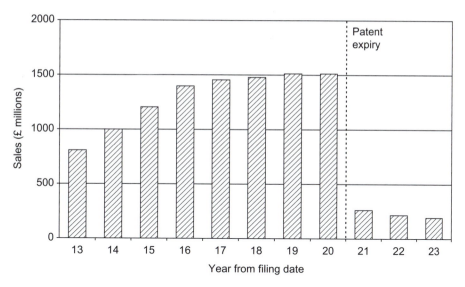

Figure 17.1 Typical sales figures for a drug through the time of patent expiry.

All of the costs of the development of a drug must be recouped before the price collapse, and patents are the main tool that a drug originator has to maintain its exclusivity and thus delay the price collapse. Patents are thus key to the viability of a drug development company as a business. It is vital that scientists working in drug discovery gain a good working knowledge of the patent system, both to avoid costly errors, and to help flag inventions when they are made. This chapter aims to give an insight into what patents are, what they can do, and how they are obtained and used.

17.2 WHAT IS A PATENT?

A patent is an exclusionary right, and it gives its owner the right to exclude others from using the invention. It is important to note that a patent does not give its owner the right to practice his own invention; there may be other patents or impediments that prevent that. Patents are one of several rights collectively referred to as "Intellectual Property". Other forms of Intellectual Property include Trade Marks (names and logos applied to goods or services), Copyright (for creative works), Registered Design Right (for the appearance of articles), confidential information (including "know-how") and regulatory data exclusivity. These other rights are also important in the pharmaceutical industry, especially Trade Marks and regulatory data exclusivity, but they generally arise late in the drug discovery process or when a medicine is already on the market. Patents are the most important rights for most organisations involved in drug discovery and development or drug manufacture.

Patents are granted for products (compounds, compositions, formulations, *etc.*) or for processes (routes of synthesis, methods of use). A patent for a product gives the owner the right to prevent others from making, using, offering for sale, or selling the invention defined by the patent claims in the country in which the patent has been granted and to prevent others from importing a product covered by the claims into that country. Where the invention relates to a process, the patent gives the owner the right to exclude others from using the claimed invention or from using, offering for sale or selling in the country in which the patent has been granted, or importing, products made by that process.

Use of a patented invention without the permission of the patent proprietor is called infringement. The exclusive patent right is enforceable in a court of law, where the patent owner sues the infringer for patent infringement. More information on infringement is given in Section 17.8.

In general, policy makers and legislators are reluctant to grant monopoly rights: they can harm competition and they interfere with the operation of the free market. Patents are an exception to this general rule, and a patent constitutes a special "deal" between an inventor and the state: in return for the inventor disclosing his invention to the public (all patent applications are published, generally 18 months after the first patent application is made for the invention), the inventor is granted a monopoly over the commercial use of the invention for a limited period (20 years in most countries). At the end of the patent term, the invention is available to the public for use. The period of exclusivity serves as an incentive to innovate and invent. This way, patents stimulate innovation and advances happen that would otherwise not happen. The net result is that technology progresses and humanity benefits.

The principles behind patents apply across all fields of technology, and they are especially important in drug development: society as a whole requires medicines for diseases that are not yet adequately controllable. Without the availability of a period of market exclusivity, companies (and indeed charities and governments) could not invest the necessary, and large, resources needed to develop the next generation of medicines.

To ensure that the monopoly granted to an inventor is fair and commensurate to the inventor's contribution, it is necessary for the scope of the patent to be accurately defined and scrutinised. The scope of a patent is defined by its claims. A patent claim is a concise and precise definition of the invention, generally in a single sentence. The claims are usually found at the end of the patent specification, under the heading "claims". When patent attorneys and lawyers speak of the claims of a patent, this is what they are referring to (rather than phrases extolling the virtues of the compounds, such as "it has now been found that the compounds of the invention exhibit surprisingly high potency"). The claims are of central importance to a patent: it is the claims that are compared with what had been known before (the "prior art") in order to determine whether the grant of the patent is justified; and it is the claims that determine what competitors are not permitted to do. More information on claims is given in Section 17.4.

In some quarters, patents are perceived as being linked to secrecy. In fact, the opposite is true: the word "patent" is derived from the Latin *patens* meaning to lie open. This refers to how patents force open disclosure of technology, which might otherwise be kept secret. The lasting record of much of humanity's technology development of the last few centuries is to be found primarily in the patent literature. Patent publications thus form an important repository of information. More information on the patent literature and how to access it is given in Section 17.10.

For a more detailed treatment of patents for medicines than can be provided in this chapter, we recommend the book by Grubb and Thomsen which, at the time of writing, is in its 5th Edition.[5]

17.3   WHAT CONDITIONS NEED TO BE FULFILLED IN ORDER FOR A PATENT TO BE GRANTED? PATENTABILITY

Patents need to be applied for *via* the national or regional patent office serving a particular country. Patent office officials, known as "examiners", carry out rigorous checks to determine whether or not all the requirements are met. If they are, then a patent is granted. In all countries, it is possible for a court to revoke a patent that has been granted, if it is later established that the requirements for patentability have not in fact been fulfilled.

Patents are granted by governments and the details vary somewhat from country to country. The following discussion gives broad-brush information applicable to most developed patent systems.

The statutes on patents set out that an invention must fulfil the following four criteria in order to be patentable. The criteria ensure that the inventor fulfils his side of the bargain in the "deal" between an inventor and the state mentioned in Section 17.2. The invention must:

- Be novel
- Be inventive
- Have industrial applicability
- Not be specifically excluded from being protectable by a patent

In addition to the invention fulfilling these criteria, the specification for the patent application must:

- Describe the invention sufficiently clearly and completely for the person skilled in the art to be able to put the invention into practice.

A patent which meets all of the requirements of the law is said to be "valid". Taking each of the above criteria in turn:

17.3.1   Novelty

An invention is novel if, at the date on which the patent application is filed, the invention is not in the state of the art. The state of the art is taken to be everything that has been disclosed to the public up to that date. Disclosures in any form must be taken into account. The state of the art thus includes not only printed publications and publications on the internet, but also oral disclosures, such as conference presentations or non-confidential discussions. Disclosure to the public can also occur by demonstrating a prototype or showing images in a video. There is no requirement for a disclosure to be generally known to workers in the field: an obscure disclosure, which might be in a foreign language, must also be taken into account.

The requirement for novelty flows from the requirement of the "deal" mentioned in Section 17.2: if the invention was already known, then the inventor is not providing anything to fulfil his side of the bargain.

An important feature of novelty is that the inventor's own disclosure of his own invention is also part of the state of the art. An inventor's own disclosure before filing a patent application can thus compromise the novelty of an invention. It is thus crucial that an invention be kept confidential until a patent application is filed. In some countries, notably the USA, the law provides a "grace period", a period in which an inventor may publish his invention without damaging his own later patent application. In the USA, the grace period is 12 months. In some other countries it is 6 months. A large number of countries (including most of Europe) have no grace period, so deliberate use of the grace period should be contemplated only with extreme caution. Controlling publications prior to filing patent applications is a vital part of achieving effective patent protection across the world.

17.3.2   Inventive Step

In addition to possessing novelty, an invention must have an inventive step. An invention has an inventive step if it is not obvious over what was in the state of the art at the patent filing date.

Referring again to the "deal" mentioned in Section 17.2 above, the public should not be prevented from doing something that is a non-inventive variant of something that was already known.

Once it has been established what the state of the art comprises, the question of novelty should be a black and white one: "was something inside the patent claims disclosed before the patent filing date or not?" The question of inventive step is generally more nuanced; in particular the answer can turn on just who should be taken into account as considering whether an advance is obvious over what was already known.

A large number of patent cases that reach the courts turn on the question of inventive step. The courts and patent offices have thus formulated tests in order to bring structure to the assessment so as to increase legal certainty for patent applicants and the public. The tests by the various authorities require a definition of the "person skilled in the art". Much has been written about this legal construct: he is taken to know everything that has ever been published or disclosed in other ways, but he has no inquisitiveness, and follows only very highlighted leads. In the case of Pfizer's patent covering the mechanism of action of Sildenafil (marketed under the trade mark Viagra®) and other cGMP phosphodiesterase inhibitors in the treatment of erectile dysfunction, the person skilled in the art was described as follows by the UK High Court judge Mr Justice Laddie:

"The question of obviousness has to be assessed through the eyes of the skilled but non-inventive man in the art. This is not a real person. He is a legal creation. [...] He is deemed to have looked at and read publicly available documents and to know of public uses in the prior art. He understands all languages and dialects. He never misses the obvious nor stumbles on the inventive. He has no private idiosyncratic preferences or dislikes. He never thinks laterally. He differs from all real people in one or more of these characteristics."[6]

Thus, the "person skilled in the art" is not a single person that could ever exist, certainly not in a research or development role. In practice, it is necessary to gather evidence from the field of the invention regarding what was generally known to teams working in the relevant area, and then to impute that knowledge into the fictitious "person skilled in the art".

The European Patent Office (EPO) has probably the most rigid approach to the assessment of inventive step. This has become known as the "Problem and Solution Approach". The approach starts with identifying what is the single disclosure in the prior art that is the closest to what is claimed in the patent in question. Next, one has to see what the difference is between that disclosure and the claimed invention. Then one has to identify what technical effect that difference has. That technical effect defines what the problem is that is considered to be solved by the invention (it may or may not be the same problem that the inventor actually thought he was working on). One then asks the question, "starting from the closest prior art disclosure, and trying to solve that problem, would it have been obvious to arrive at what is now claimed?"

Taking a chemical example: an inventor has devised a set of compounds with general formula (X) which have good activity at receptor Y. In the prior art, there is a compound (A) with structure similar to (X), but with a methyl group at a particular location where (X) has a phenyl group. The compounds (X) with the phenyl group have a higher affinity to receptor Y than compound (A) with the methyl group. One would then have to answer the question: starting from compound (A) and wishing to obtain a compound with higher affinity, would it be obvious to replace the methyl group in the compound with the phenyl group? The answer will depend on whether that replacement is known in other compounds to bring about the change in activity that has been found by the inventor.

An important factor in determining the answer is how predictable the field of technology is. If the field is highly unpredictable, then a finding that a compound has activity might very often be surprising, and hence supportive of inventive step. In a field that is unpredictable, it is all the more important that a patent applicant provides good evidence that the invention does indeed solve the problem. Generally, the evidence will be in the form of results of laboratory experiments. The better the evidence that the patent applicant can provide, the better will be the prospects of succeeding in demonstrating that the invention has an inventive step.

The EPO "Problem and Solution Approach" is perhaps the most rigid of the tests that have been developed. Slightly different tests are used in other places, sometimes giving similar results, and sometimes not. For a more extended analysis of inventive step treatment in different territories, see references.[7,8]

The structured tests are of assistance when assessing inventive step and it is important that they are considered when one is writing a patent application. At the same time, it must be remembered that they are only helpful tests rather than the ultimate law itself, and so they must not be overly relied upon. Amongst other things, the tests evolve and can be overturned. A decade or more can pass between a patent application being written and a case coming to court. That is plenty of time for the approach on a particular matter to evolve. In general, a patent applicant puts himself in the best possible position by obtaining good data showing that the invention works well. Experiments comparing the invention side-by-side with what has been done before are especially useful. This sometimes involves making a different comparison from what would be done for an academic research paper: for a research paper, the important comparator to look at might be the "Gold Standard", or the current standard clinical treatment. To support inventive step in a patent application, the comparator should be the prior art compound or method that is closest in its features to what is being claimed. That compound or method from the prior art might be obscure, or it might not have achieved mainstream acceptance for other reasons, such as cost.

17.3.3 Industrial Applicability

An invention is considered to be capable of industrial application if it is made or used in any kind of industry. The concept of "industry" is interpreted very broadly, and includes essentially any commercial enterprise. In practice, and certainly in the field of drug discovery, essentially any invention will satisfy the industrial applicability requirement. In the early days of DNA sequencing, when certain sequences were determined but their function was not known, some patent applications for DNA sequences were refused under this heading. In 2011, the UK Supreme Court determined in the case of Eli Lilly *vs.* Human Genome Sciences regarding a DNA sequence (for the gene for neurokine alpha) that HGS's identification of the gene as being a member of the TNF ligand superfamily was adequate to show that the sequence had an industrial applicability, even though the assertion had not been supported by any wet laboratory work by the time the patent application was filed.[9] The requirement forces applicants and their patent attorneys to think about real-world uses of their inventions when formulating patent claims, but in general it is rarely a factor that actually interferes with being able to obtain a patent.

17.3.4 Exclusions

Certain forms of development are specifically excluded from patentability by the patent statutes. The exclusion that most affects medicine is that most jurisdictions have a bar on the patenting of methods of medical treatment. The rationale for the exclusion is that patent law should not interfere with what a doctor can do when he or she is treating a patient.

In recognition of the fact that patents are key to the medical industry, many statutes make it clear that the exclusion does not prohibit the grant of a patent for a substance or composition for use in a method of treatment. This means that special wording must be used. In Europe, the use of a compound as a drug can be patented, but a patent claim must have the wording "compound X for use in the treatment of disease Y", rather than the more intuitive "method of treating disease Y by administering compound X". This may look like wordplay that subverts the exclusion in the statute, but that is not the case: the wording of the allowed claim makes it clear that it is the compound that is the infringement, thus making manufacturers and suppliers the potential infringers, rather than the doctor at the bedside carrying out the method.

In the USA, methods of medical treatment are not excluded from patentability, and they are regularly granted by the US Patent and Trademark Office. The US patent statute deals with this matter in a different way, by exempting from the definition of patent infringement a medical practitioner's performance of a "medical activity" on a human or animal. The same end result of allowing doctors to treat patients how they see fit without a threat of patent infringement is thus achieved in the USA in a different, and most would say more logical, way than in Europe.

In addition to methods of medical treatment, the excluded subject matter list for Europe also includes abstract concepts (discoveries, scientific theories or mathematical methods, and also methods for performing a mental act, playing a game or doing business), or a program for a computer, and inventions the exploitation of which would be contrary to public policy or morality. This latter category presents a bar to the grant of patent claims to a human being, in any stage of its development. It has recently been decided by the highest court in Europe that this includes human embryonic stem cells.[10,11] In certain areas of medical research, if patents are to be obtained, it is necessary to develop a strategy to deal with this exclusion, and the best approach will depend on just how central stem cells are to the ultimate therapeutic product. The exclusion of computer programs should not be taken at face value: over the years, case law has developed to take account of the evolution of computers from simple mathematical machines that carry out abstract calculations to integral parts of modern technological devices with real-world technical effects. In practice, many software-based inventions are now patent-eligible.

17.3.5   Clarity and Sufficiency/Reproducibility

It is necessary for a patent application text to describe the invention with sufficient clarity and detail for the person skilled in the art to carry it out without having to use any inventive skill. The reason for this requirement is again linked to the "deal" described in section 17.2 above. If the patent applicant is to deserve the 20 year period of monopoly, the patent must describe the invention in such a way that at the end of the 20 year patent term, the public has full use of the invention and can put it into effect. The patent applicant is not allowed to keep important details secret. The standard of description is similar to that required for publication of a paper in a peer-reviewed journal.

For the preparation of chemical compounds, the sufficiency requirement is generally met if the patent application contains a description of the synthesis of specific example compounds. It is important that the descriptions start from compounds that are in the literature or are commercially available. For claims to compounds used in very specific treatments, the description should include details of how the treatment is carried out.

The sufficiency requirement is particularly important in the case of certain biotechnological inventions. An antibody that has been raised by challenging the immune system of a laboratory mouse with a selected antigen may have very beneficial properties and may in principle be patentable. However, the isolation of a particular specific antibody can be a matter of chance, and a paper description alone will often not provide a skilled reader with everything he needs to carry out the invention himself. To deal with this situation, there exist specifically approved International Depository Authorities which house and maintain samples of antibodies or other biological entities and make them available to members of the public who request them.

In the USA, there is a particular requirement to disclose what the inventors consider at the time of filing the patent application to be the "best mode" of working the invention. The rationale behind the requirement is sound: it is a more rigorous onus on the patent applicant to keep his side of the deal described in Section 17.2. The patent applicant is not allowed to keep important details secret. In practice, the best mode requirement is found by many practitioners to be an overly onerous and costly feature: in the preparation of patent specifications, multiple versions of an example might be included if it not clear which might be "best"; in litigation, it has often been necessary to find and then consider a large number of internal company documents to check that a patentee has properly fulfilled the "best mode" requirements. Therefore, when the US patent law was revised in 2011 with the passing into law of the "America Invents Act", the requirement to disclose the "best mode" was retained, but failure to disclose the "best mode" ceased to be a basis for invalidating or rendering unenforceable an issued patent. The attention given to the requirement is now thus much reduced.

17.4 ANATOMY OF A PATENT SPECIFICATION

As will be apparent from the material above, patents are a specialised and complex field. Whilst it is possible for an inventor to write and file his own patent application, the drafting of a patent application is best carried out as a collaborative effort between the inventors and their patent attorney. Patent attorneys are scientists by background with specialised legal training. As a medicinal chemist, you will have significant involvement in the drafting of patent applications in your field. This section should provide some guidance as to what may be expected of you in that process.

The key part of a patent application is the "patent specification" which consists of a description and a set of claims, and in many cases figures. The claims will generally be written by the patent attorney, in close discussion with the scientists and often also a business manager. As a medicinal chemist, you will probably be involved mostly in the writing of the description. The writing of a patent specification has some features in common with the writing of an academic paper, but some important differences.

17.4.1 The Description

The description has several parts:

Introduction: this has to inform the reader of the technical field and background art. The technical field is important to direct the Patent Office Searcher to the correct area in which to carry out his search. The background does not need to be an analysis of the literature in the broad area of the invention, and it does not need to explain to the reader how the invention came about. It thus differs from the typical introduction to an academic paper. The introduction to a patent application should simply describe a small number of closest previous disclosures to show the prior art from which the invention is an advance. Ideally, this should include the "closest prior art" discussed in Section 17.3 under inventive step, but that is not always known at the time of writing. The introduction may include a description of drawbacks or shortfalls of the prior art. Care needs to be taken in describing those shortfalls, as the new invention might well be an advance but not necessarily overcome all of the shortfalls in the prior art.

Before any patent searching has been carried out, the medicinal chemist will have a better idea of the background art than the patent attorney, and the medicinal chemist might thus often write the first draft of the introduction.

Statement of Invention: after the introduction, there follows a statement of invention. This is a re-stating of the main patent claim: "The invention provides compounds of formula (X)."

That is followed by some statements setting out what is new and useful about the invention: "The compounds have high activity at receptor Y and they thus have use in the treatment of disease Z."

Detailed description: this is the meaty part of the specification. Each aspect of the invention should be described in further detail. This includes setting out fall-back positions that are narrower than the main claims, in case something comes to light that means that the broadest claim cannot be pursued to grant. For the purposes of sufficiency of disclosure, the description also needs to explain how to carry out the invention. If there are alternative ways of carrying out the invention, the detailed description should include as many of them as possible. It is often also in a patent applicant's interest to disclose features here (for example combination treatments with other drugs) so as to put them into the public domain. When the patent application publishes, that disclosure prevents a third party from applying for a patent for those features.

Worked Examples: these follow the detailed description and they describe actual preparations and experiments that have been carried out. They serve in part to substantiate the sufficiency of disclosure. They also serve to provide evidence of the beneficial properties that underpin the inventive step of the invention. Generally, the medicinal chemist is the main author of the worked examples section.

The claims: The claims define the invention: they define the monopoly that the applicant is requesting. They must be entirely clear and self-contained, and they must define the invention. It is important that they do not overlap with what is in the prior art and that everything within the claims has the beneficial features that underpin the invention. The specific worked examples will have a strong bearing on inventive step, but the applicant actually provides more to the public than the specific examples: he may be opening up a whole new field or he may be providing a new class of compounds. Commensurate to the contribution to the art, the applicant is thus generally entitled to claim his invention more broadly than the specific worked examples: close variants can also be claimed. One of the most difficult elements in the drafting of a patent application is determining the best scope of claim to apply for. Features can be generalised, but everything in the claim must have the invention's beneficial features. The generalisation is thus constrained by what can reasonably be extrapolated from the data that has been obtained. At the same time, the whole of the scope must be novel and not obvious. This requires a complicated balance to be struck.

The patent application will contain several claims, so a broad and generalised first claim can be backed up by narrower claims. A patent can also contain claims in several different categories, for example a compound and its use. Categories of claims are summarised in the next section below.

17.4.2 Types of Patent Claim

Depending on the stage of the research, an invention can arise in various elements of a new medicine:

Compound per se: if a compound has never been made before, it is possible to claim the compound (and compounds similar to it) itself ("*per se*"). That claim covers the compound in any setting, in any form and for any use. In the USA, this is called a "composition of matter" claim.

Uses in medicine (first medical use claims/methods of medical treatment): if a compound is known, but it has not previously had a use in medicine, then it is possible to claim the compound for use as a medicament (Europe, Japan and many other territories), or a method of treatment using the compound (USA).

Further uses in medicine (second medical use claims/second method of medical treatment): if a compound is known to have one use in medicine, and the compound is later found to be useful in the treatment of a second condition, then it is possible to claim the compound for use as a medicament for the treatment of that second condition (Europe), or for use in the manufacture of a medicament for the treatment of the second condition (Europe, Japan and many other territories), or a method of treatment of the second condition using the compound (USA).

Salts or crystalline forms: during drug development, new physical forms of the drug compound often need to be developed. If those have beneficial and surprising properties, then they can be claimed in a patent.

Composition claims and combinations: a particular composition with particular excipients or with a particular other drug might be developed. If such a composition provides beneficial properties, the composition can be claimed.

Dosage regimen claims: if a clinical trial shows that a particular dosing regimen brings unexpected benefits, then such a regimen can, in many countries, be the subject of a patent.

Synthetic processes: improved methods of preparing compounds can be patentable.

In general, going down this list, the value of the protection decreases. This is for a combination of two reasons: the further one goes down the list, the more opportunities there are for a competitor to work around the claim, for example by devising an alternative route of synthesis; also in general the further one goes down the list the easier it is for a third party to challenge the inventive step of a claim. A drug development team puts itself in the best overall position by having a "portfolio" of patents covering various aspects of the final product. The later-filed patents to downstream developments may be inherently less valuable than the first compound *per se* patent, but the later-filed patents have later expiry dates which give them added value in practice.

17.4.3 Case Study (a): Typical Claims in a Pharmaceutical Patent

Atorvastatin is the cholesterol-lowering statin drug originally marketed by Warner-Lambert and later by Pfizer under the name Lipitor®. Atorvastatin has the structure shown in Figure 17.2. It was first approved by the European Medicine Agency (EMA) in 1997 and by the US Food and

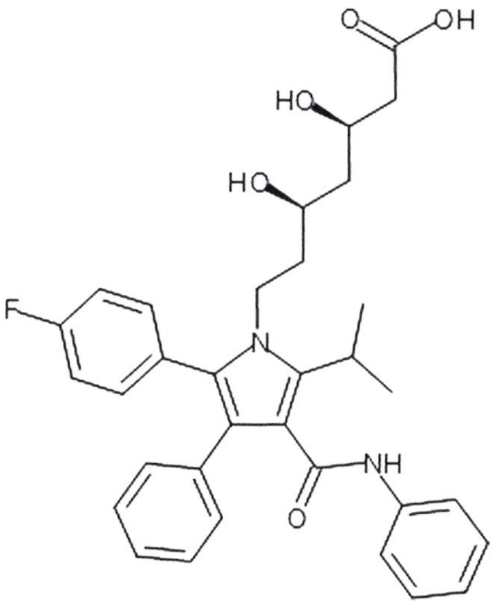

Figure 17.2   The structure of Atorvastatin, sold as Lipitor®.

Drug Administration (FDA) in 1996. It is the world's largest selling drug to date by value, with total global sales estimated to be around \$125 billion between 1996 and 2012, when it lost its patent protection.

The main patent protecting Atorvastatin in Europe was EP 0 247 633B. It expired on 29 May 2007, and its Supplementary Protection Certificate (SPC) expired on 6 November 2011 (see Section 17.7.2 below for details of SPCs). The patent specification can be accessed for free over the internet from the Espacenet Database.[12]

Atorvastatin is specifically disclosed as Example 1 in the patent. As described above, it is not the examples that determine what is protected by the patent—it is the claims. In the patent, there are eight claims, with claim 1 reading:

"1. A compound of structural formula I

wherein X is $-CH_2-$, $-CH_2CH_2-$, $-CH_2CH_2CH_2-$ or $-CH_2CH(CH_3)-$; R_1 is 1-naphthyl; 2-naphthyl; cyclohexyl; norbornenyl; 2-, 3-, or 4-pyridinyl; phenyl, phenyl substituted with fluorine, chlorine, bromine, hydroxyl; trifluoromethyl; alkyl of from one to four carbon atoms, alkoxy of from one to four carbon atoms, or alkanoyloxy of from two to eight carbon atoms; either of R_2 or R_3 is -$CONR_5R_6$; where R_5 and R_6 are independently hydrogen; alkyl of from one to six carbon atoms; 2-, 3-, or 4-pyridinyl; phenyl; phenyl substituted with fluorine, chlorine, bromine, cyano, trifluoromethyl, or carboalkoxy of from three to eight carbon atoms; and the other of R_2 or R_3 is hydrogen; alkyl of from one to six carbon atoms; cyclopropyl; cyclobutyl, cyclopentyl, cyclohexyl; phenyl; or phenyl substituted with fluorine, chlorine, bromine, hydroxyl; trifluoromethyl; alkyl of from one to four carbon atoms, alkoxy of from one to four carbon atoms, or alkanoyloxy of from two to eight carbon atoms; R_4 is alkyl of from one to six carbon atoms; cyclopropyl; cyclobutyl; cyclopentyl; cyclohexyl; or trifluoromethyl; or a hydroxy acid or pharmaceutically acceptable salts thereof, derived from the opening of the lactone ring of the compounds of structural formula I and having the formula X

where X, R_1, R_2, R_3, and R_4 are as defined above.

Claim 1 does not refer to any other claims in the patent and is called an independent claim. It covers within its general formula not only the individual compounds exemplified in the patent application, but also compounds that are similar to them. Because the compound can exist in a ring-opened and a ring-closed form, the patentee included both structures, for the avoidance of any doubt. Atorvastatin falls within the scope of claim 1 when, in formula I, X is -CH$_2$CH$_2$-, R$_1$ is phenyl substituted with fluorine in the *para* position, R$_2$ is phenyl, R$_3$ is -CONR$_5$R$_6$ where R$_5$ is phenyl and R$_6$ is hydrogen, and R$_4$ is iso-propyl."

The next claims, claims 2 to 5, each refer back to claim 1 ("A compound as defined in claim 1, wherein. . ."), and they introduce further limitations in the scope. These are referred to as "dependent claims" and they define the invention more narrowly. They are fall-back positions in case claim 1 should later be found to have a problem. Claims 3 to 5 are directed to single compounds selected from the examples. Specifically, Atorvastatin is claimed in claim 3.

Claims 6 and 7 read:

"6. A pharmaceutical composition, useful as a hypocholesterolemic agent, comprising a hypocholesterolemic effective amount of a compound in accordance with Claim 1 in combination with a pharmaceutically acceptable carrier."
"7. A method of use of a compound according to Claim 1 for the manufacture of a pharmaceutical composition as defined by Claim 6 for inhibiting cholesterol biosynthesis in a patient in need of such treatment."

Claim 6 thus relates to pharmaceutical compositions of the compounds of claim 1. The use as a hypocholesterolemic agent is specifically mentioned. That use is not actually particularly limiting on the claim: a composition that is prescribed for a different use, but that could be used as a hypocholesterolemic agent will fall within the claim.

On the face of it, claim 7 looks it is directed to a method of manufacture. In fact, it is a second medical use claim and it covers the use of the compound in treatments for inhibiting cholesterol biosynthesis. The claim is present as a fall-back in case a prior art document comes to light which discloses a compound within the scope of the patent, but for a different use.

Claim 8 is directed to a method of preparing a compound having structural formula I according to Claim 1, comprising a series of recited steps. The full claim is not reproduced here. The claim would catch a manufacturer of the compound using the claimed process.

It might seem that there is a lot of redundancy built in to the claims. Because the Atorvastatin compound was new *per se* when the patent application was filed, the compound was claimed *per se* and the patent covers the compound in any composition and for any use.

Within the compound *per se* claims (claims 1 to 5), the broadest claim (claim 1) covers a group of compounds, and would catch "me-too" copies that a competitor might want to develop. Within the claim breadth, the most important compound to protect is the compound that has been taken to market, Atorvastatin. Probably, that compound was the lead development compound at the time the patent application was filed, as it is included as Example 1. That individual compound is therefore picked out individually in claim 3 and, if the worst came to the worst, the patent could be amended to just that single compound. That cut-back patent would still protect the active ingredient from direct competition from generic manufacturers.

Despite the presence of the compound *per se* claims, it is important that the patent contains various different types of claims so that there are fall-back positions in case the compound *per se* claims should later be found to have a problem; in that situation, the additional different types of claim are necessary so that different infringers can be caught by the patent. The pharmaceutical composition claim, claim 6, would be infringed by somebody manufacturing tablets of Atorvastatin. The second use claim, claim 7, would be infringed by somebody manufacturing the compound for use in inhibiting cholesterol biosynthesis in a patient. Lastly, claim 8 would be infringed by somebody manufacturing the compound using the claimed general method.

17.4.4 Case Study (b): How Broad Should a Claim Be?

Litigation took place between Pharmacia and Merck in 2000 and 2001 first in the English High Court and then in the English Court of Appeal concerning their COX-II inhibitors. The decisions in the case can be accessed for free over the internet.[13]

In the 1990s, Pharmacia's predecessor company, G D Searle, developed the drug Celecoxib, the active substance of the medicine Celebrex®, which was the first in a new class of anti-inflammatory drug: COX-II inhibitors. These had powerful anti-inflammatory effects whilst minimising the adverse gastrointestinal effects seen with established Non-Steroidal Anti-Inflammatory Drugs (NSAIDs), such as aspirin or ibuprofen. At about the same time, Merck developed a competing anti-inflammatory drug Rofecoxib under the trade name Vioxx®, which was also a selective COX-II inhibitor. The two drugs were similar in chemical structure, as seen in Figure 17.3, but there was no suggestion that Searle's patent protecting Celecoxib described Rofecoxib.

It was clear at the time that Vioxx® and Celebrex® were potential blockbusters that could earn billions of dollars for the company successfully marketing them. If Searle were able to keep Vioxx® off the market, then it would potentially have this new market to itself. Searle had other patents relating to COX-II inhibitors.

Just before the UK launch of Vioxx®, Pharmacia/Searle (which was by now owned by Monsanto) sued Merck for patent infringement, seeking an injunction to prevent them from commercialising Vioxx®. Pharmacia/Searle's patent was EP 0 679 157B,[14] in which claim 1 was to a compound of formula I:

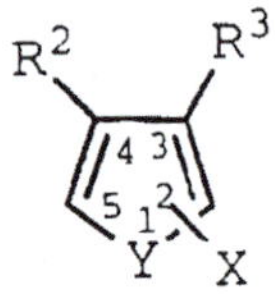

According to the claim, the definition of Y included O, the definition of X included -OH, and the definitions of R^2 and R^3 included aryl optionally substituted by a radical that could be "lower alkyl sulfonyl". Further it was a requirement that at least one of R^2 and R^3 was substituted by methylsulfonyl.

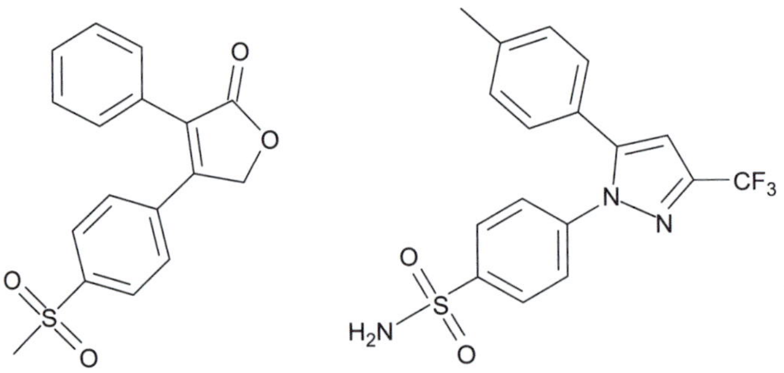

Figure 17.3   The structures of Rofecoxib, sold as Vioxx®, and Celecoxib, sold as Celebrex®.

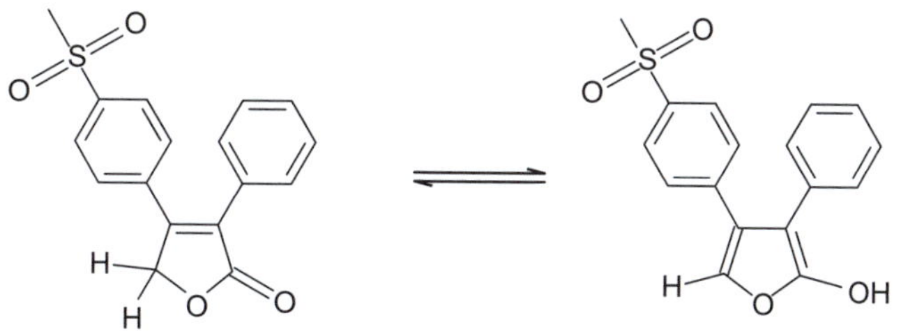

Figure 17.4 The structures of the keto and enol forms of Rofecoxib.

Pharmacia/Searle argued that this meant that the claim covered Merck's drug substance, Rofecoxib (at the time referred to by its code MK-966). They also argued that although the patent claim covered hydroxyfuran derivatives, rather than furanones, the skilled person would know that this would tautomerise to give the corresponding furanone, as shown in Figure 17.4. Pharmacia/Searle were able to make this argument because the claims had been written broadly. A broad claim is, however, susceptible to attack on its validity, and Merck exploited that in a counter-claim for invalidity of the patent.

Although there was no dispute that MK-966 was a potent and selective COX-II inhibitor, Merck showed that a significant number of close analogues of MK-966 were inactive in the screens disclosed in the patent.

As discussed in Section 17.3, it is a requirement in the patent "deal" between the applicant and the state that the claims of a patent must reflect the invention that the patentee has made. This is controlled by the requirement for inventive step (see Section 17.3.2), and by the requirement for a sufficient description of the invention (see Section 17.3.5). The fact that some of the compounds covered by the claim were not selective COX-II inhibitors (in fact some were neither COX-I nor COX-II inhibitors) meant that many of the compounds in the claims did not have an inventive step.

The judge ruled that:

"If compounds having the features of the claim may or may not possess the qualities which the patent says unify the class, it cannot be said that the claim reflects a true class at all. It is just a generalised description of a large number of chemical compounds. Such a claim is not analogous to a claim to a new principle, since the patentee has given no information, such as a structure/activity relationship, which enables the reader of the specification to draw any conclusions as to the properties of any particular compound without further experiment. All he has done is to describe the scope of the claim with spurious precision."

The patent was thus revoked for lack of inventive step.

Furthermore, the judge held that the patent was also bad for a lack of a sufficient disclosure, and he also revoked the patent for that reason:

"If the invention is a selection of certain compounds, in order to secure an advantage or avoid some disadvantage, not only must the specification contain sufficient information on how to make the compounds, it must also describe the advantage or how to avoid the disadvantage."

So, in this case, initially Pharmacia/Searle were probably pleased to have obtained grant of a patent that was so broad as to cover not only their own compound Celecoxib but also Merck's Rofecoxib. However, the breadth of the claim was also its downfall: it was not inventive across the whole of its scope and the description did not enable it to be operated across the whole of its scope.

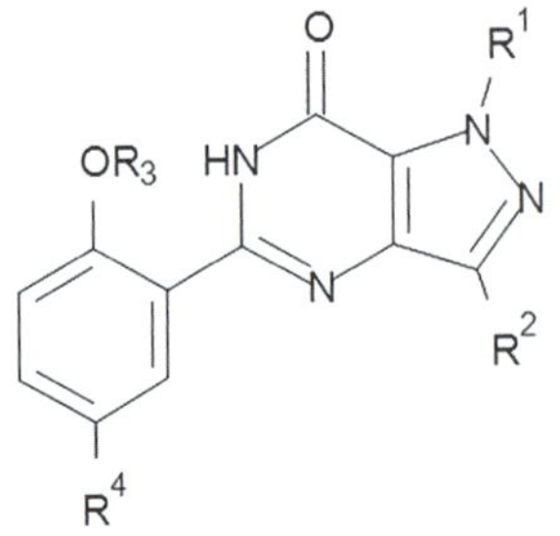

Figure 17.5 The generic structure of the group of pyrazolopyrimidinones covered by Pfizer's EP 0 463 756.

17.4.5 Case Study (c): A Second Therapeutic Use

Sometimes an invention concerns the new use of a known substance. An infamous example is based on the finding of a new therapeutic application for Sildenafil (sold under the name Viagra®), the case mentioned briefly in Section 17.3.2. In June 1991, Pfizer filed a patent application, EP 0 463 756, for a group of pyrazolopyrimidinones for use as anti-angina agents.[15] The compounds had the general formula shown in Figure 17.5.

It was said that the basis of their therapeutic effects was that the compounds of the invention exhibited selectivity for inhibition of cyclic guanidine 3',5'-monophosphate phosphodiesterases (cGMP PDEs) rather than cyclic adenosine 3',5'-monophosphate phosphodiesterases (cAMP PDEs) and, as a consequence of this selective PDE inhibition, cGMP levels were elevated. In turn, that could give rise to beneficial platelet anti-aggregatory, anti-vasospastic and vasodilatory activity, as well as potentiation of the effects of endothelium-derived relaxing factor (EDRF) and nitrovasodilators.

During its early clinical development for the treatment of angina, Sildenafil was found to be useful in the treatment of erectile dysfunction, and in May 1994 a patent application was filed claiming the use of essentially the same group of pyrazolopyrimidinones as EP 0 463 756, including Sildenafil, for the treatment of erectile dysfunction. The new patent application included not only claims to the use of the group of pyrazolopyrimidinones, but also a claim to the use of *any* cGMP PDE inhibitor, particularly a PDE$_V$ inhibitor for the treatment of erectile dysfunction.

The broad use claim was granted in Europe in patent EP 0 702 555B.[16] The claim covered any compound which is a cGMP PDE inhibitor, whether previously known or not, for this treatment. It was a particularly powerful claim, enabling Pfizer to monopolise the use of the cGMP PDE mode of action across the entire therapeutic area. That included Tadalafil which was in development by Lilly ICOS (marketed under the name Cialis®) and Vardenafil which was in development by Bayer, GSK and Schering Plough (marketed under the name Levitra®), even though Pfizer's patent did not disclose or claim compounds with those structures within its generic formula (Figure 17.6).

After it was granted, the patent was attacked by several parties, including Lilly ICOS, Bayer and Schering Plough, and the opponents succeeded in convincing the European Patent Office (EPO) to revoke the patent on the grounds of lack of inventive step: the invention claimed at that level of generality was not inventive over the prior art.

Even though Pfizer's patent covering the broad uses was revoked, it should be noted that there was nothing wrong in principle with claiming the treatment by a compound in a functionally-defined class. If the application had been filed a few years earlier, and before various publications had taken place, then the patent would probably have survived.

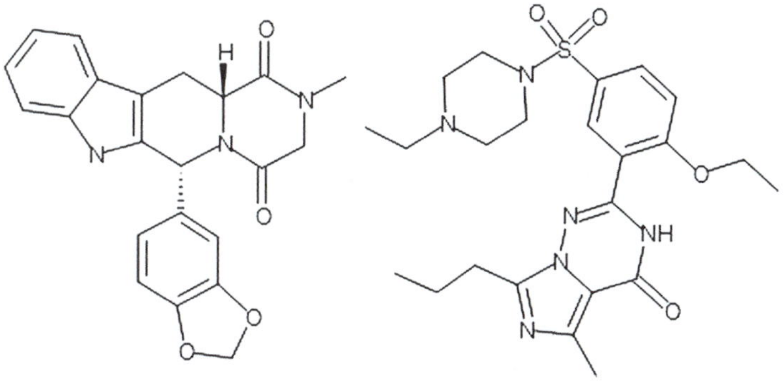

Figure 17.6   The structures of Tadalafil, sold as Cialis®, and Vardenafil, sold as Levitra®.

17.5   OWNERSHIP AND INVENTORSHIP

As patents are valuable pieces of property, it is important that they are owned by the body (generally a company) that needs to own them. This sounds nice and simple, but it is surprising how often things can go wrong, and increasingly so in the current age of spin-out companies, joint-ventures and out-sourced research.

There are small variations from country to country, but in most jurisdictions, the first owner of an invention, and hence the patent rights in it, is the inventor or the group of co-inventors. Either by virtue of an employment contract, or by operation of the national law, an invention made by an employee who is employed to make inventions belongs to his employer. That is a relatively straightforward situation, though in corporate groups, it needs to be checked if the employing entity is actually the company that needs to own the patent rights. If a contract research organisation or an independent consultant is involved in the development of a product, then it is important that the contract with them includes an obligation to assign all patent rights to the company that engages them.

Given that ownership of a patent is determined by the inventorship, it is important that inventorship is determined properly. An inventor is a person who is the "actual deviser" of the invention. That is to say the person, or group of people, who came up with the invention that is being claimed. Inventorship differs from authorship on an academic paper: supervising research and providing resources for a laboratory does not constitute inventorship; nor does carrying out the laboratory work to show that an invention works, no matter how specialised or time-consuming the laboratory work may be. A rigorous approach needs to be taken to the determination of inventorship.

The USA and some other countries (including China) have laws that restrict the ability of applicants to file patent applications for locally-devised inventions abroad. These restrictions exist to prevent inventions relevant to national security from being disclosed in other countries. Early determination of inventorship is especially important if there is any involvement by somebody resident in the USA, China or another country with national filing restrictions in place. Taking the USA as an example, if a resident of the USA is an inventor for a patent application, then the application must be first filed in the USA, or a foreign filing licence must be obtained from the US Patent and Trade Mark Office (USPTO), before the application is filed in a country other than the USA. If a foreign filing licence is not obtained, any subsequently granted US patent will be invalid and hence unenforceable unless the failure to procure the license was "through error". A foreign filing licence can be obtained from the USPTO on an expedited basis in around 3 days.

Getting inventorship or ownership wrong can be very costly. It is always best to ensure that inventorship and ownership of patents are correct from the outset. It can be problematic to sort out issues later on, particularly if a patent covers a valuable product.

17.6   THE PROCESS FOR OBTAINING A PATENT

17.6.1   The National Nature of Patents

Despite their importance to international business, patents remain matters of national law. Ultimately, therefore, it is necessary for a company to have a separate national patent for an invention in each state in which patent protection is needed. Until recently, substantive requirements for patentability were also often very different in different territories. The TRIPS agreement (Agreement on Trade Related Aspects of Intellectual Property Rights of 1994) went a long way to harmonising laws to a minimum standard. Some national differences remain, and it remains necessary for there to be a separate patent in each territory. The major downside to this fragmented national arrangement is that the patent specification needs to be translated into a large number of languages, and that litigation may need to be pursued in multiple jurisdictions.

In general, obtaining a patent in one's home market is not sufficient, and most organisations involved in medical research will want to obtain patents in several territories. Two long-standing international treaties facilitate the process of obtaining patents around the world: under the Paris Convention,[17] one can file a patent application in a first country, and then file corresponding applications in other countries within 12 months and claim for the later applications the filing date of the first country (known then as the "priority date"). The second international agreement is the Patent Cooperation Treaty (PCT). Under the PCT, one can file a single patent application for a large number of countries.[18] The PCT application can be filed in one language, in the applicant's home country and the claims are subjected to a high quality patent office search. A PCT application remains effective until 30 months from the priority date. At that time, it is necessary to decide which territories to pursue the patent in, and which not.

17.6.2   A Typical Application Process

A typical application process in the field of medicinal chemistry for a company based in the UK will thus be as follows:

- After discussion between medicinal chemists, business development manager and patent attorney, a decision is made to file a patent application.
- The application is then filed at the UK patent office. This filing date is the "priority date".
- 12 months later, if the invention remains of interest, a PCT application is filed. The PCT application may include additional material beyond what was in the original application (*e.g.* additional examples).
- A search is carried out and the search report is sent to the applicant.
- At 18 months from the priority date, the PCT application is published.
- At 30 months from the priority date, it is necessary to enter the "national phase" in each country in which a patent is wanted.
- Over the following 5 years, the application is examined by the various national patent offices. Providing no surprising unknown prior art documents come out of the woodwork, the patent should be granted in the various territories over this time scale.

Many observers are surprised that the patenting takes as long as it does: typically 4–6 years from filing to grant in the USA, 5–8 years in Europe, and 7–9 years in Japan. In almost every country, there are mechanisms by which the process can be accelerated if an applicant wants to do so. In general, it is in applicants' interests for things to go slowly: every time the case advances a step through patent office prosecution, there tend to be costs; furthermore, when the patent is granted, the scope of the claims becomes fixed. If a product is still being developed, it can be useful if the claims could be changed to accommodate changes to the product, though it should be noted that it is not possible to add new material to a patent specification once it has been filed.

It is seen from the process summary above that there are three key decision points in the process:

1) The initial decision to file, and when to do it.
2) The decision at 12 months whether to proceed with a PCT filing. If the product development is going ahead well, then the PCT application should be filed. If the invention or the programme is not yet sufficiently advanced, the first priority application can be withdrawn, and a new priority application filed. This results in the application being re-dated, so it should be checked that there has not been any relevant publication in the intervening year that will compromise the validity of the patent. The third option, if the invention is no longer of any interest, is to abandon the application altogether. Proceeding with the PCT application at 12 months sets the application on the course for publication at 18 months.
3) The decision regarding which countries to proceed with for national phase entry at 30 months. This decision is very much determined by cost: how important is the project, and for which countries are the costs justified?

17.6.3 Costs

Costs are an important consideration for any organisation, and long term budgetary planning is important for a patenting programme. Of course, the exact costs will vary from case to case depending on the complexity. Considering the process described above, the typical costs for drafting and filing a "priority application" in the medicinal chemistry area are in the region of £4000–7000. Filing a PCT application at the 12 month point typically costs around £5000, though there are page fees if the specification has more than 30 pages. If there is significant additional material, then the process of introducing that into the specification will also increase the costs.

The national phase entry at 30 months is the usually single largest cost point in the patenting process, often amounting to tens of thousands of pounds if a significant number of territories are to be covered. The 30 months national phase entry deadline has been arrived at as a compromise between providing certainty to third parties about whether there will be a patent in a particular territory, whilst allowing a patent applicant a reasonable amount of time to assess whether the invention will be sufficiently commercially successful to justify the national phase costs. In medicines research, 30 months can pass very quickly. Particularly for a small company, it is important to try and align this deadline with milestones in the development programme.

Larger corporations generally have defined sets of country lists for national phase entry and the appropriate list will be selected depending on the level of interest in the invention and its likely future value. Country listings and the associated costs might be as shown in Table 17.1.

17.6.4 The National Phase: Examination of Patent Applications

In order to make sure that a patent application fulfils the requirements for patentability and disclosure described in Section 17.3, most countries have a patent examination system in place.

Table 17.1

Group	Level of interest	Approximate costs
A	Very promising and valuable product: cover top 25 economies	£100 000 +
B	Medium level of interest, an out-licensing opportunity: cover most G8 economies and next tier, *e.g.* USA, Europe, Japan, China, India	£30 000–£60 000
C	Low level interest: USA and Europe only	£8000

After an application enters the national phase in a particular country, it goes before an examiner at the national patent office. The examiner first carries out a search of the art (in some cases the search carried out previously in the PCT phase may be sufficient), and then writes a report to the applicant setting out whether or not the application fulfils the requirements. If it does not, then the examiner's report will explain why the application cannot be allowed and the applicant is given an opportunity to amend the patent claims and to provide arguments in favour of the claims. This correspondence between the patent office examiner and the applicant and his attorney is known as "patent prosecution". It is usual that a patent specification is amended during this time. However, the applicant does not have a completely free hand when making amendments: he is not allowed to introduce additional features or limitations. All amendments must be based on information contained in the application as originally filed.

Eventually, either the patent is granted in the country concerned, or the application is refused. If an application is refused by the patent office, the applicant has the opportunity to appeal to a higher court.

The patent prosecution process takes place in parallel in each country in which the applicant has applied for a patent. Some of this work is duplicative. On many occasions, different prior art is found by different patent offices, meaning that in the end the confidence in the validity of the patent is increased.

Having said above that patent applications remain national rights, in Europe, the majority of states (including all EU states and Switzerland) are members of the European Patent Convention and patents for those countries can thus be obtained at the European Patent Office (the EPO). Examination at the EPO takes place under a single procedure in English, French or German (at the Applicant's choice). This avoids the duplication mentioned above within Europe. It is only after the "European Patent" has been granted that the patent splits into separate national rights, and the applicant needs to translate the text into local languages and decide in which European countries to have the patent in force.

At the time of writing in 2014, 25 states of the European Union have signed an agreement to create a "European Patent with Unitary Effect", a single patent that has effect across all of the 25 states involved. A second agreement was also reached between a similar set of countries to create a "Unified Patent Court", a single court in which European patents can be enforced for the whole of the territories involved. The details of the implementation of the new systems are complex and they have not yet been completed. The current best estimate for when the new systems will come into force is some time in 2017. For up to date information on the progress of this, please see the websites of the appropriate government and private sector organisations.[19]

17.7 THE PATENT AFTER GRANT

17.7.1 Maintenance

Once a patent has been granted, it is necessary to maintain it in force. In most territories, this is done by paying an annual renewal fee. There are two public policy reasons for governments insisting on annual fees: the first is that as part of the "deal" between the state and the patentee, the monopoly should only be maintained if the patentee is actually making use of his invention. If he is not using it, then the public should benefit from being able to use it before the 20 year expiry date. The second reason is that the annual renewal fees subsidise the costs of the patent office's search and examination operation. This enables the office to charge low up-front fees (thus facilitating access to the system), and it weights the cost of the patent system towards those who use it most, *i.e.* those patentees who maintain their patents in force the longest.

17.7.2   Extension of Patents

The patent system is a single system that applies to all fields of technology. It does not take any account of the relative speeds of progress, or the different levels of investment needed in different fields. The 20 year patent term applies equally to a new shape of toilet brush as to a highly complex mobile phone or a specialised cancer treatment. In recognition of the fact that before a new pharmaceutical can be marketed, extensive trials are needed in order to obtain government approval, many countries have put in place patent term extension provisions for pharmaceuticals. In many states there are similar extensions for the main other class of products that require such approval: plant protection products.

The exact mechanism for the calculation of the extension period and the method of implementation of the extension varies from country to country. In general, up to five additional years of patent-like protection can be obtained. In Europe, the extension is provided by a "Supplementary Protection Certificate" (SPC). It can be up to 5 years long, but has a maximum expiry date of 15 years from the first European marketing authorisation. It is not an extension of the whole patent; only the scope that covers the exact compound that is marketed. In the USA, the extension is called a "Patent Term Restoration" and its duration is calculated on the basis of the time that the drug spends in pre-clinical and clinical regulatory review. Again, the extension cannot be more than 5 years, but the maximum expiry date is 14 years from the US marketing authorisation, so earlier than the 15 year European expiry. There are similar arrangements in Japan, Korea, Taiwan, the Russian Federation, Israel and Australia.

These patent term extensions are of crucial importance to a drug development company. It is therefore important to consider early in the product development process whether a given product will be entitled to an extension and, if so, how long it will be.

17.7.3   Challenges to Validity

It is a surprise to some observers that grant of a patent by a patent office does not constitute a guarantee of validity. Patent office examiners can work only on the basis of the documents and evidence that they have before them; their searches are limited to written literature, and they do not have facilities to repeat experiments. Examiners will not be able to find out if an invention was disclosed non-confidentially at a meeting, or if it does not actually work. Therefore, although patent office examiners do a good job, a patent that has been granted cannot definitively be assumed to be valid. A granted patent remains open to challenge after grant, either at a national patent office or before national courts.

17.8   USE OF PATENTS

17.8.1   Infringement and Enforcement

The primary *raison d'être* of a patent is to keep competitors off the market and to preserve the patent holder's market exclusivity.

Infringement of a patent occurs if a third party uses a patented invention without permission of the patent owner. The patent owner has the legal right to stop the third party from carrying out the infringement. To assert this right, the patent holder must bring an action for infringement in the courts. The action is a civil action—in other words, it is the patent owner, rather than the state, that brings the action against the alleged infringer. In most countries, the case is heard before a technically qualified judge, although in certain countries, in particular the US, infringement actions are heard before a jury. In the UK, the judges usually have a technical degree and then many years of experience arguing patent matters in the patents courts, either on behalf of the patent owner or on behalf of the infringer.

The procedure and the costs of litigation vary tremendously between countries, with patent litigation in the US and to a somewhat lesser extent the UK, being very expensive.

In general, in the medicinal chemistry field, competitors are sophisticated companies and they know of the drug originator's patents. The existence of a patent will generally keep the competitor from launching a product until after the patent has expired (or, if the competitor considers the patent invalid, he may challenge the validity). Actual patent infringement is thus not a very common occurrence in the medicinal chemistry field. When infringement does take place, it is important to act quickly, so that the infringement can be stopped before the market price of the product drops.

In order to bring an action for patent infringement, the patent holder must identify who the infringer is, and what infringing acts they are carrying out. In a complex supply chain, this can be more difficult than one might expect: a drug being manufactured in India might be imported into the Europe by a Greek company which passes it to a UK distributor who distributes the drug to pharmacists on behalf of a UK marketing authorisation holder that is headquartered in Monaco. If we have a UK patent, then we need to identify what in this chain is an infringing act in the UK, and who is carrying out that act. Most probably in this imaginary scenario, it would be the UK distributor and marketing authorisation holder that the patentee should bring the action against.

17.8.1.1   Proving Infringement

Once the infringing acts and the infringer have been identified, it is necessary to prove that an infringement has taken place. There are two parts to this. The first is gathering evidence of what acts have taken place. This might involve making test purchases and obtaining samples, or else obtaining copies of invoices or delivery notes showing what has taken place. Then, it is necessary to show that what is being sold falls within the claims of the patent.

17.8.1.2   Claim Interpretation

In general, it is usually a straightforward matter for the court to decide that a pharmaceutical patent is infringed. Where the invention concerns a product, defined by a general chemical formula (known as a Markush formula), it will usually be sufficient to show that the alleged infringement meets the definition for the particular generic formula. Often, the substance in question is specifically named in the patent, so there is little room for argument that the claims of the patent are infringed.

But occasionally, there is real doubt as to whether or not a product is covered by a patent claim. This requires the patent claim to be interpreted to determine what its scope is. In Europe, the statutes require a claim to be interpreted in such a way as to "combine a fair protection for the patent proprietor with a reasonable degree of legal certainty for third parties."

This came to the fore in the Pharmacia *vs.* Merck case mentioned in Section 17.4.4. The broadest claim in Pharmacia /Searle's patent EP 0 679 157B[14] was to a compound of formula I:

$$R^2 \quad\quad R^3$$
$$\begin{array}{c} 4 \;\; 3 \\ 5 \;\; 2 \\ 1 \end{array}$$
$$Y \quad X$$

According to the claim, the definition of Y included O, the definition of X included -OH, and the definitions of R^2 and R^3 included aryl.

The Merck compound MK-966 was a furanone. Pharmacia argued that although the structure drawn in the claim was of hydroxyfuran derivatives, the claim should be taken as including the corresponding furanones because hydroxyfurans tautomerise to furanones in water. The product that Merck proposed to sell was a dry tablet, but the compound would inevitably come into contact with water in the body once the patient took it. Pharmacia argued that the skilled person would realise that the tautomerism would happen and would read the claim as including furanones. The structures of the keto and enol forms of Rofecoxib are shown in Figure 17.4.

To give this point credence, Pharmacia brought forward testimony from an expert from the field, an "expert witness". The purpose of an expert witness is to assist the court in understanding technical points when applying the law. In an attempt to discredit Pharmacia's view, the defendants, Merck, brought forward testimony from their own expert witness. Pharmacia called on Professor Sir Jack Baldwin FRS, then head of the Dyson Perrins Laboratory in Oxford, and Merck relied on Professor Anthony Kirby FRS, then professor of bioorganic chemistry at Cambridge. The result was a fascinating match between two heavyweights of academic mechanism chemistry bringing their knowledge and experience to bear on a legal question.

A patent document is intended to be read by a person with a reasonable technical knowledge and understanding of the area. In this particular case, the skilled person was expected to have at least an undergraduate knowledge of organic chemistry—and so was expected to be familiar with keto-enol tautomerism. It was accepted by the parties that the proportion of furanone to enol in water at physiological pH was $6.3 \times 10^{10} : 1$. Further, this overwhelming excess of keto to enol was known from undergraduate text books—so it could be expected to be part of the skilled person's "common general knowledge".

Professor Baldwin read the patent and its claims as referring to the keto tautomer as well as the expressly mentioned enol. Professor Kirby took a more literalist approach and interpreted the claim as limited to the enol form, to the exclusion of the keto form. At first instance, the judge favoured the literalist approach and held that keto forms were not included, and thus that Merck's compound did not fall within the claim.

However, Pharmacia appealed the decision to the Court of Appeal and, on appeal, this decision was overturned. The appeal judges were more convinced by Professor Baldwin's interpretation. The Court of Appeal took the view that the compounds of claim 1 are a class which are said to have activity in the body. In order to give fair protection to the patentee, it was reasonable to take into account the form in which the compound existed in the body. It was accepted that the enol form would exist in an inseparable equilibrium with the keto form, and that it was the composite that has the effect in the body. To restrict the claim to the minor tautomer would not appear to achieve any useful purpose so far as the patentees were concerned.

To construe claim 1 as covering the keto tautomer would not, in the Appeal Judges' view, prevent there being a reasonable degree of certainty for third parties. Of course third parties would appreciate that the keto form was not specifically mentioned. But they would know that that did not necessarily mean that the claim would be construed as limited to the explicitly mentioned compounds. If the contrary were the position, then claims to chemical compounds would have to be construed using too literal an approach. The skilled reader would also know that the inevitable result of using the enol was the formation of the other tautomer. They would therefore not be surprised if manufacture and sale of the other tautomer, which would form the enol in solution, would infringe. The Appeal Judges therefore concluded that claim 1 includes the keto form and so Rofecoxib infringed.

Unfortunately for Pharmacia/Searle, the broad interpretation of the claim contributed to the finding that many compounds within the claim did not work, and thus that the claim lacked inventive step. The patent was therefore revoked. So, although the Merck compound had been found to infringe the claim, Pharmacia were not able to keep Merck's compound off the market.

17.8.2 Defences and Exemptions to Infringement

If a company is accused of patent infringement, what can it do?

17.8.2.1 Defence of Non-Infringement

The first line of defence to an allegation of patent infringement is to argue that the product or method in question does not fall within the claims of the patent. That is what Merck did in the case described above. In many cases in the pharmaceutical industry, it is a generic medicine copy that is alleged to be an infringement. In that case, there is little scope for arguing non-infringement, as the generic pharmaceutical compound will necessarily have to be the same as the originator's compound, and it is quite rare that an originator's patent does not cover its compound.

That said, the values at stake in the pharmaceutical industry can prompt some generic companies to try making arguments that most medicinal chemists would think have little chance of success: in a recent case in the USA, a collective group of 8 generic pharmaceutical companies argued that their generic versions of the drug pregabalin should not be held to infringe Pfizer/Northwestern University's patent US 6,197,819 that covered the compound.[20] Pregabalin is sold by Pfizer under the name Lyrica® and it has the structure shown in Figure 17.7. It is approved for marketing as the single (S) enantiomer.

The patent claim in issue was directed to "4-amino-3-(2-methylpropyl) butanoic acid, or a pharmaceutically acceptable salt thereof." The generic companies argued that the claim did not specify that the compound is a single isomer and thus that it should only be considered infringed by a racemic mixture. The court disagreed with the generic companies and found the claim to be infringed by the single enantiomer compound.

17.8.2.2 Counterclaim for Invalidity

The next line of defence to an allegation of patent infringement is to counter-claim that the patent in question is not valid. This is what Merck did in the case described in Section 17.8.1. This approach generally starts with carrying out a thorough search of prior art that can be raised

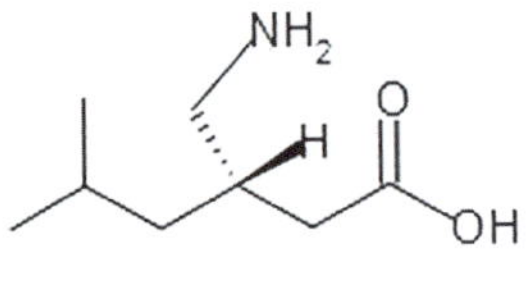

Figure 17.7   The structure of Pregabalin, sold as Lyrica®.

against the patent. Documents or other disclosures that were not taken into account by the patent office when it granted the patent can be particularly useful here.

The case then needs to be made to the court to demonstrate which of the criteria for patentability are not fulfilled by the patent. The patent holder may request that the claims in his patent be narrowed once he has seen the case put forward by the counter-claimant. Very often, the validity of the patent will turn on inventive step: was it obvious to make the change in question at the time the patent application was filed? The two parties will bring together as much evidence as they can to show that it was, or was not, obvious to do that.

17.8.2.3   Exemptions to Infringement

Certain uses of a patented invention, even though not authorised by the patent owner, are nonetheless excluded from the definition of infringement. In the UK, acts done privately and for purposes which are not commercial are not infringements. The exemption that is most relevant to medicinal chemists is the experimental use exemption: acts that are "done for experimental purposes relating to the subject-matter of the invention".

The exemption needs to be treated with some caution. It does not mean that all uses of a patented invention in experiments are exempted. For example, if the patent relates to an assay system capable of identifying new drug candidates for the treatment of a particular disease, then using such an assay without permission is almost certainly an infringement—the use is not in an experiment investigating the invention—rather it is a commercial use of the invention in an experiment, with the aim of identifying the potential new drugs.

The question often arises as to whether carrying out a clinical trial falls within the experimental use exemption. In many countries, including Germany and the USA, it is not an infringement to carry out studies and clinical trials on a patented substance, if those trials are being conducted with a view to getting regulatory approval. Such trials are taken to fall within the experimental use exemption. In the UK, the matter is not so clear cut, and all of the circumstances would have to be taken into account.

A separate exemption was introduced in Europe in 2005 (including in the UK) which permits a generic company to do the studies necessary for getting a marketing authorisation for a generic product, before patent expiry.

In summary, certain experiments are exempted from patent infringement. The exemptions are quite specific and so it is not safe to assume that an exemption will apply.

17.8.3   The Consequences of Patent Infringement

If the court finds that a defendant's use of an invention is an infringement, then the patent holder will request that the court imposes one or more remedies, generally including:

- An injunction: a court order prohibiting any further unauthorised use of the invention during the lifetime of the patent.
- Withdrawal of any product on the market and destruction of stocks.
- Payment of damages, to compensate the patent owner for its lost revenue as a result of the unauthorised use of the invention.
- Legal costs: the infringer may have to reimburse the patent owner for a substantial part of its legal costs in bringing the infringement action. Such costs can be extremely high. Each side in a major infringement action in the UK (which will usually involve a counterclaim from the infringer that the patent is invalid) will expect to incur well over £1 million in legal fees. In the US, where recovery of costs is relatively rare, each party's legal costs may well be over $10 million.

A special form of injunction merits a mention here: the preliminary injunction. Generic products are generally priced at a very substantial discount to the corresponding product of the originator. Unless it drops its price, the originator will rapidly lose market share to the generic. But if the originator drops its price, it is unlikely to be able to put it up again, even if following a successful patent infringement action, the generic product is withdrawn from the market. In these circumstances, the loss of revenue to the patent owner is impossible to quantify. Infringement actions generally take at least a year to come to court—if the generic drug were on the market for all that time, it would cause a dramatic change to the market place and a huge loss of earnings for the patent owner.

To avoid this situation, the patent owner will try to get an injunction immediately, and before the merits of the case are considered fully by the court. In the UK, an injunction obtained before a full trial is called a preliminary injunction. To be able to get a preliminary injunction, the patent holder must show the court that it has an arguable case for patent infringement and for patent validity, and that it would suffer unquantifiable damages by the infringement. In general, the patent owner must agree to compensate the alleged infringer for the loss suffered, in the event that the court eventually decides that the injunction should not have been granted.

A preliminary injunction is sometimes the most important remedy that a patent holder in the medicinal chemistry field will try to obtain.

17.8.4 Licensing

As discussed above, a patent does not give its owner the right to use his invention himself. Rather, it gives him the right to prevent others from using the invention commercially without his permission. A licence is, basically, the grant of permission to do things which, absent the licence, would be an infringement of the patent. A patentee is often said to be giving an "immunity from suit" when granting a licence.

A patentee (the licensor) and the person taking the licence (the licensee) may agree the terms of their licence to suit themselves, with one very important caveat: the terms of the licence must not be in breach of the law. Specifically, various laws exist which control monopolies and unfair competition, dealing with issues such as pricing, restrictive practices, and cross-border trade. European Competition Law, which regulates (or, arguably, deregulates) trade between the member states of the EU is a complex subject and outside the scope of this chapter, but must always be taken into account when businesses operating in the EU enter into licence agreements. In addition to EU competition law, various national laws operate in similar areas, while in the USA, these laws are called "anti-trust" laws.

Subject to this, any licence will be tailored to suit the particular situation. Common issues which arise in patent licence negotiations include the following:

The licence needs to clearly define what the licensee is allowed to do. It should define the permitted product, or process, or both. This can include everything within the scope of the claims of the patent, or it may be a narrower definition.

Subject to competition law, a licence may give permission to operate in a specific territory. For example, a patentee may wish to satisfy the whole market in his home territory, say Europe, himself, while raising revenue from licensing the technology in other territories, say the US. In this case the licence will contain a *territorial restriction*. Similarly, a patentee may wish to satisfy the whole market in a particular field of technology, but grant licences in other fields. If, for example, the invention is a pharmaceutical compound which can be used in either a cream for topical application, or in an injectable formulation for systemic use, he may choose to satisfy the whole market for one of these formulations himself, while licensing the alternative use to others. In this case, the licence will contain a *field of use restriction*.

A licence can be a bare patent licence, *i.e.* include the grant of rights under a patent and nothing else. Many licences however include other things, for example the grant of rights to use confidential *know-how* belonging to the licensor, or the requirement for the patentee to provide technical advice and assistance.

Exclusive and *non-exclusive* are common terms in licence agreements, but can be misleading, because their precise definition differs from country to country. On the UK definition, if a licensor grants a non-exclusive licence, he remains free to grant additional non-exclusive licences to others, while if he grants an exclusive licence, he agrees not to grant additional licences to others, and is not free to commercialise the invention himself. A *sole licence* is an agreement that the patentee will grant no further licences to others, but retains the right to commercialise the invention himself. A licence may or may not give the licensee the right to grant further licences, or *sub-licences*, to others.

Payment for the grant of a licence can be in many forms. It can be in the form of on-going royalty payments, or lump-sum payments, or a mixture of both, either form of payment being triggered by defined events. Possibly, no money at all changes hands: the licence may be *royalty free*. A licence in which, rather than making payments, each party grants a licence to the other, is known as a *cross-licence*.

Finally, the *termination* clause is very important. This determines under what conditions one or both of the parties can terminate the licence. Bare patent licences usually terminate at the latest when the patent terminates. If know-how is included, then a licence to use this know-how can continue for many years after patent expiry provided that the know-still remains confidential.

Numerous other issues will arise, including the handling of disputes, action and responsibilities if third parties infringe or attack the patent, warranties and indemnities given by each party to the other, and regulation of liability. The prudent licensor always includes an audit clause in a license, which gives him the right to appoint an independent auditor to inspect the books of the licensee to ensure that the correct payments are being made. Expert help in preparing a licence agreement is important.

17.8.5 The Patent Box

Until recently, a patent has served only one purpose: to prevent competitors putting their products or services on the market (or at least to control the market by licensing). That remains the primary reason for any business to obtain patents. In addition, there came into force in the UK in 2013 a government tax initiative commonly referred to as the "patent box" which now provides UK businesses with an additional reason to obtain patents.[21]

The regime allows a business to pay corporation tax at the lower rate of 10% on net profits derived from a patented product. The profits can be made anywhere in the world, but to qualify for the lower tax rate, the organisation must have a patent in the UK (or a short list of other European countries), or be an exclusive licensee. The reason for the patent being needed is to demonstrate that the product from which the profits are derived is "innovative". There is no need to enforce the patent against competitors, and it does not matter if the patent is narrow and could be worked around.

The patent box regime is designed to encourage product development work in the UK. Similar regimes are in place in a small number of other countries, for example in the Netherlands. The 2013–2014 tax year was the first year in which the UK patent box tax regime was in place and the relief is being phased in over five years to 2017. At the time of writing, it remains to be seen just what effect it will have on the UK economy. It was lobbied for by the UK pharmaceutical industry, and certainly it has the potential to help medicinal chemistry businesses in the UK and to encourage their activities.

17.9 GENERIC MEDICINES AND BARRIERS TO GENERIC COMPETITION

An originator drug development company will often be keen to keep products from other drug originator companies out of its market patch. However, as mentioned in several places in this chapter, it is the generic pharmaceutical products that are the larger threat to an originator's business. Generic medicines thus merit special discussion.

17.9.1 What is a Generic Medicine?

Although precise legal definitions vary from country to country, in general a generic medicine is a medicine which is identical to that of the originator in terms of its active principle(s) and functionally equivalent in terms of its therapeutic profile, its safety profile and pharmaceutical form. A physician or pharmacist is permitted to substitute the generic drug when fulfilling a prescription for the originator compound.

Generic pharmaceutical companies have a very different business model to originator companies. They do not carry out research and development on new active ingredients. Instead, they focus all of their efforts on being able to efficiently manufacture and distribute pharmaceuticals. Generic pharmaceutical companies face direct competition from each other and from the originators, and so they operate under pressure for product price and quality. Generic pharmaceutical companies are important to health care organisations as they facilitate serving the healthcare needs of the population within the inevitable budgetary constraints.

A generic medicine is not always precisely the same as the originator medicine: it will contain the same active ingredient, but it may, for example, be in a different salt form or a different physical form. For example, Pfizer's calcium antagonist, Amlodipine (marketed under the name Norvasc®) is marketed by Pfizer as the besylate salt. It faced generic competition from the functionally equivalent product amlodipine maleate. Similarly, GSK marketed Paroxetine (under the name Paxil®) in the hemihydrate form, which was protected by a patent. The medicine containing anhydrous paroxetine hydrochloride was considered to be essentially the same by the medicines regulators, and so the generic product with anhydrous paroxetine hydrochloride could be substituted for GSK's product with paroxetine hydrochloride hemihydrate.

In general, there are three barriers to generic competition for a medicine. The first is patents, as described in most of this chapter. Further barriers are provided by regulatory data exclusivity and synthetic challenges.

17.9.2 Regulatory Data Exclusivity

In all the major pharmaceutical markets, including USA, Europe and Japan, the sale of prescription medicines is tightly regulated. The approval of a new medicine is only permitted after Phase III clinical trials have been carried out in patients, and demonstrated a satisfactory benefit–risk balance.

It is self-evident that once the originator has demonstrated a satisfactory benefit–risk balance to secure a marketing authorisation, it is not in the public interest that authorisation of generic versions of the medicine should require the same clinical trials as the originator to come to the market. Indeed, it would be highly unethical to carry out a further clinical trial and deny a control group of patients a treatment that we now know they would benefit from. Regulatory law makers have struck a compromise between the needs of the originator pharmaceutical companies and the needs of the public by permitting companies to apply for a marketing authorisation for a generic medicine without repeating the originator's clinical trials. Instead, the competitor can rely on the safety and efficacy data generated by the originator. However, such use of the data by a competitor company is only permitted after a fixed period during which the originator has the exclusive permission to use the data. This is referred to as "data exclusivity".

The period after which this loss of "data exclusivity" occurs varies from country to country. In the USA, it is five years. So, five years after an originator's product has been approved for marketing, it is possible to make an application for a marketing authorisation for a generic version of an originator's product, on the basis only of bioequivalence studies—a so called "abbreviated new drug application (ANDA)". If the product in question is protected by a patent which the generic company considers is invalid, then the period of data exclusivity is very influential in determining when the generic company makes its challenge to the patent. A challenger will often aim to clear the way by removing the patent to coincide with the end of the data exclusivity period.

In Europe, for medicines first authorised after 2003, the competitor must wait until eight years have elapsed before it can apply for a generic marketing authorisation. Further, the marketing authorisation of the generic medicine cannot take effect until 10 years after the first authorisation of the originator's version of the medicine in Europe. In addition, if the originator gains approval for a new therapeutic indication in the first 8 years of marketing which brings significant clinical benefit over existing therapies, then the 10 year period before generic marketing can begin is extended by a further year. This combination of 8 years before a generic marketing application can be submitted, together with the further two years before marketing can begin delayed by a further one year in the event of a significant new indication is often referred to as the "8 + 2 + 1" rule for determining the earliest date when generic competition can begin in Europe.

Given that a patent expires 20 years after filing (or 25 years if an SPC-type extension is available), whilst data exclusivity runs from the time of marketing authorisation, a medicine that reaches the market only very slowly may have a data protection period that goes beyond the date of patent expiry. In this situation, the timing of generic entry is determined not by the expiry date of the patent (and any relevant patent extensions) but rather by the availability of the abbreviated regulatory procedures. An example of this is the synthetic low molecular weight heparin derivative, Fondaparinux, marketed by Sanofi and then GSK as Arixtra®. In Europe, the patent protection for the active principle expired in January 2008, but the regulatory data protection period did not expire until March 2012, 10 years after the medicine was first approved.

17.9.3   Technical Barriers

The final barrier to generic entry is the technical barrier of synthetic complexity. This is illustrated, to some extent, by Fondaparinux. Fondaparinux is marketed by GSK under the name Arixtra®. It is a pentasaccharide with the structure shown in Figure 17.8 and it is manufactured by a 57 stage process involving many difficult purification steps, as well as the introduction and removal of protecting groups.

Many thought that irrespective of the patent and regulatory barriers to generic entry, the synthetic complexity would be an insuperable barrier to overcome and that generic competition to Arixtra® would be unlikely. However, by 2011 a generic version of the medicine was developed by Alchemia and it is marketed in the USA by Dr. Reddy's Laboratories.

It is fair to say that technical difficulty is rarely a reliable barrier to generic entry onto the market, though there are occasions when generic entry is delayed by technical issues. Medicine originators should not plan for that to occur, but rather treat it as an unexpected bonus if it does.

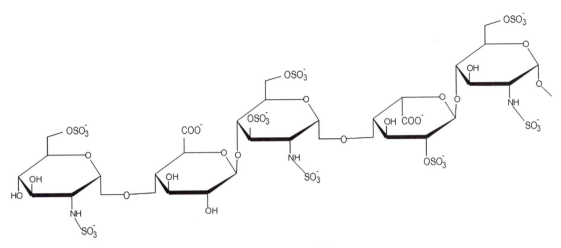

Figure 17.8 The structure of Fondaparinux, sold as Arixtra®.

17.10 PATENTS AS A SOURCE OF INFORMATION

The lasting record of much of humanity's technology development of the last few centuries is to be found primarily in the patent literature. Patent publications thus form an important repository of information, and they are available for free in paper form in various libraries and electronically over the internet. Patent specifications are quite dry and they generally do not tell as interesting a story as academic papers, but the information in them is necessarily complete and detailed, and they are well indexed and archived. They thus constitute a source of information that is highly beneficial to those engaged in medicinal chemistry research.

The free online database with the broadest international coverage is Espacenet, operated by the European Patent Office. It is found here: http://worldwide.espacenet.com

Patent publications can be located in the database using keywords, inventor names, applicant company names and various other search criteria.

The USPTO operates a similar database, confined to US patent publications. It is found here: http://www.uspto.gov/patents/process/search

Google operates a database search interface. At the time of writing, its content is not yet global (though it covers EP, JP, US and WO), but it is steadily growing. The search output is less structured than in the databases run by the patent offices, but it is perhaps more user-friendly for non-specialist users. It is found here: http://www.google.com/advanced_patent_search

To an extent, these free databases rely on the text of abstracts submitted by the patent applicants. They are thus not 100% reliable. There thus also exist pay-for-access databases, such as the Derwent World Patents Index® operated by ThomsonReuters (see http://thomsonreuters.com/derwent-world-patents-index/), and the CAplus database operated by CAS, the "Chemical Abstracts" division of the American Chemical Society (see: https://www.cas.org/content/references/patentcoverage). These databases include value-added abstracts and indexing. The commercial databases allow more complex search queries to be formed as well as often allowing the keyword searches to be carried out through the claims or full text of patents. Further examples of such databases and systems include Thomson Innovation (http://info.thomsoninnovation.com/en), STN (http://www.stn-international.de/index), PatBase (https://www.patbase.com) and Questel-Orbit (http://www.questel.com).

Chemical Structure and Chemical Name searching is possible *via* more specialised databases. The CAS Registry database, produced by CAS, is searchable by structures and sub-structures. Searching of CAS Registry is available through STN (operated jointly by CAS and FIZ Karlsruhe), with its interface optimised for the more expert searcher user, and through Scifinder (operated by CAS), which is designed for academic chemists. Entry of structure search queries in the CAS Registry database is not straightforward, and specialised training and experience is generally needed to perform a reliable search. More limited structure-based searching can be done for free using, for example, the SureChem Open database portal (https://www.surechem.com/products/open/).

The databases mentioned above will provide lists of patent documents that disclose the requested information, and allow access to the patent documents themselves. They do not provide up to date details of the status of a patent application: has it been granted? Is it still in force? Has it been abandoned? For that information, it is necessary to look in the national patent registers. These are also available for free online, at the addresses given below. The language used in many of the databases necessarily contains much specialised legal phrasing, so it is recommended to consult a patent attorney before drawing important conclusions from these registry reports:

European Patent Office:	https://register.epo.org/
UK IPO:	http://www.ipo.gov.uk/types/patent/p-os/p-find/p-ipsum
Germany:	https://register.dpma.de/DPMAregister/pat/einsteiger?lang=en
USPTO:	http://portal.uspto.gov/pair/PublicPair
Japan:	http://www.jpo.go.jp/ and http://www.ipdl.inpit.go.jp/homepg_e.ipdl

17.11   SUMMARY

We hope that this chapter has given a general, if necessarily brief, introduction to the important role that patents play in medicinal chemistry. The main lessons that we hope you will take away are:

- Patents are crucial to medicinal chemistry as they allow the massive investment in product development to be recouped.
- A patent must be applied for and it constitutes a special deal between a patent applicant and the state: in return for full disclosure of an invention, the inventor is granted a monopoly of limited duration (20 years).
- During the 20 year period, the patent does not give the applicant the right to use his invention, but it does give him the right to stop others using it commercially.
- Patent applications are examined by patent offices, but patent validity can be challenged after grant.
- Patents are enforced through the courts. Patent litigation is costly for both parties, so should only be undertaken after serious consideration. Instead of enforcing his patent, a patentee may grant a licence permitting another to use the invention in return for payment or some other benefit.

HINTS AND TIPS

- A patent is a "deal" between an inventor and the state, an exclusionary right over the commercial use of the invention for a limited period in return for the inventor disclosing his invention to the public. A patent does not give the applicant the right to use his invention, but it does give him the right to stop others using it commercially.
- To be patentable, an invention must be novel, inventive, have industrial applicability and not be specifically excluded from being protectable by a patent.

- An important feature of novelty is that the inventor's own disclosure of his own invention is also part of the state of the art. An inventor's own disclosure before filing a patent application can thus compromise the novelty of an invention. It is thus crucial that an invention be kept confidential until a patent application is filed.
- A patent claim is a concise and precise definition of the invention, generally in a single sentence. It is the claims that are compared with what had been known before (the "prior art") and it is the claims that determine what competitors are not permitted to do.
- The patent must also describe the invention sufficiently clearly and completely for the person skilled in the art to be able to put the invention into practice.
- A patent applicant puts himself in the best possible position by obtaining good data showing that the invention works well. Experiments comparing the invention side-by-side with what has been done before are especially useful. To support inventive step in a patent application, the comparator should be the prior art compound or method that is closest in its features to what is being claimed.
- One of the most difficult elements in the drafting of a patent application is determining the best scope of claim to apply for. Features can be generalised, but everything in the claim must have the invention's beneficial features.
- Patent applications are examined by patent offices, but patent validity can be challenged after grant.
- A drug development team puts itself in the best overall position by having a "portfolio" of patents covering various aspects of the final product. The later-filed patents to downstream developments may be inherently less valuable than the first compound *per se* patent, but the later-filed patents have later expiry dates which give them added value in practice.
- Patents are enforced through the courts. Patent litigation is costly for both parties, so should only be undertaken after serious consideration. Instead of enforcing his patent, a patentee may grant a licence permitting another to use the invention in return for payment or some other benefit.

Acknowledgements

The authors acknowledge the advice of David Clark of Victor Green & Company in the compilation of the list of databases in section 17.10, and the assistance of Andrew MacGowan of mac eleven in the improvement of the presentation of Figure 17.1. Many thanks are due to Sue Scott, Toby Thompson, Chloé Wildman and Matthew Fletcher of Abel & Imray for their helpful suggestions, drawing of chemical structures and thorough proofreading.

KEY REFERENCES

P. W. Grubb and P. R. Thomsen, *Patents for Chemicals, Pharmaceuticals and Biotechnology*, Oxford University Press, UK, 5th edn., 2010.

REFERENCES

1. A. Ward, "Loss of drug patents delivers more pain to AstraZeneca", *Financial Times*, 6 February 2014.
2. Reuters, *Pfizer says U.S. court invalidates Celebrex patent; generics loom*, 12 March 2014, http://www.reuters.com/article/2014/03/12/us-pfizer-celebrex-idUSBREA2B1TJ20140312.

3. S. M. Paul, D. S. Mytelka, C. T. Dunwiddie, C. C. Persinger, B. H. Munos, S. R. Lindborg and A. L. Schacht, *Nature Rev. Drug Discov.*, 2010, **9**, 203.

4. J. Mestre-Ferrandiz, J. Sussex and A. Towse, *The R&D Cost of a New Medicine*, Office of Health Economics, UK, 2012.

5. P. W. Grubb and P. R. Thomsen, *Patents for Chemicals, Pharmaceuticals and Biotechnology*, Oxford University Press, UK, 5th edn., 2010.

6. Lilly Icos LLC (Petitioner) v. Pfizer Ltd (Respondent), *[2000] EWHC Patents 49*, 2000, http://www.bailii.org/ew/cases/EWHC/Patents/2000/49.html.

7. Trilateral Initiative of the European Patent Office, the Japan Patent Office and the United States Patent and Trademark Office, *Comparative Study on Hypothetical/Real Cases: Inventive Step/Non-obviousness*, 2008, http://www.trilateral.net/projects/worksharing/study/cases2.pdf.

8. L. Xiang, *Journal of Intellectual Property Law & Practice*, 2013, **8**, 539.

9. Human Genome Sciences Inc (Appellant) v Eli Lilly and Company (Respondent), *[2011] UKSC 51*, 2011, http://www.bailii.org/uk/cases/UKSC/2011/51.html.

10. Brüstle v Greenpeace eV, *Court of Justice of the European Union Case C-34/10*, 2011, http://curia.europa.eu/juris/liste.jsf?language=en&num=C-34/10.

11. E. Callaway, *Nature*, 2011, **478**, 441.

12. EP 0 247 633 B, Warner-Lambert Co., 1991, http://worldwide.espacenet.com/publicationDetails/originalDocument?CC=EP&NR=0247633B1&KC=B1&FT=D&ND=4&date=19910130&DB=EPODOC&locale=en_EP.

13. (a) Monsanto & Ors v. Merck & Co Inc & Anor (Defendant), [2000] EWHC Patents 154 First instance High Court Decision: www.bailii.org/ew/cases/EWHC/Patents/2000/154.html; (b) Pharmacia Corporation & Ors v. Merck & Co Inc & Anor, [2001] EWCA Civ 1610 [2002] ENPR 10, [2002] RPC 41, Court of Appeal Decision: www.bailii.org/ew/cases/EWCA/Civ/2001/1610.html.

14. EP 0 679 157 B, G. D. Searle & Co. and The Monsanto Company, 1997: http://worldwide.espacenet.com/publicationDetails/originalDocument?FT=D&date=19971119&DB=EPODOC&locale=en_EP&CC=EP&NR=0679157B1&KC=B1&ND=4.

15. EP 0 463 756 A, Pfizer Ltd, filed: 7 June 1991, http://worldwide.espacenet.com/publicationDetails/originalDocument?FT=D&date=19971119&DB=EPODOC&locale=en_EP&CC=EP&NR=0679157B1&KC=B1&ND=4.

16. EP 0 702 555 B, Pfizer Limited & Pfizer Research and Development Company, N.V./S.A., Date of Grant: 11 March 1998, http://worldwide.espacenet.com/publicationDetails/originalDocument?FT=D&date=19980311&DB=EPODOC&locale=en_EP&CC=EP&NR=0702555B1&KC=B1&ND=4.

17. (a) Paris Convention for the Protection of Industrial Property, of 1883, as last revised at Stockholm, 1967, and as amended, 1979. *Paris Convention for the Protection of Industrial Property*, http://www.wipo.int/treaties/en/text.jsp?file_id=288514; (b) *Contracting Parties > Paris Convention (Total Contracting Parties: 175)* http://www.wipo.int/treaties/en/ShowResults.jsp?lang=en&treaty_id=2.

18. (a) *PCT – The International Patent System* http://www.wipo.int/pct/en/; (b) *The PCT now has 148 Contracting States* http://www.wipo.int/pct/en/pct_contracting_states.html.

19. (a) *Unified Patent Court*, http://www.unified-patent-court.org/; (b) *The Unitary Patent and Unified Patent Court*, http://www.ipo.gov.uk/pro-types/pro-patent/p-policy/pro-p-upandupc.htm; (c) *Bristows, Unified Patent Court* http://www.bristowsupc.com/.

20. *US. Pat.*, US 6,197,819 B1, Northwestern University, 2001, http://worldwide.espacenet.com/publicationDetails/originalDocument?CC=US&NR=6197819B1&KC=B1&FT=D&ND=3&date=20010306&DB=EPODOC&locale=en_EP.

21. HM Revenue and Customs, *The Patent Box: Technical Note and Guide to the Finance Bill 2012 clauses*, 2012, http://www.hmrc.gov.uk/budget-updates/march2012/patent-box-tech-note.pdf.

CHAPTER 18

The Modern Drug Discovery Process

MARK C. NOE

Pfizer Worldwide Research and Development, Eastern Point Road, Groton, CT 06340, USA
E-mail: Mark.c.noe@pfizer.com

18.1 INTRODUCTION

Drug discovery began largely as an empirical science, exploiting observations of pharmacological effects in cells or whole animals, utilizing lead matter inspired by natural products, dyes and their synthetic analogues. Advances in biochemistry and cell biology allowed results in animal models to be put into context of activity against isolated enzymes or cell receptors. This *in vivo* phenotypic screening approach provided one key advantage: it focused the medicinal chemist on lead matter with drug-like physicochemical properties because reasonable solubility, metabolic stability and tissue distribution are required to produce an observation of biological activity in this context. However, it also suffered from some major drawbacks. A rigorous understanding of molecular pharmacology and the molecular basis for target interactions was often elusive or came late in the lead optimization process, limiting opportunities for mechanistic differentiation. As a result, medicinal chemistry design was a largely empirical process, involving a systematic survey of conservative substitutions from the initial lead structure. The pace of screening was slow, limiting the speed with which new design ideas could be tested.

As the fields of molecular biology, structural biology, computational chemistry, drug metabolism and drug safety evolved, the drug discovery process became more rational. Drug discovery projects could be started with an underlying hypothesis of how modulating a specific biomolecular target could impact disease. Medicinal chemistry design was guided by a richer understanding of how compounds interact with their biological targets and the influence of physicochemical properties and structural features on absorption, metabolic stability, tissue distribution and drug safety. Structure-based drug design and quantitative models for activity and selectivity (Quantitative Structure–Activity Relationships or QSAR) created a more informed approach to drug design. Lead molecules and clinical candidates could be more systematically evaluated for attrition risks, building *in vitro*–*in vivo* correlations for permeability, metabolic

The Handbook of Medicinal Chemistry: Principles and Practice
Edited by Andrew Davis and Simon E Ward
© The Royal Society of Chemistry 2015
Published by the Royal Society of Chemistry, www.rsc.org

stability, cellular efflux, drug interactions, cellular toxicity and cardiovascular risk due to ion channel block.

The modern drug discovery process involves a hypothesis-oriented approach to target selection and medicinal chemistry design. The massive array of knowledge derived from the human genome project and the association of disease states with specific genetic variations provides a more solid foundation for drug discovery projects underpinned by a specific molecular biology rationale. Advances in bioinformatics and systems biology enable a richer understanding of how targets are interconnected through cellular signaling networks and the multiple points of intervention to produce a cellular phenotypic response. The emerging fields of chemical biology and the development of advanced techniques in protein mass spectrometry enable target deconvolution for phenotypic screening hits in cellular assays.[1,2] Advanced biophysical techniques, such as surface plasmon resonance and microcalorimetry, provide information on target binding kinetics and thermodynamics.[3,4] Hydrogen–deuterium exchange mass spectrometry and advanced NMR techniques allow evaluation of target dynamics and the differential effects of compounds on protein conformational flexibility.[5] New developments in chemoinformatics enable the medicinal chemist to draw project specific inferences across large compound data sets and to test the predictive power of *in silico* models for potency, selectivity, ADME properties and safety. Future advances in medicinal chemistry and biology will only further enhance the sophistication of the drug discovery process, and the combination of technologies to drive new hypotheses in project and compound design represents another dimension of creativity by which the medicinal chemist can impact drug discovery.

18.2 HYPOTHESIS GENERATION

Most modern drug discovery programs begin with identifying a specific biomolecular target for which pharmacological intervention is believed to be beneficial for treating disease. The underlying target selection rationale is based on connecting knowledge of cell biology, clinical disease biomarkers and/or human genetic evidence implicating the target itself or other members of its biological pathway to the disease. Research programs are becoming more reliant on human genetic evidence that modulating a particular biomolecular target will be effective in treating disease, as this evidence reduces the likelihood of Phase 2 clinical trial failure due to weak association of the target with the disease. An example of this principle is the CCR5 antagonist Maraviroc: CCR5 is a chemokine receptor that is associated with HIV virulence. Specifically, individuals who lack a functional CCR5 receptor are strongly protected from HIV infection, while those who have a heterozygous mutation have a reduced likelihood of infection with HIV.[6] This negative association suggested that a CCR5 antagonist might represent an effective treatment for HIV infection. Importantly, individuals with these genetic mutations are otherwise healthy, implying that CCR5 antagonism might be well tolerated. These genetic findings formed the underpinning of the Maraviroc drug discovery program.

Once a biological target has been selected based on its postulated role in the disease state, small molecule druggability assessments can be used to judge the feasibility of a traditional drug discovery program. In cases where the target has a known ligand, substrate or cofactor, the structure of that ligand can sometimes give clues on druggability. For instance, the ATP binding site is a common target for protein kinase inhibitors because of the diverse array of heterocycles that mimic the adenine ring of ATP. Target bioinformatic analysis can be used to determine homology of the drug target to other proteins known to be modulated by drug-like small molecules, providing an inference of druggability (the likelihood that a low MW compound will potently modulate the target). If a crystal structure has been determined for the proposed biomolecular target, the binding pocket proposed for screening can be characterized for shape, hydrophobicity and solvent accessible surface area in characterizing its druggability.[7,8] In some cases, a small screening campaign can be conducted using a random or targeted subset of a corporate screening collection or a fragment collection, and the resultant target hit rate can be used to infer druggability.[9] If the protein target is determined to be druggable, the program usually progresses to screening in an attempt to find lead matter. If, however, target druggability is predicted to be low, then alternative strategies for modulating the target could be considered (see Figure 18.1). For example, if the target resides in a signaling pathway, proteins that are upstream or downstream of the target could be considered to modulate the signal. Alternatively, modulating the biological half-life of the target by intervening in its production or catabolic processes is another possibility. An example of where this approach was used successfully is sitagliptin, a DPP-IV inhibitor that prolongs the half-life of glucagon-like peptide (GLP-1), an important incretin hormone for glucose-dependent insulin secretion, by interfering with the protease that is responsible for its destruction *in vivo*.[10] Peptide analogues of GLP-1 are validated compounds for managing diabetic hyperglycemia; however, these agents are not orally bioavailable and must be administered by injection. Inhibition of DPP-IV was identified as a viable alternative to produce an orally bioavailable therapeutic agent indirectly acting on the GLP1 receptor through inhibiting the destruction of its endogenous agonist.

The newly emerging field of chemical biology offers vast opportunity for the medicinal chemist to play a prominent role in understanding how biomolecular targets interact with cellular pathways. Chemical biology is underpinned by the fundamentals of organic and analytical chemistry, which are applied to interrogating biological systems. One example is the development of "bump and hole" kinase interrogation technology wherein the gatekeeper residue of a target kinase is mutated to enable selective binding of a modified ATP analogue bearing an enlarged purine ring and a thiophosphate tag. This functionally active system transfers stable thiophosphates to target proteins, enabling direct identification of the kinase's phosphorylation substrates.[11] Another example is the tag and modify strategy for site-selective protein modification, which enables the study of cellular processes regulated by post-translational modification. A diverse array of chemistry enabled through the *in situ* conversion of cysteine to dehydroalanine enables site-selective incorporation of sugars, terpenes and

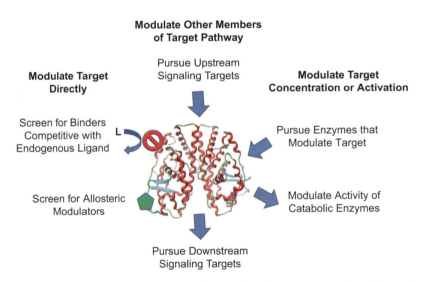

Figure 18.1   Different strategies for modulating the activity of a target protein—through directly interacting with it, influencing other members of its biological pathway or altering concentration of its activated form within the biological system.

phosphorylated functionality to mimic different manifestations of post-translational protein modification.[12]

Established techniques in chemical biology also enable the capture, purification and detection of proteins that interact directly with the target compound or whose expression is influenced indirectly by its mechanism of action. Chemical biology therefore facilitates whole cell phenotypic screening in the modern era of target-based drug discovery. One exemplification of this principle is phenotypic screening against a desired pathway biomarker, which has the advantage of identifying modulators of an entire biochemical pathway in one cell-based assay. Modern chemical proteomic approaches that combine new technology in organic chemistry, mass spectrometry and bioinformatics enable target identification in a more efficient manner than was possible in the era of classical phenotypic screening.[1,13,14] Bio-orthogonal chemistry technologies ("Click" Chemistry) for *in situ* ligation of capture ligands (often biotin or desthiobiotin) or fluorescent tags allow the manipulation of test compounds covalently bound to their target proteins (either through a photoactivatable crosslinker or electrophilic handle) to produce isolable and detectable complexes (see Figure 18.2).[15,16] It is important to choose the appropriate bioorthogonal ligation strategy, recognizing that some metal catalysts may not be compatible with experiments conducted in live cells, and the efficiency of ligation chemistry can be influenced by the local protein environment around the ligation functionality. If fluorescent tags are used, the cell lysate is run on an electrophoretic gel, and slices containing fluorescent bands are excised. If biotin tags are used, a streptavidin bead or resin is used to isolate proteins bound to the test compound. Once captured, the amino acid sequence of the target protein is obtained using mass spectrometry after trypsin digestion. Because non-specific binding can produce artifactual results, a control experiment is needed wherein the lysate is treated with both the labeled test compound and a large excess of its unlabeled parent. In this case, proteins that specifically bind to the test compound are eliminated from detection, allowing nonspecific effects to be determined.[13] Remarkable advances in mass spectrometer efficiency and analytical speed have delivered rapid and accurate determination of peptide sequence on minute quantities of protein. The vast array of protein sequence information in bioinformatic databases then allows the identity of each target protein to be rapidly determined from the amino acid sequence of its peptide fragments.[17]

Once a chemical tool has been identified, its effect on cellular pathways can be determined through gene expression array methodology, to examine effects on mRNA expression across a subset of the genome, or whole cell proteomic analysis using SILAC (Stable Isotope Labeling of Amino Acids in Cell Culture – see Figure 18.3).[14,18,19] Because transcriptional changes do not necessarily result in altered protein expression levels, proteomic methods offer a more direct

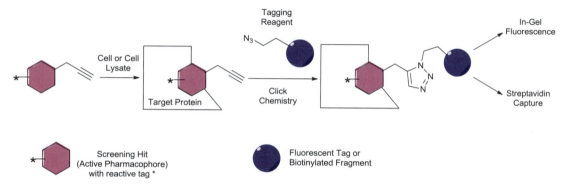

Figure 18.2 Tagging strategy using click chemistry to enable chemical proteomic-based target identification for phenotypic screening hits.

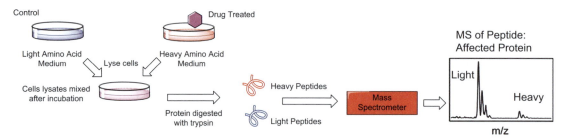

Figure 18.3 Schematic depicting SILAC analysis to determine the effect of drug treatment on protein expression within cells.

observation of functional consequences from compound treatment. The advantage of the SILAC method is that it provides a chemically identical internal standard for protein expression. In the SILAC experiment, isotopically labeled arginine and lysine are added to cell culture media for one set of cells—these amino acids are incorporated into proteins produced by the drug-treated cells. Since tryptic digestion occurs at amino acid sequences bearing lysine or arginine, peptide fragments will be differentially labeled for mass spectrometric detection relative to the control cells, which are grown in media containing unlabeled amino acids. Upon completion of drug incubation, the cells are lysed, and the heavy (drug treated) cell lysate is mixed with the light (control treated) cell lysate. The combined cell lysates are subjected to trypsin digestion followed by mass spectral analysis. The heavy and light peptides corresponding to a particular protein are then quantified and compared with the ratio for other proteins, giving an indication of differential expression modulated by the drug. By identifying proteins whose expression is upregulated or downregulated, the pathway effects of target modulation by the drug molecule can be better understood.

18.3   LEAD IDENTIFICATION

Once a biological hypothesis has been developed, the next step is to identify chemical leads that will enable early medicinal chemistry efforts. In the case of target-based screens, a direct binding or functional assay is developed that allows screening large compound collections for leads of moderate affinity. It is essential for the medicinal chemist to understand the mechanics of this target-based assay, including the method of detecting compounds that interact with the target (optical, fluorescence resonance energy transfer (FRET), radioligand displacement) and the presence of coupling enzymes that may be used to detect cofactor depletion (for instance, luciferase, which is used to detect ATP depletion in kinase assays). Most assay artifacts are driven by interference with these detection methods and manifest themselves when secondary assays not designed for high throughput screening are used to validate hits.[20]

Of course, one also needs to recognize the inherent limitations of SAR translation from a simplified biochemical assay relative to the complex environment of the cell. Assays configured to detect a binding event will give no information on functional activity for hits, which could be inhibitors/antagonists, activators/agonists, allosteric modulators or have no functional activity whatsoever—depending on where they actually bind to the protein and how they bias its conformation. Also, biochemical assays do not account for cellular penetration or efflux, which will influence the activity of hits against intracellular targets. Finally, biochemical assays may be conducted with truncated protein targets produced for ease of expression, purification and manipulation in the assay. Because such reagents lack large segments of the full-length protein that may be involved in regulation of its activity, structure–activity relationships may not translate to a native whole cell system.

It is also important to consider the binding parameters of the target for specific cofactors or substrates that interact with the binding site of interest, along with the intended concentration of those reagents in the screening assay relative to their endogenous concentration in cells relevant to the disease state. A classic example is working with ATP-competitive kinase inhibitors: the millimolar endogenous ATP concentration is very difficult to overcome for kinases that have a strong affinity for ATP ($K_m = 50$ µM for some kinases that operate in saturation), and therefore a kinase assay using low ATP concentrations will be needed to detect unoptimized inhibitors. However, one needs to bear in mind that inhibitors discovered under such conditions are unlikely to show functional activity in cells, where ATP concentrations are much higher.

In some cases, it is desirable to pursue an allosteric target modulation approach for reasons of needing to achieve better selectivity or to work under a wide range of endogenous cofactor/substrate concentrations. While one could take the approach of characterizing compounds for direct binding competition, either through mechanistic enzymology or ligand displacement assays, it is often better to design the screen properly up-front to bias detection for allosteric inhibitors. Typically, a high concentration of substrate or cofactor is used to swamp the protein's binding site and prevent detection of competitive binders. In other cases, the protein is modified with a fluorescent tag that detects motion induced by the presence of an allosteric modulator, providing definitive evidence of binding interactions outside its active site which perturb protein conformation. Such screens have been effectively implemented for kinases, but are undoubtedly more broadly applicable.[21]

The typical target-based drug discovery campaign begins with a high throughput screen of a large compound collection—typically in the order of 1–3 million compounds. Considering the fact that this number represents a vanishingly small percentage of the $>10^{40}$ possible compounds in the universe of drug-like small molecules, it is remarkable that such screening campaigns ever produce hits.[22] However, it is important to bear in mind that most corporate compound collections represent a heavily biased area of chemical space because the compounds in these collections are derived from natural product or small molecule leads that have already been proven to interact with human targets—often with only minor structural modifications.

The concept of diversity oriented synthesis was recently introduced to complement chemical space associated with typical corporate small molecule collections. Diversity oriented synthesis (DOS) emphasizes highly efficient modular coupling reactions that expand chemical diversity on the basis of richly populated functionalized monomer collections.[23,24] Compounds from DOS libraries generally have a greater number of stereocenters and saturated heterocyclic rings than those from corporate compound collections emphasizing aromatic heterocycle scaffolds and therefore provide complementary chemical space for screening. Many diversity oriented synthesis libraries emulate natural product pharmacophores—either by virtue of compounds being derived from natural product scaffolds, incorporating natural product motifs which are presented in novel molecular contexts or possessing richly adorned arrays of functionalized stereocenters that better emulate natural product chemical space.[25] Because these natural product structural features evolved to interact with proteins, these libraries have produced novel bioactive compounds against a variety of challenging targets.

While many drug discovery efforts have focused in traditional "Rule of 5" chemical space (being defined by Lipinski's Rule of 5, which focuses on matter having MW < 500, clogP < 5, ≤5 hydrogen bond donors and ≤10 nitrogen or oxygen atoms), there is an increasing realization that these guidelines can be inappropriately restrictive for challenging biomolecular targets.[26] Exceptions to the Lipinski rules include certain natural products like cyclosporine A, macrolide antibiotics and the recently approved small molecules aliskiren and argatroban. While operating in this chemical space increases the risk of poor cellular permeability or excretion, it may be necessary for protein targets with large peptidic endogenous ligands or for modulating protein-protein interactions. This issue has prompted a revisiting of peptide leads.

Peptide drug discovery is fraught with complications ranging from instability to gut and serum peptidases, cellular impermeability and the difficult attainment of secondary structure with short, linear peptides. Some of these limitations have been addressed through conformational constraint—either by cyclizing the peptide or introducing a short intramolecular "staple" that nucleates secondary structural features, such as helicity.[27,28] Some of these conformational constraints also improve permeability, as in the example of conotoxin peptide knots.[29] Presumably this permeability improvement results from reduced molecular volume due to the conformational constraints and masking hydrogen bond donors and acceptors through reinforcing intramolecular hydrogen bonding interactions. While the principles for oral peptide delivery are still elusive, these advances potentially enable injectable, topical or inhalable peptide discovery programs for extracellular targets and establish a foundation for further developments in peptide drug delivery.

The cost to screen an entire corporate collection is significant, with the typical high throughput screening assay costing in excess of $1 million when accounting for assay development time, screening expenses and depreciation of capital equipment required to screen large libraries. As such, many drug discovery efforts substitute focused screening campaigns using a subset of the larger collection. That subset might consist of compounds biased toward chemical space already known to hit other members of that target family (*e.g.* kinase subsets or GPCR subsets), increasing the probability of finding valid hits. This approach suffers from the risk of limiting possible lead matter to that which is already known to hit similar targets and the attendant concerns regarding lack of novelty and selectivity. Alternatively, a diversity subset derived from random sampling of chemical space represented in the full library might be used in an initial screen for lead compounds or to sample the full corporate file's ability to produce viable leads.

Unbiased screening methods to detect compounds that bind to a target protein without regard to its functional activity can provide ultrahigh throughput screening methods at significantly lower cost than biochemical assay screening. One example is affinity selection mass spectrometry (ASMS – see Figure 18.4).[30,31] In this technique, mass encoded libraries with very high compression (sometimes in excess of 100 compounds per well) are used to provide substrate for lead identification. The target protein is incubated with these compound mixtures, and a rapid separation process is applied where the protein and any small molecules bound to it are separated from nonbinding small molecules. This separation process is critical to the success of an ASMS campaign and typically relies on centrifugal filtration through a semiporous membrane to elute non-binders or size exclusion chromatography. Each of these methods has its limitations: centrifugal filtration requires multiple washes to remove nonbinders, limiting the ability to detect weak binders or compounds with low ionization efficiency due to dilution. Size exclusion chromatography precludes detection of compounds with fast off rates, because they dissociate from the protein before it is eluted from the column. Once the protein complex is separated from nonbinding small molecules, it is denatured, and the binders are detected by mass spectrometry.

Another approach for detecting binders is biophysical screening applied to fragment libraries. Fragment based drug discovery has been extensively reviewed and offers several advantages to screening more fully elaborated leads (see Figure 18.5).[32–36] The principal advantage of fragment based drug discovery is better coverage of chemical space. The number of theoretically possible compounds increases exponentially with molecular weight—therefore, by limiting the size of molecules in a compound collection, one can get better (although still far from complete) coverage with a small compound collection.[37] Moreover, many of these fragments are commercially available or exist in well-stocked corporate collections, representing an

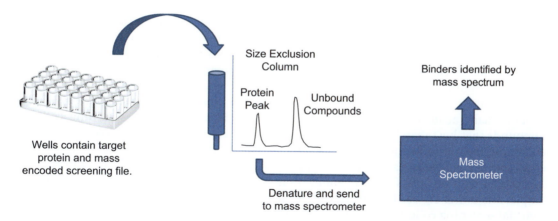

Figure 18.4 Schematic representation of typical Affinity Selection Mass Spectrometry screening workflow using size exclusion chromatography to separate protein–compound complex from non-binders.

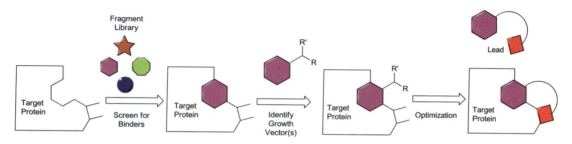

Figure 18.5 Pictorial representation of fragment based screening and hit optimization processes to derive lead matter from small fragment collections.
Figure adapted from reference 36.

easily-obtained source of chemical diversity. Fragment libraries are typically constructed with a few key principles: (1) absence of functional groups that might react with the target protein, (2) high solubility to enable detection at millimolar concentrations, (3) low molecular weight (MW < 250) to enable elaboration while retaining oral permeability, (4) synthetic flexibility through incorporating functionalizable handles and (5) molecular diversity.[38,39]

Fragment screening is typically performed using biophysical assay techniques, such as Saturation Transfer Difference NMR (STD-NMR) or Surface Plasmon Resonance (SPR), and X-ray crystallography, although there are many variants of these techniques employed.[40] The basic principle of STD-NMR relies on energy transfer from irradiated methyl groups on the protein to small molecules, coupled with the slower relaxation time of small molecules relative to protein complexes. Once irradiation occurs, magnetization is transferred from the protein to the bound small molecule ligand. If that ligand dissociates before relaxation occurs, its magnetic energy persists for a longer time and can be detected. Surface Plasmon Resonance relies on subtle refractive index changes that occur when a small molecule binds to an immobilized protein. This method is often implemented in a flow based system, where compound and buffer solutions are passed by the immobilized protein, producing a refractive index response that is plotted on a sensorgram. The SPR method offers the added advantage of being able to determine binding affinity and kinetics.[41] Once identified, fragment hits can be rapidly elaborated to more potent leads through adding appendages, often guided by X-ray crystal structures of the fragment bound to its target.

In silico approaches have also been used to identify lead compounds.[42] To enable target based screening methods, an X-ray crystal structure of the protein (or at a minimum, a homology model) is required. From this structure, compounds from a virtual screening collection are docked into the target binding site and scored based on their protein interaction energies. High scoring compounds represent the virtual screening hit set. Alternatively, a target pharmacophore is defined for the desired binding pocket, and this 3D pharmacophore is screened computationally against 3D pharmacophores for ligands in the virtual screening collection. Compounds that best match the target pharmacophore are then tested in biological assays for their ability to bind and modulate the target. It is also possible to conduct a ligand based pharmacophore screen in cases where a small molecule lead already exists for a target. Ligand based models can be built from the endogenous ligand or substrate for the protein target, for example, and either 2D or 3D pharmacophore searches can be subsequently conducted against the virtual screening collection. As in the target based method, hits are subsequently confirmed through biochemical assay prior to progression.

After identifying the chemical space serving as potential lead matter with which to start the drug discovery program, the medicinal chemistry team must understand the intellectual property landscape relevant to those leads. The first consideration is whether the lead compounds are represented specifically or generically in issued claims within other patent applications. If that is the case, the drug discovery team must consider whether lead optimization efforts are likely to evolve the chemical equity outside the scope of those claims—otherwise there will be issues with freedom to operate (the ability to commercialize the drug substance without interfering on another inventor's patent claims). If the lead series is not specifically or generally claimed in another patent application, then the discovery team will need to demonstrate novelty, non-obviousness and utility of the drug substance beyond what is otherwise known in the literature. These hurdles can generally be met through demonstrating the convergence of target potency and selectivity, appropriate pharmacokinetic profile and safety of the drug substance—each of which will be hallmarks of the lead optimization effort—due to the challenges associated with integrating all of these favorable properties into a single compound. For these reasons, patent attorneys should be involved once leads have been identified and lead optimization begins to ensure that the intellectual property landscape and its implications are understood.

18.4   LEAD VALIDATION AND OPTIMIZATION

Once lead matter has been identified, it is important to conduct validation screens that confirm specific binding interactions to the drug target, to gauge SAR depth around the lead series, and to characterize the chemical matter for drug-like ADME and safety properties. One of the most basic and useful pieces of information used in lead validation is the mechanism by which the compound modulates its target. This information not only builds confidence in lead matter validity, but also highlights potential selectivity issues and differentiation from other lead matter that may be known for the target. If the drug target is an enzyme, mechanistic enzymology can readily determine whether the inhibitor is competitive, noncompetitive or uncompetitive with substrate or cofactor. Radioligand displacement assays can be used for receptor-based targets to determine orthosteric or allosteric modulation mechanisms relative to the labeled ligand probe. In the event that a radiolabeled ligand is not available, biophysical studies can be used as a surrogate to determine competitive binding. Often STD NMR offers a fast and reliable answer for ligands whose dissociation rate from the target is rapid. In this experiment, either the test compound or the ligand/substrate is titrated against the other to determine whether there is a direct effect on the STD signal, which could signal a competitive binding event.

The kinetic and thermodynamic parameters of compound interaction can be determined using common biophysical techniques and provide helpful information for judging the quality of the lead and potential PK/PD disconnects that might be seen within the series. In any of these experiments, it is important to use high quality purified protein reagent, preferably from a construct similar to that used for biochemical assays to ensure relevance of the results. Isothermal titration calorimetry (ITC) is useful for determining the enthalpy and entropy of binding.[43] There are two principal drawbacks to this technique: first, it requires large quantities of purified protein target; second, the energetics of protein–ligand interactions can be strongly influenced by buffer composition and other experimental conditions. Nonetheless, there is evidence to suggest that enthalpically dominated protein–ligand interactions are beneficial lead quality attributes relative to entropically dominated binding energetics.[44] It is also helpful to understand binding kinetics: k_{on} (on rate) and k_{off} (off rate), as this information has implications for assay design and understanding potential PK/PD disconnects.[45] Surface plasmon resonance (SPR) is often used to determine binding kinetics, its only principal disadvantage being the requirement for immobilizing the drug target on a biosensor chip. Target immobilization is typically accomplished through biotinylating solvent-exposed lysines and capture on a streptavidin sensor. The immobilized target is validated through comparing K_d values measured by SPR with those measured by solution biochemistry techniques. A final biophysical technique that is useful for characterizing protein ligand interactions is to measure thermal stabilization of the protein target by the test compound. Every protein target has a melting temperature, which is the temperature at which the tertiary and secondary structure of the protein begins to change, resulting in different optical properties that can be measured.[4] Compounds having tight binding interactions with their protein target typically stabilize its bound conformation, resulting in a higher melting temperature in thermal denaturation studies. While this technique does not provide direct insights into binding energetics, one can determine binding constants and validate binding interactions with relatively small amounts of purified protein.

The ultimate biophysical technique for determining protein–ligand interactions is x-ray crystallography. While this technique can provide high resolution atomic coordinates for directly visualizing protein–ligand interactions, it suffers from three limitations: it requires crystallized protein, which can be difficult to procure, it produces a time-averaged picture of the protein–ligand complex, and the conformation of protein subunits can be influenced by crystal lattice contacts. The information produced using this technique directly enables structure based drug design and can provide some insight into regions of the protein that are very rigid and those that are more flexible. These crude measurements of protein flexibility can be performed through examining disorder that might be present in the crystal structure and noting large temperature factors that are present in backbone carbon and nitrogen atoms. It is best to study several crystal structures of different protein–ligand complexes to get a sense of binding pocket flexibility. Significant changes can be seen for proteases, where the regulatory flap region can exist in several conformations depending on properties of the bound inhibitor. Likewise, the ATP-binding site of kinases can exist in several conformations, most notably through movement of the regulatory DFG loop, depending on the structure of the kinase inhibitor. Computational tools to construct 3D ligand and binding pocket pharmacophores can be developed and used for enabling scaffold hopping and alternative lead structure searches.

Protein structures are dynamic, adopting several conformations in solution both in the apo and bound state. Understanding protein dynamics can provide useful information for understanding how the target might be differentially modulated by different compounds, informs strategies for attaining selectivity against homologous targets (by understanding differential accessibility of the conformation competent to bind the lead) and can be used to understand binding kinetics data.[46,47] Protein dynamics can be computed using molecular force fields and a starting structure, typically obtained from x-ray crystallography. An NMR structure can also be used to produce protein dynamics information through understanding the range of structures consistent with distance restraints (produced through NOE observations) and torsional constraints (produced from coupling constant values). NMR structures require that all amino acid resonances be assigned, which can be a laborious task and requires multiply labeled proteins to be produced for NMR studies. Sometimes rich information can be obtained from a singly labeled amino acid, such as phenylalanine or cysteine bearing a fluoroalkyl substituent. These techniques were recently used to characterize DFG loop dynamics in kinases and GPCR conformations on binding ligands with different functional properties.[48,49]

Another technique for studying protein dynamics is hydrogen–deuterium exchange mass spectrometry (HDX). In this experiment, the exchange of hydrogen for deuterium on the backbone NH groups of a protein is a probe for dynamic motion. Solvent exposed and rapidly moving residues are more likely to exchange than residues that are conformationally constrained and on the interior of the protein. Purified protein is incubated with test compound in D_2O for a fixed period of time, then cooled to 4 °C and treated with acid to stop the exchange process. Peptic digestion followed by mass spectral analysis of the peptide fragments produces a map of exchangeable protons which can be superimposed on an x-ray crystal structure to get a dynamic picture of the protein. This technique has been used to characterize GPCR dynamics and the differential effect of PPAR antagonists.[50,51] The advantage of HDX is its speed and the lack of prior structural assignment as is required for NMR. However, it is a low resolution technique that requires crystallographic information to provide maximum benefit and is also limited by the completeness of mass spectral coverage for the protein.

18.5   IMPORTANT CONSIDERATIONS FOR OPTIMIZING POTENCY

Once valid lead matter has been identified, it needs to be optimized for potency, selectivity, ADME properties, safety and biopharmaceutical properties. These topics will be covered in subsequent sections. The challenge of medicinal chemistry is to optimize what are many times interdependent properties that do not move in the same favorable direction against each of these endpoints. As such, holistic measurements of compound quality have been introduced to help medicinal chemists focus on achieving the most efficient target–ligand interactions. These measurements are directed at the two most common culprits detracting from lead quality: molecular size and lipophilicity.[52] Ligand efficiency is calculated by dividing the number of non-hydrogen atoms into the free energy of binding for the drug molecule to its target, and returns a value of kcal/heavy atom, which provides an indicator of binding efficiency based on size.[53] Medicinal chemists typically optimize for LE values of >0.3 kcal/mol/heavy atom. Lipophilic efficiency is calculated as pKi – log D, and it has been shown that highly optimized compounds have LipE values >5.[54] LipE provides an indication of how well polar functionality on the drug molecule complements the biomolecular target and seeks to minimize the amount of lipophilicity required to produce a target potency value (see Figure 18.6).

Computational tools for understanding ligand–target energetics are commonly used in structure based drug design to optimize potency and selectivity. These techniques typically rely on a molecular mechanics force field for speed and can be run by medicinal chemists on the typical desktop computer. Algorithms have been developed that consider many of the crucial free energy terms for ligand binding, including enthalpic interactions from hydrogen bonding, ionic pairing, cation–pi interactions and van der Waals interactions, along with ligand desolvation. These interactions have been reviewed extensively elsewhere, and the section below will focus on less appreciated protein–ligand interactions.[55]

The role of water in the protein binding pocket has more recently become a significant design consideration. By creating a polar dielectric environment, bound water molecules have

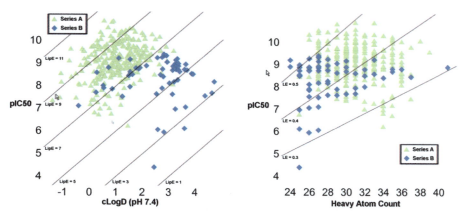

Figure 18.6   Plots of potency *vs.* calculated lipophilicity to illustrate separation of compounds based on lipophilic efficiency (LipE) values and potency *vs.* number of heavy atoms to illustrate ligand efficiency (LE) values for different series within a drug discovery project.

important effects on intramolecular hydrogen bonds that may be crucial for stabilizing the bound conformation of a protein. On displacing these water molecules with a ligand, these structural hydrogen bonds (dehydrons) become further stabilized due to the resulting lower dielectric constant of their environment, resulting in a binding energy enhancement. This dehydron concept has been successfully applied to structure based drug design.[56] Another approach is to consider the energetics of the water molecules themselves to determine whether they should be retained or displaced from the active site. Commercially available applications, such as WaterMap™ (Schrodinger), enable the medicinal chemist to estimate the enthalpy and entropy of a bound water molecule.[57] These applications require a high resolution crystal structure of the protein target, and an easy measure of their accuracy is whether they predict the presence of crystallographically detected water molecules. Waters that are energetically favorable with a high enthalpy of binding are generally targets for making a hydrogen bond with a donor or acceptor on the inhibitor, as direct displacement would require a highly optimized hydrogen bond to be made with the protein. Waters that are energetically unfavorable are targeted for displacement by the inhibitor.

Hydrogen bonding is a major interaction in reinforcing secondary structure of biological macromolecules such as proteins and DNA. Hydrogen bonding also contributes to the enthalpic interactions between small molecule drugs and their biomolecular target. Hydrogen bonds form between an electropositive hydrogen atom that is attached to a much heavier electronegative atom (such as nitrogen or oxygen)—denoted the hydrogen bond donor (HBD) and a hydrogen bond acceptor (HBA), which in proteins is typically the oxygen atom of a peptide carbonyl, or a polar amino acid side chain. Halogen atoms also have the potential to interact as a donor group with peptide carbonyl oxygens in forming a halogen bond. This interaction is due to a unique property of carbon–halogen bonds wherein an electropositive crown (also known as a sigma hole) forms due to polarization of the halogen along its covalent bond. A Lewis basic group such as oxygen, nitrogen or sulfur can interact with the sigma hole. The typical distance for a halogen bond is *ca.* 3 Å between the donor and acceptor atoms, which is similar to the distance between the heavy atoms involved in a hydrogen bond. Halogen bonds have the potential to serve as complementary intermolecular interactions to hydrogen bonds (see Figure 18.7). A survey of crystallographic data for structures containing both hydrogen and halogen bonds to the same carbonyl oxygen revealed that the bond angle between the hydrogen bond and the halogen bond is typically 88° ($+/-$ 14°).[58] Moreover, the halogen bond has an increased propensity to lie in the plane perpendicular to the amide bond. *Ab initio* calculations

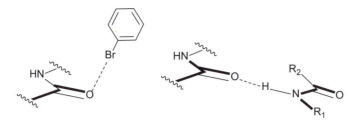

Figure 18.7   Differences in preferred geometries for halogen bonds relative to hydrogen bonds in protein–ligand interactions.

suggest that halogen and hydrogen bonds are energetically independent due to the existence of two sets of independent electronegative potentials, allowing for two sets of energetically independent electrostatic interactions. An optimum halogen bond has been calculated to be worth *ca.* 3–4 kcal/mol in binding energy.

18.6   IMPORTANT CONSIDERATIONS FOR ABSORPTION, DISTRIBUTION, METABOLISM AND EXCRETION

It is often desirable to administer drugs *via* the oral route either for patient convenience or out of practical necessity (in the case of regularly administered drugs). In order to achieve successful oral delivery of a drug, medicinal chemists must consider biopharmaceutical properties such as solubility, dissolution rate and stability to enzymes and pH excursions in the gastrointestinal tract as well as permeability. Aqueous solubility is influenced both by lipophilicity and solid state energetics. Lipophilicity or hydrophobicity drives down intrinsic water solubility by definition. Similarly, highly stable crystalline solids require a lot of energy to break up the crystalline lattice, which is required for dissolution. It is important for preclinical oral PK work to be performed on crystalline as opposed to amorphous material because oral exposure may be overestimated with amorphous drug substance. Therefore, chemists typically optimize compounds to deliver crystalline solids with melting point between 120 °C and 225 °C and intrinsic aqueous solubility of > 100 µg/mL (preferably > 250 µg/mL for high dose compounds) at the pH range relevant to absorption in the ileum and jejunum (pH 6.0–7.5). Thermodynamic solubility measurements made under physiologically relevant conditions can then provide a theoretical maximum absorbable dose based on the additional parameters of permeability, GI residence times, particle size and dissolution kinetics.[59]

Permeability is largely influenced by ionization state and lipophilicity, and often the parameters that are optimum for permeability work against those that are optimum for aqueous solubility. The majority of drugs are absorbed through passive transcellular processes, meaning that diffusion through the phospholipid-based cell membrane governs flux into the enterocyte. As such, the medicinal chemist must consider: (1) the overall lipophilicity of the drug molecule, which governs partitioning into the cell membrane, (2) the molecular size, which governs the amount of membrane displacement that must occur for absorption, and (3) the presence of hydrophilic functionality, which governs desolvation effects that must occur for plasma membrane partitioning. In the late 1990's, Lipinski and co-workers derived a simple mnemonic for predicting the probability of oral absorption called the "Rule of 5", which states that oral drugs generally have MW < 500, clogP < 5, no more than 5 hydrogen bond donors and no more than 10 N or O atoms.[26] One can envision how this statistically derived mnemonic overlaps with the physicochemical principles described for permeability and biopharmaceutics; however, several additional principles have been advanced to aid the design of orally bioavailable drugs. Molecular size can be described as a function of both gyration radius and molecular flexibility.[60]

These properties are not perfectly correlated with molecular weight, as certain heavy atoms have smaller size relative to isobaric functional groups containing only C, H, N and O atoms. Additionally, topological polar surface area (TPSA) is used as a complement to analyzing number of hydrogen bond donors or acceptors, as TPSA provides the opportunity to differentiate polar functional groups based on their exposed hydrophilic surface and hydrogen bonding potential, which is not equal for all functional groups.[60] Higher dimensional (3D) polar surface area calculations also offer the possibility of considering intramolecular hydrogen bonds, which mask the availability of these functional groups for solvation and reduce the desolvation penalty for lipid membrane partitioning. Finally, sp^3 atom content can influence solid state stability and therefore aqueous solubility.[61] The fraction of sp^3 hybridized atoms is also important to consider for achieving pharmacological selectivity, as protein binding pockets typically have significant 3-dimensional character that offers an additional opportunity for engineering selectivity through the appropriate introduction of chiral centers in a drug molecule. Of course, this principle also has to be balanced with not introducing significant synthetic difficulty that makes subsequent drug development impractical.

Medicinal chemists must also consider the ability of a drug to access its biomolecular target within the diseased tissue. Two key parameters influencing tissue distribution are cell permeability and binding to plasma proteins. The free drug hypothesis states that it is only the unbound fraction of drug that is available for interaction with its target, diffusion into tissues and clearance (see Figure 18.8).[62] Albumin is the major plasma protein involved in drug binding and has particular affinity for acidic compounds. Because it is present in high concentrations and possesses several drug binding sites, albumin binding typically is not saturable at physiological drug concentrations, although this assumption should be checked by measuring plasma protein binding at different concentrations representing the extremes of expected peak to trough plasma concentration excursions. Alpha-1-acid glycoprotein is another major plasma protein involved typically in the binding of basic drug molecules. Drugs bind differentially to proteins and phospholipids within tissues as well; this phenomenon, coupled with tissue-specific active transport, can influence concentration gradients between plasma and certain tissues and creates the potential for tissue-based drug reservoirs at equilibrium.[63] Asymmetric tissue distribution typically only has pharmacological consequences when free drug concentrations are different in the target tissue relative to plasma (*i.e.* when active transport is involved), as the absence of active transport implies that free drug concentrations in plasma and in tissues will otherwise be equal despite differences in total drug concentrations.

One tissue where free drug distribution is typically restricted is the brain and central nervous system components. The tight junctions within the capillary endothelial cells surrounding CNS tissues means that drugs must enter through passive diffusion, often against a negative

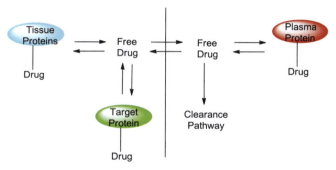

Figure 18.8 Effect of protein binding equilibria on tissue distribution, target binding and clearance of drug substances.

concentration gradient created by efflux pumps, such as P-glycoprotein (Pgp) and Organic Anion Transporters (OATP). This blood–brain-barrier creates a significant hurdle for the discovery of drugs against CNS targets. Until recently, the physicochemical property ranges for drugs to enable distribution into CNS tissue were thought to be limited to lipophilic compounds preferentially possessing a weakly basic center. However, a systematic study of brain penetration for compounds across a broader range of physicochemical properties resulted in the development of *in silico* tools with good predictive power for CNS penetration.[64] These tools are based on a multiparameter optimization (MPO) wherein specific physicochemical properties found to be statistically significant in driving brain penetration are given functional ranges and weighting factors for influencing the predicted brain penetration of a compound.[65,66] This tool has expanded the accessible chemical space for CNS penetrant compounds to include kinase and phosphodiesterase (PDE) inhibitors, classes of compounds for which it was once thought to be difficult to achieve physiologically meaningful brain concentrations.

Drug distribution can be directly observed through whole body autoradiography experiments, although their requirement of radioactively labeled drug typically means these studies are conducted at the candidate selection stage of the drug discovery process. These studies involve the use of whole animal radioimaging, wherein the distribution of drug as a function of time can be directly seen at the whole organ level. Because these techniques use stable radioisotopes (^{3}H is common) and are non-invasive, images can be taken at several points post-dose to investigate organs where drug accumulation might be observed and to study organs involved in clearance. Because the presence of radiolabel is not dependent on metabolic stability of the drug molecule, observation of the radiolabel does not necessarily imply presence of the drug itself, but possibly also of its metabolite(s). Therefore, one should assess drug levels in tissues directly using bioanalytical mass spectrometry to confirm the presence of unchanged drug and the possibility of metabolites also concentrating in tissues.

The human body has developed several mechanisms for eliminating xenobiotic compounds from systemic circulation. The purpose of all metabolic clearance methods is to make the molecule more polar so that it can be excreted, and it is common for more than one metabolite to be produced from a given compound. Common mechanisms for metabolism include oxidation reactions catalyzed by cytochrome P450 enzymes, aldehyde oxidase and flavin-dependent monoamine oxidase. The most common drug-metabolizing cytochrome P450 enzymes include CYP 1A1, 2C9, 2D6, 3A4 and 3A5. Each of these enzymes catalyzes aerobic oxidation of xenobiotics. They are NADPH dependent, possess a heme moiety that performs the oxidation chemistry and have a defined binding pocket recognizing certain target types and functionality.[67] In general, increasing lipophilicity will drive more rapid CYP-mediated metabolism. Xenobiotic metabolizing CYP isoforms are predominantly found in the liver, although some CYP isoforms are also active in lung and in the small intestine. CYP mediated metabolism is typically measured based on compound half-life in the presence of liver microsomes containing NADPH as a reductive cofactor. The most important technique to reduce CYP-mediated metabolism is to reduce overall lipophilicity. In addition, compounds possessing acidic functionality tend to be less susceptible to CYP-mediated metabolism. Blocking sites of metabolism with a fluorine or a methyl group gives mixed results, with metabolism sometimes being shifted to another part of the molecule. The advantage of CYP-mediated metabolism is that it can be readily scaled from *in vitro* intrinsic clearance to predicted *in vivo* clearance by considering physiological parameters for the species of interest. It is also possible to predict CYP-mediated metabolism using *in silico* statistical models that are trained on large data sets and compare similarity of structural and physicochemical properties for the test compound relative to nearest neighbors in the data set to derive a predicted intrinsic clearance. Statistical models typically return a predicted clearance value along with a confidence metric based on similarity of compounds in the data set to the test compound. Additionally, crystal structures exist for various CYP isoforms, which allow one to derive structurally-based predictive tools for CYP interactions.[68]

Aldehyde oxidase and xanthine oxidase are cytosolic molybdenum-containing, FAD dependent enzymes that contribute to hepatic metabolism. These enzymes are responsible for C–H oxidation adjacent to the nitrogen atom of six-membered heterocyclic compounds, oxidation of iminium ions produced by CYP oxidation of carbon atoms adjacent to amines and reduction of certain 5-membered heterocycles (see Figure 18.9). Unlike CYP-mediated metabolism, the predictive tools for AO metabolism and scaling *in vitro* intrinsic clearance to *in vivo* clearance are less well refined because of interspecies differences in AO isoform expression and an incomplete understanding of how these enzymes function.[69] Typical strategies for avoiding AO-mediated metabolism are to try isosteric replacements for the subject heterocycle, place bulky substituents on or near the heterocyclic ring to discourage oxidation or to place blocking substituents at the site of metabolism.[70]

Beyond oxidative metabolism, hepatocytes possess several enzymes that are responsible for conjugative reactions such as glucuronidation and sulfation. The glucuronidation reaction is catalyzed by UDP-glucuronidyl transferase (UGT) and attaches a molecule of glucuronic acid to the drug substance at the anomeric center.[71] Typical sites of conjugation include carboxylic acids, phenols, alcohols and amines. Common strategies for circumventing glucuronidation include eliminating the functional group that is conjugated, placing steric bulk proximal to the conjugated group or placing polarity proximal to the conjugated group. The latter two strategies are aimed at reducing enzymatic efficiency through interfering with substrate binding. Like aldehyde oxidase, it is challenging to scale UGT-mediated clearance from *in vitro* clearance assessments to *in vivo* clearance predictions.

Drug substances can also be excreted—either through the biliary excretion pathway, in which case they are ultimately eliminated in feces, or through the urinary excretion route. The biliary excretion pathway is influenced by transporters, such as ABCG2, which act on a variety of amphipathic lipid soluble drugs, and MRP2, which acts predominantly on drug conjugates such as glucuronides.[63] An important phenomenon with hepatobiliary excretion is the potential for enterohepatic recirculation, wherein the excreted substance is subsequently reabsorbed through the intestine back into the systemic circulation. In some cases, this reabsorption process requires enzymatic hydrolysis of drug conjugates, such as glucuronides, which are

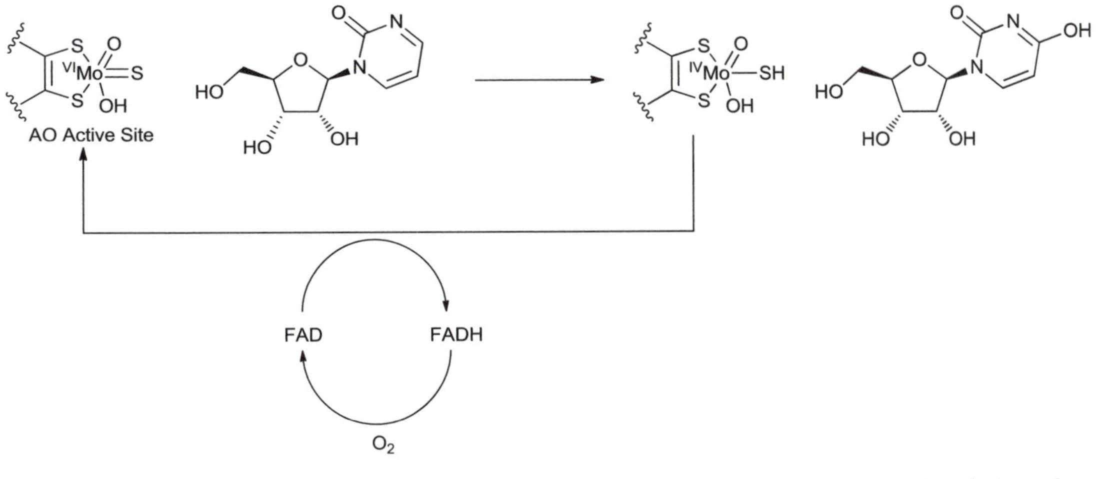

Figure 18.9 Aldehyde-oxidase mediated metabolism of zebularine as an illustration of oxidation of carbon atoms adjacent to nitrogen in electron deficient aromatic heterocycles.

cleaved prior to reabsorption through the gastrointestinal tract. Enterohepatic recirculation typically manifests itself in a second hump on a plasma concentration *vs.* time pharmacokinetic plot and a high volume of distribution, reflecting the residence of drug in the gastrointestinal depot. Renal excretion involves the processes of glomerular filtration, which is a passive diffusion of free drug into the tubular lumen. This process is augmented by active tubular secretion in cases where the drug molecule possesses functionality that is recognized by transporters present in the proximal renal tubule, such as MRP2, Pgp and organic cation transporters. A subsequent reabsorption process occurs wherein unionized drug can be reabsorbed in both the proximal and distal regions of the renal tubule. The net reabsorption of water and electrolytes by the kidney drives a concentration gradient required for reabsorption of xenobiotics. Therefore, renal clearance is influenced by the glomerular filtration rate (which is a function of kidney health), affinity for renal excretion transporters and pH of the urine, which affects the unionized fraction of drug available for reabsorption.

18.7   PK/PD RELATIONSHIPS INFLUENCING DESIGN

As compounds become better optimized for potency, selectivity and pharmacokinetics, it becomes important to understand the relationship between plasma drug concentrations and efficacy *in vivo*. This relationship between pharmacokinetics (PK: drug concentration as a function of time post-dose) and pharmacodynamics (PD: efficacy as a function of time post dose) has significant implications for which ADME parameters are targeted for optimization as the lead series evolves closer to enabling drug candidate selection. While *in vivo* PK/PD studies are often more intensive in terms of animals and resource investment relative to more typical dose-titration efficacy studies, they provide much clearer guidance to the medicinal chemist as to how half-life and clearance should be optimized. There are several instances where concentration-effect relationships observed in a static concentration *in vitro* assay do not accurately reflect observed PK-PD relationships *in vivo*, and often times these situations are characterized by kinetic parameters for drug binding and distribution. For example, in cases where target off rate is slow relative to pharmacokinetic elimination rate, the pharmacological efficacy will persist for longer than observed drug levels in plasma. Drugs that covalently modify their target, such as aspirin (cyclooxygenases), selegiline (monoamine oxidase-A) and clopidogrel (P2Y12) are examples of compounds that exhibit prolonged efficacy relative to their pharmacokinetic half-life.[72,73] This effect presents a unique benefit in that it can offset issues associated with high clearance or short half-life of the drug *in vivo*.[74] The construction of a PBPK (Physiologically Based Pharmacokinetic) model integrating target binding kinetic parameters with pharmacokinetic variables governing the rate of compound influx and elimination from the relevant tissue compartment can inform the relevance of optimizing for slow target offset to overcome clearance issues.[75] Conversely, delayed tissue distribution or system based properties, such as target turnover rate, can produce an observed lag between maximal drug concentrations in plasma and maximum pharmacological effect.[76] Examples of drugs showing delayed onset of action due to slow target binding kinetics or delayed tissue distribution include buprenorphine and morphine.[72] Receptor pharmacology can also demonstrate complexities in that responses can be amplified depending on cellular signaling feedback mechanisms or receptor association upon drug binding.[74] Dynamic concentrations of drug can be simulated *in vitro* using hollow-fiber technology.[77] Here, cells to be studied are placed in a hollow fiber bordered by a semi-permeable membrane that allows small molecules to pass but precludes exchange of proteins and cells. The hollow fiber is placed in a media chamber, and pumps are used to administer drug (simulating absorption) or flush the system with media (simulating clearance). Cells can be harvested from the hollow fiber and analyzed, or the fluid within the fiber studied for biomarkers. Hollow fiber technology has been used to study PK/PD relationships for anti-infectives and oncology drugs, and it should also be applicable to other disease areas.

It is often tempting to target a "no regrets" dose for clinical study—sustaining free drug concentrations that represent significant multiples of the target IC_{50} or IC_{90} for the entire dosing interval. This approach is often used where there is not sufficient confidence in existing *in vivo* disease models or in disease areas where clinical failure in Phase 2 due to lack of efficacy is a frequent occurrence. However, this strategy has some limitations: it places a significant burden on the medicinal chemist to optimize ADME and safety beyond what would otherwise be required and it exposes patients to unnecessarily high drug concentrations in clinical trials where efficacy might otherwise be achieved at lower doses. Most importantly, if not informed by diligently run PK/PD studies, it leaves the medicinal chemist in the dark about which strategy should be used for optimizing PK/PD and biopharmaceutical properties. It is therefore important that chemists partner with biologists to understand PK/PD relationships in relevant disease models as early as possible in a drug discovery project.

Drug discovery teams have powerful tools to measure engagement of the biomolecular target at the tissue site of action. Positron Emission Tomography (PET) imaging provides a non-invasive, medium-resolution view of target occupancy using low concentrations of a target-specific PET radiotracer probe. Typical PET studies look for displacement of the PET probe by unlabeled drug molecule, where quenching of the PET image signals complete target occupancy by the unlabeled drug.[78] The advantages of PET are its resolution at the tissue-structure level and the ability to study target occupancy as a function of time in whole animals. The principal disadvantages of PET are its requirement for a highly potent and specific radiolabeled probe that generally has a short radioisotope half-life (^{11}C or ^{18}F) and the fact that the target must be located in solid tissue. Therefore, synthetic incorporation of the PET label typically is done as the final or penultimate step in PET probe synthesis, which typically occurs at the imaging site and is used within hours of preparation.

A complementary strategy to PET imaging is the use of mechanistic biomarkers to investigate target engagement. With this strategy, examination of a biochemical pathway provides insights into which secreted or cytosolic proteins would have altered expression as a result of a drug modulating the desired biomolecular target. For example, inhibition of Janus Activated Kinases (JAKs) reduces phosphorylation of STATs, downstream signaling proteins that are the immediate targets of phosphorylation by JAK. By measuring the level of phosphorylated STAT proteins in control and drug-treated cells (or tissue samples from efficacy studies), one can infer the level of JAK signaling inhibition.[79] The relationship between downstream target modulation and disease efficacy from preclinical PK/PD studies can be correlated with clinical mechanistic PK/PD information to set dose for longer term efficacy trials. The advantage of mechanistic biomarker studies is that they measure functional modulation of the target (as opposed to just binding). Their disadvantages are the fact that one must assume biomarker modulation is an effect of drug target engagement (as opposed to secondary pharmacology), that target engagement must be measured *ex vivo* (requiring tissue extraction and homogenization) and that they may not provide resolution at the tissue structure level.

The rapidly increasing power of protein mass spectrometry technology provides the opportunity to directly measure target occupancy using competitive activity based proteomic profiling (cABPP).[80] Here, animals are treated with test drug followed a few hours later by dosing with an activity based probe incorporating a click-chemistry enabled handle for protein capture or imaging. Tissues are harvested, cells are then lysed and analyzed through in-gel fluorescence. Occupancy of the drug target binding site impairs labeling by the activity based probe, resulting in a reduction in signal for that protein in the tissue from drug treated animals relative to controls. These results can then be correlated back to the pharmacokinetic profile of the compound to understand concentration effects on target occupancy. In order to produce a signal, the drug molecule needs to effectively compete with an activity based probe that co-valently modifies the drug target. Therefore, this technique only works for tight binding drug compounds with slow dissociation kinetics. Also, the drug target must belong to an enzyme family for which general activity based proteomic profiling probes are available.[81] Its power is in enabling a direct examination of drug target occupancy that complements what can be achieved with PET or mechanistic biomarker investigations.

18.8 DRUG SAFETY

The safety and tolerability profile of a drug can be one of the most challenging properties to optimize due to the multifactoral nature of drug toxicity. The first question that must be considered is whether the underlying mechanism of drug action could cause safety problems. There are many examples of mechanism based toxicity that have resulted in discontinuation of drug candidates—some of the best known examples are dose limiting emesis with PDE4 inhibitors,[82] musculoskeletal side effects associated with MMP inhibition[83] and valvulopathies due to 5HT2B activity.[84] Drug discovery teams will typically attempt to determine mechanism-based toxicity through genetic knockout studies. In the case of embryonic gene knockouts, results must be interpreted with caution due to developmental compensation or toxicities. These issues can be circumvented using conditional knockouts, where the gene of interest is rendered non-functional after treating the organism with a stimulus that alters expression of the target gene. Molecular approaches exploiting RNA interference or short hairpin RNA technology can also provide mechanistic safety inferences, although delivery challenges make *in vivo* studies using these technologies difficult. A conceptually simple approach is to test multiple series of lead compounds in short-duration animal toxicology studies, inferring that safety issues shared among the compounds could be mechanism based. This approach suffers from several challenges, including the need for multiple tool compounds (which likely require some degree of pharmacology and PK optimization from screening hits), the required inference that activity against the primary target is the only shared pharmacology contributing to observed toxicities, and variability in dose–exposure and tissue distribution profiles between the compounds that could influence toxicity. However, the ability to get an early view of possible chemotype and target-based toxicities make this approach a commonly used method for assessing drug safety early in drug discovery projects.[85]

In addition to primary mechanism based toxicity, chemotype factors such as promiscuous secondary pharmacology, presence of reactive functionality, and property based toxicities such as phospholipidosis or crystallization within tissues—either with the parent drug molecule or its metabolites—can all result in safety or tolerability issues. As a result, medicinal chemists have developed a number of design guidelines to avoid chemotype-based toxicity. Among the earliest developed guidelines are structural alerts, which are chemical functionality known to be associated with safety issues in drug molecules (see Figure 18.10).[86] These structural alerts typically have some basis in chemical reactivity of the underlying functional group (for example, electrophiles such as epoxides, Michael acceptors and 2-fluoro pyridines) or of a likely metabolite (for example, naked thiophenes, which are epoxidized by cytochrome P450, or catechols, which are oxidized to quinones *in vivo*). Despite some of these concerns, there has been a recent resurgence in covalent inhibitors, recognizing the prevalence of covalent target inhibition mechanisms currently represented in the human pharmacopeia and the fact that many of these structural alert guidelines are formulated around highly reactive functionality (often generated by metabolism in the liver or other metabolically active tissue) that is not target specific.[87] As such, it is important to bear in mind that presence of a structural alert does not guarantee a toxicity issue, but it may result in a higher likelihood of problems as the compound is developed further. One must also consider the intended dose of the drug substance, as even

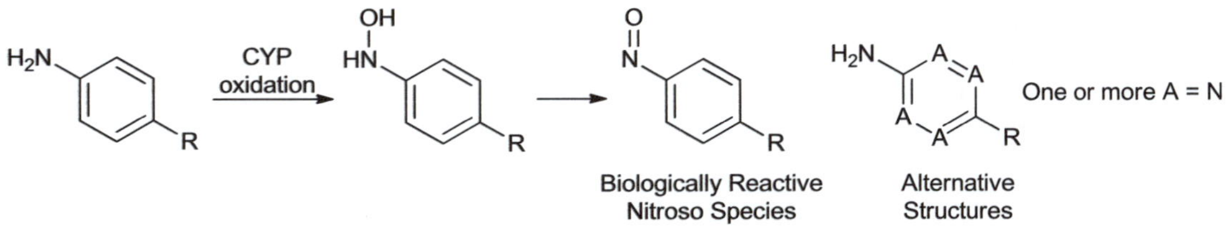

Figure 18.10 Example of a structural alert: Anilines are oxidized by CYP enzymes to produce potentially mutagentic aryl nitroso species and hydroxylamines.

reactive functionality can be tolerated if concentrations are low enough to be effectively handled by endogenous detoxification mechanisms.

In recent years, guidelines for safety optimization have moved beyond functional group alerts to include consideration of molecular properties as well. These property-based guidelines are largely based on TPSA and lipophilicity, which can be inferred to influence metabolism, tissue binding, phospholipidosis and secondary pharmacology. According to these guidelines, the probability of encountering chemotype-based toxicity of any kind is lowest if clogP < 3 and TPSA > 75 and is highest when clogP > 3 and TPSA < 75.[88,89] These safety trends are further magnified if a basic center is present on the molecule. Lipophilic molecules tend to have higher target promiscuity and greater susceptibility to oxidative metabolism. Lipophilic bases are known pharmacophores for phospholipidosis and hERG channel binding, which is associated with cardiac arrhythmogenicity. Lipophilic compounds and their metabolites are also more susceptible to biliary excretion and therefore have the potential to be implicated in biliary cholestasis, leading to liver enzyme leakage and hepatotoxicity. When combined with structural alerts, property guidelines provide significant opportunity to enrich for compounds with a safer profile in design.

It is becoming more recognized that drug toxicity can often be described at the organelle level, with major mechanisms of subcellular toxicity contributing to detrimental effects on cell health at the tissue level and concomitant effects on organ damage. Mitochondrial toxicity—through inhibiting protein synthesis, affecting integrity of mitochondrial DNA, imparting oxidative stress or interfering with metabolism—is a common mechanism for toxicity at the sub-cellular level.[90] Various tissues have different requirements for basal mitochondrial function and reserve mitochondrial capacity; interfering with mitochondrial function therefore would be expected to have effects in tissues with high requirements for mitochondrial function, such as heart and liver. The endoplasmic reticulum (ER) is another target organelle for drug toxicity; it plays a critical role in several cell functions, such as protein folding, steroid biosynthesis, regulating calcium storage and membrane lipid synthesis. Since the ER hosts a number of enzymes that metabolize xenobiotics, it is a proximal target for toxic metabolites. Interference with ER function in protein folding or calcium homeostasis triggers an ER stress response that can either lead to cell recovery or cell death.[91]

The other major mechanism of drug toxicity is direct effects on DNA or chromosomal integrity. These effects are typically studied through *in vitro* assays designed to select for mutations or observe chromosomal aberrations. Because these assays are generally conducted at high compound concentrations (often to the limit of cytotoxicity), it is important to ensure that the material being tested is highly pure, often with no single impurity exceeding one percent. Genetic toxicity associated with direct DNA binding or reactivity typically manifests itself in a positive Ames assay, as a result of inducing mutant bacteria that are capable of surviving on special nutrient-deficient media. Chromosomal aberrations (induced by a variety of processes, including interference with topoisomerases) are typically identified by studying effects on replicating lymphocytes or CHO-cells and identifying extra chromosomes or chromosomal fragments. Direct DNA damage or clastogenic (chromosomal breakage) behavior typically results in immediate discontinuation of further development due to potential for oncogenic effects. Aneugenic behavior (alteration in the number of chromosomes) can sometimes be addressed by determining a therapeutic window of significant multiple relative to efficacious plasma levels.

Because of the diverse mechanisms of toxicity observed in drug substances, batteries of *in vitro* safety assays can be combined with known or assumed properties of the compound to assess the potential for toxicity issues. In one example, the combination of *in vitro* cytotoxicity assays and measurements of covalent binding burden for a set of human drugs demonstrated reasonable correlation with observations of idiosyncratic drug toxicity.[92] Other groups have combined *in vitro* measurements of covalent binding burden with human dose to rationalize observations of idiosyncratic toxicity for sets of marketed drug substances.[93] While none of these methods provides perfect sensitivity or specificity, the significant downstream cost of safety-based drug attrition requires medicinal chemists to be vigilant in minimizing potential safety risks at the design level and understanding a compound's potential safety risks early in the drug discovery process.

18.9   THE PRECLINICAL STAGE: PREPARING FOR FIRST IN HUMAN STUDIES

As compounds become better optimized for target potency, selectivity, ADME properties and safety, drug discovery efforts move toward compound selection, wherein a clinical candidate is chosen for progression to human studies. This is a crucial stage of the drug discovery process, because once a development candidate is selected, it is impossible to change its intrinsic properties through further rounds of design. Attributes and issues that have been characterized need to be managed, and those that are not already known remain only to be discovered as the compound progresses through clinical trials. The preclinical stage also marks a dramatic increase in the costs and time to characterize attributes of the compound. Toxicology studies require large amounts of bulk drug, clinical studies require material synthesized under GMP (Good Manufacturing Practices) conditions, clinical PK assays need to be developed under GLP (Good Laboratory Practices) conditions, and more extensive characterization of pharmaceutical properties occurs. The focus of the medicinal chemist turns to enabling some of these more advanced studies—typically through revisiting the drug's synthesis, solid state properties and formulation options in collaboration with development chemists. This section will review the medicinal chemist's involvement in these areas.

None of the advanced studies mentioned above can occur without an adequate supply of material—often requiring bulk compound on >100 gram scale. Clinical studies will require significantly more bulk compound depending on the projected dose and duration of therapy, and attention will shift to preparing kilograms to metric tons of bulk as the compound advances through the various phases of clinical testing. Because the primary objective of the discovery synthesis is to enable rapid production of diverse analogues, it typically is not suitable for preparing large preclinical and clinical lots. Process syntheses need to consider overall efficiency of the synthetic route, measured both in yield as well as throughput for each step. Throughput-limiting issues are addressed either through reaction optimization or identifying different methods for effecting the same synthetic transformation. Examples of some common throughput-limiting issues include the use of environmentally unsustainable reagents or solvents, highly exothermic reactions, chromatographic chiral separations, high energy intermediates or reagents that create process safety issues and high catalyst loading. Process chemistry considerations, therefore, require the synthetic chemist to adopt a different perspective when analyzing the synthetic route and developing a strategy that offers speed, practicality, low cost and use of green (environmentally sustainable) processes to enable preparation of clinical bulk supplies. Ideally, some of these considerations are incorporated into synthetic routes that evolve before the candidate selection stage such that clinical progression is not slowed due to the need for extensive reworking of the synthetic route at this stage.

As the synthetic process is being optimized for scaling the API (active principal ingredient), formulation work begins to assess the qualities of the drug substance itself and to develop a drug product suitable for administration *via* the desired route.[59] In many cases, the API is produced as a crystalline solid, which must be characterized for stability, melting temperature and hygroscopicity. It is rarely acceptable to progress an amorphous solid to clinical development, because the amorphous form represents a metastable solid state that typically has higher aqueous solubility than the crystalline form. As such, much of the oral pharmacokinetic data generated during Phase I clinical studies would need to be re-evaluated upon identification of a crystalline form. Crystalline solids exist as a single polymorphic form, which is defined by the crystalline lattice that constitutes the solid form of the compound and defines its biopharmaceutical properties (such as solubility and stability). Typically, polymorph switches produce a more stable polymorphic form that is accompanied by lower solubility, which can have adverse effects on absorption. It is therefore important that the most stable polymorphic form be identified early in the development process. Typically, material science research groups will screen hundreds of different crystallization conditions in an attempt to identify different polymorphic forms of the compound. Another common technique is to triturate the crystalline solid for extended periods of time in different solvents to see if an alternative polymorph can be identified—its relative stability being characterized by its melting point. If the compound is ionizable, a salt screen is typically done to ensure the most appropriate salt counterion is selected—based on solubility, stability and hygroscopicity of the crystalline form. Robotic systems are typically used for this purpose, and the salt screen is directed by a list of pharmaceutically acceptable counterions.[94,95]

Most drugs are administered orally, and biopharmaceutical data produced during the discovery phase will determine parameters such as particle size that will influence dissolution kinetics and fraction of the dose that is absorbed. In the event that the drug product is for parenteral administration, the IV dosage form must be considered. Many IV products are supplied in solution or as a lyophilized powder for reconstitution. Similar solid form characterization is performed for parenterally administered compounds, with the added requirement that solution stability and solubility be significantly higher for a parenteral dosage form because the IV drug product will exist in solution for a more extended period of time. Typical solubility requirements for a parenteral dosage form are >1 mg/mL to keep dosage volumes low. There is some added flexibility in adjusting the pH and storage conditions of the IV solution to optimize these properties. In addition, IV products must be sterile-filtered, which adds complexity to the manufacturing process.

While conventional formulations offer the most straightforward path to drug development, several emerging technologies enable drug discovery teams to pursue compounds once thought to be undevelopable due to suboptimal biopharmaceutical properties. Controlled release technologies enable drug compound to be released more slowly or in specific parts of the GI tract. Controlled release formulations typically involve coating a tablet or capsule with material that dissolves only at a specific pH or in the presence of certain localized digestive enzymes. The effect of these formulations is to release the drug more slowly, leading to a PK profile with a blunted C_{max} and prolonged half-life, or to spare certain regions of the GI tract from drug exposure.

In cases where solubility is limited, nano-suspensions may be utilized to improve solubility and enhance absorption. One method involves the production of SEDDS (Self Emulsifying Drug Delivery Systems), which are used to improve the solubility of lipophilic drugs. The SEDDS formulation consists of the drug compound, an oily excipient and a surfactant with a high hydrophilic/lipophilic balance.[96] The SEDDS formulation produces an emulsified suspension with particle sizes approaching 50–100 nm. The advantages of SEDDS technology include its broad applicability, ability to formulate in both liquid and solid dosage forms (depending on the type of excipients used) and low cost. In addition, SEDDS formulations can also provide some protection from GI enzymes to address GI stability issues. Some of the disadvantages of SEDDS include practical limitations on dose size (because of the requirement for excipients and surfactants in the dosage form), variable incorporation of the drug into the lipid nanoparticle, and more complex absorption profile involving lipoprotein formation and absorption into the intestinal lymphatic system.[96] The appearance of SEDDS formulations in marketed drug products is still rare: cyclosporin was the first drug marketed using this technology, which delivered improved bioavailability relative to conventional formulation.[97] There is also a SEDDS formulation of the antiviral agent Norvir. However, the majority of SEDDS formulations are confined to drugs currently in clinical development.

One of the most important objectives of preclinical testing is to assess the safety profile of the drug substance and determine its therapeutic index, the reversibility of any toxicology findings and the monitorability of adverse events. This work is critical to informing whether the compound can advance to clinical testing and if its safety profile is sufficiently differentiated from other therapies that are used in clinical practice. The standard battery of assays to support preclinical safety studies include *in vitro* assays for genetic toxicity, *in vitro* assays for cardiovascular safety, *in vitro* assays for broad secondary pharmacology screening, *in vivo* assays for general organ toxicity and *in vivo* cardiopulmonary assays. Genetic toxicology assays were covered briefly in the lead optimization section. The preclinical genetic toxicity assays are run under GLP conditions and are reported to regulatory authorities as part of the Investigational New Drug (IND) application. The *in vitro* assay for cardiovascular safety is typically a hERG (human Ether-à-go-go) channel patch clamp assay to determine whether the compound has pharmacology pertinent to this ion channel. Inhibition of hERG channel function is associated with prolongation of the QTc interval, which can lead to fatal cardiac arrhythmias. Therefore, sufficient therapeutic margin must be established using *in vitro* assays and follow-up *in vivo* cardiovascular safety studies to look for prolongation of the QTc interval. Several reviews have been written on hERG channel blockage and the structural features and physicochemical properties associated with hERG inhibition.[98] Finally, broader receptor, kinase and ion channel pharmacology screens are conducted to determine any polypharmacology associated with the drug candidate and potential side effects that could be associated with activity against targets having known physiological function.

The principal aim of *in vivo* safety studies is to determine how well the drug substance is tolerated *in vivo* and whether there are any significant end-organ toxicities that could preclude further development. These studies are initially conducted in one rodent and one non-rodent species for two weeks duration. One of the most important activities in planning for *in vivo* safety studies is to characterize the pharmacokinetics of the compound at high doses *via* the clinically intended route of administration. Safety studies should target substantial multiples ($>10\times$) of the intended clinical exposure (both C_{max} and AUC) at the highest doses such that adequate margin is given for intersubject PK variability and the potential for higher clinical exposures than predicted from preclinical PK studies. The endpoint for these safety studies is clinical chemistry from blood and possibly urine to look for any functional organ damage, followed by histopathology on major organs. A reversibility arm may be included to determine whether any safety findings are reversible upon discontinuation of drug exposure—typically this is accomplished by having a second group of animals that receive drug for the two week period and then are maintained for 2–4 weeks without drug exposure prior to sacrifice and analysis. Proper attention to physicochemical properties, managing structural alerts and gaining a good understanding of PK/PD relationships for efficacy (such that the compound does not need to be overdosed) is critical for maximizing survival of the compound through safety studies.

18.10 CLINICAL STUDIES—ASSESSING PK, SAFETY AND EFFICACY

The traditional route for drug development involves three phases of clinical testing. The first phase (Phase 1) seeks to determine clinical PK and tolerability of the compound. Typically these studies use small numbers of healthy volunteers and use simple formulations (suspension of drug powder). Phase 1 studies begin with a single ascending dose (SAD) of compound and are designed to test the drug through its projected efficacious dose and up to the maximum allowable dose based on preclinical safety study results. Upon completion of the SAD arm, a multiple ascending dose (MAD) study is used to assess the compound's PK and tolerability under steady state conditions. The duration of dosing is always limited by the preclinical toxicology study duration and margins. According to FDA and ICH guidelines, clinical studies up to 2 weeks in duration must be supported by 2 week rodent and non-rodent toxicity studies. Clinical trials with durations between 2 weeks and 6 months must be supported by toxicity studies in rodents and non-rodents of at least the same duration as the intended clinical trial.[99] The typical output of Phase 1 studies is clinical chemistry from blood and urine samples, patient observations and PK of the drug substance and its metabolites in blood and possibly urine. It is possible to run Phase 1 studies under different regulatory paradigms (exploratory IND or under microdosing conditions) that offer more expedited testing with shorter duration toxicology studies.[100] These Phase 1 protocols are typically used to quickly answer PK questions, potentially with the objective of selecting one compound from several candidates. It is important for the drug discovery team to carefully consider the merits and issues associated with these alternative regulatory paradigms—the speed of gaining an early answer on PK is typically offset by the requirement to redo safety studies with longer duration and/or limitations on how high the compound can be dosed in early clinical studies.

Following Phase 1 clinical studies, the compound is typically subjected to longer duration preclinical toxicology studies, which follow the same principles as the initial preclinical toxicology studies. Durations of 30 days, 90 days and 6 months are used at different stages of clinical testing depending on the intended duration of drug treatment for the targeted disease. Typically therapeutic indexes drop with longer duration studies, so it is important to begin clinical testing with compounds in which there is a high degree of confidence in safety. Phase 2 clinical testing involves actual disease patients and aims to establish efficacy of the drug and tolerability in patients. The first stage of Phase 2 testing (Phase 2a) typically looks for some mechanistic evidence of target modulation through biomarker evaluation. A mechanism based biomarker and a disease based biomarker are typically used for this purpose. Provided adequate target modulation and disease biomarker effects are produced, the compound advances to Phase 2b testing, where some evidence of disease modification is typically sought to demonstrate clinical proof of concept. It is at this stage that the biological mechanism is actually tested for disease relevance. Here, it is crucial that the endpoint and controls (active or placebo) are correctly selected, as the proof of concept decision marks a substantial inflection in resource investment as the compound advances to Phase 3 studies.

The final stage of clinical testing for marketing authorization is the Phase 3 development program. Here, the drug candidate is studied in a broad range of patients to establish efficacy relative to current standard of care with sufficient statistical power to show differentiation (or non-inferiority in the case of some medicines, such as anti-infectives, where a placebo controlled trial is unethical). Phase 3 studies are designed to produce licensed indications for the drug substance and are typically constructed to evaluate specific types of disease. The other purpose of Phase 3 studies is to establish the safety database that will support drug registration. This objective often requires thousands (sometimes >10 000) patients, and the cost and resources involved requires high confidence in safety and efficacy prior to trial initiation. The final safety studies (reproductive toxicity and carcinogenicity) are typically conducted late in

Phase 2 or in Phase 3 to support registration. Following completion of Phase 3 studies, a New Drug Application (NDA) is filed.

18.11 CONCLUSIONS

Medicinal chemistry is a unique discipline in drug discovery due to the centrality of chemistry to the entire discovery and development pathway. Medicinal chemists are typically involved in the hypothesis generation and modality selection stages—based on their understanding of disease biology and druggability of different targets within the cellular system of interest. Because medicinal chemists design and prepare the actual drug substance, their understanding of drug-like physicochemical properties, molecular interactions, organic synthesis and reactivity of functional groups is critical to the success of the drug discovery program. In that way, medicinal chemists "own" the compound's properties and pharmacological properties, because they are responsible for designing them in (or out) of the compound. A broad array of technologies has advanced rapidly over the last decade to help with this process, including a plethora of biophysical methods to understand protein–ligand interactions, computational tools for structure based drug design, *in silico* property predictions, mechanistic safety assays and *in vitro* ADME assays to understand toxicity, clearance, absorption and tissue distribution. The partnership among chemists, biologists, toxicologists and drug metabolism/disposition scientists is crucial to defining the properties that are sought in a drug molecule and designing the experiments to determine whether they have been attained. As the drug candidate is selected and moves toward development, medicinal chemists are also involved in partnerships with process chemists to develop scalable manufacturing processes and support clinical studies. Their knowledge of the disease area makes medicinal chemists valuable partners in designing clinical trials and determining development strategy—often with an eye toward determining what properties need to be understood early to define objectives for backup candidate seeking programs. Finally, medicinal chemists also play an important role in defending intellectual property through participating in patent strategy and prosecution. The field of medicinal chemistry continues to evolve in scope and sophistication, and the role of medicinal chemists in drug discovery is bounded only by their creativity and interest.

HINTS AND TIPS

- Human genetic evidence is becoming more important for establishing confidence in rationale for drug targets. Well characterized single nucleotide polymorphisms or other mutations affecting both functional activity of a proposed drug target and incidence or severity of a particular disease increase confidence that the pharmaceutical intervention will be safe and effective for treating that disease.
- The ability of a biological target to be effectively modulated by small molecule drugs, also referred to as druggability, can be assessed computationally and experimentally. Computational approaches typically require structural information about putative binding sites on the target and assess size, shape, hydrophobicity and other aspects of the pocket. Experimental methods typically involve assessing hit rates with fragment libraries or screening subsets.
- Screening strategies incorporating two or more orthogonal methods for hit validation (for example a biochemical assay for functional activity and a biophysical assay for binding) minimize the impact of screening artifacts that produce false hits in a single assay.

- Obtaining mechanistic information on how a lead compound modulates target activity can provide useful insights into potential selectivity issues and differentiation of chemical equity based on its molecular mode of action.
- Molecular size and lipophilicity are important properties to manage while focusing on lead optimization. Excess molecular size and lipophilicity can erode selectivity, aqueous solubility and metabolic stability leading to significant pharmacokinetic and safety issues.
- It is important to understand the energetics of structural water molecules in a binding site when considering how to optimize a lead for potency. Computational tools such as WaterMap™ can inform whether it is better to displace a water molecule or make a hydrogen bond to it based on its binding thermodynamics.
- An optimal halogen bond has been calculated to be worth ca. 3–4 kcal/mol in binding energy.
- It is important for preclinical oral PK work to be performed on crystalline as opposed to amorphous material because oral exposure may be overestimated with amorphous drug substance due to its increased solubility. However, drugs are rarely developed as amorphous solids due to the inherent instability of the amorphous state.
- Chemists typically optimize compounds to deliver crystalline solids with melting point between 120 °C and 225 °C and intrinsic aqueous solubility of >100 µg/mL (preferably >250 µg/mL for high dose compounds) at the pH range relevant to absorption in the ileum and jejunum (pH 6.0–7.5).
- For most drugs, only the unbound fraction (that which is not bound to plasma proteins) is available for interaction with its biological target or for metabolism. Since these two parameters produce opposite effects on dose, medicinal chemists should focus on reducing unbound clearance and improving intrinsic potency to lower the projected therapeutic dose. Plasma protein binding in itself is typically not a productive parameter to influence for lowering dose.
- The most important technique to reduce CYP-mediated metabolism is to reduce overall lipophilicity. Compounds with acidic functionality typically tend to have lower CYP-mediated metabolism.
- Blocking sites of metabolism with a fluorine or a methyl group gives mixed results, with metabolism sometimes being shifted to another part of the molecule.
- Crystal structures exist for various CYP isoforms, which allow one to derive structurally-based predictive tools for CYP interactions.
- Common strategies for circumventing glucuronidation include eliminating the functional group that is conjugated, placing steric bulk proximal to the conjugated group or placing polarity proximal to the conjugated group.
- When considering molecular properties, the probability of encountering chemotype-based toxicity of any kind is lowest if clogP < 3 and TPSA > 75 and is highest when clogP > 3 and TPSA < 75.
- As a more stable polymorphic form of a drug substance is usually accompanied by lower solubility, which can have adverse effects on absorption, it is important that the most stable polymorphic form be identified early in the development process.

Acknowledgements

The author acknowledges Dr Justin Montgomery for providing Figure 18.6, Drs Veerabahu Shanmugasundaram, Adam Gilbert and Michael Shapiro for their critical review of this manuscript and Drs Bruno Hancock, Kieran Geoghegan and Jane Withka for helpful discussions.

KEY REFERENCES

A. L. Hopkins, C. R. Groom and A. Alex, *Drug Discovery Today*, 2004, **9**, 430–431.

C. Bissantz, B. Kuhn and M. Stahl,*J. Med. Chem.*, 2010, **53**, 5061–5084 (see erratum in 6241).

D. A. Smith, L. Di and E. H. Kerns, *Nat. Rev. Drug Discovery*, 2010, **9**, 929–939.

T. Kenakin, *ACS Chem. Biol.*, 2009, **4**, 249–260.

J. Gabrielsson, H. Dolgos, P.-G. Gillberg, U. Bredberg, B. Benthem and G. Duker, *Drug Discovery Today*, 2009, **14**, 358–372.

A. F. Stepan, D. P. Walker, J. Bauman, D. A. Price, T. A. Baillie, A. S. Kalgutkar and M. D. Aleo, *Chem. Res. Toxicol.*, 2011, **24**, 1345–1410.

REFERENCES

1. U. Rix and G. Superti-Furga, *Nat. Chem. Biol.*, 2009, **5**, 616–624.
2. K. Wierzba, M. Muroi and H. Osada, *Curr. Opin. Chem. Biol.*, 2011, **15**, 57–65.
3. W. Huber and F. Mueller, *Curr. Pharm. Des.*, 2006, **12**, 3999–4021.
4. G. A. Holdgate and W. H. J Ward, *Drug Discovery Today*, 2005, **10**, 1543–1550.
5. Y. Tsutsui and P. L. Wintrode, *Curr. Med. Chem.*, 2007, **14**, 2344–2358.
6. P. Dorr, M. Westby, S. Dobbs, P. Griffin, B. Irvine, M. Macartney, J. Mori, G. Rickett, C. Smith-Burchnell, C. Napier, R. Webster, D. Armour, D. Price, B. Stammen, A. Wood and M. Perros, *Antimicrob. Agents Chemother.*, 2005, **49**, 4721–4732.
7. T. A. Halgren, *J. Chem. Inf. Model.*, 2009, **49**, 377–389.
8. E. B. Fauman, B. K. Rai and E. S. Huang, *Curr. Opin. Chem. Biol.*, 2011, **15**.
9. F. N. B. Edfeldt, R. H. A. Folmer and A. L. Breeze, *Drug Discovery Today*, 2011, **16**, 284–287.
10. A. E. Weber and N. Thornberry, *Annu. Rep. Med. Chem.*, 2007, **42**, 95–109.
11. J. J. Allen, M. Li, C. S. Brinkworth, J. L. Paulson, D. Wang, A. Hubner, W.-H. Chou, R. J. Davis, A. L. Burlingame, R. O. Messing, C. D. Katayama, S. M. Hedrick and K. M. Shokat, *Nat. Methods*, 2007, **4**, 511–516.
12. J. M. Chalker, G. J. L. Bernardes and B. G. Davis, *Acc. Chem. Res.*, 2011, **44**, 730–741.
13. M. Raida, *Curr. Opin. Chem. Biol.*, 2011, **15**, 570–575.
14. R. E. Moellering and B. F. Cravatt, *Chem. Biol.*, 2012, **19**, 11–22.
15. E. M. Sletten and C. R. Bertozzi, *Acc. Chem. Res.*, 2011, **44**, 666–676.
16. J. S. Cisar and B. F. Cravatt, *J. Am. Chem. Soc.*, 2012, **134**, 10385–10388.
17. D. N. Perkins, D. J. C. Pappin, D. M. Creasy and J. S. Cottrell, *Electrophoresis*, 1999, **20**, 3551–3567.
18. S.-E. Ong, M. Schenone, A. A. Margolin, X. Li, K. Do, M. K. Doud, D. R. Mani, L. Kuai, X. Wang, J. L. Wood, N. J. Tolliday, A. N. Koehler, L. A. Marcaurelle, T. R. Golub, R. J. Gould, S. L. Schreiber and S. A. Carr, *Proc. Natl. Acad. Sci. U. S. A.*, 2009, **106**, 4617–4622.
19. S.-E. Ong, B. Blagoev, I. Kratchmarova, D. B. Kristensen, H. Steen, A. Pandey and M. Mann, *Mol. Cell. Proteomics*, 2002, **2002**, 376–386.
20. N. Thorne, D. S. Auld and J. Inglese, *Curr. Opin. Chem. Biol.*, 2010, **14**, 315–324.
21. J. R. Simard, S. Kluter, C. Grutter, M. Getlik, M. Rabiller, H. B. Rode and D. Rauh, *Nat. Chem. Biol.*, 2009, **5**, 394–396.
22. P. Gribbon and A. Sewing, *Drug Discovery Today*, 2005, **10**, 17–22.
23. S. L. Schreiber, *Nature*, 2009, **457**, 153–154.
24. S. Dandapani and L. A. Marcaurelle, *Curr. Opin. Chem. Biol.*, 2010, **14**, 362–370.
25. D. Morton, S. Leach, C. Cordier, S. Warriner and A. Nelson, *Angew. Chem., Int. Ed.*, 2008, **48**, 104–109.

26. C. A. Lipinski, F. Lombardo, B. W. Dominy and P. J. Feeney, *Adv. Drug Delivery Rev.*, 1997, **23**, 3–25.
27. L. K. Henchey, A. L. Jochim and P. S. Arora, *Curr. Opin. Chem. Biol.*, 2008, **12**, 692–697.
28. L. Thorstholm and D. J. Craik, *Drug Discovery Today: Technol.*, 2012, **9**, e13–e21.
29. R. J. Clark, J. Jensen, S. T. Nevin, B. P. Callaghan, D. J. Adams and D. J. Craik, *Angew. Chem., Int. Ed.*, 2010, **49**, 1–5.
30. A. Annis, C.-C. Chuang and N. Nazef, in *Mass Spectrometry in Medicinal Chemsitry*, ed. K. T. Wanner and G. Hofner, Wiley-VCH, Weinheim, Germany, 2007, pp. 121–156.
31. H. Zehender and L. M. Mayr, *Expert Opin. Drug Discovery*, 2007, **2**, 285–294.
32. A. G. Coyne, D. E. Scott and C. Abell, *Curr. Opin. Chem. Biol.*, 2010, **14**, 299–307.
33. G. Chessari and A. J. Woodhead, *Drug Discovery Today*, 2009, **14**, 668–675.
34. D. E. Scott, A. G. Coyne, S. A. Hudson and C. Abell, *Biochemistry*, 2012, **51**, 4990–5003.
35. M. Baker, *Nat. Rev. Drug Discovery*, 2013, **12**, 5–7.
36. R. A. E. Carr, M. Congreve, C. W. Murray and D. C. Rees, *Drug Discovery Today*, 2005, **10**, 987–992.
37. A. R. Leach and M. M. Hann, *Curr. Opin. Chem. Biol.*, 2011, **15**, 489–496.
38. W. F. Lau, J. M. Withka, D. Hepworth, T. V. Magee, Y. J. Du, G. A. Bakken, M. D. Miller, Z. S. Hendsch, V. Thanabal, S. A. Kolodziej, L. Xing, Q. Hu, L. S. Narasimhan, R. Love, M. E. Charlton, S. Hughes, W. P. v. Hoorn and J. E. Mills, *J. Comput.-Aided Mol. Des.*, **25**, 621–636.
39. S. M. Boyd and G. E. D. Kloe, *Drug Discovery Today: Technol.*, 2010, **7**, e173–e180.
40. A. Larsson, A. Jansson, A. Aberg and P. Nordlund, *Curr. Opin. Chem. Biol.*, 2011, **15**, 482–488.
41. K. Retra, H. Irth and J. E. v. Muijlwijk-Koezen, *Drug Discovery Today: Technol.*, 2010, **7**, e181–e187.
42. P. Ripphausen, B. Nisius, L. Peltason and J. Bajorath, *J. Med. Chem.*, 2010, **53**, 8461–8467.
43. E. Freire, *Drug Discovery Today: Technol.*, 2004, **1**, 295–299.
44. E. Friere, *Drug Discovery Today*, 2008, **13**, 869–874.
45. R. A. Copeland, D. L. Pompliano and T. D. Meek, *Nat. Rev. Drug Discovery*, 2007, **5**, 730–739.
46. H. Huang, H. Ji, H. Li, Q. Jing, K. J. Labby, P. Martasek, L. J. Roman, T. L. Poulos and R. B. Silverman, *J. Am. Chem. Soc.*, 2012, **134**, 11559–11572.
47. J. D. Dujrant, C. A. F. de Oliveira and J. A. McCammon, *Chem. Biol. Drug Disc.*, 2011, **78**, 191–198.
48. J. J. Liu, R. Horst, V. Katritch, R. C. Stevens and K. Wüthrich, *Science*, 2012, **335**, 1106–1110.
49. M. Vogtherr, K. Saxena, S. Hoelder, S. Grimme, M. Betz, U. Schieborr, B. Pescatore, M. Robin, L. Delarbre, T. Langer, K. U. Wendt and H. Schwalbe, *Angew. Chem., Int. Ed.*, 2006, **45**, 993–997.
50. G. M. West, E. Y. T. Chien, V. Katritch, J. Gatchalian, M. J. Chalmers, R. C. Stevens and P. R. Griffin, *Structure*, 2011, **19**, 1424–1432.
51. Y. Hamuro, S. J. Coales, J. A. Morrow, K. S. Molnar, S. J. Tuske, M. R. Southern and P. R. Griffin, *Protein Sci.*, 2006, 1883–1892.
52. M. M. Hann, *Med. Chem. Commun.*, 2011, **2**, 349–355.
53. A. L. Hopkins, C. R. Groom and A. Alex, *Drug Discovery Today*, 2004, **9**, 430–431.
54. T. Ryckmans, M. P. Edwards, V. A. Horne, A. M. Correia, D. R. Owen, L. R. Thompson, I. Tran, M. F. Tutt and T. Young, *Bioorg. Med. Chem. Lett.*, 2009, **19**, 4406–4409.
55. C. Bissantz, B. Kuhn and M. Stahl, *J. Med. Chem.*, 2010, **53**, 5061–5084 (see erratum in 6241).
56. A. Fernandez, *Structure*, 2005, **13**, 1829–1836.
57. L. Wang, B. J. Berne and R. A. Friesner, *Proc. Natl. Acad. Sci. U. S. A.*, 2011, **108**, 1326–1330.
58. A. R. Voth, P. Khuu, K. Oishi and P. S. Ho, *Nature Chemistry*, 2009, **1**, 74–79.

59. R. G. Strickley, *Annu. Rep. Med. Chem.*, 2008, **43**, 419–451.

60. D. F. Veber, S. R. Johnson, H.-Y. Cheng, B. R. Smith, K. W. Ward and K. D. Kopple, *J. Med. Chem.*, 2002, **45**, 2615–2623.

61. F. Lovering, J. Bikker and C. Humblet, *J. Med. Chem.*, 2009, **52**, 6752–6756.

62. D. A. Smith, L. Di and E. H. Kerns, *Nat. Rev. Drug Discovery*, 2010, **9**, 929–939.

63. I. L. O. Buxton, in *Goodman and Gilman's Pharmacological Basis of Therapeutics*, ed. L. L. Brunton, J. S. Lazo and K. L. Parker, McGraw Hill, New York, 2006, pp. 1–39.

64. T. T. Wager, R. Y. Chandrasekaran, X. Hou, M. D. Troutman, P. R. Verhoest, A. Villalobos and Y. Will, *ACS Chem. Neurosci.*, 2010, **1**, 420–434.

65. T. T. Wager, X. Hou, P. R. Verhoest and A. Villalobos, *ACS Chem. Neurosci.*, 2010, **1**, 435–449.

66. T. T. Wager, A. Villalobos, P. R. Verhoest, X. Hou and C. L. Shaffer, *Expert Opin. Drug Discovery*, 2011, **6**, 371–381.

67. F. P. Guengerich, in *Enzyme Systems That Metabolize Drugs and Other Xenobiotics*, ed. C. Ioannides, John Wiley and Sons, Chichester, West Sussex, UK, 2002, pp. 33–67.

68. M. J. DeGroot, *Drug Discovery Today*, 2006, **11**, 601–605.

69. D. C. Pryde, D. Dalvie, Q. Hu, P. Jones, R. S. Obach and T.-D. Tran, *J. Med. Chem.*, 2010, **53**, 8441–8460.

70. D. C. Pryde, T.-D. Tran, P. Jones, J. Duckworth, M. Howard, I. Gardner, R. Hyland, R. Webster, T. Wenham, S. Bagal, K. Omoto, R. P. Schneider and J. Lin, *Bioorg. Med. Chem. Lett.*, 2012, **22**, 2856–2860.

71. K. W. Bock, in *Enzyme Systems That Metabolize Drugs and Other Xenobiotics*, ed. C. Ioannides, John Wiley and Sons, Chichester, West Sussex, UK, 2002, pp. 281–318.

72. B. J. Pleuvry, *Anesth. Intens. Care*, 2008, **9**, 372–373.

73. J. M. Herbert, D. Frehel, E. Vallee, G. Kieffer, D. Gouy, Y. Berger, J. Necciari, G. Defreyn and J. P. Maffrand, *Cardiovasc. Drug Rev.*, 1993, **11**, 180–198.

74. T. Kenakin, *ACS Chem. Biol.*, 2009, **4**, 249–260.

75. G. Dahl and T. Akerud, *Drug Discovery Today*, 2013, **18**, 697–709.

76. J. Gabrielsson, H. Dolgos, P.-G. Gillberg, U. Bredberg, B. Benthem and G. Duker, *Drug Discovery Today*, 2009, **14**, 358–372.

77. J. J. S. Cadwell, *Drug Discovery Dev.*, 2010.

78. V. J. Cunningham, C. A. Parker, E. A. Rabiner, A. D. Gee and R. N. Gunn, *Drug Discovery Today: Technol.*, 2005, **2**, 311–315.

79. M. E. Quaedackers, W. Mol, S. S. Korevaar, E. A. F. J. v. Gurp, W. F. J. v. Ijcken, G. Chan, W. Weimar and C. C. Baan, *Transplantation*, 2009, **88**, 1002–1009.

80. A. Adibekian, B. R. M. J. W. Chang, K.-L. Hsu, K. Tsuboi, D. A. Bachovchin, A. E. Speers, S. J. Brown, T. Spicer, V. Fernandez-Vega, J. Ferguson, P. S. Hodder, H. Rosen and B. F. Cravatt, *J. Am. Chem. Soc.*, 2012, **134**, 10345–10348.

81. B. F. Cravatt, A. T. Wright and J. Kozarich, *Annu. Rev. Biochem.*, 2008, 77, 383–414.

82. A. Robichaud, F. D. Tattersall, I. Choudhury and I. W. Rodger, *Neuropharmacology*, 1999, **38**, 289–297.

83. J. T. Peterson, *Cardiovasc. Res.*, 2006, **69**, 677–687.

84. C. S. Elangbama, L. E. Jobb, L. M. Zadroznyb, J. C. Bartona, L. W. Yoona, L. D. Gatesa and N. Slocuma, *Exp. Toxicol. Pathol.*, 2008, **60**, 253–262.

85. J. A. Kramer, J. E. Sagartz and D. L. Morris, *Nat. Rev. Drug Discovery*, 2007, **6**, 636–649.

86. A. F. Stepan, D. P. Walker, J. Bauman, D. A. Price, T. A. Baillie, A. S. Kalgutkar and M. D. Aleo, *Chem. Res. Toxicol.*, 2011, **24**, 1345–1410.

87. J. Singh, R. C. Petter, T. A. Ballie and A. Whitty, *Nat. Rev. Drug Discovery*, 2011, **10**, 307–317.

88. M. P. Edwards and D. A. Price, *Annu. Rep. Med. Chem.*, 2010, **45**, 381–391.

89. J. D. Hughes, J. Blagg, D. A. Price, S. Bailey, G. A. DeCrescenzo, R. V. Devraj, E. Ellsworth, Y. M. Fobian, M. E. Gibbs, R. W. Gilles, N. Greene, E. Huang, T. Krieger-Burke, J. Loesel, T. Wager, L. Whiteley and Y. Zhang, *Bioorg. Med. Chem. Lett.*, 2008, **18**, 4872–4875.

90. K. B. Wallace, *Trends Pharmacol. Sci.*, 2008, **29**, 361–366.

91. A. E. Cribb, M. Peyrou and S. Muruganandan, *Drug Metab. Rev.*, 2005, **37**, 405–442.

92. R. A. Thompson, E. M. Isin, Y. Li, L. W. K. Page, I. Wilson, S. S. B. Middleton, S. Stahl, A. J. Foster, H. Dolgos, R. Weaver and J. G. Kenna, *Chem. Res. Toxicol.*, 2012, **25**, 1616–1632.

93. S. Nakayama, R. Atsumi, H. Takakusa, Y. Kobayaski, A. Kurihara, Y. Nagai, D. Nakai and O. Okazaki, *Drug Metab. Dispos.*, 2009, **37**, 1970–1977.

94. C. Saal and A. Becker, *Eur. J. Pharm. Sci.*, 2013, **49**, 614–623.

95. S. M. Berge, L. D. Bighley and D. C. Monkhouse, *J. Pharm. Sci.*, 1977, **66**, 1–19.

96. B. Tang, G. Cheng, J.-C. Gu and C.-H. Xu, *Drug Discovery Today*, 2008, **13**, 606–612.

97. S. Gibaud, *Expert Opin. Drug Delivery*, 2012, **9**, 937–951.

98. C. Jamieson, E. M. Moir, Z. Rankovic and G. Wishart, *J. Med. Chem.*, 2006, **49**, 5029–5046.

99. *Guidance for Industry – M3(R2) Nonclinical Safety Studies for the Conduct of Human Clinical Trials and Marketing Authorization for Pharmaceuticals*, US Dept. of Health and Human Services, US Food and Drug Administration, Center for Drug Evaluation and Research, Center for Biologics Evaluation and Research, 2010.

100. I. R. Wilding and J. A. Bell, *Drug Discovery Today*, 2005, **10**, 890–894.

CHAPTER 19

Target Validation for Medicinal Chemists

PAUL BESWICK*[a] AND KEITH BOWERS[b]

[a] Translational Drug Discovery Group, University of Sussex, Brighton, East Sussex, BN1 9QJ, UK
[b] Biopta Ltd., Weipers Centre, Garscube Estate, Bearsden Road, Glasgow, G61 1QH, UK
*E-mail: p.beswick@sussex.ac.uk

19.1   INTRODUCTION

Target Validation is a frequently used term in the arena of drug discovery and to different people it can mean very different things. An individual scientist's definition of the term often depends on their area of speciality and where their work is focussed along the drug discovery process. There are those who believe that target validation is the initial step in drug discovery and is the process of associating a gene with a disease.[1] In contrast there are scientists who believe that a target is only truly validated when modulation has been demonstrated to show clinical efficacy.[2] These are just two extreme definitions, both at key stages of the drug discovery process; early target identification and clinical proof of concept (PoC). In this chapter target validation will be reviewed in the context of the path between target identification and PoC. It will be demonstrated that target validation is a stepwise, often pragmatic process, which aims to configure experiments with the tools and resources available but one that also recognises the need to have a clear view of the ultimate goal of a robust PoC study.

Improved target identification, selection and validation are essential for the success of future drug discovery. There are many published case histories describing successful preclinical target validation studies which have correctly identified high value targets and modulation of which has delivered drugs. However, there are a disturbing number of cases where clinical modulation of the target has failed to produce a significant effect, despite a large volume of highly encouraging preclinical validation data. One such example is the case of selective antagonists of the substance P receptor NK1.[3] In the case of this target a comprehensive and highly convincing preclinical data package was generated suggesting a high chance of clinical success. The biological mechanism lay at the centre of what was believed to be a key pain signalling pathway. Highly potent, selective, brain penetrant molecules were produced which showed excellent pharmacokinetics and efficacy in a range of preclinical models.[4] In clinical studies compounds also demonstrated excellent pharmacokinetics and brain penetration, and achieved a high level

The Handbook of Medicinal Chemistry: Principles and Practice
Edited by Andrew Davis and Simon E Ward
© The Royal Society of Chemistry 2015
Published by the Royal Society of Chemistry, www.rsc.org

of receptor occupancy.[5] They failed, however, to demonstrate any significant analgesic efficacy in numerous clinical trials.[6,7]

An unequivocal explanation for the failure of NK-1 antagonists in clinical pain studies still remains elusive and is the subject of recent publications.[8,9] This story illustrates that despite having a large volume of supportive data, true target validation can only be achieved in patients.

The issue of attrition in clinical PoC studies was first highlighted nearly 10 years ago[10] and a recent survey suggested that the situation is actually getting worse rather than better[11] with an increasing number of new entities failing to demonstrate efficacy in proof of concept studies. It is therefore imperative that the drug discovery community improves its target selection process and identifies methodologies which more confidently predict clinical efficacy. Target validation is central here, and therefore is one of the most important aspects of modern drug discovery. Fortunately in recent years, due to a combination of scientific advances and retrospective analyses, new target validation tools have become available and improved strategies have been proposed, thus suggesting that there is a significant chance that the current high rate of late stage clinical attrition will reduce in the coming years.

The topic of target validation has been thoroughly reviewed, and there are many excellent articles available.[12,13] This chapter will aim to build on the content of these reviews and provide a contemporary and instructive summary of target validation practices.

19.2 TARGET VALIDATION—DEFINITION AND CONTEXT

Before describing techniques employed by the biologist to support target validation it is worth defining the term, and then to convey the information which techniques provide with regard to these definitions, within the context of the drug discovery process. The majority of this chapter will focus on target based drug discovery, which is the process currently adopted by the majority of drug discovery groups and involves the initial identification of a gene or protein target and its association with a disease. Target validation is defined as the process of gathering evidence to provide confidence that modulation of a target has the potential to treat a disease. Target identification is the initial stage of this process, providing evidence linking a drug target to a disease. Frequently, target identification utilises an *in silico* approach performed by bioinfomaticians, scientists skilled at extracting relevant data from large databases of gene expression information and the scientific literature.

For the purpose of this chapter a target is defined as a protein that is linked to a biological process, which is a key driver of a disease/pathophysiological state. The premise is that altering the function of the protein (be it stimulation or inhibition, a receptor, enzyme, transporter) will have a positive effect on the pathology associated with a disease.

Following target identification, workers seek tools to facilitate the process of early target validation *in vitro*. Typically tools are small molecules which modulate the target—either to inhibit target activity where over activity or expression is associated with the disease phenotype, or stimulate it where the converse is true. More recently alternative tools have been considered and include peptides, siRNA, oligonucleotides and antibodies. The preferred tool once identified will be used to investigate the effect of target modulation initially in *in vitro* assays, typically using cells which express the target and can be either recombinant systems, native cells or diseased tissue. Observation that the tool is able to elicit the desired response is considered by many groups to be evidence of early target validation and the data used to support further work.

The reader may be familiar with the generic stages of the drug discovery and development process which is depicted in Figure 19.1.

It represents, simplistically, a sequence of events starting with target identification, through a set of compound identification processes (high throughput screening, lead optimisation) to early

Figure 19.1 Depiction of the stages of typical drug screening program from target identification to clinical trials.

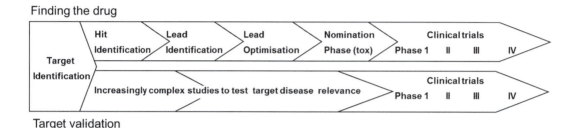

Figure 19.2 Target validation experiments and drug screening programs are often run as parallel but overlapping activities.

toxicological testing and then into clinical testing. Progression through the stages represents an increase in confidence, and therefore commitment, to the project in terms of resources; each stage getting closer to producing a molecule fit for clinical testing and potentially a medicine. Target validation is a concept that usually develops over the lifetime of a project. To 'validate' a target then represents the confidence sought, necessary to progress the project through these phases.

The level of target validation required before committing to a hit finding campaign is frequently less than that required to commit to a clinical trial. Moving from the beginning of the process to the end sees a growing level of target validation data.

This increase in data collection mirrors the progress in the ultimate aim of the project, to produce a drug. Thus, a clearer depiction of the overall drug discovery process is that depicted in Figure 19.2.

19.3 KEY QUESTIONS ASKED IN TARGET VALIDATION AND TECHNIQUES EMPLOYED

Examples of experiments that populate the lower section of Figure 19.2 can be split into themes, which answer the following key questions:

Is the target, or are its ligands, physically present in the disease tissue of interest?

- Gene expression analysis (GEA) shows DNA/RNA levels present and potential correlation with disease progression.
- Antibodies with immunohistochemistry show target associated with pathological features of disease (target on the cell types at the site of disease).

Is the target linked functionally to a biological process that is implicated in the disease and does modulation of the target have the desired effect on that biological process? Examples from a variety of disease settings include:

- Target causes bronchoconstriction in human airways for asthma.
- Target drives cell proliferation for cancer.
- Target drives cell chemotaxis of white blood cells for inflammatory diseases.

Use of compounds blocks these functions—particularly important if the target is 'activated' by non-selective agonists. For example interleukin-8 (IL-8) will cause chemotaxis by activating both CXCR1 and CXCR2 receptors. A selective antagonist would be required to show which one of these receptors (targets) is important for a particular cell type.

Other tools to inhibit the target mechanism such as siRNA or oligonucleotides—particularly useful for early targets where no suitable small molecules are available.

Does modulation of that biological process alter the disease/pathophysiological process?

These experiments are designed to determine if target modulation can in turn modulate a mechanism linked to a disease process, and determine if target modulation can potentially modulate the disease pathology. Importantly they attempt to reproduce the complexity of the biological processes involved and can, for example, demonstrate if other mechanisms are involved which can compensate for a specific intervention. Thus for some targets no net effect is observed in these more challenging assays. The assays include human tissue assays, *in vivo* animal models and early clinical tests.

- Human tissue assays *in vitro*—examples are chosen from inflammatory diseases:
  - Sputum from asthmatic patients causes chemotaxis of neutrophils *in vitro* and can be inhibited by selective modulators of a number of targets which show efficacy in treating asthma. Despite the myriad of proteins in asthmatic sputum clinically efficacious, target specific ligands block this effect.
  - Skin biopsies for psoriatic patients release protein factors into culture media. This 'conditioned media' with its complex, disease-relevant protein mix causes effects in a secondary bioassay (T-cell chemotaxis maybe) that is inhibited by target specific intervention and has proved to be predictive of clinical efficacy.
- *In Vivo* Models

  A number of *in vivo* model types are employed in target validation studies. When considering animal models it is important to highlight that those which accurately represent human disease are very powerful assets in target validation, however many are not predictive of the human situation and it is therefore equally important to validate the model before drawing solid conclusions from the data generated. Predictive models can also serve as pharmacodynamic models useful in selecting optimal compounds for further evaluation, and for predicting clinically effective doses. Animal models commonly used in target validation exercises can be divided into three categories summarised below:
  - Animal models that reproduce a biological process believed to be relevant in a particular disease. For example tracheal instilled lipolysaccharide (LPS) induced neutrophilia as a model of neutrophil inflammation in the lung.
  - Animal models that represent the disease/pathophysiologic process more closely; sometimes called tertiary models or disease models. The term disease model is controversial as it implies that a disease can be replicated in an animal, which is not the case, however these models still represent systems with complex biological disease-like processes that represent to a greater or lesser degree a process that occurs in clinical disease states. For example collagen-induced arthritis (CIA) in the mouse is a model of T-cell mediated inflammation in the joint; a standard model in which to test potential drugs for the treatment of rheumatoid arthritis, which modulate T-cell pathways.
  - Models using knock out (KO) mice, where the target is removed through genetic manipulation rather than by pharmacological intervention. These models have proved particularly useful in early target validation studies, where no tool compounds exist, and allow workers to observe the phenotype of mice lacking the relevant target gene.

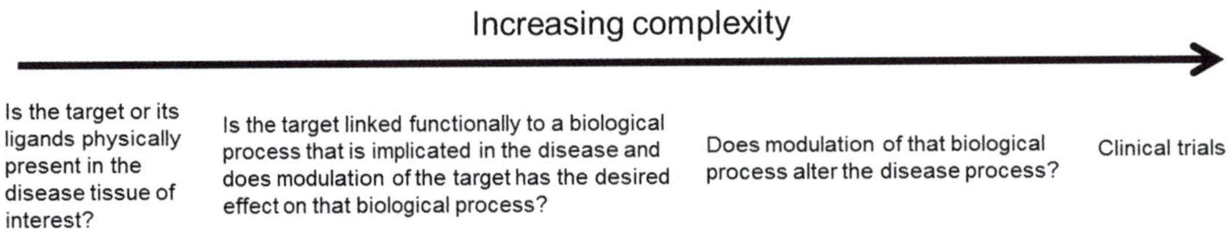

Increasing complexity

Is the target or its ligands physically present in the disease tissue of interest?

Is the target linked functionally to a biological process that is implicated in the disease and does modulation of the target has the desired effect on that biological process?

Does modulation of that biological process alter the disease process?

Clinical trials

Figure 19.3   The increase in confidence of a target is gained by investigations using increasing complex models systems.

Data from the general phenotype of the KO are useful in determining if there are any safety related issues with inhibiting a particular target. Additionally, if these mice are treated with agents which induce disease pathology, and exhibit a reduced response this would suggest that target modulation has the potential to modify disease phenotype. In recent years workers have realised that data from such studies is not always accurate, hence it is important to perform additional validation studies to give further confidence.

- Clinical outcomes and regulatory biomarkers
  - The ultimate test of target validation is through clinical testing. This is worthy of mention but beyond the scope of this chapter

There is an increasing level of complexity in each of the assay systems as a project progresses down the above lists as depicted in Figure 19.3.

19.4   EXAMPLES OF TARGET VALIDATION STUDIES AND DATA INTERPRETATION

As the complexity increases it becomes increasingly demanding to link the end point measurement of an assay to the particular target mechanism. This is of high importance as for most biological processes there are multiple mechanisms which mediate the same response. For example, consider the potential of inhibiting the enzyme caspase 1 to treat an inflammatory disease such as rheumatoid arthritis. Caspase 1 was initially identified as a key enzyme involved in the production of the inflammatory cytokine interleukin 1β (IL1β) by human monocytes. IL1β in turn was implicated as a mediator in the pathology and progression of rheumatoid arthritis. However other proteins also control production of this cytokine and additionally there are other mediators of the disease pathology. Thus there are a number of key questions that need to be addressed, most importantly does reducing IL1β production have a positive effect on disease pathology and does inhibition of caspase 1 produce a reduction of IL1β sufficient to offer a potential treatment.

The first stage in the target validation process was to demonstrate that inhibiting caspase 1 blocked the production of IL1β by activated monocytes (the biological process believed relevant to the disease), the second stage of the process was to show that a tool inhibitor was effective at reversing inflammatory response and associated pathology in an animal model, and that this reversal corresponded to a reduction in IL1β levels *in vivo* (a model of disease pathology). The final and ultimate stage of target validation was to demonstrate efficacy in patients with rheumatoid arthritis and that this efficacy was associated with a reduction of IL1β levels (clinical outcome). In this abbreviated example at each stage the mechanism has been demonstrated to be directly linked to responses observed and illustrates the need to carefully plan experiments which determine the potential role of a target in a disease at each stage.

Viewing target validation in this progressive manner with an emphasis on increasing complexity reveals a number of important facets about how data should be interpreted at each stage. Pragmatically it divides target validation into manageable units that parallel the resource and practical commitment that is justified by the other aims of the project, notably compound quality. In the above description a distinction has been made between biological process, disease/pathophysiological processes and clinical outcomes. The following example illustrates the distinction in more detail, keywords being highlighted in bold.

For the treatment of COPD the biomarker of **clinical outcome** is the measurement of 'force expiratory volume 1 second (FEV1)'; defined by regulators and known to be directly linked to the breathing capabilities of the patient. The complex disease processes underlying the decline in FEV1 that occurs as COPD progresses are not fully understood and remain the focus of much current research. One hypothesis is that inhibiting connective tissue breakdown, for example, is a **disease/pathophysiological process** that should be targeted. At the cellular level neutrophils can be postulated as a key cell type which produce the enzymes that cause this breakdown and hence migration (chemotaxis) of neutrophils would be the **biological process** one may want to block. The **target** would then be the protein entity that drives chemotaxis of neutrophils. There are many to choose from. The key points are summarised in Figure 19.4.

Thinking this way helps in the design of key target validation experiments within the project context.

Considering one of the potential targets highlighted, the chemokine receptor CXCR2, (which regulates production of the inflammatory cytokine interleukin-8 (IL8)); a key early target validation experiment would be to block CXCR2 activity and, if IL8-induced chemotaxis is inhibited, this demonstrates that the CXCR2 receptor can mediate neutrophil chemotaxis. This is far from definitive target validation, but provides encouragement for further studies. More complex models for blocking CXCR2 mediated neutrophil migration into a lung and then measuring the effect of tissue breakdown would be required for the next stages of validation, such as an *in vivo* model of LPS-induced lung injury.[14] A positive readout in this model would then provide further confidence in the role of CXCR2 for lung disease.

However it is known that other disease processes (and targets) exist that can drive the pathophysiology of COPD, such as mucus production, and this raises another potential role of target validation studies—to choose the preferred mechanism when there are several implicated in disease pathology. In this case an alternative target validation pathway to discover treatments for to treat COPD could be that as depicted in Figure 19.5.

In such a situation it is important to plan key experiments which are able to closely reproduce the role of a target in a disease and ideally quantify its contribution relative to others under consideration, thus allowing a data driven selection of the optimal target. The nature of these experiments is beyond the scope of this chapter but it is worth highlighting this extension of target validation to inform target selection decisions.

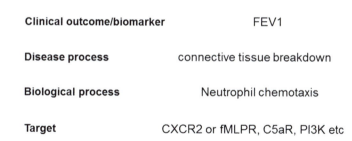

Clinical outcome/biomarker	FEV1
Disease process	connective tissue breakdown
Biological process	Neutrophil chemotaxis
Target	CXCR2 or fMLPR, C5aR, PI3K etc

Figure 19.4   Separating the biology of a disease into levels of complexity.

Clinical outcome/biomarker	FEV1	Exacerbation rate
Disease process	connective tissue breakdown	Mucus production
Biological process	Neutrophil chemotaxis	Goblet hyperplasia
Target	CXCR2 or fMLPR, C5aR, PI3K etc	

Figure 19.5   Any disease will have a number of processes and possible clinical outcomes that can be targeted.

19.5   WIDER CONSIDERATION OF A TARGET—MODE OF MODULATION

So far only the basic definition of a target has been considered; that of a novel protein playing a pivotal role in a disease.

In addition to there being a variety of different classes of biological target currently available there are multiple modes of action with which a particular ligand can modulate the target. For example in the case of antagonists of G protein-coupled receptors it is possible to inhibit receptor function in a variety of modes which include competitive inhibition, non-competitive inhibition and allosteric modulation among others. The mode of modulation may affect the pharmacology in terms of magnitude, duration and selectivity of the pharmacological response. Therefore it is critical that the optimal mode of modulation is understood as part of any target validation study. If this is not the case then there is a significant chance that an incorrect decision will be made regarding the role of the target in the disease. This is well illustrated by the detailed mechanism of actions studies performed to understand the optimal profile for an inhibitor of the glycine transporter GlyT1 which have recently led to a successful positive PoC and a potential breakthrough in the treatment of schizophrenia.[15]

In the voltage gated ion channel field mode of action is highly important in achieving a sufficient window between efficacy and target related side effects. Use-dependent sodium channel blocker anti-convulsants such as lamotrigine are safe and effective because they have a significantly greater affinity for the inactivated state of the channel over the open/closed states thus allowing selective inhibition of hyperactive neurons over normally functioning cells in conditions such as epilepsy.[16] Given the key role that sodium channels play in many physiological processes, tonic inhibition would not have been an acceptable mechanism to achieve an adequate safety profile. The majority of sodium channel blockers in clinical use today were discovered by phenotypic *in vivo* screening, thereby effectively 'bypassing' detailed *in vitro* mechanistic studies. Such an approach is not acceptable in modern drug discovery and therefore it is important to determine the most appropriate mechanism both for efficacy and safety at an early stage.

The example of N-type calcium channel blockers described in Section 19.6 is interesting as the inhibitor ziconotide, currently on the market, is a tonic blocker which completely inhibits N-type channel function and therefore has to be administered by injection directly into the spine to avoid adverse cardiovascular effects due to inhibition of peripheral N-type calcium channels. Use-dependent inhibitors are currently in clinical evaluation which only inhibit spinal neurons firing at high frequency associated with aberrant pain signalling and can be safely administered orally. Workers from Abbott have recently published their target validation work which addresses the key question as to the potential for analgesic efficacy with a use dependent inhibitor.[17] In this study they were able to show that their use-dependent inhibitor is able to inhibit neurotransmitter release from stimulated spinal neurons at predicted therapeutic concentrations in a similar manner to ziconotide, thus giving confidence that modulating the channel in a different manner will retain efficacy.

The understandings of G-protein receptor interactions have evolved in the past 10 years to embrace the concept of ligand bias. It is now accepted that G-protein receptors are coupled to multiple intracellular pathways; G-proteins (Gs, Gi) and beta arrestin as examples. Individual molecules can interact with the receptor protein to stabilise a plethora of conformation states that signal through intracellular pathways to different degrees. This now allows the differentiation of molecules based upon their ability to stimulate or inhibit one pathway and not another. This is being exploited to change the potential safety profile of a compound (or compound class) even though the physical target protein is common. Again, a target validation approach of the correct balance of properties is required to be confident that the molecule will be a successful drug. Building on this complexity is the concept of receptor dimerisation; the ability for individually defined receptors to bind together in complexes and, importantly, change the pharmacology of the new entity. Although not a new concept in the realms of *in vitro* cell biology with over expressing systems, new evidence is suggesting that this is a physiologic and pathophysiologic relevant phenomenon.[18]

Using the examples above, integrated with the more traditional view of target validation reveals the need to see validation of a target in all its guises; from relevance of biological target, how that target is influenced by a molecule (mode of action) and its impact on the biology, how the target behaves in its native environment and the impact of any dual approaches.

19.6 TOOLS USED IN FUNCTIONAL TARGET VALIDATION

As described in the previous section an early component of current target validation practice is to identify a molecular tool which modulates the protein target in the desired mode. Most commonly these tools are small molecules. It is imperative that these tools are of high quality so that the data generated gives sufficient confidence to make decisions about the target. Unfortunately the literature is full of studies which have been performed with inadequate tools and the conclusions drawn are not valid. It is perhaps surprising that this issue was only recently addressed in detail.[19] The author highlights that until 2010 there was no general consensus as to the requisite properties for a tool compound (or chemical probe as he refers to them) and that much of the published target validation data uses poor quality molecules and the conclusions that are proposed are incorrect. He illustrates this with the example of staurosporine. There are over 8000 publications of studies with this molecule and a significant number draw conclusions about specific kinases which are not valid, as staurosporine is a highly promiscuous inhibitor.

In his paper Frye[19] then proposes a series of generic criteria that a tool compound should possess in order to generate meaningful data including adequate potency, selectivity, cell permeability, physicochemical properties and to be freely available to the scientific community. It is important to consider the whole profile and not be seduced by potency alone as a highly potent ligand for an intracellular target has little value if it is not cell permeable or precipitates from the assay buffer due to poor solubility. The five key attributes and associated properties that a tool molecule should possess proposed by Frye are as follows.

1. Molecular profiling. A quality chemical probe has sufficient potency *in vitro* and selectivity data to confidently associate its profile *in vitro* to its cellular or *in vivo* profile.
2. Mechanism of action. A quality chemical probe has sufficient mechanistic data *versus* its intended molecular target to enable interpretation of its qualitative and quantitative effect (dose dependency) on a target-dependent action in either a cell-based assay or a cell-free assay that recapitulates a physiologic function of the target.
3. Identification of the active species. A quality chemical probe has sufficient chemical and physical property data to permit utilisation in *in vitro* (here defined as biochemical) and

cell-based assays with interpretations of results attributed to its intact structure or a well-characterised derivative.

4. Proven utility as a probe. A quality chemical probe has sufficient cellular activity data to confidently address at least one hypothesis about the role of the molecular target in a cell's response to its environment.

5. Availability. A quality chemical probe is readily available to the academic community with no restrictions on use

Interestingly, and perhaps importantly, Frye does not attempt to propose numerical values for these criteria as they may well be target dependent. These rules provide an excellent guide for the identification of a tool compound. This paper has been cited many times since its publication and subsequent authors have suggested further developments.[20,21]

At the point a new target of interest is identified it is highly unlikely that tool compounds fitting the criteria previously discussed will exist, therefore rather than using a suboptimal compound and potentially generating equivocal data it is now increasingly common to seek a non-small molecule tool. Examples of tools that have been used in recent studies include peptides,[22] siRNA,[23] oligonucleotides[24] and antibodies.[25] One clear limitation with all of these alternatives is poor cell penetration but recent advances have enabled intracellular delivery in some cases;[22] however, for extracellular targets they represent an excellent alternative.

Peptides have enjoyed a resurgence of interest in recent years, largely due to the development of new technologies to allow their production. Peptides offer the advantage of high potency and selectivity and despite advances in technology they are best suited to extracellular targets. An excellent example of the use of a peptide to validate a target is in the discovery of the analgesic ziconotide which is a selective inhibitor of the N-type calcium channel (Cav2.2).[26] ω-Conotoxin MVIIA is a naturally occurring peptide isolated from a marine snail and is a highly potent and selective blocker of the N-type calcium channel. It was used to demonstrate that the channel inhibition in isolated neurones inhibited firing of sensory neurones and when tested showed efficacy *in vivo* in a range of pain models. Subsequently a closely related analogue entered clinical evaluation where it demonstrated excellent analgesic efficacy and was eventually launched onto the market as ziconotide.

This is an extreme example but illustrates the power of peptides in target validation (http://www.youtube.com/watch?v=N8IRlBig8zY).

RNA interference (RNAi), also called post transcriptional gene silencing (PTGS), is a biological process in which RNA molecules inhibit gene expression, typically by causing the destruction of specific mRNA molecules.[23,27] This area has been extensively described in the molecular biology chapter and therefore an extensive definition will not be repeated in this chapter. It is however worth highlighting the fact that this technology offers a highly selective method of modulating gene expression and therefore enables scientists to study the effect that the expression (or deletion) of a gene has on the phenotype of a cell, thereby allowing an insight into the influence of a single gene on a phenotype and therefore confidence in the role of the target gene in a disease setting.[28]

When using siRNA and indeed interventions where the target protein is removed (*e.g.* KO mice) it should be appreciated that pharmacological intervention is not the same as removing the protein for a number of reasons. Proteins can have multiple functions, for example both as enzymes and scaffold proteins. A pharmacological inhibitor may remove only one of these effects and leave the other, a case not mimicked by siRNA or KO cells. Additionally there are cases where receptors have multiple functions, ligands only modulate specific functions and not others; again not a subtlety that can be mimicked by protein removal. These scenarios should be considered when attempting to translate results to clinical efficacy.

Another powerful use of this technique is in target identification, and one area where this is of particular value is in phenotypic screening which is discussed in the following section. Phenotypic screening involves testing compounds for their ability to affect a particular biological response without prior knowledge of the target. Once active molecules are identified then it is important to identify their mechanism of action. Comparing the phenotypic response of an active compound with the effects of RNAi can be a very effective method of identifying targets.

Antibodies have been described in detail in the chapter Molecular Biology chapter but it is worth highlighting here that they have proved particularly useful in target validation studies with more complex protein targets such as ion channels and G protein-coupled receptors (GPCRs). These are often highly complex proteins and are composed of multiple distinct subunits. Particularly in the case of ion channels which frequently exist as multisubunit, heteromeric proteins, and the subunit composition can vary depending on the tissue. It is extremely important to know the exact nature of the channel which is involved in the disease process of interest and to determine if a particular subunit (in certain cases) is to be targeted for optimal efficacy and safety. An excellent example of this was described by Rhodes *et al.*[29] in their work on voltage-gated potassium channels in the CNS They were able to use antibodies to demonstrate that targeting the β sub-unit of the Kv1.4 channel would be an optimal strategy for identification of a safe efficacious epilepsy treatment. Their validation data was subsequently confirmed by the identification of selective compounds which demonstrated an excellent preclinical profile *in vivo*.

19.7 EXPERIMENTS COMMONLY CONDUCTED FOR TARGET VALIDATION

Building on the methodologies described in Section 19.3 this section explores the benefits and limitations that should be considered at each stage.

19.7.1 Presence of Target and/or Target Pathway

Target expression in the form of its gene expression or mRNA levels are an early indication of target association with disease, particularly if analysed with regard to disease tissue *vs.* normal healthy control or further, correlating with degrees of pathologic progression. Methodology for this is covered in the molecular biology chapter (Chapter 9) as are other benefits to these techniques. Points of caution are that gene expression changes are not associated with all disease pathologies. So care should be taken not to conclude negatively if no changes are seen. Similarly, when performing Gene Expression Analysis (GEA) on disease tissue, samples taken from a disease biopsy contain a mixed population of cells. A false positive for target gene expression being elevated can exist if the population of cells simply changes; infiltration of white blood cells in an inflamed biopsy for example. The mRNA expression thus becomes a marker for cell diversity rather than innate increased expression in one particular cell type. This focus on genetic studies to support the concept of patient stratification leading to personalised medicine should not be lost on the target validation biologist. Understanding patient variation in disease phenotypes is a key part of that disease understanding that allows the correct pathway to be mapped from cellular systems to complex models. Measurement of target protein or indeed the modulators of the target (agonist levels to support the role of a target receptor) are valuable supporting evidence for a targets role. Processing of tissue (disease tissue ideally) to reveal protein levels with either antibody blotting on gels (Western blotting) is one approach. Antibody staining for a target on histology slices from disease tissue (Immunohistochemistry) is arguably a step further in target validation as it associates the target with particular cells and areas of pathology. Staining for multiple proteins and cellular substructures provides information on co-localisation; studies designed to support the linkage of the target to known cellular/biological processes and diseases processes. Imaging techniques are suitably advanced to allow quantification of the images from these studies. However, advances in technology are yet to supplant the eyes of an experienced pathologist for the interpretation of sections from disease; bringing clinical experience to early target validation.

The application of the technologies described above (and those under functional studies below) additionally has powerful utility in the understanding of the disease processes in animal model studies.

19.7.2 *In Vitro* Functional Models for Target Validation

Performing studies *in vitro* has great advantages in that they can be configured and manipulated to answer specific questions about a target. They allow the researcher to have a greater control over the environment of the system, and tool compounds can be more easily applied and have less stringent criteria than those used for testing *in vivo*. There are a myriad of assays that can be run *in vitro* for any particular target. It is important to determine the most appropriate assay to perform to answer a particular target validation question. Thinking in the terms introduced above, target, biological process, disease process and clinical outcome, can help. Assays should generate supporting data to link the target to the biological process, then the target and biological process to the disease process. There are then two aspects to testing *in vitro*; choice of functional readouts and the types of tissue used. The optimal functional measurement in terms of target validation is that which most accurately substantiates the role of the putative target in the disease. This is often distinct from assays used to routinely screen compounds. Measurement of intracellular calcium or cAMP are standard assays for the screening of compounds but only a very specific target validation question can be answered with this technology. For example, they can be very useful bioassays to show that bioactive material relating to a target is present in diseases supernatants *i.e.* linkage of the target to the disease. Measures that have a more disease relevant endpoint are most appropriate to address target validation issues: *e.g.* for cancer targets, cell proliferation for tumour growth or endothelial cell tube formation for angiogenesis (formation of new blood vessels).

In terms of choice of tissue, where possible human tissue is preferred; from single cells, cell suspension through to tissue slices and organ biopsy sized pieces in culture. The choice is driven by the particular disease setting and the degree of complexity required validating your target. Cell systems that have had targets cloned into them are by definition artificial and only have utility in target validation for answering very specific questions about mode of action of a compound. Fresh, intact, functional human tissue assays aim to bridge the gap between cell-based studies *in vitro*, animal studies *in vivo* and clinical trials. Such tissues offer advantages over simpler cell-based models, avoid species differences and have the potential to reflect the diverse patient population. For example, they maintain important cell-to-cell relationships in a 3-D structure which many single cell-based assays lose. There is currently a large focus on alternative methods for providing complex organ-like assays, using cell-lines and stem cells to reconstruct miniaturised organs—termed organs on a chip. The reader is referred to the 'Organs on Chips' themed section at RSC Publishing (http://pubs.rsc.org/en/journals/article-collectionlanding?sercode=ib&themeid=7bb6b005-9960-4088-87fb-d62ae72e6e77).

However, reconstructed or engineered 3-D tissues produced from stem cells or as reconstructed organoids often fail to reflect the actual disease phenotype and diversity of responses found in healthy and diseased tissues obtained directly from patients. These systems will have the throughput to allow screening programmes to be conducted but their utility in validating targets relies on them reflecting truly the behaviour of the organ they are attempting to mimic.

Much progress is being made to improve the access to fresh human tissue. This is ceasing to be a rate limiting step for many systems. For projects targeting mechanisms in white blood cells for example, then human tissue is readily available. In contrast with targets for CNS disorders it is more difficult to obtain human brain tissue.

If using non-human tissue, a knowledge of the differences in target biology between humans and other species is important. Numerous examples exist, showing differences in human and other species tissue responsiveness. Human tissue from a diseased-like state represents the most relevant system to test a target validation hypothesis. The target is in a natural environment, any interaction between pathways within the cells is conserved; receptor dimerisation, balance of converging pathways *etc*. For human organ culture experiments there is a complex mixture of interacting cells as there would be in the human body. For many disease states these human tissues are available and most target validation experiments need only be conducted periodically in the lifetime of a project. However, it is always advisable to confirm activity of lead compounds in these systems.

19.7.3 *In Vivo* Models for Target Validation

Expanding on the concept that the confidence in a target will become greater if it is shown to modulate a more complex model, the ultimate complexity outside the clinical setting is to test a target's utility in an animal model. However, that extrapolation can only be made if indeed that complexity is actually reproduced in the animal as it is in man. There is much current debate regarding the validity of animal models in the drug discovery community. Animal models themselves vary in complexity according to the question being asked in the model, and can range from the simpler pharmacodynamic models that allow a distinct pathway to be activated, to models that intend to mimic a disease process and have a read-out that is directly relevant to the clinical setting. Even these most complex models fall short in mimicking the complexity of the human disease in both pathway engagement and, for many diseases, chronicity. They can reproduce patterns of pathophysiologic processes that are similar to disease states and therefore do have utility. Choice of the correct model to represent a disease process is important. For example, there are a number of possible *in vivo* models of joint diseases, for both osteoarthritis and rheumatoid arthritis. The publication of the output from a recent steering group has performed the valuable exercise of 'positioning them on a pathogenesis map whereby model selection is determined by the specific aspect of disease to be studied.'[30] Animal model studies serve two functions in a drug discovery program; to understand the behaviour of the compound being tested and to give confidence in the utility of the target mechanism (target validation). Arguably the simpler pharmacodynamic models, being less complex, will have less utility for target validation but if a step wise approach is adopted for gaining confidence in a target then they may be valuable. For example, a model in which LPS is instilled into a rat lung. This causes a neutrophil movement into the lung and is therefore a model used to look at mechanisms for inhibiting pathways involved in this system. The fact that a particular selective compound inhibits this effect would give a degree of confidence for the target's prominent role in neutrophil biology. Failure to work would lead to concern that the mechanism under investigation is not a major mediator of the response and therefore may trigger further investigations or depending on the data available that the mechanism is not appropriate to provide a treatment for the disease. As for all experiments, understanding the system and its components are vital for data interpretation and extrapolation (translation) to man For example, the difference in distribution of chemokine receptors in rat and human. Work indicates cell expression in rats and mice is different from man in many cases. The value of this model for testing a compound's distribution and behaviour in an animal is still valid but one would need to consider the translation of these data in a target validation context.

Paramount to data interpretation is the quality of the tools used in these animal models. Building on Section 19.6 and the quality required for any experimentation, testing in any animals requires even greater criteria. Three key points are:

- Species differences. The compound to be tested in the animal has the appropriate activity at the target in that species and ideally the same activity at the human target.
- Appropriate drug metabolism and pharmacokinetics (DMPK) of the molecule to allow enough 'coverage' of the target for the required duration of the experiment.
- Selectivity. A tool should have sufficient selectivity for the target under investigation to allow adequate blood concentration can be achieved allowing selective target modulation.

Very often animal model data is published without reference to any of these points and so the interpretation of literature data should be conducted with caution.

Further information about the importance of a target mechanism can be gained from treating the animal either prophylactically or therapeutically. Prophylactic dosing is when a compound is administered prior to the disease-mimicking insult to the animal. In this case efficacy is deemed as being the abrogation or delaying of any effect. Therapeutic dosing is when a compound is administered after the insult and the effect is to halt or even reverse progression of the disease readout. This later effect is closer to the situation seen in clinical dosing, hence the term therapeutic, and is therefore deemed to be more convincing.

Chemical tools are not the only form of intervention that can be used for target validation purposes. The use of KO mice (or indeed Knock-in) can be a favoured option but relies on the availability of suitable mouse models. The rate of breeding of mice and the ease of genetic manipulation is the deciding factor for using mice but limitations exist. For example the mouse collagen-induced arthritis model is a favoured model for mimicking T-cell mediated joint damage as a model for the processes in human rheumatoid arthritis. However, the biology of this model is often only exhibited in particular strains of mice.[31] Cognisant of this the authors Labasi *et al.*,[32] wishing to test the role of the purinergic receptor P2X$_7$ using KO mice, only available in a mixed strain of mice, were careful enough to look at collagen-antibody-induced arthritis rather than collagen-induced arthritis, a subtly different insult to the animal. Both these approaches and that of other molecular biology approaches (siRNA) are powerful tools and reviewed in greater detail in the chapter on molecular biology. The level of confidence gained from the interpretation of any animal model is very much dependent on the target and disease. Conditional KO animals are also available. The ability to switch off target genes after an animal's development stages, just as a therapeutic agent would, can alleviate some concerns over adaptation of the organism (mouse). However, the issues still remain that this may not mimic drug–target interaction.

Negative allosteric modulators, ligand biasing compounds, receptor dimerisation may all infer a change to a target proteins biology that has a subtlety that is missed by genetic approaches. Again, using this term 'platform of evidence' to gain confidence in your target validation approach should lead you to use other supporting evidence, to complement (if not completely replace in some cases) animal model data; specifically data generated from the *ex vivo* studies using human systems, described above.

Alternatively one can use nature's provision of human KO to help validate your target, *i.e.* individuals who either naturally lack the target or the target in a normal form due to mutation. Their phenotype can give greater confidence in the role of that target in a disease. Two recent examples include: firstly, a subset of individuals that lack surface expression of the chemokine receptor CCR5 and showed tolerance to the HIV virus. Based on this observation small molecules CCR5 inhibitors were developed which protected animals from developing AIDS when infected by the HIV virus, a candidate compound subsequently demonstrated clinical efficacy and is launched onto the market as Maraviroc.[33] Secondly, natural human mutations in of the gene SCN9A, encoding the voltage-gated sodium channel $Na_v1.7$, have led to the linkage of this target to the perception of pain.[34] It has been demonstrated that individuals who lack the SCN9A gene or who have loss of function mutations have reduced sensitivity to pain, conversely individuals possessing gain of function mutations show increased sensitivity to painful stimuli. These observations have led to successful drug discovery programmes and compounds which produced molecules with an encouraging preclinical profile and are currently under clinical evaluation.

19.8 SINGLE TARGET *VS.* MULTIPLE TARGETS

One conclusion from the many recent clinical failures is that modulation of certain targets individually is not sufficient to produce a clinical effect. However, it has been proposed that modulation of some of these targets together with a second additional target may produce efficacy, or alternatively molecules which selectively modulate two targets may have additional benefits over fixed dose combinations. An example of the latter in the respiratory area is the development of the MABAs which are molecules in which a selective β2 adrenergic agonist and a muscarinic M3 antagonist are connected by a chemical linker and retain the potency of the two individual parts. Clinical data recently published suggested that these molecules are at least as efficacious as administering the same dose of both agents alone and may have an improved side-effect profile. Additionally, as these are both inhaled drugs, the dual compound represents a more convenient option for the patient with the additional possibility of combining with a third agent such as an anti-inflammatory agent.

A number of groups have also investigated the possibility of molecules with dual pharmacology such as the combined H1 and H3 antihistamines published by GSK,[35] this is a chemically very challenging exercise and relies on targets having pharmacophores which overlap.

The arena of combinations is a complex one and the key challenge for the target validation scientist is to demonstrate that the dual activity is significantly superior to a fixed dose combination or indeed a single agent.

19.9 PHENOTYPIC SCREENING

Phenotypic screening is in many ways the complete opposite of target based drug discovery which has been the focus of this chapter so far. There are many ways to define phenotypic screening but a simple definition is 'screening for a biological response without prior knowledge of a protein target.'

This approach is the mainstay of many anti-infective programmes, for example where there are few clearly defined targets, particularly in the area of neglected diseases. Interestingly, it was not that long ago that this approach was the mainstay of the pharmaceutical industry and successfully led to the discovery of many marketed drugs. Even though this approach has not been widely employed for several decades a recent survey revealed that 37% of drugs approved by the FDA between 1999 and 2008 were discovered *via* phenotypic screening.[36]

Given the current poor success rate in clinical proof of concept studies the validity of target based drug discovery has been questioned and consequently there has been a renewed interest in phenotypic screening. The major advantage of phenotypic screens is that they can target any protein (or other entity, such as a lipid or nucleic acid) in its biological context, without the *a priori* need to know the target. This means that, in addition to either enzyme inhibitors or receptor agonists or antagonists, small molecule hits from phenotypic screens could, for example, act as allosteric inhibitors or could ablate protein–protein interactions. Essentially in phenotypic screening focus has shifted from concentrating on how effectively molecules interact with a predefined target to how effectively they elicit or inhibit a biological response.

There are, however, a number of considerations to be taken into account before embarking on such an approach and Eggert offers a number of recommendations:[37]

(1) The workers should have a good understanding of the phenotype which will form the basis of the assay. Simple phenotypic assays such as cell death do not normally make good assay formats.

(2) The assay design needs to be robust, in a very similar manner to any screening assay care must be taken to ensure activity is not dependent on plate position, that an acceptable signal to noise ratio is achievable and that 'off-phenotype effects' such as cytotoxicity have been considered. Availability of positive a negative controls are also important, very often high quality small molecule ligands may not be available and in such instances RNAi can be very useful.

(3) A target identification strategy should be in place early in the screening schedule. Target identification early in a phenotypic screening programme is important, often challenging and designing the most appropriate screening assay can often assist this process. There are a number of target identification strategies available, some involve *in silico* approaches and will be covered in the next section, but an alternative again is the use of RNAi. By comparing the phenotypic effect of a compound in a cell with the behaviour of a cell in which a particular gene has been silenced can often offer insights as to potential targets.

(4) If the screen is to be performed by a specialist service provider then the choice of screening organisation is of high importance for a number of reasons, in particular it is advisable to work with groups who are experienced in this area and the size and nature of compound collection is also important. In contrast to target based drug discovery it is not necessary to have a vast screening library, often 50 000 compounds is sufficient; however, as with all screening campaigns, molecule quality is of high importance.

(5) Unlike the majority of target based screens high hit rates are often encountered in phenotypic screens and often screening scientists are faced with the task of hit prioritisation. Two key criteria are recommended here: (a) To prioritise hits based on 'most interesting phenotype' which are always the most potent compounds but those demonstrating the most interesting phenotype specific to the cellular and disease setting. (b) To avoid false positives, as with all screens. False positives will inevitably be identified through this approach and it is important to identify them early to avoid wasting time. Standard approaches are employed here, *i.e.* confirming the purity of hits and re-confirming with fresh batches of material.

(6) Finally it is important to profile hits in secondary assays. A panel of such assays are recommended for such an approach and are key to both determining that a phenotype is specific and can help in determining the molecular target.

It will be interesting to see the degree of success of this new found interest in phenotypic screening over the coming years, many group are investing considerable time and resources in this approach, but success can only be measured in terms of clinical efficacy.

19.10   SERENDIPITOUS TARGET VALIDATION

On occasions serendipitous discoveries have resulted in targets being validated for indications for which they were not originally intended for. Perhaps the best described is the discovery of the Sildenafil (Viagra) for the treatment of erectile dysfunction.[38] The initial project objective was to develop selective inhibitors of phosphodiesterase V (PDEV) that would increase tissue levels of cGMP, and that could be beneficial for the treatment of hypertension and other cardiovascular indications. After an extensive discovery effort UK 92480 (sildenafil) was identified as a clinical candidate which possessed a promising preclinical profile. However, initial clinical results were disappointing and the compound failed to meet clinical endpoints when tested in patients with coronary heart disease. During the study one of the side effects reported was a slight increase in erectile function and this led the Pfizer group to investigate this further. As a result the group demonstrated that by inhibiting PDEV, sildenafil potentiated the natural activity of nitric oxide (NO) and improved erectile function in conditions where NO release or smooth muscle relaxation are impaired in the penis. The remainder of the story is now well known and through this serendipitous discovery in a failed clinical study a completely new indication was identified for which the compound eventually provided an effective treatment.

This is not a unique example and there are other examples of such serendipitous target validation. Clearly by their very nature serendipitous discoveries cannot be designed but this story demonstrates the value of considering all observations in clinical studies and that side effects are not always negative.

Molecules which have reached advanced stages of clinical testing have had many of the development risks discharged and therefore given positive clinical data have a high chance of reaching the market in a relatively short time frame. In recent years repositioning initiatives have been established by many companies in which new or alternative indications are actively sought for compounds that were originally designed for a different condition.

19.11   *IN SILICO* TARGET VALIDATION

There has been a significant increase in the amount of biological data available to drug discovery scientists in recent years through a number of initiatives such as the human genome project, increased efforts in structural biology and the growth of molecular biology. Combined with the development of sophisticated chemoinformatics and bioinformatics packages this has led to a greater understanding of disease pathways. Whilst no single target can be truly validated by purely *in silico* techniques, computational techniques have had a significant impact on a number of areas of target selection and validation. This section will focus on three: (1) target identification in phenotypic screening, (2) pathway analysis for target selection and validation, and (3) drug repositioning.

Phenotypic screening generates large amounts of functional data on small molecules without prior knowledge of the target; one approach to target identification is Chemogenomics. Chemogenomics comprises a systematic relationship between targets and ligands that are used as target modulators in living systems such as cells or organisms. In recent years, data on small molecule bioactivity relationships have become increasingly available, and consequently so have the number of approaches used to translate bioactivity data into knowledge. *In silico* target prediction tools can suggest likely biological targets of small molecules *via* data mining in target-annotated chemical databases.[39] The principle behind these tools is the generation of a so-called biological fingerprint which links a specific phenotypic response to a series of interaction between a compound and protein targets. A recent example of this is the 'Target Hunter' database (www.cbligand.org/TargetHunter).[40] This approach has proved highly effective in the identification of many targets following phenotypic screening campaigns and is currently in widespread use.

The vast amount of data currently available combined with advances in computer software now allows the modelling of complex biological networks which are relevant to the clinical situation (*i.e.* it is possible to model a 'virtual patient') and predict the outcome of a drug.[41]

An excellent example of this is the potential of IL-5 blockade in treating asthma. The original hypothesis was that by inhibiting the effect of IL-5 would cause a reduction in eosinophilia and thereby reduce airway inflammation thus alleviating the symptoms of asthma. Modelling studies predicted that whilst IL-5 blockade would cause a significant reduction in eosinophilia, this would have little effect on the symptoms of asthma due to the presence of other redundant inflammatory pathways. This prediction was subsequently proved correct in clinical evaluation of an experimental anti-IL-5 antibody therapy.[42] There are an increasing number of such examples thus raising confidence that where sufficient pathway data is available for complex biological networks computers are able to predict clinical outcomes, thus validating (or invalidating) targets.

19.12 CONCLUSION

It is clear from all of the statistics that the success rate of experimental therapeutics in clinical proof of concept studies has fallen in recent years, and many authors have suggested reasons for this decrease, indeed some have proposed that the reason is due to poor target validation. In the context of traditional definitions of the term this may indeed be true, where organisations synthesised prototype compounds with predefined potency for the target and tested them in traditional animal models, many of which were designed for different indications and then on seeing positive results assumed the target was validated and progressed to clinical evaluation.

It is also evident that groups who continue with this approach will continue to face a high chance of failure in expensive clinical testing.

The modern drug discovery scientist has available a wide range of targets to choose from (which are not expressed equally in the general population or often in patient populations, nor do the targets always share the same function across species). Additionally, there has been a recognition of the number of alternate mechanisms with which to modulate these targets. Furthermore, there is a continual broadening of the 'tool box' of potential reagents and methodologies available for target validation experiments, to provide confidence or otherwise that the correct target has been chosen and is appropriate for the disease. If used wisely and continually along the path from target identification to PoC then the chances of success should increase. However if the tools and methodologies are not used wisely and negative data ignored then a project will fail.

Today's drug discovery scientist should therefore not think of target validation as a single key experiment, but as an ongoing part of programme strategy which begins with target identification and is not complete until the definitive clinical study.

HINTS AND TIPS

1) Target validation is a continual process which starts when an initial link is identified between a protein and a disease process and ends with the ultimate experiment—a clinical proof of concept study.
2) Experiments should be carefully planned and should always reflect the clinical situation as closely as possible. Data from preclinical *in vivo* models should be interpreted with caution, they do not always accurately reflect the clinical situation and can lead to incorrect conclusions.
3) Careful tool compound selection is critical. There is little value in conducting an experiment with a suboptimal tool. If a suitable small molecule is not available workers should consider alternatives such as peptide, antibodies or siRNA.

4) Recent genomic studies represent potentially powerful sources of high value target validation data, however the scientist needs to carefully study the evidence to eliminate possible false leads such as those which may have arisen from developmental factors.

5) In recent years a number of new technologies have become available to assist the drug discovery scientist with target validation studies (*in silico* techniques, novel tools), almost certainly further developments will be made in the near future. All available technologies should be considered for incorporation in target validation studies when relevant.

KEY REFERENCES

R. M. Plenge, E. M. Scolnick and D. Altshuler, *Nat. Rev. Drug Discovery*, 2013, **12**, 581–594
F. Cong, A. K. Cheung and S. M. Huang, *Annu. Rev. Pharmacol. Toxicol.*, 2012, **52**, 57–78.
S. V. Frye, *Nat. Chem. Biol.*, 2010, **6**, 159–161.

REFERENCES

1. M. Sioud, *Methods Mol. Biol.*, 2007, **360**, 1–12.
2. A. Hayes and J. Hunter, *Br. J. Pharmacol.*, 2012, **167**, 1395–1397.
3. M. K. Herbert and P. Holzer, *Anaesthetist*, 2002, **51**(4), 308–319.
4. J. M. Humphrey, *Curr. Top. Med. Chem. (Sharjah, United Arab Emirates)*, 2003, **3**(12), 1423–1435.
5. K. Van Laere, J. De Hoon, G. Bormans, M. Koole, I. Derdelinck, I. De Lepeleire, R. Declercq, S. M. Sanabria Bohorquez, T. Hamill, P. D. Mozley, D. Tatosian, W. Xie, Y. Liu, F. Liu, P. Zappacosta, C. Mahon, K. L. Butterfield, L. B. Rosen, M. G. Murphy, R. J. Hargreaves, J. A. Wagner and C. R. Shadle, *Clin. Pharmacol. Ther.*, 2012, **92**(2), 243–250.
6. S. Boyce and R. G. Hill, in *Proceedings of the 9th World Congress on Pain*, ed. M. Devor, IASP Press, Washington, USA, 2000, pp. 313–324.
7. R. A. Dionne, *Curr. Opin. Cent. Peripher. Nerv. Syst. Invest. Drugs*, 1999, **1**(1), 82–85.
8. D. Borsook, J. Upadhyay, M. Klimas, A. J. Schwarz, A. Coimbra, R. Baumgartner, E. George and W. Z. Potter, *Drug Discovery Today*, 2012, **17**(17–18), 964–973.
9. R. Hill, *Trends Pharmacol. Sci.*, 2000, **21**(7), 244–246.
10. I. Kola and J. Landis, *Nat. Rev. Drug Discovery*, 2004, **3**, 711–716.
11. J. Arrowsmith and P. Miller, *Nat. Rev. Drug Discovery*, 2013, **12**, 569.
12. R. M. Plenge, E. M. Scolnick and D. Altshuler, *Nat. Rev. Drug Discovery*, 2013, **12**, 581–594.
13. F. Cong, A. K. Cheung and S. M. Huang, *Annu. Rev. Pharmacol. Toxicol.*, 2012, **52**, 57–78.
14. R. W. Chapman, M. Minnicozzi, C. S. Celly, J. E. Phillips, T. T. Kung, R. W Hipkin, X. Fa and D. Rindgen, *J. Pharmacol. Exp. Ther.*, 2007, **322**(2), 486–493.
15. J. S. Albert, in *Targets and Emerging Therapies for Schizophrenia*, ed. J. S. Albert and M. W. Wood, John Wiley & Sons, Inc. (Chichester United Kingdom), 2012, pp. 233–254.
16. H. Tao, A. Guia, B. Xie, D. SantaAna, G. Manalo, J. Xu and A. Ghetti, *Assay Drug Dev. Technol.*, 2006, **4**(1), 57–64.
17. V. E. Scott, *Biochem. Pharmacol.*, 2012, **83**(3), 406–418.
18. G. Milligan, *Br. J. Pharmacol.*, 2009, **158**(1), 5–14.
19. S. V. Frye, *Nat. Chem. Biol.*, 2010, **6**, 159–161.

20. P. Workman and C. Collins, *Chem. Biol.*, 2010, **17**, 561–577.

21. M. E. Bunnage, C. Piatnitski and L. H. Jones, *Nat. Chem. Biol.*, 2013, **9**, 195–199.

22. R. A. Myers, L. J. Cruz, J. E. Rivier and B. M. Olivera, *Chem. Rev.*, 1993, **93**(5), 1923–1936.

23. K. K. Jain, *Drug Discovery Today*, 2004, **9**(7), 307–309.

24. M. F. Taylor, *Expert Opin. Ther. Targets*, 2001, **5**(3), 297–301.

25. M. Visintin, M. Quondam and A. Cattaneo, *Methods*, 2004, **34**(2), 200–214.

26. M. Essack, V. Bajic and J. A. C. Archer, *Marine Drugs*, 2012, **10**, 1244–1265.

27. W. Xiong, J. Zhao, Y. Zhang, J. Wang, Y.-X. Zheng, Q.-C. He, L.-H. Zhang, D. M. Zhou, *Medicinal Chemistry of Nucleic Acids*, eds. L.-H. Zhang, Z. Xi, J. Chattopadhyaya, John Wiley & Sons Inc., Hobken NJ, USA, 2011, 390–404.

28. J. Kurreck, *Expert Opin. Biol. Ther.*, 2004, **4**(3), 427–429.

29. K. J. Rhodes and J. S. Trimmer, *J. Gen. Physiol.*, 2008, **131**(5), 407–413.

30. T. L. Vincent, R. O. Williams, R. Maciewicz, A. Silman and P. Garside, *Rheumatology*, 2012, **51**(11), 1931–414.

31. P. H. Wooley, H. S. Luthra, J. M. Stuart and C. S. David, *J. Exp. Med.*, 1981, **154**(3), 688–700.

32. J. M. Labasi, N. Petrushova, C. Donovan, P. Lira, M. M. Payette, W. Brissette, J. J. Wicks, L. Audoly and C. A. Gabel, *J. Immunol.*, 2002, **168**, 6436–6445.

33. N. Ray and R. W. Doms, *Curr. Top. Microbiol. Immunol.*, 2006, **303**, 97–120.

34. J. J Cox, F. Reimann, A. K. Nicholas, G. Thornton, E. Roberts, K. Springell, G. Karbani, H. Jafri, J. Mannan, Y. Raashid, L. Al-Gazali, H. Hamamy, E. M. Valente, S. Gorman, R. Williams, D. P. McHale, J. N. Wood, F. M. Gribble and C. G Woods, *Nature*, 2006, **444**(7121), 894–898.

35. P. Daley-Yates, C. Ambery, L. Sweeney, J. Watson, A. Oliver and B. McQuade, *Int. Arch. Allergy Immunol.*, 2012, **158**(1), 84–98.

36. D. C. Swinney and J. Anthony, *Nat. Rev. Drug Discov.*, 2011, **10**, 507–519.

37. U. S. Eggert, *Nat. Chem. Biol.*, 2013, **9**, 206–209.

38. S. Campbell, *Clin. Sci.*, 2000, **99**, 255–260.

39. A. Bender, D. W. Young, J. L. Jenkins, M. Serrano, D. Mikhailov, P. A. Clemons and J. W. Davies, Comb, *Chem. High Throughput Screening*, 2007, **10**(8), 719–731.

40. L. Wang, *AAPS J.*, 2013, **15**(2), 395–406.

41. E. Minch and I. Vacheva, in *In Silico Technologies in Drug Target Identification and Validation*, eds. D. Leon and S. Markel, CRC Press, Florida, USA, 2006, 195–223.

42. A. K. Lewis, T. Paterson, C. C. Leong, N. Defranoux, S. T. Holgate and C. L. Stokes, *Int. Arch. Allergy Immunol.*, 2001, **124**, 282–286.

CHAPTER 20

Lead Generation

MARK FURBER,* FRANK NARJES AND JOHN STEELE

AstraZeneca R&D, Mölndal, Sweden
*E-mail: mark1.furber@astrazeneca.com

20.1 INTRODUCTION

20.1.1 What Do We Mean By 'Lead' And 'Lead Generation'?

In its simplest definition lead generation is the discovery of one or more lead chemical series on which to base a drug discovery program, whereas a lead can be defined as a molecule or chemical series which possesses biochemical and physicochemical properties suitable for optimisation to a candidate drug, and is the immediate end result of the lead generation process. Lead generation is often itself split into two activities, an initial hit-finding phase and a separate hit-to-lead phase; both will be considered in this chapter, but with a particular emphasis on the earlier hit-finding phase. A hit can best be defined as an active molecule or series whose activity and target engagement has been confirmed but whose scope and promise for further development is largely unknown, information that a hit-to-lead campaign hopes to provide.

As a phase in the drug discovery process lead generation represents the earliest point at which knowledge-based decisions can steer the course of a discovery program, with impact through to clinical development and drug launch. Alongside target validation it arguably represents the most critical step in the discovery process in the sense that decisions made at this very early stage of the project will influence many of the questions and the ability to answer those questions at all subsequent stages of the process. More specifically, the choice of assays employed to characterise actives or hits and the molecular properties of the emergent lead series, will define both the main hurdles that will need to be overcome and the limits to what is achievable in the progression to a candidate drug.[1,2]

The properties of what are generally seen as desirable lead series have evolved over time, and lead-like simplicity in terms of lower molecular weight and lipophilicity has emerged as a paradigm to facilitate optimisation of screening hits.[3-5] This is predicated on the observation that property optimisation often results in elevation of both. Since neither high molecular weight nor high lipophilicity are favourable for oral drug delivery, a low molecular weight more

The Handbook of Medicinal Chemistry: Principles and Practice
Edited by Andrew Davis and Simon E Ward
© The Royal Society of Chemistry 2015
Published by the Royal Society of Chemistry, www.rsc.org

polar hit/lead series should have many advantages. This is often captured in widespread use of ligand efficiency (LE) and lipophilic ligand efficiency (LLE) terms which try to capture the binding efficiency per heavy atom, or unit of molecular weight, or as a function of lipophilicity.[6] LE and LLE are variously defined as:

$$LE = \frac{\Delta G, \, pK_i, \, pK_d, \, pIC_{50}}{No. \, of \, heavy \, atoms} \text{ or } LE = \frac{pK_i, \, pK_d, \, pIC_{50}}{Molecular \, weight \, (kDa)}$$

$$LLE = pIC_{50}(pK_i) - \log P(\log D)$$

In practice, free energy of binding is calculated as $\Delta G = - RT(logK_i)$, $\Delta G = - RT(logK_d)$ or $\Delta G = - RT(logIC_{50})$ and logP/logD values are often calculated. Each of these terms has drawbacks and many other descriptors of binding efficiency have been described (*e.g.* size-independent ligand efficiency SILE, ligand efficiency dependent lipophilicity LELP), but LE and LLE remain the most commonly used. A critique and useful reference source for some for these descriptors, including LE and LLE, is available.[7]

Several representative examples of lead/drug pairs with associated LE and LLE values are provided in Table 20.1. Within this limited set of illustrative examples, but also within much larger comparison sets,[6] analyses of the binding efficiencies of drugs and their leads illustrate how potency and molecular size both increase on progressing from hits and leads to successful drugs, as generally does lipophilic ligand efficiency (LLE). In contrast, ligand efficiency (LE) can increase or it can decrease. Observation that LE can often decrease as part of any optimisation program, wherein multiple properties need to be balanced, perhaps highlights the importance of identifying lead series with good LE as an important goal for the LG phase.

As such LE, LLE and other binding efficacy indices form an important component of a *combined* overall assessment of hits and leads encompassing properties such as selectivity and propensity for metabolism in *in vitro* or *in vivo* systems.

20.1.2 The Process of Lead Generation and How The Industry Has Evolved Over Recent Years

Historically, drugs were discovered from natural sources (*e.g.* traditional therapies) or through serendipitous discovery.[8] Later, chemical libraries of synthetic small molecules, natural products or extracts were screened in intact cells or whole organisms to identify substances having a desirable therapeutic effect in a process traditionally described as 'classical pharmacology'. With greater understanding of largely protein-based molecular targets and a growing ability to manipulate such targets through advances in molecular biology, the lead generation process has become increasingly dependent on screening processes using *in vitro* biological systems to identify hit and lead series. Parallel advances in robotics greatly facilitated this process and allowed high throughput screening (HTS) of rapidly developing corporate and commercial compound collections, and HTS remains today a central (but by no means the only) driver of lead generation. Ultra high throughput screening has also developed in recent years as a significant tool for lead generation based on very large (typically >1 billion compound) DNA-tagged small molecule libraries or phage display peptide libraries.[9,10] Alongside advances in molecular biology came advances in the ability to engage protein-based molecular targets through use of monoclonal antibodies and related 'biologics'. Whilst outside the scope of this chapter, a lead generation approach can also be defined for large molecules or biologics, but the details of the process are significantly different and the reader is directed to recent reviews.[11,12]

HTS has occupied a central role in many research organisations and continues to do so given its broad applicability and tolerance of sometimes poor knowledge or understanding of the

Table 20.1 Representative drug–lead pairs comparing LE($-RT\log K_i$/no. heavy atoms) and LLE descriptors.

Lead structure	Drug structure	Drug name	Target	LE (LLE) lead	LE (LLE) drug	K_i μM lead	K_i μM drug
		Alvimopan	μ-opioid receptor	0.60 (2.9)	0.40 (5.8)	0.08	0.0008
		Sunitinib	VEGF kinase	0.55 (4.6)	0.33 (5.6)	0.39	0.08
		Argatroban	Thrombin	0.19 (4.6)	0.29 (8.1)	1000	0.032
		Oseltamivir	Influenza Neuraminidase	0.47 (8.4)	0.61 (10.2)	6.3	0.001

target or natural ligand(s). But equally, many other techniques hold an important place in the lead generation armoury, including fragment-based screening, phenotypic screening, *de novo* design, virtual screening and scaffold hopping. Each of these is discussed further in this chapter, but it should be recognised that no single hit finding approach will guarantee success and only through appropriate integration of approaches can success rates be maximised. Indeed, the need to tackle traditionally more difficult targets and open up more of the genome to small molecule hit finding approaches has increased the need to explore multiple parallel hit finding strategies. Recent examples include the search for inhibitors of the aspartyl protease β-secretase (BACE1) for treatment of Alzheimer's disease, and heat shock protein 90 (HSP90) as antitumour agents.[13,14] The first generation of BACE1 inhibitors were peptidomimetic transition state analogs, which suffered from problems such as low oral bioavailability, poor blood–brain barrier permeability and susceptibility to P-glycoprotein transport. To discover novel non-peptide inhibitors, a variety of lead finding methods were used, including conducting the HTS at high compound concentration of 100 µM instead of the usual 10 µM, the use of encoded compound libraries and fragment-based NMR or X-ray screening.[15–17] HSP90 is a molecular chaperone, which is responsible for activation and stabilisation of several oncoproteins in cancer cells. A phenotypic screen identified natural products such as Geldanamycin as inhibitors, and subsequently structure-based design, based on the X-ray structure of Geldanamycin bound to HSP90, as well as virtual screening and fragment-based methods have been used to find a wide variety of different chemotypes.[14,18] These two target examples are also used to illustrate the fragment based lead generation approach to lead generation in Section 20.2.2.4.

20.1.3   Issues Faced and Resolutions

Despite advances in many areas of the lead generation process over recent years, it is possible to identify significant sources of uncertainty or failure in the lead generation process. Perhaps the two most significant of these are the contrasting situations of finding too few or no hits for a target, and finding too many hits. Each can be wasteful of resource, but the latter can be particularly wasteful if there is an underlying cause which leads to choosing inappropriate hits and subsequently lead series.

20.1.3.1   *Inability to Identify Suitable Small Molecule Hits or Leads for Some Types Of Target*

It has been estimated that 60% of small molecule drug discovery projects fail due to lack of suitable leads.[19] The accuracy of this number can be debated, but failure to generate leads is a significant risk in any drug discovery project and the probability of this risk becoming a reality can usually be assessed at target identification. To a significant extent, the risk of success/failure for any target-based lead finding exercise (druggability) can be estimated based on precedent and knowledge that protein family sequence underlies protein folding, topology and binding site architecture. Information on known ligands as well as information obtained from analysis of target protein topology, fragment–protein docking and fragment screening can all be incorporated to give an assessment of target tractability, yet such assessments are never absolute and the need to focus on disease-relevant targets necessarily drives attempts to extend the boundaries of small-molecule druggable space. Consider the human genome, sequencing of which has invited speculation on the size of the druggable genome. Various estimates[20,21] put this number at approximately 3000 genes from a total human genome comprising approximately 30 000 genes. Of course, druggable does not equate to disease-relevance and similar estimates have been made for the number of disease-relevant human genes and this number has likewise been placed at approximately 3000 genes. It is the overlap between these two numbers, between druggable and disease relevant space, estimated to be 600–1500 small-molecule human drug targets, that defines what is therapeutically relevant druggable space.[20] As with many descriptors, the assignment of whether a target is likely to be druggable or not is defined by information available at the time and, with many disease-relevant targets being poorly tractable for small molecules, pressure to identify new means to tackle such targets increases (see Section 20.2). Pressure to succeed with poorly tractable targets necessarily brings higher risk of failure which needs to be balanced against potential reward and the overall balance of risk of an organisation's lead generation portfolio. In part, this has driven development of both computational assessments of 'druggability' or 'ligandability'[22] as well as experimental methods of determining tractability. Computational target druggability assessments require target protein 3D structural information but can provide a valuable insight into the likelihood of finding ligands for novel targets or targets where little or no ligand information is available.[22] Experimental fragment-based screening can also prove a valuable indicator of target druggability, supporting the inclusion of fragment-based screening for such targets (*e.g.* NMR-based fragment screening) prior to execution of a more resource and costly HTS. Based on the number and affinity of hits, targets then can be prioritised accordingly.[23]

20.1.3.2 Identification of Too Many Hits—False Positives

To a large extent this problem is confined to the HTS approach to lead generation wherein a HTS output delivers such a wide range of compound structural types engaging the target that the investigators begin to question the assay used and the data fidelity in terms of promiscuity. Also, given corporate compound collections of the order of 10^6 molecules, hit rates beyond 1% become difficult to triage through a staged series of experiments. Often this type of problem is one of compound interference which manifests as false positives and is a consequence of the assay adopted and the counterscreens put in place to remove them. The term *false positive* can describe compounds which interfere with the assay format such that whilst the readout of the assay indicates a positive hit, the mechanism of action is unrelated to that intended and the compounds might not engage the intended target protein at all. Such compound interference can be especially difficult to identify if it is reproducible and concentration-dependent, characteristics generally attributed to compounds with genuine activity. *False positive* also encompasses compounds which by virtue of their chemical structure are promiscuous in their binding to target proteins. Both scenarios can lead to a high hit rate depending on the quality of the assay cascade, the availability of biophysical screens, the quality of the compound library (including purity) and the nature of the protein target. Examples of several types of compound interference (fluorescent compounds, redox compounds, aggregators, luciferin mimics, reactive compounds, metal chelators) will be given to illustrate why false positives are still one of the key problems in the HTS approach to lead generation.[24]

20.1.3.2.1 Fluorescent Compounds.
Many HTS assays rely on fluorescence-based detection systems that can be subject to assay interference producing false positives. Compound fluorescence can directly interfere with many biochemical and cell-based assays simply by interference in the detection systems used, by absorbing and emitting light at excitation or emission wavelengths relevant to the assay technology. Excitation and emission wavelength usually depends on the degree of conjugation and, in practice, assays that rely on excitation at relatively short wavelengths ($\lambda_{ex} \sim 350$ nm) with detection of fluorescence in the blue spectral region ($\lambda_{em} = 450–495$ nm) tend to produce more false positives due to compound fluorescence than assays using red shifted fluorophores. By one analysis, a 50% false positive hit

rate can be expected in some fluorescent assays, so it is important to introduce measures to remove such fluorescent false positives.[25] One frequently used method is to introduce a time delay between excitation of the fluorophore and detection of the emitted light, so called TRFRET.[26] This is made possible by the use of specific fluorescent lanthanides that have the unusual property of emitting over long periods of time (measured in milliseconds) after excitation, when most standard fluorescent dyes (*e.g.* fluorescein) emit within a few nanoseconds of being excited. This results in greatly reduced sensitivity to fluorescent interference and, combining such an output with evaluation of hits in an orthogonal assay measuring a different output, is usually enough to support a high confidence in hit quality.

20.1.3.2.2 Redox Cycling Compounds. Compounds that readily undergo redox cycling can also produce high hit rates in HTS assays for oxidase enzymes or for target proteins containing catalytic cysteine residues, or where the reduced form of cysteine is important for activity of the target protein. As an example, prior to a HTS to identify inhibitors of the cysteine protease Caspase-8 a pre-screen of 20 000 compounds identified a 1% hit rate, and from this set of compounds 85% of the hits were identified as acting through a redox cycle mechanism.[27] Caspase-8 is maintained in an active form by inclusion of a reducing agent in assay and storage buffers (*e.g.* dithiothreitol DTT, tris(2-carboxyethyl)phosphine TCEP). However, some compound classes undergo redox cycling in the presence of O_2 and a reducing agent to generate H_2O_2, which can indirectly modulate target protein activity. Equally, where the target protein does not contain a catalytically or structurally important cysteine, but where the enzyme functions as an oxidase, such redox cycling can interfere with assay readouts based on H_2O_2 detection and produce a high false positive rate. A high throughput assay has been described to identify and eliminate redox cycling compounds, and selected structural types identified as redox actives are illustrated (Figure 20.1).[28]

20.1.3.2.3 Aggregators. Aggregation, a process in which small molecules associate to form much larger particles capable of indiscriminately associating with protein targets, is another frequent cause of high hit rates and false positives. As an example, a screen of 70 563 small 'drug-like' molecules against β-lactamase identified 1274 inhibitors, 1204 (95%) of which were shown to function in this particular assay as aggregate-based inhibitors.[29] The implication from such studies is that at the higher concentrations used in HTS primary assays (5–10 μM), 1–2% of drug-like molecules behave in this way. Such false positives can usually be removed by use of detergent in either the primary assay or in a counterscreen, they can also be predicted by descriptor-based computational methods and screened for in aqueous media by NMR methods.[30] Nonetheless, the problem can still be easily missed. A good example is provided by the identification of hit compound (1) and development into lead structure (2) as inhibitors of the protease Cruzain (targeting Chagas' disease) (Figure 20.2).[31] Despite interpretable SAR, apparently competitive non-covalent inhibition and a 350-fold potency increase over the initial hit, the series had actually been optimised against

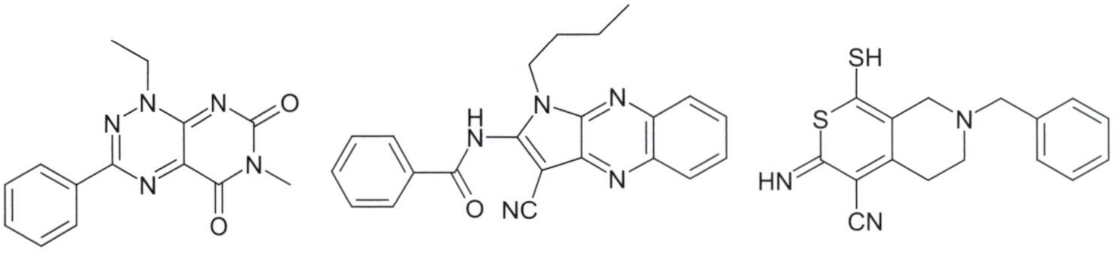

Figure 20.1 Selected examples of redox active compounds detected as false positives.[28]

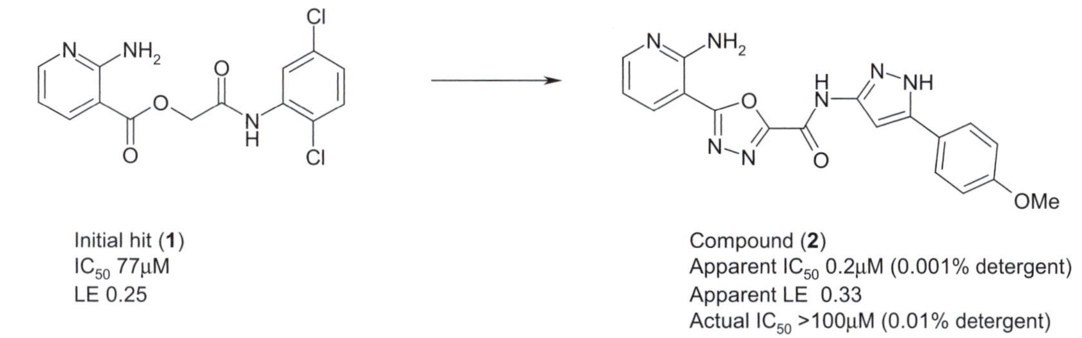

Figure 20.2 Misidentification of a hit series due to aggregation phenomena.[31]

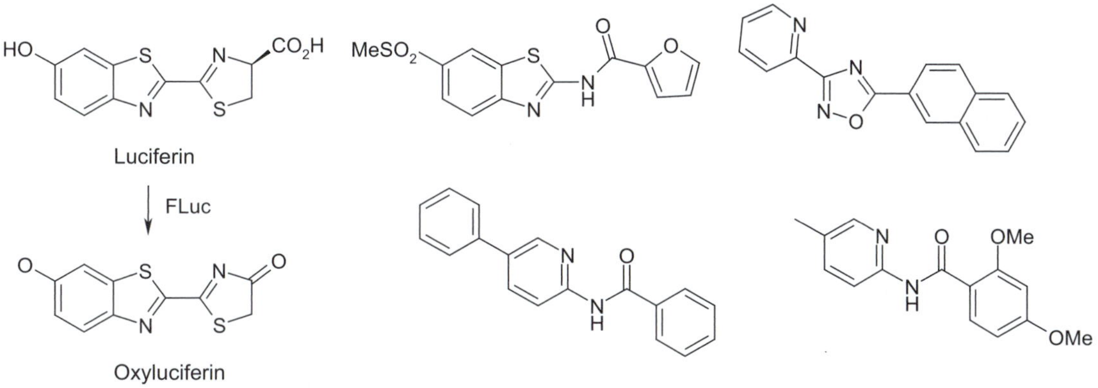

Figure 20.3 A simplified representation of the enzymatic reaction catalysed by FLuc alongside selected examples of compounds interfering with luciferase assays through structural resemblance to luciferin.[33–35]

aggregation. In this case, initial studies in the presence of detergent had shown activity but proved misleading. Failure to show any activity versus the protozoa combined with very steep dose response curves (a possible indicator of aggregation) stimulated further investigation of the mechanism and repeating the studies in the presence of 10-fold higher detergent concentration removed all activity.

20.1.3.2.4 Luciferase Inhibitors. Luciferase-based reporter gene assays are commonly used in high throughput format and can also be subject to compound interference. Firefly luciferase (FLuc), an enzyme which converts luciferin to oxyluciferin, is itself susceptible to inhibition by small molecules and such compounds can appear as activators or inhibitors of the cell-based assay depending on affinity for FLuc, FLuc concentration, detection system and whether inhibition is competitive or non-competitive.[32] In a retrospective analysis of cell based luciferase reporter assays as many as 60% of the hits were ascribed as false positives.[33] Compounds found to inhibit FLuc include 1,2,4-oxadiazoles, benzothiazoles and compounds structurally similar to the natural FLuc substrate luciferin (Figure 20.3).[33] This can lead to potentially large numbers of false positives being identified as plausible and attractive series for lead identification thus resulting in wasted expense and resource.

20.1.3.2.5 Promiscuous Compounds. False positives also arise from compound classes which show promiscuous binding behaviour. Sometimes this is due to chemical reactivity

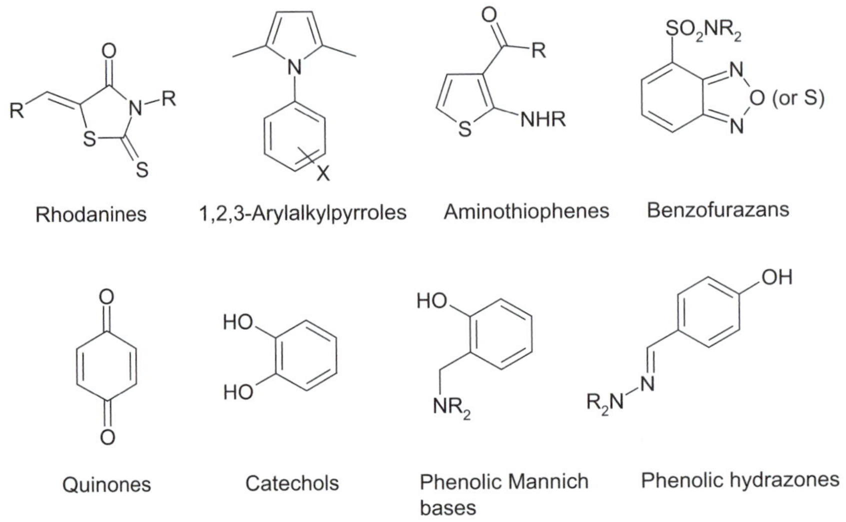

Figure 20.4   Selected examples of compound classes showing pan assay interference through protein reactivity.[36]

and formation of covalent bonds to target proteins. How indiscriminately such compounds react can be difficult to predict and can depend on multiple factors. What is clear, however, is that some structural types recur repeatedly in HTS outputs and whilst they should not necessarily be dismissed outright, the investigator needs to be aware of potential selectivity problems that might be encountered further into the lead generation process and beyond (Figure 20.4). A much publicised example is that of a class of compound called 'rhodanines'. In a well-cited publication on pan assay interference compounds (PAINS)[36] the authors quote 800 literature references to rhodanines showing biological effects. An assessment of this class of molecule recently concluded that '*compounds possessing a rhodanine moiety should be considered very critically despite convincing data generated in biological assays.*' In addition to lack of selectivity, unusual structure–activity relationship profiles and safety and specificity problems mean that rhodanines are generally not optimisable.[37] Whilst an extreme example, other functional groups present in 'frequent hitter' molecules and whilst it is often possible to identify some frequent hitters by obviously undesirable structural features or molecular properties, other compounds will escape such initial detection. A number of groups have published on use of substructure filters and rule-based filtering of compound types showing promiscuous activity across large numbers of high throughput screens and assay formats.[38,39] Such filters or rules should be used as a guide to the likely difficulty optimising compounds with such functionality and not as absolute rules. For example, a compound class which inhibits a target protein by covalent interaction might be of more or less interest depending on the disease indication and on the reversibility of the covalent interaction (and the likely selectivity that can be both achieved and tolerated for a particular target and indication).

20.1.3.2.6   Metal Impurities and Metal Chelators.   Metal ions can form a dual source of false positives in screening campaigns. In the first instance, where a metal ion plays an active role in protein structure and/or function, compounds which bind metals can indirectly and indiscriminately target such functions and produce what would generally be described as false positive hits. As such, it is important to consider the involvement of metals in target protein function and assess the likelihood of identifying such false positives. Where such liabilities exist a counterscreen can be included to remove such hits. For example, in a screen for inhibitors of homocitrate synthase which utilises an active site manganese (Mn^{2+}) or magnesium (Mg^{2+}) for catalysis, a counterscreen in the presence of 20 μM $MnCl_2$ was able to effectively remove metal chelators *e.g.* 8-hydroxyquinolines.[40] Secondly, inorganic metal contaminants can themselves provide a source of false positives in HTS screening which are not frequently identified/detected by routine purity checks and which resynthesis may or may not resolve. In one recent report, a survey of 175 HTS campaigns carried out at Roche identified 41 screens that were susceptible to inhibition by low levels of zinc impurities in screening compounds.[41] Typically such contaminants were observed to produce false positive signals in the low μM range and thus close to common selection criteria. As a counterscreen for such false positives, screening in the presence of a chelator can be effective at removing compounds contaminated by metals.[41]

Such examples, chosen to illustrate the potential blind alleys that a HTS output can direct the investigator down, serve only to illustrate the importance of critical assessment of any screening output. They illustrate the importance of functional group evaluation and potential for assay interference, as well as the importance of secondary assessment to demonstrate true target engagement and, if so desired, reversible stoichiometric binding. Many such techniques are available, including NMR based methods and surface plasmon resonance based assay systems *e.g.* BiaCore™. For example, NMR techniques have been used successfully to triage HTS outputs for fluorescent false positives[42] and aggregating compounds[30] measuring chemical shift perturbations in standard one dimensional ^{1}H-NMR experiments, as well as to remove compounds which are active by virtue of their intrinsic chemical reactivity (ALARM NMR[43]) using ^{1}H/^{13}C-HSQC and ^{1}H/^{15}N-HSQC NMR to measure chemical shift perturbation and line-broadening for a reactive cysteine probe (human La antigen protein).[43]

20.2 HIT IDENTIFICATION: HOW DO WE FIND A START POINT?

20.2.1 Strategy—What Are We Trying To Do?

In choosing a hit finding strategy for a particular biological target or phenotype, consideration is usually first given to the available proprietary or publically available ligand information to steer the course of a lead generation strategy. If suitable small molecule lead structures are already available the principle concerns might centre on novelty, mechanistic understanding and property optimisation, and screening for new hits might take a secondary or redundant role. Indeed, manipulation of endogenous ligands, natural products and published small molecules/competitor compounds has a long history and continues to result in successful leads and drugs. This approach to lead generation has been reviewed extensively,[1,2,8] but one particular strategy for discovering novel lead structures—lead hopping—is discussed in Section 20.2.2.7. For novel or less explored targets, this is often not the case and available ligands either do not exist or are deemed unsuitable as leads. In this case, a screening approach must be adopted and this may take two basic forms: *target-based* (*e.g.* HTS, fragment screening, DNA-encoded library screening, substrate/ ligand-based approaches, *in silico* virtual screening) and *phenotype-based*. Each approach has its own advantages and disadvantages.[44,45] In the target-based approach a specific biological hypothesis is tested through identification of molecules (small molecules or biologics) acting at a single protein target. Assays are generally adaptable to a high-throughput format and SAR is, in principle, understandable and sufficient to drive potency. Also, the growth in available *apo*-protein and ligand-bound protein structural information enables adoption of fragment screening as well as virtual screening approaches. The hit finding part of this chapter will focus on target-based screening but the disadvantages of this approach need to be recognised and have been extensively analysed in recent years.[44,45] These disadvantages derive from our variable understanding of disease mechanisms, sometimes resulting in compounds that have little or no clinical effect. Phenotypic screening, on the other hand, evaluates compounds at the cell or whole organism level and requires no initial understanding of molecular mechanism of action or specific protein target. Indeed compounds identified by this method might work through novel targets and/or through more than one target protein. Advantages and disadvantages of the approach are discussed at the end of this section.

Often a target will be considered of sufficiently high interest that more than one lead generation approach will be adopted. For example, where the target protein can be solubilised and crystallised a combination of target-based HTS plus fragment screen may be exploited, and in some cases multiple approaches might be justifiable, as already exemplified for BACE1 and HSP90.

20.2.2   Target-Based Approaches

20.2.2.1   Assays

Of critical importance to any target-based lead generation campaign is the ability to identify and establish biological assays with appropriate throughput, quality, reproducibility, cost and pharmacological relevance. Principally these assays, in combination, must be able to unequivocally establish the desired level and type of activity at the target under investigation. They might also give preliminary indications of selectivity if a project has identified key closely related targets over which it wishes to achieve selectivity. As already discussed (Section 20.1), efficient separation of true actives from compounds working by unwanted mechanisms *e.g.* interference with the assay detection systems, is crucial to the hit finding approach.

Key primary-screen components of a robust screening assay cascade include:[46]

- Counterscreens—to exclude off-target activity often at near-neighbour proteins in the genome (*e.g.* JAK2 when seeking a JAK1/3 inhibitor).
- Orthogonal screens—which replicate the same biology in an assay in which a single component is switched (*e.g.* the detection system).
- Biophysical or direct-binding screen—to confirm target engagement (*e.g.* NMR or label free technologies such as EPIC™ and BiaCore™).

A primary assay should translate a specific biomolecular process into an observable parameter, typically a measurement of radioactivity, photon absorption or photon emission. Of these, photon emission (fluorescence and luminescence) is by far the most widely adopted assay technology due to its wide applicability, sensitivity and suitability for high throughput automation. TRFRET, FLIPR™, AlphaScreen® and bioluminescent reporter assays are examples of very sensitive and widely used fluorescence or luminescence based assays with applicability to a broad spectrum of target classes including GPCRs, enzymes, ion channels and NHRs. Although scintillation proximity assays (SPA) and other radioactive methods have also been successfully applied in compound screening, they have become less popular due to the undesirability of handling radioactive reagents and waste, coupled to the ready availability of non-radioactive alternatives.

20.2.2.2   HTS

HTS is a central part of the hit finding strategy for many research organisations and is able to find hit series for many targets. Success rates strongly depend on target class but success rates of ca 50% have broadly been observed across multiple target classes.[45] Looking beyond hits and leads to marketed drugs derived from HTS origins, one analysis indicates 19 of 58 drugs of known origin and approved between 1991 and 2008 derived from a HTS hit finding approach.[6]

20.2.2.2.1 Screening Platforms. A screen is an optimised assay format that is sufficiently robust to be automated to generate 100 000–2 000 000+ data points; high throughput screens are typically biochemical or cell-based assays run in 384 or 1536-well microtitre plate formats. With increasing automation and assay volume miniaturisation the definition of HTS has evolved from a typical capacity of 100 96-well plates per day in the late nineties to recent standard capacities of several hundred 1536-wellplates per day, allowing complete corporate compound collections to be screened in 1–2 weeks. Alongside this commitment to automation and screening technology is a need for efficient processes for handling large amounts of assay data. HTS triaging to define key series has been facilitated by data handling and display software such as TIBCO Spotfire™ integrated with in-house or commercial software to cluster hit sets based on structural similarities and to calculate and correlate physical properties (*e.g.* clog P) with potency.

20.2.2.2.2 Targeted or Diverse Compound Collections. In deciding whether to screen a targeted or a diverse small-molecule library, consideration must be given to the desired outcome and to the available compound collection and its content, as well as to screening capacity. Successful diversity screening requires significant investment in sample acquisition, screening technology/automation and data manipulation software. Yet in spite of this complexity, screening of large diverse libraries remains the dominant Pharma lead generation approach due to the promise of novelty and the consequent IP freedom it potentially offers; it can be applied to unfiltered corporate or commercial compound collections or computationally pre-filtered libraries in which an attempt is made to remove unattractive chemical features in advance of screening.[47] Targeted screening, based on clustering targets into families on the hypothesis that similar ligands should bind to similar targets, nonetheless has many advantages since it reduces the screening investment per hit series.[48] Indeed, for some target classes the need for diversity screening is reduced by the accumulated knowledge of the target family as a whole and targeted screening is often the first option for such targets. Kinase targets are classic examples of this, since focussed libraries of ATP-binding site inhibitors are readily available. Targeted screening can also be aided by computational methods, particularly where three-dimensional structural information can be used with 3D-docking algorithms, or where ligand information is available to generate 3D pharmacophore models, both of which can be used to select compounds from large collections, commercial databases or virtual compound libraries.[49]

It should be noted that the true diversity of small-molecule screening collections of the order of 10^6 molecules is actually very low considering the theoretical size of small-molecule chemical space. Estimates of the number of possible carbon-based small-molecule drug molecules vary but a recent estimate places this at 10^{60},[50] whilst the CAS registry contains just 72 million unique substances, illustrating the scale of the difference.[50] Even with the advent of billion-compound libraries synthesized on DNA tags, the ability to explore a significant portion of chemical space remains a major challenge, but what is clear is that the proportion of this very large hypothetical chemical space that needs to be covered to sustain a lead generation campaign is demonstrably within synthetic achievability.

20.2.2.3 DNA-Encoded Libraries (DEL) and Ultra-High Throughput Affinity Selection

The DEL approach was first theoretically described in 1992 by Brenner and Lerner and involves the construction of very large chemical libraries tagged to DNA.[51] Such libraries are constructed by alternating parallel combinatorial synthesis and are encoded in a tagged oligonucleotide sequence constructed in parallel and containing all the information required for identification of the attached small molecule. As such, it has analogy to antibody phage display techniques wherein antibodies are linked to phage particles containing the gene for the attached antibody. DEL libraries are then subjected to an affinity selection process, typically using immobilised target protein, after which non-binders are removed by washing steps and binders amplified by polymerase chain reaction (PCR) and identified through their encoding DNA by some form of high throughput sequencing technique.

20.2.2.4 Fragments

Fragment based lead generation is grounded in theoretical and observable higher hit rates for small (MW < 250) fragments compared to larger molecules typically found in HTS collections,[52,53] combined with a demonstrated ability to develop small fragment hits into drug-like leads utilising 3D structural information of the bound ligands. The lower affinity of such starting fragments requires more sensitive assays but the lower molecular weight of hits allows greater scope to optimise properties. As a caution, assay robustness and many of the causes of false positives (*e.g.* aggregation, as discussed in the introduction section for general screening) can be acute when testing compounds at high concentration (up to 1 mM). A recent review discusses these problems especially in relation to fragment based lead generation.[54] Furthermore, solubility must be considered in fragment library design to reduce the risk of false negatives arising from precipitation.

Fragment-based approaches have evolved to the extent that they are frequently considered standalone or alongside HTS for soluble targets where 3D protein structural information is available, or feasible. Recent reviews are available[52,53] in addition to an excellent blog[55] with links to practical information, recent literature and a list of fragment-derived clinical candidates (currently standing at 25), including one approved drug Vemurafenib arising from a fragment based approach.

An example from Merck's β-secretase program (BACE-1) is shown in Figure 20.5. NMR-based fragment screening identified a thioisourea hit (1) with K_d 15 μM determined by ^{15}N–^1H NMR chemical shift perturbation. This was evolved through use of X-ray structural data from the thioisourea liganded BACE-1 into a series of iminohydantoins[55] and iminopyrimidones[56] of which the more potent lead (2) is representative. A multiparameter lead optimisation program delivered compound (3) an advanced potent oral candidate with good CNS penetration (Figure 20.5).[57]

A second example derives from the work of Astex researchers on the heat shock protein HSP90.[53,58] A library of 1600 fragments was screened against HSP90 using NMR methods to

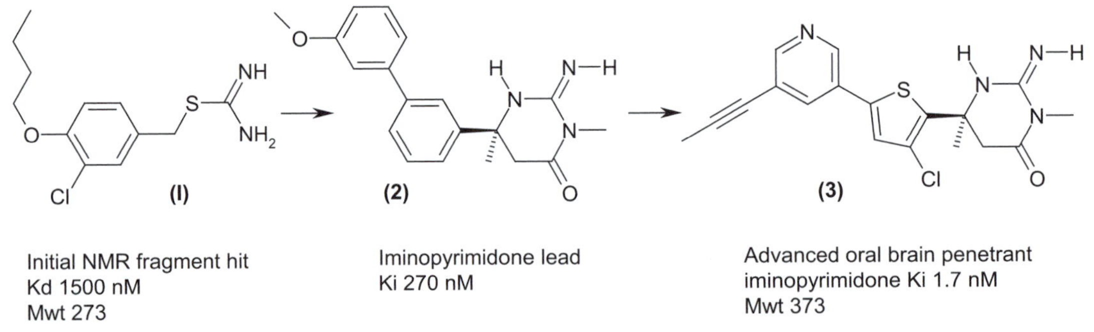

Initial NMR fragment hit
Kd 1500 nM
Mwt 273

Iminopyrimidone lead
Ki 270 nM

Advanced oral brain penetrant
iminopyrimidone Ki 1.7 nM
Mwt 373

Figure 20.5 Development of hit and lead BACE-1 inhibitors to brain penetrant iminopyrimidone (3).[56]

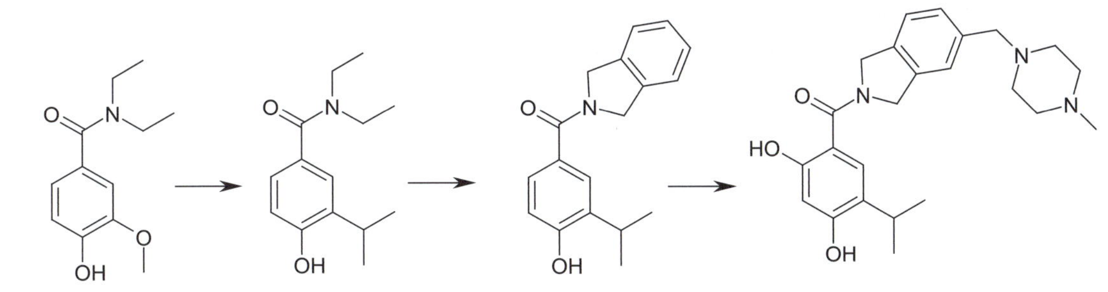

Fragment hit (4)			AT13387
K_d 790 μM	K_d 7 μM	K_d 0.068 μM	K_d 0.00071 μM
Mwt 223	Mwt 235	Mwt 281	Mwt 410
LE 0.26	LE 0.41	LE 0.47	LE 0.42
		IC_{50}(cell) 17 μM	IC_{50}(cell) 0.048 μM

Figure 20.6 Progression of a weak fragment hit to a 48 nM cell IC_{50} HSP90 clinical candidate (AT13387).[58]

identify hits in competition with ADP, allowing identification of compounds with affinity for the nucleotide binding site. The 125 fragments were progressed to crystallography experiments, from which 26 co-crystal structures were obtained. Isothermal titration calorimetry (ITC) was used to determine dissociation constants (K_d) for all compounds. Through structure-based design utilising crystallographic data, one chemical series based on catechol (**4**) led to the clinical development compound AT13387 showing both a ligand efficiency (LE) improvement and a 10^6 fold potency improvement over the initial fragment hit (Figure 20.6).[58]

20.2.2.4.1 ***Screening Platforms.*** The typically weak affinity of small molecular weight fragments has led to widespread adoption of highly sensitive biophysical screening techniques as the basis for fragment based lead generation approaches—including NMR, surface plasmon resonance (SPR), mass spectrometry, isothermal titration calorimetry (ITC) and electrophoresis methods, often in combination and frequently with guidance from X-ray crystallography.[52,53] Fragment screening libraries are small in comparison to HTS and typically comprise several hundred to a few thousand individual compounds. Indispensable to any fragment screening campaign is the ability to experimentally determine the binding mode of compounds through generation of 3D structural information of fragments bound to target protein using X-ray or NMR methods. Whilst advantageous for HTS approaches based on corporate compound collections, 3D structural information becomes essential when dealing with small, weakly bound fragments since the distance to final drug candidate is usually larger in terms of structural changes that need to be introduced, and small fragments by their very nature are often demanding medicinal chemistry start points without guidance from structural information.

20.2.2.4.2 ***Collections and Properties.*** Fragments are usually defined as compounds with molecular weight less than 250, but definitions vary and higher molecular weight limits, *e.g.* 300, have sometimes been used in setting the constraints for library composition. Irrespective of the precise constraint used, keeping fragments small allows greater freedom in the subsequent steps of property optimisation due to the commonly held and demonstrable view that lead optimisation invariable adds molecular weight. Keeping fragment lipophilicity low is beneficial for the same reason and a rule of 3 ($MW \leq 300$, $clogP \leq 3$, the number of hydrogen bond donors is ≤ 3, the number of hydrogen bond acceptors is ≤ 3) can be considered a useful maxim.[59] Incorporation of 3D diversity is equally important to fragment library composition since, in spite of the smaller size of fragment screening libraries ($<10^4$), hypothetical fragment space (of all possible carbon-based fragments MW <250) is greatly reduced and potentially allows increased diversity coverage through appropriate fragment library design.

20.2.2.5 *Target-Based De Novo Design*

Target-based *de novo* design relies on knowledge of the 3D structure of a biological target or a pharmacophore model to identify molecular structures that might bind to the target protein. It relies on a medicinal/computational chemist with the help of *de novo* molecular design software to build molecules within the constraints of a binding pocket or pharmacophore model by assembling small fragments or individual atoms. As such, it is an extremely difficult undertaking, not least due to difficulties presented in modelling interactions between flexible molecules and protein targets which themselves also do not present a static and fixed conformation, but also because of the presence of enclosed water molecules that can often play an important role in ligand recognition. The designed compounds also need to be chemically stable and readily accessible since the inherent approximations in the methods rarely justify lengthy synthesis. As a consequence, the number of true examples of leads generated by *de novo* design is rather small. Nonetheless, *de novo* design still represents an approach to generate new leads and, with growing numbers of tools available to aid the chemist, can represent a viable approach to design leads, especially where chemical space needs to be screened rationally. As a first step in such a process, the binding pocket or pharmacophore and its key interaction sites need to be identified and extracted from 3D structural information, then various seed fragments or building blocks are added to the structure and their structures optimised to identify lowest energy conformations. Fragments are then 'grown' or 'linked' to build larger ligands with higher predicted scoring functions (predictions of binding affinity). More sophisticated computational tools allow the entropic effects of bound water displacement (solvation) to be incorporated into the estimation of binding. Various *de novo* design programs are available to enable this type of approach *e.g.* SPROUT, TOPAS/Flux, Skelgen, BREED, Fragment Shuffling, SQUIRRELnovo, and several reviews can be found for more detailed information on the approach.[60,1]

20.2.2.6 *In Silico Virtual Screening*

In silico virtual screening is a computational technique to search large compound collections as well as virtual compound libraries to identify structures with the greatest potential to bind to a biological target. The approach can be applied in a structure- or ligand-based mode. In the structure-based mode, compound libraries are docked into a target protein and a scoring function is applied to give a calculated estimate of binding energy, and the hits are then ranked by this scoring function to identify the most promising examples for experimental testing. In the ligand-based mode, known ligand structures are used to generate a pharmacophore model which is used to score the virtual screening collection based on matching of steric and electronic properties. The approach suffers some of the same difficulties of *de novo* design but, when operated in a mode wherein the library is restricted to corporate or commercial compound collections, this removes problems associated with chemical space, synthesis and some stability concerns, and has been more successfully applied to generate leads. Virtual screening methods, pitfalls and successful applications have been recently reviewed.[61]

20.2.2.7 Lead Hopping

In the hit identification phase it is common to consider lead hopping (scaffold hopping) where screening or literature ligands are available. The approach aims to identify the central core of the hit series and bring one or more beneficial changes, and it can be guided by structural information when available or be driven in a more pragmatic manner dictated by the need for chemical novelty, structural simplicity and synthetic accessibility, or a need to move away from an inherently undesirable structural feature or toxophore present in the core of the ligand. A scaffold hop can increase binding affinity through additional interaction with the target protein or through reduced entropy wherein the change results in a more rigid scaffold. It can also be used to modulate physicochemical properties including lipophilicity and solubility. A very simple example might be the replacement of an indole core with a benzimidazole to lower lipophilicity and improve solubility, but examples can be significantly more complicated than this. When literature leads are taken as a start point, the ligands are often fairly advanced compounds and in finding new hit series the compounds that are produced tend to higher molecular weight than typical hits selected from screening, but when successful this approach frequently generates more potent compounds than are normally produced by HTS, simply because the start point on which the scaffold hopping approach was based was already part-optimised. A good example is the PPARδ agonist (6) produced through a scaffold hop from the known PPARδ agonists (5) and LC1765 (Figure 20.7). In this example a combination strategy was used, involving not only a core switch but also a switch of the side chain. Other benzazepine isomers or combinations were only weakly active, but this particular modification gave a novel hit series and a lead compound (7).[62]

A number of software programs are now commercially available that can help with generation of replacement scaffolds.[63]

20.2.2.8 Covalent Inhibitors

Many researchers innately avoid covalent compounds because of selectivity and toxicity fears, even though many successful marketed drugs operate through a covalent mechanism

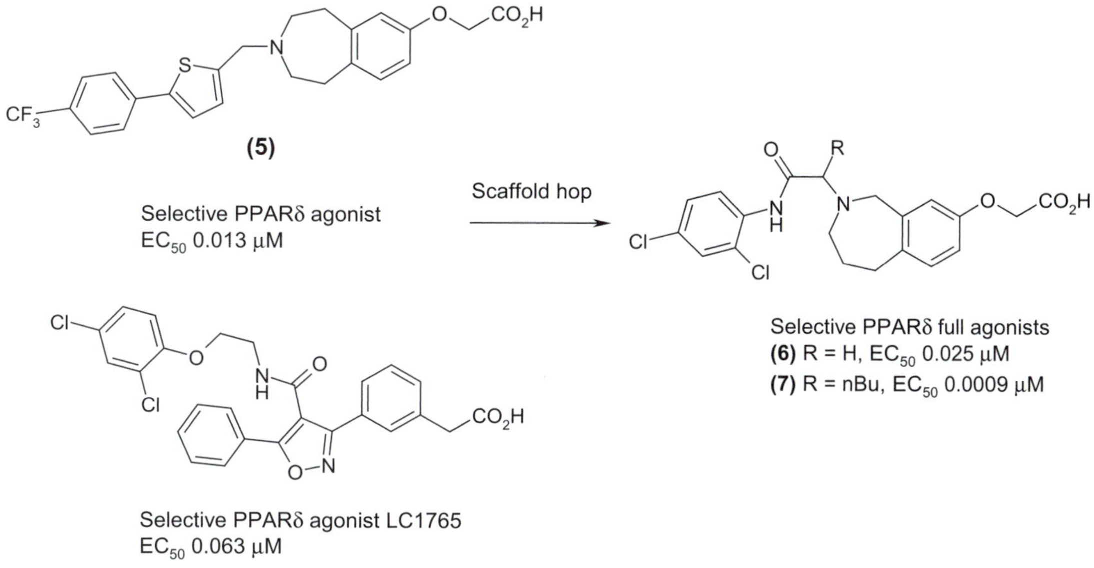

(5)

Selective PPARδ agonist
EC_{50} 0.013 μM

Scaffold hop

Selective PPARδ full agonists
(6) R = H, EC_{50} 0.025 μM
(7) R = nBu, EC_{50} 0.0009 μM

Selective PPARδ agonist LC1765
EC_{50} 0.063 μM

Figure 20.7 An example of a lead hop from compounds (5) and LC1765 to selective PPARδ agonists (6) and (7).[62]

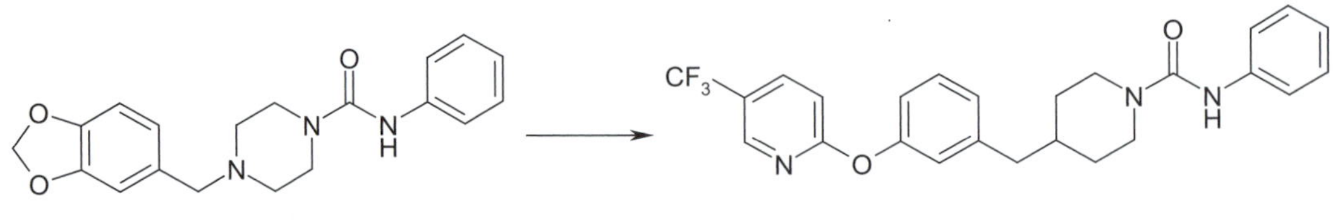

Figure 20.8   Initial covalent hit leading to a highly selective and potent FAAH inhibitor.[65]

(*e.g.* Plavix, Nexium, Prevacid and, historically, aspirin and penicillin antibiotics).[64] Invariably, these covalent drugs arose not by design but by retrospective investigation of their mechanism of action and no conscious choice was made to select for hits with a covalent binding mode. In many target-based hit finding programs irreversible covalent binders are rejected due to an understandable assumption that they would bind irreversibly to multiple proteins, RNA or DNA. This is factually the case for many such compounds, but nonetheless many examples can be found wherein high selectivity is achieved despite the irreversible nature of the inhibition.[65] One example is the fatty acid amide hydrolase (FAAH) inhibitor PF3845 (Figure 20.8). Whilst not showing structural features classically associated with irreversible covalent inhibition, the normally stable urea reacts covalently with Ser-241 of FAAH in the active site to inhibit the enzyme in an effectively irreversible manner. In both *in vitro* studies and *in vivo* protein pull-down experiments PF-3845 and analogues showed very high selectivity for FAAH.[66]

In considering covalent inhibition one should at the same time consider the dose and the reversibility/irreversibility spectrum. For compounds that have intrinsically low chemical reactivity such that they can bind and selectively form a covalent bond, reversibly or irreversibly, the stoichiometry of drug to target could ideally approach 1 : 1 and thus the delivery dose would be low and the associated selectivity and toxicity concerns would be small. Furthermore, many covalent inhibitors can operate in a reversible sense. The covalent bond they form is reversibly formed (*e.g.* Michael addition or addition of a cysteine thiol to an activated nitrile) and it is only the rate at which this covalent bond forms and breaks that differentiates the covalent inhibitor from a traditional high affinity non-covalent ligand. This reversibility can also mitigate an otherwise poorer selectivity for the target protein.[67]

20.2.3   Phenotype-Based Approaches (Phenotypic Screening)

Phenotypic screening will identify compounds that modulate in some desirable manner the observable characteristics (phenotype) of a cell, tissue or organism. Once compounds are shown to demonstrate a desirable effect, an attempt can be made to identify the biological target. Historically this is how drugs were discovered before target-based approaches came to the forefront. The approach can identify novel target proteins when coupled with genomic or chemical proteomic target deconvolution approaches and offers a potentially more reliable translation to disease pathology, although cell based (or even whole organism-based) phenotypic assays must still be shown to be relevant in the human disease. Disadvantages of the approach can include lower throughput screening capacity and the need to identify the molecular target to drive property optimisation. Even though target identification methods have improved significantly in recent years *via* chemical genetics and chemical proteomics,[68] target deconvolution remains a significant hurdle. Of course, historically drugs could reach the market before the target was identified (*e.g.* aspirin and more recently rapamycin[69]), but increased regulatory requirements and the need for clinical target-engagement biomarkers make this extremely challenging. One recent analysis indicates that the phenotypic screening approach has been more successful than target-based approaches in delivering first-in-class small molecule drugs to market,[44] but comparisons often polarise opinions and in reality each approach has its own merits and can provide many examples of successful outcomes,[44,45] in addition to failures.

20.3 HIT-TO-LEAD

The hit-to-lead stage of the early drug discovery process is variously defined, but in many organisations relates more closely to the subsequent lead optimisation phase (LO) than it does to any prior hit identification, in process and in outcomes. Initial profiling of hits is usually carried out as an intrinsic part of hit identification to allow selection of defined series, but then a more detailed hit-to-lead investigation of series properties and SAR is required to establish whether one or more chemical series has a reasonable probability of delivering a candidate drug; thus determining if it is worth committing the resource of a full LO project to the target and the newly discovered chemical equity. Prosecution of hit-to-lead has been recently reviewed in some detail,[1,2] but the aim of this phase is to produce compounds and compound series with greater potency and selectivity, and with adequate PK properties to support *in vivo* studies.

20.3.1 Strategy—What Are We Trying To Do?

From a medicinal chemistry perspective, the aim of a hit-to lead program is not to identify a candidate drug. The main goals should be:

- To establish which, and how many, of the hit series have the required properties to sustain an LO program.
- To identify obvious weaknesses in the hit series and show the potential for these weaknesses to be resolved during the course of LO.
- To establish that a series doesn't have flat SAR and is not positively correlated to an undesirable property such as high lipophilicity, or low solubility.
- To increase potency and selectivity.

The hit-to lead phase will involve hit series expansion through a target design and synthesis cycle to answer specific questions of the series and will explore the wider chemical scope to define the series boundaries for activity. This is because the end product of the lead generation phase as a whole is to deliver one or more structural classes of compound that will be primed for property optimisation specifically focused on delivery of a candidate drug.

20.3.1.1 *Right Cascade And Compound Profiling—What Criteria Do You Set And How Do You Monitor Progress?*

Because of the need to explore SAR around each of the hit series and to explore diverse chemistry to determine SAR boundaries within each series, a screening cascade needs to be in place in the hit-to-lead phase that can deal with the desired throughput of compounds; lower than in the hit finding phase but sufficient to support a design-make-test cycle. Any screening cascade might also incorporate secondary assays to further explore activity, as well as additional selectivity and cross-species assays. A lead target profile can be defined in advance which recognises the need to balance multiple properties as an objective for the phase, and to clarify the objectives of the subsequent LO phase. The number of parameters that can be addressed is difficult to define but the LO transition can become a distant prospect if the sole positive quality in a lead series is potency.

A typical lead target profile for an oral project might appear as follows:[2]

- Potency:
  - $pIC_{50} > 7$ at isolated biochemical assay target.
  - $pIC_{50} > 6$ in functional cell assay.
  - Estimated/measured whole blood potency $pIC_{50} > 5.0$.
  - $pIC_{50} > 6$ in a biological effect assay measuring activity in a relevant disease cell.
- Selectivity:
  - > 30-fold selectivity over closely related targets
  - > 300-fold selectivity *vs.* chronic (24 h) cellular toxicity
- Species crossover:
  - < 10-fold variance *vs.* human
- DMPK and physicochemical properties:
  - Lead-like physicochemical properties: MW < 450, logD < 3. 5, solubility > 10 μM, H-bond donors ≤ 4, H-bond acceptors ≤ 8.
  - Intrinsic clearance < 25 μL/min/mg in oxidising *human* microsomes or < 10 μL/min/10^6 cells in *human* hepatocytes
  - Intrinsic clearance < 40 μL/min/mg in oxidising *rat* microsomes or < 15 μL/min/10^6 cells in *rat* hepatocytes.
  - Rat Cl < 50% of liver blood flow for neutrals and bases
  - Rat Cl < 15% of liver blood flow for acids
  - IV elimination half-life > 1 h (rat)
  - Oral bioavailability (rat) > 20%
  - Cyp inhibition: $IC_{50} > 10$ μM for 4 of 5 isoforms.
  - Chemical stability at pH 7.4 > 100h
- *In vivo* activity (PD model):
  - Pharmacological hypothesis supported by activity in an *in vivo* model.
- Safety:
  - Broad secondary pharmacology screening at 10 μM.
  - QT liability assessment: minimum 30-fold selectivity over hERG.
  - Preliminary genotoxicity evaluation: 2-strain AMES negative.
- Intellectual property:
  - An assessment has been made of the patent space around lead series and the potential to identify novel chemical equity in the area.

The last bullet is noteworthy in that, at this stage, the chemical structure and therefore the potential novelty of the final candidate drug is not known. This will be defined by the lead optimisation process. If exploration of chemical scope in the hit identification and hit-to-lead phases is able to provide some definition to the boundaries of SAR, then the project should be able to make a good assessment of whether the existing patent literature significantly restricts the scope of any LO program or not. The option to make an early IP filing exists, but this risks creating prior art complexity later in the discovery phase.

20.3.1.2   Decision Making Process

Any decision making process must satisfy many demands. A good small molecule lead generation project will deliver a biological target that has been validated and shown to have disease relevance, but it will also deliver a target that can be successfully modulated by small molecules.

As such, the ability to identify appropriate small molecule chemical equity is just one part of the lead generation exercise, but a critical one nonetheless since many lead generation programs fail for lack of such compounds. Assuming that small molecule series can be found, then lead target profiles such as the one described above provide a means to benchmark progress and control resource to achieve delivery.

20.3.1.3 Success Stories

Three examples will be given. One from a HTS-derived lead generation program leading to the discovery of the DPP4 inhibitor Januvia™ (Sitagliptin),[70] the second from a fragment-based lead generation program which delivered the BRAF inhibitor Zelboraf (Vemurafenib),[71] and the third from the hepatitis C (HCV) field where most hit finding strategies failed but a substrate modification approach proved successful, delivering Telaprevir, Boceprevir and Simeprevir, with many more inhibitors in clinical trials.[72–74] Each of these marketed drugs represents a success story in its own right, presenting learning which can only be adequately obtained by reading the full story behind their respective discoveries, but serve to illustrate how low potency hits can be transformed into potent oral, marketed drugs through judicious choice of hit and lead series.[70–74]

Merck's DPP4 inhibitor Januvia™ (Sitagliptin) provides a good example of a HTS-driven lead generation program which was able to capitalise on knowledge gained from a parallel DPP4 in-licensing program identifying key DPP8/9 selectivity requirements that could be built into the lead generation program. Three hit series were identified and each was investigated, but it was the piperazine series exemplified by (9) that ultimately delivered the highly selective inhibitor Sitagliptin with high selectivity versus DPP8/9, a > 600 fold increase in potency and significantly improved bioavailability (Figure 20.9).

Plexxicon's BRAF inhibitor Zelboraf (Vemurafenib, PLX4032) is also a good example because it has delivered the first marketed drug from a fragment-based lead generation program. The drug was approved in 2011 for the treatment of late-stage melanoma and the target in this case was a mutated form of BRAF present in human cancers. Plexxicon initiated a fragment based lead generation campaign using a 20 000-fragment library MW range 150–350 to screen at high concentration (200 µM) against a panel of kinase enzymes, and identified a generic 7-azaindole hinge binding fragment with several putative sites of substitution to optimise potency and selectivity (Figure 20.10). Compound libraries based on substitution of the 3, 4 and 5-positions of the starting fragment were then screened against oncogenic BRAF and, by generating cocrystal X-ray structural data, an optimisation program led to the discovery of PL4720 and subsequently the development compound PLX4032 based on improved pharmacokinetic properties.

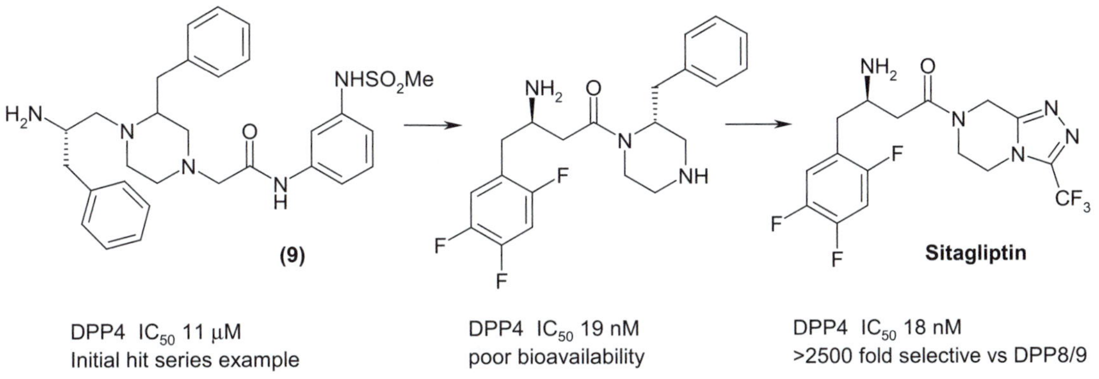

| DPP4 IC$_{50}$ 11 µM | DPP4 IC$_{50}$ 19 nM | DPP4 IC$_{50}$ 18 nM |
| Initial hit series example | poor bioavailability | >2500 fold selective vs DPP8/9 |

Figure 20.9 The discovery of Sitagliptin from a HTS derived hit series.[70]

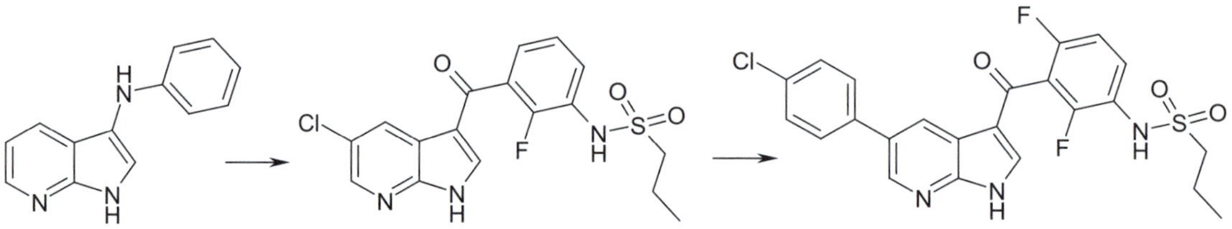

PLX4720 PLX4032

IC_{50} 100 μM IC_{50} 13 nM IC_{50} 31 nM

Figure 20.10 The discovery of Zelboraf (Vemurafenib, PLX4032) from a fragment-based lead.[71]

Asp-Asp-Ile-Val-Pro-Cys-OH K_i = 79 μM

Ac-Asp-Glu-Met-Glu-Glu-Cys-OH K_i = 0.6 μM

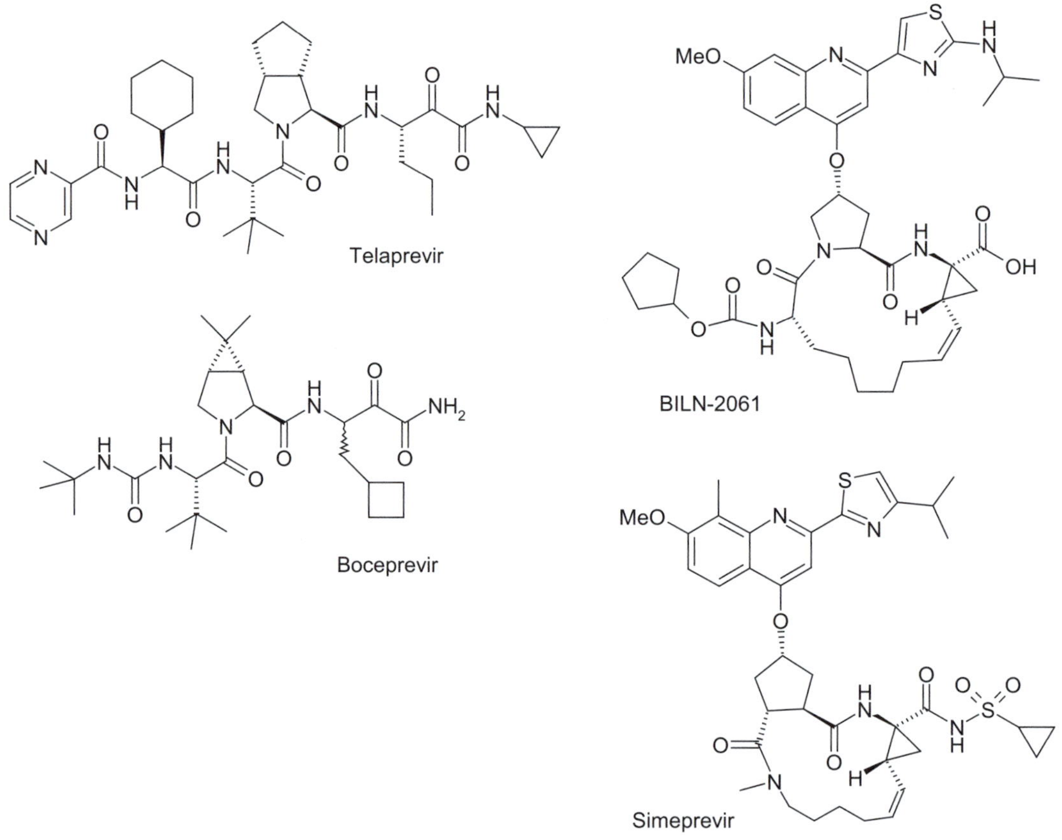

Figure 20.11 The discovery of hepatitis C virus (HCV) protease inhibitors from substrate peptides.[73,74]

Finally, the discovery of hepatitis C virus (HCV) protease inhibitors Telapravir, Boceprevir and Simeprevir provides an example where multiple HTS approaches, virtual screening and other approaches failed to deliver tractable hits, but where a substrate based approach proved ultimately successful (Figure 20.11). This protease together with its co-factor NS4A is essential for replication of the virus, and the first X-ray crystal structure revealed a shallow active site and hydrophobic substrate binding groove which interacts with its minimum decapeptide substrates, spanning P6–P4'. Protease-substrate interactions are more reminiscent of protein–protein interactions and are governed mostly by electrostatic interactions, thus explaining why conventional hit finding strategies failed.[72] Despite these challenges, there are now three approved drugs and 11 protease inhibitors in clinical trials.[73] All of these originate from the discovery that hexapeptide carboxylic acids, based on the substrate peptides, are inhibitors of the protease. These peptidic and hydrophilic leads have been transformed in several years of chemistry efforts and structure-based design to a series of reversible covalent inhibitors, such as Telapravir and Boceprevir, or to reversible acid based compounds such as BILN-2061 or Simeprevir.[73,74]

20.4 SUMMARY

Successful lead generation depends on many factors with no one technique or approach in isolation able to guarantee success. Indeed, for some targets, all approaches might fail and this is an inherent risk in any lead generation exercise. Nonetheless, success can be linked to judicious choice and combination of hit-finding approaches as well as to the process of evaluating hits and developing them into leads. Chemical space is inherently large with much of it far removed from drug-like space, and in any lead generation campaign many molecular classes will be identified that are unlikely to be drug candidates yet have the potential to waste large amounts of resource on the way to discovering this fact. Successful lead generation is as much about removing these blind alleys as it is about generating leads and should at least attempt to look beyond the criteria of a hit and a lead and consider the end goal of a marketable drug. Although this is a very difficult task and many problems cannot be identified at this early stage, some certainly can be and will help steer decision making at the lead generation stage.

Success rates for any hit finding approach are related to target class and this is certainly true of all of the approaches discussed in this chapter. It is also true that the much described R and D productivity challenge arising from decreasing numbers of successful drug approvals will inevitably increase the drive of lead generation for some pharma companies into target areas traditionally regarded as poorly tractable from a small molecule perspective, and this will impact overall lead generation success rates as well as the need to identify new hit-finding approaches. The area of protein–protein interactions is an obvious one that provides many highly disease-relevant targets but in which success rates remain low for small molecules. Nonetheless, the scientific challenge and potential rewards for success provide plenty of stimulus to succeed and approaches such as ultra-high throughput affinity selection of DNA-encoded libraries[51] and covalent inhibitor design[64] offer hope that adoption of new approaches might raise success rates in such areas.

HINTS AND TIPS

- The identification and establishment of biological assays with appropriate throughput, quality, reproducibility, cost and pharmacological relevance is of critical importance to any target-based lead generation campaign.
- Success in lead generation can be linked to judicious choice and combination of hit-finding approaches as well as to the process of evaluating hits and developing them into leads.
- Efficient separation of true actives from compounds working by unwanted mechanisms *e.g.* interference with the assay detection systems is crucial to the hit finding approach.

- Lead-like simplicity in terms of lower molecular weight and lipophilicity has emerged as a paradigm to facilitate optimization of screening hits.
- Pressure to succeed with poorly tractable targets needs to be balanced against potential reward and the overall balance of risk of an organization's lead generation portfolio.

KEY REFERENCES

Lead Generation Approaches in Drug Discovery, ed. Z. Rankovic and J. Morphy, Wiley & Sons, Inc., Hoboken, New Jersey, 2010.

A. M. Davis, D. J. Keeling, J. Steele, N. P. Tomkinson and A. C. Tinker, *Curr. Top. Med. Chem.*, 2005, **5**, 421–439.

E. Perola, *J. Med. Chem.*, 2010, **53**, 2986–2997.

REFERENCES

1. *Lead Generation Approaches in Drug Discovery*, ed. Z. Rankovic and J. Morphy, Wiley & Sons, Inc., Hoboken, New Jersey, 2010.

2. A. M. Davis, D. J. Keeling, J. Steele, N. P. Tomkinson and A. C. Tinker, *Curr. Top. Med. Chem.*, 2005, **5**, 421–439.

3. S. J. Teague, A. M. Davis, P. D. Leeson and T. I. Oprea, *Angew. Chem. Int. Ed.*, 1999, **38**(24), 3743–3778.

4. M. M. Hann, A. R. Leach and G. Harper, *J. Chem. Inf. Comput. Sci.*, 2001, **41**, 856–64.

5. T. I. Oprea, A. M. Davis, S. J. Teague and P. D. Leeson, *J. Chem. Inf. Comput. Sci.*, 2001, **41**, 1308–1315.

6. E. Perola, *J. Med. Chem.*, 2010, **53**, 2986–2997.

7. M. D. Schultz, *Bioorg. Med. Chem. Lett.*, 2013, **23**(21), 5980–5991.

8. G. M. Cragg and D. J. Newman, *Biochim. Biophys. Acta.*, 2013, **1830**(6), 3670–3695.

9. L. Mannocci, M. Leimbacher, M. Wichert, J. Scheuermann and D. Neri, *Chem. Commun.*, 2011, **47**, 12747–12753.

10. M. A. Clark, *Curr. Opin. Chem. Biol.*, 2010, **14**(3), 396–403.

11. H. H. Shih, in *Development of antibody-based therapeutics*, ed. M. A. Tabrizi, G. G. Bornstein, S. L. Klakamp, Springer, New York, 2012, pp. 9–32.

12. Antibody lead generation, Chapters 6–13 in *Antibody engineering: methods and protocols (Methods in molecular biology)*, ed. B. K. C. Lo, Humana Press, Totowa, New Jersey, 2004, **248**, pp. 117–244.

13. G. Probst and Y.-Z. Xu, *Expert Opin. Ther. Patents*, 2012, **22**, 511–540.

14. H. J. Patel, S. Shanu Modi, G. Gabriela Chiosis and T. Taldone, *Expert Opin. Drug Discovery*, 2011, **6**, 559–587.

15. (a) J. C. Barrow, S. R. Stauffer, K. E. Rittle, P. L. Ngo, Z. Yang, H. G. Selnick, S. L. Graham, S. Munshi, G. B. McGaughey, M. K. Holloway, A. J. Simon, E. A. Price, S. Sankaranarayanan, D. Colussi, K. Tugusheva, M. T. Lai, A. S. Espeseth, M. Xu, Q. Huang, A. Wolfe, B. L. Pietrak, P. Zuck, D. A. Levorse, D. J. Hazuda and J. P. Vacca, *J. Med. Chem.*, 2008, **51**, 6259–6262; (b) S. J. Stachel, T. G. Steele, A. Petrocchi, S. J. Haugabook, G. McGaughey, M. K. Holloway, T. Allison, S. Munshi, P. Zuck, D. Colussi, K. Tugasheva, A. Wolfe, S. L. Graham and J. P. Vacca, *Bioorg. Med. Chem. Lett.*, 2012, **22**, 240–244.

16. C. A. Coburn, S. J. Stachel, Y.-M. Yue-Ming, Li, D. M. Rush, T. G. Steele, E. Chen-Dodson, M. K. Holloway, M. Xu, Q. Huang, M.-T. Lai, J. DiMuzio, M.-C. Crouthamel, X.-P. Shi,

V. Sardana, Z. Chen, S. Munshi, L. Kuo, G. M. Makara, D. A. Annis, P. K. Tadikonda, H. M. Nash, J. P. Vacca and T. Tong Wang, *J. Med. Chem.*, 2004, **47**, 6117–6119.

17. Y. Gravenfors, J. Viklund, J. Blid, T. Ginman, S. Karlström, J. Kihlström, K. Kolmodin, J. Lindström, S. von Berg, F. von Kieseritzky, C. Slivo, B.-M. Swahn, L.-L. Olsson, P. Johansson, S. Eketja, J. Fälting, F. Jeppsson, K. Strömberg, J. Janson and F. Rahm, *J. Med. Chem.*, 2012, **55**, 9297–9311.

18. M. Sgobba and G. Rastelli, *ChemMedChem*, 2009, **4**, 1399–1409.

19. D. Brown and G. Superti-Furga, *Drug Discovery Today*, 2003, **8**(23), 1067–1077.

20. A. L. Hopkins and C. R. Groom, *Nat. Rev. Drug. Disc.*, 2002, **1**, 727–730.

21. A. P. Russ and S. Lampel, *Drug Discovery Today*, 2005, **10**(23/24), 1607–1610.

22. P. Schmidtke and X. Barril, *J. Med. Chem.*, 2010, **53**, 5858–5867.

23. F. N. Edfeldt, R. H. A. Folmer and A. L. Breeze, *Drug Discovery Today*, 2011, **16**, 284–287.

24. R. Sink, S. Gobec, S. Pecar and A. Zega, *Curr. Med. Chem.*, 2010, **17**, 4231–4255.

25. A. Simeonov, A. Jadhav, C. J. Thomas, Y. Wang, R. Huang, N. T. Southall, P. Shinn, J. Smith, C. P. Austin, D. S. Auki and J. Inglese, *J. Med. Chem.*, 2008, **51**, 2363–2371.

26. F. Degorce, A. Card, S. Soh, E. Trinquet, G. P. Knapik and B. Xie, Curr, *Chem Genomics*, 2009, **3**, 22–32.

27. G. K. Smith, D. G. Barrett, K. Blackburn, M. Cory, W. S. Dallas, R. Davis, D. Hassler, R. McConnell, M. Moyer and K. Weaver, *Arch. Biochem. Biophys.*, 2002, **399**(2), 195–205.

28. P. A. Johnston, K. M. Soares, S. N. Shinde, C. A. Foster, T. Y. Shun, H. K. Takyi, P. Wipf and J. S. Lazo, *Assay Drug Dev. Technol.*, 2008, **6**(4), 505–518.

29. B. Y. Feng, A. Simeonov, A. Jadhav, K. Babaoglu, J. Inglese, B. K. Stoichet and C. P. Austin, *J. Med. Chem.*, 2007, **50**, 2385–2390.

30. S. R. LaPlante, R. Carson, J. Gillard, N. Aubry, R. Coulombe, S. Bordeleau, P. Bonneau, M. Little, J. O'Meara and P. L. Beaulieu, *J. Med. Chem.*, 2013, **56**, 5142–5150.

31. R. S. Ferreira, C. Bryant, K. K. H. Ang, J. H. McKerrow, B. K. Stoichet and A. R. Renslo, *J. Med. Chem.*, 2009, **52**(16), 5005–5008.

32. N. Thorne, D. S. Auld and J. Inglese, *Curr. Opin. Chem. Biol.*, 2010, **14**(3), 315–324.

33. D. S. Auld, N. Thorne, D. T. Nguyen and J. A. Inglese, *ACS Chem. Biol.*, 2008, **3**, 463–470.

34. L. H. Heitman, J. P. D. van Veldhoven, A. M. Zweemer, K. Ye, J. Brussee and A. P. Ijzerman, *J. Med. Chem.*, 2008, **51**, 4724–4729.

35. D. S. Auld, N. T. Southall, A. Jadhav, R. L. Johnson, D. J. Diller, A. Simeonov, C. P. Austin and J. Inglese, *J. Med. Chem.*, 2008, **51**, 2372–2386.

36. J. B. Baell and G. A. Holloway, *J. Med. Chem.*, 2010, **53**, 2719–2740.

37. T. Tomasic and L. P. Masic, *Expert Opin. Drug Discovery*, 2012, **7**(7), 549–560.

38. J. Che, F. J. King, B. Zhou and Y. Zhou, *J. Chem. Inf. Mod.*, 2012, **52**, 913–926.

39. R. F. Bruns and I. A. Watson, *J. Med. Chem.*, 2012, **55**, 9763–9772.

40. S. L. Bulfer, T. J. McQuade, M. J. Larsen and R. C. Trievel, *Anal. Biochem.*, 2011, **410**(1), 133–140.

41. J. C. Hermann, Y. Chen, C. Wartchow, J. Menke, L. Gao, S. L. Gleason, N.-L. Haynes, N. Scott, A. Petersen, S. Gabriel, B. Vu, K. M. George, A. Narayanan, S. H. Li, H. Qian, N. Beatini, L. Niu and Q.-F. Gan, *ACS Med. Chem. Lett.*, 2013, **4**, 197–200.

42. P. J. Hajduk and D. J. Burns, *High Throughput Screening*, 2002, **5**, 613–622.

43. J. R. Huth, R. Mendoza, E. T. Olejniczak, R. W. Johnson, D. A. Cothron, Y. Liu, C. G. Lerner, J. Chen and P. J. Hajduk, *J. Am. Chem. Soc.*, 2005, **127**, 217–224.

44. D. C. Swinney and J. Anthony, *Nat. Rev. Drug Discovery*, 2011, **10**, 507–519.

45. R. Maccaron, M. N. Banks, D. Bojanic, D. J. Burns, D. A. Cirovic, T. Garyantes, D. V. S. Green, R. P. Hertzberg, W. P. Janzen, J. W. Pasley, U. Schopfer and G. S. Sittampalam, *Nat. Rev. Drug Discovery*, 2011, **10**, 188–195.

46. J. W. Noah, *Int. J. High Throughput Screening*, 2010, **1**, 141–149.

47. J. G. Cumming, A. M. Davis, S. Muresan, M. Haeberlein and H. Chen, *Nat. Rev. Drug Discovery*, 2013, **12**, 948–962.

48. M. J. Valler and D. Green, *Drug Discovery Today*, 2000, **5**(7), 286–293.

49. K. H. Bleicher, H.-J. Böhm, K. Muller and A. I. Alanine, *Nat. Rev. Drug. Discovery*, 2003, **2**, 369–378.

50. A. M. Virshup, J. Contreras-Garcia, P. Wipf, W. Yang and D. N. Beratan, *J. Am. Chem. Soc.*, 2013, **135**(19), 7296–7303.

51. L. Mannocci, M. Leimbacher, M. Wichert, J. Scheuermann and D. Neri, *Chem. Comm.*, 2011, **47**, 12747–12753.

52. M. Baker, *Nat. Rev. Drug Discovery*, 2013, **12**, 5–7.

53. C. W. Murray, M. L. Verdonk and D. C. Rees, *Trends Pharm. Sci.*, 2012, **33**(5), 224–232.

54. B. J. Davis and D. A. Erlanson, *Bioorg. Med. Chem. Lett.*, 2013, **23**, 2844–2852.

55. Practical Fragments: http://practicalfragments. blogspot. se.

56. Z. Zhu, Z-Y. Sun, Y. Yuanzan, J. Voigt, C. Strickland, E. M. Smith, J. Cumming, L. Wang, J. Wong, Y.-S. Wang, D. F. Wyss, X. Chen, R. Kuvelkar, M. E. Kennedy, L. Favreau, E. Parker, B. A. McKittrick, A. Stamford, M. Czarniecki, W. Greenlee and J. C. Hunter, *J. Med. Chem.*, 2010, **53**, 951–965.

57. A. W. Stamford, J. D. Scott, S. W. Li, S. Babu, D. Tadesse, R. Hunter, Y. Wu, J. Misiaskzek, J. N. Cumming, E. J. Gilbert, C. Huang, B. A. McKittrick, L. Hong, T. Guo, Z. Zhu, C. Strickland, P. Orth, J. Voigt, M. E. Kennedy, X. Chen, R. Kuvelkar, R. Hodgson, L. A. Hyde, K. Cox, L. Favreau, E. Parker and W. Greenlee, *ACS Med. Chem. Lett.*, 2012, **3**, 897–902.

58. A. J. Woodhead, H. Angove, M. G. Carr, G. Chessari, M. Congreve, J. E. Coyle, J. Cosme, B. Graham, P. J. Day, R. Downham, L. Fazal, R. Feltell, E. Figueroa, M. Frederickson, J. Lewis, R. McMenamin, C. W. Murray, M. A. O'Brien, L. Parra, S. Patel, T. Phillips, D. C. Rees, S. Rich, D-M. Smith, G. Trewartha, M. Vinkovic, B. Williams and A. J.-A. Woolford, *J. Med. Chem.*, 2010, **53**(16), 5956–5969.

59. M. Congreve, R. Carr, C. Murray and H. Jhoti, *Drug Discovery Today*, 2003, **8**(19), 876–877.

60. C. A. Nicolaou, C. Kannas and E. Loizidou, *Mini Rev. Med. Chem.*, 2012, **12**, 979–987.

61. S. Kar and K. Roy, *Expert Opin. Drug Discovery*, 2013, **8**(3), 245–261.

62. C. A. Luckhurst, M. Ratcliffe, L. Stein, M. Furber, S. Botterell, D. Laughton, W. Tomlinson, R. Weaver, K. Chohan and A. Walding, *Bioorg. Med. Chem. Lett.*, 2011, **21**, 531–536.

63. H. Sun, G. Tawa and A. Wallqvist, *Drug Discovery Today*, 2012, **17**(7/8), 310–324.

64. L. Guterman, *Chem. Eng. News*, 2011, **89**(36), 19–26.

65. D. S. Johnson, E. Weerapana and B. F. Cravatt, *Future Med. Chem.*, 2010, **2**(6), 949–964.

66. D. S. Johnson, K. Ahn, S. Kesten, S. E. Lazerwith, Y. Song, M. Morris, L. Fay, T. Gregory, C. Stiff, J. B. Dunbar Jr, M. Liimata, D. Beidler, S. Smith, T. K. Nomanbhoy and B. F. Cravatt, *Bioorg. Med. Chem. Lett.*, 2010, **19**(10), 2865–2869.

67. A. J. Wilson, J. K. Kerns, J. F. Callahan and C. J. Moody, *J. Med. Chem.*, 2013, **56**(19), 7463–7476.

68. F. Cong, A. K. Cheung and S-M. A. Huang, *Ann. Rev. Pharm. Toxicol.*, 2012, **52**, 57–78.

69. J. Heitman, N. R. Movva and M. N. Hall, *Science*, 1991, **280**(5369), 1603–1607.

70. N. A. Thornberry and A. E. Weber, *Curr. Top. Med. Chem.*, 2007, **7**, 557–568.

71. J. Tsai, J. T. Lee, W. Wang, J. Zhang, H. Cho, S. Mamo, R. Bremer, S. Gillette, J. Kong, N. K. Haass, K. Sproesser, L. Li, K. S. M. Smalley, D. Fong, Y.-L. Zhu, A. Marimuthu, H. Nguyen, B. Lam, J. Liu, I. Cheung, J. Rice, Y. Suzuki, C. Luu, C. Settachatgul, R. Shellooe, J. Cantwell, S.-H. Kim, J. Schlessinger, K. Y. J. Zhang, B. L. West, B. Powell, G. Habets, C. Zhang, P. N. Ibrahim, P. Hirth, D. R. Artis, M. Herlyn and G. Bollag, *Proc. Nat. Acad. Sci. U. S. A*, 2008, **105**(8), 3041–3046.

72. F. Narjes, U. Koch and C. Steinkühler, *Expert Opin. Invest. Drugs*, 2003, **12**, 153–163.

73. M. P. Manns and T. von Hahn, *Nature Rev. Drug Discovery*, 2013, **12**, 595–610.

74. N. Goudreau and M. Llinàs-Brunet, *Expert Opin. Invest. Drugs*, 2005, **14**, 1129–1144.

Lead Optimisation: What You Should Know!

STEPHEN CONNOLLY*[a] AND SIMON E. WARD*[b]

[a] Scientific Leader, Medicinal Chemistry, Almirall R&D, 08980 Sant Feliu de Llobregat, Barcelona, Spain; [b] Professor of Medicinal Chemistry and Director of Translational Drug Discovery Group, University of Sussex, Brighton, BN1 9QJ, United Kingdom
*E-mail: steve.connolly@almirall.com; simon.ward@sussex.ac.uk

21.1 THE ROLE OF LEAD OPTIMISATION

The identification, clinical evaluation and ultimate marketing of a drug molecule is a long and complicated process, as outlined in this book, and one that to many looking from outside may seem to be so risky and convoluted that it is easy to see why so many drug discovery programs fail.[1,2] To help investors and governance groups measure success, and to allow appropriate functional expertise to be involved at the right time, the drug discovery process is normally split up in to smaller steps. The segment called lead optimisation is critical to the endeavour because this is the step that takes the output from lead identification, likely a number of series of molecules with a range of properties, into a single candidate drug molecule whose molecular properties are fixed. Value is created in the optimisation phase as development candidates represent tangible assets.

21.1.1 What Is Obtained From Lead Identification: Assessing the Series

The start of lead optimisation usually begins with one or more series of compounds with a range of properties that give some confidence that they could be optimised to the final molecule. Many of these properties are generic, but some are more specific, depending, for instance, on whether the program is aimed at oral or topical treatment, whether once-a-day treatment is required or a more frequent dosing regimen is acceptable, or whether CNS access is required or should be avoided *etc*. Each individual project will have a set of project-specific properties that will need to be optimised. Selectivity against closely related targets may be necessary, as well as generic selectivity against targets that cause common issues, *e.g.* hERG (human cardiac potassium channel) or phospholipidosis risk. Metabolic stability should be in line with the desired dosing regimen. If the target is intracellular this will place additional

The Handbook of Medicinal Chemistry: Principles and Practice
Edited by Andrew Davis and Simon E Ward
© The Royal Society of Chemistry 2015
Published by the Royal Society of Chemistry, www.rsc.org

Table 21.1 Typical example of an oral candidate profile.

Potency	Primary Binding IC_{50}	<5 nM
	Cell Potency IC_{50}	<20 nM
	Whole blood potency	<1 µM
Selectivity	Versus key related targets	>100-fold
	Versus key generic liabilities, *e.g.* hERG, phospholipidosis	>30 µM
	Versus large panel of diverse targets	$<50\%$ activity at 10 µM spot testing
Pharmacokinetics	Stability in Microsomes and Hepatocytes	Clint < 10 µl/min/mg (mics) and <3 µl/min/10^6 cells (heps)
	In vivo rat Pharmacokinetics	T1/2 >2 hrs and Bioavailability $>30\%$
	In vivo dog Pharmacokinetics	T1/2 >4 hrs and Bioavailability $>30\%$
	Predicted human Pharmacokinetics	Predicts T1/2 suitable for desired dosing regimen *e.g.* 12–25 hrs for once-a-day
Physical Chemistry	Solubility	High enough for *in vivo* dosing in safety studies *e.g.* >50 µM
	Stability	Stable in solution
Chemistry	Synthesis	Capable of being scaled-up
	Patents	Novel to secure patent, and patent filed
Toxicology	Selectivity	No strong activity in general screening panels
	Selectivity	No strong activity in key toxicology screens *e.g.* hERG, cell toxicity
	Reactivity	No activity in reactivity screens *e.g.* cyanide, glutathione trapping
	In vivo Toxicology	Good margin with maximal tolerated dose (MTD) at multiples of predicted therapeutic dose

demands on properties to ensure good cell penetration. If the compound acts on a receptor it may be required to be a full or a partial agonist, an antagonist or an allosteric modulator. If the compound is competitive with a natural ligand, a high concentration of the ligand may provide an additional challenge. Many such questions need to be posed, and answered before the journey begins. In addition, the series should be novel and patentable, and will require advantages over existing or competitor compounds. In short, the properties of a candidate molecule will need to fulfil a large range of demands, both specific and generic, and be tailored to the treatment pattern envisaged, or demanded by the disease to be treated. This collection of properties that the final single molecule should be aimed at is commonly termed the candidate profile. See Table 21.1 for an example candidate profile.

21.1.2 What Does Lead Optimisation Deliver to Development: Meeting the Candidate Profile

At the outset of lead optimisation the project should have a well-considered candidate drug profile that it will aim for. This is not a 'mindless checklist'; some demands that are not reached may be compensated by other properties that exceed the minimum standards set. For instance, some deficit in potency may be compensated for by an excellent half-life. The rationale for setting a candidate profile up-front is to ensure that the properties of the final molecule will provide a compound that is of sufficient potency and selectivity, and whose pharmacokinetics allow a reasonably low dose to highly interact with the target over the whole of the dosing period, without affecting undesired targets that will cause side-effects.

An ideal drug candidate will have sufficient:

- POTENCY, such that the minimum concentration (C_{min}) required to be maintained for efficacy is low (*e.g.* less than 50 nM).
- HALF-LIFE, such that the dose required to maintain C_{min} does not produce a very much higher maximal concentration (C_{max}) *e.g.* a $t_{1/2}$ of 24 h will give a $C_{max}:C_{min}$ ratio of 2
- SELECTIVITY, such that toxicity (undesired pharmacology) is not seen until very much higher plasma levels than C_{max} are achieved (*e.g.* levels > 10-fold above C_{max} to give a safety margin of > 10)

This is illustrated in Figure 21.1.

The minimum concentration required for efficacy (C_{min}) should be in line with that found in animal models (allowing for cross species potency differences), while the human phar-macokinetics should be predicted from that measured in at least two animal species, usually rat and dog (allowing for interspecies scaling). Throughout the project the profile of the candidate may change with additional information. It may become tougher if a competitor compound emerges with an excellent property profile, or easier if no once-a-day compound appears and twice-a-day dosing becomes acceptable. New criteria may be added such as regulators demanding better selectivity, or increased margins against toxicity. This target profile should be a living document that captures the set of properties against which the project is aimed. But equally, if well-crafted from the outset, the profile should not change so much that the ongoing optimisation is continually set towards new directions. A well thought through profile will act as a goal that the project team members will jointly focus on to deliver a suitable compound which meets those properties. Thus, the candidate profile becomes a key set of numbers to be attained.

It is also important to remember that, while the candidate profile may already contain quite a few demanding properties, when reaching the later stages of optimisation with a range of

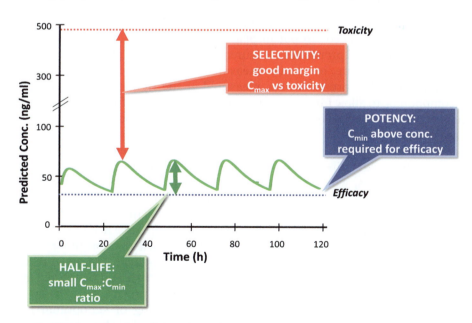

Figure 21.1 Idealised graph of candidate drug plasma profile. Graph demonstrates how both potency and half-life govern the margin to unwanted side-effects (with thanks to Dr John Dixon for generating the graphic).

compounds close to meeting the candidate profile, there will be a further extended range of properties that need to be satisfied. This extended range of criteria may not form part of the optimisation process, but need to be tested before the final molecule is proposed as a candidate, and taken into the expensive development phase. Thus, the profile of the candidate drug should encompass a set of properties, which if satisfied, will allow the clinical concept to be tested appropriately with a suitable molecule, by the envisaged administration.

21.1.3   The Process of Optimisation

Once the candidate profile is written the properties of the lead molecules identified should be measured against it to assess which properties need to be optimised. If the hit to lead phase has gone well the lead compounds should have many of the desired properties already built in, but normally some of the required properties will be present in some molecules, and other properties in other molecules. Usually it is not an easy job to combine all the required properties into a single molecule. The project will need to decide what the main issues remaining in the compounds are, and these will vary from project to project. Lead compounds do not usually have the potency required by the candidate, or if they do then it may be accompanied with an undesired property. For instance it is common for early compounds that have high potency to also have high lipophilicity, leading to poor pharmacokinetics and physicochemical characteristics, while other examples in the same series may have lower lipophilicity and good pharmacokinetics, but much lower potency. This potency–lipophilicity relationship is well described, and it is the ratio of these two properties, described as Lipophilic Efficiency (variously LLE[3] or LipE[4]) that often needs to be improved (see CCR8 case study below). Understanding the key deficiencies of the 'lead compounds' from the hit to lead campaign and the ability to change these undesired properties without adversely affecting any of the other good properties is key to making progress in the optimisation phase.

21.1.4   Screening Cascade

Once the candidate profile is written, and accepted, and the deficiencies of the lead compounds are understood, the project needs to put in place an appropriate screening plan so that compounds can have the required properties measured, and progressed towards the desired profile. Naturally not all properties are, or can be, measured at once, so the screening plan must be organised in order that key properties are measured at the right time, and on a reasonable number of compounds, to allow filtering of the best compounds for further testing, in particular into the more expensive or time consuming tests, such as *in vivo* potency in a disease model, or rat pharmacokinetics.

Note that many properties on the extended list of measurements required for potential candidate compounds might already have been performed on earlier compounds in the series, *i.e.* forerunner compounds, which did not meet the candidate profile but were progressed earlier in order to check that the series itself has no show-stopping activities, *i.e.* properties that will kill the whole series. However, it is important that the actual potential candidate compounds are tested, because if nominated as a candidate all properties need to be verified on the actual candidate drug molecule. It should also be emphasised that profiling a wider set of properties is not merely stamp collecting, but has evolved to evaluate and avoid risks of failure later in the development process. The extended testing cascades should be refined and tailored to the individual issues and challenges present in each unique project. A particular example of this pressure testing of series is early safety toxicology testing of suboptimal members of the series in order to avoid failures late in the screening cascade. This proactive early safety toxicology testing was prompted by expensive and resource consuming failures of molecules in late safety testing. The reported data[5,6] shows that such safety front-loading has reduced the attrition due to toxicity, and a greater proportion of late stage failures are now due to insufficient efficacy. Whilst any later stage attrition is disappointing, reduced attrition due to toxicity at least allows more compounds to be progressed to man to test the concept of inhibiting the target in the disease, whereas late stage toxicity failures are both costly, and mean the compounds fail prior to mechanism testing in man.

In putting together the screening cascade the project leader should ensure that all tests are in place to ask, and answer, key questions. Are the screens relevant to the disease, and route of administration? Is each assay in the right place in the cascade to enable fast filtering of compounds with the desired properties? Is the availability and throughput for each assay fit for the place it occupies in the screening cascade? Compounds will stall in the progression sequence if a key assay delivers results too slowly for the screening plan, or runs too infrequently. The screening cascade is critical to the project in providing the right data at the right time to allow the project to make critical decisions about compounds, to stop weaker compounds and progress stronger compounds. It also helps the project to identify the drivers of the key properties, and allows the project to run the iterative cycle rapidly. In short a well-planned screening cascade organises the discovery phase efficiently, and enables the project to rapidly identify a candidate drug with the right property profile. An example lead optimisation screening plan is shown in Figure 21.2.

21.1.5   Decision-Making in the Screening Cycle

Once a suitable screening plan is in place the project must decide how best to run the plan. This will depend on many factors. How many compounds will be produced? Which tests are available to the project, and how often do they run? Which project members are best placed to evaluate the test results, and choose which compounds to progress further? What is the critical data that will allow a compound to be progressed, and which value needs to be met? How these and other questions around the screening plan are answered will depend on company trends and to some degree on the project leader's own experience. However a project is set up there needs to be a way for the project to decide which compounds to progress to the next level, or to analyse results and decide on actions to progress the project. To avoid prolonged debate around each individual compound it may be useful to set pre-determined progression criteria for the main decision-making assays.

21.1.6   Progression Criteria

Throughout the optimisation phase the project should have the candidate profile in mind as the ultimate goal. However, at early stages, when the molecular properties are still some distance away, it may make sense to have looser progression criteria for moving a compound through the screening plan in order to understand all the molecule properties in a holistic way. For instance some compounds in a series with insufficient potency should be advanced into rodent pharmacokinetics with a view to testing if the PK of the series is a strength or weakness. As the optimisation progresses and the properties are advanced toward those desired of the candidate drug the progression criteria should be tightened, and ultimately coalesce with the candidate profile.

The project will need to decide which data to collect automatically on all molecules made *e.g.* primary potency, solubility, *in vitro* metabolism, measured log D. Some of this data can be key to analysis of series to determine what are the drivers of specific properties, and where limits *i.e.* progression criteria, need to be drawn. On the other hand many properties are now well predicted, and it may not make sense to collect all properties on all compounds, just because it is possible.

Equally, in these days of property prediction,[7] it makes no sense to make molecules with properties that are predicted to be bad. All compounds should be prepared in order to clearly

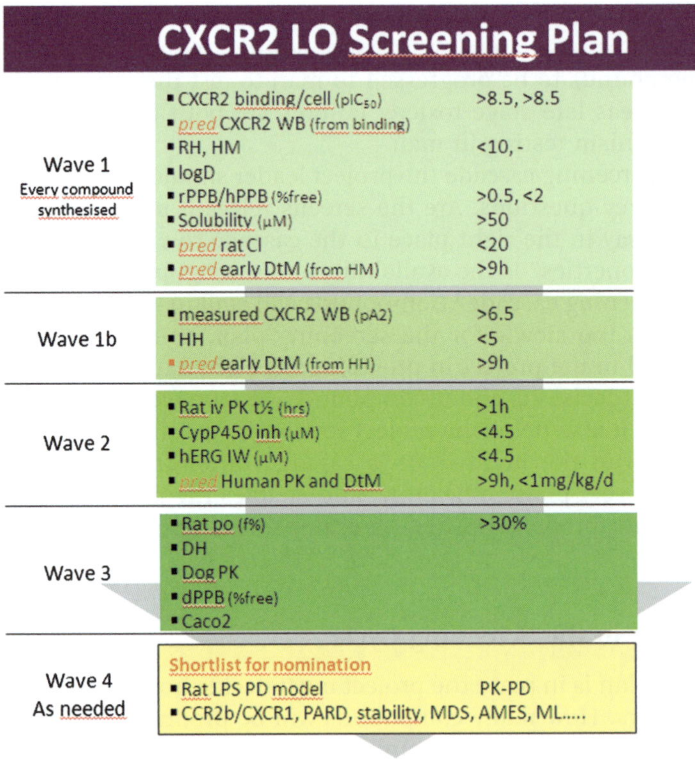

Figure 21.2   Example of a screening cascade as used in the CXCR2 project at AstraZeneca. Note, the table is shown in original form to indicate the complexity and flow of a typical screening plan, but in this case simplified with a variety of local acronyms such as pred: predicted, WB: whole blood potency, RH: *in vitro* rat hepatocyte metabolism, HM: *in vitro* human microsomal metabolism, rPPB: rat plasma protein binding, hPPB: human plasma protein binding, Cl: clearance, DTM: dose to man (predicted from *in vitro* clearance and assuming a certain volume of distribution estimated from earlier members in the series), HH: *in vitro* human hepatocyte metabolism, PK: pharmacokinetics, iv: intravenous, po: oral, DH: *in vitro* dog hepatocyte metabolism, PD: pharmacodynamics in an appropriate disease model, PARD: pharmaceutical property evaluation, MDS indicates extended target screening, ML: mouse lymphoma mutagenesis.

address a question. Small, well-chosen arrays of each sub-type are appropriate to explore property space, but the chemistry should be driven by molecules made to test a theory, and not just because one could easily synthesise 500 simple analogues.

21.1.7   Predicted Properties

As mentioned previously the properties of a compound are fixed, formulation aside. Indeed it should be recognised that the properties of a molecule are fixed from the moment of conception. Many properties are easily predicted (though not always correctly) and prior to synthesis the key properties should be predicted and the molecule made only if the predictions are in an acceptable region. It is the defining of acceptable that may be problematic. Some will argue against rigidity in selecting compounds for synthesis solely on predicted properties, however, judicious choice of what to make can avoid difficult issues later. If a compound predicts to be completely insoluble why would you make it?

21.1.8 The Use of Colour to Simplify Decision-Making

Once the project is in full throttle many decisions will need to be made about the range of compounds that enter, and travel through, the screening plan. This can be simplified by the use of colour in the results database. Looking at all the results for a compound, with the key results having their progression criteria coloured in a traffic light fashion, can make it easy to see when a compound passes all the relevant criteria and should be progressed. It is also very easy to see when a compound passes all criteria in the screening plan, and is a potential candidate molecule. However, caution must be applied: the use of colour is powerful, and should be used carefully with appropriate judgment, especially at the boundaries of criteria. Nonetheless, the use of colour can streamline decision-making remarkably. For examples of the use of colour in assessing the strength of compounds in the lead optimisation process see Figures 21.3 and 21.4. Understanding when a possible candidate has been found is a key step for an optimisation project because it may change the nature of the following synthesis and testing from a hunting phase, to a selection phase involving scale-up and a more detailed examination of a relatively small number of more focussed compounds.

21.2 LEAD OPTIMISATION—THE PRACTICALITIES

21.2.1 Quality of Start Point Is of Paramount Importance

In the same way as the most important decision in the drug discovery process is the selection of the correct target, the most important decision in the lead optimisation process is the very first decision, namely the selection of the chemical starting point itself. Whilst this is an obvious statement, there are a number of factors which drive initiation of medicinal chemistry activities, which can lead to optimisation starting on either poorly validated hits, or on molecules with a low probability of success of being able to be optimised into drug candidates. Given the resource that can be required to optimise from hit to lead to candidate and the momentum that can build up within a live project, it is vital that the resource is used effectively and applied along the correct trajectory.

Back | Layout ▼ | Columns... | Highlight ▼ | Number of Columns: 9 | ◄◄ | ►► ►►| | Print...

Compound Name	IT06371 B2 pEC50	IT06371 B2 Int Act	IT07288 B1 pEC50	IT07288 B1 Int Act	Ratio B2f:B1f	IT06359 GPT pEC50	IT06400 MEASURED GP Onset (min)	A CORRECTED Human Onset (min)
AZ10086610	6.4	0.77	5.2	0.8	15	6.7	3	2
IT06384 A1 bind pIC50	Ratio B2f:A1b	IT06275 B1 bind p IC50	Ratio B2f:B1b	IT06273 B2 bind pIC50	Ratio B2f:B2b	IT06276 D2 bind pIC50	Ratio B2f:D2b	Amount (mg)
<5.0	?25	<5.0	?25	5.3	14			70832.7
IT05855 CL (ml/min/kg)	IT05855 Vz (l/kg)	IT05855 t½ (h)	IT06295 GPT DRC ED80 (µg/kg)	IT06295 K+ effect (%)	IT06171 RHep (µl/min/1E6)	IT06172 HHep (µl/min/1E6)	IT05887 HMic (µl/min/mg)	IT03423 hERG iw IC50 (µM)
36	2.3	0.7	34	NV	12		<3	
IT06446 gp PB (% free)	IT06447 hum PB (% free)	IT06441 rat PB (% free)	IT06034 Cyp1A2 pIC50	IT06036 Cyp2C9 pIC50	IT06037 Cyp2C19 pIC50	IT06038 Cyp2D6 pIC50	IT06035 Cyp3A4 pIC50	IT06056 hERG b pIC50
85	77	65	<4.5	<4.5	<4.5	<4.5	<4.5	
desOH pred logD	IT06433 LogD	ACD LogD (7.4)	ACD LogP	IT06436 LogDmem	F.W.	IT06204 Stability pH7.4 Comments		
0.5	<-1.5	-1.80	0.02	<1	239.3			

Figure 21.3 Example of progression using colour to simplify decision-making—early chart. Example views of key data in a long-acting β-agonist project where the starting point was a low potency compound, with poor selectivity, and a high free fraction (related to a systemic side-effect in this case). The property weaknesses are clearly highlighted.

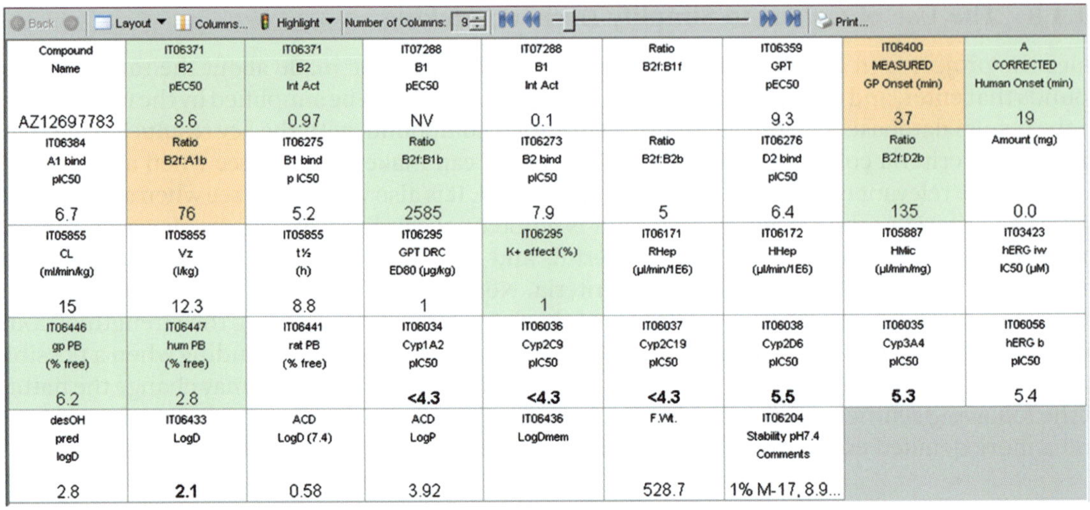

Back	Layout ▾	Columns...	Highlight ▾	Number of Columns: 9	⏮ ⏪ —⌐	⏩ ⏭	Print...	

Compound Name	IT06371 B2 pEC50	IT06371 B2 Int Act	IT07288 B1 pEC50	IT07288 B1 Int Act	Ratio B2f:B1f	IT06359 GPT pEC50	IT06400 MEASURED GP Onset (min)	A CORRECTED Human Onset (min)
AZ12697783	8.6	0.97	NV	0.1		9.3	37	19

	IT06384 A1 bind pIC50	Ratio B2f:A1b	IT06275 B1 bind pIC50	Ratio B2f:B1b	IT06273 B2 bind pIC50	Ratio B2f:B2b	IT06276 D2 bind pIC50	Ratio B2f:D2b	Amount (mg)
	6.7	76	5.2	2585	7.9	5	6.4	135	0.0

IT05855 CL (ml/min/kg)	IT05855 Vz (l/kg)	IT05855 t½ (h)	IT06295 GPT DRC ED80 (µg/kg)	IT06295 K+ effect (%)	IT06171 RHep (µl/min/1E6)	IT06172 HHep (µl/min/1E6)	IT05887 HMic (µl/min/mg)	IT03423 hERG iw IC50 (µM)
15	12.3	8.8	1	1				

IT06446 gp PB (% free)	IT06447 hum PB (% free)	IT06441 rat PB (% free)	IT06034 Cyp1A2 pIC50	IT06036 Cyp2C9 pIC50	IT06037 Cyp2C19 pIC50	IT06038 Cyp2D6 pIC50	IT06035 Cyp3A4 pIC50	IT06056 hERG b pIC50
6.2	2.8		<4.3	<4.3	<4.3	5.5	5.3	5.4

desOH pred logD	IT06433 LogD	ACD LogD (7.4)	ACD LogP	IT06436 LogDmem	F.W.	IT06204 Stability pH7.4 Comments		
2.8	2.1	0.58	3.92		528.7	1% M-17, 8.9...		

Figure 21.4   Example of progression using colour to simplify decision-making—later chart. After some cycles of compound synthesis and testing a compound with improved properties in all areas was found that subsequently became a candidate drug.

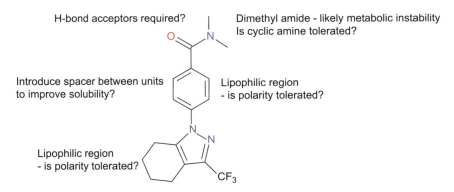

Figure 21.5   Examples of early hit to lead questions. Lead obtained for AMPA receptor positive allosteric modulator project. Initial questions posed leading to first wave of analogues prepared and/or purchased.

Assuming the lead generation is complete and the output appropriately validated, data should be generated to enable an appropriate selection of hit matter for a chemical start point. Typically this should include data which is generated computationally and in high throughput *in vitro* assays as well as examination by an experienced medicinal chemist. Additional early *in vitro* toxicology, wider selectivity and *in vivo* pharmacokinetics are informative at this stage, but their accessibility will be driven by the facilities available.

Following this analysis, two common scenarios emerge:

1. A specific go/no-go criterion can be defined for an individual series. Typically this could be to see if an undesirable structural feature can be removed *e.g.* removal of a carboxylic acid group to achieve CNS penetration or removal of a toxicophore such as a Michael acceptor.
2. A broad optimisation strategy is devised encompassing optimisation of several properties. The compounds required to be made should test specific hypotheses and should only be made if essential to answer that question (see Figure 21.5[8]).

21.2.2 Starting Lead Optimisation: Identifying the Weaknesses

As with all aspects of drug discovery, the principle of 'the more information the better' applies and often, at the start of lead optimisation, there is insufficient data within the chemical class to allow effective use of QSAR/similar tools described in earlier Chapters. Consequently, from any established lead, there are several initial avenues to gather additional data:

1. As above, the lead molecules should be characterised as extensively as resource/budget/facilities allow in order to maximise the chances of addressing key issues early on. Ideally this would include information on the kinetics and thermodynamics of the protein–ligand interaction.
2. If possible, structural data of the lead bound to the protein should be generated by X-ray crystallography. If unavailable, a computational model of the ligand bound to a homology model of the protein may provide some structural insight.
3. The structure–activity relationships for the lead should be developed by purchasing all related analogue structures (so-called 'SAR by catalogue').
4. Focussed small library sets (10–20 compounds) can be prepared around synthetically tractable areas of the molecule to generate additional SAR. Specifically, this would seek to generate a number of pair-wise comparisons to understand:
   - Which features of the molecule are required for binding to the target?
   - Which physical property (measured or calculated) is driving the parameter in need of improvement?
   - If the project is in an area of pre-existing research, then clearly literature mining and pharmacophore generation from published SAR can drive new analogue design.

The output of this initial investigation should allow the medicinal chemist to prepare plans for the classical iterative cycle of design–synthesise–evaluate/test–analyse allowing full lead optimisation to begin (see Figure 21.6).

21.2.3 Formulating a Strategy for Full Lead Optimisation

The principle challenge in lead optimisation is to recognise that the process requires optimisation of multiple parameters in parallel, and that a number of these parameters may be driven in opposing directions by the same underlying properties. The clearest example of this

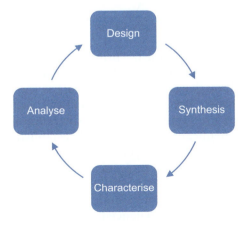

Figure 21.6   Design cycle.

is lipophilicity, for which it is apparent that affinity for the target can nearly always be achieved by increase in lipophilicity. However, it is equally apparent that increases in lipophilicity are associated with a reduction in overall drug-likeness and an increased chance of attrition.[5]

To this end, it is imperative to identify the underlying cause of the issue (*e.g.* low solubility, poor distribution *etc.*) that requires optimisation, such as specific structural features, functional groups or physicochemical properties. These observations will have been reinforced by the pairwise comparisons and focussed library sets described above. The less data that is available, the more important it is to ensure that the analogues proposed cover a wider range of physicochemical space to be able to derive future hypotheses. Subsequent analogues should be used to refine these hypotheses, and data should be analysed for simple correlations between properties (*e.g.* P450 3A4 inhibition and clog P; hERG affinity and pKa *etc.*) as well as multi-parametric principal component analyses to identify trends between weighted combinations of several parameters.

However, in addition to these general principles, there are some additional common issues that are encountered in lead optimisation, for which a selection are discussed below:

21.2.4 Strategies to Optimise Common Parameters in Early Lead Optimisation

21.2.4.1 *Target Affinity or Potency vs. Lipophilicity*

Potency is very commonly the principal driver in early lead optimisation. When pursued in isolation from other molecular properties it has been blamed for the observed trends in increasing molecular weight and/or lipophilicity leading to compound-related attrition.[3] There are many hurdles to overcome during the process of drug discovery and development, and it is clear that the balance of biological *vs.* physicochemical properties is the key to a successful lead optimisation strategy. Furthermore, given the unpredictable and challenging path to identify and then develop a drug candidate, it is imperative that medicinal chemists apply their skills to modulate the parameters and properties that lie within their control.

Recognising the inherent challenge of balancing multiple parameters, a number of tools and indices have been adopted to assist the medicinal chemist in early lead optimisation. The most simple of these are the various modes of binding or ligand efficiency, which is most easily considered as a ratio of binding affinity or potency against a descriptor of molecular size (Equation (21.1)):

$$\text{Ligand efficiency (LE)} = \frac{\Delta G}{n\,Heavy\,Atoms} \quad \text{or} \quad \frac{pXC_{50}}{n\,Heavy\,Atoms} \quad \text{or} \quad \frac{pXC_{50}}{MWt} \qquad (21.1)$$

Calculating the ligand efficiency alongside the primary screening potency for each molecule is a powerful means of ensuring the efficiency of the optimisation process.[9,19] Furthermore, analysis of ligand efficiencies through optimisation campaigns indicates that they generally decrease, reinforcing the need to start with the most efficient chemical start point. In the same way that the physical properties of a molecule are fixed at the point of conception of the idea, the overall ligand efficiency of a chemical series is generally established from the ligand efficiency of its parent hits. Indeed, data supports the premise that the maximum ligand efficiency that can be achieved decreases with increasing molecular weight. This can be rationalised by considering that an ideal fit of ligand to protein is less likely as molecular complexity increases. The optimisation from leads to both Gefitinib and Sunitinib in Table 21.2 illustrates the usual path, for which increasing molecular size was balanced with introduction of solubility and other properties leading to an overall decrease in efficiency of binding.

Table 21.2 Example ligand efficiencies from lead to candidate drug. LE calculated as pK_i/HAC, where HAC = number of heavy, *i.e.* non-hydrogen, atoms. Table adapted from content within An Analysis of the Binding Efficiencies of Drugs and Their Leads in Successful Drug Discovery Programs[9].

Lead			Drug	Drug name		
	LE	pK_i	Drug	Drug name	LE	pK_i
Leads optimised to drugs with decrease in LE						
	0.43	7.8		gefitinib (EGFR Tyr kinase)	0.25	7.6
	0.4	6.4		sunitinib (VEGF-R2 kinase)	0.26	7.1
Leads optimised to drugs with increase in LE						
	0.13	3.0		argatroban (thrombin)	0.22	7.5
	0.28	5.4		zanamivir (influenza neuraminidase)	0.44	9.7

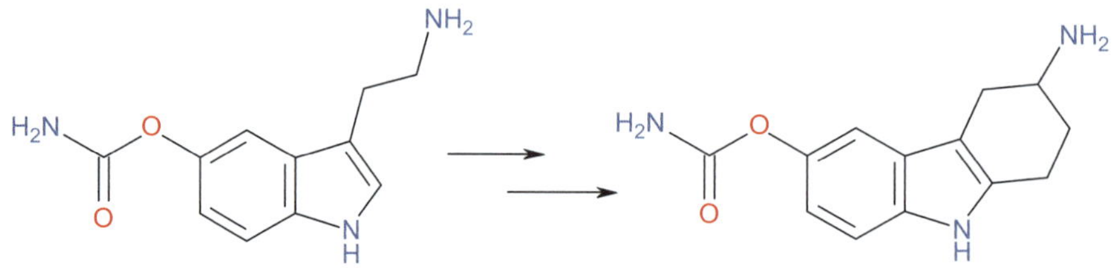

Figure 21.7 Conformational restriction during lead optimisation.

Notable exceptions to this can occur, however, a consideration of the first LE equation above should indicate that this requires increased ΔG for the same molecular size, *i.e.* increase in ΔH or decrease in ΔS. The former can be achieved by identifying additional polar or ionic protein–ligand interactions (as illustrated by argatroban and zanamivir in Table 21.2). However, it must be stressed that the design of additional polar interactions is challenging, even with the structural information provided by X-ray crystal structures of liganded proteins. A decrease in ΔS can be achieved by freezing the conformation of the ligand into its bioactive conformation to minimise entropic energy loss on binding, such as in Figure 21.7.

In addition to the observations above, it is apparent that on average the potency of a drug is superior to that of its parent lead, however the ligand efficiency remains constant. This means that that the overall molecular size increases to achieve greater potency. Earlier chapters have stressed the importance of controlling lipophilicity in drug design as it plays a major role in off-target and non-specific protein binding, phospholipidosis, solubility, toxicity and countless other parameters. Together, this requires the medicinal chemist to be particularly aware of the impact of the design strategy on lipophilicity, and as such, the lipophilic ligand efficiency index (Equation (21.2)) is helpful:

$$\text{Lipophilic ligand efficiency (LLE)} = \text{pXC}_{50} - \text{clog P (or clog D at pH 7.4)} \qquad (21.2)$$

The common and significant increase in lipophilicity over the course of an optimisation programme has been recognised as a probable contributor to current candidate attrition rates, although it is a parameter that can be controlled during the optimisation phase with scrutiny of both potency and lipophilicity of proposed and prepared molecules. The challenge to medicinal chemists is to design an optimisation path which improves potency and drug-like properties (potentially including an increase in molecular weight) but which does not overly increase lipophilicity. One way to do this is to optimise LLE. For instance, if a typical oral candidate has a potency of 1–10 nM, or in other words a pIC_{50} of 8–9, and a typical log D of 2 for acceptable physicochemical properties, then the LLE of such a compound should be in the region of 6–7. Since early lead compounds will have lower potency, and possibly higher lipophilicity, starting LLEs are usually much lower *e.g.* in the region of 2–3.

21.2.4.2   Case Study: Use of LLE to Avoid the Lipophilicity Trap in CCR8

CCR8 is a G-protein coupled receptor that is found on T-cells, and in skin, which responds to a single chemokine receptor, I-309. It has been implicated in T-cell homing to lung and skin when activated in disease situations, and has been a target for a number of pharmaceutical companies.

A high throughput screening (HTS) campaign at AstraZeneca gave a number of hit compounds, the most interesting of which is shown labelled as Compound 1 in Figure 21.8. In line with many HTS outputs compound **1** showed fairly weak potency in a radio-labelled I-309

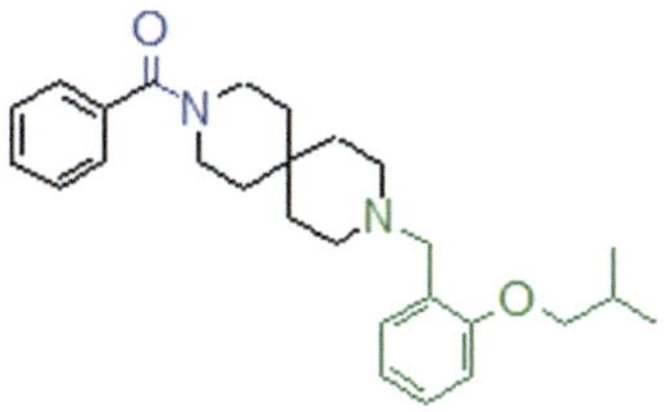

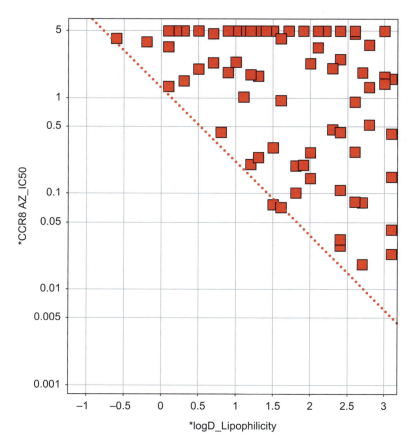

	Compound 1
CCR8 IC$_{50}$ (µM)	0.15
logD	3.1
Human Mic Clint	91
hERG IC$_{50}$ (µM)	0.5
Cyp2D6 IC$_{50}$ (µM)	5.5

Figure 21.8 Initial lead for CCR8 project.

Figure 21.9 Plot of lipophilicity *vs.* potency for example CCR8 project.

competition binding assay, and had high lipophilicity, leading to a number of issues, most notably high *in vitro* metabolic clearance and potent hERG activity.

As is normal in the lead optimisation phase a large number of simple analogues were rapidly synthesised and tested. This resulted in an initial conclusion that potency could be enhanced by increasing lipophilicity but this also made *in vitro* clearance and hERG inhibition worse. In contrast reducing lipophilicity did reduce clearance and activity against hERG, but also reduced potency against CCR8. This is a classic case of a *lipophilicity trap*.

This can be easily seen in Figure 21.9 which shows potency plotted against log D, where a diagonal *leading edge* is clearly outlined. Since the slope of this line is −1 it is clear that potency is directly correlated to lipophilicity. Addition of lipophilicity increases potency accordingly, and reduction of lipophilicity decreases potency. Compounds that are off the leading edge to the right have increased lipophilicity with reduced potency (compared to the leading edge compounds) and thus their additional lipophilicity is detracting from potency (*i.e.* making a worse interaction).

If lipophilicity needs to be reduced to a more reasonable level, *e.g.* log D ∼ 2, in order to control metabolism and unwanted activities that are related to lipophilicity, like hERG (and Cyp2D6 and others), it is clear that the potency–lipophilicity correlation needs to be broken.

How can a project break out of such a trap? There are a number of possible ways, including:

1. *Introducing polar substituents* that increase potency. Since the initial correlation shows adding polarity will reduce potency the newly added polar substituents must do more than just reducing lipophilicity, they must make positive binding interactions with the receptor.
2. *Introducing polar substituents that* do not affect potency. It is possible to add a polar group that does not interact with the target, but points outwards into solvent, thus altering the lipophilicity of the overall molecule without interacting with the target protein.
3. *Adding lipophilic substituents that add more potency than that expected by the initial correlation, i.e.* lipophilic groups that bind strongly in a lipophilic region of the receptor (note however that this strategy further increases the overall molecule lipophilicity, which ultimately still has to be reduced to a level required for good physicochemical properties).
4. *Add conformational rigidity* to a flexible part of the structure so that the rigid structure binds in the productive mode (*i.e.* increase potency without increasing lipophilicity).
5. *Removing lipophilic groups* that do not add to potency (if they exist).

The molecular journey taken by the CCR8 project[10,11] to find a candidate drug with high potency and lower lipophilicity, leading to excellent metabolic stability (and correspondingly excellent *in vivo* rat pharmacokinetics) and high margins to hERG is outlined in Table 21.3 to show one way in which the lipophilicity trap can be overcome, and to illustrate that LLE is an important parameter that can drive molecular properties into drug-like space.

The main features of the starting point **1** (see Table 21.3) are a spiro-piperidine core containing on one side a lipophilic benzamide group, and on the other side a basic tertiary amine group with an *N*-benzyl substituent, itself further substituted by a lipophilic *ortho*-alkoxy group. It can be surmised that the basic amine is key to binding, while the amide also likely makes an interaction with the receptor. The remaining interactions all appear to be lipophilic.

Initial investigations showed that adding more lipophilicity, as in the chlorobenzyl compound **2**, could improve potency slightly but only worsened other properties. Introduction of a 4-pyridyl amide group, as in **3**, at this position did not change potency much, but markedly reduced log D. Binding to hERG and *in vitro* clearance improved only two-fold. Nonetheless, it should be recognised that this change was *a significant jump forward* as it demonstrated that potency could be maintained in a much less lipophilic molecule. This is clearly seen in the improvement in LLE. Importantly, it also indicated that the pyridine lone pair was making a key interaction with the receptor, and strongly suggested that other polar groups could be acceptable in this region. The persistent poor hERG and *in vitro* clearance figures, despite the lower log D, may indicate that there were other, specific reasons for these poor properties that remained to be addressed.

Next an *ortho*-methoxyphenoxy group was introduced onto the *N*-benzyl moiety, as in compound **4**, which greatly improved potency, which was now into the nanomolar region, whilst keeping log D at the same place, so improving LLE further. Disappointingly hERG and *in vitro* clearance worsened appreciably, and this may be seen as a backwards step, however it did centre attention on the *ortho*-bis-ether moiety and possible interactions it might have with the receptor.

Table 21.3 Discovery path taken by the CCR8 project that led to a candidate drug.

No.	Structure	CCR8 IC$_{50}$ (μM)	log D	hERG IC$_{50}$ (μM)	Cl*int* Hum Mic (μl/min/mg)	LLE pIC$_{50}$-log D
1		0.15	3.1	0.5	91	3.7
2		0.13	3.7	0.5	108	3.2
3		0.10	1.8	1.1	60	5.2
4		0.012	1.8	0.05	89	6.1
5		0.012	1.2	21	73	6.7
6		0.004	0.8	8.8	17	7.6
7		0.003	1.0	>30	18	7.5
8		0.007	1.3	>30	<10	6.9

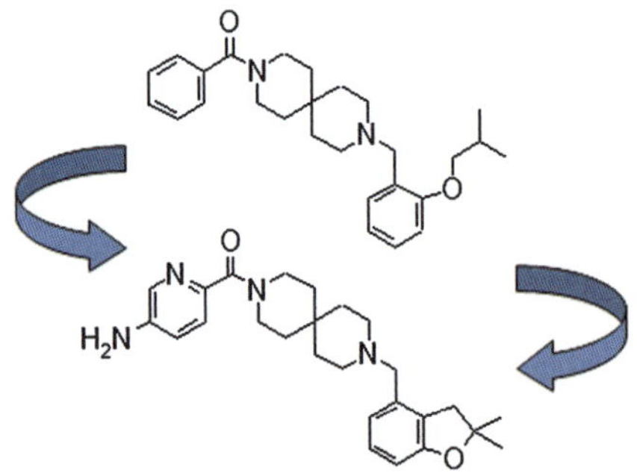

	Compound 1	Compound 8
CCR8 IC$_{50}$ (µM)	0.15	0.007
logD	3.1	1.3
Human Mic Clint	91	<10
hERG IC$_{50}$ (µM)	0.5	>30
Cyp2D6 IC$_{50}$ (µM)	5.5	>30
pIC$_{50}$	6.8	8.15
LLE (pIC$_{50}$-logD)	3.7	6.85

Figure 21.10   Summary of structural and property changes from lead compound to candidate drug.

Focusing on replacements for the *ortho*-methoxyphenoxy group we next discovered that a dihydrobenzofuran group, as in **5**, could replace the larger *N*-benzyl group, retaining potency whilst reducing log D further and vastly improving hERG, though clearance remained high. The gem-dimethyl substituents on the ring were crucial as removing these reduced potency 1000-fold (data not shown).

Rationalising that facile oxidation of the pyridine ring N-atom was responsible for the high *in vitro* metabolic clearance, amino groups were introduced to the pyridine ring to produce a range of aminopyridines, such as **6** and **7**, which were potent, with low log D, had excellent hERG selectivity and much lower clearance values. Further metabolism investigations showed that **6** and **7** were oxidised on the exposed ring benzyl CH$_2$-group, and transposing the furan ring O-atom, as in **8**, blocked benzylic oxidation and gave a compound that had high potency, excellent selectivity against hERG, and was very stable *in vitro*. Subsequently compound **8** was also shown to have excellent *in vivo* rat pharmacokinetics and was selected as a candidate drug to enter development.

If we examine what has happened in this process of lead optimisation to candidate drug (see Figure 21.10) we can see that we have improved potency around 20-fold, whilst reducing lipophilicity >60-fold giving an overall improvement of approximately 1200-fold.

This 1200-fold improvement in LLE was achieved by two main tactics. Firstly, on the benzamide side SAR investigations found that a lipophilic phenyl group could be replaced by a polar aminopyridine with no loss of activity, suggesting that the polar heterocycle makes positive interactions with the receptor. Secondly the flexible isobutoxy group on the *N*-benzyl amine side of the molecule can be tied up into a rigid bicycle improving potency and reducing log D. It is likely the ether oxygen makes a productive interaction with the receptor, and that the *gem*-dimethyl substitution on the dihydrobenzofuran causes the ring to pucker, reducing oxygen conjugation with the ring, and making it a better H-bond acceptor. In summary, it was possible in only a few steps outlined above to convert a highly lipophilic HTS hit with poor potency and poor properties into a much less lipophilic candidate drug with high potency, and excellent properties. The improvement in LLE at each point in the path allowed true progress to be observed, while specific compound issues still need to be tackled separately. However, it is worth noting that only the successful path is illustrated here; at each point there were many equally plausible routes to take and a much larger number of compounds than shown here were made in order to discover the right sequence.

21.2.5   Drop-Off in Cellular Potency

One commonly observed issue is that the activity seen in isolated protein assays is not maintained through to cellular systems. The shift to a considerably more complex biological system inevitably introduces many other variables, and one objective of the target validation was to ensure that the targeted protein–ligand interaction is sufficient to drive the required biological effect in a whole organism. Nonetheless, two common parameters can often contribute to lower cellular than isolated protein efficacy. Firstly, permeability, which is required for the lead molecule to access intracellular protein targets or penetrate pathogenic invading organisms/cells. Clearly a molecule which is a potent inhibitor of a target mitochondrial enzyme but which has low passive permeability will be unlikely to exert its required effect in cells. Furthermore, active efflux from the cell may result in even lower intra-cellular concentrations. Secondly, non-specific protein binding can restrict the levels of free drug which can bind to the target protein.

Fortunately these parameters are easily accessed by a range of low-to-high throughput techniques in early lead optimisation. In particular, two models of permeability are commonly employed:

1. *PAMPA* (parallel artificial membrane permeability) comprises an artificial hexadecane membrane, across which the test compound can equilibrate, providing a derived apparent permeability for each compound (P_{app})
2. *Caco-2* (colon adenocarcinoma cell line) comprises a confluent monolayer across which the test compound can equilibrate, and which importantly includes a number of transport proteins (additionally, transporter inhibitors, such as the P-gp inhibitor verapamil, can be included to assess transporter efflux). Compound permeation is measured by LC-MS/MS and P_{app} values are calculated. This model is used as a surrogate of intestinal absorption but clearly gives information additionally on general cellular permeability.

Whereas the PAMPA assay gives direct information about the intrinsic permeability of a compound the Caco-2 assay, being an intact human cell system, gives additional information about the possible efflux of the compound by inherent transporters. Flow from apical to basolateral (AB) is governed by permeability, while flow from basolateral to apical (BA) includes both permeability and transporter effects.

Since plasma binding controls the free concentration in the plasma compartment, and hence the free concentration in all other compartments (if these are at equilibrium, but they are not always) it might seem useful to increase the free concentration available to engage the target. However, competing processes, such as metabolism, are increased, and in general optimisation of plasma binding (alone) for efficacy is not fruitful.[12]

21.2.6 Selectivity

It is important early in the drug discovery screening cascade to identify those protein targets that are likely to present undesirable off-target activities. These can be proteins that are closely related to the target protein or other proteins that are identified during wider screening. For the latter, it is important to characterise lead exemplars from each chemical series through a wide panel of enzymes, transporters, receptors and ion channels to identify potential off-target liabilities which can then be included in the screening cascade. Clearly the specific design strategy required to achieve selectivity will depend on the profile being sought, however a few general principles apply:

- *Conformational restriction.* In general, the more degrees of freedom open to a molecule, the greater the likelihood it will be able to interact with a wider range of biological targets.
- *Lower clog P.* In general terms, there is a correlation of increasing clog P with increasing promiscuity, and in particular log P > 3 gives increased risk of off-target activity.[13]

- *Use the literature*. If an undesirable off-target activity is uncovered through cross-screening, this will inevitably be against a target for which there exists at least one of protein structure, homology model and/or developed SAR. All this information can be used to guide design of new molecules which will not bind to the undesired protein.

Additionally, a number of companies have shared data across this cross screening panel set to identify the core set of cross screening targets required.[14] This table is included as Appendix 21.1.

21.2.6.1 Selectivity vs. hERG

In addition to the general selectivity mentioned above, considerable attention is focussed on ensuring molecules do not interact with the hERG potassium ion channel and cause potentially lethal disruption to cardiac rhythms. All drug discovery projects assess hERG binding at an early stage and, whilst the interpretation and follow-up studies are discussed in more detail elsewhere in this book, it is useful to review the main strategies for avoiding hERG inhibition. Optimisation away from hERG blockade can be achieved by lowering log P, reducing the pKa of a basic nitrogen (or introducing a zwitterion) or by standard structural steric or electronic shape modification.[5,15] An analysis of the literature reported in 2008[16] identified that for molecules with clog P > 3, the most successful strategy was to reduce log P, with an average reduction of $0.8 \times$ hERG binding per log unit of clog P reduction. However, for molecules with clog P < 3, then structural modifications such as to reduce pKa, introduce steric shielding of basic nitrogen, disrupt potential π-stacking interactions or replace basic amines with alternative solubilising groups is more successful. It should be stressed that not all hERG blockers are lipophilic molecules containing basic tertiary amines, however analysis of a wide compound data set[17] identified that to have a 70% chance of achieving a hERG $IC_{50} > 10$ μM needs log D < 3.3 for a neutral compound and log D < 1.4 for a basic molecule.

Apart from its lethal effects on the heart, another difficulty with hERG inhibition is that historically selectivity against this channel *in vivo* has been overestimated by measuring selectivity *in vitro*. It is apparent that effects on heart rhythm *in vivo* can occur at low levels of hERG inhibition, around IC_{20}, rather than the oft-quoted IC_{50}. Thus, margins to hERG activity are required to be higher than to other off-target activities.

21.2.6.2 Selectivity vs. Cytochrome P450 Inhibition

Inhibition of P450 enzymes is typically screened initially against the five major metabolising isoforms. As the project nears nomination of pre-clinical candidates, it will also be important to understand a wider panel of inhibition potential against specific additional isoforms and subtypes as well as understanding which P450 enzymes are responsible for the metabolism of the parent drug (to predict future drug–drug interactions). Typically, molecules should have no or low inhibition of these enzymes; significant inhibition will lead to additional clinical drug–drug interactions during development and potentially restrict its label for administration. Each P450 enzyme has different, although overlapping, substrate requirements listed in Table 21.4 and advances in the structural determinations of these enzymes, coupled with the ability to use homology models to dock in leads, has allowed rational progress to design out a particular P450-driven interaction.

Given that the major role of Phase I metabolism is to introduce polar functionality to enable Phase II conjugation and elimination, it should not be surprising that the cytochrome P450 (CYP) enzymes have broad affinity for lipophilic molecules. Therefore, probably the more

Table 21.4 List of major P450 enzymes and their typical and example substrates.

CYP isoform	% of total CYP	Typical substrate	Example substrate
1A2	13%	Planar Weak base	Caffeine
2C9	>5%	Weak acid	Ibuprofen
2C19	>5%	Neutral or weak base 2–3 H-bonding groups	Imipramine
2D6	2%	H-bond acceptor 5 Å from metabolic site	Imipramine
3A4	30–70%	Promiscuous	Terfenadine

commonly successful is the strategy of lowering clog P and/or altering the overall electronic structure (dipole moment *etc.*) of the molecule. Again, analyses of physicochemical data suggest that for the CYP isoforms below clog P < 3 reduces the likelihood of significant P450 inhibition. Furthermore, many specific functional groups/substructures are known to interact with CYP isoforms, and so can be avoided to reduce risk of P450 inhibition or interaction. A list of these groups is included as Appendix 21.2.

21.2.7 Solubility

The most common complaint of screening biologists is that they have been given 'brick dust' to test, and poor solubility can also lead to false negatives in panel screening, and difficulties in increasing dose in safety studies. As such, optimisation of solubility is a key consideration (later chapters deal with possibilities in formulation and salt/polymorphic forms). Three main strategies are commonly employed:

1. *Introduction of polar or ionisable groups* such as in Figures 21.11 and 21.12. This list is intended to be illustrative rather than exhaustive, however the options for introducing neutral, polar or ionisable groups (with awareness of their pKa's) should be clear to any trained organic chemist.

   By example, the lead optimisation of the tyrosine kinase inhibitor Gefitinib included a late stage optimisation step in which a solubilising motif was appended to the quinazoline ring at a position known to be tolerant to substitution. However, caution should be expressed over introduction of both acidic and basic groups, which increase risk of hepatotoxicity, and both hERG inhibition and phospholipidosis, respectively.

2. *Reduction in clog P*—either by introduction of polar functionality as above, or by removal of lipophilic groups. As a useful mnemonic, solubility decreases approximately 10-fold when clog P increases by 1 unit. For over 45 000 compounds binned into solubility classes, high solubility was less likely once clog P > 3.[17]

3. *Disruption of crystal packing*. For example by introducing substituents to reduce planarity (such as Figure 21.13). As another useful mnemonic, solubility decreases by approximately 10× when melting point increases by 100 °C, or more specifically using the general equation (21.3) reported:[18]

$$\text{Log Sol (aq.)} = 0.5 - \log P - 0.01(m - 25) \tag{21.3}$$

where m = melting point in °C.

This means that a molecule with a melting point of 150°C needs to have log P < 3.25 to achieve aqueous solubility > 100 μM.

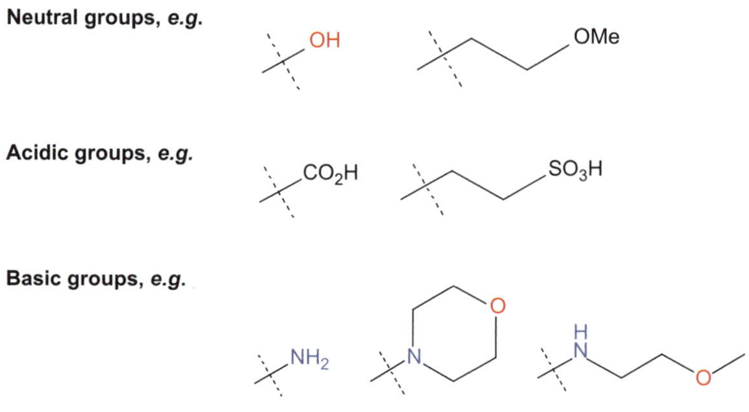

Neutral groups, *e.g.*

Acidic groups, *e.g.*

Basic groups, *e.g.*

Figure 21.11   Example structures of solubilising functionality.

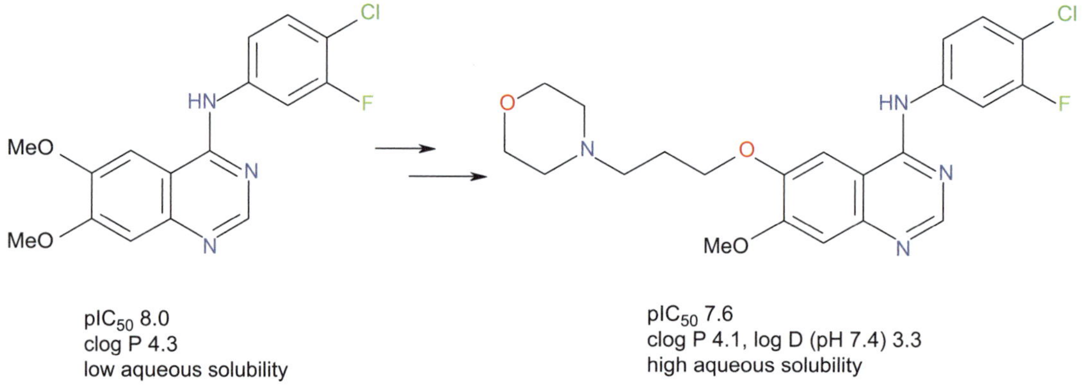

pIC$_{50}$ 8.0
clog P 4.3
low aqueous solubility

pIC$_{50}$ 7.6
clog P 4.1, log D (pH 7.4) 3.3
high aqueous solubility

Figure 21.12   Solubilising strategy to reach tyrosine kinase inhibitor Gefitinib.

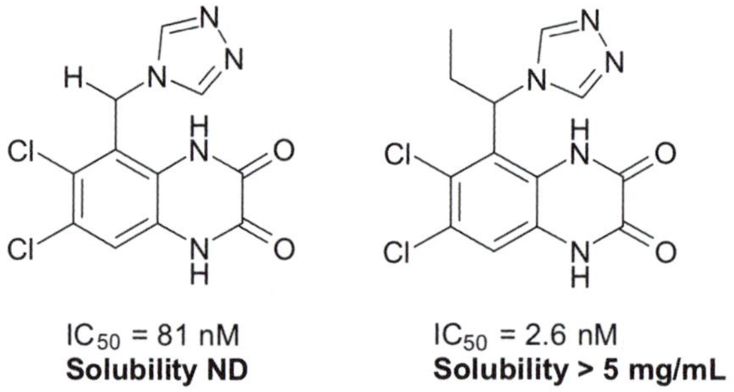

IC$_{50}$ = 81 nM
Solubility ND

IC$_{50}$ = 2.6 nM
Solubility > 5 mg/mL

Figure 21.13   Solubility increase achieved with disruption of crystal packing.

21.2.8   Metabolism

21.2.8.1   *Relevance of Test System*

A commonly encountered problem early in lead optimisation is the need to make a molecule more metabolically stable. Metabolic instability is identified both using *in vitro* and *in vivo* experiments and, as discussed in the screening cascade section, it is essential that the *in vitro* screening model is an appropriate surrogate for the *in vivo* profile. The example below illustrates this requirement. In the search for a clinical dopamine D3 receptor antagonist, the

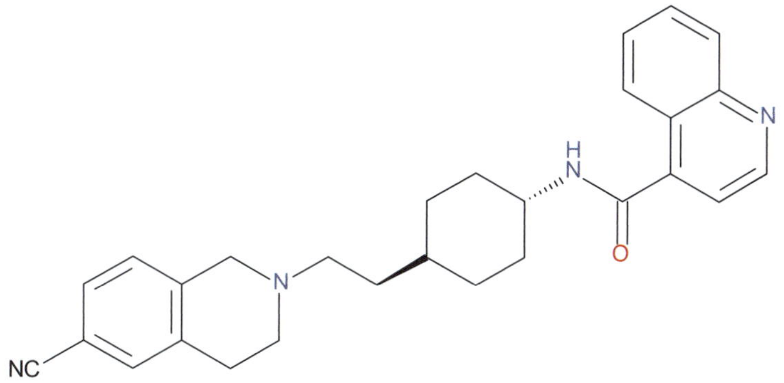

Figure 21.14 Structure of D3 antagonist.[20]

Table 21.5 Evaluation of D3 antagonist candidate drug in cross-species pharmacokinetics.

	Microsomes CLi mL/min/g	Liver homogenate CLi mL/min/kg	CLb (%LBF) mL/min/kg	Fpo %
Rat	1.4	<0.6	20 (85%)	35
Dog	<0.6	<0.6	14 (31%)	43
Monkey	1.7	9.9	58 (44%)	2
Human	1.3	45	ND	ND

%LBF is %liver blood flow.

molecule under investigation was found to have good metabolic stability in microsomal preparations leading to moderate turnover *in vivo* and moderate bioavailability in both rats and dogs (Figure 21.14). Human microsomal turnover was also low and so it was presumed that acceptable PK would be observed in man. However, a confirmatory primate PK study found the molecule to have high clearance and low bioavailability in contradiction to the microsomal stability data, creating uncertainty around the risk of progressing to clinical studies. Analysis of *in vitro* systems with more metabolic competency, *i.e.* hepatocytes, then fractions of liver homogenates, identified that the latter had a good correlation with all *in vivo* data hitherto generated (see Table 21.5). Performing the same protocol with human liver homogenates showed the molecule was rapidly turned over and so would have led to a wasted clinical progression. Microsomes are only capable of Phase I metabolism, largely oxidative, and thus predict *in vivo* clearance well only when the *in vivo* metabolism is largely oxidative. Hepatocytes can perform both Phase I and Phase II metabolism, the latter involving conjugative metabolism, and thus better predict the whole metabolic process *in vivo*.

It is worth noting that in those cases where even hepatocyte metabolism does not predict *in vivo* clearance well, that other clearance pathways, *e.g. via* renal or biliary excretion, may account for the increased total clearance observed.

21.2.8.2 Strategies to Improve Metabolic Stability

A significant proportion of early lead optimisation can be occupied by the need to improve the metabolic stability of the lead molecules. There are many approaches to achieve this, the most commonly used being:

1. *Reduction in lipophilicity*. Analysis of a set of molecules from Pfizer identified[16] that acceptably low *in vitro* metabolic turnover was more likely for molecules with

log D<3, which can be rationalised by assuming that polar compounds are less likely to undergo Phase I metabolism. However, renal clearance of unchanged parent molecule increases up to 50% for compounds that are very polar, *i.e.* with log D in the range 0 to −1.

2. *Use of classical or non-classical bioisosteres.* This can be a direct replacement of a substituent which is the site of metabolism, for example replacing a hydrogen with a fluorine, or methyl with chlorine to block an oxidative Phase I metabolism, or replacing a metabolically labile peptide linkage with a non-classical isostere (examples in Figure 21.15;[21] readers are referred to the papers on tetrazoles[22] and oxadiazoles[23] as isosteres).

3. *Steric shielding.* Recognising that most metabolic processes are driven by enzymes operating as bio-catalysts, we can learn from synthetic organic chemistry and appropriately

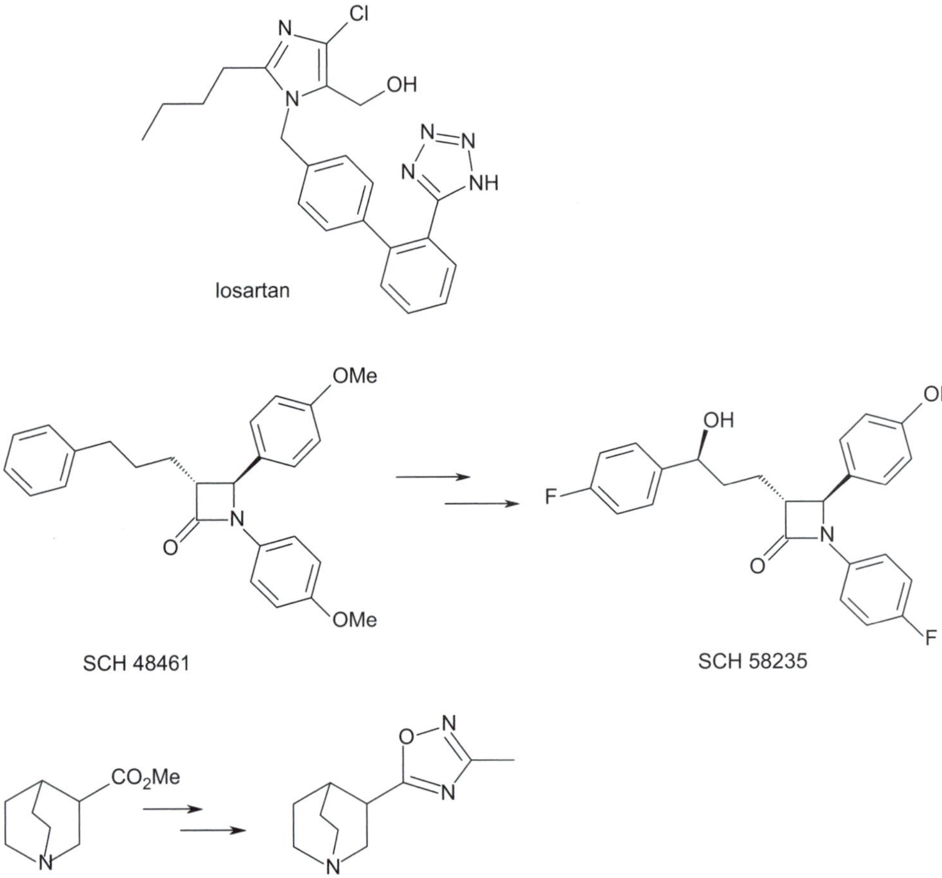

Figure 21.15 Examples of metabolic blocking strategies. (i) Replacement of a carboxylic acid with a tetrazole as achieved in the hypertension drug losartan; (ii) various strategies implemented during optimisation of SCH 48461 to SCH 58235 including both blocking oxidative metabolism by introduction of a fluorine substituent and also preparation of the oxidised metabolite as a more stable entity. The combination of changes led to a decrease in ED$_{50}$ from 2.2 mg/kg/day to 0.04 mg/kg/day driven by improved pharmacokinetic exposure; (iii) replacement of a labile ester group with an isosteric oxadiazole motif.

modify our lead (or starting material) by introducing steric hindrance to slow the rate of reaction.

4. *Use metabolite rather than parent.* Progressing through lead optimisation it is necessary to know not only how much metabolic turnover is occurring but also what metabolites are being formed. Later in the process we also need to understand which are the major metabolising enzymes (to be aware of both drug–drug interaction potential in the clinic and of P450 genetic polymorphisms that exist in the wider population) and whether preclinical efficacy and toxicology species produce the same metabolite profile as man (for appropriate toxicology evaluation). Commonly, simple metabolites also have affinity for the target protein, and sometimes this allows the metabolite to be progressed rather than the parent (see example in Figure 21.15), and also avoids the complexity of producing biologically active metabolites which would have to be extensively characterised and understood.

21.2.8.3 Absorption

As for many of the preceding parameters, an analysis of 232 drugs by Topliss indicated that to achieve oral bioavailability (F_{po}) of greater than 80%, essentially all drugs had log D of between -2 to $+3$.[24] Further studies have suggested that optimal absorption is achieved at the range log D 1–3, although specifically for rat F_{po}, a combination of PSA and rotatable bonds gave a stronger correlation.[25]

21.2.8.4 CNS Penetration

In addition to the barriers to achieve high levels of absorption there is an additional barrier involved with achieving high free concentrations of drug in the CNS. In particular, many recent analyses have reiterated the need to focus on *free* concentration of drug in the brain and the ratio of the unbound fraction in brain *vs.* plasma rather than the traditional approach of simply measuring total brain concentrations and deriving brain/blood ratios. Additionally the CSF drug concentration can sometimes be a surrogate for unbound drug concentrations in the brain, but these data can be misleading, particularly for drugs which are actively transported (P-gp at blood–CSF barrier pumps into CSF in contrast to P-gp in blood–brain barrier).

In broad terms, to achieve appropriate distribution into the CNS, the optimal range of lipophilicity has been analysed to lie within the range logP 2–3/log D 1–3, maintaining a polar surface area PSA < 75 Å. There are many exceptions to these approximations, but these criteria generally give good passive permeability as well as minimising possible interactions with efflux proteins.[26]

21.2.9 Toxicity and Phospholipidosis

Whilst it is clearly challenging to give specific guidance on how to avoid a variety of toxicities which can halt progression of a molecule or series there are nonetheless some general observations that can be made:

1. *Aim for low overall dose.* Targeting a clinical dose of less than 1 mg/kg reduces the impact of toxicological issues, in particular hepatotoxicity. Even for reactive metabolites a total dose of <10 mg has been reported to be acceptable.[27]
2. *Avoid functional groups/substructures known to cause specific toxicities.*[14,28]

3. *Reduce lipophilicity.* Analysis of Pfizer data indicated that logP < 3 and PSA > 75 Å gives the lowest risk of non-specific toxicities.[16]
4. *Be aware of risk of phospholipidosis.* The combination of a basic nitrogen and lipophilicity is often required either for mode of action, or designed to balance potency and solubility. However, there is an increased risk of phospholipidosis. In other words, a molecule with a basic centre of pKa 9 needs logP < 3 and, in general, singly positively charged molecules require logP < 2.75 for lower overall risk of phospholipidosis.[28]

21.2.10   Rules and Guidelines

To simplify the seemingly intractable difficulties involved in multi-parametric optimisation, a number of guidelines and mnemonics are commonly used. All however must be approached with caution and not followed slavishly.

21.2.10.1   Lipinski Rule of 5

The forerunner of all guidelines was the 'Rule of 5' proposed by Chris Lipinski,[29] which states that for a molecule to achieve good oral bioavailability it should possess at least three of the four following characteristics:

Hydrogen bond acceptors $\leq$ *10*
Hydrogen bond donors $\leq$ *5*
clogP $\leq$ *5*
Molecular weight $\leq$ *500*

   Many groups use variants of these criteria, with some focussing on graphical representations such as the Golden Triangle plots of molecular weight *vs.* clog D (Figure 21.16).[30]

21.2.10.2   Astex Rule of 3

Since then, a number of groups have advocated stricter guidelines, particularly in the early stages of lead optimisation. Another widely used set was the 'Rule of 3' proposed by Astex workers to deal specifically with the chemical space occupied by fragments, although this is commonly applied as a general rule for attractive properties for hits.[31] Adapted from the Lipinski guidelines above, these criteria require:

Hydrogen bond acceptors $\leq$ *3*
Hydrogen bond donors $\leq$ *3*
clogP $\leq$ *3*
Molecular weight $\leq$ *300*
TPSA $\leq$ *60 Å^2*

21.2.10.3   Pfizer MPO

To balance the seemingly impossible task of achieving good CNS penetration (clog P 2–3; TPSA $\leq$ 75 Å^2 in chemical space with low risk of unpredictable *in vivo* toxicity (clog P < 3; TPSA > 75 Å^2), Pfizer have published on a multi-parametric optimisation tool which uses clog P, clog D at pH 7.4, molecular weight, TPSA, HBD and pKa to create a means of prioritising molecules for synthesis with an improved chance for CNS penetration. An active version of this calculator is available.[32]

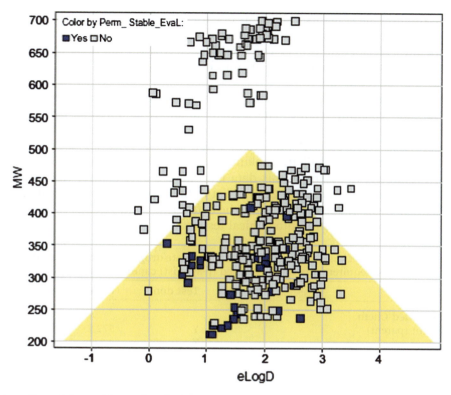

Figure 21.16   Plot of the Golden Triangle demonstrating combined *in vitro* permeability and clearance trends across MW and log D.

When aiming to optimise a series of lead molecules towards a candidate drug it is clear that many parameters require to be optimised in parallel, often in opposing directions. However, as described above, many of the parameters are dependent upon lipophilicity, which is a key property to control.

21.3   THE END GAME: CHOOSING THE CANDIDATE DRUG

21.3.1   Shortlisting

Once a project has identified a compound that meets the candidate profile it becomes a possibility for nomination as a candidate drug. Usually, once one compound is found that fits the screening plan view of a candidate molecule, the project will be able to rapidly synthesise a range of close analogues that all fit, or come close to this standard. These compounds will all have subtly different properties. How does the project choose between them? One way is to assemble these compounds into a shortlist, and test all these compounds in a stricter fashion to define the best overall compound. At this point there are two main things to check before committing to extensive animal safety studies.

Firstly, the key biochemical results (binding potency, cell potency, iso-enzyme potency, hERG potency, *etc.* should be repeated enough times that the average result is secure, and will not change, *i.e.* normally n = 6 repetitions will suffice if the results are not too variable. This is particularly important if selectivity criteria are set, *e.g.* 100-fold selectivity over a related receptor, since the selectivity number is the quotient of two biochemically derived numbers, and even small changes to the primary numbers on retesting can turn a selectivity of 200-fold to one of 50-fold overnight! This is not as unlikely as one might imagine since the screening plan will naturally select the compounds with the best selectivities on first testing. However, due to the inherent variability of biochemical screens the best selectivities can arise from the best combination of outliers. Beware! On retesting the numbers will 'regress to the mean' and ultimately the selectivity may be reduced.

Secondly, the project will need to test the shortlisted compounds in a variety of other assays, not on the normal screening plan, to check their suitability as potential development candidates. Dependant on project and company this extended list of tests can be quite long, including simple tests such as CYP induction, whole animal tests for CNS effects, detailed predictions of human predicted pharmacokinetics, safety toxicology studies and many others. One way for the project to assemble this data is to have a checklist for the data required at candidate drug nomination (see Table 21.6). A list of this type is also very useful to review and update regularly thereafter, and for preparing the data package that will go to governance to get approval for taking the compound into clinical testing.

Table 21.6 Example of a checklist for data required at candidate drug nomination. Note that this example contains the requirements for an inhaled (β-agonist) candidate drug.

Measurement	Test done	Data obtained
Phys Chem. and Med Chem.		
Molecular Weight (parent)	Y	572.2
log D	Y	2.1
ACDLOGP	Y	3.2
Solution Stability (24 h, pH 2,7,10)	Y	>264 h, >264 h, >264 h
Glutathione stability	Y	>100 h
Extended Stability (7 day)	ongoing	
Human PPB (%free) 5 μM	Y	4.2%
Guinea pig PPB (%free) 5 μM	Y	14.5%
Dog PPB (%free) 5 μM	Y	1.90%
Rat PPB (%free) 5 μM	Y	14.3%
Human CACO AB/BA (efflux)	Y	0.3/11 (efflux 37)
PAMPA	NA	
pKa	Y	7.7, 9.2
Solubility	Y	>64 μM
Patent filed	Y	filed 6th June
Scale up route for 150 g in place	Y	25 g synthesis completed
Radio-labelled compound synthesised	ongoing	
Elemental analysis (final form)	ongoing	
DMPK		
Rat PK	Y	Cl 30 Vss 3 $t_{1/2}$ 7.5 F0.3%
Dog PK	Y	Cl 11 Vss 0.9 $t_{1/2}$ 12 F0.5%
GP PK	NA	
Rat intratracheal PK	NA	
CYP inhibition	Y	all $5<6$ μM
CYP induction (human hep)	N	to check if necessary
CYP Time dependent inhibition	ongoing	
Dog Plasma Stability	Y	>1500 min
Human Plasma Stability	Y	>1500 min
Rat Plasma Stability	Y	>1500 min
Blood : plasma ratio (hu, dog, rat)	Y	0.8, 0.9, 0.8
Human Heps CL_{int}	Y	24
Human Mics CL_{int}	Y	124
Rat Heps CL_{int}	Y	57
Dog Heps CL_{int}	Y	21
Met ID Rat	Y	oxidation
Met ID dog	NA	
Met ID Human	Y	oxidation

Table 21.6 (*Continued*)

Measurement	Test done	Data obtained
Dose to man prediction (<15 µg/kg/day)	Y	1 µg/kg/day
Human half-life prediction	Y	17 h
Human Duration Prediction	Y	compatible with once-a-day
C_{max} arterial prediction	Y	0.1 nM
C_{max} venous prediction	Y	0.025 nM
Human lung S9 metabolism	Y	stable
Rat Bile duct cannulation (cold)	NA	
Dog Bile duct cannulation (cold)	NA	
Rat Bile duct cannulation (Radiolabel)	ongoing	await radio-labelled compound
Dog Bile duct cannulation (Radiolabel)	N	to check if required
PPB on radiolabel (hu, rat, dog)	ongoing	await radio-labelled compound
Met ID on Radiolabel (hu, Rat, dog)	ongoing	after 7 day repeat dosing
Rat oral dosing for F%	Y	F 0.5%
Dog oral dosing for F%	Y	F 0.5%

Biology

Measurement	Test done	Data obtained
β2 Receptor binding ($n = 6$) pIC_{50}	Y	8
β2 Functional agonism ($n = 6$) pIC_{50} and IA	Y	8.2, 0.99 ($n = 16$)
Selectivity binding/binding (α_1, β_1, D_2)	Y	479, >1260, 923
β_1 functional and IA (selectivity β_2/β_1)	Y	<5.8, 0.9 (>274)
α_1 functional and IA	Y	6.2, 0.35
D2 functional and IA	Y	NA, 0.0
MDS Pharma (AZ core)	Y	no hits at 1 µM
GP trachea potency pIC_{50}	Y	8.8 ($n = 5$)
Onset GP/mins	Y	20 ($n = 7$)
Onset predicted human (min)	Y	25
Efficacy intra tracheal (i.t) (GP bronchoconstriction)	Y	4 µg/kg
Efficacy inhaled (GP bronchoconstriction)	Y	0.03 mg/mL
Effect on plasma K^+	Y	1% at 10-fold ED_{80}
Duration GP i.t.	Y	24 h
Duration GP inhaled	Y	24 h
GP β_2 potency pIC_{50}	Y	7.4
Dog β_2 potency pIC_{50}	NA	
hERG Ionworks IC_{50}	Y	>33 µM
NaV1.5 IC_{50}	Y	>33 µM

Safety

Measurement	Test done	Data obtained
Inhaled irritancy testing	Y	clean
In silico predictions	Y	clean
AMES	Y	clean
Mouse Lymphoma	ongoing	
Phospholipidosis (IC_{50}, IA, toxic dose)	Y	45 µM, 0.08, >250 µM
GP Monophasic Action Potential	Y	clean
Reactive Metabolite Screening	NA	

Final form

Measurement	Test done	Data obtained
Final salt chosen	Y	fumarate
XRPD	Y	highly crystalline
DSC	Y	not hygroscopic
GVS	Y	single melting point at 185 °C
Colour	Y	white
Slurry to confirm Thermodynamic form	Y	done
Micronisation	Y	80% yield

PR&D

Measurement	Test done	Data obtained
Manufacturability	Y	no issue
Final Salt Selection	NA	at Campaign 3
Cost of Goods	Y	no issue
Scale-up safety review	Y	no issue

The list of properties required in Table 21.6 is meant to be illustrative, and each project would need to tailor the requirements for its own needs (project, route of administration, governance and institution requirements). However, use of such a checklist can help a project assemble a complex data package, check what data is already available, what is ongoing and what remains to be done.

21.3.2 Scale-Up and Safety Testing

Shortlisting is also an effective way of preparing the scale-up chemists for the synthesis of the larger quantities that are needed during the extensive shortlist testing, as well as alerting them to any need for process development before even larger scale synthesis. The scale-up may require several phases, preparing for example batches for multi-day rat and dog safety testing, 1 month GLP standard rat and dog safety toxicology testing and the large batch required for first time into man (FTiM) Phase 1 clinical testing.

All being well, the extended testing in the shortlisting phase will select the best compound from the shortlist (or at least leave at least one compound left to choose!) for nomination as a candidate drug. Hopefully the compound will come through the safety toxicity testing in rat and dog cleanly, or with sufficient margins to give confidence that dosing in man will be acceptable. The package of data from this phase, and from previous work, will be put together in a nomination document that can be presented to governance to seek approval to progress into a clinical testing phase. The scale-up and process development work has been done to allow the large-scale synthesis of a suitably formulated product to be in place. The clinical and commercial plans have been worked up. The lead optimisation phase of the project is finished. Isn't it?

21.3.3 Back-Up Approaches

The nature of drug discovery is an inherently difficult enterprise. Despite the best efforts of all in the industry to bring forward high quality candidate drugs into clinical testing, avoiding many of the known toxicities by early screening where it is possible, nevertheless, the industry in general still suffers from high late-stage attrition. While the fraction of late-stage attrition due to late-appearing toxicity, or sub-optimal pharmacokinetics has reduced in recent times, it still forms a part of clinical failure. Once a project has selected a candidate drug, it has been normal practice to search for a back-up compound that could replace the leading molecule if that happened to fail in the clinic for toxicity or PK reasons. Because of the long timescales for feedback from the first compound (CD1) in the clinic many back-up compounds (CD2, CD3, *etc.*) have been made without knowledge of any issue with the frontrunner. Therefore, chemists, in an effort to anticipate problems developing with CD1, searched for a back-up CD2 from a different structural class, rationalising that a different structure would be most likely to avoid any problems that might arise from the first compound. Thus projects often made two or more candidate drugs with different structures which progressed into clinical testing in a sequential manner, a so-called CD family. Recently this strategy has been less favoured for a number of reasons:

1. The structural differences between CD1 and CD2 were often not as large as might have been hoped for at the outset, and the second compound failed for the same reason as the first.
2. Even when the structure was from a different series the second compound still failed for the same reason, because the deficiency was due to a common substituent, or a common physicochemical parameter.

3. It is costly and inefficient to have a number of clinical candidates aimed at the same target all going through clinical testing, even if staggered. Especially if they do not actually serve to replace a failure of CD1.

It appears that the CD family strategy often fails. However, it is also clear that the first molecule progressed into the clinic often does not become the drug, and only after a number of candidate drugs have been through the cycle that the learning of these late-stage experiments can inform the project of how to prepare improved molecules that can go on to become successful drugs.

An alternative strategy might be to find CD1, then move resource off the project (to work on other targets). Only if, and when, CD1 has an issue, would resource then be reapplied to discover a new improved CD2 without the issue found in CD1. This alternate strategy saves resource, but may lose time (and dissipate experience).

It is clear that nowadays, to have a successful candidate drug, the organisation not only has to have the right molecule, but it should have the right animal data predictive of the disease state, the right safety profile, and the right clinical plan with the right biomarkers, aimed at delivering real benefit to the right patient segment, delivered at the right cost. All of this has to be in place for the project's proposal to get its candidate drug accepted to progress into the next phase, that of clinical testing, where the organisation can have confidence of producing a meaningful clinical effect in patients.

HINTS AND TIPS

- Multi-objective optimisation can be aided by some software tools like STARDROP (www.optibrium.com/stardrop/).
- Set a candidate drug profile to guide the optimisation process. It's a living document which can be evolved as new data emerges.
- A screening cascade should be established that can deliver a candidate drug matching the required drug profile.
- Using colour coding in results tables simplifies decision making for compound progression.
- Breaking the potency–lipophilicity "correlation" is a common challenge in most LO programs.
- The properties of a molecule are fixed as soon as you have thought of it! So if you can predict the properties, and the molecule is predicted to be poor...why make it?
- When trying to remove an off-target liability, the literature is a great source of inspiration.
- Targeting a low dose reduces the impact of toxicological issues, as it on average improves safety margins. Targeting a dose <1 mg/kg/day is typical for an oral project. A dose <10 mg total can mitigate concerns over reactive metabolites, as the body burden is considered acceptable.
- Lead optimisation should focus on achieving the maximum concentration of drug within the biophase of the target. Optimisation should focus on increasing solubility, membrane permeability and reducing clearance and not on plasma protein binding as free drug concentration changes are counterbalanced by commensurate changes in drug clearance.

APPENDIX 21.1

Table 1　Recommended targets to provide an early assessment of the potential hazard of a compound or chemical series. From the following Article: *Reducing safety-related drug attrition: the use of in vitro pharmacological profiling.*[33]

Targets (gene)	Hit rate[a]		Main organ class or system	Effects		Refs[b]
	Binding	Functional or enzymatic		Agonism or activation	Antagonism or inhibition	
G protein-coupled receptors						
Adenosine receptor A$_{2A}$ (*ADORA2A*)	High	Low (agonist)	CVS, CNS	Coronary vasodilation; ↓ in BP and reflex; ↑ in HR; ↓ in platelet aggregation and leukocyte activation; ↓ in locomotor activity; sleep induction	Potential for stimulation of platelet aggregation; ↑ in BP; nervousness (tremors, agitation); arousal; insomnia	57
α$_{1A}$-adrenergic receptor (*ADRA1A*)	High	Low (agonist); high (antagonist)	CVS, GI, CNS	Smooth muscle contraction; ↑ in BP; cardiac positive ionotropy; potential for arrhythmia; mydriasis; ↓ in insulin release	↓ in smooth muscle tone; orthostatic hypotension and ↑ in HR; dizziness; impact on various aspects of sexual function	58
α$_{2A}$-adrenergic receptor (*ADRA2A*)	High	Low (agonist); medium (antagonist)	CVS, CNS	↓ in noradrenaline release and sympathetic neurotransmission; ↓ in BP; ↓ in HR; mydriasis; sedation	↑ in GI motility; ↑ in insulin secretion	59
β$_1$-adrenergic receptor (*ADRB1*)	Medium	NA	CVS, GI	↑ in HR; ↑ in cardiac contractility; electrolyte disturbances; ↑ in renin release; relaxation of colon and oesophagus; lipolysis	↓ in BP; ↓ in HR; ↓ in CO	60
β$_2$-adrenergic receptor (*ADRB2*)[‡]	High	Medium (agonist); medium (antagonist)	Pulmonary, CVS	↑ in HR; bronchodilation; peripheral vasodilation and skeletal muscle tremor; ↑ in glycogenolysis and glucagon release	↓ in BP	61
Cannabinoid receptor CB$_1$ (*CNR1*)	Medium/ high	Medium (antagonist)	CNS	Euphoria and dysphoria; anxiety; memory impairment and poor concentration; analgesia; hypothermia	↑ in weight loss; emesis; depression	62
Cannabinoid receptor CB$_2$ (*CNR2*)	Medium	Medium (agonist)	Immune	Insufficient information	↑ in inflammation; ↓ in bone mass	63

Target			System			Ref.
Cholecystokinin A receptor (*CCKAR*)	Low/medium	NA	GI	↓ in food intake; gallbladder contraction; pancreatic enzyme secretion; ↑ in GI motility; activation of dopamine-mediated behaviour	↑ in development of gallstones	64
Dopamine receptor D$_1$ (*DRD1*)‡	Medium/high	Medium (antagonist)	CVS, CNS	Vascular relaxation; ↓ in BP; headaches; dizziness; nausea; natriuresis; abuse potential	Dyskinesia; parkinsonian symptoms (tremors); anti-emetic effects; depression; anxiety; suicidal intent	65
Dopamine receptor D$_2$ (*DRD2*)‡	Medium/high	Medium/high (agonist); medium (antagonist)	CVS, CNS, endocrine	↓ in HR; syncope; hallucinations; confusion; drowsiness; ↑ in sodium excretion; emesis; ↓ in pituitary hormone secretions	Orthostatic hypotension; drowsiness; ↑ in GI motility	66
Endothelin receptor A (*EDNRA*)	Low	NA	CVS, development	↑ in BP; aldosterone secretion; osteoblast proliferation	Teratogenicity	67
Histamine H$_1$ receptor (*HRH1*)‡	High	Very high (antagonist)	CVS, immune	↓ in BP; allergic responses of flare, flush and wheal; bronchoconstriction	Sedation; ↓ in allergic responses; ↑ in body weight	68
Histamine H$_2$ receptor (*HRH2*)	High	Low (agonist)	GI, CVS	↑ in gastric acid secretion; emesis; positive inotropy	↓ in gastric acid secretion	69
δ-type opioid receptor (*OPRD1*)	Medium/high	NA	CNS, CVS	Analgesia; dysphoria; psychomimetic effects; cardiovascular effects; convulsion	↑ in BP; ↑ in cardiac contractility	70
κ-type opioid receptor (*OPRK1*)‡	High	Medium (agonist and antagonist)	GI, CNS, CVS	↓ in GI motility; ↑ in urinary output; sedation and dysphoria; confusion; dizziness; ↓ in locomotion; tachycardia	Insufficient information	71
μ-type opioid receptor (*OPRM1*)‡	High	Medium (agonist and antagonist)	CNS, GI, CVS	Sedation; ↓ in GI motility; pupil constriction; abuse liability; respiratory depression; miosis; hypothermia	↑ in GI motility; dyspepsia; flatulence	72
Muscarinic acetylcholine receptor M$_1$ (*CHRM1*)	High	Low (agonist); high (antagonist)	CNS, GI, CVS	Proconvulsant; ↑ in gastric acid secretion; hypertension; tachycardia; hyperthermia	↓ in cognitive function; ↓ in gastric acid secretion; blurred vision	73
Muscarinic acetylcholine receptor M$_2$ (*CHRM2*)‡	High	Low (agonist); medium (antagonist)	CVS	↓ in HR; reflex; ↑ in BP; negative chronotropy and inotropy; ↓ in cardiac conduction (PR interval); ↓ in cardiac action potential duration	Tachycardia; bronchoconstriction; tremors	74

Table 1 (Continued)

Targets (gene)	Hit rate[a] Binding	Hit rate[a] Functional or enzymatic	Main organ class or system	Effects Agonism or activation	Effects Antagonism or inhibition	Refs[b]
Muscarinic acetylcholine receptor M$_3$ (**CHRM3**)	High	NA	GI, pulmonary	Bronchoconstriction; ↑ in salivation; GI and urinary smooth muscle constriction	Constipation; blurred vision; pupil dilation; dry mouth	75
5-HT$_{1A}$ (**HTR1A**)	Medium/high	Low (agonist); medium (antagonist)	CNS, endocrine	↓ in body temperature; reduced REM sleep; ↑ in ACTH; cortisol and growth hormone secretion	Potentially anxiogenic	76
5-HT$_{1B}$ (**HTR1B**)	High	High (agonist); medium (antagonist)	CVS, CNS	Cerebral and coronary artery vasoconstriction; ↑ in BP	↑ in aggression	77
5-HT$_{2A}$ (**HTR2A**)‡	Very high	Low/medium (agonist); medium/high (antagonist)	CVS, CNS	Smooth muscle contraction; platelet aggregation; potential memory impairments; hallucinations; schizophrenia; serotonin syndrome	Insufficient information	78
5-HT$_{2B}$ (**HTR2B**)	High/very high	Low (agonist); high (antagonist)	CVS, pulmonary, development	Potential cardiac valvulopathy; pulmonary hypertension	Possible cardiac effects, especially during embryonic development	79
Vasopressin V$_{1A}$ receptor (**AVPR1A**)	Medium	High	Renal, CVS	Water retention in body; ↑ in BP; ↓ in HR; myocardial fibrosis; cardiac hypertrophy; hyponatraemia	Insufficient information	80
Ion channels						
Acetylcholine receptor subunit α1 or α4 (**CHRNA1** or **CHRNA4**)‡	Medium/high	Low (opener); very high (blocker)	CNS, CVS, GI, pulmonary	Paralysis; analgesia; ↑ in HR; palpitations; nausea; abuse potential	Muscle relaxation; constipation; apnoea; ↓ in BP; ↓ in HR	81
Voltage-gated calcium channel subunit α Cav1.2 (**CACNA1C**)‡	NA	Medium/high (blocker)	CVS	Insufficient information	Vascular relaxation; ↓ in BP; ↓ in PR interval; possible shortening of QT interval of ECG	82

Target			Organ system			Ref
GABA_A receptor α1 (rat cortex) BZD site (**GABRA1**)‡	Medium	NA	CNS	Anxiolysis; muscle relaxation; ataxia; anticonvulsant; abuse potential; sedation; dizziness; depression; anterograde amnesia	Seizure (when used as a BZD antidote)	83
Potassium voltage-gated channel subfamily H member 2; hERG (**KCNH2**)	High	High	CVS	Insufficient information	Prolongation of QT interval of ECG	84
Potassium voltage-gated channel KQT-like member 1 (**KCNQ1**) and minimal potassium channel MinK (**KCNE1**)	NA	Low	CVS	Atrial fibrillation	Long QT syndrome; potential hearing impairment, deafness and GI symptoms	85
NMDA receptor subunit NR1 (**GRIN1**)‡	Low/medium	Medium (blocker)	CNS	Psychosis (schizophrenia-like); hallucinations; delirium and disoriented behaviour; seizures; neurotoxicity	Insufficient information	86
5-HT3 (**HTR3A**)‡	Medium	Very high	GI, endocrine	Emesis; gastric emptying; hyperglycaemia; possible ↑ in HR	Constipation; dizziness	87
Voltage-gated sodium channel subunit α Nav1.5 (**SCN5A**)	NA	High	CVS	Insufficient information	Slowed cardiac conduction; prolonged QRS interval of ECG	88
Enzymes						
Acetylcholinesterase (**ACHE**)	NA	High	CVS, GI, pulmonary	Insufficient information	↓ in BP; ↓ in HR; ↑ in GI motility (↓ at high doses); bronchoconstriction; ↑ in respiratory secretions	89

Table 1 (*Continued*)

Targets (gene)	Hit rate[a] Binding	Hit rate[a] Functional or enzymatic	Main organ class or system	Effects Agonism or activation	Effects Antagonism or inhibition	Refs[b]
Cyclooxy-genase 1; COX1 (*PTGS1*)	NA	Medium	GI, pulmonary, renal	Insufficient information	Gastric and pulmonary bleeding; dyspepsia; renal dysfunction	90
Cyclooxy-genase 2; COX2 (*PTGS2*)[‡]	NA	Medium/high	Immune, CVS	Insufficient information	Anti-inflammatory activity; anti-mitogenic effects; myocardial infarction; ↑ in BP; ischaemic stroke; athero-thrombosis	91
Monoamine oxidase A (*MAOA*)[‡]	NA	Medium	CVS, CNS	Insufficient information	↑ in BP when combined with amines such as tyramine; DDI potential; dizziness; sleep disturbances; nausea	92
Phosphodies-terase 3A (*PDE3A*)	NA	High	CVS	Insufficient information	↑ in cardiac contractility; ↑ in HR; ↓ in BP; throm-bocytopaenia; ventricular arrhythmia	93, 94
Phosphodies-terase 4D (*PDE4D*)[‡]	NA	Very high	CNS, immune	Insufficient information	Anti-inflammatory activ-ities; antidepressant-like activities; emesis; vascu-litis and arteritis; pos-sible thymus atrophy	95, 96
Lymphocyte-specific pro-tein tyrosine kinase (*LCK*)	NA	Medium/high	Immune	T cell activation	T cell inhibition; SCID-like immuno-deficiency	97

Target	Hit rate		Therapeutic area			Ref.
Transporters						
Dopamine transporter (**SLC6A3**)	High/very high	NA	CNS	Insufficient information	Addictive psychostimulation; depression; parkinsonism; seizures; dystonia; dyskinesia; acne	98
Noradrenaline transporter (**SLC6A2**)‡	High/very high	NA	CNS, CVS	Insufficient information	↑ in HR; ↑ in BP; ↑ in locomotor activity; constipation; abuse potential	99
Serotonin transporter (**SLC6A4**)‡	High	NA	CNS, CVS	Insufficient information	↑ in GI motility; ↓ in upper GI transit; ↓ in plasma renin; ↑ in other serotonin-mediated effects; insomnia; anxiety; nausea; sexual dysfunction	100
Nuclear receptors						
Androgen receptor (**AR**)	Medium	Medium	Endocrine	↑ in prostate carcinoma; oedema; androgenicity in females; ↑ in muscle mass; ↑ in hostility; sleep apnoea; liver complications	↓ in spermatogenesis; impotence; gynecomastia, mastodynia; ↑ in breast carcinoma	101, 102
Glucocorticoid receptor (**NR3C1**)	Medium	Medium	Endocrine, immune	Immunosuppression; hyperglycaemia; insulin resistance; muscle wasting; ↑ in body weight; ↑ in BP; ↓ in plasma potassium and arrhythmia; osteoporosis; glaucoma	Hypoglycaemia	103

[a]Hit rates were determined at 10 µM. 'Low' corresponds to to <1% hit rate; 'medium' corresponds to 1–5% hit rate; 'high' corresponds to 5–20% hit rate; 'very high' corresponds to >20% hit rate.

[b]Targets that were included in the panels of all four companies.

[§]The references cited are key references giving details of some of the main adverse drug reactions (ADRs) for each target, but not all of the ADRs listed are mentioned in the cited publications.

APPENDIX 21.2

Table 1 Known Drug Transporters along with Their Natural Substrates and a Subset of the Identified Drug Substrates. Table adapted from *Rational Approaches to Improving Selectivity in Drug Design.*[34]

Transporter	Natural substrate	Drug substrates
PEPT1	dipeptides, tripeptides	ampicillin, temocapril, enalapril, midodrine, valacyclovir
PEPT2	dipeptides, tripeptides	amoxicillin, cefadroxil, cefaclor, bestatin, valganciclovir
OCT1	organic cations	zidovudine, acyclovir, ganciclovir, metformin, cimetidine
OCT2	organic cations	memantine, metformin, propranolol, cimetidine, quinine
OAT1	organic anions	quinidine, pyrilamine, verapamil, valproate, cephaloridine
OAT2	organic anions	zidovudine, tetracycline, salicylate, methotrexate, erythromycin
OATP-A	organic anions	fexofenadine, rocuronium, enalapril, temocaprilat, rosuvastatin
OATP-B	organic anions	pravastatin, glibenclamide, atorvastatin, fluvastatin, rosuvastatin
OATP-C	organic anions	benzylpenicillin, rifampicin, cerivastatin, pitavastatin, methotrexate

KEY REFERENCES

P. D. Leeson and B. Springthorpe, *Nat. Rev. Drug Discovery*, 2007, **6**, 881.

J. Bowes, A. J. Brown, J. Hamon, W. Jarolimek, A. Sridhar, G. Waldron and S. Whitebread, *Nat. Rev. Drug Discovery*, 2012, **11**, 909.

T. T. Wager, X. Hou, P. R. Verhoest, A. Villalobos, *ACS Chem. Neurosci.*, 2010, **1**, 435.

REFERENCES

1. I. Kola and J. Landis, *Nat. Rev. Drug Discovery*, 2004, **3**, 711.
2. B. Munos, *Nat. Rev. Drug Discovery*, 2009, **8**, 959.
3. P. D. Leeson and B. Springthorpe, *Nat. Rev. Drug Discovery*, 2007, **6**, 881.
4. M. P. Edwards and D. A. Price, *Annu. Rep. Med. Chem.*, 2010, **45**, 381.
5. P. D. Leeson and J. R. Empfield, *Annu. Rep. Med. Chem.*, 2010, **45**, 393.
6. R. A. Thompson, E. M. Isin, Y. Li, R. Weaver, L. Weidolf, I. Wilson, A. Claesson, K. Page, H. Dolgos and J. G. Kenna, *Chem.-Biol. Interact.*, 2011, **192**, 65.
7. J. G. Cumming, A. M. Davis, S. Muresan, M. Haberlein and H. Chen, *Nat. Rev. Drug Discovery*, 2013, **12**, 948.
8. S. E. Ward, M. Harries, L. Aldegheri, N. E. Austin, S. Ballantine, E. Ballini, D. M. Bradley, B. D. Bax, B. P. Clarke, A. J. Harris, S. A. Harrison, R. A. Melarange, C. Mookherjee, J. Mosley, G. Dal Negro, B. Oliosi, K. J. Smith, K. M. Thewlis, P. M. Woollard and S. P. Yusaf, *J. Med. Chem.*, 2011, **54**, 78.
9. E. Perola, *J. Med. Chem.*, 2010, **53**, 2986.
10. I. Shamovsky, S. Connolly, L. David, S. Ivanova, B. Norden, B. Springthorpe and K. Urbahns, *J. Med. Chem.*, 2008, **51**, 1162.

11. I. Shamovsky, C. de Graaf, L. Alderin, M. Bengtsson, H. Bladh, L. Börjesson, S. Connolly, H. J. Dyke, M. van den Heuvel, H. Johansson, B.-G. Josefsson, A. Kristoffersson, T. Linnanen, A. Lisius, R. Männikkö, B. Norden, S. Price, L. Ripa, D. Rognan, A. Rosendahl, M. Skrinjar and K. Urbahns, *J. Med. Chem.*, 2009, **52**, 7706.

12. X. Liu, C. Chen and C. E. Hop, *Curr. Top. Med. Chem.*, 2011, **11**, 450.

13. M. P. Gleeson, *J. Med. Chem.*, 2008, **51**, 817.

14. J. Bowes, A. J. Brown, J. Hamon, W. Jarolimek, A. Sridhar, G. Waldron and S. Whitebread, *Nat. Rev. Drug Discovery*, 2012, **11**, 909.

15. Y. Kawai, S. Tsukamoto, J. Ito, K. Akimoto and M. Takahashi, *Chem. Pharm. Bull.*, 2011, **59**, 1110.

16. J. D. Hughes, J. Blagg, D. A. Price, S. Bailey, G. A. Decrescenzo, R. V. Devraj, E. Ellsworth, Y. M. Fobian, M. E. Gibbs, R. W. Gilles, N. Greene, E. Huang, T. Krieger-Burke, J. Loesel, T. Wager, L. Whiteley and Y. Zhang, *Bioorg. Med. Chem. Lett.*, 2008, **18**, 4872.

17. P. Gleeson, G. Bravi, S. Modi and D. Lowe, *Bioorg. Med. Chem.*, 2009, **17**, 5906.

18. Y. Ran, N. Jain and S. H. Yalkowsky, *J. Chem. Inf. Comput. Sci.*, 2001, **41**, 1208.

19. A. L. Hopkins, G. M. Keserü, P. D. Leeson, D. C. Rees and C. H. Reynolds, *Nat. Rev. Drug Discovery*, 2014, **13**, 105.

20. N. E. Austin, S. J. Baldwin, L. Cutler, N. Deeks, P. J. Kelly, M. Nash, C. E. Shardlow, G. Stemp, K. Thewlis, A. Ayrton and P. Jeffrey, *Xenobiotica*, 2001, **31**, 677.

21. S. B. Rosenblum, T. Huynh, A. Afonso, H. R. Davis Jr., N. Yumibe, J. W. Clader and D. A. Burnett, *J. Med. Chem.*, 1998, **41**, 973.

22. R. J. Herr, *Bioorg. Med. Chem.*, 2002, **10**, 3379.

23. B. S. Orlek, F. E. Blaney, F. Brown, M. S. G. Clark, M. S. Hadley, J. Hatcher, G. J. Riley, H. E. Rosenberg, H. J. Wadsworth and P. Wyman, *J. Med. Chem.*, 1991, **34**, 2726.

24. F. Yoshida and J. G. Topliss, *J. Med. Chem.*, 2000, **43**(13), 2575.

25. D. F. Veber, S. R. Johnson, H. Y. Cheng, B. R. Smith, K. W. Ward and K. D. Kopple, *J. Med. Chem.*, 2002, **45**, 2615.

26. Y. Lai, K. E. Sampson and J. C. Stevens, *Comb. Chem. High Throughput Screening*, 2010, **13**, 112.

27. R. A. Thompson, E. M. Isin, Y. Li, L. Weidolf, K. Page, I. Wilson, S. Swallow, B. Middleton, S. Stahl, A. J. Foster, H. Dolgos, R. Weaver and J. G. Kenna, *Chem. Res. Toxicol.*, 2012, **25**, 1616.

28. A. J. Ratcliffe, *Curr. Med. Chem.*, 2009, **16**, 2816.

29. C. A. Lipinski, *Drug Discovery Today*, 2004, **1**, 337.

30. T. W. Johnson, K. R. Dress and M. Edwards, *Bioorg. Med. Chem. Lett.*, 2009, **19**, 5560.

31. M. Congreve, R. Carr, C. Murray and H. Jhoti, *Drug Discovery Today*, 2003, **8**, 876.

32. T. T. Wager, X. Hou, P. R. Verhoest and A. Villalobos, *ACS Chem. Neurosci.*, 2010, **1**, 435.

33. J. Bowes, A. J. Brown, J. Hamon, W. Jarolimek, A. Sridhar, G. Waldron and S. Whitebread, *Nature Reviews Drug Discovery*, 2012, **11**, 909.

34. D. J. Huggins, W. Sherman and B. Tidor, *J. Med. Chem.*, 2012, **55**(4), 1424.

CHAPTER 22

Pharmaceutical Properties—the Importance of Solid Form Selection

ROBERT DOCHERTY* AND NICOLA CLEAR

Pharmaceutical Sciences, Pfizer Global R&D, Ramsgate Road, Sandwich, Kent, CT13 9NJ, UK
*E-mail: Robert.Docherty@pfizer.com

22.1 INTRODUCTION AND CONTEXT

The selection of the commercial solid form is one of the key milestones in the development of any new chemical entity (NCE). It is critical not only from an active pharmaceutical ingredient (API) standpoint but also from a drug product performance and stability perspective. The regulatory landscape associated with the solid form of the API and dosage form development has already been described.[1,2] The issues associated with the formation of an unexpected solid form have also been well documented and range from delays in the progression of a candidate through development (batch reworks, additional bioequivalence studies) through to product withdrawal in the most challenging circumstances. The most famous example of this is the antiretroviral drug ritonavir (Norvir) where a more stable, less soluble polymorph appeared in production reducing the dissolution of the capsule product.[3] This is not an isolated incident and as highlighted in a recent review there have, over the last couple of decades, been a number of product recalls due to ambiguous product performance as a result of unexpected solid state transformations.[4] The importance of the solid state features of a NCE to intellectual property has also been described extensively.[5,6] Progress of automation technologies for solid form screening and the emergence of structural informatics which allow development scientists to search and identify the solid form with optimal properties have also been reported.[7,8]

In 1987 the Nobel Prize for chemistry was awarded to Cram, Lehn and Pedersen for their work on supramolecular chemistry. Since then publications[9–12] have charted the evolution of pharmaceutical materials science. Pharmaceutical materials science has emerged as a critical linkage between medicinal chemistry and pharmaceutical development, with solid form and particle engineering being core elements linking the final steps of the API design to drug product attributes. Whilst increasing interest in the crystallisation of pharmaceutical entities

The Handbook of Medicinal Chemistry: Principles and Practice
Edited by Andrew Davis and Simon E Ward
© The Royal Society of Chemistry 2015
Published by the Royal Society of Chemistry, www.rsc.org

within academia has resulted in substantial progress over the last decade, the challenge for the medicinal chemist and pharmaceutical scientist in tackling the crystallisation of highly complex chemical entities remains a significant one because:

- Increasing molecular complexity results in a complicated solid form space that needs to be understood and evaluated (salts, co-crystals, polymorphs, hydrates and solvates).
- Multiple conformational degrees of freedom (*i.e.* molecules with a large number rotatable bonds) can result in complex solid form structures and consequently significant barriers to crystallisation.
- Different solid forms may have different chemical or physical stabilities and biopharmaceutical properties.
- Anisotropic external particle morphologies with different crystal faces exhibit different surface chemistry, and interactions with solvents and impurities hinder crystallisation and stable solid state structure formation.

Physical and chemical properties of a new chemical entity that impact product performance and product robustness are strongly influenced by the solid state structure of the drug substance as shown in Figure 22.1. The formation of different solid state structures (salt, co-crystal and polymorph) provides the pharmaceutical scientist an opportunity to eliminate undesirable properties by switching to an alternative stable crystal morphology, thus enabling a rapid and successful development program. Product performance can only be assured when the NCE is delivered to the patient in a chemically and physically stable solid form. In this chapter we will attempt to link new cutting edge academic progress to the best current industrial practices that medicinal chemists and pharmaceutical scientists can apply in selecting the optimal solid form, along with the pharmaceutical properties that enable the rapid advancement of a new chemical entities (NCE) to a medicines.

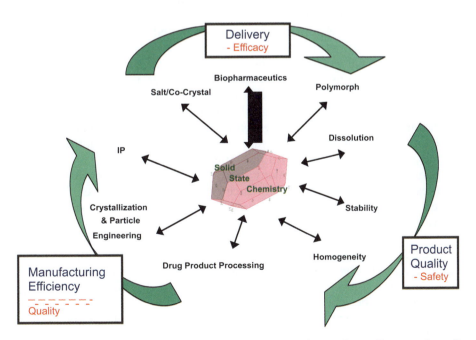

Figure 22.1　The importance of the solid state in impacting product safety, efficacy and quality.

22.2 SOLID STATE CHEMISTRY

22.2.1 Crystallography

Crystals may simply be considered as three-dimensional (3-D) repeating patterns of atoms or molecules. As with any other pattern, they can be described by defining:

  (i) The item to be repeated (the motif).
  (ii) The way in which it is repeated (symmetry operations).

Extending this general concept to crystal structures of organic materials, the motif is the molecule, the lattice describing the scheme of repetition is now a 3-D array and the unit cell is the smallest repeating unit within this 3-D structure. The unit cell is fully described by six lattice parameters, comprising three lengths of the unit cell (a, b and c) with the three inter-axial angles (α, β and γ). Unit cells range from the highly symmetrical cubic (a = b = c and $\alpha = \beta = \gamma = 90°$) through to the lower symmetry monoclinic and triclinic systems (a $\neq$ b $\neq$ c and $\alpha \neq \beta \neq \gamma$) favoured by drug molecules.[13]

So far we have considered the unit cells in terms of their basic shape and relative dimensions. It is also possible to further refine these systems in terms of the symmetry elements which they possess. These elements represent various combinations of rotation, translation and inversion and together they define the full 3-D arrangement of atoms or molecules in a given structure. The symmetry exhibited by a unit cell is thus reflected in both the arrangement of molecules in the internal crystal structure and in the physical and chemical properties of the resulting macroscopic crystal. Symmetry is evident in properties such as crystal growth rates and crystal shape and surface chemistry. Figure 22.2 shows two unit cells of paracetamol along the crystallographic b-direction. Each unit cell contains four paracetamol molecules.[14]

The crystallographic planes that define the external growth morphology of the 'as grown' crystal, and the repeating layers within the crystal bulk itself, can be expressed through their Miller indices (hkl). These indices define the orientation of the surface (or layer) in relation to the crystallographic unit cell. Each surface (layer) is designated with three numbers, these being the inverse of the intersection of that face with the three crystallographic axes a, b and c. The (011) face as shown in the paracetamol morphology in Figure 22.2 cuts the b and c axes one unit cell from the origin and runs parallel to the a-axis. Miller indices are important to the pharmaceutical scientist as they provide a link between the modern structural crystallography of X-ray diffraction and classical morphological crystallography of shape and habit. This allows the process chemist or pharmaceutical scientist to link the internal molecular structure to the chemical functionality on the external surface structure. Figure 22.2 (bottom) shows the observed morphology for paracetamol[14] with the Miller indices labelled.

The most common solid state chemistry analysis technique is Powder X-ray diffraction (PXRD), which examines the angular dependence of X-rays when scattered from crystal lattice planes. This technique provides a definitive fingerprint of the solid state structure and probes the crystallinity and structural integrity of the packing arrangement.

  • The unit cell dimensions and corresponding repeat layers in the bulk, as defined by the Miller indices, govern the angular occurrence of the peaks in the PXRD trace.
  • The position of the molecular species within the unit cell and Miller planes govern the relative intensity of these peaks.
  • The angular width of the peaks is roughly proportional to the quality and perfection of the crystal lattice.

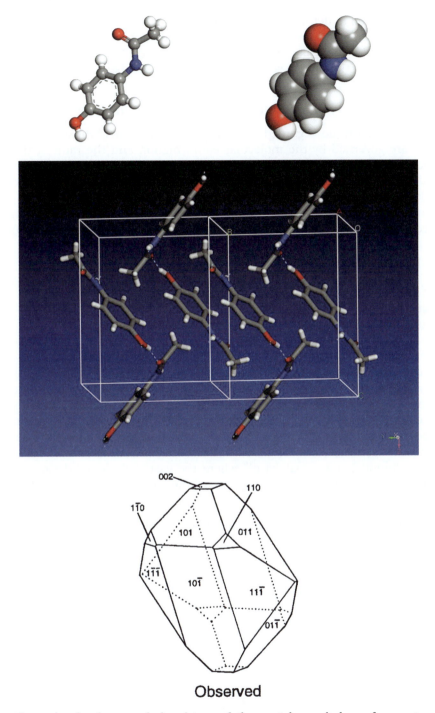

Observed

Figure 22.2 The molecule, the crystal chemistry and the crystal morphology of paracetamol. The molecular structure of paracetamol (top) is shown in ball and stick and space-fill. The crystal packing of paracetamol (middle) contains one unit cell in the a and c directions with two unit cells of paracetamol along the crystallographic b-direction. Each unit cell contains four paracetamol molecules.[14] The crystal morphology (bottom) shows the observed crystal faces labelled with the corresponding Miller indices.

PXRD is routinely used to define the form (polymorph or salt) being isolated, the consistency of the crystallinity of the solid form being produced and also as a baseline for the solid state structure for any solubility measurement being determined. The background to and use of PXRD in the characterisation of the solid state has been described elsewhere.[15]

22.2.2   Crystal Chemistry and Crystal Packing of Drug Molecules

Molecules can essentially be regarded as impenetrable systems whose shape and volume characteristics are governed by the molecular conformation and the radii of the constituent atoms. The atomic radii are essentially exclusion zones in which no other atom may enter except under special circumstances, such as bonding. Figure 22.2 (top) shows a comparison between a ball and stick and van der Waals (space-fill) representation of paracetamol.

The structures and crystal chemistry of molecular materials are often classified into different categories according to the type of intermolecular forces present.[16] A number of factors are of particular importance in assessing the influence of intermolecular bonding on the physico-chemical properties of organic solids[17–19] including:

- The size and shape of the molecular entities that make up the structure.
- The strength of the intermolecular interaction.
- The distance over which the interaction exerts an influence.
- The extent to which the interaction is directional or not.

Organic molecules in general and drug molecules in particular are found in only a limited number of low symmetry crystal systems. The general uneven shapes of molecular structures tend to result in unequal unit cell parameters. A further consequence of their unusual shape is that organic molecules prefer to adopt space groups which have translational symmetry elements, as this allows the most efficient spatial packing of the protrusions of one molecule into the gaps left by the packing arrangements of its neighbours. These tendencies are reflected in an analysis of the Cambridge Crystallographic Database[19] where the vast majority of the organic structures reported prefer the lower symmetry triclinic, monoclinic and orthorhombic crystal systems.[20,21]

22.2.3   Intermolecular Interactions, Crystal Packing (Lattice) Energies

In order to understand the principles which govern the wide variety of solid state properties and structures of drug molecules it is important to describe both the energy and direction of interactions between molecules. As a result of the pioneering work in the development of atom–atom intermolecular potentials,[22,23] it is now possible to interpret inter-molecular packing effects in organic crystals in terms of interaction energies. The basic assumption of the atom–atom method is that the interaction between two molecules can be considered to simply consist of the sum of the interactions between the constituent atom pairs.

$$E_{latt} = \frac{1}{2} \sum_{k=1}^{N} \sum_{i=1}^{n} \sum_{j=1}^{n'} V_{kij} \qquad (22.1)$$

The lattice energy E_{latt} (often referred to as the crystal binding or cohesive energy) for molecular materials can be calculated by summing up all the interactions between a designated central molecule and all the surrounding molecules. Hence, if there are n atoms in the central molecule and n' atoms in each of the N surrounding molecules then the lattice energy can be calculated using Equation 22.1.[22,23]

Each atom–atom interaction pair (V_{kij}) consists of an attractive and repulsive dispersive interaction which can be described by a van der Waals representation, together with an electrostatic interaction and in some cases (particularly for pharmaceuticals) a hydrogen bonding potential. The former two are broadly speaking undirected interactions whilst the latter is not. On a per atom basis, the H-bond is much stronger than say a dispersive interaction but the latter involve many more atomic interactions and so, for molecules such as pharmaceuticals where the molecular weight is relatively high, contributions from the undirected van der Waals interaction tend to dominate the lattice energy.

The use of these potentials has been validated by comparing the theoretical values against the known crystal structures and experimentally measured lattice energies (sublimation enthalpies).[24,25] A particular advantage of the calculated energy is that it can be broken down into specific interactions along particular crystallographic directions and further partitioned onto the constituent atom–atom and/or group contributions. This is the key link between the intrinsic molecular structure and the crystal packing, allowing a profile of the important interactions to be built up within families of compounds. This approach permits the discussion between the medicinal chemist and the pharmaceutical scientist in optimising the design of molecular features, and with the pre-formulation scientist working on the optimisation of the physical properties for the intended dosage form. A number of papers have highlighted the impact of this increased understanding in recent years, including the design of features to disrupt crystal packing and therefore enhance solubility.[26–28]

22.2.4 Crystallisation Solubility, Supersaturation and the Metastable Zone

The crystallisation process can be viewed as a two-step process involving the dissolution of the NCE, and then changing some attribute of the system, such as temperature, solubility or solvent content to induce crystallisation. At a given temperature and pressure there is a maximum amount of solute that can dissolve in a given amount of solvent. When this maximum is reached, the solution is said to be saturated. The amount of solute required to make a saturated solution at a given condition is the solubility.[29]

During crystallisation the molecules must de-aggregate from their solvated state and self-assemble, aligning certain structural elements such as conformation and intermolecular packing in order to enable the production of a stable 3-D ordered crystallographic array of molecules. This highly time-dependent event makes crystallisation essentially a kinetically driven process. The time required for crystallisation to proceed depends on a driving force called supersaturation. A solution in which the solute concentration exceeds the equilibrium saturation at a given temperature is known as a supersaturated solution. Supersaturated solutions are metastable, implying that crystallisation will ultimately occur, albeit after time has elapsed. Every solution has a maximum limit that it can be supersaturated to before it becomes unstable and crystallisation spontaneously occurs.[29] The region between the saturation curve and this unstable boundary is called the metastable zone, and it is within this that all crystallisation activities normally occur. If we plot concentration versus temperature behaviour we find three regions, as shown in Figure 22.3:

- A stable or undersaturated region where crystal growth is not favoured.
- A metastable region where the solution is supersaturated to a degree and where crystallisation will take place after a time.
- An unstable region where the solution is more supersaturated and where spontaneous crystallisation with no time delay is expected.

The crystalline form in which a material is obtained can potentially be controlled through manipulation of three main aspects of polymorphic behaviour: nucleation, crystal growth and

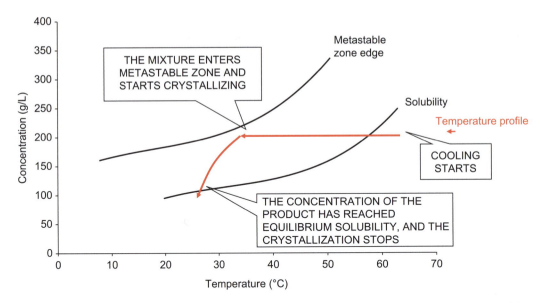

Figure 22.3   An overview of the Metastable Zone. The region below the solubility curve is stable and undersaturated. The region above the metastable zone curve is unstable and spontaneous crystallisation will occur. The region in the middle is the meta-stable zone. The red line represents a potential cooling curve which initially generates supersaturation to induce nucleation and then controlled growth.

phase transformations. The nucleation process involves the formation of aggregates of molecules, some of which reach critical size at which point they become stable and subsequently grow to form crystals. In a polymorphic system, the situation can be more complex. It is assumed that a number of different types of aggregate may exist in solution. Each type of aggregate is connected by competing equilibria and may subsequently develop into one or more different polymorphic forms. Nucleation is seen by many leading academics as the key to crystal engineering.[30,31]

22.2.5   Pharmaceutical Properties and the Solid State

Structural diversity in pharmaceutical solids arises from both the molecular conformations available from drug structures and from the broad range of intermolecular interactions available, as described in Section 22.2.2. These broad ranges of structural types represent both a challenge to control and ensure a robust product but also an opportunity to engineer a solid form that can enable a drug candidate with sub-optimal properties to move forward into development and ultimately to a patient. These different structural types are captured in Figure 22.4 for paracetamol.

Although there is a continuum of structural diversity it might be worth considering a few distinct aspects to context the discussion on physical property impact.

- *Amorphous through to crystalline.* Amorphous forms of pharmaceutical solids are chemically and physical metastable with respect to their crystalline counterparts. They present an opportunity in that they can be more soluble than the crystalline version but are generally less stable chemically. Managing an amorphous form through development is a major undertaking given the inherent propensity for chemical instability and the potential for crystallisation impacting the bioavailability on longer term stability.

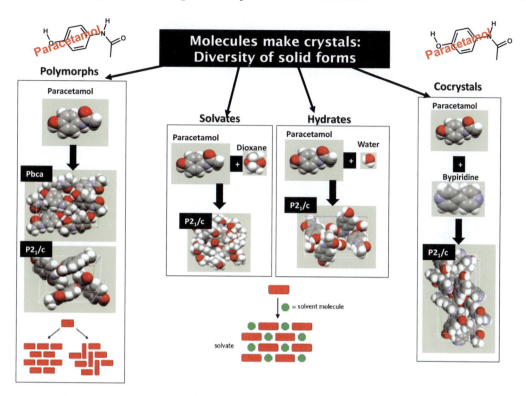

Figure 22.4 The solid form landscape for paracetamol including polymorphs, solvates and co-crystals. The schematic links the geometrical features discussed in section 22.2.1 (red bricks in different arrangements) with the crystal chemistry of drug molecules described in section 22.2.2 (hydrogen bonds and molecular packing). The propensity to form these diverse set of structures is a combination of the molecular shape packing (density) and the degree of satisfaction of different intermolecular interaction patterns (lattice energy).
(Reproduced with permission from Prof. William Jones, University of Cambridge.)

- *Free base through co-crystal to salt.* By manipulating the solid state packing through the addition of co-formers, properties of the solid such as melting point, lattice energy and solubility can be changed. The difference with a salt is that there is a proton transfer and this of course makes a difference to the structure of the molecule, including conformation and electronic structure. This changes both the solid state packing and the solvation properties of the drug species. This enables a greater change in solubility, but recent evidence suggests these structural changes result in greater tendencies for solvate formation in salt structures relative to the comparable co-crystal.

In Section 22.3 we will describe the current screening and selection practices used in industry to attempt to optimise the physical form and properties at the medicinal chemistry pharmaceutical development interface. In Section 22.5 we will describe the evolution of these processes towards physical form design.

22.2.6   Polymorphism, Thermodynamic Stability and Solubility

Despite considerable debate in the scientific literature a comprehensive definition of the term polymorphism is by no means straightforward. It can be simply defined as the existence of a compound in at least two different crystal structures. While this definition is not comprehensive, it is sufficient for the discussion here. The use of the term pseudopolymorphism to describe the relationship between such solids and solvates is misleading. Solvates and non-solvated solids differ in composition and, therefore, should not be considered to be polymorphs.

In general for molecular materials it is their intrinsic desire to closely pack effectively in the solid state that is the single biggest driving force towards selected structural arrangements. For complex molecular materials such as drug molecules there will be notable exceptions where the need to form complex H-bonding arrangements will override this desire.[16] Weaker interactions such as C–H::::::O=C bonds and other polar interactions are probably not primary drivers in the packing arrangements adopted by drug molecules but will tend to be optimised within potential arrangements. Crystallisation and the properties of the solid state are a result of molecular recognition processes on a grand scale and polymorphism is due to different balances of these subtle intermolecular interactions.[9]

The formation and behaviour of polymorphs is determined by the relative free energy of the different structures and the kinetic barriers to their formation and inter-conversion. The polymorph with the lowest free energy (lowest lattice energy) under a given set of conditions will be the most stable. Particular properties such as density, melting point, solubility and mechanical properties can all be impacted by different solid state structures.

The influence of polymorph stability on solubility can be understood by considering two polymorphs A and B, where B is the more stable form. The solubility of a given molecule is ultimately a balance between the energy of solvation (how much the molecule likes to be in a solvent environment) and the lattice energy (how much the molecule likes to be in the solid state).[32,33] For a given drug candidate the energy of solvation is a molecular property and constant in a given solvent. Given B is thermodynamically more stable at room temperature it will have a better (lower) lattice energy (stronger packing) and hence will have a lower relative solubility. Differential scanning calorimetry (DSC)[15] is a thermal method to determine melting point and enthalpy of melting (sometimes referred to as fusion enthalpy). The solid state structure that has the larger enthalpy of fusion for a given compound tends to be the most stable polymorphic form. Published data for two polymorphs of chloramphenicol palmitate[34] show that Form A is the most stable, it has the highest melting point (by 6 °C), the greatest heat of fusion (which is a surrogate measurement of the lattice energy) and less than half the solubility of form B. It should be noted that the energy difference between the polymorphs of 3.8 kcal/mol is not unusual and neither is the resultant two fold change in solubility (see Section 22.3.3).

22.3   INDUSTRY PRACTICES

22.3.1   Salt Screening and Selection

The predominant reason for screening for salts of NCEs is to overcome any undesirable chemical or physical properties of the free acid/base. Such properties may include poor solubility in bio-relevant media, exposure limiting dissolution rate, chemical instability and poor mechanical properties. In addition, salts typically exhibit good crystallinity due to the additional electrostatic intermolecular interactions compared to the free acid/base, which assists achieving good isolation, purification and stability. In the case of enantiomeric compounds, optical resolution can be achieved by diastereoisomeric salt formation in order to obtain high enantiomeric purity. An overview of acceptable salts and development factors that influence this selection process has been covered elsewhere.[35]

A summary of the key development criteria are summarised in Table 22.1.

Table 22.1 Properties which may be assessed during solid form selection for a standard oral dosage form. These are examined in a systematic approach as described in Table 22.2. In the case of other dosage forms the balance of the criteria maybe refined. For a parenteral dosage form the solubility and chemical stability in solution of the API may over ride other criteria. For a dry powder inhalation dosage from compatibility with lactose is likely to be a key selection criteria.

Safety—generally stable under conditions of isolation, purification and storage	Salt is precedented for route of administration, frequency of use and dose Chemical Stability (*e.g.* hydrolysis, oxidation, photolysis) Excipient Compatibility Thermal Behaviour (decomposition, phase transitions) Purity—good purge of key impurities through final step Tendency to form unwanted solvates
Efficacy—timely and complete dissolution of dose administered	Aqueous Solubility Common Ion Effect on Solubility pH Solubility Profile Crystallinity Melting Point Particle Size distribution and crystal habit and dissolution rate
Quality—ease of manufacture to ensure consistent product attributes.	Hygroscopicity—no change of hydrate state under storage conditions Degree of polymorphism/landscape—ease to secure desired form in API manufacture Physical stability to milling, micronisation, and compaction Absence of corrosiveness Acceptable powder flow for tablet and capsule production

Most pharmaceutical compounds contain acidic or basic functional groups and the majority of these will ionise allowing salt screening and selection. It is estimated that around half of all NCEs are utilised in dosage forms as salts. A wide variety of organic and inorganic anions and cations are used in forming salts of drug compounds. While final selection of the most suitable salt former involves assessment of an extensive range of physical and chemical properties, the primary concern is that potential counter-ions must not exhibit any adverse physiological effects, so the safety of the counter-ion selected in the proposed dosage form regime is of critical importance. The most frequently used salts and their precedence in different delivery routes has already been described.[36] When designing a salt screen there are some important factors to consider. Each drug molecule is different so routinely screening all counter-ions could prove wasteful. In the literature, guidance is provided for salt formation with pharmaceutical compounds such that there should be a pKa difference (ΔpKa) of 2 to 3 units[35] between the compound and its potential salt former. Constraining the selection of counter-ions by using this approximation ignores the impact that solvent and temperature can have on potential salt forming reactions.

When designing a salt screen for multi-basic or multi-acidic compounds and counter-ions, the potential stoichiometries should be considered. Even for mono-basic or mono-acidic compounds multiple salt stoichiometries may be achieved. Similarly, when screening for salts of a chiral API, unique solid forms may be achieved by the use of single enantiomers salt formers. For example, a compound may form a salt with L-tartaric acid that meets all development selection criteria, whereas the D-tartrate salt is non-ideal with poor crystallinity and hygroscopicity. The best counterion to use will be dependent on the NCE features, dosage form type and likely dose. The range of counter-ions and frequency of use is captured in Figure 22.5 and recent reviews have examined in detail trends in this area.[37,38]

The physical properties of the candidate salts must be assessed—a task which requires comprehensive characterisation of selected salts and their polymorphs or hydrates.

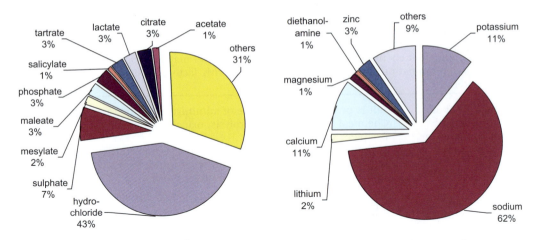

Figure 22.5 The relative usage of the most commonly utilised acidic and basic counterions.

Table 22.2 Multi-tier evaluation of potential salt forms.

	Properties	Techniques
Tier 1	Crystallinity Crystal form	Optical microscopy, Powder XRD
Tier 2	Thermal properties and thermal behaviour (decomposition, phase transitions, desolvation)	Differential Scanning Calorimetry (DSC), Thermogravimetric Analysis (TGA)
	Hygroscopicity, hydrate/solvate formation	Dynamic Vapour Sorption (DVS)
Tier 3	Polymorph and hydrate screening Aqueous and pH solubility Chemical stability testing (*e.g.* hydrolysis, oxidation, photolysis) Accelerated physical stability testing	Powder XRD, Raman microscopy, DSC Powder XRD of solubility residues HPLC Powder XRD, DSC
Tier 4	Humidity/temperature induced changes in crystal form Influence of processing conditions (*e.g.* milling, micronisation, compaction) on solid form Compatibility with excipients	Environmental Powder XRD Powder XRD, DSC, DVS HPLC

The characterisation and assessment of the physical form of salts is extremely demanding in terms of experimental effort.[36] While it is desirable to screen a range of salts, it is impractical to attempt characterisation of all aspects of every potential salt form. For this reason, salt selection is typically carried out using a tiered approach. After characterisation at each level, a decision is made as to whether the salt proceeds to the next level of assessment. The range of potential salts is, therefore, progressively narrowed down. The depth of characterisation increases from one level to the next. The number of levels required typically depends upon the number of candidate salts. In this way, a substantial number of salt forms can be investigated while minimising experimental effort. Table 22.2 shows a schematic representation of a possible multi-level salt selection process drawing down upon the desirable properties of the candidate salts.

The importance of having a robust salt selection process from a development standpoint was recently highlighted through the concerns of the partial breakdown of the salt of prasugrel back to its free base and the potential impact that had on the bioavailability and efficacy of the product.[39]

A recent analysis shows a trend away from using a small selection of counterions towards a much broader variety of ions. This trend started in the 1990s and has increased in momentum in recent years.[37]

22.3.2 Co-crystals Screening

Co-crystals are defined as neutral multi-component systems having extended molecular networks formed through strong H-bonding patterns. Components of a pharmaceutical co-crystal include at least one API and one or more ligand (co-former), all of which are neutral and solid at room temperature and atmospheric pressure.[40] The appeal of co-crystals is that they offer a pathway for altering the solid-state properties of non-ionisable API where salt options may be limited. Co-crystals provide opportunities to:

- Deliver crystalline material and avoid an amorphous final form when salts are not an option.
- Reduce the requirement for sophisticated formulation technology as co-crystals can be progressed largely within the same operational development paradigm as salts.
- Provide additional purity control options for the final form.
- Enhance solid form stability, dissolution rates and oral bioavailability of the NCE.

Co-crystal formation between an API and ligand(s) relies on complementary, non-covalent interactions such as H-bonds, van der Waals, π–π stacking and electrostatic interactions. Research on hydrogen bond motifs led to guidelines for the design of molecular assemblies.[16–18] These rules can also be applied to the targeted design of co-crystals, while taking broader aspects into consideration such as crystallisation kinetics and thermodynamic properties. As well as traditional solution based approaches to making co-crystals other areas such as eutectic based thermal screening and liquid assisted grinding are proving fruitful.[41]

The ability of co-crystals to demonstrate different solubilities, dissolution rates and stabilities has also been described. The fluoxetine HCl : succinic acid (2 : 1) co-crystal shows an approximate three-fold increase in intrinsic dissolution rate relative to fluoxetine HCl. However, fluoxetine HCl : fumaric acid (2 : 1) co-crystal has a similar dissolution rate to the API alone, whereas the dissolution rate for fluoxetine HCl : benzoic acid (1 : 1) is approximately half of the API dissolution rate.[42] Other examples include co-crystals of itraconazole with diprotic carboxylic acids, which achieve and sustain 4- to 20-fold higher drug concentrations than crystalline itraconazole during aqueous dissolution.[43] Co-crystals remain an area of intense research activity. Whilst they are unlikely to address solubility challenges associated with NCE's where solubility is limited by solvation there is increasing evidence that co-crystals may have other attractive development attributes with lower tendency to form solvates/hydrates.[40,41]

22.3.3 Polymorph Screening

The importance of polymorphism to the pharmaceutical industry can, in the views of most experts, be traced back to reviews by McCrone[44] and Byrn.[45] The temporary withdrawal of the protease inhibitor ritonovir just over a decade ago[3] highlighted the potential impact of the appearance of a more stable polymorph on solubility, dissolution and ultimately bioavailability. The recent appearance of a new polymorph in the rotigotine transdermal system[46] resulted in snowflake patterns on the patch. The subsequent regulatory discussion and concerns about the efficacy of the patch highlighted the need for continued vigilance in screening for polymorphs.[4]

The possible impact on product manufacturing robustness, product development timelines and the potential need for repeated clinical and stability studies means that most pharmaceutical companies have incorporated solid form selection and screening programs and practices within their development strategies. In this section we will outline the key elements of screening practices and the appropriate timing of these activities during drug development. For the medicinal chemists/development scientist the key questions are:

- What form will my molecule crystallise in?
- How often do these new polymorphs, solvates appear?
- What will be the impact on biopharmaceutics (*e.g.* solubility)?
- How many experiments do I need to carry out to ensure robustness?

These questions are difficult to give definitive answers to, but Figure 22.6 consolidates some data based around recent reviews in this area.[47–49] The summary suggests that the vast majority of pharmaceutical compounds exhibit multiple polymorphs. It also shows that hydrates are more prevalent for salts, and polymorphs are more likely for non-salts. A review of some key publications on high-throughput solid form screening suggests that 52 500 crystallisations on 51 new API's identified 155 new solid forms. In general, polymorph pair solubility ratios are equal to or less than three-fold changes but there are instances of greater than five-fold solubility changes, especially for highly complex drug molecular structures.[49]

In preclinical development only a limited amount of material is available. Early screening is focused on the definition of the form being used and how to ensure consistent delivery of that form with the enabling chemistry. Limited screening will be targeted towards an early awareness of the potential forms accessible from the salts that have been identified. A selection of techniques is used to provide a fingerprint of the solid state chemistry. As a candidate progresses into the early clinical studies there is a commensurate increase in the effort to find polymorphs and solvates of the API and to link this emerging information to the product design and overall development strategy. This will involve a variety of crystallisation experiments and physical stability investigations of the bulk drug in different humidity and temperature ranges.

In recent years the concepts of 'stable form' screens have emerged with increasing importance especially at this point in development. These 'stable form' screens, sometimes called 'low energy' screens, are essentially slurries in solvents with suitable solubility to facilitate the transformation from a given metastable solid state structure to a potentially more stable structure.[50] This solvent mediated phase transformation involves two steps. The dissolution of the metastable polymorph to form a solution supersaturated with respect to the stable form is then followed by the nucleation and growth of the new more stable phase from this solution.

The rate of such solvent mediated transformations is driven by solubility in the selected solvents. The number of solvents used, their solubilities, temperature and the duration of the slurry are often company practice dependent. Whilst there is no definitive standard industry practice the slurry experiment has become commonplace as it can rapidly help identify the most stable form and therefore baseline the clinical exposure from a bioavailability perspective. A number of publications have articulated these principles but have also highlighted that these slurries are not a panacea for polymorph screening as there are caveats that can reduce the effectiveness of the screen.[50,51]

- Small solubility differences between polymorphs may limit the transformation.
- A lack of solubility/stability in preferred solvents may limit the screen design.
- Solvation of a particular molecular functionality might limit the H-bonding that could be adopted, preventing a more stable structure emerging.
- Impurities that could inhibit the more stable form appearing may be present in the early clinical batches.

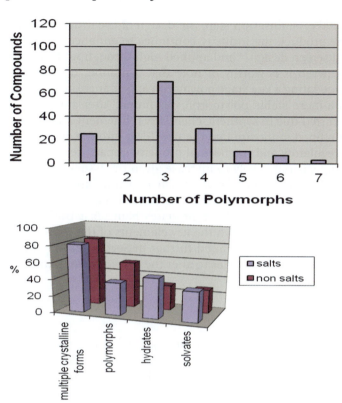

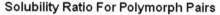

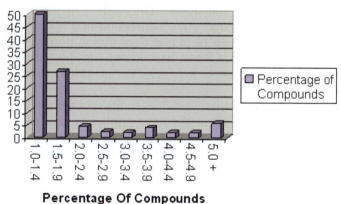

Figure 22.6 The number of compounds with a given number of polymorphic forms. The frequency of forms found for salts and non-ionisable materials and the histogram plot of solubility ratios for 180 polymorph pairs.
Based on an amalgamation of data from references.[47–49]

In the later stages of development, screens are designed to find all known polymorphs to ensure a comprehensive understanding of the solid form space and the potential impact on the commercial manufacturing process. Screens underwrite final process solvent variations, temperature excursions and the impact of impurities. Given the pivotal importance of the correct polymorph on the final crystallisation step, the design of the drug product, and the stability of both the API and the product, the industry continues to enhance screening practices through both better informed screen design[52] and applied high-throughput technologies.[7] Whilst much progress is being achieved it is worthy of note that a number of recent papers highlight the challenges of accommodating a late stage polymorph appearance on a fast moving candidate.[53,54] In both these cases a more stable polymorph was found, after small changes in the impurity profile of the API occurred due to synthetic chemistry refinement and optimisation. The clinical plan then had to be refined to incorporate the new form into the overall development program. Given the potential regulatory and intellectual property impact, research on product processing induced transformations is only likely to increase given the issues recently highlighted.[4,39,46]

The exact timing of when to do this work can be candidate/dosage form dependent and dictated by the overall development strategy. An inhalation candidate is likely to have these challenges addressed earlier as would a candidates being fast tracked in the oncology therapeutic area due to a desire to lock down the key elements of the commercial formulation prior to pivotal clinical studies. This is discussed further in Section 22.4.3 and the opportunities to apply newer structural based technologies to alter the development paradigm will be considered in Section 22.5.

22.3.4 Hydrate Screening

As with polymorphs, it is known that a hydrated solid form of a drug can have a potentially significant negative impact upon bioavailability.[48] It is also possible for the presence of water in the crystal structure to lead to chemical and physical instability, and this may result in undesirable chemical reactivity with the excipients within a solid dosage form.

In a previous section we identified Dynamic Vapour Sorption as a key technique for assessing API hygroscopicity examining water update as a function of relative humidity.[15] This approach may also be used to screen for potential hydrate formation, identifying the key temperatures and humidity for anhydrous–hydrate transitions. The method may not be relied upon to prepare hydrates for all new drug candidates since conversion to a hydrated form may be a kinetically slow process. Also, the anhydrous form may need dissolution and recrystallisation steps in order to incorporate water molecules into a new, low-energy, hydrate crystal structure. Even relatively high humidity such as 90% RH may be insufficient to rapidly ensure this rearrangement.

One way to try to overcome these kinetic barriers to hydrate formation is to recrystallise or slurry in water. In theory this would seem the most applicable approach, however, the drug molecule may have insufficient aqueous solubility to make this practicable. An alternative approach is to utilise organic solvent-water mixtures, of known water activity (α_w), to enhance the solubility of the drug candidate. Equation (22.2) relates water activity to relative humidity:

$$\text{Relative Humidity} = \alpha_w \times 100 \tag{22.2}$$

Recrystallisation or slurry equilibration[55] in solvent systems of high water activity such as $\alpha_w = 0.90$ would be a typical approach. The formation of a thermodynamically stable hydrate by this technique is reported to have a high success rate.[56] In addition, equilibration at multiple temperatures for these slurry conversions allows a broad picture of the anhydrous–hydrate phase diagram[57] to be produced, which is a useful tool when assessing the risk of isolation of the incorrect phase, or a solid form change during downstream processes.

For the case where an anhydrous form is developed, if hydrated forms are identified, it is important to consider their potential impact upon bioavailability, stability or other physicochemical properties that may be affected by an anhydrous–hydrate conversion. In contrast, if a hydrate is to be the commercial solid form developed then this is also true for a hydrate–anhydrous conversion. In such cases it is important to take a product overview ensuring the secondary processing and stability aspects have been considered, as well as the control of hydrate state during API production.

This section has highlighted that the solid state form of an NCE can have a major effect on drug product safety, efficacy and performance. Only through a comprehensive screening process will the solid form landscape of a new NCE be fully understood, and an optimal solid form selected to enable rapid product design.[58]

22.4 INTEGRATION WITHIN THE EARLY CLINICAL PHASES OF DEVELOPMENT

22.4.1 The Changing Drug Product Design Paradigm

Over the last decade, through embracing both academic advances and technology initiatives, significant progress has been made in defining relationships between the NCE properties and the formulation design aspects of new products. Examples of established progress include:

- API particle size distributions and content uniformity.[59]
- API particle size distributions and flow,[60] mechanical properties.[61,62]
- API particle size and dissolution.[63]
- Crystal brittleness, milling behaviour and particle size reduction.[64]

Models have been built that allow the pharmaceutical scientist to explore the impact of particle size variation on dissolution rate and bioavailability. The Biopharmaceutics Classification System (BCS)[65] is used to define classes of compounds based on the solubility and permeability of the compounds. Permeability is a molecular property, but solubility and dissolution rate are related to the internal structure (salt and polymorph) and particle size distribution/surface area. More recently this concept was built upon where the interplay between permeability and solubility has been refined through the use of simulated gastric fluids for measuring solubility and this has brought greater definition to the impact of API particle attributes on drug efficacy.[66] These relationships, combined with institutionalised corporate knowledge of formulation design practices have opened up the potential of a fully integrated holistic product design process.[67]

22.4.2 Different Requirements for Dosage Form Types

In reality the factors which influence formulation selection (Figure 22.7) are wide ranging and encompass both scientific consideration of the compounds physical properties, the proposed dosage form, the stage of development and overall project investment strategy. The majority of the industry would agree that the lowest energy/most stable crystalline solid form is the preferential form for drug product development, and this is largely true independent of the route of administration or dosage form type.

Although the greatest numbers of drug therapies are presented as oral medications, non-oral drugs are developed for a variety of reasons; to overcome high first pass metabolism, reduce side effects and or enhance local efficacy. Examples include steroids which are delivered locally in smaller doses through inhalation, nasally and topically, and can decrease inflammation without inducing harmful side effects seen by higher systemic concentrations. Non oral dosage forms most commonly exist as solutions, suspensions or powders and can be delivered by different routes into the body.

Factors Affecting Formulation Selection

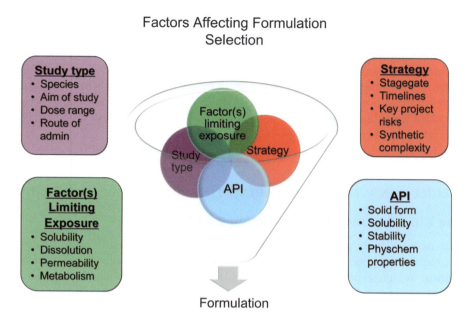

Study type
- Species
- Aim of study
- Dose range
- Route of admin

Strategy
- Stagegate
- Timelines
- Key project risks
- Synthetic complexity

Factor(s) Limiting Exposure
- Solubility
- Dissolution
- Permeability
- Metabolism

API
- Solid form
- Solubility
- Stability
- Physchem properties

Formulation

Figure 22.7   Factors affecting formulation selection.

For solution formulations a key consideration during drug development is to ensure the drug remains fully dissolved, as well as physically and chemically stable over the drug product shelf life. The formulator is interested in ensuring the API will not precipitate out to a more stable (lower energy) form, which could ultimately limit the dose achievable in a set volume. Therefore, a thorough understanding is required of the physical chemistry of the molecule, followed by the careful selection of buffer and counter-ion components, all to ensure less soluble *in situ* salts do not form/precipitate in the formulation. A common situation is the Cl^- ion effect, whereby a free base molecule can combine with a Cl^- ion *in vivo* to form a less soluble species that precipitates.

Suspension product performance relies on particles retaining the initial characteristics (size, shape, surface chemistry and amorphous content) over the product shelf life. Changes in particle size distribution within the suspension formulation can be due to dissolution from small particles and re-crystallisation on to larger particles (Ostwald ripening), or agglomeration due to primary particles interacting. These are common issues for suspension based products that can be linked back to solid form attribute definition (*e.g.* surface chemistry, amorphous content and crystallinity).

Inhaled drug delivery commonly sees the API blended with a carrier excipient (usually lactose), to bulk out the dosage form as doses are typically <1 mg. The API powder is then dispersed from the carrier at the point of administration *via* an inhaler device. The API solid form and particle properties (size distribution, the amount of fines, amorphous content, surface roughness, lactose binding energy) all have the potential to impact the respirable fraction and hence the efficacy of the product.

For both powder and suspension non-oral products, particle size reduction through the use of milling or micronisation is frequently employed to create suitable properties for drug product performance. Different solid forms of the same API will behave differently during size reduction. This can be related to solid form properties including melting point, compressibility and brittle fracture index.[64] The end result maybe different degrees of amorphous content, which in turn could impact product performance and the shelf life achievable. Thus selecting the solid form which is more amenable to such particle processing conditions can be a strategy to facilitate rapid compound development.

22.4.3 Integration of Enabling Formulation Strategies Within Development Paradigms

Drug development is a technically complex, inter-disciplinary, high risk, high cost business operating within an increasingly challenging environment. The industry employs a variety of different drug development investment paradigms in order to manage R and D costs whilst keeping a healthy product pipeline. During drug development with each new piece of information the project team continually assess the candidate progression usually against 3 key factors:

- Confidence in efficacy.
- Confidence in safety.
- Confidence in development.

The relative ranking of a given molecule against others in the company's own pipeline and compared to competitor products will have a significant bearing on the speed and costs employed in development. A first in class molecule in an area of unmet medical need with high confidence in safety and efficacy is likely to secure an accelerated development path especially once proof of concept in the actual patient population is achieved. In this situation a company will often complete the screening and characterisation studies to identify the commercial solid form nomination and drug product formulation ahead of proof of clinical concept readout, accepting the risk that this investment could be redundant should the molecule fail in these clinical studies. This development paradigm is also often seen for diseases with a high mortality rate and few existing effective therapies, *e.g.* certain cancers and viral infections. In the opposite situation where confidence in safety and/or efficacy are unproven for a new mechanism of action, companies have taken probe molecules/formulations into the clinic to establish a rapid readout with minimal solid form/formulation investment. These molecules are known from the outset to have some major development flaws, *e.g.*, poor pharmacokinetics, a known side effect, non-commercial relevant formulation. In these situations, although a crystalline form would still be preferable to achieve synthetic purity, providing the formulation has guaranteed stability over the shelf life needed for the clinical study even an amorphous form can be employed.

Influencing the physicochemical properties and hence solid form options for a molecule is a delicate balance and trade-off between optimisation of potency, selectivity and pharmacokinetics. The optimum time for the pharmaceutical scientist to engage with a project team and start to assess confidence in development is before lead series selection. Here a vast range of structural options are still in play and a combination of rapid solubility and stability screening can help direct the chemist towards more favourable series. Engagement in understanding solid form (and if crystalline material can be synthesised) should ideally start before predictive pharmacokinetic studies commence, especially if the API is low solubility (BCS class 2 or 4) and is destined for oral delivery. Although the most stable form may not be identified at this point, the ability to conduct an *in vivo* exposure study with a suspension of crystalline API versus a solution will provide a good indication of the solubility and or dissolution challenges faced in molecule progression. In parallel to completing efficacy, safety and pharmacokinetic studies, the pharmaceutical scientist will assess the compounds' "developability" for both early clinical and commercial requirements. Depending on the intended route of administration, dosage form type and projected dose, the API synthetic complexity, solubility, stability, solid form and

Table 22.3 The compatibility of API physical and molecular properties with oral low solubility enabling formulation approaches.

Formulation Options	API physical/molecular properties
Co-solvent systems	Utility generally increases with increasing log P and decreasing melting point
Micellisation	
S-SEDDS (supersaturable self-emulsifying drug delivery system)	LogP 1–4, solubility in ethanol, PEG
SEDDS (self-emulsifying drug delivery system)	LogP > 4, solubility in triglycerides
Complexation	Rings with low substitution or aliphatic chains without nearby bulky groups
cyclodextrins	
Particle size reduction	
Nanosuspension	Crystalline, $T_m > 125\ °C$, low aqueous solubility
micronisation	Crystalline, $T_m > 125\ °C$
pH adjustment	Acids of pKa < 9; Bases of pKa > 4
(*in situ* salts)	
Solid Dispersion	$T_m < 220\ °C$; $T_g > 70\ °C$, logP 2–8, solubility in methanol/acetone > 10 mg/mL

drug product processability of the molecule will be assessed. For an oral compound, a highly crystalline solid form, with appropriate solubility versus dose, a melting point greater than 125 °C and non-hygroscopic (<3% moisture uptake over 0–90% RH) would give the pharmaceutical scientist some confidence in ease of further development at the pre-clinical compound selection stage. However in reality this is often not achievable due lack of appropriate API, time or solid form complexity and usually a risk is carried forward into early development. One of the most common reasons for lack of crystalline material prior to first in human studies is a lack of purity. The purity required for pre-clinical safety and efficacy studies is often only 95–98%, whereas these levels of impurities can often have a negative impact on crystal growth making even the best crystallisation attempts futile (see Section 22.1 and Section 22.2.3).

Due to the general increase in more potent lipophilic molecules in development, there has been a parallel growth in enabling formulations to address poor solubility. Table 22.3 shows which molecular physical and chemical properties are amenable to which technology.

Traditionally a trial and error approach combining different *in vitro* dissolution models and *in vivo* studies has been employed to select the most suitable enabling formulation to enhance oral exposure. The science of biopharmaceutical modelling now focuses on reducing the process of drug absorption down to a series of theoretical equations linking drug and physiological parameters.[67] Drug absorption comprises four main processes: dissolution, GI transit, nucleation and permeation. It is the equilibrium solubility of the most stable solid form in representative GI media which is critical to establish should valuable insights be gained from this modelling.

For ionisable compounds salts are one of the simplest and most common means of enhancing dissolution and overcoming solubility limited exposure. As the salt dissolves it creates a microclimate pH at the solid surface, this in turn creates a more favourable environment for more drug to dissolve, resulting in transiently supersaturated solutions in the GI tract. The extent of any enhanced exposure is then a result of the fraction absorbed versus the time before the compound precipitates back to the free form from the supersaturated state. This situation of nucleation and precipitation *in vivo* can exist for:

- Salt to free form.
- Amorphous form to crystalline form.
- Co-crystal to free form.
- Anhydrous to hydrate form.

The exact mechanisms are not well understood, but classical nucleation theory can be used to simulate these effects in biopharmaceutical modelling.

There are a number of enabling technologies that can be employed to move a drug forward when solid form design is not successful or possible. These include

- Particle size reduction
  - When considering whether particle size reduction will provide suitable enhancement the rule of thumb is that the particle size diameter should be less than the solubility in µg/mL. To achieve particle sizes <15 µm specialised micronisation equipment would be needed. The importance of particle size distributions on dissolution and on oral absorption has been discussed elsewhere.[68,69]
- Self-emulsifying drug delivery systems (SEDDS)
  - In some cases NCE's can have properties that allow them to be dissolved in surfactants or mixtures of surfactants, lipids and co-solvents. Upon contact with water these mixtures form micelles and emulsions.[70,71] This enhances the solubility and the percentage absorbed.
- Solid dispersions
  - Solid dispersions are amorphous API composites stabilised by common excipient polymers. They can significantly increase the oral absorption of low solubility drugs.[72,73] Whilst the *quasi* stable form of these amorphous (high energy) solids can provide enhanced exposure, from an oral dosage form perspective the metastable nature of this state can present challenges in terms of reproducibility and longer term physical stability.
- Supersaturable formulations
  - This is described as a spring and parachute concept, salts, co-crystals or self- emulsifying drug delivery system (SEDDs) formulations are used to create the initial supersaturation or spring, this state is then maintained for longer *in vivo* by the parachute, usually a second excipient that is a crystallisation inhibitor.[74,75]
- Pro-drugs
  - Pro-drug strategies that improve solubility are not that common because of the additional synthetic complexity and costs. Despite this there are a number of examples where this has been applied successfully.[76]

A comprehensive overview of these and other strategies to address low solubility in drug discovery and development has recently been published.[77]

22.5   FUTURE OUTLOOK

22.5.1   Solid Form Design

ICH Q9[78] Quality Risk Management guidelines define risk as a combination of the probability of occurrence and the severity of the impact. The ICH Q6a guidelines[79] consolidate this risk framework into a decision tree on polymorphs. The first two decision points on this framework remain the key questions that need to be addressed by the development scientist.

- Decision Point 1: PROBABILITY—"can different polymorphs be formed?"
- Decision Point 2: IMPACT—"do the forms have different properties? (*e.g.* solubility)."

In this section we consider new computational/structural approaches to describe the probability of new forms, and consolidated institutional knowledge to try and quantify the impact of a potentially different structure. By understanding the structural chemistry and the bio-pharmaceutics the risk can be quantified with greater rigour, and experimental plans shaped accordingly.

In attempting to get a greater definition of probability of a new form appearing, tools ranging from quantum chemistry analysis,[80] H-bonding statistics,[81] and full polymorph prediction,[82,83] can be applied either individually or in combination. Recent developments in theoretical chemistry mean that from molecular knowledge, 3D structure optimization can occur and charges can be visualized on the van der Waals surface. These can be used to quantitatively describe the relative strength of the hydrogen bond donors and acceptors.[80] The Cambridge Crystallographic Data Centre (CCDC) has for the last 40 years consolidated organic crystal structure information and distilled this knowledge into tools and software that is routinely applied to drug design.[18] Recently there has been an enhanced effort in the application of these sorts of tools in crystal engineering and solid form design. The Logit model[81] carries out a statistical analysis of hydrogen bonding patterns for a given structure. It identifies potential hydrogen bonding patterns which are compared to those in known crystal structures in order to rationalise the physical form stability.

The *ab initio* generation of reliable solid state structural details and properties through computational methods based only on molecular descriptors remains a major scientific goal. The methods being developed for structure prediction usually involve the stages of generating, clustering and refining trial structures. Final refinement of the potential structures is carried out minimising the lattice energy (see Section 22.2.2) with respect to the unit cell dimensions (a, b, c, α, ß and γ). Despite the inherent difficulties, predictions from first principles have been the subject of much elegant investigation through the last decade with increasing application to pharmaceutical compounds.[84] Proponents of these methods have now become so confident in their approaches that they are prepared to engage in blind tests to assess the predictability of their methods.[85] Figure 22.8 shows a typical energy density

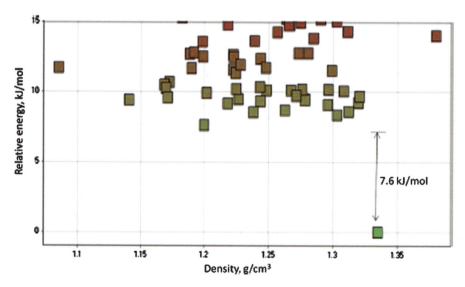

Figure 22.8   The packing energy/density plot for a polymorph prediction run. Each dot represents a potential crystal structure. The red blocks represent high energy forms. The most stable forms are those with the lowest energy and highest density highlighted in green. This demonstrated that the current from is the thermodynamically stable and distinct from the other possible structures.
Reprinted with permission from.[86] Copyright 2013 American Chemical Society.

diagram from a polymorph prediction run. Through comparing the computationally predicted solid form landscape with the experimentally known solid form data on a new candidate, it is possible to guide the solid form screening work needed. If the current solid form is consistent with the thermodynamically stable form and energetically distinct (green square in Figure 22.8) then confidence in the current form should be high and experimental screening work minimal. If the current solid form is more consistent/closely matched with those less stable structures (orange square in Figure 22.8) then this is a situation where more expansive experimental screening should; be undertaken to ensure solid form space has been effectively explored. This is a practical example of the application of these tools and technologies to allow the rapid acceleration of an oncology candidate to the patient post promising clinical results.[86]

22.5.2 Particle Design to Enable Clinical Studies

Whilst, traditionally the solid-form selection process has focused on two main factors; that is achieving an appropriate degree of product stability and bio-availability, increasing emphasis is also being focused on selecting solid forms which display optimal physical attributes for drug product processing (*i.e.* mechanical behaviour, surface properties and particle size, shape). Given the aforementioned perspective, it is worthy of note that there are elegant computational technologies emerging as foundation elements for the development and manufacture of advanced pharmaceutical particulate product. Recent examples illustrate predicting both crystal surface/solvent interactions and the enhancement of the solubility as a function of particle size reduction and morphological change.[87,88]

22.5.3 Solvation Crystal Packing Balance for Low Solubility Candidates

During recent years there has been tremendous effort made in the field of predicting the aqueous solubility of crystalline drug molecules.[89] From a pharmaceutical perspective, the crystalline solid is usually the solid-state of choice when developing a drug into a usable product and it would be of great value to be able to accurately predict the intrinsic solubility of crystalline drug molecules. This would improve the quality of the selection of compounds for synthesis and *in vitro* and *in vivo* testing, as well as improve our understanding of how structural variations change the solubility. Despite these significant efforts a unified theoretical approach that provides a definitive accurate and comprehensive approach to predicting solubility has proven elusive.

There have however been a number of semi empirical attempts to probe the changes in solubility as a function of structural changes in NCE structures in specific classes[26–28] as well as systematic approaches to looking at matched molecular pairs to determine improved solubility as a function of small structural changes and inferred crystal packing disruption.[90] The tools described in this chapter provide capabilities that can allow until now an unprecedented deconstruction of the importance of molecular solvation and crystal packing on solubility.[91] Recent work has shown a systematic experimental approach to examine key thermodynamic functions such as sublimation and hydration properties as a function of structural modifications[92] and a comprehensive computational approach to lattice energy estimation from molecular descriptors.[93] Two recent papers highlight the potential of these approaches and the scientific bridging across the two communities. The first paper optimises the solubility of a BCS class 4 antibiotic drug using structural modifications to disrupt the crystal lattice which was limiting the solubility[94] and the second uses co-crystals to optimise the dissolution rate of a psychotropic drug with known dissolution challenges.[95]

22.6 CONCLUDING REMARKS

Crystallisation is the final step of the API manufacture and so from a regulatory perspective must be both controlled and reproducible. In particular, it must provide API of a suitable quality in terms of both purity and the appropriate physical properties for robust dosage form design and processing. In recent years, a greater interest in the latter aspect has resulted in an emphasis of the link between the emerging NCE structure and the solid form in early development being considered in a more integrated fashion. This has the benefits of streamlined API solid form selection, rapid commercial product design and IP creation and product protection.

In this chapter we have attempted to bridge the new cutting edge academic progress to the best industrial practices that medicinal chemists and pharmaceutical scientists can apply to rapidly advance NCEs. The physical characteristics of the API solid form ultimately have the potential to affect the safety, efficacy and manufacturability of the product being designed and manufactured. The emergence of a range of computational tools, coupled to state of the art characterisation technologies, has allowed a greater range of desirable particle attributes to be accessed and understood in term of product performance. Harnessing this capability to new sophisticated small scale materials testing, and institutionalised product design rules, has led to the creation of a 'design by first intent' strategy for solid forms with tailored physicochemical attributes and functionality.

HINTS AND TIPS

- The lowest energy/most stable crystalline solid form is the preferential form for drug product development.
- Thus selecting the solid form which is more amenable to particle processing conditions (including melting point compressibility and brittle fracture) can be a strategy to facilitate rapid compound development.
- One of the most common reasons for lack of crystalline material prior to first in human studies is a lack of purity.
- The slurry experiment has become commonplace and can rapidly help identify the most stable crystalline form in the preclinical phase or early development.
- When considering whether particle size reduction will provide suitable enhancement in absorption, the rule of thumb is that the particle size diameter should be less than the solubility in µg/mL.

KEY REFERENCES

R. J. Davey, J. ter Horst and S. L. M. Schroeder, *Angew. Chem. Int. Ed.*, 2013, **52**, 2.
P. G. Stahly, *Cryst. Growth Des.*, 2007, 7, 1007.
P. T. A. Galek, F. H. Allen, L. Fábián and N. Feeder, *CrystEngComm.*, 2009, **11**, 2634.

REFERENCES

1. W. H. DeCamp, *Am. Pharm. Rev.*, 2001, **4**(3), 70.
2. S. R. Byrn, R. Pfeiffer and J. G. Stowell, *Am. Pharm. Rev.*, 2002, **5**(3), 92.

3. S. R. Chemburker, J. Bauer, K. Deming, H. Spiwek, K. Patel, J. Morris, R. Henry, S. Spanton, W. Dziki, W. Porter, J. Quick, P. Bauer, J. Donaubauer, B. A. Narayananan, M. Soldani, D. Riley and K. McFarland, *Org. Process Res. Dev.*, 2000, **4**, 413.
4. D. Erdemir, A. Y. Lee and A. S. Myerson, *Curr. Opin. Drug Discovery Dev.*, 2007, **10**, 746.
5. Glaxo Inc. and Glaxo Group Limited versus Novopharm Limited, No 5:94-CV-527-BO(1) 931 F. Supp. 1280 and 96-1466 DCT 94-CV-527.
6. W. Cabri, P. Ghetti, G. Pozzi and M. Alpegiani, *Org. Process Res. Dev.*, 2007, **11**, 64.
7. R. Storey, R. Docherty and P. D. Higginson, *Am. Pharm. Rev.*, 2003, **6**(1), 100.
8. A. Thayer, *Chem. Eng. News*, 2007, **85**, 17.
9. *The Crystal as a Supramolecular Entity*, ed. G. R. Desiraju, John Wiley and Sons, Chichester, England, 1997.
10. C. C. Sun, *J. Pharm. Sci.*, 2009, **98**(5), 1671.
11. K. Chow, H. Y. H. Tong, S. Lum and A. H. L. Chow, *J. Pharm. Sci.*, 2008, **97**, 2855.
12. B. H. Hancock and J. Elliot, *MRS Bulletin*, 2006, **31**, 869.
13. 'An Introduction to Crystallography', F. C. Phillips, Third Ed., Longmans, London, 1963.
14. G. Nichols and C. J. Frampton, *J. Pharm. Sci.*, 1998, **87**, 684.
15. *Physical Characterization of Pharmaceutical Solids*, ed. H. G. Brittain, Dekker, New York, 1995.
16. *Molecular Crystals*, ed. J. G. Wright, Cambridge University Press, Cambridge, 1987.
17. M. C. Etter, *Acc. Chem. Res*, 1990, **23**(4), 120.
18. R. Taylor and O. Kennard, *Acc. Chem. Res.*, 1984, **17**, 320.
19. R. Taylor and O. Kennard, *J. Chem. Soc.*, 1982, **104**, 5063.
20. R. Docherty and W. Jones, in *Organic Molecular Solids: Properties and Applications*, ed. W. Jones, CRC Press, London, 1997, Chapter 3, pp. 113–148.
21. C. P. Brock and J. D. Dunitz, *Chem. Mater.*, 1994, **6**, 1118.
22. *Molecular Crystals and Molecules*, ed. A. I. Kitaigorodsky, Academic Press, New York, 1973.
23. D. E. Williams, *J. Chem. Phys.*, 1966, **45**, 3770.
24. S. Lifson, A. T. Hagler and P. Dauber, *J. Am. Chem. Soc.*, 1979, **101**, 5111.
25. F. A. Momany, L. M. Carruthers, R. F. McGuire and H. A. Scherega, *J. Phys. Chem.*, 1974, **78**, 1595.
26. J. S. Scott, A. M. Birch, K. J. Brocklehurst, A. Broo, H. S. Brown, R. J. Butlin, D. S. Clarke, O. Davidsson, A. Ertan and K. Goldberg, *J. Med Chem.*, 2012, **55**, 5361.
27. M. Ishikawa and Y. Hashimoto, *J. Med Chem.*, 2011, **54**, 1539.
28. A. P. Hill and R. J. Young, *Drug Discovery Today*, 2010, **15**, 648.
29. *From Molecules to Crystals*, ed. R. Davey and J. Garside, Oxford University Press, Oxford, 1998.
30. R. J. Davey, K. Allen, N. Blagden, W. I. Cross, H. F. Lieberman, M. J. Quayle, S. Righini, L. Seton and G. J. Tiddy, *CrystEngComm*, 2002, **4**, 1.
31. R. J. Davey, J. ter Horst and S. L. M. Schroeder, *Angew. Chem. Int. Ed.*, 2013, **52**, 2.
32. *Solubility Behavior of Organic Compounds*, ed. D. J. W. Grant and T. Higuchi, Wiley Interscience, New York, 1990.
33. G. L. Perlovich, N. N. Strakhova, V. P. Kazachenko, T. V. Volkova, V. V. Tkachev, K. J. Schaper and O. A. Raevsky, *Int. J. Pharm. (Amsterdam, Neth.)*, 2008, **349**, 300.
34. A. Koda, S. Ito, S. Itai and K. Yamamoto, *Yakuzaigaku*, 2000, **60**, 43.
35. *Handbook of Pharmaceutical Salts*, ed. P. H. Stahl and C. G. Wermuth, Wiley-VCH, Weinheim, 2002.
36. S. M. Berge, L. D. Bighley and D. C. Monkhouse, *J. Pharm. Sci.*, 1977, **66**, 1.
37. G. S. Paulekuhn, J. B. Dressman and C. Saal, *J. Med Chem.*, 2007, **50**(26), 6665.
38. R. J. Bastin, M. J. Bowler and B. J. Slater, *Org. Process Res. Dev*, 2000, **4**, 427.

39. http://www.fda.gov/ohrms/dockets/ac/09/slides/2009-4412s1-01-FDA.pdf and http://www.fda.gov/ohrms/dockets/ac/09/briefing/2009-4412b1-01-FDA.pdf.

40. C. B. Aakeroy, M. E. Fasulo and J. Desper, *Mol. Pharmaceutics*, 2007, **4**, 317.

41. T. Friscic, A. V. Trask, W. Jones and W. D. S. Motherwell, *Angew. Chem. Int. Ed.*, 2006, **45**, 7546.

42. S. L. Childs, L. J. Chyall, J. T. Dunlap, V. N. Smolenskaya, B. C. Stahly and P. G. Stahly, *J. Am. Chem. Soc.*, 2004, **126**(41), 13335.

43. J. F. Remenar, S. L. Morissette, M. L. Peterson, B. Moulton, M. J. MacPhee, H. R. Guzman and O. Almarsson, *J. Am. Chem. Soc.*, 2003, **125**, 8456.

44. J. Haleblian and W. C. McCrone, *J. Pharm. Sci.*, 1969, **58**, 911.

45. *The Solid State Chemistry of Drugs*, ed. S. R. Byrn, R. R. Pfeiffer and J. G. Stowell, Academic Press, New York, 1982.

46. European Medicines Agency, London, 18 June 2008, Doc. Ref.: EMEA/265069/2008 Rev. 1.

47. P. G. Stahly, *Cryst. Growth Des.*, 2007, **7**, 1007.

48. M. Pudipeddi and A. T. M. Surajaddin, *J. Pharm. Sci.*, 2005, **5**, 94.

49. R. Docherty and K. Pencheva, (Unpublished Results).

50. G. C. H. Young and D. J. W. Grant, *J. Pharm. Sci.*, 2001, **90**, 1878.

51. J. M. Miller, B. M. Collman, L. R. Greene, D. W. Grant and A. C. Blackburn, *Pharm. Dev. Tech.*, 2005, **10**, 291.

52. N. Blagden and R. J. Davey, *Cryst. Growth Des.*, 2003, **3**, 873.

53. S. Desikan, R. L. Parsons, W. P. Davis, J. E. Ward, W. J. Marshall and P. H. Toma, *Org. Process Res. Dev.*, 2005, **9**, 933.

54. M. Prashad, P. Sutton, R. Wu, B. Hu, J. Vivelo, J. Carosi, P. Kapa and J. Liang, *Org. Process Res. Dev.*, 2010, **14**, 876.

55. H. Zhu and D. J. W. Grant, *Int. J. Pharm. (Amsterdam, Neth.)*, 1996, **139**, 33.

56. Y. Cui and E. Yao, *J. Pharm. Sci.*, 2008, **97**, 2730.

57. M. D. Ticehurst, R. A. Storey and C. Watt, *Int. J. Pharm. (Amsterdam, Neth.)*, 2002, **247**, 1.

58. S. M. Reutzel and G. A. Stephenson, in *Pharmaceutical Stress Testing: Predicting Drug Degradation*, ed. S. W. Baertschi, K. M. Alasante and R. A. Reed, Informa Healthcare, 2005, Chapter 10, pp. 254–285.

59. S. H Yalkowsky and S. Bolton, *Pharm. Res.*, 1990, **7**, 962.

60. M. P. Mullarney and N. Leyva, *Pharm. Technol.*, 2009, **33**, 126.

61. P. Narayan and B. C. Hancock, *Mat. Sci. Eng. A*, 2003, **A355**, 24.

62. C. C. Sun, H. Hou, P. Gao, C. Ma, C. Medina and J. Alvarez, *J. Pharm. Sci.*, 2009, **98**(1), 239.

63. K. Johnson and A. C. Swindell, *Pharm. Res.*, 1996, **13**, 1795.

64. L. Taylor, D. G. Papadopoulos, P. G. Dunn, A. C. Bentham, N. J. Dawson, J. C. Mitchell and M. J. Snowden, *Org. Process Res. Dev.*, 2004, **8**, 674.

65. G. L. Amidon, H. Lennernas, V. P. Shah and J. R. A. Crison, *Pharm. Res.*, 1995, **12**, 413.

66. J. M. Butler and J. B. Dressman, *J. Pharm. Sci.*, 2010, **99**(12), 4940.

67. K. Sugano, *Expert Opin. Drug Metab. Toxicol.*, 2009, **5**, 263.

68. W. I. Higuchi and E. N. Hiestand, *J. Pharm. Sci.*, **196**(52), 67.

69. D. M. Oh, R. I. Curl, C. S. Yong and G. L. Amidon, *Arch. Pharm. Res.*, 1995, **18**, 427.

70. C. W. Poulton, *Adv. Drug Delivery Rev.*, 1997, **25**, 47.

71. W. N. Charman, *J. Pharm. Sci.*, 2000, **7**, 967.

72. B. C. Hancock and G. Zografi, *J. Pharm. Sci.*, 1990, **7**, 902.

73. C. Luener and J. Dressman, *Eur. J. Pharm. Biopharm.*, 2000, **50**, 47.

74. P. Gao, R. D. Rush, W. P. Pfund, T. Huang, J. M. Bauer, W. Morozowich, M. Kuo and M. J. Hagema, *J. Pharm. Sci.*, 2003, **92**, 2386.

75. P. Gao, M. E. Guyton, T. Huang, J. M. Bauer, K. J. Stefanski and Q. Lu, *Drug Del. Ind. Pharm.*, 2004, **30**, 221.

76. P. Ettmayer, G. L Amidon, B. Clemant and B. Testa, *J. Med. Chem.*, 2004, **47**, 2393.

77. H. D. Williams, N. L. Trevaskis, S. A. Charman, R. M. Shanker, W. N. Charman, C. W. Pouton and C. J. H. Porter, *Pharmacol. Rev.*, 2013, **65**, 315.

78. ICH Q9, *Quality Risk Management*, http://www.ich.org/ (last accessed January 2011).

79. ICH Q6a, *Specifications: Test Procedures And Acceptance Criteria For New Drug Substances And New Drug Products: Chemical Substances*, http://www.ich.org/ (last accessed August 2014).

80. Y. A. Abramov and K. Pencheva, in *Chemical Engineering in the Pharmaceutical Industry: R&D to Manufacturing*, ed. D. J. am Ende, John Wiley & Sons Inc., Hoboken, NJ, USA, 2010, Chapter 25, pp. 477–490.

81. P. T. A. Galek, F. H. Allen, L. Fábián and N. Feeder, *CrystEngComm.*, 2009, **11**, 2634.

82. H. R. Karfunkel and R. J. Gdanitz, *J. Comput. Chem.*, 1992, **13**, 1771.

83. A. Gavezzotti, *Acc. Chem. Res.*, 1999, **27**, 309.

84. R. S. Payne, R. C. Rowe, R. J. Roberts, M. H. Charlton and R. Docherty, *J. Comp. Chem.*, 1999, **20**(2), 262.

85. G. M. Day, T. G. Cooper, A. J. Cruz-Cabeza, K. E. Hejczyk, H. L. Ammon, S. X. M. Boerrigter, J. S. Tan, R. G. Della Valle, E. Venuti, J. Jose, S. R. Gadre, G. R. Desiraju, T. S. Thakur, B. P. van Eijck, J. C. Facelli, V. E. Bazterra, M. B. Ferraro, D. W. M. Hofmann, M. A. Neumann, F. J. J. Leusen, J. Kendrick, S. L. Price, A. J. Misquitta, P. G. Karamertzanis, G. W. A. Welch, H. A. Scheraga, Y. A. Arnautova, M. U. Schmidt, J. van de Streek, A. K. Wolf and B. Schweizer, *Acta Crystallogr., Sect. B: Struct. Crystallogr. Cryst. Chem.*, 2009, **65**, 107.

86. Y. A. Abramov, *Org. Process Res. Dev.*, 2013, **17**, 472.

87. R. B. Hammond, K. Pencheva and K. J. Roberts, *Cryst. Growth Des.*, 2007, **7**, 875.

88. R. B. Hammond, K. Pencheva, K. J. Roberts and T. Auffret, *J. Pharm. Sci.*, 2009, **98**(12), 4589.

89. P. Jain and S. H. Yalkowsky, *Int. J. Pharm. (Amsterdam, Neth.)*, 2010, **385**, 1.

90. A. G. Leach, H. D. Jones, D. A. Cosgrove, P. W. Kenny, L. Ruston, P. MacFaul, M. J. Wood, N. Colclough and B. Law, *J. Med. Chem.*, 2006, **49**, 6672.

91. C. M. Wassvika, A. G. Holmen, C. A. S. Bergstrom, I. Zamora and P. Artursson, *Eur. J. Pharm. Sci.*, 2006, **29**, 294.

92. G. L. Perlovich *Mol. Pharmaceutics*, 2014, **11**, 1–11.

93. M. Salahinejad, T. C. Le and D. A. Winkler, *J. Chem. Inf. Model.*, 2013, **53**, 223.

94. U. Tehler, J. H. Fagerberg, R. Svensson, M. Larhed, P. Artursson and C. A. Bergstrom, *J. Med. Chem.*, 2013, **56**, 2690.

95. T. Zhang, Y. Yang, H. Wang, F. Sun, X. Zhao, J. Jia, J. Liu, W. Guo, X. Cui, J. Gu and G. Zhu, *Cryst. Growth Des.*, 2013, **13**, 5261.

CHAPTER 23

The Chemical Development and Medicinal Chemistry Interface

DAVID LATHBURY*[a] AND DAVID ENNIS[b]

[a] Albany Molecular Research Incorporated, 26 Corporate Cir, Albany, NY 12203, United States;
[b] AstraZeneca, Charter Way, Silk Road Business Park, Macclesfield, Cheshire, SK10 2NX, UK
*E-mail: David.Lathbury@amriglobal.com

23.1 WHAT'S THE INTERACTION TRYING TO ACHIEVE?

The hope of all medicinal chemists is that, one day, their molecule will actually make it all the way to approval and ultimately treat patients. This being the case, at some stage, their chemistry will be handed over to a chemical development group and the quality of that hand over can have a big influence on the speed of the early development program and therefore the commercial value of the target or new chemical entity (NCE).

With this in mind, we will try and convey the types of considerations and thoughts that go through the mind of the chemical development scientist when confronted with a new molecule. Although early on, supply of the active pharmaceutical ingredient (API) is on everyone's mind, the chemical development scientist is thinking both short and long term.

The first questions should be strategic ones, *i.e.* is there anything "peculiar" about the molecule that might be a show stopper from the API view which may ultimately halt the compound's development. Secondly, how big is the technical task, how closely will the API rate of supply meet the aspirations of the overall development project? As none of the other development departments can do much without an adequate supply of API, it is generally a good idea to keep the Chemical Development activities a phase ahead of the other departments. This often doesn't happen but is a relatively cheap way to de-risk many projects.

In terms of assessing the current and/or future potential routes of synthesis, use of criteria based on the "SELECT" criteria[1] are a good starting point. Initially it is very difficult to be definitive as to whether a NCE is developable in the longer term. With nearly 50 years' experience in chemical development between us, we have only seen two projects stopped at nomination into preclinical development due to cost of goods. Both were high dose antibiotics (0.75 to 1.5 g per day) with low targeted cost of goods ($2000/kg) and both in excess of 30 steps

The Handbook of Medicinal Chemistry: Principles and Practice
Edited by Andrew Davis and Simon E Ward
© The Royal Society of Chemistry 2015
Published by the Royal Society of Chemistry, www.rsc.org

long. Most times even the most difficult problems can be solved. So in the vast majority of cases, as the list of possible stumbling blocks are being identified, focus turns to rapid API supply. This is where medicinal chemistry can really help.

In Section 23.4 we will cover the information that Chemical Development would find useful and hopefully convince the reader that much of it is available "for free" if the medicinal chemist is thinking ahead. However, this prompts the obvious question, if it is valuable and easy to get, why is it often not provided?

23.2 WHY DON'T MEDICINAL CHEMISTS THINK AHEAD TO CHEMICAL DEVELOPMENT?

In the early to mid-1980's, before the arrival of high throughput screening (HTS), several grams of the new chemical entity (NCE) were often required for even the early biological assays. This meant that most medicinal chemists often carried out reactions on a 100 g scale and hence issues with respect to work up and isolation were more frequently identified and often solved before reaching Chemical Development. Mundane techniques such as crystallization were common place. There was much more connectivity between Medicinal Chemistry and Chemical Development in terms of the day to day job. Today, a few grams of material can get you a long way towards compound selection. Advances in purification technology mean that crystallization is seldom practiced in medicinal chemistry and the scale of reactions normally carried out has reduced tremendously.

This, along with the pressure of working to deadlines, can often drive "acceptance" of "poor" chemistry from a naïve view, taking comfort in the fact that the synthesis had been used before, however difficult.

These changes are understandable and to a large degree inevitable. However, the growth size of R&D departments and the use of more external contract research organization services, for scale-up and bulk chemical synthesis, make the process much more complex. It means that the need for a good technical transfer procedure from Medicinal Chemistry to Chemical Development is more important now than ever before. The final point in this section is that, as you will see, much of the information needed is collected contemporaneously, and if observation are not made or recorded, it's difficult if not impossible to recreate them. Therefore thinking about the needs of Chemical Development has to be well established if this is to be successful.

23.3 WHAT CONSTITUTES A GOOD SYNTHESIS?

Chemical development scientists are often asked to define what a good manufacturing route looks like. A good review of the "SELECT" (Safety, Environment, Legal, Economy, Control, Throughput) criteria, that need to be met to obtain a commercial route, has been written by Butters *et al.*[1] The criteria are outlined below:

- *Safety*: Any Safety (*e.g.* thermal and explosivity issues) and Health (exposure) issues are manageable.
- *Environment*: Process waste is minimized, and process meets current and anticipated future environmental regulations. Process Mass Intensity (PMI) (a measure of kg of waste, including solvent and reagents per kg of product produced) is a good measure of this.[2]
- *Legal*: Route has freedom to operate and can be protected by process patents.
- *Economy*: Route meets long-term cost target. Number of steps is generally a good measure of this. The shorter and more convergent a process, the more cost effective it is likely to be.

- *Control*: All process steps are reproducible and tolerant (within defined limits) of variation of process parameters (robust). Process contains stable intermediates to allow flexibility in production planning. All starting materials are available (or can be made available) in bulk, supply is assured and quality can be established (specifications).
- *Throughput*: Convergent processes and short cycle times that allow parallel and efficient manufacturing. This is especially important for high volume products.

However, the synthesis developed in Medicinal Chemistry is unlikely to meet these criteria mainly due to the fact that a medicinal chemistry synthesis is designed to give diversity *vs.* a chemical development focus is on a single compound.

Once lead optimization starts to focus on one or two compounds, the criteria converge normally to the shortest most convergent synthesis the medicinal chemist can devise. What is not often appreciated is that Chemical Development's unit of capacity is number of chemical steps not the number of drug candidates.

The object of this chapter is not to get Medicinal Chemistry to do Chemical Development's job. Irrespective of how your particular organization is structured more efficient chemistry will speed up the late stages of discovery. Medicinal Chemistry and Chemical Development are both staffed with synthetic organic chemists so this is an exercise where both departments can equally contribute.

Whether Medicinal Chemistry has any accountability in the route development can be answered with a straight forward question, as to whether the effort will speed up the late lead optimization phase or not. If it does then we would argue that this is very much down to Medicinal Chemistry with help from Chemical Development. If not then the responsibility lies clearly with Chemical Development.

The two examples given below (Schemes 23.1–23.4) look fairly dramatic; however, we could easily have included others. In the majority of cases, synthetic routes are reduced by 20–30%. The faster we can achieve these reductions, the faster the overall development will be, along with substantial cost reduction.

The original synthesis outlined in Scheme 23.1 is not atypical. It was designed to examine specific structure–activity relationships and the advances in chiral separations enabled the resolution to be accomplished late on in the synthesis. For a few hundred milligrams the route was acceptable, and was more than adequate to support the majority of the lead optimization phase.

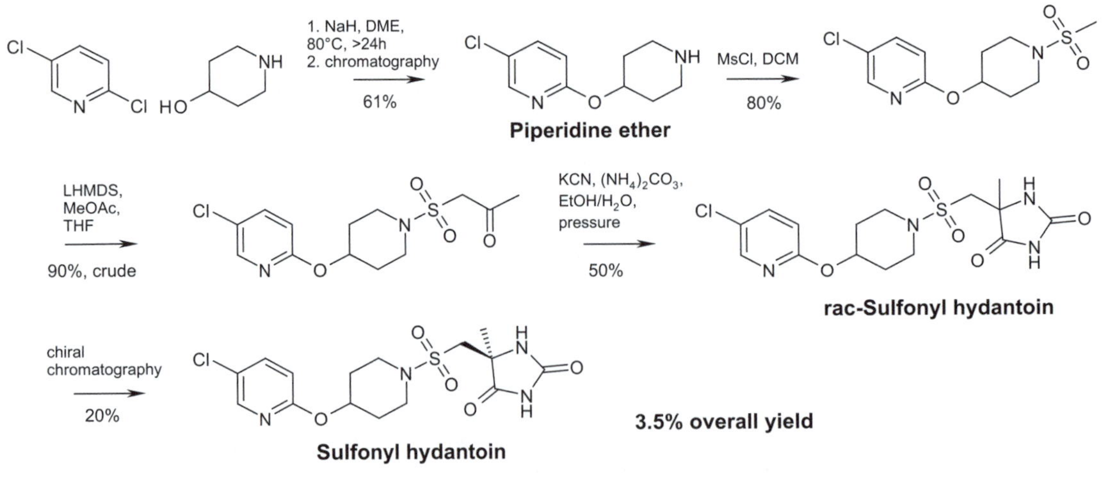

Scheme 23.1   The first synthesis of potential candidate drug "sulfonyl hydantoin" (Route A).

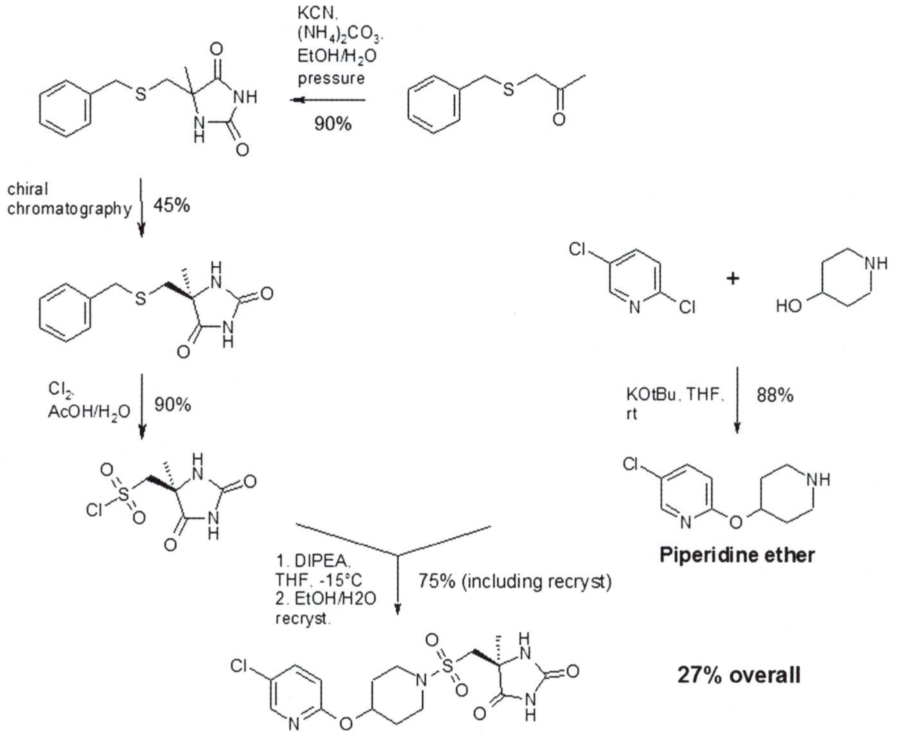

Scheme 23.2 Early route change to "sulfonyl hydantoin" (Route B).

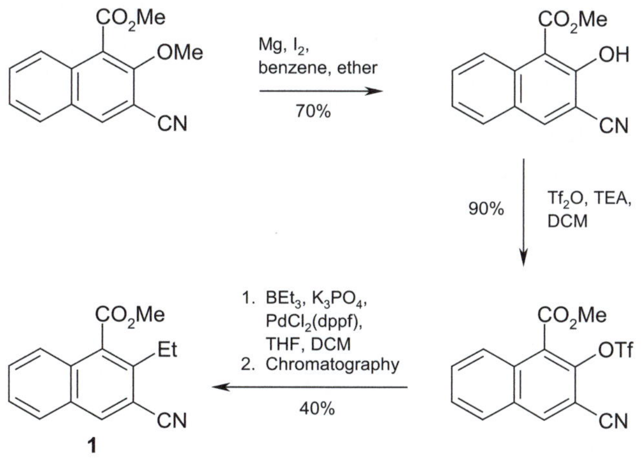

Scheme 23.3 Original route for the preparation of naphthalene intermediate (**1**).

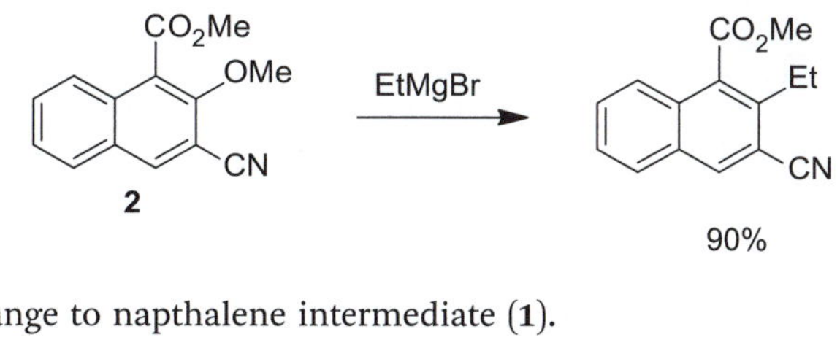

Scheme 23.4 Route change to napthalene intermediate (**1**).

However, as the project moved towards candidate selection the 250–300 g amount of API required would mean the first stage of the synthesis would need to be carried out on a kg scale and that Phase I supply would require almost 30 kg of the early stage. Again these amounts would not present insurmountable challenges but the question should always be, "can chemistry do better?" Nearly always the answer is yes, but early on in development, the inefficient route may still be the fastest way. However, the question should always be asked and an analysis be carried out. As you will see, one can often save time and even early in development save large sums of money by improving the route of synthesis.

In this particular case, the accountability for preparing material for the 7 day non-rodent toxicology studies lay with Medicinal Chemistry; however, the team was sufficiently interested in improving the quality of the synthetic route (given low solubility of the racemic sulfonyl hydantoin of approximately 1 mg/mL, essentially precluding the original sequence as a viable method) that they were willing to examine alternatives. Indeed within 2 weeks of starting the work, a significantly better alternative shown in Scheme 23.2 was developed.

By taking advantage of the obvious, sulfonamide disconnection the synthesis became much more convergent thereby both increasing yield, from 3.5% to 27%, and significantly shortening the delivery time.

Moving the resolution step earlier also gave a dramatic improvement. Indeed a rough estimate of cost would suggest that in terms of speed, the dog safety material was prepared over 2 months earlier by changing the chemistry and the cost savings achieved in the first GMP campaign were of the order of $700K (see Figure 23.1). There have been many initiatives in many companies trying to reduce the cost or the speed of early drug development. However, efficient chemistry is the single biggest factor, and the one discussed the least.

A second example shown in Scheme 23.4 demonstrates an even more stark improvement where by the initial overall yield was increased > three-fold with a single trial reaction. In this case, a heroic effort was made to prepare 200 g of a key intermediate (**1**) to support the lead optimization development programme. This took the best part of 2 months, a result of low overall yields, special containment requirements for sensitive reagents, and poor chemical selectivity, necessitating extensive use of chromatographic purification.

Stepping back, the synthetic chemist recognized the opportunity to introduce the ethyl group in a "one-step" process, by reaction of **2** with EtMgBr. Why wasn't this attempted sooner?

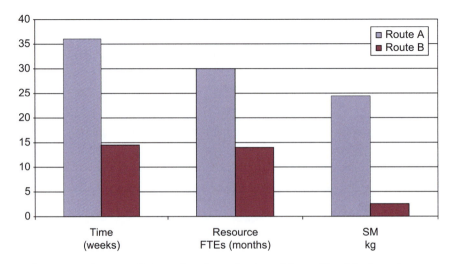

Figure 23.1 Advantages of more efficient chemistry developed for sulfonyl hydantoin.

All reagents were readily available! Yes, there may be potential chemoselectivity issues with such an approach, but the gain could have significantly reduced preparation times, and hence sped up the LO program. With the pressure to progress the program, people were happy to repeat what was done previously, rather than try something new.

In actual fact, the reaction of **2** with EtMgBr was tried in the Chemical Development laboratories, and the attempt gave a 90% yield of the desired product **1** with remarkable selectivity. It subsequently took 1 day to prepare 500 g of material.

The consequence of this was earlier candidate drug nomination (several months!) and as described in Figure 23.2: increased speed of manufacture of kg quantities to support the First GLP toxicology and first time in man (FTIM) studies; reduced resources to achieve this (productivity); and reduced quantities of starting material (reduced lead times and cost). These outcomes will be of value to any organization.

In the simple example below (Scheme 23.5), the reaction of the *N*-Boc protected hydroxy piperidine with NaH was a very slow reaction, and chromatography was required to obtain

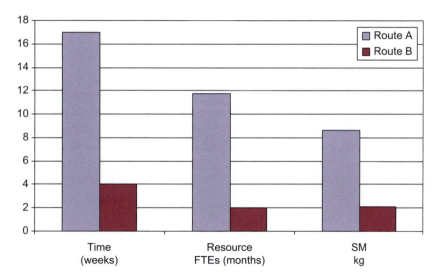

Figure 23.2 Metrics showing value of the route change.

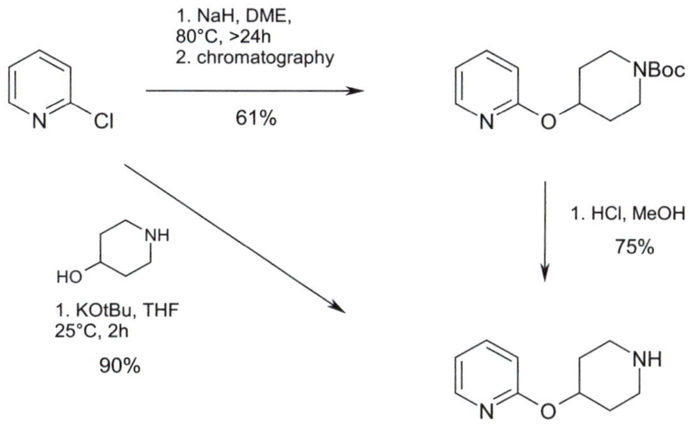

Scheme 23.5 Rapid route change to piperidine ether.

appropriate quality material for onward processing. Successful scale-up was limited to 50 g batches. Use of KOtBu as base gave a dramatic improvement in reaction, allowing for rapid progress of material supply and hence speeding-up the early development programme.

All of these examples demonstrate two key principles. If there are obvious opportunities to improve an early discovery synthesis they will often be surprisingly quick to evaluate.

Secondly, these early changes can have quite a dramatic effect on reducing time lines and cost. When we reflect on the numerous projects worked on over the years in Chemical Development, the biggest issues were created when we were not brave enough to introduce early route change.

So be brave early on. If it does not work, as Ed Grabowski[3] said, you do not have to tell anyone. This is known as the Grabowski axiom and is as true today as it was 10 years ago.

23.4 WHAT MEDICINAL CHEMISTRY CAN DO AT NO/LITTLE EXTRA COST TO HELP CHEMICAL DEVELOPMENT

There are a few basic items that if carried out on a regular basis will significantly improve the technical transfer and can speed up early safety and clinical supplies of API.

23.4.1 Recording Experimental Data

The key to any valid experiment is the ability to repeat it, and this requires an appropriate level of experimental write-up. From experience over many years, the amount of time that can be wasted because of a poorly recorded experiment is scandalous. It is particularly important when transferring knowledge between different groups. The introduction of a quality standard to ensure the appropriate level of information/observation is documented pays off many times.

The following is an example of a checklist established for technical transfer of knowledge used at AstraZeneca over the years:

The Chemistry
- Typed experimental details of current discovery process. Many institutions have a system of electronic lab note books which can be very helpful in this regard.
- Indication of scale operated, range of yields observed. Not just the highest achieved. If you a have a range of say 30% without an adequate explanation, this highlights an area of investigation that chemical development will have to carry out.

Experience on scale-up
- Did things go according to plan?
- Did reaction time, yield, quality of output, match earlier results on smaller scale?
- Any indications of instability of intermediates/API at any stage. Simply leaving samples at ambient for a few days to week or so and then reanalyzing them can give valuable information.
- Any Health & Safety issues, (exothermic reactions, for example). At small scale these temperature rises can be small, but can be serious on scale up. The same is true for issues with gas evolution.

Route Development
- Other routes investigated which were unsuccessful or less attractive than current route.
- Other ideas not investigated.

Analytical

- All existing analytical methods (TLC, HPLC, GC *etc.*) for all stages.
- Any available information on impurities and any insight as to how they might be formed or how easily they can be removed.

Intellectual Property/Literature

- Current patent situation on compound and route.
- Relevant literature references.

Raw Materials

- Availability of key raw materials. This is often the rate limiting step for early phase drug substance.
- Any experience of outsourcing?

Bulk Drug solid state issues (from early preformulation studies)

- Bulk drug stability.
- Salt form.
- Information on hydrates, solvates and appropriate crystallization conditions.
- Information on physical properties *e.g.* DSC, XRD, any indication of polymorphism.

All the above are important and can speed up the chemical development process. In particular, samples of advanced intermediates can really speed up the development of the final stages of the synthesis.

23.5   WHAT ARE THE TELL-TALE SIGNS OF POTENTIAL ISSUES?

As discussed previously, in the Medicinal Chemistry department, the initial route is typically designed to maximize diversity. For this reason, it is unlikely to be the "best" commercial route to the selected Candidate Drug. There are some fairly obvious signs that should alert the medicinal chemist, that there may be scale up or rapid supply issues.

Typical issues with initial routes are:

- *Long linear sequence.* For example, with a 10 step route, even with 80% yield per step, the overall yield is only 10%. This becomes much more significant when requiring 50–250 g quantities (dose range finding studies) and >2 kg of material for the first GLP (FGLP) toxicology and first time in man (FTIM) studies.
- *Low/unpredictable yields.* This can be especially significant later on with the synthesis where the value of the later intermediates is obviously much larger and the implications of late stage failure on project time lines will be more significant.
- *Poor availability of raw materials (long lead times).* This can be especially significant if larger quantities are needed, based on throughput (see Schemes 23.1 and 23.2), and it is becoming a bigger issue as it is much harder to determine where the material is actually being made.
- *Non-scalable reagents.* There are not many reagents that would be truly classed as "non-scalable", but significant effort may be required to facilitate use, *e.g.* diazomethane. Or the reagents just aren't available on any sensible scale.
- *Unstable intermediates.* These become more significant on scale-up. If your intermediates are decomposing within a few hours then start to think how this might be addressed if you

need several kg. Continuous processing (flow chemistry) is increasingly coming to the rescue here but it still has to be applied on a case by case basis.

- *Safety concerns.* For example, intrinsically energetic reagents/reactions and occupational toxicity considerations.

However, as the reader has by now concluded, none of the above will count for much unless there is a good and open dialogue between the medicinal chemistry and chemical development scientists. Even if they are not in the same company one shouldn't allow the traditional "silo mentality" that pervades many companies to filter the above information. Medicinal chemists have a particular expertise as do Chemical Development; the ideal situation is when we can utilize both for the good of the project. The better the communication, the better the outcome will be.

23.6 HOW BEST TO DEAL WITH THE ABOVE ISSUES?

Over the last 25 years, companies have adopted many models to try and accelerate development times or reduce cost.

Whatever one decides to do, it has to be consistent with the scientific challenge posed by the particular project. In that sense there are probably no universal solutions other than think about what you are going to do and what the longer term impacts are likely to be, before you do them.

However we can share our experiences of some of the options we've seen and adopted over the years.

In terms of a demarcation, the interaction works best when both Medicinal Chemistry and Chemical Development are actively involved. Having Medicinal Chemistry accountable for delivery of the first 150 g or so of API seems to have been the most successful strategy for ensuring this happens. It follows, therefore, that a small scale up group in discovery often works well. In some organizations, rotating staff between the departments has been a very successful way of improving the longer term relationship. A further point about Medicinal Chemistry scale up groups is that they should have a good level of analytical support. This is an area that is often neglected by Medicinal Chemistry and when things go wrong on scale up is often high on the list of contributory factors. In addition it should be staffed with some of the best synthetic chemists and not be seen as a second class role.

The challenge for any group of this type is to see that they create real value, otherwise it's probably cheaper nowadays to outsource this activity. The value can be measured in terms of speed, or whether the processes emerging from such a group are significantly improved from the initial discovery processes.

In terms of outsourcing in general, again there are no hard and fast rules. Some companies outsource the entire development of the API whilst others do much more in house. If one thinks about what is needed in the long term, the company or group that is ultimately responsible for the commercial manufacture of the API has to have much knowledge and experience of the final process as possible. Therefore any outsourcing strategy has to ensure this happens. In our view this should mean that instead of outsourcing activities based simply on the concept of the earlier the phase of the programme the more amenable to outsourcing it is, a bit more thought should be given.

There are two major sources of chemical development attrition. Firstly, is the molecule likely to make it all the way to market? This is common for all other development departments. However, the second question is whether the current process being studied for a given molecule is likely to be the final commercial one. If it isn't, it is probably a much better idea to tactically outsource a process on a mid-phase project where the process is unlikely to be viable long term, rather than outsource an earlier phase project where the process is.

So there is a common theme running though chemical development that also applies to medicinal chemistry scale up groups which we would state as getting good chemistry established as soon as possible, and tactically using outsourcing to support that objective. The use of kilo labs with, for example, large rotary evaporators and chromatography, can be of great use to enable the introduction of more efficient chemistry quickly without having to solve all the work up and isolation issues associated with larger scale plant production. The down side of course, is that the same technology can be used to support poor chemistry way beyond its useful life.

The popular concept of "fit for purpose" has been much misused in our view. In its broadest sense it can be a valuable tool, but if used without thinking, it normally becomes a mandate for mediocrity and can inflict long term damage to a departments scientific capabilities.

Of course there are times when you have to just run through a poor synthesis to get material, however some of the examples given above hopefully show that this strategy is used too often. In many cases it actually takes more time and it's always more expensive.

23.7   FINAL THOUGHTS

Molecules aren't intelligent. How they behave isn't driven by their importance or where they fit in a company portfolio. It is driven simply by the inherent scientific challenges they pose. Any strategy therefore can only be general and exceptions have to be made from time to time.

However, the quicker one can introduce effective chemistry into any programme, the faster it will go and the cheaper it will be. Time spent working on the interface between Medicinal Chemistry and Chemical Development will be hugely beneficial whether that is within the same company or externally. Lastly although the API is the fuel for the development engine, ultimately, it's the quality of the manufacturing process that will be the most valuable output.

Getting a molecule's chemical development off to a good start is one of the best investments one can make.

HINTS AND TIPS

- As most development activities are dependent on an adequate supply of API, keep the chemical development activities a phase ahead of the other departments.
- Think ahead to chemical development, be brave and introduce early route changes.
- The quicker one can introduce effective chemistry into any programme, the faster it will go and the cheaper it will be.
- Be alert to tell-tale signs of potential future problems (safety concerns, long linear sequence, poor/variable yields/reagent availability *etc.*).

KEY REFERENCES

For a general introduction to Chemical Development the reader should see:

N. Anderson, *Practical Process Research and Development: A Guide for Organic Chemists*, Academic Press, Waltham, MA, USA, 2012.
C. Thomson and G. Robinson, *On Chemistry, 5 On Process Chemistry*, SCI-Ink Ltd, Oxford, 2013.

See also a guide to synthetic route change:

D. Ennis, A. Harris and D. Lathbury, *Chemistry Today*, 2012, **30**(2).

For an overview of FDA guidance documents see:

Food and Drug Administration web site, Guidance documents, in particular, the section on Investigational New Drug (IND) Application, http://www.fda.gov/downloads/Drugs/.../Guidances/ucm074980.pdf.

REFERENCES

1. M. Butters, D. Catterick, A. Craig, A. Curzons, D. Dale, A. Gilmore, S. P. Green, I. Marziano, J. Sherlock and W. White, *Chem. Rev.*, 2006, **106**, 3002–3027.
2. C. Jimenez-Gonzalez, C. S. Ponder, B. Broxterman and J. B. Manley, *Org. Process Res. Dev.*, 2011, **15**, 912–917.
3. Personal communication from Dr. Ed Grabowski, former senior leader at the Merck process group at Rahway, USA.

CHAPTER 24

Project Management

PAULINE STEWART-LONG

11 Brands Hill Avenue, High Wycombe, Bucks, HP13 5PZ, UK
E-mail: paulinesl@btinternet.com

24.1 INTRODUCTION

As a medicinal chemist working in drug discovery you are likely to be part of a project team and it will help to understand how you can best contribute to the team and the project. You may also be leading a sub-project team which will need managing. For both roles an understanding of basic project management principles and techniques will be invaluable and help you to achieve your objectives. This chapter provides an introduction to those principles and techniques but cannot be a comprehensive guide; for those interested in developing project management skills, further reading and practical training is recommended.

24.1.1 What is a Project?

A project is a temporary endeavour to create a unique product or service[1] with a definite beginning and a definite end and differs from routine business in that it is a unique not repetitive operation. Some tasks within the project may be considered routine by the individuals conducting them. For example, a certain type of toxicology study may only be done to support a particular and unique type of project so is not routine business. However, a test assay in a manufacturing process is truly repetitive business operation.

Project Management is the application of skills and techniques to deliver projects effectively and efficiently to time, quality and budget. It is a strategic competency for companies, enabling them to link project delivery and business goals—and thus better compete in their particular market sector.

The Project Manager is rarely involved directly in the activities that make up the project but has a role to maintain progress and ensure integration of the various team members and their contributions so that the ultimate goal is delivered. However, in early drug discovery there is rarely an individual assigned as the Project Manager from a central function except perhaps in the large pharmaceutical companies. It is more common in small companies, biotech

The Handbook of Medicinal Chemistry: Principles and Practice
Edited by Andrew Davis and Simon E Ward
© The Royal Society of Chemistry 2015
Published by the Royal Society of Chemistry, www.rsc.org

companies, academic groups and not-for-profit organisations for someone to have a dual role for managing the project whilst also delivering drug substance or developing assays for the same project.

In drug development the unique product of the project is a drug, a device or a diagnostic, plus the documentation to support marketing approval and reimbursement from the regulatory authorities. The target product is likely to change during the initial phases as technology is discovered and project teams have to cope with regular failures and changes in scope. The regulated and rigid nature of some critical development activities adds to making planning both difficult and essential in this industry. Unexpected technical problems will occur in a drug development project and lines of research will fail no matter how well it is planned and managed. As many as 46% projects fail because of lack of efficacy, closely followed by animal toxicity and adverse effects in man.[2] Detection of early warning signals is one of the main aspects of project management to avoid repeating expensive studies and to terminate poor performing projects. It is a peculiarity of the industry that it is an achievement and not a failure to terminate a project early.

Irrespective of company size – startup, medium or large company – projects need plans which are visible to both the decision makers and the staff working in the laboratories. It is often stated that planning for cost and time restricts the chance of a scientific discovery but if discovery and development are not co-ordinated and activities integrated there will be a wasting of resources and consequently a reduction in the value of the product. As a member of a project team or a medicinal chemistry sub-team you will be working with experts in speciality areas some of whom may not understand how the contributions of the different functions fit together. A project manager needs good leadership, interpersonal and communication skills to manage and integrate activities across the multiple functions and team members who contribute to the whole project. Indeed as a sub-team leader those same skills will be required, as representatives from other functions may also be part of your team and you may have an influencing role rather than direct line authority over those people.

An example of a project team is given in Figure 24.1. The centre indicates the functions represented on the core team. Some of these functional members also lead sub-teams, the size of which may change as the project progresses, the size of the organisation and whether the

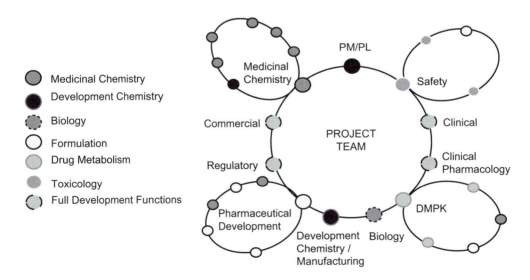

Figure 24.1 Typical drug discovery project team.

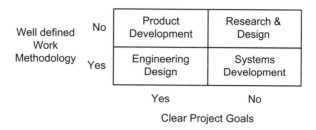

Figure 24.2 Goals and Methods Matrix (adapted from Turner).[3]

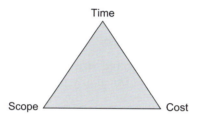

Figure 24.3 The 'Iron Triangle'.

activities are outsourced to a contract organisation. Indeed some functions such as Medicinal Chemistry may leave the team all together as the project progresses into development and drug substance activities are taken over by a Development Chemistry function.

Organisations run different types of projects for different purposes and these have been classified by Turner into four types according to how well the work methodology is defined and the clarity of the project goals.[3]

Drug research and development projects are located in the upper right quadrant of Figure 24.2 as the goals are often ill-defined at the beginning and the work methods are variable. Turner suggests this means the chances of success are reduced. Indeed when compared to an engineering project to build a bridge, with a clear goal and well defined methodology for construction, that is likely to be true. R&D personnel need to be optimistic about what can potentially be achieved. However, whilst opportunities need to be recognised and developed, too much freedom to be overly creative may increase scientific knowledge but might not lead to a new medicine. In times of financial constraint and diminishing pipelines there needs to be a fine balance between creativity and productivity.

Having a well-defined goal in mind, which may be a target product profile, reduces the risk of adding to knowledge but not delivering a product. A target profile is a guide for lead optimisation and will ideally be quite broad unless there is a clearly defined target patient population. The profile may include the route of administration (in some diseases oral is the only option), acceptable side effects (the risk:benefit ratio for childhood asthma is much lower than for a rare tumour type) and acceptable cost of good (an expensive synthesis will not be acceptable if the competition is cheap).

All projects typically operate within constraints of time, cost and scope or quality—represented by the project management or iron triangle shown in Figure 24.3, where each side represents a constraint. One side of the triangle cannot be changed without affecting the others.

The time constraint refers to the amount of time available to complete the project. The cost constraint refers to the budgeted amount of money and people available for the project. The scope constraint refers to what must be done to produce the project's end result. These three constraints are often competing. Typically increasing scope will increase both time and cost, a tight time constraint could mean increased costs and reduced scope and a tight budget could mean increased time and reduced scope.

The discipline of project management means different things to different people but is about providing the tools and techniques that enable the project team to organise their work to meet these constraints. There are different ways in which the project process and lifecycles are defined, but in essence they can be summarised in four stages beginning with initiation and planning, through execution to closure. The PMI (Project Management Institute) in the United States[4] and the APM (Association for Project Management) in the United Kingdom[5] both publish bodies of knowledge regarding what is considered to be core project management knowledge upon which this summary is based.

24.2   THE PROJECT PLANNING PROCESS

The project planning process is integrated with the drug development process. Drug development takes an average of 10 years but many projects will fail along the way and only a few will make it all the way to market. Because of the high attrition rate and the uncertainty of development where the results of an experiment or study can fundamentally change the future development of the drug, a rolling or stage gate approach is generally taken for the planning of projects with peer review and governance authorisation at each decision gate. Each phase could be considered a project in its own right but to avoid prolonging the overall development even further there needs to be a continuation of the planning process. Consequently the planning process is iterative around each phase of development so that at any point a single project plan could be in all stages of the planning process (see Robert Docherty in Chapter 22 for a description of the pharmaceutical development process). The planning process can be described in four stages.

24.2.1   Initiation

During the initiation phase of a project the scope and objectives are agreed—sometimes referred to as the Statement of Work.[6] The team is formed and the way the team will work together should be discussed. An outline or high-level plan of work may be required for formal approval to start the project, which will also contain a target product profile to direct the strategy.

24.2.2   Planning

Planning involves the whole team not just the project manager. Multiple options should be considered and the risks and trade-offs for each discussed. Assumptions and risks should be recorded and there should be commitment by the team members to the overall plan and deliverables, as well as to their own activities and timelines. Due to the large number of unknowns, detailed planning covers the near-term or current phase with a higher level plan being developed for the rest of the project. In this way resources are not committed to a tentative plan which may radically change based on information obtained in the early stages of the project.

24.2.3   Execution and Control

The execution phase is when the work takes place and actual progress is monitored and assessed against the baseline. Communication is key and there should be regular proactive updates on progress, issues and risks.

24.2.4 Close

Finally the project is closed, documentation is completed and any lessons learned are captured and shared for future benefit.

When target validation starts and the project is first created the planning cycle starts but as the project progresses through discovery towards development, planning for future phases will start before the current phase is completed. This is illustrated in Figure 24.4. For example, a project which is described as being 'in lead discovery' will be executing lead discovery activities but closing the previous target validation phase of the project, and planning for the future lead optimisation, as well as doing some early high level planning and initiation work with development groups for the candidate selection.

So what does this mean for team members and the project manager?

The project plan belongs to the team not just the project manager. As a team member it is important to contribute prior knowledge and experience which includes seeking out experiences from colleagues within medicinal chemistry. This knowledge should be shared in planning meetings and brainstorms so that the right activities are included in the initial plans with a realistic estimate of the durations so that reasonable expectations of delivery can be set.

Once a project is active and the work has started, team members have accountability for being transparent with project information and communicating to the rest of the team. Whilst there is often a tendency for a group to work away quietly solving a particular problem, the rest of the team need to know if there is an issue with delivery. For example, whether the next batch of samples will be coming through for testing or whether they can be getting on with some other work in the meantime. Good communication and transparency on progress means others can organise their work efficiently and are more likely to be available and willing to support you when needed.

At the end of a phase or the end of the project it is important for team members to contribute to capturing the lessons learnt so that they can be shared with colleagues and across the organisation and future projects can benefit from the identified best practices.

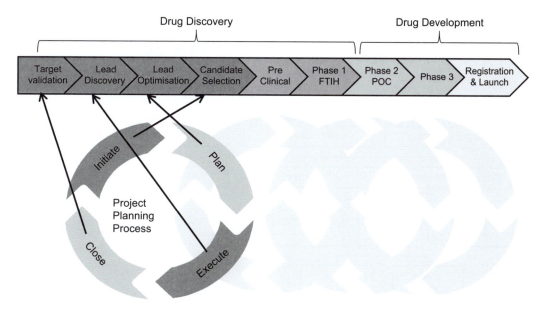

Figure 24.4 Development and planning cycle integration.

24.3   KEY PROJECT MANAGEMENT PRACTICES

The project manager role will vary according to the size and structure of the organisation within which they work and the technology which they have to support that role. However, key practices demonstrate value through cost and time reductions and can be applied in most situations.

24.3.1   Planning—Scheduling and Estimating

Planning determines what needs to be done, by whom and when. Scheduling and estimating practices ensure project plans are optimised for consistency, integrity and realistic estimations resulting in high quality plans with clarity of all activities required to achieve the project strategy, key dependencies and transparent assumptions. Good planning requires team members to focus on:

- *Content*: all activities needed to deliver the project scope and deliverables.
- *Estimates*: consistent process considering all key drivers, risks and assumptions.
- *Dependencies*: what is the activity for, what happens to the information it produces, who needs it and what is the activity dependent on before it can start.

These elements, together with milestones and intended start and finish dates, are combined to create the project schedule.

24.3.1.1   *Content—Identifying the Activities*

A work breakdown structure (WBS) as illustrated in Figure 24.5, is a way of splitting a project into distinct levels with deliverables and packages of work, with each descending level showing a greater amount of detail until the work is in manageable chunks or work packages. These work packages form the basis for scheduling and budgeting and represent the various activities that are undertaken by each functional line, such as a clinical study, a toxicology study or drug substance campaign. Many aspects of discovery and development have standard work package descriptions that can be applied to all projects although the duration and cost may vary according to the technical complexity of the particular chemical target.

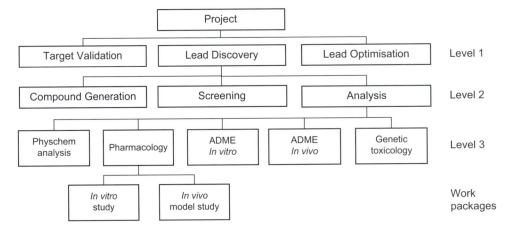

Figure 24.5   Simplified partial work breakdown structure (WBS).

24.3.1.2 Estimates—Assessing the Time and Cost

For every work package in the WBS we must establish how long it will take (days), how much it will cost (£) and how many resources it will need (people). We also need to know if there are any constraints which could affect that work package, such as availability of raw materials or completion of an analytical test. The constraints and assumptions should be documented for future reference.

Estimating requires expert judgement and an understanding of what has happened before in similar circumstances. Mathematical methods involving the statistical relationship between historical data and other variables to estimate duration are available but rarely applicable in discovery phases and will not be discussed further here.

Dates in a project plan and costs in a budget look like certainties but we never have precise information about the future. Estimates are an opinion or judgement, a rough or approximate calculation influenced by available resources, team experiences, project complexity, past performances and competition from other activities, yet the project team and management need to be aligned around realistic estimates. When estimating, it is important to understand and communicate uncertainty and three point estimating is a simple way of doing this, even if the uncertainty is not captured as additional data in the project schedule. Figure 24.6 illustrates the factors which drive three point estimates. Human nature means that some team members will be natural optimists and some will be pessimists. If optimistic dates are used throughout the plan there is a high chance the project will deliver late possibly missing an expected commercial window of opportunity. If pessimistic dates are used the project may deliver early meaning the resources for the next phase of development are not ready or able to start the work. Either way the organisation is not able to make efficient decisions about the pipeline or resources.

Ideally the plan will contain the most likely estimates and have a good chance of delivering to schedule. Start by asking what is the quickest the activity can be done if everything goes just right. This sets the optimistic date. Then consider what could go wrong and the potential delays that this could introduce. This sets the pessimistic date. Finally consider, from experience and instinct, what the actual duration is most likely to be—realistic but still challenging. That is the most likely delivery date, which should be used in the schedule but is not necessarily the arithmetic mean between the other two dates.

Common pitfalls associated with estimating include:

- Poorly defined scope of work.
- Failure to involve those who do the work in the discussions.
- Failure to assess risk and uncertainty.

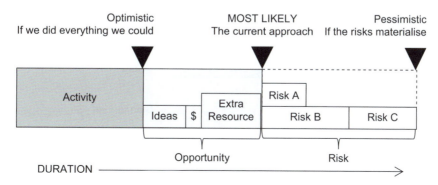

Figure 24.6 Factors driving the estimate range.

- Omission of activities.
- Rampant optimism.
- Padding (adding time to ensure delivery as promised).
- Time pressure (insufficient time to develop estimates).
- External pressure to meet pre-agreed targets.

24.3.1.3   *Dependencies—Sequencing the Activities into a Schedule*

Once the work packages or activities have been defined they can be sequenced taking into account the constraints and assumptions already identified. Sequencing can be done simply with sticky notes on paper or in a spreadsheet. If a planning software package such as Microsoft Project is available then the dates are calculated from the activity durations entered, the critical path can be automatically identified from the logical links applied, and updates are easily managed.

Most work packages or activities are sequenced so that as one finishes the next one starts—Finish to Start (FS) sequence. Figure 24.7 also shows how two or more activities may share a start if there are synergies in starting them together—Start to Start (SS) sequence—or, more rarely, two or more may share a finish when there is a desire to phase a second activity as late as possible—Finish to Finish (FF) sequence. These sequences are illustrated in Figure 24.7.

Logic is also applied to the sequencing through dependencies. Mandatory or hard logic is a dependency that cannot be broken, is unmoveable or unavoidable. An example is drug product that cannot be made until drug substance has been made, or a clinical study that cannot start until ethical approval is obtained. Discretionary or soft logic relates to experience and optimisation of the plan for resources and budget. For example the start of one study may not be dependent on the completion of another study but that relationship may be desirable. The plan is typically visualised as a Gant chart which clearly shows the time aspect of the schedule, as in the top part of Figure 24.8, or as a PERT chart, which makes the logic clearer but loses clarity on the time element as shown in the lower part of Figure 24.8.

The critical path is the path or route through the schedule network on which any delay in the activities will impact the project finish date. It is the longest of all paths but the shortest time in which the whole project can be completed. The critical path is usually depicted in red as shown in Figure 24.9 so the activities requiring most attention to avoid delay can be easily identified.

The term 'float' is used to describe 'the amount of time that a schedule activity may be delayed from its early start date without delaying the project finish date.' Activities on the critical path have zero float since to achieve the overall project timeline there is no flexibility in when they can occur. Activities not on the critical path have by definition some flexibility or 'free float' around their timing, which will not delay the start of any immediately following activities or impact the overall duration of the project.

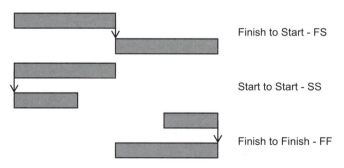

Figure 24.7   Sequencing activities.

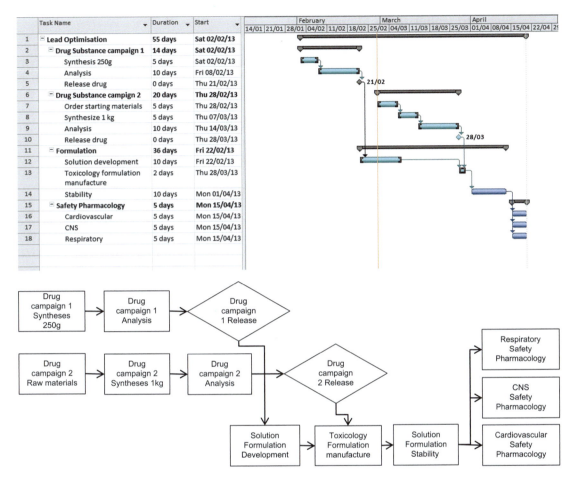

Figure 24.8   Example simple Gant and PERT Charts.

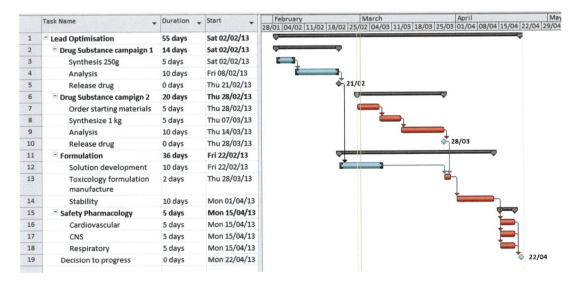

Figure 24.9   Example critical path.

24.3.2   Risk and Opportunity Management

Risk is defined as 'an uncertain event or condition that, if it occurs, has a positive or a negative effect on a business or project objective.' A risk with a positive effect is an opportunity and one with a negative effect is a threat. These are both referred to as 'risks', and should both be considered in the same project management process, which is described below. A risk that becomes a reality is an issue.

Risk and Opportunity management practices ensure identification, assessment, management and mitigation of project risks together with transparent communication to stakeholders. Projects are inherently uncertain, so risk management is important for completing them efficiently and predictably. Indeed several authors regard risk management as a 'key part of project management'[7,8] although the life science industry has historically lagged behind other industries in this capability.[9]

Risk management takes time but has the potential to reduce the time spent fire fighting and managing issues. Understanding the risks increases the reliability of estimates, gives greater assurance on the likelihood of delivery and helps provide information to support decision making. Rework following an issue is generally expensive in terms of time and money whereas mitigation activities to reduce risks are often much cheaper. Figure 24.10 illustrates this process.

24.3.2.1   Identification

The whole team should participate in a workshop to identify the risks and opportunities at the initiation of the project and when the proposal for the next phase is being developed. Several techniques are available—brainstorming, review of risk registers and lessons learned reports from previous projects, checklists of common risks and discussions with experts such as line heads, therapy area heads and other senior managers. Cultural issues can affect the ease with which the team will identify risks. Scientists are often reluctant to raise risks as they are optimistic that they will discover something and development functions don't want to be seen as making excuses for potentially not delivering. Technical risks based on the inherent characteristics of the molecules may be forthcoming but identifying operational risks to the delivery according to plan often requires more skilful facilitation.

Risks, both threats and opportunities, have:

- A cause.
- A definable event.

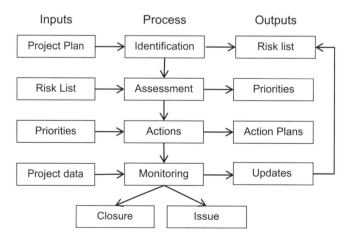

Figure 24.10   Risk and opportunity management process.

- A consequence or impact.
- A probability of occurrence.

If a risk is described properly with the cause, event and impact this ensures that it is linked to a factual event or cause and has a quantifiable impact on the project. Drug discovery is inherently risky and it is important to avoid recording all generic risks. For example, potential gender differences in pharmacokinetics are a generic risk unless there is animal pharmacology data with the class of compound which indicates specific gender differences. John Bartlett[10] recommends using meta language to ensure the three aspects are included in the description.

Because of . . . (CAUSE) . . . there is a risk that . . . (EVENT) . . . resulting in . . . (IMPACT).

Example risk threat statements:

Because of low solubility of drug substance at low pH *there is a risk of* reduced absorption after oral dosing *resulting in* inadequate exposure for efficacy in humans at commercially viable doses.

Because of the chemical structure containing 'x' *there is risk that* reactive metabolites are produced *resulting in* idiosyncratic adverse events which are not acceptable in asthma therapy.

Example risk opportunity statement:

Because of the very low molecular weight compounds being active there is an *opportunity for* brain barrier transport *resulting in* potential for additional clinical indications.

The risks and opportunities should be captured in a register developed in a spreadsheet, or a specific software package such as Predict, @Risk or Xactium, ready for the next stage of the process. Figure 24.11 is an example using Microsoft Excel.

24.3.2.2 Assessment

Assessment is a key stage in defining the risk as a product of the likelihood of occurrence and the impact should it occur. Assessing the risks and scoring the impact and probability allows them to be prioritised and for subsequent action planning to be carried out on the most important as indicated in Figure 24.12. Involvement of the team ensures consistency in interpretation and understanding across the range of risks. Both probability and impact are scored against a standard table developed by the organisation. Three or five point scales are utilised for low or very low through to high or very high ratings. Probability is a subjective assessment based on experience and generally there is insufficient historical data for a statistical approach. Impact can usually be quantified in terms of cost or time or for later phase projects, peak sales. Impact scores may be adjusted for different projects in advance of the assessment as commercial drivers need to be taken into account. If a project is targeting a valuable commercial opportunity which is time bound by competitor activities then impacts on time will be more important than impacts on cost and a very high impact may be a delay of just one month whereas in a unique but niche clinical indication, impacts on cost may be more important than time. The ultimate very high impact is that the drug is not developable and the project is terminated.

Once the risks have been assessed they can be plotted on a Heat Map or Risk Severity Grid as shown in Figure 24.13. This helps to identify those severe or key risks that require active management, and those risks that just require monitoring because either their probability of occurrence or likely impact is lower. The resulting 9 or 25 severities (which depends on whether a 3- or

RISK REGISTER

Date of entry	Risk No.	Description	Probability	Impact	Severity	Key Risk?	Expected Date of occurrence	Strategy	Actions	Action Cost	Severity post action	Action Owner
21/01/2013	1	Because of low solubility at low pH there is a risk of reduced absorption after oral dosing in the oral ADME studies resulting in inadequate exposure in humans at viable doses	VH	H	VH/H	Y	15/04/2013	Mitigate	Further development of chemical series find compounds with increased solubility	£10k	L/L	J Smith
21/01/2013	2	Because of	VL	L	VL/L	N	03/04/2013	Accept	None	£0		F Bloggs

Figure 24.11 Example risk register.

PROBABILITY	
VH	>94%
H	>70%
M	>30%
L	>5%
VL	<5%

IMPACT		
	DELAY	COST
VH	<1 week	<£10k
H	1-4 weeks	<£50k
M	>1 month	>£50k
L	>3 months	>£100k
VL	>6 months	>£250k

Figure 24.12 Probability and impact tables.

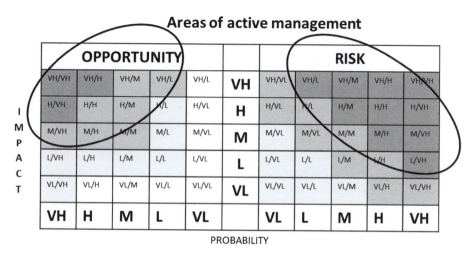

Figure 24.13 Risk and opportunity severity grid—heat map.

5-point scale is used) can be given a numerical value to assist prioritisation and reporting. Action planning is aimed at reducing the severity of the risk and moving it towards the bottom left corner of the grid. Action planning for opportunities is aimed at realisation of the benefit.

24.3.2.3 Actions

The first stage of action planning is to assign an owner on the project team who is responsible for managing the risk or opportunity on a day to day basis, developing the action plans, assessing their effectiveness and communicating back to the project team. Action plans are created for the most important risks and opportunities bearing in mind that the cost of intervention may outweigh the cost of the potential impact with little change on the probability of the risk occurring, or in the case of an opportunity the cost may outweigh the potential benefit.

Risk response strategies take into account the severity and the potential for changing the outcome.

- AVOID: eliminate the threat usually by eliminating the cause.
- ACCEPT: do nothing and accept the consequences should the risk occur.
- MITIGATE: reduce the expected impact or probability of occurrence.
- TRANSFER: shift some or all of the threat to a third party.

Mitigation plans usually involve additional resources and require approval for the cost-benefit before the additional activities are added to the schedule. When an action plan is put in place to mitigate a risk there should be a reassessment of the risk in light of the new activities to see if the desired reduction in probability and/or impact has been achieved.

Risk contingency plans may also be developed to describe what would need to happen should the risk event actually occur. Such plans should include the scope to deal with the event, any costs and resources required and the schedule of the activities. The activities themselves would not be added to the overall project schedule until the action plan is activated. Depending on local practices it may be possible to get pre-approval for the contingency spend confirming it is agreed in principle and avoiding a delay to get authorisation should the plan need to be activated.

Opportunity planning enables the team to decide on actions to maximise the probability of an opportunity coming into existence and maximising the impact should it do so.

- EXPLOIT: maximise the probability of the opportunity occurring.
- ENHANCE: maximise the impact on the project from the opportunity.
- IGNORE: do nothing but continue monitoring.
- SHARE: action taken by a stakeholder best placed to maximise the benefits for the company if not the project.

If a decision is taken to exploit or enhance the opportunity the team will need to invest time and potentially money. Maximisation plans will add to project scope but also add benefits and will require authorisation before implementation.

Opportunity realisation plans may also be developed to describe what would need to be done should the opportunity actually happen. Such plans should include the scope to deal with the event, any costs and resources required and the schedule of the activities. The activities themselves would not be added to the overall project schedule until the action plan is activated. As with risk contingency plans it may be possible to get pre-approval for the contingency spend confirming it is agreed in principle although further authorisation is likely before the plan can be put into action.

24.3.2.4 *Monitoring*

The team must take an active role in the ongoing risk monitoring for a project by being an active risk owner and regularly asking a series of questions.

- Is the due date approaching?
- Is a risk becoming an issue?
- Have the mitigation actions been successful?
- Do we need to do anything else?
- Has anything changed in the plan that I need to review for risks?
- Have any new risks occurred?
- Can I manage within the sub-team or does it have a wider impact and need to be raised at the project team?
- Do I need to generate new action plans?

Effective communication of the uncertainty around a project, which is largely subjective, is critical but challenging in a scientific environment where stakeholders are focused on facts and data. The risk register is not a communication tool but risk heat map shown in Figure 24.12 can be utilised for plotting the key risks to quickly demonstrate the level of uncertainty on a particular project.

24.3.2.5   Closure

By monitoring the risks the team will know when the time for the risk to occur has passed and at that point the risk should be closed on the risk register if it has not occurred or transferred to an issues log it has indeed become a reality. Completion of the risk register in this way allows for a review of the outcome of identified risks and an assessment of the success of mitigation activities. Lessons learned can be documented so that future teams can learn from past mistakes.

24.3.3   Project Control

Project control should not be viewed with suspicion or as policing a project team as it can offer a competitive advantage. The primary aim is to deliver what the team has committed to and maintaining control means minimising the difference between where the project actually ends up and where you thought it would end up which is reflected in the baseline plan. Project control practice describes the processes which ensure that a project delivers the objectives and deliverables that it was set up to achieve in the most efficient manner. It involves the schedule and budget baseline, capturing progress and detecting variance from plan and taking corrective action within the agreed thresholds or by seeking approval for changes from the governing bodies as appropriate. See Figure 24.14.

The baseline is a snapshot of the plan as it stands at the point of approval by the governing body and should contain milestones, activities, logic, assumptions, costs, risks and critical path. The threshold is the extent to which the plan can vary from baseline, be it time and/or budget, before a formal re-approval process needs to be instigated for a plan revision. If the governing body uses thresholds to empower the teams they would be agreed at the same time as the baseline is taken. At approval of a revised plan a new baseline is taken.

As well as providing clarity on the 'contract' between the team and the governing body and empowering teams to truly manage the plan within thresholds, baselines are valuable for capturing learnings and comparing what was estimated with what actually happened. In that way estimates provided for future projects should be more realistic.

Effective control is achieved through detailed and regular monitoring of progress and collection of 'actual' information on a weekly or monthly basis, for example invoice payments, time taken to complete an activity. This information enables the identification of critical variances from baseline and the ability to take corrective action and forecast future events. This is

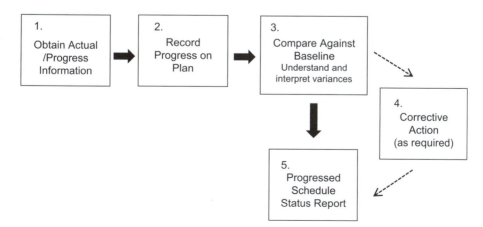

Figure 24.14   Control cycle.

underpinned by good communication across the team and precise data. Team members reporting that they are 'on schedule', 'doing OK' or 'a bit more and then I'll be done' does not provide enough information for the project manager to maintain the plan.

Control information capture includes:

- Activity start date.
- Activity finish date.
- Remaining duration to an estimated finish date.
- Actual cost.

The project plan is constantly adjusted to reflect the day to day reality and the captured information may reveal that the schedule is going off track. In many cases the first course of action is to try and maintain the original plan and not all changes should be automatically accepted of accommodated. Recovery may be possible by adjusting future activities in the project schedule rather than attempting to deal with the immediate problem. If neither of these approaches works then the team needs to consider adding resources, accepting partial deliverables, renegotiating cost and schedule targets, reducing scope or investigating alternative work methods. The trade-offs and risks will need to be reassessed too.

24.3.4 Stakeholder Management

A stakeholder can be defined as 'anyone who can affect what you are trying to achieve.'

Whilst some stakeholders may be shared across all functional groups and influence the project as a whole, there will be some stakeholders closer to home whose main influence will be on a particular functional group *e.g.* the head of medicinal chemistry who assigns resources or the procurement manager responsible for ordering raw materials in a timely manner. As research is increasingly funded and conducted in different partnerships and alliances it is important to also manage the external stakeholders, the charity funders, the academic institutes and the contract organisations.

Stakeholder analysis is a logical process involving analysis, mapping to needs and action planning including assignment of team members to particular roles. Good people skills are required to actively manage stakeholders and influence them to ensure project success.

Whilst some stakeholders need to be champions for the project, offering practical support or giving approval for funding, many need just to be co-operative to allow the project work to progress. Stakeholders can be also be obstructive; prioritising other work above the project in question. Some may even oppose the project as a matter of scientific principle and provide opposition and delay in decision making forums. As a team member, sub-team leader or a project manager it is vital to the success of the project to identify these stakeholders, understand their needs and opinions and have a plan to manage them. An example template is shown in Figure 24.15.

Most projects have regular meetings. Executive management often require monthly status review meetings and the team will need to meet weekly, bimonthly or monthly. However, meetings need not be long. They must have a purpose, an agenda, address risks and be action orientated. Core team members should attend but ad hoc members will be invited to share their expertise and support decision making. Technology such as Sharepoint and Dropbox can be used for sharing information to avoid long presentations of data. Meetings should not be held unnecessarily but equally it is not good practice to wait until a set meeting to raise issues and risks with other team members or stakeholders. Poor communications cause bottlenecks and executives do not like to be surprised by bad news in meetings.

	Stakeholder	Success Criteria	Actions to be taken	Outcome	Owner
Need as a resource					
Directly / Indirectly affected					
Need to be supportive					
Observing progress					

Figure 24.15 Example stakeholder action plan.

24.4 YOUR ROLE AS A PROJECT TEAM MEMBER

The role of project team members is to collectively plan and execute a project and it is important that teams are fully functioning with clear roles, have a consistency of approach where necessary and are following a defined framework against which the best practices can be applied to deliver the project in the best way for the organisation.

The project manager is the driver of the process but each team member is a technical expert, the voice of the line function on the project team and the voice of the team back into the functional group. Good team members are proactive communicators and have a big picture, project focussed view.

Features of fully functioning teams:

- Individual members are committed to the team as well as to the project; they look out for one another, support each other through the highs and lows and want the team to achieve success.
- Everybody knows what needs to be done, how, by whom and why. Accountabilities are agreed and an integrated plan has been developed.
- Team members take time to check in with each other, are aware of their own communication style and the preferences of others and use technology to assist not hinder communications.
- Stakeholders have been identified and are being managed by the team.

24.4.1 Team Charters

A project team charter is a statement of the scope, objectives and participants of the project. The exact structure will vary according to organisation and there may be two documents. One detailing the project scope, strategy, plan, budget *etc.* which may be called a Project Plan, a Project Strategy document, a Project Charter or perhaps a Statement of Work. The second document, a Project Team Charter, is focused on how the team will work together to deliver the project. It provides an outline of the project deliverables, a delineation of roles and responsibilities for those deliverables, methods and frequency of communication, identifies the main stakeholders and defines the authority of the project manager.

High performing teams will have a structured conversation about how team members will work together, communicate and make decisions. Understanding how individual team members like to communicate can be a useful team building activity and prevent communication breakdown in times of project stress. The scope and the objectives of the project need to be defined so each member is clear about their roles, the expectations of them by the rest of the team and the priority of the activities they are accountable to deliver. Summarising the agreements from that conversation in a charter like the one in Figure 24.16 means they are easily accessible for periodic review and for sharing with new team members. If there is a dysfunctional team member the charter can be used to remind them about what the team agreed on how they would behave in a non-confrontational way.

PROJECT NAME:		
CHARTER DATE (first prepared):	**CURRENT VERSION & DATE:**	
BUSINESS UNIT:	**PROJECT MANAGER:**	**TA PROJECT SPONSOR:**
PROJECT TEAM DYNAMICS & BEHAVIOUR		
Required Behaviour / Actions:	1. Team members will read all briefing documents prior to team meetings 2. Team members will communicate regularly between meetings and not wait until formal meetings to raise risks or issues with the Project Manager 3. If team members cannot attend team meetings a suitable, well briefed and empowered deputy will be provided 4. Sponsor to communicate to SVPs of all line functions about the overall priority of the project	
PROJECT SCOPE & OBJECTIVES		
Objectives	The project primary objectives are: 1. ..... 2.	
Priorities	The priorities within the project are: 1. .... 2.	
Phase Completion	Completion of Candidate Selection	
Phase Completion Timing	October 2013	
Key Asumptions s	......	
Project Dependencies	......	
Project Risks	See Project Risk Register	
PROJECT TEAM MEMBERSHIP		
Project Manager Medicinal Chemistry Development Chemistry Pharmaceutical Development Safety Assessment	DMPK Clinical Pharmacology Regulatory Commercial	
PROJECT GOVERNANCE		
Governing Body	Discovery Development Board	
Project Time & Budget Thresholds	+/- 1 month on major milestones +/- 3% approved budget	
Next Formal Review	Commit to Candidate stage gate	
Project Manager:	Date	
Sponsor:	Date	

Figure 24.16   Example project charter.

> **HINTS AND TIPS FOR TEAM MEMBERS**
>
> - Take time to understand your role and the planning process in your organisation.
> - Seek to understand the project plan and how your work fits with others.
> - Ensure you know the governance structure and where decisions are made. Identify your functional representative on the governing bodies.
> - Contribute to all aspects of the project—the plan, the stakeholders *etc.*
> - Endeavour to understand the communication style preferences of team members.

24.5 SUMMARY

This chapter has provided an overview of how projects are can be managed in the life sciences industry and an introduction to key practices. This industry has been slower than many in adopting good project management practice for two reasons—lack of financial constraint during the boom years and a belief that it could constrain scientific discovery. The adoption within drug discovery where there is most uncertainty is relatively recent in many organisations but the value of focusing effort, integrating activities and enabling teams to deliver more efficiently is now recognised. The experience of project teams in running projects should be captured and transferred to other teams and later projects just as scientific knowledge is published and shared.

Good project management practice can be implemented at both the project and sub-project level and adoption of scheduling, risk management, project control and stakeholder management can reap dividends for the company, the team and the individual.

KEY REFERENCES

T. Kennedy, *Pharmaceutical Project Management*, ed. T. Kennedy, Marcel Dekker, New York, 1998.

J. Bartlett, *Managing Risks for Projects and Programmes*, Project Manager Today Publications, Hook, UK, 2nd edn, 2008.

H. Kerzner, *Project Management: a systems approach to planning, scheduling and controlling*, John Wiley & Sons, New York, 10th edn, 2009.

SUGGESTED FURTHER READING

K. Schwalbe, *An Introduction to Project Management, Third Edition: With Brief Guides to Microsoft Project 2010 and @task*, Kathy Schwalbe LLC, USA, 3rd edn, 2010.

J. Ferraro, *Project Management for Non-Project Managers*, AMACON Division American Management Association, New York, NY., 1st edn, 2012.

P. Taylor, *The Lazy Project Manager*, Infinite Ideas Limited, Oxford, UK, 1st edn, 2009.

REFERENCES

1. Project Management Institute, *A Guide to the Project Management Body of Knowledge (PMBOK Guide)*, Project Management Institute, Upper Darby, PA., 4th edn, 2008, pp. 5–6.
2. T. Kennedy, in *Pharmaceutical Project Management*, ed. T. Kennedy, Marcel Dekker, New York, 2nd edn, 1998, pp. 12–13.

3. J. R. Turner, *The Handbook of Project Based Management: Improving the Processes for Achieving Strategic Objectives*, McGraw-Hill, London, 2nd edn, 1999, pp. 25–26.

4. Project Management Institute, *A Guide to the Project Management Body of Knowledge (PMBOK Guide)*, Project Management Institute, Upper Darby, PA. 4th edn, 2008, pp. 1–468.

5. Association for Project Management, *APM Body of Knowledge*, Association for Project Management, High Wycombe, UK, 5th edn, 2006, pp. 1–258.

6. H. Kerzner, *Project Management: a systems approach to planning, scheduling, and controlling*, John Wiley & Sons, New York, 10th edn, 2009, pp. 426–433.

7. H. Kerzner, *Project Management: a systems approach to planning, scheduling, and controlling*, John Wiley & Sons, New York, 10th edn, 2009, pp. 743–745.

8. S. J. Simister, *The Wiley Guide to Managing Projects*, John Wiley & Sons, New York, 1st edn, 2004, p. 30.

9. P. Harpum and T. R. Dunson, *Portfolio, Program and Project Management in the Pharmaceutical and Biotechnology Industries*, ed. P. Harpum, John Wiley & Sons, New York, 1st edn, 2010, pp. 136–137.

10. J. Bartlett, *Managing Risk for Projects and Programmes*, Project Manager Today Publications, Hook, UK, 2nd edn, 2008, pp. 20–21.

CHAPTER 25

Clinical Drug Development

MAARTEN KRAAN

AstraZeneca R&D Mölndal, SE-431 83 Mölndal, Sweden
E-mail: Maarten.Kraan@astrazeneca.com

25.1 INTRODUCTION

Clinical research on the effects of therapeutic interventions is probably as old as medicine itself. In documents of the old Greek masters, such as Hippocrates, specific and systematic approaches to study the effects of medicines in human disease are documented. In the centuries since then, there have been evolutions and revolutions, but especially in the second half of the 20th century systematic studies of the benefits and risks of interventions in human disease have developed rapidly.

Currently, clinical trials in pharmaceutical research are primarily designed to acquire data on the safety/tolerability and efficacy of pharmaceutical interventions related to health/disease. Clinical trials can be used for drugs, devices, and diagnostics but in the context of this book it will be limited to drugs only.

The clinical part of drug development has evolved dramatically over the past decades with hallmarks of success and breakthrough, but also across the industry, ever declining success rates. This decline in success rates of programs of potential medicines entered into investigations in man has been associated with significant increases in cost, duration and complexity.

Significant effort has been put into understanding of the various items contribution to the current challenges such as cycle time, probability of technical success/risk, value and cost.

Drug development can now take as much 25 thousand patient years, over 15 years, and can have a significant post registration commitment, increasing even further the amount of data generated on a therapeutic intervention. These changes have been driven by changes in demands by patients, physicians, regulators and payors to various extents. In this chapter the choices to be made in a modern clinical development program are described.

The Handbook of Medicinal Chemistry: Principles and Practice
Edited by Andrew Davis and Simon E Ward
© The Royal Society of Chemistry 2015
Published by the Royal Society of Chemistry, www.rsc.org

25.2   TYPES OF CLINICAL TRIALS

Clinical trials can be sorted in various ways; one is by the intentions of the investigator.

In an *observational* trial the aim is primarily to observe specific elements of disease, with or without active treatment applied. Data are generally generated on the outcome of the disease or symptom over a certain period of time.

In an *interventional* trial the investigator applies a specific treatment or intervention. There is usually a control group that can be with or without treatment. When treatment is given to the control group this can be standard of care, or a so-called active comparator. In the case of an active comparator the investigator uses a specific second drug intervention to compare against the investigative compound. To increase the rigour of the trial and hence the confidence in the data when no active treatment is given to the control group, the patient is given a fake dose without active ingredient, a so-called placebo.

Within the interventional trial there are some more relevant topics; when the patient is aware of what treatment he/she receives it is an open label investigation, a further maturation of the interventional trial is blinding of the groups involved, this implies the patients are not aware of which treatment arm they are exposed to. This will improve the reliability and reproducibility of the investigation since the patient is not aware of whether he or she receives active treatment or the control treatment; blinding has been demonstrated to significantly influence the results of a study. For further standardization the investigator and staff can be blinded for the treatment as well; this is usually done by pre-packaging the medication so not allowing visual identification of the doses or by using an independent pharmacist where local preparation is required. This is called double blind. When the study treatment is blinded and the two comparative treatments are very different, for example comparison of an intra-venous with an oral treatment, a double dummy is applied; in that case there is an active and a placebo which are identical in color, shape, taste and overall appearance enabling maximal control of the bias that either patient, investigator or staff can identify what treatment is administered to the patient.

To further control the variability between the various relevant clinical and demographic parameters patients can be randomized. By assigning them randomly to the treatment options, the variations between the various groups can be controlled. In the case that the investigator expects that there is a very relevant parameter that significantly can influence the outcome, he can choose to stratify the patients; this stratification will assign patients to *strata* supporting a better analysis of the data obtained.

So the best controlled trial is prospective, randomized, placebo controlled and double dummy randomized.

25.2.1   Example of a Clinical Development Program of a Small Molecule for the Treatment of Rheumatoid Arthritis (RA)

The single ascending dose (SAD) and multiple ascending dose (MAD) trial are performed in healthy volunteers at a single specialized site in the USA and the pharmacokinetic profile is confirmed as a once a day dose with a good fit on the modelled curve. There were no major drug related safety observations made in the dosing groups up to three multiples of the modelled target dose. For the pharmacodynamic profile an assay of target engagement was used which demonstrated a 95% coverage of the target within a 24 hour period at the 20 mg/day dose, this was pre-specified as the target dose. Because of the use of healthy volunteers no clinical activity parameters have been included but there was a blood sample repository built to enable testing/validation of various biomarkers. In the MAD there was a requirement for 2 weeks dosing modelled to get to steady state.

Next, a 200 patient two arm double blind randomized single dose *versus* placebo controlled Phase IIa trial was performed in RA patients that did not respond to the current standard of care (maximally tolerated dose of methotrexate (MTX) with use of a non-steroidal anti-inflammatory drug (NSAID) and a maximal daily dose of steroids of 10 mg/day) and had active disease defined per a disease activity score (DAS-28) above a certain threshold. The per patient duration of exposure was 3 months driven by the period in which a relevant clinical change can be expected. The primary endpoint was a statistical difference in clinical response (DAS good response) when active drug was compared to placebo, the secondary endpoint was a set of alternative measures to monitor clinical parameters and a set of biomarkers that were identified in relation with the target and RA. There was a standard panel against which safety was monitored. The primary and secondary endpoints were met and Proof of Concept was declared.

A four arm 400 patient double blind randomized three dose *versus* placebo controlled Phase IIb dose ranging trial was started where a similar framework of clinical, safety and biomarker data were used. The doses were chosen as lowest doses—significantly below expected efficacious dose, the middle dose around the efficacious dose and the highest significantly over. The middle dose was aimed to be the target dose. The numbers of patients per arm were chosen to allow discrimination on the primary endpoint (clinical efficacy) *versus* the placebo group. Safety was monitored and revealed a signal in blood pressure that increased with dose. The primary endpoint was met for the two highest doses including the target dose (middle dose). There was a Safety extension up to 1 year; this was open label because placebo control is not ethically accepted. The data generated were used to build the safety database and allow patients to stay on treatment when proven or perceived to be effective.

There were a series of Phase III programs, two dose arms, and five trials, all with 1 year duration and with extension/compassionate use programs. Given the needed duration of 1 year and exposure deemed unethical for 1 year, a placebo arm could not be used. The numbers of patients in these trials were driven by a series of parameters: The number of patients exposed for at least 1 year needed for the safety database was 1000 patients. The five trials studied the various relevant patient groups, two trials in MTX failures (primary population so two separate but similar trials in design) one trial of TNF failures, and one trial of MTX naive/early RA. All these trials now included a structural damage claim as co-primary endpoint next to the signs and symptoms endpoint. A fifth trial was designed with alternative imaging technology (MRI). All trials included a framework to generate data relevant for payer evidence and pricing needed for the commercialization later on.

In situations where the forward/prospective generation of a new data set is not needed/possible or warranted, existing data can be used. The type of trial that investigates data that are already generated is called retrospective. The challenge with retrospective trials is that the data used can be of variable quality or the circumstances under which they are acquired were subject to bias; this can significantly influence the outcome/conclusions of the trial without the investigator being aware of it.

A third approach is to categorize trials towards the projected outcome of the trial.

Prevention trials will measure the difference between two groups where an intervention is aimed to actually prevent the occurrence of either a disease or alternatively of a sign or symptom of a disease. An example is the assessment whether the use of statins interfering with the lipid metabolism is effective in the prevention of stroke or myocardial infarction. Another example is the trials that resulted in the vaccination for human papilloma virus in young female adults to prevent against cervical cancer.

Screening trials are designed to detect disease in a certain population, a good example is the screening for cervical and breast cancer trials.

Closely related are diagnostic trials where the aim is to test the ability of a method to detect a certain diagnosis and measure the effect of a (biological) probe on detecting a disease state or diagnosis. An example is the use of PAP smears to demonstrate the sensitivity and specificity to predict cervical cancer; this example also demonstrates the sequential use where a screening trial to demonstrate the value of the procedure is followed by a diagnostic trial to optimize the method used.

The most traditional clinical trial of all is the treatment trial. Here one or more interventions aimed at interfering with a disease state are compared.

Quality of life trials are designed to measure changes in the impact of a disease state on the quality of day to day living, these trials are used for example to assess the ability to have a normal life after a myocardial infarction or cerebrovascular accident. Data generated can subsequently be used to demonstrate the medical intervention that prevents such a condition.

Compassionate use trials or expanded access trials provide partially tested, unapproved medicines to a small number of patients who have no other realistic options. Usually, this involves a disease for which no effective therapy exists, or a patient who has already attempted and failed all other standard treatments. Usually, case-by-case approval must be granted by both the regulatory bodies on the applicability and the pharmaceutical company on a patient-by-patient basis for such exceptions.

25.3 PHASES OF DRUG DEVELOPMENT

Phase 0—This is traditionally a study in patients where no pharmaceutical intervention or a registered product is used. Patients are either subject to standard of care or to a drug that is already registered to be used in that disease and the application is done within the label as approved by the health authorities. The study readout can be almost anything but most widely used it is a symptom, biomarker or any other signal of biological relevance.

With the growing realization in the medical community that disease heterogeneity will have a direct effect on how pharmaceuticals will work on human disease there is, with increasing frequency, a strong need/demand to study pathways and biosignals in humans. In the preclinical phase of drug development there is still a frequent use of animal models touching upon elements of disease. However, we have learned in recent history that these have a limited translatability to human disease, hence there is a need to translate these signals in humans or to identify them. The purpose of these exploratory clinical trials in Phase 0 is to study the disease to be targeted, with the aim to better understand variable/s influencing outcome or response or to develop biomarkers *etc.*

Phase 1—The aim of this phase is to demonstrate the safety and tolerability of the investigational products and consequently this is the phase that humans are exposed to the investigational product for the first time. This can be done in either healthy volunteers or in patients. The choice is usually made based upon the expected profile of the drug, where especially tolerability drives the choice to go directly into patients (oncology). Alternatively the choice to go into patients can be driven by the fact that the target is only expressed in diseased humans. The disadvantage of going directly into patients is the presence of co-morbidities and other treatments.

The first study of a new investigational product is usually a single rising dose program. Based upon the pre-clinical data a dose escalating program is built where the purpose is to find the maximally tolerated dose based upon clinical determinants. Although the route of administration can be both enteral and parenteral, over the last decade, the route chosen is usually the route preferred for the product when the product is commercialized after approval by the health authorities and captured on the package insert or label.

After completion of the single rising dose program a multiple rising dose program usually follows. Exceptions can generally be found in programs with large molecules where half-life can be up to 21 days, hence a single administration results in a durable exposure. On designing such a program the rules of a rising multiple dose are followed, these imply that the projected maximal dose is subject to higher scrutiny.

Objectives are usually safety and tolerability but there is usually a strong component of clinical pharmacology.

Phase II—In this phase the aim is to demonstrate that the investigational product has a meaningful effect on the disease targeted. The questions addressed are best captured in the following principles: 1) Proof of Mechanism, 2) Proof of Principle, and 3) Proof of Concept. These questions are usually answered in the first set of experiments or Phase IIa. After these principles are addressed there is commonly a Phase IIb where the primary aim is to establish a dose.

Phase III—In this phase the investigational product has proven to be able to influence the targeted disease and needs to demonstrate a robust safety/benefit profile in the target population. Generally this phase is used to generate the data that establish the label of the product, *i.e.* the data that are used by the sponsor to negotiate with the various health authorities what claims can be made towards what effect the medicine will have upon the targeted disease. In the past decade a second stream of studies has also been started in this phase to demonstrate the value for health care providers such as insurance companies or governmental bodies, such as the National Institute for Health and Care Excellence (NICE) in the UK, which are trying to distribute limited resources towards those health interventions that drive most value.

Many choices need to be made and options need to be considered here. This phase is dominated by the four cornerstones, risk, time, value and cost. In weighing these four cornerstones, trade-offs need to be made, where changes in any one of these elements commonly impacts the others. For example if a certain risk identified in Phase II is mitigated, there is usually an increase in cost and a delay in the timeline which shortens the time on the market with patent protection *i.e.* value.

Phase IV—This is the phase after the right for commercialization has been established. In this phase two dominant themes dictate the data that need to be generated. 1) The need of the commercial organization to make sure the potential of the now approved drug is utilized and 2) the increasing need of regulators to understand the benefit/risk profile when the drug is exposed to a wider population, not limited by stringent inclusion and exclusion criteria of the clinical development programmes.

25.3.1 Example of Timelines

The total duration of the example clinical development RA program from IND preparation to achieving approval in the USA was 134 months or a little over 11 years. If we exclude periods of regulatory preparations and regulatory review it is 106 months, almost 9 years.

Preparation for the Phase I program documentation took 3 months, submission window 1 month, completion of the seven arm SAD was done in 2 months and completion of the six arm MAD was 4 months, reporting *etc.* took 1 month making this period 11 months.

Preparation of the Phase IIa trial took 6 months, period between first patient in and last patient out was 1 year, completion of documentation took 2 months making it a total of 18 months.

Preparation of the Phase IIb trial took 6 months, the period between first patient in and last patient out was 18 months and the completion of documentation took 3 months making it a total of 30 months.

The design of the Phase III program took 9 months, enrolling and completing all trials took 36 months.

Preparation for filing took place 6 months after database lock of the last trial and file was accepted by all regulatory authorities. Japan and major other markets were filed in the next 6 months followed by a third wave in the next 6 months serving most of the world. Review of the dossier took on average 18 months and a label was granted in all countries filed.

25.4   BASIC STATISTICAL CONSIDERATIONS/PRINCIPLES

Statistical methods have become an integral part of modern clinical trials. This is driven by the complexity and size of the datasets generated, consequently the demand on rigorous statistical design and support increases with the phase of development.

Randomization has been discussed before and is used to make sure potential not equally distributed characteristics do not interfere with the assignment of the comparators to each other.

Powering is the technique where, based upon the predicted effect size and the assumed variability, the size of the trial is calculated.

Adaptive design is the technique where there is a flexibility built into the design that allows the investigation to proceed depending on a predetermined outcome.

Superiority/inferiority are terms used to describe the preset tolerability for difference between two comparators. With a superiority design the expected outcome is that one treatment will demonstrate a benefit over the comparator arm, whilst in an inferiority domain the base assumption is the there is no real difference between the two treatments.

Stratification is a method used to control the heterogeneity of the disease/population studied. Before the start of the trial specific subgroups of patients are identified, and *via* the design, it is warranted that these groups are balanced within the trial during recruitment and analysis.

25.4.1   Examples of Statistical Considerations

Randomization—In a placebo controlled two arm randomized trial with four centers aiming to enroll 36 patients, a total of 36 blocks containing four numbers each is generated (144 numbers). Each number has a code for treatment or placebo. In this way there will always be a balanced number of patients to each arm irrespective of the number of patients per site. Of course block size and numbers will vary. The assignment of treatment code is usually done *via* a central randomization center which will identify the treatment code to be used at randomization.

Stratification—In a RA trial it is known that there are two main populations, patients with high inflammatory load measured by CRP (C-reactive protein) and those with low inflammatory load. The investigator wants equal numbers of both in the trial and creates two *strata*, one below a cut off and one above the cut off. On enrolment, the investigational sites will identify this with the inclusion/exclusion criteria. Half of the randomization blocks are pre-assigned to each one of the *strata* for each of the site. When one *stratum* is full the investigators are notified. This again requires a central control of assignment of randomization codes.

Adaptive design—This is used to accelerate the timelines between phases and allows the investigator to use the data from completed cohorts (for instance dose cohorts) and to skip or add additional cohorts. In Phase II it can be used to collapse the proof of concept trial into the dose finding trial without having to do the dose used for the POC again.

Interim analysis—This is used to take a snapshot at a usually pre-defined point in the trial to test the data against the primary endpoint before the trial is fully completed. A specific use is the futility analysis where a predefined dataset at a predefined time point is used to determine whether the trial is likely to meet the primary endpoint.

Power—At design, the investigator determines the willingness to accept errors (usually either false positive or false negative outcomes) and this risk definition is used to calculate the number of subjects in each arm. On average a clinical trial will be called of good power when the number of patients in each arm allows a solid interpretation reflecting the documented response in the placebo groups (preferably nil) and the variability in the parameters used for the endpoints (especially the primary endpoint).

Superiority design—When the trial is designed to prove one treatment is better than the comparator.

Non-inferiority design—When the trial is designed to demonstrate the tested treatment is similar to the comparator.

25.5 TARGET PRODUCT PROFILE

Very early in the development of an investigational product the development team captures the profile of the aspirational drug in a target product profile or TPP. This TPP describes the targeted disease and, if applicable, the sub-segment of this disease with regard to the disease phenotype, currently available treatment options/unmet medical need and the various specifications the future drugs is thought to meet.

25.6 STUDY PROTOCOL

Every clinical investigation requires a study protocol. This captures the primary, secondary and exploratory hypothesis/endpoints that the study is aimed to test. Good protocols test only one primary hypothesis. The disease is specified and there are inclusion and exclusion criteria specified to make sure the patients/subjects are meeting the criteria and that they are protected against the associated risks.

Patient Informed Consent, expressed at the language level of an 8 year old, is essential to protect the interests and safety of the study subjects.

Over the last decades a set of criteria has been established by the regulatory bodies that have identified and validated specific elements of many disease states that are related to a meaningful change in that disease when patients are subject to treatment, these are called validated endpoints. When a relation with disease activity of prediction is established but not yet rigorous validated it is called a surrogate endpoint.

Independent Review Boards are essential for a review of the protocol against the Good Clinical Practice (GCP) guidelines and local practices. Independent Data Review boards are generally put into place to independently monitor the safety and efficacy of the investigational product. On design of clinical programs margins are steered towards the efficacy needed and the tolerability that do not support continuation of the protocol at hand.

25.7 HEALTH AUTHORITIES AND ETHICAL CONSIDERATIONS

Clinical trials are closely supervised by appropriate regulatory authorities against a well-documented framework of mandatory submissions, the first submission is commonly the request for start of dosing in humans and examples are the opening of an Investigational Dossier (IND) with the Food and Drug Agency (FDA) in the USA and Investigation of Medicinal Product Dossier (IMPD) with the European Medicines Agency (EMA) in Europe. Other health authorities have developed their own criteria or use an approval by one of the major health authorities as reference. At the end of Phase III, when all relevant data are collected, the sponsor submits a final dossier for review at these agencies. In the interim, a variety of communications are possible with the agencies; further details can be found on their respective websites. In addition, many regulatory bodies have established guidelines for many diseases in which they describe the type of information they are seeking around requests for approval and the right to commercialize. Examples are recognized endpoints, acceptable surrogate markers for efficacy and/or safety, and numbers of patients required to justify the claims the sponsor wants to make based upon the clinical data generated. During the clinical development phase there is a well-structured schedule of interactions according to which the sponsor of the investigation communicates with the health authorities. In addition to the dossier with regulatory bodies all interventional or observational studies on patients must be approved by a supervising ethics committee before permission is granted to run a clinical investigation. Generally, this body is called the Institutional Review Board (IRB). IRBs can be located at the local investigator's hospital or institution; the alternative is a central (independent/for profit) IRB.

To warrant generally agreed ethical conduct, the International Committee for Harmonization (ICH) has developed guidelines for Good Clinical Practice (GCP) that are widely adopted as a framework to conduct clinical investigations. The guidelines aim to ensure the "rights, safety and wellbeing of trial subjects are protected."

25.7.1   Regulatory Examples

Communication with the health authorities and ethical committees is generally done in a well-defined framework frequently supported by guidance documents (available on US FDA, EMA, and Japan Health websites).

Before commencing investigations in humans, IND and IMPD packages are completed and submitted for review with a usually defined review period. There is the opportunity for consultation with the health authorities *via* a pre-IND and pre-IMPD meeting which can be face to face or written communications. For Japan and most other countries a similar framework exists. After approval of the package, investigations in humans can be started. There is an ongoing communication with all regulatory authorities on the safety observations made during the entire program and they can withdraw the authorization for investigation in humans based upon these reports. More and more frequently, sponsors install an independent Data Monitoring Board of experts to monitor the safety/benefit risk actively.

At the end of the Phase II program an end of Phase II meeting is usually scheduled to review the generated data and get agreement on the design of the Phase III program and the potential label claim.

After completion of the Phase III program the dossier is filed (NDA/IMPD or similar).

Health authorities have defined clinical endpoints where significant changes in a test population are considered to be reflective of a meaningful effect on the disease studied. These are called validated endpoints and are used to design the trials in various phases (but especially Phase III trials) and are mandatory for registration and subsequent commercialization.

In some situations or diseases these clinical endpoints are very difficult to test. In these situations there are sometimes surrogate endpoints defined that usually are not clinical determinants but laboratory or imaging parameters.

Investigators must obtain the full and informed consent of the participating human subjects (one of the IRB's main functions is to ensure potential patients are adequately informed about the benefits and risks of the clinical investigation they are asked to participate in). If the patient is unable to consent for him/herself, researchers can seek consent from the patient's legally authorized representative. An additional general rule is that when material is collected for genetic testing or that could be used for such purposes a separate informed consent is obtained.

Given that most investigations will collect and document information that is considered private there are rules and regulations to warrant the privacy of study subjects. In the USA researchers must understand and follow the federal patient privacy (HIPAA) law and good clinical practice.

25.8   INVESTIGATIONAL BROCHURE

This is a document generated by the sponsor of the study on the investigational product that functions as a fact base of all the core data generated with the investigational product irrespective of the source of the data. It is generally first used at the time of the submission of the IND/IMPD or equivalent and matures with the drug product throughout the entire life of the molecule. It is a core document in many of the packages that are used to get IRB approval and to inform study staff, investigators and alike.

25.9   STUDY TEAMS

The composition of study teams varies between the different phases of development, but is also strongly influenced by the type of trial planned/executed and the various companies and anchored in their structure, culture and historical performance. Generally speaking, early development programs tends to be in a single country with a relatively small number of subjects (<100) requiring smaller teams whereas later stage clinical programs tends to be at least on a regional scale and Phase III programs generally tend to be multinational/on a global scale with very sizable patient populations of up to multiple thousands. Generally speaking, teams consist of physicians, clinical research associates, statisticians, data managers and operational support. Hence, in early phase trials these teams are usually small but with the increase in size and complexity they can become very sizable for the Phase III pivotal/registration trials. In Phase IV they tend to become smaller again.

HINTS AND TIPS

1. The best possible clinical experiment is a prospective, randomized, placebo controlled, double blind, double dummy controlled clinical trial.
2. Double blinding is when neither the patient nor the investigator/staff know whether active or control (placebo) treatment is used.
3. Double dummy means that the active and placebo are identical in color, type, shape, taste and appearance.
4. Patients can be randomized, *i.e.* assigned randomly to treatment options. Within the randomization they can be stratified, *i.e.* assigned to groups based on underlying characteristic that will be relevant to treatment response.
5. Phase 0—This is traditionally a study in patients where no pharmaceutical intervention or a registered product is used. Patients are either subject to standard of care or to a drug that is already registered to be used in that disease and the application is done within the label as approved by the health authorities.
6. Phase I trials are to demonstrate safety and tolerability.
7. Phase II trials are for demonstration of impact on disease and dose finding.
8. Phase III trials are to generate a robust risk/benefit profile in the target disease population. The data package produced is used to negotiate with health authorities and to demonstrate the value of the product.
9. Phase IV is carried out after commercialization rights to understand the risk/benefit profile in the wider population.

10. Important to ensure studies are appropriately powered to be able to determine effect—*i.e.* statistically designed with appropriate patient numbers to demonstrate differences between the treatment arms or groups.
11. The start of a clinical development program is generally by submission of an IND to FDA or IMPD to EMA.

KEY REFERENCES

US FDA, general http://www.fda.gov/default.htm

US FDA, drugs http://www.fda.gov/Drugs/default.htm

US FDA, guidance documents http://www.fda.gov/Drugs/GuidanceComplianceRegulatory Information/Guidances/default.htm

EMA, general http://.ema.europa.eu

Pharmaceuticals and Medical Devices Agency, Japan http://www.pmda.go.jp/english

Pharmaceuticals and Medical Devices Agency, Japan, guidance site http://www.pmda.go.jp/english/service/regulation.html

CHAPTER 26

Aleglitazar: A Case Study

PETER MOHR

Roche Pharmaceutical Research & Early Development, Small Molecule Research,
Roche Innovation Center, Basel, Switzerland
E-mail: peter.mohr@roche.com

26.1 THE HISTORY OF DIABETES

Diabetes has been an awful plague to mankind for millennia. The Greek physician Arataeus from Cappadocia described it in the first century—quite appropriately—as liquefaction, as melting down of flesh and limbs into urine. He refers to one of the key symptoms: without treatment, diabetic patients are literally starving to death. At the beginning of the 20th century, they were still often termed "living skeletons". Diabetes was, fortunately enough, a relatively rare disease, but turned out always to be fatal. With the best available treatment, which was sticking to a very strict diet, patients could survive one year, or two at best. This changed all of a sudden when F. G. Banting, together with J. J. R. Macleod, J. B. Collip and C. H. Best, discovered insulin in 1921, and revolutionised therapy. Not surprisingly, it spread like wildfire that a new treatment had become available in Toronto. Patients from all over the world travelled to Canada to look for healing. Fred Banting was worshiped like a saint when people saw, for the first time in history, patients, often children, recover from diabetic coma. Not surprisingly, the Nobel Prize was awarded to him and Macleod in 1923. For a stimulating and detailed account, the excellent book by Michael Bliss is recommended to the interested reader.[1]

But times have changed dramatically; nowadays, late-onset type 2 diabetes is predominant and has become a major and ever-increasing health burden. The International Diabetes Federation estimates that more than 371 million people are suffering from this ailment, and the outlook is dismal! Figure 26.1 summarises the estimated sobering figures for the year 2012.

All experts agree that the most important culprit is our sedentary lifestyle. But fundamentally changing one's lifestyle is definitely more challenging than the daily swallowing of one or several pills. Not surprisingly, therefore, many pharmaceutical companies embarked decades ago upon the search for safe and efficacious treatments. This endeavour turned out to be successful, although it has to be stated that the available armamentarium to fight this disease is still limited. Table 26.1 provides an overview of the currently available therapeutic options, with

The Handbook of Medicinal Chemistry: Principles and Practice
Edited by Andrew Davis and Simon E Ward
© The Royal Society of Chemistry 2015
Published by the Royal Society of Chemistry, www.rsc.org

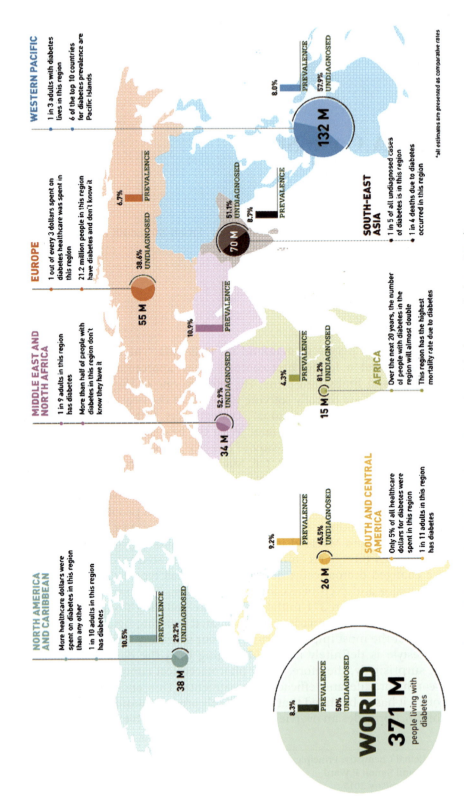

Figure 26.1 Estimated prevalence of diabetes in 2012 (reproduced from ref. 2 with the permission of IDF).

Table 26.1 Currently used therapeutic agents to treat type 2 diabetes.

Drug class	Insulin	Sulfonylureas	Glinides	Biguanides	Glucosidase Inhibitors	Thiazolidine-diones	DPP-IV inhibitors	GLP-1 analogues	SGLT2 inhibitors
Representative example	Insulin	Glibenclamide	Repaglinide	Metformin	Acarbose	Pioglitazone	Sitagliptin	Liraglutide	Dapagliflozin
Target	Insulin receptor	SU-receptor	SU-receptor	Largely unknown	α-Glucosidase	PPARγ	Dipeptidyl peptidase IV	GLP-1 receptor	Sodium glucose cotransporter 2

one example chosen from each a class of agent. Many of them were identified and developed without a thorough understanding of the disease pathology or knowing their molecular target. Even today, the mechanism of action of metformin—a cheap, efficacious but surprisingly safe drug—is not yet fully elucidated.

The World Health Organization defines diabetes according to the following diagnostic criteria: fasting plasma glucose ≥ 7.0 mmol/L (126 mg/dL) or 2 hr plasma glucose (venous plasma glucose 2 h after ingestion of 75 g oral glucose load) ≥ 11.1 mmol/L (200 mg/dL).[3] Typically, diabetes develops over a long period of time, and many patients, seemingly without symptoms, are only diagnosed several years after an impaired glucose tolerance has set in. But it should be borne in mind that an elevated glucose level is a biomarker rather than a disease of its own, and its devastating effects are triggered by manifold, complex, not fully understood mechanisms. They lead to the development of diabetes-specific microvascular pathology in the retina, in the renal glomerulus and peripheral nerves. And as a consequence, diabetes is still a leading cause of blindness, (at least in the developed world), end-stage renal disease and a variety of debilitating neuropathies. In addition, diabetes is also associated with accelerated atherosclerotic macrovascular damage affecting arteries that supply the heart, the brain and lower extremities.[4] As a result, patients with diabetes suffer from a much higher risk of myocardial infarction, stroke and limb amputation. In conclusion, control of elevated glucose levels is still a cornerstone of today's therapy, and the FDA states in their guidelines: ''For purposes of drug approval and labelling, final demonstration of efficacy should be based on reduction in HbA1c (*i.e.*, HbA1c is the primary endpoint of choice, albeit a surrogate), which will support an indication of glycaemic control.''[5] (HbA1c stands for glycated haemoglobin and provides a reliable measure for the average glucose concentration over the previous two weeks.) But within the very same document, the agency also raised the bar for approval by insisting on a robust assessment of cardiovascular safety. All drugs represented in Table 26.1 lower glucose in one way or another. Whereas insulin is just a replacement therapy, if this peptide hormone is no longer produced in sufficient quantity by the pancreas, sulfonylureas and glinides boost insulin secretion from the beta-cells, biguanides block hepatic glucose production, α-glucosidase inhibitors reduce glucose uptake in the intestine, and SGLT inhibitors increase the elimination of glucose *via* kidney and urine. DPP-IV inhibitors indirectly, and GLP-1 analogues directly, trigger insulin release from the pancreas. Finally, the glitazones, which activate the peroxisome proliferator-activated receptor (PPAR)-γ improve insulin sensitivity by controlling the expression of genes that are responsible for production, transport and metabolism of glucose and lipids.[6]

26.2   THE PEROXISOME PROLIFERATOR-ACTIVATED RECEPTORS

The peroxisome proliferator-activated receptors (PPARs) exist in three isoforms: PPAR-α, PPAR-δ, sometimes also called PPAR-β, and PPAR-γ. They are ligand-dependent transcription factors and belong to the superfamily of nuclear hormone receptors (NHRs) involved in crucial physiological functions. 48 members are encoded by the human genome[7] and most of them are built according to the general blueprint schematically represented in Figure 26.2.

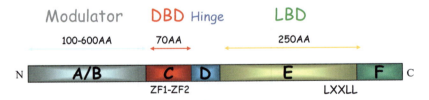

Figure 26.2   Schematic representation of nuclear hormone receptors; for details, see text.

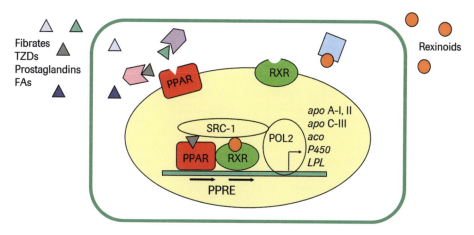

Figure 26.3 Schematic mechanism of action of PPARs (TZD = thiazolidinedione; FA = fatty acid).

The N-terminus, often called the A/B domain, is highly variable and exhibits ligand-independent (constitutively functional) transactivation activity. The A/B domain's sequence and length are highly variable between different receptors and receptor subtypes. Binding to the hormone response elements (HREs) in their target promoters is mediated through the DNA-binding domain (DBD) or C domain. Comprising two zinc fingers, the DBD is the most conserved region within the NHR superfamily. NHRs bind to DNA as heterodimers, homodimers or monomers, depending on the class. PPARs, like thyroid, retinoid, vitamin D and most orphan receptors, bind to DNA as a heterodimer with the retinoid-x-receptor (RXR). Adjacent to the DNA binding domain we find the D or hinge domain, which has an ill-defined function. It is thought to allow for conformational changes in the protein structure following ligand binding. The ligand binding domain (LBD) or E domain varies substantially between NHRs, but they all share a common structure of 11–13 α-helices organised around a hydrophobic binding pocket. Ligand-dependent activation requires the presence of activation function 2 (AF-2), located at the extreme C terminus of the NHR. Upon binding of the respective ligands—a fibrate, thiazolidinedione (TZD), prostaglandin or a fatty acid to the PPAR, and a rexinoid to RXR—the dimeric complex interacts with a PPAR response element within a target promoter. Conformational changes are induced, cofactors recruited, corepressors released, and as a consequence, the enzyme RNA polymerase 2 starts transcribing a given gene into messenger RNA (Figure 26.3).

PPAR-α is highly expressed in tissues with high rates of mitochondrial fatty acid oxidation, such as liver, heart, muscle, kidney and cells of the arterial wall (such as monocyte-derived macrophages, smooth muscle and endothelial cells), and it is activated by fibrates, fatty acids and eicosanoids, 15d-PGJ2 (15-deoxy-delta prostaglandin-J2), and oxidised fatty acids. PPAR-γ is found in adipose tissues, is activated by fatty acids or their derivatives, and plays a pivotal role in insulin sensitivity, adipogenesis and placental function. PPAR-δ is found in most tissues, is only weakly activated by fatty acids, prostaglandins and leukotrienes, and has no known physiologically relevant ligand.[8]

26.3 PPAR PROGRAMME AT ROCHE BASEL

Roche Basel inherited the PPAR programme from colleagues in Penzberg, *via* the acquisition of Boehringer-Mannheim in 1997. This event coincided with the strategic decision to transfer all antibacterial and dermatological activities to the newly founded spin-off company Basilea Pharmaceutica and to build in-house, from scratch, a strong department dedicated to metabolic

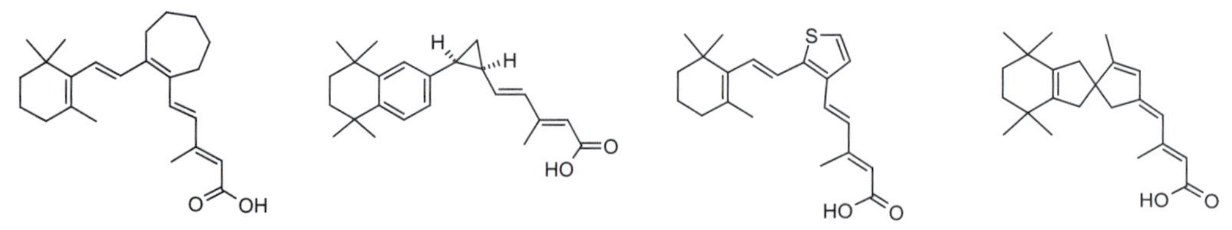

RO0257386-000 RO0482250-000 RO0256943-000 RO0487025-000

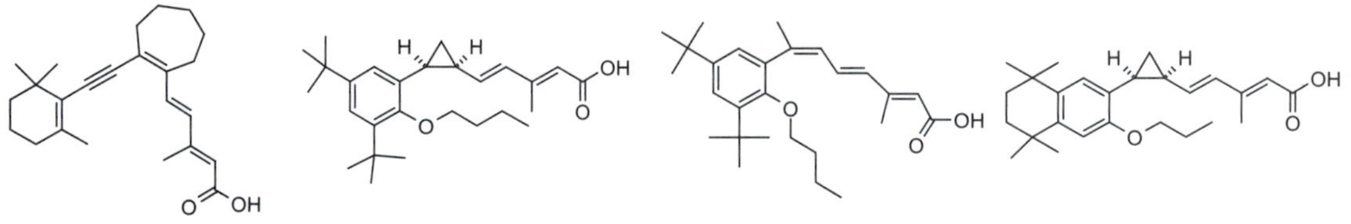

RO0620373-000 RO0640039-000 RO0265405-000 RO0717127-000

Figure 26.4 RXR agonists (top row) and antagonists (bottom row) from Roche.

research. We already had a lot of expertise in NHR research, due to our pioneering work with retinoids that culminated in the successful development of Roaccutane®, Tigasone® and Neo-Tigasone®, and our recent programmes with vitamin D derivatives, although it has to be stated that the former compounds were identified in the 1970s and 1980s without knowing their molecular target. Later, in the 1990s, we identified subclass-selective retinoids for retinoic acid receptor-α (RAR-α), retinoic acid receptor-β (RAR-β), and retinoic acid receptor-γ (RAR-γ), but also potent agonists and antagonists for the retinoid X receptors (RXRs). Figure 26.4 provides a compilation of representative RXR ligands that Roche chemists had designed and synthesised in-house. Since RXRs form heterodimers with PPARs, as stated above, it seemed obvious to test a few in models of diabetes and obesity. Although they turned out to be active, the observed effect was moderate at best, and the project was not pursued any further. In spite of immense efforts over more than two decades, just two RXR agonists, bexarotene (LGD1069) and alitretinoin (9-*cis*-Retinoic Acid), have reached the market, but they are prescribed only for niche indications (cutaneous T cell lymphoma the former, chronic hand eczema the latter).[9]

The race to discover PPAR agonists in the late 1990s, was much more successful. A search in Chemical Abstracts Service for the two terms "PPAR" and "patent" reveals no less than 3755 references (Figure 26.5); although not all of them reveal chemically novel ligands, they give testimony to a huge endeavour from a plethora of pharmaceutical companies.

PPAR-γ is a master regulator of efficient energy storage and translates nutritional and metabolic stimuli into the expression of genes. When we embarked on our endeavour to identify a PPAR-α/γ-co-agonist, three PPAR-γ ligands were already on the market, namely rosiglitazone (Avandia®) from GSK, pioglitazone (Actos®) from Takeda/Eli Lilly and troglitazone (Rezulin®) from Daiichi Sankyo/Park Davis. The latter, however, had to be withdrawn in 2000 due to an unacceptable, sometimes fatal, liver toxicity. All thiazolidinediones (TZDs) were developed and marketed as racemic mixtures since the chiral centre was supposed to be configurationally unstable, under physiological conditions. RO2052349-000 (edaglitazone) and its putative back-up compound, RO2060297-000, had been discovered by Boehringer-Mannheim and became part of the Roche portfolio in 1998. Both were not only extremely similar with respect to their structure, they also exhibited a similar profile towards the receptor, and behaved, as most but not all of the TZDs, as pure PPAR-γ agonists. Accordingly, and due to the more labour-intensive and expensive synthesis of the thiophene analogue, development of RO2060297 was dropped relatively early (December 2001) (Figure 26.6).

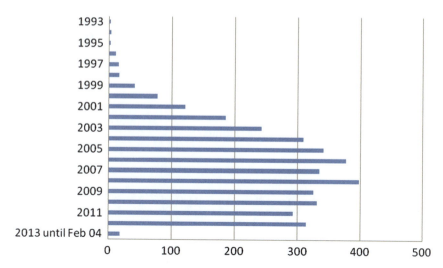

Figure 26.5 PPAR-related patents found within CAS.

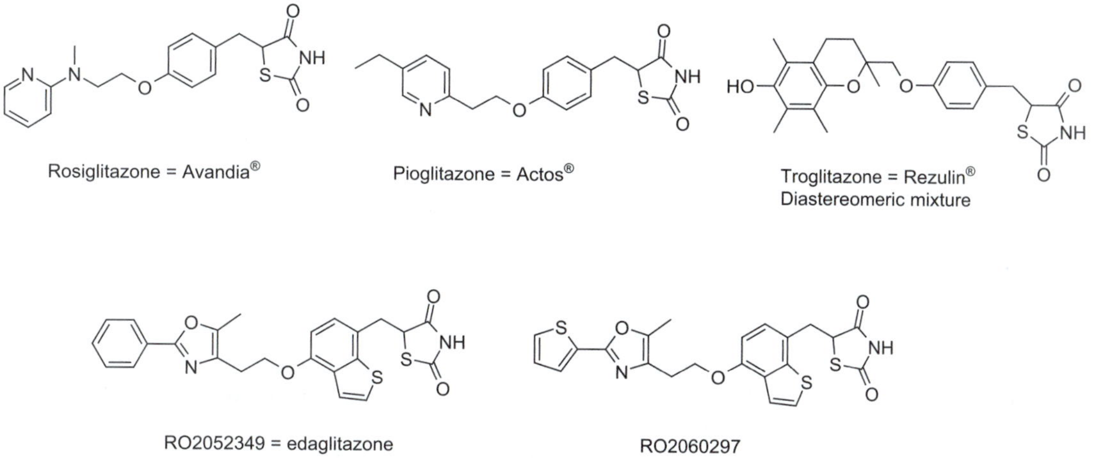

Rosiglitazone = Avandia®

Pioglitazone = Actos®

Troglitazone = Rezulin®
Diastereomeric mixture

RO2052349 = edaglitazone

RO2060297

Figure 26.6 Marketed PPAR-γ agonists and Boehringer/Roche development compounds from the TZD class.

Edaglitazone, on the other hand, was pursued until Phase III, where it ultimately succumbed to a slow and silent death. It failed for several reasons (not an atypical phenomenon these days), among them the lacklustre outcome of the carcinogenesis study, lack of clear superiority with respect to the putative competitors Actos® and Avandia® (which would have become generic around the date of launch, hence no longer a clear business case), and an increasingly unfavourable sentiment for PPARs in general. The latter was mainly due to the mind-boggling attrition rate all PPAR ligands in the clinics had suffered to date (see below).

PPAR-α is mainly expressed in liver, kidney and heart, and is thought to be the key regulator of mitochondrial and peroxisomal β-oxidation. In addition, PPAR-α regulates the expression of genes coding for lipoproteins (*e.g.* ApoAI, ApoAII and ApoCIII).[8] The clinical proof of concept for PPAR-α agonism comes from the fibrate class of compounds (*e.g.*, clofibrate, bezafibrate and fenofibrate [see Figure 26.7]) that improve the lipid profile, albeit rather modestly (weak LDL and triglyceride decrease, and moderate HDL increase). Although, again, it has to be stated that

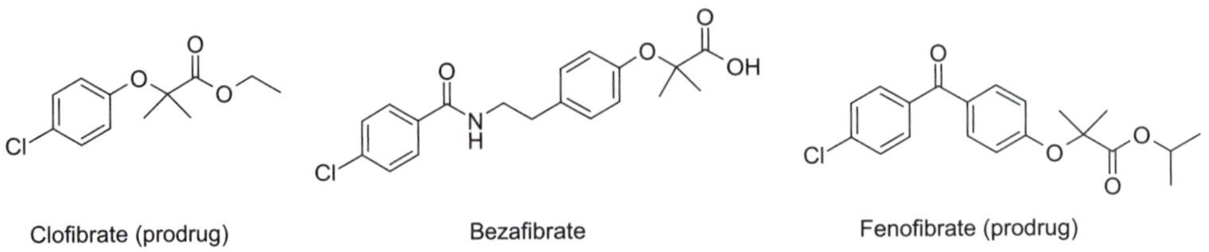

Figure 26.7   Clinically selected PPAR-α agonists (Fibrates).

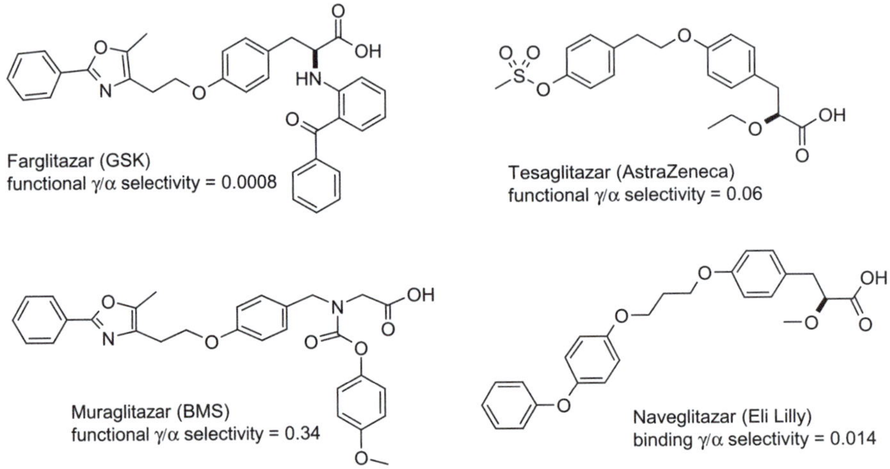

Figure 26.8   Some PPAR co-agonists that made it to advanced clinical development and their reported receptor selectivity in favour of the γ-isoform.

the mechanism of action was discovered several decades after first clinical use. Therefore, our strategy, as well as that of our competitors, was to combine the fuel-storing activities of PPAR-γ with the fuel-burning effect of PPAR-α in one molecule. This we hoped would simultaneously improve insulin sensitivity (γ effect) and lipid profile (α effect). Our target product profile (TPP) stated treatment of type 2 diabetes in monotherapy as a primary indication, but also suggested combination therapy with sulfonylureas or the other insulin secretagogues, metformin, or insulin. Combining both effects in one molecule, and hence in one pill, seemed not only advantageous with respect to patient compliance, it also avoided the necessity to adjust bioavailability and to synchronise the pharmacokinetic (PK) behaviour. In addition, it would reduce the drug load, and last but not least, it would simplify clinical development.

26.4   A JUMP-START

The synthetic chemistry of the PPAR co-agonist project started at the end of June 2000 in our laboratory. Our mission was to discover a potent, well-balanced PPAR-α/γ co-agonist with excellent physicochemical properties, outstanding PK behaviour and devoid of any off-target activity, preferably within a couple of months. Speed was of the utmost importance because several co-agonists from competitors had already entered clinical development (see Figure 26.8).

At the beginning—in order to rationally guide our design—we studied the X-ray structure of known competitor molecules bound to the human PPAR ligand-binding domain. Figure 26.9 shows the AstraZeneca compound tesaglitazar bound to PPAR-α. The binding pocket exhibits a

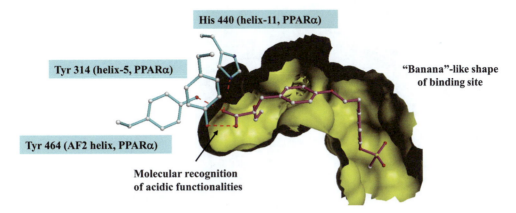

Figure 26.9 Tesaglitazar bound to human PPAR-α.

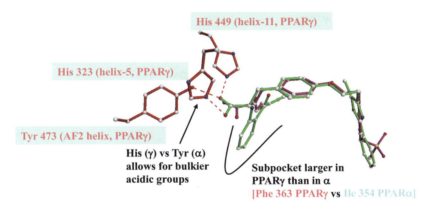

Figure 26.10 Comparison of farglitazar and tesaglitazar bound to human PPAR-γ.

"banana"-like shape; therefore, all decent ligands must be able to adopt an energetically favourable conformation with a corresponding kink to smoothly fill out the cavity. The acidic warhead, present in all potent PPAR ligands, forms three hydrogen bonds to His 440, Tyr 314 and Tyr 464. A fourth hydrogen bond to serine in the foreground is omitted for clarity. Figure 26.10 illustrates some key differences between the α and γ receptor by overlaying tesaglitazar with farglitazar from GSK in the latter. Whereas the "eastern wings" are almost superimposed, two subtle but pivotal differences in the acidic head group are clearly visible: 1) one tyrosine is replaced by another histidine, allowing bulkier head groups to be accommodated. This is the structural reason why TZDs always show more or less pronounced selectivity in favour of the γ receptor; this group is simply too large to snugly fill the available limited space. 2) The sub-pocket where the ethoxy residue of tesaglitazar fits in is definitely larger in PPAR-γ.

This difference nicely explains why farglitazar turned out to be a not very balanced co-agonist; it is one of the strongest γ binders ever described in the literature but shows only moderate activity at the α receptor. The isoleucine residue (Ile 354) in PPAR-α, much bulkier than phenylalanine in PPAR-γ (Phe 363), reduces the available space, and the large benzophenone moiety of farglitazar is therefore simply too voluminous for a perfect fit.

The X-ray structure of edaglitazone became another source of inspiration. A closer look at the picture, presented in Figure 26.11, clearly illustrates how well the benzothiophene linker fills the central cavity in the PPAR-γ ligand-binding domain. And since this part of the binding site is

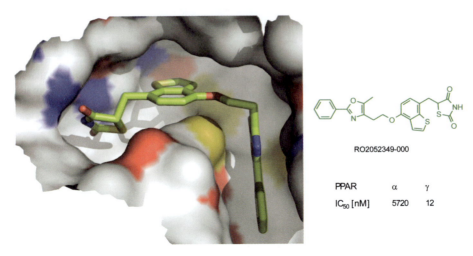

Figure 26.11 X-ray structure of edaglitazone bound to human PPAR-γ ligand-binding domain and its *in vitro* binding activity.

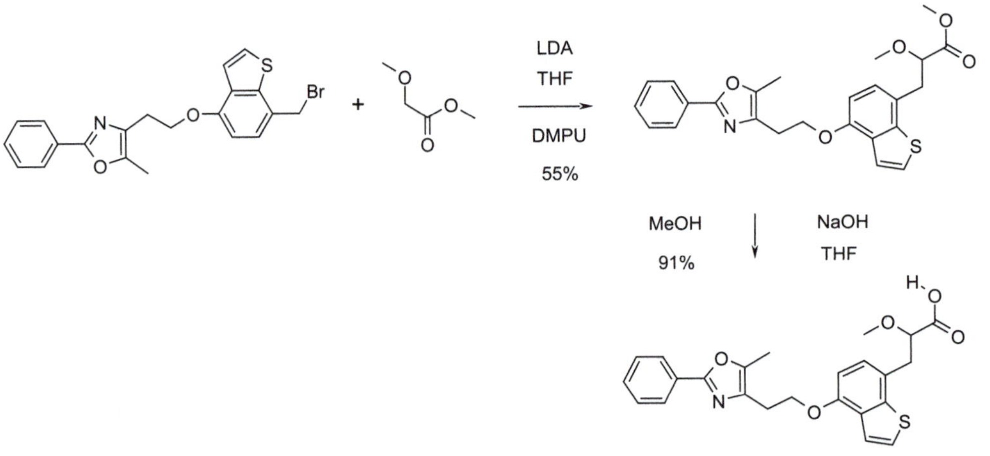

Figure 26.12 First synthesis of racemic aleglitazar in July 2000.

quite similar with respect to size, shape and polarity in both receptors, we were tempted to incorporate this building block into novel putative co-agonists. Fully in line with this optimistic prediction, this simple idea turned out to be very rewarding.

Only 2 weeks after the start of the experimental chemistry, we registered RO0717122, aleglitazar in its racemic form, in our internal Roche database; without realising of course, what an interesting molecule we had in our hands. Capitalising on an advanced building block available from process research (later banned due to the extreme toxicity of an unavoidable side product (bis(bromomethyl)-derivative) formed during its manufacture)—we prepared our first PPAR-α/γ ligand in just two steps (Figure 26.12).

When we submitted our first handful of putative ligands for screening to Markus Meyer, the project leader in biology, who had joined Roche Basel from Boehringer-Mannheim, we got really excited when the results came back. Three molecules had been tested, and all three turned out to be potent PPAR-γ ligands exceeding edaglitazone with respect to potency! Enthusiastically, we focused all our efforts on the new PPAR project and started systematically exploring the structure–activity relationship (SAR). We also got functional data back—confirming that our molecules were indeed full agonists. Shortly thereafter, Markus established the PPAR-α assays; and again the results met our boldest expectations: the majority of our molecules exhibited exactly the profile we had envisaged—they turned out to be full agonists at both receptors, and many showed a balanced, almost equipotent activity at both PPAR-α and -γ. The PPAR-δ data, on the other hand, became available much later, because this receptor subtype had first to be cloned, and at that point in time, this off-target profiling was not our top priority. The limited data generated later illustrated that many of our ligands were "pan-binders"; however, they had very little affinity for PPAR-δ and exhibited only weak transactivation activity, if any.

Since at this point in time our project was still in the exploratory stage (later, the resources in chemistry were gradually increased to four fully equipped laboratories, each staffed with one PhD chemist and two technicians), we varied only some obvious parameters in a first round of optimisation. We studied the influence of size, shape and nature of the alkoxy side-chain, we replaced the benzothiophene building block with naphthalene, and we also varied the phenyloxazole moiety.

26.5 STRIVING FOR OPTIMISATION

Table 26.2 summarises our first SAR study. Two trends seem obvious: the PPAR-γ affinity increased with size and lipophilicity of the side-chain. For example, with n-hexyl, we were already down to 2 nM affinity in the racemic series. For PPAR-α, the opposite was true; methoxy and ethoxy analogues were the most potent ligands for this particular subtype. Fully in line with predictions from modelling, bulkier side-chains did not fit perfectly into the smaller sub-pocket of this receptor. The transactivation data, although always to be interpreted with a pinch of salt, as data obtained with living cells are intrinsically less robust and reliable, followed a similar pattern. Furthermore, all ligands with a short side-chain tended to exhibit slightly greater agonistic potency. As a note of caution, it should be emphasised that even the figures for binding affinity, although they had been generated with utmost care, are not carved in stone. Radiotracers used for the competition experiment were intrinsically unstable and decomposed over time, which led to a "time-dependency" of the data. More importantly, in the beginning we had

Table 26.2 First SAR generated with respect to binding (IC_{50}) and transactivation (EC_{50}) of racemic α-alkoxypropionic acids.

Structure	RO-Number	R	$IC_{50}\alpha$ (μM)	$IC_{50}\gamma$ (μM)	$EC_{50}\alpha$ (μM) (% Effect)	$EC_{50}\gamma$ (μM) (% Effect)
	RO0717122-000	CH_3	0.072	0.046	0.103 (168)	0.059 (131)
	RO0717241-000	CH_3CH_2	0.022	0.007	0.027 (164)	0.018 (100)
	RO0722324-000	$CH_3CH_2CH_2$	0.175	0.004	0.075 (123)	0.012 (64)
	RO0727465-000	$CH_2CH(CH_3)_2$	0.239	0.003	0.048 (181)	0.004 (63)
	RO0727678-000	n-Hexyl	0.120	0.002	0.249 (115)	0.004 (85)
	RO0727682-000	n-Butyl	0.214	0.003	0.032 (142)	0.008 (60)
	RO4418746-000	i-Propyl	n. d.	0.004	0.120 (83)	0.010 (49)
	RO4499092-000	CH_2CH_2CH-$(CH_3)_2$	0.243	0.008	0.969 (76)	0.128 (82)
	RO0724715-000	CH_3	0.061	0.113	0.131 (159)	0.123 (98)
	RO0722898-000	CH_3CH_2	n. d.	0.030	n. d.	n. d.
	RO0726037-000	$CH_3CH_2CH_2$	0.069	0.007	0.087 (56)	0.016 (81)
	RO0728373-000	n-Butyl	0.139	0.007	0.061 (171)	0.067 (95)
	RO4512437-000	$CH_2CH(CH_3)_2$	n. d.	0.042	0.346 (61)	0.038 (92)

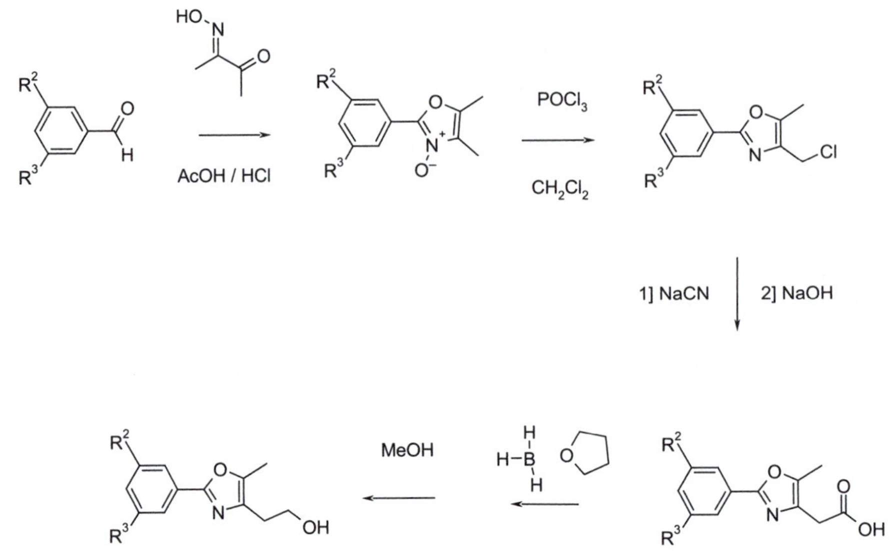

Figure 26.13 Synthesis of various oxazole building blocks following the procedure described in reference 10.

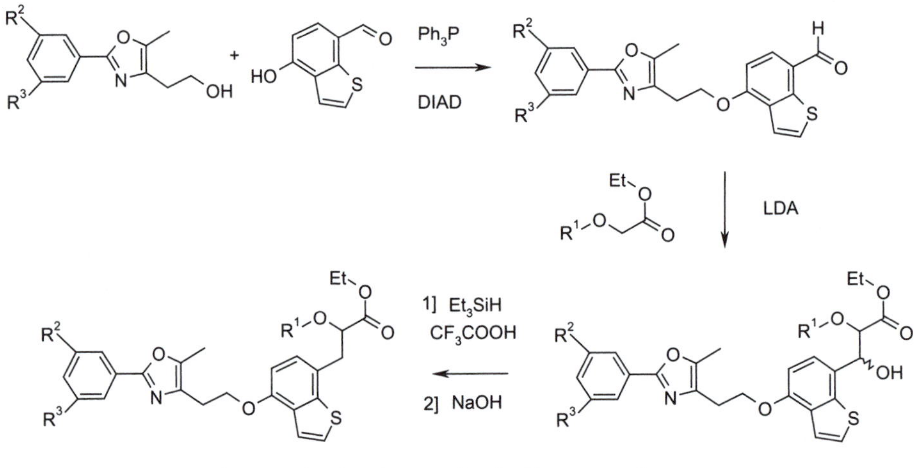

Figure 26.14 Second-generation synthesis of racemic aleglitazar analogues.

used labelled farglitazar to determine PPAR-α binding IC_{50}s which due to its modest affinity was not particularly well suited as radiotracer for the competition experiment.

In a next round of optimisation, we began to vary, more or less systematically, the substitution pattern of the phenyl-oxazole moiety. The corresponding building blocks were synthesised (see Figure 26.13) based on the then current literature, capitalising on a reaction step whose precise mechanism still remains elusive. The mechanistic explanation proposed by Goto *et al.*[10] is not fully convincing, but the procedure was nonetheless reliable and offered ample scope for the preparation of a plethora of different 2-(2-aryl-5-methyl-oxazol-4-yl)-ethanol building blocks.

In the meantime, we had also developed a new synthetic access to our lead series, based on an Aldol approach that allowed more flexibility, gave significantly better yields, and avoided the use of hazardous intermediates (Figure 26.14). The appropriated phenyloxazole building block was condensed with readily available 4-hydroxy-benzo[b]thiophene-7-carbaldehyde under Mitsunobu conditions, followed by LDA-mediated coupling with the respective α-alkoxy-acetic acid ester.

The synthesis end game comprised of ionic deoxygenation of the *syn/anti* mixture, typically with triethylsilane and trifluoroacetic acid, and subsequent hydrolysis with aqueous base.

As can be seen at first glance in Table 26.3, additional substituents contributed little to the affinity towards the PPARs. In the benzothiophene as well as in the naphthalene series that we prepared for the sake of comparison, the "naked" phenyloxazole seemed close to a local optimum. Thus, phenyloxazole was the building block of choice with respect to ligand efficiency as well as lipophilic efficiency (LiPE).[11] *Ortho* substituents, particularly large ones, actually seemed to be detrimental (probably because they forced the phenyl ring out of the oxazole plane).

In general, the benzothiophene analogues exhibited a slightly more favourable profile. The absolute potency was in general better, and the physicochemical properties, including lipophilicity (logD) and solubility, were definitely more drug-like.

In Table 26.4, we present some analogues within the α-ethoxy series. A similar pattern emerged: decorating the phenyloxazole moiety did not generate better compounds. The molecular size increased without creating any added value. Again, as a rule of thumb, we concluded that *ortho* substituents larger than fluorine were to be avoided, and that most of the compounds were already clearly γ-biased, hence not optimally balanced.

Encouraged by these exciting results, it soon became clear that we should synthesise our ligands in optically pure form sooner, rather than later. Based on available X-ray data, and also from the absolute configuration of competitor compounds, *e.g.* the AstraZeneca ligand AZ242 = tesaglitazar = Galida®, it was easy to predict which enantiomer had to be prepared

Table 26.3a SAR around the phenyloxazole part—benzothiophene series.

Structure	RO-Number	R	IC$_{50}$α (μM)	IC$_{50}$γ (μM)	EC$_{50}$α (μM) (% Effect)	EC$_{50}$γ (μM) (% Effect)
	RO0717122-000		0.072	0.046	0.103 (168)	0.059 (131)
	RO4399985-000		n. d.	0.021	0.058 (243)	0.052 (90)
	RO4403338-000		n. d.	0.056	0.190 (193)	0.171 (79)
	RO4507960-000		n. d.	4.950	n. d.	n. d.
	RO4507977-000		n. d.	0.729	0.260 (118)	0.348 (101)
	RO4508834-000		n. d.	0.384	0.316 (189)	0.274 (85)
	RO4509580-000		0.294	0.916	0.435 (101)	1.615 (64)
	RO4510847-000		n. d.	6.470	n. d.	n. d.

Table 26.3b SAR around the phenyloxazole part—naphthalene series.

Structure	RO-Number	R	$IC_{50}\alpha$ (μM)	$IC_{50}\gamma$ (μM)	$EC_{50}\alpha$ (μM) (% Effect)	$EC_{50}\gamma$ (μM) (% Effect)
	RO0724715-000		0.061	0.113	0.131 (159)	0.123 (98)
	RO4387190-000		0.211	0.083	0.286 (126)	0.788 (66)
	RO4387620-000		0.052	2.750	0.261 (78)	1.502 (42)
	RO4499561-000		0.432	0.802	0.123 (114)	0.716 (78)
	RO4499564-000		n. d.	0.400	0.225 (134)	0.489 (58)
	RO4503694-000		n. d.	10.000	n. d.	n. d.
	RO4509577-000		n. d.	3.000	n. d.	n. d.
	RO4510850-000		n. d.	1.810	0.485 (135)	0.881 (59)
	RO4512653-000		n. d.	5.950	n. d.	n. d.

(the development of Galida was stopped in spring 2006, after four Phase III studies had been completed, based on an unfavourable risk/benefit ratio [concerns about elevations in serum creatinine and an associated decrease in glomerular filtration rate]). All biological activity should reside in the (S)-enantiomer (Figure 26.15).

Whereas AstraZeneca had initially relied on a classical separation of diastereomeric amides[12] we opted directly for an enantioselective synthesis. Evans' boron-enolate methodology appeared particularly appealing to us,[13] and we could indeed prepare RO0728804-000 by this technology with a decent yield and very high enantiomeric excess (Figure 26.16). The enol-borinate, prepared *in situ* by treatment with dibutylboron triflate and triethylamine, added to the advanced aldehyde intermediate as predicted with good stereocontrol; traces of other diastereomers formed could easily be removed by chromatography or converged to the very same product upon ionic reduction of the second chiral centre, again by treatment with triethylsilane/trifluoroacetic acid. Routine crystallisation delivered the anticipated product with 99% enantiomeric excess. Aleglitazar, our clinical candidate, was thus registered in March 2001, 8 months after launching the project. This Aldol route also proved suitable for scale-up, and the first batches for toxicity studies ($\sim$50 g) were prepared in our lab, following the very same synthetic scheme. Needless to say, countless other PPAR ligands, including tesaglitazar for direct comparison, were synthesised, capitalising on this very reliable and robust method.

Table 26.4 SAR in the α-ethoxy series.

Structure	RO-Number	R	$IC_{50}\alpha$ (μM)	$IC_{50}\gamma$ (μM)	$EC_{50}\alpha$ (μM) (% Effect)	$EC_{50}\gamma$ (μM) (% Effect)
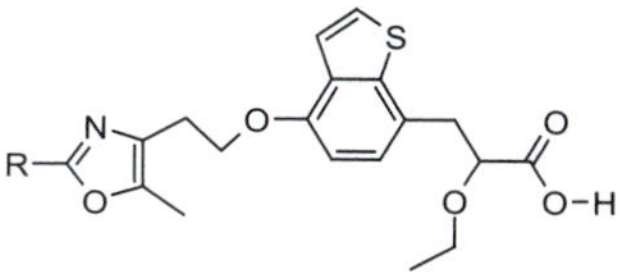	RO0717241-000	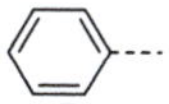	0.022	0.007	0.027 (164)	0.018 (100)
	RO4403857-000	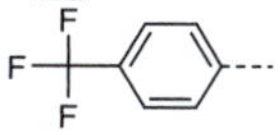	n. d.	0.778	0.017 (119)	0.706 (92)
	RO4431482-000	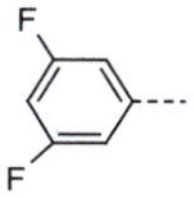	n. d.	0.024	0.259 (119)	0.044 (82)
	RO4431934-000	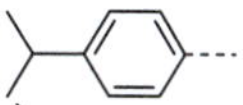	0.007	0.002	0.048 (86)	0.006 (45)
	RO4432900-000	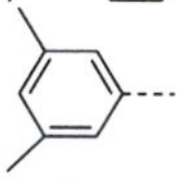	n. d.	0.002	0.008 (70)	0.004 (43)
	RO4432904-000	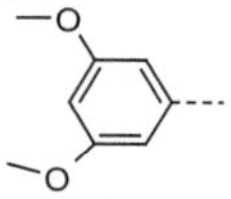	n. d.	0.003	0.017 (122)	0.004 (38)
	RO4507982-000	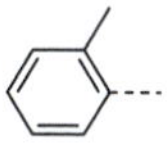	n. d.	0.182	0.209 (215)	0.124 (100)
	RO4508836-000	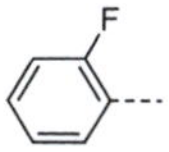	n. d.	0.076	0.203 (129)	0.070 (102)
	RO4510654-000	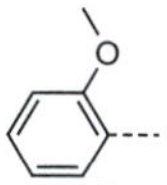	n. d.	0.264	0.102 (54)	0.171 (43)
	RO4510660-000	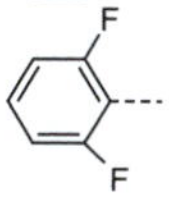	n. d.	0.128	0.008 (38)	0.091 (100)
	RO4510849-000	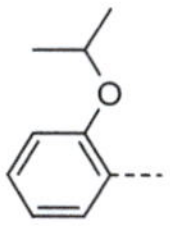	n. d.	3.110	n. d.	n. d.
	RO4516298-000	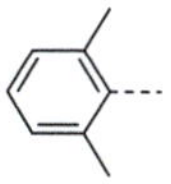	n. d.	5.420	n. d.	n. d.
	RO4543343-000	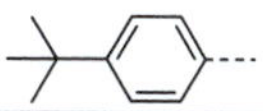	n. d.	0.004	0.187 (138)	0.008 (26)

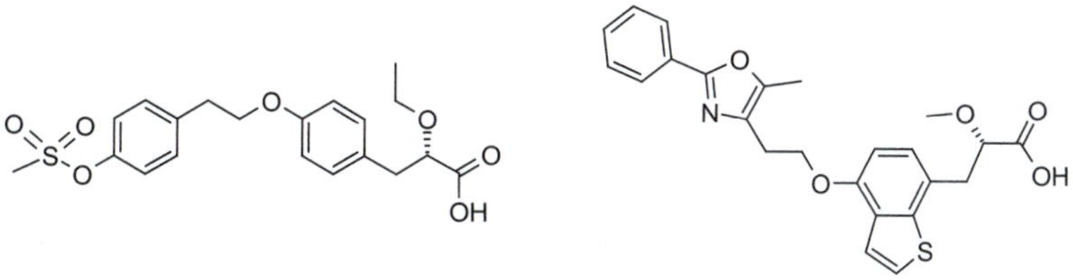

Tesaglitazar = Galida = AZ242 = RO4433150-000

Aleglitazar = RO0728804-000

Figure 26.15 Structure of tesaglitazar and aleglitazar.

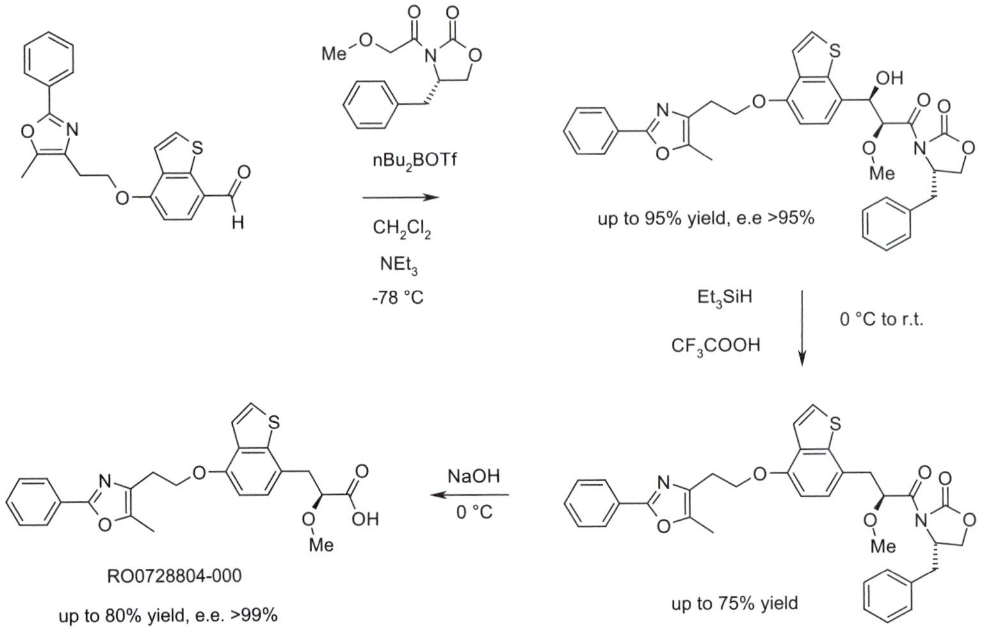

Figure 26.16 Medicinal chemistry synthesis of aleglitazar.

Table 26.5 provides some SAR data within the (*S*)-methoxy series, varying only the phenyl-oxazole moiety. The same pattern as before emerged: there was little to gain, but also little to lose, by incorporating other aryl-oxazoles; not-too-bulky lipophilic residues in the *para*-position slightly boosted potency but at the expense of increased molecular volume; or to phrase it positively: the receptors were rather tolerant within this part of the binding pocket. Electron-withdrawing as well as electron-donating substituents were found to be acceptable. Needless to say, naphthalene analogues behaved similarly but didn't offer any clear advantage.

Table 26.6 provides an update of the impact of the α-alkoxy chain length. Again, fully in line with the structural analysis of the respective receptor subpocket alluded to above, increasing the chain length brought the activity at PPAR-γ down to 1 nM, but, unfortunately, the affinity to the α-receptor was simultaneously eroded. Only the somewhat smaller 1-butenyl residue was fairly well accommodated.

26.6 IN-DEPTH PROFILING

What we call nowadays a multi-dimensional optimisation (MDO) tool, for fast profiling of new compounds, was not yet established as a routine process. However, due to promising properties, a few typical representatives were soon submitted for *in vitro* metabolic clearance studies; and to our delight, they turned out to be extraordinarily stable. Table 26.7 shows one of the first data sets we had generated with some racemic derivatives. We note in passing that a few farglitazar analogues were profiled as well. It is beyond the scope of this case study to discuss their properties in detail but suffice to say that they turned out to be—as expected—much less balanced with respect to receptor affinity and to suffer from lacklustre physicochemical properties. In addition, these hybrid molecules didn't match the extreme potency of farglitazar at the γ receptor which, in our hands, exhibited an IC$_{50}$ value of 1 nM. Farglitazar was, at that point in time, supposed to be the most advanced and most serious competitor molecule. However, its profile was definitely not well balanced, binding at least two orders of magnitude weaker to the

Table 26.5 SAR of the phenyloxazole part in the (*S*)-methoxy series.

Structure	RO-Number	R	IC$_{50}\alpha$ (μM)	IC$_{50}\gamma$ (μM)	EC$_{50}\alpha$ (μM) (% Effect)	EC$_{50}\gamma$ (μM) (% Effect)
	RO0728804-000		0.038	0.019	0.050 (156)	0.021 (67)
	RO4368972-000		0.053	0.023	0.002 (109)	0.053 (132)
	RO4391770-000		n. d.	0.008	0.043 (?)	0.011 (61)
	RO4399407-000		0.045	0.002	0.020 (151)	0.010 (53)
	RO4399408-000		0.048	0.003	0.061 (196)	0.023 (44)
	RO4431929-000		0.047	0.027	n. d.	n. d.
	RO0732244-000		0.228	0.033	0.205 (404)	0.165 (89)
	RO0729412-000		0.034	0.073	0.061 (164)	0.003 (37)
	RO4402361-000		0.010	0.027	0.047 (128)	0.054 (56)
	RO4389640-000		0.364	0.019	0.407 (130)	0.107 (79)

Table 26.6 SAR of the α-alkoxy residue; the impact of the chain length.

Structure	RO-Number	R	IC$_{50}\alpha$ (μM)	IC$_{50}\gamma$ (μM)	EC$_{50}\alpha$ (μM) (% Effect)	EC$_{50}\gamma$ (μM) (% Effect)
	RO0728804-000	Me	0.038	0.019	0.050 (156)	0.021 (67)
	RO0728811-000	Et	0.053	0.021	0.027 (109)	0.021 (63)
	RO0732129-000	nPr	0.393	0.002	0.013 (204)	0.004 (25)
	RO0731091-000	nBu	0.162	0.001	0.017 (428)	0.063 (85)
	RO0732673-000	CF$_3$CH$_2$	0.060	0.003	0.555 (179)	0.151 (119)
	RO4389983-000	CH$_2$CH$_2$CHCH$_2$	0.021	0.002	0.027 (108)	0.032 (79)

α receptor, and most physicochemical properties far from optimal—the molecule was much too lipophilic and, consequently, almost insoluble. The announcement at the end of 2001 that development was stopped after having reached Phase III studies didn't really come as a big surprise to us! (Alternative indications were later explored, *e.g.*, hepatic fibrosis.[14])

Studies in hepatocytes, where phase II metabolic processes also operate, corroborated the excellent metabolic stability of RO0717122-000 across different species (Table 26.8). The maximal achievable bioavailability (MAB, calculated assuming that intrinsic clearance is the only limiting factor) was close to 100%. Soon after, a PK study confirmed that these favourable *in vitro* properties also translated *in vivo*. In rat, low clearance, reasonable half-life, decent oral

Table 26.7 Metabolic stability in rat and human microsomes for selected racemic PPAR co-agonists.

RoNo:	Structure	Rat		Human	
		Cl_{int} (µl/min/mg prot.)	max ach.bioav.	Cl_{int} (µl/min/mg prot.)	max ach.bioav.
RO-71-8825/000		21 (medium) 38 (medium)	58% 44%	12 (medium) 15 (medium)	58% 51%
RO-71-7122/000		12 (low) 14 (low)	71% 68%	12 (medium) 4 (low)	58% 78%
RO-72-4715/000		11 (low) 9 (low)	73% 77%	6 (low) 0 (low)	72% 99%
RO-72-5858/000		53 (medium) 56 (medium)	36% 35%	4 (low) 12 (medium)	80% 57%

Table 26.8 Metabolic stability of racemic aleglitazar in rat and human hepatocytes.

RoNo:	Structure	Rat hepatocytes		Human hepatocytes	
		Cl_{int} (ml/min/mio cells)	MAB (%)	Cl_{int} (ml/min/mio cells)	MAB (%)
RO-71-7122/000		2.2 (low) 1.6 (low)	86% 89%	0.2 (low) 0.1 (low)	97% 98%

bioavailability and little inter-animal variability were observed in the first and all following experiments (Figure 26.17). CYP interaction was only later studied with the pure enantiomer. There was a liability with respect to the isoform CYP-2C9, but taking into account the high potency and, hence, low doses later applied, this was really not a serious issue (Figure 26.18).

Encouraged by all these favourable findings, representative analogues were soon submitted for the first *in vivo* pharmacodynamic studies. Some were performed in house; others were outsourced to the group of Prof. Auwerx in Strasbourg. As a screening model, we chose the well-established *db/db* mouse; this leptin deficient diabetic mutant mouse exhibits elevation of plasma insulin, hyperglycemia and obesity. As anticipated, RO0728804-000 did produce, after treatment during 4 days, a robust lowering of glucose and triglyceride levels, the typical profile of a decent insulin sensitiser. Figure 26.19 provides some screening data in the benzothiophene (red) and naphthalene (blue) series, in direct comparison with rosiglitazone (RO0641588-000). Aleglitazar is definitely not the most potent compound tested in this model, but this is not at all surprising. Most of the effects are PPAR-γ driven, and as we have seen above, RO0732129-000 is clearly more active at this receptor subtype. Please note that some derivatives (RO0727678-000, RO0726037-000 and RO0728373-000) were submitted as racemates.

Sustained and strong glucose lowering was definitely an indispensable prerequisite but certainly not sufficient to qualify a PPAR co-agonist as a clinical candidate suitable for

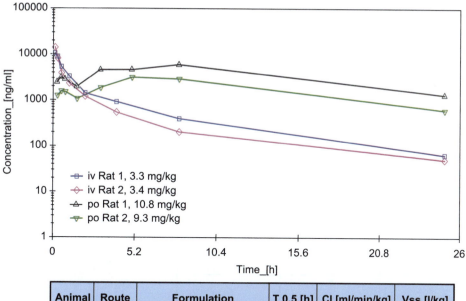

Animal	Route	Formulation	T 0.5 [h]	Cl [ml/min/kg]	Vss [l/kg]
Rat 1	i.v.	Solution (PEG 400:NaCl 0.9%, 4:6)	5.73	3.31	1.02
Rat 2	i.v.	Solution (PEG 400:NaCl 0.9%, 4:6)	6.66	4.42	1.22
Mean		Solution (PEG 400:NaCl 0.9%, 4:6)	**6.20**	**3.87**	**1.12**

Animal	Route	Formulation	Tmax [h]	Cmax [ng/ml]	T 0.5 [h]
Rat 1	p.o.	Microsuspension	8.0	5953.8	9.6
Rat 2	p.o.	Microsuspension	5.0	3087.3	8.1
Mean		Microsuspension	**6.5**	**4520.6**	**8.9**

Figure 26.17 PK of racemic aleglitazar in male Wistar rat at 3 mg/kg i.v. and 10 mg/kg p.o.

CYP	n1 [μM]	n2 [μM]	Mean [μM]
3A4	23.24	15.75	19.49
2C9	4.52	3.51	4.02
1A2	>50	>50	>50
2D6	>50	>50	>50
2C19	10.35	14.12	12.24

Figure 26.18 Interaction of aleglitazar with cytochrome P450 isoforms (measured in duplicates).

development. Favourable effects on the lipid profile, that is, significant lowering of triglycerides and a concomitant increase of HDL, if possible, accompanied by a decrease in LDL, were also a must. Aleglitazar combined both effects in a single molecule. Figure 26.20 demonstrates the robust effect in an Oral Glucose Tolerance Test (OGTT) in *db/db* mice, in comparison with the

	glucose	insulin	TG			glucose	insulin	TG
vehicle	0	0	0		vehicle	0	0	0
RO0641588-000 = Rosiglitazone	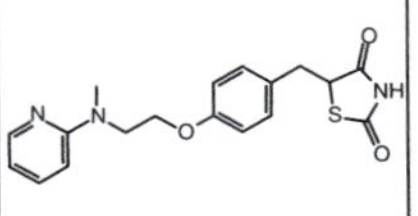 −6.81	27.67	−10.08		RO0727678-000	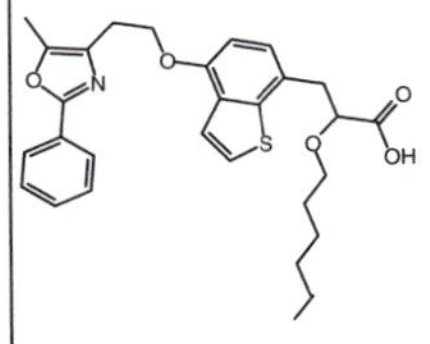 −25.05	−24.51	−36.55
RO0728804-000	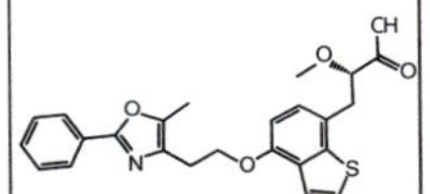 −24.92	−8.70	−55.88		RO0730062-000	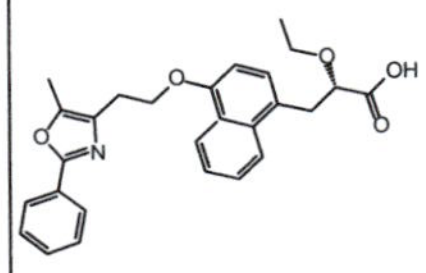 −32.40	−18.58	−44.96
RO0728811-000	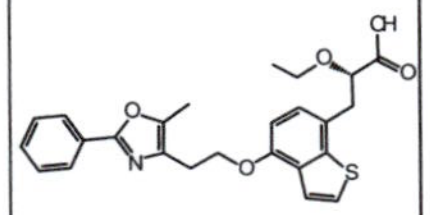 −23.84	2.37	−57.56		RO0726037-000	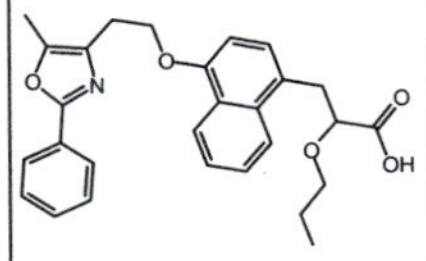 −25.59	−47.83	−44.54
RO0732129-000	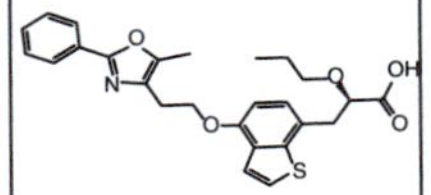 −44.17	16.21	−52.10		RO0728373-000	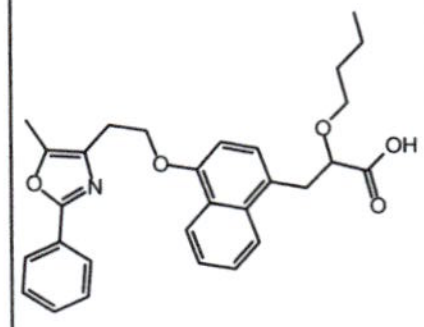 −17.09	34.78	−13.03
RO0731091-000	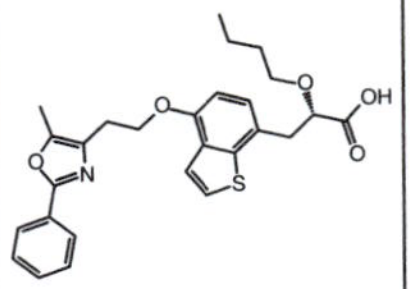 −31.12	−34.39	−38.24					

Figure 26.19 Aleglitazar and a few analogues in the *db/db* model; changes *vs.* control [%], 4d, at 3 mg/kg.

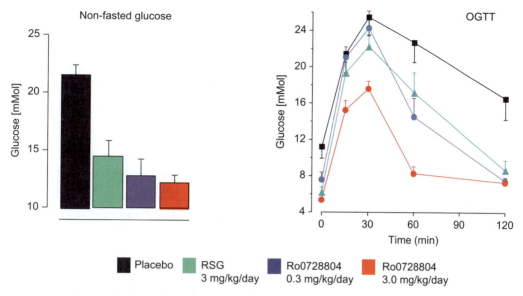

Figure 26.20 Effect of aleglitazar (RO0728804) and rosiglitazone (RSG) on glucose tolerance (reproduced from ref. 15 with the permission of Elsevier Limited).

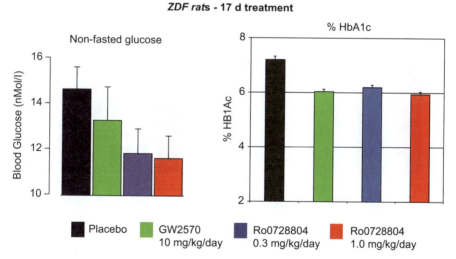

Figure 26.21 Efficacy of aleglitazar (RO0728804) and farglitazar (GW2570) on glucose lowering; see text for explanation.

pure γ agonist rosiglitazone (RSG) at a dose of 3 mg/kg/day, and the associated non-fasted glucose levels. Aleglitazar clearly outperformed the reference compound. Figure 26.21 shows data for glucose lowering and HbA1c (glycosylated haemoglobin, a biomarker for the average glucose concentration of the last couple of weeks) in *fa/fa* Zucker (ZDF) rats. This time, aleglitazar was compared with farglitazar at a dose of 10 mg/kg/day, and although the latter was found to be ten times more potent at the receptor *in vitro*, this didn't translate into better efficacy *in vivo*. Aleglitazar seems to have reached already a plateau at 0.3 mg/kg/day.

One, but perhaps not the only explanation, is certainly the excellent PK behaviour of RO0728804. The results of a third diabetes model are summarised in Figure 26.22: again,

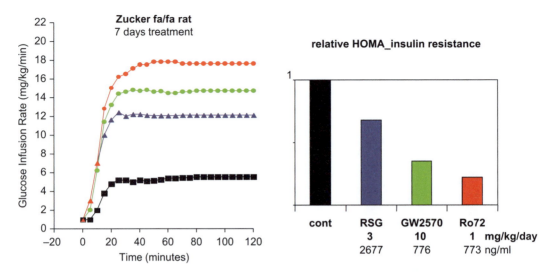

Figure 26.22 Efficacy of aleglitazar (RO72), rosiglitazone (RSG) and farglitazar (GW2570) on insulin sensitisation (reproduced from ref. 15 with the permission of Elsevier Limited; see text for explanation.

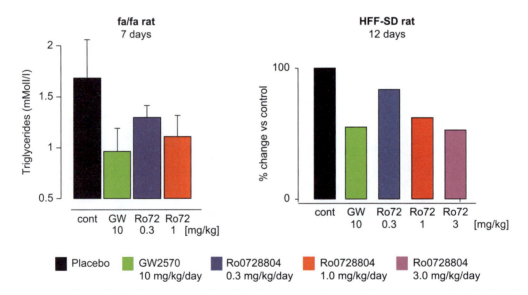

Figure 26.23 Effect of aleglitazar on triglycerides in two murine models as compared with farglitazar (GW2570).

aleglitazar clearly exceeded the other PPAR agonists in a clamp study (left side), which translated into a much-improved relative HOMA-IR index (Homeostasis Model of Assessment – Insulin Resistance, calculated from fasting glucose and fasting insulin) (right side). The numbers below the respective doses refer to the plasma levels and corroborate again the favourable pharmacokinetics of RO0728804. Switching now to lipid models, the picture looked slightly less exciting. In our *in vivo* rodent models (*fa/fa* Zucker rat and high-fat fed Sprague Dawley (HFF-SD) rat), aleglitazar behaved as anticipated and exhibited a clear effect on triglycerides, definitely not inferior to our seemingly most advanced competitor, farglitazar (GW2570), which was administered as positive control. Figure 26.23 illustrates this finding with a nicely dose-dependent pharmacodynamic effect. (At this point in time, we had not generated any murine receptor data.

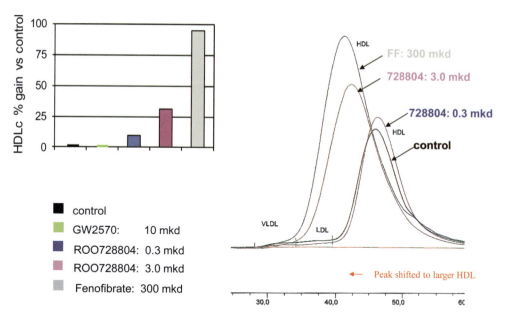

Figure 26.24 Effect of aleglitazar, farglitazar (GW2570) and fenofibrate (FF) on HDL in human ApoAI-transgenic mice; mkd = mg/kg/day (reproduced from ref. 15 with the permission of Elsevier Limited).

As realised much later, aleglitazar and its structural neighbours show a distinct species difference towards PPAR-α—they are much more potent at the human receptor than at the mouse or rat receptor, typically a factor of 50 (see Figure 26.36). Therefore, it is not surprising that we discovered later agonists from other subclasses that performed much better in these murine models. Fortunately enough, and thanks to our ignorance, this didn't hamper or delay the development of aleglitazar.) Aleglitazar at 3 mg/kg was at least equipotent to farglitazar (green bar) at 10 mg/kg, irrespective of the model used (*fa/fa* Zucker or HFF-SD rat). The same holds true for the effect on HDL and LDL in human ApoAI-transgenic mice: aleglitazar showed the anticipated behaviour. The results, compiled in Figure 26.24, illustrate the effect in comparison with fenofibrate, a poor but selective PPAR-α agonist, and again with farglitazar, respectively, whose relative effect compared with placebo was, in our hands, found to be zero! Fenofibrate, on the other hand, showed a much stronger HDL increase, albeit at a 100–1000 times higher dose.

The FPLC trace (Fast Protein Liquid Chromatography) on the right hand side illustrates that not only the area under the curve increases under the treatment with PPAR-α agonists (or α/γ co-agonists), but that the peaks are slightly "left-shifted" towards larger particles.

With really promising data in hand and encouraged by our management, eager to come up with a clinical candidate as soon as possible, we progressed quickly. For the initial toxicology studies we prepared the first 50 g batch in our own medicinal chemistry laboratory, relying on the boron enolate chemistry shown above. The physicochemical properties were definitely in a drug-like range. Polymorph A—the most stable one out of three identified so far—is transformed into polymorph B upon heating beyond 80 °C; the latter melts at 153 °C. Log D (1-octanol/buffer; pH = 7.4) was found to be 1.18, and the acid pKa found to be 3.36. Solubility was of course, as expected for a carboxylic acid, strongly pH-dependent: 0.015 mg/mL in a phosphate buffer of pH 7—still acceptable taking into account the low anticipated daily dose of 0.15 mg per patient—but already 8.0 mg/mL at pH 9 and 32 mg/mL at pH 11.

Permeability was high in the parallel artificial membrane permeability assay (PAMPA) $(2.21 \times 10^{-6}$ cm/s), as well as in the Caco-2 model $(34 \times 10^{-6}$ cm/s). Ames, Micronucleus Test (MNT) and chromosomal aberration tests turned out to be negative, as were the human

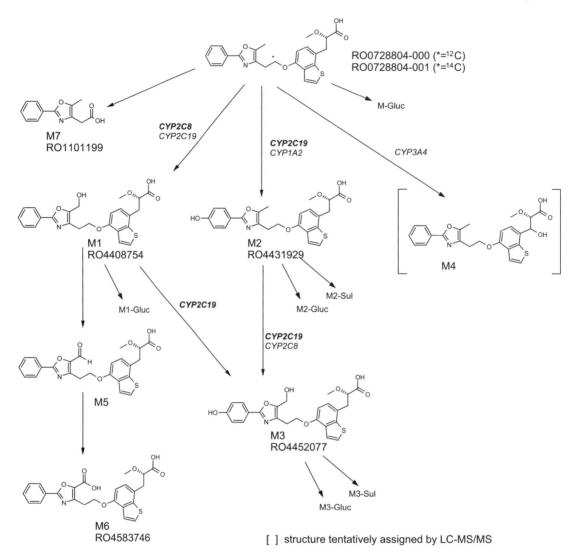

Figure 26.25   Metabolic pathway of aleglitazar.

ether-a-go-go related gene (hERG) and the phototox assays: again, very positive results without liabilities. Metabolic studies revealed quite a complex pattern, but to rely on several routes of elimination is not intrinsically negative. Several oxidative metabolites were formed in different species in microsomal incubations, with M1 (methyl-oxazole oxidised to hydroxymethyl-oxazole, see Figure 26.25) being the major metabolite in most cases.[16] In hepatocytes, phase II metabolites (glucuronides, sulphates) were also observed. The acid metabolite M6 appeared to be more readily formed in human *in vitro* systems than in animals. Consistent with this finding, the latter was identified as the main metabolite in man after multiple dosing in a clinical Phase I study, present at approximately 100% of the parent. Both could be prepared either by biotransformation or Evans' established Aldol methodology (Figure 26.26). Whereas M6, a diacid, was found to be inactive in both binding and transactivation assays, the alcohol M1 (monoacid) exhibited a very weak but significant affinity; the IC_{50}s were determined to be 5.0 μM (PPAR-γ) and 1.6 μM (PPAR-α). Thus, they were not thought to contribute significantly to the pharmacological effects seen during treatment with aleglitazar.

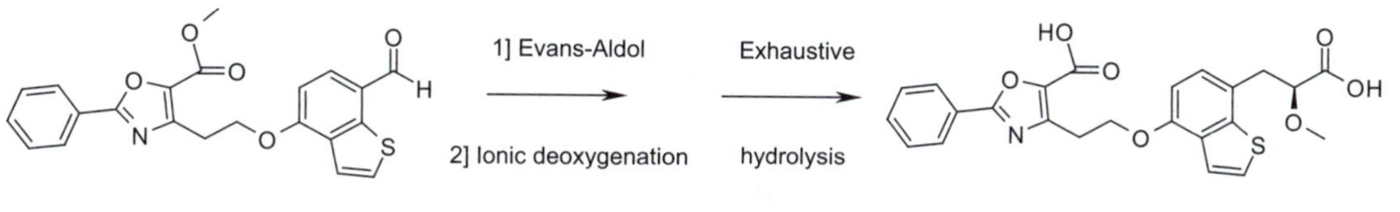

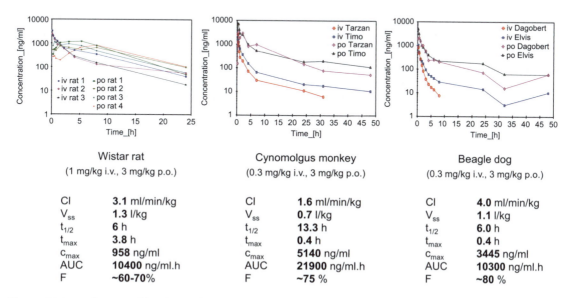

Figure 26.26 Chemical synthesis of metabolite M6.

	Wistar rat (1 mg/kg i.v., 3 mg/kg p.o.)		Cynomolgus monkey (0.3 mg/kg i.v., 3 mg/kg p.o.)		Beagle dog (0.3 mg/kg i.v., 3 mg/kg p.o.)
Cl	**3.1 ml/min/kg**	Cl	**1.6 ml/min/kg**	Cl	**4.0 ml/min/kg**
V_{ss}	**1.3 l/kg**	V_{ss}	**0.7 l/kg**	V_{ss}	**1.1 l/kg**
$t_{1/2}$	**6 h**	$t_{1/2}$	**13.3 h**	$t_{1/2}$	**6.0 h**
t_{max}	**3.8 h**	t_{max}	**0.4 h**	t_{max}	**0.4 h**
c_{max}	**958 ng/ml**	c_{max}	**5140 ng/ml**	c_{max}	**3445 ng/ml**
AUC	**10400 ng/ml.h**	AUC	**21900 ng/ml.h**	AUC	**10300 ng/ml.h**
F	**~60-70%**	F	**~75 %**	F	**~80 %**

Figure 26.27 Pharmacokinetic studies of aleglitazar in rat, monkey and dog.

The excellent properties translated directly into very favourable pharmacokinetic behaviour. As already seen above for the racemate, homochiral aleglitazar exhibited an almost ideal PK profile. As can be gleaned from Figure 26.27, all parameters measured in rat, dog and monkey were within a very promising range: low clearance, low volume of distribution, high bioavailability and a reasonable half-life.

As required by regulatory agencies, toxicological studies by necessity should be performed at doses that exceed the expected exposure levels in humans (see below) and show signs of toxicity. Toxicological studies in rats and monkeys showed a profile similar to that of other PPAR-α and PPAR-γ agonists. The main target organs for toxicity were red blood cells (RBC), liver, heart, kidney, ovaries, testes, bone marrow and adipose tissue. Overall, the severity of observed effects followed a shallow dose–response curve and, fortunately, all findings turned out to be fully reversible after cessation of treatment. Single doses are almost "non-toxic", a typical phenomenon for PPAR ligands. The maximum tolerated dose after single-dose administration in the monkey was considered to be 100 mg/kg. The corresponding C_{max} and area under the curve (AUC) values were found to be stunningly high: 111 000 ng/mL and 1 070 000 ng·h/mL, respectively, in males, and 89 500 ng/mL and 875 000 ng·h/mL, respectively, in females. A slightly increased body weight gain was seen in rats at doses ≥ 0.03 mg/kg/day, but not in monkeys. This effect in rats was considered to be due to increased adipogenesis, a known effect of PPAR-γ agonists, also found as a typical side effect in man under treatment with rosiglitazone or pioglitazone.

Our X-ray experts managed to solve no fewer than 26 PPAR crystal structures in-house, with a diverse set of ligands, among them the two shown below representing, on the one hand, aleglitazar bound to PPAR-γ and the receptor co-activator fragment SRC-1 HD1 (Figure 26.28), and on the other the complex with PPAR-α (Figure 26.29). The pictures nicely illustrate the crucial hydrogen bonds the carboxylate forms to 2 histidines, 1 tyrosine and 1 serine in the former case, and to 1 histidine, 2 tyrosines and 1 serine in the latter, respectively. Furthermore, they show the smooth space-filling fit of the very same molecule into both cavities, although the OCH₂CH₂-spacer has to adopt a slightly different conformation. The following overlay (Figure 26.30) again highlights the subtle differences within the binding cavity of the two receptor isoforms.

Retrospectively, modelling can even rationalise some peculiar SAR observations. As alluded to above, the PPAR-α pocket hosting the alkoxy side-chain is narrower than its counterpart in PPAR-γ, thus tolerating only relatively "slim" residues. Therefore, it is not surprising that 1-butenoxy accommodates quite well and gives rise to a potent and rather balanced co-agonist, albeit still slightly biased towards γ, whereas the bulkier n-butyloxy fits worse and leads to significant erosion of the α affinity, although it exhibits outstanding affinity to the γ receptor (Figure 26.31, see also Table 26.6).

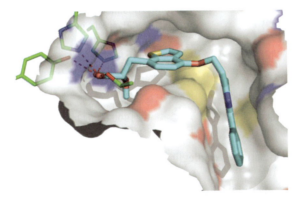

Figure 26.28 X-ray structure of aleglitazar bound to hPPAR-γ ligand-binding domain and the receptor coactivator fragment SRC-1 HD1 (reproduced from ref. 15 with the permission of Elsevier Limited.

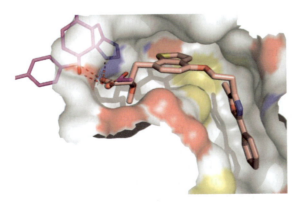

Figure 26.29 X-ray structure of aleglitazar bound to hPPAR-α ligand-binding domain (reproduced from ref. 15 with the permission of Elsevier Limited).

**Overlay of the PPARα and PPARγ RO0728804
co-crystal structures**

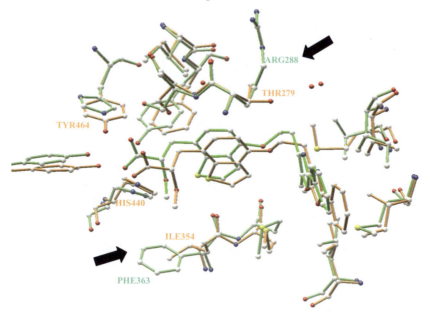

PPARα is shown in orange and PPARγ in green

Figure 26.30   Overlay of aleglitazar bound to hPPAR-α and hPPAR-γ.

Differences in Binding Pocket: PPARγ vs α

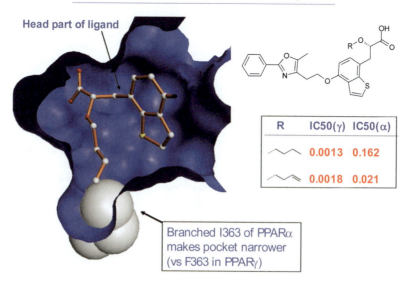

R	IC50(γ)	IC50(α)
⁀⁀⁀	0.0013	0.162
⁀⁀≈	0.0018	0.021

Head part of ligand

Branched I363 of PPARα
makes pocket narrower
(vs F363 in PPARγ)

Figure 26.31   Modelling based rationale for different affinity pattern of 1-butenyloxy- and n-butyloxy-analogue.

The whole PPAR team worked in quite a tough environment. On one hand, we were engaged in a race with a great number of competitors, seemingly significantly ahead of us. On the other, we were pushed by our management to come up with an in-depth, profiled clinical candidate as soon as possible. Nevertheless, we also pursued other avenues generally followed in a small-molecule medicinal chemistry programme and tried to identify decent PPAR ligands from completely different structural classes. High throughput screening of the Roche compound library didn't reveal any useful starting points; a few hits, like RO0925821-000, turned out to be covalent binders, an experience GSK[17] as well as Merck[18] had already faced with L-764406 and GW9662, respectively (Figure 26.32). In all three cases, the strongly nucleophilic SH group of a cysteine attacks a reactive chloride, forming an irreversible covalent bond.

Other endeavours turned out to be more rewarding. Figure 26.33 illustrates, by means of a generic structure, all the systematic variations we had tried, once the synthetic resources had been significantly increased; and it provides some rough rules of thumb in form of a general SAR. By systematic variation, many potent PPAR ligands with quite promising properties were identified. It goes without saying that not all permutations could be tried (a recurrent challenge for medicinal chemists) and that the different parts of the molecule cannot be varied independently of each other; a consequence of the old wisdom that the properties of the ensemble do not simply represent the sum of the individual components.

Whereas a full discussion of the SAR is clearly beyond the scope of this case study, suffice it to say that one representative, RO4543941-000, was long considered as a putative back-up compound since it produced greater effects than aleglitazar in murine lipid models[19] (see below, however, for an explanation of this relative weakness of aleglitazar). RO4552952-000, on the other hand, cleverly designed in the hope of capitalising on the structurally more differentiated

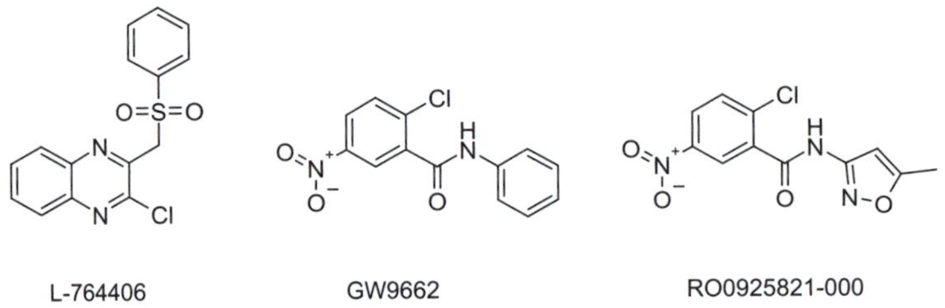

Figure 26.32   Covalently binding PPAR ligands.

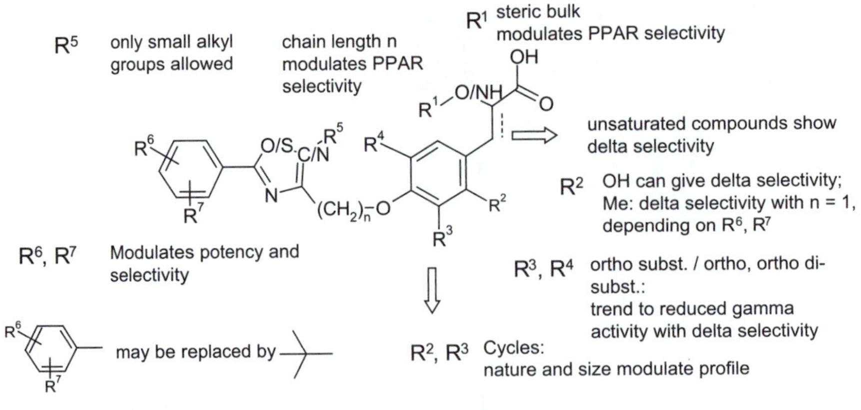

Figure 26.33   Generic SAR of the Roche PPAR programme.

modelling-inspired scaffold, finally made it as a clinical candidate, internally called PPAR(2) (Figure 26.34).[20] However, due to unexpected mortality and severe toxicity in a 4-week dose range-finding study in CD1 mice, development of this compound was stopped before entry into humans, as these findings were not known to be PPAR class effects, and no plausible hypothesis for the mechanism of toxicity could be identified.

26.7 CLINICAL DEVELOPMENT

After successfully passing two milestones, internally referred to as RDC1 (8 November 2001, Clinical Candidate Selection Endorsement) and RDC2 (16 January 2003, Entry into Human), development progressed smoothly. Efficacy studies in pre-diabetic Rhesus monkeys, performed in Barbara Hansen's laboratory, provided very convincing data that turned out to be a key turning point in the project. Even after administering tiny doses of 0.03 mg/kg/day, all glycaemic parameters improved dramatically: fasting insulin went down by 58%, fasting plasma glucose was reduced by 15%, and the insulin resistance index improved by 60% (Figure 26.35). The same holds true with respect to the antilipidaemic effects mediated predominantly by the α receptor: triglycerides tumbled by 89%, LDL decreased by 41% and ApoB by 57%, whereas HDL levels more than doubled.[21] Good potency at human PPAR-α translated into the desired effect in

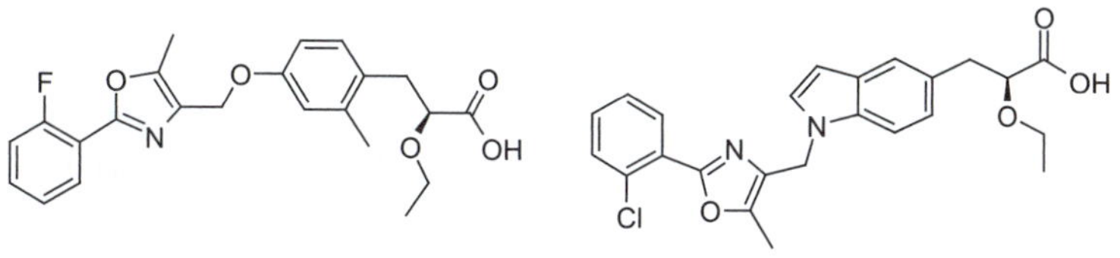

RO4543941-000   RO4552952-000

Figure 26.34 Former PPAR co-agonist back-up compounds.

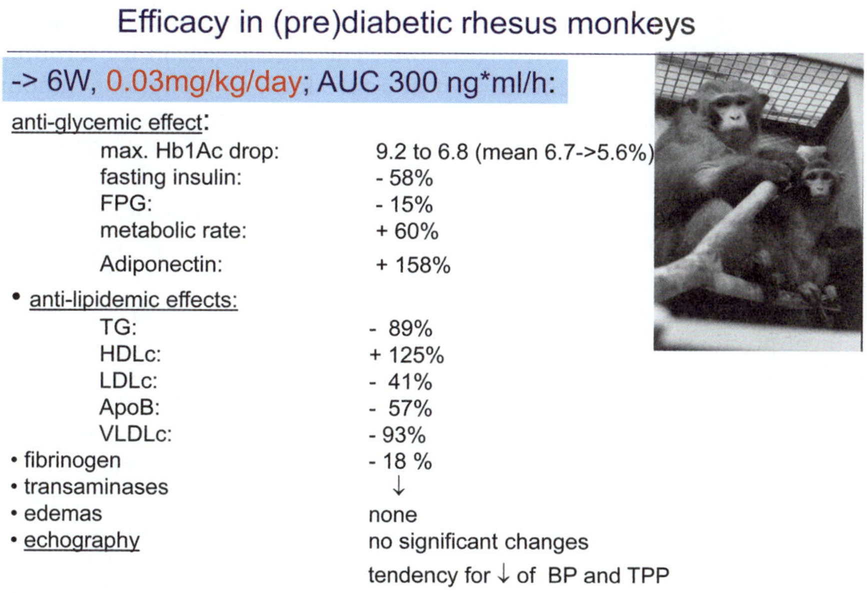

Efficacy in (pre)diabetic rhesus monkeys

-> 6W, 0.03mg/kg/day; AUC 300 ng*ml/h:

anti-glycemic effect:

max. Hb1Ac drop:	9.2 to 6.8 (mean 6.7->5.6%)
fasting insulin:	- 58%
FPG:	- 15%
metabolic rate:	+ 60%
Adiponectin:	+ 158%

• anti-lipidemic effects:

TG:	- 89%
HDLc:	+ 125%
LDLc:	- 41%
ApoB:	- 57%
VLDLc:	- 93%
• fibrinogen	- 18 %
• transaminases	↓
• edemas	none
• echography	no significant changes
	tendency for ↓ of BP and TPP

Figure 26.35 Pharmacodynamic effect of aleglitazar in rhesus monkeys.

primates, and the relative activities at the two targets, as we had hoped for with our concept of a balanced co-agonist, were in a close-to-optimal ratio. Last but not least, we were very pleased to see that no animal developed oedema—a typical side effect reported for many PPAR-γ agonists. A trend towards lowering blood pressure was also noticed—another beneficial finding.

In the meantime, our molecular biologists had cloned the mouse and rat receptors. This helped to solve an old enigma and to reconcile seemingly contradictory results in different animal species. Aleglitazar, and all the other co-agonists with a central bicyclic spacer (benzothiophene, naphthalene) studied, exhibit a peculiarity: they are roughly 50 times weaker at the murine PPAR-α than at the human one (Figure 26.36). Please note that the absolute value of the EC_{50} in man differs slightly from the data given in Table 26.5, reflecting the inter-experimental variation. Taking this modest EC_{50} of roughly 2.3 μM into account, all the effects seen with aleglitazar in our different mouse and rat lipid models are still rather impressive and reflect the excellent bioavailability and pharmacokinetic behaviour. In monkeys, however, like in man, full α effects are triggered with extremely low doses! Frankly, we had initially planned to escalate the dose later, but with 30 μg/kg/day, we almost cured the animals of diabetes!

PK data in man generated during the Phase I single ascending dose (SAD) study also turned out to be very convincing (Figure 26.37). Single doses of up to 30 mg were well tolerated in healthy male volunteers. No treatment-related clinically significant changes were seen in the safety/tolerability parameters (including renal markers). The C_{max} and AUC inter-subject variability of RO0728804 was low, ranging from 12% to 40%. Last but not least, plasma levels were proportional to the applied doses over 3.5 orders of magnitude (0.01–30 mg). Metabolites M1 (RO4408754-000, 8–10% of parent AUC) and M6 (RO4583746-000, 70–120% of parent AUC) were detected as main metabolites. The former is, as mentioned above, at least 100 times less potent then aleglitazar itself—the latter is almost devoid of any PPAR-related activity (Figure 26.38). In other words, Phase I data provided a clear go for Phase II!

Nonetheless, dark clouds soon loomed on the horizon. The attrition rate of PPAR co-agonists in clinical studies was indeed mind-boggling. One competitor project after another was discontinued, which did not bode well for us, although, as a notorious optimist, one could argue that we now had the chance to become first in class on the market. "Classical" TZDs like isaglitazone = netoglitazone (J&J/Mitsubishi) or KRP297 from Merck/Kyorin (for the respective structures see Figure 26.39) were never considered as serious threats since they show, not unexpectedly, a similar clinical profile to rosiglitazone. The official reason for termination of the latter at the end of 2003 was the identification of a rare form of malignant tumours in mice. Farglitazar-derived insulin sensitisers like GW409544 were also of less concern; they are not well α/β-balanced and/or their physicochemical properties are too unfavourable. This holds true for the first-generation non-TZDs like DRF-2725 (Dr. Reddy's Laboratories/Novo Nordisk) or DRF-4158 (Dr. Reddy's Laboratories/Novartis, structure not disclosed). Finally, Dr Reddy's Laboratories also had to give up on balaglitazone, a much-hyped TZD, purported to be a partial γ agonist, since they couldn't find a co-development partner after the first Phase III data generated in 2010.[22] Rivoglitazone was pursued by Daiichi Sankyo until recently; in a 26-week study it showed a clinical profile similar to high doses of pioglitazone.[23] Unanimously, we considered AZ242 = tesaglitazar = Galida® from AstraZeneca and BMS-298585 = muraglitazar = Pargluva® from Bristol-Myers Squibb as the strongest competitors. Tesaglitazar's development was dropped after some Phase III studies, judging the risk/benefit ratio not attractive enough to proceed having seen elevations in serum creatinine and an associated decrease in glomerular filtration rate (GFR). However, complete reversal of serum creatinine increase was not demonstrated for all patients upon treatment discontinuation. This finding and the greater than expected exposure in subjects with severely impaired renal function (GFR 10–30 mL/min/ 1.73 m^2) have been explained by interconversion of the acyl glucuronide that inhibits its own elimination (*via* saturable renal excretion) further increasing its plasma levels.[24,25] As a result,

ERN		alpha-human		alpha-rat		alpha-mouse	
		EC$_{50}$ (μm)	fold_stimulation @ 2×10^{-6} M	EC$_{50}$ (μm)	fold_stimulation @ 2×10^{-6} M	EC$_{50}$ (μm)	fold_stimulation @ 2×10^{-6} M
RO0728804-000-001	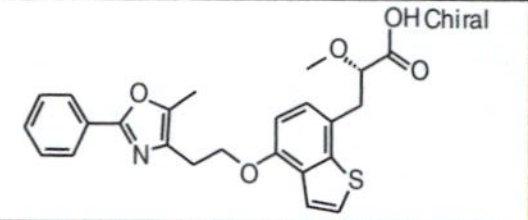	0.029	22	2.26	31	2.34	50

Figure 26.36 *In vitro* species difference of aleglitazar towards PPAR-α.

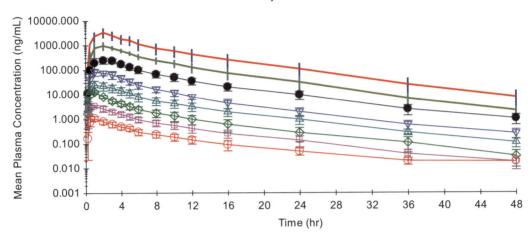

Figure 26.37 Aleglitazar PK data, generated during the Phase I SAD study.

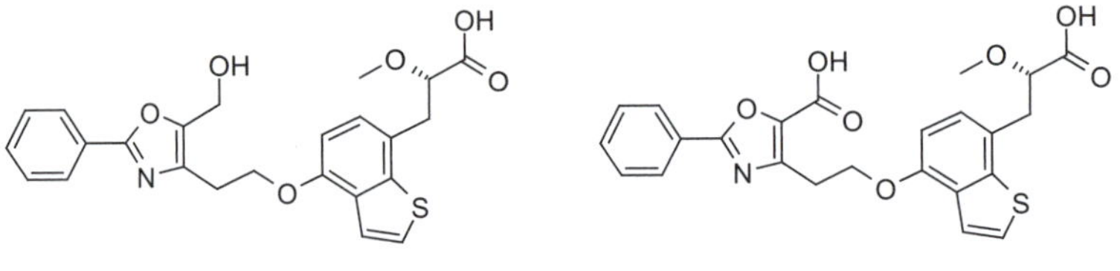

M1=RO4408754-000   M6=RO4583746-000

Figure 26.38 Main metabolites of aleglitazar in man.

the acyl glucoronide is excreted into bile and undergoes hydrolysis to tesaglitazar by β-glucuronidase in the gut.[24] Tesaglitazar is then reabsorbed perpetuating parent drug circulation and thus sustaining the rise in serum creatinine. Muraglitazar was abandoned in May 2006 when the FDA had asked for additional, costly and time-consuming clinical outcome studies with respect to cardiovascular safety following an independent analysis of the muraglitazar data revealed occurrence of death, myocardial infarction, or stroke in 1.47% of patients treated with muraglitazar with a combined relative risk of 2.23, significantly ($p = 0.03$) higher than in patients administered pioglitazone or placebo.[26] However, these failures are not isolated; naveglitazar, licensed from Ligand, was discontinued in Phase II by Lilly in the same year—2006. AZD-6610, supposed to be a PPAR-α agonist/PPAR-γ partial agonist, whose structure was only recently released, was also dropped in 2007. AVE-0847, another dual insulin sensitiser in development by Genfit and Sanofi-Aventis was given up in spring 2008, officially due to prioritisation of the product portfolio. To conclude this (not comprehensive) sad and rather long list, LBM-642 (cevoglitazar) was discontinued by Novartis in 2007. Consequently, not too many PPAR ligands remain under active development. According to information from the company, Genfit-505, a selective PPAR modulator with preferential action on the PPAR-α receptor family, is still actively pursued; in 2012, Phase II clinical trials were initiated for the treatment of non-alcoholic steatohepatitis.

Another compound, designed and synthesised by Pfizer in the mid-90s, also caught our attention early on (Figure 26.40). It shows not only an intriguing structural similarity to aleglitazar, comprising a bicyclic heteroaromatic core (albeit the relative orientation of the two exit

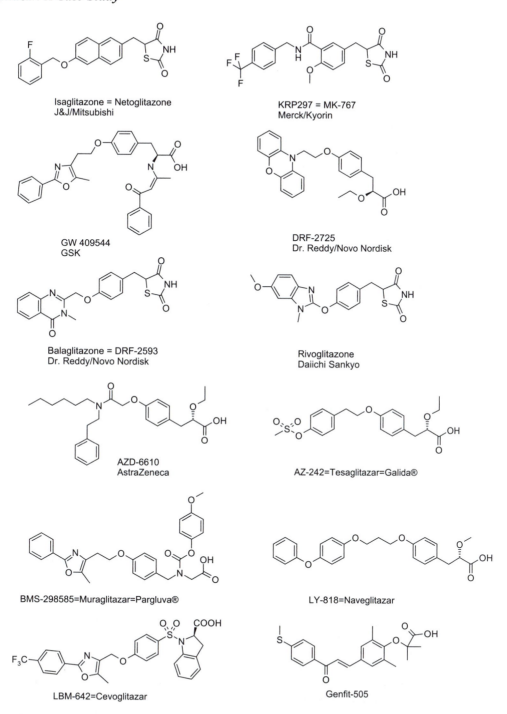

Figure 26.39 PPAR ligands no longer in clinical development and structure of Genfit-505.

vectors looks different), its *in vitro* receptor profile is also not too far away from our molecule. We prepared RO0732012-000 from scratch[27] and corroborated its dual agonism, although slightly biased towards γ. According to grapevine information it was stopped—either before or during clinical studies—due to unspecified toxic effects. It could well be that this molecule was

Code	Structure	Transactivation		Binding	
		alpha-human	gamma-human	alpha-human	gamma-human
		EC_{50} (µM)	EC_{50} (µM)	EC_{50} (µM)	EC_{50} (µM)
RO0728804-000	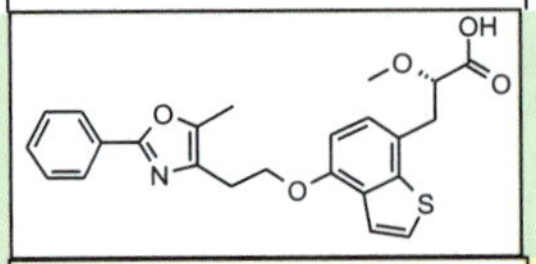	0.05	0.021	0.038	0.019
RO0732012-000	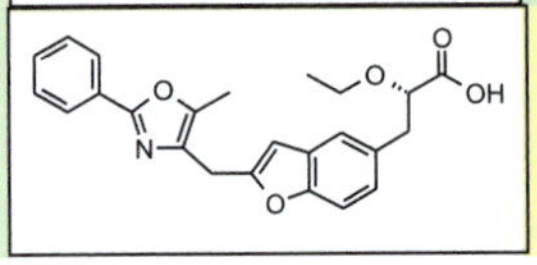	0.069	0.016	0.18	0.002

Figure 26.40   *In vitro* comparison of aleglitazar with an ''old'' dual agonist comprising a benzofuran scaffold from Pfizer.

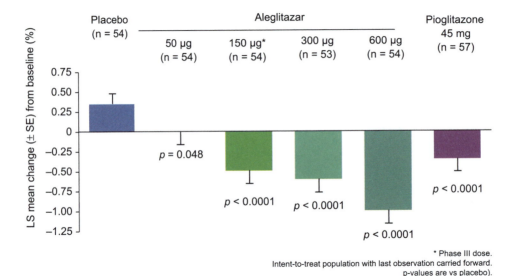

Figure 26.41   Effect of aleglitazar and Actos® on HbA1c (reproduced from ref. 28 with the permission of Elsevier Limited).

simply discovered too early, when nobody had experience with the toxicity pattern of PPAR co-agonists, which deterred pharmaceutical companies from developing such compounds. But it also cannot be excluded that RO0732012 is flawed with severe off-target toxicity or suffers from other unacceptable properties like the formation of reactive metabolites. As far as we are aware, only a few PPAR agonists remain in clinical development, amongst them the above-mentioned compound from Genfit purported to be a PPAR modulator.

Early development of aleglitazar progressed fast and smoothly. Based on the very favourable outcome of Phase I SAD and multiple ascending dose (MAD), it entered Phase II without delay. And what really showed promise were the very encouraging data generated in the so-called SYNCHRONY study, summarised below.

The compilation in Figures 26.41–26.45 presents some highlights of this randomised, dose-ranging study, which tested the effect of aleglitazar on cardiovascular disease risk factors (glucose and lipids) in 332 patients with type 2 diabetes.[28] For all relevant parameters (HbA1c

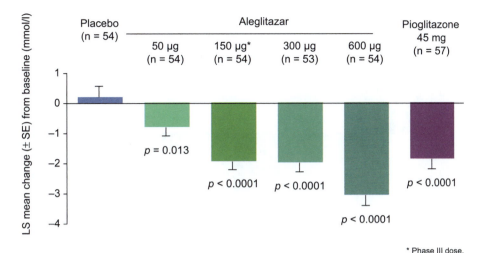

Figure 26.42   Effect of aleglitazar and Actos® on fasting plasma glucose (reproduced from ref. 28 with the permission of Elsevier Limited).

[Figure 26.41], fasting plasma glucose [Figure 26.42], triglycerides [Figure 26.43], LDL [Figure 26.44], HDL [Figure 26.45], and many more [data not shown]), aleglitazar produced a robust dose-dependent effect. The dose of 0.15 mg was, with respect to all parameters, found to be at least as efficacious as open label 45 mg of the comparator Actos® (pioglitazone). Concerning the antilipidaemic effects, in particular HDL-C and triglycerides, the efficacy seen with this very small dose was already reaching the maximal achievable effect (HDL-C roughly + 25%, triglycerides − 30%). Efficacy is one key prerequisite, and sufficient safety margins another! With respect to (mechanism-based) side effects (body weight gain, haemoglobin decrease, oedema) the outcome again compared favourably with pioglitazone, with the exception of elevations in serum creatinine that were subsequently shown to be reversible in the SESTA-R and AleNephro studies (see below). Figure 26.46 illustrates the data for body weight gain—the anticipated, dose-dependent modest increase. Figure 26.47 summarises the influence on serum creatinine, indicating a trend towards a dose-dependent increase of creatinine; however, at 0.15 mg, the levels are still within the range of placebo-treated patients. In the meantime, the FDA approved Phase III clinical trials for aleglitazar when results of the carcinogenicity study were considered by FDA/EMA to be acceptable to continue long-term Phase III trials. Thus, in February 2010, almost 9 years after the first milligrams had been prepared in the laboratory and after a long, tortuous, tedious journey with many ups and downs, aleglitazar entered into Phase III when the first patient of the AleCardio study was recruited (see Figure 26.48 for the trial design). This double-blind, randomised, parallel, two-arm study was to evaluate the potential of aleglitazar to reduce cardiovascular morbidity and mortality and the long-term safety and tolerability profile of aleglitazar compared with placebo on top of standard of care in patients with recent acute coronary syndrome (ACS) and type 2 diabetes mellitus. Patients had been randomised to receive either aleglitazar or placebo once daily as oral doses. The study was to last until at least 950 events occur, and time on study treatment would be for at least 2.5 years.[29]

In September 2012, aleglitazar's renal safety data were disclosed based on a dedicated Phase IIb study, AleNephro. This 52-week study met its primary endpoint and showed non-inferiority for the change in estimated glomerular filtration rate after 52 weeks of double-blind treatment plus 8 weeks of off-treatment follow-up versus pioglitazone, confirming reversibility of the on-treatment change in renal serum creatinine and eGFR changes. These findings were similar to

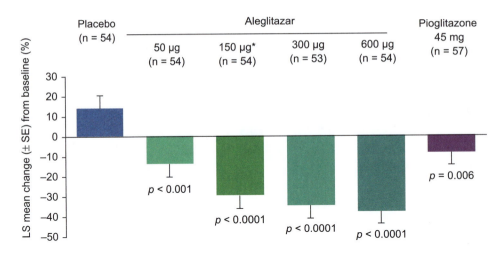

Figure 26.43   Effect of aleglitazar and Actos® on triglycerides (reproduced from ref. 28 with the permission of Elsevier Limited).

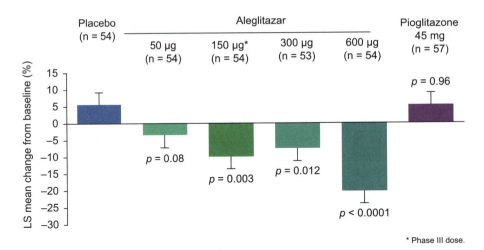

Figure 26.44   Effect of aleglitazar and Actos® on LDL-C (reproduced from ref. 28 with the permission of Elsevier Limited).

results from fibrate studies such as FIELD and ACCORD-Lipid. In these trials, although fibrate therapy was associated with non-progressive and reversible increases in serum creatinine and decreases in eGFR, a reduction in albuminuria was also observed, indicating beneficial effects on renal function.[30,31] The outcome of AleNephro was an important milestone supporting the development of aleglitazar and boded well for the future development programme. Based on these findings, aleglitazar was progressed to evaluation in a large clinical development programme including the 19 000-patient AlePrevent (NCT01715818) cardiovascular outcomes trial in patients with stable cardiovascular disease and either type 2 diabetes or pre-diabetes, as well as the AleGlucose (NCT01691755) programme, a set of glycaemic control trials to further characterise the impact of aleglitazar on glycaemic control in patients with type 2 diabetes.

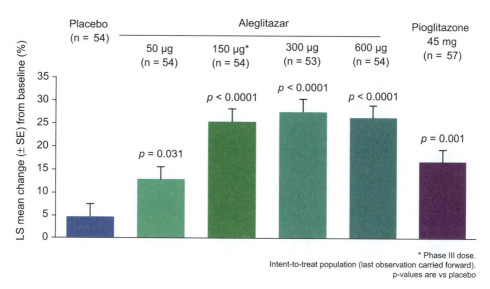

Figure 26.45 Effect of aleglitazar and Actos® on HDL-C (reproduced from ref. 28 with the permission of Elsevier Limited).

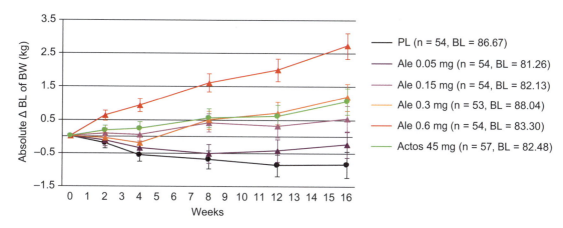

Figure 26.46 Effect of aleglitazar and Actos® on body weight gain (reproduced from ref. 28 with the permission of Elsevier Limited).

26.8   TECHNICAL PROCESS

Although the Active Pharmaceutical Ingredient (API) demand for aleglitazar, due to its high potency translating into a very low dose, will remain moderate, even in the case of a successful launch, a reliable, robust, economic, safe and environmentally friendly manufacturing process is indispensable. It is beyond the scope of this case study to discuss all the different avenues that have been explored by the Roche kilo laboratory, process research, and process development teams, respectively. Suffice it to show here two sample reaction schemes. Figure 26.49 illustrates the elegant preparation of the benzo[b]thiophen-4-ol key intermediate based on the Pd(0)-catalysed insertion of carbon monoxide followed by cyclisation and aromatisation. Starting with thiophene-2-carbaldehyde, it could be obtained in 76% overall yield. Ensuing regioselective formylation (not shown) delivers a bifunctional key building block. Figure 26.50, on the other hand, summarises the synthetic endgame under good manufacturing practice conditions from the

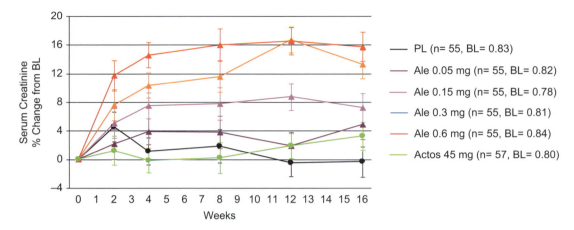

Figure 26.47   Effect of aleglitazar and Actos® on serum creatinine (reproduced from ref. 28 with the permission of Elsevier Limited).

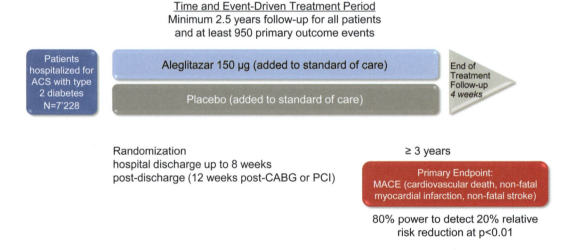

Figure 26.48   Design of the Phase III study AleCardio.

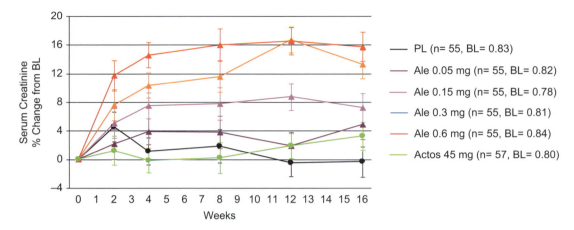

Figure 26.49   Large-scale synthesis of the benzo[b]thiophen-4-ol building block.

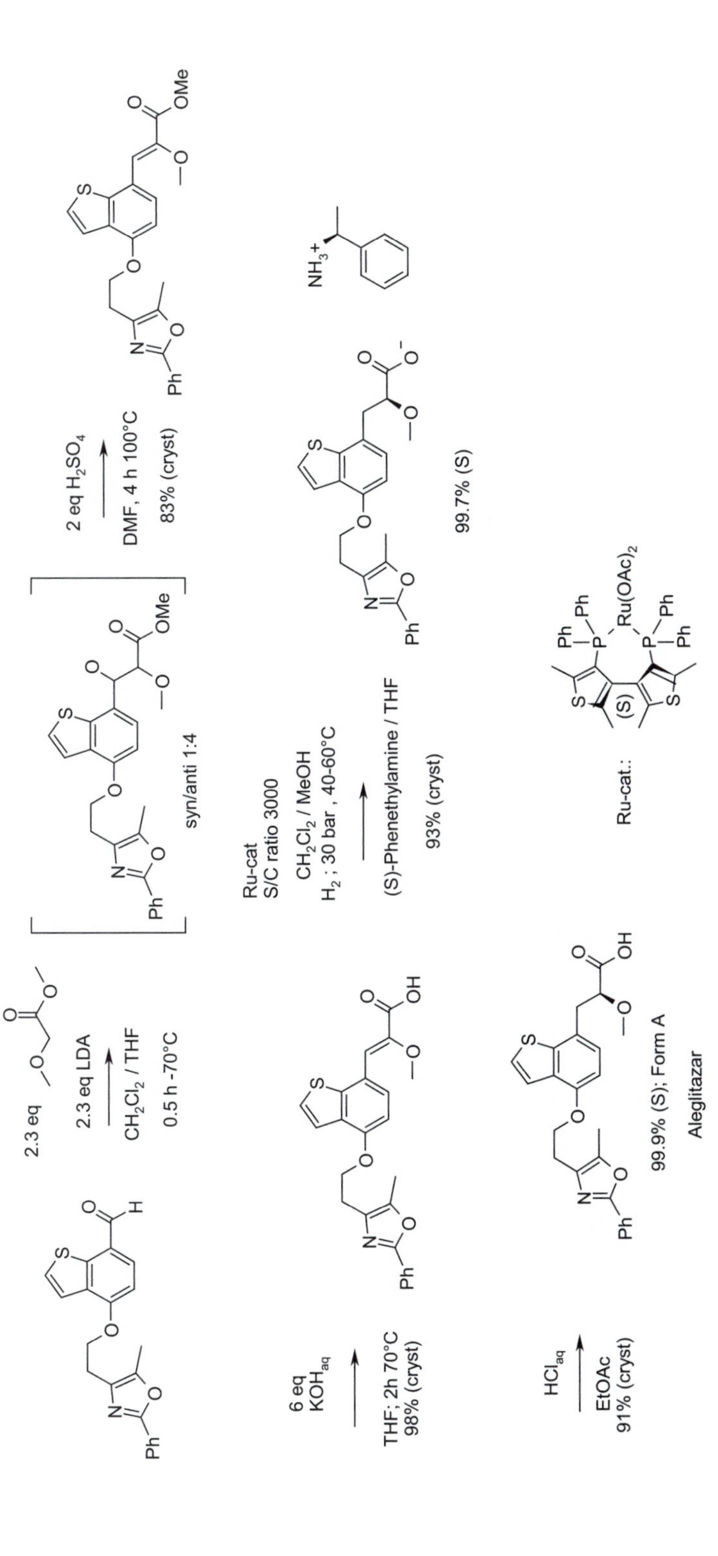

Figure 26.50 Final large-scale assembly of aleglitazar.

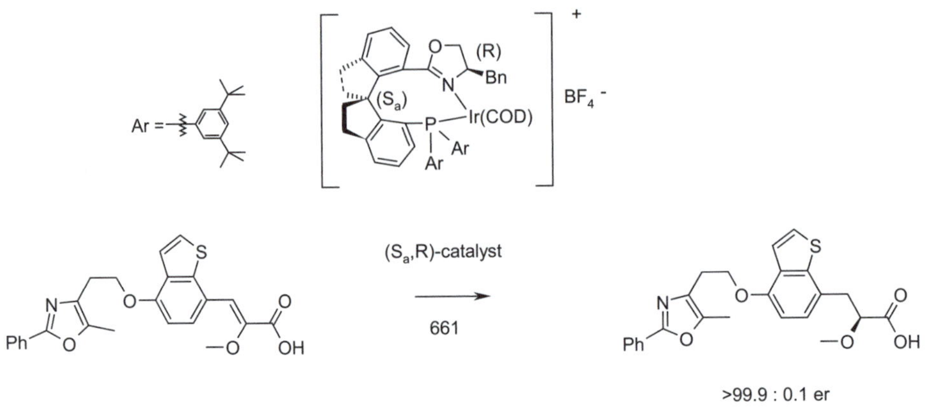

Figure 26.51 Recently optimised asymmetric hydrogenation.

beginning up to the scale of 6.7 kg of final product for the Phase I and II trials. In this highly optimised version, methoxy-acetic acid methyl ester was condensed with the very same aldehyde already used in the synthesis of edaglitazone (see also Figure 26.16). Acidic treatment under rather harsh conditions stereoselectively yielded the Z-ester, which was saponified to the crystalline acid, suitable for enantioselective hydrogenation. It occurred at high substrate-to-catalyst ratio of 3000 in the presence of an atropisomeric chiral ruthenium catalyst in a decent selectivity of 97/3; the enantiomeric excess could be further improved to 99.7/0.3 by crystallisation of the corresponding phenethylamine salt. Liberation of the free acid eventually generated homochiral aleglitazar. In a recent modification with the sophisticated Ir-catalyst 661 (Figure 26.51) the enantioselectivity could be significantly improved to 99.9/0.1, obviating the need for the additional purification step.

26.9 EPILOGUE

We are often confronted with the simple but legitimate question: what is the difference between aleglitazar and all the other PPAR activators? Why should the former succeed while all the others have failed? There are probably no simple answers to these daunting questions. A key element is certainly the high potency, which translates into a very small dose of only 0.15 mg per day per patient. This definitely reduces the risk of any non-PPAR-related toxicity, although we have so far not identified any off-target actions for aleglitazar. Another important aspect includes the excellent physicochemical properties and the almost ideal pharmacokinetic behaviour.

A third element is without doubt the well-balanced nature of aleglitazar, as shown in Figure 26.52, by direct head-to-head comparison with rosiglitazone and pioglitazone by means of novel receptor constructs (the absolute EC_{50} values are therefore not identical to those in Table 26.5). The former activates both receptors to an almost identical extent, at clinically relevant doses. Rosiglitazone, on the other hand, behaves like a pure γ agonist, whereas pioglitazone exhibits some weak α activity—possibly responsible for the subtle HDL-C increase seen in man. However, the two curves are significantly shifted away from each other; definitely not an ideal situation if one would like to capitalise on balanced agonism at both PPAR-α and PPAR-γ.

Why did this project—retrospectively analysed—turn out to be successful (at least so far, the ultimate proof, *i.e.*, launch, is still lacking)? As most often in life, it is not one single reason which made the difference. Certainly, the chosen target is "druggable". As Roche and a plethora of competitor companies have convincingly demonstrated, it is quite easy to come up with potent ligands. In addition, the targets are, to some degree, already validated, inasmuch as PPAR-α agonists as lipid-modifying agents (the famous fibrates) and PPAR-γ agonists as insulin

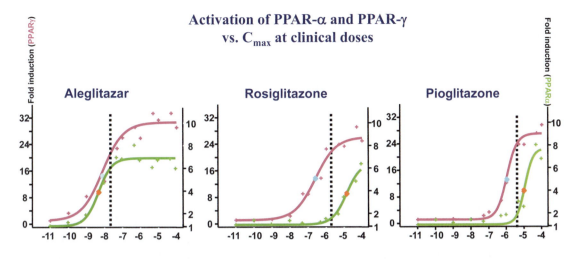

Figure 26.52 *In vitro* comparison of aleglitazar with rosiglitazone and pioglitazone (reproduced from ref. 32 with the permission of John Wiley and Sons).

sensitisers (rosiglitazone, pioglitazone) belong to the classical armamentarium of the physician; the combination within one molecule, however, is a new paradigm. Interference with a NHR is—I firmly believe—in general, a powerful but intrinsically risky therapeutic concept: one receptor directly modulates at a very high level the downstream expression of possibly hundreds of genes. And one can safely assume that this will imply significant pharmacological effects (whether to the patient's advantage or not, remains to be seen in the clinic).

We tried hard to understand the failure of other coagonists and to design our molecules based on starting points that are not flawed with nasty or toxic properties. In our particular case, we had, in addition, to face the problem of speed, since we started the race significantly behind our competitors. We can also phrase it positively: retrospectively, really good ideas are often very simple ones. Aleglitazar is a highly potent molecule (daily dose in Phase III only 0.15 mg per day per patient); nevertheless, it has a modest size (MW = 437), hence excellent ligand efficiency, doesn't violate any Lipinski rules, and really emphasises the strength of small molecules. I'm convinced that small molecules have to be really small otherwise they lose their intrinsic advantage over biologicals. This trait of aleglitazar also translates into outstanding physicochemical and pharmacokinetic behaviour. In the past, it often proved to be the case that medicinal chemists managed to boost the activity at a desired target by attaching a lot of lipophilic grease; however, the price to pay often turned out to be unacceptably high, since physicochemical properties deteriorate in parallel, and activities at undesired off-site targets increase as well leading to molecular promiscuity. Striving for decent lipophilic efficiency should always be a guiding principle for a medicinal chemist. And last but not least, one has to successfully capitalise on serendipity and luck!

NOTE ADDED IN PRESS

Following a recent regular safety review of the AleCardio Phase III trial, aleglitazar's development program was halted due to lack of efficacy in CV outcomes indicating no CV benefit, but showing PPAR-related class side effects in the post-ACS population. The data from the AleCardio trial will be further analysed to fully understand the findings and will be made available in dedicated publications.

HINTS AND TIPS

- Not all SAR permutations could be tried—choices have to be made.
- Different parts of the molecule cannot always be varied independently of each other—the properties of the ensemble do not simply represent the sum of the individual components.
- Be aware of variability in your primary assays, particularly cell assays—be careful of making high resolution SAR decisions within the noise of the assay.
- Be aware that species differences in affinity can occur, which may explain unexpected drop or gain in efficacy in preclinical models. Ideally potency in the required species should be obtained to rationalize PKPD, rather than assume crossover.
- High potency, leading to low dose, reduces the risk of off-target toxicity, and is a worthwhile generic design paradigm.
- A druggable and clinically validated target de-risks drug development, although differentiation becomes the new hurdle.
- Simple design concepts, including potency, ligand efficiency and good drug-like properties, were a guiding paradigm that led to development success relative to other less efficient and drug-like candidates from other companies.
- Striving for decent lipophilic efficiency should always be a guiding principle for a medicinal chemist.

Acknowledgements

This project has been a joint effort of many talented and highly motivated scientists and technicians. All their valuable contributions are much appreciated and gratefully acknowledged!

KEY REFERENCES

R. S. Rosenson, R. S. Wright and M. Farkouh, *American Heart Journal*, 2012, **164**, 672–680.
M. P. Menéndez-Gutiérrez, T. Röszer and M. Ricote, *Current Topics in Medicinal Chemistry*, 2012, **12**, 548–584.
A. J. Krentz, M. B. Patel and C. J. Bailey, *Drugs*, 2008, **68**, 2131–2162.

REFERENCES

1. M. Bliss, *The Discovery of Insulin*, University of Chicago Press, Chicago, 2007, 25th Anniversary Edition.
2. http://www.idf.org/sites/default/files/5E_IDFAtlasPoster_2012_EN.pdf.
3. http://whqlibdoc.who.int/publications/2006/9241594934_eng.pdf.
4. W. T. Cade, *Phys. Ther.*, 2008, **88**, 1322–1355.

5. http://www.fda.gov/downloads/Drugs/.../Guidances/ucm071624.pdf.

6. D. M. Nathan, J. B. Buse, M. B. Davidson, E. Ferrannini, R. R. Holman, R. Sherwin and B. Zinman, *Diabetes Care*, 2009, **32**, 193–203.

7. P. Germain, B. Staels, C. Dacquet, M. Spedding and V. Laudet, *Pharmacol. Rev.*, 2006, **58**, 685–704.

8. B. Desvergne and W. Wahli, *Endocr. Rev.*, 1999, **20**, 649–688.

9. M. I. Dawson and Z. Xia, *Biochim. Biophys. Acta*, 2012, **1821**, 21–56.

10. Y. Goto, M. Yamazaki and M. Hamana, *Chem. Pharm. Bull*, 1971, **19**, 2050–2057.

11. M. P. Edwards and D. A. Price, *Annu. Rep. Med. Chem.*, 2010, **45**, 381–391.

12. Initially claimed in WO9962871 (Astra, 1999), cf. also WO2001040169 (AstraZeneca, 2001).

13. J. R. Gage and D. A. Evans, *Org. Synth*, 1990, **68**, 83–87.

14. J. McHutchison, Z. Goodman, K. Patel, H. Makhlouf, M. Rodriguez-Torres, M. Shiffman, D. Rockey, P. Husa, W. L. Chuang, R. Levine, M. Jonas, D. Theodore, R. Brigandi, A. Webster, M. Schultz, H. Watson, B. Stancil and S. Gardner, *Gastroenterology*, 2010, **138**, 1365–1373.

15. A. Bernardeau, J. Benz, A. Binggeli, D. Blum, M. Boehringer, U. Grether, H. Hilpert, B. Kuhn, H. P. Marki, M. Meyer, K. Puntener, S. Raab, A. Ruf, D. Schlatter and P. Mohr, *Bioorg. Med. Chem. Lett.*, 2009, **19**, 2468–2473.

16. S. Sturm, M. Seiberling, I. Weick, A. Paehler, C. Funk and T. Ruf, *Clin. Ther*, 2012, **34**, 420–429.

17. L. M. Leesnitzer, D. J. Parks, R. K. Bledsoe, J. E. Cobb, J. L. Collins, T. G. Consler, R. G. Davis, E. A. Hull-Ryde, J. M. Lenhard, L. Patel, K. D. Plunket, J. L. Shenk, J. B. Stimmel, C. Therapontos, T. M. Willson and S. G. Blanchard, *Biochemistry*, 2002, **41**, 6640–6650.

18. A. Elbrecht, Y. Chen, A. Adams, J. Berger, P. Griffin, T. Klatt, B. Zhang, J. Menke, G. Zhou, R. G. Smith and D. E. Moller, *J. Biol. Chem.*, 1999, **274**, 7913–7922.

19. U. Grether, A. Bénardeau, J. Benz, A. Binggeli, D. Blum, H. Hilpert, B. Kuhn, H. P. Märki, M. Meyer, P. Mohr, K. Püntener, S. Raab, A. Ruf and D. Schlatter, *ChemMedChem.*, 2009, **4**, 951–956.

20. B. Kuhn, H. Hilpert, J. Benz, A. Binggeli, U. Grether, R. Humm, H. P. Märki, M. Meyer and P. Mohr, *Bioorg. Med. Chem. Lett.*, 2006, **16**, 4016–4020.

21. B. C. Hansen, X. T. Tigno, A. Bernardeau, M. Meyer, E. Sebokova and J. Mizrahi, *Cardiovasc. Diabetol*, 2011, **10**, 7–14.

22. R. Agrawal, P. Jain and S. N. Dikshit, *Mini-Rev. Med. Chem.*, 2012, **12**, 87–97.

23. H. S. Chou, K. E. Truitt, J. B. Moberly, D. Merante, Y. Choi, Y. Mun and A. Pfuetzner, *Diabetes, Obes. Metab.*, 2012, **14**, 1000–1009.

24. B. Hamren, H. Ericsson, O. Samuelsson and M. O. Karlsson, *Br. J. Clin. Pharmacol.*, 2008, **65**, 855–863.

25. B. Hamren, K. P. Ohman, M. K. Svensson and M. O. Karlsson, *J. Clin. Pharmacol.*, 2012, **52**, 1317–1327.

26. S. E. Nissen, K. Wolski and E. J. Topol, *JAMA, J. Am. Med. Assoc.*, 2005, **294**, 2581–2586.

27. B. Hulin, L. S. Newton, D. M. Lewis, P. E. Genereux, E. M. Gibbs and D. A. Clark, *J. Med. Chem.*, 1996, **39**, 3897–3907.

28. R. R. Henry, A. M. Lincoff, S. Mudaliar, M. Rabbia, C. Chognot and M. Herz, *Lancet*, 374, 126–135.

29. A. M. Lincoff, J. C. Tardif, B. Neal, S. J. Nicholls, L. Ryden, G. Schwartz, K. Malmberg, J. B. Buse, R. R. Henry, H. Wedel, A. Weichert, R. Cannata and D. E. Grobbee, *Am. Heart J*, 2013, **166**, 429–434.

30. A. Keech, R. J. Simes, P. Barter, J. Best, R. Scott, M. R. Taskinen, P. Forder, A. Pillai, T. Davis, P. Glasziou, P. Drury, Y. A. Kesaniemi, D. Sullivan, D. Hunt, P. Colman, M. d'Emden, M. Whiting, C. Ehnholm and M. Laakso, *Lancet*, 2005, **366**, 1849–1861.

31. The ACCORD Study Group, *N. Engl. J. Med.*, 2010, **362**, 1563–1574.

32. M. Dietz, P. Mohr, B. Kuhn, H. P. Maerki, P. Hartman, A. Ruf, J. Benz, U. Grether and M. B. Wright, *Chem. Med. Chem.*, 2012, **7**, 1101–1111.

CHAPTER 27

Lessons Learned From the Discovery and Development of Lapatinib/Tykerb

G. STUART COCKERILL*[a] AND KAREN E. LACKEY[b]

[a] School of Medicine, Pharmacy and Health, Wolfson Research Institute, Queens Campus, Durham University, Stockton-on-Tees, TS17 6BH, UK; [b] Center for Drug Discovery at Medical University of South Carolina, 171 Ashley Ave, Charleston, SC 29425, USA
*E-mail: stuart.cockerill@durham.ac.uk

27.1   INTRODUCTION

To be pioneering in the drug discovery field, one must evolve one's knowledge base as new findings emerge and adapt the project strategy along the way. The significance of the genetic drivers of certain tumour types was realised in the 1980's, but required years to determine how to modify the aberrant cellular signalling that caused the unchecked proliferation of tumour cells. The concept of blocking kinase activity caused by protein over-expression or constitutive activation was hotly debated in the early days of oncogene discoveries. However, now the background, structure and mechanism of the protein kinase target family, also known as the "kinome", has been reviewed on multiple occasions.[1] For the purposes of this chapter, the multiplicity of kinase targets, the conserved nature of the ATP binding site and the ability to modulate the selectivity profile of kinase inhibitors through the utilisation of subtle but key binding site changes across this kinome has been described over the last 15 to 20 years. When the Wellcome Foundation initiated its interest in the ERBB2 receptor tyrosine kinase in the 1990's, none of this information was available and the project teams involved in the successful progression of the project acquired this knowledge as the project progressed. The discovery of lapatinib/Tykerb is a testimony to this multidisciplinary team's ability to contribute to a new scientific field while creating a targeted medicine.

The ERBB or (HER) family of receptor tyrosine kinases had been identified as a cancer target in the 1980's due to the clinical influence of the oncogene on the disease free and overall survival of breast cancer patients.[2] A general scheme for the signalling pathway is shown in Figure 27.1. Interestingly, whilst the overall protein structure was similar, it became clear that the four members of the ERBB family display differences in the autophosphorylation docking sites, in substrate specificity, and in the potency of the kinase activity. By way of example, ERBB3 was found to lack

The Handbook of Medicinal Chemistry: Principles and Practice
Edited by Andrew Davis and Simon E Ward
© The Royal Society of Chemistry 2015
Published by the Royal Society of Chemistry, www.rsc.org

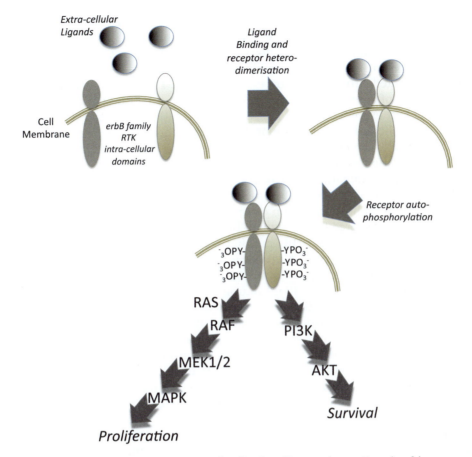

Figure 27.1 Schematic rendition of the ERBB family signalling pathway. Tyrosine kinase receptors are membrane spanning receptors containing an extracellular cysteine rich ligand binding domain, a hydrophobic membrane spanning region and intracellular domain that contains the kinase function. Peptide ligands bind to the extracellular region and the receptors dimerise. Subsequent phosphorylation, both of the kinase domain by the ERBB kinase itself (autophosphorylation) and down stream signalling proteins (including a series of kinases), leads to proliferation and/or survival outcomes for the cell.

kinase activity and ERBB2 had and still has no known ligand, but has the highest transforming capability.[3] These ERBB family receptors are best described as consisting of a cysteine rich extracellular ligand binding domain, a hydrophobic membrane-spanning region and an intracellular domain containing the kinase function.[4] Following a specific extracellular ligand binding event, the homo- or hetero-dimerisation of ERBB receptors leads to the activation of the intrinsic tyrosine kinase activity *via* autophosphorylation, or hetero-dimerisation and transactivation-autophosphorylation of ERBB2. The formation of the ERBB family dimers activate downstream ERK1/2 MAP kinases and PI3K/AKT kinase survival pathways.[4,5] Subsequent rounds of aberrant cellular proliferation are a product of these activated signalling cascades and lead to tumour formation.

27.2 PROJECT SCREENING STRATEGY AND ASSAYS EVOLVED OVER TIME BUT *IN VITRO* AND *IN VIVO* ASSAY ALIGNMENT WAS CONSTANT

The initial interest within the project was focused on inhibiting both EGFR and ERBB2 signalling with the latter being significantly less characterised. In order to evaluate small molecules for potency and selectivity, the project team created kinase enzyme assays and cellular assays that would allow the team to evaluate ERBB inhibition specifically as compared to inhibition of cell proliferation caused by downstream or non-specific interactions.

Initial studies directed at enzyme assay design for EGFR were carried out using receptor/kinase directly isolated from EGFR over expressing A-431 cell preparations utilising a radiometric measure of autophosphorylation. In the case of the ERBB2 enzyme assay, the team was able to clone and express a partial construct of the ERBB2 receptor. Since this construct contained the intracellular kinase portion of the receptor and had functional activity, the team was able to specifically target the protein with small molecules by means of a radiometric substrate phosphorylation assay.

In order to be able to evaluate activity of compounds within cells, two cell line clones were established so that compounds could be assessed for their specific intracellular mechanistic inhibition of ERBB2 autophosphorylation as opposed to other downstream inhibition events. These cell lines were engineered from a human breast cell line and were designated as HB4-erb, an ERBB2 over-expressing line and HB4-ras that contained mutant *RAS* (downstream of the desired signal inhibition) that served as a pathway selectivity control (HB4-erb is an abbreviated name for BH4a.C5.2 and HB4-ras is an abbreviated name for HB4a.C4.2).[9] Thus compounds could be specifically evaluated for ERBB inhibition in otherwise identical cells. While this work was ongoing, further target validation was added to the scientific literature by Brandt and colleagues, describing the prognostic relevance of ERBB oncogenes in several cancers.[6]

As described above, in the early stages of the project a rather unsatisfactory unpurified version of the assay system was used and was able to identify active EGFR and ERBB2 inhibitors, but the SAR interpretations were less robust. Different catalytic properties and substrate kinetics exist for the three kinase active members of the ERBB family.[16] Subsequent purification of these enzymes and optimisation of biochemical and kinetic parameters allowed accurate SAR comparisons from the EGFR, ERBB2 and ERBB4 assay results to be made.

One benefit of the Glaxo/Burroughs Wellcome merger in 1995 was the increased kinase profiling capability of the combined company, Glaxo Wellcome. Project compounds were routinely screened in the ERBB family kinase assays and also in a panel of over 30 kinase enzyme assays that included c-SRC, c-RAF1, VEGFR2, TIE2, and CDK2. At this stage in the project's life, lead compounds were understood to be competitive with ATP, evaluation of compounds therefore was carried out in these selectivity assays at K_m concentration of ATP thus allowing a valid comparison of IC_{50} values.

As the project progressed, further assays came to the fore that caused the compound evaluation scheme to evolve. In the latter stages of the project this scheme is summarised in Figure 27.2. There was an increased ability to screen compounds for ERBB2 and EGFR inhibition in a range of tumour types whose proliferation was driven by the over expression of these receptors. A panel of six parallel cell assays to provided information regarding the effectiveness of the inhibitors to block cell proliferation in breast (BT474[7,8] and HB4-erb, ERBB2 driven), head and neck (HN5, EGFR driven), gastric (N87, ERBB2 driven) tumour lines. Foreskin fibroblast (HFF) control lines of normal cells or HB4-ras discussed previously remained an integral part of the scheme. Three of the cell lines could be grown as xenografts in a mouse model of solid tumours to aid in the understanding of the *in vivo* efficacy of advanced compounds. Particularly, the faster growing gastric tumour N87 line allowed a more rapid turnaround time for evaluation of compounds *in vivo* compared to the very slow growing BT474 breast cell line. In order to distinguish the anti-tumour activity of compounds, it was important to be able to evaluate them multiple times to accurately assess their efficacy, thus a faster growing line afforded more data.

Dual inhibitors for example would be expected to be efficacious in both EGFR and ERBB2 driven cell lines, whilst the selective EGFR TK inhibitors that were advanced in the literature were significantly more effective in the HN5 line as compared with BT474, as expected.

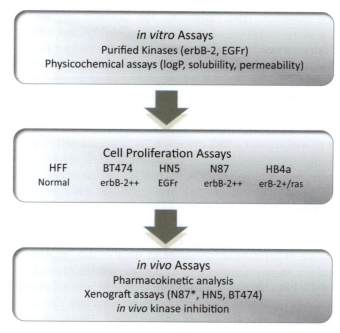

Figure 27.2 Compound evaluation scheme. Compounds progressed through initial ERBB2 and EGFR enzyme assays, physicochemical evaluation before cell proliferation studies in a range of ERBB2 and EGFR driven tumour cell lines. The final stage of evaluation involved pharmacokinetic and *in vivo* efficacy studies in specific xenografts in mice.

Selection criteria for compound progression changed as the project team grew in knowledge. In the earlier stages, flexibility on the exact criteria for enzyme potency and kinase selectivity was employed, so compounds that achieved a nominal cell activity ($IC_{50} < 500$ nM) and rat pharmacokinetic characteristics criteria progressed into the human xenograft mouse models (bioavailability > 10%, with plasma concentration > 500 nM over an 8 hr exposure period). As the medicinal chemistry understanding of the SAR of enzyme inhibition and selectivity was well honed, the cell assays became the key decision making assays because the pathway selectivity, cell permeability, and compound properties were simultaneously reflected in the assay outcomes. As the project moved closer towards the drug candidate selection stage, the desired profile to achieve efficacy became better understood. Specifically:

- Potent inhibition of ERBB2 and EGFR enzyme activity with IC_{50} values less than 0.02 μM was necessary.
- Selectivity *versus* a panel of kinases, with an expectation that it was necessary to be greater than 100-fold selective for the Type 1 receptor kinases compared with all other kinases tested.
- Activity in the tumour lines with the desired IC_{50} values less than 0.2 μM and greater than 15-fold selectivity between the *ERBB2* driven cancer cell line over the *Hb4-ras* cell line.
- Another selectivity parameter was that greater than 50-fold inhibition of cell proliferation in the tumour lines *versus* the HFF normal cell line needed to be demonstrated.
- Efficacy was expected in all of the tumour cell lines, based on the dual inhibition profile, and this efficacy was required to demonstrate anti-tumour activity in the same cell lines grown as subcutaneous xenograft models. To achieve this goal, compounds were required to demonstrate rat oral bioavailability in excess of 20% and to show no in-life toxicity during *in vivo* evaluations.

It was important to establish that inhibiting the desired target(s) led to an efficacious response *in vivo*, *i.e.* a demonstration of systemic availability and efficacy. Human xenograft mouse models using relevant cell lines whose growth was driven by ERBB family signalling were chosen at the time due to the lack of other viable ways to test multiple compounds in an anti-tumour activity assay. Early human tumour xenograft mouse studies were performed on less potent compounds that demonstrated cell activity in EGFR and ERBB2 driven lines consistent with the desired mechanism. Inhibition of tumour growth could be observed in this higher throughput, rapidly growing xenograft and the better compounds were progressed into the slower, more difficult to maintain, ERBB2 driven, BT-474 breast cell line xenograft model. Early studies were at relatively high doses in these xenograft models for these compounds based on the achievement of cover over multiples of the *in vitro* IC_{50} value in cell lines for the compounds.

Two tumour lines, BT474 (ERBB2 +++) and HN5 (EGFR +++), were grown as sub-cutaneous human xenograft models to determine the *in vivo* effectiveness of inhibiting tumour growth and became the mainstay of the evaluation process. Many compounds were evaluated in each model at two doses (30 and 100 mg/kg, b.i.d., 21 days) after the tumours reached a standard size in duplicate experiments.[10,11] After 21 days of dosing, tissue samples were saved for analysis of clinical chemistry parameters and liver, gastrointestinal, kidney, and cardiac pathology since one of the key criteria for candidate selection was a differential in terms of these clinical chemistry parameters. It was also important to correlate the observed tumour growth inhibition to the inhibition of receptor TK phosphorylation. Due to the goal of dual inhibition, the phosphorylation of both receptors was required to be inhibited with treatment whilst the expression level of the protein remained unchanged. The methods involved the establishment of the subcutaneous xenograft models; again, for both the BT474 and the HN5 tumour lines, have been reported.[12] Animals were treated orally for five doses (b.i.d.), the treated tumours excised and the inhibition of phosphorylated tyrosine levels were measured compared with the tumours of the untreated animals. The advantages of this *in vivo* assay are that smaller amounts of compound were needed and that mechanism of action information was obtained in significantly less time as compared with the efficacy studies described above.

The therapeutic index refers to the difference between where efficacy and anti-tumour activity are observed as compared with toxicity. In traditional anti-cancer therapeutic agents that have cytotoxic mechanisms, the expectation is that toxicity is observed at all levels of efficacy, but the risk/benefit is the determining factor in a fatal disease. For example, the body weight loss of a traditional cytotoxic mechanism would need to be measured alongside the efficacy measurements, and parameters would be set for an "acceptable" weight loss. The dual inhibition of ERBB2 and EGFR was termed targeted therapy and should avoid this conundrum of mechanism based toxicity at the therapeutic doses. Our project sought to create a therapeutic index whereby there was no toxicity observed at the efficacious doses, and preferably provided a margin of at least ten fold above the therapeutic levels of toxicity free range.

27.3 THE IDENTIFICATION OF LEAD COMPOUNDS INVOLVED A "SCREEN ALL" STRATEGY AND RAPID PROGRESSION TO EVALUATION IN ANIMAL MODELS

In the early 1990s, pharmaceutical companies were in the process of establishing high throughput screening technologies that included hundreds of thousands of compounds. The identification of dual EGFR and ERBB2 tyrosine kinase inhibitor starting points did not benefit from this emerging technology, but rather derived from a focused screening approach based on published literature of compounds that were found to bind in the ATP binding pocket of kinase enzymes. Many of these known inhibitors were based on natural products and were not selective kinase inhibitors. The screening of known kinase chemotypes was initiated on a modest scale.

There was a limited company collection available at that time, but staurosporines, their aglycones and bis-indolymaleimides derived from other projects in the company were screened, as well as coumarin analogues. Many of these compounds exhibited modest levels of activity but were viewed as uninspiring hits due to a whole raft of issues; synthetic tractability, gross insolubility and non-selectivity even amongst the then known and assayable kinome. The project progress benefited from the landmark discovery at Parke Davis of PD153035, wherein selective EGFR activity was demonstrated for the first time.[13] A representation of the relative selectivities of staurosporine and PD153035 is shown in Figure 27.3.

Due to the amenability of the anilino-quinazoline to be readily functionalised and to modulate EGFR activity, sets of novel compounds were synthesised based on this substructure. Initial examples contained either an unsubstituted quinazoline or the 6,7-dimethoxy variant. Multiple variations of substitutions on the aniline were prepared where the compound design was based on generating SAR data as no protein crystal structure or literature information was available for ERBB2 inhibition. While in advance of Lipinski's guidelines[14] and the advent of ligand efficiency design principles,[15] compounds synthesised at this stage were low molecular weight (<400) with appropriate clogP values for elaboration in lead optimisation. Additionally, compounds were prepared in the related quinoline series but were found to be inferior to their quinazoline analogues for inhibiting ERBB family kinases, although they were useful for other kinase targets in the screening panel. Importantly, all compounds synthesised for the inhibition of the target, including intermediates, were screened in the enzyme and cellular assays. One such intermediate compound, 4557W, was synthesised in the approach to the target phenol compound **1** and was discovered to possess the desired target profile (Figure 27.4).

There were very clear trends observable amongst this group of compounds. Potency in both the ERBB2 assay and EGFR assays was much enhanced with the 6,7-dimethoxy substitution when compared with the unsubstituted quinazoline scaffold. The 4-aniline substitution pattern provided a more subtle set of relationships and the SAR remained consistent throughout the drug discovery effort. For example, substituents such as halogens, particularly in the 3-position of the aniline, **2** (representative data for ERBB2 shown in Figure 27.5), provided modest EGFR and ERBB2 activity. It was found the ERBB2 potency could be retained with the inclusion of a larger substituent such as the 4-benzyloxy as shown for 4557W.[16] Notable amongst other modifications of this larger substituent were the "tied back" analogues, benzimidazole **3**, indole

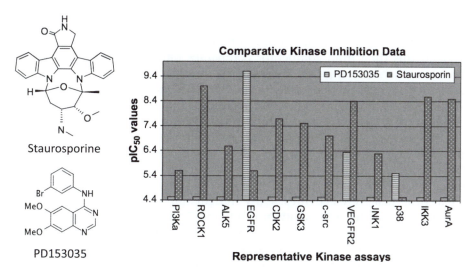

Figure 27.3   Relative selectivities of PD153035 and staurosporine.

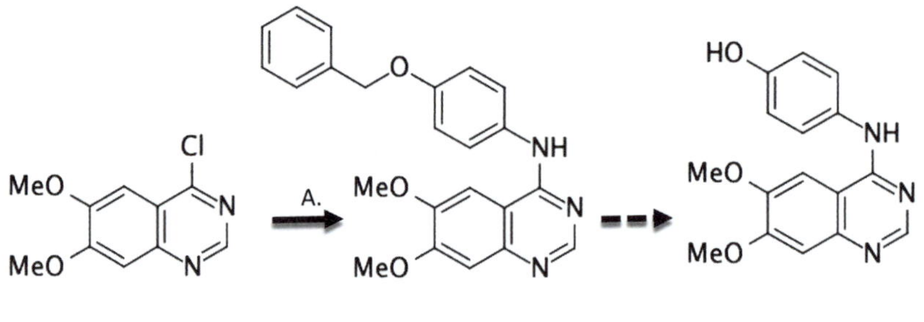

Figure 27.4   Attempted synthesis of phenol 1. (A) 4-Benzyloxyaniline, acetonitrile heat.

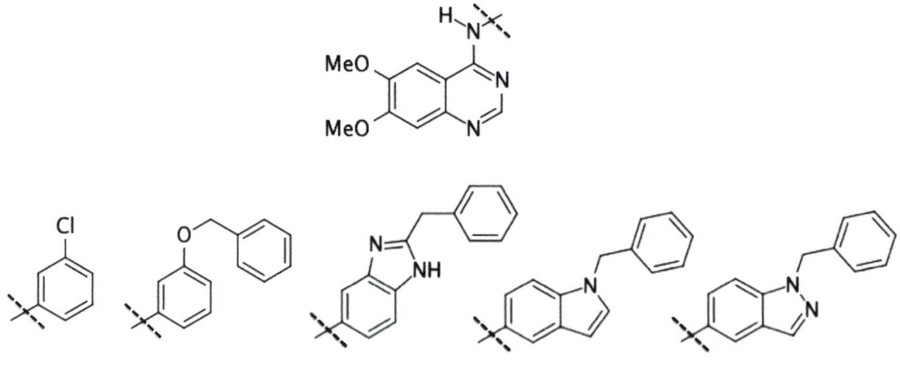

Compound	2	4557W	3	4	5
C-erbB-2 IC$_{50}$ (µM)	0.1	0.08	0.49	0.001	0.01

Figure 27.5   Early representative ERBB2 inhibitors.

4 and indazole **5**. These findings were important in staying focused on identifying novel compounds with potent ERBB2 inhibition. Information gleaned from literature and scientific presentations at this time from competitor companies only described examples of selective EGFR inhibitors. The structural and biochemical differentiation provided by the benzyloxy group, despite its physical property issues, *i.e.* increased molecular weight and lipophilicity, strongly supported the retention of this functionality at this stage of the project.

It was important to establish that inhibiting the desired target(s) led to an efficacious response *in vivo* because dual inhibition of receptor kinases were unproven, new therapeutic opportunities. In early anti-cancer drug discovery, this process of validating the target for the intended disease indication in animal studies involved a proof of principle or mechanism study. It is essential to perform these studies as early as possible to rationalise the continued investment in drug discovery activities. Early human tumour xenograft mouse studies were performed for **4** because the cell activity in EGFR and ERBB2 driven lines was consistent with the desired mechanism, albeit not very potent. The *in vivo* efficacy evaluation strategy was biased towards ERBB2 by utilising the N87 gastric cell line as described in the previous section. Subsequently the compound was progressed into the BT-474 breast cell line xenograft model at relatively high doses. 4557W was shown to be efficacious in both these xenograft models.

4557W was shown to be a selective inhibitor of the ERBB family kinases over the existing panel of over 30 kinases. Combining the data for the lead compound's kinase profile with the cell activity in relevant tumour lines, the dual EGFR/ERBB2 tyrosine kinase inhibition concept proved feasible. These results were a key achievement in the project, and more extensive medicinal chemistry on multiple related scaffolds ensued. This tool compound, served a valuable purpose by establishing the *in vivo* models and the confidence to move forward and as such, the project became high priority for the oncology research area.

27.4 PARALLEL EVALUATION IN POTENCY AND PHARMACOKINETIC ASSAYS ALLOWED A RAPID EVALUATION OF LEAD COMPOUNDS

Modifications to the central core quinazoline now became a feature of the medicinal chemistry campaign to search for improved potency and more extensive SAR. This approach was intended to reduce logP values and improve solubility, a possible counterpoint to the lipophilic nature of the benzyl group. A binding hypothesis was used at this stage, based upon a rationalisation of the SAR generated in the hit to lead phase of the project. The importance of the pyrimidine core and the aniline N-H suggested that it functioned as a hydrogen-bonding region and directed the team towards modifications of the putative variable region shown in Figure 27.6. A range of heteroatoms were investigated to provide alternatives to the phenyl ring of the quinazoline, amongst these were thienopyrimidines exemplified by **6** shown in Figure 27.6 and regioisomers of the pyridopyrimidine system **7**.

The program strategy utilised several assays in parallel thus required the chemists to make sufficient quantity of the first batch of each novel compound in order to perform enzyme, cell and pharmacokinetic evaluations. Since the team lacked access to a raft of *in vitro* permeability, stability and solubility assays, an LC-MS based *in vivo* cassette pharmacokinetic rat assay assumed a primary testing position.[17] Selection criteria as described above were not rigidly

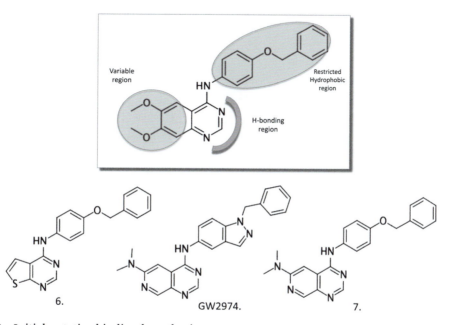

Figure 27.6 Initial putative binding hypothesis.

enforced because it was more important to develop SAR and to expand the science for ERBB family inhibition.

An exceptional compound to appear from this exercise was the benzyl indazole GW2974[18] shown in Figure 27.6. It was one of the first compounds to exhibit an IC_{50} value under 100 nM in the HB4-ERB, the c-ERBB2 over-expressing line cell line, and appeared potent in the other relevant tumour cell lines, HN5 and BT474 with IC_{50} values at or below 0.50 µM (Figure 27.7). In contrast, the control lines, HFF and HB4-ras, have IC_{50} values in excess of the maximum concentration tested of 30 and 24.8 µM, respectively. Another characteristic of GW2974, and indeed the series subsequently, was its selectivity for the ERBB family of kinases as no appreciable activity was observed for the SRC, map kinase and cyclin dependant kinase families. The cell selectivity demonstrating that the efficacy correlated to the mechanism of action was achieved as measured by comparing the ERBB driven *vs.* the ras driven HB4a results. GW2974 possessed moderate clearance in rats, marmosets and dogs; a variable intravenous half life in these species (18–44 minutes) and a moderate volume of distribution (0.19 L/kg in rats and 0.8 L/kg in dogs) equivalent to or slightly in excess of whole body volume. Mean plasma levels on repeat dosing in rats and mice showed that levels above the IC_{50} for the compound could be maintained at doses of 20 mg/kg twice daily.

Significant inhibition of average tumour growth at 10 mg/kg and complete inhibition of average tumour growth at 30 mg/kg of GW2974 after 21 days of twice daily dosing was also observed in the subcutaneous xenograft models of the same tumour lines, BT474 and HN5, and the compound appeared well tolerated in animals as evidenced by lack of body weight loss and in life observations.[16] Additionally, extended observation of a treatment group dosed at 30 mg/kg for 5 days followed by a dose reduction to 10 mg/kg for another 15 days, showed a sustained efficacious response whereby the tumours did not re-grow.[18] The inhibition of ERBB2 and EGFR tyrosine phosphorylation was measured to be greater than 90% both *in vitro* and *in vivo* after treatment with GW2974, while no effect on protein levels was observed.

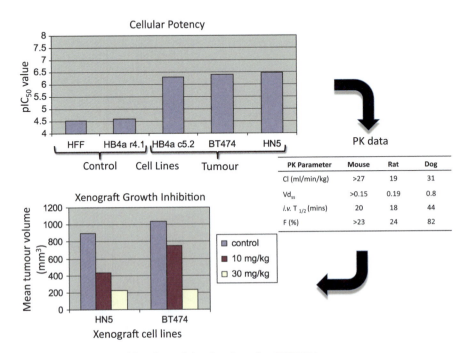

Figure 27.7 *In vitro,* pharmacokinetic and *in vivo* data for GW2974.

The inhibition of receptor autophosphorylation that was observed in tissue culture translated into inhibition or tumour cell proliferation and growth in mouse models.

The efficacy and mechanism of action of GW2974 allowed the compound progress into pre-clinical development. Several issues were uncovered at this stage. The pharmacokinetics were evaluated more extensively and found that plasma exposure did not directly correlate with dose. The compound appeared to saturate metabolism at higher dose, an effect associated with N-demethylation. Variably active metabolites were formed, both single and double demethylation of the 6 position dimethylamino group. In addition the inherent insolubility of the compound posed formulation and handling issues. Finally, a number of hematological and lymphoid effects precluded the initial definition of a no effect level. GW2974 could have been a first generation, effective dual ERBB2/EGFR TK inhibitor for a selective set of advanced cancer patients whose tumours are driven by this mechanism. Although the issues could have been addressed, the further development of GW2974 was halted with the strategy now focused on the task to deliver a dual inhibitor that possessed both an improved physicochemical profile, structural novelty an improved therapeutic index.

The drug discovery efforts and data that were generated during the optimisation phase that led to the progression of GW2974 as a first drug candidate selection for the program provided a plethora of meaningful tools and information. The fact that GW2974 failed pre-clinical toxicity studies could not have been predicted from the efficacy data, but provided the program team with an outstanding tool compound to develop more refined assay systems, and served as a minimum efficacy benchmark for a new drug candidate selection effort. In this way, the program team had a standard for all of the desired mechanism and efficacy characteristics of an ideal drug candidate.

27.5 THE USE OF BINDING HYPOTHESES AND SYNTHETIC TRACTABILITY PROVIDED ACCESS TO NOVEL STRUCTURAL SPACE BY EXPLORING A "PUTATIVE VARIABLE REGION"

Attention was now paid to the proposed variable region as shown in Figure 27.8. The chemistry-led design process progressed in two phases and focussed upon the additional ring and chain variations to the quinazoline/pyridopyrimidine nuclei as this appeared the most likely way to provide the desired balance of novelty and improved physical properties. The use of palladium-catalysed ring coupling methodologies was utilised extensively[21] in this exercise.

Dual ERBB2/EGFR TK potency, good rat pharmacokinetic properties, and cellular activity were discovered in the 6-heteroaryl quinazolines such as **8** when combined with the moderately basic methylsulfonylethyl-aminomethylene side chain and this functionality became the foundation of the medicinal chemistry strategy for achieving the desired product profile. The seven aryl substituted quinazoline series were in general less potent and were in addition significantly less tractable synthetically.

Some highlights of this work[20] are worth mentioning in respect of the subsequent effort directed at the identification of lapatinib. The key SAR observed for amine containing side chains linked through a 6-furanyl quinazoline was that the orientation of substituent on the furan ring affected the cellular activity. A marked increase in cellular activity was seen for the 5' position orientation as shown in Table 27.1. While these observations were empirically determined, after the discovery of lapatinib, a ligand bound crystal structure revealed a dramatic protein movement that could explain this finding.[27]

A large number of amine chains were investigated; a range of groups could be tolerated, notably morpholine and thiomorpholine groups, as well as number of alkyl substituted

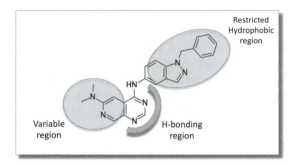

Figure 27.8 Compounds prepared as part of the exploration of the putative variable region.

Table 27.1 Substituted furan chain position SAR.

Chain Positon	Average ERBB2/EGFR IC$_{50}$ (μM)	Average Tumour IC$_{50}$ (μM,HN5/BT474/N87)
none	0.074	2.98
5$'$	0.027	0.60
4$'$	0.023	2.68
3$'$	0.068	3.81

sulfonylethylamines (Figure 27.8). The methylsulfonyl methylaminomethyl group emerged as the most interesting group due to its relative metabolic stability, ease of synthesis as well as cellular potency benefits.

Table 27.2 Lead substituted heteroaryl compounds.

Compound	Avge. ERBB2/EGFR (IC$_{50}$ µM)	Avge Tumour IC$_{50}$ (µM)
8	0.065	1.3
9	0.027	0.6

Lead compounds from this effort were thiazole quinazoline **8** and furanyl quinazoline **9** (Table 27.2). Compound **8** exhibited good linear, multi species pharmacokinetics but only moderate cell potencies and animal model efficacy was achieved. Compound **9** exhibited very good cell efficacy but only very moderate pharmacokinetics in rats. Despite this pharmacokinetic profile, **9** (GW0277) demonstrated desirable activity in mouse xenograft models and became a second significant compound for the project, being extensively investigated in the pre-clinical phase. Although not understood, perhaps some correlation can now be drawn (in hindsight) with efficacy dependent upon a pseudo-irreversible mechanism of inhibition of tumour growth.

27.6 SYNTHETIC DESIGN AND COMPOUND SYNTHESIS QUANTITY ALLOWED THE RAPID EVALUATION OF COMPOUNDS

The demands on synthetic groups in the lead optimisation phase of projects can be viewed as divergent. There is a need to produce numbers of compounds in sufficient quantity to allow the acquisition of meaningful levels of data at a pace that allows momentum to be developed and retained within the project. Although the synthesis of a minimum quantity of compound and the synthesis of numbers of diverse analogues are opposing tensions within a lead optimisation phase, the approach taken within the team at GlaxoWellcome was to design convergent syntheses that utilised advanced intermediates that could be synthesised at scale. Although this approach necessitated commitment within the team to solve significant synthetic issues, the pay off was that there was a constant supply model established, both in terms of intermediate and final compound. This allowed rapid design cycles to be established and promising compounds to be evaluated even to the point of early stage *in vivo* pharmacokinetic studies. Although this output and productivity varied with time, for significant periods of the project lifetime, the team was producing 40–50 mg quantities of compound at first synthesis.

Examples are shown in the schemes below. The pyrido-pyrimidine scaffold represented a key achievement for the team (Scheme 27.1). Substantial effort was directed initially at the synthesis of the key intermediates A and B at a scale up to 50g quantities. Most specifically the control of the first stage Curtius rearrangement and subsequent carboxylation at scale were notable achievements

This constant supply model was repeated in the quinazoline series where the generation of large quantities of key intermediates facilitated both the production of numbers of compounds in this series (Scheme 27.2) and also allowed the team to deal with synthetic challenges presented both by the synthesis of the 6-thiazole containing compounds shown in Scheme 27.3. Significantly, in this latter case the non-availability of the thiazole boronic acid or tin reagents

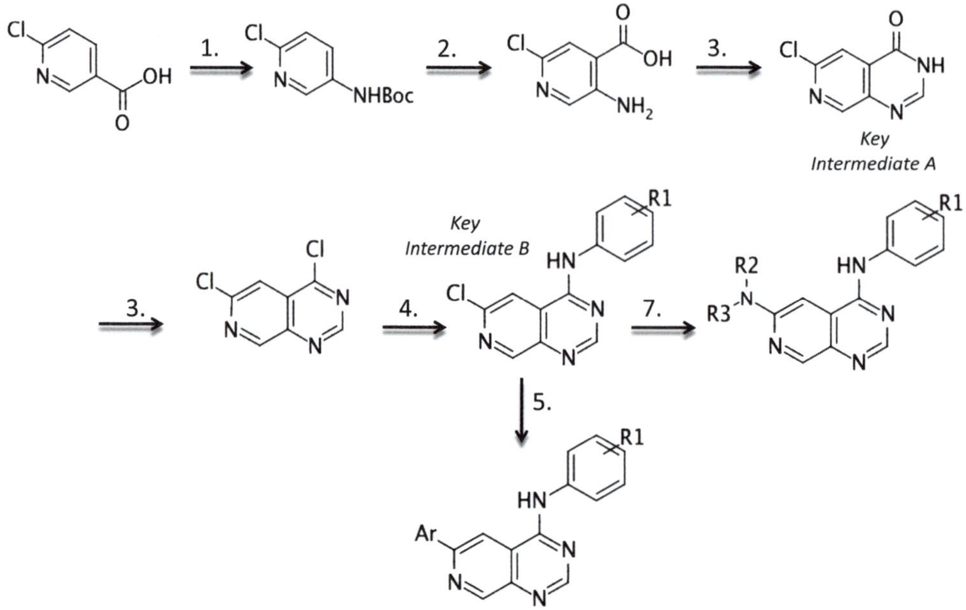

Scheme 27.1 Pyridopyrimidine synthetic strategy. (1) $(PhO)_2PON_2$, Et_3N, ^tBuOH, Δ, 83%. (2) ^nBuLi then CO_2. (3) $CH(NH)NH_2$, HOAc. (4) $PhNH_2$, CH_3CN, Δ, 90%. (5) "Benzyloxyphenyl" NH_2, CH_3CN, Δ, 90%. (6) 2-FuranSnBu$_3$, Pd(0), AgO. (7) Me_2NH, EtOH, CH_3CN.

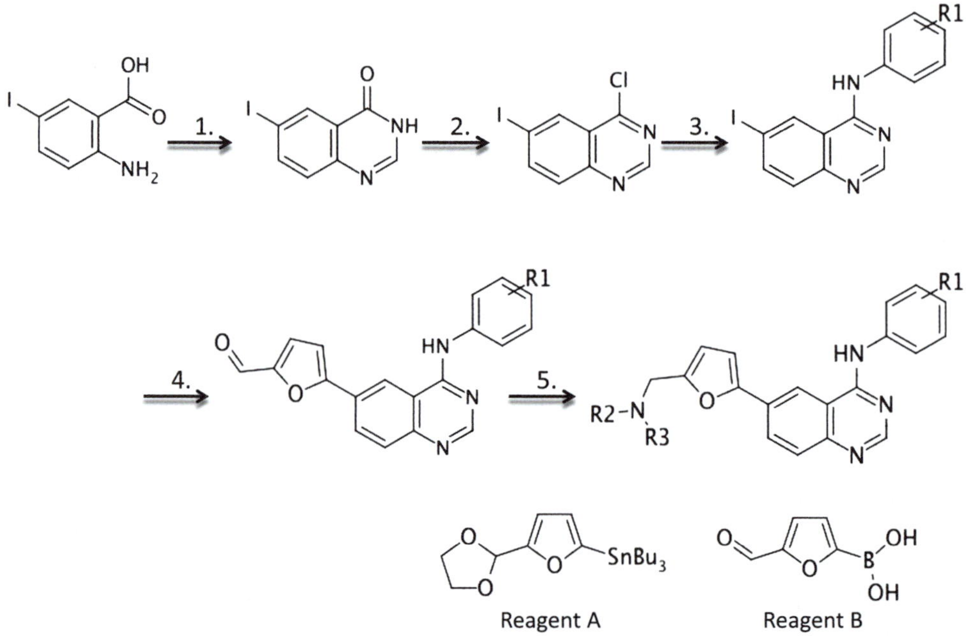

Scheme 27.2 Synthesis of heteroaryl quinazolines. (1) $CH(=O)NH_2$, D. (2) $SOCl_2$, DMF(cat). (3) 3-Chlorobenzyloxy-2-fluorophenylNH$_2$, ^iPrOH, D. (4) Reagent A or B, $Pd(OAc)_2$, PPh_3, Et_3N, DMF. (5) $MeS(O)_2CH_2CH_2NH_2$, HOAc, $Na(AcO)_3BH$.

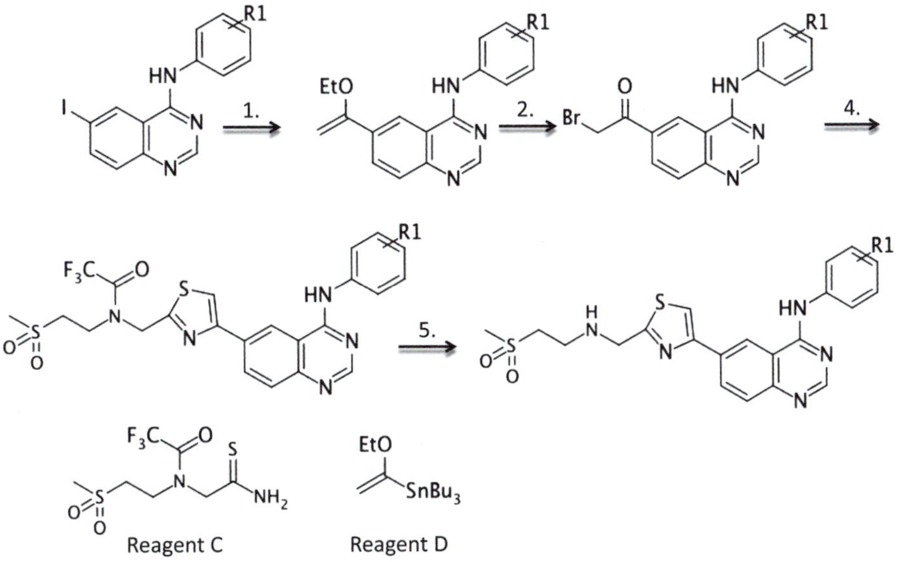

Scheme 27.3 Synthesis of substituted thiazole quinazolines. (1) Reagent B, Pd(PPh$_3$)CCl$_2$. (2) Br$_2$, CH$_2$Cl$_2$. (3) Reagent C or D, DMF, Δ. (4) (i) NaOMe, MeOH. (ii) 4 M, HCl.

enforced construction of the thiazole ring system on the quinazoline core *via* bromo-ketone/thioamide cyclisation methodology.[22]

27.7 DATA RE-EVALUATION AND RIGOROUS SELECTION CRITERIA AT THIS STAGE LED TO THE IDENTIFICATION OF GW2016

The project had reached a critical decision point as viable drug candidates had been discovered, but with features uncovered in the project progression preventing further development. In addition there remained bias in the scientific community regarding the feasibility of discovering a selective, and therefore presumed non-toxic kinase inhibitor. Considerable resource had been dedicated to the dual ERBB2/EGFR inhibitor project, so a way was needed to determine if a drug could be discovered that met the criteria for an orally bioavailable, potent dual inhibitor drug to treat solid tumours driven by the ERBB family of receptor tyrosine kinases. At this point in the drug discovery efforts there was no access to a crystal structure of EGFR, as is currently state-of-the art in the field.[23] Earlier design work was guided by using p38 crystal structures as a surrogate that created a docking model for the quinazoline and pyrido-pyrimidine series with a binding mode that did not always correlate to the enzyme SAR.

All of the data generated to date in the project, compiled and re-analysed. A calculated index for the SAR of dual inhibition or pan-ERBB family inhibition was derived from a weighted mean of pIC$_{50}$s for ERBB family inhibition, used with kinase enzyme and cellular proliferation assay data.[24] Although data for >3000 compounds were included, there was an apparent lack of correlation between the enzyme profile and the desired cellular activity. To investigate this a subset of compounds were tested in multiple developability assays (*e.g.* solubility in multiple solvents, cell permeability assays, protein binding measures, *in vitro* metabolic stability), however the physical properties of the compounds did not explain the apparent lack of correlation between the cellular and enzyme activity. For example, a number of compounds with relatively poor cell potency compared to their enzyme potency were found to be cell permeable. Although the lack of correlation between enzyme and cell data remained an issue, this rigorous data analysis suggested a synthetic strategy based on dual ERBB2/EGFR enzyme inhibition and this approach was implemented.

By this stage in the project, the compound evaluation pathway had improved (Scheme 27.4), chiefly through the incorporation of purified enzyme assays that allowed accurate SAR comparisons from the EGFR, ERBB2 and ERBB4 assay results. It is important to note that the general principles of using the cellular panel of assays and pharmacokinetic evaluations remained the same. The cell based assays included HN5 (EGFR+++), BT474 (ERBB2+++), and N87 (ERBB2++ and EGFR++),[7,8] and HFF (control). The original HB4a. assays that differentiated the ERBB2 signalling from downstream-driven ras signalling were also critical in the selection of the optimal drug candidate.

With the tumour lines, BT474 (ERBB2) and HN5 (EGFR) grown as subcutaneous human tumour xenograft models, compounds were evaluated in each model at two doses (30 and 100 mg/kg, b.i.d., 21 days) after the tumours reached a standard size in duplicate experiments with eight animals per study group.[19] An *in vivo* kinase inhibition assay was established to correlate the observed tumour growth inhibition to the inhibition of receptor TK phosphorylation without affecting the protein expression level. Animals were treated orally, the treated tumours were excised and the inhibition of phosphorylated tyrosine levels were measured and compared with the tumours of the untreated animals. A limited PK protocol was developed to assess the compounds in a standardised procedure to understand some SAR in the pharmacokinetic properties. A truncated protocol of using four time points and two animals per time point predicted full pharmacokinetic profile for oral dosing. The calculated values for the area under the curve (AUCs) for the plotted data were used to assess the compounds. Trends for improved oral bioavailability could be seen even with relatively small numbers of compounds.

Approximately 70 compounds were synthesised in this phase of the project, and six highly functionalised quinazolines and pyrido-pyrimidine compounds shown in Figure 27.9 were chosen for early toxicity studies. The 22 distinguishing candidate selection criteria included efficacy parameters (cellular and *in vivo*), biometabolism parameters (time of drug exposure over

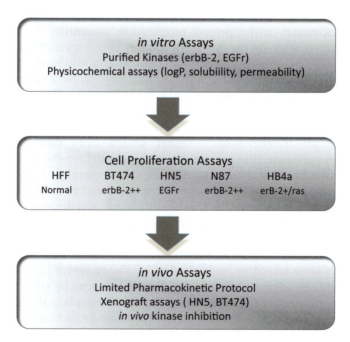

Scheme 27.4   Candidate evaluation pathway.

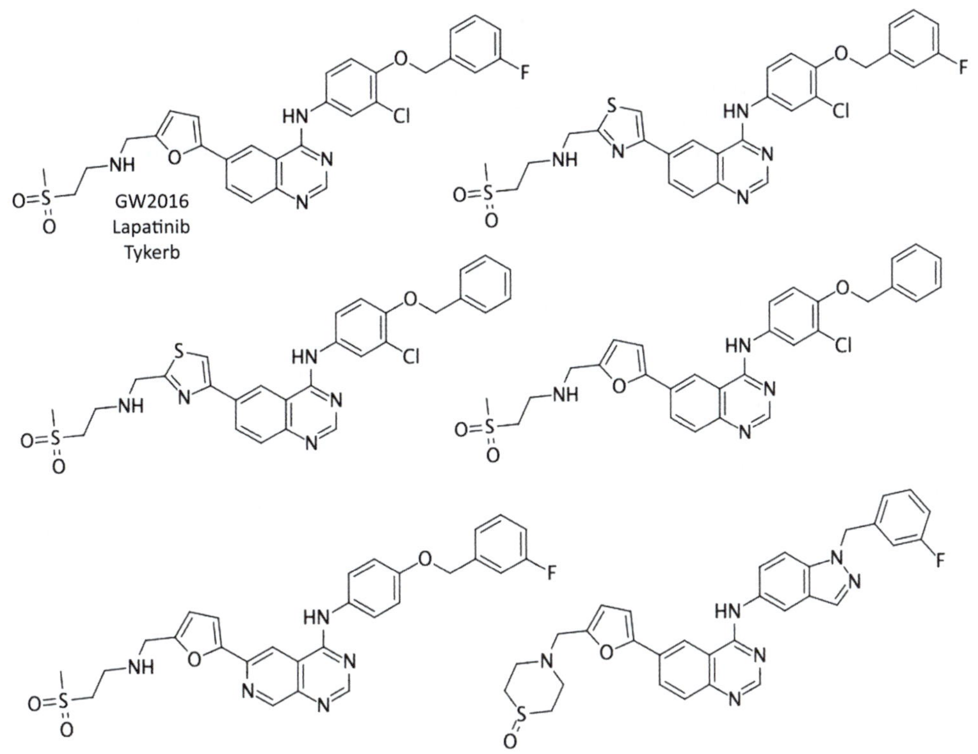

Figure 27.9 Top six derivatives evaluated for candidate selection.

IC$_{50}$ or IC$_{90}$ levels, percent oral bioavailability, p450 enzymes), toxicity measurements (cellular, cardiovascular, 7-day rat studies, Ames test) and chemical issues (cost of goods, scale-ability). Blood samples were collected at the end of 21-day dosing from all of the animals receiving 100 mg/kg b.i.d. of compound in our anti-tumour evaluations for the following clinical chemistry parameters: haemolysis, albumin, alkaline phosphatase, serum glutamic-oxaloacetic transaminase, blood urea nitrogen, cholesterol, total protein, glucose, sodium, potassium, and chloride.[10] While the general appearance and lack of body weight loss demonstrated that the top six compounds were well tolerated, varied results were obtained in the clinical chemistry analysis, and only compounds with no effects on these selected clinical chemistry parameters were considered for drug development.

27.8 LAPATINIB, GW2016

Lapatinib is potent on two of the ERBB family members with enzymatic IC$_{50}$ values against ERBB2 and EGFR receptor tyrosine kinases of 9 and 10 nM respectively, with greater than an order of magnitude loss in activity for ERBB4.[19] A small molecule–kinase interaction map was created for lapatinib (GW2016) by Ambit researchers using an ATP site-dependent competitive binding assay in a panel of 119 kinases and demonstrated a very clean profile.[25] The average range of IC$_{50}$ values obtained for lapatinib in tumour cell lines that had type I receptor expression was ~50 to 125 nM with an average cellular selectivity of 100-fold.

Lichtner and co-workers reported the cellular effects of quinazolines and 4,5-dianilino-phthalimides, two classes of potent EGFR TK inhibitors, and found that the quinazoline's cellular efficacy was due to a novel mode of action even though both classes of compounds bind in the ATP-site with similar potency.[26] The quinazoline inhibitors affected the ligand binding properties by stabilising the ligand/receptor/inhibitor complex resulting in potent cellular activity, while the dianilinophthalimides did not. Studies were performed with lapatinib to determine if there was a similar explanation for its effectiveness in the pre-clinical models. The inhibitor off-rates were evaluated using an EGFR enzyme reactivation procedure and Tarceva[TM] was found to have a rapid off-rate ($t_{1/2} < 10$ minutes) whereas after preincubation with lapatinib, there was a significantly slower off-rate ($t_{1/2} = 300$ minutes).[27]

A similar dissociation rate was observed with lapatinib using ERBB2. The crystal structure of EGFR bound to lapatinib revealed an inactive-like conformation in contrast to the published active-like structure with Tarceva.[23a] The differences in the ligand bound structures included the shape of the ATP site (closed *versus* open conformation), the position of the C helix (large back pocket *versus* intact Glu738-Lys721 salt bridge), the conformation of the COOH-terminal tail (partially blocking the ATP cleft), the conformation of the activation loop (A-loop similar to ones found in inactive structures *versus* ones found for active structures), and the hydrogen bonding pattern with quinazoline scaffold (water mediated interaction with Thr830 *versus* Thr766) (Figure 27.10). To determine if the kinetics affected cellular activity, HN5 tumour cells were treated for 4 hours with lapatinib, and the receptor phosphorylation was analysed at multiple time points after washout. The slow off-rate found for lapatinib in the enzyme reaction correlated with the observed, prolonged signal inhibition in tumour cells based on receptor tyrosine phosphorylation measurements.

Table 27.3 Comparison of lapatinib and Tarceva binding data.

Compound	ERBB2 K_i (nM)	EGFR K_i (nM)	Dissociation $T_{1/2}$ (mins)	EGFR activity post compound washout
Lapatinib	13 +/− 1	3.0 +/− 0.2	300	15% @72h
Tarceva	870 +/− 90	0.4 +/− 0.1	<10	100% @24h

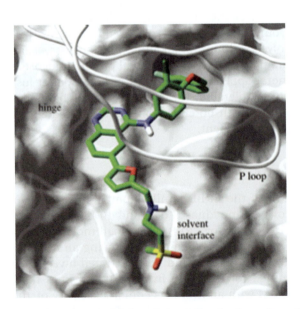

Figure 27.10 A view through the *N*-terminal lobe of EGFR for the bound conformation of lapatinib as determined by X-ray crystallography.[23b]
Image created with Maestro from Schrodinger LLC.

Blencke and co-workers identified a critical mutation of Thr766 in the EGFR kinase domain as leading to a dramatically reduced sensitivity to the 4-anilinoquinazoline PD153035 and linked it to the mechanism of tumour resistance.[28] It is possible that lapatinib's lack of an interaction at this amino acid will lead to lower mutation rates or simply to a drug that retains inhibition despite the mutation. The slow binding kinetics may offer increased signal inhibition in the tumour, thus leading to greater potential to affect the tumour growth rates or effectiveness of co-dosing with other chemotherapeutic agents. The exciting components of these ligand-bound kinase structures are the insightful links they may provide for enhancing efficacy *via* a parameter other than simple potency (*i.e.* binding to and stabilising different conformations) or in patient selection for optimal therapeutic outcomes.

Lapatinib inhibits not only baseline activation of both ERBB2 and EGFR, but also interrupts downstream activation of ERK1/2 MAP kinases and AKT.[29] The inhibition of AKT by lapatinib was associated with a 23-fold increase in apoptosis compared with vehicle controls. Lapatinib was also found to inhibit the signal transduction in the presence of saturating concentrations of epidermal growth factor (EGF), in tumour cell lines that over-express Type I receptors as well as stimulated tumour lines that do not over-express EGFR by measuring the p-Tyr, p-ERK, and p-AKT levels. These effects on downstream cell signalling markers in a range of tumour types combined with the inability of EGF to reverse the effects of Lapatinib would suggest a broader tumour type application of the drug. Specifically tumour types that remain dependent upon EGFR/ERBB2 for cell survival and growth signals where low levels of EGF can activate these receptor heterodimers. Many more cellular activity and mechanism of action evaluations were performed in order to understand the scope of a dual ERBB2/EGFR TK inhibitor. We were confident that lapatinib had the desired *in vitro* properties of efficacy and selectivity.

Lapatinib demonstrated reproducible tumour growth inhibition of $34 \pm 28\%$ (30 mg/kg b.i.d.) and $101 \pm 20\%$ (100 mg/kg b.i.d.), in the HN5 xenograft model with regression (defined as $>25\%$ reduction in tumour volume) in 33% of the treated animals.[26] In the BT474 model, inhibition of $42 \pm 35\%$ (30 mg/kg b.i.d.) and $94 \pm 18\%$ (100 mg/kg b.i.d.), was observed with 10% of the treated animals with regressions. The level of ERBB2 phosphotyrosine in tumour excised after therapy in the 100 mg/kg treatment, was reduced by 93% in the BT474 model and 85% in the HN5 model and occurred in a dose dependent manner for other treatment groups. Lapatinib was not toxic at this dose and activity/safety profiles observed in the xenograft models suggested that lapatinib could also be safely combined with standard chemotherapy.[30] We focused on the correlation of efficacy and mechanism of action, and disregarded the high doses needed in the xenograft models due to the inherent limitations of the system.

27.8.1 Clinical Studies

Safety and tolerability was demonstrated in Phase I and Phase II human clinical studies, with healthy volunteers as well as cancer patients.[31] The safety studies were designed to prepare for the long term usage anticipated in early disease and cancer preventive settings, thus the need for a drug candidate with a large therapeutic window.

Clinical responses were observed in heavily pre-treated Phase I patients with metastatic diseases in both EGFR driven and ERBB2 over-expressing solid tumours in several cancer types that included breast, non-small cell lung, bladder, and head and neck.[32] Biomarkers of signal inhibition[33] were used throughout the drug discovery program to determine maximal biological effects and patient selection (Table 27.4).

Translational research studies by Molina and co-workers comparing results in pre-clinical models with Phase I clinical trial data demonstrated encouraging results in combination

Table 27.4 Early clinical trial data for lapatinib.

Treatment	Phase	Patients Included	Summary of Responses
Lapatinib + Capecitabine	I	45 advanced solid tumours	1 CR, 4 PR, 2 SD
Lapatinib + trastuzumab	II	48 HER2 + breast cancer	1 CR, 5 PR, 10 SD
Lapatinib	II	138 HER2 + breast cancer	0 CR, 33 PR, 71 SD

CR = complete response, PR = partial response, SD = stable disease.

therapy of lapatinib and Topotecan, a topoisomerase I inhibitor.[34] They used the rationale that survival, signalling and transporter mechanisms disrupt the effectiveness of Topotecan. They found that the combination enhances Topotecan accumulation and toxicity in relevant tissue culture and xenograft mouse models in a mechanistically consistent manner. The Phase I data suggested that the combination of lapatinib (1250 mg/day for 28 days) and Topotecan (3.2 mg/m^2 i.v. on days 1, 8, and 15) is safe and tolerable. It appears that the reduction of Topotecan below the accepted MTD of 4.0 mg/m^2 when dosed alone was due to the observation that lapatinib reduced the clearance of Topotecan through transporter mechanisms. Stable disease was observed in 46% of the patients and the authors reported that the results warranted a Phase II trial of Topotecan plus lapatinib combination in ovarian cancer.

Another important comparison has been made by Ulhoa-Cintra, Greenberg and Geyer in their review of the emerging role of small molecule compounds after the success of trastuzumab, a revolutionary antibody drug that targets the extracellular domain of ERBB2.[35] Even with trastuzumab's successes, resistance occurs in a large majority of advanced breast cancers, making complementary mechanisms (*e.g.* the intracellular kinase domain) an important component of potential cancer therapies. A dose range from 500–1600 mg/day of lapatinib was evaluated in heavily pre-treated metastatic solid tumour, where partial responses were observed in patients with trastuzumab-resistant metastatic breast cancer.

Data summarised by Medina and Goodin provided a good overall picture of the *in vivo* effects of lapatinib.[36] The standard, FDA approved dose for lapatinib is 1250 mg/day which affords a steady state Cmax of 2.4 ug/mL and an AUC of 36.2 μg·h/mL, and the exposure appeared to nearly double when administered with food. Lapatinib seemed to have similar pharmacokinetics in paediatric patients, although a much smaller patient number was evaluated. Lapatinib interacts with CYP3A4, and co-administration with ketoconazole increased the levels of lapatinib by over three-fold. A small sampling of the early clinical trial data is in Table 27.3.

The Phase 3 clinical trial evaluating the efficacy of the combination of lapatinib and capecitabine in advanced breast cancer patients that failed several prior regimens of cancer therapy demonstrated a significant advantage over capecitabine therapy alone by the interim analysis of the clinical plans. Specifically, the Time To Progression (TTP) was 4.4 months for capecitabine alone, and 8.4 months in the combination therapy suggesting a 51% improvement. The trial was halted and patients were offered the combination therapy. Further follow up data from the trial showed an overall response rate of 24% for patients on lapatinib + capecitabine and 14% for those treated with capecitabine alone. The dominant side effects were usually diarrhoea and rash.[36] In a phase II study of lapatinib monotherapy in breast cancer patients that had brain metastases, >20 % reduction in CNS tumour volume occurred in 46 out of 241 patients and >50% reduction occurred in 19 out of 241 patients. These results are encouraging for the potential of lapatinib in combination therapies for this unmet medical need.

Tykerb (trade name for lapatinib) was first approved for use in breast cancer in March 2007 in the United States, and approvals followed in more than 20 countries over the next two years. Ongoing clinical trials in multiple cancers, in multiple combinations seek to define the patients that will most benefit from a dual EGFR and ERBB2 tyrosine kinase inhibitor.[37]

27.9 CONCLUSION

The drug discovery effort for lapatinib was taking place just as personalised cancer treatments began to emerge from striking advances in understanding oncogenes and their mechanisms of actions. The project required careful attention to link mechanism of action studies, dual inhibition of EGFR/ERBB2, to all observed efficacy and selectivity. In the early work, the data for the inhibition of the targets did not correlate with the desired cellular efficacy, and thus called into question whether it was the right concept for the generation of an effective drug. By performing a multi-parameter analysis linking the enzyme profile to a cellular outcome (potent on tumour cells and inactive on normal or control lines), the data could be reviewed as index analysis for the ability of each series to reach the desired profile. A subset of the data that fits the desired correlation was used to drive the design of the final sets of synthetic compounds. From these results, the most effective dual ERBB2/EGFR inhibitors were found to have a remarkably slow, but reversible, off-rate suggesting the prolonged inhibition of the signalling was beneficial for blocking tumour proliferation. Lapatinib was thereby discovered as the most effective dual EGFR/ERBB2 tyrosine kinase inhibitor.

When studied in multiple models of *in vivo* xenograft studies, tumour growth was inhibited in a dose-dependent manner, and the excised tumours showed blocked autophosphorylation consistent with the tumour inhibition. Subsequent crystal structure studies of lapatinib bound to the kinase domain of ERBB2 showed a unique binding interaction as compared with the leading EGFR selective drug candidates (inactive *versus* active forms, respectively). Enzyme kinetic studies revealed a very slow off-rate for lapatinib, further supporting why this subset of quinazoline derivatives was effective and differentiated from other potent enzyme inhibitors. Lapatinib was selected for the full development process and was dosed to humans in January 2001 in the first clinical trial in healthy volunteers. In March of 2007, Tykerb was launched as an anti-cancer treatment for breast cancer in combination with Xeloda[34] in the United States, and approvals followed in more than 20 countries over the next two years. Ongoing clinical trials in multiple cancers, in multiple combinations seek to define the patients that will most benefit from a dual EGFR and ERBB2 tyrosine kinase inhibitor.[37]

HINTS AND TIPS

- Be prepared to modify progression criteria as the project progresses and where possible test assumptions embodied by these criteria.
- In targets belonging to enzyme families like kinases, test lead compounds in standardised assays as soon as possible.
- In standardising assay formats for the evaluation of inhibitor activity, utilise K_m values for the binding of substrates when directly inhibiting a substrate conversion process. Thereby establish valid selectivity profiles within the enzyme family.
- Employ a parallel evaluation strategy, in terms of potency and pharmacokinetic optimisation, to provide a rapid evaluation of compounds and increase project progression.
- The synthesis of target compounds in the lead optimisation phase at the scale of 40–50 mg allows a rapid evaluation of *in vitro* potency and *in vivo* pharmacokinetic properties.
- Utilise convergent synthetic schemes with an intermediate stockpiling strategy to facilitate this synthetic approach.
- Utilise clinically relevant assay types early in the process. Specifically the use of cell lines directly derived from clinical samples rather than relying merely on artificially engineered cell lines.

- Establish cellular selectivity profiles between diseased (tumour) cell lines and normal cell lines and retain these as intrinsic selection criteria.
- Retain a strong link between *in vitro* evaluation data and *in vivo* efficacy read outs. When possible evaluate the causes of lack of efficacy *in vivo* and utilise the information in the development of progression criteria.

Acknowledgements

The scientists that contributed to the discovery of Tykerb are Karen Affleck, Patricia Allen, Krystal Alligood, Perry Brignola, Zongwei Cai, Malcolm Carter, Renae Crosby, Scott Dickerson, Stephen Frye, Micheal Gaul, Cassandra Gauthier, Tona Gilmer, Bob Griffin, Steve Guntrip, Yu Guo, Robert Harris, Nelson Johnson, Amanda Jowett, Barry Keith, Blaine Knight, Michael Luzzio, Robert Mook, Robert Mullin, Doris Murray, Martin Page, Kim Petrov, James Onori, Nelson Rhodes, David Rusnak, Robert Shaw, Lisa Shewchuk, Archie Sinhababu, Debby Smith, Kathryn Smith, Jeremy Stables, Colin Stubberfield, Neil Spector, Sarva Tadepalli, Peter Topley, Dana Vanderwall, Jim Veal, Ann Walker, Paul Wissel, Edgar Wood, Yue-Mei Zhang.

KEY REFERENCES

P. Fedi and S. A. Aaronson, *Signaling Networks and Cell Cycle Control: The molecular Basis of Cancer and Other Diseases*, ed. J. S. Gutkind, Humana Press, Inc., Totowa, New Jersey, 2000, Chapter 2, pp. 27–38.

E. R. Wood, A. T. Truesdale, O. B. McDonald, D. Yuan, A. Hassell, S. H. Dickerson, B. Ellis, C. Pennisi, E. Horne, K. Lackey, K. J. Alligood, D. W. Rusnak, T. M. Gilmer and L. A. Shewchuk, *Cancer Res.*, 2004, **64**(18), 6652.

P. J. Medina and S. Goodin, *Clin. Ther.*, 2008, **30**(8), 1426.

REFERENCES

1. (a) G. Manning, G. D. Plowman, T. Hunter and S. Sudarsanam, *Trends Biochem. Sci.*, 2002, **27**(10), 514; (b) P. Fedi and S. A. Aaronson, in *Signaling Networks and Cell Cycle Control: The molecular Basis of Cancer and Other Diseases*, ed. J. S. Gutkind, Humana Press, Inc., Totowa, New Jersey, 2000, Chapter 2, pp. 27–38; (c) G. J. Clark, J. P. O'Bryan and C. J. Der, in *Signaling Networks and Cell Cycle Control: The molecular Basis of Cancer and Other Diseases*, ed. J. S. Gutkind, Humana Press, Inc., Totowa, New Jersey, 2000, Chapter 12, pp. 213–230; (d) K. Parang and G. Sun, *Curr. Opin. Drug Discovery Dev.*, 2004, **7**(5), 617; (e) J. Tang, L. M. Shewchuk, H. Sato, M. Hasegawa, Y. Washio and N. Nishigaki, *Bioorg. Med. Chem. Lett.*, 2003, **13**(18), 2985.
2. J. R. Sainsbury, J. R. Farndon, G. K. Needham, A. J. Malcolm and A. L. Harris, *Lancet*, 1987, **329**(8547), 1398–1402.
3. I. Alroy and Y. Yarden, *FEBS Letts.*, 1997, **410**, 83.
4. D. J. Riese and D. F. Stern, *BioEssays*, 1998, **20**(1), 41.
5. N. Moghal and P. W. Sternberg, *Curr. Opin. Cell Biol.*, 1999, **11**(2), 190.
6. B. Brandt, U. Vogt, C. M. Sclotter, C. Jackisch, R. Werkmeister, M. Thomas, M. von Eiff, U. Bosse, G. Assmann and K. S. Zänker, *Gene*, 1995, **159**, 35.
7. H. Modjtahedi, K. Affleck, C. Stubberfield and C. Dean, *Int. J. Clin. Oncol.*, 1998, **13**, 335.

8. F. Pasleau, M. Grooteclaes and R. Gol-Winkler, *Oncogene*, 1993, **8**, 849.

9. R. A. Harris, T. J. Eichholtz, I. D. Hiles, M. J. Page and I. O'Hare, *Int. J. Cancer*, 1999, **80**(3), 477.

10. B. R. Keith, P. P. Allen, K. J. Alligood, R. M. Crosby, K. Lackey, T. M. Gilmer and R. J. Mullin, *Proc. Am. Assoc. Cancer Res.*, 92nd Annual Meeting, 2001.

11. R. J. Mullin, K. J. Alligood, P. P. Allen, R. M. Crosby, B. R. Keith, K. E. Lackey, T. M. Gilmer, R. J. Griffin, D. M. Murray and S. M. Tadepalli, *Proc. Am. Assoc. Cancer Res.*, 92nd Annual Meeting, 2001.

12. D. W. Rusnak, K. Affleck, S. G. Cockerill, C. Stubberfield, R. Harris, M. Page, K. J. Smith, S. B. Guntrip, M. C. Carter, R. J. Shaw, A. Jowett, J. Stables, P. Topley, E. R. Wood, P. S. Brignola, S. H. Kadwell, B. R. Reep, R. J. Mullin, K. J. Alligood, B. R. Keith, R. M. Crosby, D. M. Murray, W. B. Knight, T. M. Gilmer and K. E. Lackey, *Cancer Res.*, 2001, **61**(19), 7196.

13. D. W. Fry, A. J. Kraker, A. L. McMichael, A. Ambroso, J. M. Nelson, W. R. Leopold, R. W. Connors and A. J. Bridges, *Science*, 1994, **265**(5175), 1093.

14. C. A. Lipinski, F. Lombardo, B. W. Dominy and P. J. Feeney, *Adv. Drug Delivery Rev.*, 1997, **23**(1–3), 3.

15. A. L. Hopkins, C. R. Groom and A. Alex, *Drug Discovery Today*, 2004, **9**(10), 430.

16. P. S. Brignola, K. Lackey, S. H. Kadwell, C. Hoffman, E. Horne, H. L. Carter, J. D. Stuart, K. Blackburn, M. B. Moyer, K. J. Alligood, W. B. Knight and E. R. Wood, *J. Biol. Chem.*, 2002, **277**(2), 1576.

17. J. E. Shaffer, K. K. Adkison, K. Halm, K. Hedeen and J. Berman, *J. Pharm. Sciences*, 1999, **88**(3), 313.

18. G. S. Cockerill, C. Stubberfield, J. Stables, M. Carter, S. Guntrip, K. Smith, S. McKeown, R. Shaw, P. Tapley, L. Thomsen, K. Affleck, A. Jowett, D. Hayes, M. Willson, P. Woollard and D. Spalding, *Bioorg. Med. Chem. Lett.*, 2001, **11**(11), 1401.

19. D. W. Rusnak, K. Lackey, K. Affleck, E. R. Wood, K. J. Alligood, N. Rhodes, B. Keith, D. M. Murray, R. J. Mullin, W. B. Knight and T. M. Gilmer, *Mol. Cancer Therapeutics*, 2001, **1**(2), 85.

20. K. G. Petrov, Y.-M. Zhang, M. Carter, G. S. Cockerill, S. Dickerson, C. Gauthier, Y. Guo, D. W. Rusnak, R. A. Mook, A. L. Walker, E. R. Wood and K. E. Lackey, *Bioorg. Med. Chem. Lett.*, 2006, **16**, 4686.

21. (a) G. S. Cockerill and K. Lackey, *US Pat.*, WO2001004111, 2001; (b) M. C. Carter, G. S. Cockerill, S. B. Guntrip, K. E. Lackey and K. J. Smith, *US Pat.*, WO 9935146, 1999; (c) G. S. Cockerill, M. C. Carter, S. B. Guntrip and K. J. Smith, *US Pat.*, WO9802434, 1998.

22. M. D. Gaul, Y. Guo, K. Affleck, G. S. Cockerill, T. M. Gilmer, R. J. Griffin, S. Guntrip, B. R. Keith, W. B. Knight, R. J. Mullin, D. M. Murray, D. W. Rusnak, K. Smith, S. Tadepalli, E. R. Wood and K. Lackey, *Bioorg. Med. Chem. Lett.*, 2003, **13**(4), 637.

23. (a) J. Stamos, M. X. Sliwkowski and C. Eigenbrot, *J. Biol. Chem.*, 2002, **277**, 46265; (b) Y. M. Zhang, S. Cockerill, S. B. Guntrip, D. Rusnak, K. Smith, D. Vanderwall, E. Wood and K. Lackey, *Bioorg. Med. Chem. Lett.*, 2004, **14**, 111.

24. K. E. Lackey, *Curr. Top. Med. Chem.*, 2006, **6**(5), 435.

25. M. A. Fabian, W. H. Biggs, D. K. Treiber, C. E. Atteridge, M. D. Azimioara, M. G. Benedetti, T. A. Carter, P. Ciceri, P. T. Edeen, M. Floyd, J. M. Ford, M. Galvin, J. L. Gerlach, R. M. Grotzfeld, S. Herrgard, D. E. Insko, M. A. Insko, A. G. Lai, J.-M. Lelias, S. A. Mehta, Z. V. Milanov, A. M. Velasco, L. M. Wodicka, H. K. Patel, P. P. Zarrinkar and D. J. Lockhart, *Nat. Biotechnol.*, 2005, **23**(3), 329.

26. R. B. Lichtner, A. Menrad, A. Sommer, U. Klar and M. R. Schneider, *Cancer Res.*, 2001 **61**, 5790.

27. E. R. Wood, A. T. Truesdale, O. B. McDonald, D. Yuan, A. Hassell, S. H. Dickerson, B. Ellis, C. Pennisi, E. Horne, K. Lackey, K. J. Alligood, D. W. Rusnak, T. M. Gilmer and L. A. Shewchuk, *Cancer Res.*, 2004, **64**(18), 6652.

28. S. Blencke, A. Ulrich and H. Daub, *J. Biol. Chem.*, 2003, **278**, 15435.

29. W. Xia, R. J. Mullin, B. R. Keith, L.-H. Liu, H. Ma, D. W. Rusnak, G. Owens, K. J. Alligood and N. L. Spector, *Oncogene*, 2002, **21**(41), 6255.

30. R. J. Mullin, *Abstracts of Papers, 226th ACS Natl. Mtg.*, NY, USA, 2003, American Chemical Society.

31. A. K. Bence, E. B. Anderson, M. A. Halepota, M. A. Doukas, P. A. DeSimone, G. A. Davis, D. A. Smith, K. M. Koch, A. G. Stead, S. Mangum, C. J. Bowen, N. L. Spector, S. Hsieh and V. R. Adams, *Invest. New Drugs*, 2005, **23**(1), 39.

32. N. L. Spector, W. Xia, H. Burris, E. C. Hurwitz, A. Dees, B. Dowlati, B. O'Neil, P. K. Overmoyer, K. L. Marcom, D. A. Blackwell, K. M. Smith, A. Koch, S. Stead, M. J. Mangum, L. Ellis, A. K. Liu, T. Man, M. Bremer, J. Harris and S. Bacus, *J. Clin. Oncology*, 2005, **23**(11), 2502.

33. B. Corkery, N. O'Donovan and J. Crown, *OncoTargets Ther.*, 2008, **1**, 21.

34. J. R. Molina, S. H. Kaufmann, J. M. Reid, M. Gálvez-Peralta, R. Friedman, K. S. Flatten, K. M. Koch, T. M. Gilmer, R. J. Mullin, R. C. Jewell, S. J. Felten, S. Mandrekar, A. A. Adjei and C. Erlichman, *Cancer Res.*, 2008, **14**(23), 7900.

35. A. Ulhoa-Cintra, L. Greenberg and C. E. Geyer, *Curr. Oncol. Rep.*, 2008, **10**(1), 10.

36. P. J. Medina and S. Goodin, *Clin. Ther.*, 2008, **30**(8), 1426.

37. S. R. D. Johnston and A. Leary, *Drugs Today*, 2006, **42**(7), 441.

CHAPTER 28

"Daring to be Different": The Discovery of Ticagrelor

BOB HUMPHRIES*[a] AND JOHN DIXON[b]

[a] VisionRealisation Ltd, Leicestershire, UK; [b] JD International Consulting Ltd, Leicestershire, UK
*E-mail: bob.humphries@visionrealisation.co.uk

28.1 PROLOGUE

On Christmas Eve 2010, the first pack, anywhere in the world, of a new oral anti-platelet agent, ticagrelor (BRILINTA®, BRILIQUE™), was dispensed in Blackpool. A fitting event for a transformational medicine imagined, designed and initially developed in the UK: a compound that, in the pivotal Phase III study (PLATO)[1] involving more than 18 000 patients, demonstrated that, for every 72 patients with acute coronary syndromes (ACS) treated for 12 months with ticagrelor instead of clopidogrel, one more person gets to live. This case study tries to give some feel for the Discovery story behind ticagrelor. It is a story that spans three decades, so some details are necessarily sketchy, being dependent on the authors' diminishing recall. It is, of course, a story of the science behind ticagrelor but it also illustrates that ground breaking science applied by excellent scientists does not guarantee success in our industry. Without the science there is nothing, but success also requires people and teams with imagination, vision, and persistence to generate a momentum that can withstand organisational change and shifting fashion and priorities. Importantly, success comes from an unwavering focus on why we do what we do—the belief that, if we work on the right things, do the right experiments and make the right judgements, we can make a difference to patients' lives.

This chapter is dedicated to former colleagues at the Charnwood R&D site in Loughborough, Leicestershire through the Fisons, Astra and AstraZeneca years. Be proud of the fact that, at the time of writing, a medicine that came from the innovative "Can Do" Charnwood spirit is, in Leicestershire alone, being used to treat hundreds of patients each year.

28.2 ACUTE CORONARY SYNDROMES (ACS)—A STICKY PROBLEM

In Europe, Brilique, co-administered with acetylsalicylic acid (aspirin, ASA), is indicated for the prevention of atherothrombotic events in adult patients with Acute Coronary Syndromes

The Handbook of Medicinal Chemistry: Principles and Practice
Edited by Andrew Davis and Simon E Ward
© The Royal Society of Chemistry 2015
Published by the Royal Society of Chemistry, www.rsc.org

(unstable angina, non ST elevation Myocardial Infarction [NSTEMI] or ST elevation Myocardial Infarction [STEMI]); including patients managed medically, and those who are managed with percutaneous coronary intervention (PCI) or coronary artery by-pass grafting (CABG).[2]

ACS represents a life-threatening manifestation of atherosclerosis.[3] It is usually precipitated by acute thrombosis induced by a ruptured or eroded atherosclerotic coronary plaque, with or without concomitant vasoconstriction, causing a sudden and critical reduction in blood flow.

However ACS manifests in a given patient, the underlying acute pathophysiology is driven by the activation and aggregation of platelets in one or more damaged/narrowed coronary arteries, resulting in partial or complete thrombotic occlusion of the artery and intermittent or complete interruption of the blood supply to the heart muscle. Mechanical interventions such as PCI, which involves dilatation of the artery with a balloon catheter and insertion of a stent to maintain vessel patency, reduce the risk of death or myocardial infarction in ACS patients. However, the intervention itself causes considerable disruption and damage to the blood vessel wall that can also lead to thrombotic complications. The pivotal role of platelets in this process has led to the adoption of anti-platelet strategies for both the treatment of ACS and for prevention of the complications of PCI.[3] The main mechanisms that have been targeted are visualised in the simplified schematic of platelet activation and aggregation in Figure 28.1. The cartoon highlights two mechanisms, thromboxane (TxA_2) receptor activation and $P2Y_{12}$ receptor activation (by adenosine diphosphate, ADP), the understanding of which has led to dual anti-platelet therapy with aspirin and an oral $P2Y_{12}$ antagonist becoming the mainstay of anti-platelet therapy in ACS.

Prior to approval of ticagrelor and its inclusion in ACS treatment guidelines, only indirect (pro-drug) oral $P2Y_{12}$ inhibitors were available, and clinical practice guidelines recommended dual anti-platelet treatment with aspirin and clopidogrel. However, the efficacy of clopidogrel is hampered by the slow and variable transformation of the prodrug to the active metabolite,

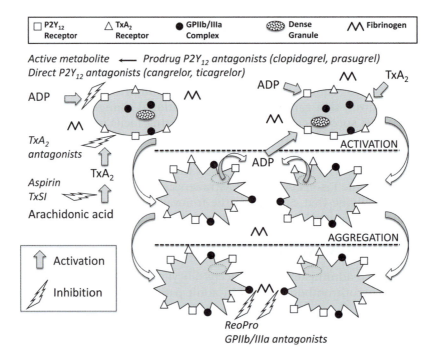

Figure 28.1 Key platelet activation pathways.

modest and variable platelet inhibition, an increased risk of bleeding, and an increased risk of stent thrombosis and myocardial infarction in patients with a poor pharmacodynamic response.[4] As compared with clopidogrel, prasugrel, another thienopyridine prodrug, has a more consistent and pronounced inhibitory effect on platelets, resulting in a lower risk of myocardial infarction and stent thrombosis, but is associated with a higher risk of major bleeding in patients with an acute coronary syndrome who are undergoing percutaneous coronary intervention (PCI).[4] Bleeding risk also reflects the fact that both clopidogrel and prasugrel are irreversible inhibitors of the $P2Y_{12}$ receptor and restoration of normal haemostasis requires generation of new platelets.

Ticagrelor provides a new therapeutic option in ACS as the first direct acting, reversible $P2Y_{12}$ receptor antagonist.[5] In addition, ticagrelor and other potent, selective $P2Y_{12}$ antagonists from the Charnwood $P2Y_{12}$ antagonist project acted as precision pharmacological tools to help explain the relative roles of $P2Y_{12}$ and $P2Y_1$ receptors on platelets,[6] and how targeting the $P2Y_{12}$ receptor can lead to profound clinical benefit.[7] Adenosine diphosphate (ADP), the endogenous agonist at the $P2Y_{12}$ receptor, is an important primary stimulus of sustained platelet activation and is also released from dense granules of platelets activated by ADP or other stimuli.[8–10] Targeting this pathway inhibits platelet activation and, consequently, platelet aggregation, dense and α-granule secretion and further pro-coagulant activity. Thus, the ADP/$P2Y_{12}$ axis plays an important role in amplifying and sustaining platelet activation initiated by other pathways, leading to stable platelet-rich thrombus generation.[11,12] Consequently, blocking the $P2Y_{12}$ receptor has important inhibitory effects on overall platelet function regardless of the initial activating stimuli.[13]

So, with pivotal, definitive clinical studies completed, ticagrelor available to patients, further studies in progress, ADP/$P2Y_{12}$ pharmacology unravelled and the shortcomings of thienopyridines fully understood, it is obvious now that the $P2Y_{12}$ receptor is an excellent therapeutic target—but, how did things look 25 years ago?

28.3 "I WOULDN'T START THERE"

At the start of any long and difficult journey into the unknown there are always plenty of reasons not to take the first step. In the case of ticagrelor, there are a host of reasons why it might never have existed (see Figure 28.2). In the late 1980s, when the story started, there was a surge of interest in anti-platelet therapy for the prevention of thrombotic events. Aspirin had transformed the outlook for patients with coronary artery disease,[14] leading to substantial efforts across the industry to identify alternative, more selective means (thromboxane (TxA2) antagonists, thromboxane synthase inhibitors (TxSI)) of inhibiting generation or effects of products of the cyclooxygenase pathway (see Figure 28.1). The other area pursued by most big Pharma active in the thrombosis field was to develop antagonists of the fibrinogen receptor (the glycoprotein IIb/IIIa complex, GPIIb/IIIa) on platelets, based on the rationale that this approach would deliver maximal anti-platelet efficacy by blocking the final common pathway in platelet aggregation, namely the cross-bridging of individual platelets by fibrinogen to form a platelet rich thrombus.[15,16] Against this background, with aspirin established, TxA_2 antagonists and TxSIs in development, GPIIb/IIIa antagonists emerging, the idea of following another single mediator approach by targeting the $P2Y_{12}$ (known then as P2T) receptor on platelets was, on the face of it, counter-intuitive, particularly since the only chemical starting point was adenosine triphosphate (ATP) for which if asked "where would you start to develop a new drug?" the answer would be "…well, I wouldn't start there." This view was easily compounded by a risk aversion to doing something different rather than follow the logic that, if everyone else is doing something then it must be the right thing—so we should do that too.

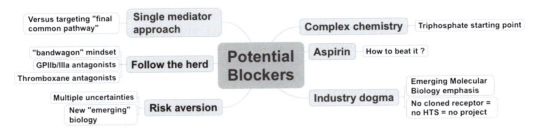

Figure 28.2   Potential blockers (real & perceived) to project initiation.

28.4   "YOU HAVE TO START SOMEWHERE…"—THE ROLE OF CANGRELOR

…and it started in the Fisons Pharmaceuticals research laboratories on the outskirts of Loughborough in the UK, which later became the Charnwood R&D site within AstraZeneca. The geographical location and resultant connections and partnerships are a big part of the ticagrelor success story.

The story started with the coming together in the late 1980s/early 1990s of two areas of research at the Charnwood site. Firstly, a naïve assessment, with no legacy in the platelet field and, therefore, no preconceptions, of where the unmet need for anti-platelet therapy might lie. Here, the geography came into play. Firstly, with the Glenfield hospital cardiology group close by, it was easy to talk to the then Professor of Cardiology, David de Bono, and obtain his insights on what would make a difference for him. At that time, the intravenous inhibitors of the platelet GPIIb/IIIa complex were becoming available for use as adjunctive therapy in coronary artery interventions.[15,16] However, it was also apparent that, with such a mechanism, there was a delicate balance between beneficial effects on thrombosis and increased bleeding risk, particularly when effects were slowly reversible. The second location factor was the proximity to Queens Medical Centre in Nottingham and the invitation to join a Platelet Discussion Group run by Professor Stan Heptinstall. This was to become a long lasting collaboration, later also involving Rob Storey who went on to become Professor of Cardiology at Sheffield and one of the lead investigators in both the cangrelor and ticagrelor clinical programmes.

From these, and other discussions, a target product profile (TPP) was identified requiring an intravenous product with rapid onset of effect, high anti-platelet efficacy and selectivity and rapid offset.

Alongside the unmet need discussions, an assessment of the mechanisms that might provide high anti-platelet efficacy led to a conclusion that blocking platelet aggregation induced by ADP held the most interest. Although, on the face of it, this was another single mediator approach, close scrutiny of the literature provided the clues that ADP could play an important amplifying role in the platelet response to most stimuli, particularly due to its presence and release from storage granules in activated platelets.[10] In addition, this was a period when clinical data were emerging for the thienopyridines ticlopidine, and later clopidogrel.[17] These orally-active compounds inhibit ADP induced platelet aggregation but in an indirect (as prodrugs), irreversible manner and, therefore, did not address the TPP we had identified.

The second area of research was a broad programme aimed at finding novel chemicals (selective receptor agonists/antagonists) to determine the importance of different subtypes of P2 receptors in human disease and to understand the therapeutic opportunities. Without this broad approach it is unlikely that the $P2Y_{12}$ antagonist project would have started— in isolation and with the challenges outlined above it is difficult to see that the substantial Medicinal Chemistry effort required would have been supported. In the event, much of the early chemistry was quite broad to support an overall exploration of the then known subtypes of P2 receptor.

The anti-platelet and P2 receptor themes centred on ADP came together with the knowledge that, while ADP was the endogenous agonist (stimulator) of the $P2Y_{12}$ receptor, uniquely located on platelets, adenosine triphosphate (ATP) was a natural antagonist at this receptor.[18] ATP was, however, a challenging starting point when attempting to design a potent, selective $P2Y_{12}$ antagonist for acute use: some properties (short half-life, high aqueous solubility) were ideally suited for this purpose; other properties (metabolic breakdown to the natural $P2Y_{12}$ agonist (ADP), lack of selectivity versus other P2 receptors) were problematic. However, Medicinal Chemistry and Pharmacological clues were there from work conducted at the Universities of Sheffield[19] and Middlesex[20,21] showing that this natural antagonist could be modified to yield stabilised ATP analogues with selectivity for the $P2Y_{12}$ receptor. Most of the pieces of the jigsaw existed—what the early $P2Y_{12}$ project within Fisons Pharmaceuticals did was to bring these pieces together: unmet medical need; tailored TPP; novel chemistry; integrative pharmacology; a strong Experimental Medicine capability—to design, deliver and develop AR-C69931MX (cangrelor),[22] the first direct acting, reversible intravenous $P2Y_{12}$ antagonist to progress into studies in patients.

Cangrelor is an important potential product in its own right, now under development by The Medicines Company, who specialise in acute care therapies. It is also an important component of the ticagrelor story—cangrelor and related compounds[23,24] removed the biological doubt and proved that targeting a "single mediator" approach could provide a broad spectrum profile. It led to a position where the $P2Y_{12}$ field was largely defined by AstraZeneca compounds, research, publications and collaborations.

So, without cangrelor, ticagrelor would not have existed. Even with the compelling evidence being generated with cangrelor, internal scepticism about being "different" remained; without it the concept of an oral direct acting $P2Y_{12}$ receptor antagonist would not have seen the light of day. In the event, innovative spirit shone through and the journey from cangrelor to ticagrelor started.

28.5 TOWARD TICAGRELOR

A crucial part of the story at this stage was the growing understanding that not all patients respond well to the thienopyridine (indirect) $P2Y_{12}$ antagonists exemplified by clopidogrel.[25] The need for metabolic conversion to an active metabolite introduced a variability not seen with cangrelor and so the concept that a direct acting $P2Y_{12}$ antagonist could fully realise the potential of the $P2Y_{12}$ antagonist mechanism, by providing more complete and consistent inhibition of platelet activation than the thienopyridines, was born. This became a differentiating thread for ticagrelor that linked basic science and preclinical data, through the Phase I and Phase II clinical trials and through, ultimately, to improved clinical outcomes in the PLATO study.

But... to rewind to the start of the oral $P2Y_{12}$ antagonist programme, here, with the pharmacological rationale gaining growing credence, the challenge lay in Medicinal Chemistry and in DMPK—how to move from a molecule such as cangrelor, with properties ideal for its intended clinical use in the acute setting, but completely incompatible with that of an orally active therapy for chronic treatment—to something that could go head-to-head with clopidogrel. This challenge was exacerbated by the fact that, at this time, the $P2Y_{12}$ receptor remained defined only by its function and its pharmacology.[26] The structure of the receptor was not known, and it was yet to be confirmed as a P2Y subtype with the "12" designation. Consequently, no cloned and expressed system (or suitable radioligand) was available to support high throughput screening (the structural identity and cloning of the $P2Y_{12}$ receptor was not published until 2001[27]—2 years after selection of ticagrelor as a candidate drug).

Faced with these challenges it would have been easy to walk away. Instead, the Project Team, strongly supported by Astra research management, maintained confidence and commitment and a belief that we could succeed. Progress was dependent on a relatively low throughput, but highly efficient and informative, functional screen (ADP induced aggregation of human washed platelets) and an empirical, hypothesis-driven synthesis/screening strategy. This approach resulted in quite long periods of small incremental steps interspersed with a handful of quite unanticipated transformational structural alterations. In all, from the inception of activities toward an oral compound to identification of the first potent, orally-bioavailable P2Y$_{12}$ antagonist represented 3 years of effort. Two years later ticagrelor was nominated as a Candidate Drug. Below, this journey is broken down into what, with the benefit of hindsight, can be seen as distinct phases of the story, each enabled by a landmark finding. Outlined for each phase are the key scientific steps taken toward the ultimate goal and highlights of the key themes, learnings and ways of working that enabled success. A common format Figure is provided summarising the following for each phase:

- The key challenge/issues faced.
- The Chemistry strategy pursued.
- The focus of the Biology (Pharmacology/DMPK).
- The key steps forward achieved.

28.5.1 Where to Start?

Where do you start when the starting point is an ATP analogue and the goal is a potent, selective, orally-bioavailable medicine? As mentioned, without access to a high throughput screen, the option of throwing away the template and finding more "druggable" hits was not available. Therefore, the only approach that could be taken was, with cangrelor as the template, to progressively explore whether the main impediments to this goal (high molecular weight, high polarity, multiple charges), and structural features not consistent with or desirable in a molecule for chronic use (adenine ring, glycosidic bond), could be modified or replaced while still retaining potency (Figure 28.3).

The first step recognised that, in cangrelor, an analogue of the endogenous P2Y$_{12}$ antagonist, ATP, the terminal phosphate group was essential for antagonism. The approach taken was, therefore, to find alternative acidic groups which could mimic the polyphosphate chain and particularly the C-phosphate unit of ATP. This initial strategy led to the discovery of a series of aspartic acid-derived di-carboxylic acids. Initially, with retention of the adenine (X = carbon) and ribose (Y = oxygen) rings, the highest attainable potency for inhibition of ADP-induced aggregation of human washed platelets was a pIC$_{50}$ of 7.0, some 300-fold less potent than cangrelor (pIC$_{50}$ = 9.4). However, we had completely replaced the triphosphate and achieved P2Y$_{12}$ antagonist potency and selectivity substantially higher than ATP itself (pIC$_{50}$ = 3.5). There then followed two of the landmark events of the ticagrelor story: introduction of triazolopyrimidine (X = N) as an isostere of purine and replacement of the ribose oxygen with carbon. These two heroic pieces of medicinal chemistry resulted in a compound with comparable P2Y$_{12}$ antagonist potency (pIC$_{50}$ = 9.5) to that of cangrelor but with considerable progression away from the triphosphate starting point, and optimism that further progress could be made. In particular, the triazolo benefit in affinity was seen in all cases and allowed previously impossible changes to be applied. Selectivity amongst P2 receptors was always maintained.

One of the key themes emerging from this phase of the project was that, with the difficulty of the chemistry, it was compound synthesis that was the rate-limiting step and the biology effort could initially be quite streamlined, predominantly focused on the primary assay for P2Y$_{12}$

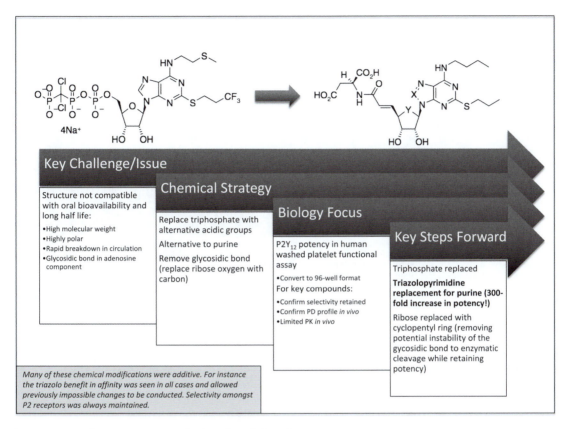

Figure 28.3 Where to start – replacing the triphosphate.

antagonist potency. This allowed the pharmacology team to complete more complex profiling of cangrelor, including building a strong differentiation story *vs.* GPIIb/IIIa antagonists in a dog model of thrombosis.[28] In turn, this built a database and expertise that subsequently helped support the oral P2Y$_{12}$ programme. However, as progress began to be made with achieving the potency goal, three things became apparent: 1) That chemistry output would need to increase if we were to deliver an optimised compound(s) in an acceptable time frame; 2) that this would require a higher capacity primary screen; and 3) that, as we moved toward more "druggable" molecules, we would need to increase capability and capacity for investigating drug metabolism and pharmacokinetics.

28.5.2 Taking Charge

Replacement of the triphosphate and the potency enhancement provided by triazolopyrimidine were encouraging steps. However, it was clear that the di-carboxylic acid compounds, although potent, were subject to rapid biliary clearance and not orally-bioavailable. Clearly, to achieve the project objective, we needed to reduce size, complexity and charge. This was indeed achieved, firstly by moving to mono-carboxylic acids and then with the landmark of achieving P2Y$_{12}$ affinity in a neutral compound (Figure 28.4), with a reasonable pIC$_{50}$ of 7.7. Importantly, this compound was not subject to biliary clearance. Instead, metabolite identification from *in vivo* DMPK experiments indicated that it was subject to hepatic metabolism. This observation moved us significantly closer to the goal of a long acting oral compound since it offered a much higher

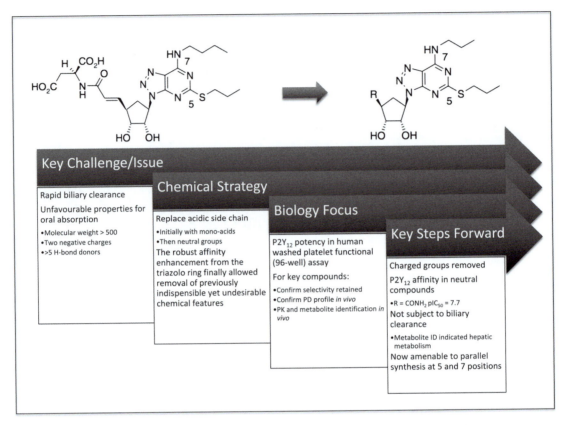

Figure 28.4 Taking charge – affinity and "normal" metabolism in neutral compounds.

likelihood of being able to make PK and dose predictions to man, based on a combination of *in vitro* metabolism data and *in vivo* metabolism and PK studies in the rat and dog.

Importantly, these less complex, neutral molecules now made it possible to increase the medicinal chemistry output, in particular by accessing the then emerging technology of parallel synthesis, enabling multiple reactions and permutations of substituent changes to be conducted in 96-well plates. In turn this was supported by conversion of the primary P2Y$_{12}$ screen to a 96-well format, initially still as a functional assay, but later as a radiolabel displacement assay, requiring identification and synthesis of a novel radioligand and development and validation of the assay. These changes exemplified another theme in the project—the flexibility and ingenuity of the team to continually evolve the key screens to adapt to the changing needs of the project and to access novel technologies, while still maintaining project delivery.

28.5.3 Parallel Universe

At this point in the project great strides had been made and considerable distance achieved from the original triphosphate starting point. However, these first neutral compounds were far from optimised—active yes—but not sufficiently so, and still with many deficits that precluded oral bioavailability and long duration of action (Figure 28.5). Here we encounter another landmark for the project—the identification, through parallel synthesis and the 96-well format primary screen, of the potency-enhancing phenylcyclopropylamine substituent that, importantly, also resulted in the first orally-active compound. This was the breakthrough that convinced the team that we were very close to candidate drug (CD) quality molecules. The growing understanding of the properties and challenges associated with the core structure also enabled refinement of the criteria required for a compound to be considered as a serious CD contender (Table 28.1).

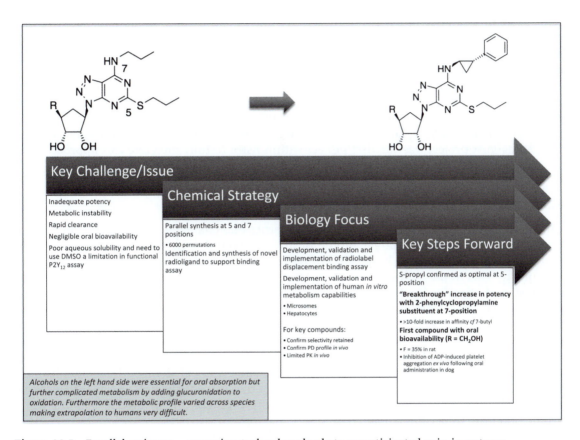

Figure 28.5 Parallel universe – emerging technology leads to unanticipated gain in potency.

Table 28.1 Outline selection criteria during prenomination phase of oral P2Y12 antagonist project.

Attribute	Criterion
Potency	$IC_{50} \leq 10$ nM *vs.* ADP-induced aggregation of human washed platelets *in vitro*. ID_{100} (inhibition of platelet aggregation *ex vivo* in dog) < 2 mg·kg^{-1} po bid.
Selectivity and specificity *in vitro*	At least 100-fold selectivity for the $P2Y_{12}$-receptor over other P2 receptor subtypes and unrelated mechanisms.
Duration	$t_{1/2}$ sufficient to support twice daily dosing in man: exact value dependent on acceptable bleeding time prolongation at predicted plasma C_{max} and interspecies scaling.
Oral bioavailability	$> 30\%$: screening target $> 30\%$ (rat); $> 50\%$ (dog).
Hypothesis testing *in vivo*	Inhibition of ADP-induced platelet aggregation *ex vivo* after oral dosing. Abolition of thrombosis in the anaesthetised dog cyclic flow reduction (CFR) model following intravenous dosing - confirmed with selected CD *via* GI route (intraduodenal).
Therapeutic index	Anti-thrombotic:bleeding time ratio better than GPIIb/IIIa antagonists.

However, there remained another twist in the tale. Namely, that the growing capability in techniques for assessing metabolism *in vitro* in liver microsome preparations and in hepatocytes from rat, dog and human, had identified significant qualitative and quantitative species differences in clearance routes and sites and extent of metabolism, making it difficult to predict, with any degree of certainty, the likely bioavailability, half-life and required dose in man. The major route of clearance in the rat of these lipophilic, neutral compounds was oxidative metabolism whereas clearance was higher as a fraction of hepatic blood flow in the dog and progressed principally *via* glucuronidation. Preliminary data from human hepatocytes suggested that human clearance was also likely to be *via* glucuronidation within the carbocycle. These observations indicated the pivotal role of *in vitro* metabolic screens and the emphasis in particular on robust *in vitro* human data. The DMPK complexities moved the Project into a phase where the predominant structure-activity information driving compound design was human *in vitro* metabolism data. There was little general precedent for this at the time, and the P2Y$_{12}$ antagonist project was the first project within Astra to fully integrate this approach into the synthesis/screening cycle and compound optimisation.

28.5.4 The Human Factor

At this stage, with growing understanding of properties and challenges of the lead compounds, it became apparent that the target profile of any compound suitable for progression would include having a predicted PK/PD profile consistent with twice daily dosing in man at a dose of <0.2 mg/kg. This in part reflected the recognised chemical complexity of the lead compounds and likely cost of goods implications, based on the knowledge available and the assumptions applied at that time. The ability to make predictions of human PK and dose with as much confidence as possible became paramount, and the main focus of the project moved to further optimisation of compounds based on improving metabolic stability in human microsomes and hepatocytes (Figure 28.6). Importantly, fluorination of the aromatic substituent not only reduced the oxidative metabolism. Despite being a remote chemical change, the metabolism on the opposite side of the molecule (glucuronidation of the primary alcohol) was also reduced by this change. This dual effect brought sufficient improvements in metabolic stability for a chronic oral drug to become a reality.

Comparison of intrinsic clearance values obtained in rat hepatocytes and microsomes *in vitro*, with clearance values obtained *in vivo*, provided support for the possibility of predicting the clearance of compounds in man *in vivo* from intrinsic clearance values obtained in human microsomes and hepatocytes *in vitro*. In addition, comparison of anti-aggregatory potency data in dog blood *in vitro* with results from combined PK/PD experiments in the dog *in vivo* supported the use of potency in human blood *in vitro* as a predictor of potency *ex vivo*. This was also supported from clinical data obtained with cangrelor. Using the predicted clearance value *in vivo*, the predicted anti-aggregatory potency (IC$_{90}$) of the compound in human blood *ex vivo* and other factors, it is then possible (Figure 28.7) to make an estimate of the dose required in man to meet the target profile (12 h full ($\geq$90%) inhibition of ADP-induced platelet aggregation measured *ex vivo*).

The use of the IC$_{90}$ value for inhibition of platelet aggregation in human blood was based on experience in the *in vivo* thrombosis model in the anaesthetised dog that was a key element of the pharmacological profiling of compounds. Developed, validated and refined throughout the course of the project, this complex model allowed a highly integrated assessment of the profile of a compound.[28] Within each experiment, the anti-thrombotic effect was evaluated against dynamic, platelet-mediated thrombosis visualised as cyclic reductions in blood flow (CFR) in the damaged, stenosed femoral artery of the anaesthetised dog. Bleeding time and ADP-induced platelet aggregation *ex vivo* were also measured. The robustness and stability of the model (experiments could be run for up to 12 hours) allowed full dose–response relationships to be

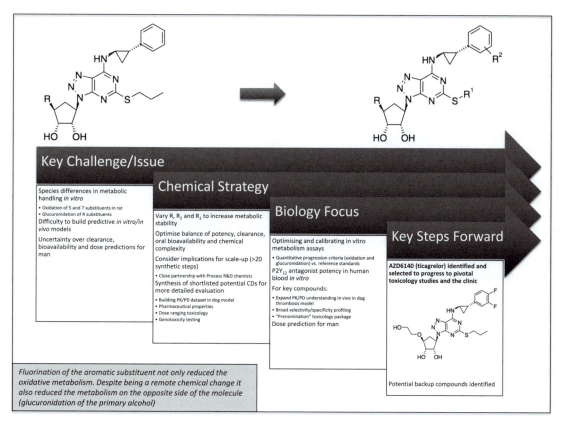

Figure 28.6 The human factor – robust *in vitro* metabolism assays enable human dose prediction.

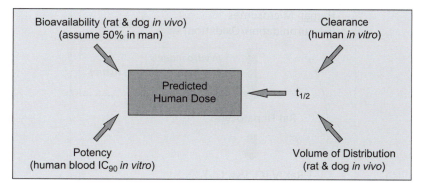

Figure 28.7 Predicting the human dose.

obtained for inhibition of thrombosis, inhibition of platelet aggregation and prolongation of bleeding time. As seen previously with cangrelor, the anti-thrombotic effect of compounds from this phase of the project was achieved with significantly less compromise of haemostasis (bleeding time prolongation) than observed with GPIIb/IIIa antagonists. For all anti-platelet agents tested, regardless of mechanism, complete inhibition of thrombosis required complete/near complete inhibition of platelet aggregation measured *ex vivo*—hence selection of the human blood IC_{90} as a component of the dose prediction.

The model also involved measurement of blood pressure, heart rate, ECG and blood flow in the non-stenosed femoral artery. Therefore, by continuing dose progression above the effective anti-thrombotic dose it was possible to also assess any potential for unwanted cardiovascular effects.

In the early stages of the project, with the focus on achieving potency in new chemical series, and reflecting the chemical complexity, it was usual for the first batch of compound synthesised to be just 10–30 mg. This was appropriate given that many compounds did not progress beyond the primary screen. However, for interesting compounds, this approach could lead to a delay of a few weeks for re-synthesis to support further progression. In the later (prenomination) final optimisation stage of the project, the majority of compounds synthesised were potent, selective, metabolically stable to varying extents, and orally bioavailable. With that in mind, tactics were changed and batch 1 of each compound was made with sufficient quantity to enable data-driven progression from primary screen, through PK and metabolism assessments and to the dog thrombosis model with no delay loop (Figure 28.8). To achieve this rapidly required very close team work and collaboration between all disciplines (medicinal chemistry, analytical and physical chemistry, drug metabolism and pharmacokinetics research, pharmaceutical sciences, pharmacology).

The efficiency that could be achieved is exemplified by experience with a close analogue of ticagrelor for which, from submission of batch 1 of the compound to availability of decision-making data from all the test systems shown took just 1 week.

28.5.5   Complexity of Science, Simplicity of Thought

Since the project was now in a rich vein of compounds and the screening load on these assays and the *in vivo* drug metabolism and pharmacokinetic assessment had become rate-limiting, tough decisions had to be made regarding compound progression. Tough decisions entailed the

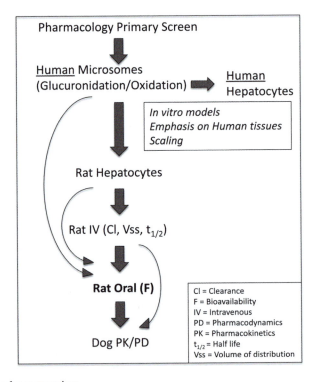

Figure 28.8   Compound progression.

whole project team being comfortable with and jointly owning pragmatic solutions, robust discussions and the making of considered judgement calls on sparse data. Examples of this highly collaborative behaviour included:

- Accepting that, if effort had been put into synthesising a batch 1 quantity sufficient for complete progression, most of the compound would not be used if it failed to progress beyond the primary screen.
- Accepting that we would make decisions to abandon series often on data on just a few representatives if the data indicated a negative trend.
- For many compounds, bypassing the rat intravenous component of progression and moving straight to oral dosing in the rat—if the PK profile after oral dosing was unfavourable that could be a stop decision and we would not necessarily go back to seek the explanation for a poor profile.

So we maintained an unrelenting focus on the ultimate goal of delivering an orally active P2Y$_{12}$ antagonist that could be progressed through safety studies and into the clinic. We only pursued compounds and data that moved the project closer to that goal. All experiments had to support a decision and all other considerations were distractions.

The ultimate achievement was identification of the compound that became ticagrelor.[5]

28.6 BITING THE BULLET

By this stage in the project it was becoming evident that clopidogrel (Plavix) in combination with aspirin would become standard of care in the ACS setting. Commitment to investment in the clinical development programme for ticagrelor would require a high level of confidence that a reversible, direct (non-prodrug) acting P2Y$_{12}$ antagonist would have a differentiated profile compared with clopidogrel. The more rapid onset and offset of effect would provide some advantages but it was clear that clinically-meaningful and reimbursable differentiation would require a significant improvement in efficacy. The scale of the PLATO study (more than 18 000 patients) demonstrates what it would ultimately take to prove this, so how could confidence be built at this early stage, even before the preclinical work to support dosing to man had commenced? The answer lay in the growing body of evidence for "poor responders" to clopidogrel, as assessed by measurement of ADP-induced platelet aggregation measured *ex vivo* following dosing of clopidogrel in healthy subjects or ACS patients. We hypothesised that these patients would remain at a high risk of thrombotic events since their platelets would still be responsive to ADP. An extension of the hypothesis was that if a direct acting P2Y$_{12}$ antagonist were to work pharmacodynamically in all subjects/patients, including those responding poorly to clopidogrel, then, in an appropriately designed and sized clinical study (such as PLATO) there would be a high level of confidence in seeing a significant improvement in efficacy across the study population. Consequently, we conducted a study in which eight healthy volunteers received the standard dose of clopidogrel (75 mg) over a period of 11 days. Blood samples were taken on days 0, 1, 2, 3 and 11. Each blood sample was split and ADP-induced platelet aggregation was measured either in the blood sample as taken or after addition *in vitro* of ticagrelor at its predicted therapeutic concentration. As illustrated in Figure 28.9, following 11 days clopidogrel treatment, in line with other studies, there was substantial variability in the degree of platelet inhibition observed. However, when ticagrelor was "spiked" into the samples, complete/near-complete inhibition was seen, even in blood from the subjects responding poorly to clopidogrel.

This small "translational" study in just eight subjects was prospectively defined as a GO/NO GO study for the project. The result represented a clear GO—we knew that, as long as adequate plasma levels of ticagrelor were achieved, we would see the same pharmacodynamic

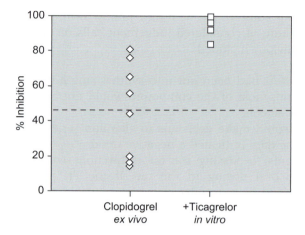

Figure 28.9   Platelet inhibition in blood samples from 8 subjects following 11 days dosing with clopidogrel with and without "spiking" with ticagrelor *in vitro*.

differentiation following oral dosing of ticagrelor in man. This was subsequently confirmed in
the Phase I and II clinical development programme.

What about the dose required to achieve this pharmacodynamic effect? How good was the
dose to man prediction? The approved dose of ticagrelor (as studied in the PLATO study) is a
loading dose of 180 mg (two tablets of 90 mg) followed by 90 mg twice a day for up to 12 months
(*i.e.*, a daily dose of 180 mg).[2] The dose prediction at the time of selection of ticagrelor as a
candidate drug was 0.2 mg/kg twice daily (14 mg daily dose based on a body weight of 70 kg)—a
five- to six-fold underestimate. Contributory factors to this difference are that the dose
prediction assumed a bioavailability of 50% (actual bioavailability 36%[2]) and a clearance of
<70 mL/min (actual clearance 362–511 mL/min with repeated twice daily dosing[29]). These
differences are probably not that surprising, given the human microsome and hepatocyte assays
were being developed and validated in parallel with project delivery, the embryonic nature of the
science of dose prediction at that time, and the number of assumptions built into the prediction
model. Importantly, as discussed in Chapter 8 of this book, information generated from the
ticagrelor clinical PK data allowed refinement of the prediction model for subsequent application to other AstraZeneca projects.

28.7   ENABLERS

As discussed above, there were certainly scientific learnings from the $P2Y_{12}$ antagonist project.
Perhaps more important though is to reflect on the human factors and ways of working that
enabled ultimate success despite all the challenges.

A characteristic of the $P2Y_{12}$ project team was the building of partnerships and ensuring
engagement with the wider project team, particularly colleagues who would be inheriting from
the Discovery team a compound with significant challenges: complex multistep synthesis; poor
solubility; potentially difficult formulation challenges; the need to consider non-standard toxicology species. One of the key factors that enabled seamless progression into early development was recognising those challenges early and building partnerships with the groups that
were going to inherit those challenges.

In particular we worked quite differently. We involved Process R&D, Formulation, Toxicology
and Experimental Medicine colleagues very early on in the project and they quickly became key
members of the extended project team. In particular this enabled the process research and
development chemists to make a head start on thinking about alternative routes and scale up.

Throughout the project lifetime, we remained true to our strategic vision and belief—we maintained momentum in the $P2Y_{12}$ antagonist field at a time when it was a difficult area to be working in. Many companies were moving out of novel antiplatelets, with one of the main reasons being that most had invested in the GPIIb/IIIa antagonist class, which, as mentioned previously, target the final common step in platelet aggregation. Targeting that mechanism proved quite successful in the acute hospital settings, with products that are providing benefit to patients. However, many of the orally active compounds with that mechanism advanced into Phase III clinical studies, but failed due to an increase in mortality[30]—for reasons still incompletely understood.

Overlaying all the Project-level factors discussed above was the critical importance of engagement and support at a senior management level. In the Astra years, Research Management had the vision to support initiation of the project, an understanding of complexity, and the patience to give the Team time and space to address complex challenges. Finally, a Phase III study of the scale and cost of PLATO study was a bold recommendation from the ticagrelor Global Product Team, and committing to it a courageous decision by the then leaders of AstraZeneca. As a result, they, and everyone associated with the $P2Y_{12}$ antagonist project, now have the satisfaction of knowing that, by daring to be different, we have been part of delivering a new medicine that, each day, is changing and saving patients' lives.

HINTS AND TIPS—INGREDIENTS FOR SUCCESS

In summary, the scientific and human ingredients for success were:

- Identification of a clear unmet need, patient population and target product profile.
- An unwavering focus on a clearly defined objective.
- Building, applying and translating the scientific knowledge of the mechanism to provide a seamless transition into the clinic using...
- ...a robust, translatable functional assay that allowed progress in the absence of what would typically now be pre-requisites for a project start (cloned receptor, high throughput screen).
- A highly integrated approach, harnessing capabilities in Medicinal Chemistry, Integrative Pharmacology, Experimental Medicine and, crucially, Drug Metabolism.
- Close teamwork enabling rapid decisions to be made, with sparse data, with an unrelenting focus on delivery.
- Building partnerships!
- "Daring to be different" and a "Can Do" attitude.
- Commitment, Courage and Conviction.

KEY REFERENCES

L. Wallentin, R. C. Becker, A. Budaj, C. P. Cannon, H. Emanuelsson, C. Held, J. Horrow, S. Husted, S. James, H. Katus, K. W. Mahaffey, B. M. Scirica, A. Skene, P. G. Steg, R. F. Storey, R. A. Harrington, for the PLATO Investigators, *N. Engl. J. Med.*, 2009, **361**, 1045.

B. Springthorpe, A. Bailey, P. Barton, T. N. Birkinshaw, R. V. Bonnert, R. C. Brown, D. Chapman, J. Dixon, S. D. Guile, R. G. Humphries, S. F. Hunt, F. Ince, A. H. Ingall, I. P. Kirk, P. D. Leeson, P. Leff, R. J. Lewis, B. P. Martin, D. F. McGinnity, M. P. Mortimore, S. W. Paine, G. Pairaudeau, A. Patel, A. J. Rigby, R. J. Riley, B. J. Teobald, W. Tomlinson, P. J. H. Webborn and P. A. Willis, *Bioorg. Med. Chem. Lett.*, 2007, **17**, 6013.

R. F. Storey, *Curr. Pharm. Des.*, 2006, **12**, 1255.

REFERENCES

1. L. Wallentin, R. C. Becker, A. Budaj, C. P. Cannon, H. Emanuelsson, C. Held, J. Horrow, S. Husted, S. James, H. Katus, K. W. Mahaffey, B. M. Scirica, A. Skene, P. G. Steg, R. F. Storey and R. A. Harrington, for the PLATO Investigators, *N. Engl. J. Med.*, 2009, **361**, 1045.
2. Ticagrelor Summary of Product Characteristics: http://www.ema.europa.eu/docs/en_GB/document_library/EPAR_-_Product_Information/human/001241/WC500100494.pdf.
3. C. W. Hamm, J.-P. Bassand, S. Agewall, J. Bax, E. Boersma, H. Bueno, P. Caso, D. Dudek, S. Gielen, K. Huber, M. Ohman, M. C. Petrie, F. Sonntag, M. S. Uva, R. F. Storey, W. Wijns and D. Zahger, *Eur. Heart J.*, 2011, **32**, 2999.
4. L. Bonello, U. S. Tantry, R. Marcucci, R. Blindt, D. J. Angiolillo, R. Becker, D. L. Bhatt, M. Cattaneo, J. P. Collet, T. Cuisset, C. Gachet, G. Montalescot, L. K. Jennings, D. Kereiakes, D. Sibbing, D. Trenk, J. W. Van Werkum, F. Paganelli, M. J. Price, R. Waksman and P. A. Gurbel, *J. Am. Coll. Cardiol.*, 2010, **56**, 919.
5. B. Springthorpe, A. Bailey, P. Barton, T. N. Birkinshaw, R. V. Bonnert, R. C. Brown, D. Chapman, J. Dixon, S. D. Guile, R. G. Humphries, S. F. Hunt, F. Ince, A. H. Ingall, I. P. Kirk, P. D. Leeson, P. Leff, R. J. Lewis, B. P. Martin, D. F. McGinnity, M. P. Mortimore, S. W. Paine, G. Pairaudeau, A. Patel, A. J. Rigby, R. J. Riley, B. J. Teobald, W. Tomlinson, P. J. H. Webborn and P. A. Willis, *Bioorg. Med. Chem. Lett.*, 2007, **17**, 6013.
6. G. E. Jarvis, R. G. Humphries and M. J. Robertson, *Leff Br. J. Pharmacol.*, 2000, **129**, 275.
7. R. G. Humphries, *Haematologica*, 2000, **85**, 66.
8. A. Gaarder, J. Jonsen, S. Laland, A. Hellem and P. A. Owren, *Nature*, 1961, **192**, 531.
9. G. V. R. Born, *J. Physiol.*, 1962, **162**, 67P.
10. K. Ugerbil and H. Holmsen, in *Platelets in Biology and Pathology*, ed. J. L. Gordon, Elsevier/North Holland, New York, 1981, Volume 2, p. 147.
11. R. T. Dorsam and S. P. Kunapuli, *J. Clin. Invest.*, 2004, **113**, 340.
12. P. A. Gurbel, K. P. Bliden, K. M. Hayes and U. Tantry, *Expert Rev. Cardiovasc. Ther.*, 2004, **2**, 535.
13. R. F. Storey, *Curr. Pharm. Des.*, 2006, **12**, 1255.
14. Antiplatelet Trialists' Collaboration, *Br. Med. J.*, 1994, **308**, 81.
15. N. S. Cook, G. Kottirsch and H. G. Zerwes, *Drugs Future*, 1994, **19**, 135.
16. T. Weller, L. Alig, M. M. Hürzeler, W. C. Kouns and B. Steiner, *Drugs Future*, 1994, **19**(5), 461.
17. CAPRIE Steering Committee, *Lancet*, 1996, **348**, 1326.
18. D. E. Macfarlane and D. C. B. Mills, *Blood*, 1975, **46**, 309.
19. G. M. Blackburn, D. E. Kent and F. Kolkmann, *J. Chem. Soc. Perkin Trans.*, 1984, **I**, 1119.
20. N. J. Cusack and S. M. O. Hourani, *Br. J. Pharmacol.*, 1982, **76**, 221.
21. N. J. Cusack and S. M. O. Hourani, *Br. J. Pharmacol.*, 1982, **75**, 397.
22. A. H. Ingall and J. Dixon, *J. Med. Chem.*, 1999, **42**, 213.
23. R. G. Humphries, W. Tomlinson, A. H. Ingall, P. A. Cage and P. Leff, *Br. J. Pharmacol.*, 1994, **113**, 1057.
24. R. G. Humphries, W. Tomlinson, J. A. Clegg, A. H. Ingall, N. D. Kindon and P. Leff, *Br. J. Pharmacol.*, 1995, **115**, 1110.
25. B. Boneu, G. Destelle, on behalf of the study group, *Thromb. Haemostasis*, 1996, **76**, 939.
26. J. L. Gordon, *Biochem. J.*, 1986, **233**, 309.
27. G. Hollopeter, H.-M. Jantzen, D. Vincent, G. Li, L. England, V. Ramakrishnan, R.-B. Yang, P. Nurden, A. Nurden, D. Julius and P. B. Conley, *Nature*, 2001, **409**, 202.
28. P. Leff, M. J. Robertson, R. G. Humphries, in *Purinergic Approaches in Experimental Therapeutics*, ed. K. A. Jacobson and M. F. Jarvis, Wiley-Liss Inc, New York, 1997, p. 203.
29. K. Butler and R. Teng, *Br. J. Clin. Pharmacol.*, 2010, **70**, 65.
30. D. P. Chew, D. L. Bhatt, S. Sapp and E. J. Topol, *Circulation*, 2001, **103**, 201.

Subject Index